Index of Process Charts
High-Level Processes

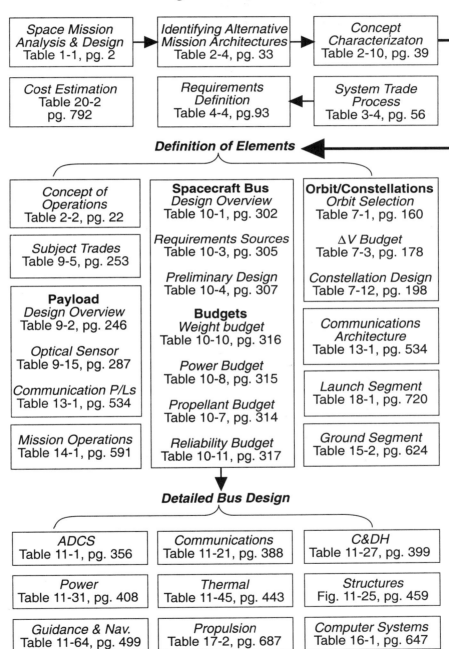

Space Mission Analysis & Design Table 1-1, pg. 2 → *Identifying Alternative Mission Architectures* Table 2-4, pg. 33 → *Concept Characterizaton* Table 2-10, pg. 39

Cost Estimation Table 20-2 pg. 792 — *Requirements Definition* Table 4-4, pg.93 ← *System Trade Process* Table 3-4, pg. 56

Definition of Elements

Concept of Operations Table 2-2, pg. 22

Subject Trades Table 9-5, pg. 253

Payload *Design Overview* Table 9-2, pg. 246

Optical Sensor Table 9-15, pg. 287

Communication P/Ls Table 13-1, pg. 534

Mission Operations Table 14-1, pg. 591

Spacecraft Bus *Design Overview* Table 10-1, pg. 302

Requirements Sources Table 10-3, pg. 305

Preliminary Design Table 10-4, pg. 307

Budgets *Weight budget* Table 10-10, pg. 316

Power Budget Table 10-8, pg. 315

Propellant Budget Table 10-7, pg. 314

Reliability Budget Table 10-11, pg. 317

Orbit/Constellations *Orbit Selection* Table 7-1, pg. 160

ΔV *Budget* Table 7-3, pg. 178

Constellation Design Table 7-12, pg. 198

Communications Architecture Table 13-1, pg. 534

Launch Segment Table 18-1, pg. 720

Ground Segment Table 15-2, pg. 624

Detailed Bus Design

ADCS Table 11-1, pg. 356 | *Communications* Table 11-21, pg. 388 | *C&DH* Table 11-27, pg. 399

Power Table 11-31, pg. 408 | *Thermal* Table 11-45, pg. 443 | *Structures* Fig. 11-25, pg. 459

Guidance & Nav. Table 11-64, pg. 499 | *Propulsion* Table 17-2, pg. 687 | *Computer Systems* Table 16-1, pg. 647

SPACE MISSION ANALYSIS AND DESIGN
Third Edition

THE SPACE TECHNOLOGY LIBRARY
Published jointly by Microcosm Press and Kluwer Academic Publishers

An Introduction to Mission Design for Geostationary Satellites, J. J. Pocha
Space Mission Analysis and Design, 1st edition, James R. Wertz and Wiley J. Larson
*Space Mission Analysis and Design, 2nd edition, Wiley J. Larson and James R. Wertz
*Space Mission Analysis and Design Workbook, Wiley J. Larson and James R. Wertz
Handbook of Geostationary Orbits, E. M. Soop
*Spacecraft Structures and Mechanisms, From Concept to Launch, Thomas P. Sarafin
Spaceflight Life Support and Biospherics, Peter Eckart
*Reducing Space Mission Cost, James R. Wertz and Wiley J. Larson
The Logic of Microspace, Rick Fleeter
Space Marketing: A European Perspective, Walter A. R. Peeters
Fundamentals of Astrodynamics and Applications, 2nd edition, David A. Vallado
Mission Geometry; Orbit and Constellation Design and Management, James R. Wertz
Influence of Psychological Factors on Product Development, Eginaldo Shizuo Kamata
Essential Spaceflight Dynamics and Magnetospherics, Boris Rauschenbakh,
Michael Ovchinnikov, and Susan McKenna-Lawlor

*Also in the DoD/NASA Space Technology Series (Managing Editor Wiley J. Larson)

Space Mission Analysis and Design
Third Edition

Edited by

James R. Wertz
Microcosm, Inc.

and

Wiley J. Larson
United States Air Force Academy

Coordination by
Douglas Kirkpatrick, *United States Air Force Academy*
Donna Klungle, *Microcosm, Inc.*

This book is published as part of the Space Technology Series, a cooperative activity of the United States Department of Defense and National Aeronautics and Space Administration.

Space Technology Library

Published Jointly by

Microcosm Press
El Segundo, California

Kluwer Academic Publishers
Dordrecht / Boston / London

Third Edition

Library of Congress Cataloging-in-Publication Data

A C.I.P. Catalogue record for this book is available from
the Library of Congress.

ISBN 1-881883-10-8 (pb) (acid-free paper)
ISBN 0-7923-5901-1 (hb) (acid-free paper)

*Cover photo of Earth from Space: View of Africa and the Indian Ocean taken in Dec. 1972, by Apollo 17,
the last of the Apollo missions to explore the Moon. Photo courtesy of NASA.*

Cover design by Jeanine Newcomb and Joy Sakaguchi.

Published jointly by
Microcosm Press
401 Coral Circle, El Segundo, CA 90245-4622 USA
and
Kluwer Academic Publishers,
P.O. Box 17, 3300 AA Dordrecht, The Netherlands.

Kluwer Academic Publishers incorporates
the publishing programmes of
D. Reidel, Martinus Nijhoff, Dr. W. Junk and MTP Press.

Sold and distributed in the USA and Canada
by Microcosm, Inc.
401 Coral Circle, El Segundo, CA 90245-4622 USA
and Kluwer Academic Publishers,
101 Philip Drive, Norwell, MA 02061 USA

In all other countries, sold and distributed
by Kluwer Academic Publishers Group,
P.O. Box 322, 3300 AA Dordrecht, The Netherlands.

Printed on acid-free paper

Table of Contents

List of Authors

Henry Apgar. Vice President, MCR International, Inc., Thousand Oaks, California. M.B.A., Northeastern University; B.S. (Electrical Engineering), Rutgers University. Chapter 20—*Cost Modeling*.

David A. Bearden. Senior Project Engineer, Corporate Business Division, The Aerospace Corporation, El Segundo, California. Ph.D., M.S. (Aerospace Engineering), University of Southern California; B.S. (Mechanical Engineering), University of Utah. Chapter 20—*Cost Modeling*.

Robert Bell. Mission Analyst, Microcosm, Inc., Torrance, California. M.S. (Aerospace Engineering), University of Southern California; B.S. (Aerospace Engineering), University of Southern California. Appendix F—*Units and Conversion Factors*.

Richard T. Berget. Program Director, BF Goodrich Aerospace, Data Systems Division, Albuquerque, New Mexico. M.S., B.S. (Electrical Engineering), University of New Mexico. Section 11.3—*Command and Data Handling*.

J. B. Blake. Director, Space Sciences Department, Space and Environment Technology Center, The Aerospace Corporation, El Segundo, California. Ph.D., M.S. (Physics), University of Illinois; B.S. (Engineering Physics), University of Illinois. Section 8.1—*The Space Environment*.

Daryl G. Boden. Associate Professor of Aerospace Engineering, U.S. Naval Academy, Annapolis, Maryland. Ph.D.(Aeronautical and Astronautical Engineering), University of Illinois; M.S. (Astronautical Engineering), Air Force Institute of Technology; B.S. (Aerospace Engineering), University of Colorado. Chapter 6—*Introduction to Astrodynamics*.

William R. Britton. Mechanisms Section Manager, Space Systems, Lockheed Martin Astronautics, Denver, Colorado. B.S. (Mechanical Engineering), Drexel University. Section 11.6—*Structures and Mechanisms*.

Robert F. Brodsky. Chief Engineer, Microcosm, Inc., Torrance, California. Adjunct Professor of Aerospace Engineering, University of Southern California. Sc.D. (Engineering), New York University; M.Aeron. Engineering, New York University; M.S. (Mathematics), University of New Mexico; B.M.E., Cornell University. Chapter 9—*Space Payload Design and Sizing*.

John B. Carraway. Principal Engineer, Jet Propulsion Laboratory, Pasadena, California. B.S. (Electrical Engineering), Massachusetts Institute of Technology. Chapter 14—*Mission Operations*.

Bruce Chesley. Small Satellite Program Manager and Assistant Professor of Astronautics, U.S. Air Force Academy, Colorado Springs, Colorado. Ph.D. (Aerospace Engineering), University of Colorado, Boulder; M.S. (Aerospace Engineering), University of Texas at Austin; B.S. (Aerospace Engineering), University of Notre Dame. Chapter 9—*Space Payload Design and Sizing*.

John T. Collins. System Engineer, Microcosm, Inc. B.S. (Aerospace Engineering), University of Illinois (UIUC); B.S. (Astronomy), University of Illinois (UIUC). Section 20.5—*FireSat Example*.

Richard S. Davies. Technical Staff, Stanford Telecommunications, Inc., Santa Clara, California. Engineer, Stanford University; M.S., B.S. (Electrical Engineering), University of Pennsylvania. Chapter 13—*Communications Architecture*.

Simon D. Dawson. Systems Engineer, Microcosm, Inc., Torrance, California. MSc (Spacecraft Technology and Satellite Communications), University College London, University of London; BSc (Hons) (Physics &European Studies), University of Sussex. Appendices; Inside Rear Pages—*Earth Satellite Parameters*.

Fred J. Dietrich. Principal Engineer, Globalstar, L.P., Palo Alto, California. Ph.D., Ohio State University; M.S. (Electrical Engineering), Purdue University; B.S. (Electrical Engineering), Missouri School of Mines. Chapter 13—*Communications Architecture*.

Peter G. Doukas. Senior Staff Engineer, Martin Marietta Astronautics Group, Denver, Colorado. B.S. (Aeronautics and Astronautics), Purdue University. Section 11.6—*Structures and Mechanisms*.

John S. Eterno. Chief Scientist, Ball Aerospace & Technologies Corp., Boulder, Colorado. Ph.D., M.S. (Aeronautics and Astronautics), Massachusetts Institute of Technology; B.S. (Aerospace Engineering), Case Western Reserve University. Section 11.1—*Attitude Determination and Control*.

Rick Fleeter. President, AeroAstro, Herndon, Virginia. Ph.D. (Thermodynamics), Brown University; M.Sc. (Astronautics and Aeronautics), Stanford University; A.B. (Engineering and Economics), Brown University. Chapter 22—*Design of Low-Cost Spacecraft*.

D. J. Gorney. Principal Director, Defense Support Program, The Aerospace Corporation, El Segundo, California. Ph.D., M.S. (Atmospheric Sciences), University of California, Los Angeles; B.S. (Physics), University of Bridgeport, Connecticut. Section 8.1—*The Space Environment*.

L. Jane Hansen. President, HRP Systems, Torrance, California. M.B.A., Pepperdine University School of Business and Management; B.S. (Applied Mathematics), California Polytechnic State University, San Luis Obispo. Chapter 16—*Spacecraft Computer Systems*.

Herbert Hecht. Chairman of the Board, SoHaR, Inc., Beverly Hills, California. Ph.D. (Engineering), University of California, Los Angeles; M.S. (Electrical Engineering), Polytechnic University of New York; B.S. (Electrical Engineering), City University, New York. Section 19.2—*Reliability for Space Mission Planning.*

Robert W. Hosken. Member of the Technical Staff, The Aerospace Corporation, El Segundo, California. Ph.D. (Physics), University of Illinois; B.S. (Electrical Engineering), Polytechnic Institute of Brooklyn. Chapter 16—*Spacecraft Computer Systems.*

Douglas Kirkpatrick. Visiting Professor, U.S. Air Force Academy, Colorado Springs, Colorado. Ph.D., University of Texas at Austin; M.S., Purdue University; B.S., U.S. Air Force Academy. Section 11.2—*Telemetry, Tracking, and Command*; Chapter 18—*Launch Systems.*

Malcolm K. Kong. Manager, Hardness & Survivability Engineering, TRW Systems & Information Technology Group, Redondo Beach, California. M.S. (Systems Engineering), West Coast University, Los Angeles; B.S. (Electrical Engineering), Purdue University. Section 8.2—*Hardness and Survivability Requirements.*

H. C. Koons. Distinguished Scientist, Space Sciences Department, Space and Environment Technology Center, The Aerospace Corporation, El Segundo, California. Ph.D. (Geophysics), Massachusetts Institute of Technology; B.S. (Physics), Massachusetts Institute of Technology. Section 8.1—*The Space Environment.*

Wiley J. Larson. Director, Space Mission Analysis and Design Program, U.S. Air Force Academy, Colorado Springs, Colorado. D.E. (Spacecraft Design), Texas A&M University; M.S. (Electrical Engineering), University of Michigan; B.S. (Electrical Engineering), University of Michigan. Editor; Chapter 1—*The Space Mission Analysis and Design Process*; Chapter 14—*Mission Operations*; Chapter 23—*Applying Space Mission Analysis and Design Process.*

Joseph P. Loftus, Jr. Assistant Director for Plans, L.B. Johnson Space Center, NASA, Houston, Texas. Sloan Fellow (Business), Stanford University; M.A. (Math and Psychology), Fordham University; B.A. (Math and Psychology), Catholic University. Chapter 18—*Launch Systems.*

Reinhold Lutz. Director, Technology Strategy, Daimler Chrysler Aerospace, Bergen, Germany. Dr.-Ing., University of German Forces; Dipl.-Ing., Technical University Munich. Chapter 9—*Space Payload Design and Sizing.*

Ronald A. Madler. Associate Professor, Embry-Riddle Aeronautical University, Prescott, Arizona. Ph.D., M.S., B.S. (Aerospace Engineering Sciences), University of Colorado, Boulder. Section 21.2—*Orbital Debris—A Space Hazard.*

James R. McCandless. Manager (Retired), Stress Analysis, Space Systems, Lockheed Martin Astronautics, Denver, Colorado. B.S. (Architectural Engineering), University of Texas. Section 11.6—*Structures and Mechanisms.*

Joseph K. McDermott. Engineering Manager, Lockheed Martin Astronautics Group, Denver, Colorado. M.E. (Engineering Management), University of Colorado; B.S. (Chemistry), Loras College. Section 11.4—*Power.*

Darren S. McKnight. Vice President, Titan Research and Technology, Reston, Virginia. Ph.D. (Aerospace Engineering), University of Colorado; M.S. (Mechanical Engineering), University of New Mexico; B.S. (Engineering), U.S. Air Force Academy. Section 21.2—*Orbital Debris—A Space Hazard.*

Robert McMordie. Technologies Manager, International Operations (Retired), Martin Marietta Astronautics Group, Denver, Colorado. Ph.D. (Mechanical Engineering), University of Washington; M.S. (Mechanical Engineering), University of Texas; B.S. (Mechanical Engineering), University of Texas. Section 11.5—*Thermal.*

Wade Molnau. Manufacturing Manager, Motorola Systems Solutions Group, Space Services and Systems Division, Scottsdale, Arizona. Ph.D., M.S., B.S. (Industrial Engineering), Arizona State University. Section 19.1— *Designing Space Systems for Manufacturability.*

Paul Nordin. NASA/TDRS Resident Office Manager at Hughes Space and Communications, El Segundo, California, employed by The Boeing Company, Seabrook, MD. Ph.D., M.A., B.A., (Nuclear Physics), University of California, Berkeley. Section 8.2—*Hardness and Survivability Requirements.*

Jean Olivieri. Teledesic Bus Production Manager, Motorola Advanced Systems Division, Chandler, Arizona. M.S., B.S. (Industrial Engineering), Arizona State University. Section 19.1— *Designing Space Systems for Manufacturability.*

Aniceto Panetti. Satellite Systems Engineer, Alenia Aerospazio Space Division, Rome, Italy. Master of Space Systems Engineering, Delft University of Technology, The Netherlands; Degree in Mechanical Engineering, Università di Roma la Sapienza, Italy. Section 11.5—*Thermal.*

Craig H. Pollock. Space Systems Engineer (Onboard Processing Design Integration and Operations Engineering), TRW Space and Defense Sector, Redondo Beach, California. M.A. (Mathematics), California State University, Long Beach; B.S. (Physics and Mathematics), University of New Mexico. Chapter 16—*Spacecraft Computer Systems.*

Emery I. Reeves. Shriever Chair Professor of Space Systems Engineering, U.S. Air Force Academy, Colorado Springs, Colorado. M.S. (Electrical Engineering), Massachusetts Institute of Technology; B.E. (Electrical Engineering), Yale University. Chapter 10—*Spacecraft Design and Sizing,* Chapter 12—*Spacecraft Manufacture and Test.*

Richard P. Reinert. Staff Consultant, Spacecraft and Mission Design Division, Ball Aerospace Systems, Boulder, Colorado. B.S. (Aeronautical Engineering), Massachusetts Institute of Technology. Chapter 2—*Mission Characterization.*

Robert L. Sackheim. Deputy Director, Propulsion Technology and Fluid Mechanics Center, TRW, Inc., Redondo Beach, California. M.S., B.S. (Chemical Engineering) Columbia University. Chapter 17—*Space Propulsion Systems*.

Thomas P. Sarafin. Consultant/President, Instar Engineering and Consulting, Inc., Littleton, Colorado. B.S. (Civil Engineering), Ohio State University. Section 11.6—*Structures and Mechanisms*.

Michael Schulz. Senior Scientist, Lockheed Martin Advanced Technology Center, Palo Alto, California. Ph.D. (Physics), Massachusetts Institute of Technology; B.S. (Physics), Michigan State University. Section 8.1—*The Space Environment*.

Chad Spalt. Production Technical Lead, Motorola Satellite Communications Group, Mobile Satellite Systems Division, Chandler, Arizona. M.S. (Industrial Engineering), Arizona State University; B.S. (Mechanical Engineering Technology), Southern Illinois University. Section 19.1— *Designing Space Systems for Manufacturability*.

Gael F. Squibb. Director for Telecommunications and Mission Operations, Jet Propulsion Laboratory, Pasadena, California. M.S. (Systems Management), University of Southern California; B.S. (Physics), Harvey Mudd College. Chapter 14—*Mission Operations*

Charles Teixeira. Chief, Systems Definition Branch, L.B. Johnson Space Center, NASA, Houston, Texas. M.S. (Mechanical Engineering), Louisiana State University; B.S. (Aeronautical Engineering), New York University. Chapter 18—*Launch Systems*.

Alan C. Tribble. Manager, Applications Development, Information Technology, Rockwell Collins, Cedar Rapids, Iowa. Ph.D., M.S., (Physics), University of Iowa; B.S. (Physics), University of Arkansas. Section 8.1—*The Space Environment*.

A. L. Vampola. Consultant, Space Environment Effects, Vista, California. Ph.D., M.S. (Physics), St. Louis University; B.S. (Mathematics and Physics), Creighton University, Nebraska. Section 8.1—*The Space Environment*.

R. L. Walterscheid. Senior Scientist, Space Sciences Department, Space and Environment Technology Center, The Aerospace Corporation, El Segundo, California. Ph.D. (Atmospheric Sciences), University of California, Los Angeles; M.S. (Meteorology), University of California, Los Angeles; A.B. (Physics), University of California, Berkeley; B.S. (Meteorology), University of Wisconsin. Section 8.1—*The Space Environment*.

Stanley I. Weiss. Visiting Professor, Massachusetts Institute of Technology and University of California Davis; Formerly Corporate VP Engineering and VP Research and Development, Lockheed Corporation; Ph.D., University of Illinois; M.S., B.S., Renesselaer Polytechnic Institute. Chapter 4—*Requirements Definition*.

James R. Wertz. President, Microcosm, Inc., Torrance, California. Ph.D. (Relativity & Cosmology), University of Texas, Austin; M.S. (Administration of Science and Technology), George Washington University; S.B. (Physics), Massachusetts Institute of Technology. Editor; Chapter 1—*The Space Mission Analysis and Design Process*; Chapter 2—*Mission Characterization*; Chapter 3—*Mission Evaluation*; Chapter 5—*Space Mission Geometry*; Chapter 7—*Orbit and Constellation Design*; Section 8.1—*The Space Environment*; Section 11.7—*Guidance and Navigation*; Chapter 23—*Applying the Space Mission Analysis and Design Process*; Appendix E—*Universal Time and Julian Dates*.

Gary G. Whitworth. Senior Engineer (Retired), Applied Physics Laboratory, Johns Hopkins University, Laurel, Maryland. B.S. (Electrical Engineering), University of Tennessee. Chapter 15—*Ground System Design and Sizing*.

Michael S. Williams. Vice President, Personal SATCOM Systems, Lockheed Martin Global Telecommunications, Reston, VA. M.S. (Electrical Engineering), University of Pennsylvania; M.B.A. (Statistics/Operations Research), Temple University; B.S. (Physics), St. Joseph's University. Chapter 4—*Requirements Definition*.

William B. Wirin. General Manager, Wirin & Associates. J.D. (University of Southern California School of Law); B.A. (Political Science), Occidental College. Section 21.1—*Law and Policy Considerations*.

Robert E. Wong. Manager, Economic Analysis, TRW Systems Engineering and Development Division, TRW, Inc., Redondo Beach, California. Ph.D. (Economics), University of Southern California; M.S. (Mathematics), University of Southern California; M.A. (Economics), University of Southern California; B.S. (Physics), Iowa State University. Chapter 20—*Cost Modeling*.

Sidney Zafran. Program Manager, TRW, Inc., Redondo Beach, California. B.S. (Mechanical Engineering), Massachusetts Institute of Technology. Chapter 17—*Space Propulsion Systems*.

SMAD I, II, and III Authors

The third edition is the end result of a substantial effort by the authors of all the SMAD editions. Consequently, we would like to acknowledge and express our thanks to all of the authors who have contributed to this series:

Henry Apgar
David A. Bearden
Robert Bell
Richard T. Berget
J. B. Blake
James E. Boatwright
Daryl G. Boden
William R. Britton
Robert F. Brodsky
John B. Carraway
Bruce Chesley
Arthur Chomas
John T. Collins
Richard S. Davies
Simon D. Dawson
Fred J. Dietrich
Peter G. Doukas
Neal Ely
John S. Eterno
Rick Fleeter
John R. Ford, Jr.
Martin E.B. France
Steven Glaseman
D. J. Gorney
L. Jane Hansen
Herbert Hecht
Robert W. Hosken
Douglas Kirkpatrick
Malcolm K. Kong
H. C. Koons
Wiley J. Larson
Joseph P. Loftus, Jr.
Reinhold Lutz
Ronald A. Madler
François Martel
James R. McCandless

Joseph K. McDermott
Darren S. McKnight
Robert K. McMordie
Wade Molnau
Ralph L. Mueller
David Negron, Jr.
Paul Nordin
Thomas P. O'Brien
Jean Olivieri
Aniceto Panetti
Craig H. Pollock
Emery I. Reeves
Richard P. Reinert
Robert L. Sackheim
Thomas P. Sarafin
Michael Schulz
Chad Spalt
Gael F. Squibb
Charles Teixeira
Merlin E. Thimlar
Alan C. Tribble
Tim Turner
A. L. Vampola
R. L. Walterscheid
Richard Warner
Stanley I. Weiss
James R. Wertz
Gary G. Whitworth
Michael S. Williams
William B. Wirin
Robert S. Wolf
Robert E. Wong
Sidney Zafran
Robert O. Zermuehlen
Harold F. Zimbelman

Preface

Space Mission Analysis and Design, known as SMAD to its many friends, has gained widespread use as a text and reference throughout the astronautics community. The purpose of the third edition of SMAD is to both update the book and make it more useful and more practical wherever possible. Some topics, such as astrodynamics and mission geometry, have changed relatively little since publication of the second edition in 1992. Here we have made minor modifications to make the material clearer and more precise. On the other hand, topics such as space computers and the design of observation payloads have been nearly completely rewritten. Because of the growing interest in "LightSats" and low-Earth orbit constellations we have added a SmallSat cost model, expanded the discussion of constellation design, and included a new section on multi-satellite manufacturing. The entire volume reflects a greater emphasis on reducing mission cost and doing more with less people and fewer resources.* Finally, the FireSat sample mission has been extended further and the appendices and end matter updated and expanded to provide greater utility as a quick reference. We hope the new edition is better and more useful to you.

As with the first two editions, the goal of the book to is allow you to begin with a "blank sheet of paper" and design a space mission to meet a set of broad, often poorly defined, objectives at minimum cost and risk. You should be able to define the mission in sufficient detail to identify principal drivers and make a preliminary assessment of overall performance, size, cost, and risk. The emphasis is on low-Earth orbit, unmanned spacecraft. However, we hope the principles are broad enough to be applicable to other missions as well. We intend the book to be a practical guide, rather than a theoretical treatise. As much as possible, we have provided physical and engineering data, rules of thumb, empirical formulas, and design algorithms based on past experience. We assume that the reader has a general knowledge of physics, math, and basic engineering, but is not necessarily familiar with any aspect of space technology.

The third edition represents an amalgam of contributions over the last decade by many engineers and managers from throughout the community. It reflects the insight gained from their practical experience, and suggests how things might be done better in the future. From time to time the views of authors and editors conflict, as must necessarily occur given the broad diversity of experience. We believe it is important to reflect this diversity rather than suppress the opinions of individual experts. Similarly, the level of treatment varies among topics, depending both on the issues each author feels is critical and our overall assessment of the level of detail in each topic that is important to the preliminary mission analysis and design process.

---

* The continuing, unrelenting demand to drive down mission cost has led to the creation a companion volume to SMAD, *Reducing Space Mission Cost* [Wertz and Larson, 1996], which addresses cost reduction in all aspects of mission design and includes 10 case studies of how the process works in practice.

The book is intended as a textbook for either introductory graduate or advanced undergraduate courses, or as a reference for those already working in space technology. It can also provide valuable supplementary material for related courses such as spacecraft design or space mission operations. We believe the book can be a key tool for payload designers who need to find out more about space mission design and for those charged with the responsibility of developing space mission requirements and specifications. Finally, we hope that it will be of use to many system engineers in this field who have a detailed knowledge of one area, but need to broaden their background or verify their understanding in related topics.

The book is meant to be read sequentially, although most of the chapters are self-contained, with references to other parts of the book as needed. For readers with specific interests, we recommend the following:

- Those concerned primarily with mission analysis and design should read Chaps. 1–9 and 19–23.

- Those concerned with spacecraft and subsystem design should read Chaps. 1, 2, 4, 8–13, and 16–23.

- Those concerned primarily with mission operations and the ground interaction should read Chaps. 1, 2, 4, and 13–16.

- Those concerned with requirements definition, logistics, and putting a space system in place should read Chaps. 1–4, 7, 9, 10, and 18–23.

- Those interested in constellation design and multi-satellite systems should read Chaps. 1–9, 13–16, and 19–23.

- Those interested in reducing mission cost and the design of low-cost missions should read Chaps. 1–3, 7–10, 12, 20–23, and the companion volume, *Reducing Space Mission Cost*.

SI (metric) units are used throughout the book. Conversions for essentially all common units are contained in Appendix F. Conversion factors and physical constants are generally given to their full available accuracy so that they can be inserted into computer programs and not considered further. As discussed in the introduction to the appendices, the values given are those adopted by the National Bureau of Standards based on a least-squares fit to the fundamental physical constants or international agreement on the definitions of various units. In the case of astronomical constants, values adopted by the International Astronomical Union are given. The most commonly used astronautical formulas and constants are in the appendices. An expanded table of space mission parameters for Earth orbits is on the inside back endleaf. For those wishing to expand that table or use it for other central bodies, the formulas used for creating it are on the preceding pages.

Leadership, funding, and support essential to updating the book were provided by numerous programs at the Air Force Space and Missile Center, Air Force Space Command, NASA Headquarters, NASA/Goddard Space Flight Center, and the Advanced Projects Research Agency. Obtaining funding to create and maintain much-needed reference material is exceptionally difficult. We are deeply indebted to the sponsoring organizations, particularly Air Force Phillips Laboratory, for their support and their recognition of the importance of projects such as this one.

The third edition of this book is the result of nearly two years of effort by a dedicated team of government, industry, and academic professionals. The Department of Astronautics, United States Air Force Academy, provided unwavering support for the project. Michael DeLorenzo, Chairman of the Department of Astronautics, provided the leadership and continuing support critical to projects of this type. Both Doug Kirkpatrick and Perry Luckett performed a detailed grammatical review in a valiant effort to prevent the rest of us from demonstrating why we became engineers rather than writers. Several graphics artists at the Academy, particularly Mary Tostanoski and Debra Porter, spent many hours developing and updating artwork. Joan Aug and Bert Reinertson cheerfully handled the huge administrative burden at the Academy. Numerous faculty members, staff, and students graciously sacrificed their time to provide assistance, review, and comments. Daryl Boden assisted with the editing and reviewing even after changing assignments to the Naval Academy. Doug Kirkpatrick managed the task for the Air Force with great skill and patience and reviewed nearly all of the material for both technical and linguistic correctness!

OAO Corporation, Colorado Springs, Colorado, provided the contract support for the project. Anita Shute at the Air Force Academy spent many hours revising drafts, creating artwork, and working all aspects of the project. Eugene deGeus of Kluwer Academic Publishers supplied substantial assistance with all aspects of the publishing activity. This was his final project at Kluwer before taking a science administration position with the Dutch government. We will miss his wisdom and guidance and wish him the best of future success.

At Microcosm, the entire analysis and publications staff worked virtually all aspects of the book (art, grammar, equation checking, technical reviews, and camera-ready copy) and suffered patiently through "the book project" as it continually absorbed great amounts of limited resources. Much of the new graphics was done by undergraduate students Karen Burnham, Paul Murata, Alan Chen, and Julie Wertz under the very capable guidance of Kevin Polk and Simon Dawson. Jennifer Burnham and Judith Neiger did much of the proofing. Robert Bell did most of the demanding task of updating units and conversion factors. John Collins created the new FireSat cost model. Wendi Huntzicker and Joy Sakaguchi created the new camera-ready copy for most of the book. Joy and Chris deFelippo did much of the new art. Finally, Donna Klungle did a truly remarkable job managing, administering, editing, reviewing, and preparing revisions, drafts, and the final camera-ready copy. Donna accomplished this with skill and good humor, while dealing with the conflicting demands of multiple authors and editors.

Arthur Cox of Lawrence Livermore National Labs and the editors of *Astrophysical Quantities* [1999] graciously permitted the use of drafts of their forthcoming volume so that we could obtain the most current values for physical quantities. We highly recommend that readers consult *Astrophysical Quantities* for solar system and astronomical parameters which are not contained here.

Every effort has been made to eliminate mathematical and factual errors. Many errors from prior editions have been found largely through readers' comments and constructive criticism. Please continue to send any errors, omissions, corrections, or comments to either editor at the addresses below. We sincerely hope that the book will be of use to you in our common goal of reducing the cost and complexity of space utilization.

Finally, one of the most exciting aspects of space mission analysis and design is that after 40 years of space exploration we have only begun to scratch the surface of the variety of important missions that can and should be done. In spite of problems, setbacks, and higher costs than any of us would like, people young and old remain excited about space. The exploration of space will take dramatic new turns in the future, from communications constellations and microgravity work now beginning to become a reality to solar power satellites, space tourism, space industrialization, and settlements on the Moon and planets which are still to be designed. We hope that this volume provides a portion of the roadmap and incentive to those who will undertake these tasks. We wish you the best of success in this endeavor.

June, 1999

James R. Wertz
Microcosm, Inc.
401 Coral Circle
El Segundo, CA 90245-4622
FAX: (310) 726-4110
jwertz@smad.com

Wiley J. Larson
Department of Astronautics
United States Air Force Academy
Colorado Springs, CO 80840-6224
FAX: (719) 333-3723
wileylarson@adelphia.net

Cox, A.N. ed. 1999. *Astrophysical Quantities*, New York: Springer-Verlag.

Wertz, James R. and Wiley J. Larson. 1996. *Reducing Space Mission Cost*. Torrance, CA and Dordrecht, The Netherlands: Microcosm Press and Kluwer Academic Publishers.

Chapter 1

The Space Mission Analysis and Design Process

James R. Wertz, *Microcosm, Inc.*
Wiley J. Larson, *United States Air Force Academy*

Space mission analysis and design begins with one or more broad objectives and constraints and then proceeds to define a space system that will meet them at the lowest possible cost. **Broad** objectives and constraints are the key to this process. Procurement plans for space systems too often substitute detailed numerical requirements for broad mission objectives. To get the most performance for the money spent, we must require of the system only what it can reasonably achieve. Thus, while our overall objectives to communicate, navigate, or observe will generally remain the same, we will achieve these objectives differently as technology and our understanding of the process and problem evolve. This chapter summarizes, and the book as a whole details, this process of defining and refining both what is to be done and what mission concept will do it at the lowest cost.

There are now a number of references available on the mission design process and the definition of mission objectives. Rechtin [1991] and Ruskin and Estes [1995] provide general discussions of this process. Shishko [1995] provides an overview from the NASA perspective and Przemieniecki [1993] gives a similar treatment for defense missions. Davidoff [1998] and Wertz and Larson [1996] discuss this process from the perspective of very low-cost missions and methods for dramatically reducing mission cost, respectively. Boden and Larson [1996] discuss the analysis and design process specifically for mission operations. Finally, Kay [1995] examines the fundamental difficulty of doing technical trades within a democratic political environment.

1.1 Introduction and Overview

Table 1-1 summarizes our approach to the space mission analysis and design process. Space missions range widely from communications, to planetary exploration, to proposals for space manufacturing, to burial in space. No single process can fully cover all contingencies, but the method in Table 1-1 summarizes a practical approach evolved over the first 40 years of space exploration.

1

Space is expensive. Cost is a fundamental limitation to nearly all space missions and is becoming more so. Consequently, this and subsequent tables reflect the assessment of each author on how things traditionally have been done **and** how they should be done differently, both to lower cost and to achieve the greatest return from the space investment.

Analysis and design are iterative, gradually refining both the requirements and methods of achieving them. Thus, we must repeat the broad process defined in Table 1-1 many times for each mission. The first several iterations may take only a day, but more detailed assessments will take far longer.

Successive iterations through Table 1-1 will usually lead to a more detailed, better-defined space mission concept. But we must still return regularly to the broad mission objectives and search for ways to achieve them at a lower cost. In defining and refining the approach, there is strong pressure to proceed to ever greater detail, and never revise a decision once it has been made. Although we must maintain orderly progress, we must also review the mission design regularly for better ways to achieve the mission objectives. Methods may change as a result of evolving technology, a new understanding of the problem, or simply fresh ideas and approaches as more individuals become involved.

TABLE 1-1. The Space Mission Analysis and Design (SMAD) Process. Tables of this type appear throughout the book. The far right column refers to sections in the book that give details of each step. See text for further explanation.

Typical Flow	Step		Section
	Define Objectives	1. Define broad objectives and constraints 2. Estimate quantitative mission needs and requirements	1.3 1.4
	Characterize the Mission	3. Define alternative mission concepts 4. Define alternative mission architectures 5. Identify system drivers for each 6. Characterize mission concepts and architectures	2.1 2.2 2.3 2.4
	Evaluate the Mission	7. Identify critical requirements 8. Evaluate mission utility 9. Define mission concept (baseline)	3.1 3.3 3.4
	Define Requirements	10. Define system requirements 11. Allocate requirements to system elements	4.1 4.2–4.4

Finally, we must document the results of this iterative process. If we wish to go back and reexamine decisions as new data becomes available, we must clearly understand and convey to others the reasons for each decision. We need this documentation for decisions based on detailed technical analyses, and, equally important, for those based on simplicity, ease of assessment, or political considerations.

This book presents many examples from real space missions. To illustrate the mission analysis and design process without being tied to existing space systems, we invented the hypothetical *FireSat* space mission. Figure 1-1 shows the broad mission statement we used to begin the process of space mission design for FireSat. We wish to stress that the parameters developed throughout the book are by no means the only possible set for FireSat, nor necessarily the best. To show how solutions may vary, Chap. 22 presents a very low-cost spacecraft as an alternative for FireSat. Our example system simply illustrates the iterative process of space mission analysis and design. Different assumptions, requirements, or proposed solutions may lead to dramatically different results.

FireSat
Mission Statement

Because forest fires have an increasing impact on recreation and commerce and ever higher public visibility, the United States needs a more effective system to identify and monitor them. In addition, it would be desirable (but not required) to monitor forest fires for other nations; collect statistical data on fire outbreaks, spread, speed, and duration; and provide other forest management data.

Ultimately, the Forest Service's fire-monitoring office and rangers in the field will use the data. Data flow and formats must meet the needs of both groups without specialized training and must allow them to respond promptly to changing conditions.

Fig. 1-1.   **Origin of the Hypothetical FireSat Mission**. FireSat is used as the primary example throughout this book.

To illustrate the broad process of Table 1-1, we will go through each of the top-level steps for the FireSat mission and indicate the type of information that needs to be developed:

In **Step 1**, we define what the mission needs to achieve. What are our qualitative goals, and why? This information should come largely from the mission statement of Fig. 1-1. We need to return to this broad goal over and over to ask whether we are doing what we set out to do.

Step 2 is significantly different. It **quantifies** how well we wish to achieve the broad objectives, given our needs, applicable technology, and cost constraints. These quantitative requirements should be subject to trade as we go along. A major error in many space-system procurements is to set requirements in concrete at a relatively early stage. An example for FireSat might be a 100 m positioning accuracy for initial fire detection. A 100 m requirement seems to be a reasonable place to start, but compared to an accuracy of 200 m, it could add tens or even hundreds of millions of dollars to the overall system cost. We might spend this extra money better in acquiring fire detection airplanes, providing more personnel on the ground, or using better fire-fighting technology. Congress, the Department of the Interior, and the Forest Service must ultimately decide how well FireSat should do and at what cost. Space mission analysis and design provides the quantitative data needed to support such decisions.

Our next step is to define and characterize a space mission to meet the objectives. **Step 3** begins this process by developing alternative mission concepts. A *mission concept* or *concept of operations* is a broad statement of how the mission will work in practice. It includes issues such as how the data will be sensed and delivered to the end user, how the mission will be controlled, and the overall mission timeline. Alternative mission concepts include, for example, conceptually distinct approaches to the problem such as the very low-cost approach defined in Chap. 22. These would also include different orbits or different wavelength bands for fire detection that would require dramatically dissimilar systems.

Step 4 defines alternate combinations of mission elements or the *space mission architecture* to meet the requirements of the mission concept. The space mission architecture is the mission concept **plus** a definition of each of the elements of the mission shown in Fig. 1-3 (Sec. 1.2). A good way to begin Step 4 is to look at the mission elements in Fig. 1-3 and consider what alternatives for each of them would best meet mission objectives.

In any real system, many things influence overall cost, performance, or the design of detailed components. However, these are influenced mainly by a relatively small number of key parameters or components, called *drivers*. Thus, there may be *cost, performance,* or *system drivers* which affect the design of the overall space system. In **Step 5** we identify the principal cost and performance drivers for each alternative mission concept. For most space missions, system drivers include the number of satellites, altitude, power, and instrument size and weight. (Sec. 2.3 gives a more detailed list.) By explicitly identifying the system drivers, we can concentrate our effort on parameters having the most impact on the design and therefore on the cost of the space mission. This method improves our chances of getting the best possible design within the available budget.

Step 6 is typically the most involved in mission design because it defines in detail what the system is and does. Here we determine the power, weight, and pointing budgets[*] and decide what to process on the ground or in space. Characterizing the mission is the most costly step because it requires the expertise of many people. Developing detail is always comforting in managing any design process but, as noted earlier, we must take care not to overdo details while characterizing the mission. System-level requirements and trades must remain our primary focus.

The next step in mission analysis and design is to evaluate the systems we have defined. Having defined and characterized alternative mission concepts, we return in **Step 7** to our initial quantitative requirements and identify the *critical requirements,*[†] that is, the key requirements principally responsible for determining the cost and complexity of the system. Recall that the *system drivers* are those defining parameters, such as altitude or payload aperture, which most strongly affect the cost, performance, and system design. System drivers are not normally system requirements. However, a critical requirement for coverage or resolution may result in altitude and aperture becoming performance or system drivers. The implication of this for mission analysis and design is that we must put substantial effort into understanding the quantitative relationship between, for example, altitude, aperture, coverage, and resolution, in order to set intelligently both the requirements (coverage and resolution) and system parameters (altitude and aperture). For FireSat, the critical requirements might be fire location accuracy, resolution, coverage, or timeliness of the data. We should concentrate on these requirements to determine how firm they are, how good we should make them, and how much we will pay for them to achieve our broad objectives. Critical requirements may differ for alternative mission concepts.

---

[*] A *budget* is a numerical list of the components of any overall system parameter. Thus, the total spacecraft *weight budget* would consist of the weights assigned to the payload instruments, the various subsystems, the propellant required, and typically some margin for growth.

[†] In the first and second editions of this book, critical requirements were called driving requirements. We changed the terminology to avoid confusion with the system drivers of Step 5.

The above questions form the basis of *mission utility analysis*, **Step 8**, in which we quantify how well we are meeting both the requirements and the broad objectives as a function of either cost or key system-design choices. We would like to provide the decision maker a single chart of potential performance vs. cost. More typically, we must settle for something less ideal, such as the percent of fires detected within 2 hours vs. the aperture of the instrument, or the delay time in detecting forest fires vs. altitude and the number of satellites in the constellation. Only the user or developer of the system can ultimately determine the goodness of these critical performance measures, called *Measures of Effectiveness* or *Figures of Merit*. Consequently, mission definition must be to some degree a joint process between those who understand the mission analysis and design process and those who eventually must use the system or justify its cost.

Having evaluated alternative designs and done a preliminary assessment of mission utility, we select one or more baseline system designs in **Step 9**. A *baseline* design is a single consistent definition of the system which meets most or all of the mission objectives. A *consistent system definition* is a single set of values for all of the system parameters which fit with each other—e.g., resolution and coverage rates which correspond to the assigned altitude, aperture, and resulting spacecraft weight. In actually designing a space system, many parameters are being defined and changed simultaneously. The baseline provides a temporary milestone against which to measure progress. It also allows us to limit the number of options which must be evaluated. Rather than looking at all possible combinations of altitude, aperture, power, and spectral band (a nearly impossible task), it is much more feasible to look at the impact of varying each of these individually relative to one or two baseline designs. As the system design matures, the baseline becomes more firm, and eventually becomes the system design. However, we should always remember that the baseline is only a starting point for the iterative trade process and should **not** be regarded as an ironclad definition of mission parameters.

Because builders of a space system work from specific requirements, we must translate the broad objectives and constraints of the mission into well-defined system requirements in **Step 10**. In **Step 11**, we *flow down* or allocate these numerical requirements to the components of the overall space mission in the same way that a budget allocates weight and power to the spacecraft's components. The final list of detailed requirements reflects how well we have done the job of space mission analysis, design, and allocation.

1.1.1 Changes in Future Space Missions

The way we analyze and design space missions is itself continually evolving. In particular, we expect major changes in this process because of increasing technological maturity, increasing use of onboard processing, and continuing emphasis on low-cost missions.

Technological limits on space exploration are giving way to those of policies, politics, and economics. Nearly any mission is technically feasible. It is well within our technical capacity to build a lunar base, mount manned explorations to Mars or other planets, create an industrial base in space, or build networks of satellites to provide truly global communications and observations. Our activity in space depends on what we can afford to do or what we choose to do. Therefore, we must carefully analyze **why** we choose to use or explore space. We must select each space mission, not just to achieve something that could not have been done before, but to achieve something that should be done or is worth doing.

A major technological change in future space missions will be increased use of onboard computers. Space system developers have been very slow to use computers because of the conservative approach to spacecraft design, long lead times in spacecraft production, and very real difficulties associated with running a computer reliably in space.* The shift to increased onboard processing is moving spacecraft toward more autonomy and increased complexity in terms of the tasks they undertake. Whether this change drives space costs up or down depends upon the government and industry's approach to autonomy and software development for space. Spacecraft may follow the example of ground systems, carrying either low-cost commercial systems or vastly more expensive but more capable special purpose systems.

We anticipate continuing emphasis on low-cost spacecraft. Small spacecraft will increase for future space missions. These could be either individual, single-purpose, small satellites or large constellations of small satellites used for communications, space-based radar, or tactical applications. Again, the community appears to be dividing into those who can build small, low-cost spacecraft and those who continue to build large, expensive systems. Creating LightSats represents a new ethic and a new way of doing business in space. If the space business is to grow and prosper as commercial aviation has, we must find a way to reduce the costs of using space. Lowering cost is the real challenge for space mission analysis and design, as well as the government and industrial groups which have created and used this process.

Finally, the mission concept and associated space mission architecture largely determine the cost, complexity, and efficiency of the overall system, This is compounded greatly when you begin to consider integrating the operational aspects of many different missions. For example, today within DoD, we have communication, navigation, signal intelligence, reconnaissance, and weather systems; each with their own mission concept and architecture. The upcoming challenge is to find ways for these systems to operate together to meet user needs.

The fundamental question is "Who are the customers, and what products or services do they require?" In trying to answer this we find ourselves dealing with information-related issues: What information is required, where, and in what form? Most customers don't care about the existence of communications, navigation, or weather satellites. They need specific information and are not interested in what systems provide it. Today's challenge is to blend the capabilities and information available from multiple systems to meet customer needs. Military people often express this as tasking, processing, interpretation, and dissemination, whereas commercial people often express the same issues as customer requests processing, formatting, and delivery.

Figure 1-3 is divided along somewhat arbitrary, functional boundaries. We need to find ways to dissolve these artificial boundaries and create cost-effective solutions to our customer's information needs. For example, instead of trying to integrate the separate systems discussed above, we might consider having multimission payloads and spacecraft that have the ability to gather intelligence information, weather, and provide navigation using one payload—multimission payloads.

An alternative to creating multimission payloads is to divide the architecture differently by placing all sensors on one space asset, processing capability on another

---

* Space computers are far more susceptible than ground computers to *single-event upsets* caused by the high-radiation environment or randomly occurring cosmic rays. To protect against this damage, we must design computers specifically for use in space, as described in Chap. 16.

and using existing or proposed communications links to move the information around. A third alternative might be to use a series of low-cost LightSats each doing a separate function, but in such a way that the end results can be easily and directly integrated by the user's equipment on the ground.

These examples provide a slightly different perspective which is difficult for many organizations, both industrial and government, to adopt because we think and organize functionally—launch, spacecraft, operations, and so on. Being able to functionally decompose our missions and divide them into workable pieces has been one of the reasons, for our success. On the other hand, if we think only functionally it may cause significant problems. We must also think horizontally and create systems that can be integrated with other space and ground systems to create capabilities that are greater than the sum of their parts. As always, our goal is to meet the total user needs at minimum cost and risk.

1.2 The Space Mission Life Cycle

Table 1-2 illustrates the life cycle of a space mission, which typically progresses through four phases:

- *Concept exploration*, the initial study phase, which results in a broad definition of the space mission and its components.

- *Detailed development*, the formal design phase, which results in a detailed definition of the system components and, in larger programs, development of test hardware or software.

- *Production and deployment*, the construction of the ground and flight hardware and software and launch of the first full constellation of satellites.

- *Operations and support*, the day-to-day operation of the space system, its maintenance and support, and finally its deorbit or recovery at the end of the mission life.

These phases may be divided and named differently depending on whether the *sponsor*—the group which provides and controls the program budget—is DoD, NASA, one of the many international organizations, or a commercial enterprise. The time required to progress from initial concept to deorbiting or death of the space asset appears to be independent of the sponsor. Large, complex space missions typically require 10 to 15 years to develop and operate from 5 to 15 years, whereas small, relatively simple missions require as few as 12 to 18 months to develop and operate for 6 months to several years.

Procurement and operating policies and procedures vary with the sponsoring organization, but the key players are the same: the space mission operator, end user or customer, and developer. Commercial space missions are customer driven. The main difference between users and customers is that customers usually pay for a service, whereas users receive services that others pay for. *Operators* control and maintain the space and ground assets, and are typically applied engineering organizations. *End users* receive and use the products and capability of the space mission. They include astronomers and physicists for science missions, meteorologists for weather missions, you and me for communication and navigation missions, geologists and agronomists for Earth resources missions, and the war fighter for offensive and defensive military

TABLE 1-2. Space Program Development Phases. Every space program progresses through the top-level phases. Subphases may or may not be part of a given program. The time required to complete the process varies with the scope. (Diagram courtesy R. Bertrand.)

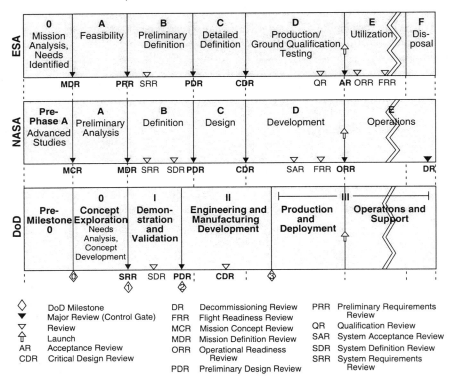

space missions. The *developer* is the procuring agent, be it government or a commercial enterprise, and includes the contractors, subcontractors, and government development, and test organizations. The operators and users must generate technically and fiscally responsible requirements; the developer must provide the necessary product or capability on time and within changing political and funding constraints.

Three basic activities occur during the Concept Exploration Phase (see Fig. 1-2). Users and operators develop and coordinate a set of broad needs and performance objectives based on an overall concept of operations. At the same time, developers generate alternative concepts to meet the perceived needs of the user and operating community. In addition, the sponsor performs long-range planning, develops an overall program structure, and estimates budgetary needs and available funding to meet the needs of the users, operators, and developers. In order to be successful in producing and deploying a new space capability, the four key players in this activity must closely integrate their areas of responsibility.

This book emphasizes the concept exploration phase which further divides into needs analysis and concept development, as detailed in Table 1-3. The goal during concept exploration is to assess the need for a space mission and to develop affordable alternatives that meet operator and end-user requirements. The *Needs Analysis* is a

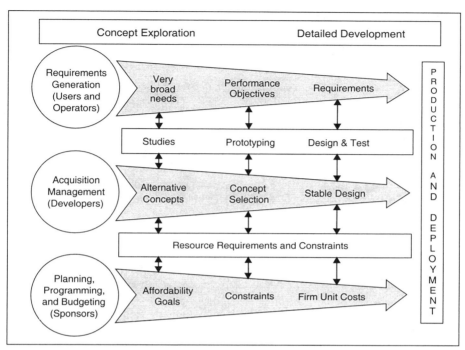

Fig. 1-2. Organizations and People that Play Key Roles in Space Missions. During concept exploration users and operators provide requirements, developers create the design of the mission and systems, while sponsors provide funding.

continuing process which culminates in a new program start. Operators and end users develop potential mission requirements based on the considerations shown in the left-hand column of Table 1-3. The process is different for each organization, but at some point a new program begins with a set of mission objectives, concept of operations, and desired schedule. In DoD, the *Mission Needs Statement* documents this information and becomes part of the planning, programming, and budgeting system [Defense Systems Management College, 1990]. If approved, the program receives funding and proceeds to concept development.

At the *Program Initiation* milestone, the funding organization commits to proceeding through concept development. The program will receive different levels of scrutiny depending on its scope, political interest, and funding requirements. In DoD, programs classed as *major programs* receive the utmost attention at the highest levels of the Defense Department. Various components of the military use distinct criteria to identify major programs [Defense Systems Management College, 1990]. A DoD critical requirements program is "major" if it requires more than $200 million in research, development, test, and evaluation funds or more than $1 billion in production costs. Programs that require participation by more than one component of the armed forces or have congressional interest may also be classified as major programs.

During *Concept Development* the developer must generate alternative methods of meeting the operator's and end user's needs. This procedure includes developing and assessing different concepts and components for mission operations, as well as

TABLE 1-3. Further Breakdown of Concept Exploration Phase. During concept exploration the operator and end users define their needs and requirements and pass them to the developing organization for concept development. A basic premise of this book is that the operator, user, and developer should work together to create realistic and affordable mission objectives and requirements that meet user needs.

Concept Exploration and Definition	
Needs Analysis	**Concept Development**

Generate potential requirements based on
  Mission objectives
  Concept of operations
  Schedule
  Life-cycle cost and affordability
  Changing marketplace
  Research needs
  National space policy
  Long-range plan for space
  Changing threats to national defense
  Military doctrine
  New technology developments
  Commercial objectives
	Reassess potential requirements generated during needs analysis
	
	Develop and assess alternative mission operations concepts
	
	Develop and assess alternative space mission architectures
	
	Estimate
	  performance supportability
	  life-cycle cost produceability
	  schedule funding profiles
	  risk return on investment

estimating the factors shown in the right-hand column of Table 1-3. The information becomes part of an overall system concept. High-level managers in the user, operator, and development communities evaluate whether the concepts, initial mission objectives, and potential requirements meet the mission's intentions. If the program satisfies the need at a reasonable cost, it passes the *Requirements Validation* milestone and proceeds into the Detailed Development Phase.

This book provides the technical processes and information necessary to explore concepts for many space missions. Table 1-3 identifies a major concern that can undermine the entire process: in many cases, users and operators analyze the needs and formulate mission requirements apart from the development community. Then they pass these requirements "over the wall" without negotiating. The developer often generates alternatives without the operators and users. These autonomous actions produce minimum performance at maximum cost.

> **To explore a concept successfully, we must remove the walls between the sponsor, space operators, users or customers, and developers and become a team.**

A good team considers the mission's operations, objectives, and requirements as well as the available technology to develop the best possible mission concept at the lowest possible life-cycle cost.

All space missions consist of a set of *elements* or *components* as shown in Fig. 1-3. The arrangement of these elements form a *space mission architecture*. Various organizations and programs define their mission elements differently, although all of the elements are normally present in any space mission.

The *subject* of the mission is the thing which interacts with or is sensed by the space payload: moisture content, atmospheric temperature, or pressure for weather missions; types of vegetation, water, or geological formations for Earth-sensing missions; or a rocket or intercontinental ballistic missile for space defense missions. We must decide

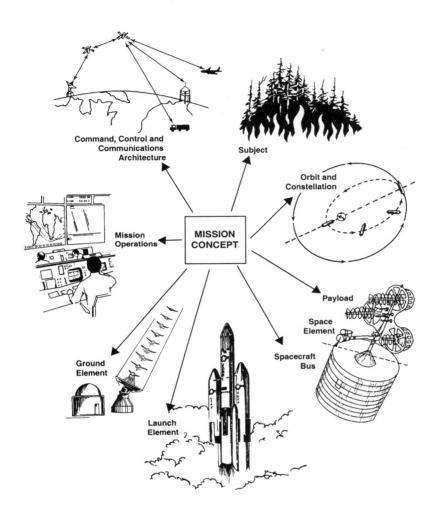

Fig. 1-3.   Space Mission Architecture. All space missions include these basic elements to some degree. See text for definitions. Requirements for the system flow from the operator, end user, and developer and are allocated to the various mission elements.

what part of the electromagnetic spectrum to use in order to sense the subject, thus determining the type of sensor as well as payload weight, size, and power. In many missions, we may trade off the subject. For example, if we are trying to track a missile during powered flight, the subject could be the rocket body or exhaust plume, or both.

For communications and navigation missions the subject is a set of equipment on the Earth or on another spacecraft, including communication terminals, televisions, receiving equipment for GPS navigation, or other user-furnished equipment. The key parameters of this equipment characterize the subject for these types of missions.

The *payload* consists of the hardware and software that sense or interact with the subject. Typically, we trade off and combine several sensors and experiments to form the payload, which largely determines the mission's cost, complexity, and effectiveness. The subsystems of the *spacecraft bus* support the payload by providing orbit and attitude maintenance, power, command, telemetry and data handling, structure and rigidity, and temperature control. The payload and spacecraft bus together are called the *spacecraft, space segment,* or *launch vehicle payload.*

The *launch system* includes the launch facility, launch vehicle and any upper stage required to place the spacecraft in orbit, as well as interfaces, payload fairing, and associated ground-support equipment and facilities. The selected launch system constrains the size, shape, and mass of the spacecraft.

The *orbit* is the spacecraft's path or trajectory. Typically, there is a separate initial parking orbit, transfer orbit, and final mission orbit. There may also be an end-of-life or disposal orbit. The mission orbit significantly influences every element of the mission and provides many options for trades in mission architecture.

The *communications architecture* is the arrangement of components which satisfy the mission's communication, command, and control requirements. It depends strongly on the amount and timing requirements of data to be transferred, as well as the number, location, availability, and communicating ability of the space and ground assets.

The *ground system* consists of fixed and mobile ground stations around the globe connected by various data links. They allow us to command and track the spacecraft, receive and process telemetry and mission data, and distribute the information to the operators and users.

Mission operations consist of the people, hardware, and software that execute the mission, the mission operations concept, and attendant policies, procedures, and data flows. Finally, the *command, control, and communications (C^3)* architecture contains the spacecraft, communications architecture, ground segment, and mission operations elements.

1.3 Step 1: Definition of Mission Objectives

The first step in analyzing and designing a space mission is to define *mission objectives*: the broad goals which the system must achieve to be productive. Figure 1-4 shows sample objectives for FireSat. We draw these qualitative mission objectives largely from the mission statement. In contrast, the *mission requirements* and *constraints* discussed in Sec. 1.4 are quantitative expressions of how well we achieve our objectives—balancing what we want against what the budget will allow. Thus, whereas we may modify objectives slightly or not at all during concept exploration, we often trade requirements throughout the process. For FireSat to be FireSat, it must detect, identify, and monitor forest fires. As we trade and implement elements of the system during concept exploration, we must ensure that they meet this fundamental objective. An excellent example of the careful definition of broad mission objectives for space science missions is given by the National Research Council [1990].

Ordinarily, space missions have several objectives. Some are secondary objectives which can be met by the defined set of equipment, and some are additional objectives which may demand more equipment. Nearly all space missions have a *hidden agenda* which consists of secondary, typically nontechnical, objectives. Frequently political, social, or cultural, they are equally real and equally important to satisfy. For example, a secondary objective for FireSat could be to show the public a visible response to

FireSat Mission Objectives
Primary Objective: **To detect, identify, and monitor forest fires throughout the UnitedStates, including Alaska and Hawaii, in near real time.**
Secondary Objectives: **To demonstrate to the public that positive action is underway to contain forest fires.** **To collect statistical data on the outbreak and growth of forest fires.** **To monitor forest fires for other countries.** **To collect other forest management data.**

Fig. 1-4. FireSat Mission Objectives. Unlike *requirements*, which specify numerical levels of performance, the *mission objectives* are broad statements of what the system must do to be useful.

frequent forest fires. Third World nations produce satellites in part to show that their developing technology makes them important players in international politics. Of course, this secondary political objective for space programs has been important for many years in both the United States and the former Soviet Union. If we are to meet all of a space mission's objectives, we must identify secondary and nontechnical objectives as well as primary ones.

Multiple objectives also occur when we use a single satellite to meet different demands. For example, we may use FireSat's temperature-sensing instruments to monitor global changes in ocean temperatures. In this case, the secondary objectives could become as important as the primary ones. A second example would be adding a communications payload to FireSat to permit better communications among the distributed groups who fight forest fires. Although the primary objective usually will be quite stable, secondary objectives may shift to meet the users' needs and the redefined potential of the space mission concept.

As in the case of most of the top-level trades, we recommend strongly against numerical formulas that try to "score" how well a mission meets its objectives. We can compute probabilities for achieving some technical objectives, but trying to numerically combine the coverage characteristics of different FireSat constellations with the political impact of launching FireSat is too simplistic for effective decision making. Instead, we must identify objectives separately so we can judge how to balance alternative objectives and mission concepts.

Good mission objectives incorporate user needs and, at least indirectly, the space characteristics we are exploiting to achieve them. As stated earlier, space is expensive. If our end objective does not use one of the fundamental space characteristics, it will likely cost less to do on Earth. For example, there is little reason to manufacture low-cost consumer goods or publish books in space.

What fundamental characteristics of space make space missions desirable? Table 1-4 lists some of them with their corresponding missions. Exploring and using space serves various objectives, from extremely practical telecommunications and weather, to major scientific observatories hoping to understand the universe better, to advanced military applications and exploring and exploiting the Moon and planets. Our objectives are diverse partly because we use many different space characteristics.

For example, materials processing uses the microgravity and high vacuum of space, disregarding the spacecraft's position over the Earth. Conversely, communications or observation satellites emphasize Earth coverage as the most fundamental space characteristic to achieve their objectives.

TABLE 1-4. **Characteristics of Space Exploited by Various Space Missions.** Note the wide variety and that many are only beginning to be used. (Spacecraft acronyms are defined in the index.)

Characteristic	Relevant Missions	Degree of Utilization	Sample Missions
Global Perspective	Communications Navigation Weather Surveillance	Some are mature industries; major new advances will come with increased onboard processing	IntelSat GPS NOAA satellites DBS
Above the Atmosphere	Scientific observations at all wavelengths	Well developed; space observatories will continue to dramatically change our view of the universe	Space Telescope GRO Chandra X-Ray Observatory IUE
Gravity-free Environment	Materials processing in space	Now in infancy; may be many future applications	Industrial Space Facility ISS Comet
Abundant Resources	Space industrialization Asteroid exploration Solar power satellites	Essentially none	Space colonies Solar power satellites NEAP
Exploration of Space Itself	Exploration of Moon and planets, scientific probes, asteroid and comet missions	Initial flybys have been done; Some landings done or planned; limited manned exploration	Manned lunar or Martian bases Apollo Galileo

Table 1-4 reveals a second important feature: the varying levels of exploitation for different space characteristics. Many current missions use the global perspective of space—for telecommunications, weather, navigation, and other aspects of Earth monitoring. Space-based telecommunications will continue to grow, but it is already a major and mature industry. Satellite communications by telephone and television have become a part of everyday life and have helped to bring about a communications revolution largely responsible for our shrinking world. Equally dramatic changes are likely in the future as new applications for space-based communications and navigation continue to emerge.

In contrast to telecommunications, materials processing and precision manufacturing in gravity-free space is only in its infancy. Major strides appear possible in pharmaceutical and semiconductor devices that may bring about an entirely new industrial segment. Exploiting space's almost limitless resources is even further removed. Unlimited continuous power and huge, accessible supplies of physical materials may, in the long run, maintain an industrialized society without destroying the Earth's fragile environment. These objectives will require greater vision than those for the more fully developed areas of communications, resource mapping, and monitoring.

We see from Table 1-4 that we have either not used or only begun to use most of the major characteristics of space, so changes in future space exploration should be far larger than present development. To take practical advantage of these characteristics, we must greatly reduce the costs of exploring and exploiting space. Finding ways to lower these costs is a principal objective of this book. (See Wertz and Larson [1996].)

1.4 Step 2: Preliminary Estimate of Mission Needs, Requirements, and Constraints

Having defined the broad objectives that the space mission is to achieve, we wish to transform them into preliminary sets of numerical requirements and constraints on the space mission's performance and operation. These requirements and constraints will largely establish the operational concepts that will meet our objectives. Thus, we must develop requirements which truly reflect the mission objectives and be willing to trade them as we more clearly define the space system.

To transform mission objectives into requirements, we must look at three broad areas:

- *Functional Requirements*, which define how well the system must perform to meet its objectives.

- *Operational Requirements*, which determine how the system operates and how users interact with it to achieve its broad objectives.

- *Constraints*, which limit cost, schedule, and implementation techniques available to the system designer.

The needs, requirements, and constraints for any specific mission will depend upon the mission itself and how we implement it. For example, the mission may be a commercial venture, a government scientific program, or a crash emergency program responding to dire need. Still, most space missions develop their requirements according to the basic characteristics in Table 1-5.

Establishing top-level mission requirements is extremely difficult, depending on mission needs and on the perceived complexity or cost of meeting them. Therefore, contrary to frequent practice, we should iterate the numerical requirements many times in the design process. The first estimate of mission requirements must come from the goals and objectives combined with some view of what is feasible. Often, we can reiterate or slightly modify requirements and specifications from previous missions, thus carrying over information known from those missions. Of course, we must be prepared to trade these requirements as we develop the mission concept, thereby avoiding the problem of keeping old and inappropriate requirements.

The next step in setting up preliminary mission requirements is to look for the "hidden agenda" discussed in Sec. 1.3 and Chap. 2. This agenda contains the developer's implicit goals and constraints. For example, the FireSat mission may need to be perceived as responding quickly to public demand. Thus, an extended R&D program to develop the most appropriate FireSat satellite may not be acceptable.

As discussed further in Chap. 21, we must recognize that developing a space mission depends on political, legal, and economic elements, as well as technology. Thus, the most appropriate solution must meet mission technical requirements and the developer's political and economic goals. For example, satellite systems for a small nation may use components built in that nation or develop some new components locally, even though they would cost less if bought in other countries. In this case, we would spend more money to meet a political constraint: using the space mission to

TABLE 1-5. **Examples of Top-Level Mission Requirements**. We typically subdivide these top-level requirements into more specific requirements applicable to specific space missions.

Requirement	Where Discussed	Factors which Typically Impact the Requirement	FireSat Example
FUNCTIONAL			
Performance	Chaps. 9,13	Primary objective, payload size, orbit, pointing	4 temperature levels 30 m resolution 500 m location accuracy
Coverage	Sec. 7.2	Orbit, swath width, number of satellites, scheduling	Daily coverage of 750 million acres within continental U.S.
Responsiveness	Sec. 7.2.3, Chap. 14	Communications architecture, processing delays, operations	Send registered mission data within 30 min to up to 50 users
Secondary Mission	Chap. 2	As above	4 temperature levels for pest management
OPERATIONAL			
Duration	Secs. 1.4, 10.5.2, 19.2	Experiment or operations, level of redundancy, altitude	Mission operational at least 10 years
Availability	Sec. 19.1	Level of redundancy	98% excluding weather, 3-day maximum outage
Survivability	Sec. 8.2	Orbit, hardening, electronics	Natural environment only
Data Distribution	Chaps. 13, 15	Communications architecture	Up to 500 fire-monitoring offices + 2,000 rangers worldwide (max. of 100 simultaneous users)
Data Content, Form, and Format	Chaps. 2, 9, 13, 14	User needs, level and place of processing, payload	Location and extent of fire on any of 12 map bases, average temperature for each 30 m^2 grid
CONSTRAINTS			
Cost	Chap. 20	Manned flight, number of spacecraft, size and complexity, orbit	< $20M/yr + R&D
Schedule	Secs. 1.3, 19.1, Chaps. 2, 12	Technical readiness, program size	Initial operating capability within 5 yrs, final operating capability within 6 yrs
Regulations	Sec. 21.1	Law and policy	NASA mission
Political	Sec. 21.1	Sponsor, whether international program	Responsive to public demand for action
Environment	Secs. 8.1, 21.2	Orbit, lifetime	Natural
Interfaces	Chaps. 14, 15	Level of user and operator infrastructure	Comm. relay and interoperable through NOAA ground stations
Development Constraints	Chap. 2	Sponsoring organization	Launch on STS or expendable; No unique operations people at data distribution nodes

develop and promote national engineering resources. The technical community often sets aside nontechnical considerations and regards them as less important or less real than technical constraints. But a successful mission design must include **all** requirements and constraints placed on the system.

Finally, we reiterate that preliminary mission requirements should be established subject to later trades. Mission designers often simply try to meet the procuring group's requirements and constraints, because not meeting them appears to be a strong competitive disadvantage. Consequently, designers may not modify them, even if changes could make the system cost less or perform better for a given cost. Section 3.3 and Chap. 4 detail this process of trading on system requirements to maximize performance vs. cost.

As an example, we consider the requirement for mission duration or spacecraft lifetime, which may or may not be the same. This parameter exemplifies the difficulty of establishing requirements. The length of the mission is often indefinite. We want to detect, identify, and monitor forest fires continuously at a reasonable cost per year. In practice, however, we must develop a system that meets this need and then deploy it with an established design life and, perhaps, a replenishment philosophy. The design life of the individual FireSat spacecraft will strongly affect cost and will determine the level of redundancy, propellant budgets, and other key system parameters. In principle, we would like to obtain a graph of spacecraft cost vs. design life as shown in Fig. 1-5. We could then easily compute the total expected cost per year for different design lives, as shown by the dashed line, and the minimum spacecraft cost per year. We could also assess technological obsolescence, or the point at which we wish to replace the spacecraft because of better or cheaper technology.

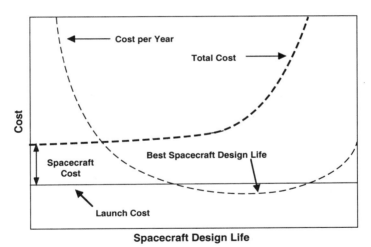

Fig. 1-5. Hypothetical Curve of Cost vs. Spacecraft Design Life. The cost per year is the total cost divided by the design life. In principle, we should use such curves to set the Spacecraft Design Life requirement. In practice, they rarely exist. See Sec. 20.5 for a Cost vs. Design Life curve for FireSat.

In practice, figures such as 1-5 are almost never done or, at best, are done qualitatively. The mission duration is normally assigned rather arbitrarily with a general perception of cost per year. Thus, there may be a push to produce spacecraft lasting 5 or 10 years because people believe these will be more economical than ones lasting only a few years. No matter how we choose the design life, we would like to go through the process described above for decisions about mission lifetime. If at all possible, it would be desirable to create a chart similar to Fig. 1-5 based on even crude estimates of spacecraft cost. Doing so provides a much stronger basis for establishing mission requirements and, e.g., determining whether we should push harder for a longer spacecraft lifetime or back off on this requirement to reduce spacecraft cost.

Having made a preliminary estimate of mission requirements and constraints, we proceed in Chap. 2 to define and characterize one or more baseline mission concepts. The issue of refining requirements and assessing how well objectives can be met is discussed in Chaps. 3 and 4.

References

Boden, Daryl G. and Wiley J. Larson. 1996. *Cost Effective Space Mission Operations.* New York: McGraw-Hill.

Davidoff, Martin. 1998. *The Radio Amateur's Satellite Handbook.* Newington, CT: American Radio Relay League.

Defense Systems Management College. 1990. *System Engineering Management Guide.* Ft. Belvoir, VA: U.S. Government Printing Office.

Kay, W.D. 1995. *Can Democracies Fly in Space? The Challenge of Revitalizing the U.S. Space Program.* Westport, CT: Praeger Publishing.

National Research Council. 1990. *Strategy for the Detection and Study of Other Planetary Systems.* Washington, DC: National Academy Press.

Przemieniecki, J.S. 1993. *Acquisition of Defense Systems.* Washington, DC: American Institute of Aeronautics and Astronautics, Inc.

Rechtin, Eberhardt. 1991. *Systems Architecting.* Englewood Cliffs, NJ: Prentice Hall.

Ruskin, Arnold M. and W. Eugene Estes. 1995. *What Every Engineer Should Know About Project Management (2nd Edition).* New York: Marcel Dekker, Inc.

Shishko, Robert. 1995. *NASA Systems Engineering Handbook.* NASA.

Wertz, James R. and Wiley J. Larson. 1996. *Reducing Space Mission Cost.* Torrance, CA: Microcosm Press and Dordrecht, The Netherlands: Kluwer Academic Publishers.

Chapter 2

Mission Characterization

James R. Wertz, *Microcosm, Inc.*
Richard P. Reinert, *Ball Aerospace Systems*

Mission characterization is the initial process of selecting and defining a space mission. The goal is to select the best overall approach from the wide range available to execute a space mission. Typically we wish to choose the lowest cost or the most cost-effective approach, and provide a traceable rationale that is intelligible to decision makers.

The initial process of mission characterization is discussed for general missions by Griffin and French [1991] and Pisacane and Moore [1994]. Elbert [1987, 1996] and Agrawal [1986] provide similar discussions for communications and geosynchronous satellites. Eckart [1996] and Woodcock [1986] discuss this process for manned missions and Wall and Ledbetter [1991] do so for remote sensing. Boden and Larson [1996] discuss initial characterization for mission operations and London [1994] provides a similar overview for launch vehicles, with a strong emphasis on reducing cost. Davidoff [1998] and Wertz and Larson [1996] discuss specific mechanisms applicable to low-cost and reduced cost missions.

The unconstrained number of mission options is huge, considering all possible combinations of orbits, launch systems, spacecraft, and mission concepts. The goal of this chapter is to prune this large number to a manageable level, without discarding options that offer significant advantages. We will do so by applying the requirements and constraints from Chap. 1 to pare down the list of alternatives. As an example, for most commercial communications applications, we would traditionally restrict ourselves to a geosynchronous orbit and only a few launch systems. However, the large number of low-Earth orbit communications constellations suggests that other options should be considered.

With requirements and constraints defined and alternative mission concepts selected, we must define each concept to the level required for meaningful comparisons. As Fig. 2-1 shows, we need to do this independently for each of the alternative mission concepts identified as "A" and "B" in the figure. Chapter 3 describes in more detail how we then evaluate the concepts, compare them in terms of cost and performance, and select one or more baselines. At the same time, we must keep track of the

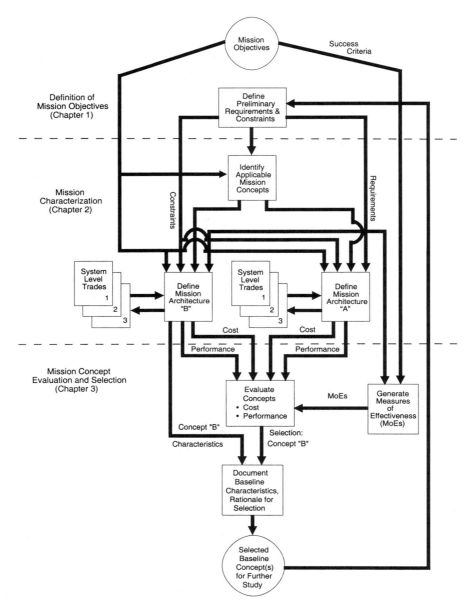

Fig. 2-1.   Concept Exploration Flow. One key for successfully implementing this flow is to iterate. Successive iterations through the flow will result in greater understanding, and can uncover critical requirements and system drivers.

element and system costs using the characteristics generated in the study and the techniques in Chap. 20. This procedure results in a *rough order of magnitude (ROM)* cost and an understanding of relative costs to support further trades and system evaluations.

In common use, "mission concept," "concept of operations," and "mission architecture" are frequently interchangeable and, at best, vaguely defined. Throughout this chapter, we wish to clearly distinguish between them. The *mission concept*, discussed in Sec. 2.1, is a broad statement of how the mission will work in practice. This should not be confused with *mission operations*, which provides the details of how people will operate and control the mission. The *mission architecture*, introduced in Sec. 2.2, is the mission concept **plus** a definition of each of the major elements of the mission.

2.1 Step 3: Identifying Alternative Mission Concepts

The broad mission concept is the most fundamental statement of how the mission will work—that is, how it gets its data or carries out the mission to satisfy the end user's needs. The mission concept as we are using it here consists of the four principal elements in Table 2-1. Notice that most of these elements are somehow associated with data or information. Except for manufacturing in space and a small number of other space payloads, most space missions are concerned fundamentally with the generation or flow of information. Thus, FireSat's mission is to generate and communicate to an end user information about forest fires. Communications satellites move data and information from one place to another. Weather, surveillance, and navigation satellites are all concerned with generating and communicating information. Thus, data flow is central to most space missions. How will FireSat determine where a fire is and how big it is? How will the system communicate that information to the fire fighter in a truck or plane? Once we answer these broad questions, we begin to understand FireSat's abilities and limits.

TABLE 2-1.   Elements of the Mission Concept of Operations. See Table 2-2 for a list of key trades and where discussed. Note that we discuss communications architecture in Sec. 13.1.

Element	Definition	FireSat Example
Data Delivery	How mission and housekeeping data are generated or collected, distributed, and used	How is imagery collected? How are forest fires identified? How are the results transmitted to the fire fighter in the field?
Communications Architecture	How the various components of the system talk to each other	What communications network is used to transmit forest fire data to the users in the field?
Tasking, Scheduling, and Control	How the system decides what to do in the long term and short term	What sensors are active and when is data being transmitted and processed? Which forested areas are receiving attention this month?
Mission Timeline	The overall schedule for planning, building, deployment, operations, replacement, and end-of-life	When will the first FireSat become operational? What is the schedule for satellite replenishment?

As Table 2-2 shows, defining the mission concept consists of defining the various options that are available and then selecting the most appropriate. Section 2.2 describes how we define options and take a first cut at the broad choices available to us. The process of selecting among them described in Sec. 3.2 is called *system trades*. Here we are interested in what these trades are and what some of the broader alternatives are to generate and transmit data. The process of defining how to transmit the data between the spacecraft and various users and controllers on the ground is called the *communications architecture* and is discussed in Chap. 13.

TABLE 2-2.   Process for Defining the Mission Concept of Operations. See Table 2-1 for definitions and FireSat example.

Step	Key Trades	Where Discussed
1. Define data delivery process for – Mission and housekeeping data	Space vs. ground processing Level of autonomy Central vs. distributed processing	Sec. 2.1.1 Chap. 13

2. Define tasking, scheduling, and control for
  – Mission and housekeeping data
  – Long term and short term
	Level of autonomy
	Central vs. distributed control
		Sec. 2.1.2

3. Define communications architecture for – Mission and housekeeping data	Data rates bandwidth Timeliness of communications	Sec. 13.1

4. Define preliminary mission timeline for
  – Concept development
  – Production and deployment
  – Operations and end-of-life
	Replenishment and
	end-of-life options
	Deployment strategy for
	multiple satellites
	Level of timeline flexibility
		Sec. 2.1.3

5. Iterate and document	N/A	N/A

The mission timeline differs from other elements of the mission concept in Table 2-1. It represents the overall schedule for developing, planning, and carrying out the mission. This defines whether it is a one-time only scientific experiment or long-term operational activity which will require us to replace and update satellites. In either case, we must decide whether the need for the mission is immediate or long term. Should we give high priority to near-term schedules or allow more extensive planning for the mission? Of course, much of this has to do with the funding for the mission: whether money is available immediately or will be available over time as we begin to demonstrate the mission's usefulness.

2.1.1 Data Delivery

Space missions involve two distinct types of data—mission data and housekeeping data. *Mission data* is generated, transmitted, or received by the mission payload. This is the basic information that is central to what the mission is all about. For FireSat, this data starts out as infrared images on a focal plane and ends up as the latitude, longitude, and basic characteristics of a forest fire transmitted to a fire fighter on the ground. The mission data has potentially very high data rates associated with it. However, the need for this data may be sporadic. Thus, FireSat may generate huge quantities of raw data during periods of time that it is passing over the forests, but there is little need for this same level of data when it is over the poles or the oceans.

Ultimately, the processed mission data may go directly to the end user or through ground stations and communication networks associated with mission operations. This will, of course, have a fundamental effect on how the mission works. In the first case, FireSat would process its imagery and send the forest fire information as it is being observed to the fire fighters in the field. In the second case, data would go instead to an operations center, where a computer system or human operators would evaluate it, compare it with previous data, and determine the location and characteristics of a forest fire. Then, the operations center would transmit this information to the fire fighters in the field. The result is about the same in both cases, but the system's abilities, limits, characteristics, and costs may be dramatically different.

In contrast to the mission data, *housekeeping data* is the information used to support the mission itself—the spacecraft's orbit and attitude, the batteries' temperature and state of charge, and the status and condition of the spacecraft's parts. Unlike the mission data, which is typically sporadic and may have huge data rates, the housekeeping data is usually continuous and at a low data rate. Continuously monitoring system performance does not require much information transfer by modern standards. In addition, rather than going to the end user, housekeeping data goes to the system monitoring and control activity within mission operations. Although housekeeping and mission data are distinct, we often need housekeeping data to make the mission data useful. For example, we must know the spacecraft's orbit and attitude to determine the ground lookpoint of the payload sensors and thereby locate the fire.

For both mission and housekeeping data, the data delivery system should be an information management-oriented process. We want to take a large amount of raw data, frequently from a variety of sensors, and efficiently transform it into the information the users need and provide it to them in a timely manner. We do not know at first whether sending FireSat data directly to the field or sending it first to a mission operations center for interpretation and analysis is the best approach. But we do know our choice will dramatically affect how well FireSat works and whether or not it is an efficient and effective system.

The principal trades associated with data delivery are:

- *Space vs. ground*—how much of the data processing occurs on board the spacecraft vs. how much is done at mission operations or by the end user?

- *Central vs. distributed processing*—is one computer talking to another computer, or does one large central computer on the spacecraft or on the ground process everything?

- *Level of autonomy*[*]—how much do people need to intervene in order to provide intelligent analysis and minimize costs?

These trades are strongly interrelated. Thus, autonomy is important by itself, but is also a key element of the space vs. ground trade. If human intervention is required (i.e., it can't be done autonomously), then the process must be done on the ground—or it must be a very large spacecraft. We will discuss each of these trades below after we have looked at the data delivery process as a whole. Autonomy is discussed in Sec. 2.1.2, because it is also critical to tasking and control.

The best way to start looking at the data-delivery problem is a *data-flow analysis*. This defines where the data originates and what has to happen to it before it gets to where it needs to go. To examine the data flow we can use a *data-flow diagram* as shown in Fig. 2-2 for the FireSat mission. A data-flow diagram lets us outline the tasks which we need to do, even though we don't understand yet how we will do most of them. For FireSat we know that we need some type of information collection, probably

---

[*] The language here can be confusing. An *autonomous operation* runs without human intervention. An *autonomous spacecraft* runs without intervention from outside the spacecraft.

a camera or imager or some other mechanism for detecting fires. As shown across the top row of Fig. 2-2, this imaging information must be digitized, probably filtered in some fashion, and transferred to a map of forest regions. We must then interpret the image to identify whether a fire exists, incorporate the results on a map, and distribute the map to the end user.

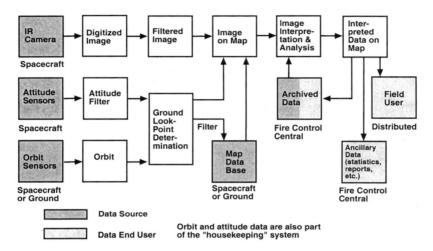

Fig. 2-2.   FireSat Data-Flow Diagram. The purpose of the data flow is to view the space mission from a data-oriented perspective. We want to know where the data comes from, what processing must be done, and where the results are used. Our principal mission objective is to provide the necessary data to the end user at minimum cost and risk.

To put the image on a map, we need to determine the spacecraft's orbit and attitude. The attitude will almost certainly be determined on board. The orbit may be determined either on board or by observations from the ground. In either case, the orbit and attitude information are combined to determine where on the ground the sensor is looking. We then select the map corresponding to the area we are looking at so we can correlate the sensor data with some physical location the fire fighters recognize.

Even though we are not certain yet how the data will be used, we can be fairly sure that our end data from FireSat will have several applications other than immediate use by the fire fighters. We will want to archive it in some central location for record-keeping and improving our capacity to analyze and interpret future data. Finally, we will sort out a large amount of ancillary data, such as statistics, reports, and forest-management information, and use it over an extended period. The need for this data does not have the real-time demand of the fire data itself.

The importance of the data-flow diagram is that it lets us see what has to happen in order to make our mission work. For FireSat, we need to combine the mission sensor with orbit and attitude information in order to make our system work in real time. However, the most difficult step is probably the one labeled "Image Interpretation and Analysis." Can an automated system quickly detect forest fires and send information directly to the user, or do we need extensive interpretation and analysis by trained people in mission operations? What type of experiments or information must we have to determine which of these is possible? Even after we have selected an approach, we should revisit it regularly to see that it still makes sense. If we decide that FireSat's real-time requirements demand data processing in a computer on board the spacecraft, we may dramatically drive up the cost because onboard processing is expensive. Our mission analysis may result in an automated FireSat which costs several times the annual budget of the Forest Service. If so, we need to reconsider whether it would be more economical to have an analyst on the ground interpret the data and then simply phone the results to an appropriate fire station. The data-flow diagram is valuable in helping to identify and track these central issues.

We will now look at two of the three principal trades associated with data delivery: space vs. ground processing and central vs. distributed processing. Section 2.1.2 and Chap. 23.3 discuss the level of autonomy.

Space vs. ground processing trades. In most earlier space missions, ground stations processed nearly all of the data because spaceborne processors could not do much. Chapter 16 describes several reasons onboard processing lags ground processing. But many onboard processors are now available with dramatically increased capacity. Consequently, a major trade for future missions is how much to process data on board the spacecraft vs. on the ground, either at a mission-operations facility or with the end user.

Section 3.2 describes how we undertake these and other system trades and compare the results. The main issues in the space vs. ground trade are as follows:

1. *Autonomy*—how independent do we want the system to be of analysis and control by a mission operator? If evaluation by people is critical, we must do much of the data processing on the ground. If autonomous processing is appropriate, it can be done on board the spacecraft, at a central ground facility, or among the end users. The level of autonomy is both a key trade in its own right and an element of the space vs. ground trade.

2. *Data latency*—how late can the data get to the end user? If we are allowed only fractions of a second, we must go to automated processes, probably on board the spacecraft. For FireSat, although we need the data in "near real time," the delays associated with sending the data to the ground for processing are not critical.

3. *Communications bandwidth*—how much data needs to be transmitted? If we have large amounts of data from a sensor, we should process and compress it as near the source as possible. Bringing down all of the FireSat imaging data and then deciding what to process further on the ground will cause an enormous communications problem and will probably drive up the FireSat mission's cost needlessly.

4. *Single vs. multiple users*—if there are a large number of end users, as would be the case for FireSat, we may be able to save considerable money by doing a high level of processing on board the spacecraft and sending the results directly down to the individual users.

5. *Location of end user*—is the "end user" for any particular data element on the ground or in space? In a space-to-space relay or a system for providing automatic orbit maintenance, the end application is in space itself. In this case, sending data to the ground for processing and then returning the results to the space system can be very complex and costly. On the ground, the complexity of the system is strongly affected by whether there is one end user at the mission operations center or multiple, scattered users, as in the case of FireSat.

Even if we choose to process data mostly in space, the basic system design should allow us to obtain or recreate selected raw data for analysis on the ground. A fully automated FireSat should have some means to record or broadcast the raw imaging data, so mission planners and analysts can evaluate how well the system is working, fix problems, and plan alternative and better techniques for later missions.

Traditionally, space software has been much more expensive than ground software. This suggests that processing on the ground is generally lower cost than processing on board the spacecraft. We believe that this will change in the future and, therefore, software cost should not be a major trade element in the space vs. ground processing trade. The cost of software is a function of what is done and how reliable we need to make it, rather than where it is done. We can choose to make highly reliable software as nearly error-free as possible for our ground systems and this software will have the high cost inherent with most previous onboard software systems. On the other hand, simple software with many reusable components can be developed economically and used on the spacecraft as well as on the ground.

The space vs. ground processing trade will be a key issue and probably a significant stumbling block for most missions in the near future. For short-lived, nontime-critical missions, it will probably be more economical to work on the ground with little automation. For long-lived missions, or time-critical applications, we will have to automate the processing and then do space vs. ground trades to minimize the operation and end-user costs. In any case, we wish to use the data flow analysis to evaluate where the data is coming from and where it will be used. If possible, we would like to minimize the communication requirements and *associate* data (e.g., attach time or position tags) as early as possible after the relevant data has been created.

For FireSat the payload sensor generates an enormous amount of data, most of which will not be useful. One way to effectively deal with large amounts of raw data on board the spacecraft is to *compress* the data (i.e., reduce the amount of data to be stored or transmitted) prior to transmitting it to the ground. The data is then recreated on the ground using *decompression* algorithms. There is a variety of methods for compressing data, both lossless and lossy. *Lossless data compression* implies that no information is lost due to compression while *lossy compression* has some "acceptable" level of loss. Lossless compression can achieve about a 5 to 1 ratio whereas lossy compression can achieve up to 80 to 1 reduction in data. Many of the methods of data compression store data only when value changes. Other approaches are based on quantization where a range of values is compressed using mathematical algorithms or fractal mathematics. By using these methods, we can compress the data to a single algorithm that is transmitted to the ground and the image is recreated based on the algorithm expansion. With the use of fractals, we can even interpolate a higher resolution solution than we started with by running the fractal for an extended period of time [Lu, 1997]. We select a method for data compression based on its strengths and weaknesses, the critical nature of the data, and the need to recreate it exactly [Sayood, 1996].

When we transmit housekeeping data we would generally use lossless compression for several reasons. First, raw housekeeping data is not typically voluminous. Second, it is important that none of the data is lost due to compression. However, when we transmit an image we might easily use lossy compression. We could either preview the image using lossy compression of we could say that the recovered image is "good enough." Alternatively, a high resolution picture may have so much information that the human eye can not assimilate the information at the level it was generated. Again, in this case a lossy compression technique may be appropriate.

In the FireSat example, we might use a sensor on board the spacecraft that takes a digital image of the heat generated at various positions on the Earth. The digital image will be represented by a matrix of numbers, where each pixel contains a value corresponding to the heat at that point on the Earth's surface. (Of course, we will need some method, such as GPS, for correlating the pixel in the image to the location on the Earth.) If we assume that the temperature at each location or pixel is represented by 3 bits, we can distinguish eight thermal levels. However, if we set a threshold such that a "baseline" temperature is represented with a 0, we might find that over many portions of the Earth, without fire, the image might be up to 70% nominal or 0. This still allows for several levels of distinction for fires or other "hot spots" on the Earth. Rather than transmit a 0 data value for each cold pixel, we can compress the data and send only those pixel locations and values which are not 0. As long as the decompression software understands this ground rule, the image can be exactly recreated on the ground. In this case, we can reduce our raw data volume to the number of hot spots that occur in any given area.

Central vs. distributed processing. This is a relatively new issue, because most prior spacecraft did not have sufficient processing capability to make this a meaningful trade. However, as discussed above, the situation has changed. The common question now is, "how many computers should the spacecraft have?" Typically, weight and parts-count-conscious engineers want to avoid distributed processing. However, centralized processing can make integration and test extremely difficult. Because integration and test of both software and hardware may drive cost and schedule, we must seriously consider them as part of the processing trade.

Our principal recommendations in evaluating central vs. distributed processing are:

- Group like functions together
- Group functions where timing is critical in a single computer
- Look for potentially incompatible functions before assigning multiple functions to one computer
- Maintain the interface between groups and areas of responsibility outside of the computer
- Give serious consideration to integration and test before grouping multiple functions in a single computer

Grouping like functions has substantial advantages. For example, attitude determination and attitude control may well reside in the same computer. They use much of the same data, share common algorithms, and may have time-critical elements. Similarly, orbit determination and control could reasonably reside in a single navigation computer, together with attitude determination and control. These hardware and software elements are likely to be the responsibility of a single group and will tend to undergo common integration and testing.

In contrast, adding payload processing to the computer doing the orbit and attitude activities could create major problems. We can't fully integrate software and hardware until after we have integrated the payload and spacecraft bus. In addition, two different groups usually handle the payload and spacecraft bus activities. The design and manufacture of hardware and software may well occur in different areas following different approaches. Putting these functions together in a single computer greatly increases cost and risk during the integration and test process, at a time when schedule delays are extremely expensive.

Another problem which can arise from time to time is incompatible functions, that is, activities which do not work well together. One example would be sporadic, computationally-intensive functions which demand resources at the same time. Another example occurs when the initial processing of either spacecraft bus or payload sensors may well be an interrupt-driven activity in which the computer is spending most of its time servicing interrupts to bring in observational data. This could make it difficult for the same computer to handle computationally-intensive processing associated with higher-level activities. This can be accommodated either by having the functions handled in separate computers or using a separate I/O processor to queue data from the process with a large number of interrupts.

Finally, we must consider the groups who oversee different activities. Integration and test of any computer and its associated software will be much more difficult if two distinct groups develop software for the same computer. In this case, significant delays and risks can occur. This does not necessarily mean, however, that elements controlled by different groups cannot be accommodated in the same computer. One approach might be to have two engineering groups be responsible for development of specifications and ultimately for testing. The detailed specifications are then handed over to a single programming group which then implements them in a single computer. This allows a single group to be responsible for control of computer resources. Thus, for example, the orbit control and attitude control functions may be specified and tested by different analysis groups. However, it may be reasonable to implement both functions in a single computer by a single group of programmers.

2.1.2 Tasking, Scheduling, and Control

Tasking, scheduling, and control is the other end of the data-delivery problem. If the purpose of our mission is to provide data or information, how do we decide what information to supply, whom to send it to, and which resources to obtain it from? Many of the issues are the same as in data delivery but with several key differences. Usually, tasking and control involve very low data rates and substantial decision making. Thus, we should emphasize how planning and control decisions are made rather than data management.

Tasking and scheduling typically occur in two distinct time frames. *Short-term tasking* addresses what the spacecraft should be doing at this moment. Should FireSat be recharging its batteries, sending data to a ground station, turning to look at a fire over Yosemite, or simply looking at the world below? In contrast, *long-term planning* establishes general tasks the system should do. For example, in some way the FireSat system must decide to concentrate its resources on northwestern Pacific forests for several weeks and then begin looking systematically at forests in Brazil. During concept exploration, we don't need to know precisely how these decisions are made. We simply wish to identify them and know broadly how they will take place.

On the data distribution side, direct downlink of data works well. We can process data on board, send it simultaneously to various users on the ground, and provide a low-cost, effective system. On the other hand, direct-distributed control raises serious problems of tasking, resource allocation, and responsibility. The military community particularly wants distributed control so a battlefield commander can control resources to meet mission objectives. For FireSat, this would translate into the local rangers deciding how much resource to apply to fires in a particular area, including the surveillance resources from FireSat. The two problems here are the limited availability of resources in space and broad geographic coverage. For example, FireSat may have limited power or data rates. In either case, if one regional office controls the system for a time, they may use most or all of that resource. Thus, other users would have nothing left. Also, FireSat could be in a position to see fires in Yosemite Park and Alaska at the same time. So distributed control could create conflicts.

For most space systems, some level of centralized control is probably necessary to determine how to allocate space resources among various tasks. Within this broad resource allocation, however, we may have room for distributed decisions on what data to collect and make available, as well as how to process it. For example, the remote fire station may be interested in information from a particular spectral band which could provide clues on the characteristics of a particular fire. If this is an appropriate option, the system must determine how to feed that request back to the satellite. We could use a direct command, or, more likely, send a request for specific data to mission operations which carries out the request.

Spacecraft Autonomy. Usually, high levels of autonomy and independent operations occur in the cheapest and most expensive systems. The less costly systems have minimal tasking and control simply because they cannot afford the operations cost for deciding what needs to be done. Most often, they continuously carry on one of a few activities, such as recovering and relaying radio messages or continuously transmitting an image of what is directly under the spacecraft. What is done is determined automatically on board to save money. In contrast, the most expensive systems have autonomy for technical reasons, such as the need for a very rapid response (missile detection systems), or a problem of very long command delays (interplanetary missions). Typically, autonomy of this type is extremely expensive because the system must make complex, reliable decisions and respond to change.

Autonomy can also be a critical issue for long missions and for constellations, in which cost and reliability are key considerations. For example, long-duration orbit maneuvers may use electric propulsion which is highly efficient, but slow. (See Chap. 17 for details.) Thruster firings are ordinarily controlled and monitored from the ground, but electric propulsion maneuvers may take several months. Because monitoring and controlling long thruster burns would cost too much, electric propulsion requires some autonomy.

As shown in Fig. 2-3, autonomy can add to mission reliability simply by reducing the complexity of mission operations. We may need to automate large constellations for higher reliability and lower mission-operations costs. Maintaining the relative positions between the satellites in a constellation is routine but requires many computations. Thus, onboard automation—with monitoring and operator override if necessary—will give us the best results.

With the increased level of onboard processing available, it is clearly possible to create fully autonomous satellites. The question is, should we do so or should we continue to control satellites predominantly from the ground?

Three main functions are associated with spacecraft control: controlling the payload, controlling the attitude of the spacecraft and its appendages, and controlling the spacecraft orbit. Most space payloads and bus systems do not require real-time control except for changing mode or handling anomalies. Thus, the FireSat payload will probably fly rather autonomously until a command changes a mode or an anomaly forces the payload to make a change or raise a warning. Autonomous, or at least semi-autonomous payloads are reasonable for many satellites. There are, of course, exceptions such as Space Telescope, which is an ongoing series of experiments being run by different principal investigators from around the world. In this case, operators control

Traditional Approach

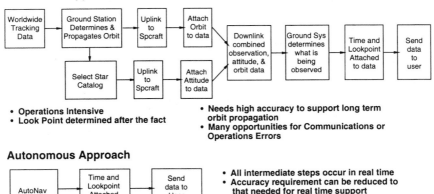

- Operations Intensive
- Look Point determined after the fact

- **Needs high accuracy to support long term orbit propagation**
- **Many opportunities for Communications or Operations Errors**

Autonomous Approach

- **All intermediate steps occur in real time**
- **Accuracy requirement can be reduced to that needed for real time support**
- **"Direct to User" data flow is both feasible & economical**

Fig. 2-3.   Comparison of Traditional vs. Autonomous Approach to Satellite Navigation.
Use of autonomous operations may significantly reduce mission complexity and thereby increase reliability.

the payload, but we would use some automatic operations to save money or to make the operator's job easier.

Controlling the attitude of the spacecraft and its appendages is done autonomously on board for nearly all satellites. Controlling the attitude from the ground is too expensive and too risky. The attitude control system on board most spacecraft provides various attitude control modes and can work over extended periods with little or no intervention from the ground.

Ground control has remained strongest in orbit maintenance and control, in which virtually all thruster firings intended to change the orbit are set up and enabled by ground command. This ground control will probably continue whenever large rocket engines produce orbit maneuvers such as when a kick stage moves the satellite from a parking orbit into a geosynchronous transfer orbit. Once in their operational orbit, however, many satellites either leave the orbit entirely uncontrolled or simply maintain the orbit at a given altitude or within a given window. In this case, low-thrust propulsion is both feasible and desirable because it is much less disturbing to the normal spacecraft environment. Low-thrust orbit maneuvers have been used on geosynchronous spacecraft for a long time so normal satellite operations can continue during the course of these stationkeeping maneuvers.

With low-thrust propulsion and current technology for autonomous navigation, autonomous orbit control is cheap, easy, and inherently less risky than autonomous attitude control. If the attitude control system stops working for even a short time, the spacecraft can have various potential problems, including loss of power, loss of command, and pointing of sensitive payloads toward the Sun. In contrast, if we lose low-thrust orbit control for a while, nothing disastrous happens to the spacecraft. The spacecraft proceeds in its orbit drifting slowly out of its predefined position. This is easily detected and corrected by the ground, assuming that the orbit control system didn't fail completely.

The major problem facing autonomous orbit control and, therefore, with autonomous satellites as a whole, is tradition. The ground does it mostly because it has always been done that way. However, there are some signs of change. Both UoSAT-12 and EO-1 are planning experiments in autonomous orbit control and several of the low-Earth orbit communications constellations have baselined autonomous orbit control to minimize both cost and risk.

Current satellite technology allows us to have fully autonomous, low-cost satellites. Autonomy can reduce cost and risk while enabling mission operations people to do what they do best—solve problems, handle anomalies, and make long-term decisions. We believe fully autonomous satellites, including autonomous orbit maintenance, will come about over the next decade as lower costs and risks, validated by on-orbit experiments, begin to outweigh the value of tradition.

2.1.3 Mission Timeline

The mission timeline is the overall schedule from concept definition through production, operations, and ultimately replenishment and end of life. It covers individual satellites and the whole system. Table 2-3 lists the mission timeline's main parts and where they are discussed. Notice that two distinct, potentially conflicting, demands can drive planning and production. One is the demand for a particular schedule or time by which the system must be operational. Thus, a Halley's Comet mission depends on launching a satellite in time to rendezvous with the comet. On the other hand, funding constraints frequently slow the mission and cause schedule gaps which add both further delays and cost. Of course, funding constraints can affect much more than timelines. They can determine whether we will do a mission, as well as its scope.

TABLE 2-3.   Principal Elements of the Mission Timeline. Key milestones in the mission or project timeline can have a significant effect on how the space system is designed and operated.

Element	Typically Driven By	Where Discussed
Planning and Development	Funding constraints System need date	Sec. 1.2, Chap. 1
Production	Funding constraints Technology development System need date	Chap. 12
Initial Launch	Launch availability System need date	Chap. 18
Constellation Build-up	Production schedule Launch availability Satellite lifetime	Sec. 7.6.1
Normal Mission Operations	Planned operational life Satellite lifetime (planned or failure constrained)	Chap. 14
Replenishment	Production schedule Launch availability Satellite lifetime (planned or failure constrained)	Sec. 19.1
End-of-Life Disposal	Legal and political constraints Danger to other spacecraft	Sec. 21.1

If the mission involves a constellation of satellites, a key timeline driver is the need to have the full constellation up for most of the satellite's lifetime. If a single satellite will last 5 years and we need a constellation of 50, we'll never get a full constellation with a launch rate of 5 per year. If having the full constellation is important, we must deploy the initial constellation within 20–25% of an individual satellite's lifetime. This schedule allows some margin for almost inevitable stretch-out as difficulties arise during the mission. If the constellation must remain complete, we need to plan for regular replenishment of satellites. We can replenish on a predefined timeline as satellites wear out or become technically obsolete, or we can respond to on-orbit failures or other catastrophic events which "kill" a particular satellite.

Two areas of the mission timeline typically do not receive adequate attention during concept exploration: performance with less than a full set of satellites while building up the constellation, and end-of-life disposal. In a constellation of satellites we would like to increase performance levels as we add satellites. If FireSat is a constellation, we want to achieve some protection from fires with the first satellite launch and more with each added launch until all satellites are in place. As described further in Sec. 7.6, designers of constellations often concentrate only on the full constellation's performance. However, the period of time before the constellation is brought fully into place can frequently be long and may well be a critical phase since a large fraction of the funding has been spent and yet full capability has not been achieved. Thus it is particularly important for constellation design to take into account the problem of performance with less than a full set of satellites. In addition, we want graceful degradation, so a satellite failure will still allow strong performance until we replace the failed satellite. These issues are important during concept exploration because they may significantly affect the design of the constellation and the entire system.

There is now growing concern with disposal of satellites after their useful life in orbit. We have already recognized this need for geosynchronous satellites because the geosynchronous ring is rapidly filling. But it is also very important for low-Earth orbit constellations in which debris and spent satellites left in the pattern can threaten the remaining constellation. Again, we must address this issue early in concept definition.

2.2 Step 4: Identifying Alternative Mission Architectures

A *mission architecture* consists of a mission concept plus a specific set of options for the eight mission elements defined in Sec. 1.2. Although we need all of the elements to define and evaluate a mission architecture, some are more critical than others in determining how the space mission will meet its objectives. Typically, we define a mission architecture by specifying the mission concept plus the subject, orbit, communications architecture, and ground system. These provide a framework for defining the other elements. Alternatively, we may define the architecture by specifying a unique approach to mission operations or a unique payload which then drives the definition of the remaining elements

Our goal is to arrive at a set of candidate architectures for further evaluation large enough to encompass all approaches offering significant advantages, but small enough to make the more detailed definition and evaluation manageable. Table 2-4 summarizes the mechanism for doing this, which we describe below.

Step A. Identify the mission elements subject to trade. We begin by examining our basic mission concept and each of the eight mission elements in light of the requirements and constraints from Sec. 1.4 to determine which have more than one option.

TABLE 2-4.   **Process Summary for Identifying Alternative Mission Architectures.** This highly creative endeavor can have a significant impact on mission cost and complexity.

Step	Where Discussed
A. Identify the mission elements subject to trade.	Table 2-5
B. Identify the main options for each tradeable element.	Table 2-6
C. Construct a trade tree of available options	}Fig. 2-4,
D. Prune the trade tree by eliminating unrealistic combinations.	}Table 2-7
E. Look for other alternatives which could substantially influence how we do the mission.	Chap. 22

Usually this step greatly reduces the number of tradeable elements. Table 2-5 summarizes this process for FireSat. The FireSat mission has multiple options that will affect
not only cost but also performance, flexibility, and long-term mission utility. Thus, for
this mission we should carry through several different options so the decision-making
audience can understand the main alternatives.

TABLE 2-5.   **Selecting FireSat Elements Which can be Traded.** Many options exist for FireSat,
not all of which are compatible with each other.

Element of Mission Architecture	Can be Traded	Reason
Mission Concept	Yes	Want to remain open to alternative approaches
Subject	No	Passive subject is well defined
Payload	Yes	Can select complexity and frequencies
Spacecraft Bus	Yes	Multiple options based on scan mechanism and power
Launch System	Cost only	Choose minimum cost for selected orbit
Orbit	Yes	Options are low, medium, or high altitude with varying number of satellites
Ground System	Yes	Could share NOAA control facility, use dedicated FireSat facility, or direct downlink to users
Communications Architecture	No	Fixed by mission operations and ground system
Mission Operations	Yes	Can adjust level of automation

Table 2-5 lists one of the options as "Cost only," meaning that the trade depends
mainly on cost and only secondarily on how or how well the mission is accomplished.
An example would be the launch system, for which the main concern normally is what
launch vehicle will get the spacecraft into orbit at the lowest cost. Still, these trades
may be important in selecting the mission concept. For example, a major increase in
the launch cost may outweigh being able to use a smaller number of identical satellites
in a higher orbit.

Step B. Identify the main options for each tradeable element. Although in theory we
have almost an unlimited number of options, we normally draw them from a limited
set such as those in Table 2-6. Thus, we first choose options that apply to our mission
and then look for special circumstances which may lead us to consider alternatives not
listed in the table.

TABLE 2-6.   Common Alternatives for Mission Elements. This table serves as a broad checklist for identifying the main alternatives for mission architectures.

Mission Element [Where Discussed]	Option Area	Most Common Options	FireSat Options
Mission Concept [Sec. 2.1]	Data delivery	Direct downlink to user, automated ground processing, man-in-the-loop processing and transmission	Direct downlink or through mission control
	Tasking	Ground commanding, autonomous tasking, simple operations (no tasking required)	Simple operation or ground commands
Controllable Subject [Secs. 2.3, 13.4, 22.3]	Selection	Standard ground stations, private TV receivers, ship or aircraft transceivers, special purpose equipment	N/A [See Sec. 9.3]
	Performance	EIRP, G/T (See 13.3 for definitions)	
	Steering	Fixed or tracking	
Passive Subject [Sec. 2.3]	What is to be sensed	Subject itself, thermal environment, emitted radiation, contrast with surroundings	Heat or visible light; chemical composition; particles
Payload [Chaps. 9, 13] (some items may not apply, depending on mission type)	Frequency	Communications: normal bands Observations: IR, visible, microwave Radar: L, C, S bands, UHF	IR, visible
	Complexity	Single or multiple frequency bands, single or multiple instruments	Single or multiple bands
Payload	Size vs. sensitivity	Small aperture with high power (or sensitivity) or vice versa	Aperture
Spacecraft Bus [Chap. 10]	Propulsion	Whether needed; cold gas, monopropellant, bipropellant	Determined by definition of payload and orbit
	Orbit control	Whether needed, onboard vs. ground	
	Navigation	Onboard (GPS or other) vs. ground-based	
	Attitude determination and control	None, spinning, 3-axis; articulated payload vs. spacecraft pointing; actuators and sensors	
	Power	Solar vs. nuclear or other; body-mounted vs. 1- or 2-axis pointed arrays	
Launch System [Chap. 18]	Launch vehicle	SSLV, Atlas, Delta, STS, Titan, Pegasus, Ariane, other foreign	Determined by definition of spacecraft and orbit
	Upper stage	Pam-D, IUS, TOS, Centaur, integral propulsion, other foreign	
	Launch site	Kennedy, Vandenberg, Kourou, other foreign	
Orbit [Chaps. 6, 7]	Special orbits	None, geosynchronous, Sun-synchronous, frozen, Molniya, repeating ground track	Single GEO satellite, low-Earth constellation
	Altitude	Low-Earth orbit, mid-altitude, geosynchronous	
	Inclination	0°, 28.5°, 57°, 63.4°, 90°, 98°	
	Constellation configuration	Number of satellites; Walker pattern, other patterns; number of orbit planes	Min. inclination dependent on altitude

TABLE 2-6.   Common Alternatives for Mission Elements. (Continued) This table serves as a broad checklist for identifying the main alternatives for mission architectures.

Mission Element [Where Discussed]	Option Area	Most Common Options	FireSat Options
Ground System [Chap. 15]	Existing or dedicated	AFSCN, NASA control center, other shared system, dedicated system	Shared NOAA system, dedicated system
Communications Architecture [Chap. 13]	Timeliness	Store and dump, real-time link	Either option
	Control and data dissemination	Single or multiple ground stations, direct to user, user commanding, commercial links	1 ground station; commercial or direct data transfer
	Relay mechanism	TDRSS, satellite-to-satellite crosslinks, commercial communications relay	TDRS or commercial
Mission Operations [Chap. 14]	Automation level	Fully automated ground stations, part-time operations, full-time (24-hr) operations	Any of the options listed
	Autonomy level	Full ground command and control, partial autonomy, full autonomy (not yet readily available)	

Steps C and D. Construct and prune a trade tree of available options. Having identified options we next construct a *trade tree* which, in its simplest form, is a listing of all possible combinations of mission options. Mechanically, it is easy to create a list of all combinations of options identified in Step B. As a practical matter, such a list would get unworkably long for most missions. As we construct the trade tree we need to find ways to reduce the number of combinations **without** eliminating options that may be important.

The first step in reducing the number of options is to identify the system drivers (as discussed in Sec. 2.3) and put them at the top of the trade tree. The *system drivers* are parameters or characteristics that largely determine the system's cost and performance. They are at the top of the trade tree because they normally dominate the design process and mandate our choices for other elements, thus greatly reducing our options.

The second step in reducing options is to look for trades that are at least somewhat independent of the overall concept definition or which will be determined by the selection of other elements. For example, the spacecraft bus ordinarily has many key options. However, once we have defined the orbit and payload, we can select the spacecraft bus that meets the mission requirements at the lowest cost. Again, although bus options may not be in the trade tree, they may play a key role in selecting workable mission concepts because of cost, risk, or schedule.

The third tree-pruning technique is to examine the tree as we build it and retain only sensible combinations. For example, nearly any launch vehicle above a minimum size will work for a given orbit and spacecraft. Because cost is the main launch-vehicle trade, we should retain in the trade tree only the lowest-cost launch vehicle that will fulfill the mission. This does not mean that we will ultimately launch the spacecraft on the vehicle listed in the trade tree. Instead, we should design the spacecraft to be compatible with as many launch vehicles as possible and then select the vehicle based on cost (which may well have changed) when we are deciding about initial deployment.

Steps C and D produce a trade tree such as the one for FireSat in Fig. 2-4. Our goal is to retain a small number of the most promising options to proceed to more detailed definition. For each option we will have selected most, though not necessarily all, of the elements shown in Table 2-7 for options 1 and 6. Of course we should reevaluate the trade tree from time to time as the system becomes more completely defined.

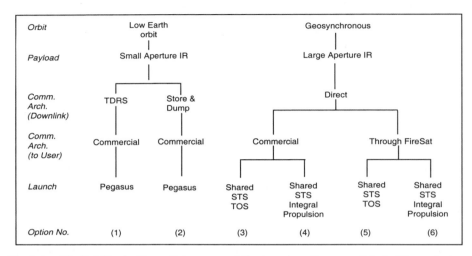

Fig. 2-4. **FireSat Trade Tree.** Only some of the launch options are listed. Other mission elements are largely independent of the trade tree options. The goal is to create drastically different options for comparison.

TABLE 2-7. **Two of the Six Preliminary FireSat Mission Concepts.** See the trade tree in Fig. 2-4 for the other options.

Element	Option 1	Option 6
Mission Concept	IR detection of fires with results put on a map and transmitted	IR detection of fires with results put on a map and transmitted
Subject	Characteristics defined by the specification	Characteristics defined by the specification
Payload	Small-aperture IR	Large-aperture IR
Spacecraft Bus	Small, 3-axis	Mid-size, 3-axis
Launch System	Pegasus	STS, integral propulsion
Orbit	Low-Earth, 2 satellites, 2 perpendicular polar planes	Geosynchronous, 1 satellite centered over west coast of U.S.
Ground System	Single, dedicated ground station	Single, dedicated ground station
Communications Architecture	TDRS data downlink; commercial links to users	Direct to station; results relayed to users via FireSat
Mission Operations	Continuous during fire season, partial otherwise	Continuous during fire season, partial otherwise

Step E. Look for other alternatives. Defining alternative architectures cannot be purely mechanical. For nearly any mission, we may find new and better ways of doing anything the basic elements do. A new, low-cost launch vehicle may dramatically change the available design alternatives. Alternative definitions of the subject or user may allow major performance improvements or cost reductions. For example, as an alternative mission concept we could detect forest fires by using small sensors spread throughout the forests being monitored and simply use the satellite system to interrogate the sensors and provide the data to the users. Chapter 22 further explores this low-cost alternative. The key point is that alternatives nearly always exist. We must look carefully for them and be willing to revise normal requirements and constraints to meet our fundamental mission objectives.

2.3 Step 5: Identifying System Drivers

System drivers are the principal mission parameters or characteristics which influence performance, cost, risk, or schedule **and** which the user or designer can control. For example, the temperature at which a forest fire burns may heavily influence how easily it can be detected; however, this is beyond the system designer's control and, therefore, is not a system driver. Correctly identifying the key system drivers is a critical step in mission analysis and design. Misidentifying system drivers is one of the most common causes of mission analysis error. For example, we may focus a lot of time and effort on getting the most coverage for an orbit when the system's ultimate performance depends mainly on data rates or cloud cover.

Table 2-8 lists the most common system drivers for space missions, along with what limits them, what they limit, and where they are discussed. The table helps us ensure that we do not overlook system drivers. In identifying these drivers we must clearly determine whether we are looking for drivers of *performance*, *cost*, *risk*, or *schedule*. These may be the same or different. To identify system drivers, we:

1. *Identify the Area of Interest*
   Explicitly identify the area of interest, typically performance, cost, risk, or schedule.

2. *Identify Parameters Which Measure the Area of Interest*
   Define numerical parameters which measure the identified area of interest. (See Sec. 3.4 on measures of effectiveness and performance parameters for more details on how to do this.) The important point is to find parameters which genuinely measure the goal rather than ones which simply are easy to compute.

3. *Develop First-Order Algorithms*
   Develop a formula or algorithm to express the first-order estimate for the value of the parameter identified above. This could include either system algorithms as defined in Sec. 3.1, or unique algorithms for the identified parameter. (See Table 2-9 for the FireSat example.)

4. *Examine the Factors*
   Examine each of the factors in the expression identified above. Those which can be adjusted and which have the strongest effect on results are the system drivers.

5. *Look for Possible "Hidden Drivers"*
   Examine each of the first-order algorithms for implicit variables or for factors affecting more than one characteristic. For example, altitude will influence the ground resolution of a given instrument, the area covered by the field of view, and the spacecraft's velocity relative to the Earth. Therefore, it will more strongly influence effective area search rates than a single formula may show.

TABLE 2-8.   Common System Drivers. System drivers can frequently be identified by examining the parameters in this list.

Driver	What Limits Driver	What Driver Limits	Where Discussed
Size	Shroud or bay size, available weight, aerodynamic drag	Payload size (frequently antenna diameter or aperture)	Chaps. 9, 10
On-orbit Weight	Altitude, inclination, launch vehicle	Payload weight, survivability; largely determines design and manufacturing cost	Sec. 10.4.1
Power	Size, weight (control is secondary problem)	Payload & bus design, system sensitivity, on-orbit life	Secs. 10.2, 11.4
Data rate	Storage, processing, antenna sizes, limits of existing systems	Information sent to user; can push demand for onboard processing	Sec. 13.3
Communications	Coverage, availability of ground stations or relay satellites	Coverage, timeliness, ability to command	Sec. 7.2, Chap. 13
Pointing	Cost, weight	Resolution, geolocation, overall system accuracy; pushes spacecraft cost	Sec. 5.4
Number of Spacecraft	Cost	Coverage frequency, and overlap	Secs. 7.2, 7.6
Altitude	Launch vehicle, performance demands, weight	Performance, survivability, coverage (instantaneous and rate), communications	Secs. 3.3, 7.1, 7.4, 7.6
Coverage (geometry and timing)	Orbit, scheduling, payload field of view & observation time	Data frequency and continuity, maneuver requirements	Secs. 5.2, 7.2
Scheduling	Timeline & operations, decision making, communications	Coverage, responsiveness, mission utility	Sec. 3.2.4, Chap. 14
Operations	Cost, crew size, communications	Frequently principal cost driver, principal error source, pushes demand for autonomy (can also save "lost" missions)	Chap. 14

The way we have defined our particular problem, or which parameters are available to us, may affect our list of system drivers. Thus, defining system drivers depends in part on the physical and technical nature of the problem and in part on the constraints imposed on the mission analyst. Usually, we want to make these constraints explicit, so we will know which variables are available for adjustment and which are assumed to be given. Table 2-9 shows the major performance drivers for FireSat.

2.4 Step 6: Characterizing the Mission Architecture

Once we have established alternative mission concepts, architectures, and system drivers, we must further define the mission concepts in enough detail to allow meaningful evaluations of effectiveness. For concept exploration, the steps in this process correspond to the space mission elements. Figure 2-5 illustrates the sequence of activities and shows schematically the major interactions between the steps, as well as primary trade study areas and their interactions with main elements of the process. The steps are described below and summarized in Table 2-10.

TABLE 2-9. **Identification of Performance Drivers for FireSat.** First-order algorithms are given to allow us to estimate the performance drivers. Definition of performance drivers may change as we create more detailed definitions of the system and system algorithms. Comparison of columns two and three shows that the performance drivers may depend on the mission concept used.

Key Parameters	First Order Algorithm (Low-Earth Orbit)	First Order Algorithm (Geosynchronous)	Performance Drivers
Observation Frequency	(Number of spacecraft)/ 12 hr	Scan frequency	Number of spacecraft for low orbit
Time Late	Onboard storage delay + processing time	Communications + processing time	Storage delay (if applicable)
Resolution	Distance ×[(wavelength/ aperture) + control error]	Distance ×[(wavelength/ aperture) + control error]	Altitude, aperture, control accuracy
Observation Gap	Cloud cover interval or coverage gap	Cloud cover interval or coverage gap	None (weather dominated)

TABLE 2-10. **Summary of the Concept Characterization Process.** See text for details. See Fig. 2-5 for a typical process flow.

	Step	Where Discussed
A	Define the preliminary mission concept	Chap. 2
B	Define the subject characteristics	Chap. 9
C	Determine the orbit or constellation characteristics	Chap. 7
D	Determine payload size and performance	Chap. 9, 13
E	Select the mission operations approach	
	• Communications architecture	Chap. 13
	• Operations	Chap. 14
	• Ground system	Chap. 15
F	Design the spacecraft bus to meet payload, orbit, and communications requirements	Chap. 10
G	Select a launch and orbit transfer system	Chap. 18
H	Determine deployment, logistics, and end-of-life strategies	Secs. 7.6, 19.1, 21.2
I	Provide costing support	Chap. 20
J	Document and Iterate	Chap. 20

A. Define the Preliminary Mission Concept (Chapter 2)

As described in Sec. 2.1, the key elements are data delivery; tasking, scheduling, and control; communications architecture; and mission timeline. We begin with a broad concept and refine this concept as we define the various mission elements in the steps below. (See Tables 2-1 and 2-2 for a further definition of these elements and how to define them.)

B. Define the Subject Characteristics (Chapter 9)

We can divide space missions into two broad categories. One services other system elements, typically on the user premises, such as Comsat ground stations or GPS navigation receivers. The other category senses elements that are not a part of the mission system, such as the clouds observed by weather satellites. Our first step in defining the system elements (Chap. 9) is to determine the subject's key characteristics.

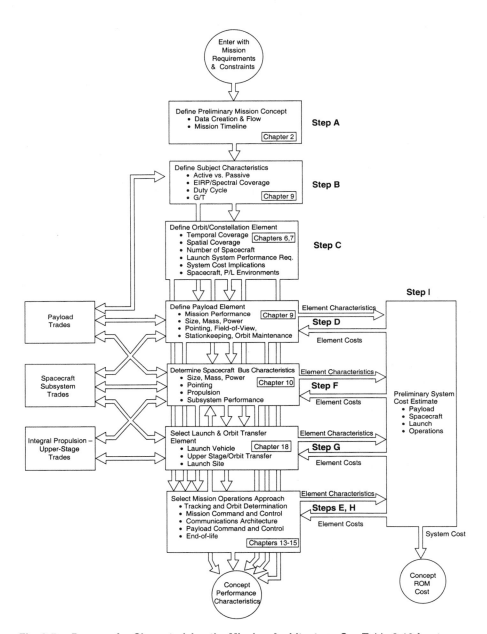

Fig. 2-5.   Process for Characterizing the Mission Architecture. See Table 2-10 for steps.

If a mission interacts with user equipment, we must define the subject characteristics either from known information for well-established services or by a trade study involving the rest of the system. The parameters for specifying passive subjects are largely the same as those for specifying user elements, except that we don't have a "receiver" to characterize, and the effective isotropic radiated power (EIRP) specification for the transmitter is replaced by definition of the object's emission intensity as a function of bandwidth. Table 2-11 summarizes the characteristics of both types of elements.

TABLE 2-11.   Summary of Main Characteristics of Space Mission Subjects. See Chap. 13 for definitions of communications parameters.

Controllable Subjects	Passive Subjects
1. Quantity	1. Quantity
2. Location or range	2. Location or range
3. Transmitter EIRP	3. Emission intensity (W/sr) as a function of frequency or spectral band
4. Receiver G/T	
5. Frequency and bandwidth	4. Needed temporal coverage (duty cycle)
6. Duty cycle	

C. Determine the Orbit and Constellation Characteristics (Chapter 7)

The mission orbit profoundly influences every part of space mission development and operation. Combined with the number of spacecraft, it determines all aspects of space-to-ground and ground-to-space sensor and communication coverage. For the most part, the orbit determines sensor resolution, transmitter power, and data rate. The orbit determines the spacecraft environment and, for military spacecraft, strongly influences survivability. Finally, the orbit determines the size and cost of the launch and delivery system.

Chapter 7 gives detailed directions for orbit design. As Table 2-12 shows, the design should include parameters for the mission and transfer orbits, propellant requirements, and constellation characteristics.

D. Determine the Payload Size and Performance (Chapters 9 and 13)

We next use the subject characteristics from Step 2 and orbit characteristics from Step 3 to create a mission payload concept. We can divide most mission payloads into six broad categories: observation or sensing, communications, navigation, in situ sampling and observations, sample return, and crew life support and transportation. More than 90% of current space-system payloads observe, sense, or communicate. Even the navigation payloads are basically communications payloads with ancillary data processing and stable time-base equipment to provide the navigation signal. Detailed directions for sizing and definition appear in Chap. 9 for observation payloads and in Chap. 13 for communications payloads. Table 2-13 summarizes the key parameters we need to specify.

System-level payload trades typically involve the user element, selecting a mission orbit, and allocating pointing and tracking functions between the payload and spacecraft elements. User element trades involve balancing the performance of the payload and elements on the user's premises to get the lowest overall system cost for a given orbit and constellation design. As an example, if a single geosynchronous spacecraft must service thousands of ground stations, as for direct broadcast TV, we would minimize the system cost by selecting a large, powerful spacecraft that can broadcast to simple and inexpensive ground stations. A system designed for trunkline communication between half a dozen ground stations uses more complex and capable ground systems and saves cost with simpler spacecraft.

TABLE 2-12.   Summary of Orbit and Constellation Characteristics. See text for discussion.

1 Altitude
2 Inclination
3 Eccentricity
4 Argument of perigee for noncircular orbits
5 ΔV budget for orbit transfer
6 ΔV budget for orbit maintenance
7 Whether orbit will be controlled or uncontrolled
8 Number and relative orientation of orbit planes (constellations)
9 Number and spacing of spacecraft per orbit plane (constellations)

TABLE 2-13.   Summary of Mission-Payload Characteristics. For multiple payloads, we must determine parameters for each payload.

1. Physical Parameters
1.1 Envelope dimensions
1.2 Mass properties
2. Viewing and Pointing
2.1 Aperture size and shape
2.2 Size and orientation of clear field of view required
2.3 Primary pointing direction*
2.4 Pointing direction range and accuracy required
2.5 Tracking or scanning rate
2.6 Pointing or tracking duration and duty cycle
3. Electrical Power
3.1 Voltage
3.2 Average and peak power
3.3 Peak power duty cycle
4. Telemetry and Commands
4.1 Number of command and telemetry channels
4.2 Command memory size and time resolution
4.3 Data rates or quantity of data
5. Thermal Control
5.1 Temperature limits (operating/non-operating)
5.2 Heat rejection to spacecraft (average/peak wattage/duty cycle)
*e.g., Sun, star, nadir, ground target, another spacecraft

Payload vs. orbit trades typically try to balance the resolution advantages of low altitudes against the fewer spacecraft needed for the same coverage at higher altitudes. The counterbalancing factor is that we need a sensor with a larger aperture and better sensitivity to obtain the same resolution at higher altitudes; the more capable sensor costs more and needs a larger spacecraft and launch system.

Payload vs. spacecraft trades usually try to meet pointing and tracking requirements at the lowest cost. At one extreme, the payload does all the pointing independently of the spacecraft attitude; an example is the use of gimballed scan platforms on the JPL Mariner MK-II spacecraft. At the opposite extreme, Space Telescope and Chandra X-Ray Observatory point the entire spacecraft with the required level of accuracy. An intermediate approach used on RME points the entire spacecraft to a lower level of accuracy, allowing the payload to do fine pointing over a limited field of regard.

E. Select the Mission Operations Approach (Chapters 13–15)

We next select and size the elements needed to support communications and control of the spacecraft and payload. Table 2-14 gives the key parameters. Typically a *mission operations control center* commands and controls the spacecraft and delivers data to the user. With rare exceptions, we would choose an existing control center, based on the user's needs, downlink data rates, and, in some cases, security considerations. Both NASA and the Air Force have existing systems. Particular institutions, such as Intelsat or Comsat, use custom systems. Most commercial operators employ system-peculiar control centers. If needed, we can interconnect most systems with different options for relaying communications. Chapter 15 details the specification, selection, and design of this element.

TABLE 2-14.   Summary of Mission Operations Characteristics.

1. Communications Architecture
1.1   Number and distribution of ground stations
1.2   Downlink and uplink path design
1.3   Crosslink characteristics, if used
1.4   Relay satellites used
1.5   Communications link budget
1.6   Space-to-ground data rates
2. Ground System
2.1   Use of existing or dedicated facilities
2.2   Required transmit and receive characteristics
2.3   Required data handling
3. Operations
3.1   Level of automation
3.2   Software lines of code to be created
3.3   Full-time or part-time staffing
3.4   Number of personnel
3.5   Amount of commanding required
3.6   Timeliness of data distribution

The *communications architecture* transfers the required mission data (payload and housekeeping data) from the spacecraft down to the mission operations control center. In addition, we must send commands back to the spacecraft, and meet other requirements such as encryption. Thus, we select the communications relay elements along with the mission control system after most payload and orbit trades are complete. Typical options are SGLS for Air Force missions or TDRSS/GSTDN with the NASA mission control centers. Custom systems are required for some applications and are commonly used for commercial missions in geosynchronous orbit. Chapter 13 describes communications architectures, and Chap. 14 treats operations.

F. Design the Spacecraft to Meet Payload, Orbit, and Communications Requirements (Chapter 10)

The spacecraft and its subsystems support the payload in the mission orbit—point it and supply power, command and data handling, and thermal control. They must be compatible with the communications architecture and mission-operations concept. These elements, along with the launch system, drive the spacecraft design. We usually choose the launch system that costs the least to place the minimum required weight in the mission or transfer orbit. Once we make this selection, the spacecraft's stowed configuration is constrained by the shroud volume of the selected vehicle or vehicles. Table 2-15 summarizes the items we need to specify while defining the spacecraft. Chapter 10 covers how we synthesize spacecraft concepts and their definition and sizing.

TABLE 2-15. Summary of Spacecraft Characteristics. See text for discussion.

1. General arrangement including payload fields of view (deployed and stowed)
2. Functional block diagram
3. Mass properties, by mission phase (mass and moments of inertia)
4. Summary of subsystem characteristics
4.1 Electrical power (conversion approach; array and battery size; payload power available, average/peak overall spacecraft power, orbit average, peak)
4.2 Attitude control (attitude determination and control components; operating modes; ranges and pointing accuracy)
4.3 Navigation and orbit control (accessing requirement, use of GPS; onboard vs. ground)
4.4 Telemetry and command (command/telemetry format; command and time resolution; telemetry storage capacity; number of channels by type)
4.5 Computer (speed and memory; data architecture)
4.6 Propulsion (amount and type of propellant; thruster or motor sizes)
4.7 Communications (link margins for all links; command uplink data rate; telemetry downlink data rates)
4.8 Primary structure and deployables
4.9 Unique thermal requirements
4.10 Timing (resolution and accuracy)
5. System parameters
5.1 Lifetime and reliability
5.2 Level of autonomy

A key spacecraft-versus-launch-system trade is the use of integral propulsion. Many commercial spacecraft ride the launch system to transfer orbit and then insert themselves into the mission orbit using an internal propulsion or an internal stage. Some DoD spacecraft, such as DSCS III and DSP, depend on a launch system with an upper stage for insertion directly into the mission orbit. They do not carry large integral propulsion subsystems. We should consider this trade whenever the spacecraft and payload cost enough to justify the reliability offered by an expensive upper stage.

Another trade between the spacecraft and launch system involves guidance of the upper stage. Often, the spacecraft control system can guide the upper stage, which may allow deletion of equipment from that stage, thereby increasing performance and lowering cost. This trade is particularly important when using three-axis-stabilized stages.

G. Select a Launch and Orbit Transfer System (Chapter 18)

The launch system and its upper stage need to deliver the spacecraft and payload to the mission orbit or to a transfer orbit from which the spacecraft can reach the mission orbit on its own. The chosen launch system usually determines the launch site. The launch site organization provides pre-launch processing, checkout, and installation to the launch system, usually on the launch pad.

Launch vehicles and upper stages may be combined in many ways to match almost any reasonable combination of payload and mission orbit. Chapter 18 details the characteristics and selection of launch systems. Selecting a launch system typically involves the trades with the spacecraft discussed above. In addition, we must decide between a single spacecraft launch and manifesting two or more spacecraft in a shared launch. In general, multiple manifesting costs less, but constrains the schedule. Finally, we should bring certain launch-system parameters to the system level design process: type of vehicle, cost per launch, and flow times for processing and prelaunch activities at the launch site.

H. Determine Logistics, Deployment, Replenishment, and Spacecraft Disposal Strategies (Sections 7.6, 19.1, and 21.2)

Logistics is the process of planning to supply and maintain the space mission over time. Whereas only military missions typically demand formal plans, the process described in Sec. 19.1 can strongly affect costs for any multi-year mission requiring extended support. Historically, most life-cycle costs have been locked in by the end of concept exploration, so we must evaluate operations, support, replenishment, and mechanisms during this phase.

Planners often overlook the sequence for building up and maintaining satellite constellations. To deploy a constellation effectively, we must create performance plateaus which allow us to deploy in stages and to degrade the system gracefully if individual satellites fail. These performance plateaus develop from the constellation design, as described in Sec. 7.6.

Section 21.2 describes the ever-increasing problem associated with *orbital debris,* consisting of defunct satellites and associated parts. Because of this problem, all new satellite designs should plan for deorbiting or otherwise disposing of satellites at the end of their useful life. In particular, satellites must be removed from areas such as the geostationary ring, where they would seriously threaten other spacecraft or any low-Earth orbit constellation.

I. Provide Costing Support for the Concept-Definition Activity (Chapter 20)

Developing costs for system elements is vital to two objectives: finding the best individual mission architecture and comparing mission architectures at the system level. Chapter 20 describes parametric, analogous, and bottoms-up methods for costing. Typically, for concept exploration, we use only the first two because we lack a detailed definition of the design. At this level, we simply want relative comparisons rather than absolute estimates, so we can accept the greater uncertainty in these methods.

References

Agrawal, Brij N. 1986. *Design of Geosynchronous Spacecraft.* Englewood Cliffs, NJ: Prentice Hall.

Boden, Daryl G. and Wiley J. Larson. 1996. *Cost Effective Space Mission Operations.* New York: McGraw-Hill.

Davidoff, Martin. 1998. *Radio Amateur's Satellite Handbook.* Newington, CT: American Radio Relay League.

Eckart, Peter. 1996. *Spaceflight Life Support and Biospherics.* Torrance, CA: Microcosm Press and Dordrecht, The Netherlands: Kluwer Academic Publishers.

Elbert, Bruce R. 1987. *Introduction to Satellite Communication.* Boston: Artech House Publishers.

Elbert, Bruce R. 1996. *The Satellite Communication Applications Handbook.* Boston: Artech House Publishers.

Griffin, M. and J.R. French. 1991. *Space Vehicle Design.* Washington, DC: AIAA.

London, J.R. 1994. *LEO on the Cheap—Methods for Achieving Drastic Reductions in Space Launch Costs.* Maxwell AFB, AL: Air University Press.

Lu, Ning. 1997. *Fractal Imaging.* San Diego, CA: Academic Press.

Pisacane, V.L. and R.C. Moore. 1994. *Fundamentals of Space Systems.* New York: Oxford University Press.

Sayood, Khalid. 1996. *Introduction to Data Compression.* San Francisco, CA: Morgan Kaufman Publishers.

Wall, Stephen D. and Kenneth W. Ledbetter. 1991. *Design of Mission Operations Systems for Scientific Remote Sensing.* London: Taylor & Francis.

Wertz, James R. and Wiley J. Larson. 1996. *Reducing Space Mission Cost.* Torrance, CA: Microcosm Press and Dordrecht, The Netherlands: Kluwer Academic Publishers.

Woodcock, Gordon R. 1986. *Space Stations and Platforms.* Malabar, FL: Orbit Book Company.

Chapter 3

Mission Evaluation

James R. Wertz, *Microcosm, Inc.*

3.1 Step 7: Identification of Critical Requirements
3.2 Mission Analysis
The Mission Analysis Hierarchy; Studies with Limited Scope; Trade Studies; Performance Assessments
3.3 Step 8: Mission Utility
Performance Parameters and Measures of Effectiveness; Mission Utility Simulation; Commercial Mission Analysis and Mission Utility Tools
3.4 Step 9: Mission Concept Selection

Chapter 2 defined and characterized alternative concepts and architectures for space missions. This chapter shows how we evaluate the ability of these options to meet fundamental mission objectives. We address how to identify the key requirements which drive the system design, how to quantify mission performance, and how to select one or more concepts for further development or to decide that we cannot achieve the mission within current constraints or technology.

Although essentially all missions go through mission evaluation and analysis stages many times, there are relatively few discussions in the literature of the general process for doing this. Fortescue and Stark [1995] discuss the process for generic missions; Przemieniecki [1993, 1994] does so for defense missions; and Shishko [1995] provides an excellent overview for NASA missions. Kay [1995] discusses the difficulty of doing this within the framework of a political democracy and Wertz and Larson [1996] provide specific techniques applicable to reducing mission cost.

The key mission evaluation questions for FireSat are:

- Which FireSat requirement dominates the system design or is the most difficult or expensive to meet?

- How well can FireSat detect and monitor forest fires, and at what cost?

- Should the FireSat mission evaluation proceed, and if so, which alternatives should we pursue?

We must readdress these questions as we analyze and design the space mission. By addressing them when we first explore concepts, we cannot obtain definitive answers. But we can form the right questions and identify ideas, parameters, and requirements we should be monitoring throughout the design. More extensive discussions of this systems engineering process are provided by Rechtin [1991] and the *System Engineer-*

ing Management [Defense Systems Management College, 1990]. The *NASA Systems Engineering Handbook* [Shishko, 1995] provides an excellent and detailed account of the process used by NASA. Przemieniecki [1990a, b] provides a good introduction to mathematical methods associated with military programs and has an associated software package. Other software packages intended specifically to support mission evaluation include *Satellite Tool Kit (STK)* from Analytical Graphics (1998), the *Mission Utility/Systems Engineering module (MUSE)* from Microcosm (1998), and the Edge product family from Autometric (1998).

3.1 Step 7: Identification of Critical Requirements

Critical requirements are those which dominate the space mission's overall design and, therefore, most strongly affect performance and cost[*]. For a manned mission to Mars, the critical requirements will be clear: get to Mars all of the required mass to explore the planet and return, and maintain crew safety for a long mission in widely varying environments. For less ambitious space missions, we cannot establish the critical requirements so easily. Because we want to achieve the best performance at minimum cost, we need to identify these key requirements as early as possible so they can be a part of the trade process.

Table 3-1 lists the most common critical requirements, the areas they typically affect, and where they are discussed. There is no single mechanism to find the critical requirements for any particular mission. Like the system drivers discussed in Sec. 2.3, they may be a function of the mission concept selected. Consequently, once we establish the alternative mission concepts, we usually can determine the critical requirements by inspection. Often, concept exploration itself exposes the requirements which dominate the system's design, performance, and cost. One approach to identification of critical requirements is as follows:

1. *Look at the principal performance requirements.* In most cases, the principal performance requirement will be one of the key critical requirements. Thus, for FireSat, the requirements on how well it must detect and monitor forest fires would normally be principal drivers of the system design.

2. *Examine Table 3-1.* The next step is to look at the requirements list in Table 3-1 and determine which of these entries drive the system design, performance, or cost.

3. *Look at top-level requirements.* Examine each of the top-level requirements established when we defined the mission objectives (Sec. 1.3) and determine how we will meet them. For each, ask whether or not meeting that requirement fundamentally limits the system's design, cost, or performance.

4. *Look for hidden requirements.* In some cases, hidden requirements such as the need to use particular technologies or systems may dominate the mission design, and cost.

---

[*] Critical requirements should be distinguished from *system drivers* (as discussed in Sec. 2.3), which are the defining mission parameters most strongly affecting performance, cost, and risk. The goal of mission engineering is to adjust both the critical requirements (e.g., coverage and resolution) and the system drivers (e.g., altitude and aperture) to satisfy the mission objectives at minimum cost and risk.

TABLE 3-1.   **Most Common Critical Requirements**. See text for discussion.

Requirement	What it Affects	Where Discussed
Coverage or Response Time	Number of satellites, altitude, inclination, communications architecture, payload field of view, scheduling, staffing requirements	Secs. 7.2, 13.2
Resolution	Instrument size, altitude, attitude control	Sec. 9.3
Sensitivity	Payload size, complexity; processing, and thermal control; altitude	Secs. 9.5, 13.5
Mapping Accuracy	Attitude control, orbit and attitude knowledge, mechanical alignments, payload precision, processing	Sec. 5.4
Transmit Power	Payload size and power, altitude	Secs. 11.2, 13.5
On-orbit Lifetime	Redundancy, weight, power and propulsion budgets, component selection	Secs. 6.2.3, 8.1.3, 10.4, 19.2
Survivability	Altitude, weight, power, component selection, design of space and ground system, number of satellites, number of ground stations, communications architecture	Sec. 8.2

For most FireSat approaches, resolution and coverage are the principal critical requirements, and we could find them by any of the first three options listed above. The critical requirements depend on a specific mission concept. For the low-cost FireSat of Chap. 22, they are coverage and sensitivity. Resolution no longer concerns us because the sensing is being done by ground instruments whose positions are known well enough for accurate location.

3.2 Mission Analysis

Mission analysis is the process of quantifying the system parameters and the resulting performance. A particularly important subset of mission analysis is *mission utility analysis*, described in Sec. 3.3, which is the process of quantifying how well the system meets its overall mission objectives. Recall that the mission objectives themselves are not quantitative. However, our capacity to meet them should be quantified as well as possible in order to allow us to make intelligent decisions about whether and how to proceed. *Mission requirements*, introduced in Chap. 1 and discussed in more detail in Chap. 4, are the numerical expressions of how well the objectives must be met. They represent a balance between what we want and what is feasible within the constraints on the system and, therefore, should be a central part of the mission analysis activity. In practice, mission analysis is often concerned with how and how well previously defined mission requirements can be met. In principle, mission analysis should be the process by which we define and refine mission requirements in order to meet our broad objectives at minimum cost and risk.

A key component of mission analysis is documentation, which provides the organizational memory of both the results and reasons. It is critical to understand fully the choices made, even those which are neither technical nor optimal. We may choose to apply a particular technology for political or economic reasons, or may not have enough manpower to investigate alternatives. In any case, for successful analysis, we must document the real reasons so others can reevaluate them later when the situation may be different. Technical people often shy away from nontechnical reasons or try to justify decisions by exaggerating their technical content. For example, we may choose for our preliminary FireSat analysis a circular orbit at 1,000 km at an inclination of 60 deg because this is a good mid-range starting point. If so, we should document this as the reason rather than trying to further justify these parameters. Later, we or others can choose the best altitude and inclination rather than having to live by choices for which there is no documented justification.

3.2.1 The Mission Analysis Hierarchy

I like to think of the mission analysis process as a huge electronic spreadsheet model of a space system. On the left side of the spreadsheet matrix are the various parameters and alternatives that one might assess, such as power, orbit, number of satellites, and manning levels for ground stations. Along the bottom row are the system's quantitative outputs, indicating its performance, effectiveness, cost, and risk. The matrix itself would capture the functional relationships among the many variables. We would like to wiggle any particular parameter, such as the diameter of the objective in a detector lens or the number of people assigned to the ground station, and determine the effect on all other parameters. In this way, we could quantify the system's performance as a function of all possible variables and their combinations.

Fortunately for the continuing employment of mission analysts, the above spreadsheet model does not yet exist.* Instead, we analyze as many reasonable alternatives as possible so we may understand how the system behaves as a function of the principal design features—that is, the system drivers. This approach does not imply that we are uninterested in secondary detail, but simply recognizes that the mission analysis process, much like the space system we are attempting to analyze, is ultimately limited in both cost and schedule. We must achieve the maximum level of understanding within these limits.

If the resources available for concept exploration are limited, as is nearly always the case in realistic situations, then one of the most critical tasks is to intelligently limit the scope of individual analyses. We must be able to compute approximate values for many parameters and to determine at what level of detail we should reasonably stop. In practice, this is made difficult by the continuing demand for additional detail and depth. Thus, we must be able to determine and make clear to others what elements of that detail will significantly affect the overall system performance and what elements, while important, can reasonably be left to a more detailed design phase.

We use two main methods to limit the depth of analysis in any particular area. The first is to clearly identify each area's system drivers by the methods in Sec. 2.3 and to concentrate most of the mission analysis effort on these drivers. The second is to clearly identify the goal of the system study and to provide a level of detail appropriate to that goal. This approach leads to a *mission analysis hierarchy*, summarized in Table 3-2, in which studies take on increased levels of detail and complexity as the activity progresses. The first three types of studies are meant to be quick with limited detail and are not intended to provide definitive results. The last three are much more complex ways to select an alternative to provide the best system performance.

---

* The Design-to-Cost model at JPL [Shishko, 1996] and similar models being developed throughout the aerospace community are attempting to automate this basic design process of evaluating the system-wide implication of changes. In due course, system engineers may become technologically obsolete. Much like modern chess players, the challenge to future system engineers will be to stay ahead of the computer in being creative and innovative.

TABLE 3-2. **The Mission Analysis Hierarchy.** These help us decide how much detail to study during the preliminary design phase.

Analysis Type	Goal	
Feasibility Assessment	To establish whether an objective is achievable and its approximate degree of complexity	Quick, limited detail
Sizing Estimate	To estimate basic parameters such as size, weight, power or cost	Quick, limited detail
Point Design	To demonstrate feasibility and establish a baseline for comparison of alternatives	Quick, limited detail
Trade Study	To establish the relative advantages of alternative approaches or options	More detailed, complex trades
Performance Assessment	To quantify performance parameters (e.g., resolution, timeliness) for a particular approach	More detailed, complex trades
Utility Assessment	To quantify how well the system can meet overall mission objectives	More detailed, complex trades

3.2.2 Studies with Limited Scope

The first three types of analyses in Table 3-2 provide methods for undertaking a quick-look assessment. They provide limited detail, but can frequently be done quickly and at low cost. Consequently, these quick-look assesments are important in any situation which is funding-limited. We will outline these methods very briefly here. However, nearly the entire book is devoted to the process of making initial estimates, which is the basic goal of limited scope studies. We want to be able to understand whether or not a particular project is feasible, and to get some idea of its size, complexity, and cost. Doing this requires that we be able to make numerical estimates and undertake limited studies in order to develop insight into the nature of the problem we are trying to solve.

The biggest difficulty with limited scope studies is the tendency to believe that they are more accurate than they really are. Thus it is not uncommon to use a feasibility assessment or point design to establish the requirements for a mission in such detail that in practice the point design becomes the only alternative which can meet them. As long as we recognize the limited scope of these studies, they have a valuable place in the mission analysis activity and represent one of the most important tools that we can use to understand the behavior of the system we are designing.

Feasibility Assessment. The simplest procedure in the mission analysis hierarchy is the *feasibility assessment*, which we use to establish whether a particular objective is achievable and to place broad bounds on its level of complexity. Frequently, we can do a feasibility assessment simply by comparison with existing systems. Thus, we are reasonably convinced that FireSat is feasible because most FireSat tasks could be performed by existing Earth resources satellites. Similarly, it is feasible to land a man on the Moon and return him safely to Earth because we have done so in the past.

We can also determine whether a particular goal is feasible by extrapolating our past experience. Is it feasible to send people to Mars and bring them back safely? Here we need to look at the principal differences between a Mars mission and a lunar mission. These differences include a longer flight time and higher gravity and, therefore, higher lift-off velocity required to leave Mars. These factors make the job more challenging and possibly more expensive than going to the Moon, but there is nothing about the Mars mission which makes it inherently impossible. Getting to Mars is feasible. The problem is being able to do so at modest cost and risk.

The third method of providing a feasibility assessment is to provide a very broad design of how such a mission might be accomplished. For example, in the 1970s, Gerard O'Neill of Princeton University proposed building large space colonies at the Lagrange points between the Earth and the Moon [O'Neill, 1974]. No mission of this scope had ever been undertaken, and it certainly was not a straightforward extrapolation of any of our normal space experience. O'Neill and his colleagues proceeded to establish the feasibility by developing a variety of alternative designs for such space colonies [Richard D. Johnson and Charles Holbrow, 1977]. While the work done was far in excess of a simple feasibility assessment, it clearly established that such colonies were feasible and gave at least an estimate of the scope of the problem.

Sizing Estimate. The purpose of the *sizing estimate* is to provide an estimate of basic mission parameters such as size, weight, power, or cost. We can do sizing estimates in much the same manner as the feasibility assessment: by analogy with existing systems. Thus, if we are aware of an Earth observation system which has resolution and information characteristics comparable to what we believe are needed for FireSat, we can use these parameters to give us an initial estimate of the FireSat parameters.

We can provide a quantitative estimate of key mission parameters by scaling the parameters from existing missions or payloads in order to obtain estimates of the component sizes for our particular mission. This scaling process is described in Sec. 9.5 for space payloads, and in Sec. 10.5 for the spacecraft as a whole. The process of sizing by scaling existing equipment is an extremely powerful approach to estimating what it will take to achieve mission objectives. It is of use not only during the conceptual design process, but throughout the hardware design definition and development phases to evaluate the system design as it evolves. If scaling existing systems leads to the suggestion that a particular component should be twice as heavy as the current design suggests, this gives us reason to look very closely at the current design and to try to determine whether or not any factors have been overlooked. We assume that designers of previous systems did a reasonable job of optimizing their system. If the current design is significantly different, either better or worse, then we would like to understand the reasons for these differences. This is a good way to gain confidence in the design process as we proceed.

As the design proceeds, more and more accurate sizing estimates come from the scaling process. We proceed by breaking down the system into components and sizing individual components based on scaling estimates with prior systems. Thus, we may initially estimate the system as a whole divided into a spacecraft and ground station. As the design becomes more detailed, we will break down the spacecraft into its relative components and estimate the size, weight, and power of each element based upon scaling from prior systems or engineering estimates of the new system to be built. Similarly, we initially size the ground station by comparison with existing systems and eventually by building a list of all the ground system components and undertaking similar sizing estimates for each component. As introduced in Chap. 1, this process of creating a list of components and estimating parameters for each is known as budgeting and is described in more detail in Sec. 10.3.

Point Design. A *point design* is a design, possibly at a top level, for the entire system which is capable of meeting the broad mission objectives. We refer to it as a point design if we have not attempted to optimize the design to either maximize performance or minimize weight, cost, or risk. The point design serves two basic purposes. It demonstrates that the mission is feasible, and it can be used as a baseline for comparison of alternatives. Thus, if we can establish a point design for FireSat that meets mission objectives with a spacecraft that weighs 500 kg and costs $50 million, then we can use this as a comparison for later systems. If other systems cost more, weigh more, and do not perform as well, then we will abandon those alternatives in favor of the original baseline. If we continue to optimize the design so that the cost and risk decrease, then we will let the baseline evolve to take into account the new design approaches.

A point design is valuable because we can do it quickly and easily. There is no need to optimize any of the parameters associated with the design unless it is necessary to do so to meet mission objectives. This gives us a sense of whether it will be easy or hard to meet the mission objectives and what are likely to be the most difficult aspects. One of the biggest problems in a point design is taking it too seriously at a later stage. We are always likely to regard work which we have done as representing the best approach, even though we may not have been aware of alternatives. The key issue here is to make use of point designs but at the same time to recognize their limitations and to continue to do trades to reduce overall cost and risk and to look for alternative approaches to meet mission objectives.

3.2.3 Trade Studies

Deciding whether to proceed with a mission should be based on a strawman system concept or point design which shows that the mission objectives can be met within the assigned constraints. Of course, the point design may not be the best solution, and we would ordinarily consider a number of alternatives. The system trade process evaluates different broad concepts to establish their viability and impact on performance and cost. We then combine the system trade results with the mission utility analysis described in Sec. 3.3 to provide input for concept selection.

System trades consist of analyzing and selecting key parameters, called *system drivers*, which determine mission performance. We use these parameters to define a mission concept and mission architecture which can then be used for performance analysis and utility analysis as described in Sec. 3.3. The key system trades are those that define how the system works and determine its size, cost, and risk. Typically, the key system trades will be in one of the following major areas:

- Critical requirements
- Mission concept
- Subject
- Type and complexity of payloads
- Orbit

Table 3-3 shows typical examples of areas in which there are key system trades for representative missions. For essentially all missions, specification of the critical requirements will be a key system trade. For the FireSat mission, the subject is probably the heat from the fire itself and the payload is probably an IR sensor. Thus, the principal system trades are probably the mission concept, the resolution and coverage requirements, and the orbit. For a mission such as the Space Telescope, the orbit is of marginal importance and the subject is moderately well defined, if only very poorly known. Here the principal trades will be the resolution and pointing requirements, the payload, and the mission concept. Communications satellite systems are normally in geosynchronous orbit with a well defined concept of operations. Here the only real trade is with the required traffic load, the subject, and the size and complexity of the payload.

Truly innovative approaches—those that really change how we think about a problem—typically involve finding a new option among these key system trades. Motorola's Iridium program and subsequent low-Earth orbit communications constellations represent a new way of thinking about using satellites for communications. These have a very different concept of operations and different orbit from traditional systems. Similarly, Chap. 22 presents an innovative approach to thinking about FireSat that provides a totally different concept of operations and type of payload. Innovative solutions are never easy to come by. To try to find them, a good place to start is with the key system trade areas given in Table 3-3.

TABLE 3-3. Representative Areas for Key System Trades. Although these system trades are critical, we can't expect numerically precise answers to our system design problem.

Trade Area	Where Discussed	FireSat	Space Telescope	Communications Satellite
Critical Requirements	Chap. 3	Yes	Yes	Yes
Mission Concept	Chap. 2	Yes	Yes	No
Subject	Chap. 9	No	No	Yes
Payload Type and Complexity	Chaps. 9, 13	No	Yes	Yes
Orbit	Chap. 7	Yes	No	No

We cannot normally do system trades in a straightforward numerical fashion. Choosing a different concept of operations, for example, will result in changes in most or all of the mission parameters. Consequently, the fact that Option A requires twice the power of Option B may or may not be critical, depending on the orbit and number of satellites for the two options. We need to look at the system as a whole to understand which is better.

The best approach for key system trades is a utility analysis as described in Sec. 3.3. We use the utility analysis to attempt to quantify our ability to meet mission objectives as a function of cost. We then select the option which fulfills our objectives at the lowest cost and risk. As described in Sec. 3.4, this is still not a straightforward numerical comparison, but does have quantitative components.

The simplest option for system trades is a list of the options and the reasons for retaining or eliminating them. This allows us to consider the merits and demerits at a high level without undertaking time-consuming trades. This, in turn, allows our list to be challenged at a later date. We should go back to our key system trades on a regular basis and determine whether our assumptions and conclusions are still valid. It is this process of examination and review that allows us to use technical innovation and new ideas. It is a process that must occur if we are to drive down the cost of space systems.

The alternative to simply articulating trade options or conducting a complex mission utility analysis is a system trade in which we make a quantitative comparison of multiple effects. This can be particularly effective in providing insight into the impact of system drivers. For the purpose of trade studies, system drivers generally divide into two categories—those for which more is better and those with multiple effects. By far the easier to deal with are the "more is better" drivers, for they simply require us to ask: "What is the cost of achieving more of the commodity in question?" For example, in a space-based radar, added power improves performance but costs more money. Thus, the designer will want to understand how much performance is available for how much power. A second example is coverage. For virtually any Earth-oriented system, including our FireSat example, more coverage means better performance at higher cost. Increasing coverage ordinarily means adding satellites or, perhaps, increasing a single satellite's coverage by increasing its altitude or the range of its sensors. Therefore, we often do a coverage trade considering performance vs. number of satellites, substituting the latter for cost. Assessing performance as a function of power or coverage may take considerable work, but it is relatively easy to present the data for judging by the funding organization, the users, or other decision makers.

System drivers and critical requirements which cause multiple effects demand more complex trade studies. Pushing parameters one way will improve some characteristics and degrade others. In trades of this type, we are looking for a solution which provides the best mix of results. Examples of such trade studies include instrument design, antenna type, and altitude. Each antenna style will have advantages and disadvantages, so we must trade various possible solutions depending upon the end goals and relative importance of different effects.

In trades with multiple effects, selecting the correct independent parameter for each trade is critical. Consider, for example, selecting either a reflector or a phased-array antenna for a space-based radar [Brookner and Mahoney, 1986]. From the radar equation, we know that a principal performance parameter for a radar is the antenna aperture. All other things being equal, larger antennas will provide much better performance. Thus, for our example, we might choose to compare reflector and phased-array antennas of equal aperture. On this basis, we would choose the phased array because its electronic steering makes it more agile than a reflector antenna, which must be mechanically steered. But our choice becomes more complex when we recognize that weight typically limits large space structures more than size does. Generally, we can build a reflector larger than a phased array for a given weight. Based on weight, a reflector may have considerably more power efficiency and, therefore, be a better radar than a phased-array system. Thus, we would have to trade the better performance of a larger reflector vs. the better agility of a smaller phased array. Depending upon the application, the results may be the same as for an aperture-based trade or reverse. The important point is the critical nature of selecting the proper independent variable in system trades. To do so, we must find the quantities which inherently limit the system being considered. These could be weight, power, level of technology, cost, or manufacturability, depending on the technology and circumstances.

Table 3-4 summarizes the system trade process for parameters with multiple effects. Typically the trade parameter is one of our system drivers. We begin by identifying what performance areas or requirements affect or are affected by the trade parameter. For example, the altitude of the spacecraft will have a key effect on coverage, resolution, and survivability and will be limited by launchability, payload weight, communications, and radiation. We next assess the effect in each of these areas and document and summarize the results, generally **without** trying to create a numerical average of different areas. Figure 3-1 shows this step for FireSat. We use the summary to select the parameter value and a possible range. Although the process is complex and may not have a well defined answer, it is not necessarily iterative unless we find that the results require fundamental changes in other system parameters.

TABLE 3-4. System Trade Process for Parameters with Multiple Effects. The example is the altitude trade for the FireSat mission. See also Fig. 3-1.

Step	FireSat Example	Where Discussed
1. Select trade parameter (typically a system driver)	Altitude	Sec. 2.3
2. Identify factors which affect the parameter or are affected by it	Coverage	Sec. 7.2
	Deployment strategy (coverage evolution)	Sec. 7.6
	Orbit period	Secs. 6.1, 7.2
	Time in view	Sec. 7.2
	Eclipse fraction	Sec. 5.1
	Response time	Sec. 7.2
	Number of spacecraft needed	Secs. 7.2, 7.6
	Launch capability	Sec. 18.2
	Resolution	Sec. 9.3
	Payload weight	Sec. 9.5
	Radiation environment	Sec. 8.1
	Survivability	Sec. 8.2
	Jamming susceptibility	Secs. 8.2, 13.5
	Communications	Secs. 13.1, 13.2
	Lifetime	Secs. 6.2.3, 8.1.5
3. Assess impact of each factor	Can launch up to 1,800 km Best coverage above 400 km Resolution—lower is better Survivability not an issue	
4. Document and summarize results	Launch Coverage Resolution Survivability	Fig. 3-1
5. Select parameter value and possible range	Altitude = 700 km 600 to 800 km	discussed in text

Altitude trades are perhaps the most common example of a trade in which multiple influences push the parameter in different ways. We would normally like to move the satellite higher to achieve better coverage, better survivability, and easier communications. On the other hand, launchability, resolution, and payload weight tend to drive the satellite lower. The radiation environment dictates specific altitudes we would like to avoid, and the eclipse fraction may or may not play a crucial role in the altitude trade. We must assess each of these effects and summarize all of them to complete a trade study. One possible summary is a numerically weighted average of the various outcomes, such as three times the coverage in square nautical miles per second divided by twice the resolution in furlongs. Although this provides a convenient numerical answer, it does not provide the physical insight or conceptual balance needed for intelligent choices. A better solution is to provide the data on all of the relevant parameters and choose based on inspection rather than numerical weighting.

The FireSat altitude trade provides an example of trading on parameters with multiple effects. For FireSat, neither survivability nor communications is a key issue, but coverage does push the satellite upward. On the other hand, payload weight and good resolution tend to push the satellite lower. Figure 3-1 shows the results of a hypothetical FireSat altitude trade. Notice that each parameter has various possible outcomes. Altitudes above or below a certain value may be eliminated, or we may simply prefer a general direction, such as lower altitude providing better resolution. Based on these results, we select a nominal altitude of 700 km for FireSat and a possible range of 600 to 800 km. This selection is not magic. We have tried to balance the alternatives sensibly, but not in a way that we can numerically justify.

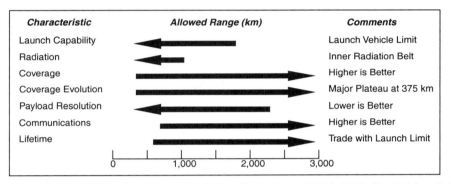

Fig. 3-1.   Results of FireSat Altitude Trade. See Table 3-4 and Table 7-6 in Sec. 7.4 for a list of trade issues. Political constraints and survivability were not of concern for the FireSat altitude trade.

3.2.4 Performance Assessments

Quantifying performance demands an appropriate level of detail. Too much detail drains resources away from other issues; too little keeps us from determining the important issues or causes us to assess the actual performance incorrectly.

To compute system performance, we use three main techniques:

- System algorithms
- Analogy with existing systems
- Simulation

System algorithms are the basic physical or geometric formulas associated with a particular system or process, such as those for determining resolution in diffraction-limited optics, finding the beam size of an antenna, analyzing a link budget, or assessing geometric coverage. Table 3-5 lists system algorithms typically used for space mission analysis. System algorithms provide the best method for computing performance. They provide clear traceability and establish the relationship between design parameters and performance characteristics. Thus, for FireSat, we are interested in the resolution of an on-orbit fire detector. Using the formula for diffraction-limited optics in Chap. 9, we can compute the achievable angular resolution from the instrument objective's diameter. We can then apply the geometric formulas in Chap. 5 to translate this angular resolution to resolution on the ground. This result gives us a direct relationship between the altitude of the FireSat spacecraft, the size of the payload, the angles at which it works, and the resolution with which it can distinguish features on the ground.

TABLE 3-5. Common System Algorithms Used for Quantifying Basic Levels of Performance. These analyses use physical or geometrical formulas to determine how system performance varies with key parameters.

Algorithm	Used For	Where Discussed
Link Budget	Communications and data rate analysis	Sec. 13.3.6
Diffraction-limited Optics	Aperture sizing for optics or antennas; determining resolution	Sec. 9.3
Payload Sensitivity	Payload sizing and performance estimates	Secs. 9.4, 9.5
Radar Equation	Radar sizing and performance estimates	[Cantafio,1989]
Earth Coverage, Area Search Rates	Coverage assessment; system sizing; performance estimates	Secs. 5.2, 7.2
Mapping and Pointing Budget	Geolocation; instrument and antenna pointing; image sensing	Sec. 5.4

System algorithms are powerful in that they show us directly how performance varies with key parameters. However, they are inherently limited because they presume the rest of the system is designed with fundamental physics or geometry as the limiting characteristic. For FireSat, resolution could also be limited by the optical quality of the lens, by the detector technology, by the spacecraft's pointing stability, or even by the data rates at which the instrument can provide results or that the satellite can transmit to the ground. In using system algorithms, we assume that we have correctly identified what limits system performance. But we must understand that these assumptions may break down as each parameter changes. Finding the limits of these system algorithms helps us analyze the problem and determine its key components. Thus, we may find that a low-cost FireSat system is limited principally by achieving spacecraft stability at low cost. Therefore, our attention would be focused on the attitude control system and on the level of resolution that can be achieved as a function of system cost.

The second method for quantifying performance is by comparing our design with existing systems. In this type of analysis we use the established characteristics of existing sensors, systems, or components and adjust the expected performance according to basic physics or the continuing evolution of technology. The list of payload instruments in Chap. 9 is an excellent starting point for comparing performance with existing systems. We could, for example, use the field of view, resolution, and integration time for an existing sensor and apply them to FireSat. We then modify the basic sensor parameters such as the aperture, focal length, or pixel size, to satisfy our mission's unique requirements. To do this, we must work with someone who knows the technology, the allowable range of modifications, and their cost. For example, we may be able to improve the resolution by doubling the diameter of the objective, but doing so may cost too much. Thus, to estimate performance based on existing systems, we need information from those who understand the main cost and performance drivers of that technology.

The third way to quantify system performance is simulation, described in more detail in Sec. 3.3.2. Because it is time-consuming, we typically use simulation only for key performance parameters. However, simulations allow much more complex modeling and can incorporate limits on performance from multiple factors (e.g., resolution, stability, and data rate). Because they provide much less insight, however, we must review the results carefully to see if they apply to given situations. Still, in complex circumstances, simulation may be the only acceptable way to quantify system performance. A much less expensive method of simulation is the use of commercial mission analysis tools as discussed in Sec. 3.3.3.

3.3 Step 8: Mission Utility

Mission utility analysis quantifies mission performance as a function of design, cost, risk, and schedule. It is used to (1) provide quantitative information for decision making, and (2) provide feedback on the system design. Ultimately, an individual or group will decide whether to build a space system and which system to build based on overall performance, cost, and risk relative to other activities. As discussed in Sec. 3.4, this does not mean the decision is or should be fundamentally technical in nature. However, even though basic decisions may be political, economic, or sociological, the best possible quantitative information from the mission utility analysis process should be available to support them.

Mission utility analysis also provides feedback for the system design by assessing how well alternative configurations meet the mission objectives. FireSat shows how this process might work in practice. Mission analysis quantifies how well alternative systems can detect and monitor forest fires, thereby helping us to decide whether to proceed with a more detailed design of several satellites in low-Earth orbit or a single larger satellite in a higher orbit. As we continue these trades, mission analysis establishes the probability of being able to detect a given forest fire within a given time, with and without FireSat, and with varying numbers of spacecraft. For FireSat, the decision makers are those responsible for protecting the forests of the United States. We want to provide them with the technical information they need to determine whether they should spend their limited resources on FireSat or on some alternative. If they select FireSat, we will provide the technical information needed to allow them to select how many satellites and what level of redundancy to include.

3.3.1 Performance Parameters and Measures of Effectiveness

The purpose of mission analysis is to quantify the system's performance and its ability to meet the ultimate mission objectives. Typically this requires two distinct types of quantities—performance parameters and measures of effectiveness. *Performance parameters,* such as those shown in Table 3-6 for FireSat, quantify how well the system works, without explicitly measuring how well it meets mission objectives. Performance parameters may include coverage statistics, power efficiency, or the resolution of a particular instrument as a function of nadir angle. In contrast, *measures of effectiveness* (MoEs) or *figures of merit* (FoMs) quantify directly how well the system meets the mission objectives. For FireSat, the principal MoE will be a numerical estimate of how well the system can detect forest fires or the consequences of doing so. This could, for example, be the probability of detecting a given forest fire within 6 hours, or the estimated dollar value of savings resulting from early fire detection. Table 3-7 shows other examples.

TABLE 3-6. Representative Performance Parameters for FireSat. By using various performance parameters, we get a better overall picture of our FireSat design.

Performance Parameter	How Determined
Instantaneous maximum area coverage rate	Analysis
Orbit average area coverage rate (takes into account forest coverage, duty cycle)	Simulation
Mean time between observations	Analysis
Ground position knowledge	Analysis
System response time (See Sec. 7.2.3 for definition)	Simulation

TABLE 3-7. Representative Measures of Effectiveness (MoEs) for FireSat. These Measures of Effectiveness help us determine how well various designs meet our mission objectives.

Goal	MoE	How Estimated
Detection	Probability of detection vs. time (milestones at 4, 8, 24 hours)	Simulation
Prompt Knowledge	*Time late* = time from observation to availability at monitoring office	Analysis
Monitoring	Probability of containment	Simulation
Save Property and Reduce Cost	Value of property saved plus savings in firefighting costs	Simulation + Analysis

We can usually determine performance parameters unambiguously. For example, either by analysis or simulation we can assess the level of coverage for any point on the Earth's surface. A probability of detecting and containing forest fires better measures our end objective, but is also much more difficult to quantify. It may depend on how we construct scenarios and simulations, what we assume about ground resources, and how we use the FireSat data to fight fires.

Good measures of effectiveness are critical to successful mission analysis and design. If we cannot quantify the degree to which we have met the mission objectives, there is little hope that we can meet them in a cost-effective fashion. The rest of this section defines and characterizes good measures of effectiveness, and Secs. 3.3.2 and 3.3.3 show how we evaluate them.

Good measures of effectiveness must be

- Clearly related to mission objectives
- Understandable by decision makers
- Quantifiable
- Sensitive to system design (if used as a design selection criterion)

MoEs are useless if decision makers cannot understand them. "Acceleration in the marginal rate of forest-fire detection within the latitudinal coverage regime of the end-of-life satellite constellation" will likely need substantial explanation to be effective. On the other hand, clear MoEs which are insensitive to the details of the system design, such as the largest coverage gap over one year, cannot distinguish the quality of one system from another. Ordinarily, no single measure of effectiveness can be used to quantify how the overall system meets mission objectives. Thus, we prefer to provide a few measures of effectiveness summarizing the system's capacity to achieve its broad objectives.

Measures of effectiveness generally fall into one of three broad categories associated with (1) discrete events, (2) coverage of a continuous activity, or (3) timeliness of the information or other indicators of quality. Discrete events include forest fires, nuclear explosions, ships crossing a barrier, or cosmic ray events. In this case, the best measures of effectiveness are the rate that can be sustained (identify up to 20 forest fires per hour), or the probability of successful identification (90% probability that a forest fire will be detected within 6 hours after ignition). The probability of detecting discrete events is the most common measure of effectiveness. It is useful both in providing good insight to the user community and in allowing the user to create additional measures of effectiveness, such as the probability of extinguishing a forest fire in a given time.

Some mission objectives are not directly quantifiable in probabilistic terms. For example, we may want continuous coverage of a particular event or activity, such as continuous surveillance of the crab nebula for extraneous X-ray bursts or continuous monitoring of Yosemite for temperature variations. Here the typical measure of effectiveness is some type of coverage or gap statistics such as the mean observation gap or maximum gap under a particular condition. Unfortunately, Gaussian (normal probability) statistics do not ordinarily apply to satellite coverage; therefore, the usual measure of average values can be very misleading. Additional details and a way to resolve this problem are part of the discussion of coverage measures of effectiveness in Sec. 7.2.

A third type of measure of effectiveness assesses the quality of a result rather than whether or when it occurs. It may include, for example, the system's ability to resolve the temperature of forest fires. Another common measure of quality is the timeliness of the data, usually expressed as time late, or, in more positive terms for the user, as the time margin from when the data arrives until it is needed. Timeliness MoEs might include the average time from ignition of the forest fire to its initial detection or, viewed from the perspective of a potential application, the average warning time before a fire strikes a population center. This type of information, illustrated in Fig. 3-2, allows the decision maker to assess the value of FireSat in meeting community needs.

3.3.2 Mission Utility Simulation

In analyzing mission utility, we try to evaluate the measures of effectiveness numerically as a function of cost and risk, but this is hard to do. Instead, we typically use principal system parameters, such as the number of satellites, total on-orbit weight, or payload size, as stand-ins for cost. Thus, we might calculate measures of effectiveness as a function of constellation size, assuming that more satellites cost more money. If we can establish numerical values for meaningful measures of effectiveness as a function of the system drivers and understand the underlying reasons for the results, we will have taken a major step toward quantifying the space mission analysis and design process.

Recall that mission utility analysis has two distinct but equally important goals—to aid design and provide information for decision making. It helps us design the mission by examining the relative benefits of alternatives. For key parameters such as payload type or overall system power, we can show how utility depends on design choices, and therefore, intelligently select among design options.

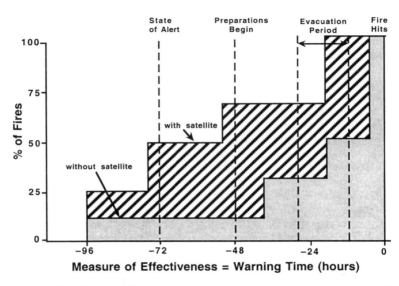

Fig. 3-2.   Forest Fire Warning Time for Inhabited Areas. A hypothetical measure of effectiveness for FireSat.

Mission utility analysis also provides information that is readily usable to decision makers. Generally those who determine funding levels or whether to build a particular space system do not have either the time or inclination to assess detailed technical studies. For large space programs, decisions ultimately depend on a relatively small amount of information being assessed by individuals at a high level in industry or government. A strong utility analysis allows these high-level judgments to be more informed and more nearly based on sound technical assessments. By providing summary performance data in a form the decision-making audience can understand, the mission utility analysis can make a major contribution to the technical decision-making process.

Typically, the only effective way to evaluate mission utility is to use a mission utility simulation designed specifically for this purpose. (Commercial simulators are discussed in Sec. 3.3.3.) This is **not** the same as a *payload simulator*, which evaluates performance parameters for various payloads. For FireSat, a payload simulator might compute the level of observable temperature changes or the number of acres that can be searched per orbit pass. In contrast, the *mission simulator* assumes a level of performance for the payload and assesses its ability to meet mission objectives. The FireSat mission simulator would determine how soon forest fires can be detected or the amount of acreage that can be saved per year.

In principle, mission simulators are straightforward. In practice, they are expensive and time consuming to create and are rarely as successful as we would like. Attempts to achieve excessive fidelity tend to dramatically increase the cost and reduce the effectiveness of most mission simulators. The goal of mission simulation is to estimate measures of effectiveness as a function of key system parameters. We must restrict the simulator as much as possible to achieving this goal. Overly detailed simulations require more time and money to create and are much less useful, because computer time and other costs keep us from running them enough for effective trade studies. The simulator must be simple enough to allow making multiple runs, so we can collect statistical data and explore various scenarios and design options.

The mission simulation should include parameters that directly affect utility, such as the orbit geometry, motion or changes in the targets or background, system scheduling, and other key issues, as shown in Fig. 3-3. The problem of excessive detail is best solved by providing numerical models obtained from more detailed simulations of the payload or other system components. For example, we may compute FireSat's capacity to detect a forest fire by modeling the detector sensitivity, atmospheric characteristics, range to the fire, and the background conditions in the observed area. A detailed payload simulation should include these parameters. After running the payload simulator many times, we can, for example, tabulate the probability of detecting a fire based on observation geometry and time of day. The mission simulator uses this table to assess various scenarios and scheduling algorithms. Thus, the mission simulator might compute the mission geometry and time of day and use the lookup table to determine the payload effectiveness. With this method, we can dramatically reduce repetitive computations in each mission simulator run, do more simulations, and explore more mission options than with a more detailed simulation. The mission simulator should be a collection of the results of more detailed simulations along with unique mission parameters such as the relative geometry between the satellites in a constellation, variations in ground targets or background, and the system scheduling or downlink communications. Creating sub-models also makes it easier to generate utility simulations. We start with simple models for the individual components and develop more realistic tables as we create and run more detailed payload or component simulations.

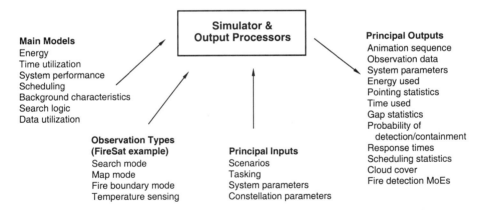

Fig. 3-3.   Results of FireSat Altitude Trade. See Table 3-4 and Table 7-6 in Sec. 7.4 for a list of trade issues. Political constraints and survivability were not of concern for the FireSat altitude trade.

Table 3-8 shows the typical sequence for simulating mission utility, including a distinct division into data generation and output. This division allows us to do various statistical analyses on a single data set or combine the outputs from many runs in different ways. In a constellation of satellites, scheduling is often a key issue in mission utility. The constellation's utility depends largely on the system's capacity to schedule resource use appropriately among the satellites. At the end of a single simulation run, the system should collect and compute the statistics for that scenario, generate appropriate output plots or data, and compute individual measures of effectiveness, such as the percent of forest fires detected in that particular run.

TABLE 3-8. **Typical Sequence Flow of a Time-Stepped Mission Utility Simulation.** Following this sequence for many runs, we can create statistical measures of effectiveness that help us evaluate our design.

Phase I — Data Generation
Advance time step
Compute changes in target or background characteristics
Update satellite positions
Update viewing geometry parameters
Schedule observations or operations
Compute pointing changes
Compute and save performance statistics
Update satellite consumables
Save data for this time step
Go to next time step
Phase II — Output Generation and Statistics Collection
Compute scenario statistics
Compute measures of effectiveness for the individual run
Prepare output plots and data for the individual run
Phase III — Monte Carlo Runs
Set new scenario start time
Repeat Phase I and II
Collect multi-run statistics
Compute statistical measures of effectiveness
Prepare Monte Carlo output plots and data

The next step is to run more simulations using new start times or otherwise varying the conditions for the scenarios. Changing the start times alters the relative timing and geometry between the satellites and the events they are observing, thus, averaging results caused by these characteristics. Collecting statistics on multiple runs is called a *Monte Carlo simulation.* For example, we might average the percentage of forest fires detected over different runs with different timing, but on the same scenario, to estimate the overall probability of detecting forest fires—our ultimate measure of effectiveness. The system simulator should accumulate output statistics and prepare output plots over the Monte Carlo runs.

Frequently, in running mission simulations, we must choose between realistic and analytical scenarios. Realistic scenarios usually are too complex to help us understand how the system works but are still necessary to satisfy the end users. On the other hand, simple scenarios illuminate how the system is working but do not show how it will work in a real situation. The best answer is to use simple scenarios for analysis and realistic scenarios to assess mission performance. In the FireSat example, we might begin by studying a single satellite to determine how it behaves and then expand to a more complex simulation with several satellites. We might also start evaluating a multi-satellite constellation by looking at its response to a simple situation, such as one fire or a small group of uniformly distributed fires. This trial run will suggest how the system performs and how changes affect it. We can then apply this understanding as we develop more realistic simulations.

A related problem concerns using a baseline scenario to compare options and designs. Repeating a single scenario allows us to understand the scenario and the system's response to it. We can also establish quantitative differences by showing how different designs respond to the same scenario. But this approach tends to mask characteristics that might arise solely from a particular scenario. Thus, we must understand what happens as the baseline changes and watch for chance results developing from our choice of a particular baseline scenario.

Finally, mission simulations must generate usable and understandable information for decision makers—information that provides physical insight. Two examples are strip charts of various system characteristics and animated output. A *strip chart* plot is similar to the output of a seismograph or any multi-pin plotter, in which various characteristics are plotted as a function of time. These characteristics might include, for example, whether a particular satellite is in eclipse, how much time it spends in active observation, and the spacecraft attitude during a particular time step. Plots of this type give a good feel for the flow of events as the simulation proceeds.

A valuable alternative for understanding the flow of events is looking at an animation of the output, such as a picture of the Earth showing various changes in the target, background, and observation geometry as the satellites fly overhead. Thus, as Fig. 3-4 illustrates, an animated simulation of FireSat output could be a map of a fire-sensitive region with areas changing color as fires begin, lines showing satellite coverage, and indications as to when fires are first detected or when mapping of fires occurs. Animation is not as numerical as statistical data, but it shows more clearly how the satellite system is working and how well it will meet broad objectives. Thus, mission analysts and end users can assess the system's performance, strengths and shortcomings, and the changes needed to make it work better.

3.3.3 Commercial Mission Analysis and Mission Utility Tools

Creating a mission utility simulation for your specific mission or mission concept is both time consuming and expensive. It is not uncommon for the simulation to be completed at nearly the same time as the end of the study, such that there is relatively little time to use the simulation to effectively explore the multitude of options available to the innovative system designer.

In my view, the single largest step in reducing software cost and risk is the use of commercial, off-the-shelf (COTS) software. The basic role of COTS software in space is to spread the development cost over multiple programs and reduce the risk by using software that has been tested and used many times before. Because the number of purchasers of space software is extremely small, the savings will be nowhere near as large as for commercial word processors. Nonetheless, reductions in cost, schedule, and risk can be substantial. Most COTS software should be at least 5 times cheaper than program-unique software and is typically 10 or more times less expensive. In addition, COTS software will ordinarily have much better documentation and user interfaces and will be more flexible and robust, able to support various missions and circumstances.

The use of COTS software is growing, but most large companies and government agencies still develop their own space-related software for several reasons. One of the

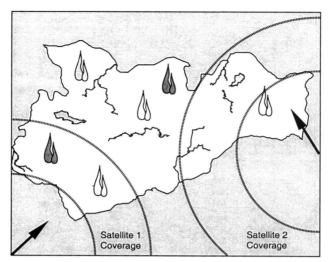

Fig. 3-4.   Hypothetical Animation Output for FireSat Mission Utility Simulator. Color displays are very valuable for animation sequences because we need to convey multiple parameters in each frame.

best ways to develop and maintain expertise is to create your own systems and models. Thus, organizations may want to support their own software group, particularly when money is tight. Also, it's hard to overcome the perception that it costs less to incrementally upgrade one's own system than to bear the cost and uncertainty of new COTS tools. In this trade, the "home built" systems often don't include maintenance costs. Finally, customers often don't know which COTS tools are available. Professional aerospace software doesn't appear in normal software outlets, advertising budgets are small, and most information is word-of-mouth through people already in the community. Despite these substantial obstacles, many organizations are now using COTS software in response to the strong demand to reduce cost.

In order to use COTS tools to reduce space system cost, we need to change the way we use software. We need to adapt to software not being exactly what we want, look for ways to make existing software satisfy the need, or modify COTS software to more closely match requirements. This is a normal part of doing business in other fields. Very few firms choose to write their own word processor, even though no single word processor precisely meets all needs. Instead, they choose one that most closely matches what they want in terms of functions, support, and ease of use. We should use the same criteria for COTS space software. In addition, we need to set realistic expectations concerning what COTS software can do. Clearly, we can't expect the low prices and extensive support that buyers of globally marketed commercial software enjoy. We have to adjust our expectations to the smaller market for space-related software, which means costs will be much higher than for normal commercial products. Maintenance and upgrades will ordinarily require an ongoing maintenance contract. Within the aerospace community, a standard arrangement is for a maintenance and upgrade contract to cost 15% of the purchase price per year.

Using COTS software and reusing existing noncommercial software requires a different mindset than continuously redeveloping software. We need to understand both the strengths and weaknesses of the relatively small space commercial software industry. Because the number of copies sold is small, most space software companies are cottage industries with a small staff and limited resources. We shouldn't expect space-software developers to change their products at no cost to meet unique needs. For example, it would be unrealistic to expect a vendor of commercial software for low-Earth orbit spacecraft to modify the software for interplanetary missions at no cost, because few groups will buy interplanetary software. On the other hand, the small size of the industry means developers are eager to satisfy the customers' needs, so most are willing to work with their customer and to accept contracts to modify their products for specific applications. This can be far less expensive than developing software completely from scratch.

There is a hierarchy of software cost, going from using COTS software as is, to developing an entirely new system. In order of increasing cost, the main options are

1. Use COTS software as sold

2. Use COTS software libraries

3. Modify COTS software to meet specific program needs (modification may be done by mission developer, prime contractor, or software developer)

4. Reuse existing flight or ground software systems or modules

5. Develop new systems based largely on existing software components

6. Develop new systems from scratch using formal requirements and development processes

This hierarchy contains several potential traps. It may seem that the most economical approach would be for the prime contractor or end-user to modify COTS software to meet their needs. However, it is likely that the COTS software developer is in a better position to make modifications economically and quickly. Although the end-users are more familiar with the objectives and the mission, the software developer is more familiar with the organization and structure of the existing code.

Secondly, there is frequently a strong desire to reuse existing code. This will likely be cheaper if the code was developed to be maintainable and the developers are still available. On the other hand, for project-unique code developed with no requirement for maintainability, it may be cheaper, more efficient, and less risky simply to discard the old software and begin again.

Commercial mission analysis tools fall into three broad categories, each of which is described below. Representative examples of these tools are listed in Table 3-9.

Generic Analysis Systems. These are programs, such as *MatLab*™, which are intended to allow analysis and simulation of a wide variety of engineering and science problems. They typically cost a few hundred to several thousand dollars and can dramatically reduce the time needed to create simulations and analyze the results. Because these are generic tools, specific simulation characteristics are set up by the user, although subroutine libraries often exist. Thus, we will need to create orbit propagators, attitude models, environment models, and whatever else the problem dictates. We use this type of simulation principally for obtaining mathematical data and typically not for animation.

Low-Cost Analysis Programs. These are programs intended for a much wider audience such as the amateur astronomy or space science community. However, when carefully selected and used appropriately, they can provide nearly instant results at very low cost. The programs themselves cost a few hundred dollars or less, are

TABLE 3-9. Commercial Space Mission Analysis and Design Software. New versions are typically released roughly annually. Because of the very small size of the space market, commercial space software both enters and leaves the marketplace on a regular basis.

Product	Publisher	Approx. Cost	Purpose
Dance of the Planets	Arc Science Simulations	$250	Amateur visual and gravitational model of the solar system useful for interplanetary mission design
Edge	Autometric	$5,000 +	Professional mission analysis system; many modules; can be customized
EWB	Maxwell Labs	High	Professional tool for space mission trade studies; used for Space Station
MUSE module	Microcosm	$6,500	MIssion Utility/Systems Engineering tool; evaluates figures of merit; can be customized by user
ORB	AIAA	< $100	Orbit analysis tool included with the book *Spacecraft Mission Design*; primarily interplanetary
Orbit Works	ARSoftware	$700	Orbit analysis, pass geometry, related tools; used by many ground operations groups
SMAD Software	KB Sciences	$500	10 software modules that implement equations in the *SMAD* book
Satellite Tool Kit, STK	Analytical Graphics	(*)	Professional mission analysis system; many modules

* Base program is free; modules range from $2,000 to $30,000.

immediately available from mail-order retailers, and can be run within a few hours of receiving them. A typical program in this category is *Dance of the Planets*, developed by Arc Science Simulations, for simulating astronomical events and allowing amateur space enthusiasts to create simulations of solar system events and obtain views from spacecraft which they define. A key characteristic of this program is that it creates simulations by integrating the equations of motion of celestial objects, thus allowing the user to define an interplanetary spacecraft orbit and determine its interaction with various celestial bodies. While less accurate than high-fidelity simulations created after a mission is fully funded, this type of tool can produce moderately accurate results quickly and at very low cost.

A second type of system used by amateurs consists of data sets, such as star catalogs, and the associated programs used to access and manipulate the data. For example, the complete *Hubble Guide Star Catalog*, created for the Space Telescope mission and containing over 19 million stars and nonstellar objects, is available on two CD-ROMs for less than $100. Smaller star catalogs contain fewer entries, but typically have much more data about each of the stars. All of the electronic star catalogs can be read and star charts created by any of the major sky plotting programs, again available off-the-shelf for a few hundred dollars.

Space Mission Analysis Systems. These are professional engineering tools created specifically for the analysis and design of space missions. Prices are several thousand dollars and up. These tools can create very realistic simulations, including data generation, animation, user-defined figures of merit, and Monte Carlo simulations. One of the most widely used tools in this category is *Satellite Tool Kit (STK)*, developed by Analytical Graphics, which provides a basic simulation capability and a variety of add-on modules for animation generation, orbit determination and propagation, coverage analysis, and mission scheduling. The *Mission Utility/Systems Engineering Module (MUSE)* by Microcosm allows the evaluation of a variety of generic figures of merit (such as coverage or timeliness) and provides Monte Carlo simulation runs to create statistical output. *MUSE* is intended specifically to allow the user to define new figures of merit to allow the rapid creation of mission-specific simulations. The *Edge* product family by Autometric provides very high-fidelity animation of a variety of space missions and is intended to be adapted by either the company or the user to become a mission-specific simulation. Each of the tools in this category can provide high-fidelity simulations at a much lower cost than creating systems from scratch.

3.4 Step 9: Mission Concept Selection

This section is concerned not with the detailed engineering decisions for a space mission, but with the broad trades involved in defining the overall mission—whether to proceed with it and what concept to use. Decisions for space missions fall into three broad categories: (1) go/no-go decision on proceeding with the mission; (2) selection of the mission concept; and (3) detailed engineering decisions, which are generally described throughout this book.

In principle, the go/no-go decision depends on only a few factors, the most important of which are:

- Does the proposed system meet the overall mission objectives?

- Is it technically feasible?

- Is the level of risk acceptable?

- Are the schedule and budget within the established constraints?

- Do preliminary results show this option to be better than nonspace solutions?

In addition to the above technical issues, a number of nontechnical criteria are ordinarily equally or more important in the decision-making process:

- Does the mission meet the political objectives?

- Are the organizational responsibilities acceptable to all of the organizations involved in the decision?

- Does the mission support the infrastructure in place or contemplated?

For example, a mission may be approved to keep an organization in business, or it may be delayed or suspended if it requires creating an infrastructure perceived as not needed in the long term. The mission analysis activity must include nontechnical factors associated with space missions and see that they are appropriately addressed.

The top-level trades in concept selection are usually not fully quantitative, and we should not force them to be. The purpose of the trade studies and utility analysis is to make the decisions as informed as possible. We wish to add quantitative information to the decisions, not quantify the decision making. In other words, we should not undervalue the decision-maker's judgment by attempting to replace it with a simplistic formula or rule.

Table 3-10 shows how we might try to quantify a decision. Assume that a system costs $500 million, but an improvement could save up to $300 million. To save this money, we could use option A, B, or C. Option A would cost $35 million, but the probability of success is only 70%; B would cost $100 million with 99% probability of success; C would cost $200 million with a 99.9% probability of success.

TABLE 3-10. **Mathematical Model of Hypothetical Decision Process (costs in $M).** Numerically, we would choose B or A′ if it were available. Realistically, any of the choices may be best depending on the decision criteria.

| Current Cost | | | | | $500M | |
| Potential Savings if Improvement is Successful | | | | | $300M | |
Option	Cost of Improvement	Probability of Success	Total Cost if Successful	Total Cost if Failed	Expected Total Cost	Expected Savings
A	35	70%	235	535	325	175
B	100	99%	300	600	303	197
C	200	99.90%	400	700	400.3	99.7
A′	35	80%	235	535	295	205

Which option should we select? The table gives the cost if successful, the cost if the improvement fails, and the expected values of both the cost and net savings. By numbers alone, we would select option B with an expected savings of $197 million. However, reasonable and valid cases can be made for both A and C. In option A, we risk only $35 million, and, therefore, are minimizing the total cost if the improvement succeeds or if it fails. In fact, the $600 million cost of failure for option B may be too much for the system to bear, no matter the expected savings. Option C provides a net savings of "only" $100 million, but its success is virtually certain. Although savings for this option are less dramatic, it does provide major savings while minimizing risk. In this case, we may assume the cost to be a fixed $400 million, with failure being so unlikely that we can discount it. Option B, of course, balances cost and risk to maximize the expected savings.

Suppose, however, that option A had an 80% probability of success as shown in A′, rather than the original 70% probability. In this case, the expected savings of A′ would increase to $205 million, and would make it the preferred approach in pure expectation terms. However, most individuals or groups faced with decisions of this sort are unlikely to change from option B to A′ based solely on the increase in estimated probability to 80%. Their decisions are more likely to depend on perceived risk or on minimizing losses. Using nonmathematical criteria does not make the decisions incorrect or invalid, nor does it make the numerical values unimportant. We need quantitative information to choose between options but we do not have to base our decisions exclusively on this information.

As a second example, we can apply the results of utility analysis to concept selection for FireSat. In particular, the number of satellites strongly drives the cost of a constellation. If we select the low-Earth orbit approach for FireSat, how many satellites should the operational constellation contain? More satellites means better coverage and, therefore, reduces the time from when a fire starts until it is first detected. Consequently, one of our key parameters is the *time late*, that is, the time from when a fire starts until the system detects its presence and transmits the information to the ground. Figure 3-5 plots the hypothetical time late vs. the number of satellites for

FireSat. The details of such plots will depend on the latitude under consideration, swath coverage, altitude, and various other parameters. However, the characteristic of increasing coverage with more satellites eventually reaches a point of diminishing returns. This will normally be true irrespective of the coverage assumptions.

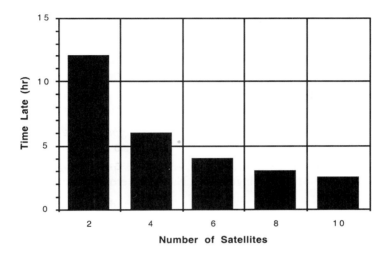

Fig. 3-5. **Hypothetical Coverage Data for FireSat.** See text for definitions and discussion. As discussed in Sec. 7.5, satellite growth comes in increments or plateaus. These are assumed to be 2-satellite increments for FireSat.

If we assume an initial goal for time late of no more than 5 hours, we see from the plot that a system of 6 satellites can meet this goal. Alternatively, a 4-satellite system can achieve a time late of 6 hours. Is the smaller time late worth the increased number of satellites and the money to build them? Only the ultimate users of the system can judge. The additional warning may be critical to fire containment and, therefore, a key to mission success. However, it is also possible that the original goal was somewhat arbitrary, and a time of **approximately** 5 hours is what is really needed. In this case, fire-fighting resources could probably be used better by flying a 4-satellite system with 6 hours time late and applying the savings to other purposes. Again, mission utility analysis simply provides quantitative data for intelligent decision making.

Of course, we must remember that the number of FireSat satellites will depend not only on the utility analysis but also on politics, schedules, and resources. The public must see FireSat as an appropriate response to the problem, as well as an efficient use of scarce economic resources compared to, for example, more fire fighters. In addition, a satellite system may serve several missions, with multiple mission criteria and needs. Just as we cannot apply only one criterion to some system drivers, we may not be able to balance numerically the several criteria for mission selection. Instead, the developers, operators, and users must balance them using the insight gained from the system trades and mission utility analysis.

Having undertaken a round of system trades, evaluated the mission utility, and selected one or more baseline approaches, we are ready to return to the issue of system requirements and their flow-down to various components. Chapter 4 treats this area, which is simply the next step in the iterative process of exploring concepts and defining requirements.

References

Brookner, Eli, and Thomas F. Mahoney. 1986. "Derivation of a Satellite Radar Architecture for Air Surveillance." *Microwave Journal.* 173–191.

Cantafio, Leopold J., ed. 1989. *Space-Based Radar Handbook.* Norwood, MA: Artech House.

Defense Systems Management College. 1990. *System Engineering Management Guide.* Ft. Belvoir, VA: U.S. Government Printing Office.

Fortescue, P., and J. Stark. 1995. *Spacecraft Systems Engineering (2nd Edition).* New York: John Wiley & Sons.

Johnson, Richard D., and Charles Holbrow, eds. 1977. *Space Settlements, A Design Study.* NASA SP-413. Washington, DC: National Aeronautics and Space Administration.

Kay, W.D. 1995. *Can Democracies Fly in Space? The Challenge of Revitalizing the U.S. Space Program.* Westport, CT: Praeger Publishing

O'Neill, Gerald. 1974. "The Colonization of Space." *Physics Today.* 27:32–40.

Przemieniecki, J.S. 1990a. *Introduction to Mathematical Methods in Defense Analysis.* Washington, DC: American Institute of Aeronautics and Astronautics.

————. 1990b. *Defense Analysis Software.* Washington, DC: American Institute of Aeronautics and Astronautics.

————. 1993. *Acquisition of Defense Systems.* Washington, DC: American Institute of Aeronautics and Astronautics, Inc.

————. 1994. *Mathematical Models in Defense Analysis.* Washington, DC: American Institute of Aeronautics and Astronautics, Inc.

Rechtin, Eberhardt. 1991. *Systems Architecting.* Englewood Cliffs, NJ: Prentice Hall.

Shishko, Robert. 1995. *NASA Systems Engineering Handbook.* National Aeronautics and Space Administration.

Shishko, Robert, et al. 1996. "Design-to-Cost for Space Missions." Chapter 7 in *Reducing Space Mission Cost*, James R. Wertz and Wiley J. Larson, eds. Torrance, CA: Microcosm Press and Dordrecht, The Netherlands: Kluwer Academic Publishers.

Chapter 4

Requirements Definition

Stanley I. Weiss, *Massachusetts Institute of Technology*
Michael S. Williams, *Lockheed Martin Global Telecommunications*

An early adage in systems engineering was "requirements before analysis, requirements before design." This emphasizes the importance of defining and developing requirements as the front-end process for system design, development, and deployment. Regardless of size and complexity, and whatever the formality and scope of this process, it should follow the general pattern described in this chapter.

All requirements must begin with succinct but well defined user and customer mission needs, focusing on the critical functional and operational requirements, without unnecessarily constraining or dictating the design. Section 4.1 shows that the requirements derived from these mission needs and progressively allocated to lower levels of the design are central to meeting a program's performance commitments. Section 4.2 describes the process of analyzing requirements and budgeting performance. As we derive functions and the associated performance requirements, we must document them to provide the basis for developing, producing, deploying, and operating the system, as well as a referencable history governing the development. Section 4.3 shows the role of requirements documentation. Finally, Sec. 4.4 summarizes a brief step-by-step method of establishing requirements for typical space mission programs.

This traditional approach to systems engineering is to first define the requirements and then design the system to meet those requirements at minimum cost and risk. More recently a number of authors and organizations have advocated "trading on requirements" as a formal process intended to provide a compromise between what the user wants and what the buyer can afford. This process is discussed in detail by Wertz and Larson [1996].

4.1 Role of Requirements in System Development

To this point, the book has dealt with the mission analysis and concept development process which ideally drives the system design. The mission objectives and system concepts we have adopted have involved five basic measures: (1) required performance, (2) cost, (3) development and deployment schedule, (4) implicit and explicit constraints, and (5) risk. The same measures continue to apply during the entire system engineering process, from concept to implementation. Through this process, we decompose and allocate the central system-derived requirements (sometimes expressed as system specifications) to individual segments or system elements, interfaces between these as well as interfaces external to the system. To define the total system, therefore, users, customers, system engineers, and segment developers must constantly interact. Although we initiate the process in a "top-down" fashion, we typically must continually reconcile system level requirements with technology and lower-level design development.

A healthy tension often exists between the user and development communities. Developers may consider the user wedded to current operational approaches and insensitive to how over-specified requirements constrain design. Users often believe that developers favor new technology and ignore the practical needs associated with operating a system and exploiting the mission data. Thus, the developer may establish mission requirements without consulting the user, or the user may produce "non-negotiable stone tablets" and carry them down from the mountain too late or too over-specified for actual use. Because both sides have valid concerns, however, they must cooperate from the start in developing the mission's operational requirements. We may implement this cooperation through so-called IPTs (Integrated Product Teams) involving both users/customers and developers.

Typically, developers wanting to build as soon as possible drive prematurely toward low-level detail. Sometimes they underemphasize the original mission drivers —requirements which dominate performance, cost, and schedule risk. Customers often constrain system development with overly specific requirements at levels below the critical requirements that determine most of a program's cost and risk. While the level of formality and detail may vary depending upon system maturity, complexity, and size, critical requirements must remain in the forefront during design, development, and validation of the system.

Overzealous requirements can also find their way into mission statements. For example, a user may specify the scan rate and swath width under payload and coverage performance. Clearly, these constraints on sensor design and constellation are inappropriate in this case, prior to establishing a system which meets the key requirements, i.e., timely data with enough accuracy and resolution. Specifications on launch rate, launch responsiveness, and spacecraft reliability are also common. But so long as a system meets availability and maximum outage needs, the developer should be able to allocate requirements for reliability, maintenance, and replacement. Mission requirements concerning launch, operation, or maintenance may establish the design domain, but not dictate the design. On the other hand, the user must also be a party to the system design as it converges, to identify design characteristics likely to produce operational problems.

Table 1-5 in Sec. 1.4 shows essential requirements for the FireSat mission. These requirements neither dictate nor impose needless constraints on design, but they do specify what is essential to perform the mission and operate the system. The table contains enough information to derive the specific design characteristics with sufficient controls on the user's essential requirements. Also, the table includes no unverifiable terms or goals such as "maximize," "sufficient," or "optimize," because these words have no quantifiable interpretations. Requirements which we are asked to implement only if no "impact" results, are in fact goals and we cannot treat them as design drivers. Every meaningful requirement bears cost and will have an impact.

Constraints are those requirements for a system which we cannot trade, usually under any circumstances. They may pertain to performance when levels of capability of a system must have a certain value to be useful. One example is the necessity for a resolution level of an optical or RF signal, above which the desired information could not be derived or would not be sufficiently better than existing systems to justify new development. A related, fixed requirement could also be coverage and timeliness of data, clearly a major consideration for FireSat. Another might be cost—a constraint increasingly important to the financial success of a new mission. Thus, if a cost ceiling of N millions could not be met for a new development, the feasibility, design attributes or method of achieving a mission would be directly affected. The term "design to cost" applies directly to a cost constraint. Schedule may also be a constraint, and many technically worthwhile projects get scrubbed because developers could not solve some problems soon enough to be competitive—this is often called a "time to market" constraint. Others, but by no means all, include environmental and safety issues, legal and political mandates, fixed asset usage, involvement of geographically distributed or foreign offset contractors.

An alternative view of "goals" vs. "requirements" is that the former represent design margin. Any firm requirement must result in a level of margin in the design, and we can regard the "goal" as specifying the desired margin. As the design matures, the margin represents the trade space available to decision-makers. The user must ultimately decide whether the additional performance is worth its associated incremental cost.

Designers often focus on performance areas, such as operating the payload and distributing the mission data, and underemphasize the more mundane requirements, such as availability and accommodation to the external environment. Yet these can be critical to cost and risk. For example, availability can demand increased component reliability and therefore raise development costs. It can drive maintenance concepts, including replenishment and on-orbit support. It can also affect production time, especially for critical components. Likewise, ignoring external interfaces can produce a system design without the external support needed to deploy and operate the mission.

When space systems perform more than one mission, planners must develop requirements which account for each mission. For example, the IR surveillance payload on FireSat may serve other users with its performance in IR imaging and radiometric measurement. If the increased cost and risk are acceptable, their requirements could lead to more payload bands, added coverage, and added distribution requirements. That is why we must establish all valid missions early in requirements definition, or we should incorporate accommodations for new missions in future upgrades to a system's capabilities.

While we must address system requirements throughout all aspects of the development cycle, the role and characteristics of requirements change in each development phase. Consequently, we should use specific structure and language early in the process without premature detail. Table 4-1 shows how the requirements converge during system development. Concept development must continue to reflect the driving requirements, including internal and external interfaces. Top level or mission requirements drive early activities—developing the system concept and assessing technology. We must be prepared to modify these as the concepts and design mature and cause re-evaluation.

TABLE 4-1.   Evolution of Requirements Activities and Products. Each development phase tends to focus on specific requirement and design considerations.

Needs Analysis
• Defining mission requirements
• Defining environment
• Identifying mission drivers and constraints
• Technology programs

Concept Development
• Identifying critical driving requirements and associated risks
• Developing operations and design concepts
• Cost estimates
• Functional analysis and major interfaces
• System studies and simulations
• Prototyping and assessing technology

Concept Validation
• Tailored system and segment definitions
• Preliminary internal interface requirements
• Preliminary system standards
• Preliminary requirements flowdown
• Integrated system validation including test planning
• Transition planning
• Validating technology

Design and Implementation
• Detailed requirements flowdown
• Developing formal design documentation and interface control
• Integrating and testing the system
• Demonstrating and verifying the system
• Test procedures and reports

During concept development, we normally carry forward and evaluate many design options, so we need to specify and document requirements in critical areas in a flexible fashion. We generally don't require formal specifications complying with acquisition standards and serving as the legal basis for the system until full-scale development. At that point, we need to have solved the critical program risk areas. Until then, however, there are no set prescriptions for the requirements products other than what the program finds applicable and workable.

We should, of course, recognize that the spectrum of valid approaches for requirements development and application is broad. Significant differences exist among NASA, DoD, ESA, NASDA and other development agencies, as well as their contractors, and even among locations within the same organization. For example, all

NASA organizations conduct Phase A and Phase B studies which result ultimately in a Request for Proposal, including top-level specifications. But they vary widely in their approaches to conducting these studies and their requirements products. For DoD organizations, the rituals of MIL-STD-499 have often overwhelmed arguments based on unique program needs, and requirements become over-detailed and over-formalized too early. In full-scale development, most of the requirements activities center on integrating program interfaces (inter-segment and external to the system) and resolving ways of carrying out specific requirements at segment level. Solving major system issues at this point can be expensive and risky. Usually, we freeze requirements once the system passes into production. Rarely can a program afford to accept changes at this point, opting far more often to accept limits to the system as designed or to defer the change to a later upgrade.

We often hear that requirements drive technology programs, but in fact, new technologies frequently make systems possible. For example, improvements in bandwidth for communications processing have permitted greater use of real-time data downlinks. But relying on new technologies or production abilities can be risky. New technologies which allow us to reduce design specifications for power, weight, and volume can improve system performance and cost. We must, however, monitor the technology and production base and carry backup plans, in case program risk management demands changes to basic design requirements and interfaces to reallocate performance.

Although the success of every program hinges on performance, cost, and schedule, cost is typically the most constraining. One reaction to cost emphasis is the *design-to-cost* practice by which a fixed dollar amount affects possible design solutions. Thus, progressive design development may, under cost limitations, cause review of requirements, with attendant trades between cost and performance. This has clearly been a factor in the design and functions of the International Space Station (ISS). We can do much to control program costs while analyzing requirements. For instance, over-specified requirements may be "safe," but evaluation of necessary design margins early via close interaction between the developer and the requirements specifier permits us to make timely trades.

As discussed earlier, defining requirements without attending to production and operational support is also costly. Thus, with every major decision, we must consider which performance option meets essential requirements while minimizing cost.

Sometimes, standardizing can reduce costs and improve operability. For example, particularly in the commercial communications industry, use of a "standard bus" or basic vehicle can yield lower costs for many programs. We sometimes call this process "platform-based design." In addressing approaches to standardization, however, we must always consider trade-offs between reduced cost and increased development risk.

As shown in Chaps. 1–3, mission development is an iterative process. Although each stage seems to cascade forward without hesitation, each requires significant feedback and adjustments. Typically, most of the feedback occurs between adjacent phases of development. However, some situations may demand feedback across multiple phases, such as when an element design falls short on a particular requirement and causes a change in the design and operations concept, and possibly a change to the original schedule.

An aside on requirements and cost control is imperative here. Solutions to constraining cost (e.g., design-to-cost specification, imposed standardization) are difficult to implement in truly innovative space systems. In fact, well-intentioned approaches early in the design cycle may result in serious cost growth later in design and operation. But this difficulty in explicit cost control does not imply we should avoid the challenge. The growth in cost from the early estimates performed during Concept Development is typically driven by a few controllable problems. First, not fully accounting for all elements of cost in these early estimates is common. Frequently, not consulting with designers and manufacturers who will develop the system and the operators who will control the system results in misunderstanding cost or missing elements of cost. Second, overspecifying the system inhibits trades which we can focus on cost reduction. Finally (and probably the most prevalent problem), heavy and uncontrolled changes to requirements as the system proceeds through latter stages of design can create major growth in cost due to constant redesign and related material and time waste. Worse, the loss of a fully understood system baseline becomes more likely and potentially very costly later in the program. The process of defining and flowing down requirements affects cost more than any other program activity.

Then, too, on several occasions, customer requirements accepted without rational challenge have led to unjustifiable project costs and, in two well-documented cases, eventually caused cancellation. One of the authors once had the opportunity to convince a customer that a new requirement that was inserted after program start would not enhance the mission; millions of dollars were saved and the customer's belief in our integrity was solidified.

4.1.1 Quality Function Deployment—A Tool for Requirements Development

While there are several structured approaches to developing requirements from the customer/user needs, the most commonly used tool is *Quality Function Deployment*, or QFD. Its application is not product limited; we also use it in developing of requirements for processes and services.

Quality Function Deployment derives from three Japanese words or characters meaning (1) quality or features, (2) function or mechanization, and (3) deployment or evaluation. Symbolically we define the combination as "attribute and function development." It involves a series of matrices organized to define system characteristics and attributes and can be applicable over multiple areas of decomposition. The first level, connecting customer needs or requirements to technical attributes or requirements, we often called the *House of Quality* and configure it in its simplest form as in Fig. 4-1. We often call the left hand column the "*Whats*" (at this first level, this is called the "voice of the customer") and we call the horizontal attributes the "*Hows*." This relationship will become apparent as the "*Hows*" define the means for fulfilling the "*Whats*."

Weightings are applied to the "what" side of the matrix and are usually graded in three levels to help establish priorities of needs and related technical attributes. While of value in trading requirements, the primary use at this stage should be to define trade space.

Figure 4-2 shows a simplified application to FireSat. Referring to Table 1-5 and illustrating with only a few of the identified mission needs, an abbreviated matrix shows some five needs and six relevant attributes. Note the conflicts between competing satellite orbits which could potentially satisfy key requirements. This suggests carrying out extensive analysis and trades. Note also the relative priorities emphasizing technical attributes which assure timely coverage.

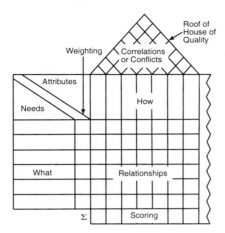

Fig. 4-1.   Abbreviated House of Quality.

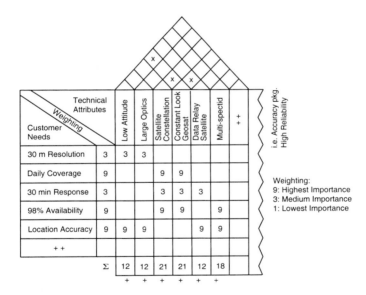

Fig. 4-2.   Simplified QFD for FireSat Mission. This limited set of needs and responding technical attributes would be expanded significantly for the complete design process.

Recalling the discussion on constraints (at the end of Sec. 4.1), we understand that the customer needs that the system cost no more than $20 million per year of operation; that is a constraint and all needs and technical attributes must meet this criterion. Thus, while it is a fixed requirement, we may leave it off the customer needs column of the QFD so as not to overbalance weighted scoring. If it were stated as having a target cost of $20 million or less, we might trade that figure and put it on the left hand side of the matrix. As a next stage use of QFD, the technical attributes developed in the top level would then become the requirements or "what" (left side) of the QFD matrix, with definitive characteristics, such as specific orbits, coverage per pass or unit time and top reliability level which we would derive to satisfy the set of specified technical requirements. Figure 4-3 illustrates this progression.

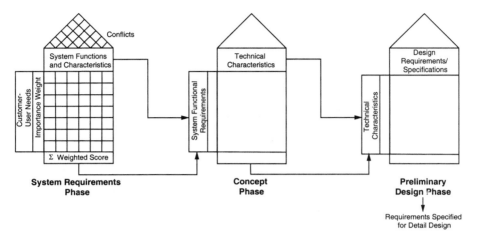

Fig. 4-3.   Progression of QFD Process. Illustrated is the derivation of successive "What" aspects from previous levels' responsive "Hows."

Thus the QFD is a structured means for a design team to address customer needs and to develop the consequent design characteristics to satisfy them. It also serves to sustain the trail of requirements derivation and provides a means for analyzing the impact of changes to requirements at any level. And since we can link the technical attributes responsive to needs, to functions of the system, there is a logical translation to functional analysis via functional flow diagrams and thence architecture and interface definitions.

As an added note regarding understanding the customer, I know of several satellite projects that have had little success as commercial ventures because the contractor's designers established requirements based on their own interpretation of potential customer needs. This was also the cause of a major military satellite contract loss to the competition due to inaccurately presumed knowledge of customer's desires. The voice of the customer must be heard before fixing a design.

4.2 Requirements Analysis and Performance Budgeting

We must decompose every system requirement into progressively lower levels of design by defining the lower-level functions which determine how each function must be performed. *Allocation* assigns the function and its associated performance requirement to a lower level design element. Decomposing and allocating starts at the system level, where requirements derive directly from mission needs, and proceeds through segment, subsystem, and component design levels. This process must also ensure closure at the next higher level. *Closure* means that satisfying lower-level requirements ensures performance at the next higher level, and that we can trace all requirements back to satisfying mission needs. This emphasizes the iterative character of the requirements development process.

Figure 4-4 shows how a single mission need—the FireSat geopositioning error—flows through many levels of design. Errors in the final mission data depend on many sources of error in the processing segments for space, mission control, and mission data.

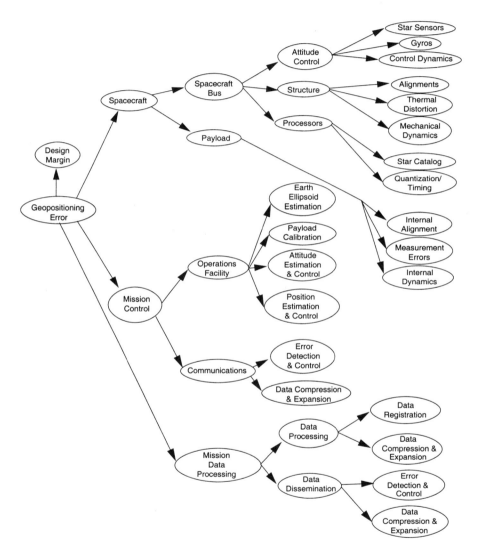

Fig. 4-4.   Allocation from Mission Requirements through Component Design. Understanding the sources contributing to top-level requirements is essential.

Two important observations are necessary. First, the system encompasses more than the spacecraft, and errors come from numerous segments. The accuracy of the geolocated object in a FireSat image is driven by much more than the spacecraft's pointing capability. Second, while the number of error sources is large, they are not all equal. Many are predictable and relatively constant—star catalogs and Earth ellipsoid estimates. Others are more variable, but small and not significant drivers for cost or technology. The remaining errors are those which are most amenable to cost-performance-risk trade-offs and need the greatest level of attention during requirements flowdown and validation.

4.2.1 Functional Analysis

The simplest way to represent functions—or actions by or within each element of a system—is through a functional-flow block diagram. As Fig. 4-5 shows, we define the topmost or first level functions of a system in the sequence in which they occur. Successive decomposition permits identifying how a system works at each level before proceeding to lower levels. For example, to address sensor misalignment three levels down in the functional flow (Function 4.4.4 in Fig. 4-5), it is necessary to consider the production (1.0) and integration (2.0) phases, which require manufacture and validation within reasonable tolerances.

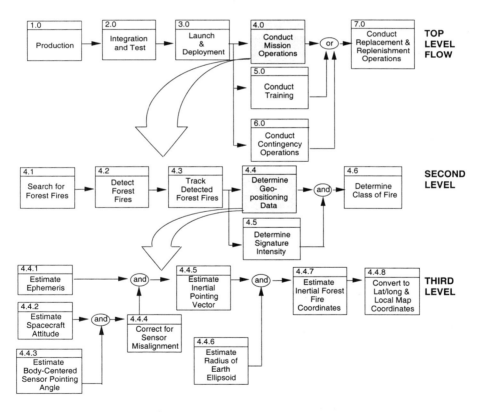

Fig. 4-5.   Functional Flows Generating Geopositioning Information for FireSat Mission. The functional flow defines what is to be done in a hierarchical framework. Additional features can be added to the representation (e.g., data interfaces, control sequences) using different diagramming techniques.

Once we establish the top-level functions and sequences, we can decompose and analyze each function throughout the remaining layers of the flow. For example, determining geopositioning data (Function 4.4 in Fig. 4-5) for FireSat requires a sequence of actions, from estimating key spacecraft and payload parameters to deriving local Earth coordinates. Functional decomposition, regardless of how formalized, is necessary in allocating design characteristics at each level of system architecture (the organization of system elements into a structured hierarchy). This organization permits us to allocate performance budgets together with other budgets affecting cost and risk.

We can also use functional flow diagrams to depict information or data flow, as well as control gates governing function sequencing. Information may include interface data flowing between functions, control relationships showing what must happen before another function can begin, or data sources and destinations.

In applying these techniques, we may use manual methods, particularly for simple systems, for top-level mission descriptions, but CASE (computer-aided system engineering) tools facilitate diagramming decompositions and maintaining traceability. But as with other computer applications, the software for developing diagrams and maintaining support databases does not drive the analysis. In fact, the functional framework which evolves is often a compromise among estimates of performance, cost, schedule, and the risk associated with each decision. (McClure [1988] and INCOSE Sixth Annual Proceedings [1996] provides an interesting discussion of support tools and techniques.)

4.2.2 Initial Performance Budgets

Analyzing requirements leads eventually to hierarchically organized performance metrics and budgets for the interactive development segments. The iterative process starts with budgets derived using analysis, simulation, known design or test data, and a large measure of experience. We should note that in the development of requirements and derived functions, mission drivers must be the primary drivers.

Experience or related reference missions are especially important in developing the initial performance budgets to meet system performance requirements. In the example of Fig. 4-5, the geopositioning accuracy reflects this. The major trade-offs and focus for validating performance therefore reside in how accurately the system can estimate and control position and attitude, and we must evaluate the options considered against not only performance requirements but also cost and schedule.

Figure 4-6 illustrates the combined effect of spacecraft attitude and position errors on the geopositioning estimate's accuracy in locating fires. In this simple example, three broad options are possible. The first option gives a very loose spacecraft position budget, which permits only limited support from the GPS and/or remote tracking stations. However, it requires a tight attitude budget, which is likely to create problems for both the space and mission control segments. Though payload sensitivity and resolution drive the selection of the FireSat orbit envelope, using a low-Earth orbit could severely affect attitude accuracy because of atmospheric drag. A higher altitude would reduce drag, but produce even tighter pointing tolerances. Thus, two main costs make this a poor budget option: the satellite's subsystem for controlling attitude and the potentially taxing calibration which the mission control segment must perform on the attitude sensors.

At the other extreme, leaving the attitude budget loose and tightly estimating spacecraft position can have risks if a full GPS constellation is not in operation. Using GPS

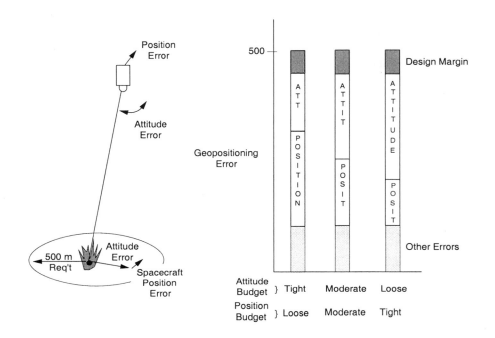

Fig. 4-6.   Typical Options in Error Budgets for Attitude and Position. Variations in attitude and ephemeris accuracy requirements have implications on allocation and attendant design risk. A balance of cost, performance, and implementation risk must enter the evaluation of options. Details of mapping budget development are given in Sec. 5.4.

risks degraded performance without a full constellation. Resorting to remote tracking stations or other sources of information can require excessive response times. A third option allows some risk for both attitude and position error budgets, but balances that risk against the cost of achieving the required geopositioning accuracy.

Table 4-2 lists the elements we would normally budget with the chapter and paragraph where we discuss each element. Budgeted items may come directly from requirements such as geolocation or timing, or they may be related to elements of the overall system design such as subsystem weight, power, or propellant.

Timeline budgets at the system level are also typical mission drivers. For FireSat, tight timelines for tip-off response and data distribution will require developing an initial budget. We must define and decompose all functions necessary to meet this timeline, as well as define their allocation and control sequences (functions which cannot start without completion of others and potential data hand-offs). Simulation will help us estimate delays in processing and communication. Applying experience or data from related systems provides some calibration. But this initial budget is just that, since as the design process progresses, we will introduce changes from design iterations among different levels.

It is, however, extremely important to recognize the nature of initial design budgets. They are typically developed by system engineers with a broad understanding of the

TABLE 4-2. **Elements Frequently Budgeted In Space Mission Design.** Primary budgets are directly related to mission requirements or ability to achieve the mission (e.g., weight). These primary requirements then flow down into secondary budgets.

Primary	Secondary	Where Discussed
Weight	Subsystem weight	Secs. 10.3, 10.4
	Power	Secs. 10.3, 10.4, 11.4
	Propellant	Secs. 10.3, 10.4, 17.4
Geolocation or System Pointing Errors	Pointing & Alignment	Secs. 5.4, 10.4.2, 11.1
	Mapping	Sec. 5.4
	Attitude Control	Secs. 4.2, 10.4.2, 11.1
	Attitude Determination	Secs. 4.2, 10.4.2, 11.1
	Position Determination	Secs. 4.2, 6.1
Timing	Coverage	Secs. 5.2, 7.2
	Communications	Sec. 13.1
	Operations	Sec. 14.2
	Processing	Sec. 16.2.1
Availability	Reliability	Secs. 10.5.2, 19.2
	Operations	Sec. 14.2
Cost	Development cost	Sec. 20.3
	Deployment cost	Sec. 20.3
	Operations and maintenance cost	Sec. 20.3

system and its elements. But the details of new technology and lower-level design studies can and should result in adjustments to these budgets as experts familiar with specific subsystem and component design review the initial allocations. A key aspect of the system design is a robust initial allocation (i.e., one which can tolerate changes at subsequently lower design levels) and adaptable to iterations as noted previously. Just as it is important to involve representatives of all affected levels of design in the development of the initial budgets, it is also important to recognize the iterative nature and that a system solution which minimizes total cost and risk may impose more stringent demands on certain aspects of lower-level designs than others. The process of reconciling the imposed costs and allocated risks involves a high degree of negotiation.

Table 4-3 shows how the response timeline may affect the space and ground segments of the system. While it may seem desirable to assign responsibility for a specified performance parameter to a single segment, we must evaluate and integrate critical system parameters across segments. For example, FireSat must respond quickly to tip-offs in order to provide the user timely data on suspected fires. This single response requirement alone may define the size and orbit envelope of the satellite constellation to ensure coverage when needed. Thus, time budgets for the following chain of events will be critical to the mission control segment's performance:

- Formulating the schedule for pass & time intervals
- Developing and scheduling commands to the spacecraft
- Developing and checking constraints on the command load
- Establishing communications with the spacecraft

TABLE 4-3.　**Impact of Response-Time Requirement on FireSat's Space and Ground Segments.** The assumed requirement is for fire data to be registered to a map base and delivered to a user within 30 min of acquisition.

Impact on Space Segment
Spacecraft constellation accessibility to specified Earth coordinates
Command load accept or interrupt timelines
Communication timelines to ground segments
Satellite availability
Impact on Ground Segment
Time to determine and arbitrate satellite operations schedule
Manual interrupt of scheduled operations
Command load generation and constraint checking time
Availability of mission ground segments and communications
Image processing timelines
Image sorting and distribution timelines

The space and ground segment budgets may involve interrupting current command loads, maneuvering the spacecraft, collecting the mission data, establishing communications links scheduled from the ground, and communicating the mission data. Mission data processing must receive, store, and process the mission data, sort it by user or by required media, and send it to the user. We must consider all of these activities in establishing budgets to meet the system requirement of delivering specified data and format within 30 min of acquiring it.

Requirements Budget Allocation Example

Pointing budget development, described in Sec. 5.4, is a problem on space missions using pointable sensors. Another common budget example is the timing delay associated with getting mission data to end users. It can be a critical requirement for system design, as is the case of detecting booster plume signatures associated with ballistic missile launches. In that case, coverage (i.e., the time from initiation of a launch to initial detection) as well as the subsequent transmission, processing, distribution, and interpretation of the detection, is time critical. Because of the severe coverage requirement, geosynchronous satellites with sensitive payloads and rapid processing are needed.

The FireSat mission does not require timing nearly as critical as missile detection, but clearly the detection of forest fires is a time-sensitive problem. Figure 4-7 shows both the timeline and the requirements budget associated with it. For FireSat's Earth coverage (i.e., Time Segment 1), it would be ideal to provide continuous surveillance using a geosynchronous satellite. However, cost and ground resolution favor a low-Earth orbit implementation which results in Time Segment 1 being three to six hours, depending principally on the number of satellites in the constellation.

Once detection occurs, a series of shorter timeline events must occur to achieve the 30-minute requirement for Time Segment 2. The system may need to validate each detection to minimize the number of false alarms transmitted to the ground for processing. This may impose design specifications for onboard detection processing and additional payload "looks." The time spent downlinking the data after validating a detection could have a significant impact on the communications architecture that assures rapid acquisition of the required links. The availability of direct or relay links

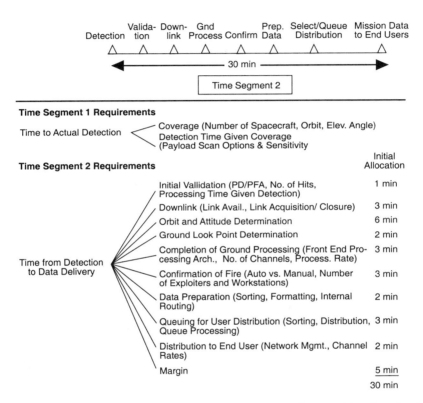

Fig. 4-7. **Mission Data Timeline and Requirements Budget.** The actual time from the detection of a fire to distribution of the time-urgent data is related both to coverage and specific timing requirements.

to meet this timeline is a significant cost driver, potentially replacing a "store and dump" approach appropriate for purely scientific missions.

Once the ground system receives the data it must process the data to format it, perform orbit, altitude, and ground-look-point determination, and then extract the relevant fire-detection data. A short time requirement here will likely demand real-time processing and a substantial capacity to support real-time operation. Identification and subsequent confirmation of a fire prior to broader dissemination may drive either a high performance pattern-matching process or manual processing in a time-critical fashion. Once the system confirms a fire, the data must be registered and prepared for distribution to appropriate end users. This preparation may involve merging it with standard data sets to support evaluating the fire at a later time. The data processing system must also queue the data for distribution over a network. Priorities and protocols may drive the management of input queues and network routing. Figure 4-7 shows the initial allocations for the components of Time Segment 2.

This example punctuates two critical activities: First, the components of a timeline must follow the step-by-step functional flow described in 4.2.1. The functions themselves may be strictly sequential or capable of being processed in parallel to shorten timelines. Functional representation diagrams and support tools (e.g., built-in simulation) can ease this evaluation. Second, there are numerous performance-cost trade-offs at each decision point which dictate the time-budget allocations. The objective is to meet the highest level requirement while equally sharing the potential performance risk and cost associated with meeting each derived requirement.

4.2.3 Refining and Negotiating the Performance Budgets

System engineers must thoroughly understand how to develop and define requirements, then allocate and negotiate budgets associated with them. Failure to meet key budgets can lead to major system problems. Early definition permits the iterative process of adjusting allocations, margins and even operations well before major cost or schedule penalties occur.

Performance budgeting and validating key system requirements is the iterative process, as shown in Fig. 4-8. Before the process can actually start, however, the specific performance parameter and associated requirement statement must be clear and traceable to the mission need. The Quality Function Deployment methodology and several tools make this possible by maintaining the link between the need and the technical requirement in traceable documentation. Vague, inconsistent, or unquantifiable requirements too often lead to inaccurate understanding, misinterpretation and/or exploitation. This applies especially to critical areas of system performance which without early and thorough interaction and/or prototype testing can become expensive and program-threatening later. We should also note that the iterative process includes negotiation and re-negotiation of budgets based upon evidence from the design process and the discovery of errors and "injustices" in the initial allocation.

We know of several programs in which major difficulties have resulted from conflict among requirements. One case involved the difference between operational availability of ground stations with that of the satellites in a system. Another involved the selection of the launch vehicle before a design concept was established, the requirements for the latter driving the mass far beyond the booster's lift capability. And in a third case, the changes in a customer's program management introduced new requirements for a payload which invalidated the flowdown of the original project requirements. The response to this required both data and persuasiveness, the latter being unfortunately insufficient until serious problems arose in the systems design.

An aside is worthwhile at this point on the issue of requirements-level vs. design-level budgeting. The system-level design is a logical integration or synthesis of segment designs. Defining functions and their performance requirements and those interfaces requiring support lays the framework for deciding "how" to design each segment. For FireSat, this relates to the accuracy of the geolocation and the allocation to segments of ephemeris, attitude, and other contributions. The "how" relates to space segment hardware decisions such as whether to use star sensor or gyro performance to achieve the required attitude accuracy. But such decisions affect mission operations which must then schedule star sensor calibration and gyro alignment so the spacecraft

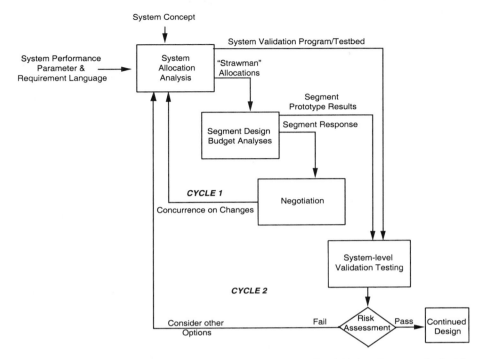

Fig. 4-8.   Iterative Performance Budgeting and Validation. The initial performance budget is a point of departure for later evaluation of segment design and validation testing. The design will converge to a validated value which may require system-level budget adjustments to match the changes.

can meet its requirements. Thus, both procedural and data interfaces must be identified and documented. In addition, mission operations must take into account the spacecraft's system for attitude control to keep from misusing it when scheduling operations. Likewise, in scheduling calibration or alignment, the mission operators must know the sensor payload and other electronic package performance characteristics to prevent accidental maneuvers. System and segment specifications provide system and interface requirements, and lower-level specifications provide design requirements, but in fact, the system engineer and segment designers must interact at all design levels.

It should be noted that initial budget estimates almost never correspond with design considerations at lower levels. Clearly, the early budgets are starting points for negotiation and budget adjustment, to reconcile early system allocations with segment estimates of design and performance. As we reconcile requirements, we should document them in a requirements reference which changes only with full traceability and visibility for all stakeholders such as segment designers, system engineers, and program managers. (A number of tools may be useful: these include the previously noted QFDs, plus software packages ranging from Excel to QSS DOORS and Ascent Logic's RDD 100.) Each of these critical performance parameters matches an established system and segment budget. These budgets can and normally do change as developers proceed on the design and validate performance.

At this stage of budgeting, design margin becomes an issue; specifically, how much is reasonable to keep, who knows where it is, and who has the authority to adjust it? Typically, margin is statistical (e.g., two-sigma error requirements), so as it cascades to various levels of design it can produce significant overdesign and cost. Design engineers can complicate appropriate adjustment by keeping margin at lower tiers of design, where it tends not to be visible or usable for reallocation. Here prescription cannot substitute for judgment. Sometimes, margins can provide robustness against on-orbit failures, but can also cause problems. For example, too much margin in communication links could actually saturate receivers. Key system requirements must also have margins, which we can trade or allocate downward, so as to permit meeting realistic performance and reliability with minimum risk.

Once the first cycle of interactions between system and segments personnel has established the best controlled estimate of key performance budgets we must continue to test the design we are developing. Configurations should be validated via simulations or prototypes. These early exercises in system integration are important in developing a consensus that continues through the initial design phase.

At all times a baseline of common requirements must support this process of analyzing and estimating performance requirements, interacting and negotiating with segment implementors, and validating the key performance drivers early in the design phase. The validation exercises use many specific scenarios or point situations to evaluate performance. Meeting performance budgets in these point situations is comforting, but not sufficient. Scenarios designed to stress one aspect of system performance may not provide adequate coverage of other aspects. The converging, controlled system requirements captured in requirements documents, interfaces, and standards are often the only reference for system functions and performance. The requirements documentation must match the phase of system development in maturity, but it must always reflect the results of analyses, performance budget negotiations, and validation exercises—faithfully, openly, and quickly.

4.3   Requirements Documentation and Specifications

In dealing with criteria for requirements documents, we should note the references governing much of today's systems engineering practice in the aerospace industry. With the deletion of most military standards in the United States as contractual requirements, internal documents most often establish and govern system design and engineering practices. These documents, however, are based largely upon either the previously controlling MIL-STD-499 or its successor (not issued but available in final draft) 499b, or newer civil organization standards. These include the Electronics Industries Association (in conjunction with the International Council on Systems Engineering) EIA/IS 632, Systems Engineering and the Institute of Electrical and Electronic Engineering (IEEE) Trial Use standard, and Application and Management of the Systems Engineering Process, IEEE 1220. Most recently, the International Standards Organization (ISO) moved to incorporate systems engineering in the growing body of international standards and to develop ISO Standard 15288, System Life Cycle Processes, which can serve as a framework for activities in the increasingly global context of the aerospace industry. All of these documents place mission and requirements development and management at the head of system design processes.

Effective requirements documents must be consistent and complete relative to the maturity of the system in its development cycle. *Consistency* means that we should write a performance requirement only once, keep it in a logical place, and position it appropriately in the hierarchy of established requirements. *Completeness* depends on the system's phase of development, but it means that at every level of requirement we must ensure that we satisfy the next higher level. As an example of completeness, the geolocation requirement of the FireSat mission should address all error sources both in segment and interface documents.

Requirements Traceability

Requirements must be rigorously traceable as we develop, allocate and decompose or derive them. While computer support tools exist which link and show dependencies and derivations among requirements, the complexity of the product should govern the form of the documentation. This form can range from notebooks for small projects to sophisticated database systems. We must base every design and decision task on requirements, and trade studies at any level must take into account all related require-ments, while considering the impact of changes throughout the product (and system) architecture. Any indexing method will suffice, so long as it permits traceability upwards as well as across all elements. Requirements documents should specify this tracing method, however, and the basis for derived requirements must be clearly iden-tifiable. The specific requirements document may be purely electronic, possibly using the database features of the computer tools. Whatever the documentation form, it must have a concise entry for every requirement. Each entry should index the documents and specific paragraphs from which we traced the requirement. Where analysis pro-duced a derived requirement, it should reference the specific technical memo or report showing the results of the analysis for future use.

Traceability emphasizes the need for effective requirements to be unambiguous and verifiable to avoid misinterpretation and exploitation. Words such as "optimize" or "minimize" in specifications cannot govern the design, and they defy verification.

We should note that requirements reviews are necessary corollaries to design reviews and issues must have the same weight as design issues in readiness decisions by a program to proceed to its next step of development. We sometimes call these "gates." Requirements assessments at such review points are critical. They may iden-tify the need to reassess project drivers, including:

- Accelerate or emphasize particular design areas

- Relax design budgets

- Reallocate requirements

- Consider operational work-arounds

- Acknowledge a program slip

- Revise funding profiles

Requirements documentation notionally falls into nine classes (Fig. 4-9). These are often designated as *specifications*. The figure also shows descriptive or supporting documents which need to be current with the requirements baseline.

Based on mission needs, analyses, and validation exercises, the *system require-ments document* (usually called "system requirements specification") should cover every relevant aspect of what the system should do (*functions*) and how well it should do it (*performance requirements*). It should address every aspect of system perfor-mance. Since ideally system requirements are the basis for segment requirements, they

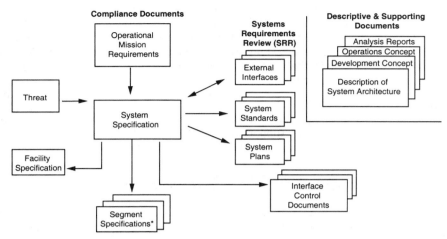

Fig. 4-9.   Requirements Documentation Hierarchy or *System Specification*.

should come before the latter. However, once segments are defined, there may be trade-offs required at the system level in response to cost, interface issues, performance limitations, or schedules related to segment designs.

We should note that among the system plans derived from requirements are test plans which will reflect validation and verification of these requirements in qualification and acceptance processes. These characteristically are reflections from test specifications which identify objectives, environments and levels of assembly at which tests are to be performed.

It should be remembered that requirements specifications, at system and lower levels, are potentially subject to change. Therefore, they should be designated, "preliminary" prior to reviews at each stage of design. During formal design phases, while requirements may have to be traded, the specifications must, like design documents, be subject to rigorous change control.

In addition, when requirements specifications at a top level govern more than one system segment, tailoring to accommodate the specific character of a segment may be appropriate. This is particularly so with requirements not directly associated with system performance.

Interface Management

Often, developers overlook or assume external interfaces in the early stages of system development, but they must be carefully considered with the total system architecture. Internal to the system, documenting interfaces between segments, usually through *interface control documents* or *ICDs*, is the key to integrating and maintaining relationships between these segments. The system level ICD may be referred to or included in the system specification. Documents covering critical interfaces, such as the spacecraft-to-ground segment for FireSat, can become highly complex. Two guidelines are important in developing and refining interface documents. Each document normally covers only two segments, although multiple elements within segments may require consideration of relationships with other segments. In general, we should avoid designs necessitating such added complexity.

In all cases, we must document these agreements at every level of design, usually in ICDs. At the system level, project managers or system engineers control these, while internal to segments, this is the responsibility of individual element leaders (see Chap. 16). Although the content and format of interface documents vary significantly with products and organizations, elements always addressed include physical and data or signal interfaces and interactions. Thus pin connections and message formats clearly must be defined in interface documents; but the characteristics of gyro drift and star sensor performance (such as nonlinearities of the transfer function, output axis coupling or star sensor noise) require the same definition level so that the mission ground station can correctly calibrate them.

4.4 Summary: The Steps to a Requirements Baseline

We have commented that we cannot prescribe a single means for establishing requirements. This chapter does, however, present guidelines for establishing a requirements baseline in approximately sequential order. This baseline is a reference not only for establishing the premises for functional design, but also a means of continually assessing the impact of design decisions on requirements validation. We can predetermine some requirements, such as constraints on a system. (One example could be the requirement to use existing NASA ground facilities.) We must recognize that requirements can and do change and that flexibility in the design process is necessary to accommodate such change, as in the need to iterate the relationships among design, functions and requirements. Documentation is also a critical aspect of the requirements process, for sustaining the baseline reference, as well as providing the translation for system development of the mission objectives.

TABLE 4-4.   Steps to Developing a Requirements Baseline.

1. Identify the customer and user of the product or services. A customer may be a procuring agent but not the ultimate user and both must be understood.

2. Identify and prioritize customer/user objectives and needs for the mission to be accomplished.

3. Define internal and external constraints.

4. Translate customer/user needs into functional attributes and system characteristics. Quality Function Deployment is one tool to do this.

5. Establish functional requirements for system and provide for decomposition to elements.

6. Establish functional flow and representative for its performance of functions.

7. Translate functional attributes into technical characteristics which will become the requirements for the physical system.

8. Establish quantifiable requirements from all the above steps.

9. Through the use of block diagrams expressing interfaces and hardware/software/data relationships for the system level.

10. From the architecture expressed by step 9 at the system level, decompose the functional requirements and characteristics sets to successive lower levels, i.e., the next level defining the basis of the elements of the system.

11. At all the steps above, iteration with preceding activities is necessary both to test the assumptions made and to reconcile higher levels of requirements and functional implementation.

In the steps which relate to determining requirements, every requirement must have at least the following three components: first, "what" the system is to do (the function); second, "how well" it is to perform the function (*performance requirement*); last, how we verify the requirement (*verification*). This last component should be of particular concern to us early in the requirements development process, and we should translate it into a verification and validation plan which will govern the quality and qualification test programs.

Table 4-4 lists ten steps to establishing a requirements baseline in the early phase of a development program. It emphasizes activities concerned with analyzing and validating system requirements versus the design of segments, subsystems, or components. These activities produce a hierarchical baseline of requirements which lead to allocation throughout a decomposed system.

References

Blanchard, B. and W. Fabrycky. 1990. *Systems Engineering*. Englewood Cliffs, NJ: Prentice-Hall, Inc.

Chubb, W.B., and G.F. McDonough. 1991. *System Engineering Handbook* MSFC-HDBK-1912. Huntsville, AL: NASA Marshall Space Flight Center.

Defense Systems Management College. 1990. *System Engineering Management Guide (3rd Edition)*. Ft. Belvoir, VA: Defense Systems Management College.

Electronic Industries Association. 1998. *Standard 632*. EIA.

Institute of Electrical and Electronic Engineering. 1996. *Standard 1220*. IEEE

International Council on Systems Engineering. 1998. *Systems Engineering Handbook*. INCOSE.

International Council on Systems Engineering. 1996. "Systems Engineering Practices and Tools." Proceedings of Sixth Annual Symposium.

Lugi, V. Berzins. 1988. "Rapidly Prototyping Real-Time Systems." *IEEE Software*. September: 25–36.

McClure, C. 1988. *Structured Techniques, The Basis for CASE*. Englewood Cliffs, NJ: Prentice Hall, Inc.

Shishko, R. and R.G. Chamberlain. 1992. *NASA Systems Engineering Handbook*. Washington, DC: NASA Headquarters.

Wertz, James R. and Wiley J. Larson. 1996. *Reducing Space Mission Cost*. Torrance, CA: Microcosm Press and Dordrecht, The Netherlands: Kluwer Academic Publishers

Chapter 5

Space Mission Geometry

James R. Wertz, *Microcosm, Inc.*

Much spaceflight analysis requires knowing the apparent position and motion of objects as seen by the spacecraft. This type of analysis deals predominantly, though not entirely, with directions-only geometry. We want to know how to point the spacecraft or instrument, or how to interpret the view of a spacecraft camera or antenna pattern. Two formal mechanisms for dealing with directions-only geometry are unit vectors and the celestial sphere. Unit vectors are more common in most areas of analysis. However, the celestial sphere provides greatly improved physical insight which can be critical to the space mission designer. Consequently, we first introduce the basic concept of using the celestial sphere for directions-only geometry and then apply the concept to space mission geometry as seen from either the Earth or the spacecraft. Finally, we develop a methodology for drawing up spacecraft mapping and pointing budgets.

To begin any formal problem in space mission geometry, we must first select a coordinate system. In principle, any coordinate system will do. In practice, selecting the right one can increase insight into the problem and substantially reduce the prospect for errors. The most common source of error in space geometry analyses is incorrectly defining the coordinate systems involved.

To define a *coordinate system* for space applications, we must first specify two characteristics: the location of the center and what the coordinate system is fixed with respect to. Typically, we choose the Earth's center as the coordinate system center for problems in orbit analysis or geometry on the Earth's surface; we choose the spacecraft's position for problems concerning the apparent position and motion of objects as seen from the spacecraft. Occasionally, coordinates are centered on a specific spacecraft instrument when we are interested not only in viewing the outside world but also in obstructions to the field of view by other spacecraft components. Typical ways to fix a coordinate system are with respect to inertial space, to the direction of the Earth or some other object being viewed, to the spacecraft, or to an instrument on the spacecraft. Table 5-1 lists the most common coordinate systems in space mission analysis and their applications. These are illustrated in Fig. 5-1. If you are uncertain of the coordinate system to select, I recommend beginning problems with the following:

- Earth-centered inertial for orbit problems
- Spacecraft-centered local horizontal for missions viewing the Earth
- Spacecraft-centered inertial for missions viewing anything other than the Earth

TABLE 5-1.   Common Coordinate Systems Used in Space Applications. Also see Fig. 5-1.

Coordinate Name	Fixed with Respect to	Center	Z-axis or Pole	X-axis or Ref. Point	Applications
Celestial (Inertial)	Inertial space[*]	Earth[†] or spacecraft	Celestial pole	Vernal equinox	Orbit analysis, astronomy, inertial motion
Earth-fixed	Earth	Earth	Earth pole = celestial pole	Greenwich meridian	Geolocation, apparent satellite motion
Spacecraft-fixed	Spacecraft	Defined by engineering drawings	Spacecraft axis toward nadir	Spacecraft axis in direction of velocity vector	Position and orientation of spacecraft instruments
Local Horizontal[‡]	Orbit	Spacecraft	Nadir	Perpendicular to nadir toward velocity vector	Earth observations, attitude maneuvers
Ecliptic	Inertial space	Sun	Ecliptic pole	Vernal equinox	Solar system orbits, lunar/solar ephemerides

[*] Actually rotating slowly with respect to inertial space. See text for discussion.
[†] Earth-centered inertial coordinates are frequently called *GCI (Geocentric Inertial)*.
[‡] Also called *LVLH (Local Vertical/Local Horizontal)*, *RPY (Roll, Pitch, Yaw)*, or *Local Tangent Coordinates*.

Unfortunately, the inertial coordinate system which everyone uses, called *celestial coordinates*, is not truly fixed with respect to *inertial space*—that is, the mean position of the stars in the vicinity of the Sun. Celestial coordinates are defined by the direction in space of the Earth's pole, called the *celestial pole*, and the direction from the Earth to the Sun on the first day of spring, when the Sun crosses the Earth's equatorial plane going from south to north. This fundamental reference direction in the sky is known as the *vernal equinox* or *First Point of Aries*.[*] Unfortunately for mission geometry, the Earth's axis and, therefore, the vernal equinox precesses around the pole of the Earth's orbit about the Sun with a period of 26,000 years. This *precession of the equinoxes* results in a shift of the position of the vernal equinox relative to the fixed stars at a rate

---

[*] The position of the vernal equinox in the sky has been known since before the naming of constellations. When the zodiacal constellations were given their current names several thousand years ago, the vernal equinox was in Aries, the Ram. Consequently the zodiacal symbol for the Ram, ♈, or sometimes a capital T (which has a similar appearance), is used for the vernal equinox. Since that time the vernal equinox has moved through the constellation of Pisces and is now slowly entering Aquarius, ushering in the "Age of Aquarius."

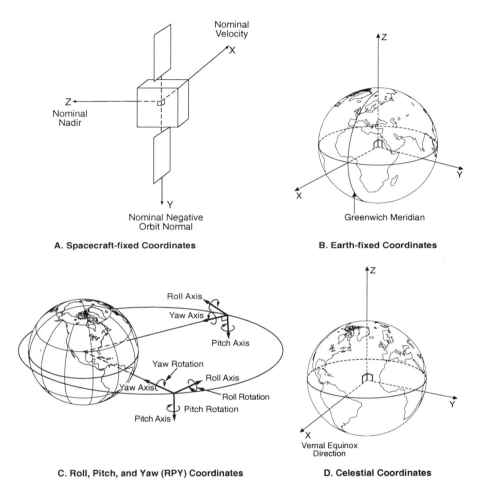

Fig. 5-1.   Coordinate Systems in Common Use. See Table 5-1 for characteristics.

of 0.014 deg/yr. Because of this slow drift, celestial coordinates require a corresponding date to accurately define the position of the vernal equinox. The most commonly used systems are *1950 coordinates, 2000 coordinates*, and *true of date*, or *TOD*. The latter coordinates use the same epoch as the orbit parameters and are traditionally used for spacecraft orbit analysis. The small corrections required to maintain TOD coordinates are conveniently done by standard computer subroutines. They are important for precise numerical work, but are not critical for most problems in mission analysis.

Once we have defined a coordinate system, we can specify a direction in space by a *unit vector*, or vector of unit magnitude, in that direction. While a unit vector will have three components, only two will be independent because the magnitude of the vector must be one. We can also define a unit vector by defining the two coordinates of its position on the surface of a sphere of unit radius, called the *celestial sphere*, centered on the origin of the coordinate system. Clearly, every unit vector corresponds to one and only one point on the celestial sphere, and every point on the surface of the sphere corresponds to a unique unit vector, as illustrated in Fig. 5-2. Because either representation is mathematically correct, we can shift back and forth between them as the problem demands. Unit vector analysis is typically the most convenient form for computer computations, while the celestial sphere approach provides the geometrical and physical insight so important to mission analysis. In Fig. 5-2 it is difficult to estimate the X, Y, and Z components of the unit vector on the left, whereas we can easily determine the two coordinates corresponding to a point on the celestial sphere from the figure on the right. Perhaps more important, the celestial sphere allows us easily to represent a large collection of points or the trace of a moving vector by simply drawing a line on the sphere. We will use the celestial sphere throughout most of this chapter, because it gives us more physical insight and more ability to convey precise information in an illustration.

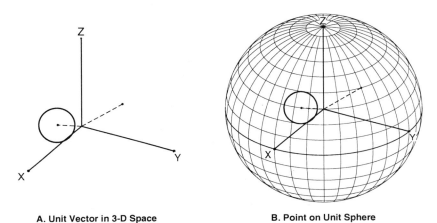

A. Unit Vector in 3-D Space   **B. Point on Unit Sphere**

Fig. 5-2.   Alternative Representations of Unit Vectors. In (B) it is clear that the small circle is of 10 deg radius centered at (15°, 30°) and that the single vector is at (60°, 40°). In (A) even the quadrant is difficult to estimate. Also note that the body of the unit vectors from the center of the sphere can be omitted since any point on the sphere implies a corresponding unit vector. Thus, the 3-dimensional pointing geometry is reduced to a 2-dimensional representation. See Fig. 5-5 for the definition of notation.

5.1 Introduction to Geometry on the Celestial Sphere

The *celestial sphere* is an imaginary sphere of unit radius centered on the observer, used to represent directions in space. It comes from classical observational astronomy and is far older than almost any other modern astronomical concept. The compelling image of the bowl of the sky at night makes it easy to think of stars and planets moving on a fixed, distant sphere. We now know, of course, that their distances from us are vastly different. But the concept of watching and computing the position and motion of things on the unit celestial sphere remains a very valuable contribution of classical astronomy to modern spaceflight analysis. Unfortunately, relatively few modern references are available. By far, the most detailed treatment is provided by Wertz [2001]. Green [1985], and Smart [1977] provide information on spherical astronomy.

Figure 5-3 illustrates the use of the celestial sphere to represent directions to objects in space. These objects may be very close, such as other components of the spacecraft, or very far, such as the surface of the Earth, the Sun, or stars. Although we will drop the observer from illustrations after the first few figures, we always assume that the observer is at the center of the sphere. Having become familiar with the idea of the observer-centered celestial sphere, we can easily work with points and lines on the sphere itself, ignoring entirely the unit vectors which they represent

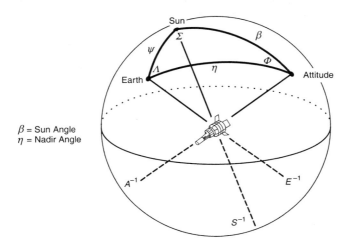

Fig. 5-3.   Use of Celestial Sphere to Represent Direction of Objects in Space. The sides of the triangle are arc lengths. The angles of the triangle are rotation angles.

Points on the celestial sphere represent directions in space, such as the direction to the Sun, the Moon, or a spacecraft axis. The direction opposite a given direction is called the *antipode*, or *antipoint*, and frequently has a "−1" superscript. Thus, S^{-1} is the direction opposite the Sun, and is called the *antisolar point*. *Nadir* is the direction to the center of the Earth. The direction opposite nadir is called the *zenith*. Points on the sphere may represent either directions to real objects or simply directions in space with no object associated with them, such as the direction parallel to the axis of the Earth (the celestial pole) or parallel to the +Z-axis of a spacecraft coordinate system.

A *great circle* on the celestial sphere is any circle which divides the sphere into two equal hemispheres. Any other circle on the sphere is called a *small circle*. Any part of a great circle is called an *arc* or *arc segment* and is equivalent to a straight line segment in plane geometry. Thus, the shortest path connecting two stars on the celestial sphere is the *great circle arc* connecting the stars. Two points which are not antipoints of each other determine a unique great circle arc on the celestial sphere.

Given 3 points on the sky, we can connect them with great circle arc segments (ψ, η, and β on Fig. 5-3) to construct a *spherical triangle*. The angles Λ, Σ, and Φ at the vertices of the spherical triangle are called *rotation angles* or *dihedral angles*. The lengths of arc segments and size of rotation angles are both measured in degrees. However, as illustrated in Fig. 5-4, these are distinctly different measurements. The arc length represents the side of the spherical triangle, and is equal to the angular separation *between* 2 points seen on the sky. The rotation angle, which is always measured *about* a point on the sphere, represents the angle in a spherical triangle, and is equal to the dihedral angle between 2 planes. For example, assume that we see the Earth,

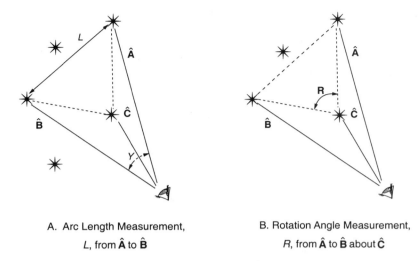

A. Arc Length Measurement,
L, from $\hat{A}$ to $\hat{B}$

B. Rotation Angle Measurement,
R, from $\hat{A}$ to $\hat{B}$ about $\hat{C}$

Fig. 5-4.   Distinction between Arc Length and Rotation Angle Measurements.

Moon, and Sun on the spacecraft sky. The arc length between the Sun and the Moon is the angular separation between them as measured by the observer. The rotation angle about the Earth between the Sun and the Moon is equal to the angle between 2 planes. The observer, Earth, and Sun form the first plane, and the observer, Earth, and Moon form the second. Both types of angles are important in mission geometry problems, and we must clearly understand the distinction between them. Table 5-2 lists the properties of these two basic measurement types.

As shown in Fig. 5-5, the +X-axis is normally toward the reference point on the equator, and the +Z-axis is toward the positive or North Pole. The great circles through the poles and perpendicular to the equator are called *meridians*. The meridian through any point on the sphere determines the *azimuth coordinate* of that point. *Azimuth* is the equivalent of longitude on the Earth's surface, and is measured along the equator. The azimuth is also equivalent to the rotation angle measured counterclockwise about the pole from the reference point to the point in question. The second coordinate which determines the position of any point on the sphere is the *elevation* or latitude component. It is the arc-length distance above or below the equator. The *co-latitude* or *co-elevation* is the arc length from the pole to the point in question. Small circles at a constant elevation are called *parallels*. Because a parallel of constant elevation is not a great circle (except at the equator), the arc length along a parallel will not be the same as the arc-length separation between two points. As Table 5-3 shows, several spherical coordinate systems in common use have special names for the azimuth and elevation coordinates.

The following equations transform the azimuth, Az, and elevation, El, to the corresponding unit vector coordinates (x, y, z):

$$x = \cos(Az)\cos(El) \tag{5-1a}$$

$$y = \sin(Az)\cos(El) \tag{5-1b}$$

$$z = \sin(El) \tag{5-1c}$$

TABLE 5-2.   Properties of Arc Length and Rotation Angle Measurements.

Characteristic	Arc Length Measurement	Rotation Angle Measurement
Solid Geometry Equivalent	Plane angle	Dihedral angle
How Measured in 3-D Space	Between 2 lines	Between 2 planes
How Measured on Sphere	Between 2 points	About a point or between 2 great circles
Unit of Measure	Degrees or radians	Degrees or radians
Component in Spherical Triangle	Side	Angle
Unit Vector Equivalent	Arc cos $(\hat{\mathbf{A}} \cdot \hat{\mathbf{B}})$	Arc tan $[\hat{\mathbf{C}} \cdot (\hat{\mathbf{A}} \cdot \hat{\mathbf{B}}) / ((\hat{\mathbf{A}} \cdot \hat{\mathbf{B}}) - (\hat{\mathbf{C}} \cdot \hat{\mathbf{A}})(\hat{\mathbf{C}} \cdot \hat{\mathbf{B}}))]$
Examples	Nadir angle Sun angle	Azimuth difference Rotation about an axis
How Commonly Expressed	"Angle from **A** to **B**" or "Arc length between **A** and **B**"	"Rotation angle from **A** to **B** about **C**"

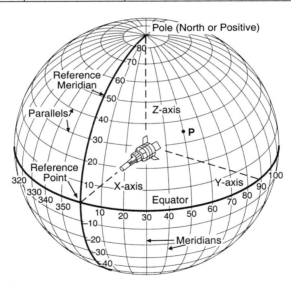

Fig. 5-5.   Definition of a Spherical Coordinate System on the Unit Sphere. The point **P** is at an azimuth of 50 deg and elevation of 35 deg, normally written as (50°, 35°).

Similarly, to transform from unit vectors to the corresponding spherical coordinates, use

$$Az = \text{atan2}\ (y/x) \tag{5-2a}$$

$$El = \text{asin}\ (z) \tag{5-2b}$$

TABLE 5-3.   Coordinate Names in Common Spherical Systems.

Coordinate System	Azimuth Coordinate	Elevation Coordinate (Z-axis)	Applications
Celestial Coordinates	Right ascension	Declination	Inertial measurements, Astronomy
Earth-fixed	Longitude	Latitude	Earth applications
Spacecraft-fixed	Azimuth or clock angle	Elevation	Spacecraft measurements, attitude analysis
Local Horizontal	Azimuth	Elevation*	Directions relative to central observer
Ecliptic Coordinates	Celestial longitude	Celestial latitude	Planetary motion

* Also used are *zenith angle* = angle from point directly overhead to point in question = 90 deg minus elevation angle; and *nadir angle* = angle at the observer from the center of Earth to point in question = 90 deg plus elevation angle.

where atan2 is the software function with output defined over 0 deg to 360 deg and the asin function is evaluated over –90 deg to +90 deg.

Spherical geometry is distinctly different from plane geometry in several ways. Most fundamental is that parallel lines do not exist in spherical geometry. This can be seen by thinking of 2 meridians which are both perpendicular to the equator of a coordinate system, but which ultimately meet at the pole. All pairs of great circles either lie on top of each other or intersect at 2 points 180 deg apart.

Another concept in spherical geometry that is different than plane geometry is illustrated in Fig. 5-6, in which we have constructed a spherical triangle using the equator and two lines of longitude. The intersection of the longitude lines with the equator are both right angles, such that the sum of the angles of the triangle exceed 180 deg by an amount equal to the angle at the pole. The sum of the angles of any spherical triangle is always larger than 180 deg. The amount by which this sum exceeds 180 deg is called the *spherical excess* and is directly proportional to the area of the spherical triangle. Thus, small triangles on the sphere have a spherical excess near zero, and are very similar to plane triangles. Large spherical triangles, however, are very different as in Fig. 5-6 with 2 right angles.

A *radian* is the angle subtended if I take a string equal in length to the radius of a circle, and stretch it along the circumference. Similarly, if I take an area equal to the square of the radius, and stretch it out on the surface of a sphere (which requires some distortion, since the surface of the sphere cannot fit on a flat sheet), the resulting area is called a *steradian*. Since the area of a sphere is $4\pi r^2$, there are 4π steradians in a full sphere, and 2π steradians in a hemisphere. These units are convenient for area problems, because in any spherical triangle, the spherical excess, expressed in radians, is equal to the area of the triangle expressed in steradians. In general, the area, A, of any spherical polygon (a figure with sides which are great circle arcs) expressed in steradians is given by:

$$A = \Sigma - (n-2)\pi \qquad (5\text{-}3)$$

where n is the number of sides, and Σ is the sum of the rotation angles expressed in radians.

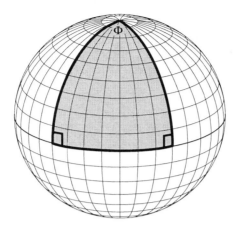

Fig. 5-6.   The Sum of the Angles in a Spherical Triangle is Always Greater than 180 deg.
The amount by which the sum exceeds 180 deg is called the *spherical excess* and is proportional to the triangle area.

Figure 5-7 shows a variety of spherical triangles. Note that all of the triangle sides are great circle arcs. Figure 5-7A is a nearly plane triangle, for which the sum of the angles is approximately 180 deg and plane geometry is a close approximation. Figure 5-7B is called a *right spherical triangle* because the angle at B is a right angle. Just as plane right triangles have particularly simple relationships among the angles and sides, right spherical triangles also have exceptionally simple relationships between the sides and angles. These are expressed by *Napier's rules* which are written out in Appendix D. Right spherical triangles are common in mission geometry problems, and provide simple, straightforward solutions for many problems associated with angular measurements.

There is a second type of special and particularly simple spherical triangle shown in Fig. 5-7C. Here, side A-B has an arc length of 90 deg. This is called a *quadrantal spherical triangle*. An equally simple set of rules apply to the relationship among the angles and sides in quadrantal spherical triangles. These are also summarized in Appendix D. Between them, right and quadrantal spherical triangles provide solutions to most problems encountered in mission analysis.

Figure 5-7D shows an *obtuse isosceles triangle* with two equal rotation angles larger than 90 deg. Clearly, this cannot exist in plane geometry. A similar strange triangle for plane geometry is Fig. 5-7E which shows an *equilateral right triangle* in which all three angles and all three sides are 90 deg. This triangle represents 1/8 of the surface of the celestial sphere, and has an area of 0.5π steradians, which can be seen either by examination or from the spherical excess rule. Finally, Fig. 5-7F shows a very large spherical triangle. Note that this triangle is remarkably similar in appearance to the small triangle in 5-7A. This is because the triangle can be thought of either as a small triangle with three angles of approximately 60 deg, or as a very large one, with three angles of approximately 120 deg. That is, the area between A, B, and C can be thought of either as the inside of a small triangle or as the outside of a large one which covers all of the surface of the sphere except for the small area between A, B, and C. In this large spherical triangle, the rules of spherical geometry still apply and can be used in very convenient fashions as described further in Wertz [2001].

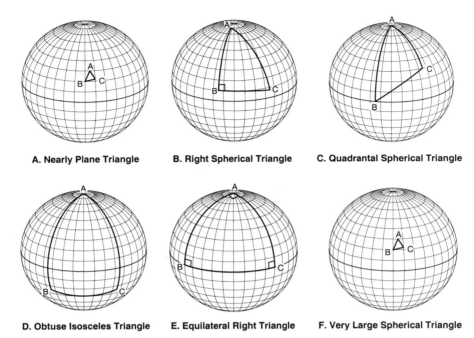

A. Nearly Plane Triangle	B. Right Spherical Triangle	C. Quadrantal Spherical Triangle

D. Obtuse Isosceles Triangle	E. Equilateral Right Triangle	F. Very Large Spherical Triangle

Fig. 5-7.   Types of Spherical Triangles. See text for discussion.

In plane geometry, we can make triangles larger or smaller and maintain the same relative proportions. In spherical geometry, this is not the case. A spherical triangle is uniquely specified by either 3 sides **or** 3 rotation angles. Additional details on elementary spherical geometry are in Appendix D, which includes references to several standard books.

Because spherical triangles approach plane triangles in the limit of small size and because most analysts are much more familiar with plane geometry, they tend to use plane geometry approximations even when it is entirely inappropriate. An example would be a geometry problem dealing with the surface of the Earth as seen from nearby space. Figure 5-8 shows the differences between plane geometry and spherical geometry, using the example of a right spherical triangle with one 45-deg rotation angle. Here both the length of the hypotenuse **and** the other rotation angle are a function of the size of the triangle. For a spacecraft in geosynchronous orbit, the apparent Earth radius is 8.7 deg and, from the figure, the differences between plane and spherical geometry will be of the order of 0.1 deg. If this amount does not matter for a particular problem, then the plane geometry approximation is fine. Otherwise, we should use spherical geometry. In low-Earth orbit, the angular radius of the Earth is 60 deg to 70 deg, so plane geometry techniques give not only incorrect numerical answers but results which are conceptually wrong. Thus, we should avoid plane geometry approximations for problems involving spacecraft in low-Earth orbits or nearly any type of precision pointing.

We will illustrate computations on the celestial sphere with two examples—the duration of eclipses in a circular low-Earth orbit, and the angle between any spacecraft

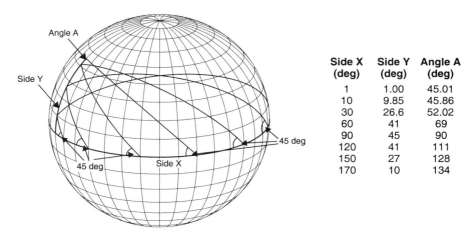

Side X (deg)	Side Y (deg)	Angle A (deg)
1	1.00	45.01
10	9.85	45.86
30	26.6	52.02
60	41	69
90	45	90
120	41	111
150	27	128
170	10	134

Fig. 5-8. **Succession of Right Spherical Triangles with One 45 deg Rotation Angle**. As spherical triangles become larger, they become less and less similar to plane triangles. In the plane geometry approximation $y = x$ and $A = 45$ deg. On the sphere tan $y =$ sin x tan 45 deg and cos $A =$ cos x sin 45 deg.

face and the Sun. We can use the latter either to calculate Sun interference or for thermal analysis. In both cases, once we choose the correct coordinate system, we can easily develop exact formulas using spherical techniques.

Example 1. Analyzing Eclipses for a Low-Earth Orbit

The first example is a satellite in a circular low-Earth orbit at altitude $H = 1,000$ km and inclination $i = 32$ deg. We wish to determine the eclipse fraction for any date and also the maximum and minimum eclipses over a year. Figure 5-9 shows four different views of the geometry of this problem. In Fig. 5-9A we have drawn the geometry in the "normal" fashion, with a Cartesian coordinate system centered on the Earth and vectors to the spacecraft and to the direction of the Sun. Although we could apply this coordinate representation to our problem, it gives us no particular insight. Also, to use this coordinate representation without further analysis, we would have to simulate the orbit and do a large number of trials to sample various eclipse durations throughout the year.

Figure 5-9B provides more information by plotting the system on the unit celestial sphere centered on the spacecraft. It shows the celestial equator, the orbit plane, and the *ecliptic*, which is the path of the Sun over the year. The light dashed circles are the outline of the disk of the Earth as seen by the spacecraft as the Earth's center moves along the spacecraft orbit path. An eclipse will occur whenever the disk of the Earth crosses in front of the Sun. The circle centered on the orbit pole is tangent to the disk of the Earth throughout the orbit. Any object within this circle can be seen at all times and will not be eclipsed by the Earth. The ecliptic does not pass through this circle of no eclipse. Consequently, for the illustrated orbit, eclipses will occur at all times of the year with no eclipse-free season. That is, no time exists when an eclipse does not occur during some part of the orbit. (For the time being, we are ignoring the rotation of the satellite's orbit due to various perturbing forces. This orbit rotation may make the associated arithmetic more complex; it does not change the basic argument.) Although

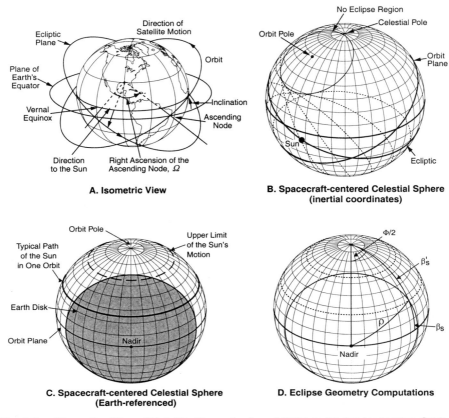

A. Isometric View

B. Spacecraft-centered Celestial Sphere (inertial coordinates)

C. Spacecraft-centered Celestial Sphere (Earth-referenced)

D. Eclipse Geometry Computations

Fig. 5-9.   Alternative View of Satellite Geometry for a 1,000-km, 32-deg Inclination Orbit.

we have gained some additional insight from this figure, it would still be awkward to compute the eclipse duration for any particular geometry.

Figure 5-9C illustrates the same geometry in a celestial coordinate system centered on the spacecraft, in which the orbit plane is the equator and we hold the direction to the Earth fixed along the +X-axis. In this coordinate frame the Earth's disk is the fixed shaded circle. Because one axis is always facing the Earth, this coordinate frame rotates once per orbit in inertial space about the orbit pole. Thus any objects approximately fixed in inertial space, such as the stars, Sun, or Moon, will appear to rotate once per orbit about the orbit pole. The heavy, solid line shows a typical path of the Sun in one orbit. Again, an eclipse will occur whenever the path of the Sun goes behind the disk of the Earth. We now have enough insight to understand what is happening throughout the year and to develop straightforward formulas for the eclipse fraction under any conditions.

In any one orbit, the Sun will move along a small circle path and the duration of the eclipse will be the fraction of the small circle behind the disk of the Earth. For the orbit illustrated, the eclipse covers 113 deg of azimuth. Thus, the eclipse will last for 113 deg/360 deg = 32% of the orbit period or about 33 min for the 105 min, 1,000 km orbit [see Eq. (7-7) for the orbit period]. As the Sun moves along the ecliptic through-out the year, it will move slowly up and down on the globe plot. The angle of the Sun above or below the orbit plane, β_S, goes from a maximum of 55 deg above the orbit plane to 55 deg below the plane. (55 deg is the sum of the assumed inclination of 32 deg and the angle between the ecliptic and the Earth's equator of 23 deg). By inspection, the maximum eclipse occurs when the Sun is in the orbit plane. It covers 120 deg of azimuth or 35 min for this orbit. Again by inspection, we see that the minimum eclipse occurs when the Sun is at the upper or lower limit of its range. It covers approximately 60 deg of azimuth. Thus an eclipse will occur on every orbit, with a minimum eclipse duration of about half the maximum, or 17 min. However, we can also see by inspection that eclipses near minimum duration will occur only when the Sun is quite close to its extreme range limit. Most of the time, the eclipse duration will be close to the maximum. Consequently, if we wish to assume an average eclipse for analysis purposes, we should take a value close to the maximum eclipse rather than one midway between the maximum and minimum values.

The geometry of Fig. 5-9C allows us easily to compute the eclipse fraction for any given Sun angle conditions. Specifically, Fig. 5-9D shows a *quadrantal spherical triangle* (that is, having one side = 90 deg) between the orbit pole, nadir, and the point at which the Sun is on the Earth's horizon. Let ρ be the angular radius of the Earth, β_S be the angle of the Sun above the orbit plane, and $\Phi/2$ be half of the rotation angle corresponding to the eclipse duration. From the rules for quadrantal triangles (Appendix D) we find immediately that

$$\cos (\Phi/2) = \cos \rho / \sin \beta_S' = \cos \rho / \cos \beta_S \qquad (5\text{-}4a)$$

The duration of the eclipse in a circular orbit, T_E, is then

$$T_E = P (\Phi/360 \text{ deg}) \qquad (5\text{-}4b)$$

where P is the orbit period from Eq. (7-7).

For our example, $\rho = 60$ deg, β_S has been chosen to be 25 deg, and, therefore, $\Phi = 113$ deg and $T_E = 33$ min as expected. Eq. (5-4) provides the eclipse fraction for any Sun geometry involving circular orbits and an approximate check for orbits which are not precisely circular. By adjusting ρ appropriately, we can use the same equation to determine the time the Sun will be a certain number of degrees above or below the Earth's horizon. This example shows how we develop physical insight and simple formulas by using global geometry to analyze mission geometry problems.

Example 2. Sun Angle Geometry

We can extend the straightforward computations of the preceding example to determine the angle of the Sun relative to any arbitrary face on the spacecraft as the spacecraft goes around in its orbit. This helps us analyze thermal effects and assess possible Sun interference in the fields of view of various instruments.

We assume the spacecraft is flying in a traditional orientation for Earth-referenced spacecraft with one axis pointed toward nadir and a second axis toward the orbit pole. The computational geometry is in Fig. 5-10, which is geometrically identical to Fig. 5-9. In this coordinate system, fixed on the spacecraft, the normal to a given spacecraft face is represented by a point, N, on the celestial sphere. N remains fixed in spacecraft coordinates, as does the orientation of nadir and the orbit pole. The Sun moves once per orbit along a small circle of radius β_S'. This radius remains essentially fixed for a single orbit. γ is the angle from N to the orbit pole. The rotation angle ΔAz

is the azimuthal difference between the Sun and N. It varies uniformly once per orbit from 0 to 360 deg. If I is the incident energy on the face with area A, K is the solar constant in the vicinity of the Earth = 1,367 W/m^2, and β is the angle between the Sun and the normal, N, to the face, then at any given moment when the Sun is shining on the face:

$$I = AK \cos \beta \qquad (5\text{-}5)$$

and, from the law of cosines for sides,

$$\cos \beta = \cos \gamma \cos \beta_S' + \sin \gamma \sin \beta_S' \cos (\Delta Az) \qquad (5\text{-}6)$$

By inspection, the maximum and minimum angles between the Sun and N are:

$$\beta_{max} = \beta_S' + \gamma \quad \text{and} \quad \beta_{min} = |\beta_S' - \gamma| \qquad (5\text{-}7)$$

β_S' and γ are both constants for a given orbit and spacecraft face, whereas ΔAz changes throughout the orbit.

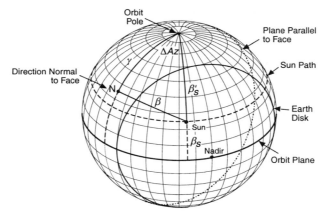

Fig. 5-10. **Geometry for Computation of Sun Angle on an Arbitrary Spacecraft Face.** N is the unit vector or direction normal to the face. As the spacecraft moves in its orbit, the apparent position of the Sun moves along the dashed line and the arc length β between the Sun and the normal to the face undergoes a sinusoidal oscillation.

Equations (5-5) to (5-7) apply to either circular or elliptical orbits. If the orbit is circular with period P and angular frequency $\omega = 2\pi/P$, then Eqs. (5-5) and (5-6) can be integrated directly to determine the total energy, E, incident on the face between azimuths Az_1 and Az_2:

$$E_{Az_1 \text{ to } Az_2} = (AK/\omega) \, [(\Delta Az_2 - \Delta Az_1) \cos \gamma \cos \beta_S'$$
$$+ (\sin \Delta Az_2 - \sin \Delta Az_1) \sin \gamma \sin \beta_S' \,] \qquad (5\text{-}8)$$

In Eq. (5-8), the 0 azimuth is in the direction of N, and the angles are in radians. In a full orbit the Sun will shine on the face except for two periods: (A) during eclipse and (B) when $\beta > 90$ deg and, therefore, the Sun is on the "back side" of the face. From Eq. (5-4a) the conditions for eclipse are

$$Az_{eclipse} = Az_0 \pm \text{arc} \cos (\cos \rho / \sin \beta_S') \qquad (5\text{-}9)$$

where Az_0 is the azimuth of nadir relative to N. For condition (B) we use a quadrantal triangle with $\beta = 90$ deg to determine

$$Az_{back} = \pm \arccos[-1/(\tan\gamma \tan\beta'_S)] \tag{5-10}$$

The problem now reduces to determining whether conditions (A) or (B) or both will occur and the relative order of the azimuth limits.

As an example, consider Figs. 5-9 and 5-10, for which $\rho = 60$ deg, $\omega = 0.0010$ rad/sec, and $\beta_S = 25$ deg. We assume the spacecraft face has an area of 0.5 m² with its normal vector at an azimuth of −75 deg from nadir ($Az_0 = 75$ deg) and an elevation of 35 deg above the orbit plane ($\gamma = 55$ deg). From Eqs. (5-8) and (5-9), the azimuth limits are: $Az_{1eclipse} = 18.5$ deg, $Az_{2eclipse} = 131.5$ deg, $Az_{1back} = 109.0$ deg, and $Az_{2back} = 251.0$ deg. Therefore, the total energy input on the face F over one orbit is between the azimuths of $Az_1 = 251.0$ deg and $Az_2 = 18.5$ deg.

Example 3. *Solar Radiation Intensity*

We can continue to extend our example to look at the average solar radiation input, I_{avg}, on any spacecraft face over an orbit. This is given by

$$I_{avg} = A\,K\,F \tag{5-11}$$

where, as before, A is the area of the face and $K = 1{,}367$ W/m² is the solar constant in the vicinity of the Earth. F is the time average fraction of the surface area projected in the direction of the Sun and must lie between 0 and 1. If the attitude of the spacecraft is inertially fixed, then the angle, β, between the unit vector, $\hat{S}$, in the direction of the Sun and the unit vector, $\hat{N}$, normal to the face remains constant over an orbit. If there is no eclipse during a given orbit, then

$$F = \hat{N} \cdot \hat{S} = \cos\beta \quad \text{(inertially fixed, no eclipse)} \tag{5-12}$$

which gives the same result as using Eq. (5-5) directly.

For a 3-axis-stabilized, nadir-oriented spacecraft in a circular orbit, the angle β will oscillate sinusoidally as illustrated previously in Fig. 5-10. The amount of sunlight on the face will depend on both γ, the angle from the orbit normal to $\hat{N}$, and on β'_S, the angle from the orbit normal to the Sun. If $(\gamma + \beta'_S) \leq 90$ deg, then the face will always be in sunlight and, assuming no eclipse:

$$F = \cos\gamma \cos\beta'_S \quad \text{(nadir fixed, full sunlight, no eclipse)} \tag{5-13a}$$

If $|\gamma - \beta'_S| \geq 90$ deg, then the surface will always be shaded and, of course:

$$F = 0 \quad \text{(nadir fixed, continuous shade)} \tag{5-13b}$$

If neither of the above conditions hold, then the face will be shaded part of the time and in sunlight part of the time. In this case, we integrate the instantaneous fraction of the surface area projected in the direction of the Sun by the instantaneous ΔAz, starting and ending when β, the Sun angle, is 90 deg. Dividing by 2π gives the average F over one orbit:

$$F = \frac{1}{2\pi}\int_{-\phi_{90}}^{\phi_{90}} \cos\beta\,d(\Delta Az) = \frac{1}{2\pi}\int_{-\phi_{90}}^{\phi_{90}}\left[\cos\gamma\,\cos\beta'_S + \sin\gamma\,\sin\beta'_S\,\cos(\Delta Az)\right]d(\Delta Az)$$

$$= (\Phi_{90}\cos\gamma\cos\beta'_S + \sin\Phi_{90}\sin\gamma\sin\beta'_S)\,/\,\pi$$
$$\text{(nadir fixed, partial shade, no eclipse)} \tag{5-13c}$$

where Φ_{90} is expressed in radians and

$$\cos\Phi_{90} = -1\,/\,(\tan\gamma\,\tan\beta'_S) \tag{5-14}$$

The quantity Φ_{90} is the value of ΔAz at which $\beta = 90$ deg, i.e., when the transition occurs between shade and sunlight.

Consider our previous example for which $\beta_S' = 65$ deg and $\gamma = 55$ deg. In this case, Eq. (5-13c) is applicable and, from Eq. (5-14), $\Phi_{90} = 109.1$ deg. This means that the Sun will shine on the face in question whenever the azimuth of the Sun is within 109.1 deg of the azimuth of the face. From Eq. (5-13c), $F = 0.370$. This means that, without eclipses, the average solar input on the face is 37.0% of what it would be if the Sun were continuously shining normal to the face. If the face in question has a surface area of 0.5 m^2, then the average solar input over an orbit with full sunlight is $(0.5) (1,367) (0.370) = 253$ W.

The nadir-oriented satellite above is spinning at one rotation per orbit in inertial space. Thus, all of the above formulas can be applied to spinning spacecraft with the interpretation that I_{avg} is the average solar radiation input over one spin period, γ is the angle from the spin axis to the face in question, and β_S' is the angle from the spin axis to the Sun.

In practice there are two principal corrections to the above formulas:

- F is reduced by eclipses
- The effective F is increased by reflected or emitted radiation from the Earth

For either an inertially fixed spacecraft or a spinning spacecraft, the effect of eclipses is simply to reduce F by the fraction of the orbit over which the spacecraft is in eclipse:

$$F = F_0 \ (1 - \Phi / 360 \text{ deg}) \tag{5-15}$$

where F_0 is the noneclipse value of F determined from Eqs. (5-12) or (5-13) and Φ is the eclipse fraction from Eq. (5-4).

For Earth-oriented spacecraft the situation is more complex, because the solar input depends on the orientation of the Sun relative to both the Earth and spacecraft face being evaluated. Let η be the angular distance from $\hat{N}$ to nadir and ρ be the angular radius of the Earth. If the face in question is sufficiently near zenith (opposite the direction to the center of the Earth) that $\eta - \rho \geq 90$ deg, then $F = F_0$ and F will not be reduced by eclipses. For this condition, any eclipses which occur will happen when the face is shaded. Alternatively, consider what happens when $|\gamma - \beta_S'| \approx 90$ deg. In this case, there is only a small portion of the orbit when the Sun shines on the face. Let Φ be the eclipse fraction from Eq. (5-4) and Φ_{90} the azimuth defined above at which the transition from sunlight to shade occurs. If $\Phi/2 \geq$ the larger of $|Az_N \pm \Phi_{90}|$, then the spacecraft will be in eclipse when the Sun is in a position to shine on the face. In this case F will be reduced to 0. For conditions in between these two extremes, F will be between 0 and its noneclipse value. Specific values will need to be evaluated numerically using Eqs. (5-4) and (5-8).

The heat input from both reflected and emitted radiation from the Earth increases the effective value of F. It is significantly more complex to compute than the effect of eclipses because of the extended size of the disk of the Earth and the variability in the intensity of reflected radiation. However, reasonable upper limits for radiation from the Earth are:

- 475 W/m^2 for reflected solar radiation (*albedo*)
- 260 W/m^2 for emitted IR radiation (*thermal radiation*)

Section 11.5 provides additional details on how to compute thermal inputs to the spacecraft.

5.2 Earth Geometry Viewed from Space

The most common problem in space mission geometry is to determine the relative geometry of objects on the Earth's surface as seen from the spacecraft. One example is to use the given coordinates of a target on the Earth to determine its coordinates in the spacecraft field of view. Another is to determine the intercept point on the surface of the Earth corresponding to a given direction in spacecraft coordinates.

To begin, we determine ρ, the angular radius of the spherical Earth as seen from the spacecraft, and λ_0, the angular radius measured at the center of the Earth of the region seen by the spacecraft (see Fig. 5-11). Because we have assumed a spherical Earth, the line from the spacecraft to the Earth's horizon is perpendicular to the Earth's radius, and therefore

$$\sin \rho = \cos \lambda_0 = \frac{R_E}{R_E + H} \qquad (5\text{-}16)$$

and

$$\rho + \lambda_0 = 90 \deg \qquad (5\text{-}17)$$

where R_E is the radius of the Earth and H is the altitude of the satellite.

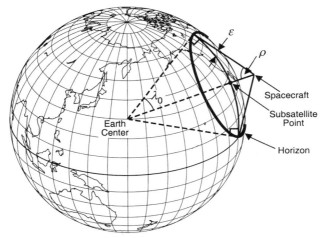

Fig. 5-11. **Relationship Between Geometry as Viewed from the Spacecraft and from the Center of the Earth.** See also Fig. 5-12.

Thus, the Earth forms a small circle of radius ρ on the spacecraft sky, and the spacecraft sees the area within a small circle of radius λ_0 on the surface of the Earth. The distance, D_{max}, to the horizon is given by (see Fig. 5-13 below):

$$D_{max} = [(R_E + H)^2 - R_E{}^2]^{1/2} = R_E \tan \lambda_0 \qquad (5\text{-}18)$$

The spherical-Earth approximation is adequate for most mission geometry applications. However, for precise work, we must apply a correction for oblateness, as explained in detail by Liu [1978] or Collins [1992]. The Earth's oblateness has two distinct effects on the shape of the Earth as seen from space. First, the Earth appears somewhat oblate rather than round, and second, the center of the visible oblate Earth is displaced from the true geometric center of the Earth. For all remaining computations in this section, we will use spherical coordinates both on the Earth and in the spacecraft frame. Computationally, we can treat oblateness and surface irregularities as simply the target's altitude above or below a purely spherical Earth. That the Earth's real surface is both irregular and oblate is immaterial to the computation, and, therefore, the results are exact.

We wish to find the angular relationships between a target, **P**, on the surface of the Earth, and a spacecraft with subsatellite point, *SSP*, also on the surface of the Earth, as shown in Fig. 5-12. We assume that the subsatellite point's latitude, Lat_{SSP} and longitude, $Long_{SSP}$, are known. Depending on the application, we wish to solve one of two problems: (1) given the coordinates of a target on the Earth, find its coordinates viewed by the spacecraft, or (2) given the coordinates of a direction relative to the spacecraft, find the coordinates of the intercept on the surface of the Earth. In both cases, we determine the relative angles between *SSP* and **P** on the Earth's surface and then transform these angles into spacecraft coordinates.

Given the coordinates of the subsatellite point $(Long_{SSP}, Lat_{SSP})$ and target $(Long_P, Lat_P)$, and defining $\Delta L = |\ Long_{SSP} - Long_P\ |$, we wish to find the azimuth, Φ_E, measured eastward from north, and angular distance, λ, from the subsatellite point to the target. (See Fig. 5-12.) These are given by

$$\cos \lambda = \sin Lat_{SSP} \sin Lat_P + \cos Lat_{SSP} \cos Lat_P \cos \Delta L \quad (\lambda < 180 \text{ deg}) \quad (5\text{-}19)$$

$$\cos \Phi_E = (\sin Lat_P - \cos \lambda \sin Lat_{SSP})/(\sin \lambda \cos Lat_{SSP}) \quad (5\text{-}20)$$

where $\Phi_E < 180$ deg if **P** is east of *SSP* and $\Phi_E > 180$ deg if **P** is west of *SSP*.

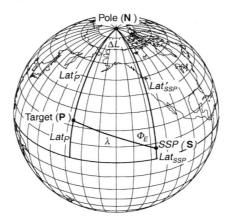

Fig. 5-12.   Relationship Between Target and Subsatellite Point on the Earth's Surface.

Alternatively, given the position of the subsatellite point $(Long_{SSP},\ Lat_{SSP})$ and the position of the target relative to this point (Φ_E, λ), we want to determine the geographic coordinates of the target $(Long_P,\ Lat_P)$:

$$\cos Lat_P' = \cos \lambda \sin Lat_{SSP} + \sin \lambda \cos Lat_{SSP} \cos \Phi_E \quad (Lat_P' < 180 \text{ deg}) \quad (5\text{-}21)$$

$$\cos \Delta L = (\cos \lambda - \sin Lat_{SSP} \sin Lat_P) / (\cos Lat_{SSP} \cos Lat_P) \quad (5\text{-}22)$$

where $Lat_P' \equiv 90$ deg $- Lat_P$ and **P** is east of *SSP* if $\Phi_E < 180$ deg and west of *SSP* if $\Phi_E > 180$ deg.

We now wish to transform the coordinates on the Earth's surface to coordinates as seen from the spacecraft. By symmetry, the azimuth of the target relative to north is the same as viewed from either the spacecraft or the Earth. That is,

$$\Phi_{E\ spc} = \Phi_{E\ surface} = \Phi_E \quad (5\text{-}23)$$

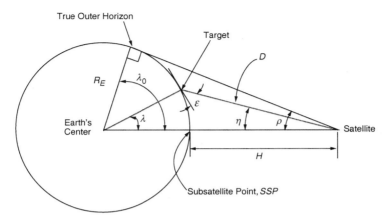

Fig. 5-13.   Definition of Angular Relationships Between Satellite, Target, and Earth Center.

Generally, then, the only problem is to find the relationship between the *nadir angle*, η, measured at the spacecraft from the subsatellite point (= nadir) to the target; the *Earth central angle*, λ, measured at the center of the Earth from the subsatellite point to the target; and the *grazing angle* or *spacecraft elevation angle*, ε, measured at the target between the spacecraft and the local horizontal. Figure 5-13 defines these angles and related distances. First, we find the angular radius of the Earth, ρ, from

$$\sin \rho = \cos \lambda_0 = \frac{R_E}{R_E + H} \qquad (5\text{-}24)$$

which is the same as Eq. (5-16). Next, if λ is known, we find η from

$$\tan \eta = \frac{\sin \rho \sin \lambda}{1 - \sin \rho \cos \lambda} \qquad (5\text{-}25)$$

If η is known, we find ε from

$$\cos \varepsilon = \frac{\sin \eta}{\sin \rho} \qquad (5\text{-}26\text{a})$$

Or, if ε is known, we find η from

$$\sin \eta = \cos \varepsilon \sin \rho \qquad (5\text{-}26\text{b})$$

Finally, the remaining angle and side are obtained from

$$\eta + \lambda + \varepsilon = 90 \text{ deg} \qquad (5\text{-}27)$$

$$D = R_E (\sin \lambda / \sin \eta) \qquad (5\text{-}28)$$

Figure 5-14 summarizes the process of transforming between spacecraft coordinates and Earth coordinates.

As an example, consider a satellite at an altitude of 1,000 km. From Eq. (5-16), the angular radius of the Earth $\rho = 59.8$ deg. From Eqs. (5-17) and (5-18), the horizon is 30.2 deg in Earth central angle from the subsatellite point and is at a line-of-sight

First, compute the angular radius of Earth, ρ

$$\sin \rho = \cos \lambda_0 = R_E / (R_E + H) \qquad (5\text{-}24)$$

To compute spacecraft viewing angles given the subsatellite point at ($Long_{SSP}$, Lat_{SSP}) and target at ($Long_P$, Lat_P), and $\Delta L \equiv |\ Long_{SSP} - Long_P\ |$

$$\cos \lambda = \sin Lat_{SSP} \sin Lat_P + \cos Lat_{SSP} \cos Lat_P \cos \Delta L \quad (\lambda < 180 \text{ deg}) \qquad (5\text{-}19)$$

$$\cos \Phi_E = (\sin Lat_P - \cos \lambda \sin Lat_{SSP}) / (\sin \lambda \cos Lat_{SSP}) \qquad (5\text{-}20)$$

$$\tan \eta = \sin \rho \sin \lambda / (1 - \sin \rho \cos \lambda) \qquad (5\text{-}25)$$

To compute coordinates on the Earth given the subsatellite point at ($Long_{SSP}$, Lat_{SSP}) and target direction (Φ_E, η):

$$\cos \varepsilon = \sin \eta / \sin \rho \qquad (5\text{-}26a)$$

$$\lambda = 90 \text{ deg} - \eta - \varepsilon \qquad (5\text{-}27)$$

$$\cos Lat_P{}' = \cos \lambda \sin Lat_{SSP} + \sin \lambda \cos Lat_{SSP} \cos \Phi_E \quad (Lat_P{}' < 180 \text{ deg}) \qquad (5\text{-}21)$$

$$\cos \Delta L = (\cos \lambda - \sin Lat_{SSP} \sin Lat_P) / (\cos Lat_{SSP} \cos Lat_P) \qquad (5\text{-}22)$$

Fig. 5-14.   Summary of the Process of Transforming Between Spacecraft Viewing Angles and Earth Coordinates. Equation numbers are listed in the figure and variables are as defined in Figs. 5-11 and 5-12.

distance of 3,709 km from the satellite. We will assume a ground station at Hawaii (Lat_P = 22 deg, $Long_P$ = 200 deg) and a subsatellite point at Lat_{SSP} = 10 deg, $Long_{SSP}$ = 185 deg. From Eqs. (5-19) and (5-20), the ground station is a distance λ = 18.7 deg from the subsatellite point, and has an azimuth relative to north = 48.3 deg. Using Eqs. (5-25) and (5-28) to transform into spacecraft coordinates, we find that from the space-craft the target is 56.8 deg up from nadir (η) at a line of sight distance, D, of 2,444 km. From Eq. (5-27), the elevation of the spacecraft as seen from the ground station is 14.5 deg. The substantial foreshortening at the horizon can be seen in that at ε = 14.5 deg we are nearly half way from the horizon to the subsatellite point (λ = 18.7 deg vs. 30.2 deg at the horizon).

Using these equations, we can construct Fig. 5-15, which shows the Earth as seen from 1,000 km over Mexico's Yucatan Peninsula in the Gulf of Mexico. The left side shows the geometry as seen on the surface of the Earth. The right side shows the geometry as seen by the spacecraft projected onto the spacecraft-centered celestial sphere. As computed above, the maximum Earth central angle will be approximately 30 deg from this altitude such that the spacecraft can see from northwestern South America to Maine on the East Coast of the U.S. and Los Angeles on the West Coast. The angular radius of the Earth as seen from the spacecraft will be 90 – 30 = 60 deg as shown in Fig. 5-15B. Because the spacecraft is over 20 North latitude, the direction to nadir in spacecraft coordinates will be 20 deg south of the celestial equator. (The direction from the spacecraft to the Earth's center is exactly opposite the direction from the Earth's center to the spacecraft.)

Even after staring at it a bit, the view from the spacecraft in Fig. 5-15B looks strange. First, recall that we are looking at the spacecraft-centered celestial sphere **from the outside**. The spacecraft is at the center of the sphere. Therefore, the view for us is reversed from right-to-left as seen by the spacecraft so that the Atlantic is on the left and the Pacific on the right. Nonetheless, there still appear to be distortions in the view. Mexico has an odd shape and South America has almost disappeared. All of this

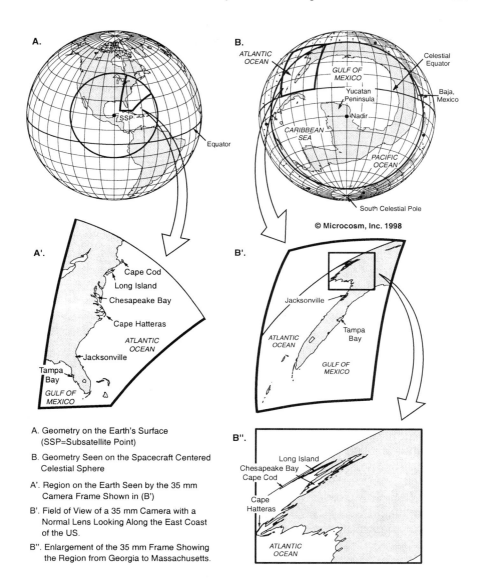

A. Geometry on the Earth's Surface
   (SSP=Subsatellite Point)

B. Geometry Seen on the Spacecraft Centered
   Celestial Sphere

A'. Region on the Earth Seen by the 35 mm
   Camera Frame Shown in (B')

B'. Field of View of a 35 mm Camera with a
   Normal Lens Looking Along the East Coast
   of the US.

B''. Enlargement of the 35 mm Frame Showing
   the Region from Georgia to Massachusetts.

Fig. 5-15.   Viewing Geometry for a Satellite at 1,000 km over the Yucatan Peninsula at 90 deg W longitude and 20 deg N latitude. See text for discussion. [Copyright by Microcosm; reproduced by permission.]

is due to the very strong foreshortening at the edge of the Earth's disk. Notice for example that Jacksonville, FL, is about halfway from the subsatellite point to the horizon. This means that only 1/4th of the area seen by the spacecraft is closer to the subsatellite point than Jacksonville. Nonetheless, as seen from the perspective of the spacecraft, Jacksonville is 54 deg from nadir, i.e., 90% of the way to the horizon with 3/4ths of the visible area beyond it.

The rectangle in the upper left of Fig. 5-15B is the field of view of a 35 mm camera with a 50 mm focal length lens (a normal lens that is neither wide angle nor telephoto). The cameraperson on our spacecraft has photographed Florida and the eastern seaboard of the US to approximately Maine. The region seen on the Earth is shown in Fig. 5-15A′ and 5-15B′ and an enlargement of a portion of the photo from Georgia to Maine is shown in Fig. 5-15B″. Note the dramatic foreshortening as Long Island and Cape Cod become little more than horizontal lines, even though they are some distance from the horizon. This distortion does not come from the plotting style, but is what the spacecraft sees. We see the same effect standing on a hilltop or a mountain. (In a sense, the spacecraft is simply a very tall mountain.) Most of our angular field of view is taken up by the field or mountain top we are standing on. For our satellite, most of what is seen is the Yucatan and Gulf of Mexico directly below. There is lots of real estate at the horizon, but it appears very compressed. From the spacecraft, I can point an antenna at Long Island, but I can not map it. We must keep this picture in mind whenever we assess a spacecraft's fields of view or measurement needs.

Thus far we have considered spacecraft geometry only from the point of view of a spacecraft fixed over one point on the Earth. In fact, of course, the spacecraft is traveling at high velocity. Figure 5-16A shows the path of the subsatellite point over the Earth's surface, called the satellite's *ground trace* or *ground track*. Locally, the ground trace is very nearly the arc of a great circle. However, because of the Earth's rotation, the spacecraft moves over the Earth's surface in a spiral pattern with a displacement at successive equator crossings directly proportional to the orbit period. For a satellite in a circular orbit at inclination i, the subsatellite latitude, δ_S, and longitude, L_S, relative to the ascending node are

$$\sin \delta_S = \sin i \sin (\omega t) \tag{5-29}$$

$$\tan (L_S + \omega_E t) = \cos i \tan (\omega t) \tag{5-30}$$

where t is the time since the satellite crossed the equator northbound, $\omega_E = 0.004\ 178\ 07$ deg/s is the rotational velocity of the Earth on its axis, and ω is the satellite's angular velocity. For a satellite in a circular orbit, ω in deg/s is related to the period, P, in minutes by

$$\omega = 6/P \leq 0.071 \text{ deg/s} \tag{5-31}$$

where 0.071 deg/s is the maximum angular velocity of a spacecraft in a circular orbit. Similarly, the ground track velocity, V_g, is

$$V_g = 2\pi R_E/P \leq 7.905 \text{ km/s} \tag{5-32}$$

where $R_E = 6,378$ km is the equatorial radius of the Earth. For additional information on the satellite ground trace and coverage, taking into account the rotation of the Earth, see Chap. 8 of Wertz [2001].

Fig. 5-16B shows the *swath coverage* for a satellite in low-Earth orbit. The swath is the area on the surface of the Earth around the ground trace that the satellite can observe as it passes overhead. From the formulas for stationary geometry in Eqs. (5-24) to (5-27), we can compute the width of the swath in terms of the Earth central angle, λ. Neglecting the Earth's rotation, the area coverage rate, ACR, of a spacecraft will be

$$ACR = 2\pi (\sin \lambda_{outer} \pm \sin \lambda_{inner})/P \tag{5-33}$$

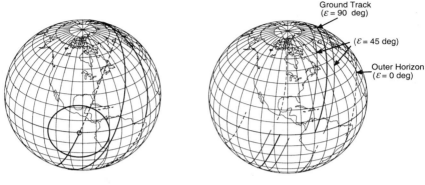

A. Satellite Ground Track.   **B. Swath Coverage for Satellite Ground Track in (A), for Several Grazing Angles, ε.**

Fig. 5-16.   Path of a Satellite Over the Earth's Surface. A swath which goes from horizon to horizon will cover a very large area, although we will see most of this area at very shallow elevation angles near the horizon.

where λ_{outer} is the effective outer horizon, λ_{inner} is the inner horizon, the area on the Earth's surface is in steradians, and P is the orbital period of the satellite. The plus sign applies to horizons on opposite sides of the ground trace and the minus sign to both horizons on one side, that is, when the spacecraft looks exclusively left or right. For a swath of width 2λ symmetric about the ground trace, this reduces to

$$ACR = (4\pi/P)\sin\lambda \tag{5-34}$$

Alternatively, this can be expressed in terms of the limiting grazing angle (or elevation angle), ε, and angular radius of the Earth, ρ, as

$$ACR = (4\pi/P)\cos(\varepsilon + \arcsin(\cos\varepsilon\sin\rho)) \tag{5-35}$$

Because the curvature of the Earth's surface strongly affects the ACR, Eqs. (5-33) to (5-35) are **not** equal to the length of the arc between the effective horizons times either the velocity of the spacecraft or the velocity of the subsatellite point.

5.3 Apparent Motion of Satellites for an Observer on the Earth

Even for satellites in perfectly circular orbits, the apparent motion of a satellite across the sky for an observer on the Earth's surface is not a simple geometrical figure. If the observer is in the orbit plane, then the apparent path of the satellite will be a great circle going directly overhead. If the observer is somewhat outside of the orbit plane, then the instantaneous orbit will be a large circle in three-dimensional space viewed from somewhat outside the plane of the circle and projected onto the observer's celestial sphere.

Because the apparent satellite path is not a simple geometrical figure, it is best computed using a simulation program. Available commercial programs include *Satellite Tool Kit* (1990), *Orbit View* and *Orbit Workbench* (1991), *Orbit II Plus* (1991), and *MicroGLOBE* (1990), which generated the figures in this chapter. These programs also work with elliptical orbits, so they are convenient—along with the appropriate formulas from this chapter—for evaluating specific orbit geometry. Unfortunately, a simulation does not provide the desired physical insight into the apparent motion of satellites. Neither does it provide a rapid method of evaluating geometry in a general case, as is most appropriate when first designing a mission. For these problems, we are interested in either bounding or approximating the apparent motion of satellites rather than in computing it precisely. After all, the details of a particular pass will depend greatly on the individual geometrical conditions. Approximate analytic formulas are provided by Wertz [1981, 2001]. For mission design, the circular orbit formulas provided below for satellites in low-Earth orbit and geosynchronous orbit work well.

5.3.1 Satellites in Circular Low-Earth Orbit

We assume a satellite is in a circular low-Earth orbit passing near a target or ground station. We also assume that the orbit is low enough that we can ignore the Earth's rotation in the relatively brief period for which the satellite passes overhead.[*] We wish to determine the characteristics of the apparent satellite motion as seen from the ground station. Throughout this section we use the notation adopted in Sec. 5.2. Figure 5-17 shows the geometry. The small circle centered on the ground station represents the subsatellite points at which the spacecraft elevation, ε, seen by the ground station is greater than some minimum ε_{min}. The nature of the communication or observation will determine the value of ε_{min}. For communications, the satellite typically must be more than 5 deg above the horizon, so $\varepsilon_{min} = 5$ deg. The size of this circle of accessibility strongly depends on the value of ε_{min}, as emphasized in the discussion of Fig. 5-15. In Fig. 5-17 we have assumed a satellite altitude of 1,000 km. The dashed circle surrounding the ground station is at $\varepsilon_{min} = 0$ deg (that is, the satellite's true outer horizon), and the solid circle represents $\varepsilon_{min} = 5$ deg. In practice we typically select a specific value of ε_{min} and use that number. However, you should remain aware that many of the computed parameters are extremely sensitive to this value.

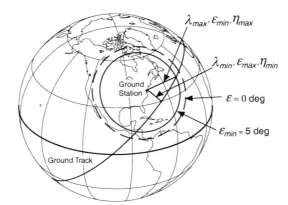

Fig. 5-17.   Geometry of Satellite Ground Track Relative to an Observer on the Earth's Surface.

---

[*] See Chap. 9 of Wertz [2001] for a more accurate approximation which takes the Earth's rotation into account.

Given a value of ε_{min}, we can define the *maximum Earth central angle*, λ_{max}, the *maximum nadir angle*, η_{max}, measured at the satellite from nadir to the ground station, and the *maximum range*, D_{max}, at which the satellite will still be in view. These parameters as determined by applying Eqs. (5-26a) to (5-28) are given by:

$$\sin \eta_{max} = \sin \rho \cos \varepsilon_{min} \tag{5-36}$$

$$\lambda_{max} = 90 \text{ deg} - \varepsilon_{min} - \eta_{max} \tag{5-37}$$

$$D_{max} = R_E \frac{\sin \lambda_{max}}{\sin \eta_{max}} \tag{5-38}$$

where ρ is the angular radius of the Earth as seen from the satellite, that is, $\sin \rho = R_E/(R_E + H)$. We call the small circle of radius λ_{max} centered on the target the *effective horizon*, corresponding in our example to $\varepsilon_{min} = 5$ deg, to distinguish it from the *true* or *geometrical horizon* for which $\varepsilon_{min} = 0$ deg. Whenever the subsatellite point lies within the effective horizon around the target or ground station, then communications or observations are possible. The duration, T, of this contact and the maximum elevation angle, ε_{max}, of the satellite depends on how close the ground station is to the satellite's ground track on any given orbit pass.

As described in Chap. 6, the plane of a spacecraft's orbit and, therefore, the ground track, is normally defined by the inclination, i, and either the right ascension, Ω, or longitude, L_{node}, of the ascending node. Except for orbit perturbations, Ω, which is defined relative to the stars, remains fixed in inertial space while the Earth rotates under the orbit. On the other hand, L_{node} is defined relative to the Earth's surface and, therefore, increases by 360 deg in 1,436 min, which is the rotation period of the Earth relative to the stars. (Again, orbit perturbations affect the exact rotation rate.) Because of this orbit rotation relative to the Earth, it is convenient to speak of the *instantaneous ascending node* which is L_{node} evaluated at the time of an observation or passage over a ground station. For purposes of geometry it is also often appropriate to work in terms of the *instantaneous orbit pole*, or the pole of the orbit plane at the time of the observation. The coordinates of this pole are

$$lat_{pole} = 90 \text{ deg} - i \tag{5-39}$$

$$long_{pole} = L_{node} - 90 \text{ deg} \tag{5-40}$$

A satellite passes directly over a target or ground station (identified by the subscript *gs*) on the Earth's surface if and only if

$$\sin (long_{gs} - L_{node}) = \tan lat_{gs} / \tan i \tag{5-41}$$

There are two valid solutions to the above equation corresponding to the satellite passing over the ground station on the northbound leg of the orbit or on the southbound leg. To determine when after crossing the equator the satellite passes over the ground station for a circular orbit, we can determine μ, the arc length along the instantaneous ground track from the ascending node to the ground station, from

$$\sin \mu = \sin lat_{gs} / \sin i \tag{5-42}$$

Again, the two valid solutions correspond to the northbound and southbound passes.

Figure 5-17 defines the parameters of the satellite's pass overhead in terms of λ_{min}, the minimum Earth central angle between the satellite's ground track and the ground station. This is 90 deg minus the angular distance measured at the center of the Earth from the ground station to the instantaneous orbit pole at the time of contact. If we know the latitude and longitude of the orbit pole and ground station, gs, then the value of λ_{min} is

$$\sin \lambda_{min} = \sin lat_{pole} \sin lat_{gs} + \cos lat_{pole} \cos lat_{gs} \cos (\Delta long) \qquad (5\text{-}43)$$

where $\Delta long$ is the longitude difference between gs and the orbit pole. At the point of closest approach, we can compute the minimum nadir angle, η_{min}, maximum elevation angle, ε_{max}, and minimum range, D_{min} as

$$\tan \eta_{min} = \frac{\sin \rho \sin \lambda}{1 - \sin \rho \cos \lambda_{min}} \qquad (5\text{-}44)$$

$$\varepsilon_{max} = 90 \text{ deg} - \lambda_{min} - \eta_{min} \qquad (5\text{-}45)$$

$$D_{min} = R_E \left(\frac{\sin \lambda_{min}}{\sin \eta_{min}} \right) \qquad (5\text{-}46)$$

At the point of closest approach, the satellite is moving perpendicular to the line of sight to the ground station. Thus, the *maximum angular rate* of the satellite as seen from the ground station, $\dot{\theta}_{max}$, will be

$$\dot{\theta}_{max} = \frac{V_{sat}}{D_{min}} = \frac{2\pi (R_E + H)}{P D_{min}} \qquad (5\text{-}47)$$

where V_{sat} is the orbital velocity of the satellite, and P is the orbit period.

Finally, it is convenient to compute the *total azimuth range*, $\Delta\phi$, which the satellite covers as seen by the ground station, the *total time in view, T*, and the azimuth, ϕ_{center}, at the center of the viewing arc at which the elevation angle is a maximum:

$$\cos \frac{\Delta\phi}{2} = \frac{\tan \lambda_{min}}{\tan \lambda_{max}} \qquad (5\text{-}48)$$

$$T = \left(\frac{P}{180 \deg} \right) \arccos \left(\frac{\cos \lambda_{max}}{\cos \lambda_{min}} \right) \qquad (5\text{-}49)$$

where the arc cos is in degrees. ϕ_{center} is related to ϕ_{pole}, the azimuth to the direction to the projection of the orbit pole onto the ground by

$$\phi_{center} = 180 \text{ deg} - \phi_{pole} \qquad (5\text{-}50)$$

$$\cos \phi_{pole} = (\sin lat_{pole} - \sin \lambda_{min} \sin lat_{gs}) / (\cos \lambda_{min} \cos lat_{gs}) \qquad (5\text{-}51)$$

where $\phi_{pole} < 180$ deg if the orbit pole is east of the ground station and $\phi_{pole} > 180$ deg if the orbit pole is west of the ground station. The maximum time in view, T_{max}, occurs when the satellite passes overhead and $\lambda_{min} = 0$. Eq. (5-49) then reduces to:

$$T_{max} = P (\lambda_{max} / 180 \text{ deg}) \qquad (5\text{-}52)$$

If satellite passes are approximately evenly distributed in off-ground track angle, then the average pass duration is about 80% of T_{max} and 86% or more of the passes will be longer than half T_{max}.

Table 5-4 summarizes the computations for ground station coverage and provides a worked example. Note that as indicated above, T is particularly sensitive to ε_{min}. If, for example, we assume a mountain-top ground station with $\varepsilon_{min} = 2$ deg, then the time in view increases by 15% to 14.27 min. Figure 5-18 shows samples of several ground tracks for satellites in a 1,000 km orbit.

TABLE 5-4.   Summary of Computations for Ground Station Pass Parameters. We assume the following parameters: orbit pole at $lat_{pole} = 61.5$ deg, $long_{pole} = 100$ deg; Hawaii ground station at $lat_{gs} = 22$ deg, $long_{gs} = 200$ deg; minimum allowable elevation angle $\varepsilon_{min} = 5$ deg. The result is a typical pass time-in-view of about 12 min.

Parameter	Formula	Eq. No.	Example
Earth Angular Radius, ρ	$\sin \rho = R_E / (R_E + H)$	5-16	$\rho = 59.8$ deg
Period, P	$P = 1.658\ 669 \times 10^{-4} \times (6{,}378.14 + H)^{3/2}$	7-7	$P = 105$ min
Max Nadir Angle, η_{max}	$\sin \eta_{max} = \sin \rho \cos \varepsilon_{min}$	5-36	$\eta_{max} = 59.4$ deg
Max Earth Central Angle, λ_{max}	$\lambda_{max} = 90$ deg $- \varepsilon_{min} - \eta_{max}$	5-37	$\lambda_{max} = 25.6$ deg
Max Distance, D_{max}	$D_{max} = R_E (\sin \lambda_{max} / \sin \eta_{max})$	5-38	$D_{max} = 3{,}202$ km
Min Earth Central Angle, λ_{min}	$\sin \lambda_{min} = \sin lat_{pole} \sin lat_{gs}$ $+ \cos lat_{pole} \cos lat_{gs} \cos (\Delta long)$	5-43	$\lambda_{min} = 14.7$ deg
Min Nadir Angle, η_{min}	$\tan \eta_{min} = (\sin \rho \sin \lambda_{min}) / (1 - \sin \rho \cos \lambda_{min})$	5-44	$\eta_{min} = 53.2$ deg
Max Elevation Angle, ε_{max}	$\varepsilon_{max} = 90$ deg $- \lambda_{min} - \eta_{min}$	5-45	$\varepsilon_{max} = 22.1$ deg
Min Distance, D_{min}	$D_{min} = R_E (\sin \lambda_{min} / \sin \eta_{min})$	5-46	$D_{min} = 2{,}021$ km
Max Angular Rate, $\dot\theta_{max}$	$\dot\theta_{max} = [(2\pi (R_E + H)] / (P\, D_{min})$	5-47	$\dot\theta_{max} = 12.6$ deg/min
Azimuth Range, $\Delta\phi$	$\cos (\Delta\phi /2) = (\tan \lambda_{min} / \tan \lambda_{max})$	5-48	$\Delta\phi = 113.6$ deg
Time in View, T	$T = (P / 180$ deg$) \cos^{-1} (\cos \lambda_{max} / \cos \lambda_{min})$	5-49	$T = 12.36$ min

5.3.2 Satellites in Geosynchronous Orbit and Above

An important special case of the satellite motion as seen from the Earth's surface occurs for geostationary satellites, which hover approximately over one location on the Earth's equator. This will occur at an altitude of 35,786 km, for which the satellite period is 1,436 min, equaling the Earth's sidereal rotation period relative to the fixed stars. Chapter 6 describes the long-term drift of geostationary satellites. We describe here the apparent daily motion of these satellites as seen by an observer on the Earth.

For convenience, we assume the observer is at the center of the Earth and compute the apparent motion from there. The detailed motion seen from a location on the Earth's surface will be much more complex because the observer is displaced relative to the Earth's center. (See Wertz [2001].) But the general results will be the same, and the variations can be computed for any particular location.

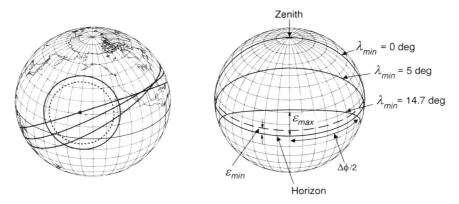

Fig. 5-18. **Motion of a Satellite at 1,000 km as Seen on the Earth and by an Observer on the Surface of the Earth.** See text for formulas.

Orbit inclination and eccentricity are the principal causes of the apparent daily motion of a geosynchronous satellite. These two effects yield different-shaped apparent orbits, which can cause confusion if the source of the apparent motion is not clearly identified. As Fig. 5-19A shows, the inclination of the orbit produces a figure eight centered on the equator, as seen by an observer at the Earth's center. The half-height, h_{inc}, and half-width, w_{inc}, of the figure eight due to an inclination, i, are given by

$$h_{inc} = \pm i \tag{5-53}$$

$$\tan \omega_{inc} = \frac{1}{2}\left(\sqrt{\sec i} - \sqrt{\cos i}\right) \cong \tan^2(i/2) \tag{5-54}$$

where the approximation in the second formula applies to small i. The source of this figure eight or *analemma* is the motion of the satellite along its inclined orbit, which will alternately fall behind and then catch up to the uniform rotation of the Earth on its axis.

The second factor which causes a nonuniform apparent motion is a nonzero eccentricity of the satellite orbit. An eccentricity, e, causes an East-West oscillation, ω_{ecc}, of magnitude

$$\omega_{ecc} = \pm\left(\frac{360 \ \text{deg}}{\pi}\right)e \approx \pm(115 \ \text{deg})e \tag{5-55}$$

In general, the inclination and eccentricity motions are superimposed, resulting in two possible shapes for the motion of the geosynchronous satellite as seen from the Earth. If the nonzero inclination effect dominates, then the satellite appears to move in a figure eight. If the eccentricity effect is larger than the inclination effect, then the apparent motion is a single open oval, as shown in Fig. 5-19B.

For satellites above geosynchronous orbit, the rotation of the Earth on its axis dominates the apparent motion of the satellite. Consequently, it is most convenient in this case to plot the motion of the satellite relative to the background of the fixed stars.

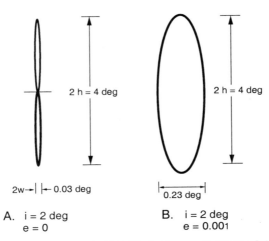

Fig. 5-19.   Apparent Daily Motion of a Satellite in Geosynchronous Orbit.

In this coordinate frame, we can handle the motion relative to the fixed inertial background just the same as we do the apparent motion of the Moon or planets. Many introductory texts on celestial mechanics treat this issue. See, for example, Roy [1991], Green [1985], or Smart [1977], or Wertz [2001].

5.4 Development of Mapping and Pointing Budgets

Nearly all spacecraft missions involve sensing or interaction with the world around them, so a spacecraft needs to know or control its orientation. We may conveniently divide this problem of orientation into two areas of pointing and mapping. *Pointing* means orienting the spacecraft, camera, sensor, or antenna to a target having a specific geographic position or inertial direction. *Mapping* is determining the geographic position of the *look point* of a camera, sensor, or antenna. Satellites used only for communications will generally require only pointing. Satellites having some type of viewing instrument, such as weather, ground surveillance, or Earth resources satellites, will ordinarily require both pointing ("point the instrument at New York") and mapping ("determine the geographic location of the tall building in pixel 2073").

The goal of this section is to develop *budgets* for pointing and mapping. A budget lists all the sources of pointing and mapping errors and how much they contribute to the overall pointing and mapping accuracy. This accuracy budget frequently drives both the cost and performance of a space mission. If components in the budget are left out or incorrectly assessed, the satellite may not be able to meet its performance objectives. More commonly, people who define the system requirements make the budgets for pointing and mapping too stringent and, therefore, unnecessarily drive up the cost of the mission. As a result, we must understand from the start the components of mapping and pointing budgets and how they affect overall accuracy. In this section we will emphasize Earth-oriented missions, but the same basic rules apply to inertially-oriented missions.

The components of the pointing and mapping budgets are shown in Fig. 5-20 and defined in Table 5-5. Basic pointing and mapping errors are associated with spacecraft navigation—that is, knowledge of its position and attitude in space. But even if the position and attitude are known precisely, a number of other errors will be present. For example, an error in the observation time will result in an error in the computed location of the target, because the target frame of reference moves relative to the spacecraft. A target fixed on the Earth's equator will rotate with the Earth at 464 m/s. A 10-sec error in the observation time would produce an error of 5 km in the computed geographic location of the target. Errors in the target altitude, discussed below, can be a key component of pointing and mapping budgets. The instrument-mounting error represents the misalignment between the pointed antenna or instrument and the sensor or sensors used to determine the attitude. This error is extremely difficult to remove. Because we cannot determine it from the attitude data alone, we must view it as a critical parameter and keep it small while integrating the spacecraft.

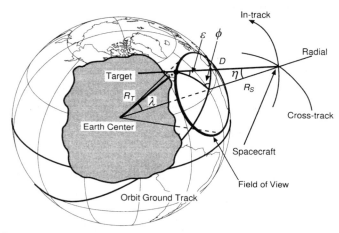

Fig. 5-20.   Definition of Pointing and Mapping Error Components.

Pointing errors differ from mapping errors in the way they include inaccuracies in attitude control and angular motion. Specifically, the pointing error must include the entire control error for the spacecraft. On the other hand, the only control-related component of the mapping error is the angular motion during the exposure or observation time. This short-term *jitter* results in a blurring of the look point of the instrument or antenna.

As discussed earlier in Sec. 4.2.2, we may achieve an accuracy goal for either pointing or mapping in many ways. We may, in one instance, know the position of the spacecraft precisely and the attitude only poorly. Or we may choose to allow a larger error in position and make the requirements for determining the attitude more stringent. In an ideal world we would look at all components of the pointing and mapping budgets and adjust them until a small increment of accuracy costs the same for each component. For example, assume that a given mission requires a pointing accuracy of 20 milliradians, and that we tentatively assign 10 milliradians to attitude determination and 5 milliradians to position determination. We also find more accurate attitude would cost $100,000 per milliradian, whereas more accurate position would cost only $50,000 per milliradian. In this case we should allow the attitude accuracy to degrade and improve the position-accuracy requirement until the cost per milliradian is the same for both. We will then have the lowest cost solution.

TABLE 5-5.   Sources of Pointing and Mapping Errors.

SPACECRAFT POSITION ERRORS:

Δl	In- or along-track	Displacement along the spacecraft's velocity vector
ΔC	Cross-track	Displacement normal to the spacecraft's orbit plane
ΔR_S	Radial	Displacement toward the center of the Earth (nadir)

SENSING AXIS ORIENTATION ERRORS (In polar coordinates about nadir):

$\Delta \eta$	Elevation	Error in angle from nadir to sensing axis
$\Delta \phi$	Azimuth	Error in rotation of the sensing axis about nadir

Sensing axis orientation errors include errors in (1) attitude determination, (2) instrument mounting, and (3) stability for mapping or control for pointing.

OTHER ERRORS:

ΔR_T	Target altitude	Uncertainty in the altitude of the observed object
ΔT	Clock error	Uncertainty in the real observation time (results in uncertainty in the rotational position of the Earth)

In practice we can seldom follow the above process. For example, we cannot improve accuracy continuously. Rather, we must often accept large steps in both performance and cost as we change methods or techniques. Similarly, we seldom know precisely how much money or what level of performance to budget. In practice the mission designer strives to balance the components, often by relying on experience and intuition as much as analysis. But the overall goal remains correct. We should try to balance the error budget so that incrementally improving any of the components results in approximately comparable cost.

A practical method of creating an error budget is as follows. We begin by writing down all of the components of the pointing and mapping budgets from Table 5-5. We assume that these components are unrelated to each other, being prepared to combine them later by taking the root sum square of the individual elements. (We will have to examine this assumption in light of the eventual mission design and adjust it to take into account how the error components truly combine.) The next step is to spread the budget equally among all components. Thus, if all seven error sources listed in Table 5-5 are relevant to the problem, we will initially assign an accuracy requirement for each equal to the total accuracy divided by $\sqrt{7}$. This provides a starting point for allocating errors. Our next step is to look at normal spacecraft operations and divide the error sources into three categories:

(A) Those allowing very little adjustment

(B) Those easily meeting the error allocation established for them, and

(C) Those allowing increased accuracy at increased cost

Determining the spacecraft position using ground radar is a normal operation, and the ground station provides a fixed level of accuracy. We cannot adjust this error source without much higher cost, so we assign it to category (A) and accept its corresponding level of accuracy. A typical example of category (B) is the observation time for which an accuracy of tens of milliseconds is reasonable with modern spacecraft clocks. Therefore, we will assign an appropriately small number (say 10 ms) to the accuracy associated with the timing error. Attitude determination ordinarily falls into

TABLE 5-6. Mapping and Pointing Error Formulas. ε is the grazing angle and *lat* is the latitude of the target, ϕ is the target azimuth relative to the ground track, λ is the Earth central angle from the target to the satellite, D is the distance from the satellite to the target, R_T is the distance from the Earth's center to the target (typically $\sim R_E$, the Earth's radius), and R_S is the distance from the Earth's center to the satellite. See Fig. 5-20.

Error Source	Error Magnitude (units)	Magnitude of Mapping Error (km)	Magnitude of Pointing Error (rad)	Direction of Error
Attitude Errors:[1]				
Azimuth	$\Delta\phi$ (rad)	$\Delta\phi D \sin \eta$	$\Delta\phi \sin \eta$	Azimuthal
Nadir Angle	$\Delta\eta$ (rad)	$\Delta\eta\, D / \sin \varepsilon$	$\Delta\eta$	Toward nadir
Position Errors:				
In-Track	ΔI (km)	$\Delta I\, (R_T/R_S) \cos H$ [2]	$(\Delta I / D) \sin Y_I$ [5]	Parallel to ground track
Cross-Track	ΔC (km)	$\Delta C\, (R_T/R_S) \cos G$ [3]	$(\Delta C / D) \sin Y_C$ [6]	Perpendicular to ground track
Radial	ΔR_S (km)	$\Delta R_S\, \sin \eta / \sin \varepsilon$	$(\Delta R_S / D) \sin \eta$	Toward nadir
Other Errors:				
Target Altitude	ΔR_T (km)	$\Delta R_T / \tan \varepsilon$	—	Toward nadir
S/C Clock	ΔT (s)	$\Delta T\, V_e \cos (lat)$ [4]	$\Delta T\,(V_e / D) \cos(lat)$ $\cdot \sin J$ [7]	Parallel to Earth's equator

Notes:
(1) Includes attitude determination error, instrument mounting error, stability over exposure time (mapping only), and control error (pointing only). The formulas given assume that the attitude is measured with respect to the Earth.
(2) $\sin H = \sin \lambda \sin \phi$.
(3) $\sin G = \sin \lambda \cos \phi$.
(4) $V_e = 464$ m/s (Earth rotation velocity at the equator).
(5) $\cos Y_I = \cos \phi \sin \eta$.
(6) $\cos Y_C = \sin \phi \sin \eta$.
(7) $\cos J = \cos \phi_E \cos \varepsilon$, where $\phi_E =$ azimuth relative to East.

category (C). Here we might have a gravity gradient-stabilized system accurate to a few degrees with no attitude determination cost at all, an horizon sensor system accurate to 0.05–0.10 deg, or a much more expensive star sensor system accurate to better than 0.01 deg (see Sec. 11.1).

This process allows us to balance cost between the appropriate components and to go back to the mission definition and adjust the real requirements. For example, achieving a mapping accuracy of 100 m on the ground might triple the cost of the space mission by requiring highly accurate attitude determination, a new system for determining the orbit, and a detailed list of target altitudes. Reducing the accuracy requirement to 500 m might lower the cost enough to make the mission possible within the established budget constraints. This is an example of trading on mission requirements, described in Chaps. 2 to 4. Requirements trading is extremely important to a cost-effective mission, but we often omit this in the normal process of defining mission requirements.

To carry out this trade process, we need to know how an error in each of the components described in Table 5-5 relates to the overall mapping and pointing errors.

Table 5-6 gives formulas relating the errors in each of the seven basic components to the overall error. Here the notation used is the same as in Fig. 5-20. For any given mission conditions, these formulas relate the errors in the fundamental components to the resulting pointing and mapping accuracies. Table 5-6 provides basic algebraic information which we must transform into specific mapping and pointing requirements for a given mission. The general process of deriving these requirements is given below. Representative mapping and pointing budgets based on these formulas are given in Table 5-7

TABLE 5-7.   Representative Mapping and Pointing Error Budgets. See Figs. 5-21 and 5-22 for corresponding plots.

Source	Error in Source	Error Budgets			
		Mapping Error (km)		Pointing Error (deg)	
		$\varepsilon = 10$ deg	$\varepsilon = 30$ deg	$\varepsilon = 10$ deg	$\varepsilon = 30$ deg
Attitude Errors:					
Azimuth	0.06 deg	2.46	1.33	0.051	0.045
Nadir Angle	0.03 deg	8.33	1.78	0.030	0.030
Position Errors:					
In-Track	0.2 km	0.17	0.17	0.002	0.005
Cross-Track	0.2 km	0.16	0.17	0.004	0.007
Radial	0.1 km	0.49	0.15	0.002	0.003
Other Errors:					
Target Altitude	1 km	5.67	1.73	—	—
S/C Clock	0.5 sec	0.23	0.23	0.005	0.008
Root Sum Square		10.39	2.84	0.060	0.055

Defining Mapping Requirements

The errors associated with mapping depend strongly on how close to the horizon we choose to work. Working in a very small region directly under the spacecraft provides very poor coverage but excellent mapping accuracy and resolution (see Fig. 5-21). On the other hand, working near the horizon provides very broad coverage but poor mapping accuracy. Thus, we must trade resolution and mapping accuracy for coverage. The mapping accuracy for a particular mission depends on the spacecraft's elevation angle at the edge of the coverage region. In almost all cases the mapping accuracy will be much better looking straight down, and the limiting accuracy will be closest to the horizon. To assess satellite coverage, we look at the satellite's swath width. That is, we assume the spacecraft can work directly below itself and at all angles out to a limiting spacecraft elevation angle as seen from a target on the ground.

Accuracy characteristics as a function of elevation angle are more complex because they involve combining several terms. A sample plot of mapping error as a function of the spacecraft's elevation angle for a satellite at 1,000 km is in Fig. 5-22. This figure is based on the equations in Table 5-6.

The total mapping error is the root sum square of the individual components. Generally, uncertainty in target altitude and in attitude determination contribute most to errors in mapping accuracy. In most cases improving other factors will have only a

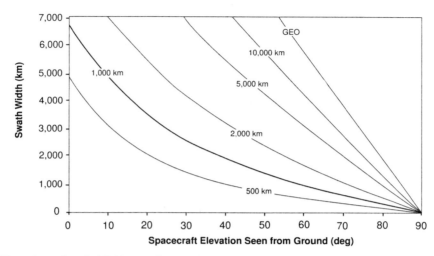

Fig. 5-21. **Swath Width vs. Spacecraft Elevation Angle for a Spacecraft at Various Altitudes.** Note that the swath width increases dramatically at small elevation angles.

second-order effect. Consequently, determining target altitude and spacecraft attitude are high priorities in assessing a mission's mapping performance and cost.

The uncertainty in target altitude typically contributes most to determining a geographic location on the Earth. The oblateness of the Earth has the largest effect on target altitude. It causes a variation in distance from the center of the Earth of approximately 25 km between the poles and the equator. But we can account for this factor analytically at very low cost, so it does not usually add to the error. The next plateau is for airplanes, clouds, or other atmospheric features. The uncertainty in target altitude at this level will typically be 10 km or larger unless we have some a priori estimate of the altitude. For features on the Earth's surface, the uncertainty in target altitude reduces to approximately 1 km, unless data analysis includes a detailed map of target altitudes. Figure 5-22 incorporates this 1 km error in target altitude as the dominant source of error. Thus, for example, for FireSat to have a mapping error of less than 1 km would require one of two arrangements. The spacecraft could work only very near nadir and therefore have very poor coverage. Alternatively, it could include the elevation of the target region as a part of data reduction, therefore requiring the use of a very large data base and making the data processing more complex.

The second principal contributor to mapping error is the uncertainty in attitude determination, which varies widely over the following cost plateaus:

Accuracy Level (deg)	Method
~10	Gravity gradient spacecraft, no attitude determination
~2	Magnetometer only
0.5	Earth sensing, no oblateness corrections
0.1	General Earth sensing
0.03	High-accuracy Earth sensing
<0.01	Star sensing

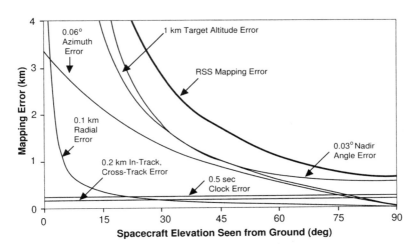

Fig. 5-22. **Mapping Error as a Function of Elevation Angle for a Spacecraft at 1,000 km Altitude.** Magnitudes of assumed error sources are marked.

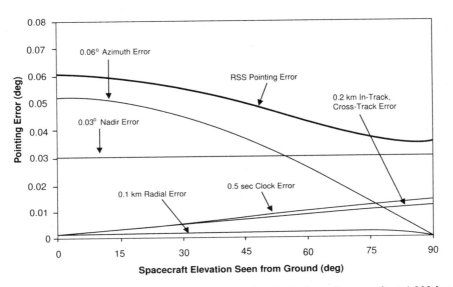

Fig. 5-23. **Pointing Error as a Function of Elevation Angle for a Spacecraft at 1,000 km Altitude.** Magnitude of assumed error sources are marked.

Using these general limits in a model such as that of Fig. 5-22 allows us to assess accuracies as a function of cost and coverage.

Defining Pointing Requirements

Unlike mapping, pointing depends only weakly on the spacecraft's elevation angle. (See Fig. 5-23.) Thus, for missions which require only pointing, working in a region near the horizon is almost as easy as pointing to a target or ground antenna at nadir. In this case the working limit on the spacecraft's elevation angle depends on other factors, such as the transmission of the atmosphere for a selected wavelength or possible obstruction by local geography. For example, ground stations ordinarily limit their elevation to approximately 5 deg above the horizon because of the reduced atmospheric transmission at lower elevation angles.

Pointing requirements normally arise from the spacecraft's housekeeping functions or from the need to point a particular instrument or antenna toward a ground target. Housekeeping requirements such as solar array pointing and orbit maneuvers ordinarily demand pointing accuracies of 0.25 to 1 deg. Consequently, for most missions, the need to point the mission sensor or antenna is more important. Here again two cases exist. If we wish to point the sensor at a single target, then we will generally try to point the center of the sensor at the target. By doing so, we establish a pointing requirement to place the target within the field of view. Specifically, if the payload sensor's field of view is four times the 3σ pointing error, then the target will lie within the field of view with a 6σ probability, or virtual certainty. For example, if the FireSat sensor has a 1 deg square field of view, an overall pointing requirement of 0.25 deg will assure that the target will be within the field of view.

In pointing we may also want to eliminate overlapping coverage. For example, if we wish to take a series of pictures, we must overlap the pictures by more than the pointing error to ensure continuous coverage of the ground. This requirement, in turn, implies that the spacing between pictures must equal the field-of-view size less the pointing error. Thus, with a large pointing error, we must accept having fewer pictures at a given time and increased resource costs in terms of time, power, and data rate for a given level of coverage. It is common to have a pointing accuracy of 10% to 20% of the field-of-view diameter. Driving the pointing under 10% of the field-of-view diameter will only slightly improve overall coverage. On the other hand, a pointing error worse than 20% of the field-of-view size can require substantial overlap, thus greatly diminishing the overall system's coverage and resource utilization.

References

Collins, Steve. 1992. "Geocentric Nadir and Range from Horizon Sensor Observations of the Oblate Earth." AIAA Paper No. 92-176 presented at the AAS/AISS Spaceflight Mechanics Meeting. Colorado Springs, CO, Feb. 24–26.

Green, R.M. 1985. *Spherical Astronomy*. Cambridge: Cambridge University Press.

Liu, K. 1978. "Earth Oblateness Modeling," in *Spacecraft Attitude Determination and Control*, ed. James R. Wertz. 98–106. Holland: D. Reidel Publishing Company.

Roy, A.E. 1991. *Orbital Motion (3rd Edition)*. New York: John Wiley and Sons.

Smart, W.M. 1977. *Textbook on Spherical Astronomy (6th Edition)*. Cambridge: Cambridge University Press.

Wertz, James R. 1981. "Global Geometry Techniques for Mission Analysis," in *Proc. Int. Symp. Spacecraft Flight Dynamics*, Darmstadt, FRG, May 18–22 (ESA SP-160, Aug. 1981).

_____. 2001. *Mission Geometry; Orbit and Constellation Design and Management*. Torrance, CA: Microcosm Press and Dordrecht, The Netherlands: Kluwer Academic Publishers.

Chapter 6

Introduction to Astrodynamics

Daryl G. Boden, *United States Naval Academy*

Astrodynamics is the study of a satellite's *trajectory* or *orbit*, that is, its path through space. The satellite's *ephemeris* is a table listing its position as a function of time. The first section below explains the terms used to describe satellite orbits, provides equations necessary to calculate orbital elements from position and velocity, and shows how to predict the future position and velocity of a satellite. We base this method on a simple, but accurate, model treating the Earth and the satellite as spherical masses. The next section discusses how forces other than the Newtonian gravitational force affect the orbit of a satellite. The third section explains maneuvering strategies used to change the satellite's orbit. The final two sections discuss the available launch times and methods for maintaining satellite orbits.

Several textbooks are available in the areas of satellite orbits and celestial mechanics. Some of the most popular are Bate, Mueller, and White [1971], Battin [1999], Danby [1962], Escobal [1965], Kaplan [1976], and Roy [1991]. More recently Chobotov [1996], Wiesel [1996], and Vallado [1997] have provided works specifically on spacecraft astrodynamics.

6.1 Keplerian Orbits

Explaining the motion of celestial bodies, especially the planets, has challenged observers for many centuries. The early Greeks attempted to describe the motion of celestial bodies about the Earth in terms of circular motion. In 1543, Nicolaus Copernicus proposed a heliocentric (Sun-centered) system with the planets following circular orbits. Finally, with the help of Tycho Brahe's observational data, Johannes Kepler described elliptical planetary orbits about the Sun. Later, Isaac Newton provided a mathematical solution for this system based on an inverse-square gravitational force.

Kepler spent several years reconciling the differences between Tycho Brahe's careful observations of the planets and their predicted motion based on previous theories. Having found that the data matched a geometric solution of elliptical orbits, he published his first two laws of planetary motion in 1609 and his third law in 1619. Kepler's three laws of planetary motion (which also apply to satellites orbiting the Earth) are:

- *First Law:* The orbit of each planet is an ellipse, with the Sun at one focus.

- *Second Law:* The line joining the planet to the Sun sweeps out equal areas in equal times.

- *Third Law:* The square of the period of a planet is proportional to the cube of its mean distance from the Sun.

6.1.1 Satellite Equations of Motion

Figure 6-1 depicts the key parameters of an elliptical orbit. The *eccentricity, e,* of the ellipse (not shown in the figure) is equal to c/a and is a measure of the deviation of the ellipse from a circle.

Isaac Newton explained mathematically why the planets (and satellites) follow elliptical orbits. Newton's *Second Law of Motion,* applied to a constant mass system and, combined with his *Law of Universal Gravitation,* provides the mathematical basis for analyzing satellite orbits. Newton's law of gravitation states that any two bodies attract each other with a force proportional to the product of their masses and inversely proportional to the square of the distance between them. The equation for the magnitude of the force caused by gravity is

$$F = -GMm/r^2$$
$$\equiv -\mu m/r^2 \tag{6-1}$$

where F is the magnitude of the force caused by gravity, G is the universal constant of gravitation, M is the mass of the Earth, m is the mass of the satellite, r is the distance from the center of the Earth to the satellite, and $\mu \equiv GM$ is the Earth's gravitational constant (= 398,600.5 km^3s^{-2}).

Combining Newton's second law with his law of gravitation, we obtain an equation for the acceleration vector of the satellite:

$$\ddot{\mathbf{r}} + (\mu r^{-3})\mathbf{r} = \mathbf{0} \tag{6-2}$$

This equation, called the *two-body equation of motion,* is the relative equation of motion of a satellite position vector as the satellite orbits the Earth. In deriving it, we assumed that gravity is the only force, the Earth is spherically symmetric, the Earth's mass is much greater than the satellite's mass, and the Earth and the satellite are the only two bodies in the system.

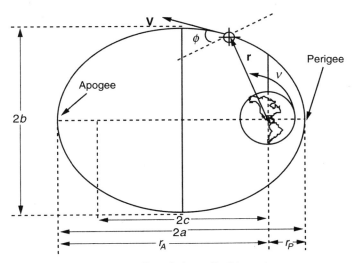

r: position vector of the satellite relative to Earth's center

V: velocity vector of the satellite relative to Earth's center

ϕ: *flight-path-angle*, the angle between the velocity vector and a line perpendicular to the position vector

a: *semimajor* axis of the ellipse

b: *semiminor* axis of the ellipse

c: the distance from the center of the orbit to one of the focii

v: the *polar angle* of the ellipse, also called the *true anomaly*, measured in the direction of motion from the direction of perigee to the position vector

r_A: *radius of apogee*, the distance from Earth's center to the farthest point on the ellipse

r_P: *radius of perigee*, the distance from Earth's center to the point of closest approach to the Earth

Fig. 6-1.   Geometry of an Ellipse and Orbital Parameters.

A solution to the two-body equation of motion for a satellite orbiting Earth is the *polar equation of a conic section*. It gives the magnitude of the position vector in terms of the location in the orbit,

$$r = a(1 - e^2)/(1 + e \cos v) \qquad (6\text{-}3)$$

where a is the semimajor axis, e is the eccentricity, and v is the polar angle or true anomaly.

A *conic section* is a curve formed by the intersection of a plane passing through a right circular cone. As Fig. 6-2 shows, the angular orientation of the plane relative to the cone determines whether the conic section is a *circle, ellipse, parabola,* or *hyperbola*. We can define all conic sections in terms of the eccentricity, *e,* in Eq. (6-3) above. The type of conic section is also related to the semimajor axis, *a,* and the specific mechanical energy, ε. Table 6-1 shows the relationships between energy, eccentricity, and semimajor axis and the type of conic section.

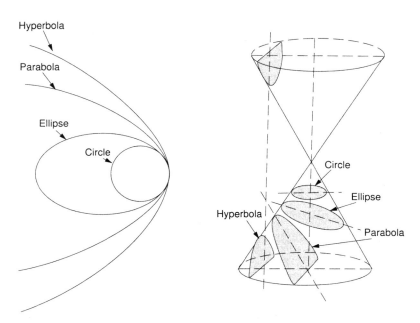

Fig. 6-2. Geometric Origin of Conic Sections. Satellite orbits can be any of four conic sections: a circle, an ellipse, a parabola, or a hyperbola.

TABLE 6-1. Conic Sections. See text for discussion.

Conic	Energy, $\mathcal{E}$	Semimajor Axis, a	Eccentricity, e
Circle	< 0	= radius	0
Ellipse	< 0	> 0	0 < e < 1
Parabola	0	∞	1
Hyperbola	> 0	< 0	> 1

6.1.2 Constants of Motion

Using the two-body equation of motion, we can derive several constants of motion of a satellite orbit. The first is

$$\mathcal{E} = V^2 / 2 - \mu / r = -\mu / (2a) \tag{6-4}$$

where $\mathcal{E}$ is the total *specific mechanical energy*, or mechanical energy per unit mass, for the system and is the sum of the kinetic energy per unit mass and potential energy per unit mass. We refer to Eq. (6-4) as the *energy equation*. Because the forces in the system are conservative, the energy is a constant. The term for potential energy, $-\mu/r$, defines the potential energy to be zero at infinity and negative at any radius less than infinity. Using this definition, the specific mechanical energy of elliptical orbits will always be negative. As the energy increases (approaches zero), the ellipse gets larger, and the elliptical trajectory approaches a parabolic trajectory. From the energy Eq. (6-4), we find that a satellite moves fastest at perigee of the orbit and slowest at apogee.

We also know that for a circle the semimajor axis equals the radius, which is constant. Rearranging the energy equation, we find the velocity of a satellite in a circular orbit.

$$V_{cir} = (\mu / r)^{1/2} \qquad (6\text{-}5)$$

$$\cong 7.905\ 366\ (R_E / r)^{1/2}$$

$$\cong 631.3481\ r^{-1/2}$$

where V_{cir} is the circular velocity in km/s, R_E is the Earth's radius in km, and r is the orbit's radius in km.

From Table 6-1, the energy of a parabolic trajectory is zero. A parabolic trajectory is one with the minimum energy needed to escape the gravitational attraction of Earth. Thus, we can calculate the velocity required to escape from the Earth at any distance, r, by setting energy equal to zero in Eq. (6-4) and solving for velocity.

$$V_{esc} = (2\mu / r)^{1/2} \qquad (6\text{-}6)$$

$$\cong 11.179\ 88\ (R_E / r)^{1/2}$$

$$\cong 892.8611\ r^{-1/2}$$

where V_{esc} is the escape velocity in km/s, and r is in km.

Another constant of motion associated with a satellite orbit is the *specific angular momentum*, **h**, which is the satellite's total angular momentum divided by its mass. We can find it from the cross product of the position and velocity vectors.

$$\mathbf{h} = \mathbf{r} \times \mathbf{v} \qquad (6\text{-}7)$$

We find that from Kepler's second law, the angular momentum is constant in magnitude and direction for the two-body problem. Therefore, the orbital plane defined by the position and velocity vectors must remain fixed in inertial space.

6.1.3 Classical Orbital Elements

When solving the two-body equations of motion, we need six constants of integration (initial conditions) for the solution. Theoretically, we could find the three components of position and velocity at any time in terms of the position and velocity at any other time. Alternatively, we can completely describe the orbit with five constants and one quantity which varies with time. These quantities, called *classical orbital elements*, are defined below and are shown in Fig. 6-3. The coordinate frame shown in the figure is the geocentric inertial frame,* or GCI, defined in Chap. 5 (see Table 5-1). Its origin is at the center of the Earth, with the X-axis in the equatorial plane and pointing to the vernal equinox. Also, the Z-axis is parallel to the Earth's spin axis (the North Pole), and the Y-axis completes the right-hand set in the equatorial plane. The classical orbital elements are:

a: *semimajor axis*: describes the size of the ellipse (see Fig. 6-1).

e: *eccentricity*: describes the shape of the ellipse (see Fig. 6-1).

---

*A *sufficiently inertial coordinate frame* is a coordinate frame that we can consider to be non-accelerating for the particular application. The GCI frame is sufficiently inertial when considering Earth-orbiting satellites, but is inadequate for interplanetary travel because of its rotational acceleration around the Sun.

i: *inclination*: the angle between the angular momentum vector and the unit vector in the **Z**-direction.

Ω: *right ascension of the ascending node*: the angle from the vernal equinox to the ascending node. The *ascending node* is the point where the satellite passes through the equatorial plane moving from south to north. Right ascension is measured as a right-handed rotation about the pole, **Z**.

ω: *argument of perigee*: the angle from the ascending node to the eccentricity vector measured in the direction of the satellite's motion. The *eccentricity vector* points from the center of the Earth to perigee with a magnitude equal to the eccentricity of the orbit.

v: *true anomaly*: the angle from the eccentricity vector to the satellite position vector, measured in the direction of satellite motion. Alternately, we could use *time since perigee passage, T*.

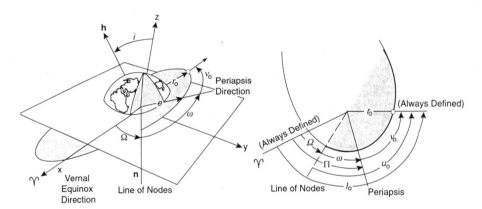

Fig. 6-3. **Definition of the Keplerian Orbital Elements of a Satellite in an Elliptic Orbit.** We define elements relative to the GCI coordinate frame.

Given these definitions, we can solve for the elements if we know the satellite's position and velocity vectors. Equations (6-4) and (6-7) allow us to solve for the energy and the angular momentum vector. An equation for the *nodal vector*, **n**, in the direction of the ascending node is

$$\mathbf{n} = \mathbf{Z} \times \mathbf{h} \qquad (6\text{-}8)$$

We can calculate the eccentricity vector from the following equation:

$$\mathbf{e} = (1/\mu)\{(V^2 - \mu/r)\,\mathbf{r} - (\mathbf{r} \cdot \mathbf{v})\,\mathbf{v}\} \qquad (6\text{-}9)$$

Table 6-2 lists equations to derive the classical orbital elements and related parameters for an elliptical orbit.

Equatorial ($i = 0$) and circular ($e = 0$) orbits demand alternate orbital elements to solve the equations in Table 6-2. These are shown in Fig. 6-3. For equatorial orbits, a single angle, Π, can replace the right ascension of the ascending node and argument of perigee. Called the *longitude of perigee*, this angle is the algebraic sum of Ω and ω. As i approaches 0, Π approaches the angle from the **X**-axis to perigee. For circular

TABLE 6-2. Computing the Classic Orbital Elements. For the right ascension of the ascending node, argument of perigee, and true anomaly, if the quantities in parentheses are positive, use the angle calculated. If the quantities are negative, the correct value is 360 deg minus the angle calculated.

Symbol	Name	Equation	Quadrant Check
a	semimajor axis	$a = -\mu/(2\varepsilon) = (r_A + r_P)/2$	
e	eccentricity	$e = |\mathbf{e}| = 1 - (r_P/a) = (r_A/a) - 1$	
i	inclination	$i = \cos^{-1}(h_Z/h)$	
Ω	right ascension of the ascending node	$\Omega = \cos^{-1}(n_X/n)$	$(n_Y > 0)$
ω	argument of perigee	$\omega = \cos^{-1}[(\mathbf{n}\cdot\mathbf{e})/(n\cdot e)]$	$(e_Z > 0)$
v	true anomaly	$v = \cos^{-1}[(\mathbf{e}\cdot\mathbf{r})/(e\cdot r)]$	$(\mathbf{r}\cdot\mathbf{v} > 0)$
r_P	radius of perigee	$r_P = a(1 - e)$	
r_A	radius of apogee	$r_A = a(1 + e)$	
P	period	$P = 2\pi(a^3/\mu)^{1/2}$ $\cong 84.489\,(a/R_E)^{3/2}\text{min}$ $\cong 0.000\,165\,87\,a^{3/2}\text{min},\ a \text{ in km}$	
ω_O	orbital frequency	$\omega_O = (\mu/a^3)^{1/2}$ $\cong 631.348\,16\,a^{-3/2}\text{rad/s},\ a \text{ in km}$	

orbits ($e = 0$), a single angle, $u \equiv \omega + v$, can replace the argument of perigee and true anomaly. This angle is the *argument of latitude* and, when $e = 0$, equals the angle from the nodal vector to the satellite position vector. Finally, if the orbit is circular and equatorial, then a single angle, l, or *true longitude*, specifies the angle between the **X**-axis and the satellite position vector.

6.1.4 Satellite Ground Tracks

As defined in Chap. 5, a satellite's ground track is the trace of the points formed by the intersection of the satellite's position vector with the Earth's surface. In this section we will evaluate ground tracks using a flat map of the Earth. Chapters 5 and 7 give another approach of displaying them on a global representation.

Although ground tracks are generated from the orbital elements, we can gain insight by determining the orbital elements from a given ground track. Figure 6-4 shows ground tracks for satellites with different orbital altitudes and, therefore, different orbital periods. The time it takes Earth to rotate through the difference in longitude between two successive ascending nodes equals the orbit's period. For *direct* orbits, in which the satellite moves eastward, we measure the change positive to the east. For *retrograde* orbits, in which the satellite moves westward, positive is measured to the west.[*] With these definitions, the period, P, in minutes is

$$P = 4(360\text{ deg} - \Delta L) \quad \text{direct orbit} \tag{6-10}$$

$$P = 4(\Delta L - 360\text{ deg}) \quad \text{retrograde orbit}$$

---

[*]This convenient empirical definition does not apply for near polar orbits. More formally, a *prograde* or *direct* orbit has $i < 90$ deg. A *retrograde* orbit has $i > 90$ deg. A *polar orbit* has $i = 90$ deg.

(For a more precise rotation rate for the Earth, use $3.98\overline{8}$ min/deg instead of 4.0 in these equations.) where ΔL is the change in longitude in degrees that the satellite goes through between successive ascending nodes. The difference in longitude between two successive ascending nodes for a direct orbit will always be less than 360 deg, and in fact will be negative for orbits at altitudes higher than geosynchronous altitude. For retrograde orbits, the difference in longitude between two successive ascending nodes (positive change is measured to the west) is always greater than 360 deg.

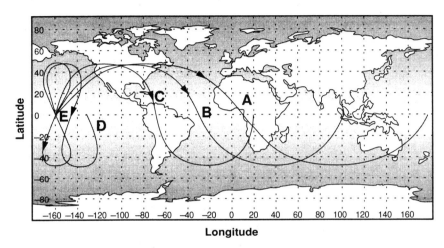

Fig. 6-4. Orbital Ground Tracks of Circular Orbits of Different Periods. (A) $\Delta L = 335°$, $P = 100$ min; (B) $\Delta L = 260°$, $P = 398$ min; (C) $\Delta L = 180°$, $P = 718$ min; (D) $\Delta L = 28°$, $P = 1,324$ min; and (E) $\Delta L = 0°$, $P = 1,436$ min.

Once we know the period, we can determine the semimajor axis by using the equation for the period of an elliptical orbit:

$$a = [(P/2\pi)^2\mu]^{1/3} \tag{6-11}$$

$$\cong 331.249\ 15\ P^{2/3}\ km$$

where the period is in minutes.

Figure 6-4 shows one revolution each for the ground tracks of several orbits with an increasing semimajor axis. The period of a *geosynchronous* orbit, E, is 1,436 min, matching the Earth's rotational motion.

We can determine the orbit's inclination by the ground track's maximum latitude. For direct orbits the inclination equals the ground track's maximum latitude, and for retrograde orbits the inclination equals 180 deg minus the ground track's maximum latitude.

The orbit is circular if a ground track is symmetrical about both the equator and a line of longitude extending downward from the ground track's maximum latitude. All the orbits in Fig. 6-4 are circular.

Figure 6-5 shows examples of ground tracks for the following typical orbits:

A: Shuttle parking orbit, $a = 6,700$ km, $e = 0$, $i = 28.4$ deg;

B: Low-altitude retrograde orbit, $a = 6,700$ km, $e = 0$, $i = 98.0$ deg;

C: GPS orbit, $a = 26{,}600$ km, $e = 0$, $i = 55.0$ deg; and

D: Molniya orbits, $a = 26{,}600$ km, $e = 0.75$, $i = 63.4$ deg, $\omega = 270$ deg.

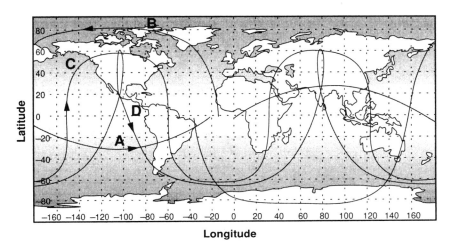

Fig. 6-5.   **Typical Ground Tracks.** (A) Shuttle parking, (B) Low-altitude retrograde, (C) GPS, and (D) Molniya orbits. See text for orbital elements.

6.1.5 Time of Flight in an Elliptical Orbit

In analyzing Brahe's observational data, Kepler was able to solve the problem of relating position in the orbit to the elapsed time, $t - t_0$, or conversely, how long it takes to go from one point to another in an orbit. To solve this, Kepler introduced the quantity M, called the *mean anomaly*, which is the fraction of an orbit period which has elapsed since perigee, expressed as an angle. The mean anomaly equals the true anomaly for a circular orbit. By definition,

$$M - M_0 \equiv n\,(t - t_0) \tag{6-12}$$

where M_0 is the mean anomaly at time t_0 and n is the *mean motion*, or average angular velocity, determined from the semimajor axis of the orbit:

$$n \equiv (\mu / a^3)^{1/2} \tag{6-13}$$

$$\cong 36{,}173.585\ a^{-3/2}\ \text{deg/s}$$

$$\cong 8{,}681{,}660.4\ a^{-3/2}\ \text{rev/day}$$

$$\cong 3.125\ 297\ 7 \times 10^9 a^{-3/2}\ \text{deg/day}$$

where a is in km.

This solution will give the average position and velocity, but satellite orbits are elliptical, with a radius constantly varying in orbit. Because the satellite's velocity depends on this varying radius, it changes as well. To resolve this problem we define an intermediate variable called *eccentric anomaly*, E, for elliptical orbits. Table 6-3 lists the equations necessary to relate time of flight to orbital position.

TABLE 6-3. Time of Flight in an Elliptic Orbit. All angular quantities are in radians.

Variable	Name	Equation	
n	mean motion	$n = (\mu/a^3)^{1/2}$ $\approx 631.348\,16\,a^{-3/2}$ rad/s	(a in km)
E	eccentric anomaly	$\cos E = (e + \cos v)/(1 + e \cos v)$	
M	mean anomaly	$M = E - e \sin(E)$ $M = M_0 + n(t - t_0)$	(M in rad) (M in rad)
$t - t_0$	time of flight	$t - t_0 = (M - M_0)/n$	($t - t_0$ in sec)
v	true anomaly	$v \approx M + 2e \sin M + 1.25e^2 \sin(2M)$	(approx.)

As an example, we find the time it takes a satellite to go from perigee to an angle 90 deg from perigee, for an orbit with a semimajor axis of 7,000 km and an eccentricity of 0.1. For this example

$$v_0 = E_0 = M_0 = 0.0 \text{ rad} \qquad t_0 = 0.0 \text{ s}$$
$$v = 1.571\,98 \text{ rad} \qquad E = 1.4706 \text{ rad}$$
$$M = 1.3711 \text{ rad} \qquad n = 0.001\,078 \text{ rad/s}$$
$$t = 1{,}271.88 \text{ s}$$

Finding the position in an orbit after a specified period is more complex. For this problem, we calculate the mean anomaly, M, using time of flight and the mean motion using Eq. (6-12). Next, we determine the true anomaly, v, using the series expansion shown in Table 6-3, a good approximation for small eccentricity (the error is of the order e^3). If we need greater accuracy, we must solve the equation in Table 6-3 relating mean anomaly to eccentric anomaly. Because this is a transcendental function, we must use an iterative solution to find the eccentric anomaly, after which we can calculate the true anomaly directly.

6.1.6 Orbit Determination

Up to this point, we have assumed that we know both the position and velocity of the satellite in inertial space or the classical orbital elements. But we often cannot directly observe the satellite's inertial position and velocity. Instead, we commonly receive data from radar, telemetry, optics, or GPS. Radar and telemetry data consists of range, azimuth, elevation, and possibly the rates of change of one or more of these quantities, relative to a site attached to the rotating Earth. GPS receivers give GCI latitude, longitude, and altitude. Optical data consists of right ascension and declination relative to the celestial sphere. In any case, we must combine and convert this data to inertial position and velocity before determining the orbital elements. Bate, Mueller, and White [1971] and Escobal [1965] cover methods for combining data, so I will not cover them here.

The type of data we use for orbit determination depends on the orbit selected, accuracy requirements, and weight restrictions on the spacecraft. Because radar and optical systems collect data passively, they require no additional spacecraft weight, but they are also the least accurate methods of orbit determination. Conversely, GPS data is more accurate but also requires a receiver and processor, which add weight. We can also use GPS for semiautonomous orbit determination because it requires no ground support. An alternative method for fully autonomous navigation is given by Tai and Noerdlinger [1989] (See Sec. 11.2.)

6.2 Orbit Perturbations

The Keplerian orbit discussed above provides an excellent reference, but other forces act on the satellite to perturb it away from the nominal orbit. We can classify these *perturbations*, or variations in the orbital elements, based on how they affect the Keplerian elements.

Figure 6-6 illustrates a typical variation in one of the orbital elements because of a perturbing force. *Secular variations* represent a linear variation in the element. *Short-period variations* are periodic in the element with a period less than or equal to the orbital period. *Long-period variations* have a period greater than the orbital period. Because secular variations have long-term effects on orbit prediction (the orbital elements affected continue to increase or decrease), I will discuss them in detail. If the satellite mission demands that we precisely determine the orbit, we must include the periodic variations as well. Battin [1999], Danby [1962], and Escobal [1965] describe methods of determining and predicting orbits for perturbed Keplerian motion.

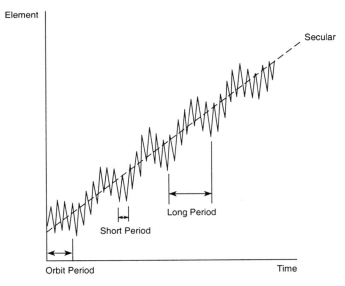

Fig. 6-6. Secular and Periodic Variations of an Orbital Element. Secular variations represent linear variations in the element, short-period variations have a period less than the orbital period, and long-period variations have a period longer than the orbital period.

When we consider perturbing forces, the classical orbital elements vary with time. To predict the orbit we must determine this time variation using techniques of either special or general perturbations. *Special perturbations* employ direct numerical integration of the equations of motion. Most common is *Cowell's method*, in which the accelerations are integrated directly to obtain velocity and again to obtain position.

General perturbations analytically solve some aspects of the motion of a satellite subjected to perturbing forces. For example, the polar equation of a conic applies to the two-body equations of motion. Unfortunately, most perturbing forces don't yield to a direct analytic solution but to series expansions and approximations. Because the orbital elements are nearly constant, general perturbation techniques usually solve directly for the orbital elements rather than the inertial position and velocity. The orbital elements are more difficult to describe mathematically and approximate, but they allow us to better understand how perturbations affect a large class of orbits. We can also obtain solutions much faster than with special perturbations.

The primary forces which perturb a satellite orbit arise from third bodies such as the Sun and the Moon, the nonspherical mass distribution of the Earth, atmospheric drag, and solar radiation pressure. We describe each of these below.

6.2.1 Third-Body Perturbations

The gravitational forces of the Sun and the Moon cause periodic variations in all of the orbital elements, but only the right ascension of the ascending node, argument of perigee, and mean anomaly experience secular variations. These secular variations arise from a gyroscopic precession of the orbit about the ecliptic pole. The secular variation in mean anomaly is much smaller than the mean motion and has little effect on the orbit; however, the secular variations in right ascension of the ascending node and argument of perigee are important, especially for high-altitude orbits.

For nearly circular orbits, e^2 is almost zero and the resulting error is of the order e^2. In this case, the equations for the secular rates of change resulting from the Sun and Moon are:

- Right ascension of the ascending node:

$$\dot{\Omega}_{MOON} = -0.003\,38\,(\cos i)\,/\,n \tag{6-14}$$

$$\dot{\Omega}_{SUN} = -0.001\,54\,(\cos i)\,/\,n \tag{6-15}$$

- Argument of perigee:

$$\dot{\omega}_{MOON} = 0.001\,69\,(4 - 5\sin^2 i)\,/\,n \tag{6-16}$$

$$\dot{\omega}_{SUN} = 0.000\,77\,(4 - 5\sin^2 i)\,/\,n \tag{6-17}$$

where i is the orbital inclination, n is the number of orbit revolutions per day, and $\dot{\Omega}$ and $\dot{\omega}$ are in deg/day. These equations are only approximate; they neglect the variation caused by the changing orientation of the orbital plane with respect to both the Moon's orbital plane and the ecliptic plane.

6.2.2 Perturbations Because of a Nonspherical Earth

When developing the two-body equations of motion, we assumed the Earth has a spherically symmetric mass distribution. In fact, the Earth has a bulge at the equator, a slight pear shape, and flattening at the poles. We can find a satellite's acceleration by taking the gradient of the gravitational potential function, Φ. One widely used form of the geopotential function is:

$$\Phi = (\mu/r)\left[1 - \sum_{n=2}^{\infty} J_n \, (R_E/r)^n \, P_n \, (\sin L)\right] \qquad (6\text{-}18)$$

where $\mu \equiv GM$ is Earth's gravitational constant, R_E is Earth's equatorial radius, P_n are Legendre polynomials, L is geocentric latitude, and J_n are dimensionless geopotential coefficients of which the first three are:

$$J_2 = \quad 0.001\ 082\ 63$$
$$J_3 = -0.000\ 002\ 54$$
$$J_4 = -0.000\ 001\ 61$$

This form of the geopotential function depends on latitude, and we call the geopotential coefficients, J_n, *zonal coefficients*. Other, more general expressions for the geopotential include sectoral and tesseral terms in the expansion. The *sectoral terms* divide the Earth into slices and depend only on longitude. The *tesseral terms* in the expansion represent sections that depend on longitude and latitude. They divide the Earth into a checkerboard pattern of regions that alternately add to and subtract from the two-body potential. A *geopotential model* consists of a matrix of coefficients in the spherical harmonic expansion. The widely used Goddard Earth Model 10B, or GEM10B, is called a "21 ×21 model" because it consists of a 21 ×21 matrix of coefficients. In order to achieve high accuracy mapping of the ocean surface and wave properties, the TOPEX mission required creating a 100 ×100 geopotential model.

The potential generated by the nonspherical Earth causes periodic variations in all of the orbital elements. The dominant effects, however, are secular variations in right ascension of the ascending node and argument of perigee because of the Earth's oblateness, represented by the J_2 term in the geopotential expansion. The rates of change of Ω and ω due to J_2 are

$$\dot{\Omega}_{J_2} = -1.5n \, J_2 \, (R_E \, / \, a)^2 (\cos i)(1 - e^2)^{-2} \qquad (6\text{-}19)$$

$$\cong -2.064\ 74 \times 10^{14} \, a^{-7/2} (\cos i)(1 - e^2)^{-2}$$

$$\dot{\omega}_{J_2} = 0.75n \, J_2 \, (R_E \, / \, a)^2 (4 - 5\sin^2 i) \, (1 - e^2)^{-2} \qquad (6\text{-}20)$$

$$\cong 1.032\ 37 \times 10^{14} \, a^{-7/2} (4 - 5\sin^2 i) \, (1 - e^2)^{-2}$$

where n is mean motion in deg/day, R_E is Earth's equatorial radius, a is semimajor axis in km, e is eccentricity, i is inclination, and $\dot{\Omega}$ and $\dot{\omega}$ are in deg/day. Table 6-4 compares the rates of change of right ascension of the ascending node and argument of perigee resulting from the Earth's oblateness, the Sun, and the Moon. For satellites in GEO and below, the J_2 perturbations dominate; for satellites above GEO the Sun and Moon perturbations dominate.

Molniya orbits are highly eccentric ($e \cong 0.75$) with approximately 12 hour periods (2 revolutions/day). Orbit designers choose the orbital inclination so the rate of change of perigee, Eq. (6-20), is zero. This condition occurs at inclinations of 63.4 deg and 116.6 deg. For these orbits we typically place perigee in the Southern Hemisphere, so

TABLE 6-4. Secular Variations in Right Ascension of the Ascending Node and Argument of Perigee. Note that these secular variations form the basis for Sun-synchronous and Molniya orbits. For Sun-synchronous orbits the nodal precession rate is set to 0.986 deg/day to match the general motion of the Sun.

Orbit	Effect of J_2 (Eqs. 6-19, 6-20) (deg/day)	Effect of the Moon (Eqs. 6-14, 6-16) (deg/day)	Effect of the Sun (Eqs. 6-15, 6-17) (deg/day)
Shuttle	$a = 6,700$ km, $e = 0.0$, $i = 28$ deg		
$\dot{\Omega}$	−7.35	−0.000 19	−0.000 08
$\dot{\omega}$	12.05	0.000 31	0.000 14
Sun-Synchronous	$a = 6,728$ km, $e \cong 0.0$, $i = 96.85$ deg		
$\dot{\Omega}$	0.986	0.000 03	0.000 01
$\dot{\omega}$	−4.890	−0.000 10	−0.000 05
GPS	$a = 26,600$ km, $e = 0.0$, $i = 60.0$ deg		
$\dot{\Omega}$	−0.033	−0.000 85	−0.000 38
$\dot{\omega}$	0.008	0.000 21	0.000 10
Molniya	$a = 26,600$ km, $e = 0.75$, $i = 63.4$ deg		
$\dot{\Omega}$	−0.30	−0.000 76	−0.000 34
$\dot{\omega}$	0.00	0.000 00	0.000 00
Geosynchronous	$a = 42,160$ km, $e = 0$, $i = 0$ deg		
$\dot{\Omega}$	−0.013	−0.003 38	−0.001 54
$\dot{\omega}$	0.025	0.006 76	0.003 07

the satellite remains above the Northern Hemisphere near apogee for approximately 11 hours/orbit. Mission planners choose perigee altitude to meet the satellite's mission constraints. Typical perigee altitudes vary from 200 to 1,000 km. We can calculate the eccentricity and apogee altitude using the semimajor axis, perigee altitude, and equations from Table 6-2.

In a *Sun-synchronous orbit*, the satellite orbital plane remains approximately fixed with respect to the Sun. We do this by matching the secular variation in the right ascension of the ascending node (Eq. 6-19) to the Earth's rotation rate around the Sun. A nodal precession rate of 0.9856 deg/day will match the Earth's average rotation rate about the Sun. Because this rotation is positive, Sun-synchronous orbits must be retrograde. For a given semimajor axis, a, and eccentricity, we can use Eq. (6-19) to find the inclination for the orbit to be Sun-synchronous.

6.2.3 Perturbations From Atmospheric Drag

The principal nongravitational force acting on satellites in low-Earth orbit is atmospheric drag. Drag acts in the opposite direction of the velocity vector and removes energy from the orbit. This energy reduction causes the orbit to get smaller, leading to further increases in drag. Eventually, the altitude of the orbit becomes so small that the satellite reenters the atmosphere.

The equation for acceleration due to drag on a satellite is:

$$a_D = -(1/2)\rho\,(C_D\,A/m)V^2 \tag{6-21}$$

where ρ is atmospheric density, A is the satellite's cross-sectional area, m is the satellite's mass, V is the satellite's velocity with respect to the atmosphere, and C_D is the drag coefficient ≈ 2.2 (See Table 8-3 in Sec. 8.1.3 for an extended discussion of C_D).

We can approximate the changes in semimajor axis and eccentricity per revolution, and the lifetime of a satellite, using the following equations:

$$\Delta a_{rev} = -2\pi\,(C_D A/m)a^2\,\rho_p\,\exp\,(-c)\,[I_0 + 2eI_1] \tag{6-22}$$

$$\Delta e_{rev} = -2\pi\,(C_D A/m)a\,\rho_p\,\exp\,(-c)\,[I_1 + e\,(I_0 + I_2)/2] \tag{6-23}$$

where ρ_p is atmospheric density at perigee, $c \equiv ae\,/\,H$, H is density scale height (see column 25, Inside Rear Cover), and I_i are Modified Bessel Functions* of order i and argument c. We model the term $m\,/\,(C_D A)$, or *ballistic coefficient*, as a constant for most satellites, although it can vary by a factor of 10 depending on the satellite's orientation (see Table 8-3).

For near circular orbits, we can use the above equations to derive the much simpler expressions:

$$\Delta a_{rev} = -2\pi\,(C_D A/m)\rho a^2 \tag{6-24}$$

$$\Delta P_{rev} = -6\pi^2\,(C_D A/m)\rho a^2 /V \tag{6-25}$$

$$\Delta V_{rev} = \pi\,(C_D A/m)\rho a\,V \tag{6-26}$$

$$\Delta e_{rev} = 0 \tag{6-27}$$

where P is orbital period and V is satellite velocity.

A rough estimate of the satellite's lifetime, L, due to drag is

$$L \approx -H/\Delta a_{rev} \tag{6-28}$$

where, as above, H is atmospheric density scale height given in column 25 of the Earth Satellite Parameter tables in the back of this book. We can obtain a substantially more accurate estimate (although still very approximate) by integrating Eq. (6-24), taking into account the changes in atmospheric density with both altitude and solar activity level. We did this for representative values of the ballistic coefficient in Fig. 8-4 in Sec. 8.1.

6.2.4 Perturbations from Solar Radiation

Solar radiation pressure causes periodic variations in all of the orbital elements. Its effect is strongest for satellites with low ballistic coefficients, that is, light vehicles with large frontal areas such as Echo. The magnitude of the acceleration, a_R, in m/s^2 arising from solar radiation pressure is

$$a_R \approx -4.5 \times 10^{-6}\,(1 + r)\,A/m \tag{6-29}$$

where A is the satellite cross-sectional area exposed to the Sun in m^2, m is the satellite mass in kg, and r is a reflection factor. ($r = 0$ for absorption; $r = 1$ for specular

---

* Values for I_i can be found in many standard mathematical tables.

reflection at normal incidence; and $r \approx 0.4$ for diffuse reflection.) Below 800 km altitude, acceleration from drag is greater than that from solar radiation pressure; above 800 km, acceleration from solar radiation pressure is greater.

6.3 Orbit Maneuvering

At some point during the lifetime of most satellites, we must change one or more of the orbital elements. For example, we may need to transfer it from an initial parking orbit to the final mission orbit, rendezvous with or intercept another satellite, or correct the orbital elements to adjust for the perturbations discussed in the previous section. Most frequently, we must change the orbital altitude, plane, or both. To change the orbit of a satellite, we have to change the satellite's velocity vector in magnitude or direction using a thruster. Most propulsion systems operate for only a short time compared to the orbital period, so we can treat the maneuver as an impulsive change in the velocity while the position remains fixed. For this reason, any maneuver changing the orbit of a satellite must occur at a point where the old orbit intersects the new orbit. If the two orbits do not intersect, we must use an intermediate orbit that intersects both. In this case, the total maneuver requires at least two propulsive burns.

In general, the change in the velocity vector to go from one orbit to another is given by

$$\Delta \mathbf{V} = \mathbf{V}_{NEED} - \mathbf{V}_{CURRENT} \qquad (6\text{-}30)$$

We can find the current and needed velocity vectors from the orbital elements, keeping in mind that the position vector does not change significantly during impulsive burns.

6.3.1 Coplanar Orbit Transfers

The most common type of in-plane maneuver changes the size and energy of the orbit, usually from a low-altitude parking orbit to a higher-altitude mission orbit such as a geosynchronous orbit. Because the initial and final orbits do not intersect (see Fig. 6-7), the maneuver requires a transfer orbit. Figure 6-7 represents a Hohmann[*] Transfer Orbit. In this case, the transfer orbit's ellipse is tangent to both the initial and final circular orbits at the transfer orbit's perigee and apogee, respectively. The orbits are tangential, so the velocity vectors are collinear at the intersection points, and the Hohmann Transfer represents the most fuel-efficient transfer between two circular, coplanar orbits. When transferring from a smaller orbit to a larger orbit, the propulsion system must apply velocity change in the direction of motion; when transferring from a larger orbit to a smaller, the velocity change is opposite to the direction of motion.

The total velocity change required for the transfer is the sum of the velocity changes at perigee and apogee of the transfer ellipse. Because the velocity vectors are collinear at these points, the velocity changes are just the differences in magnitudes of the velocities in each orbit. We can find these differences from the energy equation, if we know the size of each orbit. If we know the initial and final orbits (r_A and r_B), we calculate the semimajor axis of the transfer ellipse, a_{tx}, and the total velocity change

---

[*]Walter Hohmann, a German engineer and architect, wrote *The Attainability of Celestial Bodies* [1925], consisting of a mathematical discussion of the conditions for leaving and returning to Earth.

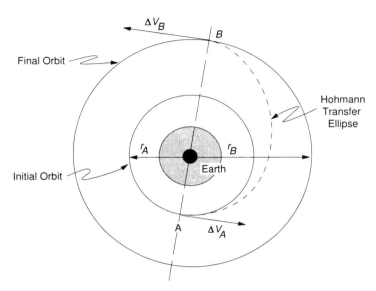

Fig. 6-7.   **Hohmann Transfer**. The Hohmann Transfer ellipse provides orbit transfer between two circular, co-planar orbits.

(the sum of the velocity changes required at points A and B) using the following algorithm. An example transferring from an initial circular orbit of 6,567 km to a final circular orbit of 42,160 km illustrates this technique.

STEP	EQUATIONS		EXAMPLE
1.	a_{tx}	$= (r_A + r_B)/2$	= 24,364 km
2.	V_{iA}	$= (\mu/r_A)^{1/2} = 631.3481(r_A)^{-1/2}$	= 7.79 km/s
3.	V_{fB}	$= (\mu/r_B)^{1/2}$ $= 631.3481(r_B)^{-1/2}$	= 3.08 km/s
4.	V_{txA}	$= [\mu(2/r_A - 1/a_{tx})]^{1/2}$ $= 631.3481\,[(2/r_A - 1/a_{tx})]^{1/2}$	= 10.25 km/s
5.	V_{txB}	$= [\mu(2/r_B - 1/a_{tx})]^{1/2}$ $= 631.3481\,[(2/r_B - 1/a_{tx})]^{1/2}$	= 1.59 km/s
6.	ΔV_A	$= |V_{txA} - V_{iA}|$	= 2.46 km/s
7.	ΔV_B	$= |V_{fB} - V_{txB}|$	= 1.49 km/s
8.	ΔV_{TOTAL}	$= \Delta V_A + \Delta V_B$	= 3.95 km/s
9.	Time of transfer $= P/2$		= 5 hr 15 min

Alternatively, we can write the total ΔV required for a two-burn transfer between circular orbits at radii r_A and r_B:

$$\Delta V_{TOTAL} \equiv \Delta V_A + \Delta V_B \tag{6-31}$$

$$= \sqrt{\mu} \left[\left| \left(\frac{2}{r_A} - \frac{1}{a_{tx}} \right)^{\frac{1}{2}} - \left(\frac{1}{r_A} \right)^{\frac{1}{2}} \right| + \left| \left(\frac{2}{r_B} - \frac{1}{a_{tx}} \right)^{\frac{1}{2}} - \left(\frac{1}{r_B} \right)^{\frac{1}{2}} \right| \right] \tag{6-32}$$

where $\sqrt{\mu}$ = 631.3481 when ΔV is in km/s and all of the semimajor axes are in km. As in step 1 above, $a_{tx} = (r_A + r_B) / 2$.

The above expression applies to any coplanar Hohmann transfer. In the case of small transfers (that is, r_A close to r_B), we can approximate this conveniently in two forms:

$$\Delta V \approx V_{iA} - V_{fB} \tag{6-33}$$

$$\Delta V \approx 0.5 \, (\Delta r / r) \, V_{A/B} \tag{6-34}$$

where

$$\Delta r \equiv r_B - r_A \tag{6-35}$$

and

$$r \approx r_A \approx r_B \qquad V_{A/B} \approx V_{iA} \approx V_{fB} \tag{6-36}$$

To make the orbit change, we divide the ΔV into two small burns are of approximately equal magnitude.

The result in Eq. (6-33) is more unusual than it might seem at first. Assume that a satellite is in a circular orbit with velocity V_{iA}. In two burns we *increase* the velocity by an amount ΔV. The result is that the satellite is higher and traveling *slower* than originally by the amount ΔV. We can best clarify this result by an example. Consider a satellite initially in a circular orbit at 400 km such that r_A = 6,778 km and V_{iA} = 7,700 m/s. We will apply a total ΔV of 20 m/s (= 0.26% of V_{iA}) in two burns of 10 m/s each. From Eq. (6-34) the total Δr will be 0.52% of 6,778 km or 35 km. Thus, the final orbit will be circular at an altitude of 6,813 km. Immediately following the first burn of 10 m/s the spacecraft will be at perigee of the transfer orbit with a velocity of 7,710 m/s. When the spacecraft reaches apogee at 6,813 km it will have slowed according to Kepler's second law by 0.52% to 7,670 m/s. We then apply the second burn of 10 m/s to circularize the orbit at 7,680 m/s which is 20 m/s slower than its original velocity. We have added energy to the spacecraft which has raised the orbit and resulted in a lower kinetic energy but sufficiently more potential energy to make up for both the reduced speed and the added ΔV.

Sometimes, we may need to transfer a satellite between orbits in less time than that required to complete the Hohmann transfer. Figure 6-8 shows a faster transfer called the *one-tangent burn*. In this instance the transfer orbit is tangential to the initial orbit. It intersects the final orbit at an angle equal to the flight-path angle of the transfer orbit at the point of intersection. An infinite number of transfer orbits are tangential to the initial orbit and intersect the final orbit at some angle. Thus, we may choose the transfer orbit by specifying the size of the transfer orbit, the angular change of the transfer, or the time required to complete the transfer. We can then define the transfer orbit and calculate the required velocities.

For example, we may specify the size of the transfer orbit, choosing any semimajor axis that is greater than the semimajor axis of the Hohmann transfer ellipse. Once we know the semimajor axis of the ellipse (a_{tx}), we can calculate the eccentricity, angular distance traveled in the transfer, the velocity change required for the transfer, and the time required to complete the transfer using the equations in Table 6-5.

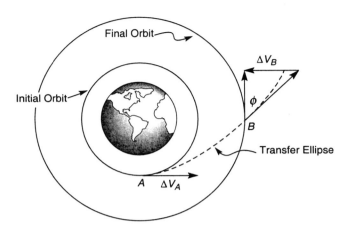

Fig. 6-8.   **Transfer Orbit Using One-Tangent Burn between Two Circular, Coplanar Orbits.**

TABLE 6-5.   **Computations for One-Tangent Burn Orbit Transfer.** See Vallado [1997], for example. Given: r_A, r_B, and a_{tx}

Quantity	Equation
eccentricity	$e = 1 - r_A / a_{tx}$
true anomaly at second burn	$v = \cos^{-1}[(a_{tx}(1 - e^2) / r_B - 1) / e]$
flight-path angle at second burn	$\phi = \tan^{-1}[e \sin v / (1 + e \cos v)]$
initial velocity	$V_{iA} = 631.3481 \, r_A^{-1/2}$
velocity on transfer orbit at initial orbit	$V_{txA} = 631.3481 \, [2 / r_A - 1 / a_{tx}]^{1/2}$
initial velocity change	$\Delta V_A = |V_{txA} - V_{iA}|$
final velocity	$V_{fB} = 631.3481 \, r_B^{-1/2}$
velocity on transfer orbit at final orbit	$V_{txB} = 631.3481 \, [2 / r_B - 1 / a_{tx}]^{1/2}$
final velocity change	$\Delta V_B = [V_{fB}^2 + V_{txB}^2 - 2 V_{fB} V_{txB} \cos \phi]^{1/2}$
total velocity change	$\Delta V_T = \Delta V_A + \Delta V_B$
eccentric anomaly at B	$E = \tan^{-1}[(1 - e^2)^{1/2} \sin v / (e + \cos v)]$
time of flight	$TOF = 0.001\,583\,913 \, a_{tx}^{3/2} (E - e \sin E)$, E in rads

Table 6-6 compares the total velocity change required and time-of-flight for a Hohmann transfer and a one-tangent burn transfer from a low altitude parking orbit to geosynchronous orbit.

Another option for changing the size of the orbit is to use a constant low-thrust burn, which results in a *spiral transfer*. We can approximate the velocity change for this type of orbit transfer by

$$\Delta V = |V_2 - V_1| \qquad (6\text{-}37)$$

where the velocities are the circular velocities of the two orbits. Following the

TABLE 6-6.   Comparison of Coplanar Orbit Transfers from LEO to Geosynchronous Orbit.

Variable	Hohmann Transfer	One-Tangent-Burn
r_A	6,570 km	6,570 km
r_B	42,200 km	42,200 km
a_{tx}	24,385 km	28,633 km
ΔV_T	3.935 km/s	4.699 km/s
TOF	5.256 hr	3.457 hr

previous example, the total velocity change required to go from low-Earth orbit to geosynchronous is 4.71 km/s using a spiral transfer. We obtain this by subtracting the results of step 3 from the results of step 2 in the above example.

6.3.2 Orbit Plane Changes

To change the orientation of the satellite's orbital plane, typically the inclination, we must change the direction of the velocity vector. This maneuver requires a component of ΔV to be perpendicular to the orbital plane and, therefore, perpendicular to the initial velocity vector. If the size of the orbit remains constant, we call the maneuver a *simple plane change* (Fig. 6-9A). We can find the required change in velocity by using the law of cosines. For the case in which V_f is equal to V_i, this expression reduces to

$$\Delta V = 2V_i \sin (\theta/2) \tag{6-38}$$

where V_i is the velocity before and after the burn, and θ is angle change required.

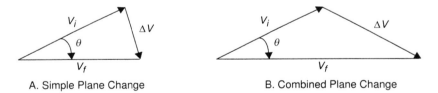

A. Simple Plane Change   B. Combined Plane Change

Fig. 6-9.   Vector Representation of Simple and Combined Changes in Orbital Plane. For the simple plane change, the magnitude of initial and final velocities are equal.

For example, the change in velocity required to transfer from a low-altitude ($h = 185$ km) inclined ($i = 28$ deg) orbit to an equatorial orbit ($i = 0$) at the same altitude is:

$$r = 6,563 \text{ km} \qquad V_i = 7.79 \text{ km/s} \qquad \Delta V = 3.77 \text{ km/s}$$

From Eq. (6-38) we see that if the angular change equals 60 deg, the required change in velocity equals the current velocity. Plane changes are very expensive in terms of the required velocity change and resulting fuel consumption. To minimize this, we should change the plane at a point where the velocity of the satellite is a minimum: at apogee for an elliptical orbit. In some cases, it may even be cheaper to boost the satellite into a higher orbit, change the orbital plane at apogee, and return the satellite to its original orbit.

Typically, orbital transfers require changes in both the size and the plane of the orbit, such as transferring from an inclined parking orbit at low altitude to a zero-inclination orbit at geosynchronous altitude. We do this transfer in two steps: a Hohmann transfer to change the size of the orbit and a simple plane change to make the orbit equatorial. A more efficient method (less total change in velocity) would be to combine the plane change with the tangential burn at apogee of the transfer orbit (Fig. 6-9B). As we must change both the magnitude and direction of the velocity vector, we can find the required change in velocity using the law of cosines:

$$\Delta V = (V_i^2 + V_f^2 - 2V_i V_f \cos\theta)^{1/2} \qquad (6\text{-}39)$$

where V_i is the initial velocity, V_f is the final velocity, and θ is the angle change required.

For example, we find the total change in velocity to transfer from a Shuttle parking orbit to a geosynchronous equatorial orbit as follows:

$r_i = 6{,}563$ km	$r_f = 42{,}159$ km
$i_i = 28$ deg	$i_f = 0$ deg
$V_i = 7.79$ km/s	$V_f = 3.08$ km/s
$\Delta V_A = 2.46$ km/s	$\Delta V_B = 1.83$ km/s
$\Delta V_{TOTAL} = 4.29$ km/s	

Completing a Hohmann transfer followed by a simple plane change would require a velocity change of 5.44 km/s, so the Hohmann transfer with a combined plane change at apogee of the transfer orbit represents a savings of 1.15 km/s. As we see from Eq. (6-39), a small plane change ($\theta \geq 0$) can be combined with an energy change for almost no cost in ΔV or propellant. Consequently, in practice, we do geosynchronous transfer with a small plane change at perigee and most of the plane change at apogee.

Another option is to complete the maneuver using three burns. The first burn is a coplanar maneuver placing the satellite into a transfer orbit with an apogee much higher than the final orbit. When the satellite reaches apogee of the transfer orbit, it does a combined plane change maneuver. This places the satellite in a second transfer orbit which is coplanar with the final orbit and has a perigee altitude equal to the altitude of the final orbit. Finally, when the satellite reaches perigee of the second transfer orbit, another coplanar maneuver places the satellite into the final orbit. This three-burn maneuver may save fuel, but the fuel savings comes at the expense of the total time required to complete the maneuver.

6.3.3 Orbit Rendezvous

Orbital transfer becomes more complicated when the objective is to rendezvous with or intercept another object in space: both the interceptor and target must arrive at the rendezvous point at the same time. This precise timing demands a phasing orbit to accomplish the maneuver. A *phasing orbit* is any orbit which results in the interceptor achieving the desired geometry relative to the target to initiate a Hohmann transfer. If the initial and final orbits are circular, coplanar, and of different sizes, then the phasing orbit is simply the initial interceptor orbit (Fig. 6-10). The interceptor remains in the initial orbit until the relative motion between the interceptor and target results in the desired geometry. At that point, we inject the interceptor into a Hohmann transfer orbit. The equation to solve for the wait time in the initial orbit is:

$$\text{Wait Time} = (\phi_i - \phi_f + 2k\pi) / (\omega_{int} - \omega_{tgt}) \tag{6-40}$$

where ϕ_f is the phase angle (angular separation of target and interceptor) needed for rendezvous, ϕ_i is the initial phase angle, k is the number of rendezvous opportunities, (for the first opportunity, $k = 0$), ω_{int} is the angular velocity of the interceptor, and ω_{tgt} is the angular velocity of the target. We calculate the lead angle, α_L, by multiplying ω_{tgt} by the time of flight for the Hohmann transfer and ϕ_f is 180 deg minus α_L.

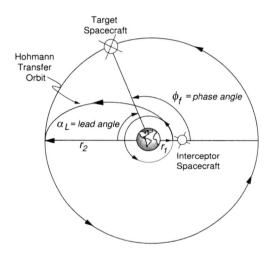

Fig. 6-10. Geometry Depicting Rendezvous Between Two Circular, Coplanar Orbits. The phase angle is the angular separation between the target and interceptor at the start of the rendezvous and the lead angle is the distance the target travels from the start until rendezvous occurs.

The total time to rendezvous equals the wait time from Eq. (6-40) plus the time of flight of the Hohmann transfer orbit.

The denominator in Eq. (6-40) represents the relative motion between the interceptor and target. As the size of the interceptor orbit approaches the size of the target orbit, the relative motion approaches zero, and the wait time approaches infinity. If the two orbits are exactly the same, then the interceptor must enter a new phasing orbit to rendezvous with the target (Fig. 6-11). For this situation, the rendezvous occurs at the point where the interceptor enters the phasing orbit. The period of the phasing orbit equals the time it takes the target to get to the rendezvous point. Once we know the period, we can calculate the semimajor axis. The two orbits are tangential at their point of intersection, so the velocity change is the difference in magnitudes of the two velocities at the point of intersection of the two orbits. Because we know the size of the orbits, and therefore, the energies, we can use the energy Eq. (6-4) to solve for the current and needed velocities.

Frequently operators must adjust the relative phasing for satellites in circular orbits. They accomplish this by making the satellite drift relative to its initial position. The *drift rate* in deg/orbit is given by

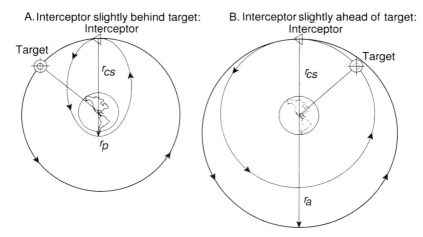

Fig. 6-11.   **Rendezvous from Same Orbit Showing the Target both Leading and Trailing the Interceptor.**

$$drift\ rate = 1{,}080\ \Delta V / V \qquad\qquad (6\text{-}41)$$

where V is the nominal orbital velocity and ΔV is the velocity change required to start or stop the drift.

The techniques described above move the target vehicle close to the interceptor. Once the two vehicles are close to each other they begin proximity operations by solving a set of relative motion equations to achieve the final rendezvous. Vallado [1997] contains an excellent discussion of the solution to the nearby relative motion problem, as addressed by the *Clohessy-Wiltshire* or *Hill's equations* of relative motion.

6.4 Launch Windows

Similar to the rendezvous problem is the launch-window problem, or determining the appropriate time to launch from the surface of the Earth into the desired orbital plane. Because the orbital plane is fixed in inertial space, the launch window is the time when the launch site on the surface of the Earth rotates through the orbital plane. As Fig. 6-12 shows, the launch time depends on the launch site's latitude and longitude and the satellite orbit's inclination and right ascension of the ascending node.

For a launch window to exist, the launch site must pass through the orbital plane. This requirement places restrictions on the orbital inclinations, i, possible from a given launch latitude, L:

- No launch windows exist if $L > i$ for direct orbit or $L > 180$ deg $- i$ for retrograde orbits.
- One launch window exists if $L = i$ or $L = 180$ deg $- i$.
- Two launch windows exist if $L < i$ or $L < 180$ deg $- i$.

The *launch azimuth*, β, is the angle measured clockwise from north to the velocity vector. If a launch window exists, then the launch azimuth required to achieve an inclination, i, from a given launch latitude, L, is given by:

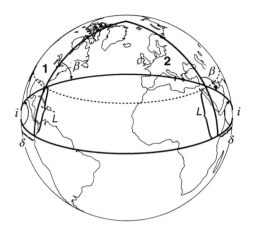

Fig. 6-12. **Launch Window Geometry for Launches near the Ascending Node (1) and Descending Node (2).** The angles shown are the orbital inclination *(i)*, launch site latitude *(L)*, and launch azimuth *(β)*.

$$\beta = \beta_I \pm \gamma \approx \beta_I \qquad (6\text{-}42a)$$

where

$$\sin \beta_I = \cos i \,/\, \cos L \qquad (6\text{-}42b)$$

and

$$\tan \gamma = \frac{V_L \cos \beta_I}{V_0 - V_{eq} \cos i} \approx \left(\frac{V_L}{V_0} \right) \cos \beta_I \qquad (6\text{-}42c)$$

where V_L is the inertial velocity of the launch site given by Eq. (6-46) below, $V_{eq} = 464.5$ m/s is the velocity of Earth's rotation at the equator, and $V_0 \approx 7.8$ km/s is the velocity of the satellite immediately after launch. β_I is the inertial launch azimuth and γ is a small correction to account for the velocity contribution caused by Earth's rotation. For launches to low-Earth orbit, γ ranges from 0 for a due east launch to 3.0 deg for launch into a polar orbit. The approximation for γ in Eq. (6-42c) is accurate to within 0.1 deg for low-Earth orbits. For launches near the ascending node, β is in the first or fourth quadrant and the plus sign applies in Eq. (6-42a). For launches near the descending node, β is in the second or third quadrant and the minus sign applies in Eq. (6-42a).

Let δ, shown in Fig. 6-12, be the angle in the equatorial plane from the nearest node to the longitude of the launch site. We can determine δ from:

$$\cos \delta = \cos \beta \,/\sin i \qquad (6\text{-}43)$$

where δ is positive for direct orbits and negative for retrograde orbits. Finally, the *local sidereal time*, *LST*, of launch is the angle from the vernal equinox to the longitude of the launch site at the time of launch:

$$
\begin{aligned}
LST &= \Omega + \delta & \text{(launch at the ascending node)} \\
&= \Omega + 180 \text{ deg} - \delta & \text{(launch at the descending node)} \qquad (6\text{-}44)
\end{aligned}
$$

where Ω is the right ascension of the ascending node of the resulting orbit.

Having calculated the launch azimuth required to achieve the desired orbit, we can now calculate the velocity needed to accelerate the payload from rest at the launch site to the required burnout velocity. To do so, we use topocentric-horizon coordinates with velocity components V_S, V_E, V_Z:

$$V_S = -V_{bo} \cos \phi \cos \beta_b \tag{6-45a}$$

$$V_E = V_{bo} \cos \phi \sin \beta_b - V_L \tag{6-45b}$$

$$V_Z = V_{bo} \sin \phi \tag{6-45c}$$

where V_{bo} is the velocity at burnout (usually equal to the circular orbital velocity at the prescribed altitude), ϕ is the flight path angle at burnout, β_b is the launch azimuth at burnout, and V_L is the velocity of the launch site at a given latitude, L, as given by:

$$V_L = (464.5 \text{ m/s}) \cos L \tag{6-46}$$

Equations (6-45c) do not include losses in the velocity of the launch vehicle because of atmospheric drag and gravity—approximately 1,500 m/s for a typical launch vehicle. Also, in Eq. (6-45c) we assume that the azimuth at launch and the azimuth at burnout are the same. Changes in the latitude and longitude of the launch vehicle during powered flight will introduce small errors into the calculation of the burnout conditions. We can calculate the velocity required at burnout from the energy equation if we know the semimajor axis and radius of burnout of the orbit [Eq. (6-4)].

6.5 Orbit Maintenance

Once in their mission orbits, many satellites need no additional orbital adjustments. On the other hand, mission requirements may demand that we maneuver the satellite to correct the orbital elements when perturbing forces have changed them. Two particular cases of note are satellites with repeating ground tracks and geosynchronous equatorial satellites, placed at an assigned longitude.

Using two-body equations of motion, we can show that a satellite will have a *repeating ground track* if it has exactly an integer number of revolutions per integer number of days. Its period, therefore, must be:

$$P = (m \text{ sidereal days}) / (k \text{ revolutions}) \tag{6-47}$$

where m and k are integers, and 1 sidereal day = 1,436.068 167 min. For example, a satellite orbiting the Earth exactly 16 times per day has a period of 89.75 min and a semimajor axis of 6,640 km.

Next we would modify the period of the satellite to account for the drift in the orbital plane caused by the Earth's oblateness (J_2). We can calculate the change in the right ascension of the ascending node, $\Delta\Omega$, because of J_2 from the two-body orbital elements. In this case the new period is:

$$P_{NEW} = P_{TWO\ BODY} + \Delta\Omega / \omega_{Earth} \tag{6-48}$$

Because we base the nodal drift on the two-body orbital elements, we must iterate to find the new orbital period and semimajor axis. Continuing with the previous example, assume a perigee altitude of 120 km and an inclination of 45 deg. In this case, we find that the compensated period is 88.20 min and the new semimajor axis is 6,563 km.

Several examples of spacecraft placed in orbits with repeating ground tracks are shown in Table 6-7.

TABLE 6-7.   Examples of Repeating Ground Tracks.

Satellite	Inclination (deg)	Semimajor Axis (km)	Revs	Days
SEASAT	108.0	7,169.0	43	3
LANDSAT 4/5	98.2	7,077.7	233	16
GEOSAT	108.05	7,162.7	244	17

The Earth's oblateness also causes the direction of perigee to rotate around the orbit. If the orbit is noncircular and the mission places limits on the altitude over specific targets, we must control the location of perigee. One possibility is to select the inclination of the orbit to be at the critical inclination (63.4 deg for a direct orbit and 116.4 deg for a retrograde orbit), so the location of perigee is fixed. If other constraints make this selection impossible, we must maintain the orbit through orbital maneuvers. We can change the location of perigee by changing the flight-path angle by an angle θ. Only the direction of the velocity vector changes, so we can find the change in velocity from the equation for a simple plane change:

$$\Delta V = 2V \sin \theta/2 \qquad (6\text{-}49)$$

A final consideration for a low-altitude orbit with repeating ground tracks is the change in the semimajor axis and eccentricity due to atmospheric drag. Drag causes the orbit to become smaller. As the orbit becomes smaller, the period also decreases, causing the ground track to appear to shift eastward. If we specify some tolerance, such as a maximum distance between the actual and desired ground tracks, the satellite must periodically maneuver to maintain the desired orbit.

We can use Eq. (6-22) to calculate the change in semimajor axis per revolution of the orbit. Given the change in the size of the orbit, we can also determine the change in the period:

$$\Delta P = 3\pi \, \Delta a / (na) \qquad (6\text{-}50)$$

If constraints exist for either the period or semimajor axis of the orbit, we can use Eqs. (6-22) and (6-50) to keep track of the period and semimajor axis until we need to correct the orbit. We can apply a tangential velocity change at perigee to adjust the semimajor axis when required. Again, we can find the current and needed velocities from the energy Eq. (6-4), because we know the size, and therefore the energy, of the two orbits.

Geosynchronous equatorial orbits also require orbital maintenance maneuvers. Satellites in these orbits drift when perturbations occur from the nonspherical Earth and from the third-body gravitational attraction of the Sun and Moon. Matching the period of a geostationary orbit with the Earth's rotational velocity results in a resonance with the J_{22} term in the geopotential. This resonance term results in a transverse acceleration, that is, an acceleration in the plane of the orbit. This causes the satellite to drift in longitude (*East-West drift*). The Sun and the Moon cause out-of-plane accelerations which make the satellite drift in latitude (*North-South drift*).

North-south stationkeeping is necessary when mission requirements limit inclination drift. If not corrected, the inclination of the orbit varies between 0 and 15 deg with a period of approximately 55 years. The approximate equations to solve for the worst-case change in velocity are:

$$\Delta V_{MOON} = 102.67 \cos \alpha \sin \alpha \text{ (m/s/yr)} \qquad (6\text{-}51)$$

$$\approx 36.93 \text{ m/s per year, for } i = 0$$

$$\Delta V_{SUN} = 40.17 \cos \gamma \sin \gamma \text{ (m/s/yr)} \qquad (6\text{-}52)$$

$$\approx 14.45 \text{ m/s/yr, for } i = 0$$

where α is the angle between the orbital plane and the Moon's orbit, and γ is the angle between the orbital and ecliptic plane.

The transverse acceleration caused by resonance with the J_{22} term results in periodic motion about either of two stable longitudes at approximately 75 deg and 255 deg East longitude. If we place a satellite at any other longitude, it will tend to orbit the closest of these two longitudes, resulting in East-West drift of up to 180 deg with periods of up to 900 days. Suppose a mission for a geostationary satellite specifies a required longitude, l_D. The change in velocity required to compensate for the drift and maintain the satellite in the vicinity of the specified longitude is:

$$\Delta V = 1.715 \sin (2 |l_D - l_s|) \qquad (6\text{-}53)$$

where l_D is the desired longitude, l_s is the closest stable longitude, and ΔV is in m/s/yr.

For example, we find the velocity change required for one year if the desired longitude is 60 deg west as

$$l_D = -60 \text{ deg} \qquad l_s = 255 \text{ deg}$$

$$\Delta V = 1.715 \text{ m/s/yr}$$

After the mission of the satellite is complete, several options exist, depending on the orbit. We may allow low-altitude orbits to decay and reenter the atmosphere or use a velocity change to speed up the process. We may also boost satellites at all altitudes into benign orbits to reduce the probability of collision with active payloads, especially at synchronous altitudes or when the satellite is part of a large constellation. Because coplanar velocity changes are more efficient than plane changes, we would normally apply tangential changes in velocity. Their magnitude would depend on the difference in energy of the two orbits. Generally, a change in altitude of a few hundred kilometers is sufficient to prevent collisions within a constellation. If we increase the altitude of the orbit for disposal, we must make sure atmospheric drag does not cause the disposal orbit to decay and return the spacecraft to the original altitude of the constellation. If we choose to deorbit our satellite, the velocity change required to deorbit a satellite in a circular orbit at initial altitude, H_i, and velocity, V, is:

$$\Delta V_{deorbit} \approx V \left[1 - \sqrt{\frac{2(R_E + H_e)}{2R_E + H_e + H_i}} \right] \qquad (6\text{-}54a)$$

$$\approx V \left[\frac{H_i - H_e}{4(R_E + H_e)} \right] \qquad (6\text{-}54b)$$

where R_E is the radius of the Earth and H_e is the perigee altitude at the end of the burn.

It is not necessary to reduce perigee altitude to 0 km. Choosing a 50 km deorbit altitude would reduce the FireSat $\Delta V_{deorbit}$ to 183 m/s. Note that we reduce only perigee in the deorbit burn. Reducing perigee to 100 to 150 km could result in several orbits over which apogee decreases before the spacecraft reenters and might not allow adequate control of the deorbit conditions.

References

Bate, Roger R., Donald D. Mueller, and Jerry E. White. 1971. *Fundamentals of Astrodynamics*. New York: Dover Publications.

Battin, Richard H. 1999. *An Introduction to the Mathematics and Methods of Astrodynamics*. New York: AIAA Education Series.

Chobotov, Vladimir A. 1996. *Orbital Mechanics (2nd Edition)*. New York: AIAA Education Series.

Danby, J.M.A. 1962. *Fundamentals of Celestial Mechanics*. New York: Macmillan.

Escobal, Pedro R. 1965. *Methods of Orbit Determination*. Malabar, FL: Robert E. Krieger Publishing Company.

Hohmann, Walter. 1925. *Die Erreich Barkeit de Himmelskörper. (The Attainability of Heavenly Body)*. NASA Technical Translation TTF44. Nov. 1960. Washington, DC: National Aeronautics and Space Administration.

Kaplan, Marshall H. 1976. *Modern Spacecraft Dynamics and Control*. New York: Wiley and Sons.

King-Hele, D. 1964. *Theory of Satellite Orbits in an Atmosphere*. London: Butterworths Mathematical Texts.

Pocha, J.J. 1987. *An Introduction to Mission Design for Geostationary Satellites*. Boston: D. Reidel Publishing Company.

Roy, A.E. 1991. *Orbital Motion (3rd Edition)*. Bristol and Philadelphia: Adam Hilger.

Tai, Frank and Peter D. Noerdlinger. 1989. *A Low Cost Autonomous Navigation System*, Paper No. AAS 89-001 presented to the 12th Annual AAS Guidance and Control Conference. Keystone, Colorado, Feb. 4–8.

Vallado, David A. 1997. *Fundamentals of Astrodynamics and Applications*. New York: McGraw-Hill Book Company.

Wiesel, William E. 1996. *Spaceflight Dynamics (2nd Edition)*. New York: McGraw-Hill Book Company.

Chapter 7

Orbit and Constellation Design

James R. Wertz, *Microcosm, Inc.*

Chapters 5 and 6 introduced the geometry and physics of spacecraft orbits, as well as formulas for computing orbit parameters. In contrast, this chapter deals with selecting or designing orbits to meet the largest number of mission requirements at the least possible cost. The orbit selection process is complex, involving trades between a number of different parameters. The orbit typically defines the space mission lifetime, cost, environment, viewing geometry, and, frequently, the payload performance. Most commonly, we must trade the velocity required to achieve the orbit as a measure of cost vs. coverage performance.

Chapter 6 lists several references on astrodynamics. Unfortunately, very few of these references contain any discussion of the orbit selection or design process. Chobotov [1996] and Vallado [1997] have some information. By far the most extensive discussion of this topic is in Wertz [2001]. Soop [1994] contains an excellent discussion of the design of geostationary orbits.

Ordinarily, spacecraft will be in various orbits during the space mission life. These could include, for example, a *parking orbit* for spacecraft checkout or storage, a *transfer orbit* to move it from the *injection orbit* where the spacecraft separated from the launch vehicle to its *operational orbit* or orbits for mission activities, and possibly a final *disposal orbit* where the spacecraft will do minimum damage when we are through using it. In early mission design, we pay most attention to defining the spacecraft's operational orbit. However, preliminary mission design needs to consider all mission phases to meet the needs of more complex missions.

7.1 The Orbit Design Process

Orbit design has no absolute rules; the method described below and summarized in Table 7-1 gives a starting point and checklist for the process.

TABLE 7-1. **Summary of the Orbit Selection Process.** See text for discussion of each step.

Step	Where Discussed
1. Establish orbit types.	Sec. 7.1
2. Determine orbit-related mission requirements.	Sec. 7.1
3. Assess applicability of specialized orbits.	Secs. 7.4, 6.3
4. Evaluate whether single satellite vs. constellation is needed.	Secs. 7.1, 7.6

5. Do mission orbit design trades
   – Assume circular orbit (if applicable)
   – Conduct altitude/inclination trade
   – Evaluate use of eccentric orbits
	Secs. 7.4, 7.5
	
	Sec. 7.4
	Sec. 7.4

6. Assess launch and retrieval or disposal options.	Secs. 2.1, 6.5, 18.3, 21.2
7. Evaluate constellation growth and replenishment (if applicable).	Sec. 7.6
8. Create ΔV budget.	Sec. 7.3
9. Document orbit parameters, selection criteria, and allowed ranges. Iterate as needed.	

Effective orbit design requires clearly identifying the reasons for orbit selection, reviewing these reasons regularly as mission requirements change or mission definition improves, and continuing to remain open to alternatives. Several different designs may be credible. Thus, communications may work effectively through a single large satellite in geosynchronous orbit or a constellation of small satellites in low-Earth orbit. We may need to keep both options for some time before selecting one.

Step 1. Establish Orbit Types

To design orbits we first divide the space mission into segments and classify each segment by its overall function. Each orbit segment has different selection criteria, so we evaluate it separately, placing it into one of the four basic types:

- *Parking Orbit*—a temporary orbit which provides a safe and convenient location for satellite checkout, storage between operations, or at end-of-life. Also used to match conditions between phases such as post-launch and pre-orbit transfer.

- *Transfer Orbit*—used for getting from place to place. Examples: transfer orbit to geosynchronous altitude; interplanetary orbit to Mars.

- *Space-referenced Orbit*—an operational orbit whose principal characteristic is being somewhere in space (specific orbit parameters may not be critical). Examples: Lagrange point orbits for space sampling and observations; orbits for celestial observations or space manufacturing.

- *Earth-referenced Orbit*—an operational orbit which provides the necessary coverage of the surface of the Earth or near-Earth space. Examples: geosynchronous satellites, low-Earth satellites for Earth resources, meteorology, or communications.

A Shuttle-launched communications satellite in geosynchronous orbit provides an example of dividing a mission into segments. Once ejected from the Shuttle, the spacecraft may briefly stay in a parking orbit near the Shuttle to provide test and checkout of spacecraft and transfer vehicle systems. The second mission segment uses a transfer orbit to move the spacecraft from its parking orbit to geosynchronous equatorial orbit. The spacecraft then enters its operational orbit in the geostationary ring where it will spend the rest of its active life. At the end of its life we must move it out of the geostationary ring to avoid a possible collision with other satellites and to free the orbital slot for a replacement. (See Sec. 21.2.) Putting the nearly dead spacecraft into a final disposal orbit above the geostationary ring requires a relatively small transfer orbit. (Going above the geostationary ring rather than below avoids collisions with other satellites in geosynchronous transfer.)

Step 2. Establish Orbit-Related Mission Requirements

For each mission segment, we define the orbit-related requirements. They may include orbital limits, individual requirements such as the altitude needed for specific observations, or a range of values constraining any of the orbit parameters. Section 7.4 discusses in detail the requirements we would follow in designing an operational orbit. Ordinarily, these multiple requirements drive the orbit in different directions. For example, resolution or required aperture tend to drive the orbit to low altitudes, but coverage, lifetime, and survivability drive the spacecraft to higher altitudes.

Selecting of parking, transfer, and space-referenced orbits is normally conceptually simpler, although it may be mathematically complex. Here the normal trade is meeting the desired constraints on a mission, such as lifetime, thermal, or radiation environments, at the lowest possible propellant cost. Section 7.5 discusses the key mission requirements for these orbits.

Step 3. Assess Specialized Orbits

In selecting the orbit for any mission phase, we must first determine if a specialized orbit applies. *Specialized orbits* are those with unique characteristics, such as the geostationary ring in which satellites can remain nearly stationary over a given point on the Earth's equator. We examine each specialized orbit to see if its unique characteristics are worth its cost. This examination precedes the more detailed design trades, because specialized orbits constrain parameters such as altitude or inclination, and thus often lead to very different solutions for a given mission problem. Consequently, we may need to carry more than one orbit into more detailed design trades.

Step 4. Choose Single Satellite or Constellation

The principal advantage of a single satellite is that it reduces cost by minimizing the mission overhead. Thus, one satellite will have one power system, one attitude control system, one telemetry system, and require only a single launch vehicle. A constellation, on the other hand, may provide better coverage, higher reliability if a satellite is lost, and more survivability. We may also need a constellation to provide the multiple conditions to carry out the mission, such as varying lighting conditions for observations, varying geometries for navigation, or continuous coverage of part or all of the Earth for a communications constellation.

To meet budget limits, we must often trade a single large satellite with larger and more complex instruments against a constellation of smaller, simpler satellites. This decision may depend on the technology available at the time of satellite design. As small satellites become more capable through miniaturized electronics (Chap. 22) and onboard processing (Chap. 16), we may be able to construct constellations of small, low-cost satellites, frequently called *LightSats*, that were not previously economically feasible. Another major cost element for large constellations is the operational problem of providing continuous navigation and control. The introduction of low-cost autonomous navigation and control (See Sec. 11.7) should promote larger constellations of small satellites in the future.

Step 5. Do Mission Orbit Design Trades

The next step is to select the mission orbit by evaluating how orbit parameters affect each of the mission requirements from Table 7-6 in Sec. 7.4. As the table shows, orbit design depends principally on altitude. The easiest way to begin is by assuming a circular orbit and then conducting altitude and inclination trades. (See Sec. 7.4 and the example in Sec. 3.3.1.) This process establishes a range of potential altitudes and inclinations, from which we can select one or more alternatives. Documenting the results of this key trade is particularly important, so we can revisit the trade from time to time as mission requirements and conditions change. If a satellite constellation is one of the alternatives, then phasing the satellites within that constellation is a key characteristic, as Sec. 7.6 describes.

Note that constellations of satellites are normally at a common altitude and inclination because the orbit's drift characteristics depend largely on these parameters. Satellites at different altitudes or inclinations will drift apart so that their relative orientation will change with time. Thus, satellites at different altitudes or inclinations normally do not work well together as a constellation for extended times.

Step 6. Assess Launch and Retrieval or Disposal Options

Chapter 18 discusses satellite launch systems in detail. The launch vehicle contributes strongly to mission costs, and ultimately will limit the amount of mass that can be placed in an orbit of any given altitude. As we define the mission early on, we must provide enough launch margin to allow for later changes in launch vehicles or spacecraft weight. Naturally, new designs require more margin than existing ones, with 20% being typical for new missions.

Although given little consideration in the past, retrieval and disposal of spacecraft have become important to mission design (Sec. 21.2). Spacecraft that will reenter the atmosphere must either do controlled reentry, burn up in the atmosphere, or break up into harmless pieces.

If the spacecraft will not reenter the atmosphere in a reasonable time, we must still dispose of it at the end of its useful life so it is not hazardous to other spacecraft. This problem is particularly acute in geosynchronous orbit where missions compete strongly for orbit slots.* As Sec. 21.2 points out, a collision between two spacecraft not only destroys them but also causes debris dangerous to their entire orbit regime.

A third option is satellite retrieval, done either to refurbish and reuse the satellite, or to recover material (such as radioactive products) which would be dangerous if they entered the atmosphere uncontrolled. Currently, the Shuttle can retrieve spacecraft

---

* Cefola [1987] gives an excellent analysis of the requirements for removing satellites from geosynchronous orbit.

only from low-Earth orbit. In the future, it would be desirable to retrieve satellites as far away as geosynchronous orbit and return them to either the Orbiter or Space Station for refurbishment, repair, disposal, or reuse.

Step 7. Evaluate Constellation Growth and Replenishment

An important characteristic of any satellite constellation is growth, replenishment, and graceful degradation. A constellation that becomes operational only after many satellites are in place causes many economic, planning, and checkout problems. Constellations should be at least partly serviceable with small satellite numbers. *Graceful degradation* means that if one satellite fails, the remaining satellites provide needed services at a reduced level rather than a total loss of service. Section 7.6 discusses further the critical question of how we build up a constellation and how to plan for graceful degradation.

Step 8. Create ΔV Budget

To numerically evaluate the cost of an orbit, we must create a ΔV budget for the orbit, as described in Sec. 7.3. This then becomes the major component of the propellant budget as described in Sec. 10.3.

Step 9. Document and Iterate

A key component of orbit or constellation design is documenting the mission requirements used to define the orbit, the reasons for selecting the orbit, and the numerical values of the selected orbit parameters. This baseline can be reevaluated from time to time as mission conditions change. Because mission design nearly always requires many iterations, we must make the iteration activity as straightforward as possible and readdress orbit parameters throughout the design process to ensure they meet all requirements.

7.2 Earth Coverage

Earth coverage refers to the part of the Earth that a spacecraft instrument[*] or antenna can see at one instant or over an extended period. The coverage available for a particular location or region is frequently a key element in mission design. In evaluating coverage, two critical distinctions must be made. First, as Fig. 7-1 shows, the *instantaneous field of view*, typically called the *FOV* or *footprint*, is the actual area the instrument or antenna can see at any moment. In contrast, the *access area* is the total area on the ground that could potentially be seen at any moment by turning the spacecraft or instrument. In the case of a truly omni-directional antenna, these two would always be the same. For most operational instruments they are not.

The second important distinction is between the area which can be seen at any one instant vs. the rate at which new land comes into view as the spacecraft and instrument move. Both are important, and either can be vital to mission success. In geosynchronous orbit, the instantaneous area is typically most important because the spacecraft is

---

[*] Throughout this section we will use *instrument* to refer to any spacecraft sensor or antenna for which we want to compute coverage.

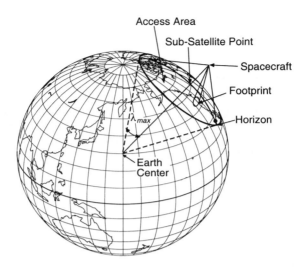

Fig. 7-1.   The Instrument *Footprint* or *FOV* is the Instantaneous Region on the Ground Be-ing Covered. The *instantaneous access area* is the total area in view.

nearly stationary relative to the Earth's surface. In low-Earth orbit, satellites are moving rapidly over the surface, so the rate at which new land appears is usually critical.

The two distinctions above lead to four key parameters for Earth coverage:

- *Footprint Area* (F_A, also *FOV area* or *instantaneous coverage area*) = area that a specific instrument or antenna can see at any instant.

- *Instantaneous Access Area* (*IAA*) = all the area that the instrument or antenna could potentially see at any instant if it were scanned through its normal range of orientations.

- *Area Coverage Rate* (*ACR*) = the rate at which the instrument or antenna is sensing or accessing new land.

- *Area Access Rate* (*AAR*) = the rate at which new land is coming into the spacecraft's access area.

For an instrument which covers all of the area available to it as the spacecraft moves along, the coverage rate and access rate will be the same. For instruments operating only part of the time or continuously selecting the region to be examined, the coverage rate and access rate may be dramatically different. Generally the access area and access rates depend only on the orbit and limiting geometry of the system, so we can easily compute them with only a minimal knowledge of the detailed system design. On the other hand, the actual area coverage rate during spacecraft operations may well depend on the spacecraft control, power, and management systems, as well as the details of mission operations.

Coverage assessment conveniently divides into two areas: first, an analytic assessment to provide approximate formulas for coverage parameters as a function of mission variables; second, numerical simulations to provide coverage Figures of Merit for more detailed studies.

7.2.1 Analytic Approximations

In this section we present analytic approximations for various Earth coverage parameters. All of the formulas here take into account the spherical surface of the Earth, but do not account for oblateness, or the rotation of the Earth underneath the orbit. These effects, in addition to those of coverage by multiple satellites, are ordinarily accounted for in numerical simulations as described in Sec. 7.2.2.

All of the formulas here are derived directly from the single-satellite geometry described in Sec. 5.2. In particular, we will use the notation developed there and summarized in Fig. 5-13 in Sec. 5.2. In this section we will parameterize coverage in terms of the *Earth central angle*, λ.* However, we can use Eqs. (5-24) to (5-28) to transform each of the formulas below into one for either the spacecraft-centered nadir angle, η, or spacecraft elevation angle, ε, seen from the ground.

As Fig. 7-1 shows, the instrument footprint is normally a beam with circular cross section substantially smaller than the access area projected onto the Earth's surface. The nomenclature and computational geometry for the footprint are in Fig. 7-2. (For instruments which see very large portions of the Earth, we can use the access area formulas below. For those which have noncircular cross sections, the logic here along with the formulas of Sec. 5.2 allows us to develop mission-specific formulas for footprint size and area.)

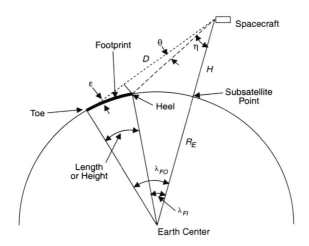

Fig. 7-2. Computational Geometry for Instrument Footprint. Note that ε is typically measured at the toe because of performance budgeting. (See also Fig. 5-13.)

The length (also called the height) of the footprint, L_F, is given by

$$L_F = K_L (\lambda_{FO} - \lambda_{FI}) \tag{7-1a}$$

$$\approx D \sin \theta / \sin \varepsilon \tag{7-1b}$$

where the variables are defined in Fig. 7-2 and, for λ expressed in degrees,

---

* λ may be thought of either as an angle at the Earth's center or as a distance measured along the Earth's surface. We will use these two views interchangeably as convenient for the problem at hand.

$$K_L = 1 \qquad \text{for length in deg}$$

$$K_L = 111.319\,543 \qquad \text{for length in km}$$

$$K_L = 60.107\,744\,7 \qquad \text{for length in nmi}$$

Note: The linear approximation given in Eq. (7-1b) is computationally convenient but can be very inaccurate, particularly near the horizon where ε is small. (For a satellite at an altitude of 1,000 km with a 1 deg diameter beam, the error in Eq. (7-1b) is 400% at $\varepsilon = 1$ deg, 10% at $\varepsilon = 15$ deg, and 1% at $\varepsilon = 60$ deg.) However, the alternative computation of Eq. (7-1a) is much less convenient. To find the footprint length for a given spacecraft elevation angle at the toe of the beam, ε, we begin by computing η and λ at the toe, then subtract the beam width, θ, from η to determine η at the heel, compute λ at the heel, and finally subtract to get the footprint length from Eq. (7-1a). An alternative that improves the approximation somewhat is to use the center rather than the toe of the beam. Because the toe represents the worst-case link budget (see Chap. 13), it is most often used for performance computations and, therefore, is commonly used for geometry calculations as well.

The footprint width, W_F, is given by

$$W_F = R_E \sin^{-1} (D \sin \theta / R_E) \qquad (7\text{-}2a)$$

$$\approx D \sin \theta \qquad (7\text{-}2b)$$

where $R_E = 6{,}378.14$ km is the radius of the Earth, θ is the beam width, and D is the distance from the spacecraft to the toe of the footprint.[*] Here the error in the approximation in Eq. (7-2b) is proportional to $1 - (W_F / \sin W_F)$[†] and is generally small relative to other errors. Thus, Eq. (7-2b) is adequate for most practical applications.

Finally, if we assume that the projection on the ground is an ellipse, then the footprint area, F_A, is given by

$$F_A = (\pi/4) L_F W_F \qquad (7\text{-}3)$$

Assuming that L_F was computed by Eq. (7-1a), the error in ignoring the curvature of the Earth in Eq. (7-3) is again proportional to $1 - (W_F / \sin W_F)$ and is negligible for most applications.

The *instantaneous area coverage rate* for the beam is defined by

$$ACR_{instantaneous} \equiv F_A / T \qquad (7\text{-}4)$$

where T is the *exposure time* or *dwell time* for the instrument. The *average area coverage rate*, ACR_{avg}, will also be a function of the *duty cycle*, DC, which is the fraction of the total time that the instrument is operating, and the *average overlap* between the footprint, O_{avg}, which is the amount by which two successive footprints cover the same area (typically about 20%):

$$ACR_{avg} = \frac{F_A \left(1 - O_{avg}\right) DC}{T} \qquad (7\text{-}5)$$

---

[*] In the case of a noncircular beam, Eq. (7-1) can be used with the beam width, θ, perpendicular to the horizon and Eq. (7-2) can be used independently with the beam width parallel to the horizon.

[†] Here W_F should be expressed in radians as seen from the center of the Earth.

Computing the instantaneous access area, *IAA*, will depend on the shape of the potential coverage area on the ground. Figure 7-3 shows several typical shapes. The most common of these is Fig. 7-3A, which assumes that the instrument can work at any point on the Earth within view for which the spacecraft elevation is above ε. This corresponds to a small circle on the Earth of radius λ centered on the current sub-satellite point. Some instruments, such as radar, cannot work too close to the subsatellite point. As Fig. 7-3B shows, these instruments have both an *outer horizon*, λ_1, and an *inner horizon*, λ_2.

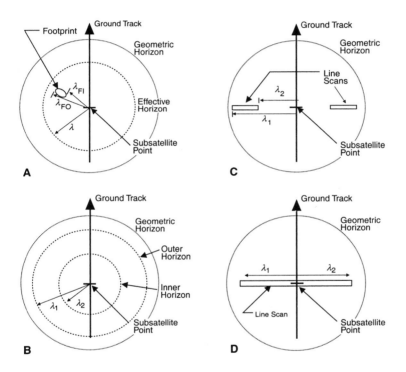

Fig. 7-3. Typical Access Areas for Spacecraft Instruments. See Table 7-2 for formulas.

For instruments with an access pattern as shown in Fig. 7-3A, the instantaneous access area, *IAA*, will be just the area of the small circle, that is,

$$IAA = K_A (1 - \cos \lambda) \tag{7-6}$$

where

$$K_A = 2\pi \approx 6.283\ 185\ 311 \qquad \text{for area in steradians}$$

$$K_A = 20{,}626.480\ 6 \qquad \text{for area in deg}^2$$

$$K_A = 2.556\ 041\ 87 \times 10^8 \qquad \text{for area in km}^2$$

$$K_A = 7.452\ 225\ 69 \times 10^7 \qquad \text{for area in nmi}^2$$

The instantaneous access areas or access lengths for the other patterns in Fig. 7-3 are given in Table 7-2, which also summarizes all of the coverage formulas for these patterns. These access area formulas **do** take into account the curved surface of the Earth and are accurate for any access area size or satellite altitude to within very small corrections for the Earth's oblateness.

TABLE 7-2. **Coverage Formulas for Patterns Shown in Fig. 7-3.** See text for definition of variables. In pattern D, the minus sign applies if λ_2 is on the same side of the ground track as λ_1. The approximation for footprint area is invalid when $\varepsilon \approx 0$. The ACR formulas for patterns C and D assume that the instrument is side-looking. P is the orbit period.

Pattern	Typical Application	Footprint Area (FA)	Instantaneous Access Area (IAA) or Length (IAL)	Area Coverage Rate (ACR)	Area Access Rate (AAR)
A	Omni-antenna, Ground station cov, General sensing	$(\pi D K_L / 4)$ $\sin\theta \times$ $(\lambda_{FO} - \lambda_{FI})$ $\approx (\pi D^2 / 4) \times$ $\sin^2\theta / \sin\varepsilon$	$IAA = K_A(1 - \cos\lambda)$	$\dfrac{F_A\left(1 - O_{avg}\right)DC}{T}$	$\dfrac{2K_A \sin\lambda}{P}$
B	Radar	As above	$IAA = K_A(\cos\lambda_2 - \cos\lambda_1)$	As above	$\dfrac{2K_A \sin\lambda_1}{P}$
C	Synthetic Aperture Radar	As above	$IAL = 2K_L(\lambda_1 - \lambda_2)$ $\approx 2K_L D \sin\theta / \sin\varepsilon$	$\dfrac{2K_A\left(\sin\lambda_1 - \sin\lambda_2\right)}{P}$	$\dfrac{2K_A\left(\sin\lambda_1 - \sin\lambda_2\right)}{P}$
D	Scanning Sensor	As above	$IAL = K_L(\lambda_1 \pm \lambda_2)$	$\dfrac{K_A\left(\sin\lambda_1 \pm \sin\lambda_2\right)}{P}$	$\dfrac{K_A\left(\sin\lambda_1 \pm \sin\lambda_2\right)}{P}$

We now wish to determine the length of time a particular point on the Earth is within the satellite access area and the access area rate at which the land enters or leaves the access area. Consider a satellite in a circular orbit at altitude H. The orbit period, P, in minutes is given by

$$P = 1.658\,669 \times 10^{-4} \times (6{,}378.14 + H)^{3/2} \qquad H \text{ in km} \qquad (7\text{-}7)$$

$$P = 4.180\,432 \times 10^{-4} \times (3{,}443.9 + H)^{3/2} \qquad H \text{ in nmi}$$

We define the *maximum Earth central angle*, λ_{max}, as the radius of the access area for the observation in question. Twice λ_{max} is called the *swath width* and is the width of the coverage path across the Earth. As shown in Fig. 7-4, the coverage for any point **P** on the surface of the Earth will be a function of λ_{max} and of the off-track angle, λ, which is the perpendicular distance from **P** to the satellite ground track for the orbit pass being evaluated. The fraction of the orbit, F_{view}, over which the point **P** is in view is

$$F_{view} = \Delta v / 180 \deg \qquad (7\text{-}8a)$$

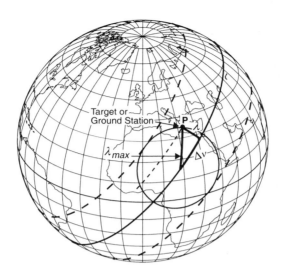

Fig. 7-4.   Earth Coverage Geometry. λ is the *off ground track angle* and 2 λ_{max} is the *swath width*. **P** is the target or ground station.

where

$$\cos \Delta v = \cos \lambda_{max}/\cos \lambda \qquad (7\text{-}8b)$$

Therefore the time in view, T, for the point **P** will be

$$T = PF_{view} = \left(\frac{P}{180 \text{ deg}}\right)\cos^{-1}\left(\frac{\cos\lambda_{max}}{\cos\lambda}\right) \qquad (7\text{-}9)$$

which is equivalent to Eq. (5-49). Note that here we use λ rather than λ_{min} for the off ground-track angle and that Δv is one half of the true anomaly range (i.e., angle along the ground track) over which the point **P** is in view by the satellite. See Fig. 5-17 in Sec. 5.3.1 for the geometry of this computation.

Finally, the area access rate as the satellite sweeps over the ground for the access area of Fig. 7-3A is

$$AAR = (2 \, K_A \sin \lambda)/P \qquad \text{(Pattern A)} \qquad (7\text{-}10)$$

Formulas for other patterns are in Table 7-2. Again note that because of the curvature of the Earth's surface, this area access rate is **not** equal to the diameter of the access area times the subsatellite point velocity.

As an example of the above computations, consider a spacecraft at 2,000 km altitude with a 1 deg diameter beam staring perpendicular to the ground track at an elevation angle of 10 deg as seen from the ground. Our linear estimate of the footprint height is 446 km from Eq. (7-1b).[*] However, from Table 7-2 we see that the true height is 355 km and therefore need to use the somewhat more complex Eq. (7-1a). From

---

[*] As indicated previously, this estimate would be substantially improved if the 10 deg elevation angle was at the center of the beam. However, we would then need to keep track of beam-center parameters for the geometry and beam-edge parameters for performance estimates.

Eqs. (5-24), (5-26), and (5-27) we determine $\lambda_{FO} = 31.43$ deg and $\lambda_{FI} = 28.24$ deg. The footprint width from Eq. (7-2a) is 77 km. From Eq. (7-3) the footprint area is 21,470 km^2. The accuracy of the area is proportional to $1 - (77/6,378) / \sin(77/6,378) = 0.002\%$. The ground track velocity is the circumference of the Earth divided by the orbit period (from Eq. 7-7) = 40,075 km/127 min = 315.6 km/min = 5.26 km/s. Multiplying this by the footprint height of 355 km gives a crude estimate of the area coverage rate of 1,867 km^2/s. Using the more accurate formula in Table 7-2 (Pattern D) and the values of λ above, we obtain a more accurate value of $ACR = 2.556 \times 10^8 \times (\sin 31.43$ deg $- \sin 28.24$ deg$) / (127 \times 60) = 1,620$ km^2/s which implies an error of 15% in the less accurate approximation.

The above formulas are in terms of off-ground-track angle, which is computationally convenient. But we often need to know the coverage as a function of latitude, L, for a satellite in a circular orbit at inclination, i. We assume that the pattern of Fig. 7-3A applies and that observations can be made at any off-track angle less than or equal to λ_{max} on either side of the satellite ground track. We also assume that L is positive, that is, in the northern hemisphere. (The extensions are straightforward for the southern hemisphere or nonsymmetric observations.) Depending on the latitude, there will be either no coverage, a single long region of coverage, or two shorter regions of coverage for each orbit as follows (See Fig. 7-5).

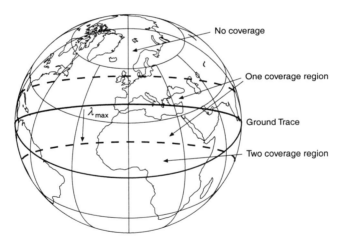

Fig. 7-5. Single Orbit Coverage is a Function of Latitude, Orbit Inclination, and Swath Width. See text for formulas.

Latitude Range	Number of Coverage Regions	Percent Coverage	
$L > \lambda_{max} + i$	0	0	(7-11a)
$i + \lambda_{max} > L > i - \lambda_{max}$	1	$\phi_1 / 180$	(7-11b)
$i - \lambda_{max} > L > 0$	2	$(\phi_1 - \phi_2)/180$	(7-11c)

where

$$\cos \phi_{1\text{ or }2} = \frac{\pm \sin \lambda_{max} + \cos i \sin L}{\sin i \cos L} \qquad (7\text{-}12)$$

where the minus sign applies for ϕ_1 and the plus sign for ϕ_2. Here ϕ is one-half the longitude range over which coverage occurs. The formula in the third column above represents the fraction of all points at a given latitude in view of the satellite during one orbit. This is approximately equal to the fraction of orbits that will cover a given point at that latitude.

As an example of the above formula, consider a satellite in a 62.5-deg inclined orbit which can see to an off-ground-track angle, $\lambda_{max} = 20$ deg. At a ground station latitude of 50 deg, the percent coverage will be 49.3%. On any orbit, 49% of the points at a latitude of 50 deg will be within view of the satellite at some time. Conversely, a given point at 50-deg latitude will be covered at some time on approximately 49% of the satellite orbits. Because there is only one coverage region, the covered orbits will occur successively during the day. If the satellite orbit period is 2 hr, then our hypothetical ground station at 50-deg latitude will typically see the satellite on 6 successive orbits followed by 6 orbits of no coverage. The number and duration of coverage passes on a given day will depend on where the ground station is located with respect to the orbit node.

As a final example, consider a satellite in a 1,000-km circular orbit at an inclination of 55 deg. From Eq. (5-24), in Sec. 5.2, $\rho = 59.82$ deg and from Eq. (7-7) the orbit period is 105 min. We assume that the satellite can make observations out to a spacecraft elevation angle of 10 deg as seen by the target, corresponding to a nadir angle $\eta_{max} = 58.36$ deg from Eq. (5-26) and maximum off-track angle, $\lambda_{max} = 21.64$ deg from Eq. (5-27). From Eq. (7-10), the potential area search rate is 1.8×10^6 km^2/min. From Eq. (7-9) a point 15 deg from the ground track will remain in view for 9.2 min. Finally, from Eqs. (7-11), a satellite in such an orbit will see 45.7% of all points at a latitude of 50 deg and 33.4% of all points at a latitude of 20 deg.

7.2.2 Numerical Simulations

The analytic formulas above provide an easy and rapid way to evaluate Earth coverage, but this approach has several limitations. It does not take into account non-circular orbits, the rotation of the Earth under the spacecraft, or possible overlapping coverage of several satellites. Although we could extend the analytic expressions, numerically simulating the coverage is a better approach for more complex situations. Any modern office computer can do a simple simulation that takes these effects into account with sufficient accuracy for preliminary mission analysis.

Analytic approximations also do not allow us to assess coverage statistics easily. For example, while we can determine the coverage time for a given orbit pass, we cannot easily compute how often we will see a given point or where regions of coverage or gaps between coverage will occur. We usually need these statistics for Earth observation applications.

Numerical simulations of coverage can become extremely complex. They may consider such activities as scheduling, power and eclipse conditions, and observability of the target or ground station. Chapter 3 briefly describes an example. In the following paragraphs we will consider two simple simulations of considerable use during preliminary mission design.

The simplest "simulation" is a ground track plot of the mission geometry, clearly revealing how the coverage works and the possible coverage extremes. Figure 7-6 shows ground trace plots for our example satellite in a 1,000-km circular orbit with a period $P = 105$ min, corresponding to approximately 14 orbits per day. The spacing, S, between successive node crossings on the equator is

$$S = \frac{P}{1,436\,\text{min}} \cdot 360\,\text{deg} \qquad (7\text{-}13)$$

The heavy circle on Fig. 7-6 represents the subsatellite points corresponding to spacecraft elevations, ε, greater than 5 deg (equivalent to $\lambda_{max} < 25.6$ deg). For the 25-deg latitude shown, we can see by inspection that we will see the point **P** on either two or three successive upward passes and two or three successive downward passes. The downward coverage passes will normally begin on the fourth orbit after the last upward coverage pass. Individual passes within a group will be centered approximately 105 min apart and will be up to $105 \times 51.2 / 360 = 14.9$ min long.

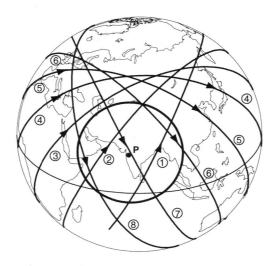

Fig. 7-6. Ground Track of 8 Successive Orbits (out of 14 per day) for a Satellite at 1,000 km. The heavy circle covers subsatellite points over which spacecraft will be at an elevation angle, ε, greater than 5 deg. See text for discussion.

The details of each day's potential observations will depend on how the orbit falls relative to the point **P**. However, the general flow will be as follows. Two or three passes of approximately 12 min each will occur 105 min apart. (Twelve minutes is estimated by inspection relative to the maximum pass duration of 15 min.) After a break of 5 hr, there will be another group of 2 or 3 passes. The process will repeat after a break of 12 hr. Though we would like to have more statistical data, the ground track analysis can rapidly assess performance, represent the coverage distribution, and crosscheck more detailed results. From this process we could, for example, generate timelines for the most and least coverage in one day.

The next step in the numerical modeling hierarchy is a *point coverage simulation*. To do this, we conceptually create a grid of points on the surface of the Earth, fly one or more spacecraft over the grid, and track the observation characteristics for each of the grid points. We can then collect and evaluate data over different geographical regions. The most common way is to collect data along lines of constant latitude and present statistical coverage results as a function of latitude. This type of simulation is an excellent way to evaluate coverage statistically. The main disadvantage is that it does not let us see the problem physically or admit the general analytic studies stemming from the formulas of Sec. 7.2.1. Thus, the best choice is to evaluate coverage by combining analytic formulas, ground trace plots, and numerical simulations.

Although the technique for the numerical point coverage simulation is straightforward, the analyst must be aware of three potential pitfalls. First, if we want to compare coverage performance at different latitudes, then we need to have grid points covering approximately equal areas over the surface of the globe. If grid points are at equal intervals of latitude and longitude, (at every 10 deg, for example), then the number of points per unit area will be much greater near the poles, thus incorrectly weighting polar data in the overall global statistics. We can easily resolve this artificial weighting by using grid points at a constant latitude spacing with the number of points at each latitude proportional to the cosine of that latitude. This covers the globe with an approximately equal number of points per unit area and properly balances the global statistics.

Our second problem is to adjust for gaps where the simulation begins and ends. Otherwise, these gaps will make gap statistics unrealistic because true gaps and coverage regions will not begin and end at the start and end points of the simulation. The easiest solution is to run the simulation long enough that start and end data have minimal impact on the statistics.

The third, and perhaps most significant, problem is that we are trying to collect statistical data on a process for which statistical distributions do not apply. Most statistical measures, such as the mean, standard deviation, or the 90th percentile assume that the data being sampled has a Gaussian or random distribution. While the distribution which we found by examining the ground track plot above was not uniform, it was also not at all important for some activities. Our estimation of above 2 to 3 passes of 12 min each, twice per day, gives an average percent coverage on the order of 1 hr/24 hr = 4%. But just collecting statistical data and concluding that the percent coverage is about 4% is remarkably uniformative. That could be the result of 1 hr of continuous coverage and 23 hr of no coverage or 2.5 min of coverage every hour. Similar problems plague all of the normal statistical measures applied to orbit analysis. The important point is:

Statistical analysis of inherently nonstatistical data, such as orbit coverage, can lead to dramatically incorrect conclusions.

Simple techniques such as ground track analysis are imperative to understand and validate the conclusions we reach.

7.2.3 Coverage Figures of Merit

Having established a simulation technique, we need to find a way to accumulate coverage statistics and to evaluate the quality of coverage. As described in Chap. 3, we can quantify coverage quality by providing a *coverage Figure of Merit* (an **appropriate** numerical mechanism for comparing the coverage of satellites and constellations). We wish to find a Figure of Merit which is physically meaningful, easy to compute in our numerical simulation, and fair in comparing alternative constellations. The most common general purpose coverage Figures of Merit are:

- *Percent Coverage*

  The *percent coverage* for any point on the grid is simply the number of times that point was covered by one or more satellites divided by the total number of simulation time steps. It is numerically equal to the analytically computed percent coverage in Eq. (7-11). The advantage of percent coverage is that it shows directly how much of the time a given point or region on the ground is covered. However, it does not provide any information about the distribution of gaps in that coverage.

- *Maximum Coverage Gap* (= *Maximum Response Time*)

  The *maximum coverage gap* is simply the longest of the coverage gaps encountered for an individual point. When looking at statistics over more than one point, we can either average the maximum gaps or take their maximum value. Thus the worldwide *mean maximum gap* would be the average value of the maximum gap for all the individual points, and the worldwide *maximum gap* would be the largest of any of the individual gaps. This statistic conveys some worst-case information, but it incorrectly ranks constellations because a single point or a small number of points determine the results. Thus, the maximum coverage gap, or *maximum response time*, is a poor Figure of Merit.

- *Mean Coverage Gap*

  The *mean coverage gap* is the average length of breaks in coverage for a given point on the simulation grid. To compute gap statistics, we must have three counters for each point on the simulation grid. One counter tracks the number of gaps. A second tracks total gap duration. The third tracks the duration of the current gap and is reset as needed. During the simulation, if no satellite covers a given point on the grid, we increment the gap length counter (3) by one time step. If the point is covered but was not covered the previous time (indicated by a value of the gap length counter greater than 0), then we have reached the end of an individual gap. We increment the counter for the number of gaps (1) by one and add the gap duration to the total gap counter (2) or incorporate it in other statistics we want to collect. The final mean coverage gap is computed by dividing the total gap length by the number of gaps. As noted above, what happens at the beginning and end of the simulation influences all statistics relating to gap distribution.

- *Time Average Gap*

  The *time average gap* is the mean gap duration *averaged over time*. Alternatively, it is the average length of the gap we would find if we randomly sampled the system. To compute the time average gap, two counters are required—one for the current gap length and one for the sum of the squares of gap lengths. During the simulation, if no satellite covers a given point on the grid, add one to the current gap length counter. If the point is covered, square the current gap length, add the results to the sum of the squares counter, and reset the current gap length counter to zero. (If the current gap length counter was previously 0, then no change will have occurred in either counter.) The time average gap is computed at the end of the simulation by dividing the sum of the squares of the gaps by the duration of the simulation.

- *Mean Response Time*

  The *mean response time* is the average time from when we receive a random request to observe a point until we can observe it. If a satellite is within view of the point at a given time step, the response time at that step will be 0.* If the point in question is in a coverage gap, then the response time would be the time until the end of the coverage gap. In principle, response time should be computed from a given time step to the end of a gap. But by symmetry we could also count the time from the beginning of the gap—a computationally convenient method with the same results. Thus the response time counter will be set to 0 if a point is covered at the current time step. We advance the response time counter by one time step if the point is not now covered. The mean response time will then be the average value of all response times for all time steps. This Figure of Merit takes into account both coverage and gap statistics in trying to determine the whole system's responsiveness. As shown below, the mean response time is the best coverage Figure of Merit for evaluating overall responsiveness.

To illustrate the meaning and relative advantages of these Figures of Merit, Fig. 7-7 diagrams a simplified coverage simulation from three satellite systems: A, B, and C. These could, for example, be three sample FireSat constellations. Our goal is to see events as quickly as possible, and therefore, minimize gaps. Constellation B is identical to A except for one added gap, which makes B clearly a worse solution than A. C has the same overall percent coverage as A, but the gaps are redistributed to create a rather long gap, making C the worst constellation for regular coverage.

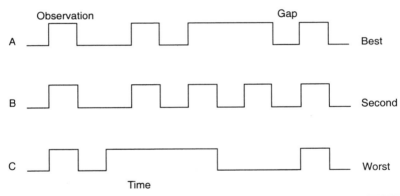

	Percent Coverage	Maximum Gap	Mean Gap	Time Average Gap	Mean Response Time
A	60	2	1.33	0.6	0.5
B	50	2	1.25	0.7	0.6
C	60	3	2.00	1.0	0.7

Fig. 7-7.   **Coverage Figures of Merit**. See text for explanation.

---

* One advantage of response time as a Figure of Merit is that delays in processing or communications (for both data requests and responses) can be directly added to the coverage response time. This results in a *total response time,* which measures the total time from when users request data until they receive it. We can also evaluate minimum, mean, and maximum total response times which have much more operational meaning than simple gap statistics but are still easy to compute.

The table below Fig. 7-7 shows the numerical values of the Figures of Merit defined above. The *percent coverage* correctly ranks constellation A better than B, but because it does not take gap statistics into account it cannot distinguish between A and C. Similarly, the *maximum gap* cannot distinguish between A and B, even though B is clearly worse by having an additional gap. In this case the maximum gap tells us which constellation is worst but cannot distinguish between two constellations which are clearly different.

The *mean gap statistic* is even more misleading. By adding a short gap to constellation B, the average length of the gaps has been decreased, and consequently, this Figure of Merit ranks constellation B **above** constellation A. (This can happen in real constellation statistics. By adding satellites we may eliminate some of the very small gaps, thus **increasing** the average gap length, even though more satellites provide more and better coverage.)

Finally, the *time average gap* and *mean response time* in the fourth and fifth columns correctly rank the three constellations in order of preference by taking into account both the percent coverage and gap statistics. Consequently, both of these are better Figures of Merit than the other three. I believe the mean response time is the stronger Figure of Merit because it provides a more useful measure of the end performance of the system and because it can be easily extended to include delays due to processing, communications, decision making, or the initiation of action. However, because each of the Figures of Merit represent different characteristics we should evaluate more than one. Specifically, I recommend evaluating mean response time, percent coverage, and maximum gap, and qualitatively (not quantitatively) weighting the results in that order, keeping strongly in mind the caveat at the end of Sec. 7.2.2.

7.3 The ΔV Budget

To an orbit designer, a space mission is a series of different orbits. For example, a satellite may be released in a low-Earth parking orbit, transferred to some mission orbit, go through a series of rephasings or alternative mission orbits, and then move to some final orbit at the end of its useful life. Each of these orbit changes requires energy. The ΔV *budget* is traditionally used to account for this energy. It is the sum of the velocity changes required throughout the space mission life. In a broad sense the ΔV budget represents the cost for each mission orbit scenario. In designing orbits and constellations, we must balance this cost against the utility achieved.

Chapter 10 shows how to develop a propulsion budget based on a given ΔV budget. For preliminary design, we can estimate the "cost" of the space mission by using the rocket equation to determine the total required spacecraft plus propellant mass, $m_f \equiv m_0 + m_p$, in terms of the dry mass of the spacecraft, m_0, the total required ΔV, and the propellant exhaust velocity, V_0:

$$m_f = m_0 e^{-(\Delta V/V_0)} \qquad (7\text{-}14)$$

This is equivalent to Eqs. (17-6) and (17-7) in Sec. 17.1, with V_0 replaced by $I_{sp}g$, where the *specific impulse*, $I_{sp} \equiv V_0/g$, and g is the acceleration of gravity at the Earth's surface. Typical exhaust velocities are in the range of 2 to 4 km/s and up to 30 km/s for electric propulsion. We can see from Eq. (7-14) that ΔV requirements much smaller than the exhaust velocity (a few hundred meters per second), will require a propellant mass which is a small fraction of the total mass. If the total ΔV required is equal to the exhaust velocity, then we will need a total propellant mass equal to $e - 1 \approx 1.7$ times the mass of the spacecraft. Propulsion systems require additional structure such as tanks, so a ΔV much greater than the exhaust velocity is difficult to achieve. It may scuttle the mission or require some alternative, such as staging or refueling.

Table 7-3 summarizes how to construct a ΔV budget. We begin by writing down the basic data required to compute ΔVs: the launch vehicle's initial conditions, the mission orbit or orbits, the mission duration, required orbit maneuvers or maintenance, and the mechanism for spacecraft disposal. We then transform each item into an equivalent ΔV requirement using the formulas listed in the table. The right-hand column shows how these formulas apply to the FireSat mission. Figure 7-8 shows the ΔV required for altitude maintenance for typical spacecraft and atmosphere parameters.

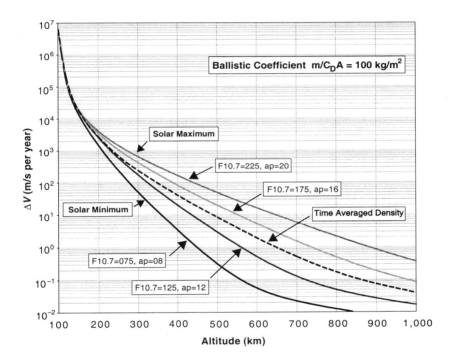

Fig. 7-8.   Altitude Maintenance ΔVs for a Ballistic Coefficient of 100 kg/m². See Sec. 8.1.3 for ballistic coefficient and atmosphere parameters. The ΔV for altitude maintenance is inversely proportional to the ballistic coefficient. The F10.7 index is in units of 10^{-22} W/(m²·Hz). Ap is an index of geomagnetic activity ranging from 0 (very quiet) to 400 (extremely disturbed).

The ΔV budget relates strongly to the propulsion requirements and to the final cost of a space mission. Yet other conditions may vary the propellant requirements relative to the nominal ΔV budget. For example, although rocket propulsion usually provides the ΔV, we can obtain very large ΔVs from a flyby of the Moon, other planets, or even the Earth itself [Kaufman, Newman, and Chromey, 1966; Meissinger, 1970]. In a fly-by, a spacecraft leaves the vicinity of some celestial body with the same velocity

TABLE 7-3. Creating a ΔV Budget. See also summary tables on inside back cover.

Item	Where Discussed	Equation Source	FireSat Example
Basic Data			
Initial Conditions	Chap. 18		150 km, 55 deg
Mission Orbit(s)	Secs. 7.4, 7.5		700 km, 55 deg
Mission Duration (each phase)	Sec. 2.3		5 yr
Orbit Maintenance Requirements	Secs. 6.2.3, 6.5		Altitude maintenance
Drag Parameters	Secs. 6.2.3, 8.1.3	Table 8-3, Fig. 8-2, inside rear endleaf	$m/C_d A = 25$ kg/m^2 $\rho_{max} = 2.73 \times 10^{-13}$ kg/m^3
Orbit Maneuver Requirements	Sec. 6.3		None
Final Conditions	Secs. 6.5, 21.2		Positive reentry
Delta V Budget (m/s)			
Orbit Transfer			
1st burn	Sec. 6.3.1*	(6-32), (6-39)	156 m/s
2nd burn	Sec. 6.3.1*		153 m/s
Altitude Maintenance (LEO)	Secs. 6.2.3, 7.3	(6-26), Fig. 7-8	19 m/s
North/South Stationkeeping (GEO)	Sec. 6.5	(6-51), (6-52)	N/A
East/West Stationkeeping	Sec. 6.5	(6-53)	N/A
Orbit Maneuvers			
Rephasing, Rendezvous	Sec. 6.3.3	(6-41)	None
Node or Plane Change	Sec. 6.3.2	(6-38), (6-39)	None
Spacecraft Disposal	Sec. 6.5	(6-54)	198 m/s
Total ΔV		Sum of the above	526 m/s
Other Considerations			
ACS & Other Requirements	Sec. 10.3		
ΔV Savings	Sec. 7.3	See text	
Margin	Sec. 10.2		Included in propellant budget

*Sec. 6.3.2 if plane change also required.

relative to the body as when it approached, but in a different direction. This phenomenon is like the elastic collision between a baseball and a bat, in which the velocity of the ball relative to the bat is nearly the same, but its velocity relative to the surrounding baseball park can change dramatically. We can use flybys to change direction, to provide increased heliocentric energy for solar system exploration, or to reduce the amount of energy the satellite has in inertial space. For example, one of the most energy-efficient ways to send a space probe near the Sun is to use a flyby of Jupiter to reduce the intrinsic heliocentric orbital velocity of Earth associated with any spacecraft launched from Earth.

A second way to produce a large ΔV without burning propellant is to use the atmosphere of the Earth or other planets to change the spacecraft's direction or reduce its energy relative to the planet. The manned flight program has used this method from the beginning to dissipate spacecraft energy for return to the Earth's surface. Mars Pathfinder used aerobraking for planetary exploration. It can also be used to produce a major plane change through an *aeroassist trajectory* [Austin, Cruz, and French, 1982; Mease, 1988].

The *solar sail* is a third way to avoid using propellant. The large, lightweight sail uses solar radiation to slowly push a satellite the way the wind pushes a sailboat. Of course, the low-pressure sunlight produces very low acceleration.

The aerospace literature discusses many alternatives for providing spaceflight energy. But experimental techniques (those other than rocket propulsion and atmospheric braking) are risky and costly, so normal rocket propulsion will ordinarily be used to develop the needed ΔV, if this is at all feasible.

The ΔV budget described in Table 7-3 measures the energy we must give to the spacecraft's center of mass to meet mission conditions. When we transform this ΔV budget into a propellant budget (Chap. 10), we must consider other characteristics. These include, for example, inefficiencies from thrusters misaligned with the ΔV direction, and any propulsion diverted from ΔV to provide attitude control during orbit maneuvers. Chapters 10 and 17 describe propulsion requirements in detail.

For most circumstances, the ΔV budget does not include margin because it results from astrodynamic equations with little error. Instead, we maintain the margin in the propellant budget itself, where we can reflect such specific elements as residual propellant. An exception is the use of ΔV to overcome atmospheric drag. Here the ΔV depends upon the density of the atmosphere, which is both variable and difficult to predict. Consequently, we must either conservatively estimate the atmospheric density or incorporate ΔV margin for low-Earth satellites to compensate for atmospheric variations.

7.4 Selecting Orbits for Earth-Referenced Spacecraft

The first step in finding the appropriate orbit for an Earth-referenced mission is to determine if a specialized orbit from Table 7-4 applies.[*] We should examine each of these orbits individually to see if its characteristics will meet the mission requirements at reasonable cost. Space missions need not be in specialized orbits, but these orbits have come into common use because of their valuable characteristics. Because they do constrain such orbit parameters as altitude and inclination, we must determine whether or not to use them before doing the more detailed design trades described below.

It is frequently the existence of specialized orbits which yields very different solutions for a given space mission problem. Thus, a geosynchronous orbit may provide the best coverage characteristics, but may demand too much propellant, instrument resolution, or power. This trade of value versus cost can lead to dramatically different solutions, depending on mission needs. For a traditional communications system, the value of providing continuous communications coverage outweighs the cost and performance loss associated with the distance to geostationary orbit. Some communications systems provide continuous coverage with a low-Earth orbit constellation as described in Sec. 7.6. In the case of FireSat, continuous coverage is not required and

---

[*] For an extended discussion see Cooley [1972] or Wertz [2001].

TABLE 7-4. **Specialized Orbits Used for Earth-Referenced Missions**. For nearly circular low-Earth orbits, the eccentricity will undergo a low-amplitude oscillation. A *frozen orbit* is one which has a small eccentricity (~0.001) which does not oscillate due to a balancing of the J_2 and J_3 perturbations.

Orbit	Characteristic	Application	Where Discussed
Geosynchronous (GEO)	Maintains nearly fixed position over equator	Communications, weather	Sec. 6.1.4
Sun-synchronous	Orbit rotates so as to maintain approximately constant orientation with respect to Sun	Earth resources, weather	Sec. 6.2.2
Molniya	Apogee/perigee do not rotate	High latitude communications	Sec. 6.2.2
Frozen Orbit	Minimizes changes in orbit parameters	Any orbit requiring stable conditions	See Chobotov [1996]
Repeating Ground Track	Subsatellite trace repeats	Any orbit where constant viewing angles are desirable	Sec. 6.5

the need for fine resolution on the ground for an IR detection system precluded a geosynchronous orbit, so its mission characteristics are dramatically different. There is no a priori way of knowing how these trades will conclude, so we may need to carry more than one orbit into detailed design trades. In any case, we should reconsider specialized orbits from time to time to see whether or not their benefits are indeed worth their added constraints.

Orbit design is inherently iterative. We must evaluate the effects of orbit trades on the mission as a whole. In selecting the orbit, we need to evaluate a single satellite vs. a constellation, specialized orbits, and the choice of altitude and inclination. For example, alternative solutions to a communications problem include a single large satellite in geosynchronous equatorial orbit and a constellation of small satellites in low-Earth orbit at high inclination.

The first step in designing mission orbits is to determine the effect of orbit parameters on key mission requirements. Table 7-5 summarizes the mission requirements that ordinarily affect the orbit. The table shows that altitude is the most important of orbit design parameter.

The easiest way to begin the orbit trade process is by assuming a circular orbit and then conducting altitude and inclination trades as described below and summarized in the table. This process establishes a range of altitudes and inclinations, from which we can select one or more alternatives. Documenting the reasons for these results is particularly important, so we can revisit the trade from time to time as mission requirements and conditions change.

Selecting the mission orbit is often highly complex, involving such choices as availability of launch vehicle, coverage, payload performance, communication links, and any political or technical constraints or restrictions. Thus, considerable effort may go into the process outlined in Table 7-1. Figure 3-1 in Sec. 3.2.3 shows the results of the altitude trade for the FireSat mission. Typically these trades do not result in specific values for altitude or inclination, but a range of acceptable values and an indication of those we would prefer. Ordinarily, low altitudes achieve better instru-

TABLE 7-5.   Principal Mission Requirements That Normally Affect Earth-Referenced Orbit Design.

Mission Requirement	Parameter Affected	Where Discussed
Coverage Continuity Frequency Duration Field of view (or swath width) Ground track Area coverage rate Viewing angles Earth locations of interest	Altitude Inclination Node (only relevant for some orbits) Eccentricity	Sec. 7.2
Sensitivity or Performance Exposure or dwell time Resolution Aperture	Altitude	Chaps. 9,13
Environment and Survivability Radiation environment Lighting conditions Hostile action	Altitude (inclination usually secondary)	Chap. 8
Launch Capability Launch cost On-orbit weight Launch site limitations	Altitude Inclination	Chap. 18
Ground Communications Ground station locations Use of relay satellites Data timeliness	Altitude Inclination Eccentricity	Chap. 13
Orbit Lifetime	Altitude Eccentricity	Secs. 6.2.3, 8.1.5
Legal or Political Constraints Treaties Launch safety restrictions International allocation	Altitude Inclination Longitude in GEO	Sec. 21.1

ment performance because they are closer to the Earth's surface. They also require less energy to reach orbit. On the other hand, higher orbits have longer lifetimes and provide better Earth coverage. Higher orbits are also more survivable for satellites with military applications. Orbit selection factors usually compete with each other with some factors favoring higher orbits and some lower.

Often, a key factor in altitude selection is the satellite's radiation environment. As described in Sec. 8.1, the radiation environment undergoes a substantial change at approximately 1,000 km. Below this altitude the atmosphere will quickly clear out charged particles, so the radiation density is low. Above this altitude are the Van Allen belts, whose high level of trapped radiation can greatly reduce the lifetime of spacecraft components. Most mission orbits therefore separate naturally into either *low-Earth orbits* (*LEO*), below 1,000 to 5,000 km, and *geosynchronous orbits* (*GEO*), which are well above the Van Allen belts. Mid-range altitudes may have coverage characteristics which make them particularly valuable for some missions. However, the additional shielding or reduced life stemming from this region's increased radiation environment also makes them more costly.

Having worked the problem assuming a circular orbit, we should also assess the potential advantages of using eccentric orbits. These orbits have a greater peak altitude for a given amount of energy, lower perigee than is possible with a circular orbit, and lower velocity at apogee, which makes more time available there. Unfortunately, eccentric orbits also give us non-uniform coverage and variable range and speed.

Eccentric orbits have an additional difficulty because the oblateness of the Earth causes perturbations which make perigee rotate rapidly. This rotation leads to rapid changes in the apogee's position relative to the Earth's surface. Thus, with most orbits, we cannot maintain apogee for long over a given latitude. As Sec. 6.2 describes, the first-order rotation of perigee is proportional to $(2 - 2.5 \sin^2 i)$ which equals zero at an inclination, $i = 63.4$ deg. At this *critical inclination* the perigee will not rotate, so we can maintain both apogee and perigee over fixed latitudes. Because this orientation can provide coverage at high northern latitudes, the Soviet Union has used such a *Molniya* orbit for communications satellites for many years. Geosynchronous orbits do not provide good coverage in high latitude regions.

Eccentric orbits help us sample either a range of altitudes or higher or lower altitudes than would otherwise be possible. That is why scientific monitoring missions often use high eccentricity orbits. As discussed in Sec. 7.6, Draim [1985, 1987a, 1987b] has done an extensive evaluation of the use of elliptical orbits and concluded that they can have significant advantages in optimizing coverage and reducing the number of satellites required.

FireSat Mission Orbit. Our first step for the FireSat mission orbit is to look at the appropriateness of the specialized orbits from Table 7-4. This is done for FireSat in Table 7-6. As is frequently the case, the results provide two distinct regimes. One possibility is a single geosynchronous FireSat. In this case, coverage of North America will be continuous but coverage will not be available for most of the rest of the world. Resolution will probably be the driving requirement.

TABLE 7-6. FireSat Specialized Orbit Trade. The conclusion is that in low-Earth orbit we do not need a specialized orbit for FireSat. The frozen orbit can be used with any of the low-Earth orbit solutions.

Orbit	Advantages	Disadvantages	Good for FireSat
Geosynchronous	Continuous view of continental U.S.	High energy requirement / No world-wide coverage / Coverage of Alaska not good	Yes
Sun-synchronous	None	High energy requirement	No
Molniya	Good Alaska coverage / Acceptable view of continental U.S.	High energy requirement / Strongly varying range	No, unless Alaska is critical
Frozen Orbit	Minimizes propellant usage	None	Yes
Repeating Ground Track	Repeating viewing angle (marginal advantage)	Restricts choice of altitude / Some perturbations stronger	Probably not

The alternative is a low-Earth orbit constellation. Resolution is less of a problem than for geosynchronous. Coverage will not be continuous and will depend on the number of satellites. None of the specialized low-Earth orbits is needed for FireSat. (A frozen orbit can be used with essentially any low-Earth orbit.) Thus, for the low-Earth constellation option, there will be a broad trade between coverage, launchability, altitude maintenance, and the radiation environment.

For low-Earth orbit, coverage will be the principal driving requirement. Figure 3-1 in Sec. 3.2.3 summarized the FireSat altitude trades and resulted in selecting an altitude range of 600 to 800 km with a preliminary value of 700 km. This may be affected by further coverage, weight, and launch selection trades. FireSat will need to cover high northern latitudes, but coverage of the polar regions is not needed. Therefore, we select a preliminary inclination of 55 deg which will provide coverage to about 65 deg latitude. This will be refined by later performance trades, but is not likely to change by much.

Zero eccentricity should be selected unless there is a compelling reason to do otherwise. There is not in this case, so the FireSat orbit should be circular. Thus, the preliminary FireSat low-Earth orbit constellation has $a = 700$ km, $i = 55$ deg, $e = 0$, and the number of satellites selected to meet minimum coverage requirements.

7.5 Selecting Transfer, Parking, and Space-Referenced Orbits

Selecting transfer, parking, and space-referenced orbits proceeds much the same as for Earth-referenced orbits, although their characteristics will be different. Table 7-7 summarizes the main requirements. We still look first at specialized orbits and then at general orbit characteristics. Table 7-8 shows the most common specialized orbits. The orbits described in this section may be either the end goal of the whole mission or simply one portion, but the criteria for selection will be the same in either case.

TABLE 7-7. Principal Requirements that Normally Affect Design of Transfer, Parking, and Space-Referenced Orbits.

Requirement	Where Discussed
Accessibility (ΔV required)	Secs. 6.3, 7.3
Orbit decay rate and long-term stability	Sec. 6.2.3
Ground station communications, especially for maneuvers	Secs. 5.3, 7.2
Radiation environment	Sec. 8.1
Thermal environment (Sun angle and eclipse constraints)	Secs. 5.1, 10.3
Accessibility by Shuttle or transfer vehicles	Sec. 18.2

7.5.1 Selecting a Transfer Orbit

A transfer orbit must get the spacecraft where it wants to be. For transfer orbits early in the mission, the launch vehicle or a separate upper stage was traditionally tasked with doing the work as described in Chap. 18. Because of the continuing drive to reduce cost, integral propulsion upper stages have become substantially more common (see Chap. 17).

TABLE 7-8. Specialized Orbits Used for Transfer, Parking, or Space-Referenced Operations.

Orbit	Characteristic	Application	Section
Lunar or Planetary Flyby	Same relative velocity approaching and leaving flyby body	Used to provide energy change or plane change	7.3
Aeroassist Trajectory	Use atmosphere for plane change or braking	Used for major energy savings for plane change, altitude reduction or reentry	7.3
Sun-synchronous	Orbit rotates so as to maintain approximately constant orientation with respect to Sun	Solar observations; missions concerned about Sun interference or uniform lighting	6.2.2
Lagrange Point Orbit	Maintains fixed position relative to Earth/Moon system or Earth/Sun system	Interplanetary monitoring; potential space manufacturing	[Wertz, 2001]

Two distinct changes can occur during transfer orbit: a change in the total energy of the satellite, and a change in direction without changing the total energy. As discussed in Sec. 6.1, the total energy of a Keplerian orbit depends only on the semimajor axis. Consequently, only transfer orbits which change the mean altitude, such as transfer from LEO to GEO, require adding energy to the satellite. Clearly, if we wish to go to an orbit with a higher energy level then we must find some process to provide the additional energy, such as rocket propulsion or a lunar or planetary flyby. If we must remove energy from the orbit, we can frequently use atmospheric drag.

A change in satellite direction without changing energy normally involves a plane change (inclination or node), although we may also change the eccentricity without changing the mean altitude. (Any small thrust perpendicular to the velocity cannot change the orbit energy, or therefore, the mean altitude.) To change the satellite orbit plane or eccentricity without changing the total energy, several options are available.

If we choose to change directions by using propulsion, then propellant requirements will typically be large; the ΔV required to change directions is directly proportional to the spacecraft velocity, which is about 7 km/s in low-Earth orbit. Fortunately, other techniques for changing the orbit plane require less energy. For example, suppose we want to shift the node of an orbit to create a constellation with nodes equally spaced around the Earth's equator or to replace a dead satellite. If the constellation is at an altitude other than that of the replacement satellite, we can use the node regression provided by normal orbit perturbations. The rate at which the node of an orbit precesses varies substantially with altitude, as described in Sec. 6.2. Specifically, if we have a final constellation at a high altitude, we can inject and leave the replacement satellite at low altitude so that the node rotates differentially with respect to the high-altitude constellation. When the satellite reaches the desired node, an orbit transfer is made with no plane change, thus using much less energy. In this case we are trading orbit transfer time for energy.

A second way to reduce the ΔV for large plane changes is to couple them with altitude changes. The net required ΔV will be the vector sum of the two perpendicular components changing the altitude and direction—substantially less than if the two burns were done separately.

Because the ΔV required to change the plane is directly proportional to satellite velocity, plane changes are easier at high altitudes where the satellite velocity is lower. That is why most of the plane change in geosynchronous transfer orbit is done at apogee rather than perigee.

If the required plane change is large, it may cost less total propellant to use a three-burn transfer rather than a two-burn transfer [Betts, 1977]. In this case, the first perigee burn puts apogee at an altitude above the ultimate altitude goal, because the apogee velocity is lower there. During a second burn, the plane change is made using a smaller ΔV and perigee is raised to the final altitude. The third burn then brings apogee back to the desired end altitude. This process is **not** more efficient for small plane changes such as those associated with launch from mid-latitudes to geosynchronous equatorial orbit. A third proposed way to make large plane changes is the aeroassist orbit described in Sec. 7.3 [Austin, Cruz, and French, 1982; Mease, 1988].

Ordinarily, we want to transfer a satellite using the smallest amount of energy, which commonly leads to using a Hohmann transfer as described in Sec. 6.3. However, as illustrated in Fig. 7-9 and described in Table 7-9, other objectives may influence the selection of a transfer orbit. For example, we can reduce the transfer time relative to a Hohmann orbit by using additional energy. These transfers are not common, but they may be appropriate if transfer time is critical as might be true for military missions or a manned mission to Mars.

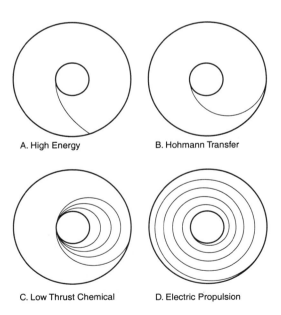

A. High Energy   B. Hohmann Transfer

C. Low Thrust Chemical   D. Electric Propulsion

Fig. 7-9.   Alternative Transfer Orbits. See Table 7-9 for characteristics.

We should also consider a low-thrust transfer, using low thrust chemical or electrical propulsion. (See Chap. 17 for hardware information.) To do a low-thrust chemical transfer with maximum acceleration of 0.05 to 0.10 g's, the satellite undertakes a series of burns around perigee and then one or two burns at apogee to reach the

TABLE 7-9.   Alternative Transfer Orbit Methods. (See Chap. 17 for discussions of hardware alternatives.)

Method	Typical Accel.	Orbit Type	ΔV	Advantages	Disadvantages
High Energy	10 g	Elliptical & hyperbolic	Table 6-5	• Rapid transfer	• Uses more energy than necessary + Hohmann disadvantages
Minimum Energy, High Thrust (Hohmann)	1 to 5 g	Hohmann transfer	Eq. (6-32)	• Traditional • High efficiency • Rapid transfer • Low radiation exposure	• Rough environment • Thermal problems • Can't use S/C subsystems • Failure unrecoverable
Low Thrust Chemical	0.02 to 0.10 g	Hohmann transfer segments	Same as Hohmann	• High efficiency • Low engine weight • Low orbit deployment & check-out • Better failure recovery • Can use spacecraft subsystems • Failure recovery possible	• Moderate radiation exposure • 3–4 day transfer to GEO
Electric Propulsion	0.0001 to 0.001 g	Spiral transfer	Eq. (6-37)	• Can use very high I_{sp} engines = major weight reduction • Low orbit deployment and check-out • Can have reusable transfer vehicle • Failure recovery possible	• 2 to 6 month transfer to GEO • High radiation exposure • Needs autonomous transfer for cost efficiency

final orbit, as illustrated in Fig 7-9C. In this case the total efficiency will approximate that of a two-burn Hohmann transfer, because all of the energy is being provided near perigee or apogee, as it is for the Hohmann transfer. With a low-thrust chemical transfer, we can deploy and check out a satellite in low-Earth orbit where we can still recover it before transferring it to a high-energy orbit where we cannot. Low-thrust transfer provides substantially lower acceleration and, therefore, a more benign environment. Also, we are more likely to be able to recover a satellite if the propulsion system fails. The principal disadvantage of low-thrust chemical transfer is that it is a very nontraditional approach. Wertz, Mullikin, and Brodsky [1988] and Wertz [2001] describe low-thrust chemical transfer further.

Another type of low-thrust transfer uses electric propulsion, with extremely low acceleration levels—at levels of 0.001 g or less [Cornelisse, Schöyer, and Wakker, 1979]. Transfer therefore will take several months, even when the motors are thrusting continuously. Consequently, as Fig. 7-9D shows, electric propulsion transfer requires spiralling out, with increased total ΔV (see Table 7-9). We need far less total propellant because of electric propulsion's high I_{sp}. Electric propulsion transfer greatly reduces the total on-orbit mass and, therefore, the launch cost. However, much of the weight savings is lost due to the very large power system required. In addition, the slow transfer will keep the satellite longer in the Van Allen belts, where radiation will degrade the solar array and reduce mission life.

Flybys or gravity-assist trajectories can save much energy in orbit transfers. Because they must employ a swing-by of some celestial object, however, missions near Earth do not ordinarily use them. Gravity-assist missions can use the Earth, but the satellite must first recede to a relatively high altitude and then come back near the Earth.* For a more extended discussion of gravity-assist missions, see Kaufman, Newman, and Chromey [1966], or Wertz [2001]. Meissinger et al. [1997, 1998] and Farquhar and Dunham [1998] have separately proposed interesting techniques for using a different orbit injection process to substantially increase the mass available (and, therefore, reduce the launch cost) for high-energy interplanetary transfers.

FireSat Transfer Orbit. We assume that FireSat will be launched into a 150-km, circular parking orbit at the proper inclination and need to determine how to get to the operational orbit of 700 km. For now, we assume some type of orbit transfer. When the spacecraft weight becomes better known and a range of launch vehicles selected, another trade will be done to determine whether it is more economical for the launch vehicle to put FireSat directly into its operational orbit.

The FireSat orbit transfer ΔV from Table 7-3 is a modest 309 m/s. It is not worth the added cost, solar array weight, or complexity for electric propulsion transfer. There is no reason for a high-energy transfer. We are left to select between a Hohmann transfer and a low-thrust chemical transfer. The Hohmann transfer is the traditional approach.

Low-thrust chemical transfer provides a more benign transfer environment and the potential for low-orbit deployment and checkout so that satellite recovery would be a possibility. The propulsion system would be lighter weight and require less control authority. We may be able to do the orbit transfer using just the mission orbit control modes and hardware which could completely eliminate a whole set of components and control logic.

For FireSat we will make a preliminary selection of low-thrust chemical transfer. This is non-traditional, but probably substantially lower cost and lower risk. Later in the mission design, the launch vehicle may eliminate this transfer orbit entirely.

7.5.2 Parking and Space-Referenced Orbits

In parking or space-referenced orbits, the position of the spacecraft relative to the Earth is unimportant except for blockage of communications or fields of view. Here the goal is simply to be in space to observe celestial objects, sample the environment, or use the vacuum or low-gravity of space. These orbits are used, for example, for space manufacturing facilities, celestial observatories such as Space Telescope and Chandra X-Ray Observatory, or for testing various space applications and processes. Because we are not concerned with our orientation relative to the Earth, we select such orbits to use minimum energy while maintaining the orbit altitude, and possibly, to gain an unobstructed view of space. For example, Sun-synchronous orbits may be

---

*Using the Earth for a gravity assist was first proposed by Meissinger [1970].

appropriate for maintaining a constant Sun angle with respect to a satellite instrument. Another example is the *parking* or *storage* orbit: a low-Earth orbit high enough to reduce atmospheric drag, but low enough to be easy to reach. We may store satellites (referred to as *on-orbit spares*) in these orbits for later transfer to other altitudes.

An interesting class of orbits which have been used for environmental monitoring and proposed for space manufacturing are *libration point orbits* or *Lagrange orbits,* named after the 18th century mathematician and astronomer, Joseph Lagrange. The *Lagrange points* for two celestial bodies in mutual revolution, such as the Earth and Moon or Earth and Sun, are five points such that an object placed at one of them will remain there indefinitely. We can place satellites in "orbit" around the Lagrange points with relatively small amounts of energy required to maintain these orbits. (For more details, see Wertz [2001].)

7.6 Constellation Design

In designing a constellation, we apply all of the criteria for designing a single-satellite orbit. Thus, we need to consider whether each satellite is launchable, survivable, and properly in view of ground stations or relay satellites. We also need to consider the number of satellites, their relative positions, and how these positions change with time, both in the course of an orbit and over the lifetime of the constellation.

Specifying a constellation by defining all of the orbit elements for each satellite is complex, inconvenient, and overwhelming in its range of options. A reasonable way to begin is by looking at constellations with all satellites in circular orbits at a common altitude and inclination, as discussed in Sec. 7.4. This means that the period, angular velocity, and node rotation rate will be the same for all of the satellites. This leads to a series of trades on altitude, inclination, and constellation pattern involving principally the number of satellites, coverage, launch cost, and the environment (primarily drag and radiation). We then examine the potential of elliptical orbits and the addition of an equatorial ring. The principal parameters that will need to be defined are listed in Table 7-10. After exploring the consequences of some of the choices, we will summarize the orbit design process in Sec. 7.6.2. A more detailed discussion is given in Wertz [2001].

No absolute rules exist. A constellation of satellites in randomly spaced low-Earth orbits is a serious possibility for a survivable communications system. The Soviet Union has used a constellation of satellites in highly eccentric Molniya orbits for decades. Various other missions may find satellite clusters useful. One of the most interesting characteristics of the low-Earth orbit communications constellations is that the constellation builders have invested billions of dollars and arrived at distinctly different solutions. For example, a higher altitude means fewer satellites, but a much more severe radiation environment (as discussed in Sec. 8.1), such that the cost of each satellite will be higher and the life potentially shorter. Similarly, elliptical orbits allow an additional degree of freedom which allows the constellation to be optimized for multiple factors, but requires a more complex satellite operating over a range of altitudes and velocities and passing through heavy radiation regimes. (See, for example, Draim [1985].) Because the constellation's size and structure strongly affect a system's cost and performance, we must carefully assess alternate designs and document the reasons for final choices. It is this list of reasons that allows the constellation design process to continue.

TABLE 7-10. Principal Factors to be Defined During Constellation Design. See Sec. 7.6.2 for a summary of the constellation design process.

Factor	Effect	Selection Criteria	Where Discussed
PRINCIPAL DESIGN VARIABLES			
Number of Satellites	Principal cost and coverage driver	Minimize number consistent with meeting other criteria	Sec. 7.6.1
Constellation Pattern	Determines coverage vs. latitude, plateaus	Select for best coverage	Sec. 7.6.1
Minimum Elevation Angle	Principal determinant of single satellite coverage	Minimum value consistent with constellation pattern	Secs. 5.3.1, 7.6.1
Altitude	Coverage, environment, launch, & transfer cost	System level trade of cost vs. performance	Secs. 7.2, 7.6.1
Number of Orbit Planes	Determines coverage plateaus, growth and degradation	Minimize consistent with coverage needs	Sec. 7.6.1
Collision Avoidance Parameters	Key to preventing constellation self-destruction	Maximize the intersatellite distances at plane crossings	Sec. 7.6.2
SECONDARY DESIGN VARIABLES			
Inclination	Determines latitude distribution of coverage	Compare latitude coverage vs. launch costs *	Secs. 7.2, 7.6.1
Between Plane Phasing	Determines coverage uniformity	Select best coverage among discrete phasing options *	Secs. 7.6.1, 7.6.2
Eccentricity	Mission complexity and coverage vs. cost	Normally zero; nonzero may reduce number of satellites needed	Secs. 7.4, 7.6.1
Size of Stationkeeping Box	Coverage overlap needed; cross-track pointing	Minimize consistent with low-cost maintenance approach	Wertz [2001]
End-of-Life Strategy	Elimination of orbital debris	Any mechanism that allows you to clean up after yourself	Sec. 6.5

*Fine tune for collision avoidance

7.6.1 Coverage and Constellation Structure

For most constellations, Earth coverage is the key reason for using multiple satellites.* A constellation can provide observations and communications more frequently than a single satellite can. Given this objective, the normal trade in constellation design is coverage as a measure of performance versus the number of satellites as a measure of cost. Thus, we normally assume that a five-satellite constellation will be less

---

* The principal alternative is the scientific satellite constellation which may, for example, want to sample simultaneously the magnetosphere and solar particle flux at various locations and altitudes.

expensive than a six-satellite one, but this assumption may be wrong. The larger constellation may be at a lower altitude or inclination and, therefore, cost less to launch or have a less harsh radiation environment. Alternatively, we may be able to have a smaller constellation with elliptical orbits, for which increased spacecraft complexity could offset the lower cost due to the number of satellites.

A principal characteristic of any satellite constellation is the number of orbit planes in which the satellites reside. Symmetry in constellation structure requires an equal number of satellites in each orbit plane. This means that an eight-satellite constellation may have either one, two, four, or eight separate orbit planes. But because moving satellites between planes uses much more propellant than moving them within a plane, it is highly advantageous to place more satellites in a smaller number of planes. Moving satellites within an orbit plane requires only a slight change in the satellite altitude. This changes the period so we can slowly rephase the satellite within the constellation, and then return it to the proper altitude to maintain its position relative to the rest. Thus, we can rephase many times using relatively little propellant. If a satellite fails or a new satellite is added to a given orbit plane, we can rephase the remaining satellites so that they are uniformly spaced. The consequence of this is to provide a significant premium to constellations which contain more satellites in a smaller number of orbit planes.

The number of orbit planes relates strongly to a coverage issue often overlooked in constellation design: the need to provide the constellation both *performance plateaus* and graceful degradation. Ideally one would like to achieve some performance level with the very first satellite launched and to raise that level of performance with each succeeding satellite. Generally, however, performance tends to come in plateaus as we put one more satellite into each orbit plane of the final constellation. If a constellation has seven orbit planes, we will achieve some performance with the first satellite, but the next major performance plateau may not come until one satellite is in each of the seven planes. We would expect this constellation to have plateaus at one, seven, fourteen, twenty-one, (and so on) satellites. Again, constellations with a small number of orbit planes have a distinct advantage over many-plane ones. A single-plane constellation produces performance plateaus with each added satellite, whereas one with two planes would have plateaus at one, two, four, six, eight, (and so on) satellites. Thus, more complex constellations will require more satellites for each performance plateau.

Frequent performance plateaus have several advantages. First, because individual satellites are extremely expensive, we may want to build and launch one or two satellites to verify both the concept and the constellation's ultimate usefulness. If a constellation is highly useful with just one or two satellites, it offers a major advantage to the system developer.

Another advantage is that coverage requirements are rarely absolute. More coverage is better than less, but we may not know at the time the constellation is designed how useful added coverage will be. For example, we may design the FireSat system for 30-min revisits, then later revise the response strategy so 45-min revisits can provide nearly equal performance. Communications constellations are normally thought of as having a very rigid requirement of continuous global coverage. Even here, however, they may want more coverage or greater redundancy over regions of high population density.

A constellation of one or two planes can be more responsive to changing user needs than a system with multiple planes can. Because we often design constellations many years before many launch, we may not be able to correctly balance performance vs.

cost. Both needs and budgets follow political constraints and economic priorities over much shorter periods than a constellation's lifetime. Thus if an eight-satellite constellation is highly useful with only six satellites, budget constraints may delay the launch of the remaining two.* At the same time, the constellation may expand to ten satellites if the first set generates substantial demand for more performance or greater capacity. This responsiveness to political and performance demands provides perhaps the largest advantage to constellations with a smaller number of orbit planes.

Finally, a smaller number of orbit planes leads to more graceful degradation. In an eight-satellite, two-plane constellation, if one satellite is lost for any reason, we may rephase the constellation at little propellant cost and thereby maintain a high performance level corresponding to a six-satellite plateau. This rephasing and graceful degradation may be impossible for constellations with a large number of orbit planes.

Another important characteristic is the orbit inclination. In principle, one could design satellite constellations with many different inclinations to get the best coverage. In practice this is extremely difficult because the rate of nodal regression for a satellite orbit is a function of both altitude and inclination. Consequently, satellites at a common altitude but different inclinations will regress at different rates, and a set of orbit planes which initially have a given geometric relationship with respect to each other will change that relationship with time. Otherwise, we would have to expend propellant to maintain the relative constellation spacing, a technique that is extremely expensive in terms of propellant and is achievable for only a short time or under unique circumstances. Thus, we usually design constellations to have all the satellites at the same inclination. A possible exception is to have all satellites at a single inclination except for a set of satellites in a 0 inclination (equatorial) orbit. Regression of the nodes is not meaningful for the equatorial orbit, so we can maintain constant relative phasing indefinitely between satellites in equatorial and inclined orbits. An example of such a constellation is satellites in three mutually perpendicular orbit planes—two polar and one equatorial.

As shown in Fig. 7-10, the spacing between satellites in a single orbit plane determines whether coverage in that plane is continuous and the width of the continuous coverage region. Assume that λ_{max} is the maximum Earth central angle as defined in Sec. 5.3.1 and that there are N satellites equally spaced at $S = 360/N$ deg apart in a given orbit plane. There is intermittent coverage throughout a swath of half-width λ_{max}. If $S > 2 \lambda_{max}$, the coverage is intermittent throughout the entire swath. If $S < 2 \lambda_{max}$, there is a narrower swath, often called a *street of coverage*, centered on the ground trace and of width $2 \lambda_{street}$, in which there is continuous coverage. This width is given by:

$$\cos \lambda_{street} = \cos \lambda_{max} / \cos (S/2) \qquad (7\text{-}15)$$

If the satellites in adjacent planes are going in the same direction, then the "bulge" in one orbit can be used to offset the "dip" in the adjacent orbit as shown in Fig. 7-11. In this case, the maximum perpendicular separation, D_{max}, between the orbit planes required for continuous coverage is

$$D_{maxS} = \lambda_{street} + \lambda_{max} \qquad \text{(moving in the same direction)} \qquad (7\text{-}16)$$

---

* One hopes that procuring agencies will not purposely select a rigid alternative to protect budgets.

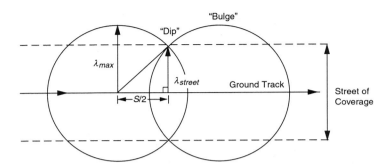

Fig. 7-10.   The "Street of Coverage" is a Swath Centered on the Ground Track for which there is Continuous Coverage.

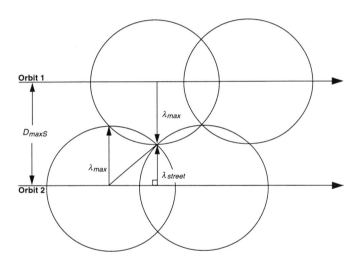

Fig. 7-11.   Coverage in Adjacent Planes. If the planes are moving in the same direction, the overlap pattern can be designed to provide maximum spacing between adjacent planes.

If the satellites are moving in opposite directions, then the bulge and dip cannot be made to line up continuously and, therefore,

$$D_{maxO} = 2\,\lambda_{street} \qquad \text{(moving opposite directions)} \qquad (7\text{-}17)$$

This leads to a polar constellation often called *Streets of Coverage,* illustrated in Fig. 7-12, in which M planes of N satellites are used to provide continuous global coverage. At any given time, satellites over half the world are going northward and satellites over the other half are going southward. Within both regions, the orbit planes are separated by D_{maxS}. Between the two halves there is a seam in which the satellites are going in opposite directions. Here the spacing between the planes must be reduced to D_{maxO} in order to maintain continuous coverage.

   This pattern clearly shows another critical characteristic of constellations—**coverage does not vary continuously and smoothly with altitude.** There are discrete

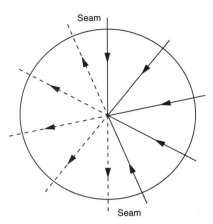

Fig. 7-12. **"Streets of Coverage" Constellation Pattern.** View seen from the north pole. Northward portions of each orbit are drawn as solid lines and southward portions are dashed. To achieve full coverage, orbit planes on either side of the seam must be closer together than the others.

jumps in coverage which depend primarily on λ_{max} which, in turn, depends on the minimum elevation angle, ε_{min}, and the altitude (See Eqs. 5-35 and 5-36). If we keep ε_{min} fixed and lower the constellation altitude, then we will reach an *altitude plateau* at which we will need to add another orbit plane, and N more satellites, to cover the Earth. The Iridium communications constellation was originally intended to have 77 satellites in a streets of coverage pattern. (The element iridium has an atomic number of 77.) By slightly increasing the altitude and decreasing the minimum elevation angle, the number of orbit planes was reduced by one, and the number of satellites required for continuous coverage was reduced to only 66. (Unfortunately, dysprosium is not a compelling constellation name.)

As the altitude changes, the fundamental constellation design changes and, consequently, the number of satellites and coverage characteristics change in steps. As a result, we cannot provide a meaningful chart of, for example, number of satellites vs. altitude without examining different constellation designs at different altitudes. While we may use this sort of chart to estimate constellation size, it would not provide realistic data for orbit design.

Requirements other than coverage can also be important in constellation design, but most are directly related to coverage issues. For example, we may need several satellites to cover a point on the ground or in space at the same time. Navigation with GPS requires that four satellites be in view with a reasonably large angular separation. A similar requirement is cross-link connectivity among satellites in the constellation. Cross-link connectivity is geometrically the same issue as overlapping coverage. At any time when the coverage of two satellites overlaps (that is, they can both see at least one common point on the ground), then the two satellites can see each other and we can establish a cross-link. Thus, forming cross-links is equivalent to the problem of multiple coverage.

Even apparently simple design problems can be very difficult, with solutions depending on various mission conditions. Perhaps the simplest constellation design problem is the question "What is the minimum number of satellites required to provide continuous coverage of the Earth?" In the late 1960s, Easton and Brescia [1969] of the United States Naval Research Laboratory analyzed coverage by satellites in two mutually perpendicular orbit planes and concluded we would need at least *six* satellites to provide complete Earth coverage. In the 1970s, J.G. Walker [1971, 1977, 1984] at the British Royal Aircraft Establishment expanded the types of constellations considered to include additional circular orbits at a common altitude and inclination. He concluded that continuous coverage of the Earth would require *five* satellites. Because of his extensive work, *Walker constellations* are a common set of constellations to evaluate for overall coverage. More recently in the 1980s, John Draim [1985, 1987a, 1987b] found and patented a constellation of *four* satellites in elliptical orbits which would provide continuous Earth coverage. A minimum of four satellites are required at any one instant to provide full coverage of the Earth. Consequently, while the above progression looks promising, the 1990s are unlikely to yield a three-satellite full Earth coverage constellation or the 2000s a two-satellite constellation.

While extensively studying regular, circular orbit patterns, Walker [1984] developed a notation for labeling orbits that is commonly used in the orbit design community and frequently used as a starting point for constellation design. Specifically, the *Walker delta pattern* contains a total of t satellites with s satellites evenly distributed in each of p orbit planes. All of the orbit planes are assumed to be at the same inclination, i, relative to a reference plane—typically the Earth's equator. (For constellation design purposes, this need not be the case. But orbit perturbations depend on the inclination relative to the equator and, therefore, the equator is the most practical standard reference plane.) Unlike the streets of coverage, the ascending nodes of the p orbit planes in a Walker pattern are uniformly distributed around the equator at intervals of 360 deg/p. Within each orbit plane the s satellites are uniformly distributed at intervals of 360 deg/s.

The only remaining issue is to specify the relative phase between the satellites in adjacent orbit planes. To do this we define the *phase difference, $\Delta\phi$,* in a constellation as the angle in the direction of motion from the ascending node to the nearest satellite at a time when a satellite in the next most westerly plane is at its ascending node. In order for all of the orbit planes to have the same relationship to each other, $\Delta\phi$ must be an integral multiple, f, of 360 deg/t, where f can be any integer from 0 to $p - 1$. So long as this condition holds, each orbit will bear the same relationship to the next orbit in the pattern. The pattern is fully specified by giving the inclination and the three parameters, t, p, and f. Usually such a constellation will be written in the shorthand notation of i: $t/p/f$. For example, Fig. 7-13 illustrates a Walker constellation of 15/5/1 at i = 65 deg. Table 7-11 gives the general rules for Walker delta pattern parameters.

While Walker constellations are important to constellation design, they are not the only appropriate options and do not necessarily provide the best characteristics for a given mission. Walker intended to provide continuous multiple coverage of all the Earth's surface with the smallest number of satellites. This plan may or may not meet all the goals of a particular program. For example, equally distributed coverage over the Earth's surface may not be the most beneficial. We may wish to provide global coverage with the best coverage at the poles, mid-latitude regions, or the equator. In these cases, we may want constellation types other than Walker orbits.

If the regions of interest do not include the poles, then an equatorial constellation may provide all of the coverage with a single orbit plane, which leads to flexibility, multiple performance plateaus, and graceful degradation. Thus, for example, if all of

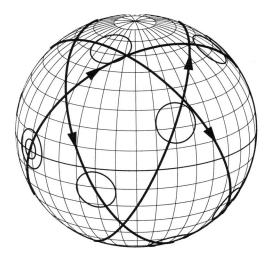

Fig. 7-13.   **A 15/5/1 Walker Constellation at 65 deg Inclination.** Circles are centered on each of the 15 satellites. The double circle is on a satellite at its ascending node.

TABLE 7-11.   **Characteristics of a Walker Delta Pattern Constellation.** See Walker [1984].

T/P/F — Walker Delta Patterns
t = Number of satellites. p = Number of orbit planes evenly spaced in node. f = Relative spacing between satellites in adjacent planes.
Define $s \equiv t/p$ = Number of satellites per plane (evenly spaced).
Define *Pattern Unit, PU* ≡ 360 deg/t.
Planes are spaced at intervals of s *PU*s in node. Satellites are spaced at intervals of p *PU*s within each plane.
If a satellite is at its ascending node, the next most easterly satellite will be f *PU*s past the node. f is an integer which can take on any value from 0 to $(p-1)$.
Example: 15/5/1 constellation shown in Fig. 7-13. 15 satellites in 5 planes (t = 15, p = 5). 3 satellites per plane ($s \equiv t/p$ = 3). PU = 360/t = 360/15 = 24 deg. In-plane spacing between satellites = $PU \times p$ = 24 × 5 = 120 deg. Node spacing = $PU \times s$ = 24 × 3 = 72 deg. Phase difference between adjacent planes= $PU \times f$ = 24 × 1 = 24 deg.

the regions of interest were within 50 deg of the equator, we would want to consider a constellation having several equatorial satellites with enough altitude to provide the appropriate coverage at the smallest spacecraft elevation angle.

If all or most regions of interest are above a given latitude, a directly polar constellation would allow all satellites to see the region of the pole on every orbit. Thus, if all targets of interest were within 50 deg of the pole, a polar constellation with a single orbit plane could provide excellent coverage. If most targets were in the polar region with lesser interest in the equatorial regions, a two-plane polar constellation could provide continuous or nearly continuous coverage of the pole while providing reduced but good coverage of the equatorial regions. One might also consider a mix of polar or high inclination satellites with some satellites at the equator to provide the added coverage needed there.

Another class of non-Walker constellations consists of two planes at right angles to each other. If both planes are perpendicular to the equator it will be a polar constellation. Although it will have substantial symmetry, it is *not* one of the Walker delta patterns. The two planes can also be tipped relative to the equator to achieve any inclination from 90 to 45 deg. Again the ascending nodes are such that they are not Walker constellations except when the inclination is 45 deg, in which case they reduce to a Walker two-plane configuration. Figure 7-14 shows examples of several non-Walker constellations.

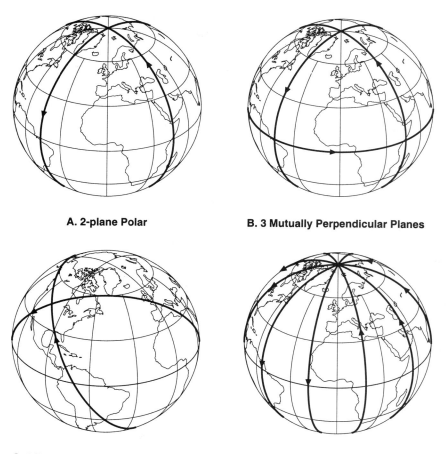

A. 2-plane Polar   **B. 3 Mutually Perpendicular Planes**

C. 2 Perpendicular Non-polar Planes   **D. 5-plane Polar "Streets of Coverage"**

Fig. 7-14.   Examples of Typical Non-Walker Constellations. All orbits are assumed to be circular.

A final example of non-Walker constellations is the Molniya orbits used for Russian communication satellites. Sections 6.2 and 7.4 describe them in more detail. As mentioned above, these constellations can fully cover high northern latitudes while requiring much less energy than circular high-altitude orbits.

7.6.2 Summary of Constellation Design

Constellation design is complex, requiring us to assess many issues and orbit characteristics. We must certainly pick a preliminary design, but this complexity demands that we document the reasons for that design and remain aware of alternatives as orbit design continues. Unfortunately, systematic reassessments of constellation design are difficult under typical budget constraints and the constellation pattern is often locked in very early in mission design.

Unfortunately, we cannot use analytic formulas to design a constellation. With numerical simulation, we can evaluate some of the Figures of Merit defined in Sec. 7.2. (That section discusses how to lay out the simulation for unbiased results.) Generally the results of such a simulation are best expressed as Figures of Merit vs. latitude for the various performance plateaus. Thus, a typical decision plot might include mean response, percent coverage, and maximum gap as a function of latitude for the various constellations being considered. Often, we must also evaluate coverage data for different instruments on board a spacecraft. Each instrument has its own coverage area and, therefore, a different swath width will apply for each principal observation type. Thus, the coverage associated with one instrument may differ dramatically from that associated with another instrument or operating mode. Alternative operating modes or instruments will likely lead us to prefer distinctly different constellation designs. We may then choose either different satellites or a compromise between the alternative instruments or modes.

Table 7-12 summarizes the constellation design process. Wertz [2001] provides a much more extended discussion of constellation design and techniques for evaluating the factors involved. Mora et al. [1997] provide an excellent chronological summary of constellation design methods. As with single-satellite orbits, we normally start by assuming circular orbits at a common altitude. Depending upon the coverage requirements, I recommend beginning with either the Walker delta orbits, the one-plane equatorial, or streets of coverage polar orbits. We should also consider elliptical orbits, either as a full constellation or to fill in missing coverage. Generally, we evaluate each constellation design for three criteria:

- *Baseline Coverage vs. Latitude*

  The coverage associated with different instruments or operating modes is best expressed as coverage vs. latitude (see Secs. 7.2.2 and 7.2.3). I regard the mean response time as the best overall measure of coverage, although percent coverage and maximum gap can also be important in some applications. We must use the maximum gap measure carefully because this single point should not be allowed to drive the design of an entire constellation as it will typically not provide the best overall performance for the cost.

- *Growth and Degradation*

  As described in Sec. 7.6.1, this is a key issue in practical constellation design. It will be different for each constellation type. In evaluating growth and degradation we should assume that rephasing within the orbit plane can be done at very modest propellant cost, and that changing orbit planes is not feasible.

TABLE 7-12. **Constellation Design Summary**. See text for discussion. See also Tables 7-11, 7-14, and 7-15 for additional details.

Step	Where Discussed
1. Establish mission requirements, particularly • Latitude-dependent coverage • Goals for growth and degradation plateaus • Requirements for different modes or sensors • Limits on system cost or number of satellites	Chap. 5 Sec. 7.2
2. Do all single satellite trades except coverage	Sec. 7.4
3. Do trades between swath width (or maximum Earth central angle), coverage, and number of satellites. • Evaluate candidate constellations for: – Coverage Figures of Merit vs. latitude – Coverage excess – Growth and degradation – Altitude plateaus – End-of-life options • Consider the following orbit types – Walker Delta pattern – Streets of coverage polar constellation with seam – Equatorial – Equatorial supplement – Elliptical	Sec. 7.6.1
4. Evaluate ground track plots for potential coverage holes or methods to reduce number of satellites.	Sec. 7.1
5. Adjust inclination and in-plane phasing to maximize the intersatellte distances at plane crossings for collision avoidance.	[Wertz, 2001]
6. Review the rules of constellation design in Table 7-13.	Table 7-13
7. Document reasons for choices and iterate.	

• *Existence of Altitude Plateaus*

   We should evaluate each constellation to see if plateaus exist in which the number of orbit planes or other key characteristics make a discrete step. Plateaus may be different for different instruments and operating modes, but usually are functions of the swath width for each instrument or operating mode.

   There are no absolute rules for choosing the proper constellation. Selection is based on the relative importance of the various factors to the owners and users of the constellation. A summary of the most common rules and the reason for them is given in Table 7-13. As with all aspects of mission design, we must document our selection, our reasons, and the coverage characteristics. It is critical to keep in mind possible alternatives and to reevaluate orbits with advances in mission definition and requirements.

   Finally, one of the most important characteristics of any constellation is collision avoidance. The reason for this is not merely the loss of the satellites which collide because we anticipate losing satellites for many reasons in any large constellation. The fundamental problem is the debris cloud that results from any satellite collision. The velocity imparted to the particles resulting from the collision is small relative to the orbital velocity. Consequently, the net effect of a collision is to take two trackable, possibly controllable satellites and transform them into thousands of untrackable

TABLE 7-13.   **Rules for Constellation Design.** While there are no absolute rules, these broad guidelines are applicable to most constellations.

Rule	Where Discussed
1. To avoid differential node rotation, all satellites should be at the same inclination, except that an equatorial orbit can be added.	Sec. 6.2.2
2. To avoid perigee rotation, all eccentric orbits should be at the critical inclination of 63.4 deg.	Sec. 6.2.2
3. Collision avoidance is critical, even for dead satellites, and may be a driving characteristic for constellation design.	Table 7-14
4. Symmetry is an important, but not critical element of constellation design.	Sec. 7.6.1
5. Altitude is typically the most important of the orbit elements, followed by inclination. Zero eccentricity is the most common, although eccentric orbits can improve some coverage and sampling characteristics.	Secs. 7.4, 7.6.1
6. Minimum working elevation angle (which determines swath width) is as important as the altitude in determing coverage.	Sec. 5.2, Fig. 5-21
7. Two satellites can see each other if and only if they are able to see the same point on the ground.	Sec. 7.6.1
8. Principal coverage Figures of Merit for constellations: • Percentage of time coverage goal is met • Number of satellites required to achieve the needed coverage • Mean and maximum response times (for non-continuous coverage) • Excess coverage percent • Excess coverage vs. latitude	Sec. 7.2
9. Size of stationkeeping box is determined by the mission objectives, the perturbations selected to be overcome, and the method of control.	[Wertz, 2001]
10. For long-term constellations, absolute stationkeeping provides significant advantages and no disadvantages compared to relative stationkeeping.	[Wertz, 2001]
11. Orbit perturbations can be treated in 3 ways: • **Negate** the perturbing force (use only when necessary) • **Control** the perturbing force (best approach if control required) • Leave perturbation **uncompensated** (best for cyclic perturbations)	[Wertz, 2001]
12. Performance plateaus for the number of orbit planes required are a function of the altitude.	Sec. 7.6.1
13. Changing position within the orbit plane is easy; changing orbit planes is hard; implies that a smaller number of orbit planes is better.	Sec.7.6.1
14. Constellation build-up, graceful degradation, filling in for dead satellites, and end-of-life disposal are critical and should be addressed as part of constellation design.	Sec. 7.6.2
15. Taking satellites out of the constellation at end-of-life is critical for long-term success and risk avoidance. This is done by: • Deorbiting satellites in LEO • Raising them above the constellation above LEO (including GEO)	Secs. 6.5, 7.6.2

particles that spread out with time **in the same orbits as the original satellites**. Because the energy is proportional to mv^2, even a small piece of a satellite carries an enormous amount of kinetic energy at orbital velocities.[*] Because the debris cloud remains in the constellation orbit, it dramatically increases the potential for secondary collisions which, in turn, continues to increase the amount of debris and the possibility of making the orbit "uninhabitable." The implication for constellation design is that we should go to great lengths to design the constellation and the spacecraft to avoid collisions, explosions, or generation of extraneous debris. Methods for doing this are summarized in Table 7-14.

TABLE 7-14.   Key Issues in Designing a Constellation for Collision Avoidance.

Approach or Issue	Comment
1. Maximize the spacing between satellites when crossing other orbit planes.	May impact phasing between planes and, therefore, coverage.
2. Remove satellites at end-of-life.	Either deorbit or raise them above the constellation, if still functioning.
3. Determine the motion through the constellation of a satellite that "dies in place."	Constellations at low altitude have an advantage.
4. Remove upper stages from the orbital ring or leave them attached to the satellite.	Do not leave uncontrolled objects in the constellation pattern.
5. Design the approach for rephasing or replacement of satellites with collision avoidance in mind.	All intersatellite motion should assess collision potential.
6. Capture any components which are ejected.	Look for explosive bolts, lens caps, Marmon clamps, and similar discards.
7. Avoid the potential for self-generated explosions.	Vent propellant tanks for spent space-craft.

References

Austin, R.E., M.I. Cruz, and J.R. French. 1982. "System Design Concepts and Requirements for Aeroassisted Orbital Transfer Vehicles." AIAA Paper 82-1379 presented at the AIAA 9th Atmospheric Flight Mechanics Conference.

Ballard, A.H. 1980. "Rosette Constellations of Earth Satellites." *IEEE Transactions on Aerospace and Electronic Systems*. AES-16:656–673.

Betts, J.T. 1977. "Optimal Three-Burn Orbit Transfer." *AIAA Journal*. 15:861–864.

Cefola, P. J. 1987. "The Long-Term Orbital Motion of the Desynchronized Westar II." AAS Paper 87-446 presented at the AAS/AIAA Astrodynamics Specialist Conference. Aug. 10.

Chobotov, V.A. ed. 1996. *Orbital Mechanics (2nd Edition)*. Washington, DC: American Institute of Aeronautics and Astronautics.

---

[*]The relative velocity of two objects in low-Earth orbit will be approximately 14 km/s $\times$ sin $(\theta/2)$, where θ is the angle of intersection between the orbits. 14 km/s times the sine of almost anything is a big number.

Cooley, J.L. 1972. *Orbit Selection Considerations for Earth Observatory Satellites.* Goddard Space Flight Center Preprint No. X-551-72-145.

Cornelisse, J.W., H.F.R. Schöyer, and K.F. Wakker. 1979. *Rocket Propulsion and Spaceflight Dynamics.* London: Pitman Publishing Limited.

Draim, John. 1985. "Three- and Four-Satellite Continuous Coverage Constellations." *Journal of Guidance, Control, and Dynamics.* 6:725–730.

————. 1987a. "A Common-Period Four-Satellite Continuous Global Coverage Constellation." *Journal of Guidance, Control, and Dynamics.* 10:492–499.

————. 1987b. "A Six-Satellite Continuous Global Double Coverage Constellation." AAS Paper 87-497 presented at the AAS/AIAA Astrodynamics Specialist Conference.

Easton, R.L., and R. Brescia. 1969. *Continuously Visible Satellite Constellations.* Naval Research Laboratory Report 6896.

Farquhar, Robert W. and David W. Dunham. 1998. "The Indirect Launch Mode: A New Launch Technique for Interplanetary Missions." IAA Paper No. L98-0901, 3rd International Conference on Low-Cost Planetary Missions, California Institute of Technology, Pasadena, CA. Apr. 27–May 1.

Farquhar, Robert W., D.W. Dunham, and S.-C. Jen. 1997. "CONTOUR Mission Overview and Trajectory Design." *Spaceflight Mechanics 1997, Vol. 95, Advances in the Astronautical Sciences,* pp. 921–934. Presented at the AAS/AIAA Spaceflight Mechanics Meeting, Feb. 12.

Karrenberg, H.K., E. Levin, and R.D. Luders, 1969. "Orbit Synthesis." *The Journal of the Astronautical Sciences.* 17:129–177.

Kaufman, B., C.R. Newman, and F. Chromey. 1966. *Gravity Assist Optimization Techniques Applicable to a Variety of Space Missions.* NASA Goddard Space Flight Center. Report No. X-507-66-373.

Mease, K.D. 1988. "Optimization of Aeroassisted Orbital Transfer: Current Status." *The Journal of the Astronautical Sciences.* 36:7–33.

Meissinger, Hans F. 1970. "Earth Swingby—A Novel Approach to Interplanetary Missions Using Electric Propulsion." AIAA Paper No. 70-117, AIAA 8th Electric Propulsion Conference, Stanford, CA. Aug. 31–Sept. 2.

Meissinger, Hans F., Simon Dawson, and James R. Wertz. 1997. "A Low-Cost Modified Launch Mode for High-C3 Interplanetary Missions." AAS Paper No. 97-711, AAS/AIAA Astrodynamics Specialist Conference, Sun Valley, ID. Aug. 4–7.

Meissinger, Hans F. and S. Dawson. 1998. "Reducing Planetary Mission Cost by a Modified Launch Mode." IAA Paper No. L98-0905, 3rd IAA International Conference on Low-Cost Planetary Missions, California Institute of Technology, Pasadena, CA. 1997.

Mora, Miguel Belló, José Prieto Muñoz, and Genevieve Dutruel-Lecohier. 1997. "Orion—A Constellation Mission Analysis Tool." *International Workshop on Mission Design and Implementation of Satellite Constellations*, International Astronautical Federation, Toulouse, France. Nov. 17–19.

Soop, E.M. 1994. *Handbook of Geostationary Orbits*. Dordrecht, The Netherlands: Kluwer Academic Publishers.

Vallado, David A. 1997. *Fundamentals of Astrodynamics and Applications*. New York: McGraw-Hill.

Walker, J.G. 1971. "Some Circular Orbit Patterns Providing Continuous Whole Earth Coverage." *Journal of the British Interplanetary Society*. 24: 369–384.

—————. 1977. *Continuous Whole-Earth Coverage by Circular-Orbit Satellite Patterns*, Royal Aircraft Establishment Technical Report No. 77044.

—————. 1984. "Satellite Constellations." *Journal of the British Interplanetary Society*. 37:559–572.

Wertz, J.R., T.L. Mullikin, and R.F. Brodsky. 1988. "Reducing the Cost and Risk of Orbit Transfer." *Journal of Spacecraft and Rockets*. 25:75–80.

Wertz, J.R. 2001. *Mission Geometry; Orbit and Constellation Design and Management*. Torrance, CA: Microcosm Press and Dordrecht, The Netherlands: Kluwer Academic Publishers.

Chapter 8

The Space Environment and Survivability

8.1 The Space Environment

Alan C. Tribble, *Intellectual Insights*
D.J. Gorney, J.B. Blake, H.C. Koons, M. Schulz, A.L. Vampola, R.L. Walterscheid, *The Aerospace Corporation*
James R. Wertz, *Microcosm, Inc.*

The near-Earth space and atmospheric environments strongly influence the performance and lifetime of operational space systems by affecting their size, weight, complexity, and cost. Some environmental interactions also limit the technical potential of these systems. They can lead to costly malfunctions or even the loss of components or subsystems [Tribble, 1995; Hastings and Garrett, 1996; DeWitt et al., 1993].

By itself, operating under vacuum-like conditions can pose significant problems for many spacecraft systems. When under vacuum, most organic materials will *outgas* —the generation of spurious molecules which may act as contaminants to other surfaces. Even before reaching orbit, particles from the atmosphere may fall onto optical surfaces and degrade the performance of electro-optical instrumentation. Because there is no practical way to clean spacecraft surfaces once the vehicle reaches orbit, maintaining effective contamination control during design and development is a significant issue for most spacecraft [Tribble et al., 1996].

Once orbit is obtained, the spacecraft is subjected to a very tenuous atmosphere [Tascione, 1994]. At lower orbits a spacecraft will be bombarded by the atmosphere at orbital velocities on the order of ~8 km/s. Interactions between the satellite and the neutral atmosphere can erode satellite surfaces, affect the thermal and electrical properties of the surface, and possibly degrade spacecraft structures.

At shuttle altitudes, ~300 km, about 1% of the atmosphere is ionized. This fraction increases to essentially 100% ionization in the geosynchronous environment. The presence of these charged particles, called the plasma environment, can cause differential charging of satellite components on the surface and interior of the vehicle. If severe, this charging can exceed breakdown electric fields and the resulting electrostatic discharges may be large enough to disrupt electronic components. More energetic space radiation, such as electrons with energies from about 200 keV to 1.5 MeV, can become embedded in dielectric components and produce electrostatic discharges in cable insulation and circuit boards. This *bulk charging* may disrupt a subsystem's signals or the operation of its devices. Even if mild, the charging may alter the electrical potential of the spacecraft relative to space and affect the operation of scientific instrumentation.

Very energetic (MeV–GeV) charged particles can be found in the trapped radiation belts, solar flare protons, and galactic cosmic rays. The total dose effects of this high-energy radiation can degrade microelectronic devices, solar arrays, and sensors. A single energetic particle can also cause *single-event phenomena* within microelectronic devices which can temporarily disrupt or permanently damage components.

Lastly, orbiting spacecraft are periodically subjected to hypervelocity impacts by 1 μm or larger sized pieces of dust and debris. If the impacting particles originate in nature they are termed micrometeoroids. If the particles are man-made they are termed orbital debris. A single collision with a large micrometeoroid or piece of orbital debris can terminate a mission. The probability of this occurring will increase significantly with the introduction of large constellations of satellites.

The subject of space environment effects is, by itself, an area of active research. The more critical of the various effects are discussed below.

8.1.1 The Solar Cycle

This subject is of particular interest because of the fact that the solar activity is seen to vary with an 11-year cycle as shown in Fig. 8-1 [NOAA, 1991]. The plot shows the *F10.7 index*, which is the mean daily flux at 10.7 cm wavelength in units of 10^{-22} $W/m^2 \cdot Hz$. The peaks in the F10.7 index are called solar maxima, while the valleys are called solar minima. Note that the variations are substantial on a day-to-day basis and that one solar maximum may have levels that vary dramatically from other solar maxima. Consequently, predicting the level at any given future time is highly uncertain. On the other hand, the average over an extended period of time is well known. As will be seen, many space environment effects are strongly dependent on the solar cycle.

8.1.2 The Gravitational Field and Microgravity*

Microgravity, also called *weightlessness, free fall,* or *zero-g,* is the nearly complete absence of any of the effects of gravity. In the microgravity environment of a satellite, objects don't fall, particles don't settle out of solution, bubbles don't rise, and convection currents don't occur. Yet in low-Earth orbit, where all of these phenomena occur, the gravitational force is about 90% of its value at the Earth's surface. Indeed, it is the gravitational field that holds the satellite in its orbit.

---

* Contributed by James R. Wertz, *Microcosm, Inc.*

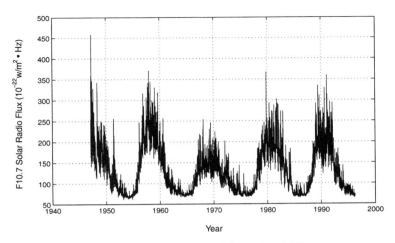

Fig. 8-1. Observed Daily Radio Flux at 10.7 cm Adjusted to 1 AU.

In Earth orbit microgravity comes about because the satellite is in *free fall*—i.e., it is continuously falling through space and all of the parts of the satellite are falling together. In a circular orbit the forward velocity of the spacecraft (tangential to the direction to the Earth) is just enough that the continual falling of the spacecraft toward the Earth keeps the satellite at the same distance from the Earth's center. Moving in a circular orbit requires a continuous acceleration toward the center.

The term microgravity is used in the space environment because, in practice, zero gravity cannot actually be achieved. Two objects traveling very near each other in orbit will not travel in quite the same path due to differences in the gravitational forces or external nongravitational forces acting on them. Two objects held side by side and dropped from a tall building will both accelerate toward the center of the Earth and, as they fall, will converge slightly toward each other. From the point of view of the objects, there is a small component of Earth gravity that pulls them toward each other. An orbiting spacecraft under the influence of atmospheric drag or solar radiation pressure will feel a very small force due to this external pressure. This force can mimic the effect of gravity, causing heavy particles in solution to settle toward the front end of a moving spacecraft. Similarly, a rotating spacecraft produces "artificial gravity" due to centrifugal force. Finally, *tidal forces,* sometimes called *gravity-gradient* forces, come about because of very small differences in the force of gravity over an extended object. For a spherical bubble drifting in orbit, the force of gravity on the lower edge of the bubble will be stronger than at the center of mass and weaker at the far edge of the bubble. This very small difference in forces results in "tides" which will distort the shape of the bubble and elongate it toward and away from the direction to the Earth.

For most practical applications, microgravity effects in low-Earth orbit can be reduced to the level of 10^{-6} g (= 1 μg). A level of 10^{-7} g can be achieved over a very small region near the center of mass of the spacecraft. Table 8-1 provides formulas for the most common forces in the microgravity environment. In this table, z is the direction toward nadir and $\ddot{z}$ is the acceleration in the nadir direction. ω is the angular velocity of the satellite in its orbit. Note that for a nadir-pointing spacecraft, the spacecraft rotates in inertial space at a rate of one rotation per orbit and thus will add to the acceleration environment due to the centrifugal force of this rotation. In low-Earth orbit, the lowest microgravity level can be achieved in an inertially-oriented spacecraft rather than a nadir-oriented spacecraft. The last row in Table 8-1 is the *coriolis force* which is an apparent sidewise force that occurs whenever objects move (e.g., fluid flow in a chemical process) in a rotating reference frame, such as a gravity-gradient stabilized spacecraft.

TABLE 8-1. Equations for Microgravity Level. ω is the orbital angular velocity. ω_{LEO} = $(\mu/a^3)^{1/2}$ = 0.00106 rad/s. x, y, and z are the distances from the spacecraft center of mass.

Source	x direction (velocity)	y direction (orbit normal)	z direction (nadir)
Aerodynamic Drag	$\ddot{x} = 0.5\,(C_D A/m)\rho a^2\omega^2$	$\ddot{y} = 0$	$\ddot{z} = 0$
Gravity Gradient	$\ddot{x} = -x\omega^2$	$\ddot{y} = -y\omega^2$	$\ddot{z} = 2z\omega^2$
Centrifugal (due to spacecraft rotation in inertial frame)	$\ddot{x} = x\omega^2$	$\ddot{y} = 0$	$\ddot{z} = z\omega^2$
Sinusoidal Vibration along X axis of frequency f and amplitude A	$\ddot{x} = A(2\pi f)^2$	—	—
Coriolis Force from material moving in the spacecraft frame	$\ddot{x} = 2\dot{z}\omega$	$\ddot{y} = 0$	$\ddot{z} = -2\dot{x}\omega$

TABLE 8-2. Microgravity Levels. Each entry gives the conditions under which a microgravity level of 1 μg will be achieved, assuming a gravity-gradient stabilized spacecraft at 700 km. c.m. = center of mass.

Source	x direction	y direction	z direction
Aerodynamic Drag	Altitude of 360 km at solar max, 260 km at solar min for $m/C_D A$ = 65 kg/m^2	—	—
Gravity Gradient	7.3 m from c.m.	7.3 m from c.m.	3.6 m from c.m.
Centrifugal	7.3 m from c.m.	—	7.3 m from c.m.
Sinusoidal Vibration	$A = 2 \times 10^{-9}$ m at 10 Hz	—	—
Coriolis Force	$\dot{z}$ = 4.6 mm/s	—	$\dot{x}$ = −4.6 mm/s

Table 8-2 uses the formulas in the previous table to compute the conditions under which 1 μg is achieved. This gives an idea of the scale over which specific microgravity values can be obtained. Microgravity levels could be substantially lowered by conducting microgravity experiments well away from the vicinity of the Earth or other large objects.

Microgravity leads to a wide variety of chemical and manufacturing processes that cannot occur on the surface of the Earth. Heavier particles in a solution do not settle out and bubbles do not rise to the surface. This allows uniform, universal mixing and permits chemical reactions to occur that could not occur on Earth because separation or weight collapse would hinder completion of the reaction or hardening of the material. Surface tension and other inter-molecular forces can take over that would otherwise be dominated by gravitational settling. Similarly, convection does not occur in space and heated materials do not churn or boil. This allows differential heating to take place to provide other chemical reactions. Because settling does not occur, very large crystals can be formed in space which have a variety of industrial applications. Containerless processing and working in the vacuum of space can lead to extremely pure chemicals for use in both pharmaceuticals and manufacturing. Microgravity is a whole new science which is just beginning to evolve.

8.1.3 The Upper Atmosphere[*]

The upper atmosphere affects spacecraft by generating aerodynamic drag lift and heat, and through the chemically corrosive effects of highly reactive elements such as atomic oxygen. The effects of aerodynamic lift and heating are important during launch and reentry. Aerodynamic drag is addressed in Sec. 6.2.3.

Drag depends on the ballistic coefficient (which is a function of atmospheric composition and temperature), velocity relative to the wind, and atmospheric density. We can either estimate the ballistic coefficient based on the configuration of the spacecraft, or bound the problem using typical upper and lower limits as shown in Table 8-3.

TABLE 8-3. Typical Ballistic Coefficients for Low-Earth Orbit Satellites. Values for cross-sectional area and drag coefficients are estimated from the approximate shape, size, and orientation of the satellite and solar arrays. [XA = cross-sectional]

Satellite	Mass (kg)	Shape	Max. XA (m²)	Min. XA (m²)	Max. XA Drag Coef.	Min. XA Drag Coef.	Max. Ballistic Coef. (kg/m²)	Min. Ballistic Coef. (kg/m²)	Type of Mission
Oscar-1	5	box	0.075	0.0584	4	2	42.8	16.7	Comm.
Intercos.-16	550	cylind.	2.7	3.16	2.67	2.1	82.9	76.3	Scientific
Viking	277	octag.	2.25	0.833	4	2.6	128	30.8	Scientific
Explorer-11	37	octag.	0.18	0.07	2.83	2.6	203	72.6	Astronomy
Explorer-17	188.2	sphere	0.621	0.621	2	2	152	152	Scientific
Sp. Teles.	11,000	cylind.*	112	14.3	3.33	4	192	29.5	Astronomy
OSO-7	634	9-sided	1.05	0.5	3.67	2.9	437	165	Solar Physics
OSO-8	1,063	cylind.*	5.99	1.81	3.76	4	147	47.2	Solar Physics
Pegasus-3	10,500	cylind.*	264	14.5	3.3	4	181	12.1	Scientific
Landsat-1	891	cylind.*	10.4	1.81	3.4	4	123	25.2	Rem. Sens.
ERS-1	2,160	box*	45.1	4	4	4	135	12.0	Rem. Sens.
LDEF-1	9,695	12-face	39	14.3	2.67	4	169	93.1	Environment
HEAO-2	3,150	hexag.	13.9	4.52	2.83	4	174	80.1	Astronomy
Vanguard-2	9.39	sphere	0.2	0.2	2	2	23.5	23.5	Scientific
SkyLab	76,136	cylind.*	462	46.4	3.5	4	410	47.1	Scientific
Echo-1	75.3	sphere	731	731	2	2	0.515	0.515	Comm.
Extrema							437	0.515	

*With solar arrays

---

[*] Contributed by R. L. Walterscheid, *The Aerospace Corporation.*

Strong drag occurs in dense atmospheres, and satellites with perigees below ~120 km have such short lifetimes that their orbits have no practical importance. Above ~600 km, on the other hand, drag is so weak that orbits usually last more than the satellites' operational lifetimes. At this altitude, perturbations in orbital period are so slight that we can easily account for them without accurate knowledge of the atmospheric density. At intermediate altitudes, roughly two variable energy sources cause large variations in atmospheric density and generate orbital perturbations. These variations can be predicted with two empirical models: the Mass Spectrometer Incoherent Scatter (MSIS) and the Jacchia models [Hedin, 1986; Jacchia, 1977].

Altitudes between 120 and 600 km are within the Earth's *thermosphere*, the region above 90 km where the absorption of extreme ultraviolet radiation from the Sun results in a very rapid increase in temperature with altitude. At ~200–250 km, this temperature approaches a limiting value, called the *exospheric temperature,* the average values of which range between ~600 and 1,200 K over a typical solar cycle. The thermosphere may also be strongly heated from geomagnetic activity, which transfers energy from the magnetosphere and ionosphere. Heating of the thermosphere increases atmospheric density because the thermosphere's expansion causes increased pressure at fixed altitudes.

Heating due to extreme ultraviolet radiation and its solar cycle variation has the greatest effect on satellite lifetimes. Geomagnetic disturbances are generally too brief to significantly affect lifetimes. Extreme ultraviolet radiation from the Sun is completely absorbed before it reaches the ground and is not measured routinely by satellite-borne instruments; consequently, its effects are unpredictable. Solar activity is monitored using such proxy indices as sunspot number and the F10.7 index which was previously discussed.

Figures 8-1 to 8-4 provide a means of estimating satellite lifetimes based on the information available to the mission designer. Figure 8-2 provides the atmospheric density as a function of altitude corresponding to various values of the F10.7 index. Densities were obtained from the MSIS atmospheric model [Hedin, 1986]. Below about 150 km, the density is not strongly affected by solar activity. However, at satellite altitudes in the range of 500 to 800 km, the density variations between solar maximum and solar minimum are approximately 2 orders of magnitude.

The large variations in density imply that satellites will decay more rapidly during periods of solar maxima and much more slowly during solar minima. This is clearly demonstrated in Fig. 8-3 which shows the altitude as a function of date for a set of 9 hypothetical satellites launched over a 6-year period. We assume that all of the satellites were launched in a perfectly circular orbit at 700 km altitude—3 in 1956 at the beginning of a solar maximum, 3 in 1959 toward the end of the solar maximum, and 3 in 1962 near the time of solar minimum. In each group one satellite had a ballistic coefficient of 20 kg/m^2, one was at 65 kg/m^2, and one at 200 kg/m^2. The histories of the 9 satellites are shown in the graph.

Several characteristics of satellite decay are easily seen in Fig. 8-3. Satellites decay very little during solar minimum, and then rapidly during solar maximum. For one satellite, each solar maximum period will generally produce larger decay than the previous maximum because the satellite is lower. It will, of course, depend on the level of the particular solar maximum. The effect of the solar maxima will also depend on the satellite ballistic coefficient. Those with a low ballistic coefficient will respond quickly to the atmosphere and will tend to decay promptly. Those with high ballistic coefficients will push through a larger number of solar cycles and will decay much

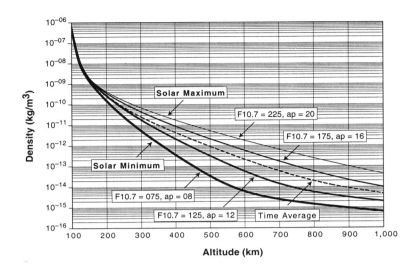

Fig. 8-2.   Density vs. Altitude for Various F10.7 Values. Note that the curves have the same shape as the altitude maintenance curves in Fig. 7-8.

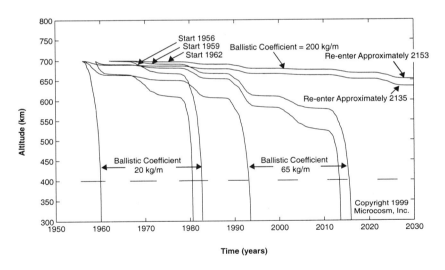

Fig. 8-3.   Altitude as a Function of Date for 9 Hypothetical Satellites Launched over a 6-Year Period. The bars at approximately 400 km altitude mark the periods of solar maxima when the Fl0.7 index was above 150.

more slowly. Note that the time for satellite decay is generally measured better in solar cycles than in years. All 9 satellites reentered during periods of solar maximum. For the range of ballistic coefficients shown, the lifetimes varied from approximately half of a solar cycle (5 years) to 17 solar cycles (190 years). Predicting where the satellite would come down would be remarkably difficult.

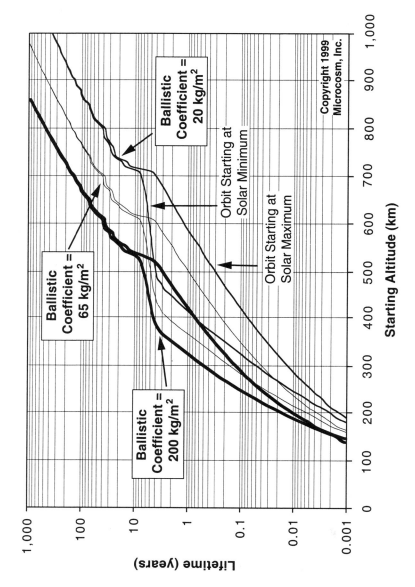

Fig. 8-4. Satellite Lifetimes as a Function of Altitude, Relationship to the Solar Cycle, and Representative Ballistic Coefficients. For each ballistic coefficient, the upper curve represents launch at the start of solar minimum when there will be a low level of decay and the lower curve represents launch at the start of solar maximum when the satellite will decay most rapidly for several years. Data generated with the *SatLife* program [1998].

Figure 8-4 summarizes satellite decay in a convenient manner for mission analysis use. The 3 sets of curves show lifetime as a function of initial circular altitude for satellites with ballistic coefficients that are low (20 kg/m^2), moderate (65 kg/m^2), and high (200 kg/m^2). The spread in the curves near the middle represents the difference between solar maximum (F10.7 = 225) and solar minimum (F10.7 = 75). At the left of the figure, below about 200 km, satellite orbits decay within a few days, the atmospheric density is largely independent of the solar cycle (see Fig. 8-11), and the upper and lower curves for each coefficient begin to merge. From there to lifetimes of about half a solar cycle (approximately 5 years) there will be a very strong difference between satellites launched at the start of solar minimum (upper curve) and those launched at the start of solar maximum (lower curve). Also note that the difference between the solar maximum and solar minimum curves is larger for satellites with a lower ballistic coefficient as we would expect. (Balloons respond to drag more than bowling balls do.) After about half a solar cycle, satellites on the upper curve of each pair will hit solar maximum and the curve will become much flatter. In contrast, those on the lower curve will hit solar minimum and will nearly stop decaying such that the curve becomes nearly vertical. This oscillatory pattern continues with a frequency of the 11-year solar cycle as can be seen in the upper portion of the curves. Finally at high altitudes and long lifetimes the curves come together because the satellite will see a large number of solar cycles and it will make very little difference when the satellite starts. of course the actual lifetime for any particular satellite will depend on both the actual F10.7 index variations and the design and attitude history of the satellite. Nonetheless, Fig. 8-13 provides an estimate of the extrema for use in mission design and can be used to estimate the lifetime for a specific satellite if the ballistic coefficient and launch date relative to the solar cycle are know.

Atomic oxygen—the predominant atmospheric constituent from ~200 km to ~600 km—is another important part of the upper atmosphere's effect on space systems. This form of oxygen can react with thin organic films, advanced composites, and metallized surfaces [Visentine, 1988], resulting in degraded sensor performance. For example, Kapton, a material commonly used for insulation and seals, erodes at a rate of approximately 2.8 µm for every 10^{24} atoms/m^2 of atomic oxygen fluence [Leger et al., 1984]. The fluence, F_0, over a time interval T, is given by

$$F_0 = \rho_N V T \qquad (8-1)$$

where ρ_N is the number density of atomic oxygen (see Fig. 8-5), and V is the satellite velocity. (See tables on inside rear cover.) In addition, chemical reactions involving atomic oxygen may produce radiatively active, excited constituents which, in turn, emit significant amounts of background radiation, create effects such as "Shuttle glow," and interfere with optical sensors.

Atomic oxygen forms when solar ultraviolet radiation dissociates molecular oxygen. Above 110 km, atmospheric constituents diffuse, and each constituent's density varies with altitude according to its *scale height*. The *scale height* of a constituent is the height change over which the density drops to 1/e of its value. In diffusive equilibrium, the scale height is inversely proportional to its molecular weight. Thus, the density of light constituents decreases less rapidly with altitude than the density of heavy constituents, and eventually the light constituents dominate the mixture of gases. Atomic oxygen is lighter than the molecular nitrogen and oxygen; therefore, near ~170 km altitude, atomic oxygen becomes the most abundant constituent.

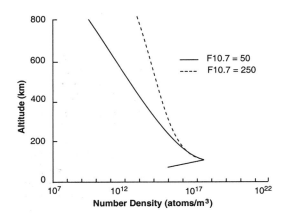

Fig. 8-5. **Altitude Profiles of Number Density of Atomic Oxygen at Solar Minimum (Solid Line) and Solar Maximum (Dashed Line).**

Figure 8-5 shows altitude profiles of atomic oxygen number density over the equator for F10.7 values of 50 and 250. These values represent the extremes for solar minimum and maximum, respectively. The profiles are based on predictions from the MSIS model [Hedin, 1986] for 3 p.m. local time at northern hemisphere equinox and for geomagnetically quiet conditions [Walterscheid, 1989]. Up to 150 km, the solar-cycle variation is small, but it increases steadily with increasing altitude. By 300 km, the number density of atomic oxygen at solar maximum becomes an order of magnitude greater than at solar minimum.

The large solar-cycle variation in atomic oxygen means spacecraft materials can be selected based on phasing the mission to the solar cycle. Since there are large differences between solar cycles, material choices made on the basis of average solar maximum conditions may be inappropriate because of the possibility of more extreme solar maximum conditions.

8.1.4 Plasmas, The Magnetic Field, and Spacecraft Charging[*]

The Earth's magnetic field is roughly dipolar; that is,

$$B(R,\lambda) = (1 + \sin^2 \lambda)^{1/2} B_0 / R^3 \qquad (8-2)$$

where B is the local magnetic field intensity, λ is the magnetic latitude, R is the radial distance measured in Earth radii (R_E), and B_0 is the magnetic field at the equator at the Earth's surface [$B_0 = B(R = R_E, \lambda = 0) = 0.30$ gauss].

As shown in Fig. 8-6, the interaction between the solar wind and the Earth's magnetic field causes magnetic field on the night side of the Earth to stretch into a very elongated structure known as the *magnetotail* (see Tsyganenko [1987] for a more complete model of the Earth's magnetic field). A thin plasma sheet bifurcates the magnetotail, which extends over 1,000 Earth radii parallel to the flow velocity of the solar wind.

---

[*] Contributed by H. C. Koons, *The Aerospace Corporation.*

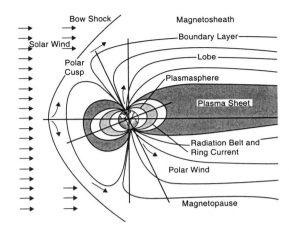

Fig. 8-6.　A Cross Section of the Earth's Magnetosphere. It shows the key plasma and energetic particle populations which respond to variations in solar wind parameters.

Through the interaction between the solar wind and the Earth's magnetic field, some of the solar wind's kinetic energy is converted to magnetic energy stored in the magnetotail. Because this energy cannot build up indefinitely, magnetic *substorms* dissipate it from time to time. These substorms produce an energized plasma (5 to 50 keV) that is injected toward the Earth. This hot plasma can extend into geosynchronous orbits, charging the surface of any spacecraft within it to high negative voltages.

The electrical potential of a spacecraft or component is measured with respect to the plasma in which it is immersed where the net current flow is zero. That is, the following currents must cancel each other: (1) the electron current from the plasma to the vehicle, (2) the ion current from the plasma to the vehicle, (3) the secondary electron current, (4) the backscattered electron current, and (5) the photoelectron current. The voltage at the component's surface also depends on the material's capacitance with respect to the surrounding materials, especially the vehicle ground.

Because materials have varying secondary emission coefficients and photoelectron currents, their equilibrium potentials also differ. An electrostatic arc occurs if the electric field exceeds the breakdown field along the surface of the material, through the material, or between adjacent materials. *Electromagnetic interference, EMI,* from such arcs can cause spacecraft to operate erratically [Robinson, 1989].

Surface charging detrimental to vehicle operation occurs mainly in orbits where electrons with energies of 10 to 20 keV dominate the electron current from the plasma to the vehicle. At low altitudes this charging occurs only at high latitudes where auroral electrons collide with the vehicle which is passing through an otherwise cold, low-density plasma [Gussenhoven et al., 1985]. For other low-altitude locations, low-energy electrons usually develop enough current to keep electric fields below breakdown levels.

In higher orbits, such as geosynchronous, surface charging occurs during magnetospheric substorms between the longitudes corresponding to midnight and dawn [Fennell et al., 1983]. We can approximate the electron-flux distribution during a substorm by summing a cold and a hot Maxwellian distribution. The cold component has a density of 0.2 cm^{-3} and a temperature of 0.4 keV; the hot component has a density of about 2.3 cm^{-3} and a temperature of 25 keV [Mullen et al., 1981]. We must design spacecraft which either keep the differential charging caused by this plasma well below breakdown potentials, or can tolerate the resulting electrostatic discharges.

Design guidelines are available to help reduce differential potentials on vehicle surfaces [Purvis et al., 1984; Vampola et al., 1985]. For example, we can select candidate materials and conductive coatings, apply numerical or analytical models using their quantifiable characteristics, and determine their differential potentials in space. If we cannot prevent discharges by selecting alternative materials, we might consider alternatives such as special filtering, cabling, or grounding. We can employ coupling models for electromagnetic interference simulation, and test the vehicle for electrostatic discharges in its flight configuration.

It is important to note that while differential charging as discussed in the preceding paragraphs is not seen in lower equatorial orbits, the spacecraft potential may be as much as 90% of the solar array voltage more negative than that of the surrounding plasma depending on the configuration of the spacecraft electrical power supply. This may be a concern on scientific missions, where nonbiased measurements of the plasma environment are desired, and may also give rise to arcing or other undesirable effects if high voltage power supplies are used [Tribble, 1995].

8.1.5 Radiation and Associated Degradation

Trapped Radiation[*]

The *Van Allen radiation belts* are a permanent hazard to orbiting spacecraft. They consist of electrons and ions (mostly protons) having energies greater than 30 keV and are distributed nonuniformly within the magnetosphere. As illustrated in Fig. 8-7, the energetic electrons preferentially populate a pair of toroidal regions centered on the magnetic shells $L \sim 1.3$ (inner zone) and $L \sim 5$ (outer zone).

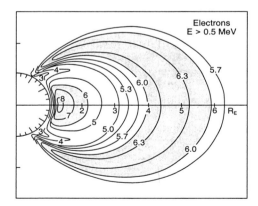

Fig. 8-7. Electron Belts of the Inner and Outer Zones. The numbers on the contours represent the $\log_{10}$ of the integral omnidirectional flux in units of particles cm^{-2} s^{-1}. The horizontal axis is the magnetic equator marked in units of Earth radii. Only electrons with energies above 0.5 MeV are included. [Adapted from Vette et al., 1966, by Schulz and Lanzerotti, 1974].

---

[*] Contributed by M. Schulz and A. L. Vampola, *The Aerospace Corporation.*

A *magnetic L-shell* is the surface generated by rotating a magnetic field line around the dipole axis. It approximately satisfies the equation

$$R = L \cos^2 \lambda \qquad (8\text{-}3)$$

where R is the distance in Earth radii from the idealized point dipole near the Earth's center, and λ is the magnetic latitude. Thus, the L value of a dipolar magnetic shell is its equatorial radius measured in Earth radii. A generalized concept of L, introduced by McIlwain [1961], takes account of the way higher harmonics of the main geomagnetic field perturb the motion of charged particles from their dipolar trajectories. Normally, we use this concept instead of the dipole L value for mapping the trapped radiation environment. Standard models of the Van Allen belts are available from the National Space Science Data Center. The model which provides inner zone ($L = 1.2 - 2.4$) electron data is AE8, which has an average energy range of 40 keV to 5 MeV [Bilitza et al., 1988]. AE8 has two forms, AE8MIN and AE8MAX, which represent the time-averaged environments during solar minimum and maximum.

Outer-zone electron fluxes vary much more over time than inner-zone fluxes. Indeed, during a major magnetic storm, the equatorial intensity at a given energy and L value may grow by several orders of magnitude (factors of 10) in less than a day. Between storms, such flux enhancements usually decay exponentially with an energy-dependent lifetime $\tau \sim 10\,E(\text{MeV})$ days at $L > 3$.

Available proton models are AP8MIN and AP8MAX, which also represent the solar minimum and maximum periods. At solar maximum, the increased atmospheric densities decrease the proton fluxes because the trapped protons collide with the atmosphere at low altitudes. The AP8 models cover the energy range from 100 keV to > 400 MeV. Figure 8-8 shows some of AP8MIN's typical contours for proton fluxes in (R, λ) space. Since the MIN model predicts slightly more flux than the MAX model, it can be used during solar maximum or as a conservative model for long-term missions.

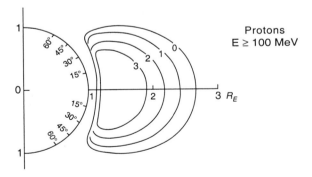

Fig. 8-8.   **High-Energy Protons in the Inner Zone.** Axes and units are the same as on Fig. 8-7. (Note a somewhat different scale.) Data is the omnidirectional flux from the AP8MIN model. Only protons with energies above 100 MeV are considered.

Using the energetic proton models (e.g., AP8) is difficult because they are organized in terms of B (local magnetic field intensity) and L. Secular variations in the Earth's magnetic field drive the normally stable energetic protons toward lower altitudes and the models do not take this into account. Thus, if we project the magnetic

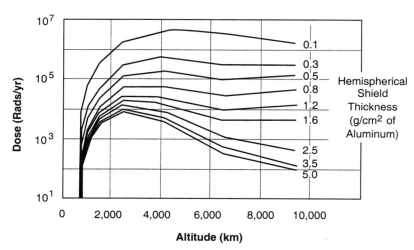

Fig. 8-9.   Radiation Dose Rates as a Function of Altitude for Low-Altitude Polar Orbits. Dose rates are shown for several shielding depths.

field more than 10–15 years into the future, our calculations will be invalid for low-altitude orbits [Konradi and Hardy, 1987]. The present and future configurations of the inner-zone proton belt probably will not differ much, provided both are described in terms of L and λ.

We can determine the appropriate amount of shielding for future spacecraft by computing the dose rate for the desired orbit as a function of shield thickness. To do so, we must apply a radiation transport computation to the time-averaged space environment. Figure 8-9 shows how various shielding depths affect radiation doses in low-altitude polar orbits. Figure 8-10 shows results for geosynchronous satellites. A *rad* is that amount of radiation which deposits 100 ergs (= 6.25×10^7 MeV) per gram of target material (100 mils of aluminum is equivalent to 0.686 g/cm²). The *total radiation dose* consists of three components: proton dose, electron dose, and *bremsstrahlung X-ray dose* produced by the interaction of electrons with the shielding material. In low-Earth orbits, energetic protons in the inner radiation belt contribute most to the total radiation dose. Radiation dose strongly depends on altitude; below 1,000 km the dose increases at approximately the 5th power of the altitude. At synchronous altitude the greater than 5 MeV proton dose is negligible and the bremsstrahlung dose dominates the electron dose for shield thicknesses greater than 1 cm.

Figure 8-10 illustrates that protons trapped near a geosynchronous orbit do not have enough energy to penetrate 10 mils of aluminum. Nevertheless, many trapped protons and heavier ions in this region of space have energies around 10–200 keV. These lower-energy ions can harm space systems differently than penetrating radiation. By depositing their energy in the spacecraft skin, the lower-energy ions can cause a temperature rise sufficient to significantly enhance the infrared background. Heat loads of up to 0.5 W/m² are possible. These same low-energy ions can degrade the effectiveness of paints and protective glass by breaking chemical bonds in their surface layers. We cannot shield against these effects.

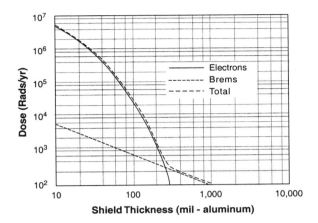

Fig. 8-10. **Radiation Dose Rate in Geosynchronous Equatorial Orbit for Various Shield Thicknesses.**

Solar Particle Events[*]

Solar particle events (SPEs) occur in association with solar flares. SPEs are rapid increases in the flux of energetic particles (~ 1 MeV to above 1 GeV) lasting from several hours to several days. On the average, only a few SPEs occur per year; however, they have important consequences for man-made systems and for man in space. For example, they degrade solar array elements, increase background noise in many types of electro-optical sensors, and cause illnesses in astronauts.

Figure 8-11 shows the typical evolution over time of a solar particle event observed near Earth. The profile depends on the time evolution of the originating solar flare, how long the energetic particles take to diffuse within the solar corona and how the particles propagate within the interplanetary medium. Protons of relativistic energies arrive at Earth within minutes after the flare's occurrence. Lower energy (~10's of MeV) protons are slowed by diffusion within the solar corona and by interactions with the interplanetary medium. After a solar particle event occurs, proton fluxes decay to background noise values over several days. The practical importance of an individual event depends on its maximum intensity, its duration, and the relative abundance of the highest-energy components and heavy nuclei.

The frequency of proton events peaks within a year or two of sunspot maximum and diminishes greatly during the few years surrounding sunspot minimum. Nevertheless, intense events can occur virtually any time within the 11-year sunspot cycle except at sunspot minimum. Table 8-4 shows when the most and fewest sunspots will occur for solar cycles 21–25.

The intensities of typical solar proton events closely follow a log-normal distribution [Jursa, 1985; King, 1974; Feynman et al., 1988]; thus, a few individual events can dominate the total proton fluence observed over a complete solar cycle. Table 8-5 shows the parameters of this distribution. For example, using the values in Table 8-5, a typical solar proton event has a fluence above 10 MeV of $10^{8.27}$ cm^{-2}, whereas an extreme (3σ) event would contribute $10^{(8.27 + 3 \times 0.59)}$ cm^{-2} = 1.1×10^{10} cm^{-2}.

---

[*] Contributed by D. J. Gorney, *The Aerospace Corporation.*

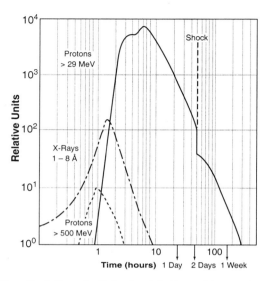

Fig. 8-11. Typical Time Evolution of a Solar Particle Event Observed on Earth.

TABLE 8-4. Years of Sunspot Maxima and Minima for Solar Cycles 21–25.

Solar Cycle	21	22	23	24	25
Sunspot Maximum	1979	1990	2001	2012	2023
Sunspot Minimum	1985	1996	2007	2018	2029

TABLE 8-5. Parameters of the Log-Normal Probability Distributions for Solar Proton Events. [King, 1974].

Energy Range (MeV)	> 10	> 30	> 60	> 100
Log Fluence (cm^{-2})	8.27 ± 0.59	7.28 ± 0.75	6.63 ± 0.95	5.77 ± 1.24
Log Peak Flux (cm^{-2} sec^{-1})	3.27 ± 0.64	2.37 ± 0.82	1.88 ± 0.78	—

Figure 8-12 shows the probability of the proton fluence (energy >10 MeV) exceeding a given value over time intervals of 1–7 years (typical of the durations of many satellite missions). Solar array outputs typically degrade by a few percent following exposure to fluences above ~10^9 cm^{-2} at energies over ~1 MeV, but actual degradation rates depend on cell type, cover glass thickness, and cell age.

Galactic Cosmic Rays[*]

Galactic cosmic rays, or GCR, are particles which reach the vicinity of the Earth from outside the solar system. The number and type of nuclei in these particles are proportional to those in solar system material. Figure 8-13 shows the energy spectrum for several elements. The sum of the curves in the lower energy portion of the figure suggests that cosmic rays undergo solar-cycle modulation.

---

[*] Contributed by J. B. Blake, *The Aerospace Corporation.*

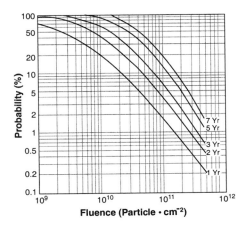

Fig. 8-12. **A Plot of the Probabilities of Exceeding Given Fluence Levels for Satellite Mission Durations of 1, 2, 3, 5, and 7 Years.** The plot represents protons with energies above 10 MeV. These values pertain to satellite missions near the solar activity maxima [Feynman et al., 1988].

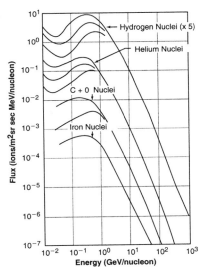

Fig. 8-13. **Energy Spectra for Several Elements of Galactic Cosmic Rays.** The energy for hydrogen has been multiplied by a factor of 5 to distinguish it from the helium curve.

Cosmic rays pose a serious hazard because a single particle can cause a malfunction in common electronic components such as random access memory, microprocessors, and hexfet power transistors. When a single passing particle causes this malfunction, we call radiation effects *single-event phenomena,* or *SEP.*

A galactic cosmic ray loses energy mainly by ionization. This energy loss depends chiefly on the square of the particle's charge, Z, and can be increased if the particle undergoes nuclear interactions within an electronic part. Thus, lower-Z (and more abundant) ions deposit as much energy in a device as less abundant, higher-Z ions. When the galactic cosmic ray leaves electron-hole pairs in a depletion region of an electronic device, the electric field in that region sweeps up the pairs. Figure 8-14 shows this process schematically.

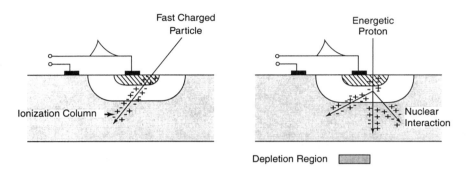

Fig. 8-14. **Schematic Showing How Galactic Cosmic Rays Deposit Energy in an Electronic Device.**

Single-event phenomena include three different effects in electronic components. The first is the so-called *bitflip*, or *single-event upset (SEU)*, which neither damages the part nor interferes with its subsequent operation. The second is *single-event latch-up (SEL)*. In this case, the part hangs up, draws excessive current, and will no longer operate until the power to the device is turned off and then back on. The excessive current drawn in the latched condition can destroy the device if the power supply cannot handle the current. When latchup demands too much current for the power supply, it may drag down the bus voltage, or even damage supply. The third effect is *single-event burnout (SEB)*. This causes the device to fail permanently.

To evaluate the frequency of single-event phenomena for a given part, we must know three things: the external environment; how the incident energy spectrum and particle intensity change as a particle passes through the spacecraft to the sensitive device; and how the electronic device responds to ionizing radiation. We find these phenomena difficult to evaluate because of the complex interactions between the radiation environment and the device's circuit elements. On-orbit failure rates can be predicted primarily for single-event upsets in memory devices, with well defined sensitive volumes, in which the galactic cosmic rays produce electron-hole pairs. A useful equation developed by Petersen [1995] expresses the upset rate R as follows:

$$R = 200 \ \sigma_L / L^2_{0.25} \tag{8-4}$$

where R is the number of upsets or errors per bit day, σ_L is the limiting cross section (sensitive area) of the device in cm^2, and $L_{0.25}$ is the *linear energy transfer* (LET) at 25% of the limiting cross section in units of MeV/mg/cm^2. If experimental cross section data is not available, but device modeling data is, then geometric data can be used in conjunction with the predicted critical charge, and now

$$R = 2 \times 10^{-10} \ abc^2 / Q_c^2 \tag{8-5}$$

where R is the number of errors per bit day, a and b are device surface dimensions in µm, c is the device depth in µm, and Q_c is the critical charge in pC. These two equations have been shown to predict upset rates in the geosynchronous orbit for solar minimum conditions with reasonable accuracy. Scale factors for estimating error rates for other orbits and other calculational methods may be found in Petersen [1995].

Single-event upset rates in complex devices such as microprocessors, or single-event latchups or burnouts in any devices, cannot be reliably predicted. We must resort to predictions based on simulated accelerator test observations and flight performance of similar devices.

Galactic cosmic rays can also generate background noise in various satellite subsystems such as star sensors, infrared detectors, and components employing charge-coupled devices. In addition to increased noise signals, these rays create spurious events which can masquerade as real signals. The spurious signals can affect satellite subsystems depending on the genuine signals' frequency of occurrence, time duration, and repetition, as well as the sophistication of the sensor system. Galactic cosmic rays are a potential source of background noise which must be taken into account when designing a satellite system. It should also be noted that, while this section addresses effects of galactic cosmic rays, similar effects are caused by high energy protons and must be considered for orbits in the range of 1,000–10,000 km altitude.

8.2 Hardness and Survivability Requirements

Paul Nordin, *The Boeing Company*
Malcolm K. Kong, *TRW Systems & Information Technology Group*

Survivability is the ability of a space system to perform its intended function after being exposed to a stressing natural environment or one created by an enemy or hostile agent. *Hardness* is an attribute defining the environmental stress level which a space system can survive. As an example, a satellite or spacecraft which can withstand an X-ray fluence of 1.0 cal/cm^2 or absorption of 10^7 rads (Si) of total dose (a rad of absorbed dose is approximately 100 ergs/g) has a hardness of that amount. (*Fluence* is the time integral of flux. *Flux* is the flow of energy per unit time and per unit cross-sectional area.)

In the aerospace industry we now consider both natural and hostile environments in the definition of hardness and survivability. Well-developed technologies, evolved over the last 35 years, make it possible to design satellites to withstand natural and modest levels of hostile environments. Although technologies for hardening against hostile military threats and for natural survival of satellites overlap, they are distinct and are usually treated separately except in the areas of survivability to total dose due to the Van Allen belts, single-event effects (SEE) caused by galactic cosmic rays and high energy protons, and space/bulk charging due to naturally occurring space plasmas. The latter phenomena must be treated synergistically in the design of satellites.

A military space system or commercial satellite must be survivable if we will need its services in times of high stress, such as a nuclear war. To do this, we must understand what may cause the system to malfunction and then design it to protect against failures. Survivability requirements include identifying the environments and their intensities and, in most cases, designing the space system so it will continue to perform its intended function for a specified time after exposure.

Commercial or scientific satellites usually do not need to be survivable to military threats, but planners must be aware that an unhardened satellite may prematurely stop operating after even very distant nuclear explosions. A slight hardening of satellites can make them much more survivable. (See Sec. 8.2.3.) The Starfish high-altitude nuclear test of July 9, 1962, illustrates the vulnerability of unhardened satellites. That test, a 1.4 megaton device at 400 km altitude above Johnston Island in the Pacific Ocean, caused the failure of several satellites when electrons became trapped in the Earth's geomagnetic field. As a result of those failures, the U.S. Joint Chiefs of Staff established hardening guidelines for all military satellites, including operational and experimental ones. Ritter [1979] discusses these guidelines.

Studies were conducted by the Defense Threat Reduction Agency (formerly the Defense Special Weapons Agency) [Webb, et al., 1995] on the possible effects of a small Third World nuclear burst (for example, 50 kT at 120 km altitude over the East Asian Peninsula) on known commercial (unhardened) satellites. Satellites considered included Hubble Space Telescope, Iridium, ORBCOMM, Globalstar, NOAA, and Nimbus. These satellites showed lifetime reductions of 67% to as large as 99%.

It is important to consider survivability from the outset of mission design. For example, if the satellite can function within a range of orbit altitudes, the highest of these is both the hardest to attack and the most expensive to reach. We should consider the system's survival in each of its life-cycle phases, including concept definition, engineering design and development, and operations in orbit. Note, however, that historically we have not hardened launch systems because of cost and weight, as well as undefined need. The main military threats against space systems are nuclear weapons, including directed energy designs such as X-ray lasers; ground- and space-based laser weapons; high-velocity-pellet (fragmentation) weapons; high-power radio frequency (microwave) weapons; homing kinetic energy weapons; and beam weapons using neutral atomic particles. We may use several approaches to make a system survivable, with hardening of the satellite as a key element. Section 8.2.4 describes these possible approaches and discusses their approximate cost and relative effectiveness.

8.2.1 The Nuclear Weapons Environment and Its Effect on Space Systems

Nuclear weapons pose the most severe threat to spacecraft or space systems. The *yield*, or explosive power, and accuracy of delivery are such that if a nuclear weapon directly attacks a spacecraft, ground station, or any other node of a space system, the node will be destroyed. Nuclear weapon yields can range from a few tons to many megatons of *TNT equivalent* (one kiloton of TNT is defined to be 10^{12} calories). Future nuclear exchanges could use yields of a few hundred kilotons to a few megatons, depending on the purpose of the specific attack and the weapon's delivery accuracy. Accurate delivery of low yields will achieve the desired kill probability, whereas less accurate delivery requires higher yields.

Approximately 80% of the energy from a nuclear weapon detonated in space appears in the form of X-rays. Other important effects include small amounts of gamma rays and neutrons, as well as small fractions in residual radioactivity and kinetic energy of bomb debris. For additional technical detail on nuclear weapons effects, see Glasstone and Dolan [1977].

X-Radiation. The X-radiation occurs because just after detonation, nuclear bomb material is at 10–100 million K. As a first approximation, the hot bomb material will very quickly radiate the energy as though it were a black body, according to the *Stefan-Boltzmann's law*:

$$E = \sigma T^4 \qquad (8\text{-}6)$$

where E is the energy in W/m^2, T is the absolute temperature in K, and σ is the Stefan-Boltzmann's constant (5.67×10^{-8} W $\cdot$ m^{-2} $\cdot$ K^{-4}). At higher black-body temperatures, more X-ray photons are emitted at higher energies. (This is *Wien's law*, which states that $\lambda_{max} T$ = constant, where λ_{max} is the wavelength at maximum intensity and T is the absolute temperature of the blackbody.)

The X-ray fluence, F_x, at a distance R from a nuclear detonation of yield Y is given by

$$F_x = f_x Y / (4 \pi R^2) \qquad (8\text{-}7a)$$

$$= 6.4 Y / R^2 \qquad (8\text{-}8a)$$

where f_x is the fraction of the energy emitted as X-rays (≈ 0.8) and in the numerical form F_x is in cal/cm^2, Y is in kilotons, and R is in km.

Neutron Radiation. One kiloton of equivalent nuclear energy arises from the fission of approximately 1.45×10^{23} nuclei. Each fission produces 2 or 3 neutrons. Approximately half of these neutrons escape during the few tens of nanoseconds of energy generation. Accordingly, the neutron fluence at a distance R cm from a nuclear detonation is given by

$$F_n = 0.5 \times 2.5 \times 1.45 \times 10^{23} Y / (4\pi R^2)$$

$$F_n = 1.4 \times 10^{12} Y / R^2 \quad \text{n/cm}^2 \qquad (8\text{-}9)$$

where Y is in kilotons, R is in km, and F_n is in n/cm^2. This equation is only approximate. The actual neutron output will depend upon the design of the nuclear weapon.

Prompt Radiation. Gamma radiation emitted during the actual nuclear burn time is *prompt radiation*, whereas gamma rays emitted after the nuclear burn time are *delayed radiation*. Prompt gammas result from fission reactions, neutron capture, and inelastic neutron scattering events occurring during intense generation of nuclear energy. The total energy and energy distribution of the prompt gamma rays depend on the nuclear weapon's specific design. To calculate preliminary survivability at range R from a nuclear burst in space, we can express the *dose*, D_γ, (energy deposited per unit mass) in silicon semiconductor material from prompt gamma radiation as

$$D_\gamma = 4 \times 10^5 Y^{2/3} / R^2 \text{ rads (Si)} \qquad (8\text{-}10)$$

where R is in km, Y is in megatons, and D_γ is in rads (Si).

Delayed Radiation. Delayed gammas, neutrons, positrons, and electrons—or residual radiation—occur when radioactive fission products decay. For about the first second after a nuclear explosion, the decay rate from residual nuclear radiation is nearly constant. Thereafter, the dose rate follows the approximate law:

$$R = R_0 (t / t_0)^{-1.2} \qquad (8\text{-}11)$$

where R is the dose rate (usually in rads/hr) at any elapsed time after the reference time, t_0, when the dose rate was R_0. The fission products causing this dose rate contain more than 300 different isotopes of 36 elements from the periodic table, so the inverse 1.2 power of time is an approximation. It is accurate to within 25% for the first 6 months after the nuclear explosion [Glasstone and Dolan, 1977].

The explosion energy rapidly disperses residual radiation. In the absence of an atmosphere and geomagnetic fields, the radioactive fission products would expand geometrically and decrease in intensity by the inverse square of the distance from the burst. However, the geomagnetic field causes the mostly ionized, radioactive weapon debris to spiral along geomagnetic field lines, in a manner similar to the charged-particle motion described in Sec. 8.1. Thus expansion of the radioactive debris will depend on the magnetic field at the nuclear event and on the magnetic field's configuration between the nuclear event and the satellite being considered. As a first approximation, we can assume a geometric expansion. A more conservative approach, however, would be to assume that the nuclear event and the satellite are on the same geomagnetic field lines. The largest possible amount of radioactive debris would then funnel from the nuclear event to the satellite.

An upper bound estimate of the delayed gamma flux due to radioactive debris, in gammas or photons per square centimeter per second, from a single nuclear burst, is given by

$$\dot{\gamma} = 9 \times 10^{15} Y / 4\pi R^2 (1 + t)^{1.2} \qquad \text{photons} / \text{cm}^2 / \text{s} \qquad (8\text{-}12)$$

where Y is yield in megatons (one-third of total yield assumed to be fission), R is distance from the burst point in kilometers, and t is time after burst in seconds [Goldflam, 1990]. This estimate applies to cases where the debris strikes and plates outer surfaces of a satellite, as well as cases where the debris is far away. Similarly, an estimate of delayed beta debris can be made by applying a one-third factor to the equation.

Both delayed gammas and betas will manifest themselves as noise spikes in electro-optical and visible sensor elements used on satellite systems (such as infrared surveillance sensors, optical/visible sensors, Earth sensors, and star trackers). The delayed gammas are a significant threat to satellites, since they can be reduced only by very thick shielding with high Z materials.

Electromagnetic Pulse (EMP). *EMP* is a secondary effect of nuclear weapon detonations. X-rays and gamma rays impinging upon the upper atmosphere create an electron flux which radiates in the RF region of the spectrum. EMP's spectral energy is mostly in the 1 MHz to 100 MHz range. As the RF energy arrives at a satellite, it will induce currents and voltages that may damage or kill the satellite if we do not design to protect it. Nominal electric field strengths, which satellites would experience, can vary from 3 to 100 V/m, depending on satellite altitude and burst location, relative geometry, and other parameters.

System-Generated EMP, SGEMP, is a phenomenon caused when X-rays and gamma rays hit a satellite or other system element, thereby creating an internal flux of electrons whose electromagnetic interactions create large currents and voltages. These large internal currents and voltages can damage sensitive components inside the satellite. Section 8.2.3 discusses how we can mitigate these effects.

Geomagnetically Trapped Radiation. Following a nuclear burst at high altitude, electrons caused by the weapon join the naturally occurring Van Allen radiation belts (Sec. 8.1). The electron flux may increase by many orders of magnitude, thus increasing the absorbed dose in unshielded materials as the satellite repeatedly traverses the Van Allen belts. To protect solid-state (silicon) electronic circuits, we normally enclose them in aluminum, with wall thickness ranging from 0.0254 cm (0.01 inch) to a centimeter or more. Aluminum shielding with a thickness of 0.1 cm and a density of 2.71 g/cm^3 corresponds to 0.27 g/cm^2. Figure 8-9 of Sec. 8.1 gives the natural dose rate in silicon, in rads per year, for polar orbits as a function of orbital altitude and for several values of aluminum shielding. The dose scales linearly with time so the curves can be used for longer or shorter durations. A polar orbit satellite will accumulate less dose than an equatorial one because the trapped radiation is essentially nonexistent at and near the geomagnetic poles of the Earth; this dose difference can be as large as a factor of 5.

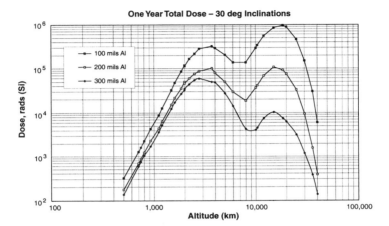

A. Natural total dose in one year, 30 deg inclination circular orbits, for three values of aluminum shielding (0.254 cm, 0.508 cm, and 0.762 cm).

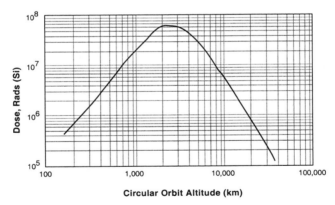

B. Nuclear-enhanced electron dose in 30 days, 30 deg inclination circular orbits, for one value of aluminum shielding (0.254 cm).

Fig. 8-15.   Nuclear-enhanced Electron Environment, 30 Days Duration. Figure assumes 30 deg inclination circular orbit and 0.1 inch aluminum shielding.

Figure 8-15B gives the dose resulting from a 30-day exposure to a nuclear weapon-enhanced electron flux for one value of aluminum shielding and as a function of orbital altitude. The dose for shorter periods can be estimated by linear scaling; however, longer periods cannot be estimated by linear scaling since the saturated environment decays rapidly. As with the natural trapped electrons, the nuclear weapon-enhanced electron flux is practically nonexistent near the north and south poles. For satellites with higher inclination orbits, i.e., greater than 60 to 70 deg, the accumulated dose is greatly reduced, compared to satellites with inclinations of zero to 60 or 70 deg.

An example will show how these calculations work in practice. Consider a satellite which must operate for 1 year in the natural environment and then operate (survive) for 1 day following a high altitude nuclear explosion which creates an electron-enhanced Van Allen belt. We assume a circular orbit, 30 deg inclination and an altitude of 6,000 km. For a wall thickness of 0.254 cm (corresponding to a shielding value of 0.69 g/cm^2), Fig. 8-15A gives a dose of about 130 krads in 1 year. Figure 8-15B gives a dose of 21 Mrads for 30 days and 700 krads for 1 day. Adding the two, we get 830 krads. The electronics must be able to function properly after accumulating a total dose of this amount. Solid-state electronics can be hardened to tolerate from a few krads to about 1 Mrad so our satellite, if hardened to 830 krads or more, would satisfy the survivability requirement of 1 year natural plus 1 day weapon enhanced.

Nuclear Weapon Effects on Materials. The X-radiation pulse lasts tens of nanoseconds, and its energy is absorbed almost instantaneously in solid material through the photoelectric effect and Compton scattering. In the *photoelectric effect*, bound electrons of the material are ejected from their atomic orbits and take on a kinetic energy equal to the difference between the energy of the incident photon and the atom's ionization energy. The incident photon disappears in the photoelectric effect, with its absorption per atom proportional to the 5th power of Z, the atomic number of the absorbing material, and inversely proportional to the 7/3 power of the incident photon's energy [Heitler, 1954]. Therefore, high-Z materials shield against X-rays more effectively, and the absorption cross section decreases dramatically for incident-photon energy from 1 to 20 keV (1 keV = 1.6×10^{-16} J).

Compton scattering is an elastic scattering event in which an electron receives some of the energy of the incident photon, and the incident photon changes direction. As a result, the photon's energy decreases and its wavelength increases [Heitler, 1954]. The cross section per atom for Compton scattering is proportional to Z and, for the range of photon energies we are interested in, is inversely proportional to the incident photon's energy. Therefore, for Compton scattering, increasing Z only slightly increases the absorption coefficient, whereas the cross section decreases moderately as the photon energy increases.

The energetic, free electrons described above can cause electronic circuits to malfunction, and their energy ultimately appears as heat in the material. In fact, the material heats rapidly enough to create shock waves which develop tensile stresses that may cause spall at its unconstrained boundaries. If the deposited energy is high enough (usually not the case at typical satellite fluence levels), the material may vaporize or melt, creating direct damage in addition to the shock waves. For spacecraft, where flux or fluence levels are low, malfunction of electronic circuits is the most likely occurrence.

Gamma rays resulting from nuclear explosions range from a few hundred keV to several MeV. In preliminary designs, we can assume that gamma rays have a mean energy of approximately 1 MeV and interact with matter primarily through Compton scattering. Because gamma radiation is very penetrating, we cannot effectively shield against it. Thus, when we wish to protect against the less penetrating, but more highly ionizing, X-radiation, we need only provide enough shielding to reduce the prompt dose to levels approximately equal to that of gamma radiation. Figure 8-18 of

Sec. 8.2.3 gives the prompt dose induced by a unit fluence of X-radiation as a function of additional shielding (the abscissa of the figure can be converted to linear dimensions by dividing by the density of the shielding material). For higher or lower fluences, linear scaling is appropriate.

Neutrons interact with material by colliding with atomic nuclei. The collisions impart energy to the atoms of the material and displace the atoms from their normal positions in the lattice. Changing the lattice structure can seriously harm solid-state electronic devices because they depend on the characteristics of the lattice for their function. At fluences greater than about 10^{12} n/cm^2, neutrons can cause solid-state devices to stop working, thus "electronically killing" a satellite.

Effects on Communications. A nuclear weapon detonated in space near the Earth interacts strongly with the atmosphere and the Earth's magnetic field. The electromagnetic energy radiated from the detonation creates large-scale ionization in the bomb material and in the atmosphere. Radioactive debris contributes *beta* particles (positrons and electrons) from radioactive decay. The ionized bomb debris and beta particles move along the lines of force of the geomagnetic field, as described in Sec. 8.1. As the magnetic field lines enter the atmosphere, the energetic particles interact with it, creating more ions and electrons. The free electrons thus created absorb and reradiate RF energy and refract the electromagnetic waves of the radio communications links between ground and satellite, creating phase and amplitude changes. These in turn reduce the signal strength in radio receivers, thus interrupting communications.

Based on the theory of electromagnetic propagation, the attenuation, a, in dB per km, is given by

$$a = 4.4 \times 10^4 N_e \nu / ((2 \pi f)^2 + \nu^2) \text{ dB/km} \qquad (8\text{-}13)$$

where N_e is the number of electrons per cm^3, ν is the frequency of collision of electrons with ions, atoms or molecules in Hz, and f is the frequency of the electromagnetic radiation in Hz. The values of these parameters are difficult, if not impossible, to obtain. However, the U.S. Defense Threat Reduction Agency (formerly the Defense Special Weapons Agency) can provide computer programs for propagation analyses in nuclear environments, assuming appropriate clearances and need to know can be established.

For space-to-ground links, we can use the form of Eq. (8-13) and the fact that the density distribution of the atmosphere is approximately exponential to infer the absorptive behavior of RF signals as a function of frequency. The form of Eq. (8-13) indicates that the absorption passes through a maximum as a function of collision frequency, ν, which is proportional to the density of air. Above approximately 80 km, the density is so low that ν is essentially zero and absorption does not occur. Below about 60 km, electrons rapidly reattach to atoms and molecules, so the low electron density again leads to small absorption. Therefore, the attenuation is at a maximum for any given radiation frequency between 60 and 80 km. In this region the attenuation varies with the inverse square of the radiation frequency. Thus, we should choose the highest communication frequency we can to minimize attenuation due to nuclear weapons environments. For a more complete treatment of nuclear effects on communications, see Mohanty [1991].

High data rate requirements for military satellites with surveillance sensor payloads and future commercial communications satellites have resulted in the use of optical/laser links in modern systems. While having many advantages over RF links, such as weight and power, optical link components are also affected by nuclear environments. Table 8-6 contains general guidelines for the effects of nuclear radiation on optical link components.

TABLE 8-6.   Radiation Effects on Optical Link Components.

Device Type	Total Dose	Neutron	Prompt Dose Rate
	Natural Van Allen belts, man-made events	Man-made events primarily	Man-Made Events with short-term irradiation times.
Optical Fibers	$\geq$ 100 krad, polymer clad silica, 20 °C, 0.85 μm: 0.02–0.5 dB/m loss (1–2 orders less loss at 1.5 μm).	> 10^{14} n/cm^2 for 0.02–0.5 dB/m loss.	Losses increase 1–2 orders, depending on dose, dose rate, wavelength and temperature. Nearly complete annealing in $\leq$ 24 hrs.
Transmitters	1–10 Mrad (up to 3.0 dB light loss) for LEDs and laser diodes, peak wavelength shifts, threshold current increases, beam pattern distorts, power loss.	10^{12}–10^{14} n/cm^2 for LEDs (threshold) 10^{13}–10^{15} n/cm^2 for laser diodes (threshold). Light output loss and peak wavelength shifts.	Ionization induced burnout at 10^9–10^{10} rads/s. Pulsed lasers turn-on delays are up to 100 ns. Power loss, wavelength shifts.
Detectors	Decrease in responsivity of 10–30% at 10 Mrad. Dark current increase of 1–2 orders at 10–100 Mrads (for Si PIN photodiodes, worse for APDs, better for AlGaAs/GaAs photodiodes)	Displacement damage thresholds of ~10^{14} n/cm^2 for Si PIN photodiodes and ~10^{12} for APDs. Dark current increases, responsivity decreases.	Dark current increases linearly up to ~10^{10} rads/s. False signal generation by radiation pulse. Upset at $\geq$ 10^7 rads/s. Burnout at $\geq$ 10^9 rads/s. APDs much more sensitive than PIN photodiodes.
Opto-modulators	Depends on device and device technology.	Depends on device and device technology.	Depends on device and device technology. Circuit upset and burnout possible.

NOTES:

Optical Fibers
- Damage worse and annealing slower for lower temperatures. Losses generally lower for increasing wavelength (to 1.5 μm)
- Polymer clad silica cores have lowest losses but losses increase below ~ 20 °C. Max dose usage of ~10^7–10^8 rads

Transmitters
- At higher temperatures, threshold current and peak wavelength increase while output power decreases for laser diodes
- LEDs have better temperature/temporal stability, longer lifetimes, greater reliability and lower cost

Detectors
- APDs are predicted to be more sensitive than PIN diodes to total dose, neutrons and dose rate
- AlGaAs/GaAs photodiodes shown to be more radiation resistant than hard PIN photodiodes

8.2.2 Other Hostile Environments

Laser Weapons. High-power lasers are being developed as potential ground-based or space-based antisatellite weapons. The flux in power per cross-sectional area from these weapons is given by

$$F = \frac{PD^2}{\pi R^2 [(1.22 \lambda Q)^2 + (JD)^2]}$$

(8-14)

where P is the average output power, D is the laser objective diameter, Q is the quality of the laser beam (dimensionless), λ is the wavelength of the laser, J is the angular jitter of the beam (in rad), and R is the range from the laser to the target. $Q = 1$ indicates a diffraction-limited weapon; laser weapons being developed will have a beam quality of 1.5 to 3.0. Both pulsed and continuous-wave lasers are in development. Equation (8-14) is for a continuous-wave laser, but is approximately correct for the average flux from a pulsed laser. The peak flux for a pulsed laser will be much higher.

For engagement ranges of several hundred km, the laser spot sizes will be several meters in diameter and will, in general, completely engulf the target satellite in laser radiation. To damage or kill a satellite at any range, the laser beam must hold steady long enough to achieve a damaging or killing level. Depending on the incident flux level and sensitivity, this dwell time could be several seconds or minutes.

Fragmentation or Pellet Weapons. The former Soviet Union operated an antisatellite weapon using fragmentation pellets that could attack satellites in low-Earth orbit [U.S. Congress, OTA, 1985]. This weapon, launched from ground locations, achieved an orbit with nearly the same elements as those of the target satellite. Hence we call it a *co-orbital antisatellite system*. Radar or optical guidance brings the weapon close to the target satellite. A high explosive then creates many small fragments which move rapidly toward the target satellite and damage or kill it by impact.

High-Power Microwave Weapons. These weapons generate a beam of RF energy intense enough to damage or interfere with a satellite's electronic systems. Their frequencies of operations range from 1 to 90 GHz, thereby covering the commonly used frequencies for command, communication, telemetry, and control of most modern satellites. A satellite's antenna tuned to receive a frequency the weapons radiate will amplify the received radiation. Thus, it could damage RF amplifiers, downconverters, or other devices in the front end of a receiver.

Neutral-Particle-Beam Weapons. Particle accelerators have been used for high-energy nuclear physics research since the early 1930s, so the technology is well developed. Weapons using this technology must be based in space because the particles cannot penetrate the atmosphere. The particles would be accelerated as negative hydrogen or deuterium ions, then neutralized by stripping an electron as they emerge from the accelerator. (The particles must be electrically neutral to avoid being deflected by the Earth's magnetic field).

8.2.3 Spacecraft Hardening

Hardening of a space system's elements is the single most effective action we can take to make it more survivable. Presently, we use hardening to prevent electronics upset or damage from nuclear-weapon effects. In the 2000s, we will see laser hardening in military satellites which must survive hostile attacks. If projected antisatellite weapons are developed and deployed, hardening will help reduce the effects of High-Power Microwave and Neutral-Particle-Beam weapons on satellites.

Figure 8-16 gives approximate upper and lower bounds on the weight required to harden a satellite to nuclear weapons effects. The technology for hardening satellites against nuclear weapons is well developed up to a few tenths of cal/cm². Above these levels, the hardening weight increases sharply, as Fig. 8-16 illustrates. Figure 8-17 gives rough upper and lower bounds on hardening costs. Comparable cost data may be found in Webb and Kweder [1998].

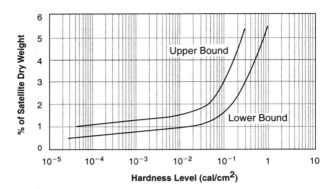

Fig. 8-16.   Weight Required to Harden a Satellite as Percent of Satellite Weight.

For X-rays with photon energy below 3 keV, shielding is very effective with almost any convenient material, such as aluminum. At higher photon energies, materials with higher atomic numbers, Z, are more effective than the low-Z materials. A commonly used shielding material is tantalum because of its availability and ease of manufacture. A satellite's external surfaces are particularly vulnerable to crazing, cracking, delamination, or micro-melting. Therefore, we must carefully select materials for these surfaces to protect functions such as thermal control, optical transmission, or reflection. In this category are covers for solar cells, optical coatings on lenses and thermal control mirrors, thermal control paints, metal platings, and optical elements made of quartz or glass. The data for typical satellite materials exposed in underground nuclear tests is, in general, classified.

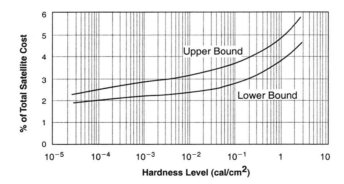

Fig. 8-17.   Cost to Harden a Satellite as a Percent of Total Satellite Cost. Costs include production cost plus proportional share of engineering costs.

Prompt dose results from penetrating X-radiation and, to a lesser extent, from prompt gamma. Typically, the X-ray prompt dose is 3 to 4 orders of magnitude larger than the dose from prompt gamma. At typical spacecraft levels, prompt dose can break the bonds of the leads on susceptible integrated circuits and can cause electronic circuits to experience burnout, latchup, and temporary upset.

In most cases, shielding of the X-radiation can reduce the internal dose to manageable levels. Figure 8-18 gives the prompt dose in silicon shielded by the basic 0.040-inch aluminum enclosure plus a range of additional tantalum shielding in g/cm^2. We can use this figure to estimate the extra tantalum weight required to shield against the prompt dose. The procedure is as follows:

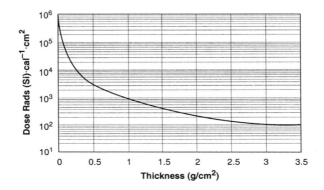

Fig. 8-18.   Prompt Dose as a Function of Additional Tantalum Shielding (Worst Case 1–15 keV Spectrum).

- Using Fig. 8-18, scale the maximum allowable prompt dose by the fluence appropriate to the system under consideration and determine the surface mass density (g/cm^2) required;
- Multiply the surface mass density required by the total area to be shielded on the satellite.

The total dose is the sum of the ionizing dose from all sources of radiation and is usually expressed in rads (Si). In almost all cases, the total dose is dominated by trapped electrons in the geomagnetic field. Figure 8-19 gives the dose in silicon as a function of thickness of shielding material in g/cm^2, normalized to an incident 1 MeV electron fluence of 10^{14} electrons/cm^2. The asymptotic nature of the dose curve for large mass densities results from the bremsstrahlung electrons produce as they stop in the shielding material. Thus, we would shield interactively for total dose and prompt dose. The prompt dose shielding also attenuates the radiation from the Van Allen belts, and the extra aluminum needed to attenuate the Van Allen belt radiation also attenuates the prompt X-radiation. For example, as Fig. 8-18 shows, an aluminum box 0.102 cm thick can reduce an external prompt dose of 3×10^8 rads (Si)$\cdot$ cal$^{-1} \cdot$ cm^2 to an internal dose of 4×10^5 rads (Si)$\cdot$ cal$^{-1} \cdot$ cm^2. We can reduce the prompt dose even further by adding more high-Z material, such as tantalum or tungsten as shown in Fig. 8-18. This high-Z material also reduces the dose caused by trapped electrons, as mentioned above.

Metals are relatively unaffected by total dose. However, total dose degrades certain properties of organic materials, beginning between 0.1 and 1 Mrad, and makes them unusable above 10 to 30 Mrad. For example, organic materials may soften, become brittle, or lose tensile strength. NASA [1980] and Bolt and Carroll [1963] give data on how the total dose affects organic materials. Figure 8-20 shows the "sure-safe" total dose capabilities for commonly used satellite materials.

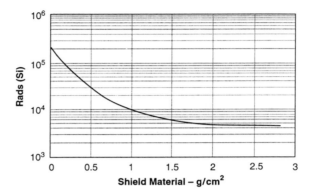

Fig. 8-19.   Dose from Trapped Electron Fluence of 10^{14} Electrons/cm^2 as a Function of Thickness of Shielding Material in g/cm^2.

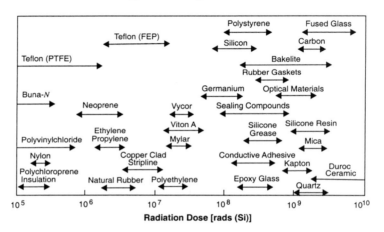

Fig. 8-20.   Total Dose Capabilities of Satellite Materials.

Silicon electronic devices suffer decreases in operating parameters such as gain, gate voltage, or lifetime of the minority carrier. We can measure these operating parameter changes as a function of dose and develop curves of radiation deratings. Thus, during circuit design, these radiation deratings are used to ensure that the devices will continue to operate satisfactorily at the design exposure level.

Total dose also includes the ionization from prompt and delayed weapon radiation, as well as neutron-generated radiation. To the total dose from man-made sources of hostile radiation, we must add radiation from the natural environments. The total dose depends on the amount of shielding, orbital parameters, and satellite life. In the absence of nuclear-weapon detonations, the total dose will normally increase linearly with time on orbit. To harden a satellite against these effects, we would use silicon-based electronic devices which tolerate the effects and shield them to the appropriate level, depending on how long we want the satellite to last. In the future, we will increase the use of electronic devices based on gallium arsenide, because gallium arsenide appears to be unusually immune to total dose effects.

Radiation hardened parts are required for all designs that must operate in nuclear weapon environments, but some commercial communications satellites can consider using radiation tolerant parts (<50 krads capability) or even commercial off the shelf (COTS) Class B type parts (10–15 krads capability), particularly if they will only operate in low-Earth orbits (less than about 1,000 km) and have orbital design lifetimes of 2 to 3 years maximum. Table 8-7 shows a comparison of typical unhardened COTS parts and hardened parts capabilities.

TABLE 8-7.   COTS and Rad Hard Parts Comparison. Rad hardening increases radiation protection significantly, thus increasing spacecraft survivability.

Characteristics	COTS	Rad Hard
Total Dose	10^3–10^4 rads	10^5–10^6 rads
Dose-Rate Upset	10^6–10^8 rads (Si)/s	>10^9 rads (Si)/s
Dose-Rate-Induced Latchup	10^7–10^9 rads (Si)/s	>10^{12} rads (Si)/s
Neutrons	10^{11}–10^{13} n/cm^2	10^{14}–10^{15} n/cm^2
Single-Event Upset (SEU)	10^{-3}–10^{-7} errors/bit-day	10^{-8}–10^{-10} errors/bit-day
Single-Event Latchup/Single-Event Burnout (SEL/SEB)	< 20 MeV-cm^2/ mg (LET)	37-80 MeV-cm^2/ mg (LET)

• COTS characteristics may vary unpredictably from lot to lot and even within a lot.
• Higher margins and more testing (screening) are required with COTS usage, which will offset lower piece part costs.
• LET is Linear Energy Transfer threshold.

Whether designing satellite electronics with RAD Hard or COTS parts, a Radiation Hardness Assurance Control Plan (RHACP) is necessary to specify radiation design requirements, parts derating methods, required design margins, parts testing requirements and the process for controlling all activities related to radiation hardness. Implementation of the RHACP will help ensure the success of the hardness design and hardness verification process. The hardware is normally hardness qualified at an appropriate level, either piece part, unit, subsystem or system, whichever is economically and technically correct.

Displacement fluence is any electromagnetic or particulate radiation which displaces atoms from their normal lattice positions. For nuclear weapons, neutron fluence is the primary cause of displacement. In the natural environment, electrons and protons are the principal contributors. The displaced atoms and their vacancies will react with the bulk material and form stable defects in the lattice structure. These defects significantly change the equilibrium-carrier concentration and minority-carrier lifetime. In silicon solar cells, these changes degrade power output. In other solid-state electronic devices, they reduce gain and increase forward voltage drop and reverse leakage currents.

We cannot harden to the neutron displacement fluence from a nuclear burst by shielding because the uncharged neutron is very penetrating and large amounts of shielding would be needed. In general, we harden to the neutron fluence by selecting devices that resist degradation by neutrons.

To protect against displacement by electrons or protons, we must shield the solid-state devices. Solar cells are shielded by a layer of fused silica, varying in thickness with the amount of shielding required. At very high ionization dose, the cover glass material darkens, reducing the solar array's power output. For solid-state devices contained inside aluminum boxes, we choose the thickness of the aluminum to stop the electrons and ignore the protons, which penetrate much less in most commonly used orbits.

Delayed beta radiation flux can also be shielded effectively in the same manner as total dose, since it is composed of electrons. In contrast, delayed gamma flux cannot be shielded easily due to its high energy content (up to about 12 MeV). As an example, a factor of 10 reduction requires about 1.1 inches (2.8 cm) of high Z material like tungsten or tantalum. Mitigation of gamma debris noise spikes in sensor systems will require heavy shielding and/or pulse suppression signal processing (such as time delay integration), or even complementary satellite tasking. Even then, the gamma noise can still be high enough to cause sensor outages lasting from seconds to minutes, depending on specific sensor performance characteristics and design. For example, a fairly robust sensor with an operational capability (noise threshold plus signal to noise ratio) of 10^9 photons/cm^2/s will be "blind" for about 34 sec, given a 1 megaton burst at 100 km away from the satellite, using Eq. (8-12).

EMP is typically in the MHz range. At satellite altitudes, EMP intensities of a few V/m can easily cause damage and upset in unhardened satellites. To prevent this, Faraday shields can keep the radiation from entering the satellite cavities. We can also use good external grounding, interconnect all conducting parts and surfaces, employ surge arrestors, and eliminate sensitive components. In addition, designing for electromagnetic compatibility, such as shielding of cables and harnesses, will reduce or eliminate much of the potential for EMP damage. Computers are particularly sensitive to EMP, as are the following components (in order of decreasing sensitivity): semiconductor diodes in microwave applications, field-effect transistors, RF transistors, silicon-controlled rectifiers, audio transistors and semiconductor diodes in power rectifier applications.

SGEMP occurs when the incident flux of photons, both X-ray and gamma ray, creates a flux of electrons inside the satellite. Some of the energetic electrons are not stopped in solid material but emerge into satellite cavities, causing currents and fields within these cavities. At representative satellite fluence levels, these electrons can generate cable injection currents of 10–100 amperes/meter of cable length and peak cavity electric fields of several hundred kilovolts/meter. The fields then couple electromagnetic energy into cables and other conductive elements in the cavity, and the sharp pulse of energy transmitted to sensitive components can make them fail.

SGEMP hardening uses the same methods as EMP hardening except for external shielding, because SGEMP generates inside the satellite. We can also treat internal surfaces with low-Z (atomic number) paints to reduce electron emission into cavities. Using specially designed low-response cables will also reduce SGEMP effects. Finally, we can protect input/output circuits and terminals with various devices—zener diodes, low-pass filters, and bandpass filters—to limit current or to clamp voltage.

The natural space phenomena causing single event upsets (SEUs) and other single-event effects (SEEs), as well as methods for predicting upset rates, were addressed in Sec. 8.1.5. Because shielding is ineffective in reducing SEEs, satellite systems must be designed to mitigate these effects, given that they will occur. Table 8-8 lists some classical approaches used in modern space system design. The extent to which these approaches are applied depends on the mission criticality, system upset specifications (allowable rates and outage times), and orbital environment expected. However, as indicated in the table, selection of acceptable parts is perhaps the single most important of all approaches for SEE mitigation, albeit not sufficient by itself. Not indicated in Table 8-8 is the effect of orbital altitude. While geosynchronous altitude is the worst case for SEEs (due to galactic cosmic rays), orbits that traverse the proton belt (elliptical orbits and those between about 1,200 km and 8,000 km altitude) will have SEEs from high energy protons, in addition to galactic cosmic rays, and the proton SEUs can be 10 times worse.

TABLE 8-8. Single-Event Effects. The effects caused by single events can be reduced by better parts, improved shielding, and process redundancy.

Approach	Comments
1. Parts Selection: • Error rate < 10^{-8} errors/bit-day • Latchup immunity to LET ≥37 MeV-cm²/mg • MOSFET SEGR/SEB immunity to LET ≥ 37 • Derate power MOSFET to 30–40% of V_{DS}	• < 10^{-10} desired • LET ≥ 60–120 desired • LET ≥ 60–80 desired • V_{DS} is rated drain to source voltage
2. Use parity and SECDED	Single error correction, double error detection
3. Use dual or redundant logic for critical functions	2 correct outputs for decision making
4. Use watchdog timers and triple modular redundancy (TMR) in spacecraft control processor.	2 out of 3 voting logic used; switching to spare processor after repeated timeouts
5. Periodic refreshing of critical memories	Periodic switchover to refreshed memory bank
6. Use of hard latches	Eliminate soft error responses
7. Design digital circuits immune to analog circuit spikes	Long response time compared to spike transient
8. Eliminate nonrecoverable system modes and failures that could result from a soft error (bit flip)	Good design practice always required to ensure no damage and recoverable modes

Note: LET is linear energy transfer threshold; SEGR/SEB is single-event gate rupture/single-event burnout.

Surface charging and resultant electrostatic discharge (ESD) due to space plasmas were addressed briefly in Sec 8.1.4, including basic design guidelines for satellite survivability. Satellites that are highly exposed to electrons (those at high altitudes, geosynchronous and highly elliptical orbits) must also be designed to survive bulk charging, in which electrons embedded in bulk dielectrics (cable dielectrics and circuit boards) and isolated conductors (such as ungrounded circuit board metallizations and spot shields on parts) build up potentials sufficient to cause discharges. Such discharges can result in anomalous upset and/or damage to electronics, much like SEUs, discussed in the preceding paragraph.

Much of the work on bulk charging is summarized by Vampola [1996], based on CRRES flight data. Mitigation approaches are indicated in Table 8-9. Designers can eliminate most bulk charging concerns simply by providing sufficient shielding to reduce both maximum current on circuit boards to less than 0.1×10^{-12} amps/cm² and maximum total integrated fluence to less than 3×10^9 electrons/cm² on ungrounded localized spot shields [Frederickson et al., 1992]. For geosynchronous satellites, this shielding is about 0.305 cm of total equivalent aluminum (which is typically provided for total dose protection).

TABLE 8-9. Bulk Charging Mitigation Approaches. Careful planning can produce adequate solutions without large investments of time and money.

Approach	Implementation
Prevention	Use leaky dielectrics and bleed-off paths with < 10^9 ohms resistance to ground (at least 2 ground paths for contiguous areas >64.5 cm^2)
	Double shielded wire harness and cables
	Adequate shielding (~0.305 cm aluminum) of circuit boards and part shields (vs. grounding of all metallizations and local part shields).
Signal Response Conditioning	Design circuits to be unresponsive to the relatively short, low level spurious ESD pulses which are typically less than 100 ns.
Circuit Hardness	Circuits should be designed for no damage by ESD pulses with energy levels up to 10 microjoules.

8.2.4 Strategies for Achieving Survivability

As described in Sec. 8.2.3 and summarized in Table 8-10, hardening is the single most effective survivability option. Table 8-11 presents other strategies for enhancing survivability. We use *redundant nodes*, also called *proliferation* or *multiple satellites*, to overlap satellite coverages. Thus, if one satellite fails, others will perform at least a part of the total mission. An attacker must use multiple attacks to defeat the space system—a costly and therefore more difficult approach for the enemy. The development of the so-called *lightsat* technology—light, inexpensive satellites performing limited functions—will support this strategy. To be effective, each node (ground station or satellite) must be separated from another node by a large enough distance to prevent a single attack from killing more than one node.

TABLE 8-10. Space Survivability Hardening Design Summary. Though the space environment is harsh, survivability can be designed into spacecraft subsystems.

Threat Type	Requirement Driver	Mitigation Design Approach
Natural Space Radiation *Enhanced Radiation from Nuclear Bursts*	Withstand total dose degradation. Minimize single-event upsets (SEU)	Radiation resistant materials, optics, detectors and electronics. Shielding at unit & part levels. Self-correcting features for SEU tolerance.
Collateral Nuclear Burst	Withstand prompt X-ray, neutron, EMP damage, minimize dose rate upsets. Tolerate induced noise due to debris.	Radiation resistant materials, optics, detectors & electronics. High Z shielding, current limiting/terminal protection. Event detection, circumvention, recovery. Sensor noise suppression. Multiple satellite coverages.
Redout	Sensor tolerance to background levels.	Processing algorithms. Multiple satellites for detection.
Ground Based Laser	Sensor tolerance to interference or damage.	2 color sensor detection, filtering and processing.
High Power Microwave and EMP	Sensor and communications tolerance to interference/damage.	Protection of detectors and circuits, processing for noise discrimination.
RF Jamming/Blackout	Communications tolerance to interference/scintillation.	Multiple links, processing, modulation and frequency choices.

Prior to the end of the Cold War, fixed ground control stations were high priority targets of ICBM-launched nuclear weapons. Therefore, satellites needed to be autonomous or capable of being controlled by multiple mobile ground control stations, or utilize a combination of the two survivability features. In the Post-Cold War era, these survivability features are less important. Nevertheless, the following principles of survivability are still relevant. Mobile ground stations are survivable because ICBMs cannot find targets whose Earth coordinates are unknown and continually changing. By deploying mobile ground stations so they are separate from one another, we allow a single nuclear weapon to kill, at most, one ground station.

TABLE 8-11.   Satellite System Survivability Options. Many options exist, each adding cost and design complexity.

Option	Cost*	Effectiveness	Features
Satellite Hardening	2–5%	Very good	Trapped electron shielding, prompt radiation shielding, latchup screening, radiation-tolerant electronics, degraded electronic parts deratings
Redundant Nodes	Cost of extra node	Good	Essential functions performed by 2 or more nodes (e.g., satellites with overlapping coverage but separated by greater than 1 lethal diameter range)
Onboard Decoys	1–10%	Good, depending upon type of threat guidance	Credible decoys simulating both radar and optical signatures of the satellite; decoys are launched when an attack is detected (detection system required)
Maneuver Capability	10–20%	Good, depending upon type of threat guidance	Thrust levels depend on satellite altitude (warning time), nature of threat, threat detection efficiency; additional satellite weight for high acceleration
Self Defense	20–40%	Very good	Kinetic energy kill homing missiles represent most likely first system
Escort Defense	Cost of 1 sat.	Very good	Kinetic kill homing missiles represent most likely first system; directed energy (e.g., high-energy laser or high-power microwave system) is future possibility
Autonomous Operations	3–8%	Provides protection against loss of ground station	Autonomous orbit control (e.g., station-keeping for geosynchronous orbits), momentum control, redundant unit control (fault detection) and substitution
Mobile Ground Control Stations	2 to 3 times cost of large grnd. stat.	Very good; provides survivable ground control station network	Multiple mobile ground control stations; while one is controlling, one is tearing down, one is setting up, and one is changing its location; survivability is achieved by physical location uncertainty.
Surv. Mobile Grnd. Term.	20–30% of fixed terminal	Very good; provides low-cost ground-control option†	Hardened against high-altitude EMP, nuclear biological chemical warfare, jamming, small arms fire. Survivability enhanced by physical location uncertainty.
Onboard Attack Reporting System	1–5%	Essential for total system survivability	System records/reports time, intensity, or direction of all potentially hostile events (e.g., RF, laser, nuclear, pellet impacts, and spoofing or takeover attempts); allows appropriate military response to hostilities

*Percent of total satellite cost.
†Survivable with min. essential com. connectivity.

Onboard systems for attack reporting tell ground-control stations that a satellite is being attacked and what the attack parameters are. Without such information, ground operators may assume a spacecraft fault or natural accident has occurred, rather than an attack. Thus, controllers could act incorrectly or fail to act when necessary. More importantly, national command authorities need timely information telling of any attack on our space assets.

Decoys are an inexpensive way to blunt an antisatellite attack. They simulate the satellite's optical or RF signature and deploy at the appropriate moment, thus diverting the attack toward the decoys. Decoys must be credible (provide a believable radar or optical simulation of the satellite) and must properly sense an attack to know the precise moment for the most effective deployment. We can also defeat a homing antisatellite by including optical or RF jammers to nullify or confuse its homing system. Such jammers weigh little and, depending on how well we know the parameters of the homing system, can be very effective.

A satellite can maneuver, or dodge, an antisatellite attack if it has thrusters for that purpose. Of course, almost every satellite has thrusters for attitude control and orbit changes. Thrusters for maneuvers are more powerful, generating higher accelerations and causing the need for stiffer, stronger solar arrays or other appendages. These extra requirements lead to weight penalties. In addition, we must supply more propellant, trading off the increased propellant weight against the increased survivability.

A satellite can defend itself against an antisatellite attack if that capability is included in the design. One possible approach is to include a suite of optical or radar sensors and small, lightweight missiles. The sensors would detect the onset of an attack, determine approximate location and velocity of the attacker, and launch the self-guided, homing missiles to kill the attacker. Of course, we would have to consider weight, power, inertial properties, and other design factors, but a self-defense system is a reasonable way to help a high-value spacecraft survive. Alternatively, we could deploy an escort satellite carrying many more missiles and being much more able to detect, track, and intercept the antisatellite attack. An escort satellite would cost more than active defense on the primary satellite, but the latter's weight and space limitations may demand it.

References

Bilitza, D., D.M. Sawyer, and J.H. King. 1988. "Trapped Particle Models at NSS-DC/WDC-A." in *Proceedings of the Workshop on Space Environmental Effects on Materials*. ed. B.A. Stein and L.A. Teichman. Hampton, VA.

Bolt, Robert O. and James G. Carroll, eds. 1963. *Radiation Effects on Organic Materials*. Orlando, FL: Academic Press.

Defense Nuclear Agency. 1972. *Transient Radiation Effects on Electronics (TREE) Handbook*. DNA H-1420-1. March 2.

DeWitt, Robert N., D. Duston, and A.K. Hyder. 1993. *The Behavior of Systems in the Space Environment*. Dordrecht, The Netherlands: Kluwer Academic Publishers.

Fennell, J.F., H.C. Koons, M.S. Leung, and P.F. Mizera. 1983. *A Review of SCATHA Satellite Results: Charging and Discharging*. ESA SP-198. Noordwijk, The Netherlands: European Space Agency.

Feynman, J., T. Armstrong, L. Dao-Gibner, and S. Silverman. 1988. "A New Proton Fluence Model for E>10 MeV." in *Interplanetary Particle Environment*, ed. J. Feynman and S. Gabriel, 58–71. Pasadena, CA: Jet Propulsion Laboratory.

Frederickson, A.R., E.G. Holeman, and E.G. Mullen. 1992. "Characteristics of Spontaneous Electrical Discharging of Various Insulators in Space Radiations." *IEEE Transactions on Nuclear Science*, vol. 39, no. 6.

Glasstone, S. and P.J. Dolan. 1977. *The Effects of Nuclear Weapons (3rd edition)*. Washington, DC: U.S. Departments of Defense and Energy.

Goldflam, R. 1990. "Nuclear Environments and Sensor Performance Analysis." Mission Research Corp. Report MRC-R-1321, October 4, 1990.

Gussenhoven, M.S., D.A. Hardy, F. Rich, W.J. Burke, and H.-C. Yeh. 1985. "High Level Spacecraft Charging in the Low-Altitude Polar Auroral Environment." *J. Geophys. Res.* 90:11009.

Hastings, D. and H. Garrett. 1996. *Spacecraft-Environment Interactions*. New York: Cambridge University Press.

Hedin, A.E. 1986. "MSIS-86 Thermospheric Model." *J. Geophys. Res.* 92:4649–4662.

Heitler, Walter. 1954. *The Quantum Theory of Radiation (3rd edition)*. Oxford: Clarendon Press.

Jacchia, L.G. 1977. *Thermospheric Temperature, Density and Composition: New Models*. Spec. Rep. 375. Cambridge, MA: Smithsonian Astrophysical Observ.

Jursa, A.S., ed. 1985. *Handbook of Geophysics and the Space Environment*, Bedford, MA: Air Force Geophysics Laboratory.

King, J.H. 1974. "Solar Proton Fluences for 1977–1983 Space Missions." *J. Spacecraft and Rockets.* 11:401.

Konradi, A. and A.C. Hardy. 1987. "Radiation Environment Models and the Atmospheric Cutoff." *J. Spacecraft and Rockets.* 24:284.

Leger, L.J., J.T. Visentine, and J.F. Kuminecz. 1984. "Low Earth Orbit Oxygen Effects on Surfaces." Paper presented at AIAA 22nd Aerospace Sciences Meeting, Reno, NV, January 9–12.

McIlwain, C.E. 1961. "Coordinates for Mapping the Distribution of Magnetically Trapped Particles." *J. Geophys. Res.* 66:3681–3691.

Mohanty, N., ed. 1991. *Space Communication and Nuclear Scintillation*. New York: Van Nostrand Reinhold.

Mullen, E.G., M.S. Gussenhoven, and H.B. Garrett. 1981. *A "Worst-Case" Spacecraft Environment as Observed by SCATHA on 24 April 1979*. AFGL-TR-81-0231, Hanscom Air Force Base, MA: Air Force Geophysics Laboratory.

National Aeronautics and Space Administration. 1980. *Nuclear and Space Radiation Effects on Materials*, NASA SP-8053. June 1980.

NOAA/National Environmental Satellite, Data and Information Service. 1991. *Monthly Mean 2800 MHz Solar Flux (Observed) Jan. 1948–Mar. 1991*. Solar-Geophysical Data prompt reports. Boulder, CO: National Geophysical Data Center.

Petersen, E.L. 1995. "SEE Rate Calculation Using the Effective Flux Approach and a Generalized Figure of Merit Approximation." *IEEE Trans. Nucl. Sci.*, vol. 42, no. 6, December 1995.

Purvis, C.K., H.B. Garrett, A.C. Whittlesey, and N.J. Stevens. 1984. *Design Guidelines for Assessing and Controlling Spacecraft Charging Effects*. NASA Technical Paper 2361.

Ritter, James C. 1979. "Radiation Hardening of Satellite Systems." *J. Defense Research* (classified Secret Restricted Data), vol. 11, no. 1.

Robinson, P.A. 1989. *Spacecraft Environmental Anomalies Handbook*. GL-TR-89-0222. Hanscom Air Force Base, MA: Air Force Geophysics Laboratory.

Schulz, M. and L.J. Lanzerotti. 1974. *Particle Diffusion in the Radiation Belts*. Heidelberg: Springer-Verlag.

Tascione, T. 1994. *Introduction to the Space Environment (2nd Edition)*. Malabar, FL: Orbit Book Company.

TRW, Inc. 1998. *Spacecraft Hardening Design Guidelines Handbook*. Vulnerability and Hardness Laboratory. September 1998.

Tribble, A.C. 1995. *The Space Environment: Implications for Spacecraft Design*. Princeton, NJ: Princeton University Press.

Tribble, A.C., B. Boyadjian, J. Davis, J. Haffner, and E. McCullough. 1996. *Contamination Control Engineering Design Guidelines for the Aerospace Community*. NASA CR 4740. May 1996.

Tsyganenko, N.A. 1987. "Global Quantitative Models of the Geomagnetic Field in the Cislunar Magnetosphere for Different Disturbance Levels." *Planet. Space Sci.* 35:1347.

U.S. Congress, Office of Technology Assessment. September 1985. *Anti-Satellite Weapons, Countermeasures, and Arms Control*, OTA-ISC-281.Washington, DC: U.S. Government Printing Office.

Vampola, A.L., P.F. Mizera, H.C. Koons and J.F. Fennell. 1985. *The Aerospace Spacecraft Charging Document*. SD-TR-85-26, El Segundo, CA: U.S. Air Force Space Division.

Vampola, A.L. 1996. "The Nature of Bulk Charging and Its Mitigation in Spacecraft Design." Paper presented at WESCON, Anaheim, CA, October 22–24.

Vette, J.I., A.B. Lucero and J.A. Wright. 1966. *Models of the Trapped Radiation Environment, Vol. II: Inner and Outer Zone Electrons*. NASA SP-3024.

Visentine, J.T. ed. 1988. *Atomic Oxygen Effects Measurements for Shuttle Missions STS-8 and 41-G,* vols. I–III. NASA TM-100459.

Walterscheid, R.L. 1989. "Solar Cycle Effects on the Upper Atmosphere: Implications for Satellite Drag." *J. Spacecraft and Rockets.* 26:439–444.

Webb, R.C., L. Palkuti, L. Cohn, G. Kweder, A. Constantine. 1995. "The Commercial and Military Satellite Survivability Crisis." *J. Defense Electronics.* August

Webb, R.C., G. Kweder. 1998. "Third World Nuclear Threat to Low Earth Orbit Satellites." Paper presented at GOMAC, Arlington, VA, 16–19 March 1998.

Chapter 9

Space Payload Design and Sizing

Bruce Chesley, *U.S. Air Force Academy*
Reinhold Lutz, *Daimler Chrysler Aerospace*
Robert F. Brodsky, *Microcosm, Inc.*

As illustrated in Fig. 1-3 in Chap. 1, the *payload* is the combination of hardware and software on the spacecraft that interacts with the *subject* (the portion of the outside world that the spacecraft is looking at or interacting with) to accomplish the mission objectives. Payloads are typically unique to each mission and are the fundamental reason that the spacecraft is flown. The purpose of the rest of the spacecraft is to keep the payload healthy, happy, and pointed in the right direction. From a mission perspective it is worth keeping in mind that fulfilling these demands is what largely drives the mission size, cost, and risk. Consequently, a critical part of mission analysis and design is to understand what drives a particular set of space payloads so that these elements can become part of the overall system trade process designed to meet mission objectives at minimum cost and risk.

This chapter summarizes the overall process of payload design and sizing, with an emphasis on the background and process for designing observation payloads such as FireSat. (Communications payloads are discussed in Chap. 13.) We begin with the flow of mission requirements (from Chap. 1) to payload requirements and the mission operations concept (from Chap. 2) to a *payload operations concept* which defines how the specific set of space instruments (and possibly ground equipment or processing) will be used to meet the end goals. We then summarize key characteristics of electromagnetic radiation, particularly those which define the performance and limitations of space instruments. Finally, we provide additional details on the design of observation payloads and develop a preliminary payload design for FireSat, which we compare with the MODIS instrument, a real FireSat payload for the Terra spacecraft in NASA Earth Observing System.

Several authors have discussed space observation payload design in detail, such as Chen [1985], Elachi [1987], and Hovanessian [1988]. More recently Cruise, et al. [1998] provides a discussion of a full range of payload design issues including optics, electronics, thermal, structures and mechanisms, and program management. In addition, a number of authors provide extended discussions of specific types of observations missions. Schnapf [1985], Buiten and Clevers [1993], and Kramer [1996] provide surveys of Earth observing missions and sensors. Huffman [1992] discusses UV sensing of the atmosphere. Meneghini and Kozu [1990] and Kidder and Vonder Haar [1995] discuss meteorology from space. Kondo [1990] and Davies [1997] discuss astronomical observatories in space. Finally, Chap. 13 provides numerous references on space communications payloads and systems.

Spacecraft missions have been flown to serve many purposes, and while virtually every mission has unique elements and fulfills some special requirement, it is nonetheless possible to classify most space missions and payloads into the following broad categories: communications, remote sensing, navigation, weapons, in situ science, and other. Table 9-1 provides a sample of missions that fall within these categories along with a primary payload and spacecraft that fits that particular mission. Many other types of space missions have been proposed or demonstrated. We include these in Table 9-1. We will introduce each of these spacecraft mission types, then focus on first-order system engineering analysis of remote sensing payloads.

Communications. The purpose of the majority of spacecraft is to simply transfer information. Communications missions range from wideband full-duplex telecommunications connectivity to one-way broadcast of television signals or navigation messages. Communications has traditionally been dominated by large geosynchronous spacecraft, but constellations of smaller spacecraft in lower orbits are emerging with alternative architectures for global coverage. New technologies are developing rapidly, including research into using lasers for spacecraft communication. A detailed discussion of communications payloads and subsystems is included in Sec. 11.2, Chap. 13, and Morgan and Gordon [1989].

Remote Sensing. Spacecraft remote sensing represents a diverse range of missions and applications. Any observation that a spacecraft makes without directly contacting the object in question is considered *remote sensing*. Imaging the Earth's surface, sounding the Earth's atmosphere, providing early warning of a ballistic missile launch, or observing the characteristic chemical spectra of distant galaxies are all remote sensing missions. Fundamentally we focus on measurements in the electromagnetic spectrum to determine the nature, state, or features of some physical object or phenomenon.

Depending on the particular mission, we can evaluate different aspects of electromagnetic radiation to exploit different characteristics of the target with respect to spatial, spectral, and intensity information content. We also evaluate this information in a temporal context that supports comparisons and cause-and-effect relationships. The types of information and sensors used to provide this information are illustrated in Fig. 9-1.

TABLE 9-1.   Types of Spacecraft Missions and Payloads.

Spacecraft Mission	Payload	Example
Communications		
Full-duplex broadband	Transceiver	Milstar, Intelsat
Message broadcast	Transmitter	DirecTV, GPS
Personal comm	Transceiver	Iridium
Remote Sensing		
Imaging	Imagers and cameras	LandSat, Space Telescope
Intensity measurement	Radiometers	SBIRS early warning,
Topographic mapping	Altimeters	Chandra X-Ray Observatory, TOPEX/Poseidon
Navigation		
Ranging	Transceiver	TDRS
Nav signal	Clock and transmitter	GPS, GLONASS
Weapons		
Kinetic energy	Warhead	Brilliant Pebbles concept
Directed energy	High-energy weapon	Space-Based Laser concept
In Situ Science		
Crewed	Physical and life sciences	Space Shuttle, Mir
Robotic	Sample collection/return	Mars Sojourner, LDEF
Other		
Microgravity Manufacturing	Physical plant and raw materials	Space Shuttle
Space power	Solar collector, converter, and transmitter	SPS
Resource utilization	Lunar soil collector and processor	Lunar Base
Tourism	Orbital hotel	Various
Space burial	Remains container	Pegasus XL

We make an additional distinction depending on the source of the electromagnetic radiation being sensed. If the instrument measures direct or reflected solar radiation in the environment, then we call it a *passive sensor*. *Active sensors*, on the other hand, emit radiation that generates a reflected return which the instrument measures. The principal active remote sensing instruments are radar and lidar.

Although our focus is on remote sensing of Earth, many scientific missions observe electromagnetic phenomena elsewhere in the universe. The physical principles of remote sensing and the categories of sensors are the same, regardless of whether the payload is looking at deep space or the planet it is circling.

Navigation. GPS, GLONASS, and other international navigation systems have demonstrated a wealth of applications for military, civilian, academic, and recreational users. As discussed in Sec. 11.7.2, GPS provides information for real-time position, velocity, and time determination. It is available worldwide on a broad range of platforms, including cars, ships, commercial and military aircraft, and spacecraft. The heart of GPS is a spread-spectrum broadcast communication message that can be exploited using relatively low-cost receivers.

Weapons. While remote sensing, communication, and navigation applications are quite mature and dominate the use of space, space-based weapons remain conceptual, occupying a small niche in the realm of space mission design. In particular, concepts

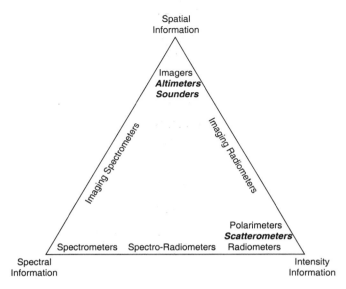

Fig. 9-1.   Electromagnetic Information Content and Sensor Types. Sensor types inside the triangle can observe the features shown outside the triangle. For example, each pixel collected by an imaging radiometer reflects both spatial and intensity information. Active instruments (such as radar) are printed in bold italic text. (Modified from Elachi [1987].)

for weapons in space became a topic of intense study and debate as part of the Strategic Defense Initiative and space-based strategic missile defense. Development of certain operational space weapons has been prohibited under the Anti-Ballistic Missile Treaty of 1972. Although some experts view widespread weaponization of space as inevitable, it has not become a stated objective of U.S. national policy [DeBlois, 1997]. Of course, space has been used to support military objectives since the dawn of artificial spacecraft [Hall, 1995; McDougall, 1985], but the vast majority of military space applications fall into the categories of remote sensing and communications.

In Situ Science. Sample collection and evaluation serves an important role in planetary and space science. Perhaps the most elaborate instance of sample collection took place in the Apollo missions when approximately 300 kg of samples from the Moon were returned to Earth for analysis. Other examples of sample collection and analysis include planetary landers (such as Viking and Mars Sojourner) and collection of solar wind particles.

Other. Exploitation of physical resources in space—either from the Moon or asteroids—has sparked innovative and imaginative concepts for augmenting Earth's limited resources or enabling human exploration of the solar system. In the nearer term, however, space-based materials processing and manufacturing are more likely to mature and exploit the characteristics of the microgravity environment (Sec. 8.1.6). Glaser et al. [1993] has done extensive studies of *satellite solar power,* i.e., generating solar power in space for use on Earth. Many authors have created designs for lunar colonies and space tourism facilities, but all require a dramatic reduction in launch cost. (See, for example, the CSTS Alliance's Commercial Space Transportation Study [1994].)

9.1 Payload Design and Sizing Process

Payload definition and sizing determines many of the capabilities and limitations of the mission. The payload determines what the mission can achieve, while the size of the payload, along with any special structural, thermal, control, communications, or pointing restrictions, will influence the design of the remainder of the spacecraft support systems.

We begin with the assumption that mission objectives are defined and the critical mission requirements are understood. This section concentrates on a top-down methodology for bounding the trade space of possible payloads and making an informed selection among them. This process is a useful guide for moving from a blank slate to a preliminary set of payloads. Iterating on the process produces a more detailed definition and more useful set of payloads that can meet the mission objectives at minimum cost and risk.

As shown in Table 9-2, the process begins with an understanding of mission requirements described in Chaps. 3 and 4. The mission requirements have a major effect on all aspects of space vehicle design, but it is frequently necessary to treat the components and subsystems separately for preliminary design and sizing. We begin with the payload because it is the critical mission element bounding spacecraft performance. Chapters 10 and 11 treat the remainder of the spacecraft systems and trade-offs involved in the overall spacecraft design.

Once the mission requirements are understood, we must determine the level of detail required to satisfy different aspects of the mission. For FireSat, varying levels of detail are required if the task is to identify the existence of a fire, assess the damage caused by fires, or characterize the combustibles in a fire. Additionally, the temporal (timeliness) demands placed on the mission could be vastly different depending on whether the data is to support long-term scientific analysis or real-time ground activity.

We summarize the basic steps in this process below and discuss them in more detail in the remainder of this chapter for remote sensing payloads and in Chap. 13 for communications payloads.

1. *Select Payload Objectives.* These objectives will, of course, be strongly related to the mission objectives defined in Chap. 1 and will also depend on the overall mission concept, requirements, and constraints from Chaps. 2, 3, and 4. However, unlike the mission objectives which are a broad statement of what the mission must do to be useful, the payload objectives are more specific statements of what the payload must do (i.e., what is its output or fundamental function). For FireSat, this is specific performance objectives in terms of identifying fires. For the space manufacturing example in the table, called WaferSat, the payload objective is a definition of the end product to be manufactured.

2. *Conduct Subject Trades.* The *subject* is what the payload interacts with or looks at. As discussed in detail in Sec. 9.2, a key part of the subject trade is determining what the subject is or should be. For a mobile communications system, it is the user's hand-held receiver. Here the subject trade is to determine how much capability to put in the user unit and how much to put on the satellite. For FireSat, we may get very different results if we define the subject as the IR radiation produced by the fire or as the smoke or visible flickering which the fire produces. In addition to defining the subject, we need to determine the performance thresholds to which the system must operate. For FireSat, what temperature differences must we detect? For WaferSat, how pure must the resulting material be? For mobile communications, how much rain attenuation

TABLE 9-2.   **Process for Defining Space Payloads.** See text for discussion. See Chap. 13 for a discussion of communications payloads.

Process Step	Product	FireSat (Remote Sensing) Example	Space Manufacturing Example	Where Discussed
1. *Use mission objectives, concept, requirements, and constraints to select payload objectives*	Payload performance objectives	Identify smoldering and flaming fires	Manufacture ultra-pure silicon wafers	Chaps.1, 2
2. *Conduct subject trades*	Subject definition and performance thresholds	Distinguish smoldering fires that are 3 K warmer than the background from flaming fires that are 10 K warmer than the background	Less than 1 ppb impurities over 50 cm square wafers	Sec. 9.2
3. *Develop the payload operations concept*	End-to-end concept for all mission phases and operating modes	Determine how end users will receive and act on fire detection data	Define user method to specify product needs, recover and use materials	Secs. 2.1, 9.4, Chap. 14
4. *Determine required payload capability to meet mission objectives [identify key characteristics of interest]*	Required payload capability	12-bit quantization of radiometric intensity in the 3–5 µm wavelength	Throughput of 5,000 wafers/day on orbit	Sec. 9.4.3
5. *Identify candidate payloads*	Initial list of potential payloads	Specifications for Sensors #1 and #2	Specifications for Factories #1 and #2	Sec. 9.6
6. *Estimate candidate payload capabilities and characteristics [mission output, performance, size, mass, and power]*	Assessment of each candidate payload	**Sensor #1** meets the sensitivity requirement but requires a data rate of 10 Mbps. **Sensor #2** can only identify flaming fires that are 10 K warmer than the background but requires a data rate of only 1.5 Mbps	**Factory #1** produces 6,000 wafers/day, weighs 80 kg, and uses 2 kW **Factory #2** produces 4,000 wafers/day (some of which will have >1 ppb impurities), weighs 100 kg, and uses 500 W	Sec. 9.5.3
7. *Evaluate candidate payloads and select a baseline*	Preliminary payload definition	Spacecraft and ground architecture based on 1.5 Mbps data rate. Adjust mission requirement to identify flaming fires only (not smoldering)	Select #1 with 1,000 wafers/day margin to be sold to reduce cost	Secs. 9.5.4, 9.6.1
8. *Assess life-cycle cost and operability of the payload and mission*	Revised payload performance requirements constrained by cost or architecture limitations	FireSat spacecraft with acceptable mission performance and cost	Payload repackaging to accommodate launch as an Ariane secondary payload on ASAP ring	Sec. 9.5.6, Chap. 20
9. *Identify and negotiate payload-derived requirements*	Derived requirements for related subsystems	Data handling subsystem requirement to accommodate payload data rate of 1.5 Mbps	ACS system to provide 140 continuous min of jitter less than ± 1 nm	Sec. 9.5.4
10. *Document and iterate*	Baseline payload design	Baseline FireSat payload	Baseline WaferSat payload	

must we be able to accommodate? These will be iterative trades as we begin to define the payload instruments and can intelligently evaluate cost vs. performance.

3. *Develop the Payload Operations Concept.* Ultimately, the data or product produced by the payload must get to the user in an appropriate form or format. How will the end user of FireSat data receive and act on the satellite data? How will the manufacturer recover the WaferSat materials and define what is to be done on the next flight? Payload operations will have a major impact on the cost of both the spacecraft and mission operations. As discussed in Chap. 15, payload operations may be done by the same facility and personnel that handle the spacecraft or, similar to the Space Telescope, may be an entirely different operations activity.

4. *Determine the Required Payload Capability.* What is the throughput and performance required of the payload equipment to meet the performance thresholds defined in Step 2? For FireSat what is the specification on the equipment needed to meet the temperature, resolution, or geolocation requirements? For WaferSat, how many wafers of what size will it produce? For mobile communications, now many phone calls or television channels must it handle simultaneously?

5. *Identify Candidate Payloads.* Here we identify the possible payloads and their specifications. For simple missions there will be a single payload instrument. For most missions, there will be multiple instruments or units which frequently must work together to meet mission requirements. Different complements of equipment may break the tasks down in different ways and may even work with different aspects of the subject. Thus, a system designed to identify the source of solar storms may have an imager and a spectrometer or a magnetometer and an instrument to map small temperature fluctuations on the photosphere or in the solar wind.

6. *Estimate Candidate Payload Characteristics.* Here we need to determine the performance characteristics, the cost, and the impact on the spacecraft bus and ground system so that we can understand the cost vs. performance for each of the viable candidate systems. Payloads will differ in their performance and cost, but also in weight, power, pointing, data rate, thermal, structural support, orbit, commanding, and processing requirements. We must know all of these impacts to conduct meaningful trades.

7. *Evaluate Candidates and Select a Baseline.* Here we examine the alternatives and make a preliminary selection of the payload combination that will best meet our cost and performance objectives. In selecting a baseline, we must decide which elements of performance are worth how much money. The payload baseline is strongly related to the mission baseline and can not be defined in isolation of the rest of the parts of the mission and what it will be able do for the end user.

8. *Assess Life-cycle Cost and Operability.* Ultimately, we want to determine mission utility as a function of cost. This process was described in detail in Chap. 3. Typically it will not be a simple cost vs. level of performance characterization. Rather it is a complex trade that requires substantial interaction with potential users and with whatever organization is funding the activity. It may become necessary at this point to relax or prioritize some of the mission requirements in order to meet cost and schedule objectives. For FireSat we may decide that only one type of fire or one geographic region will be addressed. For WaferSat we may reduce the purity, the size of the wafers, or the throughput.

9. *Define Payload-derived Requirements.* In this step we provide a detailed definition of the impact of the selected payloads on the requirements for the rest of the system (i.e., the spacecraft bus, the ground segment, and mission operations). FireSat will have power, pointing, geolocation, and data rate requirements. WaferSat may care very little about pointing and geolocation, but will have requirements on the spacecraft cleanliness levels and jitter control. These, in turn, may levy secondary requirements such as storage for onboard commands or thermal stability for pointing and jitter control.

10. *Document and Iterate.* Although this point is emphasized throughout the book, we stress again the need to document what we have decided and **why**. The "why" is critical to allowing the system trades to proceed at a future time. We can make preliminary decisions for a wide variety of reasons, but we must understand these reasons in order to intelligently continue to do payload and system trades. Like all of the space mission analysis and design process, payload definition is iterative. We will come back to the process many times as we learn more about the consequences of preliminary choices.

Figure 9-2 illustrates the conceptual process of payload sizing. At the bottom end of the curve, we need to spend a minimum amount of money to achieve any performance at all. Near minimum performance, a small amount of additional expense will substantially increase performance. At the top end of the curve, we can spend a lot of money for very small improvement. The overall payload performance per unit cost follows a straight line through the origin and whatever point on the performance vs. cost curve we are working at. Therefore, the maximum performance per unit cost occurs where a straight line through the origin is tangent to the curve.

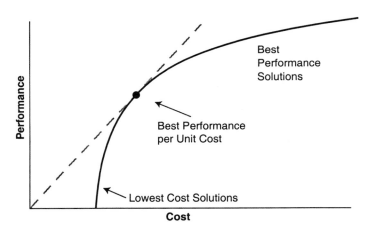

Fig. 9-2. Performance vs. Cost. The tangent point is the highest performance per unit cost.

There are good reasons for operating at any region along the curve in Fig. 9-2. To design a good payload, we must decide where along the curve our particular mission should be. At the high end we obtain the best available performance. This would be appropriate for some military or science missions, such as the Space Telescope or Chandra X-Ray Observatory. LightSats are at the bottom end of the curve. They perform modestly at very low cost. They may also be appropriate for multi-satellite, distributed systems. Large commercial activities, such as communications satellites, need the best performance per unit cost.

The key to deciding how to size our payload is to look carefully at the mission objectives, particularly the tacit rules which often imply how well we want to do. Do we need the best performance regardless of cost? Can the mission proceed only on a minimum budget? Is this a long-term, continuing, and potentially competitive activity in which performance per unit cost is critical? The answers to these questions will let us correctly size the payload and the mission to meet our mission objectives.

9.2 Mission Requirements and Subject Trades

Defining requirements and constraints for space missions occurs as described in Chaps. 1 and 2. The overall mission requirements dictate the technical performance of the payload, while the mission concepts and constraints determine the operational implementation for the mission. Frequently the technical specification and operations concept for payloads are interrelated. For example, increasing temporal resolution (revisit) may reduce the requirement for spatial resolution in an optical sensor system. We must ensure that the mission requirements capture the fundamental needs of the users without constraining the designer's ability to satisfy these requirements through alternate technical means.

For FireSat we begin with the overall mission requirement to detect, identify, and locate forest fires, then consider the level of detail needed to satisfy the mission. Often it is useful to articulate the questions that need to be answered or the decisions that need to be made based on sensor data. Possible questions for the FireSat mission planners include:

- Can a new fire be detected within 2 hours? Twenty minutes?
- What is the geographic extent of the fire?
- Can smoldering fires be distinguished from flaming fires?
- What are the primary combustibles (can fires burning organic material be distinguished from petroleum and chemical fires)?
- What direction is the fire spreading and how quickly?
- How much smoke and ash is the fire generating?
- Where is the fire burning hottest?
- At which locations would additional firefighting efforts to contain and suppress the fire be most effective?
- What other sources of information exist from air-, ground-, or space-based sources?
- If available, how might other sources of information be used?

Specific mission objectives and priorities addressed by these questions will determine the specific observables linking payload performance with mission performance. To choose a remote sensing payload, the key steps to a disciplined and repeatable design begin with determining the elements of information that we need to address the problem. We must specify the physically observable quantities that contribute to elements of information about the problem in sufficient detail to ensure they can be detected by a spacecraft payload with sufficient resolution to provide meaningful insight into the subject.

Establishing performance thresholds provides a framework for trading off performance across a number of different design features. For all missions, payload performance evaluation categories include physical performance constraints and operational constraints. Examples of physical performance constraints include limits on spatial, spectral, radiometric, and temporal resolution. Operational constraints include sensor duty cycle limits, tasking and scheduling limits on sensor time, and resource contention (inability of the sensor to view two targets of interest simultaneously).

Within each of the categories of sensor constraints, we should establish an absolute minimum threshold such that any performance that does not meet this capability is unacceptable. The minimum threshold values generally will not satisfy mission objectives, but establishing the minimum level of usefulness for the mission allows flexibility for trade-offs. At the other extreme, we should specify the desired performance to establish the performance that will fully satisfy the requirement. We can also define an intermediate value—an acceptable level of performance—to articulate a desired level of performance that will meet the bulk of mission objectives. Table 9-3 illustrates a sample of performance thresholds for the FireSat mission payload across the functional areas of resolution, quantity, timeliness, periodicity, geolocation accuracy, and completeness. These distinctions can be critical in determining the viability of a mission concept. In commercial remote sensing, for example, the range of performance requirements from minimum to desired is typically determined through extensive market analysis and business development case studies. These studies frequently identify a minimum resolution (or other performance parameter) below which a remote sensing spacecraft concept will not be profitable.

TABLE 9-3. Sample Threshold Performance Requirements for the FireSat Payload.
Desired performance represents the maximum reasonable level of performance across all design features.

	Minimum	**Acceptable**	**Desired**
Subject Characteristics	Detect presence or absence of large fires	Identify, locate, and track progress of fires	Determine thermal conditions within fires and products of combustion
Quantity	Measure existence of 1 fire	Simultaneously measure and track 7 fires	Simultaneously measure and track 20 fires
Timeliness	Report detection of fire within 6 hours	Report detection of fire within 2 hours	Report detection of fire within 20 min
Revisit Interval	Update status of fire every 2 hours	Update status of fire every 90 min	Update status of fire every 45 min
Geolocation Accuracy	Determine location of fire within ±100 km	Determine location and extent of fire within ±1 km	Determine location and extent of fire within ±100 m
Completeness	Map fires in continental U.S.	Map fires in North America and one other selectable region (e.g., Persian Gulf)	Map fires globally

We need to parameterize the mission, such as identifying and locating forest fires, in such a way that we can evaluate, size, and design candidate sensors. This parameterization involves a process of requirements analysis that focuses on matching the tasks involved in the mission with categories of discipline capabilities. If we match mission requirements with existing or probable capabilities the result is a set of potential information requirements. We then try to identify the characteristics of the subject (*signatures*) that correspond to the information requirements through a set of rules. For the FireSat example, these rules consist of the spectral wavelengths and thresholds needed to detect fires. The rules yield a set of mission observables, such as specific wavelength bands and spectral sensitivities that we need in our sensor. These observables provide the basis for the payload characteristics that comprise the baseline design to satisfy the mission. In the case of FireSat, the basic mission categories that might satisfy this mission and the corresponding information type are shown in Table 9-4.

TABLE 9-4.   Simplified Subject Trades for FireSat Mission. The information type allows for subject trades to be made among the different signatures that can be exploited to satisfy FireSat mission requirements.

Sensor Type	Information Type
Electro-optical Imager	Visible return from light or smoke cloud produced by the fire
Spectrometer	Spectral signatures from products of combustion
Radiometer	Thermal intensity

A unique signature exhibited by fires is the flickering light in a fire. This flicker has a characteristic frequency of about 12 Hz and can be exploited by processing the data stream from an electro-optical sensor to search for this frequency [Miller and Friedman, 1996]. Light flickering at this frequency produces an irritating effect on the human vision system, possibly as a survival adaptation for the species against the threat of wildfires.

There are many choices and types of sensors, more than one of which might be a candidate to perform a given mission. In the case of FireSat, it may be possible to satisfy basic mission requirements by observing a number of different phenomenologies: visible signatures associated with flame and smoke, thermal infrared signatures from the fire, spectral analysis of the products of combustion, or an algorithm combining all of these. The selection of a spacecraft payload represents the fundamental leap in determining how to satisfy mission requirements with a space sensor. In the previous section we introduced a top-down framework for considering the general problem of spacecraft design. Here we turn our attention to the payload; in particular, a methodology for determining the type of payload to employ and the physical quantities to measure.

Figure 9-3 illustrates the framework for the heart of the payload design process. The process begins with a task or mission requirement and ends with a spacecraft payload design. We have divided this process into intermediate steps to focus the effort along the way to a final design. In this section we focus on describing the process illustrated in Fig. 9-3; Sec. 9.4 provides some of the specific techniques that are employed in this process for visible and IR systems.

For the FireSat mission design, we need to identify specific signatures that would allow candidate sensors to provide viable solutions to the mission requirement. We observe physical phenomena through signatures, and we must choose which signature will provide the desired information. The specific signatures that a payload senses must be evaluated in light of the particular focus of the mission. For example, a spectrometer that is sensitive enough to detect all fires, but which cannot be used to differentiate campfires from forest fires could generate a large false alarm rate and render it operationally useless. Defining the key signatures and observables that support the information content needed to satisfy the mission determine the performance limits for the payload design.

9.2.1 Subject Trades

The objective of a space mission is typically to detect, communicate, or interact. The *subject*, as an element of the space mission, is the specific thing that the spacecraft will detect, communicate, or interact with. For GPS, the subject is the GPS receiver, For FireSat, we would assume that the subject is the heat generated by the forest fire. But other subjects are possible: light, smoke, or changes in atmospheric composition.

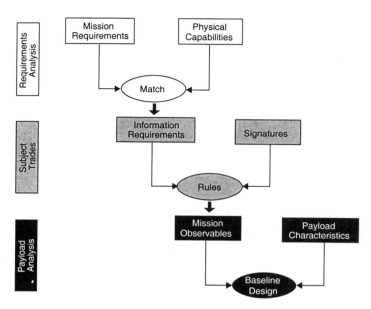

Fig. 9-3.   Process for Linking Mission Requirements to Payload Design. The process moves from mission requirements to a payload design in three steps: requirements analysis, subject trades, and payload analysis.

What we choose as the subject will dramatically affect performance, cost, and the mission concept. Thus, we must do this trade carefully and review it from time to time to ensure it is consistent with mission objectives and our goal of minimizing cost and risk.

Table 9-5 summarizes the subject-trade process. We begin by looking at the basic mission objectives and then ask what subjects could meet these objectives. To do this, we should look at what we are trying to achieve, the properties of space we intend to exploit, and the characteristics of what we are looking at or interacting with. Table 9-6 shows examples of subject trades for four representative missions. As the missions change, the nature of the subject trades will also change. For FireSat, we are looking for a well-defined subject (the forest fire), and we want to do this at minimum cost and risk. With the Space Telescope, we must ask, "What am I looking for? What am I trying to detect and how can I detect it?" For any of the science missions, we would ask, "Is the subject some distant and unknown object, or is it part of the electromagnetic spectrum I am trying to explore?"

For a space system intended to detect airplanes, the main subject trades would concern mission goals. Are the targets cooperative or noncooperative? Do we need to track over the poles? Should we track in high-density areas around airports or over the open oceans? The answers to these questions will determine the nature of the subject trades.

Perhaps the easiest subject trades are those in which the system will be interacting with a ground element that is a part of the system, such as direct broadcast television or a truck communication system. In this case, the subject trade becomes simply an issue of how much capacity should go on the spacecraft vs. how much should go in the unit on the ground.

TABLE 9-5.   Subject Trade Process. Note that the subject trades lead directly to the payload trade process as discussed in Sec. 9.2

Step	FireSat Example	Where Discussed
1. Determine fundamental mission objectives	Detect and monitor forest fires	Sec. 1.3
2. Determine what possible subjects could be used to meet these objectives (i.e., what could the system detect or interact with to meet the objectives)	Heat, fire, smoke, atmospheric composition	Sec. 9.2
3. Determine broad class of ways that the spacecraft can detect or interact with the possible subjects	Heat –> IR flame, smoke –> visual atmospheric composition –> lidar	Sec. 9.2
4. Determine if subject is passive or controllable	Initially assume passive fire detection	Sec. 9.2
5a. For controllable subjects, do trade of putting functionality at the subject, in the space system, or in the ground system	N/A	Secs. 2.1, 3.2.3
5b. For passive subjects, determine general characteristics that can be detected	Forest fire temperature range and total heat output	Sec. 9.2
6. Determine whether multiple subjects and payloads should be used	Not initially	Sec. 9.2
7. Define and document initial subject selection	IR detection of heat	N/A
8. Review selection frequently for alternative methods and possible use of ancillary subjects	See Sec. 22.3, alternative low cost for FireSat	N/A

The next step for subject trades is to determine whether the subject is controllable or passive. The system designer knows and can control characteristics of *controllable* or *active* subjects. This includes ground stations, antennas, receivers, and transmitters such as those used for ground communications, direct broadcast television, or data relay systems. Because we can control the subject, we can put more or less capability within it. Thus we might choose to have a simple receiver on the ground with a high-power, accurately pointed, narrow-beam transmitter on the spacecraft. Or we could place a sophisticated, sensitive receiver on the ground with a small, lower-cost system in space. Usually, the solution will depend on the number of ground stations we wish to interact with. If there are many ground stations, as in direct-broadcast television, we will put as much capability as possible into the satellite to drive down the cost and complexity of the ground stations. On the other hand, if there are only a few ground stations, we can save money by giving these stations substantial processing and pointing capability and using a simpler, lighter-weight, and lower-cost satellite.

Passive subjects are those in which the characteristics may be known but cannot be altered. This includes phenomena such as weather, quasars, or forest fires. Even though we cannot control the object under examination, we can choose the subject from various characteristics. We could detect forest fires by observing either the fire itself or the smoke in the visible or infrared spectrum. We could detect atmospheric composition changes or, in principle, reductions in vegetation. Thus, even for passive subjects, the subject is part of the system trades.

TABLE 9-6. Representative Subject Trades. Subject trades for the Space Telescope are particularly interesting in that a significant goal of the system is to discover previously unknown phenomena or objects.

Mission	FireSat	Airplane Detection	Truck Communications System	Space Telescope
Property of space used	Global perspective	Global perspective	Global perspective	Above the atmosphere
General object of study or interaction	Forest fires	Airplanes	Portable telecommunication centers	Distant galaxies + unknown phenomena
Alternative mission subjects	Fire (visible or IR) Smoke (visible or IR) Increased CO_2 Decreased vegetation	Skin (radar, visible) Plume (IR) Radio emissions (RF)	Current radio Current CB Standard TV New telecommunication center Cellular relay	Quasars Galaxies Planets Visible spectrum Unknown objects
Key subject trades	None—IR detection probably best choice	Radar vs. IR vs. active RF	Complexity of truck element vs. complexity of space & ground station	Is the subject known or unknown? Is it objects or spectral regions?
Comments	See low-cost alternative in Chap. 22	Need to examine goals; cooperative vs. noncooperative targets; high density vs. ocean tracking		

We do not always know whether a given mission has passive or active subjects; in some cases, we can choose either type. For example, we could detect airplanes passively with an IR sensor or radar, or actively by listening for or interrogating a transponder on the airplane. Chapter 22 summarizes an alternative for sensing forest fires by using equipment on the ground and then relaying it to space—a technique possible for various mission types. Satellites that monitor the weather or environment could do complex observations or simply collect and relay data from sensors on the ground.

The next step is to determine whether we need multiple subjects (and, probably, multiple payloads) to meet our mission objectives. Using multiple subjects at the same time has several advantages. This approach can provide much more information than is available from a single subject and can eliminate ambiguities which occur when observing only one aspect. On the other hand, multiple subjects typically require multiple payloads, which dramatically drive up the space mission's cost and complexity. Thus, a principal trade is between a low-cost mission with a single subject and single payload vs. a more expensive mission that achieves higher performance by using several payloads to sense several different subjects related to the same objective.

For FireSat, we tentatively select the heat of the forest fire as the subject of the mission, keeping in mind that this may change as the design evolves. Of course, we should make these trades as rapidly as possible because they strongly affect how the mission is done.

Finally, as always, we should document the subject selection and review it frequently during the program's early stages, looking for other possible methods and subjects. Looking for alternative subjects is perhaps the single most important way to drive down the cost of space missions. We need to continually ask ourselves, "What are we trying to achieve, including the tacit rules of the program, and how can we achieve it?"

9.3 Background

9.3.1 The Electromagnetic Spectrum

As Fig. 9-4 illustrates, the *electromagnetic spectrum* is a broad class of radiation. It includes *gamma rays* and *X-rays*, with extremely short wavelengths measured in angstroms ($\text{Å}=10^{-10}$ m), as well as visible and *infrared* (*IR*) wavelengths of 10^{-7} to 10^{-3} m and the *microwave* region from 0.1 to 30 cm. Finally, it ranges into the *radio* spectrum, with wavelengths as long as kilometers. As the figure shows, satellite systems operate over the entire spectral range. Normal wavelengths for comsats, radars, and microwave radiometers range from approximately 1 meter to 1 millimeter, whereas visual and IR systems operate from around 0.35 to 100 microns (1 *micron* = 10^{-6} m = 1 μm).

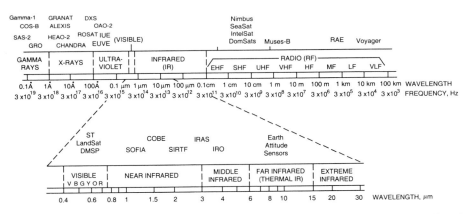

Fig. 9-4.   The Electromagnetic Spectrum. The expanded view highlights visible and infrared wavelengths frequently exploited by satellites. Sample space missions across the entire spectrum are listed above the band region.

For all electromagnetic radiation in a vacuum, the relation between the wavelength, λ, and the frequency, v,[*] is

$$c = \lambda v = 2.997\ 924\ 58 \times 10^8 \text{ m/s} \qquad (9\text{-}1)$$

where c is the speed of light. Thus, in terms of frequency, the usable electromagnetic spectrum ranges from radio waves measured in kilohertz (kHz) to gamma rays with

---

[*] Both v and f are commonly used to represent frequency. We use v throughout this chapter to avoid confusion with focal length, which is also represented by f.

frequencies in the tens of exahertz (EHz). (1 kilohertz = 1,000 cycles/s = 300-km wavelength; 1 exahertz = 10^9 GHz = 10^{18} cycles/s = wavelength of 3 angstroms or 3×10^{-10} m.)[*]

At any temperature above absolute zero (0 K), all matter continuously emits electromagnetic radiation. This is called *thermal radiation* or *blackbody radiation*. For a *perfect blackbody*, the rate of total energy emission and the energy distribution by wavelength or frequency is a function only of the temperature, T. The actual spectrum of emitted radiation from a real object will depend on the surface characteristics for small objects, such as a spacecraft, or on the atmosphere for large objects, such as the Earth or Sun. Nonetheless, in practice the blackbody energy distribution is a good starting point for analysis. The spectral energy distribution of a blackbody is given by *Planck's Law*:

$$E_\lambda = \frac{2\pi hc^2}{\lambda^5} \frac{1}{e^{ch/kT\lambda} - 1} \tag{9-2}$$

where E_λ is the energy per unit wavelength (also called the *spectral irradiance* and typically measured in $W \cdot m^{-2} \cdot \mu m^{-1}$), λ is the wavelength, h is Planck's constant $(6.626\ 075\ 5 \times 10^{-34}\ W \cdot s^2)$, T is the absolute temperature, c is the speed of light, and k is Boltzmann's constant $(1.380\ 658 \times 10^{-23}\ W \cdot s/K)$. Figure 9-5 shows typical energy distribution curves for various blackbody temperatures. When E_λ is divided by the solid angle (in steradians) leaving an extended source in a given direction, it becomes L_λ the *spectral radiance* (typical units, $W \cdot m^{-2} \cdot \mu m^{-1} \cdot sr^{-1}$).

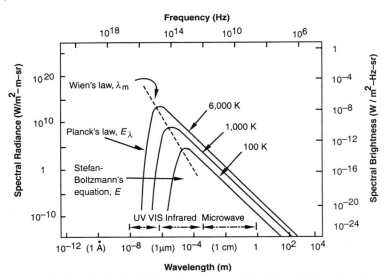

Fig. 9-5.   Planck's Blackbody Radiation Curves as a Function of Wavelength and Frequency [Chen, 1985]. Planck's law defines the shape of the curve over all frequencies, the Stefan-Boltzmann's law defines the area under the curve (the total energy emitted over all wavelengths), and Wien's displacement law defines the wavelength of maximum radiance.

---

[*] A list of all metric prefixes is at the front of Appendix F.

From Planck's Law we can derive two other important relations. First, we obtain the *Stefan-Boltzmann's Law* by integrating Eq. (9-2) over the complete spectrum, yielding the *total radiant emittance*, W_b:

$$W_b = \sigma T^4 \qquad (9\text{-}3)$$

where σ is the Stefan-Boltzmann constant, $5.670\,51 \times 10^{-8}$ W·m^{-2}·K^{-4}, and W_b is typically in W/m^2. Second, we derive *Wien's Displacement Law* by differentiating Eq. (9-2) and setting the result equal to zero. The straight-line result defines the locus of peak spectral radiance vs. temperature, as shown on Fig. 9-5 and defined by

$$\lambda_{max} = 2,898\,/T \qquad (9\text{-}4)$$

where λ_{max} is in µm when T is in K.

Remote-sensing instruments are aimed at a target on the Earth's surface or in space. Radars measure the characteristics of reflected, self-generated signals. For other sensors, an object's spectral radiance, or *brightness*, depends on its equivalent *black-body temperature*. This is the temperature of a perfect radiating body which has the same total radiance. Visual systems, which use film or solid-state detectors to form the images, can take advantage of the Sun's reflected energy, based on its blackbody temperature of about 6,000 K. Of course, without sunlight, visual images are much less distinct. On the other hand, systems using infrared and microwave radiometry measure scenes against the Earth's intrinsic thermal radiation background (corresponding to about 300 K). Thus, they can operate day or night, as well as through clouds and other atmospheric disturbances. Note, though, how much weaker the signals are in the RF bands compared to the IR ranges.

As Fig. 9-6 shows, the electromagnetic spectrum has many frequency bands for which the Earth's atmosphere is nearly opaque. We must avoid these bands if we wish to observe ground scenes. This phenomenon also allows us to sound the atmosphere and measure such interesting data as the thickness and location of cloud layers, water vapor contained in clouds, and other upper-atmospheric phenomena using the opaque bands. Clouds, rain, and snow tend to produce noise and thus attenuate signals for both communication and remote-sensing, even in the window bands.

When a sensor views an area in space, the radiation that reaches the sensor could come from a number of sources. The energy reflected directly from the target is usually the dominant feature of interest for optical remote sensing, but other emitted, reflected and scattered energy can complicate the picture. The *primary* (direct) and *secondary* (single scatter) sources of electromagnetic radiation are shown in Fig. 9-7.

The sources of radiation in Fig. 9-7 give rise to a number of different strategies for distinguishing different phenomena within the atmosphere or on the surface. For a given application, any of the sources of radiation will either be the subject being analyzed or noise to be minimized. Radiative measurements include the full complexity of all the effects on that radiation such as reflection, refraction, absorption, transmission, and scattering by material substances in solid, liquid, and gaseous phases. Distinguishing and identifying features using remote sensing techniques must take all of these vari-ables into account. As Miller and Friedman [1996] advise, "when modeling the real world, allow for some slack to represent reality."

9.3.2 Basic Telescope Optics

A brief review of physical optics and antenna theories will show that systems for gathering or transmitting optical and RF signals are exactly the same in theory—only

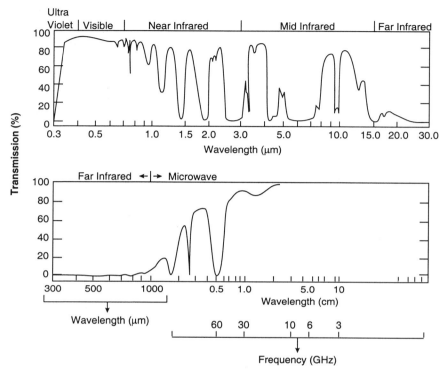

Fig. 9-6.   Transmission Characteristics of the Earth's Atmosphere. Transparent regions are referred to as *windows* in the atmosphere.

the physical hardware is different. Thus, a mirror (in visual and IR systems) and the reflector of a dish antenna (in microwave radiometry and radar) are equivalent. This section only summarizes remote-sensing instrument analysis. *The Manual of Remote Sensing* [Colwell, 1983] provides an extensive discussion, including image analysis. Seyrafi's [1985] treatment is also comprehensive, and ends with a design example of a thermal imaging system for a spacecraft.

In this discussion, we treat *reflective systems* (optical systems using mirrors) and *refractive systems* (optical lens systems) together and refer to them both as *optical systems*. Reflective and refractive systems have advantages and disadvantages, which we will discuss later.

There are several ways to describe an optical system. Parallel rays of light falling on a perfect lens will all converge at the *focal point*, whose distance from the lens is called the *focal length, f*. The focal length largely determines the length of the optical collection system and for a single lens, it is related to the lens surface's radius of curvature. The focal length of a spherical reflecting surface is one-half its radius. For a parabolic reflector whose surface is defined by the equation $z^2 = 4fy$, the quantity f is the focal length; it equals the distance from the focus to the nearest point on the reflecting surface.

In design practice, we normally determine the required focal length based on field of view and the size of the image plane. The *plate scale, s*, or length per field-of-view

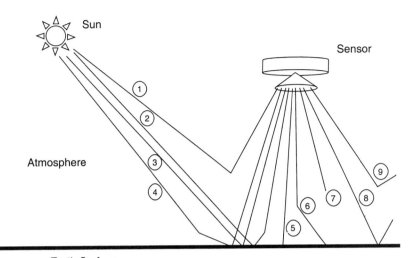

Fig. 9-7. **Sources of Radiation.** Radiation that reaches the sensor can come from a number of different sources. The diagram illustrates all direct and single-scatter radiation that reaches a space sensor. The sources of radiation are: (1) sunlight scattered by the atmosphere into the sensor; (2) sunlight reflected off the Earth and then scattered by the atmosphere into the sensor; (3) sunlight reflected off the Earth's surface; (4) sunlight scattered by the atmosphere then reflected off the Earth's surface into the sensor; (5) ground emission; (6) ground emission scattered by the atmosphere into the sensor; (7) atmospheric emission; (8) atmospheric emission reflected by the Earth's surface into the sensor; and (9) atmospheric emission scattered by the atmosphere into the sensor. [Adapted from Kramer, 1996.]

angle is given by

$$s = f \qquad \text{unit length/rad}$$
$$= 0.01745\,f \qquad \text{unit length/deg} \qquad (9\text{-}5)$$

where s and f are in the same units. The image size is a function of s and the size of the detector—ranging from a single element to a large array—employed at the focal plane. As Fig. 9-8 shows, the focal length needed to record an object or scene of radius R is given by

$$\frac{f}{h} = \frac{r_d}{R} \equiv magnification \qquad (9\text{-}6)$$

where h is the distance from the spacecraft to the object, r_d is the radius of the detector array in the image plane, and R is the radius of the object, with the image and object measured perpendicular to the line of sight. The *magnification* or *scale* $\equiv r_d /R$ is the ratio of the image size to the object size. It is ordinarily a very small number for satellites. We express the scale on the image plane as "1 cm equals x km on the ground."

We can also describe an optical element or system by its so-called *infinity F-number* or *F-stop*, often written as $f/$(read "F-stop"), F, F No., or F#. It is defined as f/D, where D is the *aperture* or diameter of the lens. Image brightness is proportional to $1/F\text{-}2$, so an F-4 lens gives an image four times brighter than an F-8 lens.

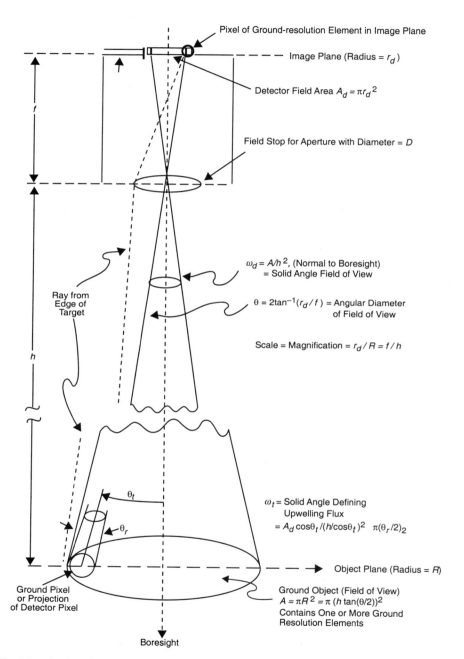

Pixel of Ground-resolution Element in Image Plane

Image Plane (Radius = r_d)

Detector Field Area $A_d = \pi r_d{}^2$

Field Stop for Aperture with Diameter = D

$\omega_d = A/h^2$, (Normal to Boresight)
= Solid Angle Field of View

$\theta = 2\tan^{-1}(r_d/f)$ = Angular Diameter of Field of View

Scale = Magnification = $r_d/R = f/h$

Ray from Edge of Target

ω_t = Solid Angle Defining Upwelling Flux
= $A_d \cos\theta_t /(h/\cos\theta_t)^2$ $\pi(\theta_r/2)_2$

Object Plane (Radius = R)

Ground Pixel or Projection of Detector Pixel

Ground Object (Field of View)
$A = \pi R^2 = \pi (h \tan(\theta/2))^2$
Contains One or More Ground Resolution Elements

Boresight

Fig. 9-8.   Optical Characteristics of a Refractive System. Note one-to-one correspondence of the ground-resolution element's size to the pixel size at the image plane. The operating wavelength is λ. As resolution elements move away from nadir, flat-Earth approximations become less precise. See Chap. 5 for additional details.

The *numerical aperture, NA,* gives the same information in another way:

$$NA \equiv \frac{1}{2F\#} = \frac{D}{2f} \tag{9-7}$$

or

$$F\# = \frac{1}{2NA} = \frac{f}{D} \tag{9-8}$$

The largest numerical aperture for optics used in air is 1. Thus, the smallest $F\#$ is 0.5.

All optical systems suffer from *aberrations,* or imperfections in the image quality, in addition to diffraction which limits the system resolution as discussed in Sec. 9.3.3. The principal optical aberrations are listed in Table 9-7. *Chromatic aberration,* or imperfections which are color or wavelength dependent, arises from various wavelengths being bent by different amounts when they pass through a lens. Consequently, only systems with at least one refractive element suffer chromatic aberration because reflective surfaces treat all wavelengths the same. (This is not absolutely true since some surfaces will reflect visible light, for example, but not X-rays. However, when reflection does occur it is independent of wavelength to first order.)

TABLE 9-7. Principal Aberrations in Optical Systems. See Table 9-8 for which of these are mitigated in various optical systems.

Chromatic Aberration = dispersion of the light due to the refractive index of a lens being a function of the wavelength. Causes different colors to focus at different distances.

Spherical Aberration = dispersion in which light from the periphery of a spherical lens or mirror is focused nearer the element than light from the center. Can be eliminated by making the optical surface parabolic, rather than spherical.

Coma = dispersion of off-axis portions of the image. (So named because in a telescope off-axis star images look like tear drops or the coma of a comet pointing toward the center of the image.)

Astigmatism = aberration in which the distorted image is asymmetric such as when light in a horizontal plane comes to a slightly different focus than light in a vertical plane. A common problem in human vision.

Distortion = when an otherwise sharp image is distorted in shape, such as when straight lines on the surface being viewed appear curved on the focal plane. A uncorrectable distortion occurs when trying to image the celestial sphere onto a flat focal plane. (See Sec. 5.1.)

Curvature of Field = when a sharp image is formed on a focal surface which isn't flat. Can be corrected in film systems by using a slightly curved focal plane.

Figure 9-9 shows the three basic types of telescopes. In each of the three, there is a corresponding refractive and reflective instrument. The aberrations which can be corrected in each type are shown in Table 9-8. The lens doublet is the classic refractive telescope lens. The doublet can be designed to eliminate spherical aberrations, coma, distortion and chromatic aberrations (see Table 9-8). In tele-optic lens systems the distance between the optical element and the focal plane is shorter than the focal length. Tele-optic lenses can eliminate spherical aberrations, coma, astigmatism, and curvature of field effects. They can also overcome chromatic aberrations. The lens triplet is the simplest refractive (spherical) optical system that theoretically allows for correction of all distortions. The price we pay for this advantage is the very high sensitivity of each of the optical elements with respect to displacement or tilt. The ray traces in

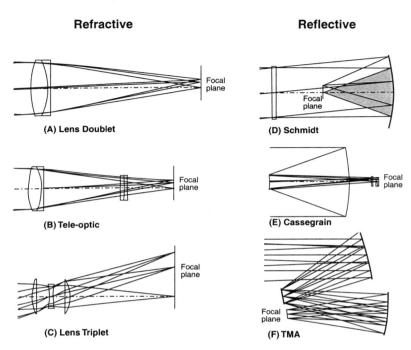

Fig. 9-9. **Basic Configurations for Refractive and Reflective Optical Systems.** Each of the reflective systems on the right is analogous to the corresponding refractive system on the left. TMA = Three-mirror anastigmatic.

TABLE 9-8. **Aberrations that can be Corrected by the Three Basic Optical Systems.** Checks indicate errors that are fully correctable and parenthetical checks indicate that corrections are possible only for dedicated design parameters. (See Table 9-7 for definitions).

Image Error	Doublet/Schmidt	TeleOptic/Cassegrain	Triplet/TMA
Lateral Chromatic Aberration	✔	✔	✔
Length Chromatic Aberration	✔	✔	✔
Spherical Aberration	✔	✔	✔
Coma	✔	✔	✔
Astigmatism		(✔)	✔
Distortion	(✔)	(✔)	✔
Curvature of Field		(✔)	✔

Fig. 9-9 indicate different locations of the image in the focal plane corresponding to various viewing angles. The lens triplet compensates for all five of the third-order aberrations: spherical aberrations, coma, astigmatism, curvature of field, and distortion. It too is free from chromatic aberrations. The same behaviors are present in the corresponding reflective systems. The Schmidt Mirror System is an all-reflective doublet, and the Cassegrain telescope is a reflective implementation of a tele-optic lens. The Three-Mirror Anastigmatic system is comparable to the lens triplet with respect to all the aberration corrections, but with an all-reflective design. Reflective optical systems generally are free from chromatic aberrations. However, reflective systems typically have a much smaller field of view than their refractive counterparts.

In reality, optical systems for space remote sensing are far more complex because the technologies for manufacturing the lenses and mirrors are limited and other effects such as thermal distortions and radiation effects can alter the performance of the instrument. Thermal distortions can limit the performance of an optical system, even if the operating temperature range is regulated within a few degrees for high performance optical systems, and cosmic radiation effects can degrade the transparency of most optical glass over time. Figure 9-10 shows the lens cross section of the high-resolution optical lens system of the German-built Modular Optoelectronic Multispectral Scanner (MOMS 2P) instrument designed to achieve 6 m resolution on the ground.

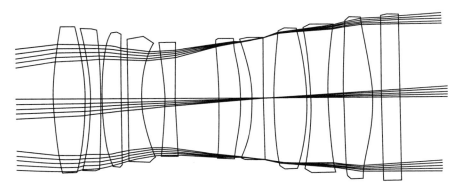

Fig. 9-10.   **Lens Cross Section of the Panchromatic Objective of the MOMS 2P Instrument.** The sensor has a focal length of 0.66 m and an aperture size of 0.15 m. The complexity of this optical system is representative of sophisticated remote sensing payloads.

9.3.3 Diffraction Limited Resolution

The *resolution* of an optical system is its ability to distinguish fine detail. In general resolution is expressed in angular terms. Thus, a telescope that can just distinguish or *resolve* two stars which are very close together is said to have a *resolving power* equal to the angular separation of the stars. For Earth observing systems we are more interested in the ability to see or resolve fine detail on the surface. Thus, for these systems resolution is commonly expressed in terms of the size of an object on the Earth that can just be distinguished from the background. To read this page requires a resolution of about 0.1 mm, whereas you may be able to distinguish a large newspaper headline with a resolution of 1 cm.

No matter how good the quality of the lens or mirror, a fundamental limitation to resolution is *diffraction*, the bending of light that occurs at the edge of the optical system. Even for a perfect optical system, diffraction causes the image of a point source of light, such as a distant star, to appear not as a point on the focal plane but as a series of concentric circles getting successively dimmer away from the center, as shown in

Fig. 9-11. This pattern is called the *diffraction disk*, the *Airy disk**, or the *point spread function*. The angular distance, θ_r, from the maximum at the center of the image to the first dark interference ring, called the *Rayleigh†* limit, or *Rayleigh diffraction criteria*, is given by

$$\theta_r = 1.22 \, \lambda \, /D \qquad (9\text{-}9)$$

where λ is the wavelength, D is the aperture diameter of the optical instrument, and θ_r is expressed in radians. The bright maximum at the center of the Airy disk, out to the first interference minimum, contains 84% of the total energy which arrives at the focal plane from a point source. For a satellite at altitude, h, the linear resolution or *ground resolution*, X, at nadir is just

$$X' = 2.44 \, h\lambda/D \qquad (9\text{-}10)$$

where we have replaced the radius from Eq. (9-9) with the diameter of the resolution element. In this expression, h can be replaced by the slant range, R_S, from Eq. (5-28), to determine the resolution away from nadir (R_S here = D in Chap. 5). Note however, that this is the resolution perpendicular to the line of sight and is made larger (i.e., worse) by $1/\sin \varepsilon$, where ε is the elevation angle at the orbital point in question, obtained from Eq. (5-26a). The ground resolution at nadir for several typical wavelengths and aperture diameters is given in Table 9-8.

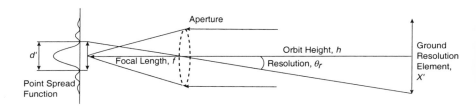

Fig. 9-11.   Point Spread Function for Imaging System with Diffraction. The optical wave front from an ideal point source on the ground is imaged as the point spread function by the optical system. The diameter of the aperture and the wavelength determine the extent of the point spread function measured by the diameter, d', of the first intensity minimum.

When we implement an optical system using a detector array, we add an additional design parameter, the *quality factor*, Q, defined as the ratio of the pixel size, d, to the diameter of the diffraction disk or point spread function, d', i.e.,

$$Q \equiv d \, / \, d' = X \, / \, X' \qquad (9\text{-}11)$$

where d' is the diameter of the first minimum in the diffraction image (i.e., twice the angular resolution), X is the ground pixel size, and X' is the ground resolution = diameter on the ground corresponding to d' on the focal plane (see Fig. 9-12). Q typically ranges from 0.5 to 2. For $Q < 1$, the pixels are smaller than the diffraction disk and resolution is limited by diffraction in the optics. This gives the best possible image

---

* Named for Sir George Airy, the British Astronomer Royal from 1835 to 1881.
† Named for Sir John Rayleigh, a 19th century British physicist and 4th recipient of the Nobel prize for physics.

resolution for a given aperture. For $Q > 1$, the resolution is limited by pixel size. This will be done if image quality is less important than aperture size, as would be the case, for example, when increased light gathering power is required. As a starting point for the design, select $Q = 1$, which allows good image quality.

From the definition of the magnification, Eq. (9-6), we have:

$$d / X = d'/ X' = f / h \tag{9-12}$$

and from the small angle approximation for the angular resolution, θ_r, we have:

$$\theta_r \approx \tan \theta_r \equiv d'/(2 f) = 1.22 \, \lambda / D \tag{9-13}$$

Combining Eqs. (9-11) to (9-13), we obtain expressions for the pixel size, d, in terms of the other basic system parameters:

$$d = d'X / X' = d'Q = (2.44 \, \lambda f / D) \, Q \tag{9-14}$$

where the parameters are defined above and, as usual, λ is the wavelength, f is the focal length, and D is the aperture diameter.

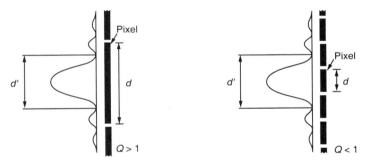

Fig. 9-12.   **Effect of Varying Quality Factor.** Different sizing of the detector pixel with respect to the point spread function is shown by varying the Quality Factor, Q. A large quality factor results in the relative sizing in the diagram on the left and a low quality factor results in the relative sizing in the diagram on the right.

TABLE 9-9.   **Diffraction-Limited Resolution.** Note that the Synthetic Aperture Radar provides resolutions similar to visual or IR systems, independent of range and wavelength for ranges up to the maximum signal-to-noise limit by synthesizing the required aperture.

		Ground Resolution = $2.44 \, h\lambda/D$		
	Aperture Size, D	Visible [$\lambda = 0.5 \, \mu m$]	IR [$\lambda = 3 \, \mu m$]	Passive Microwave [$f = 10$ GHz; $\lambda = 3$ cm]
From an orbiting space-craft at h = 900 km	1 m	1.1 m	6.59 m	65.9 km = 40.9 miles
	3 m	0.366 m = 14.4 in.	2.2 m	22 km = 13.6 miles
From a synchronous spacecraft (h = ~35,800 km)	1 m	43.7 m	262 m	2,620 km = 1,630 miles
	3 m	14.6 m	87.4 m	874 km = 543 miles
From SR-71 at h = 20 km (70,000 ft)	0.3 m	0.081 m = 3.19 in.	0.488 m	4.88 km

9.4 Observation Payload Design

The electromagnetic radiation that forms the basis of remote sensing arises as a by-product of energy being transferred from one form to another. In general, transformation processes that are more random produce wider bandwidth signatures, while a more organized process produces a more coherent return [Elachi, 1987]. For example, heat generated by a diesel motor is radiated over a wide bandwidth in the infrared spectrum, while a laser (a more organized energy transformation) generates narrow bandwidth radiation. In spacecraft remote sensing we are concerned with processing measurements from four primary spectral types.

Visible systems operate from the ultraviolet (~ 0.3 µm) to the red end of the visual spectrum (~ 0.75 µm). They offer the potential for high spatial resolution because of their short wavelengths, but can only operate in daylight because they depend on reflected sunlight.

Infrared systems operate in various bands throughout the infrared spectrum ($\sim 1–100$ µm) subject to atmospheric transmission windows. Infrared sensors can operate both day and night since the detected signal is a function of the emissivity of the scene (although the signatures will be different by day and night).

Microwave radiometers operate in the radio frequency range, chiefly at the millimeter wavelengths (20–200 GHz). Their resolution is three to five orders of magnitude worse than visible wavelength sensors with the same aperture size, but they are capable of collecting unique information over large areas. Typically, microwave sensor require extensive ground-truth calibration data to interpret the measurements.

Radar systems are active instruments which provide their own illumination of the scene in the centimeter to millimeter bands. The reflected signals can be processed to identify physical features in the scene. Radar systems can be designed to penetrate most atmospheric disturbances, such as clouds, because only larger features can reflect signals at radar wavelengths. Cantafio [1989] provides an extended discussion of space-based radar.

There are a number of different approaches for linking the fundamental physics of the Planck function to the practical design of remote sensing systems. Hovanessian [1988] treats emitted radiation as a signal to be detected and considers remote sensing essentially as a special case of antenna and propagation theory (even in the visible spectrum). Elachi [1987] begins with Maxwell's equations and focuses on the features of electromagnetic radiation, such as quantum properties, polarization, coherency, group and phase velocity, and Doppler shift to derive strategies for exploiting these features in different parts of the frequency spectrum. McCluney [1994] draws on the parallels between the science of radiometry and remote sensing in general and the science of the human eye as expressed in the literature of photometry. These references provide detailed, application-specific derivations beginning with Planck's Law. Our focus for the remainder of this chapter will be on engineering applications and rules-of-thumb to define and design remote sensing payloads.

Observation geometry, effective aperture, integration time, detector sensitivity, spectral bandwidth, transmission through the atmosphere, and ground pixel size determine the radiometric performance of an optical instrument. Depending on the spectral range, we define three basic categories of Earth observation. In the first case, the optical instrument receives reflected radiation from the surface of the Earth when it is illuminated by the Sun. The thermal emitted radiation of the Earth's surface is negligible in this case. The frequency range covered by this case includes the visible wavelength (0.4–0.78 μm), the near infrared wavelength (0.78–1 μm), and the short wavelength infrared (1–3 μm).

The second case involves optical instruments receiving emitted radiation from the surface of the Earth when the reflected radiation of the sun is negligible. This condition holds for the long wavelength infrared region (8–14 μm). The third case applies to the mid-wavelength infrared spectral region (3–5 μm) where we must consider contributions from direct and reflected sources. Figure 9-13 shows the radiance available from direct and reflected radiation sources. Corresponding to Planck's law, the thermal emitted radiance of the Earth (modeled at 290 K) increases with wavelength for the spectral region shown. The reflected radiance from the Earth's surface decreases with wavelength.

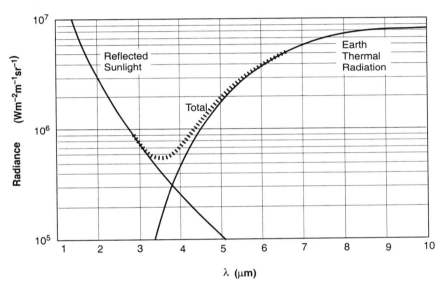

Fig. 9-13.   **Radiance from Direct and Reflected Sources.** Radiance contribution in W per square meter per meter (wavelength) per unit solid angle of reflected sunlight from the Earth and emitted radiation from the Earth as a function of wavelength. The sum is shown as a dashed line. The Sun is modeled as a blackbody with a temperature of 6,000 K, the reflection coefficient of the Earth's surface and the transmission of the atmosphere are modeled as constants for clarity.

In the visual and near IR (0.7 to 1.0 μm) bands, we resolve images produced by energy (chiefly from the Sun) reflected from the target scene rather than energy from the very limited self-emissions that occur in the visible band. But in the infrared, we see things almost entirely by their self-emission, with very little energy being reflected, particularly at night. We may use the same optical train elements—lenses, prisms, mirrors, and filters—to collect infrared energy as for visible and UV, but we must apply them differently. For example, ordinary glass is opaque to IR beyond 3 μm, whereas germanium, which is opaque in the visible band, is transparent in the 1.8 to 25-μm region. Further, we must consider atmospheric scattering caused by aerosols and particles in the air. The amount of scattered radiation is a function of the inverse fourth power of the wavelength. Thus, IR penetrates haze and dust much better than visible radiation because the IR wavelengths are four or more times those in the visible spectrum. The same phenomena explain the reddish color of the sky near dawn and sunset. At these times, shorter green, blue indigo, and violet wavelength signals are greatly attenuated as they travel farther through the atmosphere than when the Sun is overhead.

9.4.1 Candidate Sensors and Payloads

Electro-optical imaging instruments use mechanical or electrical means to scan the scene on the ground. Spacecraft in geostationary orbits perceive very little relative motion between the scene and the spacecraft, so an optical instrument needs to scan in two dimensions to form an image. A common approach for geostationary imaging spacecraft, such as ESA's meteorological spacecraft, METEOSAT, involves placing a large scan mirror in front of the instrument's optics to perform the north-south scan. Rotation of the spacecraft around a north-south axis performs the east-west scan. Three-axis stabilized spacecraft in geostationary orbits frequently use a two-axis scan mirror in front of the optics to scan the scene in two dimensions. Alternatively, we can use a two-dimensional matrix imager, which maps each picture element (pixel) in the imager to a corresponding area on the ground. Scanning the scene then becomes a process of sampling the two-dimensional arrangement of pixels in the imager.

Spacecraft in low-Earth orbits move with respect to the scene. The sub-spacecraft point moves along the surface of the Earth at approximately 7,000 m/s (see Chap. 5). This motion can replace one of the scan dimensions, so the scanning system of the optical instrument needs to perform only a one-dimensional scan in the cross-track direction. *Whiskbroom sensors* scan a single detector element that corresponds to a single pixel on the ground in the cross-track direction. Fig. 9-14A illustrates this technique. Whiskbroom scanners can also use several detectors to reduce the requirements compared to a single detector. Each detector element corresponds to a pixel on-ground (see Fig. 9-14B), and the dwell time per pixel is multiplied by the number of detector elements used.

Push broom scanners use a linear arrangement of detector elements called a line imager covering the full swath width. The name "push broom" comes from the read-out process, which delivers one line after another, like a push broom moving along the ground track. Each detector element corresponds to a pixel on-ground. Fig. 9-14C illustrates this technique. The ground pixel size and the velocity of the sub-spacecraft point define the integration time.

Step-and-stare scanners use a matrix arrangement of detector elements (matrix imager) covering a part or the full image. Each detector element corresponds to a pixel on-ground. Fig. 9-14D illustrates this technique. Step-and-stare systems can operate in two basic modes. The first mode uses integration times that are chosen as in the case of the push broom sensor for which the ground pixel size and the velocity of the sub-satellite point determine the integration time. Thus, no advantage with respect to the integration time is achieved, but a well known geometry within the image is guaranteed. We need a shutter or equivalent technique, such as a storage zone on the imager, to avoid image smear during read-out. The second mode allows a longer integration time if the image motion is compensated to very low speeds relative to the ground. We can do this by shifting the imaging matrix in the focal plane or by moving the line of sight of the instrument by other means to compensate for the movement of the sub-spacecraft point. Step-and-stare sensors require relatively complex optics if they must cover the full image size. An additional complexity is that the fixed pattern noise has to be removed from the image, since each pixel has a somewhat different responsive-

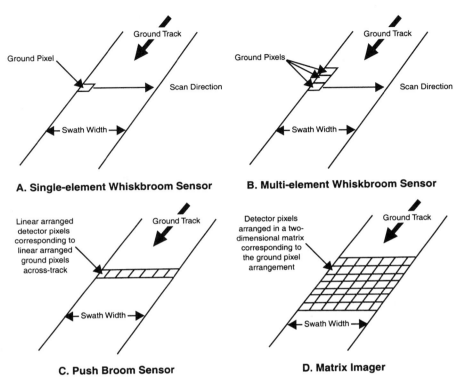

A. Single-element Whiskbroom Sensor

B. Multi-element Whiskbroom Sensor

C. Push Broom Sensor

D. Matrix Imager

Fig. 9-14.   Scanning Techniques for Electro-Optical Instruments. (A) Shows a whiskbroom scanner with a single detector element which scans one line after another across the swath. The swath width must be scanned in the time interval the sub-spacecraft point moves down one ground pixel length. (B) Shows a whiskbroom scanner with multiple detector elements which scan multiple lines across the swath at a time. The swath width must be scanned in the time interval the sub spacecraft point moves down the multiple ground pixel length. (C) Shows a push broom scanner with multiple linearly arranged detector elements which scan one line across the swath per integration time. The integration time is usually set to the time interval the sub-spacecraft point moves down one ground pixel length. (D) Shows a step-and-stare scanner with detector elements arranged in a matrix which scan the full image per integration time. The integration time is usually set to the time interval the sub-spacecraft point needs to move down one ground pixel length.

ness and dark signal. Table 9-10 summarizes the distinguishing features of optical scanning methods.

An alternate approach for capturing the scene using matrix imagers involves positioning of the scene with respect to the instrument. To image the entire scene, the instrument shifts, or "steps," to the next part of the scene (along-track and/or across-track) after the integration period. This approach is referred to as a *step-and-stare* imager. If it only covers a part of the required scene, then moderately complex (and also moderately sized) optics is required. We can use highly agile and accurate pointing mirrors in front of the instrument's optics to adjust the line of sight. For example, the Early Bird satellite avoids the complexity of large matrix imagers or sophisticated

TABLE 9-10. Comparison of Optical Sensor Scanning Methods. We list relative advantages and disadvantages of different scanning mechanisms.

Scanning Technique	Advantages	Disadvantages
Whiskbroom Scanner— Single Detector Element	High uniformity of the response function over the scene Relatively simple optics	Short dwell time per pixel High bandwidth requirement and time response of the detector
Whiskbroom Scanner— Multiple Detector Elements	Uniformity of the response function over the swath Relatively simple optics	Relatively high bandwidth and time response of the detector.
Push Broom Sensor	Uniform response function in the along-track direction. No mechanical scanner required Relatively long dwell time (equal to integration time)	High number of pixels per line imager required Relatively complex optics
Step-and-Stare Imager with Detector Matrix	Well defined geometry within the image Long integration time (if motion compensation is performed).	High number of pixels per matrix imager required Complex optics required to cover the full image size Calibration of fixed pattern noise for each pixel Highly complex scanner required if motion compensation is performed.

butting techniques of several smaller matrix imagers in favor of pointing the mirror with high dynamics and fine pointing performance.

Optical instruments for space missions usually rely on existing detector and imager designs. Custom tailoring of these detectors and imagers is common, however, to optimize the design with respect to performance and cost of the instrument. We make the distinction between detectors, which consist of one or a few radiation-sensitive elements without read-out electronics, and imagers, which usually consist of a considerable number of discrete radiation-sensitive elements combined with read-out electronics.

We must select the materials used for detector elements depending on the spectral range of the instrument being designed. The ability of detector elements to absorb photons relates directly to the energy of the incident photons (and consequently to the wavelength of the radiation as well) and the effective band gap of the material. All matter, including the detector material, generates thermal photons. Therefore, we must lower the temperature of the detector elements such that the self-generated photons do not degrade the signal-to-noise ratio of the instrument. This requirement becomes more stringent as the wavelength of the radiation being detected increases. With few exceptions, detectors and imagers have to be cooled for wavelengths in the short wave infrared (SWIR) band and longer.

For the spectral range between 400 nm and 1,100 nm, silicon detectors and imagers are used most frequently. Silicon is attractive because it is possible to combine the detector elements and read-out electronics in a single monolithic chip. We can produce line imagers with a large number of elements through this process.

Incident photons on a line imager are converted to an electrical output signal in the imager. For charge-coupled device (CCD) line imagers with read-out electronics, the process begins when incident photons are converted by each pixel (detector element) into electrons according to a conversion efficiency dictated by the characteristic spectral response. The electrons are then collected for the entire integration time for each pixel. The read-out of the charge generated—up to one million electrons per pixel—is performed via an analog shift register into a serial output port. Figure 9-15 shows a typical spectral response function for a silicon imager.

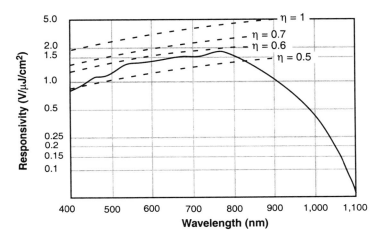

Fig. 9-15.   Spectral Response Function of a Silicon Line Imager. The spectral response is shown in terms of the output voltage resulting from illumination by energy density vs. the wavelength from 400 to 1,100 nm. The imager reaches a *quantum efficiency* (generated electrons per incident photon), denoted by η, above 50% between 400–800 nm. Outside this wavelength region the quantum efficiency drops to small values making the imager less suitable above 900 nm.

Area array imagers, or matrix CCD imagers, provide an alternative to line imagers. The principles of operation are essentially the same as line imagers. Area array imagers offer the advantage of undistorted geometry within the image. A disadvantage compared to line imagers is the possible smear effect during frame transfer. There are a variety of read-out techniques to compensate for the smear effect. Matrix imagers can also suffer from a relatively poor fill factor of pixels in the array. Table 9-11 summarizes line and matrix imager capabilities for current systems that are at least partially space qualified.

When radiometric performance of the optical instrument is paramount, we use time delay and integration (TDI) methods. TDI describes an imaging principle that uses the image motion along the rows of a matrix imager to extend the integration time. Integration time is extended by electronically shifting the integrating pixel cell synchronously to the movement along the row. The signal-to-noise ratio of this concept is improved by the square root of the number of TDI stages. The primary advantage of TDI imager systems compared to line imagers is the improved signal-to-noise ratio. The disadvantage is the increased requirement for spacecraft attitude and orbit stability (due to the required synchronization of the shifting pixel).

We classify and select infrared detectors according to their spectral band of operation and a figure of merit called specific detectivity or quantum efficiency for photon detectors. The operating temperature of the detector dictates the cooling requirements

TABLE 9-11. Characteristics of Imagers. Typical parameters for available line and matrix imager systems. *Photo response nonuniformity* is the difference between the most and least sensitive element under uniform illumination. *Dark signal uniformity* is equivalent to photo response nonuniformity, but without illumination. *Dynamic range* is the saturation exposure divided by the rms noise-equivalent exposure. Read-out speed is given in million samples per second (Msps) per output port.

Characteristic	Line Imager	Matrix Imager
Pixels	6,000 – 9,000 pixels	Up to 1,024 × 1,024 image pixels in frame transfer mode
Photo response nonuniformity	5%	5%
Dark signal nonuniformity	5%	5%
Dynamic range	10,000	5,000
Limitations on read-out speed	~10 Msps	For example, 4 ports each at 20 Msps

for the sensor focal plane. Infrared sensors often have nonnegligible time constants for response with respect to integration time. Because of technical difficulties with combining detectors and read-out structure, the total number of pixels in an IR detector array is limited in practice to several hundred.

We detect infrared wavelengths with thermal detectors or photon detectors. Thermal detectors exploit the fact that absorbed heat raises the temperature of the detector, which changes its electrical characteristics. The advantage of thermal detectors is uniform response with respect to wavelength. Thermal detectors can also be operated at ambient temperatures, although they have lower sensitivity and slower response times.

Photon detectors use absorbed photons to generate charge carriers. These systems offer the advantages of higher sensitivity and shorter time response, but they must be operated at low temperatures.

Infrared detectors are often rated by the specific detectivity, $D*$, given by

$$D* = \frac{\sqrt{A\Delta f}}{NEP} \tag{9-15}$$

where A is the detector area, Δf is the noise equivalent bandwidth, and NEP is the noise equivalent power of the detector. The factor $D*$ is strongly wavelength dependent showing its peak value at the cut-off frequency. Figure 9-16 shows the specific detectivities and the operating temperatures for infrared detectors, and Table 9-12 gives characteristics of infrared detector arrays with read-out electronics.

The selection of a detector or detector array (usually with a read-out multiplexer) is driven by several factors. The primary design issues center on maximizing detectivity in the spectral band of interest while operating at the highest possible temperature and a sufficiently small time constant. In addition, we must consider the geometry of the detector and the array as well as associated calibration issues.

9.4.2 Payload Operations Concept

In addition to the technical trade-offs in spacecraft performance, the operations concept for employing the sensor is an important consideration early in the preliminary design. We need to understand the end-to-end mission problem—not merely the physics of collection. The entire process beginning with the ultimate users or customers of the data needs to provide a feasible and efficient means to meet mission objectives.

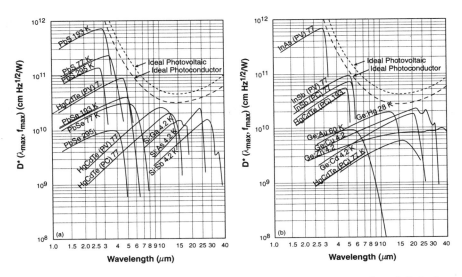

Fig. 9-16. Specific Detectivities and Operating Temperatures for Infrared Detectors. (Courtesy of Santa Barbara Research Center.) The specific detectivity, D^*, is a normalized figure of merit for the class of infrared detectors for which the noise (voltage or current) is proportional to the square root of the detector area and the electrical noise bandwidth. The values given vs. the wavelength show a sharp cut-off to higher wavelengths when the lower energy incident photons no longer generate sufficient charge carriers within the detector. Additionally, the operating mode (photo-voltaic or photo-conductive) and the operating temperature of the detector is indicated. D^* can be interpreted as the normalized inverse noise-equivalent power of the detector (with dimensions expressed as the square root of the detector area times the square root of the noise bandwidth divided by the noise-equivalent power).

TABLE 9-12. Characteristics of IR Detector Arrays with Read-out Electronics. The table shows detector characteristics based on material properties. (Source: *Photonics Spectra*, September, 1989.)

Detector Material	Usable Spectral Response Region (μm)	Operating Temperature (K)	Array Configuration	Pixel Size (μm)	Time Constant τ (μs)
InSb	3–5	77	128 × 128	100 × 100	0.020–0.200
InSb	3–5	77	1,024 × 1	100 × 100	0.020–0.200
InSb	3–5	77	64 × 2	40 × 40	0.020–0.200
HgCdTe	3–5	195	256 × 256	100 × 100	0.2–0.8
HgCdTe	3–5	195	512 × 30	100 × 100	0.2–0.8
Si:In	3–7	< 50	64 × 64	65 × 65	100
Si:In	3–7	< 50	1,024 × 1	65 × 65	100
HgCdTe	8–12	77	256 × 256	100 × 100	0.2–0.8
HgCdTe	8–12	77	1,024 × 1	100 × 100	0.2–0.8
HgCdTe	8–12	77	288 × 4	25 × 28	0.2–0.8
Si:Ga	8–17	< 30	64 × 64	75 × 75	0.1–5
Si:Ga	8–17	< 30	1,024 × 1	75 × 75	0.1–5

The nomination of a collection task, the tasking and scheduling of the sensor, the processing of the mission data, and the distribution of the data can dramatically increase the complexity of the systems engineering challenge and decrease the final accuracy of the system. Not only can physical effects such as atmospheric correction, calibration, and rectification degrade system performance, but technical effects such as quantization and data compression errors can decrease the resolution of the system from the perspective of the end user. For additional information about technical aspects of the end-to-end throughput problem for spacecraft imagery, see Shott [1997].

The concept of operations for a spacecraft system such as FireSat needs to account for the full breadth of the operational mission, including different phases of the mission and alternate operating modes. See Sec. 2.1 and Chap. 14 for a description of a mission operations concept.

For preliminary mission planning, we should pay particular attention to the projected sequence of events during each mission phase (see Activity Planning in Chap. 14). For the FireSat mission under normal operations, a sample mission timeline for normal operations includes the following steps:

1. Fire starts at some location

2. Sensor field-of-view passes over the fire

3. Signature from the fire introduced into the sensor data stream

4. Data is passed to the mission ground station for analysis (or processed on board)

5. Fire detection algorithm determines the possible presence of fire (this may be a multistage process with a preliminary, coarse fire detection process that triggers a more precise algorithm or set of measurements)

6. Generate appropriate messages indicating the presence of fire

7. Issue reports and notifications to appropriate authorities and research centers

8. Monitor the fire (this could involve switching to an alternate operating concept that tracks the progress of the existing fire and monitors surrounding areas for new outbreaks)

We should create a concept of operations for each phase of the mission and each operating mode of the spacecraft—including contingency and failure modes. This step will ensure mission success within the constraints of the operating environment. (See Chap. 14.)

9.4.3 Required Payload Capability

Frequently there are several ways to meet mission requirements. How to sort through these multiple approaches is not always obvious. The general approach we outline provides a repeatable framework for choosing a payload to satisfy a remote sensing mission. Once we select a physical phenomenology (e.g., measuring thermal infrared radiance to detect forest fires), then two things need to be established. First, the radiometric measurement levels that are needed to satisfy the information need; and second, the implications for a payload in terms of size and performance to be able to sense the required signature.

Categorizing remote sensing missions is complicated by the fact that sensors usually have multiple uses, and they can be categorized according to any number of different aspects, such as measurement technique (active or passive), event measured (such as fire or deforestation), and measurement resolution (spatial, spectral, radiometric, temporal). By way of example, however, Table 9-13 provides a small sampling of remote sensing payloads and corresponding spacecraft missions.

TABLE 9-13. Characteristics of Typical Payloads.

Purpose	Instrument Name	Size L×W×D (m)	Mass (kg)	Avg. Pwr. at 28 V (W)	Data Rate (Mbps)	Aperture (m)	Pointing Accuracy (deg)
Solar Physics	Lyman-Alpha Coronograph	2.8 × 0.88 × 0.73	250	87	13.5	—	0.003
	X-ray Telescope Spectrom.	2.7 × 1 dia.	465	30	0.4	—	0.003
	Solar Optical Telescope	7.3 × 3.8 dia.	6,600	2,000	50+	1.25	—
	Solar Magnetic Velocity Field	2 × 0.4 × 0.4	183	322	2+	—	0.003
	100 m Pinhole Camera	1 × 1 × 2	1,000	500	0.5	—	—
	Extreme UV Telescope	2.78 × 0.86 × 0.254	128	164	1.28	—	—
	Solar Gamma Ray Spectrom.	1 × 1 × 3	2,000	500	0.1	140 cm²	0.003
Space Plasma Physics	Ion Mass Spectrometer	0.5 × 0.5 × 0.4	80	334	0.01	—	1
	Beam Plasma	0.6 × 0.7 × 0.7 + two 0.7 dia. ant.	17	38	0.016	—	5
	Plasma Diagnostics	—	2,000	250	50	—	—
	Doppler Imaging Interferom.	(0.25)³	100	620	0.2	—	—
	Proton (Ion) Accelerators	6.7 × 3.4 × 3.10	500	1,500	0.256 (4.2 TV)	—	1
High Energy Astrophysics	Gamma Ray Burst	2 × 4 dia.	1,000	120	0.01	3	—
	Cosmic Ray Transition	3.7 × 2.7 dia.	1,500	230	0.10	2.70	—
	X-ray Spectrom./Polarimeter	1.6 × 1.6 × 3	2,000	300	0.03	—	0.1
	Short X-ray	1 × 1 × 3	1,000	300	0.025	1 × 3	0.1
	Hi Energy Gamma Ray Tele.	3 dia. × 4	10,000	100	0.003	3	0.1
Resources	Gravity Gradiometer	0.23 m sphere	10	1			1–2
	Synthetic Aperture Radar	2.8 × 3.7 × 1.4	808	3,000	120	8 × 2.8	2.5
	Multi-Spectral Mid-IR	1.5 × 1 dia.	800	900	30	1	0.1
	Thematic Mapper	2 × 0.7 × 0.9	239	280	85	0.406	0.08
Materials Processing	Materials Experiment Assem.	1 × 1 × 2	900	500	0	—	—
	Solidification Experiment		1,100	3,000	0.02	—	—
Life Sciences	Life Science Lab Module	7 × 4 dia.	6,800	8–25 kW	1.0	—	—
Environmental	Limb Scanning Radiometer	4.8 × 1.9 dia.	~ 800	125	0.52	—	—
	Microwave Radiometer	4 × 4 × 4	325	470	0.20	4	±0.1
	Dual Frequency Scatterom.	4.6 × 1.5 × 0.3	150	200	0.01	4.6 × 0.3	1
	Ocean SAR	20 × 2 × 0.2	250	300	120	20 × 2	0.1
	Solar Spectrum	0.4 × 0.3 × 0.6	16	60	Low	—	±3
	Doppler Imager	1.25 × 0.6 × 0.8	191	165	20	—	±3
	Photometric Imaging	1.4 × 1.4 × 0.5	147	330	0.01	—	±1.5
Comm.	TDRS Comm. Payload	2.5 × 2.5 × 1	680	715	300 (×2) + 50	4.8	~ 0.3
	DSCS III	—	550	491	—	—	—

The measurement techniques employed in sensors are tailored to provide information that can be exploited to understand the subject. We describe the fundamental information content provided by passive remote sensing instruments in terms of spatial, spectral, and radiometric resolution. We then introduce the basic types of detectors and collection techniques that are employed in the design of a remote sensing instrument.

We use various applications of imaging sensors, such as film cameras or electro-optical devices, to measure and analyze spatial features. Optical imaging in the visible spectrum is the most common approach for applications dealing with topographic mapping and photogrammetry. Other sensors that rely fundamentally on spatial measurements include sounders and altimeters (Kramer [1996] lists examples of these types of sensors).

Spatial resolution is a function of many different parameters of the remote sensing system. It is usually different from the *ground sample distance* (GSD), the distance at which the sensor spatially samples the target scene for sensors that have a correspondence between detector size and ground target size. For targets with very high contrast to the background, the spatial resolution can be finer than the GSD—usually however, it is on the order of twice the GSD. Alternatively, the spatial resolution can be characterized by the angle under which it sees the smallest target. The smallest physical features that can be discriminated using the sensor measurement characterize the limits of spatial resolution in a remote sensing system. Spatial resolution is a function of the range from the sensor to the target, the aperture and focal length of the lens, and the wavelength of the incident energy. We can characterize the spatial resolution by the angle that the sensor can resolve, or directly as the ground range in units of length for the size of the smallest object that can be discriminated.

We use spectrometers to analyze spectral content of a scene to identify the chemical composition of the objects being sensed. The spectral information received by a sensor is a composite of the spectral information emitted by all objects in the field-of-view. The spectral content reaching a potential multispectral FireSat payload, for example, will include signatures from soil, vegetation, and cities in addition to a fire that may be present. The combined spectral content of all these features may be very different from the spectrum of a forest fire by itself. Combining information from multiple spectral bands has been used successfully to differentiate key features and maximizing the utility of a sensor with a given spectral resolution.

We can use a multispectral sensor to uniquely determine the features within an image. Multispectral systems typically employ tens of bands. Hyperspectral and ultra-spectral systems employ hundreds and thousands of spectral bands, respectively. Following the example presented by Slater [1980, pp. 17–18], we consider a multi-spectral image of an area containing concrete, asphalt, soil, and grass. Figure 9-17 shows typical spectral reflectance for these four materials. If we have an imager with only three bits of radiometric sensitivity (low, medium, and high), then no single image, whether panchromatic or filtered to a particular band region could distinguish these four materials. However, a properly calibrated two-band multispectral system can uniquely resolve these four materials using bands in the 600–700 nm and 700–900 nm wavelengths. The returns for each of these materials is shown in Table 9-14.

Using multispectral sensors to study more complex scenes or to identify additional materials requires more bands or narrower bandwidths. The number of spectral bands and the bandwidth of each band determine the resolving power of a spectral sensor. We achieve higher spectral resolution is achieved by narrower bandwidths, but using narrower bandwidths tends to reduce the signal-to-noise ratio for the measurement.

We can also use intensity information contained in the electromagnetic field to extract useful information for remote sensing purposes. If we have an imager with four or more bits of sensitivity operating in the 700–900 nm wavelength, then a single

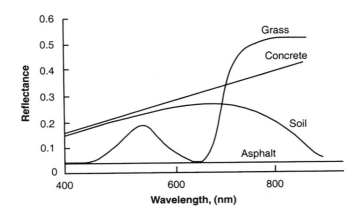

Fig. 9-17. **Typical Spectral Reflectance Curves for Grass, Concrete, Soil, and Asphalt.** A single-band low-resolution sensor cannot distinguish all 4 types, but a 2-band, low-resolution sensor can, as shown in Table 9-14 [Slater, 1980].

TABLE 9-14. **Spectral Resolution for Sample Two-Band Sensor.** Combining low-resolution sensor returns for the two-band sensor can uniquely identify the four materials from Fig. 9-17.

Material	Sensor Reading 600–700 nm	Sensor Reading 700–900 nm
Concrete	Medium	High
Asphalt	Low	Low
Soil	Medium	Low
Grass	Low	High

image could be used to distinguish all four materials represented in Fig. 9-18. The ability to distinguish the intensity of radiance at the sensor goes beyond detecting either the presence or absence of energy at a given wavelength. For example, if a FireSat sensor were being designed to detect fires based on thermal emission, then the intensity information in the thermal signature would indicate the presence of fires. The radiometric resolution of the instrument needs to allow the system to discriminate between a forest fire and a campfire, for instance.

Merely detecting the presence of thermal energy in a band region characteristic of burning biomass is not sufficient for satisfying the FireSat mission. The number of bits used to represent the intensity information for a radiometric instrument will be dictated by two practical limits [Slater, 1980, p. 19]: (1) the signal-to-noise ratio present in the data, and (2) the level of confidence we need to differentiate between two threshold signal levels.

Radiometric instruments measure the intensity of incoming energy. *Radiometers* passively measure the intensity, while *scatterometers* are active instruments that measure surface roughness by sensing the backscattered field when a surface is illuminated. *Polarimeters* measure the polarization state of a transmitted, scattered, or reflected wave. Refer to Table 9-13 to see examples of these sensor types.

The limiting factors in intensity measurement using radiometry are the signal-to-noise ratio (SNR) of the sensor and the quantization levels of the measurement device. Applications requiring high radiometric accuracy include vegetation and soil analysis studies, while applications such as sea surface temperature studies require less radiometric resolution.

Figure 9-18 shows the resolution characteristics of an example sensor across the three primary dimensions. The sensor is the MODerate-resolution Imaging Spectroradiometer (MODIS), a part of NASA's Earth Observing System. The MODIS scanning radiometer has 36 channels that have been selected to enable advanced studies of land, ocean, and atmospheric processes. We will discuss applications of the MODIS sensor to the problem of automatic fire detection in Sec. 9.6.

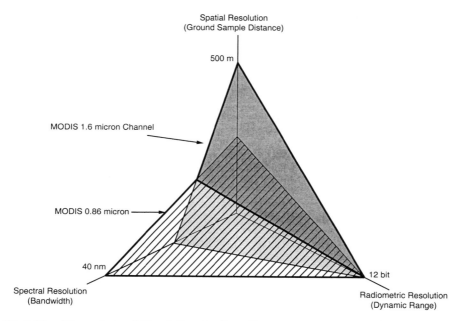

Fig. 9-18.   Dimensions of Resolution for Two Channels of the MODIS Instrument. The diagram illustrates the spatial, spectral, and radiometric dimensions of measurement resolution for MODIS. The spatial resolution of the 1.6 micron channel is 500 m at a spectral resolution of 20 nm compared to 250 m and 40 nm for the 0.86 micron channel. Both channels have 12-bit radiometric resolution. Data from King, et al. [1992].

9.5 Observation Payload Sizing

We must be able to estimate the size and main characteristics of the mission payload before completing a detailed design. We want to be able to look at several options without necessarily designing each in depth. This section provides ways to compute data rates and estimate the overall size and key parameters. In Sec. 9.6, we will apply these values to the FireSat example.

9.5.1 Signal Processing and Data Rates

Analog signal processing is very similar for CCD read-out imagers and for infrared imagers using an integrated mulitplexer. In both cases weak analog signals need to be amplified and conditioned, maintaining high dynamic range and high processing speed. Electronic signal processing at high speeds with high accuracy can become a cost and schedule driver in the development of high-resolution instruments. Typically we must incorporate massively parallel processing and tailored implementation technologies to allow high-speed, high-accuracy analog processing. Concerns about electromagnetic interference dictate that this processing be conducted as close to the focal plane assembly as possible. On the other hand, strict thermal decoupling between imagers and the heat dissipating electronics must be maintained.

For CCD imagers the dark signal typically doubles with each rise in temperature of seven degrees. Infrared imagers are usually cooled to low temperatures; therefore, we must minimize each heat leak to their cryostat to keep the cooling power low. The cooling requirement becomes more stringent as the wavelength of the radiation to be measured becomes longer. For wavelengths in the 100 μm range the detectors are typically cooled to 4 K. Even small heat leaks, such as those from the necessary electrical wiring and mechanical connection to the spacecraft, transport excess heat requiring high cooling power.

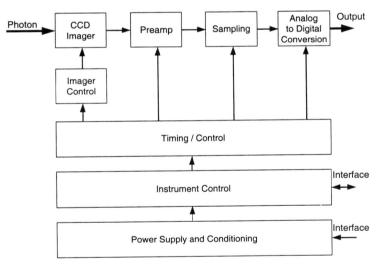

Fig. 9-19.   **Block Diagram of an Optical Instrument.** Photons are collected by the CCD imager, then amplified and processed. The primary supporting functions of power and control are also depicted. The sampling process for charge coupled device signal processing typically uses correlated double sampling to eliminate reset noise.

The block diagram in Fig. 9-19 shows the typical functional blocks found in the electronics of an optical instrument. The signal flow through the electronics begins with the detector in the focal plane of the instrument. It converts photons to analog electrical signals, which are amplified and conditioned. Any additional signal processing is performed in the digital data processing block. The analog and digital signal processing blocks operate synchronously with the read-out of the imagers. The central timing system supplies appropriate timing signals to all relevant circuit blocks to guarantee synchronous operation. Because electrical signals processed in the instrument electronics are weak, we have to pay special attention to power conditioning. Careful filtering and clean electrical grounding must be implemented to decouple digital and analog signals.

The instrument control computer manages the signal processing and timing functions performed by the instrument and it interfaces to the main spacecraft computer. Depending on the design, time-tagged commands are executed by the instrument computer or issued by the spacecraft computer.

Analog signal processing for CCD imagers usually involves correlated double sampling. This technique takes two slightly time-shifted samples of the analog signal and subtracts one from the other to extract their image-related video signal. The adaptation of this video signal to the input range of the analog-to-digital converter requires setting gain and offset parameters for the system. Digital data processing normalizes the imager pixels. Normalization requires that any non-uniformity in photo response and dark signal must be removed from each pixel. This task is frequently conducted on board the spacecraft because some of the most straightforward data compression algorithms become invalid if the pixels are not normalized. Furthermore, signal processing for normalizing pixels may be required if we use onboard calibration methods in the sensor.

We must select the process of normalizing pixels based on the characteristics of the sensor. For sensors with a linear response, one- or two-point correction is sufficient. Highly nonlinear detectors, as is often the case for infrared detectors, require n-point correction techniques. Figure 9-20 illustrates an example of *two-point correction*. The offset (dark signal) and gain factor (response) are corrected by first subtracting the individual offset value and then by dividing by the individual response for each pixel.

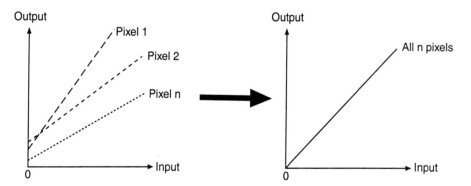

Fig. 9-20. Two-point Normalization of Pixel Response. The left diagram shows the original response functions of individual pixels. The right diagram shows the normalized response function of the same pixels, where the effects of the variations in offset values and responsivity values have been removed from the individual pixels.

High-resolution optical instruments typically generate data rates on the order of several hundred Mbps and above. To send this data stream to a ground station in real time, the system may need several parallel channels with capacities up to 100 Mbps each. If data collection is short compared to the available downlink time, a buffer memory can reduce the downlink data rate over a longer transmission time.

The trends for future systems indicate a tendency toward more onboard information extraction to decrease communication requirements. Several enabling technologies and techniques include powerful data processing hardware. Software uploads to the hardware must be possible to update and modify operational algorithms from the user community.

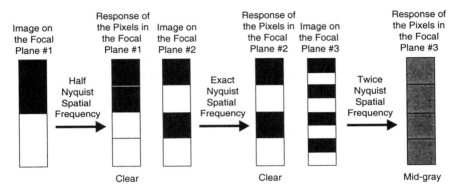

Fig. 9-21.   **Effect of Sampling Frequency on Image Quality.** The three diagrams show illumination patterns having different spatial frequencies present in the scene (left of the arrows) sampled at the same sampling frequency, that is, the same detector array (right of the arrows). The response of the pixels is shown as intensity (gray level) in the three diagrams. From left to right: Sampling an illumination pattern with half the Nyquist frequency. Sampling an illumination pattern with exactly the Nyquist frequency (the highest possible frequency that can be reconstructed without error). Sampling an illumination pattern with twice the Nyquist frequency, resulting in a pixel response of mid-gray since every pixel is illuminated over exactly half its area.

The spatial sampling process of pixels in an optical instrument is determined by the geometry of the detector elements and by the scanning principles employed for the different sensor types discussed in Table 9-8. The sampling process can introduce errors into the image data if we don't select it properly. The *Nyquist frequency* is the lowest rate at which information with a given bandwidth must be sampled to avoid errors. Figure 9-21 shows the effect of sampling three different illumination frequencies. The *spatial frequency* is defined as the inverse of the width of a black-and-white line pair, and is expressed in line pairs per meter. The sampling process is performed in all three cases by the same detector array (right of the arrows), that is, by the same spatial sampling frequency, defined as the inverse of the center distance of the sampling pixels. The maximum spatial frequency which can be reconstructed without error after the sampling/restoration process is defined as the *Nyquist frequency*, which equals half the sampling frequency, shown in the middle diagram. The response of the detector pixels is shown as intensity (gray level) in the three diagrams. On the left, sampling of an illumination pattern with half the Nyquist frequency. In the center, sampling of an illumination pattern with exactly the Nyquist frequency, showing that reconstruction of the original illumination pattern is feasible without error. (However, if the sampling pattern is shifted by half a pixel with respect to the illumination pattern, then reconstruction fails). And on the right, sampling an illumination pattern with twice the Nyquist frequency results in a pixel response of mid-gray since every detector pixel is illuminated over half its area in the example. All samples in the third case show the same constant value which corresponds to a spatial frequency of zero (line-pairs per meter) which was not present in the original scene. This creation of new frequencies is known as *aliasing*. Such frequency components cannot be removed by additional processing of the reconstructed image. Furthermore, the well known techniques to eliminate aliasing in electronic signal processing (band limiting the input signal to an electronic sampler using a low pass filter corresponding to the Nyquist condition) are not feasible when we sample images in the spatial domain. Usually we cannot use our optical system as a band-limiting low pass filter since its aperture diameter and cut-off frequency are usually defined by radiometric requirements. The result is that we usually have a certain amount of unavoidable aliasing which degrades the image quality of such systems.

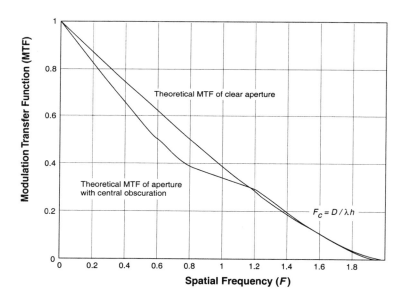

Fig. 9-22.   Modulation Transfer Function of Circular Diffraction-Limited Optical System vs. Spatial Frequency. MTF curves for a clear circular optical system is compared with one having a central obscuration (as found in on-axis reflective telescopes). The MTF can be thought of as a function dependent on the spatial frequency, *F*, which describes the modulation (contrast) function through the optical system (analogous to the frequency dependent gain of an electrical transmission block). The MTF starts at 1 for spatial frequencies near 0 and drops to 0 at the cutoff frequency, $F_c = D/\lambda\, h$.

The *Modulation Transfer Function* (MTF) is the ratio of the intensity variation of the ground scene to the intensity variation of the image at a given spatial frequency. The cut-off frequency is that spatial frequency at which the transfer function becomes zero. Figure 9-22 shows the theoretical MTF of an optical system with and without central obscuration. The theoretical MTF can be approximated by a line starting at 1 when the spatial frequency is 0 and falling to 0 at the *cut-off frequency*, $F_c = D/\lambda h$, where D is the aperture diameter, λ is the wavelength, and h is the altitude. It is the autocorrelation function of the effective aperture. In optical terms, the MTF is the absolute value of the complex Optical Transfer Function (OTF) which describes how the complex amplitudes of the optical wave front are transferred by an optical system at different spatial frequencies.

The MTF describes the transfer quality of an optical system as a function of spatial frequency. The point-spread function illustrated in Fig. 9-16 describes exactly the same properties by showing the two-dimensional intensity distribution in the focus of the optical system. The two are interrelated by the Fourier transform function.

9.5.2 Estimating Radiometric Performance

In order to estimate the radiometric performance of optical instruments in the visible or near infrared we start with the radiometric input of the Sun shown in Fig. 9-23.

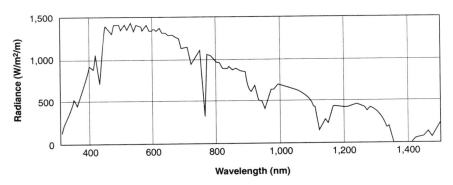

Fig. 9-23.   **Solar Radiometric Input.** Radiometric input (radiance in $Wm^{-2} m^{-1}$ [wavelength]) of the Sun at sea level as a function of the wavelength.

The integration of the spectral radiometric input over the spectral bandwidth gives the power density in the spectral band of interest. To first-order we can assume that lambertian (ideal) reflection with a constant reflection coefficient occurs at the target scene (this approximation holds for small spectral bandwidth). The area of the ground pixel resulting in back-radiated power determines the power density per solid angle. The atmosphere attenuates this radiation by a constant transmission factor (again invoking an approximation for small spectral bandwidth). The effective aperture at orbital altitude collects a very small fraction of this radiation resulting in the power at the entrance of the optics. The signal power is attenuated further by transmission through the optics, ultimately resulting in a lower power level at the detector pixel. During the integration period a certain amount of energy (power times integration period) is accumulated in each pixel. This energy is divided by the energy of one photon (which is wavelength dependent) resulting in the number of available photons per pixel. The *quantum efficiency* of the detector transforms this number of photons into the number of available electrons. These electrons comprise a charge packet and correspond to the output signal of the detector.

To fully characterize the radiometric performance of an instrument, we must also determine the signal-to-noise ratio and dynamic range. The *signal-to-noise* ratio describes the image quality at a given intensity. Due to the quantum nature of light, the number of noise electrons (temporal noise) equals the square root of the number of signal electrons. The read-out process of the imagers results in a certain number of additional noise electrons. The temporal noise is added to the read-out noise since they are statistically independent values, resulting in a total number of noise electrons to be considered for the evaluation of the signal-to-noise ratio.

The *dynamic range* of the instrument is the quotient of signal- and read-out noise electrons the sensor sees between dark and bright scenes at the given reflection coefficient of the target scene. The maximum dynamic range is the difference between the darkest and brightest possible scene. The brightest scene is typically reflection from clouds or snow.

In order to estimate the radiometric performance of optical instruments in the mid- and long-wavelength infrared spectrum, the spectral emission of the surface of the Earth must be modeled. A blackbody with an equivalent temperature of 290 K can be used for this purpose. The atmospheric transmission as a function of the wavelength is well known for a given path orientation and atmospheric characteristics. The multiplication of the spectral radiance with the atmospheric transmission results in the upwelling radiance at the sensor. The integration of it over the selected bandwidth and the multiplication with the area of the ground pixel results in the power per solid angle. After consideration of sensor altitude and effective aperture of the receiving optics the input power at the sensor's entrance aperture can be calculated. This transforms via the optical transmission factor to the input power at detector level (usually in the picoWatt region). During the integration period a certain amount of energy is accumulated per pixel. The division of that energy by the energy of one photon gives the number of photons available per pixel which is transferred by the quantum efficiency to the available number of electrons per pixel which correspond to its output signal.

To characterize the radiometric performance of an instrument with respect to the temperature resolution, we must determine the *noise-equivalent temperature difference* (NEΔT) for the instrument. The noise equivalent temperature resolution is given by the temperature difference (at scene temperature), which generates a signal equivalent to the total noise electrons at scene temperature. The NEΔT characterizes the instrument in its ability to resolve temperature variations for a given background temperature.

9.5.3 Estimating Size, Weight, and Power

We must be able to estimate the size and main characteristics of the mission payload before completing a detailed design. We want to be able to look at several options without necessarily designing each in depth. This section provides ways to compute data rates and estimate the overall size and key parameters. In Sec. 9.6.1, we will apply these values to the FireSat example.

We have looked in some detail at the design of specific observation payloads in Sec. 9.4. However, irrespective of the nature of the particular payload, we would like to estimate its size, weight, and power even before we have done a detailed design. To do so, we can use three basic methods:

- Analogy with existing systems
- Scaling from existing systems
- Budgeting by components

The most straightforward approach is to use an analogy with existing systems. To do this, we turn to the list of existing payloads in Table 9-13 in Sec. 9.4.3 or other payloads that we may be aware of which have characteristics matching the mission we have in mind. Kramer [1996] offers a very thorough list of existing sensors. We look for payloads whose performance and complexity match what we are trying to achieve and make a first estimate that our payload will have characteristics comparable to the previously designed, existing payload. While this approach is rough, it does provide a first estimate and some bounds to decide whether the approach we have in mind is reasonable.

A second approach, described in more detail below, is scaling the payload estimate from existing systems. This can provide moderately accurate estimates of reasonable accuracy if the scale of the proposed payload does not differ too greatly from current payloads. In addition, scaling provides an excellent check. Most existing payloads have been carefully designed and optimized. If our new payload is either too large or too small relative to prior ones, there should be some reason for this change in characteristics. If more detailed estimates based on detailed budgets don't scale from existing systems, we must understand why.

The most accurate process for first-order payload sizing is budgeting by components. Here we develop a list of payload components such as detectors, optics, optical bench, and electronics. We then estimate the weight, power, and number of each. This is the best and most accurate approach but may be very difficult to apply at early mission stages because we simply don't have enough initial information. Ultimately, we will size the payload with budgeting by components. We will develop budgets as outlined in Chap. 10 for each payload instrument for weight, power, and any critical payload parameters. These budgets will then help us monitor the ongoing payload development. However, even with a detailed budget estimate, it is valuable to use scaling as a check on component budgeting. Again, we wish to understand whether the components scale from existing payloads and, if not, why not.

Scaling from Existing Systems

An excellent approach for preliminary design is to adjust the parameters in Table 9-13 to match the instrument we are designing. We will scale the instruments based on aperture—a main design parameter that we can determine from preliminary mission requirements. To scale, we compute the aperture ratio, R, defined by

$$R = \frac{A_i}{A_o} \tag{9-16}$$

where A_i is the required aperture of our new instrument, and A_o is the aperture of a similar instrument (Table 9-13). We then estimate the size, weight, and power based on ratios with the selected instrument from Table 9-13, using the following:

$$L_i \approx R\,L_o \qquad L = \text{linear dimensions} \tag{9-17}$$

$$S_i \approx L_i^2 \qquad S = \text{surface area} \tag{9-18}$$

$$V_i \approx L_i^3 \qquad V = \text{volume} \tag{9-19}$$

$$W_i \approx KR^3 W_o \qquad W = \text{weight} \tag{9-20}$$

$$P_i \approx KR^3 P_o \qquad P = \text{power} \tag{9-21}$$

The factor K should be 2 when R is less than 0.5, and 1 otherwise. This reflects an additional factor of 2 in weight and power for increased margin when scaling the system down by a factor of more than 2. When the system grows, the R^3 term will directly add a level of margin. For instruments more than a factor of five smaller than those listed in Table 9-13, scaling becomes unreliable. We recommend assuming a mass density of 1 gm/cm^3 and power density of 0.005 W/cm^3 for small instruments. An example of these computations for FireSat is in Sec. 9.6.1.

9.5.4 Evaluate Candidate Payloads

Multi-attribute performance indices can be defined for comparing optical instruments with similar performance characteristics. For high-resolution spatial instruments three basic values describe the quality (corresponding to the information content) in the image. The three defining features are the signal-to-noise ratio at spatial frequency zero (high SNR corresponds to high information content), the MTF of an instrument at the Nyquist frequency (high MTF corresponds to high information content for sampling rates between zero and the Nyquist frequency), and the ground sample distance GSD (small GSD corresponds to high information content). We define a *relative quality index* (RQI) to allow straightforward quantitative comparisons with a reference instrument denoted by the suffix ref.

$$RQI = \frac{SNR}{SNR_{ref}} \frac{MTF}{MTF_{ref}} \frac{GSD_{ref}}{GSD} \qquad (9\text{-}22)$$

This relative quality index allows the designer to trade requirements with respect to each other. For example, a higher SNR can compensate for a lower MTF at the Nyquist frequency for a given GSD. Such comparisons allow for first-order insights into the relationships between complexity, performance, and cost of candidate sensors. For example, suppose we define a reference instrument to have an SNR of 512, and an MTF of 0.5 and a GSD of 25 m. If we then compute design parameters for a particular mission, we can generate a relative quality index, or score, for our design with respect to the reference instrument. For instance, if our design choices lead us to an instrument with a SNR of 705.2, a MTF of 0.47 and a GSD of 30 m, then the RQI for this system will be 108%. This index offers a straightforward method for comparing several competing sensors across three key performance measures.

9.5.5 Observation Payload Design Process

Table 9-15 contains the details of the design process for visible and infrared systems. We begin with basic design parameters such as the orbital height, minimum observation angle and ground resolution. We then compute the quantities that describe the performance of the instrument. In particular, we determine the pixel processing parameters and system data rate, the size of the optics for a given pixel size, and the radiometry of the sensor. Sample computations for the FireSat payload are given in the third column.

The data rate required for observation payloads depends on the resolution, coverage, and amplitude accuracy. With the maximum look angle, η, spacecraft altitude, h, and cross-track pixel size, X, we have to image $2\eta h / X$ pixels per swath line (cross-track). With the spacecraft ground-track velocity V_g and the along-track pixel size Y we have to scan V_g / Y swath lines in one second. If we quantify the intensity of each pixel by b bits (2^b amplitude levels) we generate a data rate, DR, of

$$DR = \frac{2\eta h}{X} \cdot \frac{V_g}{Y} \cdot b \approx \frac{S_w}{X} \cdot \frac{V_g}{Y} \cdot b \qquad \text{bits/second (bps)} \qquad (9\text{-}23)$$

where η is the maximum look angle in radians, h is the orbit altitude, V_g is the spacecraft ground-track velocity, S_w is the swath width, X is the across-track pixel dimension, and Y is the along-track pixel dimension. The approximation is good for small swath widths. The data rate can be increased by transmission overhead such as housekeeping data or coding and it can be decreased by data compression. (See Secs. 2.1.1, 13.2.2, and 15.3.2.).

TABLE 9-15. Calculation Design Parameters for a Passive Optical Sensor.

Step	Calculation	FireSat	Comments
Step 1. Define Orbit Parameters			
Define orbital altitude, h	Design parameter†	$h = 700$ km	See Table 3-4, Sec. 7.4
Compute orbit period, P	Eq. (7-7), IRC*	$P = 98.8$ min	Assumes circular orbit
Compute ground track velocity, V_g	Eq. (5-31), IRC*	$V_g = 6.76$ km/s	Assumes circular orbit
Compute node shift, ΔL	Eq. (7-13), IRC*	$\Delta L = 24.8$ deg	Function of inclination
Step 2. Define Sensor Viewing Parameters			
Compute angular radius of the Earth, ρ	Eq. (5-15), IRC*	$\rho = 64.3$ deg	Depends on orbital altitude
Compute max. distance to the horizon, D_{max}	Eq. (5-17), IRC*	$D_{max} = 3{,}069$ km	Depends on orbital altitude
Define max. incidence ang. IA, or max. Earth cen. ang. ECA_{max}	Design parameter, IRC*	$IA = 70$ deg	Adjust swath width for good coverage (Sec. 7.4)
Compute sensor look angle (= nadir angle), η	Eqs. (5-24) or (5-25b), IRC*	$\eta = 57.9$ deg	Will be less than ρ
Compute min. elev. angle, $\varepsilon = 90° - IA$	Eqs. (5-25b) and (5-26), IRC*	$\varepsilon = 20$ deg	If max. ECA_{max} given, compute ε
Compute max. Earth central angle, ECA_{max}	Eqs. (5-25b) and (5-26), IRC*	$ECA_{max} = 12.1$ deg	If ε given, compute ECA_{max}
Compute slant range, R_S	Eq. (5-27), IRC*	$R_S = 1{,}578$ km	R_S here = D in Chap. 5
Find swath width	= 2 ECA_{max}	2 $ECA_{max} = 24.2$ deg	Determines coverage
Step 3. Define Pixel Parameters and Data Rate			
Specify max. along-track ground sampling dist., Y_{max}	Design parameter	$Y_{max} = 68$ m	Based on spatial resolution requirements at ECA_{max}
Determine instantaneous field of view, $IFOV$	$IFOV = \dfrac{Y_{max}}{R_S} \cdot \dfrac{180 \text{ deg}}{\pi}$	$IFOV = 0.00245$ deg	One pixel width
Find max. cross-track pixel resolution, X_{max}, at ECA_{max}	$X_{max} = \dfrac{Y_{max}}{\cos(IA)}$	$X_{max} = 199.6$ m	Driven by resolution requirement at maximum slant range

* IRC = parameter tabulated on the Inside Rear Cover for Earth Satellites Parameters.
†Calculations are based on a circular orbit.

TABLE 9-15.   Calculation Design Parameters for a Passive Optical Sensor. (Continued)

Step	Calculation	FireSat	Comments
Step 3. Define Pixel Parameters and Data Rate (Continued)			
Determine cross-track ground pixel resolution, X, at nadir	$X = IFOV \cdot h \left(\dfrac{\pi}{180 \text{ deg}} \right)$	$X = 30$ m	Best cross-track resolution for this instrument
Determine along-track pixel resolution, Y, at nadir	$Y = IFOV \cdot h \left(\dfrac{\pi}{180 \text{ deg}} \right)$	$Y = 30$ m	Best along-track resolution for this instrument
Determine no. of cross-track pixels, Z_c	$Z_c = \dfrac{2 \cdot \eta}{IFOV}$	$Z_c = 4.7 \times 10^4$	Ground pixel size varies along the swath
Find no. of swaths recorded along-track in 1 sec, Z_a	$Z_a = \dfrac{V_g \cdot 1 \text{ sec}}{Y}$	$Z_a = 225.6$	Number of successive swaths without gaps at nadir
Find no. of pixels recorded in 1 sec., Z	$Z = Z_c \cdot Z_a$	$Z = 1.06 \times 10^7$	
Specify no. of bits used to encode each pixel, B	Design parameter	8 bits	Based on radiometric resolution requirement and dynamic range
Compute data rate, DR	$DR = Z \cdot B$	$DR = 85$ Mbps	Large number may challenge downlink capacity
Step 4. Define Sensor Integration Parameters			
Specify no. of pixels for whiskbroom inst. N_m	Design parameter	$N_m = 256$	Must be large enough to allow sufficient integration time
Find pixel integration period, T_i	$T_i = \dfrac{Y}{V_g} \cdot \dfrac{N_m}{Z_c}$	$T_i = 24.1$ μs	Integration time of each detector pixel
Find resulting pixel read-out frequency, F_p	$F_p = 1/T_i$	$F_p = 42$ kHz	
Verify detector time constant, T_{det}, is smaller than T_i	$T_{det} < T_i$	$T_{det} < T_i$	Compare with physical properties in Table 9-12.
Step 5. Define Sensor Optics			
Specify width for square detectors, d	Design parameter	$d = 30$ μm	Typical for available detectors
Specify quality factor for imaging, Q	Design parameter	$Q = 1.1$	$0.5 < Q < 2$ (Q=1.1 for good image quality)
Specify operating wavelength, λ	Design parameter	$\lambda = 4.2$ μm	Based on subject trades
Define focal length, f	$f = \dfrac{h \cdot d}{X}$	$f = 0.7$ m	Use altitude and Eq. (9-12)
Find diffraction-limited aperture diameter, D	$D = \dfrac{2.44 \lambda \cdot f \cdot Q}{d}$	$D = 0.263$ m	Eq. (9-14) equivalent

TABLE 9-15. Calculation Design Parameters for a Passive Optical Sensor. (Continued)

Step	Calculation	FireSat	Comments
Step 5. Define Sensor Optics			
Compute F-number of optics, $F\#$	$F\# = f/D$	$F\# = 2.7$	Typical range = 4–6
Compute field of view of optical system, FOV	$FOV = IFOV \cdot N_m$	$FOV =$ 0.628 deg	FOV for the array of pixels
Determine cut-off frequency, F_c	$F_c = D/\lambda h$	$F_c = 0.09$ line pairs/m	Referred to nadir
Determine cross-track Nyquist frequency, F_{nc}	$F_{nc} = 1/2X$	$F_{nc} =$ 0.017 lp/m	Referred to ground pixel resolution at nadir
Determine along-track Nyquist frequency, F_{na}	$F_{na} = 1/2Y$	$F_{na} =$ 0.017 lp/m	Referred to ground pixel resolution at nadir
Compute relative Nyquist frequencies, F_{qc} and F_{qa}	$F_{qc} = \dfrac{F_{nc}}{F_c}; \quad F_{qa} = \dfrac{F_{na}}{F_c}$	$F_{qc} = 19\%$ $F_{qa} = 19\%$	% of the cutoff frequency used for this case
Find optics PSF as a function of distance, r, from center of detector	$PSF(r) = [\lambda\, J_1\,(Z)\,/\,Z]^2$ $Z \equiv \pi\, r\, D/\lambda\, f$	See Figs. 9-25A and B	Use $-2d < r < 2d$ J_1 is the Bessel function of order 1
Find optical modulation transfer function (MTF_0) for clear circular optics	$MTF_0(F) = (2C/\pi)$ $X\left[\dfrac{\pi}{2} - \dfrac{C}{F_c\sqrt{1-C^2}} - a\sin C\right]$	See Fig. 9-25B	Use $0 \le F \le F_c$ $C \equiv F/F_c$
Step 6. Estimate Sensor Radiometry (for Nadir Viewing)			
Compute detector MTF cross-track, MTF_X, and along-track, MTF_Y	$MTF_X = [\sin(F_X)/F_X]^2$ $MTF_Y = |\sin(F_Y)/F_Y|$	See Fig. 9-25B	Use $0 \le F \le F_c$ $F_X \equiv \pi\, X\, F$ $F_Y \equiv \pi\, Y\, F$
Compute system MTF cross-track, MTF_s	$MTF_s(F) =$ $MTF_0(F) \cdot MTF_c(F)$	See Fig. 9-25B	Let F range: $0 \le F \le F_c$
Define equivalent blackbody temp. T	Design parameter	$T = 290$ K	Blackbody temperature of the Earth
Define the operating bandwidth, $\Delta\lambda$	Design parameter	$\Delta\lambda = 1.9\ \mu m$	Based on subject trades
Determine blackbody spectral radiance, L_λ,	$L_\lambda = E(\lambda)\,/\,4\pi$ $E(\lambda)$ from Eq. (9-2)		Use range $\lambda \pm \Delta\lambda\,/\,2$
Look up transmissivity, $\tau\,(\lambda)$ of the atmosphere	See Fig. 9-6	See Fig. 9-6	Evaluate operating bandwidth
Compute upwelling radiance, L_{upi}	$L_{upi}\,(\lambda) = L_\lambda\,\tau\,(\lambda)$		Total input radiance as a function of wavelength
Compute integrated upwelling radiance, L_{int},	$L_{int}\,(\lambda) = \sum_i L_{upi}\,(\lambda_i - \lambda_{i+1})$	$L_{int} =$ 0.433 W/m²/sr	Evaluate over operating bandwidth

TABLE 9-15.   Calculation Design Parameters for a Passive Optical Sensor. (Continued)

Step	Calculation	FireSat	Comments
Step 6. Estimate Sensor Radiometry (for Nadir Viewing) (Continued)			
Compute radiated power, L, from a ground pixel at nadir	$L = L_{int} \cdot X \cdot Y$	$L = 389.5$ W/sr	Total power from the ground scene that arrives at the instrument
Compute input power, P_{in}, at sensor	$P_{in} = \dfrac{L}{h^2} \cdot \left(\dfrac{D}{2}\right)^2 \cdot \pi$	$P_{in} =$ 4.3×10^{-11} W	P_{in} is power at the entrance to the optics
Define optical transmission factor τ_o	Design parameter	$\tau_o = 0.75$	Typical value for optical systems
Find input power, P_D, at the detector pixel	$P_D = P_{in} \cdot \tau_o$	$P_D =$ 3.2×10^{-11} W	Very little power arrives at each pixel
Determine available energy, E, after integration time	$E = P_D \cdot T_i$	$E =$ 7.8×10^{-16} Ws	Radiometric design challenge for FireSat
Find no. of available photons, N_p	$N_p = E \lambda / hc$	$N_p = 1.7 \times 10^4$	h is Planck's constant, c is speed of light
Define quantum efficiency, QE, of detector at λ	Design parameter	$QE = 0.5$	Typical physical property of detector material
Compute no. of electrons available, N_e	$N_e = N_p \cdot QE$	$N_e = 8.3 \times 10^3$	Evaluate for an ideal detector
Determine no. of noise electrons, N_n	$N_n = \sqrt{N_e}$	$N_n = 91$	Considers only Shott noise
Define no. of read-out noise electrons, N_r	Design parameter	$N_r = 25$	Typical value
Determine total no. of noise electrons, N_t	$N_t = \sqrt{N_n^2 + N_r^2}$	$N_t = 95$	Assumes uncorrelated noise processes
Find signal-to-noise ratio of the image, SNR	$SNR = N_e / N_t$	$SNR = 88$	Assuming signal dominates background
Determine sensor dynamic range, DR	$DR = N_e / N_r$	$DR = 332.9$	With respect to cold space
Step 7. Find the Noise-Equivalent Temperature Difference			
Recompute all the parameters in Step 6 for $\Delta T = 1$ K	$T_{new} = T + 1$ K	$N_e = 8.7 \times 10^3$	Assume scene temperature changes by 1 deg K
Determine no. of charge carriers for 1 K temp. change	$\Delta N = N_{e_{new}} - N_e$	$\Delta N = 335.8$	
Compute noise-equivalent temp. difference, $NE\Delta T$	$NE\Delta T = N_e / \Delta N$	$NE\Delta T = 0.3$ K	Temperature limit the instrument can resolve

9.5.6 Assess Life-cycle Cost and Operability of the Payload and Mission

In addition to trades between minimal and desired performance, spacecraft designs are heavily driven by cost. Several approaches have been proposed and implemented to treat cost as an independent variable. For our purposes, it is sufficient to note that trading cost and performance means it is no longer sufficient to state the mission requirements clearly and realistically. Rather, the mission requirements become involved in the iterative process of design (see Fleeter [1996]). There are several excellent descriptions of how to include cost as a system parameter rather than a given; see for example Shishko and Jorgensen [1996].

Working through the trade-offs associated with cost, performance, and requirements in this early stage of payload definition keeps payload designers focused on the best sensor characteristics to maximize mission performance and minimize cost. Designers sometimes have a tendency to want to perform a purely analytical evaluation of the costs and benefits of various design options. Unfortunately, the relative benefits of different design features are difficult if not impossible to quantify in an unambiguous and universally accepted manner. Analysis can be very useful for providing a common footing and level playing field for the different design attributes. Ultimately, however, judgments about satisfying mission objectives within cost and schedule constraints rely on human insight, adding to the difficulty and importance of this portion of the payload definition process.

Once we determine the final payload type and basic payload performance requirements, then payload final design can commence. The final payload design could be as simple as an evaluation of existing payloads that are available, or it could involve detailed design, fabrication, and testing of an entirely new instrument. The final step in the payload definition and sizing process is the decision to procure or fabricate the spacecraft payload.

Integrating a payload into a spacecraft design introduces several practical considerations for the other payload subsystems. These derived requirement can have a significant impact on the rest of the spacecraft. Table 9-16 contains an overview of some of the accommodation aspects of a payload as it impacts the other spacecraft subsystems. Resolving the impact of these requirements means we must assess the performance, cost, and technical risk of each subsystem to accommodate the payload.

9.6 Examples

We present two examples of remote sensing payload designs—one very preliminary and one very mature—to give an indication of the beginning and ending points of the design process. Sec. 9.6.1 provides an initial assessment of a payload to fulfill the FireSat mission. Sec. 9.6.2 describes features of the MODerate-Resolution Imaging Spectroradiometer (MODIS), one of the primary sensors on board the Earth Observing System EOS-AM1 spacecraft, which has a fire detection capability.

9.6.1 The FireSat Payload

To illustrate the preliminary design process for payloads, we will estimate the basic parameters for the FireSat payload developed throughout Chaps. 1–8 and earlier in Chap. 9. We cannot expect to carry out a detailed design without substantial input from an IR payload designer. Still, we would at least like to know whether the FireSat payload is the size of a shoebox or the size of a truck.

TABLE 9-16.   Impact of Remote Sensing Payloads on the Spacecraft Design. The table summarizes requirements in other elements of the spacecraft design that must be present to support a remote sensing payload.

Impact Area	Typical Requirements to Support Payload	Additional Considerations
Structure	Mount the optical instruments isostatically to the spacecraft bus Do not apply excessive forces or torques to the payload instrument Make the mounting structure or base plate for optical components stiff enough to prevent any misalignment when subjected to the forces and vibrations of launch	Carefully analyze aging of material (e.g., stress release in metal parts), humidity release, transition to micro-gravity, and acceleration forces Typical stability requirements at critical locations within the optical instrument housing are in the μm and mdeg range
Thermal	Make large opto-mechanical assemblies temperature stabilized or isothermal Operate refractive optical systems typically within a specific temperature range to achieve required performance (frequently they employ semi-active temperature control) Make reflective systems entirely from the same material which leads to a compensation of thermal effects (typically done for cryogenic optical systems)	Large reflective systems (which use Zerodur, Aluminum, or Beryllium or newly developed materials such as SiC or CsiC as materials for the mirrors) and mounting structures (which use composite materials) are temperature sensitive and may require semi-active temperature control of structure and/or mirrors Temperature gradients in optical components can severely degrade performance
External Alignment	Align the optical axis of the instrument and/or the line of sight of the pointing device with an external reference on the spacecraft. External alignments may need to be on the order of 1 arc sec	Use reference cubes to achieve alignment External alignment requires a calibrated optical bench.
Pointing	For monocular optical instruments, make the pointing requirements on the order of 0.1 to 0.01 of the swath width, typically For stereoscopic instruments, automated digital terrain mapping requires pointing knowledge of 1/5 of a pixel	Mount attitude determination sensors (e.g., star sensors) to the instrument (not the bus) to minimize the effects of thermoelasticity Do pointing by maneuvering the spacecraft or by pointing devices (such as pointing mirrors for the front of the instrument or gimbals for the entire instrument)
Assembly Integration and Verification	Optical instruments require clean rooms and clean laminar air flow benches for all integration and verification activities Clean room requirements typically range from 100 to 100,000 ppm	Cleaning optical surfaces is generally not possible During exposure to the environment, use cleanliness samples to verify the level of contamination
System Accommodation	Sensor must have an unobstructed field-of-view Sensor must have a guard cone to prevent performance degradation due to stray light Avoid pointing toward the Sun	Orient radiators and passive coolers for infrared systems to prevent interference with optical devices Calibration devices impose geometric constraints with respect to the optics of the system and the orbit

The FireSat altitude trade led to a preliminary altitude, $h = 700$ km. From this, we can determine the angular radius of the Earth, ρ:

$$\rho = \sin^{-1} (R_E/(R_E + h)) = 64.3 \text{ deg} \qquad \text{From Eq. (5-15)} \qquad (9\text{-}24)$$

A key parameter in the system design is the minimum elevation angle, ε, at which the system can work. We do not have an estimate of that yet, but we do know that IR payloads do not work well at small elevation angles. Therefore, we will tentatively assume a minimum elevation angle of 20 deg, recognizing that this may be a very critical trade at a later stage. With this assumption, we can compute the nadir angle range, η, the maximum ground-track angle or swath width, λ, and the maximum range to the target, D, from the formulas in Sec. 5.2:

$$\sin \eta = \cos \varepsilon \sin \rho \qquad \eta = 57.9 \text{ deg} \qquad \text{From Eq. (5-25a)} \qquad (9\text{-}25)$$

$$\lambda = 90 - \eta - \varepsilon = 12.1 \text{ deg} \qquad \text{From Eq. (5-26)} \qquad (9\text{-}26)$$

$$D = R_E (\sin \lambda / \sin \eta) = 1{,}580 \text{ km} \qquad \text{From Eq. (5-27)} \qquad (9\text{-}27)$$

These equations imply that the sensor on board the spacecraft will have to swing back and forth through an angle of ±57.9 deg to cover the swath. The swath width on the ground will be $2 \times 12.1 = 24.2$ deg wide in Earth-central angle, with a maximum distance to the far edge of the swath of 1,580 km. Had we been able to work all the way to the true horizon ($\varepsilon = 0$), the maximum Earth central angle would be $90 - \rho = 27.5$ deg, and the swath width would be 55 deg. Increasing the minimum elevation angle to 20 deg has very dramatically reduced the size of the available swath.

We next find the orbit period, P, and longitude shift per orbit, ΔL, (Sec. 7.2):

$$P = 1.659 \times 10^{-4} \times (6{,}378 + h)^{3/2} = 98.8 \text{ min} \quad \text{From Eq. (7-7)} \qquad (9\text{-}28)$$

$$\Delta L = 1.65 \times (360/24) = 24.8 \text{ deg} \qquad \text{From Eq. (5-17)} \qquad (9\text{-}29)$$

Therefore, at the equator, successive node crossings are 24.8 deg apart. Notice that this is slightly larger but very close to the 24.2 deg swath width which we computed above. This is an important characteristic for FireSat. It would be extremely valuable to have the swaths overlap so that every FireSat spacecraft can cover all locations on the Earth twice per day. Therefore, in designing the payload, we should work hard to maintain either the altitude or the minimum elevation angle to provide some swath overlap. Doing so could dramatically reduce the number of spacecraft required and therefore the cost of the system.

As Fig. 9-24 shows, the swath width does not need to be quite as large as the spacing between nodes along the equator. Even at the equator, it is enough to have a swath width equal to S, the perpendicular separation between the ground tracks. In Chap. 7, we selected an inclination for FireSat of 55 deg to cover up to 65 deg latitude. Consequently, we can use the spherical triangle ABC shown in the figure to compute S as follows:

$$S = \sin^{-1} (\sin 24.8 \text{ deg} \sin 55 \text{ deg}) = 20.1 \text{ deg} \qquad (9\text{-}30)$$

The perpendicular separation between the orbits at the equator is 20.1 deg. Because the swath width is 24.2 deg, we now have some overlap margin even at the equator and substantial margin at higher latitudes, which are the primary areas of interest. We could, therefore, increase the minimum elevation angle to 25 deg. This would be a reasonable option. At present, we instead choose to hold ε at 20 deg and to provide some margin on altitude and elevation angle for later payload trades.

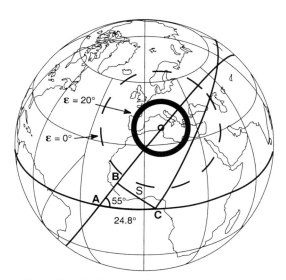

Fig. 9-24.   Computation of FireSat Ground-Track Parameters.

We next compute the required resolution and data rates for FireSat. From Table 1-5, we initially estimated the needed ground resolution as 30 m. Because this is meant to be a very low cost system, we will assume that the required resolution, θ_r, is at nadir so that from an altitude of 700 km we have:

$$\theta_r = 0.030/700 = 4.3 \times 10^{-5} \,\text{rad} = 0.00245 \,\text{deg} \qquad (9\text{-}31)$$

Had we made this requirement at the maximum slant range of 1,580 km, the required resolution would have been 0.001 deg.

Using this resolution, we can follow the procedure outlined in Table 9-15 to compute the data rate for FireSat as 85 Mbps. This data rate from the FireSat sensor is very high. However, we will be able to reduce it in many ways. We could process the data on board or, more simply, turn off the payload over the oceans or other areas where fire detection is of marginal utility. For now, we will leave the value as computed so that we remain aware of the data rate out the sensor, recognizing that this will be need to be reduced later in the system design.

We next compute mapping and pointing budgets for FireSat. We do not have a firm mapping requirement, but we do have some broad sense of what is needed. We begin, therefore, with a rough estimate of performance parameters and create the mapping error as a function of the elevation angle shown in Fig. 9-25A. In this figure, we have used a 0.1-deg nadir angle and azimuth errors corresponding to a relatively inexpensive pointing system based on an Earth sensor. We know we can go to a more expensive system if necessary. In looking at Fig. 9-25A, we see that the mapping error at our chosen minimum elevation angle of 20 deg is between 6 and 8 km. While we are not certain what our mapping requirement is, we are reasonably sure that it is smaller than 6 km. We need to locate fires more accurately than this. Note also that the accuracy has been set almost entirely by our crude attitude number of 0.1 deg.

The next most critical parameter is the 1-km error in target altitude. This means that we assume we can determine the altitude of the fire above the Earth to 1 km—a reasonable accuracy with an oblate Earth model. But significantly improving this accuracy would require carrying a map of the altitudes of all of the regions of interest. That could be very difficult, particularly in mountainous areas, and would cost a lot more money. Therefore, it is of little value to drive the error in nadir angle down below approximately 0.05 deg because it would no longer be the dominant error source. Fig. 9-25B shows the curves that we would achieve with the error in nadir angle reduced to 0.05 deg and all of the other error sources remaining the same. Now the contribution of the errors in nadir angle and target altitude are comparable, so we will use this budget to establish a preliminary mapping requirement of 5.5 km at a 20-deg elevation angle, and 3.5 km at a 30-deg elevation angle. This may still be considerably more crude than we would like, so we may need to revisit this issue.

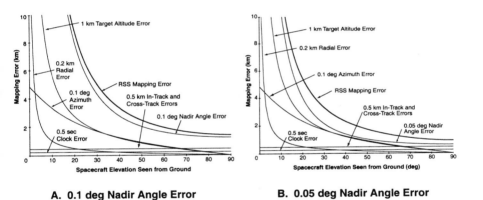

A. 0.1 deg Nadir Angle Error   B. 0.05 deg Nadir Angle Error

Fig. 9-25.   **FireSat Mapping Budget.** Reducing the nadir angle error below 0.05 deg will have relatively little impact on the overall mapping error because of the 1 km target altitude error. Compare with Fig. 5-22.

Equipped with the analysis of the mission geometry, we turn our attention to the process described in Table 9-15, these computations allow us to evaluate the optical, signal processing, and radiometric performance of the instrument. The third column in that table summarizes the results of the computations for a whiskbroom sensor design for the FireSat mission.

The example FireSat design addresses only initial feasibility of the instrument. Several challenges remain with this design and addition iterations need to be made in the context of mission requirements and constraints. The computed data rate of 85 Mbps will present a design challenge, as will the multiple pixel scanner needed to scan 256 pixels simultaneously. This will require all 256 pixels to be read out in parallel, and the signal processing will need to be designed accordingly. These features present a particularly demanding element of the initial design.

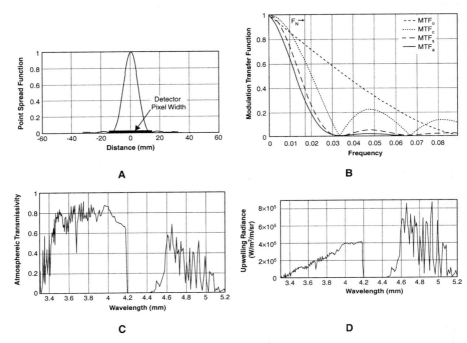

Fig. 9-26.   Sample Characteristics for Sample FireSat Sensor Described in Table 9-15 in Sec. 9.5.5. A shows the point spread function of the sensor with respect to the pixel size (shown as a horizontal bar). B shows the modulation transfer functions of the sample instrument. C illustrates optical transmission of the atmosphere over the operating wavelength of the sensor. D shows the total upwelling radiance through the atmosphere across the operating bandwidth of the sensor.

A second challenge with the design is the relatively poor signal-to-noise ratio of 88. If we consider a pixel at the limit of the field of view, then radiometric information becomes indistinguishable from noise. From this point of view our current design is at the limit of feasibility and may require changes to meet the SNR requirements. Finally the F-number of the optics of 2.7 (driven by the focal length of 0.7 m and an aperture diameter of 26 cm) is quite a demanding optical design.

We now select a "similar instrument" from Table 9-13 for our FireSat example. We appear to have two options: the Thematic Mapper or the Multispectral Mid-IR instrument. We tentatively select the Multispectral Mid-IR as our similar instrument and will scale from its fundamental parameters of 1.5 m × 1 m diameter, 800-kg weight, and 900 W power, for its 1-m aperture. We first compute the aperture ratio,

$$R = 0.26/1.0 = 0.26 \qquad \text{From (Eq. 9-16)} \qquad (9\text{-}32)$$

With this fundamental ratio, we now estimate the FireSat payload parameters as

$$\text{Size} = 0.4 \text{ m} \times 0.3 \text{ m diameter}$$

$$\text{Weight} = 2 \times 800 \times 0.26^3 = 28 \text{ kg}$$

$$\text{Power} = 2 \times 900 \times 0.26^3 = 32 \text{ W}$$

As described in Sec. 9.5.3, we have incorporated a factor of 2 to provide margin for having substantially scaled down the payload size. The estimate of the linear dimensions needs to be adjusted as well to allow for the size of the scanner which will need to be mounted in front of the sensor optics and electronics. A rough estimate of the scanner dimensions is the same size as the payload estimate. Thus, as summarized in Table 9-17, the budgeted dimensions for the optics plus scanner is 0.8 m long × 0.3 m diameter. Thus, our first guess is that the FireSat payload is a moderately sized instrument and could fit well on a small to medium-sized spacecraft.

TABLE 9-17.   Summary of FireSat Initial Parameter Estimates.

Parameter	Nominal Value	Comments
Altitude, h	700 km	Range = 600 to 800 km
Inclination, i	55 deg	Coverage to 65 deg latitude
Swath width, $2\lambda_{max}$	24.2 deg	20.1 deg required
Nadir angle range, η	±57.9 deg	
Min. elevation angle, ε	20 deg	Needs payload input
Instrument	mid-range IR scanner	Payload needs expert input
Ground resolution	30 m at nadir	Key parameter—needs trade study
Instrument resolution	4.3 x 10^{-5} rad	= 0.00245 deg
Aperture, A	0.26 m	
Size	0.4 m x 0.3 m dia	
Weight, W	28 kg	
Power, P	32 W	
Data rate, DR	85 Mbits/sec	May be limiting feature
Mapping	3.5 km @ $\varepsilon = 30°$ 5.5 km @ $\varepsilon = 20°$	

Our preliminary analysis of a small, lightweight FireSat payload shows that the mission is feasible but challenging. Several refinements and iterations on the design have the potential to result in a viable and cost-effective payload concept. To illustrate the end point of such a process, we turn our attention to MODIS, a large instrument and a mature design with a fire detection mission.

9.6.2 MODIS—A Real FireSat Example

A detailed design for a spacecraft sensor that can automatically detect fires already exists. The MODIS instrument (MODerate-resolution Imaging Spectroradiometer) on the Terra spacecraft has been designed for a comprehensive range of scientific investigations into Earth's atmosphere, oceans, and land use—much more challenging than fundamental requirements for the FireSat mission (therefore, MODIS may be overdesigned for the FireSat mission). However, the MODIS instrument represents a mature design and a sophisticated, space-based fire detection system. The features and considerations that drove the MODIS fire detection sensor and data processing algorithms offer an opportunity to inform our broader discussion of FireSat throughout this book.

The development of the MODIS sensor for Terra traces its roots to the GOES and NOAA spacecraft, and it represents at least a decade of research and design to improve the performance of the Advanced Very High Resolution Radiometer (AVHRR) flown on the NOAA series of spacecraft. The MODIS sensor on Terra is a whiskbroom, electro-optical system. Table 9-18 lists its technical characteristics and specifications. The MODIS instrument includes specific design features to capitalize on the physics of thermal detection of fires. MODIS fire products include detecting the incidence of fire, its location, emitted energy, its ratio of flaring to smoldering, and the area burned (burn scar detection). These products are important for understanding the influence of burning biomass on many atmospheric processes as well as direct and indirect effects on terrestrial ecosystems [Kaufman and Justice, 1996]. The key innovations for the fire detection algorithms include distinguishing the flaring and smoldering parts of the fire and the automatic algorithms for reporting the progress of fires.

TABLE 9-18. MODIS Instrument Characteristics. [Herring, 1997.]

Orbit	705 km, 10:30 a.m. descending node, Sun synchronous
Scan Rate	20.3 rpm cross track
Swath Dimensions	2,330 km (across track) by 10 km (along track)
Telescope	17.78 cm diam off-axis, afocal (collimated, with intermediate field stop)
Size	$1.0 \times 1.6 \times 1.0$ m
Mass	274 kg
Power	162.5 W (avg for one orbit), 168.5 W (peak)
Design Life	6 years
Quantization	12 bits
Data Rate	6.2 Mbps (avg), 10.8 Mbps (day), 2.5 Mbps (night)
Spectral Range	0.4–14.4 µm
Spectral Coverage	±55 deg, 2,330 km swath (contiguous scans at nadir at equator)
Spatial Resolution	250 m (2 bands), 500 m (5 bands), 1,000 m (29 bands) at nadir
Duty Cycle	100%

The algorithm developed for MODIS fire detection data products employs two of the 500 m resolution bands, one at 4 µm and the other at 11 µm. The algorithm is an extension of the methods developed using AVHRR. A summary of the steps in the fire processing algorithm follows [Kaufman and Justice, 1996]. (See also Sec. 16.3.)

Initialization. The algorithm eliminates pixels with potential problems due to clouds or extreme viewing angles. It corrects apparent temperature readings for atmospheric absorption (including water vapor), and estimates the background temperature for pixels containing fire.

Fire detection. The algorithm defines fire pixels based on thresholds and temperature differences between readings in the two spectral bands.

Correction. It eliminates potential false positive readings due to sun glint and consolidates fire readings from adjacent pixels to eliminate redundant reports.

Total emitted energy. It estimates the total energy based on measurements in the 4 µm channel.

Smoldering or flaming stage. It estimates the nature of the fire, namely, smoldering, flaming, or a combination of both.

The MODIS payload illustrates many of the design features of an automated fire detection system. In the context of the FireSat mission, this example provides a point design that has finalized a series of trade-offs in size, weight, power, resolution, and data rate.

References

Barnes, William L., Thomas S. Pagano, and Vincent V. Salomonson. 1998. "Prelaunch Characteristics of the Moderate Resolution Imaging Spectroradiometer (MODIS) on EOS-AM1." *IEEE Transactions in Geoscience and Remote Sensing*, vol. 36(4): 1088–1100, July.

Buiten, Henk J. and Jan G.P.W. Clevers, eds. 1996. *Land Observation by Remote Sensing: Theory and Applications*. The Netherlands: Gordon and Breach Science Publishers.

Cantafio, Leopold J., ed. 1989. *Space-Based Radar Handbook*. Boston, MA: Artech House.

Chen, H.S. 1985. *Space Remote Sensing Systems, an Introduction*. Orlando, FL: Academic Press.

Colwell, Robert N., ed. 1983. *The Manual of Remote Sensing*. Falls Church, VA: American Society and Photogrammetry and Sheridan Press.

Cruise, A.M., J.A. Bowles, T.J. Patrick and C.V. Goodall. 1998. *Principles of Space Instrument Design*. United Kingdom: Cambridge University Press.

The CSTS Alliance (Boeing, General Dynamics, Lockheed, Martin Marietta, McDonnell Douglas, Rockwell). 1994. "The Commercial Space Transportation Study." Report No. SSD94D0034, April.

Davies, John K. 1997. *Astronomy From Space: The Design and Operation of Orbiting Observation*. Chichester, England: John Wiley & Sons.

DeBlois, Bruce M. 1997. "Space Sanctuary: A Viable National Strategy." School of Advanced Aerospace Studies Manuscript, Montgomery, AL.

Elachi, Charles. 1987. *Introduction to the Physics and Techniques of Remote Sensing*. New York: John Wiley and Sons.

Fleeter, Rick. 1996. "Reducing Spacecraft Cost." Chap. 5 in *Reducing Space Mission Cost*, James R. Wertz and Wiley J. Larson, eds. Torrance, CA: Microcosm Press.

Glaser, P.E., F.P. Davidson, and K.I. Csigi. 1993. *Solar Power Satellites*. New York: Ellis Horwood, Ltd.

Hall, R. Cargill. 1995. "The Eisenhower Administration and the Cold War: Framing American Astronautics to Serve National Security." *Prologue Quarterly of the National Archives*, Vol. 27(1): 59–72, Spring.

Herring, David. 1997. *NASA's Earth Observing System: EOS AM-1*. NASA Goddard Space Flight Center.

Hovanessian, S.A. 1988. *Introduction to Sensor Systems*. Boston: Artech House.

Huffman, Robert E. 1992. *Atmospheric Ultraviolet Remote Sensing*. Boston, MA: Academic Press, Inc.

Kaufman, Yoram and Chris Justice. 1996. "MODIS Fire Products." Algorithm Technical Background Document, ATBD-MODIS-14, Ver. 2.1, Earth Observing System Document ID #2741, NASA Goddard Space Flight Center, December 3.

Kidder, S.Q. and T.H. Vonder Haar. 1995. *Satellite Meteorology*: An Introduction. San Diego: Academic Press.

King, Michael D., Yoram I. Kaufman, W. Paul Menzel, and Didier Tanré. 1992. "Remote Sensing of Cloud, Aerosol, and Water Vapor Properties from the Moderate Resolution Imaging Radiometer (MODIS)." *IEEE Transactions on Geoscience and Remote Sensing*, Vol. 30(1): 2–27, January.

Kondo, Y., ed. 1990. *Observatories in Earth Orbit and Beyond*. Boston: Kluwer Academic Publishers.

Kramer, Herbert J. 1996. *Observation of the Earth and Its Environment: Survey of Missions and Sensors (Third Edition)*. Berlin: Springer.

Lutz, Reinhold. 1998. "Design Considerations for High Resolution Earth Observation Instruments." Course notes from Technical University, Delft, Toptech Studies Program.

McCluney, Ross. 1994. *Introduction to Radiometry and Photometry*. Boston: Artech House.

McDougall, Walter A. 1997. *The Heavens and the Earth: A Political History of the Space Age*. New York: Basic Books.

Meneghini, Robert and Toshiaki Kozu. 1990. *Spaceborne Weather Radar*. Boston, MA: Artech House.

Miller, Lester John and Edward Friedman. 1996. *Photonics Rules of Thumb*. New York: McGraw-Hill.

Morgan, W. L., and G.D. Gordon 1989. *Communications Satellite Handbook*. New York: John Wiley and Sons.

Schnapf, Abraham, ed. 1985. *Monitoring Earth's Ocean, Land, and Atmosphere from Space—Sensors, Systems, and Applications*. New York: American Institute of Aeronautics and Astronautics, Inc.

Seyrafi, Khalil. 1985. *Electro-Optical Systems Analysis*. Los Angeles: Electro Optical Research Company.

Shishko, Robert and Edward J. Jorgensen. 1996. "Design-to-Cost for Space Missions," Chap. 7 in *Reducing Space Mission Cost*. James R. Wertz and Wiley J. Larson, eds. Torrance, CA: Microcosm Press and Dordrecht, The Netherlands: Kluwer Academic Publishers.

Shott, John Robert. 1997. *Remote Sensing: The Image Chain Approach*. New York: Oxford University Press.

Slater, Philip N. 1980. *Remote Sensing: Optics and Optical Systems*. Reading, MA: Addison-Wesley Publishing.

Chapter 10

Spacecraft Design and Sizing

Emery I. Reeves, *United States Air Force Academy*

10.1 Requirements, Constraints, and the Design Process

10.2 Spacecraft Configuration

10.3 Design Budgets

10.4 Designing the Spacecraft Bus
Propulsion Subsystem; Attitude Determination and Control Subsystem; Communications Subsystem; Command and Data Handling Subsystem; Thermal Subsystem; Power Subsystem; Structures and Mechanisms

10.5 Integrating the Spacecraft Design
Spacecraft Size; Lifetime and Reliability

10.6 Examples

Over the past four decades the engineering design of spacecraft has evolved from infancy to well-defined techniques supported by analysis tools, manufacturing technology, and space-qualified hardware. This chapter summarizes these techniques, with emphasis on the conceptual design of the spacecraft vehicle. The following two chapters present more detailed design and manufacturing information. To design a spacecraft, we must understand the mission, including the payload's size and characteristics, plus significant system constraints such as orbit, lifetime, and operations. We then configure a space vehicle to carry the payload equipment and provide the functions necessary for mission success. The design process shown in Table 10-1 involves identifying these functions, choosing candidate approaches for each function, and selecting the best approaches. This chapter presents design methods with rules of thumb that will help us roughly estimate the spacecraft design [Agrawal, 1986; Chetty, 1991; Griffin and French, 1991].

An unmanned spacecraft consists of at least three elements: a payload, a spacecraft bus, and a booster adapter. The *payload* is the mission-peculiar equipment or instruments. The *spacecraft bus* carries the payload and provides housekeeping functions. The payload and spacecraft bus may be separate modules, or the vehicle may be an integrated design. The *booster adapter* provides the load-carrying interface with the boost vehicle. The spacecraft may also have a propellant load and a propulsion kick stage. The *propellant*, either compressed gas, liquid or solid fuel, is used for velocity corrections and attitude control. A *kick stage,** if used, is a separate rocket motor or liquid stage used to inject the spacecraft into its mission orbit.

---

* Also called *apogee boost motor, propulsion module,* or *integral propulsion stage.*

TABLE 10-1. Overview of Spacecraft Design and Sizing. The process is highly iterative, normally requiring several cycles through the table even for preliminary designs.

Step	References
1. Prepare list of design requirements and constraints	Sec. 10.1
2. Select preliminary spacecraft design approach and overall configuration based on the above list	Sec. 10.2
3. Establish budgets for spacecraft propellant, power, and weight	Sec. 10.3
4. Develop preliminary subsystem designs	Sec. 10.4
5. Develop baseline spacecraft configuration	Secs. 10.4,10.5
6. Iterate, negotiate, and update requirements, constraints, design budgets	Steps 1 to 5

The top-level requirements and constraints are dictated by the mission concept, mission architecture, and by payload operation. For instance, the selection of orbit is intimately tied to the selected mission and payload as described in Chaps. 6 and 7. From a spacecraft design standpoint, the orbit also affects attitude control, thermal design, and the electric power subsystem. However, most of these design effects are secondary to the effect that the orbit can have on payload performance. The designer therefore selects the orbit based on mission and payload performance, and computes the required spacecraft performance characteristics such as pointing, thermal control, power quantity, and duty cycle. The spacecraft is then sized to meet these requirements. We can summarize succinctly the spacecraft bus functions: support the payload mass; point the payload correctly; keep the payload at the right temperature; provide electric power, commands, and telemetry; put the payload in the right orbit and keep it there; and provide data storage and communications, if required. The spacecraft bus consists of subsystems or equipment groups which provide these functions. Table 10-2 lists the somewhat arbitrary definitions of subsystems used here and in Chap. 11. The table also includes alternate terminology and groupings you may encounter, along with references to more detailed information. Sometimes the payload is also treated as a subsystem. Chapters 9 and 13 discuss payload design.

The *propulsion subsystem* provides thrust for changing the spacecraft's translational velocity or applying torques to change its angular momentum. The simplest spacecraft do not require thrust and hence have no propulsion equipment. But most spacecraft need some controlled thrust, so their design includes some form of *metered propulsion*—a propulsion system that can be turned on and off in small increments. We use thrusting to change orbital parameters, correct velocity errors, maneuver, counter disturbance forces (e.g., drag), control attitude during thrusting, and control and correct angular momentum. The equipment in the propulsion subsystem includes a propellant supply (propellant, tankage, distribution system, pressurant, and propellant controls) and thrusters or engines. Compressed gasses, such as nitrogen, and liquids, such as monopropellant hydrazine, are common propellants. Significant sizing parameters for the subsystem are the total impulse and the number, orientation, and thrust levels of the thrusters. Chapter 17 describes design and equipment for propulsion subsystems.

The *attitude determination and control subsystem* measures and controls the spacecraft's angular orientation (pointing direction), or, in the case of a guidance, navigation, and control system, both its orientation and linear velocity (which affects its orbit). The simplest spacecraft are either uncontrolled or achieve control by passive

TABLE 10-2. Spacecraft Subsystems. A spacecraft consists of functional groups of equipment or subsystems.

Subsystem	Principal Functions	Other Names	References
Propulsion	Provides thrust to adjust orbit and attitude, and to manage angular momentum	Reaction Control System (RCS)	Sec. 10.4.1, Chap. 17
Attitude Determination & Control System (ADCS)	Provides determination and control of attitude and orbit position, plus pointing of spacecraft and appendages	Attitude Control System (ACS), Guidance, Navigation, & Control (GN&C) System, Control System	Secs. 10.4.2, 11.1, 11.7
Communication (Comm)	Communicates with ground & other spacecraft; spacecraft tracking	Tracking, Telemetry, & Command (TT&C)	Secs. 10.4.3, 11.2
Command & Data Handling (C&DH)	Processes and distributes commands; processes, stores, and formats data	Spacecraft Computer System, Spacecraft Processor	Secs. 10.4.4, 11.3, Chap. 16
Thermal	Maintains equipment within allowed temperature ranges	Environmental Control System	Secs. 10.4.5, 11.5
Power	Generates, stores, regulates, and distributes electric power	Electric Power System (EPS)	Secs. 10.4.6, 11.4
Structures and Mechanisms	Provides support structure, booster adapter, and moving parts	Structure Subsystem	Secs. 10.4.7, 11.6

methods such as spinning or interacting with the Earth's magnetic or gravity fields. These may or may not use *sensors* to measure the attitude or position. More complex systems employ *controllers* to process the spacecraft attitude, and *actuators, torquers*, or propulsion subsystem *thrusters* to change attitude, velocity, or angular momentum. Spacecraft may have several bodies or *appendages,* such as solar arrays or communication antennas, that require individual attitude pointing. To control the appendages' attitude, we use actuators, sometimes with separate sensors and controllers. The capability of the attitude control subsystem depends on the number of body axes and appendages to be controlled, control accuracy and speed of response, maneuvering requirements, and the disturbance environment. Section 11.1 discusses design of the attitude determination and control subsystem.

The *communications subsystem* links the spacecraft with the ground or other spacecraft. Information flowing to the spacecraft *(uplink* or *forward* link) consists of commands and ranging tones. Information flowing from the spacecraft *(downlink* or *return* link) consists of status telemetry and ranging tones and may include payload data. The basic communication subsystem consists of a receiver, a transmitter, and a wide-angle (hemispheric or omnidirectional) antenna. Systems with high data rates may also use a directional antenna. The communications subsystem receives and demodulates commands, modulates and transmits telemetry and payload data, and receives and retransmits range tones—modulation that allows signal turnaround time delay and hence range to be measured. The subsystem may also provide coherence between uplink and downlink signals, allowing us to measure range-rate Doppler shifts. We size the communications subsystem by data rate, allowable error rate, communication path length, and RF frequency. Section 11.2 and Chap. 13 discuss design of the communications subsystem.

The *command and data handling subsystem* distributes commands and accumulates, stores, and formats data from the spacecraft and payload. For simpler systems, we combine these functions with the communications subsystem as a tracking, telemetry, and command subsystem. This arrangement assumes that distributing commands and formatting telemetry are baseband extensions of communications modulation and demodulation. In its more general form, the subsystem includes a central processor (computer), data buses, remote interface units, and data storage units to implement its functions. It may also handle sequenced or programmed events. For the most part, data volume and data rate determine the subsystem's size. Section 11.3 discusses subsystem design, and Chap. 16 covers computers and software.

The *power subsystem* provides electric power for the equipment on the spacecraft and payload. It consists of a power source (usually solar cells), power storage (batteries), and power conversion and distribution equipment. The power needed to operate the equipment and the power duty cycle determine the subsystem's size, but we must also consider power requirements during eclipses and peak power consumption. Because solar cells and batteries have limited lives, our design must account for power requirements at *beginning-of-life (BOL)* and *end-of-life (EOL)*. Section 11.4 discusses design of the power subsystem.

The *thermal subsystem* controls the spacecraft equipment's temperatures. It does so by the physical arrangement of equipment and using thermal insulation and coatings to balance heat from power dissipation, absorption from the Earth and Sun, and radiation to space. Sometimes passive, thermal-balance techniques are not enough. In this case, electrical heaters and high-capacity heat conductors, or *heat pipes*, actively control equipment temperatures. The amount of heat dissipation and temperatures required for equipment to operate and survive determine the subsystem's size. Section 11.5 discusses temperature control in more detail.

The *structural subsystem* carries, supports, and mechanically aligns the spacecraft equipment. It also cages and protects folded components during boost and deploys them in orbit. The main load-carrying structure, or *primary structure*, is sized by either (1) the strength needed to carry the spacecraft mass through launch accelerations and transient events during launch or (2) stiffness needed to avoid dynamic interaction between the spacecraft and the launch vehicle structures. *Secondary structure*, which consists of deployables and supports for components is designed for compact packaging and convenience of assembly. Section 11.6 discusses structural design.

10.1 Requirements, Constraints, and the Design Process

In designing spacecraft, we begin by developing baseline requirements and constraints such as those in Table 10-3. If some of the information is not available, we may need to assume values or use typical values such as those presented here or in the following chapters. For successful design, we must document all assumptions and revisit them until we establish an acceptable baseline.

To get a feel for the size and complexity of a spacecraft design, we must understand the space mission: its concept of operations, duration, overall architecture, and constraints on cost and schedule. Even if we select a mission concept arbitrarily from several good candidates, clearly defining it allows us to complete the spacecraft design and evaluate its performance.

The payload is the single most significant driver of spacecraft design. Its physical parameters—size, weight, and power—dominate the physical parameters of the

TABLE 10-3. Principal Requirements and Constraints for Spacecraft Design. These parameters typically drive the design of a baseline system.

Requirements and Constraints	Information Needed	Reference
Mission:		Chaps. 1, 2
Operations Concept	Type, mission approach	Sec. 1.4
Spacecraft Life & Reliability	Mission duration, success criteria	Secs. 1.4, 10.5, 19.2
Comm Architecture	Command, control, comm approach	Sec. 13.1
Security	Level, requirements	Secs. 13.1, 15.4
Programmatic Constraints	Cost and schedule profiles	Chaps. 1, 20
Payload:		Chaps. 9, 13
Physical Parameters	Size, weight, shape, power	Secs. 9.5, 13.4, 13.5
Operations	Duty cycle, data rates, fields of view	Secs. 9.5, 13.2
Pointing	Reference, accuracy, stability	Secs. 5.4, 9.3, 11.1
Slewing	Magnitude, frequency	Secs. 9.5, 11.1
Environment	Max and min temperatures, cleanliness	Sec. 9.5
Orbit:		Chaps. 6, 7
Defining Parameters	Altitude, inclination, eccentricity	Secs. 7.4, 7.5
Eclipses	Maximum duration, frequency	Sec. 5.1
Lighting Conditions	Sun angle and viewing conditions	Secs. 5.1, 5.2
Maneuvers	Size, frequency	Secs. 6.5, 7.3
Environment:		Chap. 8
Radiation Dosage	Average, peak	Secs. 8.1, 8.2
Particles & Meteoroids	Size, density	Secs. 8.1, 21.2
Space Debris	Density, probability of impact	Sec. 21.2
Hostile Environment	Type, level of threat	Sec. 8.2
Launch:		Chap. 18
Launch Strategy	Single, dual; dedicated, shared; use of upper kick stage	Secs. 18.1, 18.2
Boosted Weight	Launch capabilities	Secs. 18.1, 18.2
Envelope	Size, shape	Sec. 18.3
Environments	g's, vibration, acoustics, temperature	Sec. 18.3
Interfaces	Electrical and mechanical	Sec. 18.3
Launch Sites	Locations, allowed launch azimuths	Sec. 18.1
Ground-System Interface:		Chaps. 14, 15
Degree of Autonomy	Required autonomous operations	Secs. 15.4, 16.1
Ground Stations	Number, locations, performance	Secs. 15.1, 15.5
Space Links	Space-to-space link, performance	Secs. 13.3, 13.4

spacecraft. Payload operations and support are key requirements for the spacecraft's subsystems, as well. The payload may also impose significant special requirements that drive the design approach, such as cryogenic temperatures or avoidance of contamination. Fortunately, we often understand the payload's characteristics better than the spacecraft's overall characteristics in the early design phases. Thus, we can infer many important design features by understanding the payload and how it operates.

Chapters 6 and 7 show how orbital characteristics affect the mission. In spacecraft design, the orbit affects propulsion, attitude control, thermal design, and the electric power subsystem. Most of these design effects are secondary to the orbit's effect on payload performance. Therefore, we select the orbit based on mission and payload performance. Then we compute the performance characteristics needed for the spacecraft, such as pointing, thermal control, and power quantity and duty cycle. Finally, we size the spacecraft to meet these needs.

The natural space environment—especially radiation—limits two aspects of spacecraft design: usable materials or piece parts, and spacecraft lifetime. Radiation levels and dose must be considered in the design, but they do not normally affect the system's configuration or ability. Chapter 8 provides useful space environment information. However, some types of hostile (weapon) environments may affect countermeasures, configuration, shielding, or maneuvering ability.

Selecting a boost vehicle and the possible use of kick stages are central issues in designing a spacecraft. We must select a booster that can put at least the minimum version of our spacecraft into its required orbit. Chapter 18 describes available boosters, all of which have limited weight-lifting ability. In most cases, we must extrapolate published data to meet our mission requirements. Chapters 6 and 7 present the laws of orbital mechanics and the techniques of trajectory design. These include methods for computing velocity increments and guidance techniques. In some cases, the spacecraft must provide large amounts of velocity just to reach orbit or to guide the flight path. Chapter 17 presents performance characteristics for solid and liquid propulsion kick stages to implement these functions. Common nomenclature for a kick stage used to inject a spacecraft into transfer orbit is a *perigee kick motor (PKM)*, whereas a kick stage used to circularize at high altitude is called an *apogee kick motor (AKM)*.

The booster selection will also affect a spacecraft's linear dimensions. An aerodynamic cover, called a *fairing,* or *shroud,* protects the spacecraft as it travels through the atmosphere. The fairing's diameter and length limit the spacecraft's size—at least while it is attached to the booster. Chapter 18 presents the size of standard fairings for various boosters. If the on-orbit spacecraft is larger than the fairing, it must be folded or stowed to fit within the fairing and unfolded or deployed on orbit. To design an item with a large area but small intrinsic mass, such as a solar array or antenna, we make the item as light as possible, fold it and protect it during boost, and *deploy* it (unfold, pull, or stretch it into shape) on orbit. Solar cells may rest on lightweight substrates or even on film that is folded or rolled for storage. Antenna reflectors have consisted of folded rigid panels or of fabric, either film or mesh. Thus, we meet the launch vehicle's demands for a smaller spacecraft by using a stowed configuration and then deploying the spacecraft to meet the full size needed on orbit. We use weight efficiently by caging and protecting the light-weight deployables during boost.

The ground system interface determines how much ground operators and the spacecraft can interact—an important part of design. The periods of visibility between ground stations and the spacecraft limit ground control of spacecraft operations or corrections of errant behavior. Visibility periods and ground coverage issues are described in Chap. 5. If ground operations cost too much, we may want the spacecraft to operate autonomously—another major design decision.

Table 10-4 lists initial configuration decisions or trade-offs designers often face. Weight, size, and power requirements for the payload place lower limits on spacecraft's weight, size, and power. The spacecraft's overall size may depend on such payload parameters as antenna size or optical system diameter. Our approach to spacecraft design must match these dimensions and provide fields of view appropriate to the payload functions. The spacecraft must generate enough power to satisfy the payload needs as well as its own requirements. The amount of power and the duty cycle will dictate the size and shape of solar arrays and the requirements for the battery.

TABLE 10-4.   Initial Spacecraft Design Decisions or Trade-offs. Further discussion of these trades is in Sec. 10.2.

Design Approach or Aspect	Where Discussed	Principal Options or Key Issues
Spacecraft Weight	Table 10-10	Must allow for spacecraft bus weight and payload weight.
Spacecraft Power	Tables 10-8, 10-9	Must meet power requirements of payload and bus.
Spacecraft Size	Sec. 10.1	Is there an item such as a payload antenna or optical system that dominates the spacecraft's physical size? Can the spacecraft be folded to fit within the booster diameter? Spacecraft size can be estimated from weight and power requirements.
Attitude Control Approach	Secs. 10.2, 10.4.2, 11.1	Options include no control, spin stabilization, or 3-axis control: selection of sensors and control torquers. Key issues are number of items to be controlled, accuracy, and amount of scanning or slewing required.
Solar Array Approach	Secs. 10.2, 10.4.6, 11.4	Options include planar, cylindrical, and omnidirectional arrays either body mounted or offset.
Kick Stage Use	Chaps. 17, 18	Use of a kick stage can raise injected weight. Options include solid and liquid stages.
Propulsion Approach	Secs. 10.2, 10.4.1, 17.2, 17.3	Is metered propulsion required? Options include no propulsion, compressed gas, liquid monopropellant or bipropellant.

Field-of-view and pointing considerations influence how we configure the spacecraft. Instruments, sensors, solar arrays, and thermal radiators all have pointing and field-of-view requirements that must be satisfied by their mounting on the spacecraft and the spacecraft's orientation. In the simplest case, all items are fixed to the body, and control of the body's attitude points the field of view. In more complex cases, single or two-degree-of-freedom mechanisms articulate the field of view.

We must also establish how to configure the spacecraft's propulsion early in the design process. Although interaction with the Earth's gravity or magnetic field can control attitude, it cannot change the spacecraft's velocity state. If spacecraft velocity control is needed, some form of metered propulsion must be used. If we decide to use metered propulsion, we should look at using this system for such functions as attitude control or as an orbit transfer stage. Most attitude-control systems use metered propulsion to exert external torque on the spacecraft. Selecting a propulsion approach depends on the total impulse requirement, and the propulsion system's performance, as discussed in Chap. 17.

10.2 Spacecraft Configuration

To estimate the size and structure of a spacecraft, we select a design approach, develop a *spacecraft configuration* (overall arrangement) and make performance allocations to the spacecraft subsystems. We then evaluate the resulting design and reconfigure or reallocate as needed. Subsequent iterations add design detail and provide better allocations. The process of allocating design requirements involves two mutually supporting techniques. First, the allocated design requirements are dictated by considering the overall spacecraft design—a *top-down* approach. Alternatively, the allocated design requirements are developed by gathering detailed design information—a *bottom-up* approach. For instance, we may allocate 100 kg for structural weight based on 10% of the overall spacecraft weight. This is a top-down allocation. However, a detailed design of the structure may require 120 kg if aluminum is used and 90 kg if composites are used. These are bottom-up allocations, providing us with the opportunity to trade off alternatives and reallocate requirements to optimize the design. Most of the allocation methods presented in this chapter are top-down. They provide a starting point for the allocation process. However, we should use them in conjunction with bottom-up design from the more detailed information given in Chaps. 11, 16, and 17.

Figure 10-1 shows different spacecraft configurations. First, observe that each of these spacecraft has a central body or equipment compartment that houses most of the spacecraft equipment. Second, note that these spacecraft all have solar arrays either mounted on external panels or on the skin of the equipment compartment. And finally note that some of the spacecraft have appendages carrying instruments or antennas attached to the main compartment. Let's examine each of these configuration features in more detail.

Table 10-5 lists the factors called *configuration drivers* leading to the various configurations. The weight, size and shape of the payload, and the boost vehicle diameter drive the size and shape of the equipment compartment. Table 10-5 also presents rules of thumb based on analysis of a large number of spacecraft designs. This analysis shows that the average spacecraft bus dry weight (spacecraft weight excluding propellant) is approximately twice that of the payload. The minimum spacecraft bus dry weight is equal to the payload weight and is achieved only when the payload is massive and compact. At the other extreme, low-density payloads or those consisting of multiple instruments can lead to a spacecraft bus as massive as 6 times the payload. Although this is a large range of possible spacecraft bus weights, these ratios are at least a bound. Section 10.3 shows how to refine the estimate.

The spacecraft equipment compartment volume can be estimated from its weight. For 75 spacecraft launched between 1975 and 1984, the average spacecraft in launch configuration with propellant loaded and all appendages folded had a density of only 79 kg/m^3 with a maximum of 172 kg/m^3 and a minimum of 20 kg/m^3. However, appendages are usually lightweight, so the weight of the equipment compartment is only slightly less than the total spacecraft weight. We can use this experience to estimate the spacecraft size (volume and dimensions) by the steps shown in Table 10-6. We start with payload weight to obtain an estimate of spacecraft bus weight (e.g.,

A. Spin-Stabilized Spacecraft

VELA

Mission: Nuclear Detection

Orbit: Super synchronous ~107,000 km

Payload: Radiation instruments body mounted

Configuration Features:
- *Equipment compartment:* 1.6 m diapolyhedron; internal solid AKM
- *Solar array:* Body mounted solar panels
- *Appendages:* None
- *Attitude control:* Spin-stabilized, inertially pointed

Weight: 221 kg   **Power:** 90 W

EXPLORER VI

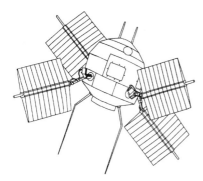

Mission: Radiation Fields Measurement

Orbit: 245 × 42,400 km 47 deg inclination

Payload: Radiation Instruments body mounted

Configuration Features:
- *Equipment Compartment:* 0.6 m dia sphere
- *Solar Array:* Four deployed paddles 2.2 m span
- *Appendages:* Whip antennas
- *Attitude control:* Spin-stabilized, inertially pointed

Weight: 64.6 kg

DSCS II

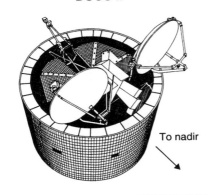

To nadir

Mission: Communications

Orbit: Geosynchronous

Payload: Communications transponder; Earth coverage horn antennas, and steerable pencil beam antennas

Configuration Features:
- *Equipment Compartment:* 3 m dia cylinder, 4.2 m long
- *Solar Array:* Body mounted on cylinder
- *Appendages:* Despun antenna platform with steerable parabolic antennas
- *Attitude control:* Spin-stabilized, spin axis normal to orbit plane

Weight: 523 kg

Power: 535 W BOL, 360 W EOL

Fig. 10-1.   Typical Spacecraft Showing Different Configuration Options. FOV = field-of-view; BOL = beginning-of-life; EOL = end-of-life.

B. 3-Axis-Stabilized Spacecraft

DSP

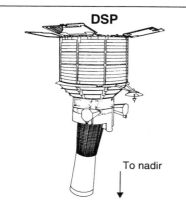

To nadir ↓

Mission: Earth Observation
Orbit: Geosynchronous
Payload: Body mounted telescope and attitude sensors
Configuration Features:
- *Equipment Compartment:* 4.5 m dia cylinder
- *Solar Array:* Body mounted on cylinder and on deployed panels
- *Appendages:* Solar panels and communications antennas
- *Attitude Control:* Cylinder axis pointed toward nadir; Body rotates slowly about cylinder axis to scan the payload FOV

Weight: 2,273 kg   **Power:** 1,274 W

LANDSAT 4, 5

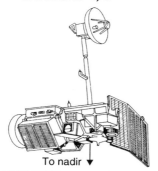

To nadir ▼

Mission: Earth Observation
Orbit: 700 km, Sun synchronous
Payload: Multispectral scanner fixed to spacecraft body with internal scanning mirror
Configuration Features:
- *Equipment Compartment:* Triangular cylinder 2 m dia, 4 m long
- *Solar Array:* Four panel deployed planar, single axis pointing control
- *Appendages:* Communication antenna
- *Attitude Control:* 3-axis control, 1 face toward nadir, 1 axis in direction of flight

Wt: 940 kg   **Power:** 990 W BOL, 840 W EOL

TDRS

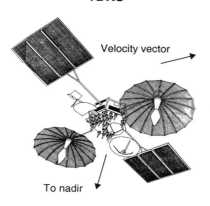

Velocity vector →

To nadir ▼

Mission: Communications
Orbit: Geosynchronous
Payload: S, K, and C band communication transponders with multiple antennas
Configuration Features:
- *Equipment Compartment:* 2.5 m hexagonal cylinder. Auxiliary compartments behind each of the large steerable antennas
- *Solar Array:* Deployed planar panels on both sides of equipment compartment. Single-axis articulation
- *Appendages:* Two 5 m steerable parabolic antennas, one 2 m steerable antenna, one 1.5 m fixed parabolic antenna
- *Attitude Control:* 3-axis control, one face toward nadir, 1-axis in direction of flight

Weight: 2,200 kg   **Power:** 1,700 W

Fig. 10-1. Typical Spacecraft Showing Different Configuration Options. (Continued)
FOV = field-of-view; BOL = beginning-of-life; EOL = end-of-life.

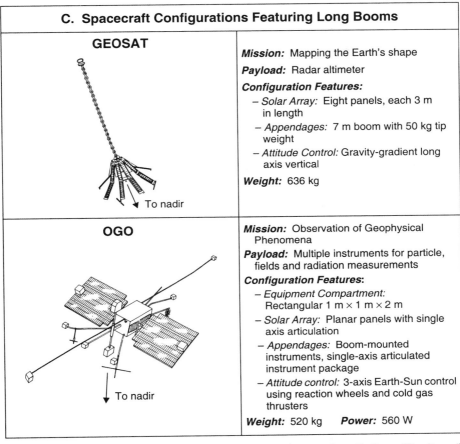

C. Spacecraft Configurations Featuring Long Booms

GEOSAT

To nadir

Mission: Mapping the Earth's shape

Payload: Radar altimeter

Configuration Features:
- *Solar Array:* Eight panels, each 3 m in length
- *Appendages:* 7 m boom with 50 kg tip weight
- *Attitude Control:* Gravity-gradient long axis vertical

Weight: 636 kg

OGO

To nadir

Mission: Observation of Geophysical Phenomena

Payload: Multiple instruments for particle, fields and radiation measurements

Configuration Features:
- *Equipment Compartment:* Rectangular 1 m × 1 m × 2 m
- *Solar Array:* Planar panels with single axis articulation
- *Appendages:* Boom-mounted instruments, single-axis articulated instrument package
- *Attitude control:* 3-axis Earth-Sun control using reaction wheels and cold gas thrusters

Weight: 520 kg   **Power:** 560 W

Fig. 10-1. Typical Spacecraft Showing Different Configuration Options. (Continued) FOV = field-of-view; BOL = beginning-of-life; EOL = end-of-life.

twice payload weight) and add an estimate of propellant weight (see Sec. 10.3) to yield spacecraft loaded weight. We now use an estimated density (e.g., 79 kg/m^3) to compute the spacecraft volume. We select an equipment compartment shape and dimensions to provide this volume, match the payload dimensions, and fit within the booster diameter.

Usually, a spacecraft in folded launch configuration is cylindrically symmetric, mounted on the booster with the axis of symmetry parallel to the thrust axis. The folded spacecraft diameter is designed to fit within the boost vehicle diameter although on rare occasions a bulbous fairing may be used to provide a slightly larger diameter. (Bulbous fairings are generally avoided because they detract from booster performance.) Since the equipment compartment diameter is approximately the same as the folded spacecraft diameter, we can use the booster diameter as an upper limit for the equipment compartment diameter and select a compartment shape to fit within this diameter. Knowing its volume, we can readily compute the compartment's length. Table 10-28 in Sec. 10.5 provides formulas for compartment parameters for a cubic

dry veh wt = payload wt + bus wt
dry   dry

TABLE 10-5. Spacecraft Configuration Drivers.

Configuration Driver	Effect	Rule of Thumb	Reference
Payload Weight	Spacecraft dry weight	Payload weight is between 17% and 50% of spacecraft dry weight. Average is 30%	Chap. 9
Payload Size and Shape	Spacecraft size	Spacecraft dimensions must accommodate payload dimensions	Chaps. 9, 10, 11, 13
Payload Power	Spacecraft power	Spacecraft power is equal to payload power plus an allowance for the spacecraft bus and battery recharging	Chap. 9, 10.2, 10.3
Spacecraft Weight	Spacecraft size	Spacecraft density will be between 20 kg/m^3 and 172 kg/m^3. Average is 79 kg/m^3	Secs. 10.2, 10.3
Spacecraft Power	Solar array area	The solar array will produce approximately 100 W/m^2 of projected area	Sec. 11.4
Solar Array Area	Solar array type	If required solar array area is larger than area available on equipment compartment, then external panels are required	Sec.11.4
Booster Diameter	Spacecraft diameter	Spacecraft diameter is generally less than the booster diameter	Sec.10.2, Table 18-3
Pointing Requirements	Spacecraft body orientation and number of articulated joints	Two axes of control are required for each article to be pointed. Attitude control of the spacecraft body provides 3 axes of control	Sec.11.1

TABLE 10-6. Estimating Spacecraft Equipment Compartment Dimensions.

Step	Procedure	Comments
1. Payload Weight	Starting point	Sec. 9.1
2. Estimate Spacecraft Dry Weight	Multiply payload weight by between 2 and 7	Average is 3.3
3. Estimate Spacecraft Propellant	Prepare a bottom-up propellant budget or arbitrarily select a weight	Normal range is 0% to 25% of spacecraft dry weight (Table 10-7)
4. Estimate Spacecraft Volume	Divide spacecraft loaded weight by estimated density	Range of density is 20–172 kg/m^3; Average is 79 kg/m^3
5. Select Equipment Compartment Shape and Dimensions	Shape and dimensions should match payload dimensions and fit within the booster diameter	In the folded configuration, spacecraft are cylindrically symmetric about the booster thrust axis. Cross-sectional shapes range from triangular to circular.

compartment using a density of 100 kg/m^3. Figure 10-1 shows examples of various spacecraft compartment shapes.

Spacecraft solar arrays are of two types: body mounted and panel mounted. Examples of both types are shown in Fig. 10-1. An array produces about 100 W/m^2 of projected solar cell area. This is unregulated power and represents an array efficiency of 7%. To use this rule of thumb, we need to estimate the total spacecraft power, as described in Sec. 10.3. However for initial configuration selection, we need only bound the power requirements and see if there is sufficient area on the equipment compartment to allow body-mounted cells. At a minimum, power must be provided for the payload. Prudent design would also make some allowance for spacecraft bus power and battery recharge power which are discussed in more detail in Sec. 10.3. If there is insufficient area on the spacecraft body for a body-mounted array, then we are forced to use panels.

Evaluation of pointing and attitude control on spacecraft configuration starts with identifying all pointing requirements (see Table 10-13) for both the spacecraft bus and payload. The process of synthesizing a control approach to meet these requirements is discussed in Sec. 10.4.2. Although we must go through this process in detail to see the full effect of pointing on the spacecraft configuration, the basic implications can be derived by the process and rules of thumb in Table 10-5. The spacecraft configuration must provide 2 axes of control for each item that is to be pointed. The spacecraft body has 3 axes so the body alone can satisfy one pointing requirement; for instance, one body axis (i.e., the yaw axis) can be pointed toward nadir by control about the other 2 axes (roll and pitch). If two items are to be pointed, then the spacecraft must be configured with at least one articulated joint between the two items. For illustration, a body-mounted antenna can be pointed toward nadir by controlling 2 axes of body attitude. A solar array can then simultaneously be pointed toward the Sun by using the third body axis and providing a single axis solar array drive to control the solar array attitude relative to the body. This approach is called *yaw steering* (see OGO, Fig. 10-1C). If the spacecraft has a second item that must point in another direction (say, an antenna that must point toward a communication relay satellite), then the configuration must provide 2 more axes of control. The TDRS spacecraft shown in Fig 10-1B has 3 separate articulated antennas with a total of 6 mechanical axes of control in addition to 3 axes of body attitude control and 1 axis of solar array control. DSP, shown in Fig 10-1B, has a body-mounted payload and uses control of two body axes to point toward nadir. The third body axis is used to scan the payload field-of-view. A communication antenna is articulated about 2 axes to point toward a ground station, and although solar array panels are used to augment the solar array area, the array is not articulated.

Spin stabilization is a particularly simple and robust method of attitude control. Satellites that employ spin stabilization are often cylindrical, such as DSCS II shown in Fig 10-1A. For such a satellite, the spin axis supplies 1 axis of control by using a motor-driven platform that is *despun* (spinning in a negative sense relative to the satellite body). This is sometimes called a *dual-spin* system. Equipment mounted on the despun platform can be articulated about additional axes to achieve further pointing capability.

The attitude control method may also interact with the solar array configuration. Solar panels may be body-fixed such as Explorer VI and DSP, or they may be articulated, as shown on OGO and TDRS in Fig. 10-1. Spin-stabilized spacecraft usually have body-fixed arrays, and 3-axis-controlled spacecraft generally use articulated panels. The power generated by a solar array is proportional to the area that intercepts sunlight (the projected area). A planar array which is pointed toward the Sun has a ratio of total array area to projected area of one. A cylindrical array which has its axis perpendicular to the Sun line has a ratio of $1/\pi$ and an array which projects equal area in all directions has a ratio of $1/4$. The method of solar array pointing control and the type of array selected therefore affect the total array area.

Sometimes a spacecraft configuration is driven to use long booms either to control spacecraft moment of inertia or to separate delicate instruments from the spacecraft body electrical fields. GEOSAT, shown in Fig. 10-1C, uses a boom to increase the moment of inertia and provide gravity-gradient attitude stabilization, a passive attitude control technique described in Sec. 11.1. OGO, also shown in Fig.10-1C, uses booms to isolate payload magnetometers from the spacecraft body.

10.3 Design Budgets

We begin allocating performance by establishing *budgets** or allocations for propellant, power, weight, and reliability. We can derive the propellant budget by estimating the propellant requirements for velocity changes (orbit correction and maintenance) and attitude control. At first, we estimate the power budget by adding the payload's power requirements to power estimates for the spacecraft bus subsystems. To derive the first weight budget for the spacecraft, we add the payload weight to estimates for the spacecraft bus, including propulsion components and power components. We make the initial reliability budget by defining the probability of achieving acceptable spacecraft performance and lifetime.

A typical propellant budget as shown in Table 10-7 contains four elements: velocity-control propellant, attitude-control propellant, margin, and residual. The velocity-correction requirement is expressed as *total velocity change* or ΔV, which is obtained from Sec. 7.3. Chapter 17 presents the rocket equation Eq. (17-21) by which we convert velocity change to propellant mass. Attitude-control propellant is used for spin stabilization and maneuvering while spinning, countering disturbance torques (including control during ΔV thrusting), attitude maneuvering, and limit cycling or oscillation. Formulas for these entries are presented in Sec. 10.4.2 and summarized in Table 10-7. *Propellant margin* is a percentage of the identified propellant requirement, typically 25% for preliminary design. 1% or 2% is unavailable propellant.

TABLE 10-7. Propellant Budget.

Elements	Reference
Velocity Correction and Control	Eqs. (17-9), (7-14), Table 7-3

Attitude Control
  Spinup and despin
  Maneuvering while spinning
  Cancelling disturbance torque
  Control during ΔV thrusting
  Attitude maneuvering
  Limit cycling
	Secs. 10.4.2, 11.1
	Eq. (10-9)
	Eq. (10-11)
	Table 10-18
	Table 10-18 and Eqs. (10-8a,b)
	Eq. (10-3)
	Eqs. (10-5) through (10-7)

Nominal Propellant	Sum of above
Margin	10%–25% of nominal
Residual	1%–2% of total
Total Propellant	Sum of above

---

* See p. 4 in Chap. 1 for the definition of a budget.

Table 10-8 outlines the three steps for estimating spacecraft power requirements. First, we prepare an *operating power budget* by estimating the power required by the payload and the spacecraft subsystems. If the spacecraft has several operating modes that differ in power requirements, we must budget separately for each mode, paying particular attention to peak power needs for each subsystem. The second step is *battery sizing*, or selecting the battery capacity appropriate to the spacecraft power requirements and battery cycle life. With size established, we can compute the battery's recharge power. The third step is accounting for power-subsystem degradation over the mission life by computing radiation damage to the solar array.

TABLE 10-8.   Steps in Preparing a Power Budget.

Step	What's Involved	Where Discussed
1. Prepare Operating Power Budget	Estimate power requirements for payload and each spacecraft bus subsystem	Chaps. 9, 11, 13
2. Size the Battery	Estimate power level that the battery must supply	Generally equal to or less than the operating power level
	Compute discharge cycle duration, charge cycle duration, and number of charge-discharge cycles	Determined by orbit selection and mission duration (Chap. 7)
	Select depth of discharge	Sec. 11.4
	Select charge rate	Sec. 11.4
	Compute battery recharge power	Sec. 11.4
3. Estimate Power Degradation Over Mission Life	Compute degradation of power system from orbital environment	Secs. 8.1, 11.4

Table 10-9 lists references that discuss operating power for the payload and the spacecraft bus subsystems, and shows typical percentages of the operating power budget devoted to each subsystem. These percentages change with the spacecraft's total power use, so I have presented ranges for a minimum (<100 W), a small (200 W), and a medium to large spacecraft (500–10,000 W). We can use these values as a starting point if we do not have more information, but Sec. 10.6 and Chap. 11 give specific examples of various power requirements.

Sections 10.4.6 and 11.3 discuss battery recharge power. At the minimum, the recharge energy must exceed the energy drawn from the battery during discharge by an amount that accounts for the efficiency of the charge-discharge process (typically 80%). Most batteries also require recharge at a minimum rate—specified as a fraction of battery capacity (typically 1/15). These two requirements translate into recharge power ranging from 7% of the discharge power for geosynchronous orbits to 60% for low-altitude orbits.

The solar array must supply enough power for operations and recharging the battery until end-of-life. The beginning-of-life power requirement must allow for degradation in the solar array. As Sec. 11.4 points out, this degradation depends on orbit altitude and radiation environment, but 30% is typical for 10 years at geosynchronous altitude. We can assume the same value for altitudes of 800 km or less. Between these altitudes the degradation is much larger.

Table 10-10 shows the build-up of spacecraft weight. As pointed out in Sec. 10.2, the ratio of spacecraft dry weight to payload weight lies in the range of 2:1 to 7:1; the payload weight is typically less than half the spacecraft dry weight and may be as little

TABLE 10-9.   Typical Power Consumption by Module or Subsystem.

Spacecraft Size Spacecraft Power Subsystem*	Minimum (< 100 W total)	% of Operating Power		References
		Small (~ 200 W)	Medium to Large (> 500 W)	
Payload	20–50 W	40	40–80	Chaps. 9, 13
Propulsion	0	0	0–5	Sec. 10.4.1, Chap. 17
Attitude Control	0	15	5–10	Secs. 10.4.2, 11.1
Communications	15W	5	5–10	Secs. 10.4.3, 11.2
Command & Data Handling	5 W	5	5–10	Secs. 10.4.4, 11.3
Thermal	0	5	0–5	Secs. 10.4.5, 11.5
Power	10–30 W	30	5–25	Secs. 10.4.6, 11.4
Structure	0	0	0	Secs. 10.4.7, 11.6

Average Power: Sum of above
Margin: 5% to 25% of power based on design maturity
Total Average Requirement for Operating Power: Total of above

* Includes conversion and line losses.

TABLE 10-10.   **Weight Budget.** The percentages shown in the right-hand column are the percent of spacecraft dry weight.

Element	Weight	Reference	Comments
Payload	$M_{P/L}$	Chaps. 9, 13	15% to 50% of M_{dry}
Spacecraft Subsystems	M_{SS}		Sum of subsystem wts
Propulsion	$M_{propulsion}$	Chap.17	
Attitude Control	M_{gnc}	Sec. 11.1	
Communications	M_{com}	Sec. 11.2	
Command & Data Handling	M_{cadh}	Sec. 11.3	
Thermal	M_{th}	Sec. 11.5	2% to 5% of M_{dry}
Power	M_{ep}	Sec. 11.4	
Structure & Mechanisms	M_{sam}	Sec. 11.6	8% to 12% of M_{inj} or 15% to 25% of M_{dry}
Margin	M_{mar}		5% to 25% of wt based on design maturity
Spacecraft Dry Weight	$M_{dry} = M_{P/L} + M_{SS} + M_{mar}$		
Propellant	M_{prop}	Table 10-7	
Loaded Weight	$M_{loaded} = M_{dry} + M_{prop}$	Chap. 17	
Kick Stage	M_{kick}	Sec. 17.3	
Injected Weight	$M_{inj} = M_{loaded} + M_{kick}$		
Adapter	$M_{adapter}$	Sec. 11.6	
Boosted Weight	$M_{boosted} = M_{inj} + M_{adapter}$	Chap. 18	

as 15% of the dry weight. Spacecraft structure weight generally falls in the range of 15% to 25% of spacecraft dry weight (see Appendix A). Spacecraft structural weight may also be estimated at 8% to 12% of injected weight (dry weight + propellant + injection stage). Spacecraft thermal subsystem weight is between 2% and 5% of spacecraft dry weight. Weight percentages for other subsystems vary widely and require more detailed investigation. (See Sec. 10.4.3, Chaps. 11 and 17.) To account for uncertainties during preliminary design we add 25% to these weights for new equipment and 5% or less for known hardware. We should hold a small allowance (1% to 2%) at the system level to account for integration hardware, such as brackets and mounting hardware, which are often overlooked.

TABLE 10-11. Preparing a Reliability Budget.

Step	Comments	Reference
1. Establish mission success criteria	The criteria should be numerical and may have multiple elements. For instance, a communication spacecraft having multiple channels for several types of service might define success as one channel of each type of service or as a total number of channels and total radiated power.	Chaps. 1, 2, 3
2. Assign numerical success probability to each element of the mission success criteria and define the method for computing success probability	This might be stated as a probability of 0.5 of operating service "A" for 5 years and a probability of 0.7 of operating service "B" for 2 years and 0.4 of operating service "B" for 7 years. For each element of the success criteria, numerical values and associated lifetimes are assigned. Several methods of evaluating success probability are available—see Chap. 19.	Sec. 19.2
3. Create the reliability budget by allocating the success probability (reliability) to each item of hardware & software	If, for instance, a system reliability of 0.6 is required, it might be allocated as: Propulsion 0.95 Comm 0.93 Structure 0.99 C&DH 0.93 Thermal 0.99 Power 0.93 ADCS 0.9 Payload 0.89	Sec. 19.2
4. Evaluate the system reliability and iterate the design to maximize reliability and identify and eliminate failure modes	Assuming independent, serial operation, hardware failure rate is generally evaluated by summing piece part failure rates. (See Chap. 19 and MIL-HDBK-17 [1991].) Failure mode analysis and elimination are discussed in Sec. 10.4. Effect of failures can be reduced and reliability raised by changing the design, selecting more reliable hardware, or adding redundant hardware and software.	Sec. 19.2

From the start of the spacecraft design we must design our hardware and software to achieve reliable operation. The process of *design-for-reliability* starts in the conceptual design phase with the determination of system reliability requirements and allocation of these requirements to the spacecraft subsystems. This is a four-part process as shown in Table 10-11. First, we establish the *mission success criteria*, which is a list of events and operations that together constitute success. Second, we assign a numerical probability to meeting each element of the mission success criteria and select a set of ground rules for computing the probability of success. Third, we allocate reliability requirements to all spacecraft hardware and software. Fourth, we evaluate system reliability and iterate the design to maximize the reliability assessment, and identify and eliminate failure modes. (See Secs. 10.5.2 and 19.2 for further discussions of reliability.)

10.4 Designing the Spacecraft Bus

10.4.1 Propulsion Subsystem

The propulsion equipment for a spacecraft includes tankage to hold the propellant, lines and pressure-regulating equipment, and the engines or thrusters. Common propellants are pressurized gas such as nitrogen, selected monopropellants such as hydrazine, and bipropellants. The pressurized feed systems typically used may be *pressure regulated* or *blow down*. Important design parameters are the number, orientation, and location of the thrusters; the thrust level; and the amount of impulse required. Chapter 17 discusses the design of propulsion subsystems and characteristics of propulsion components.

The propulsion tanks rest at or near the spacecraft's center of mass to avoid shifting of the center of mass as the propellant is used. Engines for translational control are aligned to thrust through the center of mass, whereas engines for attitude control thrust tangentially and are mounted as far away from the center of mass as possible to increase the lever arm and thus increase the torque per unit thrust. Attitude control engines which fire in the direction of flight (along or in opposition to the velocity vector) are generally used in pairs to produce a pure torque without net linear force. However, the spacecraft flight path is less sensitive to thrust at right angles to the velocity vector and single attitude control engines are sometimes used in these directions. Three-axis control requires a minimum of 6 attitude control thrusters, and many designs use 8 to 12 plus backup units for reliability.

Table 10-12 gives weight and power estimates for the propulsion subsystem. Chapter 17 gives more detailed weights and Sec. 10.6 offers examples of integrated designs. The propulsion subsystem does not use much electrical power unless it employs thrusters with heated catalyst beds, heated thrusters, or electric propulsion. Electric propulsion is rare, but heated thrusters are common. Propulsion lines and tanks must be protected from freezing, usually by thermostatically controlled guard heaters. Power for these heaters is included in the thermal subsystem. Electrically operated solenoid valves control propellant flow to the thrusters, but we account for their power in the ADC subsystem.

TABLE 10-12. Weight and Power Budget for Propulsion Subsystem. See Sec. 17.4 for specific design information.

Component	Weight (kg)	Power (W)	Comments
Propellant	Table 10-7	—	Added to overall budget in Table 10-7; **not** part of propulsion subsystem
Tank	10% of propellant weight	—	Tanks for compressed gas may be up to 50% of gas weight
Thrusters	0.35–0.4 kg for 0.44 to 4.4 N hydrazine units	5 W per thruster when firing	
Lines, Valves, & Fittings	Dependent on detailed spacecraft design	—	Example spacecraft of Sec. 10.6 used 6.8 kg (HEAO) & 7.5 kg (FLTSATCOM)

10.4.2 Attitude Determination and Control Subsystem

Attitude control requirements are based on those for payload pointing and the spacecraft bus pointing. Table 10-13 lists possible design requirements. If we must control the attitude of all or part of the payload, we have to decide whether to point the entire spacecraft or to articulate the payload or part of the payload. In the extreme case of a scanning payload fixed to the spacecraft's body, we must scan with the entire spacecraft. Similar decisions are necessary for parts of the spacecraft that need orienting in other directions such as pointing at the Sun.

TABLE 10-13. Typical Sources of Requirements for Attitude Control.

Requirement	Information Needed
Payload Requirements:	
Article to be Pointed	Entire payload or some payload subset such as antennas or a thermal radiator
Pointing Direction	Defined relative to what reference
Pointing Range	All of the possible pointing directions
Pointing Accuracy	Absolute angular control requirement
Pointing Knowledge	Knowledge of pointing direction either in real time or after the fact
Pointing Stability	Maximum rate of change of angular orientation
Slew Rate	Reorientation from one pointing direction to another in a specified time
Exclusion Zones	For example, "not within 10 deg of the Sun"
Other Requirements:	
Sun Pointing	May need for power generation or thermal control
Pointing During Thrusting	May need for guidance corrections
Communications Antenna Pointing	Toward a ground station or relay satellite

Pointing in a particular direction requires control of angular orientation about each of the 2 axes perpendicular to the pointing axis. If, for example, a payload or antenna must point toward the Earth, we need to control its attitude about 2 horizontal axes. If the payload is fixed to the spacecraft's body, these 2 axes are 2 of the 3 axes. Thus, we can use the third axis—rotation about the pointing axis—to satisfy a second pointing requirement, such as pointing one axis in the direction of flight. Table 10-14 lists types of pointing systems.

Either *spin stabilization* or *3-axis control* using sensors and torquers can be used to control the spacecraft's attitude. Spin stabilization divides into *passive spin, spin with precession control,* or *dual spin* (spin with a despun platform). Classes of 3-axis control depend on the sensor type or torquing method. Possible sensors include Earth, Sun and star sensors, gyroscopes, magnetometers, and directional antennas. Torquers include gravity gradient, magnetic, thrusters, and wheels. Wheels include variable speed *reaction wheels*; *momentum wheels*, which have a nominal nonzero speed and therefore provide angular momentum to the spacecraft; and *control moment gyros,* which are fixed-speed gimballed wheels. Table 10-15 summarizes these methods of control during thrusting and nonthrusting.

TABLE 10-14. Design Approaches for Selected Pointing Requirements.

Requirement	System
Nadir-Pointed Payload	Body-fixed payload using 2 axes of body attitude control to meet the Earth-pointing requirement. The third axis is used to point a horizontal axis in the direction of flight.
	Can also use spin-stabilized spacecraft with spin axis normal to the orbit plane and payload mounted on despun platform.
Payload Pointed in a Fixed-Inertial Direction	Body-fixed payload using 2 axes of body attitude for payload-pointing direction in inertial space. Third axis is used to keep one side toward the Sun.
Sun-Oriented Solar Array	Planar array requires 2 axes of control. May be achieved by 1 axis of body attitude and 1 rotation axis.
	Cylindrical array with array axis perpendicular to Sun line.
Communications Antenna	2-axis mechanism.

TABLE 10-15. Types of Attitude Control.

Control Mode	Type of Control
Control During Thrusting:	
Spin Stabilization with Axial Thrust	Passive spin in a fixed direction with thrust applied parallel to the spin axis.
Spin Stabilization with Radial Thrust	Passive spin in a fixed direction with thrust applied perpendicular to the spin axis in short pulses.
3-Axis Control	Attitude is sensed with sensors whose output is used to control torquers. Torquers include thrusters operated off-on or swiveled to control thrust direction.
Control While Not Thrusting:	
Spin Stabilization with Precession Control	Spin direction is controlled by applying precession torque with an off-axis thruster.
Dual Spin	Spin-stabilized with a despun platform.
3-Axis Control	Control using attitude sensors and torquers.

We use spin stabilization extensively for attitude control during kick-stage firing and for small spacecraft. Spin-stabilized satellites with a despun platform, called a dual spin system, frequently support communications payloads. In this case, we mount the payload antenna on the despun platform so we can control its pointing. If the spin axis is roughly perpendicular to the Sun line, we can mount solar cells on the spacecraft's cylindrical skin to produce electric power.

Three-axis approaches range from *passive control* using gravity gradient or magnetics to *full active control* with propulsion thrusters and wheels. Passive techniques can provide coarse control to support low-accuracy pointing requirements and simple spacecraft. The active 3-axis method gives us highly accurate pointing control, more efficient solar arrays (by allowing oriented planar arrays), and pointing of several payloads or spacecraft appendages. But active 3-axis systems are more complex and usually heavier than spin-stabilized ones, and we may need to consider both approaches carefully before deciding between them. Table 10-16 summarizes how various requirements affect this decision.

TABLE 10-16. Implication of Pointing Requirements on Attitude Control Approach.

Requirement	Implication
Control during kick-stage firing	Spin stabilization preferred in most cases
Coarse control (> 10 deg)	Spin stabilization or passive control using gravity gradient
Low-accuracy pointing (> 0.1 deg)	Either 3-axis control or dual spin
Low power requirement (< 1 kW)	Either oriented planar array or spinning cylindrical array
High-accuracy pointing (< 0.1 deg)	3-axis control
High power requirement (> 1 kW)	Oriented planar array
Multiple pointing requirements	3-axis control
Attitude slewing requirement	Dual spin with articulation mechanism or 3 axis with wheels

The *guidance and navigation* function on most spacecraft is a basic form of radio guidance. It uses ground tracking to measure the flight path (or *spacecraft ephemeris*), ground computing of desired velocity corrections, and command of the correction through the communications and command subsystems. The direction of the velocity correction is governed by the attitude control of the spacecraft body and the magnitude is controlled by engine firing time. Two elements limit performance: ground-tracking and spacecraft attitude-control during thrusting. The *Global Positioning System* (*GPS*) may provide another way to measure the flight path. Coupled with an appropriate guidance computer, a GPS receiver should be able to guide the boost phase and allow the spacecraft to navigate autonomously in orbit, at least for low altitudes. Other autonomous navigation methods are also available (see Sec. 11.7). Guiding spacecraft to intercept or rendezvous usually requires a guidance radar, a gyroscope reference assembly, and often accelerometers.

Accurate attitude control depends on the attitude sensors. Table 10-17 summarizes what present sensors can do. Each sensor class is available in either a 1- or 2-axis version. Magnetometers, Earth sensors, and Sun sensors are available in forms which use the spin motion of a spinning spacecraft to scan the sensor's field of view. Some magnetometers and Earth and Sun sensors do not require scanning, but some highly accurate Earth sensors use scanning detectors. Magnetometers apply only to altitudes below about 6,000 km because the Earth's magnetic field falls off rapidly with altitude. Uncertainty in the Earth's magnetic field and its variability with time limits the accuracy of a magnetometer. In the same way, horizon uncertainty limits an Earth sensor's accuracy. Star sensors, however, allow us to measure attitude very accurately. Most star sensors are too slow (typically several seconds) to control a spacecraft's attitude directly, so we normally use them with gyroscopes for high accuracy and rapid response. Gyroscope accuracy is limited by instrument drift, so most gyroscope systems are used in conjunction with an absolute reference such as a star sensor or a directional antenna.

After-the-fact processing can improve our knowledge of attitude. For example, we can monitor the Earth's magnetic field continuously and therefore partially correct the variable effects of magnetic sensing. Variations in the Earth's horizon tend to follow a daily cycle, so we can apply some filtering correction.

TABLE 10-17. Ranges of Sensor Accuracy. Microprocessor-based sensors are likely to improve accuracies in the future.

Sensor	Accuracy	Characteristics and Applicability
Magnetometers	1.0° (5,000 km alt) 5° (200 km alt)	Attitude measured relative to Earth's local magnetic field. Magnetic field uncertainties and variability dominate accuracy. Usable only below ~6,000 km.
Earth Sensors	0.05 deg (GEO) 0.1deg (low altitude)	Horizon uncertainties dominate accuracy. Highly accurate units use scanning.
Sun Sensors	0.01 deg	Typical field of view ±130 deg.
Star Sensors	2 arc sec	Typical field of view ±16 deg.
Gyroscopes	0.001 deg/hr	Normal use involves periodically resetting the reference position.
Directional Antennas	0.01deg to 0.5 deg	Typically 1% of the antenna beamwidth.

The attitude-control system can also produce attitude motions which combine with the attitude sensor's accuracy to affect the total control accuracy. Control systems that use thrusters alone require a *dead zone* to avoid continuous firing of the thrusters. The control system's accuracy is limited to half the dead-zone value plus the sensor accuracy. Systems which use wheels (either speed-controlled reaction wheels or gimballed constant-speed wheels) can avoid dead-zone attitude errors, so they usually can operate close to the sensor accuracy.

Torquing methods for 3-axis-controlled spacecraft include gravity gradient, magnetic, thrusters, and wheels. Spacecraft using gravity-gradient and magnetic torquing are clean and simple but do not provide high levels of control torque. We use thrusters on most spacecraft because they produce large torque and can control the spacecraft's translational velocity as well as attitude. If the spacecraft must maneuver or suffers cyclical disturbances, such as those at orbit rate or twice orbit rate (see Chap. 11), we need to use wheels. A wheel can cyclically speed up or slow down, thus producing maneuvering torque or counteracting disturbance torques. A wheel system consumes less propellant than a thruster-only system because periodic effects do not require propellant use. On the other hand, wheel systems are heavier and more complex than those without wheels. Variable-speed reaction wheels produce only limited control torque (less than 1 N·m). To obtain large values of cyclic torque, we use control moment gyros—constant-speed wheels gimballed about an axis perpendicular to the spin axis. These gyros can develop torques up to several thousand N·m. When we need more degrees of freedom or better pointing accuracy, we use mechanisms to point spacecraft appendages, such as solar arrays or directional antennas.

Estimating Torque Requirements

One important sizing parameter for the control subsystem is its torque capability. This capability, often called the *control authority*, must be large enough to counterbalance disturbance torques and control the attitude during maneuvers and following transient events such as spacecraft separation, deployment, and failure recovery. These latter events usually size the torque requirement. The separation transient is usually specified in terms of *tip-off rate*—the angular velocity, typically 0.1 to 1 deg/s, imparted to the spacecraft at release from the booster. We size the attitude control thrusters to *capture* or stabilize the spacecraft attitude before it has exceeded a specified value, as shown in Fig. 10-2A. The relation between required torque T, the tip-off rate, ω_t (in rad/s), spacecraft moment of inertia, I_s, and the maximum attitude excursion, θ_{max} (in rad), is:

$$T = \frac{1}{2}\omega_t^2\left(I_s \, / \, \theta_{max}\right) \tag{10-1}$$

The torquing ability of a thruster system, a reaction wheel, a control moment gyro, or a pointing mechanism may be set by an acceleration requirement such as that arising from an attitude slew maneuver, shown in Fig 10-2B. The torque is simply I_a where I is the moment of inertia and a is the acceleration. Sometimes the attitude maneuver is specified as a change in angle of θ in a time t_{dur}. The torque in this case is:

$$T = 4\theta I \, / \, t_{dur}^{\ 2} \tag{10-2}$$

As Fig. 10-2B shows, this is based on applying full accelerating torque for $t_{dur}/2$ and full decelerating torque for the remaining time.

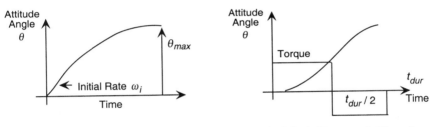

A. Attitude Capture Following Separation   B. Attitude Maneuver in Time, t_{dur}

Fig. 10-2. Estimating Torque for Attitude Capture and Maneuvering.

The control torque required to stabilize a spacecraft during velocity-correction firing of a rocket motor is the product of the rocket's thrust level and the distance that its line-of-action is offset from the spacecraft center-of-mass. This torque can be due either to thruster misalignment or cg offset (see Table 10-18).

Estimating Angular Impulse for 3-Axis Control

Another major sizing parameter for the attitude control subsystem is the angular impulse capability of its torquers. *Angular impulse* is the time integral of torque. For thruster-produced torque, the angular impulse is related to the propellant mass expended. For reaction wheels and control moment gyros, the angular impulse is related to wheel moment of inertia and speed. In all cases, angular impulse is related to control system weight.

For 3-axis control systems, we calculate angular impulse by evaluating that needed for attitude maneuvering, for counteracting the effects of disturbance torques, and for oscillation or limit cycling. We determine the angular impulse required for maneuvering from spacecraft moment of inertia and maneuver angular rate. The angular impulse required to start an attitude maneuver L_{start} is:

$$L_{start} = I_s \, \omega_{man} \tag{10-3}$$

where I_s is spacecraft moment of inertia and ω_{man} is the angular rate (rad/s) of the

TABLE 10-18. Disturbance Torques. These are vector equations where "×" denotes vector cross product and "·" denotes vector dot product. See Sec. 11.1 for simplified equations.

Disturbance	Equation	Definition of Terms
ΔV Thruster Misalignment	$s \times T$	**s** vector distance from center of mass to thrust application point **T** vector thrust
Aerodynamic Torque	$\frac{1}{2}\rho V^2 C_d A (u_v \times s_{cp})$	ρ atmospheric density C_d drag coefficient (typically 2.25) A area perpendicular to u_v V velocity u_v unit vector in velocity direction s_{cp} vector distance from center of mass to center of pressure
Gravity Gradient Torque	$\frac{3\mu}{R_0^3}\, u_e \times (I \cdot u_e)$	μ Earth's gravitational coefficient 3.986 × 10^14 m³/s² R_0 Distance to Earth's center (m) **I** Spacecraft inertia tensor u_e Unit vector toward nadir
Solar Radiation Torque	$K_s(u_s \cdot u_n) A \left[\begin{array}{c} u_s(\alpha + r_d) + \\ u_n\left\{2r_s + \frac{2}{3}r_d\right\} \end{array} \right] \times s_c$	K_s solar pressure constant 4.644 × 10^-6 N/m² s_c vector from spacecraft center of mass to area A u_n unit vector normal to A u_s unit vector toward the Sun α surface absorptivity coefficient r_s surface specular reflectance coefficient r_d surface diffuse reflectance coefficient (Note: $\alpha + r_s + r_d = 1$)

maneuver. Stopping the maneuver requires an equal amount of angular impulse of opposite polarity.

To compute angular impulse to cancel disturbance torques, we examine the disturbances affecting the spacecraft as shown in Table 10-18. These disturbance torques are all vector quantities. They can be expressed in any convenient system of coordinates, although spacecraft body coordinates are generally used. For a given spacecraft configuration, orbit, and spacecraft attitude, these torques can be computed and integrated over the spacecraft lifetime. The result is the accumulated angular impulse which, if uncontrolled, will disturb the spacecraft attitude. The control subsystem must counteract these disturbance torques by applying control torque and the control subsystem angular impulse capability must be at least equal to the disturbance angular impulse.

In addition to the disturbance torques listed in Table 10-18, the control system may be sized by the requirement to interchange momentum between spacecraft body axes (sometimes referred to as *Euler cross-coupling torque*). Numerically this torque-like effect is:

$$T_E = -\Omega \times H_s \tag{10-4}$$

and comes about when the spacecraft dynamic equations are written in rotating coordinates. T_E is the torque, Ω is the angular velocity of the coordinate system, and H_s is the spacecraft angular momentum including that due to body rotation and internal moving parts (such as reaction wheels). For a circular orbit and the dynamic equations written in a coordinate system which rotates at orbit rate, the Euler cross-coupling torques are of the same form as gravity-gradient torques and these effects are often combined.

Note that, for any single axis, the disturbance torque may have *cyclic* terms that integrate to zero over an integer number of cycles and *secular* terms that are not periodic. Also note that angular impulse required for maneuvering is all cyclic. Reaction wheels and control moment gyros can counteract cyclic torques by changing speed or direction. If we use reaction wheels or control moment gyros (CMGs), we can size them for the cyclic terms and counteract only the secular terms with thrusters. But if we are designing a system that has no wheels or CMGs, we must expel propellant to counteract all disturbances, and the angular impulse requirement is the sum of the time integrals of the absolute value of disturbance torque computed about each axis. The process of computing control system angular impulse from disturbance torques and identifying cyclic and secular components is shown in Table 10-19.

Thruster control systems operate by pulsing a thruster when the attitude error exceeds a set value known as the dead-zone limit. The thruster's design determines the length of the pulse—typically from 0.02 sec to 0.1 sec. The propellant consumption of such a system is proportional to the size of the pulses and the rate of pulse firing. A well designed control system will fire a minimum length pulse each time the dead-zone limit is exceeded. The angular velocity change, $\Delta\omega$, produced by a minimum pulse P_{min} is:

$$\Delta\omega = P_{min} / I_s \qquad (10\text{-}5)$$

where $P_{min} = T s t_{min}$, T is thrust level of the thruster, s is the lever arm through which the thruster works to produce torque, t_{min} is the minimum thruster firing time, and I_s is the spacecraft's moment of inertia. The mean angular velocity of the spacecraft while in the dead-zone is $\Delta\omega/2$ which implies that the spacecraft transverses a dead-zone of $2\theta_d$ in $4\theta_d/\Delta\omega$ seconds. Since the pulse firing time is negligible relative to the time spent in traversing the dead-zone, the average impulse rate, IR, is one minimum pulse every $4\theta_d/\Delta\omega$ seconds, or:

$$IR = P^2_{min} / (4\theta_d I_s) \qquad (10\text{-}6)$$

The total angular impulse expended during the mission, L_m, is IR times mission duration:

$$L_m = IR * \text{mission duration} \qquad (10\text{-}7)$$

The torque produced by a thruster is equal to its thrust T times its lever arm s. The time integral of thrust is linear impulse and is related to the mass of propellant used by the rocket equation (Eq. 17-6). An appropriate expression relating angular impulse to propellant use is:

$$L = s\int T dt \cong s\, I_{sp}\, g m_p \qquad (10\text{-}8a)$$

or equivalently:

$$m_p = L / (s\, I_{sp}\, g) \qquad (10\text{-}8b)$$

TABLE 10-19. Computing Control System Angular Impulse Requirements.

Step	Operation	Comments
1. Calculate disturbance torques	Use equations in Table 10-18	
2. Compute time integral of disturbance torque for each control axis	$L_x = \int T_{dx}\, dt$ $L_y = \int T_{dy}\, dt$ $L_z = \int T_{dz}\, dt$	L_x, L_y, L_z angular impulse requirements about x, y, and z control axes T_{dx}, T_{dy}, T_{dz} disturbance torques about x, y, and z control axes
3. Identify cyclic and secular components of angular impulse		Either identify cyclic components from the equations or plot each component of angular impulse to identify components
4. Size torquers		Wheels or CMGs are sized for the cyclic terms. Thrusters, if used with wheels, are sized for the secular terms.
5. If thruster control is used without wheels or CMGs, compute time integrals of the absolute value of disturbance torque	$L_x = \int \|T_{dx}\|\, dt$ $L_y = \int \|T_{dy}\|\, dt$ $L_z = \int \|T_{dz}\|\, dt$	

where L is angular impulse, t is time, I_{sp} is specific impulse, g is gravitational acceleration and m_p is propellant mass expended.

A reaction wheel produces torque by changing its speed. Its angular impulse capability is equal to its moment of inertia times its maximum speed—its total angular momentum. A control moment gyro produces torque by changing the direction of its constant-speed momentum wheel. The angular impulse capability of a CMG is its momentum (wheel mass moment of inertia times speed) times the angle through which it can be moved. The relation between momentum and weight and power for wheels and CMGs is presented in Table 10-20.

Estimating Angular Impulse for Spin Stabilization

For a spin-stabilized spacecraft, angular impulse is required for *spinup, spin maintenance*, and *spin-axis precession*. If the spacecraft is spin-stabilized for only part of its mission (such as during kick-stage thrusting), then impulse is also needed for *despin*.

TABLE 10-20.   Weight and Power of Components in an Attitude Determination and Control Subsystem. Note $T \equiv$ Torque in N·m, and $H \equiv$ angular momentum in N·m·s.

Component	Weight (kg)		Power (W)
Earth Sensor	2 to 3.5		2 to 10
Sun Sensor	0.2 to 1		0 to 0.2
Magnetometer	0.2 to 1.5		0.2 to 1
Gyroscope	0.8 to 3.5		5 to 20
Star Sensor	5 to 50		2 to 20
Processors	5 to 25		5 to 25
Reaction Wheels	$2 + 0.4 \times H$	$H < 10$	10 to 20 at constant speed;
	$5 + 0.1 \times H$	$10 < H < 100$	500 to 1,000 W/(N·m) when torquing
Control Moment Gyros	$35 + 0.05 \times H$		15 to 30 standby;
	$100 < H < 2,500$		0.02 to 0.2 W/(N·m) when torquing
Actuators (single axis)	$4 + 0.03 \times T$		1 to 5 W/(N·m)

The impulse requirement for spinup is:

$$L_{spinup} = I_{ss} \, \Omega_s \qquad (10\text{-}9)$$

where L_{spinup} is the impulse required in N·m·s, I_{ss} is the spacecraft moment of inertia about the spin axis in kg-m^2, and Ω_s is the spin speed in rad/s. Typical values of spin speed range from 0.1 rad/s for spacecraft requiring minimal spin stability to 10 rad/s for stabilization during kick motor firing. Impulse required for despin is computed with this same equation. Note that the inertia of the spacecraft during despin may be different than during spinup.

The principal merit of spin stabilization is that it is passive; that is, a spin-stabilized spacecraft will remain spinning at constant rate with its spin axis substantially fixed in inertial space. However, if the spacecraft has a thrusting mode such as kick-stage firing or thruster firing for velocity correction, both the spin speed and the spin axis orientation may be changed. Disturbance torques as presented in Table 10-18 may also make spin corrections necessary.

Variation in spin speed occurs when the thruster's axis and spacecraft center of mass are misaligned. If the offset between center of mass and the thrust axis is L_{cm}, the angular impulse, ΔH, imparted by a velocity correction ΔV to a spacecraft of mass $m_{s/c}$ is:

$$\Delta H = m_{s/c} \, \Delta V \, L_{cm} \qquad (10\text{-}10)$$

If the offset is in the direction that changes the spin speed, then impulse will be required to correct spin speed. If the offset is along the spin axis, the spacecraft angular-momentum vector will precess, thus changing the orientation of the spin axis. In either event, impulse is required to correct the unwanted change. If spin-speed correction is required, the impulse is about the spin axis and if precession is involved, it is normal to the spin axis. Typical thruster alignment tolerance is 0.1 deg. This tolerance, coupled with center of mass uncertainty, leads to typical thrust-axis to center-of-mass offset of 0.002s–0.01s where s is the distance from the thruster to the center of mass.

To change the direction of spin-axis orientation, we must precess the angular-momentum vector. This is usually done by synchronizing off-axis thruster pulses with the spin period of the spacecraft. The small increments of angular impulse imparted by properly synchronized firings add at 90 deg to the momentum vector to rotate the vector direction. Some nutation is also introduced by the thruster firings but this decays with time. The impulse required to precess a spinning spacecraft through an angle of α radians is

$$L_{precess} = I_{ss}\,\Omega_s\,\alpha \qquad (10\text{-}11)$$

where I_{ss} is the mass moment of inertia about the spin axis, Ω_s is the spin speed (in rad/s) and α is the angle of rotation of the spin vector in radians.

The total angular impulse is converted to propellant weight through the use of Eq. (10-10). Typical weight and power for attitude control components are summarized in Table 10-20 and discussed in Sec. 11.1. Actuators and wheels are available in a large number of sizes and capabilities. Their weight and power as a function of torque may be estimated by the relations given.

10.4.3 Communications Subsystem

The communications subsystem receives and demodulates uplink signals and modulates and transmits downlink signals. The subsystem also allows us to track spacecraft by retransmitting received range tones or by providing coherence between received and transmitted signals, so we can measure Doppler shift. Table 10-21 summarizes the main system considerations which drive the design of communications subsystems.

TABLE 10-21. System Considerations for Design of Communications Subsystems.

Consideration	Implication
Access	Ability to communicate with the spacecraft requires clear field of view to the receiving antenna and appropriate antenna gain
Frequency	Selection based on bands approved for spacecraft use by international agreement. Standard bands are S (2 GHz), X (8 GHz), and Ku (12 GHz)
Baseband Data Characteristics	Data bandwidth and allowable error rate determine RF power level for communications

Communication access to a spacecraft requires a clear field of view for the spacecraft antenna. It also requires sufficient received power to detect the signal with acceptable error rate. Access across many viewing angles demands an antenna with a wide beamwidth, so good spacecraft designs always include an antenna system that can receive signals over at least a hemisphere. The gain of a widebeam antenna is low—typically 0 dB for hemispheric coverage. Therefore, we must select a level of transmit power and a receiver sensitivity that allow us to detect signals with an acceptable error rate.

The spacecraft receives data consisting of commands and range tones. Command rates range from 100 bits/s to 100 kbits/s, with most systems below 1,000 bits/s. The data rate depends on mission considerations and sets the communications subsystem's bandwidth, which establishes the received power required to detect signals (Chap. 13).

For spacecraft communicating directly with ground terminals, received signal strength is not a design driver because we can set the ground terminal's transmitting power as high as necessary. However, received signal strength from a relay satellite does affect the communications subsystem's receiver sensitivity and maximum data rate. Systems that require data communications at rates greater than 1,000 bits/s normally use high-gain, directional antennas and can operate at low bit rates to allow wide-angle access when needed.

The downlink signal consists of range tones, telemetry for spacecraft status, and payload data. The baseband data is normally digital and multiplexed by frequency or time. Telemetry to report the spacecraft's status operates between 100 bits/s and 1,000 bits/s. If the downlink handles only status telemetry, or if the payload data will fit within a low-bandwidth link, we can communicate using a widebeam antenna. Data communication over a high bandwidth usually requires a high-gain, directional antenna and a low-bandwidth mode for widebeam coverage.

Table 10-22 shows how we size the communications subsystem. To do so, we must identify the data bandwidths of the uplink and downlink, select communication frequencies, prepare RF power budgets for both links (Chap. 13), and select equipment. The basic communications subsystem consists of a transmitter, a receiver, a widebeam antenna, and an RF diplexer. We may also use a high-power transmitter or a directional antenna if the data rate requires it.

TABLE 10-22. Steps in Designing a Communications Subsystem.

Step	What's Involved	Reference
1. Identify Data Rate		Payload commands and data—Chap. 9 Spacecraft bus commands and telemetry—Secs. 10.4.3, 11.3
2. Select Frequencies	Decide which of the allowed bands to use	Sec. 13.1
3. Prepare RF Power Budget	Analyze characteristics of RF links	Sec. 13.3
4. Select Equipment		Sec. 11.2

Table 10-23 shows the characteristics of a standard communications subsystem. The system operates at S-band, radiates 2 W, has a data rate of 1,000 bits/s, and weighs 5.9 kg. The transponder provides a coherent carrier response to measure range rates and retransmits ranging tones. Section 11.2 describes other equipment in communications subsystems, including transmitters with higher power and directional antennas.

TABLE 10-23. Characteristics of Communications Subsystems Using S-band.

Component	Weight (kg)	Power (W)	Comments
S-band Antenna	0.9	0	Hemispheric pattern 0 dB
Diplexer	1.2	0	
Receiver	1.8	4	Two units required for redundancy
Transmitter	2	4.4	

10.4.4 Command and Data Handling Subsystem

The *command and data handling subsystem, (C&DH)*, receives and distributes commands and collects, formats, and delivers telemetry for standard spacecraft operations (*housekeeping*) and payload operations. We usually handle housekeeping data intermittently and at rates below 1,000 bits/s. Rates for payload commanding and telemetry depend on the payload's design. They may require very high data rates (10 kb/s to 500 Mb/s) and storage of payload data.

The C&DH subsystem may include encryptors, decryptors, a sequencer or timer, a computer for data processing, and equipment for data storage. It interfaces with the communications subsystem from which it receives commands and to which it sends the formatted telemetry stream. It also delivers commands to and receives telemetry from the other spacecraft subsystems and may have similar interfaces with the payload.

The decoding of command signals is peculiar to the detailed design of the system. A typical command is a serial binary word containing a preamble, a user address, and the command word. The preamble allows the spacecraft to identify and authenticate the command. The command unit decodes the user addresses and then routes actual commands to the correct users, either by separate wires or by a data bus. Each user then decodes and executes the command.

In some cases, we need several commands to do something. If the function is time critical, we can send precursor commands and verify their receipt before sending a precisely timed execute or a time-tagged execute. A sequencer or onboard computer can execute time-tagged commands. If we need to send commands over time, we can time-tag them or use a timer-execute command followed by a sequence of timer-reset commands. If the command system fails, the command halts when the timer times out. Ordnance normally requires separate arm and fire commands.

Telemetry signals tell us about the spacecraft's health and provide operational data needed to control the spacecraft. Normally, we convert telemetry measurements to digital signals, serially multiplex them in a telemetry frame using a repetitive pattern, and transmit the frame using a main frame word for frame synchronization. We may use a main frame word as a subcommutated channel or we may supercommutate signals if we need a sampling rate higher than the frame rate. Finally, we may place the telemetry frames in packets and multiplex them with other downlink data.

Commands and telemetry signals depend on the spacecraft's operation. Each of the spacecraft's commandable functions needs a separate command. In addition, we must provide enough telemetry to define the spacecraft's state of health, as well as data for operational control.

Table 10-24 lists the steps to size the C&DH subsystem. In its simplest form, this subsystem consists of a command decoder and a telemetry multiplexer. More complex systems distribute the command decoding function by routing serial commands to user subsystems for final decoding. Telemetry multiplexing can also be distributed. High-performance subsystems use central digital computers for data processing. Table 10-25 presents typical characteristics. Chapter 11 describes more complex subsystems.

10.4.5 Thermal Subsystem

The thermal design of a spacecraft involves identifying the sources of heat and designing paths for transporting and rejecting heat, so components will stay within

TABLE 10-24. Steps in Sizing the Command and Data Handling Subsystem.

Step	What's Involved	Reference
Prepare Command List	Prepare a complete list of commands for the payload and each spacecraft bus subsystem. Include commands for each redundancy option and each commandable operation.	Secs. 10.4, 11.3
Prepare Telemetry List	Analyze spacecraft operation to select telemetry measurement points that completely characterize it. Include signals to identify redundancy configuration and command receipt.	Secs. 10.4, 11.3
Analyze Timing	Analyze spacecraft operation to identify time-critical operations, and timeliness needed for telemetry data.	Sec. 10.3, Chap. 16
Select Data Rates	Choose data rates that support command and telemetry requirements and time-critical operations.	Sec. 13.3
Identify Processing Requirements	Examine need for encryption, decryption, sequencing, and processing of commands and telemetry.	Sec. 11.3
Identify Storage Requirement	Compare data rates of payload and spacecraft to the communications subsystem's ability.	Sec. 11.3
Select Equipment	Configure the subsystem and select components to meet requirements.	Sec. 11.3

TABLE 10-25. Typical Characteristics of Basic Components for Command and Data Handling.

Component	Weight (kg)	Power (W)	Comments
Command Unit	5.0	5.4 standby 14 operating	Redundant unit, 9 user addresses capacity, 18,892 commands
Pulse Code Modulation Encoder	5.5	5.5	Redundant unit, 250 or 1,000 bits/s 64 word, 8 bit frame 5 subcommutated channels

required temperatures. The sources of heat include solar radiation, Earth-reflection and infrared radiation, and electrical energy dissipated in the electrical components. Conventional electronics operate at temperatures close to room temperature (25 °C) and will tolerate temperature variations of about ±20 °C. Battery cells, particularly nickel-cadmium cells, are more sensitive to temperature than most electronics. But they can still stand temperature ranges of 5 °C to 20 °C. We can control the temperatures of compartments for conventional electronics by coating or insulating their outer surfaces. We select these coatings to strike a balance between the heat absorbed and the heat radiated to space. The coatings include various paints and tapes, and second surface glass mirrors. The weight of such coatings is almost independent of the quantity of heat dissipated and seldom exceeds 4% of the spacecraft dry weight. The thermal coatings, particularly insulation, can close out compartment openings and may also shield components from electromagnetic radiation.

Components which have stringent temperature requirements or which dissipate large amounts of electrical power require more extensive thermal control. For example, we usually place gyros and precision oscillators in insulated compartments, or *ovens*, with active electrical heaters to control temperatures carefully. Traveling wave tubes and other elements which dissipate a lot of power concentrate their dissipation locally and produce *hot spots*. Normally, we conduct heat away from such hot spots and spread it over a thermal panel where it radiates to space. If the hot spots are less than 50 W, we can simply increase the thickness of the mounting panel so it will carry the heat away. But more intense hot spots and equipment that must meet tight thermal limits usually require *heat pipes* to move the heat and thus equalize temperatures.

The process of thermal design for spacecraft proceeds as follows. The first step is to identify heat sources and the location of radiating panels to dispose of excess heat. The heat sources include internal dissipation and external radiation from the Sun and the Earth. Often, particular surfaces of the spacecraft are not exposed or only partially exposed to the Sun or Earth. These faces are preferred locations for radiating panels and for components which dissipate large amounts of heat. The latter should be next to the radiators. We can compute the total amount of radiating area from the static-heat-balance equation for the entire spacecraft. But a component's thermal performance may differ markedly from the average.

Most power for thermal control goes to heaters that keep components from getting too cold. Heaters also compensate for imperfections in insulation or heat leaks, and are used to heat areas such as articulation joints that are difficult to insulate and dissipate little heat internally. Heaters or heater-controlled heat pipes (see Sec. 11.5) offer very tight control or control at a particular temperature value for certain components. A typical medium-sized spacecraft (1,000 W) consumes 20 W in the thermal subsystem plus any power required for special thermal control. In most cases, heaters can operate from primary power.

10.4.6 Power Subsystem

The *power subsystem* generates power, conditions and regulates it, stores it for periods of peak demand or eclipse operation, and distributes it throughout the spacecraft. The power subsystem may also need to convert and regulate voltage levels or supply multiple voltage levels. It frequently switches equipment on or off and, for increased reliability, protects against short circuits and isolates faults. Subsystem design is also influenced by space radiation, which degrades the performance of solar cells. Finally, battery life often limits the spacecraft's lifetime.

Earlier in this chapter, I described how to prepare a power budget for the spacecraft. This budget includes most of the information we need to size the power subsystem: the spacecraft's needs for operating power, storage requirements, and how the power subsystem degrades over the spacecraft's lifetime. The remaining steps to size the power subsystem are selecting a solar-array approach, sizing the array, sizing the batteries and the components that control charging, and sizing the equipment for distributing and converting power.

Solar arrays are generally planar, cylindrical, or omnidirectional. *Planar* arrays are flat panels pointed toward the Sun. Their power output is proportional to the projection of their area toward the incident sunlight. Three-axis-stabilized spacecraft normally use planar arrays. *Cylindrical* arrays appear on spin-stabilized systems in which the spin axis is perpendicular or nearly perpendicular to the Sun line. The output of a solar-cell array is nearly proportional to the amount of solar energy intercepted, and the projected area of a cylinder is $1/\pi$ times the total area. Thus, the cylindrical array should have approximately π times as many cells as a planar array with the same power rating. But temperature effects slightly favor the cylindrical array, so the actual ratio is closer to 1/2.5. If the spacecraft can receive sunlight from any aspect, then its array must have equal projected area in all directions. In other words, it must have an *omnidirectional* array. A sphere has this property, but paddles or cylinders combined with planar panels are also possible. The total area of an omnidirectional array must be approximately 4 times the projected area, so an omnidirectional array has about 4 times the area of a planar array with the same power rating.

The required area of a planar solar array is related to the required power, P, the solar constant (1,367 W/m^2), and the conversion efficiency of the solar-cell system. Although cells have had efficiencies as high as 30%, practical array designs range from 5% to 15% when taking into account the operating conditions and degradation at end-of-life. An array with an efficiency of 7% would have a required area of

$$A_a = \frac{P}{0.07 \times 1367} \approx 0.01P \tag{10-12}$$

where A_a is in m^2 and P is in watts. This area is characteristic of current arrays. The mass of a planar array with *specific performance* of 25 W/kg is

$$M_a = 0.04\,P \tag{10-13}$$

where M_a is in kg and P in watts. Current designs range from 14 to 47 W/kg at end-of- life. The high end would provide 66 W/kg at beginning-of-life. Solar arrays mounted on the spacecraft's body usually weigh less than planar arrays.

Rechargeable nickel-cadmium or nickel-hydrogen batteries are the usual devices for energy storage for unmanned spacecraft. They are available in various sizes and are highly reliable even though their performance characteristics are quite complex. The battery often represents one of the most massive components in the spacecraft. It also is very sensitive to temperature and to the use profile. Nickel-cadmium, and to a lesser extent nickel-hydrogen batteries perform best when operated between 5 °C and 20 °C. This range is both lower and more restricted than the temperature requirements for most electronic components. The battery also has complex wear-out mechanisms, thus limiting cycle life as a function of depth-of-discharge. Other variables—temperature, rate of charge, rate of discharge, and degree of overcharge—also affect cycle life but in a less well-defined way. If a battery has shallow discharge cycles, it loses capacity. To counter this tendency, most spacecraft recondition their batteries from time to time by discharging them completely.

We determine a battery's capacity from the energy it must produce (discharge power times discharge duration) and from its depth-of-discharge. We select the battery's depth-of-discharge to meet cycle life requirements. Table 10-26 gives guidelines on depths-of-discharge for nickel-cadmium and nickel-hydrogen batteries. Section 11.4 discusses these concepts in more detail.

TABLE 10-26. Allowed Battery Depth-of-Discharge vs. Cycle Life.

Cycle Life	Battery Type	Depth of Discharge
Less than 1,000 cycles	NiCd	80%
	NiH$_2$	100%
10,000 cycles	NiCd	30%
	NiH$_2$	50%

To compute a battery's capacity, we divide the discharge energy (watt-hours) by the depth-of-discharge. The ratio of battery weight to battery capacity is 30 to 40 W·hr/kg for NiCd batteries and 35 to 50 W·hr/kg for NiH$_2$. Often, several batteries operate in parallel to provide the needed capacity. By using several small batteries, we can add some redundant batteries for backup with less weight penalty than for a second large battery.

Spacecraft *primary power*—power produced by the solar array and batteries—is not well regulated (28 ± 5 V is typical). Furthermore, we must match the solar array's electrical output to the battery's charging requirements and provide switching equipment that allows the battery to supply power when needed. Section 11.4 describes various ways to meet these needs. Significant features include limiting the battery's charge rate, limiting overcharge, providing for low-impedance discharge, and providing for reconditioning. The controller or regulator must cope with the voltage swings between charge and discharge. The power control unit must isolate faults and switch to redundant units while also serving as the center of the power distribution network. An estimate of the power control unit's weight is 0.02 kg/W of controlled power.

Most electronic equipment, for both the payload and the spacecraft, requires voltage regulation tighter than that provided by the arrays and batteries. We must either regulate the primary power or convert it to *secondary power*, which we can regulate more tightly. In either case, power dissipates in the regulator or the power converters. This dissipation typically amounts to 20% of the power converted, which may be all of the spacecraft's operating power. In sizing the power subsystem, we must therefore include the weight of the power conversion equipment—typically 0.025 kg/W converted.

The power subsystem includes wiring for distribution and may have components for switching and fault isolation. The power dissipated in wiring losses and switching equipment is 2% to 5% of the operating power, and the wiring harness takes up 1% to 4% of the spacecraft dry weight. Spacecraft which must operate in high radiation environments may require shielded wire to distribute power. Table 10-27 summarizes the weight and power requirements of the power subsystem.

TABLE 10-27. Weight and Power Budget for Power Subsystem. P = required power in watts. Note that M_{dry} is used here as in Table 10-10.

Component	Weight (kg)	Power (W)	Comments
Solar Arrays	0.04 P	—	× π for cylindrical body-mounted × 4 for omnidirectional body mounted
Batteries	C/35 (NiCd) C/45 (NiH$_2$)	—	C = capacity in W·hrs
Power Control Unit	0.02 P	—	P = controlled power
Regulator/Converters	0.025 P	0.2 P	P = converted power
Wiring	0.01–0.04 M_{dry}	0.02–0.05 P	M_{dry} = spacecraft dry weight

Primary power is distributed in most unmanned spacecraft as low-voltage direct current. But for systems with power needs above 10 kW, we should consider alternating current distribution, both sine wave and square wave, at several hundred volts.

Radioisotope thermoelectric generators (RTGs) have been designed for various power levels but have been applied only to low-power needs. In practice, the units consist of a radioisotope heat source which can produce power by thermoelectrics or provide thermal energy to a rotating generator. If we use one of these units, we must dispose of excess heat during all mission phases and particularly during launch preparation and boost. We must also consider safety issues, but RTG sources are probably safer than most propellants. The design must ensure that the generator remains intact and shielded even during catastrophic launch failure.

Using rotating machines to generate primary power is another design with potential. Closed-cycle, thermal engines should be nearly twice as efficient as solar cells, and rotating generators can provide sine-wave AC power with better regulation than solar-cell designs.

10.4.7 Structures and Mechanisms

The spacecraft structure carries and protects the spacecraft and payload equipment through the launch environment and deploys the spacecraft after orbit injection. The load-carrying structure of a spacecraft is *primary structure*, whereas brackets, closeout panels, and most deployable components are *secondary structure*.

We size primary structure based on the launch loads, with strength and stiffness dominating its design. The size of secondary structure depends on on-orbit factors rather than boost-phase loads. Secondary structure only has to survive but not function during boost, and we can usually cage and protect deployables throughout this phase.

Each of the launch boosters provides maximum acceleration levels to be used for design (see Chap. 18). These acceleration levels or load factors are typically 6 g's maximum axial acceleration and 3 g's maximum lateral acceleration. These levels work for conceptual design, but some designers prefer to increase them by as much as 50% during early design phases. During preliminary sizing, we must remember that the primary structure must carry some weight, such as kick motors and propellant, which will drop away before orbit injection. Section 11.6 discusses structural design and presents methods for preliminary structural sizing.

We use cylindrical and conical shell structures and trusses for primary structure, commonly building them out of aluminum and magnesium with titanium for end fittings and high-strength attachments. Composite materials have seen limited use in primary structure to date but they will become more common. We can size primary structure by modeling it as a cylindrical beam which is mass loaded by its own weight and the spacecraft's components. The lateral load factors applied to this beam produce a moment that is a function of axial location. Compression in the extreme section of the beam carries the moment. By adding the axial load to the moment-induced, compressive load, we can estimate the critical load, which in turn sizes the primary structure (see Eq. 11-42). In these preliminary sizing calculations, we can exercise much license in assuming symmetry and in simplifying the loads. We can iterate the skin gage to withstand stress levels and check the tubular design for buckling (see Sec. 11.6.6).

We use a similar approach to size a truss-based primary structure. We reduce the truss to its simplest form by successively removing redundant members until we reach a statically determinant structure. Simply combining loading conditions allows us to size the truss members.

We must also locate and mount components on the basic, load-carrying cylinder or truss. Most electronic components have rectangular symmetry and are mounted with lugs or bosses integral to their housings. Mounting requirements include loads, good thermal contact with the mounting surface, and good electrical contact. Aluminum honeycomb is an excellent mount for components. It attaches to longeron-stringer frames to form a *semi-monocoque* (load-carrying skin) structure. Honeycomb sheets with composite faces occasionally substitute for other approaches.

Some components are not rectangular. For example, propellant tanks are normally spherical but may be elongated, have conical sections, or be toroidal. Electromechanical drives and reaction wheels are cylindrical, and control moment gyros are complex. These components include mounting provisions in their designs. Generally they have flanges, bosses, or lugs. In most cases, and particularly in the case of tanks and pressure vessels, the mounting must avoid loading the component. To do so, the mount must be statically determinant, and component loads from deflection of the mount must be minimal.

Other components of complex geometric shape, such as thrusters and connectors, may mount through brackets specifically tailored to them. Hinges and similar items are machined fittings with integral flanges or mounting bosses. We can align components by shimming, but we must be careful not to disturb thermal and electrical bonds.

A set of data on spacecraft subsystem mass is presented in Appendix A. These data show the structural mass to be approximately 20% of the spacecraft total. However these data do not include all of the injected mass (apogee kick motors carried in the spacecraft are not included). Therefore one should be careful about using these data for estimating new designs. However, structural mass of 10% to 20% of spacecraft dry mass is a reasonable starting point.

We must have an interstage structure to mount the spacecraft to the booster. This structure conforms to the booster provisions for mounting and carries loads during the boost phase. Both truss structures and conical adapters are common. Because this structure is designed for strength under high loads, it is an excellent candidate for high-strength materials and weight-efficient design. The spacecraft usually provides this interstage structure and incorporates a separation joint to release the spacecraft at orbit.

Common methods of attachment at the separation plane are marmon clamps or separation bolts. In the *marmon clamp*, the separation joint is a continuous ring held together by an annular clamp. Release of clamp tension allows the joint to separate. In the case of *separation bolts*, the joint is held by several bolts which are released by either severing the bolt or by releasing a nut. Once the separation joint is free, springs impart a small velocity increment to the spacecraft. After separation, the booster maneuvers to avoid accidental impact. For spin-stabilized spacecraft, the interstage structure may incorporate a mechanism to impart spin while ejecting the spacecraft.

The Shuttle interfaces differently from the expendable boosters. It links with its payloads at a series of hard points located along the sill of the cargo bay and along the cargo bay's centerline (*keel fittings*). The payload and its upper stages usually require a cradle or fittings to translate the loads into these hard points. Mechanisms for deploying the spacecraft may be spring-powered or motor-driven. Chapter 11 presents weight-estimating relations for motor-driven mechanisms. Spring-powered mechanisms must meet stiffness requirements, but they weigh about half as much as their motor-driven equivalents.

10.5 Integrating the Spacecraft Design

10.5.1 Spacecraft Size

If we know the spacecraft's weight and power, we can estimate its size. Most spacecraft have a main body or equipment compartment. Many also have solar panels which wrap around the compartment for launch and deploy outside the compartment on orbit.

Table 10-28 gives estimating relations based on analysis of the volume and dimensions of a number of spacecraft. These spacecraft ranged from 135 kg to 3,625 kg and represent about 15% of the U.S. spacecraft launched between 1978 and 1984 [TRW Defense and Space Systems Group, 1980–1985]. Their density ranged from 20 kg/m^3 to 172 kg/m^3, with an average of 79 kg/m^3. The spacecraft were all cylindrically symmetric, although the cross section varied from rectangular to circular. The ratio of base diameter to cube root of mass ranged from 0.16 $m/kg^{1/3}$ to 0.31 $m/kg^{1/3}$, with an average of 0.23 $m/kg^{1/3}$. The ratio of spacecraft height to cube root of mass ranged from 0.13 to 0.83, with an average of 0.39 $m/kg^{1/3}$.

TABLE 10-28. Rules for Estimating Volume, Dimension, Area, and Moments of Inertia.
$M \equiv$ spacecraft loaded mass in kg as defined in Table 10-10.

Characteristic	Estimate	Range
Volume (m^3)	$V = 0.01\ M$	0.005 to 0.05
Linear Dimension (m)	$s = 0.25\ M^{1/3}$	0.15 to 0.30
Body Area (m^2)	$A_b = s^2$	—
Moment of Inertia (kg· m^2)	$I = 0.01\ M^{5/3}$	—

Section 10.4.6 presented relations for estimating the area of a solar array. Sometimes, the required array area is smaller than the spacecraft's body area, and the body can be oriented properly relative to the Sun. In this case, we can mount the solar cells directly on the body. But high-power spacecraft usually mount the solar cells on external panels, either off to one side or symmetrically on both sides of the equipment compartment. External solar arrays greatly increase the spacecraft's moment of inertia, particularly about the axes perpendicular to the array axis. Suppose the solar array consists of two square panels, one on each side of the spacecraft, and the center of each of these panels is L_a meters from the body's center. If so, the increase in moment of inertia is approximately $L_a^2 M_a$, where M_a is the solar array weight. Table 10-29 gives an approximate expression for L_a in terms of the array area and the body dimension, s. It shows the solar array's moment of inertia relative to the spacecraft's center. External solar arrays also affect the total projected spacecraft area, which in turn influences aerodynamic drag and solar-radiation pressure. Table 10-29 summarizes estimating rules for solar-array moment of inertia and area offset. A_a is the total solar array area. We must add these inertias to the inertias of the central compartment, assuming the latter to be equal to the values for the folded spacecraft computed above.

TABLE 10-29. Rules for Estimating Area Offset and Moment of Inertia of a Solar Array.
These should be added to the body values computed in Table 10-28. See text for definition of terms.

Solar Array Area Offset (m)	$L_a = 1.5\ s + 0.5\ (A_a / 2)^{1/2}$
Solar Array Moment of Inertia (kg ? m^2)	
Perpendicular to Array Face	$I_{ax} = (L_a^2 + A_a / 12)\ M_a$
Perpendicular to Array Axis	$I_{ay} = (L_a^2 + A_a / 24)\ M_a$
About Array Axis	$I_{aa} = (A_a / 24)\ M_a$

10.5.2 Lifetime and Reliability

Reliability is a parameter under the designer's control. We should consider its potential effect on spacecraft sizing during conceptual design by examining failures from wear-out and random causes. In other words, we should identify the ways in which the spacecraft may fail and tailor the design to eliminate or limit failures to acceptable levels. This implies identifying components or functions which can wear out and designing the system so that they meet the mission's lifetime requirements. Propellant supply and battery-cycle life are examples of these components. If equipment does not wear out, we must evaluate how each part's failure affects the mission and modify the design to eliminate any single-point failures. Then, we use statistics to compute the probability of mission success and tailor the design to acceptable levels. This process is not exact, but careful attention to reliability gives us the most balanced and able system possible.

To design for reliability, we must understand what constitutes success. The more specifically and numerically we can state the success criteria, the easier we can translate these criteria into design requirements. After defining success, we should list the smallest amount of equipment or number of functions that will provide it. We can begin by placing these functions in a signal flow or block diagram. In this basic form, most functions involve only one path or set of equipment. For this reason, we sometimes call it a *single-string reliability model*. Later in the design process, we can add multiple paths or backup modes to improve the probability of success, taking care to understand both the reliability enhancement and the cost.

By understanding the functions needed for a successful mission, we understand the factors which limit mission life or threaten that success. Often a new mission depends on developing or exploiting new technology, so we need to know the technology and the factors that stress the components of our system. By reducing our knowledge to a set of specifications and applying the stresses to our design, we improve our ability to produce reliable hardware.

One of the key steps in design for reliability is to numerically predict the probability of success. To do so, we must differentiate failures from wear-out and failures from random causes. Classic reliability models depict the rate of failure when plotted against time as a "bathtub"-shaped curve. Early on, systems fail at high rates because of *infant mortality*; late in life, they fail because of wear-out. We can eliminate failures from infant mortality with careful construction, testing, and burn-in. We can avoid wear-out by understanding and eliminating the factors that cause it or by providing enough hardware to replace worn-out equipment. Between the extremes of infant mortality and wear-out, the failure rate is more or less uniform and attributed to random effects.

Wear-out shortens a mission. Random failures kill a spacecraft with accumulated effects. A successful design copes with them by providing enough backup components to cover them. Because we cannot determine when they will occur, our design must allow us to detect and correct them. Also, a good design tolerates some failures and remains useful in a degraded mode.

Searching for and identifying the ways in which equipment can fail is a basic part of design for reliability. This process, called *Failure Modes Effects and Criticality Analysis* (*FMECA*) assumes that we can identify the ways in which equipment can fail and analyze the effect. Key to this process is identifying and eliminating single-point failure modes—failures that by themselves can kill the mission. If we cannot eliminate them, we must control their probability of occurrence.

We can analyze the failure modes of our equipment in several ways. For example, the all-part method simply analyzes each of the spacecraft's parts to determine the effect of its failure. On a large spacecraft this method is a lot of work but is straightforward and easy to do. The all-part method requires us to analyze shorts and opens —systematically searching for wires or printed traces on circuit boards that can cause failure if opened or shorted together. We can also use scenarios to find potential failure modes. To do so, we simulate the spacecraft's launch, deployment, and operation to ensure that telemetry can detect failures and that the command system can correct them. This simulation normally occurs when operational procedures are being prepared, but it can more effectively detect design flaws if done earlier.

Another way to identify failure modes is the jury method. In many cases new designs do not have a lot of experience behind them, but people have had experience with similar equipment. We can poll them as part of a formal design review or in a separate meeting, thus using their experience to identify likely failure modes and probable effects.

10.6 Examples

In this section, we discuss three examples of spacecraft sizing. First we develop a preliminary estimate of the FireSat spacecraft and then review two actual systems —FLTSATCOM and HEAO-B.

The drivers for the FireSat spacecraft design are the FireSat payload design (Sec. 9-7, Table 9-15) and the orbit and ΔV requirements (Table 7-3). We will use these to get a broad estimate of the overall size, weight, and power for FireSat and then to break this down into approximate subsystem allocations. The results of the top-level process are summarized in Table 10-30. Keep in mind that these are crude estimates that allow us to begin the process of spacecraft design. We must continually evaluate and refine the requirements and resulting design and perform a variety of system trades to arrive at an acceptable, consistent design.

Our first estimate of the spacecraft mass and power come directly from the payload estimates of Sec. 9-7 (Table 9-15). As given in Table 10-5, the payload mass is between 17% and 50% of the spacecraft dry weight with an average of 30% (see also Appendix A). We know very little about FireSat at this time, so we will add margin by estimating the payload at 20% of the spacecraft mass, well below the average percentage. However, FireSat was scaled **down** from a flight unit. This implies that the bus will probably be a larger fraction of the spacecraft dry weight. Our knowledge of the weight is poor at this time because we have not yet done a preliminary weight budget. When we allocate the mass to subsystems below we will hold the margin at the system level to allow us to apply it as needed to various subsystems.

Similarly, our initial power estimate is based on the payload power of 32 W and the estimate from Table 10-9 that for moderate size spacecraft, the payload represents 40% of the spacecraft power. Our spacecraft is small with significant control and processing requirements. Therefore, we will again be conservative and assume that the payload represents only 30% of the power requirement for FireSat. Here the knowledge is very poor, because we have not yet budgeted the power and have not determined what payload duty cycle should be used—that is, should we turn the payload off over the poles and oceans? Because we will have to contend with eclipses (Sec. 5.1, Example 1), the solar array output will be estimated at 170 W to provide 110 W to the spacecraft which then provides 32 W to the payload.

TABLE 10-30. Preliminary Estimate of FireSat Spacecraft Parameters. See text for discussion. These parameters are based primarily on the payload parameters defined in Sec. 9.6.

Parameter	FireSat Estimate	Notes and References
Payload:		
Mass	28 kg	Table 9-15
Power	32 W	Table 9-15
Spacecraft:		
Dry mass	140 kg	Payload mass / 0.2; Text + Table 10-5
Average power	110 W	Payload power / 0.3; Text + Table 10-9
Solar array power	170 W	Eclipse allowance, Eq. (11-1)
Solar array design	Body-mounted omni array, 1.7 m^2 facing the Sun (8.5 m^2 total area)	Sec. 10.4.6, array on 5 non-nadir faces
Control approach	3-axis, nadir pointed	Sec. 10.4.2
Propellant:		
ΔV	28 kg	Table 7-3; Eq. (17-7)
Attitude control + residual	2 kg	7%; Secs. 10.4.2, 11.1
Margin	4 kg	15%; Table 10-7
Total propellant	34 kg	Sum of the above
Propulsion approach	Metered bipropellant (I_{sp} = 300 s); no kick stage	Text
Spacecraft Loaded Mass:	175 kg	Dry mass + propellant
Spacecraft Size and Moments:		
Volume	1.7 m^3	Table 10-28
Linear dimensions	1.4 m	Table 10-28
Body cross-sectional area	2.0 m^2	Table 10-28
Moment of inertia	50 kg · m^2	Table 10-28

In Sec. 7.5.1, we decided to try eliminating a kick stage and flying the spacecraft up using low-thrust chemical propulsion. In order to maintain reasonable efficiency, we initially assume a metered bipropellant system with an I_{sp} of 300 s (Sec. 10.4.1, Chap. 17). Using the rocket equation (Eq. 17-7), we can compute the propellant mass as 28 kg and then add small amounts for attitude control and margin as given in Table 10-30. Here our knowledge of the propellant mass as a fraction of the spacecraft mass is good, although the spacecraft mass itself is not yet well known. Because the propellant mass is small, we may choose later to go to a simpler monopropellant system or to have the launch vehicle put FireSat directly into its end orbit.

Given an approximate mass for the whole system we can estimate the size and moments of inertia from Table 10-28. This, in turn, can tell us something about the solar array configuration. We estimate the body area at 2.0 m^2 and the required solar array area at 1.7 m^2. So we can probably avoid solar panels altogether and use an omnidirectional array consisting of solar cells mounted on the non-nadir facing sides of the body. This will be compact, economical, and easy to control.

Finally, Table 10-31 presents two ways of developing a preliminary weight budget for FireSat. We can estimate the mass of each subsystem as a percentage of spacecraft dry mass or as a percentage of the payload mass. Column (1) lists the average percentage of spacecraft dry mass devoted to each subsystem based on the historical data for spacecraft listed in Appendix A. The resulting mass distribution and margin are shown in column (3). Column (2) lists the same data expressed as the average percentage of payload mass devoted to each subsystem. The resulting FireSat mass distribution is shown in column (4). We recommend using the mass distribution shown in column (4). A weight margin of at least 25% at this stage of development is appropriate. The column (3) approach resulted in a "margin" of only 11.2 kg or 8% of the spacecraft dry mass. This approach prematurely divides the available margin among the subsystems. We recommend maintaining the margin at the system level and then allocating it to the payload or other subsystems as necessary throughout the development.

TABLE 10-31. Preliminary FireSat Spacecraft Weight Budget.

Element of Weight Budget	(1)* Est. % of Spacecraft Dry Mass	(2)† Est. % of Payload Mass	(3) Est. Mass Based on Col. (1) (kg)	(4) Est. Mass Based on Col. (2) (kg)	Comments
Payload	20.0	100.0	28.0	28.0	Payload mass estimate from Table 10-30
Structures	21.0	75.0	29.4	21.0	
Thermal	4.5	16.1	6.3	4.5	
Power	30.0	107.1	42.0	30.0	
TT&C	4.5	16.1	6.3	4.5	
Att. Control	6.0	21.4	8.4	6.0	
Prop (dry)	6.0	21.4	8.4	6.0	
Margin (kg)	—	—	11.2	40.0	Note that using the approach in Col. (4) the margin is maintained at the system level, not the subsystem level
Spacecraft Dry Mass (kg)	—	—	140.0	140.0	Estimate from Table 10-30
Propellant Mass (kg)	—	—	35.0	35.0	Estimate from Table 10-30
Spacecraft Loaded Mass (kg)	—	—	175.0	175.0	Sum of spacecraft dry mass & propellant mass
Margin as % of Dry Mass	—	—	8.0%	28.6%	Margin / (Spacecraft dry mass) × 100%

* The percentages in Column (1) are the average values listed in Appendix A.
†The percentages in Column (2) are the average payload values listed in Appendix A.

To provide more detailed examples of spacecraft design, we'll look at two actual spacecraft—FLTSATCOM [Reeves, 1979] and HEAO-B [Frazier, 1981] and use our estimating techniques to describe them. Figure 10-3 shows the basic configuration of these spacecraft and their principal mission parameters.

Table 10-32 summarizes the design requirements for FLTSATCOM and HEAO-B. FLTSATCOM is a communications spacecraft that is part of a global network

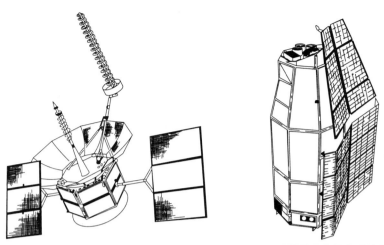

	FLTSATCOM	HEAO-B
Mission	Military communications	X-ray astronomy
Payload	Communication transponders and antennas	X-ray telescope and instruments
Size	4.9 m transmit antenna 3.10 m helical receive antenna 2.4 m hexagonal × 1 m deep electronics compartment	2.3 m octagonal × 4.7 m instrument compartment
Wt. (loaded)	927 kg	3,154 kg
Power	1,224 W	619 W
Pointing	0.25 deg accuracy	Telescope axis anywhere 1 arcmin accuracy 30 arcsec stability in 1 hr 10 deg/min slew rate
Data Rate (selectable)	1 kbps/250 bps	6.4 kbps/128 kbps 84 Mb data storage capability
Orbit	Geosynchronous, 2.5 deg inclined	540 km circular, 22.75 deg inclined
Reliability	0.267 at 5 yrs	0.81 at 1 yr

Fig. 10-3.   Mission Parameters for FLTSATCOM and HEAO-B. Note that loaded weight is mass in kg; this use continues throughout the example.

providing UHF communications between ground stations and mobile users. The communications payload equipment consists of receivers, processors, transmitters, and antennas. Its mission of global communications requires full-Earth antenna coverage. The antennas are body-fixed and oriented to point toward nadir. The orbit has a 24-hour period (geosynchronous) and initially is inclined 2.5 deg to the equator. There is no active inclination control. The launch vehicle is an Atlas-Centaur to inject the spacecraft into an elliptic transfer orbit. At apogee, a solid kick motor injects the spacecraft into geosynchronous orbit. The spacecraft uses the Air Force Satellite Control Network for command and control.

The mission of HEAO-B was X-ray astronomy. Its payload consisted of 5 X-ray instruments mounted within a telescope assembly. The mission required the telescope

TABLE 10-32. FLTSATCOM and HEAO-B Requirements and Constraints (See Table 10-3).

Space-craft	Mission	Payload	Orbit	Environ-ment	Launch System	Ground System
FLTSAT-COM	Global UHF commu-nications	Communications transponder & antennas	24-hr circular inclined 2.5 deg	Atlas-Centaur launch; 5 yrs in GEO	Atlas-Centaur with solid apogee kick motor	Air Force Satellite Control Network (SGLS)
HEAO-B	X-ray astronomy	X-ray telescope & instruments	540 km circular inclined 22.75 deg	Atlas-Centaur launch; 12 mo in LEO	Atlas-Centaur	NASA S-band GSTDN network

to be pointed anywhere and stabilized to 1 arcmin accuracy, with 30 arcsec stability over one hour. A slew rate of 10 deg/min was required. The orbit was circular at 540 km altitude inclined at 22.75 deg. HEAO-B was launched on an Atlas-Centaur in late 1978 and had a design life of one year. It interfaced with NASA's GSTDN S-band ground network.

Table 10-33 presents the key design decisions for configuring the two spacecraft. Weight dominated the FLTSATCOM design. The payload was large and complex, and mission-reliability goals required complete payload redundancy. The Atlas-Centaur's launch capability limited the weight, thus requiring tight weight control. Because the FLTSATCOM payload required over 1,000 W, designers selected an oriented planar solar array. The Atlas-Centaur's fairing diameter also limited the spacecraft's folded size. The spacecraft employs two attitude-control modes: spin stabilization for orbit injection and 3-axis stabilization for on-orbit control. All communication components are body-fixed, and the body is controlled to point the antennas toward nadir. A planar solar array is articulated about one axis to point toward the Sun. Liquid-hydrazine propulsion provides attitude and orbit control; a solid apogee kick motor injected the spacecraft into orbit.

TABLE 10-33. Initial Design Decisions for FLTSATCOM and HEAO-B (See Table 10-4).

Design Aspect	FLTSATCOM Design Approach	HEAO-B Design Approach
Spacecraft Weight and Power	Dominated by heavy weight and high-power payload; constrained by booster capability	Large margins established initially; not tightly constrained by booster capability
Spacecraft Size	Folded configuration limited by fairing; deployed configuration dominated by antenna size and solar array area	Folded configuration limited by fairing diameter; large deployables not required
Attitude Control Approach	Spin stabilization for kick motor firing; 3-axis control on orbit	3-axis control used to provide precision
Solar Array Approach	Planar-oriented array	Planar body-fixed array
Kick Stage Use	Solid AKM used for orbit insertion	Not required
Propulsion Approach	Liquid hydrazine thrusters used for attitudel	
Orbit Control	Liquid hydrazine thrusters used for attitude control	

Weight did not dominate the design for HEAO-B. Its payload instruments existed before the program developed the spacecraft to carry them. The program was able to establish large weight and power margins. The fairing diameter affected the spacecraft's size, but the instruments fit within this diameter. Power requirements were modest and a planar body-fixed array was used. The spacecraft mission required precision pointing of the payload, which led to 3-axis attitude control using gyroscopes, star sensors, and reaction wheels. Kick-stage propulsion was not required. Hydrazine monopropellant propulsion was used for attitude control.

Design Budgets

Table 10-34 presents the propellant budget, velocity increments, specific impulse, and spacecraft mass for FLTSATCOM. To derive the velocity required to insert the spacecraft into its final orbit, the designers used the methods in Chap. 7. That chapter also describes injection-error analysis which the designers used to derive the velocity increment that corrected the orbit. The ratio of this error to the AKM velocity (0.5%) is typical of current performance for solid motors. The stationkeeping velocity derives from analysis of orbit perturbations over the spacecraft's design life. The stationkeeping increment developed from an operational requirement to move the spacecraft at a rate of 15 deg longitude per day. During orbit injection FLTSATCOM spin-stabilized at 60 rpm and maneuvered through an angle of 65 deg while spinning. Eqs. (10-7) and (10-9) translate these requirements into propellant weight. During its lifetime the spacecraft uses propellant for attitude control, mainly to counter solar radiation pressure. HEAO-B needed propellant only to acquire attitude and to cancel disturbance torques (principally aerodynamics and gravity gradient; see Table 10-18). The propellant's mass was 138 kg.

TABLE 10-34. Propellant Budget for FLTSATCOM (See Table 10-4).

Element	Mass (kg)	Design Characteristics
Velocity Correction and Control		
Orbit insertion (AKM)	855.0	ΔV = 1,748.8 m/s; I_{sp} = 285.5 s
Guidance error correction	4.1	9.0 m/s; I_{sp} = 215.2 s; $M_{S/C}$ = 988 kg
Stationkeeping	6.5	14.2 m/s
Station change	52.7	115.2 m/s
Attitude Control		
Spinup and despin	6.0	Spin speed = 60 rpm; S/C inertia = 995 kg·m^2
Maneuvering while spinning	3.3	Maneuver angle = 65 deg; Inertia = 917 kg·m^2
Attitude control	8.4	I_{sp} = 100 s; lever arm = 1.25 m
Residual	1.5	
Total Propellant		
Solid	855.0	
Liquid	82.5	

Tables 10-35 and 10-36 present the power and weight budgets for FLTSATCOM and HEAO-B. HEAO-B had two normal power modes: cruise and ground pass. The table shows power requirements for both modes, together with the orbital average. The batteries were sized for 20% depth of discharge because the mission life exceeded 5,000 discharge cycles. For FLTSATCOM, payload power, including noted power conversion losses, is the main entry. The FLTSATCOM battery was large enough to support the full operating power during the maximum eclipse of 1.2 hours. The batteries operated at a maximum depth of discharge of 70% and required a recharge power of 167 W. Designers sized the array for a beginning-of-life power of 1,800 W.

TABLE 10-35. **Power Budgets for FLTSATCOM and HEAO-B (See Tables 10-6, 10-7, and 10-8).**

Subsystem	FLTSATCOM Power Budget (W)	HEAO-B Power Budget Cruise Mode (W)	HEAO-B Power Budget Ground Pass (W)
Payload	1,070*	217.0	217.0
Spacecraft Bus			
Propulsion	40	17.4	17.4
Attitude Control	33†	201.0	201.0
Communications	9‡	13.4	78.4
Data Handling	—	33.8	41.6
Thermal	17	30.2	30.2
Electric Power	56	26.0	33.7
Operating Power	1,224	538.8	619.3
Orbital Average Power	—	546.8	
Battery Recharge Power	167	244.0	
End-of-Life Power	1,391		
Beginning-of-Life Power	1,800		

* Includes 192 W power conversion losses.
† Includes guidance and navigation functions.
‡ Combined communication and command and data handling.

TABLE 10-36. **Weight Budgets for HEAO-B and FLTSATCOM (see Table 10-9).**

Subsystem	FLTSATCOM Weight Budget (kg)	HEAO-B Weight Budget (kg)
Payload	222.0	1468.0
Spacecraft Bus		
Propulsion	29.5	25.5
Attitude Control	57.7	124.0
Communications	26.3*	30.0
Command and data handling	—	41.6
Thermal	14.5	35.4
Electric power	336.0	376.0
Structure	154.0	779.0
Spacecraft Dry Weight	841.0	2868.2
Propellant	85.4†	138.0
Ballast	—	148.0
Loaded Weight	927.0	3154.2
Apogee Kick Motor	916.0	—
Injected Weight	1844.0	—
Adapter	19.5	18.1
Boosted Weight	1863.0	3172.3

* Combined communication and command and data handling.   † Includes 2.8 kg pressurant.

Subsystem Design

Table 10-37 shows characteristics of the propulsion subsystems for the example spacecraft. Each subsystem uses liquid-hydrazine propellant stored in two tanks. The tanks are pressurized with a fixed amount of nitrogen gas, so the hydrazine pressure decays as propellant burns. The tanks can cross connect through commandable isolation valves. Thrusters and propellant plumbing are redundant and able to cross connect. HEAO-B used twelve 4.4-N thrusters mounted in six dual-thruster assemblies. FLTSATCOM uses sixteen 4.4-N thrusters and four 0.44-N thrusters mounted in dual-thruster assemblies.

TABLE 10-37. Characteristics of FLTSATCOM and HEAO-B Propulsion Subsystems.

Characteristic	FLTSATCOM	HEAO-B
Propellant	Hydrazine monoprop.	Hydrazine monoprop.
Capacity	83 kg	136.4 kg
Tankage	Two 0.6 m diameter titanium	Two 0.74 m diameter titanium
Pressurant	2.8 kg nitrogen	3.6 kg nitrogen
Tank weight	14.4 kg	13.3 kg
Lines and Valves	Central isolation valve distribution assembly	Central propellant distribution module
Weight	7.5 kg	6.8 kg
Thrusters	Sixteen 4.4 N (4 roll, 4 pitch, 8 yaw and ΔV) Four 0.44 N (roll-yaw)	Twelve 4.4 N (4 pitch, 8 roll-yaw)
Power Consumption	40 W (catalyst bed and line heaters)	17.4 W
Weight	7.6 kg	5.5 kg

Table 10-38 summarizes the attitude-control requirements for the two example spacecraft (see Table 10-13). FLTSATCOM employs body-mounted antennas for its payload and orients one body face to point the antennas toward nadir. FLTSATCOM also has a planar solar array which it orients toward the Sun by combining body orientation and rotation of the array axis. Finally, FLTSATCOM has a guidance requirement for ΔV corrections. HEAO-B was required to point anywhere except close to the Sun and hold accurate pointing for long periods. It also had modest power requirements and no requirement for orbit correction.

TABLE 10-38. Attitude-Control Requirements for FLTSATCOM and HEAO-B.

Requirement	FLTSATCOM	HEAO-B
Payload Requirements		
Article to be pointed	Communication antennas	X-ray telescope
Pointing direction	Nadir	Anywhere
Pointing accuracy	0.25 deg	1 arcmin
Pointing stability	—	30 arcsec in 1 hr
Slew rate	—	10 deg/min
Exclusion	—	Within 15 deg of Sun
Other Requirements		
Sun pointing	Yaw (Z-axis) controlled to keep Y-axis normal to orbital plane	roll (X-axis) controlled to keep Sun in X-Z plane
Pointing during ΔV	Yaw (Z-axis) control as above	—

Both FLTSATCOM and HEAO-B are 3-axis-controlled spacecraft. Both spacecraft had body-mounted payloads and controlled the body's attitude to point the payloads (see Table 10-13). Both spacecraft use one axis of body attitude control to orient a planar solar array toward the Sun. FLTSATCOM, which requires more than 1 kW of electric power, orients the array about a second axis.

FLTSATCOM employs a solid apogee kick motor to inject it into orbit. Attitude is controlled during AKM firing by spin stabilization (see Tables 10-14 and 10-15). The spin direction can be rotated by using an off-axis thruster; the thruster fires pulses synchronized with the spin period. During normal operation and ΔV firing, FLTSATCOM is 3-axis stabilized. It uses Earth and Sun sensors for attitude sensing and thrusters and a reaction wheel for torque. HEAO-B did not have a ΔV mode. Under normal operation, it used gyroscopes and Sun sensors for attitude sensing with wheels and thrusters for torque. Its torqued gyroscopes slewed the reference-pointing direction, and payload star sensors allowed accurate reference and correction of gyro drift. Radiation pressure causes the main disturbance torques for FLTSATCOM. Gravity gradient and aerodynamics were the chief torques for HEAO-B. See Table 10-18 and Sec. 11.1.

Table 10-39 shows weight and power values for components of the complete guidance, navigation, and control subsystems (including redundancy) for both FLTSATCOM and HEAO-B.

TABLE 10-39. Weight and Power for the Attitude-Control Component.

	FLTSATCOM		HEAO-B	
Component	Mass (kg)	Power (W)	Mass (kg)	Power (W)
Sensors				
Sun sensors	1.5	0	1.6	—
Earth sensors	7.2	12	—	—
Gyroscopes	—	—	19.6 (6)	—
Processing Electronics	23.8	12.8	49.0	—
Reaction Wheels	11.3 (2)	8	53.8 (4)	—
Solar Array Drives	13.9 (2)	0.2 (2)	—	—
Total	57.7	33	124.0	201

FLTSATCOM employed a single system for tracking, telemetry, and command (TT&C). This spacecraft uses only 977 commands for infrequent commanding. The telemetry requirements are 178 mainframe words (0.5 s frame rate) and fewer than 1,000 subcommutated words—typical of a minimum subsystem for TT&C. A wide-beam antenna mounted on the tip of the main payload's antenna mast ensures communications access to the satellite.

The communications requirements for HEAO-B were driven by HEAO's low-altitude and low-inclination orbit, which allowed infrequent access to the spacecraft for commanding and data readout. Table 10-40 summarizes characteristics of the communications subsystem and the subsystem for command and data handling. HEAO-B's spacecraft had to maneuver and take payload data while out of sight of a ground station. The commanding subsystem included a stored command programmer which could store and execute 256 commands. Data remained on a tape recorder until the ground station read it out at communication intervals.

TABLE 10-40. Characteristics of the Subsystem for Communications and Command and Data Handling on HEAO-B.

Characteristic	Description
Communications	
Frequency	S band
Radiated power	1.0 W
Antennas	2 wide beam cross strapped
Weight	16.2 kg
Tape Recorder	
Capacity	84×10^6 bits
Data rate	6.4 kbps and 128 kbps
Weight	13.8 kg
Power (communications and tape recorder)	3.4 W cruise; 78.4 W ground pass
Command and Data Handling	
Command rate	200 bps
Stored commands	256 30 bit commands
Telemetry rate	6.4 and 128 kbps
Weight	28.4 kg
Power	33.8 W cruise; 41.6 W ground pass

The FLTSATCOM and HEAO-B components require temperatures of 5–50 °C —no challenge to the designers. Thermal balance is achieved by mounting the components on external panels of the equipment compartment and allowing the panels to radiate excess heat to space. Second-surface mirrors serve as radiators, taking advantage of their low solar absorptivity and high infrared emissivity. Areas not required for radiation are insulated. The spacecraft's interior surfaces are black to enhance internal heat transfer, and guard heaters prevent excessively low temperatures when equipment is off, or on sensitive assemblies such as propellant lines.

Table 10-35 presented power budgets for FLTSATCOM and HEAO-B. Table 10-41 shows the characteristics and components of their power subsystems. FLTSATCOM uses planar arrays oriented toward the Sun by solar-array drives and controls the body's attitude about the Z-axis (yaw control). The array produces 1,800 W at beginning of life. HEAO-B used planar solar panels which were body-mounted. They were oriented toward the Sun by X-axis (roll) attitude control which kept the Sun in the X-Z body plane. With this attitude, the Sun was up to 75 deg away from the array normal. To improve this poor illumination, designers sized the array for 1,500 W under full solar illumination.

Both FLTSATCOM and HEAO-B use NiCd batteries. The cycle life for FLTSATCOM was less than 1,000 cycles, and the three batteries operated at 70% depth of discharge. They also contain bypass electronics to allow removal of failed cells. The three HEAO-B batteries had a cycle life of over 5,000 cycles and operated at 20% depth of discharge.

Table 10-41 also shows the characteristics of the power control, switching, cabling, and conversion equipment on FLTSATCOM and HEAO-B. The FLTSATCOM spacecraft uses an unregulated bus and switches power to user subsystems in central power-control units. A central power converter provides secondary power for the spacecraft bus subsystems. HEAO-B used a regulated solar array and also switched power to users in a set of integration assemblies. These units also contained power converters.

TABLE 10-41. Characteristics and Components of the Electric-Power Subsystems.

Characteristics and Components	FLTSATCOM	HEAO-B
Solar Arrays		
Power output BOL	1,800 W BOL normal incidence 1,200 W @ 7 year 23 deg incidence	1,500 W normal incidence 613 W @ 15 deg incidence
Size	Two 2.8 m × 3.8 m panels	13.9 m^2
Weight	92.6 kg	77.1 kg
Battery		
Capacity	600 W·hr (24 A·hr)	550 W·hr (20 A·hr)
Weight	89.6 kg total	89.6 kg total
Power Control		
Type	Unregulated array	Regulated array
Weight	9.7 kg	33.1 kg
Power	12.9 W	26 W
Power Switching		
Type	Central switching assemblies (1 for S/C, 2 for P/L)	Central switching assembly (1 for S/C, 1 for P/L)
Weight	29.3 kg	35.5 kg
Power	4.3 W (S/C bus)	Included in power control
Power conversion		
Type	Central S/C bus converter; Separate P/L converters	Converter in switching assembly; Separate ACS and C&DH
Weight	8.5 kg S/C bus, 78.3 kg P/L	4.6 kg ACS, 3.3 kg C&DH
Power	14.3 W S/C bus, 192 W P/L	included in subsystems
Cabling		
Weight	72.7 kg	133.8 kg
Power	24 W	Included above

Table 10-42 summarizes the structural design of FLTSATCOM and HEAO-B. Figure 10-4 shows the structure of FLTSATCOM, and Fig. 10-5 shows the structure of HEAO-B. The HEAO-B structure was a semi-monocoque in which longerons, stiffened skin panels, and rings carried the loads. Both spacecraft were launched on the Atlas-Centaur and were designed to substantially the same boost loads. In the launch configuration, both spacecraft were about the same diameter, although HEAO-B was slightly longer. The on-orbit configurations of the spacecraft were quite different because FLTSATCOM has a large, deployable, solar array and deployed antennas. HEAO-B had essentially the same configuration on orbit as during launch. The two spacecraft had different designs for the load-carrying structure. FLTSATCOM carries loads with a central cylinder. The equipment compartment mounts around this cylinder, and the deployables stay around the outside of the compartment during boost.

Design Integration

Table 10-43 gives relations for estimating the spacecraft's size based on its weight. Equations (10-12) and (10-13) give relations for estimating solar-array area and weight based on the spacecraft's total power. The table compares the actual parameters for FLTSATCOM and HEAO-B to the estimated values.

Fig. 10-4.   **FLTSATCOM Structure**.

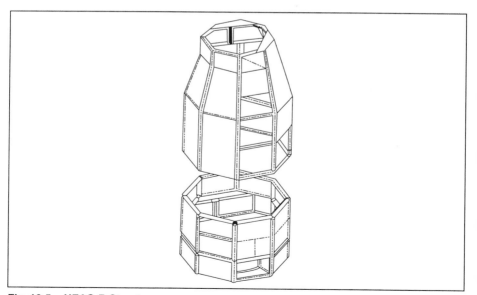

Fig. 10-5.   **HEAO-B Structure**.

TABLE 10-42.   **Structural Characteristics of FLTSATCOM and HEAO-B Designs.**

Characteristic	FLTSATCOM	HEAO-B
Weight	154 kg	779 kg
Design Approach		
Primary structure	Central cylinder	Semi-monocoque
Solar array	Aluminum honeycomb panels deployed in two wings	Aluminum honeycomb panels body mounted
Equipment compartment	Honeycomb panels mounted to longeron-stringer frame	Honeycomb panels mounted to structural frame
Deployables	Wrapped around or folded over equip compartment during boost	

TABLE 10-43.   **Estimated vs. Actual Sizes of FLTSATCOM and HEAO-B.**

Spacecraft Parameters	Units	FLTSATCOM		HEAO-B	
		Estimated	Actual	Estimated	Actual
Spacecraft Loaded Weight	kg		927		3,154
Spacecraft Power BOL	W		1800		1,500
Volume		9.27		31.5	37.8
Compartment	m³		5.0	—	—
Folded Spacecraft	m³		28.2	—	—
Linear Dimension					
Diameter	m	2.9	2.4	4.4	2.7
Length	m		5.1		6.8
Body Area	m²	8.5	12.3	19.2	18.1
Solar-Array Area	m²	18	21.5	15	13.9
Offset to Solar-Array Area	m	5.9	5.2	N/A	N/A
Moments of Inertia	kg·m²				
Folded w/ AKM		881.3	916.4	I_{min} 6,783.2	1,963.1
w/o AKM			828.5	I_{max} 6,783.2	4,860.9
Deployed I_{xx}		3453.3	3388.9		
I_{yy}		935.3	825.8		

References

Agrawal, Brij N. 1986. *Design of Geosynchronous Spacecraft*. Englewood Cliffs, NJ: Prentice-Hall, Inc.

Chetty, P.R.K. 1991. *Satellite Technology and Its Applications (2nd edition)*. New York: TAB Books.

Frazier, R.E. 1986. *HEAO Case Study in Spacecraft Design*. AIAA Professional Study Series. TRW Report 26000-100-102. Washington, DC: American Institute of Aeronautics and Astronautics.

Griffin, Michael D. and James R. French. 1991. *Space Vehicle Design*. Washington, DC: American Institute of Aeronautics and Astronautics.

Reeves, Emery I. 1979. *FLTSATCOM Case Study in Spacecraft Design*. AIAA Professional Study Series. TRW Report 26700-100-054-01.Washington, DC: American Institute of Aeronautics and Astronautics.

TRW Defense and Space Systems Group. 1980–1985. *TRW Space Log*, vols. 17–21. Redondo Beach, CA: TRW, Inc.

U.S. Department of Defense. 1991. *Reliability Prediction of Electronic Equipment*. MIL-HDBK-217 (f).

Chapter 11

Spacecraft Subsystems

This chapter provides design information for the spacecraft bus subsystems, emphasizing material most pertinent to the spacecraft engineer. It offers practical insight into the mission and interface requirements that drive how we configure spacecraft. We include first-order approximations and describe hardware to show how each subsystem works and to help estimate the subsystem's size, weight, power requirements, and eventual cost. We also reference many chapters of this book to integrate concepts and subsystems. Chapter 17 discusses the propulsion subsystem, and Chap. 13 provides much of the communications theory. Chapter 10, Agrawal [1986], Chetty [1991], and Morgan and Gordon [1989] provide insight to the theory and practice of designing spacecraft subsystems.

In the rest of this chapter we will discuss these issues and an approach for estimating the size and configuration of spacecraft subsystems.

11.1 Attitude Determination and Control
John S. Eterno, *Ball Aerospace & Technologies Corporation*

The *attitude determination and control subsystem (ADCS)* stabilizes the vehicle and orients it in desired directions during the mission despite the external disturbance torques acting on it. This requires that the vehicle determine its attitude, using sensors, and control it, using actuators. The ADCS often is tightly coupled to other subsystems on board, especially the propulsion (Chap. 17) and navigation (Sec. 11.7) functions. Additional information on attitude determination and control can be found in Wertz [1978, 2001], Kaplan [1976], Agrawal [1986], Hughes [1986], Griffin and French [1990], Chobotov [1991], and Fortescue and Stark [1992].

We begin by discussing several useful concepts and definitions, including mass properties, disturbance torques, angular momentum, and reference vectors. The mass properties of a spacecraft are key in determining the size of control and disturbance torques. We typically need to know the location of the center of mass or gravity (cg) as well as the elements of the inertia matrix: the moments and products of inertia about chosen reference axes. (See Sec. 11.6 for examples of moment of inertia calculations.) The direction of the principal axes—those axes for which the inertia matrix is diagonal and the products of inertia are zero—are also of interest. Finally, we need to know how these properties change with time, as fuel or other consumables are used, or as appendages are moved or deployed.

A body in space is subject to small but persistent disturbance torques (e.g., 10^{-4} N·m) from a variety of sources. These torques are categorized as *cyclic*, varying in a sinusoidal manner during an orbit, or *secular*, accumulating with time, and not averaging out over an orbit. These torques would quickly reorient the vehicle unless resisted in some way. An ADCS system resists these torques either passively, by exploiting inherent inertia or magnetic properties to make the "disturbances" stabilizing and their effects tolerable, or actively, by sensing the resulting motion and applying corrective torques.

Angular momentum plays an important role in space, where torques typically are small and spacecraft are unconstrained. For a body initially at rest, an external torque will cause the body to angularly accelerate proportionally to the torque—resulting in an increasing angular velocity. Conversely, if the body is initially spinning about an axis perpendicular to the applied torque, then the body spin axis will precess, moving with a constant angular velocity proportional to the torque. Thus, spinning bodies act like gyroscopes, inherently resisting disturbance torques in 2 axes by responding with constant, rather than increasing, angular velocity. This property of spinning bodies, called *gyroscopic stiffness*, can be used to reduce the effect of small, cyclic disturbance torques. This is true whether the entire body spins or just a portion of it, such as a momentum wheel or spinning rotor.

Conservation of vehicle angular momentum requires that only external torques change the system net angular momentum. Thus, external disturbances must be resisted by external control torques (e.g., thrusters or magnetic torquers) or the resulting momentum buildup must be stored internally (e.g., by reaction wheels) without reorienting the vehicle beyond its allowable limits. The momentum buildup due to secular disturbances ultimately must be reduced by applying compensating external control torques.

Often, in addition to rejecting disturbances, the ADCS must reorient the vehicle (in *slew* maneuvers) to repoint the payload, solar arrays, or antennas. These periodic repointing requirements may drive the design to larger actuators than would be required for disturbance rejection alone.

To orient the vehicle correctly, external references must be used to determine the vehicle's absolute attitude. These references include the Sun, the Earth's IR horizon, the local magnetic field direction, and the stars. In addition, inertial sensors (gyroscopes) also can be carried to provide a short-term attitude reference between external updates. External references (e.g., Sun angles) are usually measured as body-centered angular distances to a vector. Each such vector measurement provides only two of the three independent parameters needed to specify the orientation of the spacecraft. This results in the need for multiple sensor types on board most spacecraft.

Table 11-1 lists the steps for designing an ADCS for spacecraft. The FireSat spacecraft, shown in Fig. 11-1, will be used to illustrate this process. The process must be iterative, with mission requirements and vehicle mass properties closely related to the ADCS approach. Also, a rough estimate of disturbance torques (see Chap. 10) is necessary before the type of control is selected (step 2), even though the type of control will help determine the real disturbance environment (step 3).

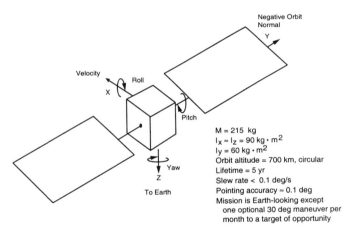

Fig. 11-1. Hypothetical FireSat Spacecraft. We use this simplified example to discuss key concepts throughout this section. See Fig. 5-1 for illustration of roll-pitch-yaw coordinates.

11.1.1 Control Modes and Requirements

Tables 11-2 and 11-3 describe typical spacecraft control modes and requirements. The ADCS requirements are closely tied to mission needs and other subsystem characteristics, as shown in Fig. 11-2. These requirements may vary considerably with mission phase or modes, challenging the designer to develop a single hardware suite for different objectives.

For many spacecraft, the ADCS must control vehicle attitude during firing of large liquid or solid rocket motors, which may be used during orbit insertion or for orbit changes. Large motors create large disturbance torques, which can drive the design to larger actuators than are needed once on station.

TABLE 11-1.   Control System Design Process. An iterative process is used for designing the ADCS as part of the overall spacecraft system.

Step	Inputs	Outputs	FireSat Example
1a. Define control modes 1b. Define or derive system-level requirements by control mode	Mission requirements, mission profile, type of insertion for launch vehicle	List of different control modes during mission (See Table 11-2) Requirements and constraints (See Table 11-3)	Orbit injection: none—provided by launch vehicle Normal: nadir pointing, < 0.1 deg; autonomous determination (Earth-relative) Optional slew: One 30 deg maneuver per month to a target of opportunity
2. Select type of spacecraft control by attitude control mode (Sec. 11.1.2)	Payload, thermal and power needs Orbit, pointing direction Disturbance environment	Method for stabilizing and control: 3-axis, spinning, or gravity gradient	Momentum bias stabilization with a pitch wheel, electro-magnets for momentum dumping, and optionally, thrusters for slewing (shared with ΔV system in navigation)
3. Quantify disturbance environment (Sec. 11.1.3)	Spacecraft geometry, orbit, solar/magnetic models, mission profile	Values for forces from gravity gradient, magnetic aerodynamics, solar pressure, internal disturbances, and powered flight effects on control (cg offsets, slosh)	Gravity gradient: 1.8×10^{-6} N·m normal pointing; 4.4×10^{-5} N·m during target-of-opportunity mode Magnetic: 4.5×10^{-5} N·m Solar: 6.6×10^{-6} N·m Aerodynamic: 3.4×10^{-6} N·m
4. Select and size ADCS hardware (Sec. 11.1.4)	Spacecraft geometry, pointing accuracy, orbit conditions, mission requirements, lifetime, orbit, pointing direction, slew rates	Sensor suite: Earth, Sun, inertial, or other sensing devices Control actuators, e.g., reaction wheels, thrusters, or magnetic torquers Data processing electronics, if any, or processing requirements for other subsystems or ground computer	1 Momentum wheel, Momentum: 40 N·m·s 2 Horizon sensors, Scanning, 0.1 deg accuracy 3 Electromagnets, Dipole moment: 10 A·m² 4 Sun sensors, 0.1 deg accuracy 1 3-axis magnetometer, 1 deg accuracy
5. Define determination and control algorithms	All of above	Algorithms, parameters, and logic for each determination and control mode	Determination: Horizon data filtered for pitch and roll. Magnetometer and Sun sensors used for yaw. Control: Proportional-plus-derivative for pitch, Coupled roll-yaw control with electromagnets
6. Iterate and document	All of above	Refined requirements and design Subsystem specification	

Once the spacecraft is on station, the payload pointing requirements usually dominate. These may require Earth-relative or inertial attitudes, and fixed or spinning fields of view. In addition, we must define the need for and frequency of attitude slew maneuvers. Such maneuvers may be necessary to:

TABLE 11-2. Typical Attitude Control Modes. Performance requirements are frequently tailored to these different control operating modes.

Mode	Description
Orbit Insertion	Period during and after boost while spacecraft is brought to final orbit. Options include no spacecraft control, simple spin stabilization of solid rocket motor, and full spacecraft control using liquid propulsion system.
Acquisition	Initial determination of attitude and stabilization of vehicle. Also may be used to recover from power upsets or emergencies.
Normal, On-Station	Used for the vast majority of the mission. Requirements for this mode should drive system design.
Slew	Reorienting the vehicle when required.
Contingency or Safe	Used in emergencies if regular mode fails or is disabled. May use less power or sacrifice normal operation to meet power or thermal constraints.
Special	Requirements may be different for special targets or time periods, such as eclipses.

TABLE 11-3. Typical Attitude Determination and Control Performance Requirements. Requirements need to be specified for each mode. The following lists the areas of performance frequently specified.

Area	Definition*	Examples/Comments
DETERMINATION		
Accuracy	How well a vehicle's orientation with respect to an absolute reference is known	0.25 deg, 3 σ, all axes; may be real-time or post-processed on the ground
Range	Range of angular motion over which accuracy must be met	Any attitude within 30 deg of nadir
CONTROL		
Accuracy	How well the vehicle attitude can be controlled with respect to a commanded direction	0.25 deg, 3 σ; includes determination and control errors, may be taken with respect to an inertial or Earth-fixed reference
Range	Range of angular motion over which control performance must be met	All attitudes, within 50 deg of nadir, within 20 deg of Sun
Jitter	A specified angle bound or angular rate limit on short-term, high-frequency motion	0.1 deg over 1 min, 1 deg/s,1 to 20 Hz; usually specified to keep spacecraft motion from blurring sensor data
Drift	A limit on slow, low-frequency vehicle motion. Usually expressed as angle/time.	1 deg/hr, 5 deg max. Used when vehicle may drift off target with infrequent resets (especially if actual direction is known)
Settling Time	Specifies allowed time to recover from maneuvers or upsets.	2 deg max motion, decaying to < 0.1 deg in 1 min; may be used to limit overshoot, ringing, or nutation

* Definitions vary with procuring and designing agencies, especially in details (e.g., 1 or 3 σ, amount of averaging or filtering allowed). It is always best to define exactly what is required.

- Repoint the payload's sensing systems to targets of opportunity
- Maneuver the attitude control system's sensors to celestial targets for attitude determination
- Track stationary or moving targets
- Acquire the desired satellite attitude initially or after a failure

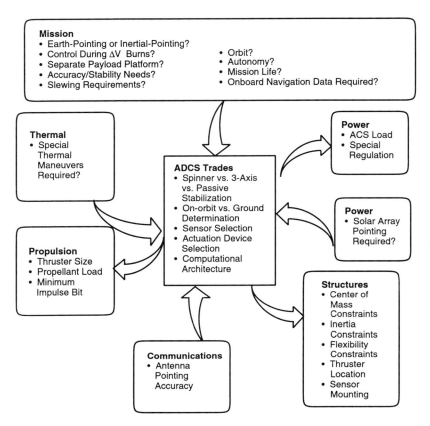

Fig. 11-2.   The Impact of Mission Requirements and Other Subsystems on the ADCS Subsystem. Direction of arrows shows requirements flow from one subsystem to another.

In most cases, we do not need to rotate the spacecraft quickly. But retargeting time may be critical for some applications. In either case, slewing mainly influences the choice and size of actuators. For example, the vehicle's maximum slew rate determines the thrusters' size or the reaction wheel's maximum torque. High-rate maneuvers may require other actuation systems, such as a second set of high-thrust reaction jets or perhaps control moment gyros.

For FireSat, we assume that the launch vehicle places us in our final orbit, with no need for ADCS control during orbit insertion. The normal pointing requirement is 0.1 deg, nadir-oriented. Attitude determination must be autonomous, providing Earth-relative knowledge better than 0.1 deg (to support the pointing requirement) while the vehicle is within 30 deg of nadir. In addition to these basic requirements, we will consider an optional requirement for occasional repointing of the spacecraft to a region of interest. We want to examine how such a requirement would influence the design, increasing the complexity and capability of the ADCS. For this option, we will assume the requirement to repoint the vehicle once every 30 days. It must repoint, or slew, up to 30 deg in under 10 min, and hold the relative nadir orientation for 90 min.

11.1.2 Selection of Spacecraft Control Type

Once we have defined the subsystem requirements, we are ready to select a method of controlling the spacecraft. Table 11-4 lists several different methods of control, along with typical characteristics of each.

TABLE 11-4. Attitude Control Methods and Their Capabilities. As requirements become tighter, more complex control systems become necessary.

Type	Pointing Options	Attitude Maneuverability	Typical Accuracy	Lifetime Limits
Gravity-gradient	Earth local vertical only	Very limited	±5 deg (2 axes)	None
Gravity-gradient and Momentum Bias Wheel	Earth local vertical only	Very limited	±5 deg (3 axes)	Life of wheel bearings
Passive Magnetic	North/south only	Very limited	±5 deg (2 axes)	None
Pure Spin Stabilization	Inertially fixed any direction. Repoint with precession maneuvers	High propellant usage to move stiff momentum vector	±0.1 deg to ±1 deg in 2 axes (proportional to spin rate)	Thruster propellant (if applies)*
Dual-Spin Stabilization	Limited only by articulation on despun platform	Momentum vector same as above. Despun platform constrained by its own geometry	Same as above for spin section. Despun dictated by payload reference and pointing	Thruster propellant (if applies)* Despin bearings
Bias Momentum (1 wheel)	Best suited for local vertical pointing	Momentum vector of the bias wheel prefers to stay normal to orbit plane, constraining yaw maneuver	±0.1 deg to ±1 deg	Propellant (if applies)* Life of sensor and wheel bearings
Zero Momentum (thruster only)	No constraints	No constraints High rates possible	±0.1 deg to ±5 deg	Propellant
Zero Momentum (3 wheels)	No constraints	No constraints	±0.001 deg to ±1 deg	Propellant (if applies)* Life of sensor and wheel bearings
Zero Momentum CMG	No constraints	No constraints High rates possible	±0.001 deg to ±1 deg	Propellant (if applies)* Life of sensor and wheel bearings

*Thrusters may be used for slewing and momentum dumping at all altitudes. Magnetic torquers may be used from LEO to GEO.

Passive Control Techniques. *Gravity-gradient control* uses the inertial properties of a vehicle to keep it pointed toward the Earth. This relies on the fact that an elongated object in a gravity field tends to align its longitudinal axis through the Earth's center. The torques which cause this alignment decrease with the cube of the orbit radius, and are symmetric around the nadir vector, thus not influencing the yaw of a spacecraft around the nadir vector. This tendency is used on simple spacecraft in near-Earth orbits without yaw orientation requirements, often with deployed booms to achieve the desired inertias.

Frequently, we add dampers to gravity-gradient spacecraft to reduce *libration* —small oscillations around the nadir vector caused by disturbances. Gravity-gradient spacecraft are particularly sensitive to thermal shocks on long deployed booms when entering or leaving eclipses. They also need a method of ensuring attitude capture with the correct end of the spacecraft pointed at nadir—the gravity-gradient torques make either end along the minimum inertia axis equally stable.

In the simplest gravity-gradient spacecraft, only two orientation axes are controlled. The orientation around the nadir vector is unconstrained. To control this third axis, a small, constant-speed momentum wheel is sometimes added along the intended pitch axis (i.e., an axis perpendicular to the nadir and velocity vectors). This "yaw" wheel is stable when it aligns with the orbit normal, and small energy dissipation mechanisms on board cause the spacecraft to seek this minimum energy, stable orientation without active control.

A third type of purely passive control uses permanent magnets on board the spacecraft to force alignment along the Earth's magnetic field. This is most effective in near-equatorial orbits where the field orientation stays almost constant for an Earth-pointing vehicle.

Spin Control Techniques. *Spin stabilization* is a passive control technique in which the entire spacecraft rotates so that its angular momentum vector remains approximately fixed in inertial space. Spin-stabilized spacecraft (or *spinners*), employ the gyroscopic stability discussed earlier to passively resist disturbance torques about two axes. The spinning motion is stable (in its minimum energy state) if the vehicle is spinning about the axis having the largest moment of inertia. Energy dissipation mechanisms on board, such as fuel slosh and structural damping) will cause any vehicle to head toward this state if uncontrolled. Thus disk-shaped spinners are passively stable while pencil-shaped vehicles are not. Spinners can be simple, survive for long periods without attention, provide a thermally benign environment for components, and provide a scanning motion for sensors. The principal disadvantages of spin stabilization are (1) that the vehicle mass properties must be controlled to ensure the desired spin direction and stability and (2) that the angular momentum vector requires more fuel to reorient than a vehicle with no net angular momentum, reducing the usefulness of this technique for payloads that must be repointed frequently.

It takes extra fuel to reorient a spinner because of the gyroscopic stiffness which also helps it resist disturbances. In reorienting a spinning body with angular momentum, h, a constant torque, T, will produce an angular velocity, ω, perpendicular to the applied torque and angular momentum vector, of magnitude $\omega = T/h$. Thus, the higher the stored momentum is, the more torque must be applied for a given ω. For a maneuver through an angle θ, the torque-time product—an indication of fuel required for the maneuver—is a constant equal to $h\theta$. Conversely, for a nonspinning vehicle with no initial angular velocity, a small torque can be used to start it rotating, with an opposite torque to stop it. The fuel used for any angle maneuver can be infinitesimally small if a slow maneuver is acceptable.

A useful variation of spin control is called *dual-spin stabilization*, where the spacecraft has two sections spinning at different rates about the same axis. Normally, one section, the rotor, spins rapidly to provide angular momentum, while the second section, the stator or platform, is despun to keep one axis pointed toward the Earth or Sun. By combining inertially fixed and rotating sections on the same vehicle, dual spinners can accommodate a variety of payloads in a simple vehicle. Also, by adding energy dissipation devices to the platform, a dual spinner can be passively stable spinning about the axis with the smallest moment of inertia. This permits more pencil-shaped spacecraft, which fit better in launch vehicle fairings. The disadvantage of dual-spin stabilization is the added complexity of the platform bearing and slip rings between the sections. This complexity can increase cost and reduce reliability compared to simple spin stabilization.

Spinning spacecraft, both simple and dual, exhibit several distinct types of motion which often are confused. *Precession* is the motion of the angular momentum vector caused by external torques such as thruster firings. *Wobble* is the apparent motion of the body when it is spinning with the angular momentum vector aligned along a principal axis of inertia which is offset from a body reference axis—for example, the intended spin axis. This looks like motion of the intended spin axis around the angular momentum vector at the spin rate.

Nutation is the torque-free motion of the spacecraft body when the angular momentum vector is not perfectly aligned along a principal axis of inertia. For rod-shaped objects, this motion is a slow rotation (compared to spin rate) of the spin axis around the angular momentum vector. For these objects, spinning about a minimum inertia axis, additional energy dissipation will cause increased nutation. For disk-shaped objects, spinning around a maximum inertia axis, nutation appears as a higher-than-spin-rate tumbling. Energy dissipation for these objects (e.g., with a passive nutation damper) reduces nutation, resulting in a clean spin.

Nutation is caused by disturbances such as thruster impulses, and can be seen as varying signals in body-mounted inertial and external sensors. Wobble is caused by imbalance and appears as constant offsets in body-mounted sensors. Such constant offsets are rarely discernible unless multiple sensors are available.

Spin stability normally requires active control, such as mass expulsion or magnetic coils, to periodically adjust the spacecraft's attitude and spin rate to counteract disturbance torques. In addition, we may need to damp the nutation caused by disturbances, precession commands, or fuel slosh. Aggravating this nutation is the effect of structural flexure and fuel slosh, which is present in any space vehicle to one degree or another. Once the excitation stops, nutation decreases as these same factors dissipate the energy. But this natural damping can take hours. We can neutralize this source of error in minutes with nutation dampers (see Sec. 11.5). We can also reduce the amount of nutation from these sources by increasing the spin rate, thus increasing the stiffness of the spinning vehicle. If the spin rate is 20 rpm, and the nutation angle is 3 deg, then at 60 rpm the nutation angle would decrease by a factor of three. We seldom use spin rates above 90 rpm because of the large centrifugal forces and their effect on structural design and weight. In thrusting and pointing applications, spin rates under 20 rpm may allow excessive nutation and are not used. However, noncritical applications, such as thermal control, are frequently insensitive to nutation and may employ very low spin rates.

Three-axis Control Techniques. Spacecraft stabilized in 3 axes are more common today than those using spin or gravity gradient. They maneuver and can be stable and accurate, depending on their sensors and actuators. But they are also more expensive and more complex. The control torques about the axes of 3-axis systems come from combinations of momentum wheels, reaction wheels, control moment gyros, thrusters, or magnetic torquers. Broadly, however, these systems take two forms: one uses momentum bias by placing a momentum wheel along the pitch axis; the other is called zero momentum with a reaction wheel on each axis. Either option usually needs thrusters or magnetic torquers as well as the wheels.

In a *zero-momentum* system, reaction wheels respond to disturbances on the vehicle. For example, a vehicle-pointing error creates a signal which speeds up the wheel, initially at zero. This torque corrects the vehicle and leaves the wheel spinning at low speed, until another pointing error speeds the wheel further or slows it down again. If the disturbance is cyclic during each orbit, the wheel may not approach saturation speed for several orbits. Secular disturbances, however, cause the wheel to drift toward saturation. We then must apply an external torque, usually with a thruster or magnetic torquer, to force the wheel speed back to zero. This process, called *desaturation*, *momentum unloading*, or *momentum dumping,* can be done automatically or by command from the ground.

When high torque is required for large vehicles or fast slews, a variation of 3-axis control is possible using *control moment gyros,* or *CMGs.* These devices work like momentum wheels on gimbals. (See Sec. 11.1.4 for a further discussion of CMGs.) The control of CMGs is complex, but their available torque for a given weight and power can make them attractive.

As a final type of zero momentum 3-axis control, simple all-thruster systems are used for short durations when high torque is needed, such as orbit insertion or during ΔV burns from large motors. These thrusters then may be used for different purposes such as momentum dumping during other mission modes.

Momentum bias systems often have just one wheel with its spin axis mounted along the pitch axis, normal to the orbit plane. The wheel is run at a nearly constant, high speed to provide gyroscopic stiffness to the vehicle, just as in spin stabilization, with similar nutation dynamics. Around the pitch axis, however, the spacecraft can control attitude by torquing the wheel, slightly increasing or decreasing its speed. Periodically, the pitch wheel must be desaturated (brought back to its nominal speed), as in zero-momentum systems, using thrusters or magnets.

The dynamics of nadir-oriented momentum-bias vehicles exhibit a phenomenon known as *roll-yaw coupling.* To see this coupling, consider an inertially-fixed angular momentum vector at some angle with respect to the orbit plane. If the angle is initially a positive roll error, then 1/4 orbit later it appears purely about the yaw axis as a negative yaw error. As the vehicle continues around the orbit, the angle goes through negative roll and positive yaw before realigning as positive roll. This coupling, which is due to the apparent motion of the Earth and, therefore, the Earth-fixed coordinate frame as seen from the spacecraft, can be exploited to control roll and yaw over a quarter orbit using only a roll sensor.

Effects of Requirements on Control Type. With the above knowledge of control types, we can proceed to select a type which best meets mission requirements. Tables 11-5 through 11-7 describe the effects of orbit insertion, payload pointing, and payload slew requirements on the selection process.

A common control approach during orbit insertion is to use the short-term spin stability of the spacecraft-orbit-insertion motor combination. Once on station, the motor may be jettisoned, the spacecraft despun using jets or a yo-yo device, and a different control technique used.

Payload pointing will influence the ADCS control method, the class of sensors, and the number and kind of actuation devices. Occasionally, pointing accuracies are so stringent that a separate, articulated platform is necessary. An articulated platform can perform scanning operations much easier than the host vehicle, with better accuracy and stability.

TABLE 11-5.   Orbit Transition Maneuvers and Their Effect. Using thrusters to change orbits creates special challenges for the ADCS.

Requirement	Effect on Spacecraft	Effect on ADCS
Large impulse to complete orbit insertion (thousands of m/s)	Solid motor or large bipropellant stage Large thrusters or a gimbaled engine or spin stabilization for attitude control during burns	Inertial measurement unit for accurate reference and velocity measurement Different actuators, sensors, and control laws for burn vs. coasting phases Need for navigation or guidance
On-orbit plane changes to meet payload needs or vehicle operations (hundreds of m/s)	More thrusters, but may be enough if coasting phase uses thrusters	Separate control law for thrusting Actuators sized for thrusting disturbances Onboard attitude reference for thrusting phase
Orbit maintenance trim maneuvers (<100 m/s)	One set of thrusters	Thrusting control law Onboard attitude reference

TABLE 11-6.   Effect of Payload Pointing Directions on ADCS Design. The payload pointing requirements are usually the most important factors for determining the type of actuators and sensors.

Requirement	Effect on Spacecraft	Effect on ADCS
Earth-pointing • Nadir (Earth) pointing • Scanning • Off-nadir pointing	• Gravity-gradient fine for low accuracies (>1 deg) only • 3-axis stabilization acceptable with Earth local vertical reference	**If gravity-gradient** • Booms, dampers, Sun sensors, magnetometer or horizon sensors for attitude determination • Momentum wheel for yaw control **If 3-axis** • Horizon sensor for local vertical reference (pitch and roll) • Sun or star sensor for third-axis reference and attitude determination • Reaction wheels, momentum wheels, or control moment gyros for accurate pointing and propellant conservation • Reaction control system for coarse control and momentum dumping • Magnetic torquers can also dump momentum • Inertial measurement unit for maneuvers and attitude determination
Inertial pointing • Sun • Celestial targets • Payload targets of opportunity	• Spin stabilization fine for medium accuracies with few attitude maneuvers • Gravity gradient does not apply • 3-axis control is most versatile for frequent reorientations	**If spin** • Payload pointing and attitude sensor operations limited without despun platform • Needs thrusters to reorient momentum vector • Requires nutation damping **If 3-axis** • Typically, sensors include Sun sensors, star tracker, and inertial measurement unit • Reaction wheels and thrusters are typical actuators • May require articulated payload (e.g., scan platform)

TABLE 11-7. Slewing Requirements That Affect Control Actuator Selection. Spacecraft slew agility can demand larger actuators for intermittent use.

Slewing	Effect on Spacecraft	Effect on ADCS
None	Spacecraft constrained to one attitude—highly improbable	• Reaction wheels, if planned, can be smaller • If magnetic torque can dump momentum, may not need thrusters
Nominal rates— 0.05 deg/s (maintain local vertical) to 0.5 deg/s	Minimal	• Thrusters very likely • Reaction wheels adequate by themselves only for a few special cases
High rates— > 0.5 deg/s	• Structural impact on appendages • Weight and cost increase	• Control moment gyros very likely or two thruster force levels—one for stationkeeping and one for high-rate maneuvers

Trade studies on pointing requirements must consider accuracy in determining attitude and controlling vehicle pointing. We must identify the most stringent requirements. Table 11-8 summarizes effects of accuracy requirements on the spacecraft's ADCS subsystem approach. Section 5.4 discusses how to develop pointing budgets.

FireSat Control Selection. For FireSat, we consider two options for orbit insertion control. First, the launch vehicle may directly inject the spacecraft into its mission orbit. This common option simplifies the spacecraft design, since no special insertion mode is needed. An alternate approach, useful for small spacecraft such as FireSat, is to use a monopropellant system on board the spacecraft to fly itself up from a low parking orbit to its final altitude. For small insertion motors, reaction wheel torque or momentum bias stabilization may be sufficient to control the vehicle during this burn. For larger motors, ΔV thruster modulation or dedicated ADCS thrusters become attractive.

Once on-station, the spacecraft must point its sensors at nadir most of the time and slightly off-nadir for brief periods. Since the payload needs to be despun and the spacecraft frequently reoriented, spin stabilization is not the best choice. Gravity-gradient and passive magnetic control cannot meet the 0.1 deg pointing requirement or the 30 deg slews. This leaves 3-axis control and momentum-bias stabilization as viable options for the on-station control as well.

Depending on other factors, either approach might work, and we will baseline momentum bias control with its simpler hardware requirements. In this case, we will use a single pitch wheel for momentum and electromagnets for momentum dumping and roll and yaw control.

For the optional off-nadir pointing requirement, 3-axis control with reaction wheels might be more appropriate. Also, 3-axis control often can be exploited to simplify the solar array design, by using one of the unconstrained payload axes (yaw, in this case) to replace a solar array drive axis. Thus, the reduced array size possible with 2 deg of freedom can be achieved with one array axis drive and one spacecraft rotation.

11.1.3 Quantify the Disturbance Environment

In this step, we determine the size of the external torques the ADCS must tolerate. Only three or four sources of torque matter for the typical Earth-orbiting spacecraft. They are gravity-gradient effects, magnetic-field torques on the vehicle, impingement

TABLE 11-8. Effect of Control Accuracy on Sensor Selection and ADCS Design. Accurate pointing requires better, higher cost, sensors, and actuators.

Required Accuracy (3σ)	Effect on Spacecraft	Effect on ADCS
> 5 deg	• Permits major cost savings • Permits gravity-gradient (GG) stabilization	**Without attitude determination** • No sensors required for GG stabilization • Boom motor, GG damper, and a bias momentum wheel are only required actuators **With attitude determination** • Sun sensors & magnetometer adequate for attitude determination at ≥ 2 deg • Higher accuracies may require star trackers or horizon sensors
1 deg to 5 deg	• GG not feasible • Spin stabilization feasible if stiff, inertially fixed attitude is acceptable • Payload needs may require despun platform on spinner • 3-axis stabilization will work	• Sun sensors and horizon sensors may be adequate for sensors, especially a spinner • Accuracy for 3-axis stabilization can be met with RCS deadband control but reaction wheels will save propellant for long missions • Thrusters and damper adequate for spinner actuators • Magnetic torquers (and magnetometer) useful
0.1 deg to 1 deg	• 3-axis and momentum-bias stabilization feasible • Dual-spin stabilization also feasible	• Need for accurate attitude reference leads to star tracker or horizon sensors & possibly gyros • Reaction wheels typical with thrusters for momentum unloading and coarse control • Magnetic torquers feasible on light vehicles (magnetometer also required)
< 0.1 deg	• 3-axis stabilization is necessary • May require articulated & vibration-isolated payload platform with separate sensors	• Same as above for 0.1 deg to 1 deg but needs star sensor and better class of gyros • Control laws and computational needs are more complex • Flexible body performance very important

by solar-radiation, and, for low-altitude orbits, aerodynamic torques. Section 8.1 discusses the Earth environment in detail, and Chap. 10 and Singer [1964] provide a discussion of disturbances. Tables 11-9A and 11-9B summarize the four major disturbances, provide equations to estimate their size for the worst case, and calculate values for the FireSat example.

Disturbances can be affected by the spacecraft orientation, mass properties, and design symmetry. For the normal FireSat orientation, the largest torque is due to the residual magnetism in the spacecraft. If, however, we use the optional 30-deg off-nadir pointing, the gravity-gradient torque increases over an order of magnitude, to become as large as the magnetic torque. Note that we use 1 deg in the gravity-gradient calculations, rather than the 0.1 deg pointing accuracy. This is to account for our uncertain knowledge of the principal axes. If the principal axes are off by several degrees, that angle may dominate in the disturbance calculations. We also note that a less symmetric solar array arrangement would have increased both the aerodynamic and solar torques, making them closer to the magnetic torque in this example.

TABLE 11-9A.   Simplified Equations for Estimating Worst-Case Disturbance Torques. Disturbance torques affect actuator size and momentum storage requirements.

Distur-bance	Type	Influenced Primarily by	Formula
Gravity-gradient	Constant torque for Earth-oriented vehicle, cyclic for inertially oriented vehicle	• Spacecraft inertias • Orbit altitude	$T_g = \dfrac{3\mu}{2R^3} \lvert I_z - I_y \rvert \sin(2\theta)$ where T_g is the max gravity torque; μ is the Earth's gravity constant (3.986×10^{14} m³/s²); R is orbit radius (m), θ is the maximum deviation of the Z-axis from local vertical in radians, and I_z and I_y are moments of inertia about z and y (or x, if smaller) axes in kg·m².
Solar Radiation	Cyclic torque on Earth-oriented vehicle, constant for solar-oriented vehicle or platform	• Spacecraft geometry • Spacecraft surface reflectivity • Spacecraft geometry and cg location	Solar radiation pressure, T_{sp}, is highly dependent on the type of surface being illuminated. A surface is either transparent, absorbent, or a reflector, but most surfaces are a combination of the three. Reflectors are classed as diffuse or specular. In general, solar arrays are absorbers and the spacecraft body is a reflector. The worst case solar radiation torque is $$T_{sp} = F(c_{ps} - cg)$$ where $F = \dfrac{F_s}{c} A_s (1+q)\cos i$ and F_s is the solar constant, 1,367 W/m², c is the speed of light, 3×10^8 m/s, A_s is the surface area, c_{ps} is the location of the center of solar pressure, cg is the center of gravity, q is the reflectance factor (ranging from 0 to 1, we use 0.6), and i is the angle of incidence of the Sun.
Magnetic Field	Cyclic	• Orbit altitude • Residual spacecraft magnetic dipole • Orbit inclination	$$T_m = DB$$ where T_m is the magnetic torque on the spacecraft; D is the residual dipole of the vehicle in amp·turn·m² (A·m²), and B is the Earth's magnetic field in tesla. B can be approximated as $2M/R^3$ for a polar orbit to half that at the equator. M is the magnetic moment of the Earth, 7.96×10^{15} tesla·m³, and R is the radius from dipole (Earth) center to spacecraft in m.
Aerody-namic	Constant for Earth-oriented vehicles, variable for inertially oriented vehicle	• Orbit altitude • Spacecraft geometry and cg location	Atmospheric density for low orbits varies significantly with solar activity. $$T_a = F(c_{pa} - cg) = FL$$ where $F = 0.5\,[\rho\, C_d A V^2]$; F being the force; C_d the drag coefficient (usually between 2 and 2.5); ρ the atmospheric density; A, the surface area; V, the spacecraft velocity; c_{pa} the center of aerodynamic pressure; and cg the center of gravity.

TABLE 11-9B.   **Example of Worst Case Disturbance Torque Estimates for FireSat.** Magnetic and aerodynamic disturbances are the largest for this small spacecraft.

Disturbance	FireSat Example
Gravity-gradient	For $R = (6,378 + 700)$ km $= 7,078$ km; $I_z = 90$ kg·m^2, $I_y = 60$ kg·m^2 and $\theta = 1$ deg (normal mode) or 30 deg (optional target-of-opportunity mode): normal: $$T_g = \frac{(3)(3.986 \times 10^{14}\, m^3/s^2)(30\, kg \cdot m^2)\sin(2\, deg)}{(2)(7.078 \times 10^6\, m)^3}$$ $$= 1.8 \times 10^{-6}\ N \cdot m$$ optional target-of-opportunity: $T_g = 4.4 \times 10^{-5}\ N \cdot m$
Solar Radiation	For a 2 m by 1.5 m spacecraft cross-section, a center-of-solar-pressure to center-of-mass difference of 0.3 m, incidence angle of 0 deg and coefficient of reflectivity of 0.6. $T_{sp} = (1,367\ W/m^2)\ (2\, m \times 1.5m)\ (0.3\, m)\ (1 + 0.6)\ (\cos 0\ deg)\ /\ (3 \times 10^8\ m/s)$ $= 6.6 \times 10^{-6}\ N \cdot m$
Magnetic Field	For $R = 7,078$ km, a spacecraft magnetic dipole of 1 A·m^{2*} and the worst-case polar magnetic field, $M = 2\ (7.96 \times 10^{15}\ tesla \cdot m^3)/(7.078 \times 10^6\ m)^3$ $= 4.5 \times 10^{-5}$ tesla $(= 0.45$ gauss) $T_m = 1 \times 4.5 \times 10^{-5} = 4.5 \times 10^{-5}\ N \cdot m$
Aerodynamics	For illustration purposes we assume a 3 m^2 surface, offset from the center of mass by 0.2 m. In a 700-km orbit the velocity is $\approx 7,504$ m/s, the atmospheric density (ρ) is $\approx 10^{-13}$ kg/m^3. For C_d, the drag coefficient, use 2.0. $F = 1/2\ [(10^{-13}\ kg/m^3)\ (2)(3\ m^2)\ (7,504\ m/s)^2] = 1.7 \times 10^{-5}\ N$ $T = FL = 1.7 \times 10^{-5}\ N\ (0.2\ m) = 3.4 \times 10^{-6}\ N \cdot m$ This is small. At a 100-km orbit, however, $\rho = 10^{-9}$ kg/m^3. This results in $T = 3.3 \times 10^{-2}$ N·m, which is significant for our small spacecraft.

* Residual magnetic dipoles can range anywhere from 0.1 to > 20 A·m^2 depending on the spacecraft's size and whether any onboard compensation is provided. On a small-sized, uncompensated vehicle, 1 A·m^2 is typical (1 A·m^2 = 1,000 pole·cm).

The other disturbances on the control system are internal to the spacecraft. Fortunately, we have some control over them. If we find that one is much larger than the rest, we can respecify it to tighter values. This change would reduce its significance but most likely add to its cost or weight. Table 11-10 summarizes the common internal disturbances. Misalignments in the center of gravity and in thrusters will show up during thrusting only and are corrected in a closed-loop control system. The slosh and operating machinery torques are of greater concern but depend on specific hardware. If a spacecraft component has fluid tanks or rotating machinery, the system designer should investigate disturbance effects and ways to compensate for the disturbance, if required. Standard techniques include slosh baffles or counter-rotating elements.

TABLE 11-10. Principal Internal Disturbance Torques. Spacecraft designers can minimize internal disturbances through careful planning and precise manufacturing which may increase costs.

Disturbances	Effect on Vehicle	Typical Values
Uncertainty in Center of Gravity (cg)	Unbalanced torques during firing of coupled thrusters Unwanted torques during translation thrusting	1 to 3 cm
Thruster Misalignment	Same as cg uncertainty	0.1 to 0.5 deg
Mismatch of Thruster Outputs	Similar to cg uncertainty	± 5%
Rotating Machinery (pumps, tape recorders)	Torques that perturb both stability and accuracy	Dependent on spacecraft design; may be compensated by counter-rotating elements.
Liquid Sloshing	Torques due to fluid motion and variation in center-of-mass location	Dependent on specific design; may be controlled by bladders or baffles.
Dynamics of Flexible Bodies	Oscillatory resonance at bending frequencies, limiting control bandwidth	Depends on spacecraft structure.
Thermal Shocks on Flexible Appendages	Attitude disturbances when entering/leaving eclipse	Depends on spacecraft structure. Worst for gravity gradient systems with long inertia booms.

11.1.4 Select and Size ADCS Hardware

We are now ready to evaluate and select the individual ADCS components.

Actuators. We first discuss the actuators, as summarized in Table 11-11, beginning with reaction and momentum wheels. *Reaction wheels* are essentially torque motors with high-inertia rotors. They can spin in either direction, and provide one axis of control for each wheel. *Momentum wheels* are reaction wheels with a nominal spin rate above zero to provide a nearly constant angular momentum. This momentum provides gyroscopic stiffness to two axes, while the motor torque may be controlled to precisely point around the third axis.

In sizing wheels, it is important to distinguish between cyclic and secular disturbances, and between angular momentum storage and torque authority. For 3-axis control systems, cyclic torques build up cyclic angular momentum in reaction wheels, as the wheels provide compensating torques to keep the vehicle from moving. We typically size the *angular momentum capacity* of a reaction wheel (limited by its saturation speed) to handle the cyclic storage during an orbit without the need for frequent momentum dumping. Thus, the average disturbance torque for 1/4 or 1/2 orbit determines the minimum storage capability. The secular torques and our total storage capacity then define how frequently angular momentum must be dumped.

The *torque capability* of the wheels usually is determined by slew requirements or the need for control authority above the peak disturbance torque in order for the wheels to maintain pointing accuracy.

For 3-axis control, at least three wheels are required with their spin axes not coplanar. Often, a fourth redundant wheel is carried in case one of the three primaries

TABLE 11-11.   **Typical ADCS Actuators.** Actuator weight and power usually scale with performance.

Actuator	Typical Performance Range	Weight (kg)	Power (W)
Thrusters *Hot Gas (Hydrazine)* *Cold Gas*	0.5 to 9,000 N* < 5 N*	Variable† Variable†	N/A† N/A†
Reaction and Momentum Wheels	0.4 to 400 N·m·s for momentum wheels at 1,200 to 5,000 rpm; max torques from 0.01 to 1 N·m	2 to 20	10 to 110
Control Moment Gyros (CMG)	25 to 500 N·m of torque	> 10	90 to 150
Magnetic Torquers	1 to 4,000 A·m²‡	0.4 to 50	0.6 to 16

*   Multiply by moment arm (typically 1 to 2 m) to get torque.
†   Chap. 17 discusses weight and power for thruster systems in more detail.
‡   For 700-km orbit and maximum Earth field of 0.4 gauss, the maximum torques would be 4.5×10^{-5} N·m to 0.18 N·m (see Table 11-9B).

fails. If the wheels are not orthogonal (and the redundant one never is), additional torque and momentum authority may be necessary to compensate for the unfavorable geometry. It is also common to use wheels larger than the minimum required in order to use a standard component.

For spin-stabilized or momentum-bias systems, the cyclic torques will cause cyclic rates, while the secular torques cause gradual divergence. We typically design the stored angular momentum, determined by spin rate and inertia of the spinning body, to be large enough to keep the cyclic motion within our pointing specification without active control during an orbit. Periodic torquing will still be required to counteract the secular disturbances. The more angular momentum in the body, the more resistant it is to external torques. An upper limit on the stored momentum, if one exists, may be defined by the fuel cost to precess this angular momentum.

For high-torque applications, *control-moment gyros* may be used instead of reaction wheels. These are single- or double-gimbaled wheels spinning at constant speed. By turning the gimbal axis, we can obtain a high-output torque whose size depends on the speed of the rotor and the gimbal rate of rotation. Control systems with control moment gyros can produce large torques about all three of the spacecraft's orthogonal axes, so we most often use them for agile (high-rate) maneuvers. They require a complex control law and momentum exchange for desaturation. Other disadvantages are high cost and weight.

Spacecraft also use *magnetic torquers* as actuation devices. These torquers use magnetic coils or electromagnets to generate magnetic dipole moments. Magnetic torquers can compensate for the spacecraft's residual magnetic fields or attitude drift from minor disturbance torques. They also can desaturate momentum-exchange systems but usually require much more time than thrusters. A magnetic torquer produces torque proportional (and perpendicular) to the Earth's varying magnetic field. Electromagnets have the advantage of no moving parts, requiring only a magnetometer for field sensing and a wire-wound, electromagnetic rod in each axis. Because they use the Earth's natural magnetic fields, they are less effective at higher orbits. We can easily specify the rod's field strength in amp·turn·m² and tailor it to any application. Table 11-12 describes sizing rules of thumb for wheels and magnetic torquers.

TABLE 11-12. Simplified Equations for Sizing Reaction Wheels, Momentum Wheels, and Magnetic Torquers. FireSat momentum wheels are sized for the baseline requirements. Reaction wheels are sized for the optional design with 30-deg slew requirement.

Parameter	Simplified Equations	Application to FireSat Example
Torque from Reaction Wheel for Disturbance Rejection	Reaction-wheel torque must equal worst-case anticipated disturbance torque plus some margin: $T_{RW} = (T_D)\,(\text{Margin Factor})$	For the example spacecraft, $T_D \approx 4.5 \times 10^{-5}$ N·m (Table 11-9). This is below almost all candidate reaction wheels. We will select a wheel based on storage requirements or slew torque, not disturbance rejection. See below.
Slew Torque for Reaction Wheels	For max-acceleration slews (1/2 distance in 1/2 time): $$\frac{\theta}{2} = \frac{1}{2}\frac{T}{I}\left(\frac{t}{2}\right)^2$$	For the 30-deg slews of the 90 kg·m^2 spacecraft (Fig. 11-1) in 10 min, this becomes: $$T = 4\theta\frac{I}{t^2} = \frac{4 \times 30\ \text{deg} \times (\pi/180\ \text{deg}) \times 90\ \text{kg·m}^2}{(600\ \text{sec})^2}$$ $$= 5.2 \times 10^{-4} \text{N·m}$$ This is also a small value.
Momentum Storage in Reaction Wheel	One approach to estimating wheel momentum, h, is to integrate the worst-case disturbance torque, T_D, over a full orbit. If the disturbance is gravity gradient, the maximum disturbance accumulates in 1/4 of an orbit. A simplified expression for such a sinusoidal disturbance is: $$h = (T_D)\frac{\text{Orbital Period}}{4}(0.707)$$ where 0.707 is the rms average of a sinusoidal function.	For $T_D = 4.5 \times 10^{-5}$ N·m (Table 11-9B) and a 700-km orbital period of 98.8 min $$h = (4.5 \times 10^{-5}\,\text{N·m})\left(\frac{98.8\ \text{min}}{4}\right)\left(\frac{60\ \text{sec}}{\text{min}}\right)(0.707)$$ $$= 4.7 \times 10^{-2} \text{N·m·s}$$ A small reaction wheel which gives us storage of 0.4 N·m·s would be sufficient. It provides a margin of > 9 in storage for the worst-case torques.
Momentum Storage in Momentum Wheel	Roll and yaw accuracy depend on the wheel's momentum and the external disturbance torque. A simplified expression for the required momentum storage is: $$T \times \frac{P}{4} = h\theta_a$$ T = torque P = orbit period h = angular θ_a = allowable momentum motion	The value of h for a 0.1 deg yaw accuracy would be $$h = \frac{(4.5 \times 10^{-5}\,\text{N·m}) \times 1482\ \text{sec}}{0.1 \times \dfrac{\pi}{180\ \text{deg}}}$$ $$= 38.2\ \text{N·m·s}$$ T_D is from Table 11-9A. For a 1 deg accuracy, we would need only 3.8 N·m·s
Momentum Storage in Spinner	Same as for a momentum wheel, but with the spin rate: $$\omega_s = \frac{h}{I}$$	For the 0.1 deg accuracy, the spin rate is: $$\omega_s = \frac{(37.3)\text{N·m·s}}{90\ \text{kg·m}^2} = 0.42\ \text{rad/sec} = 4.1 \text{rpm}$$
Torque from Magnetic Torquers	Magnetic torquers use the Earth's magnetic field, B, and electrical current through the torquer to create a magnetic dipole (D) that results in torque (T) on the vehicle: $$D = \frac{T}{B}$$ Magnets used for momentum dumping must equal the peak disturbance + margin to compensate for the lack of complete directional control.	Table 11-9B estimates the worst-case Earth field, B, to be 4.5×10^{-5} tesla. We calculate the torque rod's magnetic torquing ability (dipole) to counteract the worst-case gravity gradient disturbance, T_D, of 4.5×10^{-5} N·m as $$D = \frac{T}{B} = \frac{4.5 \times 10^{-5}\,\text{N·m}}{4.5 \times 10^{-5}\,\text{tesla}} = 1\,\text{A·m}^2$$ which is a small actuator. The Earth's field is cyclic at twice orbital frequency; thus, maximum torque is available only twice per orbit. A torquer of 3 to 10 A·m^2 capacity should provide sufficient margin.

Note: For actuator sizing, the magnitude and direction of the disturbance torques must be considered. In particular, momentum accumulation in inertial coordinates must be mapped to body-fixed wheel axes, where necessary.

Gas jets or *thrusters* produce torque by expelling mass, and are not governed by the same concerns as momentum storage devices. We consider them to be a *hot-gas system*, either bipropellant or monopropellant, when a chemical reaction produces the energy. They are a *cold-gas system* when energy comes from the latent heat of a phase change or from the work of compression without a phase change. Cold-gas systems usually apply to small spacecraft and low-impulse requirements.

Thrusters produce torques and forces that:

- Control attitude
- Control nutation
- Maneuver spacecraft over large angles
- Adjust orbits
- Control the spin rate
- Dump extra momentum from a momentum wheel, reaction wheel, or control moment gyro

Unfortunately, their plumes may impinge on the spacecraft, contaminating surfaces, and they require expendable propellant, dictating spacecraft life. An advantage is that they can provide large, instantaneous torques at any point in the orbit.

We must decide whether we need thrusters, how many we need, and where to locate them. For applications that demand fine control from the thrusters, we may have to specify the minimum impulse from a single thruster pulse—usually 20 ms or greater. Single thrust levels are usually used, unless the complication of dual or variable thrust is required.

Although the baseline FireSat spacecraft will use magnetic torquers, we illustrate the thruster sizing calculations for momentum dumping and the optional slew requirement. We will assume the thruster's moment arm is 0.5 m. Table 11-13 gives procedures and simplified equations, where applicable, for sizing thrusters and estimating propellant. Refer to Chap. 17 for a thorough discussion of propulsion subsystems.

The size of the thrusters and required propellent are small for this example. For the optional system with reaction wheels, slewing can be accomplished with the wheels, avoiding use of propellent. For the baseline momentum bias system, we would use thrusters for the optional slews, though large electromagnets could be used if thrusters were not available and maneuver time were not important.

Sensors. We complete this hardware unit by selecting the sensors needed for control. Consult Table 11-14 for a summary of typical devices, as well as their performance and physical characteristics. Note, however, that sensor technology is evolving rapidly, promising more accurate, lighter-weight sensors for future mission.

Sun sensors are visible-light detectors which measure one or two angles between their mounting base and incident sunlight. They are popular, accurate and reliable, but require clear fields of view. They can be used as part of the normal attitude determination system, part of the initial acquisition or failure recovery system, or part of an independent solar array orientation system. Since most low-Earth orbits include eclipse periods, Sun-sensor-based attitude determination systems must provide some way of tolerating the regular loss of this data without violating pointing constraints.

Sun sensors can be quite accurate (< 0.01 deg) but it is not always possible to take advantage of that feature. We usually mount Sun sensors near the ends of the vehicle to obtain an unobstructed field of view. Sun sensor accuracy can be limited by structural bending on large spacecraft. Spinning satellites use specially designed Sun sensors that measure the angle of the Sun with respect to the spin axis of the vehicle. The data may be sent to the ground for processing or used in a closed-loop control system on board the vehicle.

TABLE 11-13. Simplified Equations for Preliminary Sizing of Thruster Systems. FireSat thruster requirements are small for this low-disturbance, minimal slew application.

Simplified Equations	Application to FireSat Example
Thruster force level sizing for external disturbances: $F = T_D / L$ F is thruster force, T_D is worst-case disturbance torques, and L is the thruster's moment arm	For the worst case T_D of 4.5×10^{-5} N·m (Table 11-7) and a thruster moment arm of 0.5 m $$F = \frac{4.5 \times 10^{-5} \text{ N·m}}{0.5 \text{ m}} = 9.0 \times 10^{-5} \text{ N}$$ This small value indicates slewing rate, not disturbances, will more likely determine size. Also, using thrusters to fight cyclic disturbances uses much fuel.
Sizing force level to meet slew rates (optional zero momentum system): Determine highest slew rate required in the mission profile. Develop a profile that accelerates the vehicle to that rate, coasts, then decelerates. We calculate the thruster force from the acceleration value using the following relationships: $T = F L = I \ddot{\theta}$ Solve for F	Assume a 30-deg slew in less than 1 min (60 sec), accelerating for 5% of that time, coasting for 90%, and decelerating for 5%. $$\text{Rate}(\dot{\theta}) = 30 \text{ deg} / 60 \text{ sec} = 0.5 \text{ deg/sec}$$ To reach 0.5 deg/s in 5% of 1 min, which is 3 sec, requires an acceleration $$(\ddot{\theta}) = \frac{\dot{\theta}}{t} = \frac{0.5 \text{ deg/sec}}{3 \text{ sec}} = 0.167 \text{ deg/sec}^2 = 2.91 \times 10^{-3} \text{ rad/sec}^2$$ $$F = \frac{I\ddot{\theta}}{L} \frac{(90 \text{ kg·m}^2)(2.91 \times 10^{-3} \text{ rad/sec}^2)}{0.5 \text{ m}} = 0.52 \text{ N}$$ This is small but feasible.
Sizing force level for slewing a momentum-bias vehicle: The applied torque T is $T = FLd = h\omega$ where F = average thruster force L = moment arm d = thruster duty cycle (fraction of spin period) h = angular momentum ω = slew rate	For FireSat, allowing 10 min for a 30-deg slew, with 10% duty cycle $$= \frac{h\omega}{Ld} = \frac{(38.2 \text{ N·m·s})\left(\dfrac{30 \text{ deg}}{600 \text{ sec}}\right) \times \dfrac{\pi}{180 \text{ deg}}}{(0.5 \text{ m})(0.1)}$$ $$= 0.67 \text{ N}$$
Thruster pulse life: Develop detailed maneuver profile from mission sequence of events and determine pulse number and length for each segment	Assume example spacecraft uses thrusters only for large mission maneuvers and momentum dumping for 1 sec each wheel once a day. The large maneuver, 30 deg, in 2 axes each week includes a 3-sec acceleration pulse and a 3-sec deceleration pulse. Total Pulses = 2 pulses (start & stop) × 2 axes × 12/yr × 5 yr (maneuver) + 1 pulse × 3 wheels × 365 days/yr × 5 yr (momentum dump) = 240 + 5,475 = 5,715 pulses This is below the typical 20,000 to 50,000 pulse ratings for small thrusters.

TABLE 11-13.   Simplified Equations for Preliminary Sizing of Thruster Systems. **(Continued)** FireSat thruster requirements are small for this low-disturbance, minimal slew application.

Simplified Equations	Application to FireSat Example
Sizing force level for momentum dumping: $$F = \dfrac{h}{Lt}$$ where h = stored momentum (from wheel capacity or disturbance torque × time) L = moment arm t = burn time	For FireSat with 0.4 N·m·s wheels and 1-sec burns, $$F = \frac{0.4\,\text{N·m·s}}{(0.5\,\text{m}) \times (1\,\text{sec})} = 0.8\,\text{N}$$
Propellant: Estimate propellant mass (M_p) by determining the total pulse length, t, for the pulses counted above, multiplying by thruster force (F), and dividing by specific impulse (I_{sp}), and g as follows: $$M_p = \frac{Ft}{I_{sp}g}$$	To derive the propellant weight from the pulses above, use 3 sec for on-time for each large maneuver pulse, and 1 sec for each momentum-dump pulse at the computed force levels (actual times will change when a thruster is chosen, but the total impulse will be the same). Total impulse = I = 240 pulses 3 sec/pulse 0.52 N + 5,475 pulses × 1 sec/pulse × 0.8 N = 4,754 N·s then $$M_p = \frac{I}{I_{sp}g} = \frac{4,754\,\text{N·s}}{200\,\text{sec} \times 9.8\,\text{m/sec}^2} = 2.43\,\text{kg}$$ where an I_{sp} of 200 sec for hydrazine is a conservative estimate.

TABLE 11-14.   Typical ADCS Sensors. Sensors have continued to improve in performance while getting smaller and less expensive.

Sensor	Typical Performance Range	Wt Range (kg)	Power (W)
Inertial Measurement Unit (Gyros & Accelerometers)	Gyro drift rate = 0.003 deg/hr to 1 deg/hr, accel. Linearity = 1 to 5 × 10⁻⁶ g/g² over range of 20 to 60 g	1 to 15	10 to 200
Sun Sensors	Accuracy = 0.005 deg to 3 deg	0.1 to 2	0 to 3
Star Sensors (Scanners & Mappers)	Attitude accuracy = 1 arc sec to 1 arc min 0.0003 deg to 0.01 deg	2 to 5	5 to 20
Horizon Sensors • Scanner/Pipper • Fixed Head (Static)	Attitude accuracy: 0.1 deg to 1 deg (LEO) < 0.1 deg to 0.25 deg	1 to 4 0.5 to 3.5	5 to 10 0.3 to 5
Magnetometer	Attitude accuracy = 0.5 deg to 3 deg	0.3 to 1.2	< 1

Star sensors have evolved rapidly in the past few years, and represent the most common sensor for high-accuracy missions. Star sensors can be scanners or trackers. *Scanners* are used on spinning spacecraft. Stars pass through multiple slits in a scanner's field of view. After several star crossings, we can derive the vehicle's attitude. We use *trackers* on 3-axis attitude stabilized spacecraft to track one or more stars to derive 2- or 3-axis attitude information. The most sophisticated units not only track the stars as bright spots, but identify which star pattern they are viewing, and output the sensor's orientation compared to an inertial reference. Putting this software inside the sensor simplifies processing requirements of the remaining attitude control software.

While star sensors excel in accuracy, care is required in their specfication and use. For example, the vehicle must be stabilized to some extent before the trackers can determine where they point. This stabilization may require alternate sensors, which can increase total system cost. Also, star sensors are susceptible to being blinded by the Sun, Moon, or even planets, which must be accommodated in their application. Where the mission requires the highest accuracy and justifies a high cost, we use a combination of star trackers and gyros. These two sensors complement each other nicely: the gyros can be used for initial stabilization, and during periods of sun or moon interference in the trackers, while the trackers can be used to provide a high-accuracy, low frequency, external reference unavailable to the gyros. Work continues to improve the sample rate of star trackers and to reduce their radiation sensitivity.

Horizon sensors are infrared devices that detect the contrast between the cold of deep space and the heat of the Earth's atmosphere (about 40 km above the surface in the sensed band). Simple narrow field-of-view fixed-head types (called *pippers* or *horizon crossing indicators*) are used on spinning spacecraft to measure Earth phase and chord angles which, together with orbit and mounting geometry, define two angles to the Earth (nadir) vector. *Scanning horizon sensors* use a rotating mirror or lens to replace (or augment) the spinning spacecraft body. They are often used in pairs for improved performance and redundancy. Some nadir-pointing spacecraft use *staring sensors* which view the entire Earth disk (from GEO) or a portion of the limb (from LEO). The sensor fields of view stay fixed with respect to the spacecraft. This type works best for circular orbits.

Horizon sensors provide Earth-relative information directly for Earth-pointing spacecraft, which may simplify onboard processing. The scanning types require clear fields of view for their scan cones (typically 45, 60, or 90 deg, half-angle). Typical accuracies for systems using horizon sensors are 0.1 to 0.25 deg, with some applications approaching 0.03 deg. For the highest accuracy in low-Earth orbit, it is necessary to correct the data for Earth oblateness and seasonal horizon variations.

Magnetometers are simple, reliable, lightweight sensors that measure both the direction and size of the Earth's magnetic field. When compared to the Earth's known field, their output helps us establish the spacecraft's attitude. But their accuracy is not as good as that of star or horizon references. The Earth's field can shift with time and is not known precisely in the first place. To improve accuracy, we often combine their data with data from Sun or horizon sensors. When a vehicle using magnetic torquers passes through magnetic-field reversals during each orbit, we use a magnetometer to control the polarity of the torquer output. The torquers usually must be turned off while the magnetometer is sampled to avoid corrupting the measurement.

GPS receivers are commonly known as high-accuracy navigation devices. Recently, GPS receivers have been used for attitude determination by employing the differential signals from separate antennas on a spacecraft. Such sensors offer the promise of low cost and weight for LEO missions, and are being used in low accuracy applications or as back-up sensors. Development continues to improve their accuracy, which is limited by the separation of the antennas, the ability to resolve small phase differences, the relatively long wavelength, and multipath effects due to reflections off spacecraft components.

Gyroscopes are inertial sensors which measure the speed or angle of rotation from an initial reference, but without any knowledge of an external, absolute reference. We use them in spacecraft for precision attitude sensing when combined with external references such as star or sun sensors, or, for brief periods, for nutation damping or attitude control during thruster firing. Manufacturers use a variety of physical phenomena, from simple spinning wheels *(iron gyros* using ball or gas bearings) to *ring lasers, hemispherical resonating surfaces,* and *laser fiber optic bundles.* The gyro manufacturers, driven by aircraft markets, steadily improve accuracy while reducing size and mass.

Error models for gyroscopes vary with the technology, but characterize the deterioration of attitude knowledge with time (degrees per hour or per square-root of time). When used with an accurate external reference, such as star trackers, gyros can provide smoothing (filling in the measurement gaps between star tracker samples) and higher frequency information (tens to hundreds of hertz), while the star trackers provide the low frequency, absolute orientation information that the gyros lack. Individual gyros provide one or two axes of information, and are often grouped together as an *Inertial Reference Unit,* IRU, for three full axes. IRUs with accelerometers added for position/velocity sensing are called *Inertial Measurement Units,* IMUs.

Sensor selection. Sensor selection is most directly influenced by the required orientation of the spacecraft (e.g., Earth- or inertial-pointing) and its accuracy. Other influences include redundancy, fault tolerance, field of view requirements, and available data rates. Typically, we identify candidate sensor suites and conduct a trade study to determine the best, most cost-effective approach. In such studies, the existence of off-the-shelf components and software can strongly influence the outcome. In this section we will only briefly describe some selection guidelines.

Full 3-axis knowledge requires at least two external vector measurements, although we use inertial platforms or spacecraft angular momentum (from spinning or momentum wheels) to hold the attitude between external measurements. In some cases, if attitude knowledge can be held for a fraction of an orbit, the external vectors (e.g., Earth or magnetic) will have moved enough to provide the necessary information.

For Earth-pointed spacecraft, horizon sensors provide a direct measurement of pitch and roll axes, but require augmentation for yaw measurements. Depending on the accuracy required, we use Sun sensors, magnetometers, or momentum-bias control relying on roll-yaw coupling for the third degree of freedom. For inertially-pointing spacecraft, star and Sun sensors provide the most direct measurements, and inertial platforms are ideally suited. Frequently, only one measurement is made in the ideal coordinate frame (Earth or inertial), and the spacecraft orbit parameters are required in order to convert a second measurement or as an input to a magnetic field model. The parameters are usually uplinked to the spacecraft from ground tracking, but autonomous navigation systems using GPS are also in use (see Sec. 11.7).

FireSat sensors. The external sensors for FireSat could consist of any of the types identified. For the 0.1 deg Earth-relative pointing requirement, however, horizon sensors are the most obvious choice since they directly measure two axes we need to control. The accuracy requirement makes a star sensor a strong candidate as well, although its information needs to be transformed to Earth-relative pointing for our use. The 0.1 deg accuracy is at the low end of horizon sensors' typical performance, and we need to be careful to get the most out of their data.

TABLE 11-15. FireSat Spacecraft Control Components Selection. A simple, low-cost suite of components fits FireSat's needs.

Type	Components	Rationale
Actuation Devices	(1) Momentum Wheel	• Pitch axis torquing • Roll and yaw axis passive stability
	(3) Electromagnets	• Roll and yaw control • Pitch wheel desaturation
Sensors	Horizon Sensor	• Provide basic pitch and roll reference • Can meet 0.1 deg accuracy • Lower weight and cost than star sensors
	Sun Sensors	• Initially acquire vehicle attitude from unknown orientation • Coarse attitude data • Fine data for yaw
	Magnetometer	• Coarse yaw data

TABLE 11-16. FireSat Spacecraft Control Subsystem Characterized. The baseline ADCS components satisfy all mission requirements, with thrusters available if required.

Components	Type	Weight (kg)	Power (W)	Mounting Considerations
Momentum Wheel	Mid-size, 40 N·m·s momentum	< 5 total, with drive electronics	10 to 20	Momentum vector on pitch axis
Electromagnets	3, 10 A·m² capacity each	2, including current drive electronics	5 to 10	Orthogonal configuration best to reduce cross-coupling
Sun Sensors	4 wide-angle coarse sensors providing 4 π steradian coverage; ≈ 0.1 deg accuracy	< 1 total	0.25	Free of viewing obstructions and reflections
Horizon Sensors	Scanning type (2) plus electronics; 0.1 deg accuracy	5 total	10	Unobstructed view of Earth's horizon
Optional Thrusters	Hydrazine; 0.5 N force	Propellant weight depends on mission	N/A	Alignments and moment arm to center of gravity are critical
Magnetometer	3-axis	< 1	5	Need to isolate magnetometer from electromagnets, either physically or by duty-cycling the magnets.

We assume we also need a yaw sensor capable of 0.1 deg, and this choice is less obvious. (Often, it is useful to question a tight yaw requirement. Many payloads, e.g., antennas, some cameras, radars, are not sensitive to rotations around their pointing axis. For this discussion, we will assume the requirement is firm.) We could use sun sensors, but their data needs to be replaced during eclipses. Magnetometers don't have the necessary accuracy on their own, but with our momentum-bias system, roll-yaw coupling, and some yaw filtering, a magnetometer-Sun sensor system should work.

At this point, we consider the value of an inertial reference package. Such packages, although heavy and expensive, provide a short-term attitude reference which would permit the Earth vector data to be used for full 3-axis knowledge over an orbit. A gyro package would also reduce the single-measurement accuracy required of the horizon sensors, simplifying their selection and processing. Such packages are also useful in the control system if fast slews are required. Although nice to have, an inertial package does not seem warranted for FireSat. Table 11-15 summarizes our hardware selections.

Once the hardware selection is complete, it must be documented for use by other system and subsystem designers as follows:

- Specify the power levels and weights for each assembly

- Establish the electrical interface to the outside world

- Describe requirements for mounting, alignment, heating, or coding

- Determine what telemetry data we must process

- Document how much software we need to support equations of motion

Specific numbers depend on the vendors selected. A typical list for FireSat might look like Table 11-16, but the numbers could vary considerably with only slight changes in subsystem accuracies or slewing requirements.

ADCS Vendors. Typical suppliers for ADCS components are listed in Table 11-17.

TABLE 11-17.　ADCS Component Suppliers. Aerospace mergers can result in sudden name changes.

| Company | Sensors | | | | | Actuators | | | |
	Sun	Earth	Magnet-ometers	Star	Inertial (Gyros)	Momentum/ Reaction Wheels	CMGs	Electro-magnets	Thrus-ters
Adcole Corporation	✔								
AlliedSignal					✔	✔			
Ball Aerospace and Technologies Corp.	✔			✔					
Billingsley Magnetics			✔						
CAL Corporation				✔					
EDO (Barnes) Corp.	✔	✔							
Honeywell Space Systems					✔	✔	✔		
Ithaco Space Systems Inc.	✔	✔	✔			✔		✔	
Kearfott Guidance & Navigation Corporation					✔	✔			
Litton Industries					✔				
Lockheed Martin	✔				✔				
Kaiser Marquardt Corp.									✔
Matra Marconi Space						✔			
Meda, Inc.			✔						
Microcosm, Inc.								✔	
Primex Technologies									✔
Raytheon Systems, Inc.				✔					
Servo Corp. of America		✔							
Smiths Industries Defense Systems					✔				
Teldix GmbH						✔			
TRW									✔
Watson Industries, Inc.			✔		✔				

11.1.5 Define the Control Algorithms

Finally, we must tie all of the control components together into a cohesive system. We begin with a block diagram for a single-axis control system (See Fig. 11-3). As we refine the design, we add or modify feedback loops for rate and altitude data, define gains and constants, and fine tune the equations of motion. To do so, we need good mathematical simulations of the entire system, including internal and external disturbances. Usually, linear differential equations with constant coefficients describe the dynamics of a control system, thus allowing us to analyze its performance with the highly developed tools of linear servomechanism theory. With these same tools, we can easily do linear compensation to satisfy specifications for performance.

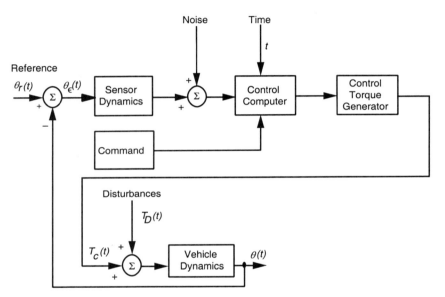

Fig. 11-3.   Diagram of a Typical Attitude Control System with Control Along a Single Axis. Control algorithms are usually implemented in an onboard processor and analyzed with detailed simulations.

We typically apply linear theory only to preliminary analysis and design. As the design matures, nonlinear effects come strongly into play. These effects may be inherent or intentionally introduced to improve the system's performance. Feedback control systems are of two kinds, based on the flow of their control signals. They are *continuous-data* systems when their control signals flow continuously without interruption; they are *sampled-data* systems when sampling occurs at equal intervals. Most modern spacecraft process data through digital computers and therefore use control systems that sample data.

Although it is beyond the scope of this handbook to provide detailed design guidance on feedback control systems, the system designer should recognize the interacting effects of attitude control system loop gain, capability of the attitude control system to compensate for disturbances, accuracy of attitude control, and control system bandwidth.

Three-axis stabilization. Different types of active control systems have different key parameters and algorithms. Frequently, 3-axis control can be decoupled into three independent axes. The most basic design parameter in each axis is its *position gain*, K_p.[*] This is the amount of control torque which results from a unit attitude error and can be expressed in N·m/deg or N·m/rad. The position gain is selected by the designer and must be high enough to provide the required attitude control accuracy in the presence of disturbances, or $K_p \geq T_D/\theta_\varepsilon$, where K_p is position gain, T_D is peak disturbance torque, and θ_ε is allowable attitude error.

The value of the position gain also determines the attitude control system bandwidth and speed-of-response. The *bandwidth* is given by $\omega_n = (K_p/I)^{1/2}$, where I is the spacecraft moment of inertia. The bandwidth defines the frequency at which control authority begins to diminish. Attitude control and disturbance rejection are effective from 0 frequency (d.c.) up to the bandwidth. *Speed of response* is approximately the reciprocal of bandwidth. Note that position gain is inversely proportional to allowable error and bandwidth is proportional to the square root of position gain. Therefore, high accuracy implies high position gain and high bandwidth. However, high bandwidth may cause bending resonances to affect control system performance.

With the relations given, the system designer can estimate required position gain from his estimates of disturbance torque and accuracy requirements. He can use this estimate to compute control system bandwidth. This allows him to specify minimum bending frequencies as discussed below.

In defining algorithms for the control system, we must also consider whether the vehicle will have flexible-body effects that can make the vehicle unstable. Spacecraft with flexible appendages such as antennas, booms, and solar panels may produce slight warping at their natural frequencies. Control torques and external-disturbance torques will cause structural vibrations, in some cases close to or within the control system's bandwidth. The lowest natural frequencies of flexible components should be at least an order of magnitude greater than the rigid-body frequencies before we can neglect flexibility. For further discussion of how structural flexibility affects the control subsystem, see Sec. 3.12 of Agrawal [1986].

Spin stabilization and momentum bias. The fundamental concept in spin stabilization is the nutation frequency of the vehicle. For a spinning body, the inertial nutation frequency (ω_{ni}) is equal to

$$\omega_{ni} = \frac{I_s}{I_T}\omega_s \qquad (11\text{-}1)$$

where I_s is the spin axis inertia, I_T is the transverse axis inertia, and ω_s is the spin frequency.

For a momentum-bias vehicle with a stable body and a momentum wheel (or a dual-spin vehicle with a stable platform and a spinning rotor), the nutation frequency is

$$\omega_{ni} = \frac{h}{I_T} \qquad (11\text{-}2)$$

---

[*] In its simplest form, a spacecraft attitude control system can be represented in s-domain as a $1/I_s^2$ plant and may be controlled by a *proportional plus derivative* (PD) *controller* where $T_c \equiv K_p\,\theta_\varepsilon + K_r\,\dot{\theta}_\varepsilon$. The position gain, K_p, controls system bandwidth and the rate gain, K_r, controls damping.

where h is the angular momentum of the spinning body. Thus, spacecraft with large inertias and small wheels have small nutation frequencies (i.e., long periods).

Attempting to move the vehicle with a bandwidth faster than the nutation frequency causes it to act more like a 3-axis vehicle. In general, we attempt to control near the nutation frequency or slower, with correspondingly small torques. In this area, the vehicle acts like a gyroscope, with the achieved angular rate, ω, proportional to the applied torque, T:

$$\omega = T/h \qquad (11\text{-}3)$$

where h is the system angular momentum.

A lower limit on control bandwidth is usually provided by the orbit rate ω_o, which for a circular orbit is

$$\omega_0 = \sqrt{\mu / r^3} \qquad (11\text{-}4)$$

where $\mu = 3.986 \times 10^{14}$ m^3/s^2 and r is the orbit radius.

Attitude determination. A full discussion of determination algorithms requires a dedicated reference such as Wertz [1978]. We will highlight only some of the basic concepts.

The basic algorithms for determination depend on the coordinate frames of interest (e.g., the sensor frames, local vertical frame, or Earth-centered inertial frame), and the geometry of the measurements, parameterized by *Euler angles* (such as roll, pitch, and yaw) or *quaternions* (which are scaled vectors for Eigen-axis rotations of coordinate frames). Inertial platforms and star sensor data usually are suited to inertial quaternions, while Earth-pointing spacecraft often use a local-vertical, aircraft-like set of Euler angles.

Simple spacecraft may use the sensor readings directly for control, while more complex vehicles or those with higher accuracy requirements employ some form of averaging, smoothing, or Kalman filtering of the data. The exact algorithms depend on the vehicle properties, orbit, and sensor types used.

FireSat algorithms. For our momentum-bias FireSat example, control separates into pitch-axis control using torque commands to the momentum wheel, and roll-yaw control using current commands to the electromagnets. The pitch-wheel desaturation commands must also be fed (at a slow rate) to the magnets. The pitch-wheel control is straightforward, using proportional-plus derivative and, optionally, integral control. The roll-yaw control design starts by using the linearized nutation dynamics of the system, and is complicated by the directional limitations of electromagnetic torque (the achievable torque is perpendicular to the instantaneous Earth magnetic field).

The nadir-oriented control system may use an Earth-referenced, aircraft-like Euler angle (roll, pitch, yaw) set, although quaternions should also be considered for their lack of singularities during off-nominal pointing. The horizon sensors directly read two of the angles of interest, pitch and roll. Yaw needs to be measured directly from Sun position (during orbit day) or from the magnetometer readings (using a stored model of the Earth's field), or inferred from the roll-yaw coupling described earlier. The magnetic field and Sun information require an uplinked set of orbit parameters, and increase the computational requirements of the subsystem. Overall, meeting the 0.1 deg yaw requirement when the Sun is not visible will be the toughest challenge facing the ADCS designer, and a form of coasting through the blackouts, without direct roll-yaw control, may be most appropriate.

11.2 Telemetry, Tracking, and Command

Douglas Kirkpatrick, *United States Air Force Academy*
Adapted from SMAD II, Sec. 11.2 "Communications," by John Ford

The *telemetry, tracking, and command* (*TT&C*) or *communications subsystem* provides the interface between the spacecraft and ground systems. Payload mission data and spacecraft housekeeping data pass from the spacecraft through this subsystem to operators and users at the operations center. Operator commands also pass to the spacecraft through this subsystem to control the spacecraft and to operate the payload. We must design the hardware and their functions to pass the data reliably for all the spacecraft's operating modes. For a discussion of how we collect and manipulate housekeeping and payload data, see Sec. 11.3, Chap. 9, and Chap. 16. Chapter 13 discusses the communication link design, and Morgan and Gordon [1989] provide a wealth of information on spacecraft communications.

The subsystem functions include the following:

- Carrier tracking (lock onto the ground station signal)

- Command reception and detection (receive the uplink signal and process it)

- Telemetry modulation and transmission (accept data from spacecraft systems, process them, and transmit them)

- Ranging (receive, process, and transmit ranging signals to determine the satellite's position)

- Subsystem operations (process subsystem data, maintain its own health and status, point the antennas, detect and recover faults.)

Table 11-18 presents specific subfunctions to accomplish these main functions. Subsystem designers must ensure that all of these functions operate reliability to accomplish the spacecraft mission.

As part of carrier tracking, most satellite TT&C subsystems generate a downlink RF signal that is phase coherent to the uplink signal. *Phase coherence* means that we transmit the downlink carrier so its phase synchronizes with the received phase of the uplink carrier. This process is the *coherent turnaround* or *two-way-coherent mode*. The coherent turnaround process creates a downlink carrier frequency precisely offset from the uplink carrier by a predefined numerical *turnaround ratio*. This is the ratio of the downlink carrier frequency to the uplink carrier frequency. This operational mode can only exist when the transmitter phase-locks to the received uplink carrier. For a given uplink signal, the downlink signal has a constant phase difference. For NASA's GSTDN-compatible transponders, the receiver downcoverts the uplink carrier, and creates a voltage such that the receiver's voltage-controlled oscillator runs at precisely 2/221 times the uplink carrier frequency. The oscillator frequency goes to the transmitter which multiplies it by a factor of 120. Therefore, the composite transmitter downlink is $120 \times 2/221 = 240/221$ times the uplink frequency, which is the turnaround ratio for NASA-compatible transponders. The turnaround ratio for transponders compatible with SGLS is 256/205. The two-way-coherent mode allows the ground station to know more exactly the downlink signal's frequency and to measure the Doppler shift, from which it computes the *range rate* or line-of-sight velocity between the spacecraft and the tracking antenna. This knowledge allows operators to

TABLE 11-18.   What a TT&C Subsystem Does. These are the functions of a communications subsystem connecting the satellite to the ground or other satellites. In a broad sense the communications subsystem receives signals from Earth or another satellite and transmits signals to Earth or another satellite.

Specific Functions
• **Carrier Tracking**
– 2-way coherent communication (downlink frequency is a ratio of the uplink frequency)
– 2-way noncoherent communication
– 1-way communication
• **Command Reception and Detection**
– Acquire and track uplink carrier
– Demodulate carrier and subcarrier
– Derive bit timing and detect data bits
– Resolve data-phase ambiguity if it exists
– Forward command data, clock, and in-lock indicator to the subsystem for command and data handling
• **Telemetry Modulation and Transmission**
– Receive telemetry data streams from the command and data handling subsystem or data storage subsystem
– Modulate downlink subcarrier and carrier with mission or science telemetry
– Transmit composite signal to the ground station or relay satellite
• **Ranging**
– Detect and retransmit ranging pseudorandom code or ranging tone signals
– Retransmit either phase coherently or noncoherently
• **Subsystem Operations**
– Receive commands from the subsystem for command and data handling
– Provide health and status telemetry to the C&DH subsystem
– Perform antenna pointing for any antenna requiring beam steering
– Perform mission sequence operations per stored software sequence
– Autonomously select omni-antenna when spacecraft attitude is lost
– Autonomously detect faults and recover communications using stored software sequence

scan fewer frequencies and thus, acquire the spacecraft more quickly. Deep-space imaging, data collection, and low-Earth orbit spacecraft best illustrate this advantage. These spacecraft typically have large volumes of data and a short field-of-view time to the ground station. To receive maximum data at the ground station on a direct down-link at the spacecraft's highest rate, operators must acquire the downlink signal in the minimum time. Also, if they use ranging for navigation, they can calculate range-rate information from the Doppler shift of the coherent signal.

Occasionally a TT&C subsystem, operating in the two-way coherent mode, loses lock on the uplink signal. At this point, the spacecraft's transmitter autonomously changes the references for the downlink carrier from the receiver's voltage-controlled oscillator to the subsystem's master oscillator. This process creates a unique downlink frequency which is no longer synchronous with the uplink carrier. This TT&C mode is two-way noncoherent communications.

For the *ranging function* (i.e., determining the *range* or line-of-sight distance), the ground station may use the ranging method of navigation to track a spacecraft.

Depending on the communication standard, the ground station modulates a pseudo-random code, tones, or both onto the command uplink signal. The TT&C subsystem's receiver detects the code or tones and retransmits them on the telemetry carrier back to the ground station. From the turnaround time of the *ranging code* or tones traveling to and from the spacecraft, the system determines the range. If the downlink carrier's phase is coherent with the uplink carrier (two-way coherent mode), we can measure the Doppler-frequency shift on the downlink carrier signal and thus obtain range-rate information. Pointing information from the ground station's directional antenna allows us to determine the satellite's azimuth and elevation angles.

Under subsystem operations, the TT&C subsystem performs antenna pointing for any antenna that requires beam steering. Closed-loop antenna pointing requires special autotracking equipment. This equipment generates error signals for the guidance, navigation, and control subsystem, so it can point the antenna. Monopulse and conical-scan systems are the most common ways of generating pointing error signals. Monopulse systems use a monopulse feed that generates difference patterns with nulls on the axis of the azimuth and elevation planes. *Conical-scan systems* rotate the received beam about its axis by a small angle. The rise and fall of the received signal amplitude per revolution indicates the pointing error. By correlating the feed position with the position where the signal is at maximum amplitude, the system generates error signals for the control subsystem to point the antenna. We can use open-loop antenna pointing when we know the spacecraft antenna's position and the direction to the receiver station.

Also under subsystem operations, the TT&C subsystem may do sequences of mission commands or respond to autonomous commands, such as putting itself in a safe mode and routing the omni-antenna to the active receiver. For certain failure scenarios, the subsystem may also execute fault-detection and recovery operations through a stored software sequence.

To create a robust TT&C subsystem, we must consider and satisfy three parts of satellite design: requirements, constraints, and regulations. The requirements come from a variety of sources and form the basis for the mission in which this subsystem plays a key role. Typically TT&C requirements include:

- Type of signals (voice, television, and data)

- Capacity (number of channels and bandwidth)

- Coverage area & ground site locations (local, regional, national, international)

- Link signal strength (usually derived from ground terminal type)

- Connectivity (crosslinks, relay ground stations, and direct links)

- Availability (link times per day and days per year, outage times)

- Lifetime (mission duration)

See Sec. 11.2.1 for a more thorough discussion of requirements.

Constraints are limits on the TT&C subsystem from various sources. Power constraints come from sizing the spacecraft and the power source (primary batteries, solar panels and secondary batteries, or radioisotope thermoelectric generator). Mass constraints arise from the mass budget, which comes from the mission design and the chosen launch vehicle. The launch vehicle generally limits the total dimensions and mass, so individual subsystems receive their allocation within those limits. The launch vehicle choice also sets the launch vibration and acoustic environment, which places constraints on the fragility of the subsystem. The interference environment further constrains the subsystem. When we choose the orbit, we also set the surrounding interference environment. The owners and developers place cost limits on the total design, which in turn limits each subsystem. These cost constraints typically determine how much new technology and subsystem margin that designers can consider. Many other constraints may arise during design, depending on the mission and the people involved.

For the TT&C subsystem, international law and regulatory agencies impact design significantly. Because all spacecraft communicate with users and operators on the ground, de-conflicting frequencies, orbital locations, and power levels are critical to civilized sharing of limited resources. So, we must apply to the regulatory bodies for:

- Desired communication frequencies (depending on the mission data rate, transmission power available, and altitude)

- Orbital assignment (further than 2 deg from a satellite with the same frequency, if geosynchronous)

- Desired power flux density on surface (depending on our receiver antennas)

The main regulatory agency enforcing standards is the International Telecommunications Union (ITU), which is now part of the United Nations. Within the ITU three bodies regulate the communication allocations: the Consultative Committee on International Telephony, the Consultative Committee on International Radiocommunications, and the International Frequency Registration Board (IFRB). The first two organizations formulate policy and set standards. The IFRB coordinates and approves frequency and orbit requests. Because these agencies are international and the number of requests is large and growing, we must plan years in advance to get approval for our communications request.

Various other bodies exist to assist organizations in coordinating and rationalizing commercial use of the radio frequency spectrum. Three of these are the International Telecommunications Satellite Consortium (INTELSAT), the European Telecommunications Satellite Consortium (EUTELSAT), and the International Maritime Satellite Organization (INMARSAT), which assist their member nations with telecommunications planning.

11.2.1 Requirements

The TT&C subsystem derives its requirements from many sources, such as (1) the mission or science objective (top-level requirements such as architecture, orbit lifetime and environment); (2) the satellite (system-level); (3) the TT&C subsystem (internal requirements); (4) other satellite subsystems; (5) the ground station and any relay satellite (compatibility); and (6) mission operations (the satellite operational modes as a function of time). From these sources come the requirements that drive the subsystem design: (1) data rates (commands and telemetry for health and status or for mission and science needs); (2) data volume; (3) data storage type; (4) uplink and downlink frequencies; (5) bandwidths; (6) receive and transmit power; (7) hardware mass; (8) beamwidth; (9) Effective Isotropic Radiated Power (*EIRP*); and (10) antenna gain/system noise temperature. Table 11-19 shows the effects of these requirements on the TT&C-subsystem design.

TABLE 11-19. TT&C Subsystem Requirements. These are typical system-level requirements that are imposed on the TT&C subsystem. See also Table 11-23. (Courtesy of TRW)

Requirement	Alternative/Considerations	Comments
Data Rates Command	4,000 bps typical, 8–64 bps deep spc	Range 2,000–8,000 bps
Health & status telemetry	8,000 bps is common	40–10,000 bps
Mission/ science	Low < 32 bps Medium = 32 bps–1 Mbps High > 1Mbps–1 Gbps	Mission dependent
Data Volume	Record data, compress data, and transmit during longer windows	Data rate × transmission duration • Shorter duration increases data rate • May require data compression
Data Storage	Solid-state recorders 128 × 10⁶ bits	Policy may dictate all data be stored that is not immediately transmitted Mission may require that data be stored then played back later
Frequency	Use existing assigned frequencies and channels Use systems that are compatible to the existing system	Policy set by FCC, ITU, National Telecommunications & Info. Admin. Refer to the atmospheric frequency absorption charts
Bandwidths	Use C.E. Shannon's theorem to calculate channel capacity; See Chap. 13, Eq. (13-24).	Primary driver is data rate Secondary driver is modulation scheme
Power	Use larger antennas, higher efficiency amplifiers Reconsider data requirements	S/C power may limit size of TT&C system transmitter
Mass	Use TWTAs for higher RF power output to reduce antenna size, reconsider data requirements	S/C TT&C system mass allocation may limit size of antennas
Beamwidth	See Tables 13-14, 13-15, and 13-16 for various antenna types, beam shapes, and beamwidths	Ground coverage area requirements or the radiation footprint on the ground Antenna gain null requirements Antenna pointing error
EIRP (Effective Isotropic Radiated Power) (Transmitter Req't)	For a constant EIRP: As antenna size (gain) increases, the transmitter power requirement decreases	EIRP (dB) = transmitter power + antenna gain – front end losses Min EIRP required = space loss + atmo loss + antenna pointing loss – receiver antenna gain – receiver sensitivity
G/T [Receiver antenna gain/receiver sys noise temp) (Receiver Req't)	See Table 13-10 for various communication system temperatures and G/Ts	G/T is the sensitivity of the receiving station and a common Figure of Merit; for an existing satellite link, a ground station can only vary its antenna gain and system noise temp to improve the system signal-to-noise ratio

Classic trade studies include size of the antenna aperture vs. transmitter power, solid-state amplifiers vs. *traveling-wave tube amplifiers* (*TWTAs*), and spacecraft complexity vs. ground complexity. If we increase an antenna's aperture size, its gain increases; therefore, we may decrease the transmitter's RF output power and still maintain the received signal strength. Unfortunately, large antenna apertures are very heavy and have narrow beamwidths (producing more stringent pointing requirements). As Chap. 13 (Eq. 13-17) shows, the beamwidth decreases with an increase in the antenna's aperture size. Depending on the frequency and gain needed, we commonly decide between solid-state and TWT amplifiers—a system-level trade. To do so, we must assess the effects of the spacecraft's total mass, solar-array size, system reliability, and antenna aperture size. Solid-state amplifiers tend to be more reliable, lighter, and smaller. TWT amplifiers have a lower technology risk (at higher gains) and a higher efficiency. We must use TWTAs when the RF output power requirement is too high at a given frequency for solid-state amplifiers, or when the solid-state amplifier efficiency is too low for our application. At today's technology level, we can design solid-state amplifiers with power levels of 65 W at UHF frequencies, 40 W at S-band, and about 20 W at EHF frequencies.

The old rule-of-thumb was to keep the satellite as simple as possible by moving all the complexity and processing to the ground. With modern processors, however, we can now do a tremendous amount of processing on a satellite. Thus, we can design for lower downlink data rates and simpler ground stations or we can collect more data while not overburdening the TT&C subsystem. The new trend is to process as much information as possible on the satellite whenever the mission or science community do not need the raw data.

At the system level, the TT&C subsystem can interface with a fixed or a mobile ground station, as well as a relay satellite. Table 11-20 lists examples of these systems. We usually select the system-level interfaces when establishing mission, satellite, and operational requirements.

TABLE 11-20. **Options for System-Level Interfaces to the TT&C Subsystem.** Shown below are several interface possibilities for a TT&C subsystem. If the interface is an existing system, we also provide the system's document number. (Courtesy of TRW)

Interface	Example of Systems	Where to Find Subsystem Parameters
Fixed Ground Station	SGLS—S-band system SDLS—Secure 44/20 GHz GSTDN—S-band 9 and 26 m NASA DSN S-, X-band 26, 34, 70 m Mission-dedicated or unique	TOR-0059 (6110-01)-3, Reissue H MIL-STD-1582 }JPL-DSN-810-5, rev D
Mobile Ground Station	GPS—Ground antenna, S-band, SGLS DMSP MK IV van, S-band AFSCN TMGS—7-m diameter, SGLS-compatible Mission-dedicated or unique • Dedicated comm trailer • Transportable tent	Aerospace Report TOR-0059 (6110-01)-3
Relay Satellite	TDRSS—S-band and Ku-band DSCS III—UHF, S-band, SHF	NASA Goddard STDN 101.2 DSCS III Interface Guide, DCA
Future Relay Sat; Sister Sat w/in Mission Constellation	MILSTAR Mission-dedicated or unique	

At the subsystem level, the TT&C subsystem interfaces directly with every subsystem except for propulsion. The interface with guidance, navigation, and control deals primarily with antenna pointing. It uses gimbals, or motorized rotary joints, that move in 1 or 2 axes to steer the spacecraft's narrow-beam antennas. Nongimbaled or fixed directional antennas rely on feed arrays for small movements of the antenna beam or on spacecraft attitude maneuvers for large movements. The World Administrative Radio Conference established antenna-pointing requirements for geostationary satellites. The pointing error must be the smaller value of 10% of the half-power (–3 dB) beamwidth or 0.3 deg. Table 11-21 summarizes the TT&C subsystem's constraints and requirements on other subsystems.

TABLE 11-21.   TT&C Subsystem Constraints and Requirements on Other Subsystems. The TT&C subsystem interfaces with all these subsystems and must reliably pass data back and forth or receive support.

Subsystem	Requirement	Constraint
Attitude Determination and Control	• Antenna pointing requirements for gimbaled antennas (gimbal degrees of freedom, amount of rotation) • Pointing requirements of the lesser of 1/10 of antenna beamwidth or 0.3 deg • Closed-loop pointing requirements (i.e., cross-links requiring autotracking)	• Spacecraft pointing and attitude knowledge for fixed antennas may impact antenna beamwidth requirements • Uncertainty in attitude and pointing estimates for the pointing loss in the link budget
Command and Data Handling	• Command and telemetry data rates • Clock, bit sync, and timing requirements • 2-way comm requirements • Autonomous fault detection and recovery requirements (ROM stored command sequence that automatically selects the backup receiver and omni-antenna) • Command & telemetry electrical interface	• Onboard storage and processing
Electrical Power Subsystem	• Distribution requirements	• Amount and quality of power, including requirements for duty cycle, average, and peak power
Structure/ Thermal	• Heat sinks for traveling wave tube amplifiers • Heat dissipation of all active boxes • Location of TT&C subsystem electronics and antennas (locate comm electronics as close to the antennas as possible to minimize RF cable loss) • A clear field of view and movement for all gimbaled antennas	• Temperature uncertainty on non-oven-controlled frequency sources resulting in frequency uncertainty
Payload	• Requirements for storing mission data • RF and EMC interface requirements (conducted emissions, conducted susceptibility, radiated emissions, radiated susceptibility) • Special requirements for modulation, coding, and decoding	• Maximum data rates for mission or science telemetry • Maximum data volume for mission or science telemetry
Propulsion	• None	• None

The interface with the subsystem for command and data handling passes spacecraft commands and telemetry, as well as the TT&C subsystem's control and reporting of health and status. The interface must allow the system to receive spacecraft commands while transmitting real-time telemetry. It also must permit safing of the subsystem and autonomous fault detection and correction.

The interface with the electrical-power subsystem controls the amount and quality of spacecraft power to the TT&C subsystem. One common design has the electrical-power subsystem deliver +28 Vdc unregulated to the transponders and other active boxes in the TT&C subsystem and +28 Vdc regulated to the TWTAs. This design requires dc-to-dc converters at each piece of equipment to provide the correct voltage changes. Another interface design centralizes the power conversion and conditioning, for the TT&C subsystem's active elements. However, because TWTAs require specific voltage levels (−1,000 Vdc, +1,000 Vdc, and +4,000 Vdc,), centralized power conversion and conditioning is not very common with those types of amplifiers.

The payload interface mainly transfers mission or science telemetry data to either the ground station or a relay satellite. To characterize this interface we must know the data rate, data volume, and any data storage requirements. This interface may have to couple signals between the payload and the TT&C subsystem and to modulate the payload telemetry.

Table 11-22 gives a design process for the TT&C subsystem. Once we state the performance parameters for TT&C and identify the ground and spaceborne equipment, we use the methods in Chap. 13 to determine overall performance of the communication links. We must iterate this process many times within the design team to attain an acceptable spacecraft weight, configuration, and performance level.

TABLE 11-22. Preliminary Design Process for the TT&C Subsystem.

Step	Comments	Where Discussed
1. Determine requirements	Range, orbit and spacecraft geometry Data rate and volume Minimum elevation angle Worst case rain conditions Bit error rate	Table 11-19
2. Select frequency	Typically an existing, assigned frequency	Sec. 11.2.2, Table 13-12
3. Determine required bandwidth	Use Shannon's theorem; primary driver is data rate	Eq. 13-24 (Chap. 13)

4. Do major subsystem trades between:
  − Receiver noise temperature
  − Receiver gain (antenna aperture)
  − Transmitter gain (antenna aperture)
  − Transmitter power
	Use link budget to trade between components	Table 13-13

5. Do major subsystem trades between the TT&C subsystem and other subsystems	Understand the satellite's sensitivity to each TT&C subsystem design feature	Table 11-21
6. Calculate performance parameters	EIRP, G/T, and margin	Tables 11-19, 13-12
7. Estimate subsystem weight and power	Use analogy with existing systems	Table 11-26
8. Document reasons for selection	Important to document assumptions	N/A

11.2.2 Designing the TT&C Subsystem

Table 11-23 lists parameters for the TT&C subsystem we should specify and monitor in addition to the system-level requirements in Table 11-19.

TABLE 11-23. Design Parameters for the TT&C Subsystem. Below are design parameters for the TT&C subsystem that are not typically specified at the system level. (See Table 11-19 for system-level parameters.)

Parameter	Comment
Antenna Sidelobe Levels	Design to minimize. Sidelobes degrade the antenna's directionality. Very high sidelobes may interfere with other antennas and receivers on the satellite. High sidelobe levels also affect security by making detection of signals more likely.
Polarization	Polarizations can be circular (right or left), or linear (horizontal or vertical). To decrease signal loss in the link, the polarizations need to be compatible. For example, the satellite antenna and ground station must both have right-circular polarization.
Frequency Stability	When we need to acquire the signal quickly, the receiver frequency must be known and stable. Thus, we specify the original receiver frequency's set point, short-term stability, temperature stability, and aging stability so we can acquire the uplink signal with little uncertainty.
Capture and Tracking Range	The *capture range* is the band of frequencies over which the uplink-carrier signal can drift from the receiver's best-lock frequency, so the receiver will still lock to the uplink signal. The *tracking range* is the band of frequencies the receiver will follow while locked to a sweeping, uplink-carrier signal without losing lock. Typically the capture range is 1% of the tracking range.
Diplexer Isolation	The *diplexer* allows us to use the same antenna for transmitting and receiving. The diplexer isolates the transmitter from the receiver. A diplexer with low isolation may require a band-reject filter between the transmitter and the diplexer.
Coupling Between Antennas	Signal and noise coupling between a transmitting antenna and a receiving antenna may cause the receiving antenna to lock onto a frequency coming from the transmitting antenna's transmitter. More commonly, broadband noise from the transmitting antenna may couple over to the receiving antenna and raise the noise floor of the receiving antenna's receiver, reducing the signal-to-noise ratio.

As Table 11-24 shows, selection criteria for TT&C subsystems fall into three categories: performance, compatibility, and other. Performance is the most important selection criterion. This subsystem's hardware must meet specifications of minimum performance to close the communication link with an acceptable signal-to-noise ratio. The *Bit Error Rate (BER)* is a Figure of Merit for the digital part of the communication link. It is the probability that a bit sent over the communication link will be received incorrectly. We typically specify this rate to be 1×10^{-6} for the command uplink and 1×10^{-5} for the telemetry downlink, depending on the nature of the data. To achieve this rate, the system must meet certain technical specifications: RF power output for the transmitter, receiver-noise figure, stable oscillator frequency, and the TT&C subsystem's front-end losses and antenna gains.

Compatibility is an important selection criterion when the TT&C subsystem must communicate with existing systems. If the TT&C subsystem must talk to the ground stations in the Space Ground Link System (SGLS), then the transponder must be compatible with SGLS. Likewise, if it must talk to the Tracking and Data Relay Satellite System (TDRSS), the cross-link transponder must be compatible with TDRSS.

TABLE 11-24. Selection Criteria for the TT&C Subsystem. Selection criteria at the subsystem level fall into three broad categories: performance, compatibility, and other. Chapter 13, Table 13-1, shows the selection process for communications architectures. (Courtesy of TRW)

Category	Criteria	Comments
Performance	• Mass • Volume	} See Table 11-26
	• Power (RF and dc)	
	• Bit error rate—Figure of Merit	Uplink = 10^{-6}, downlink = 10^{-5} are typical values (Sec. 13.3)
	• Noise figure—Figure of Merit	4 dB typical for SGLS
	• Frequency stability	Typically in parts per million
	• Insertion loss	
	• Reliability	Measured in terms of mean time between failures (Sec. 19.2)
	• Efficiency	Percent transponder-radiated power/input power
Compatibility	• Compatibility with existing systems • SGLS compatibility • TDRSS compatibility	
Other	• Technology risk • Heritage	Subjective (Sec. 20.4) Measured in terms of previous spaceflight experience

In the "Other" category, heritage is important when the schedule and budget are tight and technology risk must be low. Typically, we measure heritage in terms of previous spaceflight experience. A lot of communication hardware meets all three selection criteria. A typical SGLS transponder (1) can close the links to the ground stations with an acceptable bit error rate, (2) is compatible with the ground stations in the SGLS, and (3) has flown before.

Figure 11-4 diagrams a generic TT&C subsystem. A typical subsystem contains two transponders (for redundancy). The transmitter (downlink) path is as follows. From the left side of Fig. 11-4, two digital bit streams enter the transponder. One enters either from data storage or real-time from the payload. The other stream comes from command and data handling with telemetry data on health and status. These two data streams are modulated onto subcarriers, which are then modulated onto the carrier output. The composite signal is then amplified and routed out of the transponder and through a low-pass filter. The filter reduces second and higher-order harmonics to decrease frequency spurs and intermodulation products from the spacecraft's receiver. From the filter, the composite signal travels through a band-reject filter; a double-pole, double-throw (*transfer* or *2P2T*) RF switch; a diplexer; and finally to the antenna where it radiates to the ground station. The band-reject filter is a notch filter that attenuates frequencies coming from the transmitter and falling within the receiver's pass band. This filtering action further isolates the transmitter and receiver signals. The RF switch selects transmitter A or B and antenna A or B. The diplexer allows a transmitter and receiver to share the same antenna. It also isolates the transmitter from the receiver port at the receiver's center frequency, so the transmitter doesn't lock, jam, or damage the receiver.

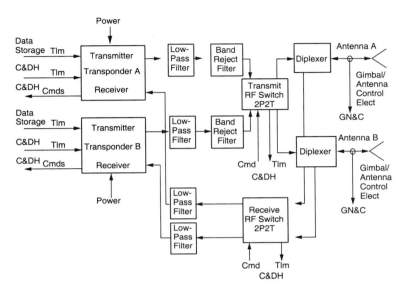

Fig. 11-4.   Block Diagram of a Generic TT&C Subsystem. This subsystem has full redundancy: two transponders with parallel transmit and receive signal paths. The diplexer allows the same antenna to be used for transmitting and receiving. The band-reject filter attenuates spurious signals originating from the transmitter at the receiver's center frequency to help the diplexer isolate the receiver from the transmitter.

The receiver's uplink path is as follows. First, we assume the equipment modulates the digital command data onto a subcarrier and further modulates the subcarrier onto the uplink-carrier frequency to form a composite uplink signal. Then, from the right of Fig. 11-4, the composite signal enters the subsystem through the antenna. The diplexer routes the composite signal to the receiver RF switch, which then selects antenna A or B and receiver A or B. The composite signal travels through the receiver's low-pass filter, which rejects unwanted transmitter harmonics and frequency spurs that may exist above the diplexer's stop band. The signal then moves into the transponder's receiver, which demodulates it and routes the digital-command bit stream to the command and data handling subsystem.

Figure 11-5 diagrams a typical transponder. The composite uplink signal enters the receiver, where the command data stream is demodulated from the carrier and subcarrier. The data stream enters the command detector, which validates the stream and forwards the data and receiver-in-lock indicator to the command and data handling subsystem.

The telemetry on the spacecraft's health and status, as well as the telemetry from the mission or science payload, enters the module that conditions it. The telemetry is modulated onto subcarriers (if applicable) and sent to the exciter/transmitter to be modulated onto the carrier. If the transponder is in the two-way-coherent mode, the transmitter generates the downlink carrier with the reference frequency (coherent drive) from the receiver's voltage-controlled oscillator. The composite downlink signal then goes to the ground station for processing.

For the ranging signals, the receiver demodulates the ranging tones or coding from the composite signal. The tones or pseudorandom noise code then moves to the trans-

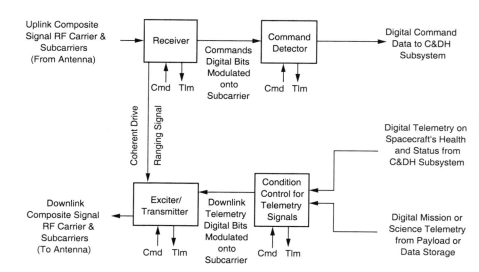

Fig. 11-5.   Block Diagram Showing How a Typical Transponder Works. The receiver is isolated from the transmitter except for the ranging signal path and the coherent drive path. The coherent drive comes from the receiver's voltage-controlled oscillator when we want the downlink carrier to be a ratio of the uplink carrier.

mitter, where it is conditioned and modulated onto the composite downlink carrier, amplified, and transmitted out of the transponder.

In a typical 3-axis-stabilized satellite, we mount omni-antennas to the top and bottom of the satellite. We mount all ground-link antennas to provide an unobstructed view of Earth and place cross-link antennas to provide an unobstructed view of the relay satellite. Historically, the TT&C subsystem's electronics are as close to the antennas as possible. If we gimbal the antennas, we must make sure the satellite's body or other appendages such as the solar arrays do not obscure the antenna's field of view.

Spin-stabilized satellites are usually cylindrical, and commonly contain a despun section. Except for omnidirectional varieties, we must mount antennas on the despun section. Frequently, we put the RF components and associated electronics on the despun section to avoid passing RF signals through rotary joints.

From a TT&C point of view, satellites in LEO or in geosynchronous orbit are similar. The primary differences are the path losses. We compensate for these losses with either antenna gain or transmitter power. An interesting case involves an antenna beam providing spot communication from a LEO spacecraft for an area of the Earth. As the satellite ranges away from the coverage area, the antenna beamwidth to cover the area decreases. The increased antenna gain resulting from the narrow beamwidth compensates for the path loss based on the inverse square law. The longer distance demands no added transmitter power.

Table 11-25 summarizes five ways we can apply a TT&C subsystem. For each application, the table specifies frequency, modulation, and common antenna characteristics.

TABLE 11-25. Attributes of Common Telemetry, Tracking, and Subsystems. This table presents the principal characteristics of two uplink/downlink systems and three cross-link systems. Each system can support various modulation schemes. The antenna characteristics for the uplink/downlink systems contain an Earth-coverage antenna for nominal operations and a hemispherical-coverage antenna for emergency operations. See Sec. 13.3 for modulation descriptions. (Courtesy of TRW)

Appli-cation	Frequency U/L	Frequency D/L	Modulation U/L	Modulation D/L	Antenna Charac-teristics	Remarks
Space Ground Link Sub-system	S-band 1.75 to 1.85 GHz	S-band 2.20 to 2.30 GHz	FSK AM PM	PCM PM FM	Earth coverage; Hemispherical coverage	SGLS Standard, refer to TOR-0059 Reissue H
Ground Space Tracking and Data Network	S-band 2.02 to 2.12 GHz	S-band 2.20 to 2.30 GHz	PCM PSK FSK	PCM PSK PM	Earth coverage; Hemispherical coverage	GSTDN is slowly phasing out. JPL-DSN is absorbing some of its assets. (See JPL-DSN-810-5 rev D)
Cross-link Within Constel-lation	W-band 60 GHz	W-band 60 GHz	Any	Any	Narrow beam 0.1 deg typical	Modulation, coding, and encryption can be customized
Cross-link to TDRSS	S-band K-band Ku-band	S-band K-band Ku-band	QPSK Spread spectrum	QPSK Spread spectrum	Narrow beam	TDRSS User Standard (See NASA Goddard STDN 101.2 TDRSS Users' Guide)
Optical	IR to UV	IR to UV	PPM/ PCM	PPM/ PCM	Telescope pencil beam	Frequency depends on device

Legend:

FSK = Frequency Shift Keying

AM = Amplitude Modulation

FM = Frequency Modulation

QPSK = Quadrature Phase Shift Keying

PSK = Phase Shift Keying

PCM = Pulsed-Code Modulation

PM = Phase Modulation

PQM = Phase Quadrature Modulation

PPM = Pulse Position Modulation

Table 11-26 contains detailed mass, power, and volume characteristics of three common TT&C subsystems: TDRSS-compatible in the S-band, generic X-band, and typical Ku-band. The information in these two tables comes from specification sheets issued by manufacturers of communication hardware, and summarizes today's technology for TT&C subsystems.

TABLE 11-26. Typical Parameters for TT&C Subsystems. Two transponders provide redundancy and two hemispherical-coverage antennas offer full coverage. The Ku-band system assumes another (S-band) system applies when the spacecraft deviates from its proper attitude. (Courtesy of TRW and L3 Communications)

Component	Qty	Mass (kg) Each	Mass (kg) Total	Power (W)	Dimensions (cm)	Remarks
Typical X-Band Communication Subsystem						
• Transponder	2	3.8	7.6		20 × 22 × 7	Generic X-band transponder
– Receiver				10.4		– 3-W RF output
– Transmitter				35.0		– Solid-state power amp
• Filters/switch diplexers.	1	1.5	1.5	0.0	10 × 22 × 4	1 set
• Antennas						
– Hemis	2	0.25	0.5	0.0	8.0 dia × 4	Circular Wave Guide
– Parabola	1	9.2	9.2	0.0	150 dia × 70	4-dBi* gain
– Waveguide	1	1.4	1.4	0.0	200 cm long	WR112
TOTAL			20.2	45.4		
Typical S-Band TDRSS User Communication Subsystem						
• Transponder	2	6.87	13.74		14 × 33 × 7	2nd generation TDRSS user
– Receiver				17.5		– 12-W RF output
– Transmitter				40.0		– Solid-state power amp
• Filters/switch diplexers	1	2.0	2.0	0.0	15 × 30 × 6	1 set
• Antennas						
– Hemis	2	0.4	0.8	0.0	9.5 dia × 13	Circular Wave Guide
– Parabola	1	9.2	9.2	0.0	150 dia × 70	4-dBi gain
– Turnstile	1	2.3	2.3	0.0	10 dia × 15	Cavity type
– Coax cables	1	0.5	0.5		1.2 dia × 150	1 set
TOTAL			28.54	57.5		
Typical Ku-Band Communication Subsystem						
• Transponder	2	4.45	8.90		17 × 34 × 9	Generic Ku-band transponder
– Receiver				4.3		– 4-W RF output
– Transmitter				20.0		– Solid-state power amp
• Filters/switch diplexers	1	1.2	1.2	0.0	8 × 19 × 4	1 set
• Antennas						
– Earth cover	1	0.5	0.5	0.0	4.0 dia × 2	Earth coverage horn
– Parabola	1	2.0	2.0	0.0	60 dia × 22	Cross-link antenna
– Waveguide	1	0.7	0.7	0.0	125 cm long	
TOTAL			13.3	24.3		

*dBi is the antenna gain relative to an isotropic radiator expressed in decibels (dB). See Sec. 13.3.5

11.3 Command and Data Handling

Richard T. Berget, *BF Goodrich Aerospace Data Systems Division*

The *command and data handling* system, *C&DH*, performs two major functions. It receives, validates, decodes, and distributes commands to other spacecraft systems and gathers, processes, and formats spacecraft housekeeping and mission data for downlink or use by an onboard computer. This equipment often includes additional functions, such as spacecraft timekeeping, computer health monitoring *(watchdog)*, and security interfaces.

While they normally provide independent functions, the combination of command and data handling into a single subsystem provides an efficient means for autonomous control of spacecraft functions. An onboard computer or microprocessor can send commands and monitor telemetry over a single interface with the C&DH system, allowing the control of multiple subsystems.

The C&DH system's size is directly proportional to spacecraft complexity. The more systems a spacecraft has, the more monitoring and configuration capability required. Reliability concerns alone may double the hardware's size if we require redundant C&DH subsystems.

The ideal C&DH system is one which has been proven on another spacecraft and which requires no modification for the mission under development. However, new missions are usually supported by systems which evolve from older designs. We make small improvements in the performance of the systems from the viewpoint of speed, power, weight, volume, or other operating parameter. In the case of a new or custom design, extensive testing must simulate the strenuous environments involved in a space launch and flight.

11.3.1 Introduction to C&DH

Figure 11-6 shows a typical command decoder. Command messages can originate from an onboard computer, uplink transponders, or a hardline test interface. An arbitration scheme is necessary for source selection which gives uplink commands priority. Commands from the computer are delayed until a time slot is available. The hardline test interface is not active during flight and when in use overrides the other command sources.

Several standards exist for command message formats. (See CCSDS 201.0-B-1, 1987.) Typically, a command consists of a synchronization code, spacecraft address bits, command message bits, and error check bits. Received commands are validated prior to execution. Validation consists of receiving synchronization code, checking command message length (correct number of bits), exactly matching the spacecraft address and any fixed-bit patterns (unused message bits), and detecting no errors in an error check polynomial code. Once the decoder validates a command, it increments a counter to record the number of executed commands. Then the message bits pass to a decoder for execution. The command decoder rejects commands that do not pass the validation criteria and it increments the command reject counter. The data handling system reads the accept and the reject counters and includes them in the downlink data to provide operational feedback.

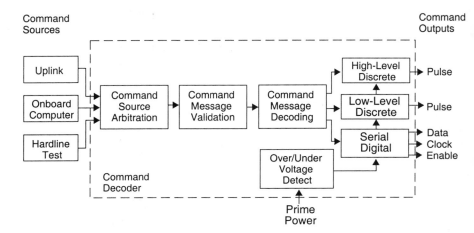

Fig. 11-6.   Command Decoder Block Diagram.

The command decoder determines command output type and the specific interface channel. A typical system provides two types of output: discrete and serial. *Discrete commands* are a fixed amplitude and a fixed pulse duration and consist of two basic types:

- *High-Level Discrete Command*: A +28 V, 10 to 100 ms pulse used to drive a latching relay coil or fire an ordnance device.

- *Low-Level Discrete Command*: An open collector or 5 V pulse typically interfacing with digital logic.

A *serial command* is a 3-signal interface consisting of a shift clock, serial command data, and a data enable used to indicate the interface is active. A portion of the received command message bits (typically 8 or 16) is sent in serial form to a user subsystem.

The terms data handling and telemetry are often used interchangeably. However, data handling is more than just telemetry. IEEE Standard 100 offers this definition of *telemetry*:

> *Telemetering (remote metering). Measurement with the aid of intermediate means that permit the measurement to be interpreted at a distance from the primary detector. The distinctive feature of telemetering is the nature of the translating means, which includes provision for converting the measurand into a representative quantity of another kind that can be transmitted conveniently for measurement at a distance. The actual distance is irrelevant.*

Data handling combines telemetry from multiple sources and provides it for downlink or internal spacecraft use. Figure 11-7 illustrates a typical spacecraft data-handling unit.

Most data handling systems are of the time-division multiplexed type. These systems sequence through their inputs in a predetermined order, then organize them in a fixed output format. Other systems process inputs as lists of data samples and/or allow random access by an onboard computer. Input signal sampling rate is determined by signal bandwidth. Sample rate must be a minimum of two times the

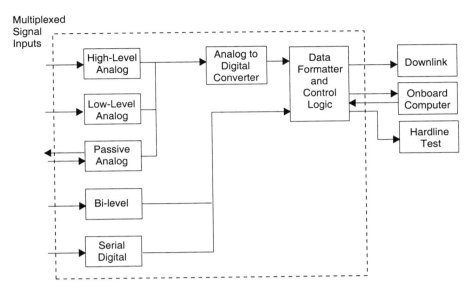

Fig. 11-7. Data Handling Unit Block Diagram.

greatest frequency component contained in the signal. (See Sec. 13.2 for a more in-depth discussion on data rates.) Data from all inputs is converted to digital form and formatted into a serial stream of continuous data for downlink. The data rate is the sum of all input sample rates plus some bandwidth for insertion of synchronization codes and a frame identification counter. IRIG Standard 106, Chap. 4, provides a detailed discussion of downlink telemetry formats and associated format terms.

The data handling system may also supply telemetry to an onboard computer. The computer sends its request to the data handling system which processes the input and returns telemetry data. This operation is interleaved with downlink telemetry gathering which is usually continuous.

Analog telemetry data comes to the data handling equipment in many forms. Often, direct transducer outputs require signal conditioning prior to conversion from analog to digital form. Data handling hardware is simplified, however, when input signals are preconditioned or fall in the general categories described below.

- *High-Level Analog*: A telemetry channel with information encoded as an analog voltage, typically in the range of 0 to 5.2 V. These are active analog inputs in that the command and data handling system does not provide measurement excitation. Data handling equipment converts this information to digital form.

- *Low-Level Analog*: A telemetry channel with information encoded as an analog voltage. The signal range is low enough to require amplification before the information is encoded into digital form. Typical gain values fall between 100 to 300. Because of the signal's low voltage range, it is subject to noise contamination and thus uses an interface in which the telemetry information is the difference between signal and reference inputs to the command and data handling system. This is differential or double-ended interface.

- *Passive Analog*: A telemetry channel with information encoded as a resistance. The command and data handling system supplies a constant current to the resistive sensor and encodes the resulting IR voltage drop into a digital word.

All analog telemetry is converted to digital form within the command and data handling system. The system determines data resolution by the number of quantization levels. More information on this topic may be found in Chap. 13.

The two most common forms of digital telemetry data are described below:

- *Bi-Level (Discrete) Input*: A telemetry channel conveying two state information (such as on/off or enable/disable). Information is encoded as voltages, but may be encoded as a resistance or the presence or absence of a signal. Typically a logic 0 = 0 to 1 V, and a logic 1 = 3 V to 5 V (or 3 V to 28 V).

- *Serial Telemetry (Digital) Interface*: A 3-signal interface used to transfer digital data from an external source to the data handling equipment. The command and data handling system provides a shift clock and an interface enable signal to control data transfer. Interface circuits may be differential line drivers or single ended. Serial rather than parallel interfaces are preferred on spacecraft, because they simplify cable design and require fewer interface circuits.

11.3.2 C&DH System Sizing Process

Table 11-27 summarizes the command and data handling subsystem estimating process. The desired output of this process is a reasonable estimate of the hardware necessary to support the mission including estimate of its size, weight, and power. Use this table in conjunction with Tables 11-28 and 11-29 to assist in estimating the system parameters in the case of unknowns. The results may then be fed back into the mission design process and adjusted as necessary. FireSat has been used as an example to illustrate the process and highlight the main points of estimating when the majority of needed inputs are unknown or flexible.

Step 1—Identify Functions to be Performed by the C&DH System. The first step in baselining the command and data handling system is to define the primary functional requirements needed to monitor and control the spacecraft. At a minimum, most missions require command processing and housekeeping data acquisition. The size and complexity of these two systems is determined by the spacecraft design, the technical requirements, and additional functions and subsystems supported.

Command Processing

There are three primary considerations for baselining a command decoder:

- The number of command output channels
- Any requirement for stored commands
- Any requirement for computer commands or ACS functions

Table 11-28 gives guidelines for channel counts and sizing. As the spacecraft design evolves, we refine the needed quantities and types of channels. Whenever possible, interfaces to the C&DH system should be standardized as a means of saving cost, but this is not mandatory. The C&DH system is an excellent place to put spacecraft functions that do not seem to fit anywhere else.

TABLE 11-27. Command and Data Handling System Baseline Process.

Step	Procedure	Issues or Data Needed	FireSat Example	Ref.

1. Identify which functions are to be performed by the C&DH system	Determine or estimate what on the spacecraft must be controlled and what must be monitored to accomplish or support the mission. Determine what tasks are to be performed on board.	**Command Processing**
		Is command processing required?
		If Yes,
		  – What is the command rate?
		  – How many channels?
		  – Is there a computer?
		  – Are stored commands needed?
		Telemetry Processing
		Is telemetry processing required?
		If Yes,
		  – How many channels?
		  – What is the housekeeping data rate?
		  – What is the payload data rate?
		  – Is a computer interface needed?
		Other
		Mission time clock needed?
		Computer watchdog needed?
		ACS functions needed?
			Yes
			
			Uncertain
			< 200
			Computer or stored commands, not both
			
			Yes
			
			< 200
			Low
			150 Mbps
			No
			
			Yes
			No
			Uncertain
				Secs. 10.3, 13.2
				
				Secs. 10.3.4, 13.2, 9.5
				
				Sec. 11.1, Chap. 16

2. Identify requirements and constraints	Determine the parameters, driven by aspects of the overall spacecraft design, which impact the C&DH system (the C&DH drivers).	Bus constraints Reliability Satellite Lifetime Radiation Environment Schedule Budget	LightSat bus 0.98 5 years Must be SEU hard 2-year development Low	Chap. 10, Secs. 19.2, 8.2
3. Determine complexity of C&DH functions	Use Table 11-28 to estimate the complexity of each function identified in step 1, applying constraints established in step 2.	Evaluate each function separately	See Table 11-28	Table 11-28
4. Determine overall C&DH level of complexity	Collect functions into command and telemetry components and determine composite complexity of each	Are other functions such as the mission time clock combined into command or telemetry (typically telemetry)?	Mission time clock to be included in telemetry component of C&DH (critical to correlate data with time of day)	
5. Estimate size, weight, and power for each component	Apply results of step 4 to Table 11-29 for command and telemetry components	Combined systems (share housing, reduce interface cabling, may share power supply)	See Table 11-29	Table 11-29

We must include the capability to store commands if we require spacecraft control when its not in view of its ground stations, or as a means of recovery if the communication link is lost. These commands may be controlled by matching a time-tag or by a simple delay counter from a controlled timing event. Stored commands of this type may be easily implemented without a general-purpose processor.

We must add an onboard computer if we require a decision-making element on the spacecraft. Once we establish the need for a computer, we can plan to use it to perform many functions including the stored command capability, attitude control algorithms, and data processing and storage. Integrating attitude control with the command system will typically add some special interface requirements for driving control elements.

Telemetry Processing

The data handling system provides the ability to acquire data for:

* Spacecraft housekeeping data (health and status)

* Feedback for onboard control of spacecraft functions

* Routing of payload or subsystem data to and from receivers and transmitters, storage or affected system controllers

The quantity of telemetry input channels required for monitoring spacecraft health is typically directly proportional to the size, complexity, and quantity of payloads and subsystems involved in meeting the primary and secondary missions. The majority of these channels are standard interfaces to temperature, pressure, and voltage transducers. Some subsystems provide the ability to monitor their own health and integrate the information into a data stream. For new subsystems, the awareness of what the data handling system can do for them may prevent an unusual design or duplication of a large amount of circuitry.

The data handling system may acquire payload or subsystem data also. Of critical importance to the system design is the quantity of data and its transfer rate. The telemetry acquired for spacecraft health is limited in speed due to the time necessary to accurately convert analog signals to digital information. If a subsystem or payload data stream exceeds 200 kbps or is greater than a few thousand bits in size, it is usually necessary to provide data buffers or to process the data in a separate section of the data handling system. Often, an interleaver may be provided to integrate and synchronize the health and payload data into a single stream.

If an onboard computer is available, it may require additional signals to perform its tasks. These signals may not be needed in the downlink telemetry format. Therefore, it is usually preferable for the computer to have the capability to request data independently of the preprogrammed downlink format. The computer may also be used to preprocess subsystem and health data to reduce the downlink bandwidth requirement. (See Chap. 16 for a discussion of onboard processing.)

Other Functions

Time. Most spacecraft designs require the availability of a *time word* (universal time, mission elapsed time, or delay) for support of attitude control, stored commanding, or data time-tagging. Several systems can provide this time, including GPS receivers, computer-maintained counters, and hardware timers. The most critical parameters for the definition of this function are:

* Time word granularity

- Stability requirement

- Acceptable uncertainty

Granularity defines the smallest increments of time maintained for use on the spacecraft or of interest on the ground. This value is usually driven by the accuracy of time needed for data time-tagging or the attitude control system. Typically this is one millisecond or one microsecond. Over specifying this value will increase the hardware required, increase the cost, and decrease the available bandwidth for data. IRIG-B time code generators transmit a 1-sec resolution time word and a 1-MHz oscillator to allow the user to create their own smaller granularity time.

The aging characteristics of the primary oscillator, which drives the timing system, determines the drift characteristics of the time word. Oscillators are typically specified by long-term and short-term stability in parts per million (ppm) over a given time. Selection of this stability determines the allowable error in the onboard time between time updates from the ground. The same stable oscillator may be used to provide other oscillator frequencies to other spacecraft subsystems. Occasionally, the stability needed by the other system may be the driving factor.

Maintaining time with the spacecraft computer is possible using internal registers and a periodic interrupt signal. However, additional uncertainty may be induced due to the nonsynchronous nature of a processor under interrupt control. Higher priority interrupts may delay the update of the time word. If other subsystems need a time base, the designer must include additional registered circuitry.

Computer Watchdog. When a spacecraft computer is used to provide decision-making capability on orbit, it is common to provide a method of determining a computer failure independent of the processor itself. This function may be integrated into the C&DH system and is usually referred to as the *watchdog timer.*

The watchdog timer ensures that the computer hardware and software functions as planned. A hardware or software anomaly could be catastrophic to the spacecraft mission if we don't provide a means of correcting the problem. Typically, this function uses one or more timers which must be reset by the onboard computer prior to timing out. The computer resets the timer by writing a specific data word to a specific address. If this is not accomplished prior to the time-out, the watchdog will execute a predetermined recovery action. The recovery may be a computer reset, interrupt, or a disable which is maintained until cleared by a ground command.

Attitude Control System Functions. Integrating attitude control functions into the C&DH system may reduce the hardware required on the spacecraft by taking advantage of C&DH circuitry that is available in other subsystems. The integration of command, telemetry and onboard computer functions allows closed-loop monitoring and control with the addition of interface channels specific to the attitude control function. These channels may be high current, high accuracy, or other special requirement interfaces. In some cases, the attitude control section provides only controlling signals, with the high power and signal conditioning circuitry integrated into the attitude control component.

Spares. As the baselining process continues, we develop an estimate of the I/O channel quantities, and use this estimate in step 3 to estimate system parameters. Unfortunately, I/O channel quantities tend to increase, as the spacecraft becomes more defined. Therefore, it is common practice to include 10% to 25% additional channels in the count for unforeseen growth requirements. We should use the channel count, including spares, to estimate system complexity in Table 11-28. This estimate must be documented carefully to prevent several increases as the concept proceeds through the various departments and levels of management involved in the spacecraft design. As always, more hardware increases the cost, size, weight, and power of the system.

Step 2—Identify Requirements and Constraints. Once the functions required by the command and data handling system have been determined, requirements and constraints imposed by external factors must be identified. We don't control these requirements and constraints and they may affect one or more aspects of the C&DH system design. Early identification and response to these issues may minimize the cost impact and design problems.

Spacecraft Bus Constraints. The physical size of a spacecraft and its design will often direct the ultimate configuration of the command and data handling system. In general, the C&DH system may be divided into three classes or architectures:

- Single-unit systems

- Multiple-unit, distributed systems

- Integrated systems

A *single-unit C&DH system* provides one unit for the command system and one unit for the telemetry system or a single unit which integrates both functions. Although the single-unit design may be simple and centralize functions, it can have a significant disadvantage on a medium to large spacecraft bus. As mentioned previously, a larger spacecraft will generally require a larger number of subsystems and associated interfaces and health monitors. A single-unit system requires every interface wire to be routed to a single physical location for monitoring and control. The result can be a wire harness that is larger than the unit itself and significantly impacts the weight budget.

Multiple-unit C&DH systems provide a potential solution to this problem and others. A multiple-unit system provides "remote" command and data handling capabilities in locations physically removed from the "central" unit. The number of remotes is driven by the spacecraft bus design or the quantity of I/O channels. One example is the design of a dual-spin satellite in which every signal must be transferred between the spinning and fixed sections of the satellite over a slip-ring interface. Slip rings limit the quantities of signals which may be practically routed and also complicate the design due to induced noise. One practical solution is to provide a remote unit on the spinning side which communicates with the central unit over a digital data bus. This allows the acquisition of hundreds of channels on the spinning side while requiring only 2 to 6 wires to pass through the slip rings.

Integrated C&DH systems typically combine command, telemetry, flight processing, and attitude control into one system. These systems tend to be small LightSat-type applications which use a single computer to monitor and control the satellite or a large high-performance system which uses multiple computers and subsystems coordinated by a central high-power processor. This type of system may provide a reduced hardware requirement and cost due to the increased capability provided by the processor. However, this system will most likely entail increased software costs associated with the increased programming requirements.

Reliability. The reliability required of the C&DH system will affect the system design in two areas: redundancy and parts quality. A low failure rate for the system provides a high confidence factor in the success of the mission. Reliability is dramatically increased by including a redundant system for all mission-critical components. Configuring a system in this manner will obviously increase the amount of hardware involved. More hardware means increasing the recurring cost, but not necessarily double the total procurement cost. Many cost items involved in manufacturing the system are fixed whether a single-string or redundant system is built.

Parts quality also affects the reliability of a system. Increasing the parts quality does not increase the amount of hardware; however, it does significantly increase cost. Specifying a Class "S" parts program (or indirectly requiring it via the reliability requirement) typically multiplies the material cost by 400% to 500%.

Radiation. The areas affected most by the radiation requirement are cost and schedule. System size and weight may be affected if we require shielding of electronic components. A radiation environment limits the part types available to the designer and system performance is typically lower due to required derating. Predicting circuit behavior is accomplished by modeling, simulation, and analysis. Environment severity may double system development time and increase parts costs by a factor of 10.

Program Constraints. The foundation of any hardware development program lies in the constraints placed upon the program to carry out the mission. In some cases, program constraints initially restrict a design so the desired mission cannot be accomplished. The budget allocated for the program will typically be the most limiting constraint in the development of the spacecraft and, in turn, command and data handling. Allocating a budget for a LightSat will clearly preclude developing a spacecraft and support systems for a national-asset satellite.

If managers define a budget, it will become the primary driver in determining the other elements of the definition process. In the case of a preliminary study, the objective may be to define the budget needed to achieve the desired mission goals. In this case, the later steps become the determinant and the dollars needed become the output of the process.

The second significant program constraint is commonly schedule. Most space-qualified electronic systems are custom designs or semicustom implementations of existing hardware. The need dates for hardware may completely determine which approach we take to develop hardware. Typical lead times for command and data handling equipment are 12 to 18 months for systems using Class B parts and 24 to 30 months for systems using Class S parts. These schedule times are almost completely driven by the lead time involved in the procurement of electronic piece parts. Therefore, a fast delivery requirement to support an urgent mission will affect the parts and reliability level of the unit.

Step 3—Determine the Complexity of C&DH Functions. Table 11-28 may be used to provide a first-order estimate of the complexity of each C&DH function. There are no absolutes in this stage of the process. The estimate is the result of the C&DH "feel" obtained by comparing known general requirements with those listed in the table. The result is a bounding of the system definition into one of three zones. We must define C&DH system drivers which may move the components between zones in the case of an unclear definition. Once a determination is made on the function complexities, steps 4 and 5 provide an estimate of the system size, weight, and power specifications. As can be seen in the FireSat example, all the requirements do not have to be defined to make a first-order estimate.

Step 4—Determine Overall C&DH Level of Complexity. Functions described as "other" are now collected into the command and telemetry components. Typically, the mission time clock is included in the telemetry component. The computer watchdog is included in the command component because a computer failure often requires spacecraft reconfiguration via the command system. ACS functions are included in both.

TABLE 11-28. System Complexity Definition. Reliability calculations assume a 5 year mission. Reliability predictions are affected dramatically by mission duration. Simple and typical reliability estimates assume single units. Complex assumes integrated command and data handling.

Requirement or Constraint	System Complexity			FireSat Requirements	FireSat Complexity Selection
	Simple	Typical	Complex		
Processing Commands:				Required	Simple-typical
CMD rates	50 cmds/s	50 cmds/s	≥ 50 cmds/s	≤ 50 cmds/s	
Computer interface	none	Computer or stored cmds (not both)	yes	Computer or stored commands needed	
Stored commands	none		not needed		
Number of channels	< 200 channels	300–500 channels	> 500 channels	< 200	
Processing of Telemetry Data:				Required	Simple
TLM rates					
Housekeeping data	500–4kbps	4–64 kbps	64–256 kbps	low	Design Driver
Payload data	none	1–200 kbps	10 kbps–10 Mbps	150 Mbps	Implies separate data link for payload
Computer interface	none	none	yes	not required	
Number of channels	< 200 channels	400–700 channels	> 500 channels	< 200	
Other:					Typical–complex
Mission time clock	none	included	included	Time required for data time-tagging and remote control	
Computer watchdog	none	included if OBC	included	ACS/autonomy requirement determined by ACS design	
ACS functions	none	none	included		
Bus Constraints	Single Unit	Single unit or multiple units	Integrated or distributed	LightSat bus (single unit)	Simple
Reliability-Class B parts					Implies redundancy and Class S parts
Single String	0.8233	0.7610	0.6983	0.98 desired	
Redundant	0.9875	0.9736	0.9496		
Reliability-Class S parts					
Single string	0.9394	0.9083	0.8285		
Redundant	0.9987	0.9964	0.9829		
Radiation Environment (total dose)	< 2 krads	2–50 krads	50K–1 Mrads	Low (< 2 krads)	Simple
Schedule (in months, after order)				Assumed to be driven by system design	24-month schedule acceptable
Class B parts	6–12	6–12	9–18		
Class S parts	9–18	9–24	9–24		

Composite complexity is determined by scoring the functions complexity and system drivers for each component (command and telemetry), giving the most weight to system drivers. The FireSat example appears to be satisfied with a "simple" system approach with one exception. The high-speed payload data requirement must be addressed to prevent this requirement from becoming the system design driver. The impact must be evaluated for the mission design. Potential solutions include a separate data link, onboard data compression to reduce bandwidth, and additional research to determine if the 150 Mbps requirement is really needed to accomplish the mission.

The reliability requirement specified for FireSat causes a significant impact on the system design and cost. To achieve the desired rating, the systems must be configured redundantly and manufactured using Class S parts. This information, its impact and alternatives, should be fed back into the mission design process as early as possible.

Step 5—Estimate Size, Weight, and Power for Each Component. The results of steps 3 and 4 may now be used in conjunction with Table 11-29 to obtain an estimate of the system parameters. The "feel" of the system or the desired design margin will determine the selected value within each zone.

TABLE 11-29. **Parametric Estimation of C&DH Size, Weight, and Power.** Peak command power will vary by command type, duration, and load. Nominal command power will be higher if ACS drivers are required.

		Simple	Typical	Complex	FireSat Baseline
Size (cm³)	Command only	1,500–3,000	2,000–4,000	5,000–6,000	3,000
	Telemetry only	1,500–3,000	4,000–6,000	9,000–10,000	3,000
	Combined systems	2,500–6,000	6,000–9,000	13,000–1,500	
Weight (kg)	Command only	1.5–2.5	1.5–3.0	4.0–5.0	2.5
	Telemetry only	1.5–2.5	2.5–4.0	6.5–7.5	2.5
	Combined systems	2.75–5.5	4.5–6.5	9.5–10.5	
Power (nominal) (W)	Command only	2	2	2	2
	Telemetry only	5–10	10–16	13–20	10
	Combined systems	7–12	13–18	15–25	

Command and data handling systems are generally conservative, evolutionary designs due to their mission-critical nature. The baselining process presented provides the mission designer with an approach to making a first-order estimate of the necessary hardware based upon previous hardware developments. This approach will provide a realistic estimate to be used in mission resource budgeting.

It is important to identify the rationale and drivers for the baseline. If possible, each specification should be allocated a rating or confidence factor to indicate if the specification is required, flexible, or merely a place holder. This information may allow the mission design team more creativity in solving a given technical problem.

11.3.3 C&DH Basics

This section is a list of details of great concern to command and data handling system design and operation. Many of these concerns are of absolute necessity when determining C&DH requirements and generating procurement specifications. Emphasis is placed on the command system because of the severity of the effects if these guidelines are not followed. Data handling basics such as data rates and the number of bits per sample are covered in Sec. 13.2.

Interfaces to other equipment must be protected so that their faults do not propagate into the command decoder.

It is paramount that no commands or any transient signals appear on command outputs during application or removal of prime power, or during under/over prime power voltage conditions.

It is a basic philosophy of command decoder designs that if the integrity of a command message is in doubt, the command is not issued. It is rejected! This is especially true when firing an ordnance device or the spacecraft is launched from a manned vehicle. It is for this reason that received command messages are not corrected, although the capability exists, using error check bits.

For safety concerns, operations such as firing ordnance, an engine, or thruster, require multiple commands configured in series forming a logical AND function. No single command causes the operation to occur. In a typical ordnance application, three commands are required: safe, arm, and fire. In this case, safe and arm are relays that enable a high level discrete command, fire. The commands must (shall) be isolated within the command decoder such that no single component or physical failure results in inadvertent function execution. To achieve this, the Hamming distance of controlling command messages must be two or greater (for isolation in the decoding scheme), and command outputs must be physically isolated to the greatest extent possible using different decoding circuits and interface connectors.

It is advised not to have any commands that turn a command decoder off during flight. In addition, there should be no commands that interrupt the uplink source to the command decoder.

In redundant applications, where command outputs are cross strapped, the interface circuits and interconnection have to be designed such that no single component or physical failure prevents the active output from functioning. Along the same lines, where telemetry inputs and serial interface outputs are cross strapped, the interface circuits and interconnections have to be designed such that no single component or physical failure prevents the interface from functioning.

The rising and falling edges of discrete command and serial telemetry outputs are often limited in frequency content so that they are not a source of noise emissions on the spacecraft.

11.3.4 A Final Note

The C&DH subsystem is often one of the last on the spacecraft to be defined. It is a tool, used to configure, control, or program the payload and other spacecraft subsystems. It is the spacecraft's senses reporting internal environment, health, and status information. C&DH equipment cannot be completely defined until the requirements of other systems have been established. The mission designer's main task is that of listing the command, telemetry and other data needs for each spacecraft system. The list must also include the rate at which commands are issued and telemetry is gathered for determination of composite data rates. Issues such as data format, encoding, and security must then be addressed. At this point it may be advantageous to stop and take an overall view of the spacecraft for other functions, which if included in the C&DH, would simplify overall design. Remember that the C&DH interfaces to nearly all spacecraft functions. Next the impact of the mission environments, duration and required reliability on the C&DH hardware is assessed. When these tasks are complete the C&DH subsystem can be fully characterized.

11.4 Power

Joseph K. McDermott, *Lockheed Martin Astronautics*

As illustrated in Fig. 11-8, the *electrical power subsystem (EPS)* provides, stores, distributes, and controls spacecraft electrical power. Table 11-30 lists typical functions performed by the EPS. The most important sizing requirements are the demands for average and peak electrical power and the orbital profile (inclination and altitude). We must identify the electrical power loads for mission operations at *beginning-of-life, BOL*, and *end-of-life, EOL*.

For many missions, the end-of-life power demands must be reduced to compensate for solar array performance degradation. The average electrical power needed at EOL determines the size of the power source. Section 10.3 shows a sample power budget that we may use to begin the sizing process. We usually multiply average power by 2 or 3 to obtain peak power requirements for attitude control, payload, thermal, and EPS (when charging the batteries). Fortunately, all the systems do not require peak power at the same time during the mission.

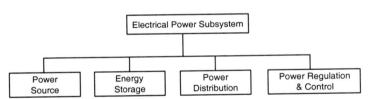

Fig. 11-8.   Functional Breakdown for the Spacecraft's Power Subsystem. We start with these four functions and must determine requirements for the hardware, software, and interfaces for each.

TABLE 11-30.   Typical Top-Level Power Subsystem Functions. Each of these functions consists of subfunctions with a myriad design characteristics which we must develop to meet mission requirements.

- Supply a continuous source of electrical power to spacecraft loads during the mission life.
- Control and distribute electrical power to the spacecraft.
- Support power requirements for average and peak electrical load.
- Provide converters for ac and regulated dc power buses, if required.
- Provide command and telemetry capability for EPS health and status, as well as control by ground station or an autonomous system.
- Protect the spacecraft payload against failures within the EPS.
- Suppress transient bus voltages and protect against bus faults.
- Provide ability to fire ordnance, if required.

Table 11-31 summarizes the power subsystem design process, which we discuss further in the following subsections, and Table 11-32 shows the principal effects of mission requirements on the power system design. We will work through the design process, beginning with the selection of a power source.

TABLE 11-31.   The Preliminary Design Process for the Power Subsystem. All of these design steps must link back to mission requirements to satisfy the owner and users. Note that derived requirements may impact previous design decisions and force designers to iterate the design process.

Step	Information Required	Derived Requirements	References
1. Identify Requirements	Top-level requirements, mission type (LEO, GEO), spacecraft configuration, mission life, payload definition	Design requirements, spacecraft electrical power profile (average and peak)	Secs. 10.1, 10.2
2. Select and Size Power Source	Mission type, spacecraft configuration, average load requirements for electrical power	EOL power requirement, type of solar cell, mass and area of solar array, solar array configuration (2-axis tracking panel, body-mounted)	Secs. 10.1, 10.2 Table 10-9 Sec. 11.4.1 Table 11-34
3. Select and Size Energy Storage	Mission orbital parameters, average and peak load requirements for electrical power	Eclipse and load-leveling energy storage requirement (battery capacity requirement), battery mass and volume, battery type	Sec. 11.4.2 Tables 11-3, 11-4, 11-38, 11-39, 11-40 Fig. 11-11
4. Identify Power Regulation and Control	Power-source selection, mission life, requirements for regulating mission load, and thermal-control requirements	Peak-power tracker or direct-energy-transfer system, thermal-control requirements, bus-voltage quality, power control algorithms	Sec. 11.4.4

TABLE 11-32.   Effects of System-Level Parameters on the Power Subsystem. Most aspects of the mission affect the power subsystem because so many other subsystems require specific power attributes.

Parameter	Effects on Design
Average Electrical Power Requirement	Sizes the power-generation system (e.g., number of solar cells, primary battery size) and possibly the energy-storage system given the eclipse period and depth of discharge
Peak Electrical Power Required	Sizes the energy-storage system (e.g., number of batteries, capacitor bank size) and the power-processing and distribution equipment
Mission Life	Longer mission life (> 7 yr) implies extra redundancy design, independent battery charging, larger capacity batteries, and larger arrays
Orbital Parameters	Defines incident solar energy, eclipse/Sun periods, and radiation environment
Spacecraft Configuration	Spinner typically implies body-mounted solar cells; 3-axis stabilized typically implies body-fixed and deployable solar panels

11.4.1 Power Sources

The power source generates electrical power within the spacecraft. Launch vehicles such as Titan IV or Delta use *primary batteries* (discussed in Sec. 11.4.2) as the power source for electrical loads because the batteries usually need to last less than an hour. But batteries alone are too massive for missions that last from weeks to years. These missions need a source that can generate power over many orbital cycles to support electrical loads and recharge the batteries.

Typically, we use four types of power sources for spacecraft. *Photovoltaic* solar cells, the most common power source for Earth-orbiting spacecraft, convert incident solar radiation directly to electrical energy. *Static power* sources use a heat source —typically plutonium-238 or uranium-235 (nuclear reactor), for direct thermal-to-electric conversion. *Dynamic* power sources also use a heat source—typically concentrated solar radiation, plutonium-238, or enriched uranium—to produce electrical power using the Brayton, Stirling, or Rankine cycles. The fourth power source is fuel cells, used on manned space missions such as Gemini, Apollo, SkyLab, and the Space Shuttle. Table 11-33 provides a comparison of various power sources.

Static power conversion uses either a thermoelectric or a thermionic concept. The most common static power source for spacecraft is the *thermoelectric couple*. This basic converter uses the temperature gradient between the p-n junction of individual thermoelectric cells connected in a series-parallel arrangement to provide the desired dc electrical output from each converter. This temperature gradient comes from slow decay of the radioactive source. The thermal-to-electric conversion efficiency for a thermoelectric source is typically 5–8%.

Thermionic energy conversion produces electricity through a hot electrode (*emitter*) facing a cooler electrode (*collector*) inside a sealed enclosure that typically contains an ionized gas. Electrons emitted from the hot emitter flow across the inter-electrode gap to the cooler collector. There they condense and return to the emitter through the electrical load connected externally between the collector and the emitter. We choose the collector and emitter temperatures for best overall system performance. In choosing the collector temperature, we try to decrease the weight and size of thermal radiators, and we choose materials based on mission life requirements. Thermionic power sources usually rely on a reactor heat source because of the high temperature required for efficient thermionic conversion. Power efficiencies for a thermionic power conversion are typically 10–20%.

In contrast to static sources, *dynamic power sources* use a heat source and a heat exchanger to drive an engine in a thermodynamic power cycle. The heat source can be concentrated solar energy, radioisotopes, or a controlled nuclear-fission reaction. Heat from the source transfers to a working fluid, which drives an energy-conversion heat engine. For a dynamic solar-power source, *the balance of energy* remains as latent and sensible heat in a heat exchanger (molten eutectic salt), which provides continuous energy to the thermodynamic cycle during eclipse periods. A dynamic power source using a nuclear reactor or plutonium-238 decay does not require thermal-energy storage because the source provides continuous heat.

Dynamic power sources use one of three methods to generate electrical power: Stirling cycle, Rankine cycle, or Brayton cycle. *Stirling-cycle engines* use a single-phase working fluid as the working medium. The thermodynamic cycle consists of two isothermal processes (compression and expansion) and two constant-volume processes (heating and cooling). Power-conversion efficiencies for Stirling engines are 25–30%. *Rankine-cycle engines* are dynamic devices that use a two-phase fluid system

TABLE 11-33. Matrix for Comparing Most Common Spacecraft Power Sources. We may use different factors to select the correct power source but specific power and specific cost are used extensively.

EPS Design Parameters	Solar Photovoltaic	Solar Thermal Dynamic	Radio-isotope	Nuclear Reactor	Fuel Cell
Power Range (kW)	0.2–300	5–300	0.2–10	5–300	0.2–50
Specific Power (W/kg)	25–200	9–15	5–20	2–40	275
Specific Cost ($/W)	800–3,000	1,000–2,000	16K–200K	400K–700K	Insufficient Data
Hardness – Natural Radiation – Nuclear Threat – Laser Threat – Pellets	Low–Medium Medium Medium Low	High High High Medium	Very high Very high Very high Very high	Very high Very high Very high Very high	High High High Medium
Stability and Maneuverability	Low	Medium	High	High	High
Low-orbit Drag	High	High	Low	Medium (due to radiator)	Low
Degradation Over Life	Medium	Medium	Low	Low	Low
Storage Required for Solar Eclipse	Yes	Yes	No	No	No
Sensitivity to Sun Angle	Medium	High	None	None	None
Sensitivity to Spacecraft Shadowing	Low (with bypass diodes)	High	None	None	None
Obstruction of Spacecraft Viewing	High	High	Low	Medium (due to radiator)	None
Fuel Availability	Unlimited	Unlimited	Very low	Very low	Medium
Safety Analysis Reporting	Minimal	Minimal	Routine	Extensive	Routine
IR Signature	Low	Medium	Medium	High	Medium
Principal Applications	Earth-orbiting spacecraft	Interplanetary, Earth-orbiting spacecraft	Inter-planetary	Inter-planetary	Inter-planetary

employing a boiler, turbine, alternator, condenser, and pump. This power-conversion cycle is essentially the same as that used to generate electricity from fossil and nuclear energy on Earth. Power-conversion efficiencies for Rankine-cycle engines are 15–20%. *Brayton-cycle engines* are dynamic devices that use a single, compressible working fluid as the working medium. The thermodynamic cycle consists of adiabatic compression and expansion stages separated and coupled by stages that add or reject heat at constant pressure. Placed after the turbine, a recuperator-heat exchanger improves the cycle's efficiency. Power conversion efficiencies for the Brayton cycle are 20–35%.

Fuel cells convert the chemical energy of an oxidation reaction to electricity. They are self-contained generators that operate continuously without sunlight, but must carry their own reactant supply, usually. The longer the mission, the larger the reactant tanks. The most popular version for space applications is the hydrogen-oxygen (referred to as "alkaline" because of the KOH electrolyte) fuel cell because of its relatively high specific power (275 W/kg on the Space Shuttle), low reactant mass (hydrogen and oxygen), and useful by-product (water).

A typical single cell produces a voltage of 0.8 Vdc. In combination, a fuel cell unit can create many kilowatts of power (each Shuttle fuel cell produces 16 kW peak or 12 kW continuous). The energy conversion efficiency can run as high as 80% for low current draws, but as current increases, the efficiency drops to 50–60%, due to activation overpotential and electrical resistance in the electrolyte solution between electrodes. However, compared with other power sources, fuel cell efficiencies are high.

The three Space Shuttle fuel cells are state-of-the-art power generators that produce all of the Shuttle electricity for the 28 Vdc bus. Their high efficiency (70%), low weight (118 kg), and excellent reliability (> 99% available) attest to their quality. Other important factors are their 15-min start-up time, instantaneous shutdown, and long lifetime (2,400 hours before refurbishment). Besides electricity, these fuel cells produce crew drinking water, at a rate of 0.36 kg/kWh, or about 104 kg a day.[*]

Research is underway to solve the short-mission limit with fuel cells, caused by carrying large reactant masses. Because the fuel-cell reaction is reversible, we can use electrolysis to create more reactants from the water by-product. To optimize each process, however, we have to use separate units for generating electricity and separating the water. Any long-duration mission could use this *regenerative* system if it had some input electricity from solar cells, nuclear generators, or other power system during periods of low electrical load.

Earth-orbiting spacecraft at low-Earth to geosynchronous orbits have usually employed photovoltaics as their power source. Often, photovoltaics were the only real candidate for these low-power missions (less than 15 kW) because solar cells were well-known and reliable. Photovoltaic sources are not attractive for interplanetary missions to the outer planets because solar radiation decreases, thus reducing the available energy from a solar array. To configure and size a solar array, we must understand cell types and characteristics; solar-array design issues, types, sizing calculations, configurations, regulation; and radiation and thermal environments. Key design issues for solar arrays include spacecraft configuration, required power level (peak and average), operating temperatures, shadowing, radiation environment, illumination or orientation, mission life, mass and area, cost, and risk. Table 11-34 shows the solar array design process.

Step 1. Mission life and the average power requirement are the two key design considerations in sizing the solar array for most spacecraft. We size a photovoltaic system to meet power requirements at EOL, with the resulting solar array often oversized for power requirements at BOL. This excess power at BOL requires coordinated systems engineering to avoid thermal problems. The longer the mission life, the larger the difference between power requirements at EOL and BOL. We usually consider photovoltaics a poor power source for missions lasting more than 10 years because of natural degradation in the solar array. Section 11.4.4 discusses how we manage excess power from the solar array. The average power requirement can be obtained from Secs. 10.1 and 10.2.

---

[*] Telephone conversation with Jay Garrows, International Fuel Cells, Inc., Oct. 98.

TABLE 11-34. Solar Array Design Process. In the FireSat example column, I_d represents inherent degradation, θ is the Sun incidence angle, L_d is life degradation, and X_e and X_d represent the efficiencies of the power distribution paths. The material following the table further explains these quantities.

Step	Reference	FireSat Example
1. Determine requirements and constraints for power subsystem solar array design		
• Average power required during daylight and eclipse	Input parameter, Secs. 10.1, 10.2	110 W during daylight and eclipse
• Orbit altitude and eclipse duration	Input parameter, end papers	700 km 35.3 min
• Design lifetime	Chaps. 2, 3	5 yr
2. Calculate amount of power that must be produced by the solar arrays, P_{sa}	Step 1 Eq. 5-5, end papers (Orbit period − T_e) Eq. 11-5	$P_e = P_d = 110$ W $T_e = 35.3$ min $T_d = 63.5$ min Assume a peak power tracking regulation scheme with $X_e = 0.6$ and $X_d = 0.8$ $P_{sa} = 239.4$ W
3. Select type of solar cell and estimate power output, P_o, with the Sun normal to the surface of the cells	*Si: $P_o = 0.148 \times 1{,}367$ W/m² $= 202$ W/m² *GaAs: $P_o = 0.185 \times 1{,}367$ W/m² $= 253$ W/m² *Multijunction: $P_o = 0.22 \times 1{,}367$ W/m² $= 301$ W/m²	Si solar cells $P_o = 202$ W/m²
4. Determine the beginning-of-life (BOL) power production capability, P_{BOL}, per unit area of the array	Table 11-35 Eq. 5-7 Eq. 11-6	$I_d = 0.77$ $\theta = 23.5$ deg (worst case) $P_{BOL} = 143$ W/m²
5. Determine the end-of-life (EOL) power production capability, P_{EOL}, for the solar array	Performance degradation Si: 3.75% per yr, GaAs: 2.75% per yr, Multijunction: 0.5% per yr Eq. 11-7 Eq. 11-8	Performance degradation is 3.75% per year $L_d = 0.826$ for 5 yr mission $P_{EOL} = 118.1$ W/m²
6. Estimate the solar array area, A_{sa}, required to produce the necessary power, P_{sa}, based on P_{EOL} an alternate approach	Eq. 11-9 Eq. 10-12†	$A_{sa} = 2.0$ m² $A_{sa} = 2.5$ m²
7. Estimate the mass of the solar array	Eq. 10-13†	$M_a = 9.6$ kg
8. Document assumptions		

* Typical demonstrated efficiencies for Si, GaAs, and multijunction solar cells are 14.8%, 18.5%, and 22%, respectively.

† Use P_{sa} in these equations.

In designing a solar array, we trade off mass, area, cost, and risk. Silicon presently costs the least for most photovoltaic power applications, but it often requires larger area arrays and more mass than the more costly gallium-arsenide cells. Programs for which mass and volume (solar array area) are critical issues may allow higher costs or technical risks. They could select a system based on gallium arsenide or some other advanced type of solar cell. Risk develops from the unproven reliability and fabrication of the photovoltaic source.

A solar array's illumination intensity depends on orbital parameters such as the Sun incidence angles, eclipse periods, solar distance, and concentration of solar energy. Tracking and pointing mechanisms on the solar array often adjust for these influences. If we mount the cells on the body of the spacecraft, we must orient them so they will generate adequate power throughout the mission.

Step 2. To estimate the solar-array area required for a spacecraft, we first determine how much power, P_{sa}, the solar array must provide **during daylight** to power the spacecraft for the entire orbit

$$P_{sa} = \frac{\left(\dfrac{P_e T_e}{X_e} + \dfrac{P_d T_d}{X_d} \right)}{T_d} \tag{11-5}$$

where P_e and P_d are the spacecraft's power requirements (excluding regulation and battery charging losses) during eclipse and daylight, respectively, and T_e and T_d are the lengths of these periods per orbit. The terms X_e and X_d represent the efficiency of the paths from the solar arrays through the batteries to the individual loads and the path directly from the arrays to the loads, respectively. The efficiency values for eclipse and daylight depend on the type of power regulation: direct energy transfer or peak-power tracking. (A description of these methods follows in Sec. 11.4.4.) For direct energy transfer, the efficiencies are about $X_e = 0.65$ and $X_d = 0.85$; for peak-power tracking they are $X_e = 0.60$ and $X_d = 0.80$. The efficiencies of the former are about 5% to 7% greater than the latter because peak-power tracking requires a power converter between the arrays and the loads.

Step 3. Table 11-35 shows the efficiencies and radiation-degradation sensitivities of three main types of cells. Gallium arsenide has the advantage of higher efficiencies, whereas indium phosphide reduces the degrading effects of radiation. Silicon solar cell technology is mature and has the advantage of lower cost per watt for most applications. Gallium arsenide and indium phosphide cost about 3 times more than silicon.

The *energy-conversion efficiency* of a solar cell is defined as the power output divided by the power input. The power input value for a planar solar array is the solar-illumination intensity (1,367 W/m²). Thus, a solar panel with a BOL efficiency of 18% will provide 246 W/m². We must be aware that reported efficiency values for solar panels often apply only to single cells. We need to identify losses inherent to panel assembly (diodes, interconnect cabling, transmission losses) to size the array adequately. We also need to note that these efficiency values often refer to **laboratory** cells and not **production** cells, which have lower average efficiencies.

To complete this step, we identify the type of solar cells and how their performance will degrade during the mission. Ideally, silicon and gallium arsenide solar cells have efficiencies of about 14.8% and 18.5%. These solar cell efficiencies give us **ideal** solar cell output performance per unit area, P_o, of 202 W/m² and 253 W/m², respectively, if the incident solar radiation (1,367 W/m²) is normal to the surface.

TABLE 11-35. **Performance Comparison for Photovoltaic Solar Cells.** Note that the stated efficiencies are for single solar cells, not solar arrays.

Cell Type	Silicon	Thin Sheet Amorphous Si	Gallium Arsenide	Indium Phosphide	Multijunction GaInP/GaAs
Planar cell theoretical efficiency	20.8%	12.0%	23.5%	22.8%	25.8%
Achieved efficiency: Production Best laboratory	14.8% 20.8%	5.0% 10%	18.5% 21.8%	18% 19.9%	22.0% 25.7%
Equivalent time in geosynchronous orbit for 15% degradation – 1 MeV electrons – 10 MeV protons	10 yr 4 yr	10 yr 4 yr	33 yr 6 yr	155 yr 89 yr	33 yr 6 yr

Step 4. Next, we must determine the **realistic** power production capability of the manufactured solar array. As shown in Table 11-36, an assembled solar array is less efficient than single cells due to design inefficiencies, shadowing and temperature variations, collectively referred to as *inherent degradation, I_d.* Solar cells are applied to a substrate, usually honeycomb aluminum, and interconnected, resulting in losses of 10% of the solar array's substrate area. This accounts for the design and assembly losses. If we configure the spacecraft well, its appendages will shadow few cells, and shadowing losses should be slight. The temperature of a typical flat solar panel receiving normal incident radiation ranges from about 67 °C in LEO to 53 °C in GEO. The reference temperature for silicon solar cells is 28 °C, with performance falling off 0.5% per degree above 28 °C. Body-mounted arrays on nonspinning spacecraft are typically about 5 °C warmer than deployed solar arrays because they can't radiate heat into deep space as efficiently.

TABLE 11-36. **Elements of Inherent Solar Array Degradation.** Although individual solar cells may have adequate efficiency, after we manufacture the solar array, these elements cause some degradation in the cumulative efficiency by the amounts indicated.

Elements of Inherent Degradation	Nominal	Range
Design and Assembly	0.85	0.77–0.90
Temperature of Array	0.85	0.80–0.98
Shadowing of Cells	1.00	0.80–1.00
Inherent Degradation, I_d	0.77	0.49–0.88

As mentioned earlier, we commonly refer to the current-voltage characteristics of a solar cell as the I-V curves. Figure 11-9 depicts a first-quadrant I-V curve for a planar array in LEO. This curve characterizes BOL and EOL performance. As the figure illustrates, the three significant points for solar-array design are:

- *Short-circuit current, I_{sc},* where voltage = 0
- *Peak-power point,* where voltage times current is maximized
- *Open-circuit voltage, V_{oc},* where current = 0

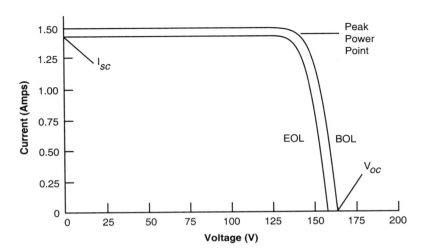

Fig. 11-9. I-V Plot for a Planar Array. The power available is simply the area under the curve.

We must also consider how temperature affects the I-V characteristics. While the spacecraft is in eclipse, the solar-array temperature can get as low as –80 °C. The highest operating temperature for an LEO spacecraft solar array is 100 °C, occurring near the end of a full Sun period during an orbit.

The operating temperature of the array is a key issue because the solar cell's performance depends on temperature. A *current-voltage,* or *I-V plot,* illustrates the performance of a solar-array cell, or the array (see Fig. 11-9). A change in the operating temperature of the solar cell or array causes three changes in the I-V curve:

- A scaling of the I-V curve along the current axis

- A translation or shifting of the I-V curve along the voltage axis

- A change in the I-V curve shape affecting the roundness of the knee region

The *temperature coefficient,* or percent degradation of performance with increasing temperature, for any solar cell depends on factors such as the type of cell and its output-power characteristics, actual operating temperature, and radiation environment. Gallium arsenide and indium phosphide have lower temperature coefficients, but higher temperature still means reduced performance. Solar arrays using gallium arsenide and indium phosphide also resist radiation better than silicon and provide greater EOL power for a given area. We must establish a profile for operating temperatures during a mission, so the photovoltaic system can generate adequate power throughout.

The peak-power point depends on the array's operating temperature at BOL and EOL. Thus, an array often provides maximum power coming out of an eclipse period because it is at its coldest operating temperature. Transient voltage excursions often occur when leaving eclipse, so we may need to clamp voltages to protect spacecraft loads. By understanding how the array's performance relates to these variables, we can get the highest output power from the array.

Usually, solar cells in series-parallel combinations make up a *solar array.* The number of series-connected solar cells in one string establishes the bus voltage required at EOL at the operating temperature; the number of parallel strings depends on the required current output. Isolation diodes mounted within the solar array typically minimize the effects of shadowing and reversed-biased solar cells.

Solar-array configurations are either planar or concentrator, and either type can be body- or panel-mounted. Most photovoltaic applications to date have employed a *planar array* in which solar cells are mounted onto a surface (typically insulated aluminum honeycomb) with an adhesive. A Kapton, Kevlar, or fiberglass sheet usually insulates the solar cell from the aluminum honeycomb support structure. *Concentrator solar arrays* increase the solar cell's output by using mirrors or lenses to focus more solar radiation on the cells.

Panel-mounted solar arrays usually apply only to 3-axis stabilized spacecraft. The panel-mounted approach tracks and points the solar array to get the best Sun incidence angle. The body-mounted approach reduces the requirements for tracking and pointing on any spacecraft (spinning or stabilized). But the less effective Sun incidence angle and increased array temperature of body-mounted cells produce a lower efficiency in orbit. Panel-mounted solar arrays are usually mounted on a boom. Deployable panel arrays are either flexible or rigid, according to the type of substrate material employed for mounting. For most spacecraft, we try to place the solar array away from the payload and other spacecraft subsystems because of the variable and often high temperature of the solar cells.

Body-mounted planar cells are typical on spinning spacecraft, which provide thermal control by radiating excess heat to space as the spacecraft spins. Body-mounted solar arrays use cells inefficiently because of higher temperature and reduced voltage. Thus, they generate lower power per unit area than a deployed, oriented panel. When solar cells are body-mounted to a spinning spacecraft, the array's total output power decreases because the cells are not always oriented toward the Sun. This decrease depends on the spacecraft's configuration and the drive mechanisms of the solar array (if any). For example, a stabilized array using Sun-tracking and pointing on two axes would fully use the solar array's surface area. But the array's reduction in output power per total surface area would be approximately π for body-mounted cells on a cylindrical, spinning spacecraft and 4 for body-mounted cells on a cubic-shaped spacecraft that does not employ active tracking. The output power decreases because not all cells are illuminated. We must trade the cost and design for the solar array's total surface area against the cost and complexity of stabilizing the spacecraft and using a drive system for the solar array.

Shadowing considerations are important because a solar cell will go into open circuit (become high resistance) when not illuminated. In a series-connected string of solar cells, the shadowing of one cell results in the loss of the entire string. Shadowing may be caused by spacecraft components such as transmitting or receiving antennas, deployment mechanisms, or structures such as the solar-array. We can reduce shadowing effects by actively pointing and tracking solar arrays on 3-axis stabilized spacecraft, using diodes, or designing series-parallel arrays. On spinning spacecraft, we must lay out solar cells so all solar cells within a string are illuminated. Diodes, which bypass groups of solar cells in a string, help prevent damage to reduce the advance effects of shadowed solar cells.

We can improve solar cell performance with coverslides, coatings, and back-surface reflectors. Coverslides provide a hermetic seal yet allow the cell to receive sunlight and reject heat. They are textured or smooth. A textured coverslide is used for body-mounted solar cells that do not actively point toward the Sun. It reflects incident solar energy back onto the solar cell, improving the overall efficiency. Smooth cover-slides are used for spacecraft whose arrays actively track and point. By decreasing reflective losses on solar cells, coatings allow cells to use more of the incident energy. Back-surface reflectors direct incident solar radiation that passes through the solar cell back through the cell again to improve overall efficiency. By reducing solar absorptance, they help the solar array manage thermal energy. Solar-cell vendors are continually improving the mechanical and thermal characteristics of coverslides, coatings, and back-surface reflectors. Thus, we must coordinate mechanical and thermal characteristics of these cells with the vendors.

At beginning-of-life, the array's power per unit area is

$$P_{BOL} = P_o I_d \cos \theta \tag{11-6}$$

where $\cos \theta$ is referred to as the *cosine loss*. We measure the *Sun incidence angle, θ,* between the vector normal to the surface of the array and the Sun line. So if the Sun's rays are perpendicular to the solar array's surface, we get maximum power. Obviously, the geometry between the array and the Sun changes throughout the mission and different solar array panels will have different geometry. We configure the solar array to minimize this cosine loss. For a flat, silicon solar array with a worst-case Sun angle of $\theta = 23.5$ deg angle between equatorial and ecliptic planes and the nominal value of I_d, the power output at beginning-of-life is 143 W/m^2.

Step 5. Radiation damage severely reduces a solar array's output voltage and current. At geosynchronous altitude, we must guard against solar-flare protons on-station, trapped electrons on-station, and trapped electrons and protons during transfer orbits. (Chapter 8 explains these terms.) Electrons and protons trapped in the Earth's magnetic field cause most degradation of solar cells. Silicon solar cells protected by coverslides lose 15% of their voltage and current (shielding assumed) when exposed to a total fluence of 10^{15} MeV equivalent electrons (4 to 5 years for a LEO spacecraft). As mission planners, we should coordinate degradation characteristics with the solar-cell manufacturer, based on the radiation environment the spacecraft will encounter. Degradation of a solar cell also depends on its design. Advanced technologies, such as indium phosphide cells, are more radiation hardened.

Next, we must consider the factors that degrade the solar array's performance during the mission. *Life degradation, L_d,* occurs because of thermal cycling in and out of eclipses, micrometeoroid strikes, plume impingement from thrusters, and material outgassing for the duration of the mission. In general, for a silicon solar array in LEO, power production can decrease by as much as 3.75% per year, of which up to 2.5% per year is due to radiation. For gallium-arsenide cells in LEO, the degradation is about 2.75% per year, of which radiation causes 1.5% per year. The actual lifetime degradation can be estimated using

$$L_d = (1 - degradation/yr)^{satellite\ life} \tag{11-7}$$

The array's performance per unit area at end-of-life is

$$P_{EOL} = P_{BOL} L_d \tag{11-8}$$

Using the FireSat example array in Table 11-34 for a 5-year mission, L_d is 82.6%, resulting in a P_{EOL} of 118.1 W/m^2. The solar-array area, A_{sa}, required to support the spacecraft's power requirement, P_{sa}, is

$$A_{sa} = P_{sa} / P_{EOL} \tag{11-9}$$

The resulting solar-array area for the example spacecraft is about 2.0 m^2. If we had used a perfectly pointed array, the BOL power would have been 155 W/m^2, resulting in an EOL power of 128 W/m^2 and an array area of 1.9 m^2. So, having to account for the cosine loss costs us 0.1 m^2 in array size and the equivalent mass.

Solar-array sizing is more difficult than it appears from the above discussion. Typically, we must consider several arrays with varying geometry. Also, the angle of incidence on the array surface is constantly changing. We must predict that angle continuously or at least determine the worst-case angle to develop an estimate of P_{EOL}.

11.4.2 Energy Storage

Energy storage is an integral part of the spacecraft's electrical-power subsystem providing all the power for short missions (< 1 week) or back-up power for longer missions (> 1 week). Any spacecraft that uses photovoltaics or solar thermal dynamics as a power source requires a system to store energy for peak-power demands and eclipse periods. Energy storage typically occurs in a battery, although systems such as flywheels and fuel cells have been considered for various spacecraft.

A *battery* consists of individual cells connected in series. The number of cells required is determined by the bus-voltage. The amount of energy stored within the battery is the *ampere-hour capacity* or *watt-hour* (ampere-hour times operating voltage) *capacity*. The design or nameplate capacity of the battery derives from the energy-storage requirements. Batteries can be connected in series to increase the voltage or in parallel to increase this current output—the net result being an increase in watt-hour capacity.

Table 11-37 lists issues to consider early in the conceptual phase of any program. Most of all, we try to provide a stable voltage for all operating conditions during the mission life because load users prefer a semi-regulated bus voltage. The difference in energy-storage voltage between end of charge and end of discharge often determines the range of this bus voltage.

TABLE 11-37. Issues in Designing the Energy Storage Capability. Energy storage usually means large batteries and we must consider all their characteristics when designing this subsystem.

Physical	Size, weight, configuration, operating position, static and dynamic environments
Electrical	Voltage, current loading, duty cycles, number of duty cycles, activation time and storage time, and limits on depth-of-discharge
Programmatic	Cost, shelf and cycle life, mission, reliability, maintainability, and produceability

Figure 11-10 highlights the charge-discharge characteristics of a spacecraft's energy-storage system. We want a flat discharge curve that extends through most of the capacity and little overcharge. Overcharging quickly degrades most batteries. We also need to match the electrical characteristics of the battery cells. Otherwise, charge imbalances may stress and degrade the batteries, resulting in a shorter life for the electrical-power subsystem.

All battery cells are either primary or secondary. *Primary battery* cells convert chemical energy into electrical energy but cannot reverse this conversion, so they cannot be recharged. Primary batteries typically apply to short missions (less than one day) or to long-term tasks such as memory backup, which use very little power. The

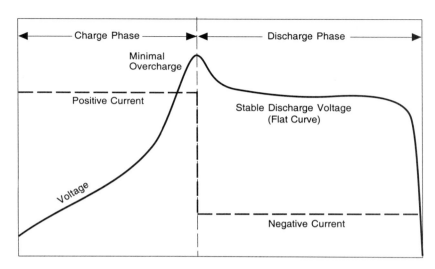

Fig. 11-10. Profile of Charge/Discharge Voltages for Batteries. Secondary batteries may cycle through this type of profile hundreds or thousands of times during their mission life. At the left edge, the voltage is low because the spacecraft just came out of eclipse where it used battery power. During the charge phase, there is positive current from the power regulator, so the battery voltage rises. In the discharge phase (in eclipse again), there is a negative current, so the battery voltage decreases.

most common batteries use silver zinc, lithium thionyl chloride, lithium sulfur dioxide, lithium monoflouride, and thermal cells. Table 11-38 highlights the applications and relative merits. It also depicts the wide ranges in each couple's specific-energy density. We cannot specify a value for specific-energy density because cells vary in design and depend on mission requirements. We must coordinate mission requirements with the battery manufacturer to specify battery performance.

TABLE 11-38. Characteristics of Selected Primary Batteries. Primary battery manufacturers can meet power requirements within these ranges of specific energy density. We must tradeoff cost and mass with capacity while ensuring mission accomplishment.

Primary Battery Couple	Specific Energy Density (W·hr/kg)	Typical Application
Silver Zinc	60 – 130	High rate, short life (minutes)
Lithium Thionyl Chloride	175 – 440	Medium rate, moderate life (< 4 hours)
Lithium Sulfur Dioxide	130 – 350	Low/medium rate, long life (days)
Lithium Monoflouride	130 – 350	Low rate, long life (months)
Thermal	90 – 200	High rate, very short life (minutes)

A *secondary battery* for energy storage can convert chemical energy into electrical energy during discharge and electrical energy into chemical energy during charge. It can repeat this process for thousands of cycles. Table 11-39 shows ranges of specific-energy density for common secondary batteries. A secondary battery provides power during eclipse periods on spacecraft that employ photovoltaics and can also level loads. Secondary batteries recharge in sunlight and discharge during eclipse. The spacecraft's orbital parameters, especially altitude, determine the number of charge/discharge cycles the batteries has to support during the mission life. A geosynchronous satellite needs to store energy for two 45-day eclipse periods per year with eclipses lasting no more than 72 min each day. The geosynchronous orbit demands few charge/discharge cycles during eclipse periods, thus allowing a fairly high (50%) depth-of-discharge. On the other hand, LEO spacecraft encounter at most one eclipse period each orbit or about 15 eclipse periods per day, with maximum shadowing of approximately 36 min. Therefore, the batteries must charge and discharge about 5,000 times each year, and the average depth-of-discharge is only 15–25%—much lower than for geosynchronous spacecraft.

TABLE 11-39. Characteristics of Selected Secondary Batteries. Though secondary batteries have much lower specific energy densities than primary batteries, their ability to be recharged makes them ideal for backup power on spacecraft powered by solar cells.

Secondary Battery Couple	Specific Energy Density (W·hr/kg)	Status
Nickel-Cadmium	25 – 30	Space-qualified, extensive database
Nickel-Hydrogen (individual pressure vessel design)	35 – 43	Space-qualified, good database
Nickel-Hydrogen (common pressure vessel design)	40 – 56	Space-qualified for GEO and planetary
Nickel-Hydrogen (single pressure vessel design)	43 – 57	Space-qualified
Lithium-Ion ($LiSO_2$, $LiCF$, $LiSOCl_2$)	70 – 110	Under development
Sodium-Sulfur	140 – 210	Under development

Depth-of-discharge (DOD) is simply the percent of total battery capacity removed during a discharge period. Higher percentages imply shorter cycle life as shown in Fig. 11-11. Once we know the number of cycles and the average depth of discharge, we can determine the total capacity of the batteries.

Figure 11-11 illustrates the relationship between average depth-of-discharge (DOD) and cycle life for secondary batteries using nickel cadmium (NiCd) and nickel hydrogen (NiH_2). Extensive data supports the predictions for both NiCd and NiH_2.

The NiCd battery is still a common secondary energy storage system for many aerospace applications. NiCd technology has been space qualified, and we have extensive databases for nearly any mission. A 28 Vdc aerospace NiCd battery usually consists of 22–23 series-connected cells. NiCd battery cells for aerospace missions have typical capacities of 5 to 100 Amp-hr.

NiH_2 technology has been the recently qualified energy storage system of choice for aerospace applications where higher specific energies and longer life are important. The three space-qualified design configurations for NiH_2 are individual pressure vessel, common pressure vessel, and single pressure vessel. The individual pressure vessel was the first NiH_2 technology used for aerospace application. Here, only a single electrochemical cell is contained within the pressure vessel. It has a working

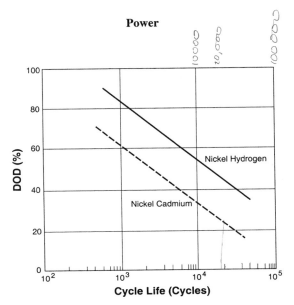

Fig. 11-11. Depth-of-Discharge vs. Cycle Life for Secondary Batteries. Increased cycle life reduces the amount of energy available from the batteries during each cycle—DOD decreases with cycle life.

terminal voltage of 1.22 to 1.25 Vdc depending upon discharge loads. The typical individual pressure vessel battery design consists of multiple cells connected in series to obtain the desired battery voltage. Cell diameters are typically 9 to 12 cm, with capacity ranges from 20 to over 300 Amp-hr. The common pressure vessel NiH$_2$ technology is very similar to individual pressure vessel, with the primary difference in the wiring connection of the internal electrode stacks. In the individual pressure vessel, the electrode stacks are all connected in parallel. In a common pressure vessel, there are two sets of electrode stacks within the pressure vessel that are series connected, yielding a working terminal voltage of 2.44 to 2.50 Vdc. This design has a higher specific-energy at the battery level since there are half as many pressure vessels and a significant reduction in cell piece-parts. Common pressure vessel NiH$_2$ technology has been space qualified in the 6 cm and 9 cm cell diameter configuration for capacities in the 12 to 20 Amp-hr range. Batteries with larger Amp-hr capacities should be qualified for aerospace application in the near future. The single pressure vessel NiH$_2$ battery is designed such that a common hydrogen supply is used by three or more series connected cells with a single pressure vessel. Each cell stack contains its own electrolyte supply which is isolated within individual cell stack containers. The key operating characteristic of this design is to allow the free movement of hydrogen within the cell stacks while maintaining cell stack electrolyte isolation. These batteries are presently available in a 12.5 cm or 25 cm diameter design.

Lithium Ion battery technology offers a significant energy density advantage and a much wider operating temperature range over NiCd and NiH$_2$ battery types. Typical cell constituents are lithium thionyl chloride, lithium sulfur dioxide, and lithium carbon monofluoride. The nominal operating voltage for a lithium ion cell is 3.6 to 3.9 Vdc, which allows us to reduce the number of cells by approximately one-third when compared to NiCd or NiH$_2$ cells. The lithium ion secondary battery system offers a 65% volume advantage and a 50% mass advantage for most present day aerospace battery applications. Lithium ion battery technology should be qualified for a planetary mission by the year 2000, with space qualification for GEO and LEO applications by the years 2005–2010.

To size a secondary battery, we must identify the parameters and apply the equation in Table 11-40. The parameter values used in the equation can vary significantly with battery type. The ideal battery capacity is the average eclipse load, P_e, times the eclipse duration, T_e. This ideal capacity must be increased to include the battery-to-load transmission efficiency, n, and the depth-of-discharge constraints. For LEO, we expect the battery's DOD to be 40–60% for NiH_2 technology, compared to 10–20% for NiCd technology. We base these expectations on the average DOD over 24 hours and assume the batteries are fully recharged at least once during this period. The number of batteries, N, may be equal to one for this calculation if you simply require a battery capacity. Two to five batteries are typical. We must have at least two (unless the battery uses redundant cells) because the spacecraft needs redundant operation with one unit failed. But more than five batteries require complex components for recharging. The secondary batteries may be required to help meet peak power loads during full Sun conditions. For some missions, the peak power loads may drive the required battery capacity rather than the eclipse load. To design the Energy-Storage subsystem, follow the steps in Table 11-40

TABLE 11-40. Steps in the Energy Storage Subsystem Design. To obtain the required battery capacity in Amp-hr, divide by the required satellite bus voltage.

Step	Consider	FireSat Example
1. Determine the energy storage requirements	• Mission length • Primary or secondary power storage • Orbital parameters 　– Eclipse frequency 　– Eclipse length • Power use profile 　– Voltage and current 　– Depth of discharge 　– Duty cycles • Battery charge/discharge cycle limits	• 5 yrs • Secondary power storage • 16 eclipses per day • 35.3 min per eclipse (T_e) • Eclipse load 110 W (P_e) 　– 26.4 V, 4.2 A (max) • 20% (upper limit) • TBD—depends on observations taken and downlinked during eclipses
2. Select the type of secondary batteries	• NiCd (space qualified) • NiH_2 (space qualified) • Li-ion (under development) • NaS (under development)	• NiCd or NiH_2—both are space-qualified and have adequate characteristics
3. Determine the size of the batteries (battery capacity)	• Number of batteries • Transmission efficiency between the battery and the load	• N = 3 batteries (nonredundant) • n = 0.90 • C_r = 119 W-hr • C_r = 4.5 Amp-hr (26.4 V bus)

Battery Capacity: $C_r = \dfrac{P_e T_e}{(DOD)Nn}$ W-hr (for battery capacity in Amp-hr, divide by bus voltage)

11.4.3 Power Distribution

A spacecraft's power distribution system consists of cabling, fault protection, and switching gear to turn power on and off to the spacecraft loads. It also includes command decoders to command specific load relays on or off. The power distribution system is a unique feature of the electrical-power subsystem and often reflects individual spacecraft loads and power-switching requirements. Power distribution designs for various power systems depend on source characteristics, load requirements, and subsystem functions. In selecting a type of power distribution, we focus on keeping power losses and mass at a minimum while attending to survivability, cost, reliability, and power quality.

Power switches are usually mechanical relays because of their proven flight history, reliability, and low power dissipation. Solid-state relays, based on power technology, which uses metal-oxide semiconductor field-effect transistors are available.

The load profile of a spacecraft is a key determining factor in the design specifications of a power distribution subsystem. Predominant spacecraft loads (radar, communications, motors, computers) may require low- to high-voltage dc (5–270 Vdc), high-voltage single-phase ac (115 Vrms, 60 Hz), or high-voltage three-phase ac (120/440 Vrms, 400 Hz)—all converted from the 28-Vdc power bus. Because the regulation requirements for these loads vary, the bus voltage may need further regulating, leveling up or down, and, possibly, inverting through dc-dc converters. Spacecraft power loads often turn on or off or otherwise vary their power consumption. Transient behavior within a load may produce noise that the distribution system translates to other loads, potentially harming working components. In addition, certain spacecraft loads require a voltage different from the bus voltage. Power converters often connect loads susceptible to noise or requiring voltage conversion to the distribution system. These converters typically isolate the load from the noise on the bus and regulate the power provided to the load against disturbances from the load and the bus. They also keep load failures from damaging the power-distribution system and provide on-off control to desired loads. Any dc-dc converter connected to the bus must dampen its electromagnetic-interference filter to keep step loads from causing excessive ringing.

We need to know the boundaries of the load profile to evaluate its effects on required bus voltage and frequency. Most spacecraft have demanded low power (< 2,000 W), so power distribution has relied on a standard, 28 V bus. This standard, with electronic parts built to match, has limited study of the best bus voltage. As power systems expand to many kilowatts, the 28 V bus may not work for power distribution because of losses in cabling and limits on mass. The harness or cabling that interconnects the spacecraft's subsystems is a large part (10–25%) of the electrical-power system's mass. We must keep harnesses as short as possible to reduce voltage drops and to regulate the bus voltage. Figure 11-12 depicts the relationship between current and cable mass.

Systems for distributing power on spacecraft have been predominantly dc because spacecraft generate direct current power. Direct-current systems will dominated throughout the 1990s. Conversion to ac would require more electronics, which would add mass to the EPS. Alternating-current power distribution applies only for high-power spacecraft, such as the International Space Station, which have many electrical loads with varying duty cycles. Even on the space station, however, recent decisions have taken planners back to dc for the entire distribution system.

Power distribution systems are either centralized or decentralized, depending on the location of the converters. The decentralized approach places the converters at each

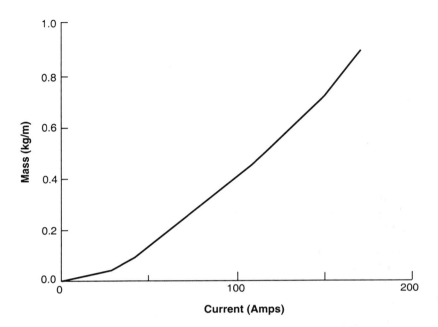

Fig. 11-12.   Cable Mass vs. Current. We must account for the cable and harness mass when designing the Power Subsystem. Operating low current (less than 30 amps) devices helps keep this mass low.

load separately, whereas the centralized approach regulates power to all spacecraft loads within the main bus. The decentralized approach implies an unregulated bus because distributed converters regulate power. A regulated power bus typically has some power converters at the load interface because electronics may require different voltages (+5, ±12 Vdc). An advantage of the centralized system is that we do not have to tailor-design the EPS for different applications. Larger spacecraft with high power levels use the decentralized distribution systems, with an unregulated bus, usually.

Fault protection within the EPS focuses on detection, isolation, and correction of faults. Its main purpose is to isolate a failed load that could eventually cause loss of the mission or the spacecraft. A failed load typically implies a short circuit, which will draw excessive power. If this condition continues, the failed load may stress cables and drain the energy-storage reserve. Typically, we would isolate these faults from the EPS bus with fuses (sometimes resettable). Most spacecraft power loads have some sort of fuse in series with the power bus to isolate faults. Of course, if the mission requires us to know where load faults occur, we can add fault-detection circuits. To design the Power Distribution subsystem, follow the steps in Table 11-41.

11.4.4 Power Regulation and Control

The energy source determines how we regulate a spacecraft's power. For example, we regulate a static or dynamic power source through the direct energy transfer method discussed below. But because most aerospace applications use solar photovol-

TABLE 11-41.   **Steps in the Power Distribution Subsystem Design.**

Step	Consider	Possibilities
1. Determine the electrical load profile	• All spacecraft loads, their duty cycles, and special operating modes • Inverters for ac requirements • Transient behavior within each load • Load-failure isolation	• Low-voltage dc: 5 V • High-voltage dc: 270 V • High-voltage 1-phase ac: 115 V_{rms}, 60 Hz • High-voltage, 3-phase ac: 120/440 V_{rms}, 400 Hz
2. Decide on centralized or decentralized control	• Individual load requirements • Total system mass	• Converters at each load—for a few special loads • Centralized converters control voltage from the main bus (no specialized power requirements)
3. Determine the fault protection subsystem	• Detection (active or passive) • Isolation • Correction (change devices, reset fuses, work around lost subsystem)	• Cable size (length and diameter) and excess current-carrying ability • Size of power storage in case of a short circuit • Location of fuses and their type

taics, we will examine power regulation emphasizing that viewpoint. Power regulation divides into three main categories: controlling the solar array, regulating bus voltage, and charging the battery.

We must control electrical power generated at the array to prevent battery over-charging and undesired spacecraft heating. The two main power control techniques, illustrated in Fig. 11-13, are a *peak-power tracker* (*PPT*) and a *direct-energy-transfer* (*DET*) subsystem. A PPT is a nondissipative subsystem because it extracts the exact power a spacecraft requires up to the array's peak power. The DET subsystem is a dis-sipative subsystem because it dissipates power not used by the loads. However, a DET subsystem can dissipate this power at the array or through external banks of shunt resistors to avoid internal power dissipation. DET subsystems commonly use shunt regulation to maintain the bus voltage at a predetermined level. Figure 11-13 depicts the main functional differences between varying PPT and shunt-regulated DET sub-systems.

A PPT is a dc-dc converter operating in series with the solar array. Thus, it dynamically changes the operating point of the solar-array source to the voltage side of the array (Fig. 11-13) and tracks the peak-power point when energy demand exceeds the peak power. It allows the array voltage to swing up to its maximum power point; then the converter transforms the input power to an equivalent output power, but at a different voltage and current. Solar-source characteristics permit us to extract large amounts of power when the array is cold (post eclipse) and at the beginning of life. A peak-power tracker replaces the shunt-regulation function by backing off the peak-power point of the arrays toward the end of the battery's charging period. Because the PPT is in series with the array, it uses 4–7% of the total power. A PPT has advantages for missions under 5 years that require more power at BOL than at EOL.

For direct energy transfer systems a *shunt regulator* operates in parallel to the array and shunts the array current (typically at the array) away from the subsystem when the

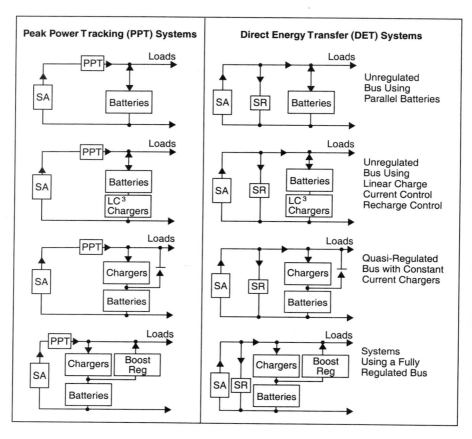

Fig. 11-13. Techniques for Power Regulation. The basic approaches are *Peak Power Tracking* (*PPT*), which places a regulator in series with the solar arrays and the load, and *Direct Energy Transfer* (*DET*), which uses a regulator in parallel with the solar arrays and load.

loads or battery charging do not need power. Power subsystems with shunt regulation are extremely efficient. They dissipate little energy by simply shunting excess power at the array or through shunt resistor banks. A shunt-regulated subsystem has advantages: fewer parts, lower mass, and higher total efficiency at EOL.

Techniques for controlling bus voltage on electrical-power subsystems fall into three categories: unregulated, quasi-regulated, or fully regulated. Figure 11-13 illustrates the main differences between these techniques. An unregulated subsystem has a load bus voltage that varies significantly. The bus-voltage regulation derives from battery regulation, which varies about 20% from charge to discharge. In an *unregulated* subsystem, the load bus voltage is the voltage of the batteries.

Quasi-regulated subsystems regulate the bus voltage during battery charge but not during battery discharge. A battery charger is in series with each battery or group of parallel batteries. During charge the bus voltage fixes at a potential several volts above the batteries. As the batteries reach full charge, the drop across the chargers decreases, but the bus voltage is still constantly regulated. The bus becomes unregulated during discharge when the voltage is about a diode drop lower than the batteries and decreases as the batteries further discharge. A quasi-regulated power subsystem has low efficiency and high electromagnetic interference if used with a peak-power tracker.

The *fully regulated* power subsystem is inefficient, but it will work on a spacecraft that requires low power and a highly regulated bus. This subsystem employs charge and discharge regulators. We can design the regulators so the charge regulator uses linear technology and the discharge regulator is a switching converter, but for best efficiency both should be converters. The advantage of this type of power subsystem is that, when we connect it to the loads, the system behaves like a low-impedance power supply, making design integration a simple task. But it is the most complex type of power subsystem, with an inherent low efficiency and high electromagnetic interference when used with a PPT or boost converter.

We can charge batteries individually or in parallel. A parallel charging system is simpler and has the lower cost, but does not allow flexibility in vehicle integration. It can also stress batteries so they degrade faster. When batteries are charged in parallel, the voltage is the same but the current and temperature are not. Because current is not rigidly controlled, one battery could receive all the available charge current, and a thermal runaway condition could result if we do not control the bus voltage from the hottest battery. Parallel batteries eventually end up balancing out, so we could use them for missions under five years. To ensure a battery life greater than five years, we should seriously consider independent chargers, such as the *linear, charge-current-control (LC^3)* design in Fig. 11-13.

Batteries usually limit the life of a spacecraft. To support a seven-year life, we must charge the batteries independently to degrade the battery as little as possible. Individual charging optimizes the battery use by charging all the batteries to their own unique limits. It also forgives battery deviations in systems with several batteries. Unfortunately, individual chargers add impedance, electronic piece parts, and thermal dissipation not present in a parallel system. To design the Power Regulation and Control subsystem, follow the steps in Table 11-40.

TABLE 11-42. Steps in the Power Regulation and Control Subsystem Design.

Step	Consider	Possibilities
1. Determine the power source	• All spacecraft loads, their duty cycles, and special operating modes	• Primary batteries • Photovoltaic • Static power • Dynamic power
2. Design the electrical control subsystem	• Power source • Battery charging • Spacecraft heating	• Peak-power tracker • Direct-energy transfer

3. Develop the electrical bus voltage control	• How much control does each load require?
	• Battery voltage variation from charge to discharge
	• Battery recharge subsystem
	• Battery cycle life
	• Total system mass
		• Unregulated
		• Quasi-regulated
		• Fully regulated
		
		• Parallel or individual charging
		    − < 5 yrs—parallel charge
		    − > 5 yrs—independent charge

11.5 Thermal

Robert K. Mcmordie, *Martin Marietta Astronautics Group*
Aniceto Panetti, *Alenia Aerospazio Space Division*

The safest systems and components used for space are the outcome of extensive laboratory experimentation and many years of on ground environment applications. This means that reliable operation of most spacecraft constituents requires a temperature range not far from room temperature. Hence the purpose of a *thermal-control subsystem* (TCS) is to maintain all the items of a spacecraft within their allowed temperature limits during all mission phases using minimum spacecraft resources. Controlling the temperature is particularly serious for alignment/pointing critical structures, batteries, hydrazine propulsion equipment, and electronic components.

In order to provide a design which meets the temperature requirements of the spacecraft, we must account for heat inputs from the Sun, the Earth, and electrical and electronic components on board the spacecraft, denoted by "*P*" in Fig. 11-14. In most instances the heat inputs are highly variable with time. For example, the mean value of direct solar heating on a spacecraft surface varies from 1,367 $W \cdot m^{-2*}$ when the Sun is normal to the surface, to zero when the spacecraft is in eclipse. The geometry of a spacecraft is generally very complex, leading to numerically complex thermal analysis. However, the thermal control subsystem accounts for only about 2 to 5% of the total spacecraft cost and about the same percentage of the dry weight.

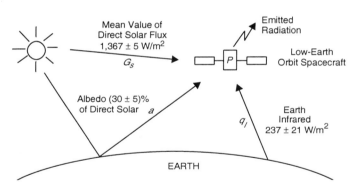

Fig. 11-14. Thermal-Radiation Environment for a Typical Spacecraft. *P* refers to the power dissipated on board the spacecraft. Values shown are taken from NASA [1982].

Requirements for a thermal-control system usually exist at several levels. Top-level system requirements define allowable temperature margins, overall testing requirements, and environmental conditions such as the flux levels for direct solar, Earth-reflected solar (*albedo*),[†] and Earth-emitted energy. Subsystem weight allocations and

---

[*] Depending on the seasonal variation of Earth's distance from the Sun, the solar flux, in Earth's orbit, can assume the following values (with reference to the Northern Hemisphere): 1,326 $W \cdot m^{-2}$ at summer solstice, 1,360 $W \cdot m^{-2}$ at autumn equinox, 1,418 $W \cdot m^{-2}$ at winter solstice, 1,381 $W \cdot m^{-2}$ at vernal equinox [NASA TM-82478, January 1983].

cost goals are requirements derived by the spacecraft project management. Finally, we base component temperature limits on supplier data.

Table 11-43 gives typical temperature ranges for spacecraft components. Silicon solar cells usually operate over a wide range; however, their efficiency increases as their temperature decreases. Normally, structures, external appendages like antennas, booms, and sensors have wide temperature limits. The exception is for structures that support equipment, such as cameras, which have extremely accurate pointing requirements. In this situation the structural elements may have a temperature variation requirement of ± 0.5 °C or less to minimize differential thermal expansion which distorts the spacecraft shape and adversely affects pointing. The temperature limits for *hydrazine*, a common monopropellant, come from its freezing point on the low end and its decomposition temperature on the high end. Because of this, the typical thermal design for hydrazine systems is to cold bias the lines and use thermostatically controlled heaters to maintain the lines above the lower limit. *Cold bias* means to design a given region of the spacecraft so that it operates below the upper temperature limit for all equipment in the region. When we do this, usually some of the equipment in the region will operate below its lower temperature limit if nothing prevents this. Therefore, we provide thermostatically controlled heaters to keep the equipment temperatures above the lower limits.

In addition to the temperature level requirement, temperature stability and uniformity requirements can play an important role. The *temperature stability* specifies the maximum allowable temperature variation over time of a given item, while *temperature uniformity* is the maximum allowable temperature gradient within the considered body. The following definitions are often used with regard to equipment temperatures:

- *Operating Temperature Range*: the temperature limits within which the equipment fulfills all specified operating performance and life requirements.

- *Switch-on Temperature Limit*: the lowest temperature for activating equipment, which ensures that it will not break upon turn-on and it will reach its operating temperature range.

- *Nonoperating Temperature Range*: the temperature limits within which the equipment, being in nonoperating mode ("off" condition), has to survive without any performance degradation once it reaches its operating temperature range.

Strong interaction exists between the thermal control subsystem and the other spacecraft subsystems. This implies that the thermal engineers need to work very closely and interactively with payload and other subsystem specialists in order to develop the most convenient spacecraft concept. This is applicable from the early definition stage, when the overall satellite architecture and subsystem layouts are defined in connection with the preliminary thermal design concepts. The process goes on during the proper development and testing phases when the detailed thermal model is developed and extensively used with a considerable exchange of information with all subsystem and testing specialists. Also during the launch, and on-orbit operation phases, the thermal engineer supports the satellite operational team providing

---

† *Albedo* is actually the percentage of solar radiation reflected off the Earth.

TABLE 11-43. Typical Operating Temperature Ranges for Selected Spacecraft Components. Temperatures in °C.[1]

Spacecraft Internal Units			
Worst case envelope[2]	0 to +40		
Telecommunications			
Payload units	−10 to +50		
Remote Sensing Payloads			
Optical sensors	+15 to +25	Infrared module[3]	−40 to +30
Radiometric units	−10 to +50	Radar units	−10 to +40
Onboard Computer	−10 to +50		
Telemetry & Command units	−10 to +50		
Electrical Power			
Batteries (NiH$_2$)	−5 to +20	Batteries (NiCd)	0 to +25
Solar Arrays	−105 to +110	Power control unit	−20 to +55
Attitude Control			
Sun & Earth sensors	−30 to +50	Magnetometer	−80 to +80
Electronic units	−10 to +55	Momentum and	
Gyro package	0 to +50	reaction wheels	−5 to +45
Propulsion			
Propellant tank, filters, valves, lines	+7 to +55	Thrusters	+7 to +65
Harness			
Spacecraft internal	−15 to +55	Spacecraft external	−100 to +100
Thermal Control			
Multilayer Insulation (MLI)	−160 to +250	Heaters, thermostats,	
Radiators	−95 to +60	heat pipes	−35 to +60
Structures			
Nonalignment critical	−45 to +65	Alignment critical[4]	+18 to +22
Mechanisms			
Pyrotechnics	−100 to +120	Electric motors	−45 to +80
Deployment hinge	−45 to +65	Solar array drive assembly	−35 to +60
Antennas			
Parabolic reflector	−160 to +95	TT&C	−65 to +95
GPS antenna	−95 to +70		

(1) For all equipment units, the reported temperatures apply to their external case or mounting interface. The temperature range for internal components of equipment is wider (e.g. a transistor can operate in the range −30 to +110 °C).

(2) This is a conservative range envelope used for the spherical spacecraft exercise, cf. Sec. 11.5.5.

(3) The detector needs a cryogenic system within the module, to maintain the range −200 to −80 °C.

(4) This range equals the temperature of the integration room where equipment is installed on the spacecraft structure and aligned with a high accuracy. The requirement for the thermal design is simply to maintain the same temperature range also during the space mission. This will assure that on-ground alignments will be maintained (duplicated) also in orbit by limiting the thermo-elastic distortion of the structure.

feedback on the proper operation of subsystems, helping to detect and resolve anomalies. The power subsystem has a great interaction with the thermal control subsystem because all the dissipated electrical energy within the spacecraft must be radiated to space. Also, batteries generally have a narrow temperature range and often require special thermal attention. If infrared detectors are part of the spacecraft payload, they can be a major challenge because they often have to operate at temperatures in the cryogenic range, 0 to 120 K. This temperature range requires a low-temperature, mechanical, refrigeration system and special, low-temperature radiators. If the mission is short, however, we can use a stored, expendable cryogen like helium for a low-temperature heat sink.

To give an idea of the interaction of the structure and the thermal subsystems, think that structural panels and thermal radiators are, in almost all cases, the same. This means that heat pipes are mounted on panels or embedded in their honeycomb. Multi-layer insulation mounting interfaces and optical solar reflectors are installed on the panel skins. Paints and tapes cover most spacecraft parts. The structural subsystem can have a critical interaction with the thermal control subsystem in the case of ultra-stable structures like optical benches, space telescopes, and military observation satellite structures. These cases require a fine thermal control to tightly limit the thermo-elastic distortion of structures. Propulsion subsystems generally require a significant amount of thermal control hardware to ensure the propellant is at the appropriate temperature and to shield those parts that can reach high temperatures. Finally, spacecraft attitude control can have a significant influence on the thermal control subsystem design. The attitude of the spacecraft determines the thermal radiation inputs from the Sun and Earth and these inputs affect radiator performance and spacecraft temperatures.

11.5.1 Principles of Heat Transfer

In this paragraph, we present the basic principles and the terminology concerning energy transportation modes. This information is needed for a good comprehension of the TCS concepts and features presented in the following paragraphs. With these concepts we can use the analytical tools, understand how certain thermal-control coatings maintain relatively low temperatures in direct sunlight, and perform thermal analyses to support preliminary mission design activities. *Heat* is a form of energy, which flows from one body to another body at a lower temperature, by virtue of the temperature difference between the bodies. *Steady state* heat transfer means that heat flows continuously at a uniform rate. This is not the case of *unsteady state* or *transient heat* transfer.

We divide *heat transfer* into three major areas: conduction, convection, and radiation. *Radiation* is energy transfer via electromagnetic waves. *Conduction* is thermal energy transfer through matter in the absence of fluid motion. *Convection* is thermal energy transfer between a flowing fluid and a solid interface.

Convection—Convection occurring between a solid surface and its adjacent fluid is described by:

$$q = h \, A \, \Delta T \tag{11-10}$$

where q is the rate of heat flow in W, h is the heat transfer coefficient in $W \cdot m^{-2} \, K^{-1}$, A is the surface area in m^2, ΔT is the temperature difference between the surface and the fluid mass in K. The heat transfer coefficient "h" depends on the geometry of the surface, the fluid's flow characteristics, and the physical properties. Determining h is the central problem of convective heat-transfer analysis. Convection can create satellite heating during satellite launch, when the launcher jettisons its fairing early to maximize the launched mass. Typically fairing separation takes place when the value of the aerodynamic heat flux, acting on a flat surface normal to the direction of flight, reaches a value lower than 1,200 $W \cdot m^{-2}$. This thermal flux can act on the spacecraft simultaneously to solar, albedo, and Earth IR fluxes. At any rate, convection plays a considerable role for the thermal control of manned vehicles or modules.

Conduction—The fundamental equation for steady-state, one-dimensional heat conduction in rectangular coordinates is:

$$Q = (kA/\Delta x)(T_1 - T_2) \tag{11-11}$$

where Q in W is the energy transfer rate, T, in K, is the temperature, k, in $W \cdot m^{-1} K^{-1}$, is the thermal conductivity, Δx, in m, is the length of the heat transfer path, and A, in m^2, is the cross-sectional area normal to the heat transfer direction. Figure 11-15 illustrates steady-state conduction for rectangular, cylindrical, and spherical coordinate systems.

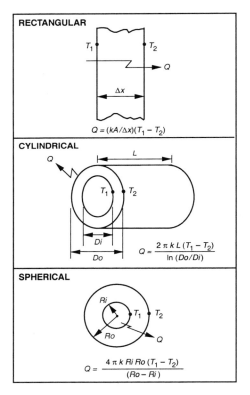

Fig. 11-15. Equations for Steady-State Conduction. These equations help us compute the amount of heat, Q, conducted through each shape, depending on the thermal conductivity, k, of the material, and the other variables.

Radiation—As described in Sec. 9.4, all matter radiates electromagnetic energy with the total radiated energy per unit time per unit surface area, q, given by the Stefan-Boltzmann's equation:[*]

$$q = \varepsilon \sigma T^4 \tag{11-12}$$

---

[*] Austrian physicist Josef Stefan established this relationship experimentally in 1879. Another Austrian physicist, Ludwig Boltzmann, developed the equation theoretically 5 years later.

where σ is the Stefan-Boltzmann's constant, 5.670 51 $\times$ 10^{-8} W·m^{-2}K^{-4}, T is the absolute temperature, and ε is the emissivity, a dimensionless number between 0 and 1. *Emissivity* is the ratio of the energy emitted by a substance to that emitted by a perfect radiator, or *blackbody*, at the same temperature. Certain black paints have emissivities very close to one. Polished gold or silver surfaces have very low emissivities in the neighborhood of 0.05.

Figure 11-16 describes surface radiation properties, in addition to emissivity. When we irradiate a transparent plate with incoming radiation, the plate's surface will reflect, absorb, or transmit the incoming radiation. The percentage of incoming radiation that the plate absorbs we define as *absorptivity*, the percentage reflected is the *reflectivity*, and the percentage transmitted is *transmissivity*. Because these are the only outcomes and they are fractions of the total irradiance, they must sum to 1.0.

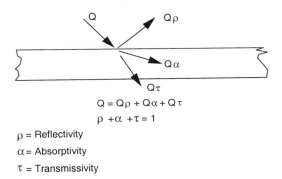

$$Q = Q\rho + Q\alpha + Q\tau$$
$$\rho + \alpha + \tau = 1$$

ρ = Reflectivity
α = Absorptivity
τ = Transmissivity

Fig. 11-16. Radiation Properties of Surfaces. Any surface irradiated by electromagnetic radiation absorbs it, α, reflects it, ρ, or transmits it, τ.

If we focus on the radiant energy emitted from a surface as a function of wavelength or *monochromatic* radiation, we use *Planck's equation*, given below, which defines the monochromatic radiation energy (energy per unit time per unit area per unit wavelength), emitted from the surface of a blackbody (emissivity = 1):

$$E_{b\lambda} = \frac{2\pi hc^2}{\lambda^5} \frac{1}{e^{ch/k\lambda T} - 1} \tag{11-13}$$

where λ is the wavelength, h is Planck's constant, T is the temperature, c is the speed of light, and k is Boltzmann's constant. (See the inside front cover for values of the physical constants.) If we include the *monochromatic emissivity* (emissivity as a function of wavelength) for a given surface in the numerator of Eq. (11-13), the resulting expression describes the monochromatic emission from the surface at the given temperature, T. If the emissivity is a constant independent of wavelength, the integration of Eq. (11-13) over all wavelengths results in Eq. (11-12), the Stefan-Boltzmann's equation. At the surface temperature of the Sun (approximately 5,800 K), the emitted energy wavelengths are concentrated between zero and 2.5 microns, with the maximum radiated power at 0.5 microns. At "room temperature" (294 K), over 98% of the emitted energy is at a wavelength greater than 5 microns, and the peak emission wavelength is at about 9.8 microns. The point is that radiation energy emitted from a surface at room temperature and solar energy occupy two very different wave bands. Solar radiation is concentrated in the short wavelengths, while room temperature emission is concentrated in the long wavelengths.

Figure 11-17 is a plot of the emissivity of white paint vs. wavelength. The curve is also the absorptivity vs. wavelength because monochromatic emissivity equals monochromatic absorptivity (this is called *Kirchoff's Law*.) The white paint properties, ε_λ and α_λ, when used with the characteristics of Planck's equation explains why a white-painted surface exposed to solar radiation in space has a relatively low temperature. We assume that the back side of the white surface (the side of the plate away from the solar input) is insulated perfectly. Therefore, the absorbed solar energy (visible) must equal the emitted (infrared) energy since these energies are the only quantities involved in the steady state energy balance on the white surface. The white paint absorbs weakly in the solar wave band where the average absorptivity is low, under 0.4. However, the white paint emits strongly in the infrared wave band where the emissivity is above 0.8. So, with little solar energy absorbed and large amounts of heat emitted, the result is the equilibrium temperature of the white surface runs relatively cool in sunlight.

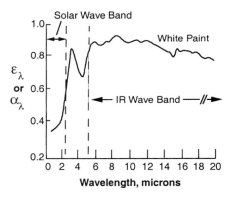

Fig. 11-17. White Paint Emissivity and Absorptivity as a Function of Wavelength. White paint properties from Siegel and Howell [1968]. Low absorptivity in the Solar Wave Band, coupled with high emissivity in the IR Wave Band, means white-painted surfaces stay relatively cool in sunlight.

Surfaces with radically different emissivity (or absorptivity) values in the solar wave band as compared to the infrared are called *selective surfaces* (such as white paint). The mean value of the absorptivity over the solar wave band is called *solar absorptivity* (α_s), while the average value of emissivity over the infrared wave band is called *IR emissivity* (ε_{IR}). Infrared emissivity is typically a weak function of surface temperature, and generally the temperature must be 30 °C or 40 °C different from room temperature to show a significant change from the room temperature value. We perform most spacecraft thermal analyses using the mean values, α_s and ε_{IR},[*] and do not require use of the monochromatic equations.

---

[*] From this point on we will drop the subscripts on emissivity, ε, and absorptivity, α. Be careful to determine the proper values before using them.

Table 11-44 provides representative values for solar absorptivity and infrared emissivity for various materials. For convenience in the analysis, we introduce the *view factor*, also called shape factor, configuration factor, or angle factor, to characterize the effects of geometry and orientation of surfaces on their radiative heat exchange. The physical significance of view factor is the fraction of the radiative energy leaving one surface element that strikes the other surface directly. We introduce the symbol F_{1-2} to indicate the view factor from surface 1, with area A_1, to surface 2, with area A_2. The view factor is a function of the size, geometry, relative position, and orientation of the two surfaces. The view factors follow the reciprocity relation, for two surfaces

$$A_1 F_{1-2} = A_2 F_{2-1} \qquad (11\text{-}14)$$

and the summation relation

$$\sum_{i=1}^{n} F_{1-i} = 1 \qquad (11\text{-}15)$$

expressing the fact that the total sum of the view factors for a given surface to all directions in the surrounding equals unity. These two relations are useful to calculate a particular view factor from the knowledge of others. View factors for systems of various geometrical arrangements are provided in Kreith and Bohn [1996] and ESA-PSS-03-108.

For heat transfer between two bodies we generally assume diffuse gray surface conditions (*diffuse* means that the reflected and emitted radiation follow Lambert's cosine law and the *gray* surfaces are those for which emissivity and absorptivity are independent of wavelength).

We generally use the following equations to compute the energy exchange rate, q (W), from one body to the other body, in a nonabsorbing medium.

For a gray surface surrounded by deep space we use:

$$q = \sigma \varepsilon_1 A_1 \, (T^4_1 - T^4_2) \qquad (11\text{-}16)$$

since $T_{deep\ space} = 4$ K, generally we assume $T^4_2 = 0$.

For two flat surfaces at a very small distance apart we use:

$$q = \cfrac{1}{\left(\cfrac{1}{\varepsilon_1} + \cfrac{1}{\varepsilon_2} - 1 \right)} \cdot \sigma \cdot A_1 \cdot \left(T_1^4 - T_2^4 \right) \qquad (11\text{-}18)$$

For coaxial long cylinders and concentric spheres we use:

$$q = \cfrac{1}{\cfrac{1}{\varepsilon_1} + \cfrac{A_1}{A_2}\left(\cfrac{1}{\varepsilon_2} - 1 \right)} \cdot \sigma \cdot A_1 \cdot \left(T_1^4 - T_2^4 \right) \qquad (11\text{-}19)$$

For the equations given above σ is the Stefan-Boltzmann constant, A is area in m^2, ε is emissivity, and T is temperature in degrees K. Note that in Eq. (11-19) the subscript 1 refers to the inside surface and 2 to the outside surface.

The book *Principles of Heat Transfer* [Kreith and Bohn, 1996] provides a systematic explanation of the basic principles of heat transfer together with a number of examples and data. *Handbook of Essential Formulae and Data on Heat Transfer for Engineers* [Wong, 1977] provides the equations of heat transfer in a compact manner together with a short explanatory text on the basic principles.

11.5.2 Thermal Control Components

Spacecraft thermal control subsystems are classified as *passive* or *active*. An example of a passive subsystem is a space radiator thermally coupled to heat-dissipating equipment by conductive paths. Active thermal-control subsystems include pumped-loop systems, heaters controlled by thermostats, and mechanical refrigerators. Examples of active and passive thermal control systems are shown in Fig. 11-18.

Thermal-control subsystems extensively use the following components and devices:

- *Materials and Coatings*—These are surfaces with special radiation or thermo-optical properties (absorptivity, emissivity, reflectivity, and transparency or transmissivity) that provide the desired thermal performance of the body. Examples are paints, high-quality mirrors, silverized plastics, and chemical conversion coatings such as anodizing. Table 11-44 provides a list of radiation properties of various materials. Particularly interesting are the following two materials:

- *Optical Solar Reflectors* (OSRs)—The type of reflective surface that produces the lowest temperature when irradiated by solar energy in space is an *optical solar reflector,* also called a *second surface mirror,* shown in Fig. 11-19. We create an optical solar reflector by mounting a highly reflective surface on a substrate and overlaying the surface with a transparent cover. For the highest possible thermal performance the reflector surface is silver and the transparent cover is quartz. The outer surface of the transparent cover partially reflects incoming radiation. The cover material partially absorbs it, but transmits most of it to the reflective surface. The reflective surface reflects the majority of the transmitted radiation back to space. Highly reflective surfaces are generally metallic, which also have relatively low emissivities. Therefore, the reflective surface does not contribute much to radiation emission. However, the outer surface of the transparent cover has a high IR emissivity and, therefore, is a major source of emission. Because the transparent cover is thin, the temperatures of the cover and the substrate are essentially the same. The overall effect is that the optical solar reflector has a high effective emissivity due to the transparent cover, but a low solar absorptivity from the highly reflecting metallic surface. Representative values for optical solar reflectors are an IR emissivity of 0.8 and a solar absorptivity of 0.15.

- *Silver-Coated Teflon*—OSRs are both fragile and costly. Silver-coated Teflon has radiation characteristics similar to OSRs. The silver-coated Teflon material does not provide as low a temperature as high-quality quartz optical solar reflectors, but is much less expensive and much more durable. In recent years, we have used silver-coated Teflon extensively on outer surfaces of spacecraft.

TABLE 11-44. Radiation Properties. All values are undegraded beginning of life.

No.	Material	Measurement Temp. (K)	Surface Condition	Solar Absorptivity, α	Infrared Emissivity ε	Absorptivity/ Emissivity Ratio	Equilibrium Temp* (°C)
1	Aluminum (6061-T6)	294	As Received	0.379	0.0346	10.95	450
2	Aluminum (6061-T6)	422	As Received	0.379	0.0393	9.64	428
3	Aluminum (6061-T6)	294	Polished	0.2	0.031	6.45	361
4	Aluminum (6061-T6)	422	Polished	0.2	0.034	5.88	346
5	Gold	294	As Rolled	0.299	0.023	13.00	482
6	Steel (AM 350)	294	As Received	0.567	0.267	2.12	207
7	Steel (AM 350)	422	As Received	0.567	0.317	1.79	187
8	Steel (AM 350)	611	As Received	0.567	0.353	1.61	175
9	Steel (AM 350)	811	As Received	0.567	0.375	1.51	168
10	Steel (AM 350)	294	Polished	0.357	0.095	3.76	281
11	Steel (AM 350)	422	Polished	0.357	0.111	3.22	259
12	Steel (AM 350)	611	Polished	0.357	0.135	2.64	234
13	Steel (AM 350)	811	Polished	0.357	0.155	2.30	217
14	Titanium (6AL-4V)	294	As Received	0.766	0.472	1.62	176
15	Titanium (6AL-4V)	422	As Received	0.766	0.513	1.49	166
16	Titanium (6AL-4V)	294	Polished	0.448	0.129	3.47	270
17	Titanium (6AL-4V)	422	Polished	0.448	0.148	3.03	251
18	White Enamel	294	Al. Substrate	0.252	0.853	0.30	20
19	White Epoxy	294	Al. Substrate	0.248	0.924	0.27	13
20	White Epoxy	422	Al. Substrate	0.248	0.888	0.28	16
21	Black Paint	294	Al. Substrate	0.975	0.874	1.12	136
22	Silvered Teflon	295		0.08	0.66	0.12	−39
23	Aluminized Teflon	295		0.163	0.8	0.20	−6
24	OSR (Quartz Over Silver)	295		0.077	0.79	0.10	−51
25	Solar Cell-Fused Silica Cover			0.805	0.825	0.98	122

Notes:
Items 1 through 21 are from Brown and Gagola [1965]
Items 22, 23, and 24 are from Ahern, Belcher, et al. [1983]
Item 25 is from NASA Jet Propulsion Laboratory [1976]
* Temperature of a flat plate located in space (radiation heat exchange with deep space), with a perpendicular solar input of 1,418 W · m⁻², no internal dissipation, and perfect insulation of its backside.

ACTIVE THERMAL CONTROL

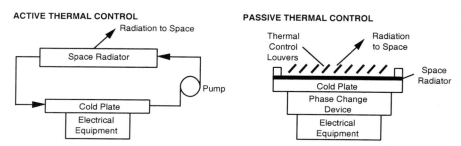

PASSIVE THERMAL CONTROL

Fig. 11-18. Arrangements for a Spacecraft's Thermal-Control Subsystem.

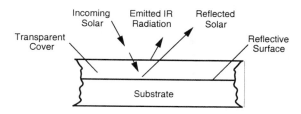

Fig. 11-19. Thermal Performance of the Optical Surface Reflector.

- *MultiLayer Insulation*—The primary spacecraft insulation is *multilayer insulation* (MLI). We use MLI blankets to minimize the radiation heat transfer from and to a spacecraft component. The MLI typical arrangement is alternate layers of aluminized Mylar or Kapton and a thin net of material, such as nylon bridal veil, Dacron, or Nomex. The net acts as a separator for the mylar layers keeping the adjacent layers from touching, which would cause a thermal conduction short across the insulation blanket. Kapton, a plastic which is much stronger than mylar, often acts as the inner and outer layers of a multilayer blanket. In Sec. 11.5.6 we will quantify the thermal performance of MLI.

- *Electrical Heaters*—These devices usually control regions that are cold biased (designed to run cold when the heaters are off). The heaters, controlled by the thermostats, bring the temperature up to the desired level. We typically apply this approach to regions which require fine temperature control. Heaters have, generally, the shape of flat sheets. They consist of a thin electrical resistor between two Kapton sheets. Heat is generated by the Joule effect. Depending on the surface on which we bond the heater, it can assume many different shapes: rectangular, square, circular, or spiral. Typical heater power densities are on the order of less than $1 \ W \cdot cm^{-2}$.

- *Thermostats*—They operate like switches to turn on/off the heaters when a component reaches its prescribed temperature limits. We bond or bolt a thermostat to the thermally controlled unit to have a *local control*. Thermostats consist of a bi-metallic, temperature-sensitive element that drives the heater switch. Their typical operative range is between –50 to +160 °C and they can perform more than 100,000 on/off cycles. Thermostats represent a simple and cheap means to control temperature set points to within ± 2 to 3 °C. For a very accurate temperature control, dedicated electronic units or the onboard computer control the heaters using the input data provided by temperature sensors. The latter is called *central control*.

- *Space Radiators*—A space radiator is a heat exchanger on the outer surface of a spacecraft that radiates waste heat to deep space even with an environmental heat input on its surface. They can be satellite structural panels or flat plates installed on the spacecraft. An example of a radiator calculation is provided in Sec. 11.5.5.

- *Cold Plates*—On *cold plates,* shown on Fig. 11-18, we mount heat dissipating equipment. A cold plate for an active thermal control subsystem uses fluid passages integral with the plate. For this type of subsystem, thermal energy dissipates as waste heat in the electrical equipment. This heat transfers across the bolted interface between the equipment and the cold plate. The fluid circulating through the cold plate then transports the waste heat to a radiator which radiates it into space. The passive subsystem shown in Fig. 11-18 combines the cold plate with the radiator. Also, for the arrangement shown, we added a phase change device and thermal control louver. The *louver* modulates heat rejection from the radiator as we discuss later in this section.

- *Doublers*—They represent the simplest type of cold plates. Doublers consist of aluminum plates connected to one or more power dissipation units, which increase the heat exchange surface area of the units. We bond doublers to the back side of thermal radiators, and interpose a filler between the doubler and the dissipating unit. We also use them to increase the thermal inertia of a unit to reduce temperature excursions during transient phases. Their typical thickness is 1 to 5 mm. Basically, they have the same function as heat pipes, but are less complex and less able to carry away heat.

- *Phase Change Devices*—A *phase change device* absorbs thermal energy using a solid-to-liquid phase change, which helps when electrical equipment has high, short bursts of power. The phase-change material, usually a paraffin, reduces temperature spikes in proportion to the amount of paraffin. During the period of heat dissipation, the paraffin absorbs the waste heat and melts. While the equipment is inactive, the phase change material cools returning the heat to the unit or another heat-dissipating device and solidifies. NASA Tech Brief B72-10464 [1972] gives detailed information concerning phase change devices.

- *Heat Pipes*—*Heat pipes* are lightweight devices used to transfer heat from one location to another. For example, a heat pipe can transfer heat dissipated in an electrical component to a space radiator. A *heat pipe* is a hermetically sealed tube with a wicking device on the inside surface. A fluid inside the tube operates by changing phases during heating and cooling phases. Heat applied at one end of the pipe causes evaporation of the liquid in the wick. The gases formed by the evaporation flow down the center of the heat pipe to the opposite end. Here, heat is transferred from the pipe, causing condensation in the wicking material. Capillary forces draw the fluid from the condenser end to the evaporator end of the heat pipe, thus completing a heat transfer loop. This loop occurs naturally when one end of the heat pipe is maintained at a higher temperature than the other. Due to the relatively high latent heat of evaporation of the heat pipe's working fluid, these devices exhibit high heat transfer rates with small temperature differences from the evaporator to the condenser end of the heat pipe.

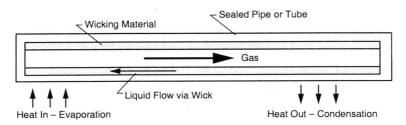

Fig. 11-20. Heat-Pipe Schematic. Heat pipes can transfer up to 1,000 times as much energy as a solid aluminum bar of the same diameter.

The heat pipe shown in Fig. 11-20 is a *fixed conductance heat pipe*. Heat pipes can also exhibit variable conductance characteristics and, therefore, the ability to automatically modulate heat transfer. We make heat pipes designed to operate at room temperature out of aluminum with ammonia as the working fluid. The material selection is critical because some combinations of working fluids and pipe materials will cause a slow chemical reaction that can severely affect the pipe performance. An 11-mm external diameter heat pipe has a mass of about 0.35 kg·m^{-1} and a heat transport capability of about 100 Wm in the range 0 to 40 °C and zero gravity. This means that the heat pipe can transfer 100 W over a 1 m distance, or 1,000 W over 10 cm. One disadvantage of heat pipes is their performance in a 1-g environment such as during testing. The orientation of the axis of the pipe relative to the gravity vector affects the fluid flow. This means that for testing on Earth, we must accurately position the pipe axis so that it is essentially horizontal. Then the heat pipe's thermal performance in space and on the Earth will be about the same. However, for some heat pipe designs, being off horizontal by a degree or less changes their performance significantly.

- *Louvers*—Louvers resemble Venetian blinds, which we position between a radiator surface and deep space, as shown in Fig. 11-18. Opening and closing the louver blades modulates the rate of heat flow to space. At a radiator temperature of 304 K with the blades wide open, the heat rejection rate is about 430 W·m^{-2}. Closed and at a radiator temperature of 283 K, the rejection rate is about 54 W·m^{-2}. One problem encountered with louvers is overheating, if the louvers inadvertently point at the Sun. Some spacecraft designs have included covers over the louvers to avoid this potential problem; however, the covers significantly decrease the thermal performance of these louvers.

- *Temperature Sensors*—To control accurately the temperature of particular equipment or portion of a spacecraft we need a number of thermometers on board. For space applications the following electrical thermometers sense temperature:

Thermistors—They are semiconductor materials that vary their resistance with temperature. They are cheap, accurate temperature sensors, having small dimensions, very small response time and power consumption. Their typical measurement range is about –50 to +300 °C. They are common on many types of spacecraft.

Resistance Thermometers—The temperature sensitive element is generally a pure platinum conductor, which varies its electrical resistance as a quasi-linear function of temperature. They are very accurate and expensive. We generally use them for temperature sensing on solar arrays. Their typical measurement range is about –260 to +600 °C.

- *Adhesive Tapes*—We use adhesive tape for manufacturing multilayer insulation or applying directly to structural, mechanism and unit surfaces to provide adequate thermo-optical properties. By using adhesive tapes we avoid obstructing movable parts with MLI. Generally these tapes consist of a Kapton substratum, acrylic or silicon adhesive and a possible aluminum coating.

- *Fillers*—Fillers increase the conductive heat transfer between contacting surfaces. This is particularly important for high power dissipation units mounted on a baseplate. Fillers consist of thin sheet materials, generally metal oxide within an elastomeric binder. Their typical thickness is 0.2 to 0.5 mm, and their conductivity is about 2 to 7 $W \cdot m^{-1} \cdot K^{-1}$. These materials fill the microscopic voids between contacting surfaces, which are due to surface roughness. They can drastically increase the heat transfer surface. Typical fillers are CHO-THERM, CALGRAPH, SIGRAFLEX.

- *Thermal Isolators*—They reduce the conductive heat transfer between contacting surfaces or objects. Generally they isolate propellant lines or other components from the spacecraft body. We interpose them under the mounting feet of some units, to isolate external appendages which prevents excessive heat leaks from the spacecraft body to the external environment. We also use them to isolate instruments. They have the shape of washers, brackets, or a more complex shape depending on the particular application. Their typical thickness is 1 to 4 mm, and in particular cases, up to 10 mm. Typical thermal isolators are low-conductivity materials like: NARMCO with a conductivity of about 0.14 $W \cdot m^{-1} \cdot K^{-1}$, fiberglass, stainless steel, titanium. Table 11-48 reports the thermal conductivity of these materials.

- *Thermoelectric Coolers*—They use the Peltier effect, i.e., an electric current induces the cooling of the junction between dissimilar metals. These devices are simple, compact, reliable, and, unlike pumped systems, have no moving parts, and don't generate any micro-vibration. This vibrationless operation is an important requirement for very stable spacecraft carrying sensitive instruments. Unfortunately their thermodynamic efficiency is modest, so we use them for local, limited cooling applications only.

- *Cryogenic Systems*—Low-noise amplifiers, superconducting equipment, and IR detectors require very low operating temperatures, in the range of –271 °C to –150 °C. We divide cryogenic systems into:

Active Refrigeration Systems—For long mission duration the most attractive cryogenic system is an active refrigeration one. These systems are generally based on the Stirling or reverse Brayton cycles. Like common refrigerators they can remove equipment waste heat at low temperature at the expense of an input electrical power. The waste heat is then rejected to space by means of a thermal radiator. Vibration control, reliability, and operating life are the critical issues of these devices.

Expendable Cooling Systems—For short duration missions or for intermittent cooling requirements, stored cryogen expendable cooling systems are used. These systems are based on a low temperature fluid or solid, generally helium, ammonia, or methane, that absorbs the heat from the spacecraft components, and transfers it to space in the form of vented gas. These systems are simple, reliable, and relatively cheap.

11.5.3 Thermal Subsystem Design

During the preliminary mission design of a spacecraft there are several thermal issues which we must address. A review of these issues will bound the thermal design problem and provide an estimate of the engineering and computer time required to develop the detailed spacecraft thermal design. Table 11-45 gives a description of the preliminary design process, and in Sec. 11.5.5 we present a numerical example of the preliminary design process.

Usually preliminary mission design indicates that we can passively control unmanned, low-Earth orbit spacecraft. A passive subsystem is generally lighter, requires less electrical power, and is less costly than an active subsystem. Thus, we start with a passive subsystem, adding other elements only as needed. We use active thermal control subsystems for manned spacecraft, for situations requiring very close-tolerance temperature control (a few degrees), and for components that dissipate large amounts of waste energy (on the order of several kilowatts).

The first step in the thermal design of a spacecraft is to determine key requirements and constraints, which include the temperature limits and power dissipation of all the spacecraft components. Next, we determine the spacecraft altitude and orientation relative to the Earth and the Sun for all mission phases so we can set the thermal boundary conditions for the mission. With the attitude and orientation we can determine the environmental heat input on the exterior surfaces.

Once we establish thermal boundary conditions, define temperature limits, and understand other requirements, we can develop a thermal design. Figure 11-21 illustrates the overall design process. The most important aspects of this process are the two generalized computer programs.

Geometric Mathematical Model (GMM)—The process sequence is to exercise the radiation program first. The program inputs are the spacecraft geometry, orientation, altitude, and surface radiation property values. The program output defines the absorbed energy on the spacecraft's external surfaces. The absorbed energy may be directly from the Sun, solar energy reflected from the Earth (albedo), or infrared energy emitted from the Earth. We also use the radiation program to define all the spacecraft surface-radiation-exchange factors, also called radiative conductors, and view factors. These factors define the radiation interaction between the spacecraft surfaces. Widely-used radiation programs are NEVADA (developed by Turner Associates Consultants), TRASYS (developed for the Johnson Space Center by Martin

TABLE 11-45.　The Preliminary Design Process for the Thermal Subsystem of an Isothermal Spacecraft.

Step	Notes	Where Discussed
1. Determine requirements and constraints	• Identify temperature limits	Table 11-43
	• Payload requirements usually dominate; batteries often have narrow temperature limits	Tables 10-8, 10-10, 10-16, 10-19, 10-21, 10-23, 10-30
	• Estimate electrical power dissipation	
	• Use power consistent with worst case hot and worst case cold conditions, if available; otherwise use orbit average power	
2. Determine diameter of sphere with surface area equal to surface area of the spacecraft	Make first-order estimates of the spacecraft thermal performance by assuming an isothermal and spherical spacecraft	See example Sec. 11.5.5
3. Select radiation surface property values	Initially assume white paint with $\alpha_s = 0.6$ and $\varepsilon_{IR} = 0.8$	Table 11-44, Sec. 11.5.6
4. Compute worst-case hot and cold temperatures for the spacecraft	For upper limit, use high side values for direct solar, albedo, and Earth IR emission; for lower limit use low value for Earth IR emission	Fig. 11-14
5. Compare worst-case hot and cold temperatures (step 4) with temperature limits (step 1)	If worst-case hot temperature is higher than required upper limit, plan to use a deployed radiator with a pumped-looped system. If the opposite is true, use body-mounted radiators	Table 11-46 example
6. Estimate required area for body-mounted radiator	Use upper temperature limit for radiator temperature, assume no environmental heat inputs, and use maximum heat dissipation	Eq. 11-23
7. Estimate radiator temperature for worst-case cold conditions	Use the area from step 6 and minimum heat dissipation	Eq. 11-23
8. Determine heater power required to maintain radiator at lower temperature limit	If temperature found in step 7 is less than lower limit, in order to compute the heater power, assume the radiator temperature is at the lower limit and the area is the value defined in step 6	Eq. 11-23
9. Determine if there are special thermal control problems	Identify components with narrow temperature ranges, high power dissipation, or low temperature requirements	Sec. 11.5.2
10. Estimate subsystem weight, cost, and power	If no special problems, use 4.5% of spacecraft dry weight, 4% of the total spacecraft cost, and heater power from step 8	Chap. 20
11. Document reasons for selections	Particularly important to document assumptions	N/A

Marietta), THERMICA (developed by Matra Marconi Space) and ESARAD (developed by ALSTOM).

Thermal Mathematical Model (TMM)—After we establish the radiation aspects of the overall problem, we assemble a heat transfer model of the spacecraft. The nodes represent physical locations on the spacecraft and are connected to each other by thermal conductors. If the model is for transient analyses, thermal capacitors are attached to the node points. This method has its conceptual origin in the thermal/electric analogy in which heat flow is equivalent to current flow and temperature to voltage.

This model accounts for all the thermal characteristics of the spacecraft and allows us to predict the spacecraft's temperatures as a function of time. Elements of this model include the spacecraft geometry, electrical power dissipation, radiation heat transfer, conduction heat transfer, and thermal characteristics of the thermal control hardware. The overall thermal model often includes results of component-level tests. Examples of component-level tests are thermal-performance tests on louvers and heat pipes. With a spacecraft thermal model, we can observe component temperatures during the mission and adjust the thermal control hardware elements of the spacecraft until we satisfy all temperature limits. Widely used thermal analyzer computer programs nclude BETA (The Boeing Company), MITAS (Martin Marietta), SINDA (developed for the Johnson Space Center by Martin Marietta), and ESATAN (developed for the European Space Agency by ALSTOM).

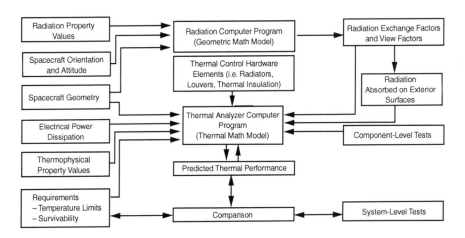

Fig. 11-21. Overall Design Process for Thermal Control of a Spacecraft.

11.5.4 Thermal Analysis Concepts

In this section we perform thermal analyses to support preliminary mission design activities. The proposed analyses are also very useful to consolidate the knowledge of the thermal control elements presented so far.

Equilibrium Temperature—We estimate the equilibrium temperature of a body in space using an energy balance equation that results from the conservation of energy: absorbed energy + dissipated energy – emitted energy = 0.

$$q_{absorbed} + q_{dissipated} - q_{emitted} = 0 \qquad (11\text{-}20)$$

For a body in space (a spacecraft), the absorbed energy is due primarily to direct and reflected solar energy and Earth-infrared emissions. The body dissipates energy via electronic components and energy leaves the body by electromagnetic radiation.

Temperature of Insulated Surfaces in Space—The equilibrium temperatures shown in Table 11-44 result from orienting a flat-plate with the Sun's rays normal to the surface. We insulate the backside of the plate so that it doesn't absorb or emit any energy. Furthermore, the plate dissipates no energy inside. The result from Eq. (11-20) is that the energy absorbed equals the energy emitted. The absorbed energy is simply the solar flux, G_s, times the projected area of the flat plate, A_p, times the absorptivity of the surface, α. We calculate the emitted energy by multiplying the Stefan-Boltzmann's equation (11-12) by the radiator area, A_r, resulting in

$$G_s A_p \alpha = \varepsilon \sigma T^4 A_r \tag{11-21}$$

For our perfectly insulated flat plate, $A_p = A_r$.

Solving Eq. (11-21) for the resulting equilibrium temperature yields

$$T = \left(\frac{G_s \alpha}{\varepsilon \sigma} \right)^{1/4} \tag{11-22}$$

where G_s is 1,418 W·m^{-2}, α is the material's solar absorptivity, ε is its infrared emissivity, and σ is the Stefan-Boltzmann's constant, 5.670 51 × 10^{-8} W·m^{-2}K^{-4}.

The radiation surface properties, α_s and ε_{IR}, are average values. We use these quantities routinely in radiation analysis. Table 11-44 gives values of these properties. Note that we use Eq. (11-21) as a general equation to estimate the temperature of insulated surfaces in space with solar heating. If the solar vector is not normal to the surface, we multiply the left side of Eq. (11-21) by cos θ. The angle θ is the angle between the surface normal and the solar vector.

Temperature of Space Radiators—We also expand Eq. (11-21) to provide a means of estimating the performance of space radiators. We accomplish this by adding the waste heat term, (Q_W/A_R), to the left side of Eq. (11-21). Q_W is the waste heat rejected by the radiator and A_R is the radiator area. The resulting equation is:

$$G_s \alpha \cos\theta + Q_W / A_R - \sigma\varepsilon \, T^4 = 0 \tag{11-23}$$

We can use this equation to estimate the heat rejection (Q_W) if the radiator operates at a given temperature, T, or, conversely, the required temperature to reject a given amount of heat.

Eq. (11-23) provides a means of estimating radiator performance for situations involving direct solar energy as the only environmental heat input. In the following paragraphs we develop equations which include the effects of reflected solar energy off the Earth's surface (albedo) and Earth emitted infrared radiation. We use this information as primary elements in the preliminary design process. The specific conditions we address are for the spacecraft directly overhead at high noon and for the spacecraft in the Earth's shadow with no view of sunlit portions of the Earth's surface. These two situations correspond to *worst-case hot* and *worst-case cold* conditions. We consider two geometries: a spherical spacecraft and a flat plate.

Solar Array/Flat Plate Analysis—The flat plate is similar to a solar array whose surface normal is parallel to the solar vector and passes through the center of the Earth.

We first consider a solar array and assume that the array is thin and isothermal; the "top" surface of the array receives direct solar energy while the "bottom" surface receives albedo and Earth IR radiation. The direct solar energy absorbed by the array in space is

$$Q_{s_a} = G_s A \alpha_t \tag{11-24}$$

where G_s is the solar constant (1,418 W·m^{-2}), A is the area of the array (the areas of the top and bottom are A), and α_t is the solar absorptivity of the solar cells on the top surface of the array. The IR energy flux at the Earth's surface, q_I, is 237±21 W·m^{-2}. Therefore, the total IR energy leaving the Earth's surface is q_I times the surface area of the Earth. The IR energy flux at the spacecraft altitude is total energy divided by area of a sphere with a radius equal to the Earth's radius plus the altitude of the spacecraft. Thus:

$$G_I = q_I(4\pi)R_E^2 / [4\pi(H + R_E)^2] = q_I R_E^2 / (H + R_E)^2 \qquad (11\text{-}25)$$
$$= q_I \sin^2\rho$$

where G_I is the Earth's IR radiation flux at the spacecraft's altitude, H, in W·m^{-2}, q_I is the Earth's IR emitted energy flux at the surface of the Earth, R_E is the Earth's radius, and ρ is the Earth's angular radius, as discussed in Chap. 5 and shown in the end papers of this book. We can obtain the same result by considering the view factor "F" for one face of an elemental plate with its normal pointed to the center of a sphere ($F = \sin^2\rho$). The absorbed Earth's infrared radiation on the bottom of the array is

$$q_{Ia} = q_I \sin^2\rho A\, \varepsilon_b \qquad (11\text{-}26)$$

where A is the area of the bottom of the array, and ε_b is the IR emissivity of the bottom surface of the array. The quantities q_I and ρ are defined above.

The equation for the solar energy reflected off the Earth (albedo) and absorbed by the bottom of the solar array is

$$Q_{Aa} = G_s a A \alpha_b K_a \sin^2 \rho \qquad (11\text{-}27)$$
$$K_a = 0.664 + 0.521\, \rho - 0.203\, \rho^2$$

where G_s is the solar constant, a is the albedo (the percentage of direct solar energy reflected off the Earth, approximately 30%), ρ is the angular radius of the Earth, A is the area of the bottom of the array, α_b is the solar absorptivity of the bottom surface of the array, and K_a accounts for the reflection of collimated incoming solar energy off a spherical Earth. In the expression for K_a ρ is in radians.

The total absorbed energy on the solar array is the sum of the direct energy absorbed on top, the Earth's infrared absorbed on the bottom, and the albedo absorbed on the bottom:

$$Q_a = G_s A\, \alpha_t + q_I A\, \varepsilon_b \sin^2\rho + G_s a A\, \alpha_b K_a \sin^2\rho \qquad (11\text{-}28)$$
$$K_a = 0.664 + 0.521\, \rho - 0.203\, \rho^2$$

The emitted radiation energy from the array is

$$Q_e = \sigma\varepsilon_b A T^4 + \sigma\varepsilon_t A T^4 \qquad (11\text{-}29)$$

where σ is the Stefan-Boltzmann's constant, A equals the area of one side of the array, ε_b is the IR emissivity of the bottom surface of the array, ε_t is the IR emissivity of the top surface of the array, and T is the absolute temperature of the array. Realizing that when it's at its equilibrium temperature, we can perform an energy balance on the solar array similar to Eq. (11-20):

$$q_{absorbed} - q_{emitted} - q_{power\ generation} = 0 \qquad (11\text{-}30)$$

For a solar array, the energy balance equation includes a power generation term since the solar cells convert solar energy directly to electrical energy. There is some dissipation associated with solar arrays (diodes, interconnect cabling, and transmission losses); however, we account for these losses in the array efficiency term. We define the *solar array efficiency*, η, as the ratio of electrical power output to incident solar radiation. Solar array efficiencies range from 0.07 to 0.2. Substituting into the energy balance equation yields

$$G_s A \alpha_t + q_I A \varepsilon_b \sin^2 \rho + G_s a A \alpha_b K_a \sin^2 \rho$$
$$- \sigma \varepsilon_b A T^4 - \sigma \varepsilon_t A T^4 - \eta G_s A = 0 \tag{11-31}$$

Solving for this worst-case hot temperature for the flat plate or solar array of any area (the area term, A, divides out of each term):

$$T_{MAX_A} = \left[\frac{G_s \alpha_t + q_I \varepsilon_b \sin^2 \rho + G_s a \alpha_b K_a \sin^2 \rho - \eta G_s}{\sigma(\varepsilon_b + \varepsilon_t)} \right]^{1/4} \tag{11-32}$$

We define the terms used in Eqs. (11-32) and (11-33) following Eq. (11-35). If the array is in the Earth's shadow and is not in view of any portion of the sunlit parts of the Earth, then we have a worst-case cold condition. For this condition there is no direct solar, albedo, or electric power generation. The equation which defines the temperature for this condition is:

$$T_{MIN_A} = \left[\frac{q_I \varepsilon_b \sin^2 \rho}{\sigma(\varepsilon_b + \varepsilon_t)} \right]^{1/4} \tag{11-33}$$

Spherical Satellite Equations—We can derive the worst-case hot temperature and worst-case cold temperature expressions for an isothermal sphere similar to the flat-plate expressions. For the sphere we assume that there is uniform energy dissipation over the surface of the sphere and there is no electrical generation on the spherical surface. Equation (11-20) gives the energy balance for the sphere and the temperature expressions are as follows:

$$T_{MAX_s} = \left[\frac{A_C G_s \alpha + AF q_I \varepsilon + AF G_s a \alpha K_a + Q_W}{A \sigma \varepsilon} \right]^{1/4} \tag{11-34}$$

$$T_{MIN_s} = \left[\frac{AF q_I \varepsilon + Q_W}{A \sigma \varepsilon} \right]^{1/4} \tag{11-35}$$

$\sin \rho = R_E / (H + R_E)$
G_s = solar constant = 1,418 W·m⁻² to 1,326 W·m⁻² (depending on the season)
q_I = Earth IR emission = 237 ± 21 W·m⁻²
a = albedo = 30% ± 5% of direct solar
Q_W = electrical power dissipation, W
σ = Stefan-Boltzmann's constant = 5.670 51 × 10⁻⁸ W·m⁻²·K⁻⁴
α_t = solar absorptivity on the top surface of the solar array
α_b = solar absorptivity on the bottom surface of the solar array
α = solar absorptivity of the sphere
ε_t = IR emissivity on the top surface of the solar array
ε_b = IR emissivity on the bottom surface of the solar array
ε = IR emissivity of the sphere
ρ = the angular radius of the Earth, $\sin \rho = R_E / (H + R_E)$ (see Chap. 5)
R_E = the radius of the Earth = 6,378,140 m

H = the altitude of the spacecraft, m

K_a = a factor which accounts for the reflection of collimated incoming solar energy off a spherical Earth

K_a = $0.664 + 0.521\,\rho - 0.203\,\rho^2$

η = solar array efficiency = 0.07 to 0.20 nominal efficiency for modern solar arrays

D = diameter of the spherical spacecraft, m

A = surface area of the satellite, m^2

A_C = $\pi \cdot D^2/4$, cross section area of the spherical satellite, m^2

F = $(1 - \cos\rho)/2$ view factor of an infinitesimal sphere viewing a finite sphere.

11.5.5 Preliminary Design Process

Spherical Satellite Analysis—Note that Eqs. (11-32), (11-33), (11-34), and (11-35) define steady-state temperatures. An actual spacecraft will likely exhibit lower maximum temperatures and higher minimum temperatures than those predicted by the equations, due to transient effects. Nevertheless, we can use these equations to provide reasonable first-order estimates of spacecraft thermal performance, and as the major element of the preliminary design process shown in Table 11-45. A numerical example of this process follows and for this example, we use the FireSat spacecraft.

The first step in the process is to establish the equipment temperature limits. Referring to Table 11-43, we find that the typical temperature range for electronics (internal units) is 0 °C to 40 °C and for nickel-hydrogen batteries the range is from –5 °C to 20 °C. Next, we estimate the electrical power dissipation of the spacecraft. If possible, we determine the power consistent with the worst-case hot condition and the power consistent with the worst-case cold condition. If this is not possible, we use the orbital average value for both conditions. Then we compute the diameter of a spherical spacecraft which has a surface area equal to the actual spacecraft. Now we can use Eqs. (11-34) and (11-35) that have been developed for an isothermal sphere to obtain first-order estimates of the spacecraft's thermal performance. For this calculation we assume a solar absorptivity of 0.6 (see Sec. 11.5.6) and an emissivity of 0.8. These values are consistent with white paint properties and represent radiation property values which are readily obtainable. Note that we have used a value for the solar absorptivity that is higher than the values shown in Table 11-44. This is to allow for UV degradation of the paint which we discuss in Sec. 11.5.6.

Table 11-46 gives the results of performance calculations for the FireSat spacecraft. Note that we use the larger values of the environmental radiation fluxes for the worst-case hot conditions and no environmental inputs or the lower values for the worst-case cold conditions. The results of the analysis are a worst-case hot temperature of +27 °C and a worst-case cold temperature of –87 °C. These values tell us that from an overall standpoint the spacecraft will be relatively easy to control thermally since the worst-case hot temperature is below the upper limit of 40 °C for the electronics. However, the 20 °C upper limit for the battery has been exceeded. This implies that we must thermally isolate the battery compartment as well as we can from the rest of the spacecraft, adopting dedicated thermal radiators and heaters. If the worst-case hot temperature had been 50 °C or 60 °C, we would know that we could not control the spacecraft using body-mounted radiators and we would need deployed radiators with a pumped-loop system probably.

TABLE 11-46.   Preliminary Thermal Performance Estimates for the FireSat Spacecraft.

No.	Item	Symbol	Value	Units	Source	Comment
1	Surface area of FireSat	A	13.5	m^2	FireSat	FireSat is 1.5-m cube
2	Diameter of sphere which equals FireSat surface area	D	2.07	m	Spherical geometry	$D = \sqrt{A/\pi}$
3	Max power dissipation	Q_w	170	W	FireSat	Given
4	Min power dissipation	Q_w	80	W	FireSat	Assumed
5	Altitude	H	700	km	Table 7-4	
6	Radius of Earth	R_E	6,378	km	Table 8-2	Rounded to 4 places
7	Angular radius of Earth	ρ	1.1221	rad	Calculated	Defined on page 110
8	Albedo correction	K_a	0.9931	—	Calculated	Defined with Eq. (11-28)
9	Max Earth IR emission at surface	q_I	258	$W \cdot m^{-2}$	Fig. 11-14	Use for worst-case hot
10	Min Earth IR emission at surface	q_I	216	$W \cdot m^{-2}$	Fig. 11-14	Use for worst-case cold
11	Direct solar flux	G_S	1,418	$W \cdot m^{-2}$	Fig. 11-14	Use maximum value
12	Albedo	a	35	%	Fig. 11-14	Use maximum value
13	Emissivity	ε	0.8	—	Table 11-44	Assume end of life white paint absorptivity
14	Absorptivity	α	0.6	—	Table 11-44	Assume white paint degrades
15	Worst case hot temperature	T_{MAX}	+27	°C	Eq. 11-34	
16	Worst case cold temperature	T_{MIN}	−87	°C	Eq. 11-35	
17	Upper temperature limit	T_U	35	°C	Table 11-43	Assume 5 °C thermal margin
18	Lower temperature limit	T_L	5	°C	Table 11-43	Assume 5 °C thermal margin
19	Radiator area based on worst case hot conditions	A_R	0.42	m^2	Eq. 11-23	Assume no solar heat input, dissipation of 170 W and radiator temp. of 35 °C
20	Radiator temperature verification for worst case cold conditions	T_R	−18.4	°C	Eq. 11-23	Assume no solar heat input, dissipation of 80 W and area of 0.42 m^2
21	Heater power required to maintain radiator at lower limit	Q_n	33.8	W	Eq. 11-23	Assume no solar heat input, area of 0.42 m^2 and radiator temp. of 5 °C

Thermal Radiator Analysis—We use Eq. (11-23) to assess the performance of a body-mounted radiator for FireSat. For line 19 of Table 11-46 we assume that the radiator temperature is 35 °C (40 °C upper equipment limit − 5 °C thermal margin), there are no environmental heat inputs, and the heat dissipation is 170 W. The radiator area required for these conditions is 0.42 m^2. The FireSat satellite is a 1.5-m cube, therefore it can easily accommodate a 0.42 m^2 body-mounted radiator. Line 20 on Table 11-46 illustrates a check on the performance of the 0.42 m^2 radiation consistent with worst-case cold conditions. For this situation we assume that the heat dissipation is 80 W, which produces a radiator temperature of –18.4 °C. This temperature is considerably below the lower limit of 5 °C (0 °C lower equipment limit + 5 °C thermal margin). Therefore, the radiator will require a thermal louver to modulate the heat flow and a thermostat and heater to maintain the temperature above the lower limit. The heater power required to maintain the radiator at 5 °C is 33.8 W (the computation on line 21 of Table 11-46). This power at the radiator is in addition to the 80 W dissipated from the electrical and electronic equipment. Note that we have not attempted to maintain the spacecraft equipment within the battery temperature limits. We will isolate the battery from the other spacecraft equipment and provide thermal control separately for this device. A reasonable approach is to cold bias the battery and thermally control the battery temperature with thermostats and heaters.

A useful thermal analysis handbook is *Spacecraft Thermal Control Design Data* ESA PSS-03-108, issue 1 [1989], which is a collection of information relevant to spacecraft thermal control design, compiled by the School of Aeronautics of the Polytechnic University of Madrid under ESA contract. ESA is also preparing, at the moment [1998], "Thermal Control Standards" ECSS-E-30-00 Part 2-1 that addresses the discipline of Spacecraft Thermal Control design, analysis, manufacturing, and verification.

11.5.6 Thermal Control Design Consideration

This section provides more information on selected TCS topics.

Thermal Capacities—Each material subjected to an input energy shows a temperature increase given by the equation:

$$Q = m\, c_p\, \Delta T/\Delta t \qquad (11\text{-}36)$$

where Q is the input (or output) energy per unit time in W, m is the mass of the system in kg, c_p is the specific heat capacity of material in J·kg^{-1}K^{-1}, ΔT is the temperature increase (decrease) in K, Δt is the duration of input (output) energy in seconds. This equation is not applicable during material change of phase. The specific heat capacity tells us how much energy we need to store in 1 kg material to increase its temperature by 1 K in 1 second. It's an intrinsic characteristic of matter. We use this equation to initially calculate, the temperature increase of power dissipating items. Thermal capacities of spacecraft components are needed to characterize the nodes of the heat transfer model when performing transient analysis. Manufacturers usually provide thermal capacitance of units and components. Values of specific heat capacities of various materials are given Table 11-47.

Thermal Conductivities—We calculate the conductive heat transfer with Eq. (11-11), considering the thermal conductivities of adopted materials that are in Table 11-48. Thermal conductivities are also used to calculate linear conductors of the heat transfer model.

MLI Effective Emissivity—Table 11-48 does not give values for multilayer insulation because we describe the thermal performance of multilayer insulation by effective emissivity rather than by conductivity. We define effective emissivity as

$$q = \sigma \varepsilon_{eff}(T_h^4 - T_c^4) \qquad (11\text{-}37)$$

where q is heat transfer through a multilayer blanket in W·m^{-2}, σ is the Stefan-Boltzmann's constant's, ε_{eff} is the effective blanket emissivity, and T_h and T_c are ab-

TABLE 11-47. Specific Heat Capacities of Materials. All data are at room temperature. (From various sources.)

Material	Specific Heat Capacity—c_p ($J \cdot kg^{-1}K^{-1}$)	Material	Specific Heat Capacity—c_p ($J \cdot kg^{-1}K^{-1}$)
Copper	390	Nylon	1,680
Aluminum Alloys	920	Carbon Fiber	840
Stainless Steel	503	Air (1 atm.)	1,006
Beryllium	1,800	Water	4,182
Magnesium	1,050	Hydrazine (liquid)	3,100
Titanium	520	Gallium Arsenide	335
Nickel	460	Ceramics	800
Tungsten	142	Silicon	712
Teflon	1,006	Kapton	1,006

TABLE 11-48. Thermal Conductivities. (From Sandia National Laboratory [1981] and Gilmore [1994]). This table provides additional properties of materials.

Material	Conductivity ($W \cdot m^{-1} \cdot K^{-1}$)	Material	Conductivity ($W \cdot m^{-1} \cdot K^{-1}$)
Copper	389	Ethylene Glycol	2.42
Aluminum Alloy 2017	164	Water	0.536
Aluminum Alloy 3003	156	Silicone	0.151
Aluminum Alloy 308	96.9	Perlite	0.0692
Aluminum Alloy 2219-0	172.9	Calcium Silicate	0.0548
Aluminum Alloy 6061-T6	167.7	Mineral Fiber Board	0.9548
Aluminum Alloy 7075-T6-T7	121	Mineral Fiber Block	0.0360
Steel 17-4PH	18	Phenolic	0.0332
Inconel 718	11.2	Glass Fiber Block	0.0317
Magnesium AZ31B H24	76.1	Urea Formaldehyde	0.0317
Titanium Ti-6Al-6V	7.26	Polystyrene	0.0288
Beryllium	150.4	Air	0.0260
Plain Carbon Steel	51.9	Polyurethane	0.0231
Stainless Steel Type 304	17.3		

solute hot and cold boundary temperatures of the MLI. We consider the temperature of the MLI as one boundary; this temperature can be assumed coincident with internal MLI layer temperature; the other boundary is the external MLI layer temperature. Other definitions may apply depending on the modeling technique used. Effective emissivity is more descriptive of the heat transfer through multilayer blankets than conductivity, since the predominant mode of heat transfer in multilayer blankets is radiation.

Figure 11-22 gives the thermal performance of multilayer insulation. An important conclusion we draw from this figure is that the effectiveness of multilayer insulation generally depends on the size of the object being insulated. Large multilayer insulation systems usually exhibit a lower effective emissivity than small systems, because edge effects, joints, seams, and penetrations (electrical wires, structural supports) represent relatively high extraneous heat-leak paths through the insulation. Generally, large multilayer insulation systems have a smaller percentage of extraneous heat-leak paths than small systems. Note in Fig. 11-22 that calorimeters and cryogenic tanks represent multilayer systems with low discontinuity densities. Most unmanned spacecraft and propulsion systems correspond to medium discontinuity densities while highly complex spacecraft and science platforms generally have high discontinuity densities.

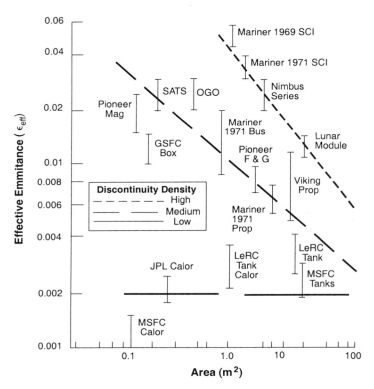

Fig. 11-22.   Performance of Multilayer Insulation. In this figure SCI stands for science scan platform, Mag for magnetometer, Prop for propellant, Calor for calorimeter, and F & G for the Pioneer Spacecraft series designations (Stimpson and Jaworski [1972]).

Figure 11-23 shows an example of typical design practices for a thermal control subsystem. In this figure BOL stands for beginning of life, and EOL stands for end of life. The TWT (traveling wave tube) shown in the figure is a high power amplifier. We require manufacturers to paint this device black to radiate as much thermal energy as possible.

Super Insulation 0
Aluminum Teflon ▲
Heaters ●
Silvered Quartz ■

Interior Finishes
CTL–1S Black Paint ε = 0.85 ⎫
CTL–1S White Paint ε = 0.85 ⎬ Except Where Noted
Vapor Deposited Gold ε = 0.03 ⎭

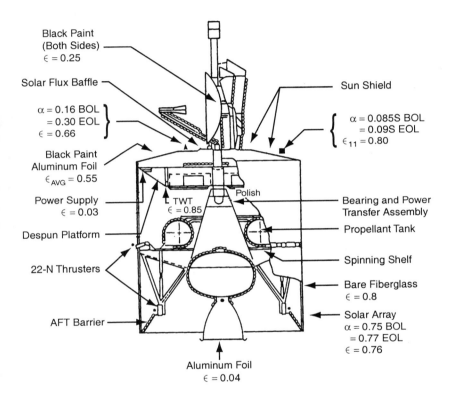

Black Paint
(Both Sides)
ε = 0.25

Solar Flux Baffle

α = 0.16 BOL ⎫
= 0.30 EOL ⎬
ε = 0.66 ⎭

Black Paint
Aluminum Foil
ε$_{AVG}$ = 0.55

Power Supply
ε = 0.03

Despun Platform

22-N Thrusters

AFT Barrier

TWT
ε = 0.85

Polish

Aluminum Foil
ε = 0.04

Sun Shield

α = 0.085S BOL
= 0.09S EOL
ε$_{11}$ = 0.80

Bearing and Power
Transfer Assembly

Propellant Tank

Spinning Shelf

Bare Fiberglass
ε = 0.8

Solar Array
α = 0.75 BOL
= 0.77 EOL
ε = 0.76

Fig. 11-23. Typical Design Practice for Thermal Control.

MLI Details—Multilayer insulation is a primary hardware component for space-craft thermal control subsystems. This material, however, requires extreme care with regard to installing, modeling, and testing. For example, we saw earlier that overlapping multilayer blankets at joints will greatly degrade the thermal performance of the insulation. If we don't account for penetrations correctly in thermal models, the analytical predictions will not match the test results. Also, test times in vacuum chambers must be long enough for the air within multilayer blankets to vent. Even though multilayer blankets are routinely perforated, it may take several hours for the trapped air to vent so the blanket performs as it would in space.

Joint Conduction—One of the elements of a thermal control subsystem that is difficult to predict is the resistance to heat flow across solid-to-solid interfaces. Thermal control subsystems often have important heat transfer paths across interfaces; for example, heat-dissipating components bolted to cold plates. The thermal performance of an interface is characterized as joint conductance and is defined by the following equation:

$$Q = h_c A \, \Delta T \tag{11-38}$$

where Q is the heat transfer rate across the interface in W, A is the interface area in m^2, ΔT is the temperature difference across the interface in K, and h_c is the joint conductance in $W \cdot m^{-2} \cdot K^{-1}$.

The difficulty in predicting joint conductance is that this property depends on the surface finish, hardness of the interface surfaces, the waviness of the interface surfaces, type and number of interface bolts, and the resulting interface pressure. A typical value for aluminum surfaces is $4,500 \, W \cdot m^{-2} \cdot K^{-1}$; however, this value can vary greatly depending on the circumstances just mentioned. Because of the difficulty in predicting contact conductance, we must measure this quantity at the component level and then use the measured values in the spacecraft thermal models. Contact conductance can be increased with the use of filler materials (cf. Sec. 11.5.2).

Degradation Effects—In Table 11-44 the values of solar absorptivity are all beginning-of-life values. This means we have not subjected the materials to the space environment, which will cause degradation of the solar absorptivity with time as a result, primarily, of UV radiation. The degradation is exponential in nature and generally there is an upper limit. For example, the silvered Teflon, aluminized Teflon, and OSR, listed in Table 11-44, have upper limit solar absorptivities of 0.241, 0.316, and 0.103, respectively. The amount of time required for the materials to approach the upper limit, or asymptotic value, is from 6,800 to 44,000 hours of solar exposure. All types of white paint generally degrade considerably in space. After only 3 years α can be in the order of 0.4 and reach up to 0.7, for some types, at end of life. In analyzing the thermal performance of a spacecraft, we must consider the beginning-of-life and end-of-life values of thermal coatings to define the temperature extremes during the entire mission. For a discussion of degradation effects and thorough lists of properties of surface finishes, see Gilmore, [1994].

Planetary Thermal Data—For planetary or deep space missions the data reported in Table 11-49 are of interest to the thermal designer.

TABLE 11-49.   Planetary Thermal Data. (From NASA [1983].)

Planet	Maximum Solar Radiation Intensity (W · m⁻²)*	Average Albedo
Mercury	9,350	0.06
Venus	2,700	0.61
Earth	1,418	0.34
Moon	1,418	0.07
Mars	600	0.15
Jupiter	52	0.41
Saturn	15	0.42
Uranus	4	0.42
Neptune	1.5	0.52
Pluto	0.9	0.16

*Rounded up figures.

Material Stability, Outgassing, and Contamination—To prevent premature thermal performance degradation and pollution of instruments or sensor heads, we must carefully analyze our subsystems' resistance to long space exposures and our thermal control systems' outgassing behavior. The thermo-optical properties of thermal control components like: MLI, paints, tapes, radiators are very sensitive to the presence of contaminants on their surfaces. So, we must minimize any performance degradation of thermal control hardware by adopting the necessary care and cleanliness control procedure during manufacturing, handling, installation, and testing.

Thermal Analyses and Related Activities—We summarize the main thermal analyses, related analyses, and other analytical activities in the following:

- *Thermal Analyses*—By the appropriate analytical model, thermal analyses provide temperature predictions at designated reference points for comparison with the requirements (i.e. consistent with the definitions provided in Table 11-43). A common algorithm is the finite difference method, which works well for nonlinear calculations. A thermal analysis may also provide interface heat fluxes, thermal gradients, temperature vs. time plots, parts temperature maps at specified times. The object of the analysis may be an entire satellite, a local model of a portion of a satellite, or a single unit/ component. The analyzed environment may be the test, launch, transfer orbit, or nominal orbital conditions.

  We often perform the analysis of electronic units by the finite element method (FEM) because a unit structural model typically already exists, and the thermal FEM works well for linear analyses since the conductive heat transfer dominates over radiation within the unit.

- *Condensed (or Reduced) Thermal Mathematical Model Preparation*— Sometimes we must integrate detailed thermal model into larger models. For instance, we may integrate an antenna model into a spacecraft model, or a spacecraft model into a launcher model. We should avoid making the simple assembly of detailed models into a huge, complex model which may not run efficiently on a computer. To avoid these problems, we must reduce the detailed models to simpler models by grouping small nodes into larger ones. The reduced model must duplicate well the temperature and heat-flux predictions for the specified reference nodes, with respect to the original detailed model.

- *Sensitivity Analyses*—We perform these to assess the impact on the temperature predictions induced by inaccuracies in the environmental, the physical, the model, as well as test facility parameters. They consist-of replacing, in the thermal model, the nominal value of the considered parameter by a new that includes the expected inaccuracy. The results provide the extent of the temperature uncertainty.

- *Thermo-Elastic and Humidity Analyses*—We consider these mechanical analyses because the final result is the outcome of the structural mathematical model. They determine structural distortions for stress and alignment error verification. The inputs are the analyzed parts' temperature maps which result from a thermal analysis. Due to the complexity of the structural architectures and relevant temperature gradients, which we usually analyze, it is not obvious which thermal case induces the worst structural distortions. For this reason we must analyze a number of worst hot, cold, and maximum gradient cases which we pass to the structural engineer.

- *Thermal Mathematical Model Correlation*—With this process we reduce the errors affecting the temperature predictions when thermal-test results are available. We do this by comparing the predicted and measured average temperatures, at convenient locations, during testing. Typically, if they differ more than 3°C, then we must increase the realism/accuracy of the thermal model.

- *Subsystem Budgets Preparation*—We request the thermal subsystem designers to establish a budget for each of the following resources: mass, size, power, energy, telemetry and telecommand channels, all along the TCS life cycle. We use these budgets to control the status of the thermal design, to provide inputs to other subsystems, and to discuss with the system authority.

Trade Studies—In this section we have described many of the hardware elements of a thermal control system. The question remains, which hardware elements should we use, or even consider, for a given spacecraft? We often answer this question in practice by performing trade studies to determine the most effective components to use for a given situation. The criteria used in trade studies usually involves: thermal performance, cost, weight, reliability, design flexibility, accessibility, testability, availability, safety, and durability. During the spacecraft design phase, a concurrent engineering approach should be applied; i.e. the thermal design should be developed in parallel with the overall spacecraft and mission design and not sequentially in order to establish the most efficient TCS concept in terms of time and money.

11.5.7 Thermal Testing

We perform thermal tests at both component, subsystem and system levels to verify the thermal design and demonstrate the proper operation of items in extreme environmental conditions. The following definitions apply to spacecraft and spacecraft components thermal testing.

Figure 11-24 makes clear the following temperature definitions.

Predicted Temperature Range—It's a temperature range defined by the minimum and maximum temperatures predicted from a thermal analysis considering conservative parameters. This means to consider undegraded thermo-optical properties, no or minimum power dissipation and environmental thermal fluxes for cold cases, and degraded thermo-optical properties and maximum power dissipation and environmental thermal fluxes for hot cases.

Design Temperature Range—(Sometimes called operating or expected temperatures). It's a temperature range defined by the predicted extreme temperatures plus thermal uncertainties. *Thermal uncertainties* traditionally range from 5 °C to 7 °C for a final analysis campaign after model correlation, to 10 °C for design definition analyses, to 15 °C for preliminary design definition analysis. They are introduced to compensate for the inevitable uncertainties due to inaccuracies in modeling, knowledge of environmental parameters, and thermal hardware characteristics. Thermal uncertainties are analogous to factor of safety in structural design. The design temperature range can either be the requirement for the termal designer or the result of his analysis.

Acceptance Temperature Range—It's specified for the operating and non operating modes of a unit and defined by the design extreme temperature range with a

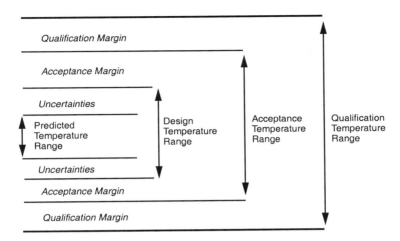

Fig. 11-24. Temperature Definitions for TCS.

± 5 °C additional environmental margin. The main purpose of tests performed at these levels is to identify any defect in materials or workmanship. We subject components to these temperature levels before accepting them as flight hardware. For example, if a component must pass flight acceptance tests at low and high temperatures of 5 °C and 50 °C, with a thermal margin of 5 °C, the spacecraft thermal control must maintain the given component, in terms of design temperatures, at 10 °C to 45 °C, 5 °C inside the flight acceptance limits.

Qualification Temperature Range—It's specified for the operating and nonoperating mode of a unit. It's a range generally 10 °C beyond the design temperatures at which components must operate on orbit. This is the extreme temperature range for which a unit is guaranteed to operate with the required performances and reliability. The main purpose of tests performed using these levels is to identify any weakness in the thermal design.

System-Level Thermal Tests—We generally require these tests as part of a spacecraft program and these tests fall into two major categories: a *solar balance test* and a *thermal vacuum test*. We conduct both tests in a vacuum chamber equipped with nitrogen-cooled walls to thermally simulate deep space.* Thermal vacuum tests are typically performed using electric heaters or infrared lamps, controlled to force spacecraft temperatures to prescribed levels. We typically perform thermal-vacuum tests to verify the workmanship of the spacecraft and provide a means of checking the operation of the spacecraft components in a simulated space environment.

We perform solar balance tests with a solar simulator and often electric heaters or infrared lamps to simulate Earth's infrared radiation. A light source (often filtered),

---

* The simulation of the deep space temperature (4 K) by means of a shroud filled with liquid nitrogen (100 K) is very efficient in spite of the apparent large temperature difference. We express the radiation heat transfer from the spacecraft to the external environment by: $Q \propto (T_s{}^4 - T_e{}^4)$, where T_s is the temperature of the spacecraft (K), and T_e is the temperature of the external environment (K). Now assuming a spacecraft temperature in the range −10 to 40 °C, and using the presented proportion, we see that the heat transfer towards the nitrogen cooled shroud is between 2 to 1% lower than the equivalent to deep space.

simulates the solar spectrum and the solar flux level expected during the mission. The purpose of balance tests is to establish experimentally the worst-case hot and worst-case cold thermal performance of the spacecraft and to validate the analytical thermal model of the spacecraft. If there are significant differences between the test results and the mathematical predictions, we try to identify the reason(s) for the differences. Experience shows that the differences can be the fault of either inaccuracies in the model or problems with the test. After we achieve close correspondence between analysis and test, we can use the thermal model with confidence in predicting the temperatures of the spacecraft for any flight situation.

Component/Subsystem Thermal Testing—We perform these tests to find faults in the design, materials, processes, workmanship, and to verify the performance of the test item. They consist of thermal (in air) or thermal-vacuum cycling where we impose convenient temperature extremes, number of cycles, dwell times, and operative modes on the item under test. Burn-in testing is used to screen electronic parts for infant mortality failures or for mass production components lot acceptance. This is generally performed as part of the thermal cycle test at ambient pressure.

Development Tests—We may require these tests, at any level, to provide an early feedback during the design or manufacturing phases. We direct them on a case by case basis and generally they are performed to provide confidence in the development of new concepts or processes.

11.5.8 Future Trends

Thermal Control Materials

Researches are in progress to produce new materials, including paints, having resistance to extreme temperatures, stable thermo-optical properties under several years of space exposure, and controlled outgassing behavior. They study materials having a very high thermal conductivity like diamond, carbon-carbon, or metal-matrix composites. The future Thermal Control design of satellite will take advantage of these new materials.

The following two tendencies, concerning satellite applications, have been envisioned in the *Satellite Thermal Control Handbook* by Gilmore et al. [1994]:

- *Large Satellites*
  The future large, high-power spacecraft will require more efficient means to reject heat. This thermal control may use large radiators that remain stowed on the spacecraft primary structure during the launch, deployed on orbit and probably steered to avoid Sun impingement. This concept implies that we develop flexible heat pipes, rotating fluid joints and high-performance pumped fluid loops.

- *Small Satellites*
  In the field of small, micro, and nano satellites, it's foreseen a considerable reduction in the size of all satellite components, and their heat rejection requirements. These satellites could be isothermal, which will benefit their reliability. Because nano satellites will concentrate all functions and subsystems in a reduced volume, this approach could lead to a high power dissipation density. Consequently we foresee developing and utilizing highly conductive materials, small-scale heat pipes, and pumped fluid loops.

11.6 Structures and Mechanisms

Thomas P. Sarafin, *Instar Engineering*
Peter G. Doukas, *Lockheed Martin Astronautics*
James R. McCandless and William R. Britton, *Lockheed Martin Astronautics*

The *structures and mechanisms subsystem* mechanically supports all other space-craft subsystems, attaches the spacecraft to the launch vehicle, and provides for ordnance-activated separation. The design must satisfy all strength and stiffness requirements of the spacecraft and of its interface to the booster. *Primary* structure carries the spacecraft's major loads; *secondary* structure supports wire bundles, propellant lines, nonstructural doors, and brackets for components typically under 5 kg.

In this section, we describe how to develop a preliminary design for a structures subsystem. We begin by considering the spacecraft's operating environments and designing the structure with overall spacecraft packaging in mind. After conducting numerous design trades, we then assess each structural member for its most likely failure modes, possible weight savings, and need for reinforcement. See Fig. 11-25 for details.

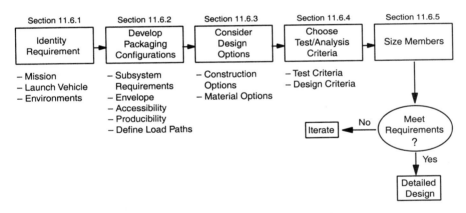

Fig. 11-25. The Preliminary Design Process for Structures and Mechanisms. We move from left to right, iterating as needed, when designing the spacecraft structure.

11.6.1 Structural Requirements

Structures must endure environments from manufacture to the end of the mission. Team members should contribute from all disciplines: engineering, manufacturing, integration, test, and mission operations. This interdisciplinary approach ensures coverage of all critical requirements—even those which seem minor. The following discussion of the Space Shuttle's external tank structure show why we should not overlook any event in the structure's lifetime.

The aluminum skin of the external tank must have a very tight manufacturing tolerance. Adding just 0.0254 mm (0.001 in) thickness to the entire shell of its forward tank for liquid oxygen adds 220 kg to tank mass. Special handling fixtures must cradle the tank's wall sections to keep them from collapsing during welding, as they cannot support their own weight. The nose section of the completed external tank experiences its most severe loads **before** launch from winds occasionally gusting against an empty and unpressurized tank. Table 11-50 lists typical mission phases and possible sources for structural requirements.

TABLE 11-50. **Typical Sources for Structural Requirements by Mission Phase.** The structural design must account for specific loads in every phase.

Mission Phase	Source of Requirements
Manufacture and Assembly	• Handling fixture or container reactions • Stresses induced by manufacturing processes (welding)
Transport and Handling	• Crane or dolly reactions • Land, sea, or air transport environments
Testing	• Environments from vibration or acoustic tests • Test fixture reaction loads
Prelaunch	• Handling during stacking sequence and pre-flight checks
Launch and Ascent	• Steady-state booster accelerations • Vibro-acoustic noise during launch and transonic phase • Propulsion system engine vibrations • Transient loads during booster ignition and burn-out, stage separations, vehicle maneuvers, propellant slosh, and payload fairing separation • Pyrotechnic shock from separation events
Mission Operations	• Steady-state thruster accelerations • Transient loads during pointing maneuvers and attitude control burns or docking events • Pyrotechnic shock from separation events, deployments • Thermal environments
Reentry and Landing (if applicable)	• Aerodynamic heating • Transient wind and landing loads

The launch vehicle is the most obvious source of structural requirements, dictating the spacecraft's weight, geometry, rigidity, and strength. The launch vehicle, selected orbit, and upper stage determine the spacecraft's allowable weight. See Table 18-4 for launch-vehicle data.

The core body structure and spacecraft adapter typically account for 10% to 20% of a spacecraft's dry weight. Appendages, component boxes, and most secondary structures apply to the weight of other subsystems. On the structures and mechanisms subsystem, we normally increase the estimated weight by 10% for fasteners and fittings. We should also add approximately 25% for weight growth to account for program additions, underestimating, and inadequate understanding of requirements. The spacecraft item most often underestimated or neglected is electronic wiring, sometimes approaching 10% of a spacecraft's dry weight. Of course, allowances for weight growth may vary for a component or subsystem, based on its design maturity and schedule risk. As subsystem designs mature, known weights replace growth estimates.

A spacecraft's size depends on choosing the *payload fairing* compatible with the launch vehicle. These protective shrouds shield the spacecraft from direct air loading and contamination. The spacecraft and its fairing have a prescribed *dynamic envelope*, or space allocation, that takes into account expected deflection and the possible addition of thermal protection blankets. The spacecraft must be rigid enough so the fairing and spacecraft do not encroach on each other's envelope. Although the Space Shuttle does not have a traditional payload fairing, its cargo bay requires a similar envelope. See Fig. 18-8 for an example.

The spacecraft's rigidity requirements specify more than maximum deflection. A launch-vehicle structure has certain natural frequencies that respond to forces from both internal (engine oscillations) and external (aerodynamic effects) sources. The launch vehicle contractor lists known natural frequencies for each launch vehicle (see Table 18-9) and describes associated axial, bending, or torsional (twisting) modes. The spacecraft structure tailored to avoid the launch vehicle's natural frequencies will experience much lower loads. Typical resonance sources to avoid include interaction between the spacecraft and the launch vehicle's control system, oscillations in the propulsion system (*pogo*), aerodynamic buffeting during ascent, and bending of the solid rocket motors.

Engine thrust during launch and ascent exposes the spacecraft to steady-state acceleration along its axis. This acceleration steadily increases as a booster depletes fuel (less mass to propel), but comes to an abrupt end, or transient, at burn-out. The acceleration resumes suddenly, with another transient, as the next stage ignites. Wind gusts and vehicle maneuvers can induce lateral transients. Transients and steady-state accelerations cause inertial loads, commonly specified as *load factors*, or multiples of weight at sea level. Table 18-8 shows typical load factors for several launch-vehicle events.

Random vibration from engines and other sources is a critical source of load. At lift-off, the major source of random vibration is acoustic noise, which radiates from the engines to engulf the vehicle. Acoustics develop from aerodynamic turbulence when the vehicle passes through the transonic portion of its flight. Structures with high surface area and low mass, including skin sections and solar array panels, respond strongly to acoustic noise.

Load factors do not express random vibration correctly. Three parameters that help describe random vibration are distribution, frequency content, and magnitude. Typically we assume that a random spectrum has a Gaussian distribution, which determines the percentage of time the vibration is within certain limits. The frequency content is most commonly expressed as *power spectral density* (*PSD*) even though "acceleration" is more precise than "power" in this application. Vibrational power in a signal is proportional to acceleration-squared. This is divided by the frequency bandwidth over which the signal was integrated, to make the quantity independent of bandwidth. Thus, PSD is in units of g^2/Hz. To illustrate the power spectral density, we use a log-log plot of g^2/Hz against frequency. The square root of the area under the curve is the time history's rms value. This value equals one standard deviation, σ, of the random acceleration. Figure 11-26 shows a random signal, its normal distribution, and a typical PSD plot.

Pyrotechnic shock, another source of load, comes from explosive separation events involving the boosters, payload fairing, and spacecraft, as well as release mechanisms for solar panels and other deployable appendages. This shock causes high acceleration and high frequency over a very short time (see Fig. 18-11). Because shock loads attenuate quickly, they seldom damage structures removed from the immediate impulse, but they may seriously harm nearby electronic components.

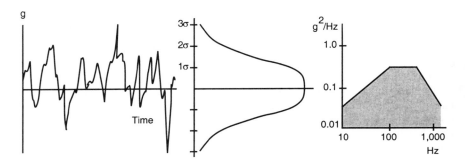

Fig. 11-26.   Random Vibration. The plot shown on the left is an example of how acceleration from random vibration would vary over time. The probability density curve (center), with the same vertical scale, describes the relative likelihood of acceleration being at a given value. This is a normal distribution, with each tick mark on the vertical scale representing a standard deviation, σ. The figure on the right is a plot of power spectral density. It describes the frequency content of the vibration and is equal to the mean-square acceleration (g^2) in a selected frequency band divided by the width, in Hz, of that band.

In actual design, we combine math models of the spacecraft and launch vehicle to do a *coupled loads analysis.* In this analysis, we drive the coupled model with *forcing functions* (force as a function of time or frequency) that are based on measured launch-vehicle environments. Before we get to this point, though, we must configure the structure, select from our design options, and roughly size the structure based on estimated design loads.

11.6.2 Packaging and Configuring the Subsystem

Designers must trade the low weight of a high-density design against the need to access individual components for testing or replacement before launch. The preliminary arrangement should account for every component in the design because the spacecraft's structure inevitably becomes heavier if it must accommodate new components. Added component mass multiplies through higher allowances for weight growth, heavier structure for support, and more onboard propellants for attitude control.

The payload and the attitude control approach most strongly influence a spacecraft's configuration, and the launch vehicle constrains it. Chapter 9 discusses payloads and their requirements. Section 11.1 discusses spin- and 3-axis stabilization to control attitude. A spin-stabilized spacecraft influences packaging most because the mass moment of inertia (MOI)* about the spin axis must be greater than any other axis to maintain stability. In 3-axis stabilized spacecraft, we must separate the magnetic torque rods and the magnetometer (device that senses the Earth's magnetic field) to preclude any magnetic interference between them.

Sensing devices always require specific fields of view and pointing accuracy. The packaging designer must locate sensors to be unobstructed by antennas or solar arrays.

---

* I is the standard notation for both mass moment of inertia and area moment of inertia. In this section, we use MOI for mass moment of inertia and I for area moment of inertia.

Advanced composite materials often help mounting-structure designs meet requirements for rigidity and thermoelastic distortion.

Communication antennas also require rigidity, thermoelastic stability, and a clear field of view. One solution is to mount sensors and antennas on an appendage that is stowed during launch, deployed on orbit for an unobstructed view, and constructed of advanced composite materials for rigidity and thermoelastic stability.

Components for command and data handling are often vulnerable to the environments of outer space, so we usually bury them in the center of the spacecraft to shield against radiation. Interfacing wire bundles also weigh less when the processor, data bus, and other control components are close together.

Propulsion subsystems include reaction-control assemblies and orbital transfer stages. By purchasing thrusters in multiaxis combinations, called *rocket engine modules*, we can reduce the number of propellant-line welds needed on site during the spacecraft's assembly. We usually place those modules on the spacecraft's periphery—far from the spacecraft's center of mass—but we must keep them from contaminating sensors, antennas, and solar array cells with propellant gases. A propulsion system with a low operating pressure helps lessen the propellant tank's weight. Another structural challenge is the need to support transfer stages, so the thrust vector remains aligned with the spacecraft's center of mass. These stages are usually heavy; so, placing them on the bottom of the spacecraft stack, near the launch vehicle interface, helps minimize structural weight.

The configuration of the power subsystem varies with power requirements and orbital conditions. For example, we must determine where to stow solar panels during launch and where to deploy them in orbit, so they do not touch or rest in the shadow of other subsystems. By using fewer folds in the panels, we can keep the deployment mechanisms simple and more reliable. Finally, batteries should be accessible for pre-launch testing or replacement and placed where they will be at their optimum temperature. Thermal control specialists can place components, select materials, and suggest surrounding structure (open truss or closed skin panels with stiffeners) to help control temperature. These measures help us avoid using other active temperature control devices.

If we configure the structure and package components at the same time, we may make the components part of the load-carrying structure. This concurrent approach may also produce better symmetry in the structure, which satisfies frequency response requirements. By using common members and joints throughout the design, we can lower fabrication costs and more easily meet weight allocations. For example, beams that make the spacecraft rigid can also support components. Establishing design routes for wire bundles and propellant lines helps avoid the inefficiencies of cutting through structure later. We should design special joints to connect members made with different materials because their varying rates of thermal expansion and contraction can be detrimental. Finally, the spacecraft adapter must transition smoothly from the spacecraft to the interface on the launch vehicle's upper stage.

11.6.3 Design Options

In designing a structure, we consider optional materials, types of structure, and methods of construction. To select from these options, we do trade studies to compare weight, cost, and risk.

A typical spacecraft structure contains metallic and nonmetallic materials. Most metals are very nearly *homogeneous*, having constant properties throughout their composition, and *isotropic*, having the same properties regardless of direction. Non-metals are usually formed with *composites*, or blends of more than one material. Composite materials are not homogeneous and are normally not isotropic. Materials are selected based on:

- Strength
- Stiffness
- Density (weight)
- Thermal conductivity
- Thermal expansion
- Corrosion resistance
- Cost

- Ductility (which can prevent cracks)
- Fracture toughness (ability to resist crack growth)
- Ease of fabrication
- Versatility of attachment options, such as welding
- Availability

By far the most commonly used metal for spacecraft structure is aluminum alloy, of which there are many types and tempers. Aluminum is relatively lightweight, strong, readily available, easy to machine, and low in raw material cost. The stiffness-to-weight ratio of aluminum is about the same as steel, but the strength-to-weight ratio is usually higher. The main advantage of aluminum over steel for flight structures is its lower density. For the same mass, an aluminum shell or plate would be thicker and thus able to carry a greater compressive load before it would buckle. If we need harder or denser materials, we normally choose steel or titanium.

Alloys are available in sheets, plates, extrusions, forgings, and castings. The primary source of material properties is MIL-HDBK-5, *Metallic Materials and Elements for Aerospace Vehicle Structures* [U.S. Air Force Materials Laboratory, 1994], which contains many properties and statistically guaranteed strengths for all commonly used aerospace metals.

One popular advanced composite is graphite-epoxy, which has graphite fibers for strength and stiffness in an epoxy matrix. Composite fabric layers normally bond together in designed fiber orientations, so they can provide properties not available in homogeneous metallic materials, including extremely high stiffness-to-weight ratios and negligible expansion and contraction resulting from temperature gradients. Other fibers in these composites include boron, Kevlar™, and glass. Graphite and boron fibers also reinforce metal-matrix composites. Techniques to manufacture and apply metal-matrix composites are presently less advanced than for epoxy-matrix materials. Tsai [1987] and MIL-HDBK-17 [1977, 1989] provide more information on composite materials.

Table 11-51 summarizes the advantages and disadvantages of the most commonly used materials in spacecraft design. Table 11-52 shows their representative properties.

Types of structures include skin panel assemblies, trusses, ring frames, pressure vessels, fittings, brackets, and equipment boxes. Sometimes only one meets objectives; but we usually have several options. We use *monocoque structures,* which are panels and shells without attached stiffening members, only if applied and reacted loads are spread out rather than concentrated. A *semimonocoque shell* has lightweight, closely spaced stiffening members (*stiffeners*) that increase its buckling strength. *Skin-stringer structures* have longitudinal members (*stringers*) and lateral members (*frames*) to accept concentrated loads and skin to spread those loads out and to transfer shear. A *truss* is an assembly that remains stable under applied concentrated loads with its structural members loaded only axially. A *sandwich structure* is a panel or shell

TABLE 11-51.   **Advantages and Disadvantages of Commonly Used Materials.**

Material	Advantages	Disadvantages
Aluminum	• High strength vs. weight • Ductile; tolerant of concentrated stresses • Easy to machine • Low density; efficient in compression	• Relatively low strength vs. volume • Low hardness • High coefficient of thermal expansion
Steel	• High strength • Wide range of strength, hardness, and ductility obtained by treatment	• Not efficient for stability (high density) • Most are hard to machine • Magnetic
Heat-resistant	• High strength vs. volume • Strength retained at high temperatures • Ductile	• Not efficient for stability (high density) • Not as hard as some steels
Magnesium	• Low density—very efficient for stability	• Susceptible to corrosion • Low strength vs. volume
Titanium	• High strength vs. weight • Low coefficient of thermal expansion	• Hard to machine • Poor fracture toughness if solution treated and aged
Beryllium	• High stiffness vs. density	• Low ductility & fracture toughness • Low short transverse properties • Toxic
Composite	• Can be tailored for high stiffness, high strength, and extremely low coefficient of thermal expansion • Low density • Good in tension (e.g., pressurized tanks)	• Costly for low production volume; requires development program • Strength depends on workmanship; usually requires individual proof testing • Laminated composites are not as strong in compression • Brittle; can be hard to attach

constructed of thin face sheets separated by a lightweight core; this form of construction efficiently adds bending stiffness and stability. Section 15.3 of Sarafin [1995] provides guidance for selecting the above types of structures.

We can attach structural elements with adhesive bonds, welds, or mechanical fasteners. But regardless of the selected structure type and method of attachment, much of the structural subsystem's weight will be in the fittings used to transfer load from one member to another.

Most composite material structures have metal end fittings or edge members attached by bonding, but the bond's strength depends on the process and workmanship. Normally, we must select a proper bonding process through development testing. We can use bolts instead; however, local stress concentrations around the fasteners can cause failure at load levels much lower than a composite material can otherwise sustain. Welding is also possible for most aluminum alloys, but heat from welding can lower material strength near welds by more than 50%. If we need stiffness more than strength, we may choose welding over mechanical joints. As with bonding, welding processes require strict development, control, and testing.

The strength of mechanical fasteners, such as rivets and bolts, is very dependable as a result of process controls, inspections, and frequent sample testing. But to fully

TABLE 11-52. Design Properties for Commonly Used Metals [MIL-HDBK-5G, 1994]. The design allowable stresses given here are statistically guaranteed for at least 99% of all material specimens. Strengths shown are for the *longitudinal* direction (rolling or extrusion direction) of the material; strengths are usually lower in the *long-transverse* (across width) and *short-transverse* (through thickness) directions.

Material Alloy and Form	ρ 10^3 kg/m^3 (lb/in^3)	F_{tu} 10^6 N/m^2 (10^3 lb/in^2)	F_{cy} 10^6 N/m^2 (10^3 lb/in^2)	E 10^9 N/m^2 (10^6 lb/in^2)	e %	α 10^{-6}/°C (10^{-6}/°F)
Aluminum						
2219-T851 1″ Plate	2.85 (0.103)	420 (61)	320 (47)	72 (10.5)	7	22.1 (12.3)
6061-T6 Bar	2.71 (0.098)	290 (42)	240 (35)	68 (9.9)	10	22.9 (12.7)
7075-T73 Sheet	2.80 (0.101)	460 (67)	380 (55)	71 (10.3)	8	22.1 (12.3)
Steel 17-4PH H1150z Bar	7.86 (0.284)	860 (125)	620 (90)	196 (28.5)	16	11.2 (6.2)
Heat-Res. Alloy A-286 2″ Bar	7.94 (0.287)	970 (140)	660 (95)	201 (29.1)	12	16.2 (9.0)
Inconel 718 4″ Bar	8.22 (0.297)	1280 (185)	1,080 (156)	203 (29.4)	12	12.2 (6.8)
Magnesium AZ31B H24 Sheet	1.77 (0.064)	270 (39)	165 (24)	45 (6.5)	6	25.4 (14.1)
Titanium Ti-6Al-4V Annealed Plate	4.43 (0.160)	900 (130)	855 (124)	110 (16.0)	10	8.8 (4.9)
Beryllium AMS 7906 Bar	1.85 (0.067)	320 (47)	—	290 (42)	2	11.5 (6.4)

ρ = Density

F_{tu} = *Allowable Tensile Ultimate Stress,* the highest uni-axial tensile stress a material can sustain before rupturing.

F_{cy} = *Allowable Compressive Yield Stress,* the compressive stress that causes a permanent deformation of 0.2% of the specimen's length.

E = *Young's Modulus,* a.k.a. *Modulus of Elasticity,* the ratio of stress to strain (length change divided by original length) in the linear elastic range (see Sec. 11.6.6).

e = *Elongation,* a measure of ductility, equal to the percentage change in length caused by plastic deformation prior to rupture.

α = *Coefficient of Thermal Expansion,* a measure of strain per degree temperature change. Values shown for α are at room temperature.

develop fastener strength, we must provide adequate fitting thicknesses, fastener spacing, and edge distances. The torque value for installing a tension fastener must provide a preload that will maintain stiffness and preclude *fatigue,* which is a failure resulting from cracks that form and grow because of cyclic loading. A locking feature, typically a deformed thread in the nut or insert, will prevent a threaded fastener from backing out when the structure vibrates. Many similar guidelines help us design a dependable structural joint.

11.6.4 Structural Design Philosophy and Criteria

To develop a structure light enough for flight, and to keep spacecraft affordable, we must accept some risk of failure. Material strengths vary because of random, undetectable flaws and process variations, and loads depend on unpredictable environments. Random variables affect the adequacy of most structures, such as a dam whose load depends on how much it rains; but for space missions, we must accept a higher probability of failure than for most other types of structures. Launch loads are affected by many different random variables, such as acoustics, engine vibrations, air turbulence, and gusts, and we seldom have enough data to confidently model the probability distributions of these variables.

Because of loads uncertainty, we cannot accurately quantify the structural reliability of a spacecraft. We can approximate it, however, and we can develop design criteria that will provide acceptable reliability. Let us work backwards from a subsystem-level reliability to understand how conservative our design approach should be for an individual structural part.

If we select a goal for structural reliability of 99% (we probably should aim higher), which means there is a 1% chance of a mission-ending structural failure, we must design each structural element to much higher reliability. If the structure has 1,000 parts whose failure would jeopardize the mission, and if their chances of failure are independent, each must have 99.999% reliability ($0.99999^{1000} = 0.99$, from probability theory). This explains why design criteria may appear so conservative. To achieve appropriate reliability, many programs use the following ground rules:

- Use a design-allowable strength for the selected material that we expect 99% of all specimens will equal or exceed.

- From available environmental data, derive a design limit load equal to the mean value plus three standard deviations. This means there will be 99.87% probability that the limit load will not be exceeded during the mission, assuming the load variability has a Gaussian distribution. Because data will be limited, we can only approximate the true probability level of the design load; but "3-sigma" remains the goal. (Some programs aim for 99% probability instead of 3σ.)

- Multiply the design limit load by a factor of safety, then show that the stress level at this load does not exceed the corresponding allowable strength.

- Test the structure to verify design integrity and/or workmanship, to correlate analytical models, and to protect against human errors.

Table 11-53 summarizes the criteria used to design space structures.

Space programs use different factors of safety, but most recognize the need to balance the factors with the type of structure and scope of testing. Factors of safety are highest for pressure vessels and for structures we will not test. For most other structures, a contractor will be able to choose from several test options. Table 11-54 shows the test options for an unmanned launch. If personnel safety is at risk, as for a Shuttle launch or during ground handling, we use higher factors.

TABLE 11-53. **Terms and Criteria Used in Strength Analysis.** We design space structures to meet specified or selected criteria for preventing yield and ultimate failures. A *yield failure* is one in which the structure suffers permanent deformation that degrades the mission; *ultimate failure* is rupture or collapse. Two factors of safety—one for yield and one for ultimate—typically apply to a structural assembly, and depend on the selected test option (Table 11-54). For each structural member, the allowable load, the design load, and the margin of safety each have two values, one for yield and one for ultimate.

Term	Definition
Load Factor	A multiple of weight on Earth, representing the force of inertia that resists acceleration. The load factor applies in the direction opposite that of the acceleration. For example, an object under an acceleration of 5 g, where g is the gravitational acceleration, has a load factor of −5; if that object weighs 500 N (a mass of 51 kg), the force it exerts on its support structure is −2,500 N.
Limit Load (or design limit load)	The maximum load expected during the mission or for a given event, at a specified or selected statistical probability (typically 99% for expendable launch vehicles and 99.87% for launches with humans aboard). The **load** can be acceleration, load factor, force, or moment.
Allowable Load or Stress	The highest load or stress a structure or material can withstand without failure, based on statistical probability (usually 99%; i.e., only 1% chance the actual strength is less than the allowable).
Factor of Safety, FS	A factor applied to the limit load to obtain the design load for the purpose of decreasing the chance of failure.
Design Load	Limit load multiplied by the yield or ultimate factor of safety; this value must be no greater than the corresponding allowable load.
Design Stress	Predicted stress caused by the design load; this value must not exceed the corresponding allowable stress.
Margin of Safety, MS	A measure of reserve strength: $$MS = \frac{\text{Allowable load (or stress)}}{\text{Design load (or stress)}} - 1 \geq 0 \text{ to satisfy design criteria}$$

TABLE 11-54. **Typical Test Options and Factors of Safety for Missions without Humans Aboard.** [DoD-HDBK-343]. See Table 11-53 for definitions and use.

Option	Design Factors of Safety	
	Yield	Ultimate
1. Ultimate test of dedicated qualification article (1.25 × limit)	1.0	1.25
2. Proof test of all flight structures (1.1 × limit)	1.1	1.25
3. Proof test of one flight unit of a fleet (1.25 × limit)	1.25	1.4
4. No structural test	1.6	2.0

Typically, a program will vary the test option for different components depending on type of structure, material or method of construction, weight criticality, perceived test difficulty, program schedule, and planned quantity of flight articles.

We should proof test (option 2) each flight article if strength is significantly affected by workmanship or process variations. The test conditions should apply the critical stresses in all load directions possible during the mission. Proof tests should also include the effects of predicted temperature changes during the mission, which can be significant in certain types of structures. Proof testing is the only viable option for advanced composite structures and bonded joints, unless we can become confident that our processes are controlled well enough to keep strength variations low. We also proof test all pressure vessels.

If strength is relatively insensitive to workmanship, such as for most mechanically fastened metals, we can choose any of the four test options. When building many articles, we can save money by testing just one (option 3), but this option carries a weight penalty because of its high factors of safety. Alternatively, we can build a dedicated qualification article and use the lower factors of safety that go with the "Ultimate Test" (option 1). With this approach, we have three additional benefits:

1. We can shorten the flight-article development schedule by conducting the test program in parallel.

2. We can use the test article as a pathfinder for launch-site operations.

3. The impact of test failure is not as severe.

We must weigh these advantages against the cost of building the test article, for which we must use the same processes as for the flight structure.

Option 4 ("No Test") can be risky, and we must use it with caution. The space industry has no design codes, such as for many commercial structures, and relies heavily on testing to verify structural integrity. Without a test, a critical analysis error or oversight could lead to a mission-ending structural failure, even with high factors of safety. However, with caution, we can confidently use this option for relatively simple structures. The "no test" option may be most cost effective for structures designed for stiffness rather than strength.

To provide confidence the structure will survive multiple loading cycles, we also perform fatigue analysis. Fatigue is a much greater concern for aircraft than for most primary structures in a spacecraft because launch is of such short duration. But if stresses are high enough, it does not take many cycles for a material to fatigue, and launch is not the only event that can cause fatigue damage. Ground testing and transportation can significantly degrade service life. Structures sensitive to high-frequency vibrations or on-orbit thermal cycling are particularly susceptible to fatigue.

All materials have internal defects, most of which are microscopic. The number, sizes, and locations of these defects all contribute to high variability in fatigue life between specimens of the same material. In a fatigue analysis, we compare the number of cycles at the limit stress level with the test-determined average number of cycles at which failure occurs. We account for variability in a material's fatigue life by multiplying the predicted number of load cycles over the mission life by a *scatter factor* of four [Rolfe and Barsom, 1977].

To account for the possibility of a large pre-existing flaw in a critical location of a structural part, we can establish a *fracture control* program. This includes inspections of raw materials and fabricated parts for defects, special handling procedures for critical parts, and *fracture-mechanics safe-life analysis*. This analysis, which is a semi-empirical method of predicting crack growth and part life, is more conservative than fatigue analysis because we assume an initial crack exists at the location of peak stress. We set the size of the assumed initial crack equal to the minimum our inspection methods can reliably detect. MIL-HDBK-5G provides fatigue and fracture mechanics data for most metals.

11.6.5 Preliminary Sizing of Structural Members

To size the structural members of a spacecraft, we consider stiffness, strength, and weight. We will rarely find a design in the first iteration that is acceptable for all three. Before the design is final, we will perform many iterations that also consider fatigue life, cost, and changes in subsystem requirements.

Stiffness—*Flexibility* is a measure of how much a structure deflects under unit load. *Stiffness* is a measure of force required to cause a unit displacement. (For a single-degree-of-freedom system, stiffness is the inverse of flexibility.) A structure's mode shapes and natural frequencies of vibration depend on its stiffness and mass properties. We discussed typical considerations for stiffness in Sec. 11.6.1.

We can estimate the primary frequencies of a stowed spacecraft by representing it with an *equivalent beam*, which simulates mass properties and core stiffness, then using simple beam-frequency equations provided in Sec. 11.6.6. As the design evolves, we construct a *finite element model* (a mathematical representation of the structure) to obtain more accurate predictions of mode shapes and frequencies. For a given mode of vibration, most finite element software can identify the locations in a structure that have the most *strain energy*, which is the energy absorbed when a structure deforms under load. Reinforcing the areas with high strain energy is the most efficient way to stiffen a structure.

A structural assembly is usually more flexible than predicted by a math model because of local flexibility in mechanical attachments. Thus, even if our model demonstrates the design is adequate, we may find out during testing that the structure doesn't meet stiffness requirements. It usually doesn't cost much in weight to stiffen a joint—the key is being aware of stiffness in the design of attachments. We should also not cut a stiffness requirement too close before verifying it by test.

Strength—We can use various methods to predict distributions of internal loads, depending on the structure's complexity and the scope of our analysis. *Free-body diagrams* show applied load, which in preliminary design equals weight multiplied by load factor at the center of gravity, and the reactions necessary for static equilibrium. With these diagrams, we can easily determine member loads in a *statically determinate* structure, which has just one solution for member loads that satisfies equilibrium. Finite element analysis is the most efficient method of predicting loads in a *statically indeterminate* structure, which has redundant load paths.

To have adequate strength, the structure must not rupture, collapse, or deform such that its function is impaired. Primary structural members made of ductile materials seldom rupture in tension for two reasons: (1) Most structures are statically indeterminate, and ductile materials will stretch enough prior to failure for loads to redistribute. (2) Tension is an easy mode of failure to assess for a member of constant cross-section and is seldom overlooked. Instead, tension members fail most often at their attachments: fittings, welds, fasteners, and adhesives.

Stability is a structure's resistance to collapsing under compression. Compressive failures are the most sudden and catastrophic, and they are often the hardest to predict. An overall instability failure of a column is called *buckling*. This is the kind of failure we would expect if we pushed on the ends of a long, slender rod. The load at which a column buckles decreases with the square of its length. *Crippling* is a compression failure that starts with local buckling of thin-walled flanges or webs in a member's cross-section.

We often design panels in skin-stringer structures to buckle under compressive loads, with shear being transferred by diagonal tension. This is a common practice for weight-critical structures and is not catastrophic if we design the rivets, stringers, and frame members properly. Diagonal tension will induce lateral loads in edge members that cause compression and bending, as discussed in Bruhn [1973].

We also must check that structural elements do not *yield*, or take on permanent deformations that can jeopardize the mission. Yielding is a characteristic of all structural materials except those that are perfectly brittle (no ductility). Other deformations can be detrimental as well, such as shifting in mechanical joints, so we must assess them as well.

When assessing rupture and collapse, we use *design ultimate loads*, which are limit loads multiplied by the ultimate factor of safety. We use the yield factor of safety to assess permanent deformations. The onset of compressive yielding will often be followed by collapse because of reduced stiffness, so we should ensure there will be no compressive yielding at design ultimate loads.

Weight—Designers of flight structures quickly develop an instinct for meeting requirements with the lightest structure. Throughout preliminary design, the configuration and loads will change, and we will have to increase the sizes of many structural elements. We will also find elements that are unnecessarily heavy, but we won't always change the design. As the design becomes more detailed, weight optimization becomes increasingly difficult and complicates production.

At each design iteration, we compare a component's weight with its allocation. Once the allocation has been met, we focus our attention on other issues. The best design will seldom be the lightest design—it will be the one that is optimal for the system, considering performance, reliability, and cost.

11.6.6 Structural Mechanics and Analysis

A part made from a solid material will change shape as force is exerted on it. *Mechanics of materials* is the term used to describe how materials respond to applied forces and other environments. The most basic term in mechanics of materials is *stress*, σ, which is the load, P, in a member, divided by its cross-sectional area, A, (Fig. 11-27).

$$\sigma \equiv \frac{\text{Load}}{\text{Area}} \equiv \frac{P}{A} \qquad (11\text{-}39)$$

Typical units for stress are N/m^2 and lb/in^2 or psi.

Strain, ε, is a dimensionless measure of deformation for a given load. In Fig. 11-27 the bar's length, L, is increased by ΔL in response to the axial load, P.

$$\varepsilon \equiv \frac{\Delta L}{L} \qquad (11\text{-}40)$$

Solids experience some thinning when elongated under an axial load. *Poisson's ratio*, ν, which describes this phenomenon, is the ratio of lateral-to-axial strain.

$$\nu \equiv \frac{\varepsilon_{lateral}}{\varepsilon_{axial}} \qquad (11\text{-}41)$$

Poisson's ratio for metals lies in the range of 0.28 to 0.33.

The *stiffness* of a material is the relationship of its stress to strain for a given load.

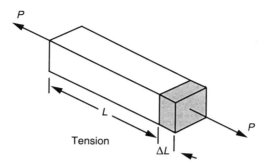

Fig. 11-27. Stress and Strain. Stress is load, *P*, divided by area, *A*. Strain is the change in length, Δ*L*, divided by the initial length, *L*.

We express it as the *modulus of elasticity** or *Young's modulus, E*:

$$E \equiv \frac{\sigma}{\varepsilon} \qquad (11\text{-}42)$$

Values for *E* were shown in Table 11-52. Metals typically start by exhibiting a linear relationship between stress and strain. Strain in this region is termed *elastic* because it will return to zero after the load is removed.

Beyond a stress called the *proportional limit* (normally assumed to be the same value in tension and compression), a material's stress/strain curve is no longer linear. In other words, if we design our structure such that its material is stressed above the proportional limit, linear methods of analysis would no longer apply. This can be risky because so many of our methods of analysis are based on the assumption of linearity; any other assumption would make loads analysis, in particular, so cumbersome it would be impractical. Inelastic effects influence structural stability more than anything else. An effective guideline for preliminary design is to keep the design ultimate compressive stress below the material's proportional limit.

Above the *elastic limit*, which is often indistinguishable from the proportional limit but can be higher, the material will undergo residual strain (*plastic strain*), which remains after the load is removed. Convention has defined the *yield stress* to be the stress that would cause the material to have a residual strain of 0.2%. Although the material actually begins to yield at the elastic limit, such initial yielding is often not noticeable in a structural assembly. For design, we commonly use the traditionally defined yield stress, based on the assumption that 0.2% permanent strain would not be detrimental. To design to this value, we need to make sure our structure would still function properly if it sustained the corresponding permanent deformation.

A material that can yield substantially before rupturing is termed *ductile*. Ductile materials can survive local concentrations of strain without failing, resist crack formation, and allow parts to be shaped through hammering and bending. Conversely, *brittle* materials, such as ceramics, do not deform plastically before rupturing. In designing with brittle materials, we must make sure the local concentration of strain around a discontinuity, such as a drilled hole, does not cause an elastic stress that exceeds the

---

* *Elasticity* is the characteristic of a material to return to its original dimensions after an applied force is removed.

rupture stress (*ultimate stress*). Otherwise, a crack would form and possibly grow uncontrollably until the part ruptures. Brittle materials also are not as resistant to impact loads; the area under a material's stress/strain curve indicates how much energy it can absorb before it ruptures. Figure 11-28 shows representative stress/strain curves for ductile and brittle materials. Refer to Table 11-52 for statistically guaranteed design stresses for commonly used alloys.

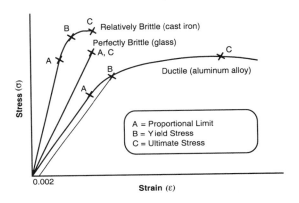

Fig. 11-28. Representative Stress/Strain Curves for Ductile and Brittle Metals. We generate curves such as these from uni-axial tensile tests. The slope of the linear region is the modulus of elasticity. When the material is unloaded, even when stressed above the proportional limit, stress is again proportional to strain according to the elastic modulus. The yield stress is the stress that causes a permanent strain of 0.002.

Beams are very common structural members. We characterize a beam by how it is supported. Examples are described in Table 11-55 and can occur in various combinations.

TABLE 11-55. Beam Examples.

Name	Constraints	Examples
Cantilevered	One end constrained against translation and rotation; other end free	Diving board
Simply Supported	Both ends constrained against translating, but free to rotate	Plank placed across a stream for hikers to cross
Rigidly Supported	Both ends constrained against translation and rotation	Floor joists
Continuous Support	Beam's entire span is supported	Railroad track, ski

Loads on beams may be concentrated forces, distributed weights or pressures, or rotational loads, called *bending moments*. Figure 11-29 shows the symbols commonly used for beam characteristics. Figure 11-30 includes examples of sketches called *free-body diagrams*, showing beams in static equilibrium. Beams are said to be in *static equilibrium* when they fully react all applied forces and moments—a very important prerequisite for static structural analysis.

Description	Symbol	Schematic	Dimension
Concentrated axial load	P or F W (if weight)		Force
Lateral, or shear load	V		Force
Concentrated reaction	R		Force (equal, opposite P)
Uniform, distributed load	w		Force per unit length
Varying, distributed load	w(x)		Force per unit length
Applied bending moment, reactive bending moment	M		Force times distance

Fig. 11-29.   Beam Symbols and Schematics.

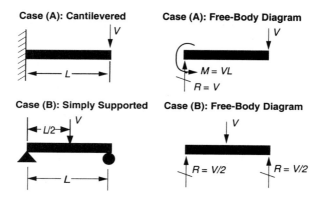

Fig. 11-30.   Common Beam Cases with Associated Free-Body Diagrams. The "brick" wall or rigid left-hand support in case (A) can be replaced with the bending moment, *M*, equal to *V* times *L*.

Some of the common beam equations and relationships can be explained by use of the cantilevered beam example of Fig. 11-31A. The distributed lateral load, *w(x)*, places the beam in a state of bending and shear. We can see evidence of the bending from the fact that the deformed beam is no longer straight (Fig. 11-31B). The *shear* is the lateral force transmitted along the beam's length. The shear reaction, *R*, at the fixed end must be equal and opposite to the sum of *w(x)* or the beam would no longer be in equilibrium.

The beam's internal shear forces and bending moments can be expressed as functions of the applied load *w(x)*. The local variation in the shear force equals the load at any point along the beam.

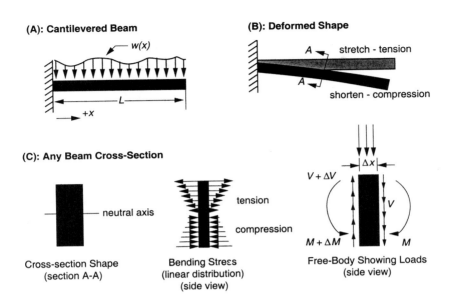

Fig. 11-31.   **Bending and Shear in a Cantilevered Beam.** Bending stresses vary linearly, peaking in the parts of the section that are farthest from the *neutral axis* (centroidal bending axis). Shear stresses vary nonlinearly and are maximum at the neutral axis. Magnitudes of both bending and shear stresses vary for different cross-sections.

$$\frac{dV}{dx} = w(x) \qquad (11\text{-}43)$$

The shear force also relates to the change in bending moment along the beam.

$$V = \frac{dM}{dx} \qquad (11\text{-}44)$$

When the applied force is continuous so that V can be differentiated, the following is also true:

$$\frac{dV}{dx} = \frac{d^2 M}{dx^2} \qquad (11\text{-}45)$$

The force $w(x)$ in Fig. 11-31 is neither tensile nor compressive as it is not applied along the beam's axis. However, from the shape of the deflected beam, we can see that the upper surface of the beam is stretched; this material is in tension. Likewise, the bottom material is shortened and is in compression. The tensile and compressive stresses are necessary to react the applied bending load. The bending moment increases for sections of the beam closer to the fixed end. For any individual cross-section of the beam, the tensile and compressive stresses are maximum at the upper and lower surfaces. Provided the maximum stress remains below the proportional limit, these stresses vary linearly for parallel surfaces inward from the extremities, finally reaching zero at a line called the *neutral axis* (Fig. 11-31C). Shearing stresses vary nonlinearly along the cross-section and, unlike bending stress, reach a maximum at the neutral axis. Extensive beam equations for stress, deflection, and reactions to applied loads are in Roark and Young [1975].

We can quantify a beam's ability to resist bending loads using the *second moment of area of a cross-section*. It is usually referred to as **area** *moment of inertia, I,* (or *moment of inertia for the section*) and should not be confused with **mass** *moment of inertia* used in control system analysis. The area moment of inertia about an arbitrary axis is

$$I_{axis} = \int_{Area} y^2 dA$$

(11-46)

where y is the distance from the centroid to the infinitesimal area, dA. Figure 11-32 presents values of I for several commonly used sections. For boxes and tubes, we find the I for a section by subtracting the I of the "hole" from the total.

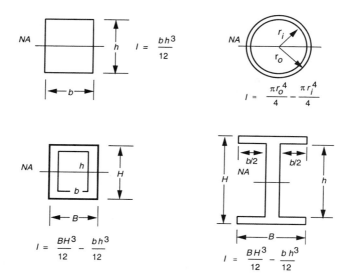

Fig. 11-32.   Common Sections and Their Centroidal Moments of Inertia. For boxes and tubes, we find the I for the section by subtracting the inner I from the outer I. For a very thin annulus, $I = \pi r^3 t$.

For calculations of I with respect to an axis other than the neutral axis, we use the parallel axis theorem.

$$I \text{ (any parallel axis)} = I \text{ (neutral axis)} + Ad^2$$

(11-47)

where A is the cross-sectional area, and d is the distance between the two parallel axes. The parallel axis theorem allows us to find the area moment of inertia for complex sections, such as the I-beam in Fig. 11-33. Note that the area moment of inertia increases for a reference axis other than the neutral axis.

The value for *bending stress*, σ_b, is given in Eq. (11-48) for a point on a symmetric cross-section at a distance, c, from the neutral axis. Use of Eq. (11-48) assumes that

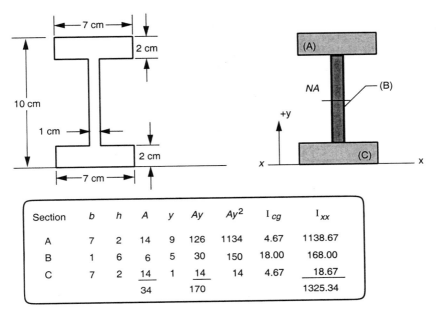

Section	b	h	A	y	Ay	Ay^2	I$_{cg}$	I$_{xx}$
A	7	2	14	9	126	1134	4.67	1138.67
B	1	6	6	5	30	150	18.00	168.00
C	7	2	14	1	14	14	4.67	18.67
			34		170			1325.34

Step 1) Calculate each part's I. $(I = bh^3 / 12)$

Step 2) Find section's neutral axis. $y_{NA} = \Sigma Ay / \Sigma A = 170 / 34 = 5.00$

Step 3) Calculate I for axis x-x. $I_{xx} = \Sigma I_{cg} + \Sigma Ay^2 = 1325.34$

Step 4) Equate I at the neutral axis. $I_{NA} = I_{xx} - A(y^2_{NA})$

$\qquad\qquad\qquad\qquad\qquad = 1325.34 - (34)(25) = 475.34 \text{ cm}^2$

Fig. 11-33. Method for Finding Neutral Axis and Moment for Inertia for Complex Sections.

cross-sections remain planes after bending so that stresses will increase linearly away from the neutral axis. To predict stress values above the material's proportional limit, we would use inelastic methods [Bruhn, 1973].

$$\sigma_b = \frac{Mc}{I} \qquad (11\text{-}48)$$

When a column under axial compression suddenly deflects laterally, or bows, we say that it *buckles*. Such an occurrence is usually catastrophic. Theoretically, a linear-elastic column in compression will buckle at a *critical*, or *Euler buckling load*, P_{cr}, given by

$$P_{cr} \equiv \frac{\pi^2 EI}{(L')^2} \qquad (11\text{-}49)$$

where L' is an *effective length*, dependent on the column's end conditions as shown in Fig. 11-34. This equation applies only if the axial stress at buckling (P_{cr}/A) does not exceed the material's proportional limit. Otherwise, we would replace E in this equation with E_t, the *tangent modulus*, which is the slope of the stress/strain curve at the operating stress level (the buckling stress, in this case). Premature column buckling can also occur as a result of imperfect geometry and local buckling of flanges or webs in the column. See Sarafin [1995] or Bruhn [1973] for details.

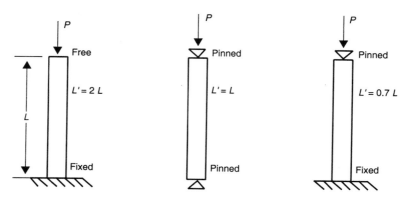

Fig. 11-34.   Effective Lengths for Columns with Different End Conditions. The square of a column's effective length, L', is inversely proportional to the force that would cause the column to buckle elastically. Conceptually, the effective length is the length over which the buckled shape would approximate that of a buckled column with pinned ends (center figure). For example, if the cantilevered column shown at left were to buckle, its free end would deflect laterally, while its fixed end would not translate or rotate. This shape is the same as half the shape of a buckled pinned-end column, so L' for the cantilever is 2L.

The elastic buckling stress, σ_{cr}, for curved skin panels in compression is given as

$$\sigma_{cr} = \frac{k\pi^2 E}{12\left(1 - v^2\right)} \left(\frac{t}{b}\right)^2 \tag{11-50}$$

where t is panel thickness, b is panel width, v is Poisson's ratio (Eq. 11-41) and k is a geometric coefficient. Figure 11-35 graphs values of k for curved panels where r is the radius of curvature and is used to compute the independent variable on the graph.

We can quickly evaluate combined axial, lateral, and bending loads on a thin-wall cylinder using the *equivalent axial load*, P_{eq} (Fig. 11-36):

$$P_{eq} \equiv P \pm \frac{2M}{R} \tag{11-51}$$

where M and R are defined on the figure.

The basis for P_{eq} is that bending stress will be greatest at the two points farthest from the cylinder's neutral axis (one point in tension, the other in compression). Because lateral and bending loads can usually come from any direction (wind or drag), this peak stress can occur at any point. Therefore, we must size the cylinder for the load that would create this peak stress along the cylinder's circumference. P_{eq} is an axial load on a cylinder that would result in a uniform stress equal to a peak stress created by a combination of an axial load and bending moment.

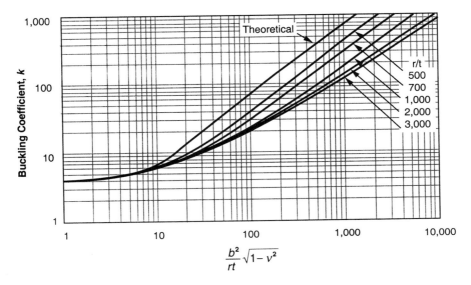

$$\frac{b^2}{rt}\sqrt{1-v^2}$$

Fig. 11-35. **Coefficients of Axial Compressive Buckling for Long Curved Plates**. The circumferential dimension is measured as an arc.

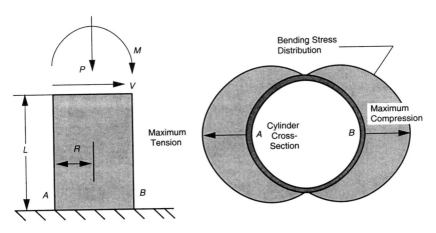

Fig. 11-36. **Equivalent Axial Load**. The cylinder at the left can be exposed to axial (P), shear (V), or bending loads (M). Note that a bending moment at the base of the cylinder could be created either by the applied bending moment, M, or the lateral load, V, applied somewhere above the base (lateral load times moment arm).

Pressure vessels are composed of doubly-curved shells such as spheres or ellipsoids. See Fig. 11-37. By *doubly-curved*, we mean that the surface geometry can be defined when two radii of curvature are known. Typical names for these radii are the longitudinal or *meridional radius of curvature*, designated R_m, and the circumferential or *hoop radius of curvature*, R_h, In all cases, these radii are measured perpendicular to the shell, not the central axis of symmetry. Figure 11-38 illustrates this double curvature.

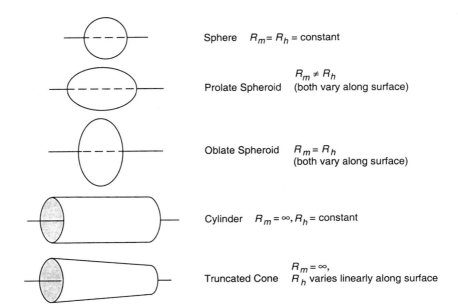

Sphere $R_m = R_h$ = constant

$R_m \neq R_h$
Prolate Spheroid   (both vary along surface)

Oblate Spheroid $R_m = R_h$
(both vary along surface)

Cylinder $R_m = \infty, R_h$ = constant

$R_m = \infty,$
Truncated Cone R_h varies linearly along surface

Fig. 11-37.   **Characteristics of Doubly-Curved Shells.** In each case, the horizontal line is the longitudinal axis of symmetry.

Shells may buckle under compressive loads. The equation for the elastic cylinder buckling stress, σ_{cr}, is

$$\sigma_{cr} = 0.6\gamma \frac{E_t}{R} \text{ (material } v = 0.3) \tag{11-52}$$

where γ is a *reduction factor* used to correlate theory to test results. Thin-shell buckling is very sensitive to minor imperfections in shape, so γ can be as low as 0.15 if the thin shell is badly dented. The reduction factor depends on a geometric parameter, φ, for cylinders.

$$\varphi = \frac{1}{16} \sqrt{\frac{R}{t}} \quad \text{(for } \frac{R}{t} < 1,500 \text{ and } \frac{L}{R} < 5 \text{)} \tag{11-53}$$

$$\gamma = 1.0 - 0.901(1.0 - e^{-\varphi}) \tag{11-54}$$

Note the caveats for Eq. (11-53), where R is the radius, t is thickness, and L is the length of the shell. If σ_{cr} is greater than the material's proportional limit, we must apply other inelastic buckling methods [NASA, 1975].

If the shell is a pressure vessel with internal pressure, p, the meridional stress is

$$\sigma_m = \frac{pR_h}{2t} \tag{11-55}$$

and the hoop stress is

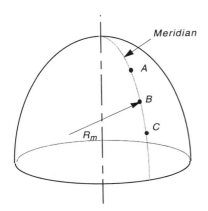
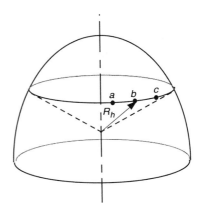

(A) Meridional or Longitudinal Radius of Curvature

- Measured perpendicular to shell
- Defines curvature of arc ABC
- Center need not lie on axis of symmetry
- Can vary along the meridian

(B) Hoop or Circumferential Radius of Curvature

- Measured perpendicular to shell
- Center must lie on axis of symmetry
- Will be constant along curve abc for a shell of revolution (defines a cone whose edge is perpendicular to the shell)
- Can vary along the meridian

Fig. 11-38.   Radii of Curvature for Doubly-Curved Shells. Together, the meridional and hoop radii of curvature fully describe the geometry of a doubly curved shell.

$$\sigma_h = \frac{pR_h}{2t}\left(2 - \frac{R_h}{R_m}\right) \qquad (11\text{-}56)$$

Note that for a sphere, $R_m = R_h$ and $\sigma_m = \sigma_h$. For a cylinder, $R_m = \infty$ and the hoop stress is twice the meridional stress.

Mass moment of inertia (MOI) is a measure of a solid's tendency to resist rotational forces. Rotational inertia depends on mass distribution and varies with the axis of revolution selected as a reference. The MOI for a solid will always be smallest for an axis passing through its center of mass[*] (Fig. 11-39).

---

[*]For an arbitrary mass with an orthogonal coordinate system (x, y, z) located at its center of mass, the moment of inertia about, say, the x-axis is

$$MOI_{xx} = \int_{volume}\left(y^2 + z^2\right)\rho dV = \int_{mass}\left(y^2 + z^2\right)dm$$

where y and z are the distances from the x axis in the y and z directions to the elemental volume, dV, and ρ is the density of the material. Using the parallel axis theorem, the *MOI* about an axis parallel to x is

$$MOI_{x'x'} = MOI_{xx} + (l_y^2 + l_z^2)m$$

where l_y and l_z are the distances from the x to x' axis in the y and z directions.

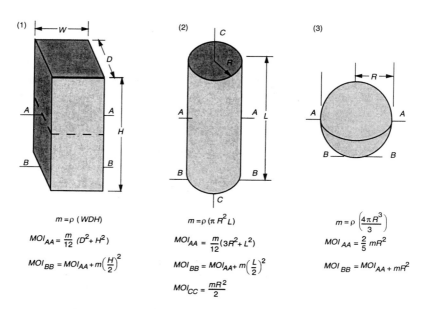

$$m = \rho \, (WDH)$$

$$MOI_{AA} = \frac{m}{12} \, (D^2 + H^2)$$

$$MOI_{BB} = MOI_{AA} + m\left(\frac{H}{2}\right)^2$$

$$m = \rho \, (\pi R^2 L)$$

$$MOI_{AA} = \frac{m}{12}(3R^2 + L^2)$$

$$MOI_{BB} = MOI_{AA} + m\left(\frac{L}{2}\right)^2$$

$$MOI_{CC} = \frac{mR^2}{2}$$

$$m = \rho \left(\frac{4\pi R^3}{3}\right)$$

$$MOI_{AA} = \frac{2}{5} \, mR^2$$

$$MOI_{BB} = MOI_{AA} + mR^2$$

Fig. 11-39. **Equations for the Mass and Moment of Inertia for Some Common Solids.** For a solid of uniform density, ρ, the mass is the product of ρ and volume. We use the parallel axis theorem to find inertias at axes B-B. $MOI_{BB} = MOI_{AA} + md^2$ where d is the distance from the A-A to B-B axes.

Sandwich structure consists of a lightweight, shear-resistant *core* bonded to outer *face sheets* (Fig. 11-40). A sandwich panel acts like an I-beam. The faces correspond to the top and bottom flanges of the beam and resist in-plane bending, tension, and compression. The core acts like the I-beam's web and carries shear and out-of-plane loads, while providing support for the faces.

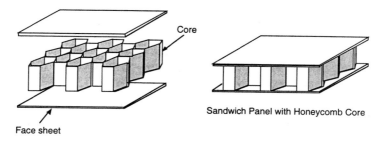

Fig. 11-40. **Sandwich Panel with Honeycomb Core.** By separating thin face sheets with a lightweight core, we efficiently increase the bending strength and stiffness of a panel or shell. Increasing the bending stiffness raises the buckling strength.

Face sheets and cores can be of nearly any metallic or composite material. The core is usually formed into corrugations or honeycomb cells built from thin strips called *ribbons*. Core properties are not isotropic, as stiffness in line with ribbons is greater than transverse stiffness. To maintain the structural integrity of a sandwich, we must make sure the adhesive bond between the core and face sheets is consistent. See Bruhn [1973] for detailed sandwich analysis. Figure 11-41 shows how sandwich structure can be stiffer than skin-only designs.

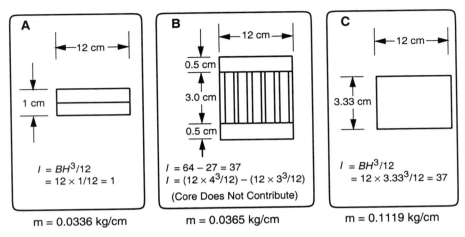

Fig. 11-41. **Comparison of Sandwich and Monocoque Construction.** Thin face sheets have little bending stiffness, as indicated by the small value of I, (A). The bending stiffness is increased by separating the faces with a low density core, (B). A monocoque wall thickness of 3.33 cm is required to obtain bending stiffness equal to the sandwich panel at three times the mass, (C). The masses per unit thickness shown are per cm using aluminum with a density of 2,800 kg/m³ and 80 kg/m³ for the face sheets and honeycomb core, respectively.

The deflection, δ, and natural frequencies, f_{nat}, of simple beams are shown in Fig. 11-42 for axial and lateral applied loads [Roark and Young, 1975]. When considering only its first natural, or fundamental frequency, a structure can be idealized as a single-degree-of-freedom spring-mass system.

The spring is the structure. We can assume an equivalent beam to represent a spacecraft with a *natural frequency*,

$$f_{nat} = \frac{1}{2\pi}\sqrt{\frac{k}{m}} \qquad (11\text{-}57)$$

where m is mass and k = stiffness = load/deflection, also called a spring constant. We find the spring constant, k, using

$$k = \frac{mg}{\delta} \qquad (11\text{-}58)$$

where δ is the deflection and g is acceleration due to gravity.

Description of a Typical Spacecraft Structure

Figure 11-43 shows the Magellan spacecraft configuration and locations of major subsystems.

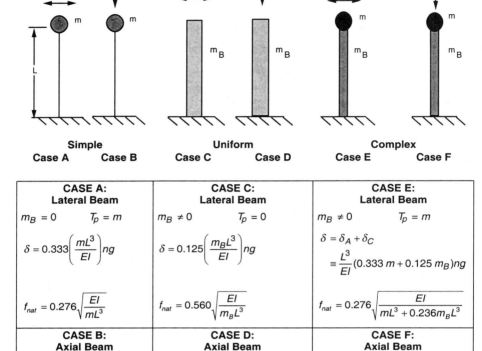

CASE A: Lateral Beam	CASE C: Lateral Beam	CASE E: Lateral Beam
$m_B = 0$ $T_p = m$	$m_B \neq 0$ $T_p = 0$	$m_B \neq 0$ $T_p = m$
$\delta = 0.333\left(\dfrac{mL^3}{EI}\right)ng$	$\delta = 0.125\left(\dfrac{m_B L^3}{EI}\right)ng$	$\delta = \delta_A + \delta_C$ $\equiv \dfrac{L^3}{EI}(0.333\,m + 0.125\,m_B)ng$
$f_{nat} = 0.276\sqrt{\dfrac{EI}{mL^3}}$	$f_{nat} = 0.560\sqrt{\dfrac{EI}{m_B L^3}}$	$f_{nat} = 0.276\sqrt{\dfrac{EI}{mL^3 + 0.236 m_B L^3}}$
CASE B: **Axial Beam**	**CASE D:** **Axial Beam**	**CASE F:** **Axial Beam**
$m_B = 0$ $T_p = m$	$m_B \neq 0$ $T_p = 0$	$m_B \neq 0$ $T_p = m$
$\delta = \dfrac{mL}{AE}ng$	$\delta = 0.5\left(\dfrac{m_B L}{AE}\right)ng$	$\delta = \delta_B + \delta_D \equiv \dfrac{L}{AE}(m + 0.5\,m_B)ng$
$f_{nat} = 0.160\sqrt{\dfrac{AE}{mL}}$	$f_{nat} = 0.250\sqrt{\dfrac{AE}{m_B L}}$	$f_{nat} = 0.160\sqrt{\dfrac{AE}{mL + 0.333\,m_B L}}$

n = load factor
g = gravitational acceleration
m_B = mass of the beam (uniformly distributed)
T_p = tip mass

I = area momentum of inertia of the beam's cross-section
E = the modulus of the elasticity
A = cross-sectional area of the beam

Fig. 11-42. Beam Deflections, f_{nat}, and Natural Frequencies, δ. We can estimate the natural frequencies and deflections of beams for both axial and lateral or bending loads. In cases E and F, the values of m and m_B are different from previous cases.

The Magellan structures subsystem consists of the following (excludes cabling and pyrotechnics):

- Spacecraft to inertial upper stage (IUS) adapter
- Solid rocket motor module and spacecraft adapter
- Ten-sided bus that houses major elements of the electronics subsystem

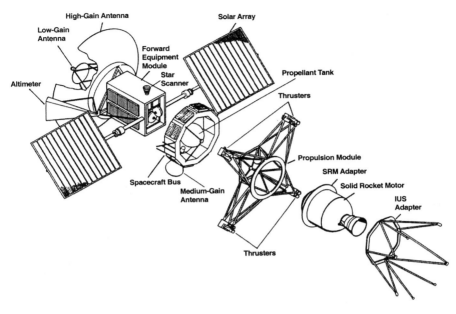

Fig. 11-43.   Magellan Spacecraft.

- Forward equipment module for radar sensor components
- Solar array support and substrate structure with rotation and deployment mechanism
- Altimeter radar mounting structure
- Cover and support structure for radar equipment
- Propulsion module structure

Some of the Magellan structures, such as the IUS and adapters on the propulsion module, lent themselves to the use of truss or strut members. The truss member design loads derive from preliminary load factors, the mass distribution, and the vehicle geometry. All 12 truss members are graphite-epoxy tubes with machined titanium end fittings. Four members are 10.2 cm in diameter. They were sized as beam-columns to withstand axial loading and bending while partially supporting solar panels during launch. The ultimate, equivalent axial load is 102,300 N. The other eight members are 8.9 cm in diameter and sized for 93,410 N.

The adapter cone for the Solid Rocket Motor is a honeycomb structure that transfers the motor's thrust loads from 24 equally spaced bolts at the forward end to four large bolts that attach to the truss for the IUS at the aft end. The forward equipment module consists of a tubular framework covered with thin shear panels. The frames are welded 2219 aluminum alloy 5.08-cm^2 tubing with wall thicknesses varying from 1.27 to 3.81 mm. Numerous equipment boxes are attached to the tubes using threaded fasteners for easy removal. The shear panels enclose the framework except where an equipment box forms an effective load path. The tube-member sizes result from beam-column analysis, including transient and acoustic loads.

11.6.7 An Example Problem

The following example parallels the process for sizing the spacecraft structure and booster adapter in Table 11-56. Structural size and mass are driven by either strength or stiffness requirements. We can begin the process by sizing the structure to meet load requirements and check the resulting natural frequency, or we can begin with a frequency requirement, size the structure, and check strength. Equations are provided to do either. Most short, heavy spacecraft are strength driven and long, lighter spacecraft or assemblies are stiffness driven. Any design with very thin skin or stringer sections can be sensitive to stability failures.

TABLE 11-56. Process for Estimating Size and Mass of the Spacecraft Structure.

Step	Description	References
1	Select a structural approach by identifying the type of structure (monocoque, semimonocoque), shape of the structure, and arrangement of components and load paths.	Chaps. 9, 10, Sec. 11.6.2
2	Estimate mass distribution for all equipment and the structure, including the booster adapter.	Sec. 11.6.8
3	Estimate size and mass of structural members using information from steps (1) and (2) and the axial and bending frequencies for the selected booster. Iterate this structural design as required.	Chap. 18, Sec. 11.6.8
4	Apply combined design loads (axial, lateral, and bending) and determine member loads.	Sec. 11.6.7
5	Compute the structural capability and compare with the applied loads to determine the margin of safety. Iterate the design as required to obtain the necessary margin of safety.	Secs. 11.6.7, 11.6.8

To illustrate the process and some of the more useful analysis methods, we have shown sizing calculations below for the simple example cylinder in Fig. 11-44. This trade study compares monocoque (skin only) and skin-stringer designs of the lightest cylinder that meets representative requirements described in Table 11-57.

Option 1—Monocoque

Sizing for Rigidity to Meet the Natural Frequency Requirement

This cylinder has uniform thickness and, by definition, no ring or longitudinal stiffeners. Using Eq. (11-57), we will find the minimum shell thickness that meets the natural frequency requirements. With $f_{nat} = 25$ (axial) and 10 (lateral), $E = 71 \times 10^9$ N/m^2, $m_B = 2,000$ kg (a weight of 19,614 N or 4,410 lb), and $L = 10$ m, we can solve to find the required cylinder A and I.

Axial Rigidity: for axial rigidity, Eq. (11-57) takes the form of case D in Fig. 11-42.

$$25 = 0.250\sqrt{\frac{AE}{m_B L}} = 0.25\sqrt{\frac{A(71 \times 10^9)}{(2,000)(10)}} \qquad (11-59)$$

from which the required A is 28.17 cm^2 and the required thickness, t, is = 0.045 cm. Lateral Rigidity: here Eq. (11-57) takes the form of case C in Fig. 11-42.

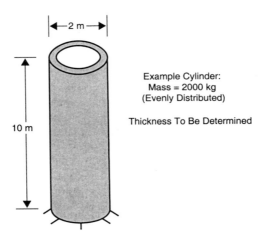

Fig. 11-44. Structural Idealization for the Example Problem. In this problem, we will idealize the spacecraft in launch configuration as a cantilevered cylinder, with all the mass of the spacecraft uniformly distributed. This is often a good starting assumption for initial sizing.

TABLE 11-57.   **Example Problem Requirements.**

Geometry:		
Cylinder Length = 10 m	Cylinder Diameter = 2 m	Distributed Mass = 2,000 kg

Requirements:

Envelope: Assume the spacecraft fits within a required launch-vehicle-fairing envelope (found in Table 18-7 or Fig. 18-8). Also assume that satisfying rigidity requirements will keep the spacecraft's deflection from violating the fairing's dynamic envelope.

Mass: Assume the 2,000 kg is the total spacecraft mass, including an allocation for structure.

Load Factors: Axial = 2.5 (steady-state) + 4.0 (transient) = 6.5, Lateral = 3.0 (Representative load factors can be found in Table 18-9.)

Rigidity: The first axial frequency of the spacecraft must be above 25 Hz. The first lateral (bending) frequency must be above 10 Hz. (See Table 18-9 for typical values.)

Pressure: An internal venting pressure similar to Fig. 18-9 has a maximum value of 6,899 Pa.

Factors of Safety: 1.25 (ultimate) and 1.10 (yield) according to Option 2 of Table 11-54.

Material Properties:	7075 aluminum is chosen.	
Young's Modulus	E	71×10^9 N/m²
Poisson's Ratio	ν	0.33
Density	ρ	2.8×10^3 kg/m³
Ultimate Tensile Strength	F_{tu}	524×10^6 N/m²
Yield Tensile Strength	F_{ty}	448×10^6 N/m²

$$10 = 0.560 \sqrt{\frac{EI}{m_B L^3}} = 0.56 \sqrt{\frac{(71 \times 10^9) I}{(2,000)(10)^3}} \quad (11\text{-}60)$$

from which the required cylinder area moment of inertia, I, is 8.982×10^5 cm^4 and the required thickness, $t = I/(\pi R^3)$ is 0.286 cm. The bending mode requirement is much more critical. For a 0.286-cm thickness, the cylinder's cross-sectional area is 180 cm^2.

Applied and Equivalent Axial Loads

By multiplying the spacecraft weight by the load factors, we can derive the limit or maximum expected loads. See Table 11-58 for the example cylinder limit loads.

TABLE 11-58. Cylinder Applied Loads. The distance is measured from the base to the cylinder's center of mass. Load factors are from Table 11-57.

Type of Load	Weight (N)	Distance (m)	Load Factor	Limit Load
Axial	19,614	—	6.5	127,500 (N)
Lateral	19,614	—	3.0	58,840 (N)
Bending Moment	19,614	5	3.0	294,200 (N·m)

With a bending moment arm of 5 m (the center of mass location is at the cylinder mid-length), we can find the equivalent axial load using Eq. (11-51):

$$P_{eq} = P_{axial} + \frac{2M}{R} = 127,500 + \frac{(2)(294,200)}{1.0} = 715,900 \, N \qquad (11\text{-}61)$$

$$\text{Limit load} \times \text{Ultimate Factor of Safety} = \text{Ultimate Load} \qquad (11\text{-}62a)$$

$$\text{or } 715,900 \times 1.25 = 894,900 \text{ N.} \qquad (11\text{-}62b)$$

Sizing for Tensile Strength

The equation for axial stress, σ, is $\sigma = P/A$. To size the cylinder for tensile strength, we use the ultimate P_{eq} load = 894,900 N, and the material's allowable stress, $F_{tx} = 524 \times 10^6$ N/m^2, and use $A = 2\pi Rt$ to solve for the required thickness.

$$524 \times 10^6 = \frac{(894,900)}{2\pi(1.0)t} \qquad (11\text{-}63)$$

$$t_{req'd} = 0.0272 \text{ cm} \qquad (11\text{-}64)$$

Although we won't show you here, we must check for yield conditions in the same way, using a factor of safety of 1.10 with limit load and $F_{ty} = 448 \times 10^6$ N/m^2.

Sizing for Stability (Compressive Strength)

We must now size the cylinder for stability [Ref. Eqs. (11-52) and (11-53)], using the cylinder thickness required for bending stability. The cylinder must withstand an ultimate $P_{eq} = 894,900$ N.

$$\varphi = \frac{1}{16}\sqrt{\frac{R}{t}} = \frac{1}{16}\sqrt{\frac{1.0}{0.00286}} = 1.17 \qquad (11\text{-}65)$$

$$\gamma = 1.0 - 0.901(1.0 - e^{-\varphi}) = 0.379 \qquad (11\text{-}66)$$

The equation for cylinder buckling stress is

$$\sigma_{cr} = 0.6\gamma \frac{E_t}{R} = (0.6)\,(0.379)\frac{(71\times10^9)\,(0.00286)}{1.0} \quad (11\text{-}67)$$

$$= 46.16\times10^6 \text{ N/m}^2$$

Note that if σ_{cr} were greater than the material's proportional limit, we would use additional methods for inelastic buckling. With the cylinder's cross-sectional area, $A = 180$ cm^2, the critical buckling load is

$$P_{cr} = A\,\sigma_{cr} = (0.0180)(46.16\times10^6) \quad (11\text{-}68)$$

$$= 830{,}900 \text{ N (ultimate)}$$

Thus, the cylinder is not adequate because the applied ultimate load is greater than the critical buckling load. Structural integrity is often shown in terms of the *margin of safety* (*MS*), defined as

$$MS = \frac{\text{Allowable Load or Stress}}{\text{Design Load or Stress}} - 1.0 \quad (11\text{-}69)$$

and must be greater than or equal to zero. For the stability conditions (ultimate),

$$MS = \frac{830{,}900}{894{,}900} - 1.0 = -0.07 \text{ (7\% negative margin of safety)} \quad (11\text{-}70)$$

Results for a small increase in thickness are shown in Table 11-59.

TABLE 11-59.   **Summary of Sizing the Monocoque Cylinder for Stability.** This table summarizes our initial sizing attempt and the first (and final) iteration for an equivalent axial load of 894,900 N.

Iteration	Thickness (cm)	γ	σ_{cr}	Area (cm^2)	P_{cr}	MS
Initial	0.286	0.379	46.16×10^6	180.0	830,900	-0.07
First	0.295	0.384	48.27×10^6	186.0	898,000	$+0.00$

Internal Pressure

We can find the hoop stress in the cylinder by using Eq. (11-56) with $R_m = \infty$:

$$\sigma_h = \frac{pR_h}{t} = \frac{(6{,}899)\,(1.0)}{(0.00295)} = 2.34\times10^6 \, N/m^2 \text{(limit)} \quad (11\text{-}71)$$

$$= 2.92\times10^6 \text{ N/m}^2 \text{ (ultimate)}$$

From Eq. (11-55), we see that the meridional (longitudinal) pressure stress is half this value.

Although these stresses are small, we must combine them with stresses from load factors when sizing for tensile strength. The pressure and load factors must be time-consistent (for example, do not combine lift-off loads with venting pressures that occur later in the ascent). In the case of stability, internal pressure can strengthen a shell. We can increase the reduction factor, γ, slightly to account for the stiffening effect of the internal pressure. Lateral shear will tend to lower the buckling load.

Calculating the Mass

The mass of the cylinder is the product of the density, ρ, and volume, $2\pi R t L$.

$$m = \rho \, 2 \, \pi \, R \, t \, L = (2.8 \times 10^3)(2)(\pi)(1.0)(0.002\ 95)(10.0) \qquad (11\text{-}72)$$

$$= 519 \text{ kg}$$

Any fasteners, attachments, and access doors would increase this mass somewhat, making allowances for material lost in drilled holes and cut outs.

Summary of Monocoque Options

The driving requirements for the monocoque cylinder are bending rigidity and compressive stability, which represent actual design conditions. Please note that the calculation for first natural frequency depends on a crude assumption of equally distributed mass. In this example, we want only to illustrate methods and clarify the need for iterative design. In an actual design, we would know the mass distribution and use computerized techniques to get a more realistic weight for the structure.

If we break the cylinder into several assemblies, such as an adapter on the bottom with a spacecraft bus on top, we could analyze each section separately. For cylinder sections closer to the base, P_{eq} loads increase. Thus, we would want to analyze different sections for varying types of construction, each with its own applied loads. In this example, we could assume that the spacecraft adapter occupies the bottom 2 m of the cylinder, resulting in a preliminary mass of $519 \times 2/10 = 103.8$ kg.

Option 2—Skin-Stringer

Suppose we stiffen the cylinder with 12 longitudinal members, called *stringers*, and 11 circumferential rings, or *frames*. The cylinder's circumference is 6.28 m, so the 30-deg stringer spacing results in a stringer spacing of 0.5236 m, measured along the curved surface. The frames separate the cylinder into 10 sections, or *bays*, each with a height of 1.0 m. Figure 11-45 identifies the stringers by number.

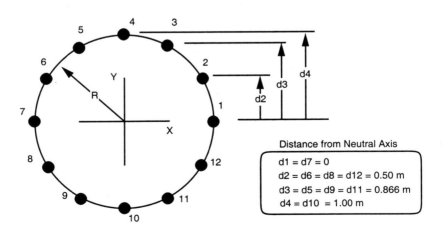

Fig. 11-45.   Stringer Arrangement and Geometry.

It is reasonable to assume that the presence of stiffening stringers and rings in this design allows us to reduce the skin's thickness. A designer's initial concern with a thinner skin is buckling; the concern is real and we will indeed check for this mode of failure. In addition, thin, external surfaces with large surface areas are also susceptible to the acoustic environment. Acoustically driven loads are based on many factors, including:

- The launch vehicle's acoustic environment

- Location of the structure within the payload fairing, or shroud

- Whether acoustic blankets are used to help diminish noise within the shroud

- Type of structure (as we said, large and thin surfaces are more affected)

- Whether the structure is an external or internal payload surface

- Boundary conditions of the surface edges

- Whether the surface is flat or curved

- The first resonant frequency of the surface (depends on size, shape, thickness, material's modulus of elasticity, and edge boundary conditions).

The calculations for acoustic loads are cumbersome; see Sec. 7.7 of Sarafin [1995] for an example of one technique. We will assume a starting standard gage skin thickness of 0.127 cm is adequate against acoustic noise for our design.

First, we must choose whether to design the skin to help sustain load or whether to allow it to buckle, forcing the stiffeners to take on more of the burden. In this example, we will design the skin not to buckle, as is usually done when performing preliminary sizing analysis. Chapter C11 of Bruhn [1973] provides details on how to analyze buckled skin.

Stiffness

Again, let's first size for stiffness. We already know from calculations for Option 1 that we need a skin thickness of 0.045 cm to meet the axial frequency requirement of 25 Hz. Therefore, the 0.127-cm-thick skin alone will be adequate for axial rigidity. In the bending case, the required area moment of inertia, I, of the cylinder's cross-section is 8.98×10^5 cm^4. The skin will satisfy part of this:

$$I_{skin} = \pi \, R^3 \, t = \pi (1.0)^3 (0.001 \; 27) = 4.00 \times 10^5 \text{ cm}^4 \tag{11-73}$$

Therefore, the contribution to I from the 12 stringers must equal the remainder:

$$I_{str} = 8.98 \times 10^5 - 4.00 \times 10^5 = 4.98 \times 10^5 \text{ cm}^4 \tag{11-74}$$

We can calculate the I of the 12 stringers in the cylinder using the parallel axis theorem, $I_{xx} = \Sigma \, (I_{cm} + Ad^2)$. We can ignore the I_{cm}, or I about each stringer's center of mass, because it will be very small compared to its Ad^2 term. Therefore, the I of the stringer system is a function of stringer cross-sectional area, A, and d, the distance from the cylinder's neutral axis (Table 11-60).

Therefore, $I_{str} = 4.98 \times 10^5$ cm$^4 = A \times 60,000$ cm^2. This results in a required cross-sectional area of each stringer of 8.32 cm^2. The cylinder area combines the skin and twelve stringers for a total area of 180.00 cm^2. Note that both the skin and stringers must contribute to overall I to meet this requirement. When we allow skin to buckle,

TABLE 11-60.   Calculations for Moment of Inertia Based on Stringer Area.

Stringer No.	d (cm)	d^2 (cm)^2	ΣA	$\Sigma A d^2$
1, 7	0	0	2 A	0
2, 6, 8, 12	50	2,500	4 A	$A \times$ 10,000 cm^2
3, 5, 9, 11	86.6	7,500	4 A	$A \times$ 30,000 cm^2
4, 10	100	10,000	2 A	$A \times$ 20,000 cm^2
			Total	$A \times$ 60,000 cm^2

we can consider only the stringers and small sections of skin near the stringers, called *effective skin* [Bruhn, 1973].

Panel Stability

Equation (11-50) is used to determine the compressive buckling stress for the skin panel:

$$\sigma_{cr} = \frac{k\pi^2 E}{12(1 - v^2)}\left(\frac{t}{b}\right)^2 = 0.923kE\left(\frac{t}{b}\right)^2$$

$$= (0.923)(55)(71 \times 10^9)\left(\frac{0.00127}{0.5236}\right)^2$$

$$= 21.2 \times 10^6 \text{ N/m}^2 \tag{11-75}$$

where $k = 55$ (from Fig. 11-35), v (Poisson's ratio) = 0.33, $r = 1$ m, $t = 0.127$ cm, and $b = 0.5236$ m (the spacing between stringers). The buckling load, $P_{cr} = \sigma_{cr} \times$ area = (21.2×10^6) $(0.0180$ m$^2) = 381,000$ N. The resulting margin of safety, *MS*, is

$$MS = \frac{381,000}{894,000} - 1 = -0.57 \tag{11-76}$$

The negative margin of safety points out the inadequacy of the design, so we must add thickness to keep the panel from buckling, resulting in the values shown in Table 11-61. When we increase the thickness like this to prevent panel buckling, we can decrease the area of the stringers, with the goal of achieving the same total area and moment of inertia needed for bending stiffness. With a skin thickness of 0.195 cm, this means the required stringer area is 4.78 cm^2.

Table 11-61 summarizes the estimation of mass for the skin-stringer option, using a mass density of 2,800 kg/m^3. Note we've included an extra 25% to account for ring frames, which are needed to stabilize the stringers and fasteners; this is simply an estimate. Note also that, in a real sizing exercise such as this, we would need to confirm the feasibility of only 4.78 cm^2 area for each stringer. To keep a stringer from buckling as a column between ring frames, we need to design its cross section to have a relatively large moment of inertia. A common strategy for doing this at low mass is to use a thin-walled I- or C-section. However, we might find that, to provide the needed area moment of inertia with an area of 4.78 cm^2, we would need to make the flanges and webs so thin that they could not carry the design load without buckling locally. Chapter 8 of Sarafin [1995] explains how to assess column buckling and local buckling for thin-walled structural members.

TABLE 11-61.   Skin-Stringer Cylinder Skin Panel Sizing for Stability.

Iteration	Thickness (cm)	k	σ_{cr} (N/m²)	Area (cm²)	P_{cr} (N)	MS
Initial	0.127	55*	21.2×10^6	180.0	381,000	−0.57
First	0.195	55	49.8×10^6	180.0	896,400	+0.00

* The fact that the value of *k* is 55 for both the initial try and first iteration is coincidental. Note that the values for *k/t* and *Z* differ between the two cases.

TABLE 11-62.   Skin-Stringer Cylinder Mass Calculation.

Part	Thickness (cm)	Area (cm²)	Length (m)	Volume (m³)	Mass (kg)
Skin	0.195	122.55	10.0	0.1225	343.0
Stringers	—	12 (4.78)=57.4	10.0	0.0574	160.7
Subtotal					503.7
25% Extra for ring frames and fasteners					125.9
TOTAL					629.6

Skin-Stringer Option Summary

Strength and stability determine the structural design. With diagonal tension designs, where we allow skin to buckle, we must calculate additional stringer and frame loads. Solid and finite-element modeling is so prevalent in today's industry that computers are important even to the first stage of the design. Still, we must not substitute computers for a thorough knowledge of the various modes of failure and the limits of our assumptions.

Cylinder geometry is key in both the monocoque and skin-stringer examples. We cube the length in the equation for bending rigidity, and use the radius prominently in calculating stability. Table 11-63 illustrates the effects of making the cylinder shorter and wider. Note that the resized cylinder has the same internal volume as the original case, so we must verify that the launch vehicle's shroud can handle the new size.

TABLE 11-63.   Cylinder Sizing Summary.

Geometry (m)	Option	Condition	Skin t (cm)	Stringer A (cm²)	Mass (kg)
L=10 R=1	M	Rigidity	0.286	—	503
		Stability	0.295	—	519
	SS	Rigidity	0.127	8.32	630
		Stability	0.195	4.78**	630
L=9.5 R=1.026	M	Rigidity	0.227	—	389
		Stability	0.289	—	495
	SS	Rigidity	0.127*	5.37	486
		Stability	0.209	0.97**	486

*Thickness required to satisfy assumed acoustics environment
**Stringers sized to accommodate bending rigidity requirement
M = Monocoque, SS = Skin-stringer

Mass decreases with a shorter cylinder, but the stability requirement becomes more critical as the radius increases. More iterations with a shorter cylinder and additional stringers would be appropriate. A more massive example would make the skin-stringer design more attractive.

11.6.8 Mechanisms and Deployables

Aerospace mechanisms can be divided into high- and low-cyclic applications. The former, such as antenna gimbals or solar array drives, require frequent or constant articulation. The latter restrain a payload on launch or retrieval, or they propel the payload to the deployed or restored position. Figures 11-46 and 11-47 show examples of these mechanisms. The design is complete only when principles of mechanics and environmental considerations lead to a producible and testable spacecraft. The most challenging requirements for mechanisms are those that demand precision pointing and a long operating life.

Fig. 11-46. High-Cyclic Mechanism, Rotary Actuator Assembly and Components. An example of an aerospace mechanism requiring precision pointing (motor driven).

Requirements. Typical spacecraft requirements for aerospace mechanisms are as follows:

* High-cyclic mechanisms
  —Antenna pointing and tracking
  —Solar array pointing and tracking
  —Attitude control reaction wheels
  —Boom extensions

* Low-cyclic mechanisms
  — Antenna launch retention
  — Antenna deployment
  — Solar array retention
  — Solar array deployment
  — Contamination cover removal
  — Spacecraft/launch vehicle separation

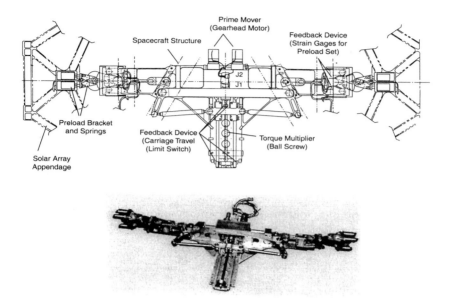

Fig. 11-47. Low-Cyclic Mechanism, Solar Array Retention Mechanism. An example of an aerospace appendage retention mechanism.

The MIL-A-83577 [1988] specification for moving mechanical assemblies gives us important technical guidance. The functional requirements for the mechanisms derive from mission requirements and resolve into torques or forces and operating rates. An operating rate profile, as shown in Fig. 11-48, establishes the payload articulation or deployment rate. This profile determines the maximum angular acceleration, α. Once we have determined the payload moment of inertia, MOI, we can compile the mechanism's operating torque, $T = \alpha \, (MOI)$. For rough torque sizing, we can add a 20% friction torque to the operating torque. The constant-speed part (s_2) of the operating rate profile, represents the mechanism operating torque because there is no acceleration during this phase. With the two operating points known, we can generate a torque-speed curve (see Fig. 11-49). This linear curve establishes the stall torque and theoretical no-load speed for the mechanism. When these mechanism-performance characteristics are arithmetically manipulated by the mechanical advantage of a gear train, the new performance characteristics represent the principal motor requirements. With the mechanism's stall torque now known, we can do first-order approximations of the mechanism parameters using Fig. 11-50.

As an example, a solar array with moment of inertia, MOI, must be deployed by rotating from a stowed position to a locked position in time, t. This time period involves accelerating the array to a maximum rate, s_1, then decelerating to the lock. Therefore, the operating torque (operating point 1) equals moment of inertia, MOI, times acceleration $(s_1 \div t/2)$. In the absence of other running friction data, we can assume operating point 2 is 20% of operating point 1. Extrapolating to a stall torque (assume 200 N·m) lets us use Fig. 11-50. If the mechanism had a 200 N·m stall torque, we can see that the mechanism mass will be about 18 kg, require 90 W of power, and have a volume of about 7,800 cm³. As a guideline, mechanisms should have a 100% torque margin to provide for uncertainties of friction, payload inertia growth, and thermal effects.

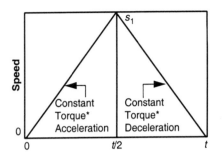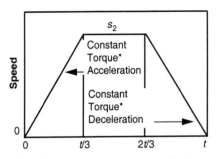

*Resulting from constant current applied to motor.

*Resulting from constant current applied to motor.

Fig. 11-48.   Typical Operating Rate Profiles and Derived Accelerations. The acceleration is calculated by dividing speed by the time increment.

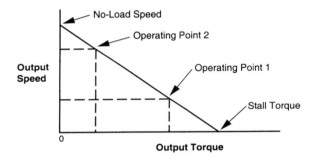

Fig. 11-49.   Derivation of Mechanism Stall Torque. Linear extrapolation of operating points establishes stall torque and no-load speed.

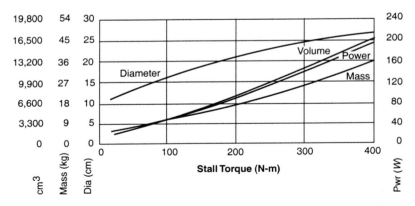

Fig. 11-50.   Actuator Characteristics Based on Stall Torque Requirements. Empirical data based on wide range of aerospace mechanisms.

Although weight, power, and volume are usually the three major spacecraft system parameters, we must not severely constrain the mechanism's weight. The mechanism design should be robust to withstand stall torques and to maintain its structural stiffness over a wide range of temperatures. The mechanism is not a major power consumer. Low-cyclic mechanisms operate only a few times in the mission. High-cyclic mechanisms draw high currents during the acceleration phase of the duty cycle, a phase that is generally 10% of its operating life. Volume constraints will dictate the design process. Also, requirements for mechanical and electrical interfaces will influence the mechanism's volume and structure. The mechanism will also produce its own requirements for torques or forces, operating rates, structural stiffness, operating life and histogram (torque/cycle matrix), and environments. The mechanism must withstand the launch and derived vibration tests, which will influence the strength and stiffness requirements. The mechanism must operate in orbit, where the thermal-vacuum environment will influence the selection of materials, lubricants, and coatings. It will also create thermally induced loads caused by difference in coefficients of thermal expansion of selected structural materials.

For more information on space mechanisms, see Conley [1998], Sarafin [1995], and Mil-A-83577 [1988].

11.7 Guidance and Navigation

James R. Wertz, *Microcosm, Inc.*

We use *navigation** and *orbit determination* interchangeably to mean determining the satellite's position and velocity or, equivalently, its orbital elements as a function of time. Similarly, we use both *guidance* and *orbit control* to mean adjusting the orbit to meet some predetermined conditions. For satellites, orbit control has two important subsets. *Orbit maintenance* refers to maintaining the orbital elements but not the timing of when the satellite is at a particular location in the orbit. *Stationkeeping* refers to maintaining the satellite within a predefined box, which includes maintaining both the in-track position and the other orbital elements. *Altitude maintenance* is an example of orbit maintenance in which occasional thruster firings are used to overcome drag and keep the orbit from spiraling downward. *Geosynchronous stationkeeping* maintains the satellite in a box over one place on the Earth. Stationkeeping in low-Earth orbit includes *constellation maintenance*, in which each satellite is maintained in a moving box defined relative to the rest of the satellites in the constellation.

The satellite *ephemeris is* a tabular listing of the position and possibly the velocity as a function of time, usually in electronic form. It is important to distinguish the satellite ephemeris from the *solar ephemeris*, which lists the relative positions of the

---

* The origin of the terminology causes some confusion, particularly when reading older sources or references not associated with satellites. *Navigation* traditionally referred to determining how to get a craft where we wanted it to go. The term *guidance* was introduced with rockets and missiles to mean computing the steering commands needed to make the rocket go where we wanted it to (thus, a *guided* missile); *control* meant carrying out these steering commands to adjust the vehicle's direction of flight. Thus, an intercept missile would have a *guidance and control (G&C)* system, and a space plane or interplanetary spacecraft would have a *guidance, navigation, and control (GN&C)* system. However, for spacecraft we use *navigation* to mean orbit determination, *guidance* to mean orbit control, and *control system* as a shortened form of attitude control system.

Earth and Sun as the Earth travels in its orbit, and *lunar* and *planetary ephemerides*, which provide similar data for other bodies in the solar system.

There are two types of orbit determination, differentiated by timing. *Real-time orbit determination* provides the best estimate of where a satellite is at the present time and may be important for spacecraft and payload operations, such as accurate pointing at some target. *Definitive orbit determination* is the best estimate of the satellite position and orbital elements at some earlier time. It is done after gathering and processing all relevant observations. *Orbit propagation* refers to integrating the equations of motion to determine where a satellite will be at some other time. Usually orbit propagation refers to looking ahead in time from when the data was taken and is used either for planning or operations. Occasionally orbits will be propagated backward in time, either to determine where a satellite was in the past or to look at historical astronomical observations in the case of comets or planets.

Traditionally, ground stations from around the world provide tracking data to a mission-operations center. When all data is available, definitive orbit determination provides the best estimate of the orbit. This is used to process the payload data for science or observation missions. The best estimate of the orbit is then propagated forward for real-time operations (such as star catalog selection or maneuver timing) and further forward for mission planning.

In 1983 NASA launched the first *Tracking and Data Relay Satellite, TDRS*, to begin replacing the worldwide ground tracking network.* TDRS provides the same functions as the traditional ground-station network. As the name implies, it tracks low-Earth orbiting satellites and relays data between the satellite and the TDRS ground station in White Sands, NM. As described in Sec. 11.7.2, GPS, GLONASS, and other more autonomous systems are also becoming operational, so orbit determination for future systems will differ significantly from what it has been in the past.

We can think of orbit determination and control as analogous to attitude determination and control. The ADCS subsystem (Sec. 11.1) measures and maintains the spacecraft's orientation about its center of mass. Similarly, the guidance and navigation function, perhaps better thought of as the *Orbit Determination and Control Subsystem, or ODCS*, measures and maintains the position of the spacecraft's center of mass. Both systems deal with spacecraft dynamics and both have the multiple functions of acquisition, determination, maintenance, and maneuver control.

11.7.1 System Definition Process

Major changes are occurring in the guidance and navigation arena. Traditionally, this has been exclusively a ground-operations activity. However, with the introduction of GPS and advanced onboard computers, several options now exist for *autonomous navigation*—determining the orbit on board the satellite in real time. We also have the capability to perform autonomous orbit maintenance and control, so the orbit determination and control function will change significantly. Even if we ultimately choose a completely traditional approach, we should evaluate new techniques which may reduce cost and risk for a particular space mission.

Table 11-64 summarizes the process of defining the orbit determination and control function. Each of the steps is described below. Section 11.7.2 then discusses the

---

* The second TDRS was lost in the Challenger accident in 1986, so the two-satellite operational constellation was not complete until 1988.

principal alternatives for navigation systems, and Sec. 11.7.3 describes the alternatives for maintaining and controlling the orbit. The implementation of these in hardware and software is discussed in Sec. 11.7.4

TABLE 11-64.   **Process for Defining the Guidance and Navigation Subsystem.** See text for discussion of each step.

Step	Principal Issues	Where Discussed
1. Define navigation and orbit-related top-level functions and requirements	Mapping and pointing Scheduling Constellation or orbit maintenance Rendezvous or destination requirements	Secs. 1.4, 2.1, 4.2, 7.1
2. Do pointing and mapping trades to determine preliminary navigation (position) accuracy requirements	What payload functions will the navigation data be used for? Payload data processing (mapping) Payload pointing	Sec. 5.4
3. Determine whether orbit control or maintenance is needed	Geosynchronous stationkeeping Constellation stationkeeping Altitude maintenance Maintaining orbit elements Mid-course corrections	Chap. 7, Sec. 11.7.3
4. If yes, do trade on autonomous vs. ground-based orbit control	Is reduced operations cost and risk worth introducing a nontraditional approach?	Secs. 2.1.2, 11.7.1, 11.7.3
5. Determine where navigation data is needed	Is it needed only at ground station for mission planning and data evaluation? Is it needed on board (orbit maintenance, Sun vector determination, payload pointing, target selection)? Is navigation (or target location) data needed by several end users who may get information directly from the spacecraft?	Sec. 2.1.1
6. Do autonomous vs. ground-based navigation trade	Does reduced operations cost and risk justify a nontraditional approach? Is there a need for real-time navigation data?	Secs. 2.1.1, 11.7.1
7. Select navigation method	See Sec. 11.7.2 for main options	Sec. 11.7.2
8. Define G&N system requirements	Top-level requirements should be in terms of **what** is needed (mapping, pointing, constellation maintenance, level of autonomy), not **how** the mission is done	Secs. 11.7.1, 11.7.4

Step 1. Define top-level functions. We want to determine the key mission objectives which require either navigation information or orbit maintenance and control. Typically, we think of maintaining the satellite in a specialized orbit over the life of the mission. Examples include a geostationary slot, Lagrange point orbit, or repeating ground track orbit. We may also need stationkeeping to maintain the relative positions

between spacecraft in a constellation. Orbit control, but not orbit maintenance, is needed to reach a particular destination such as rendezvous with another spacecraft, landing on the Moon, or insertion into a particular geosynchronous slot.

Navigation has two basic purposes. It allows us to maintain and control the orbit, just as attitude determination is used for attitude control. Thus, any requirement for orbit control will ordinarily result in a corresponding requirement for navigation. We may also need navigation information to process data from the payload. Although some science missions may actually use position data (e.g., mapping of the magnetic field or particle flux density), it is usually only part of payload pointing and mapping. Irrespective of any orbit control, we often need to point an antenna or instrument at some location or to define where an instrument is looking on the surface of the Earth. Typically, this results in a more stringent navigation requirement than for purely operational purposes.

Step 2. Do pointing and mapping trades. Because these are typically the most stringent requirements, we must do the pointing and mapping trades described in Sec. 5.4 to obtain a preliminary estimate of the needed accuracy. In most cases, the pointing and mapping requirement can be met by trading between navigation and attitude accuracy (see Fig. 4-6), so navigation trades will frequently need to be performed in conjunction with attitude trades.

Step 3. Determine need for orbit control. At the system level, we must decide whether we need to maintain or control the orbit. If we don't, we may be able to save money and weight by eliminating the propulsion subsystem. But if we need a propulsion system anyway, this hardware can often easily handle orbit maintenance and control. At the same time, maintaining the orbit may significantly extend the mission life and thereby reduce the cost per year.

Step 4. Do autonomous orbit control trade. If we must control the orbit, then we need to determine whether to do so from the ground or autonomously on board the spacecraft. Traditionally, the ground station has controlled the orbit. In most cases, this remains the best approach if the orbit control activity is nonrecurring and communications with the satellite are straightforward, as in the case of transfer to geosynchronous orbit. Here, orbit control needs are well defined, traditional ground techniques are available, and it is needed only once during the mission. Thus, an autonomous, onboard system would probably cost more than we would save from lower operations expenses. However, autonomous orbit maintenance and control can reduce life-cycle cost and risk for many missions (see Sec. 11.7.3). Over the next decade, I expect it to become as common as autonomous attitude control is on today's spacecraft.

Step 5. Determine where navigation data is needed. Section 2.1.1 describes how to do a data flow analysis to determine where data comes from and where it is needed. If we choose not to design the system to minimize communications and data flow, our decision must be justifiable. For example, we may choose to avoid the nonrecurring cost of putting the processing where it would be most efficient. An extreme would be a small LightSat with a single ground station which performs all of the data evaluation, system control, and mission planning. In this case, it is probably easiest to do the navigation on a small computer at the ground station. At the other extreme would be a satellite communicating with many distributed users, each of whom needs to know either the satellite position or derivative information such as the ground look-point location. In this case, the navigation data is probably needed on board the spacecraft, although we could navigate from the ground and uplink the results. The third possible use of navigation data is by the spacecraft itself for payload functions such as pointing or target selection and for real-time identification. We can also use navigation data for spacecraft control functions such as orbit maintenance, star catalog selection, or determining where the Sun is relative to the spacecraft for accurate attitude determination.

Step 6. Do autonomous vs. ground-based navigation trade. We must determine whether reducing the long-term, recurring operations cost and risk justifies the cost of autonomous navigation. Tradition strongly supports ground navigation. This provides a greater level of comfort to some customers and end users who are concerned primarily that there be no deviation from prior procedures. Irrespective of technical issues, there is a level of programmatic risk associated with any nontraditional solution. The issue of autonomous navigation clearly falls in this area, although semi-autonomous navigation, such as TDRS and GPS, is gaining acceptance.

A key question is whether we need real-time navigation data, either on board the satellite or for the end user. If we must provide navigation and payload data simultaneously, autonomous navigation is highly desirable. The alternative would be to navigate on the ground and uplink the solutions for use by the spacecraft or to send them to the end user. This approach makes reliable communications critical. If the work is done on the ground, then there are two possibilities. It can be done either in real time, using real-time data from the spacecraft, or it can be done using older data which is propagated forward to produce a real-time ephemeris. In the past we used older data because real-time data was not available. TDRS, GPS, GLONASS, and various autonomous navigation systems now allow us to use real-time data which does not need to be as accurate as data for propagated solutions.

Step 7. Select a navigation method. Section 11.7.2 summarizes the alternative spacecraft navigation methods and their advantages and disadvantages.

Step 8. Define requirements for the guidance and navigation system. We should define the top-level requirements in terms of what is needed rather than how it is to be done. Thus, requirements should be expressed in terms of mapping, pointing, constellation maintenance, and level of autonomy rather than the specific navigation method to be used. While we will go through detailed trades to select the best navigation method, we should focus on mission objectives to define requirements. This allows later trades which may either be more detailed or use new information.

11.7.2 Orbit Determination Systems

There are three elements to the orbit determination problem: (1) the source and type of data, (2) the algorithms for modeling the orbit, and (3) the computer program which processes the observations. The second and third elements are well established and will be described only briefly here. I will concentrate on the alternative sources of data and the advantages and disadvantages of each in an orbit determination system.

The analytical methods for orbit determination are complex but well understood. They are summarized briefly in Chap. 6 and discussed in detail in several modern reference works. Vallado [1997], Battin [1999], Chobotov [1991], Escobal [1965], Noton [1998], and Roy [1991] provide extensive discussions of orbit determination and orbit propagation methods.

Generally, the various algorithms used for orbit determination are implemented in a small number of large and complex software systems. The major orbit determination systems are used for multiple space programs. Perhaps the most frequently used is the *Goddard Trajectory Determination System, GTDS,* used by NASA to process data for nearly all low-Earth orbit satellites [Long et al.,1989]. NORAD and others use a similar system for tracking spacecraft based on radar observations. JPL's *Deep Space Network (DSN)* uses a unique and remarkably accurate system to track interplanetary spacecraft [Jordan, 1981; Miller et al., 1990]. It solves simultaneously for the orbits of the interplanetary probes and the planets and satellites which they approach. For many mission analysis purposes, Chap. 6 (or any book on astrodynamics) contains sufficient information to construct an elementary orbit propagator. Highly precise orbit propagators are now commercially available.

The observations used for orbit determination can be obtained by tracking from the ground, tracking from space, or from autonomous or semi-autonomous systems on the spacecraft. Each of these approaches is described below. Table 11-65 summarizes their relative advantages and disadvantages.

TABLE 11-65.   Advantages and Disadvantages of Alternative Navigation Methods. (See also Table 11-66.)

System	Advantages	Disadvantages
Ground Tracking	Traditional approach Methods and tools well established	Accuracy depends on ground-station coverage Can be operations intensive
TDRS Tracking	Standard method for NASA spacecraft High accuracy Same hardware for tracking and data links	Not autonomous Available mostly for NASA missions Requires TDRS tracking antenna
Global Positioning System (GPS); GLONASS	High accuracy Provides time signal as well as position	Semi-autonomous Depends on long-term maintenance and structure of GPS Orbit only (see text for discussion) Must initialize some units
Microcosm Autonomous Navigation System (MANS)	Fully autonomous Uses attitude-sensing hardware Provides orbit, attitude, ground look-point, and direction to Sun	First flight test in 1993 Initialization and convergence speeds depend on geometry
Space Sextant	Could be fully autonomous	Flight-tested prototype only— not a current production product Relatively heavy and high power
Stellar Refraction	Could be fully autonomous Uses attitude-sensing hardware	Still in concept and test stage
Landmark Tracking	Can use data from observation payload sensor	Still in concept stage Landmark identification may be difficult May have geometrical singularities
Satellite Crosslinks	Can use crosslink hardware already on the spacecraft for other purposes	Unique to each constellation No absolute position reference Potential problems with system deployment and spacecraft failures
Earth and Star Sensing	Earth and stars available nearly continuously in vicinity of Earth	Cost and complexity of star sensors Potential difficulty identifying stars

Ground-Station Tracking

This is the traditional way to obtain data for orbit determination. We either track the spacecraft's telemetry signals or use radar tracking from a site not associated with the spacecraft. In both cases, the principal data used for orbit determination are *range* and *range rate*—that is, the distance from the ground station to the satellite and the satellite's line-of-sight velocity during the overhead pass. Angular measurements are also available at times but are typically far less accurate than range or range-rate measurements.

Accurate orbit determination using ground-station data ordinarily requires a number of passes. We may accumulate data from multiple passes over a single ground station, or may receive data at a central location from multiple ground stations around the world. In either case, data from a number of passes goes to one place for processing through a large system such as GTDS, described above. Ground-based systems necessarily operate on historical data and therefore will use propagated orbits for real-time operations and mission planning. Accuracies achievable with ground-based tracking vary with a spacecraft's orbit and the accuracy and amount of data. However, 3σ accuracies typically range from several kilometers for low-Earth orbits to approximately 50 km for geosynchronous orbit.

TDRS

The *Tracking and Data Relay Satellite, TDRS*, has now replaced NASA's worldwide ground-tracking network. A major advantage of this system is that the two operational TDRS satellites can provide tracking data coverage for 85% to 100% of most low-Earth orbits. (TDRS does not work for satellites in geosynchronous orbit.) The system collects mostly range and range-rate data from the TDRS satellite to the satellite being tracked. Angular information is available, but is much less accurate than the range and range-rate data. If atmospheric drag effects on a satellite are small, TDRS can achieve 3σ accuracies of about 50 m. This is considerably better than most ground-tracking systems. Another way to track from space is to use satellite-to-satellite or crosslink tracking as described below.

Spacecraft Autonomous Navigation

As summarized in Table 11-66, manufacturers have developed a number of autonomous navigation systems for spacecraft. Determining the orbit on board is technically easy with the advent of advanced spacecraft computers and higher-order languages. The principal problem is to provide orbit determination that is reliable, robust, and economical in terms of both cost and weight. A number of systems which can do this now exist—autonomous orbit determination is clearly feasible but becoming less important with the increasing use of GPS for navigation in low-Earth orbit. Wertz [2001] and Chory et al. [1986] describe alternative methods of autonomous navigation on board satellites. Table 11-65, earlier in the section, gives the advantages and disadvantages of the primary alternatives.

Autonomous navigation is inherently real-time. Thus, definitive orbit solutions and payload data are available simultaneously, which means that we can generate ground look-points or target positions and immediately associate them with the payload data. In addition, measurements can be less accurate than those for systems that work on old data, because solutions propagated forward in time lose accuracy. For example, to do accurate orbit maneuvers without autonomous navigation, we need a greater accuracy from a definitive solution based on old data that must be propagated forward to meet

TABLE 11-66. Alternative Navigation Methods. (See also Table 11-65.)

System	Basis	Status	Determines	Typical Accuracy (3σ)	Operating Range	Comments	Manufacturer
Global Positioning System (GPS)	Network of navigation satellites	Operational	Orbit*	15 m–100 m in LEO	LEO only	Semi-autonomous	Motorola, Rockwell
Microcosm Autonomous Navigation System (MANS)	Observations of Earth, Sun, and Moon	Flight tested in 1993	Orbit, attitude, ground look point, Sun direction	100 m–400 m in LEO (using only Earth, Sun and Moon)	LEO to GEO, lunar and planetary orbits	Can use other instruments (GPS receiver, star sensor, IMUs) to improve accuracy	Microcosm
Space Sextant	Angle between stars and Moon's limb	Flight tested	Orbit and attitude	250 m	LEO to GEO	Not being actively marketed for space at the present time	Lockheed Martin
Stellar Refraction	Refraction of starlight passing through the atmosphere	Proposed, some ground tests done	Orbit and attitude	150 m–1 km	Principally LEO	Could use attitude sensor data	
Landmark Tracking	Angular measurements of landmarks	Proposed, observability conditions are uncertain	Orbit and attitude	Several kilometers	Principally LEO	Could, in principle, use observation payload data	
Satellite Crosslinks	Range and range rate or angle measurements to other satellites in a constellation	Proposed; may be used on communications constellations	Orbit†	Theoretically as good as 50 m	Principally LEO	Operation with less than full constellation can be a problem; has no absolute position reference	
Earth and Star Sensing	Observe direction and distance to Earth in inertial frame	Proposed	Orbit and attitude	100 m–400 m in LEO	LEO to GEO, planetary orbits	Similar to MANS with higher accuracy and availability	

* Attitude determination using GPS receivers has been demonstrated. Multipath limits accuracy to ~ 0.3 to 0.5 deg.

† Could, in principle, be used for attitude determination as well.

real-time needs. With real-time systems, highly accurate orbit propagation is less critical, although we will still need some forward propagation for prediction and planning.

GPS and GLONASS

The *Global Positioning System*, also called *GPS* or *Navstar*, is a system of navigation satellites funded by the U.S. Department of Defense and intended explicitly to allow position determination by very small receivers anywhere on or near the Earth's surface. Extensive discussions of GPS and its applications, including signal structure and processing algorithms, are provided by Parkinson and Spilker [1996], Leick [1995], Kaplan [1996], and Hofmann-Wellenhof [1997]. GPS receivers are now readily available and their use is becoming widespread in airplanes, ships, ground vehicles, and military equipment. The system provides a moderate accuracy signal (50 m–100 m) for general navigation and a high-accuracy coded signal (15 m) for military applications. Commercial GPS receivers are now available for spacecraft, and are gaining in popularity in low-Earth orbit [Wertz, 2001; Chory et al., 1986; Anthony, 1992; Parkinson and Gilbert, 1983; Porter and Hite, 1984].

GPS receivers use signals from four different GPS satellites to solve simultaneously for the three components of the observer's position and the time. This can be done several times, providing position and velocity data which determines the orbit elements. The GPS constellation is in a 12 hour orbit at approximately half-geosynchronous altitude. Because the GPS antennas are designed to provide signals only in a cone covering the Earth's surface, coverage drops off rapidly with altitude, even for satellites in low-Earth orbit [Wertz, 1999]. Nonetheless, both analytic and experimental studies have been done on using GPS for navigation in orbits as high as geosynchronous using the spillover of the beam beyond the edge of the Earth's disk [Chao, et al., 1992].

The GPS signal can also be used to solve for the attitude of the vehicle on which the receiver is located. This is done by using multiple GPS antennas which are a known distance apart and which are attached to a rigid element of the vehicle. By measuring the phase difference between the signal from one GPS satellite arriving at two antennas, the GPS receiver serves as an interferometer measuring the angle between the line of sight to the GPS satellite and the line joining the two antennas. The wavelength of the GPS carrier signal is about 20 cm. Therefore, the accuracy of the attitude is limited by both the long wavelength and multi-path effects which cause confusion in the identification of the signal coming directly from the GPS satellite. In practice, spacecraft have been able to achieve on-orbit attitude accuracies on the order of 0.3 to 0.5 deg.

A number of practical difficulties have prevented GPS receivers from developing substantial operational utility in space for attitude determination. One problem is that multi-path effects can cause difficulties in some geometries and the GPS constellation, by nature, will eventually present most geometrical circumstances to the spacecraft. Orbit determination is an activity which can be done intermittently without harming system performance, but attitude determination must be continuous if we are to avoid a major failure. Consequently, the potential lack of availability of four GPS satellites for even a short period due to either geometrical circumstances or the outage of one or more satellites is a major concern for a spacecraft which depends on GPS for attitude determination. Thus, GPS-based attitude sensors will probably serve principally as backup, or would require backup systems to prevent major anomalies.

The *global navigation satellite system*, GLONASS, is a Russian space-based navigation system that provides 3-D position, velocity determination and time dissemination on a worldwide basis. GLONASS is very similar to GPS. It consists of a 24 satellite constellation at approximately half geosynchronous altitude and provides accuracies very similar to those of GPS. There are a number of manufacturers of GLONASS receivers, some of which are combined GPS/GLONASS receivers.

GLONASS is operated by Russia's Ministry of Defense. Like GPS, it was initiated in the mid-1970s with military design goals. Also like GPS, the civilian applications became apparent rapidly and the system is now in use for both civilian and military purposes. While the end results are very similar, the GLONASS signal structure is significantly different than that of GPS. Both GLONASS and GPS are available for use by spacecraft, and many satellite manufacturers are considering the use of either or both systems for onboard determination of position, velocity, time, and sometimes attitude. For further information on GLONASS, see, for example, Leick [1995] and Kaplan [1996].

Space Sextant

The *Space Sextant* was developed and flight tested in the late 1970s as a means of autonomous navigation by accurately measuring the angle between a star and the limb of the Moon [Martin Marietta Aerospace, 1977; Booker, 1978]. The Space Sextant provides both orbit and attitude information and can work over a very large regime, including geosynchronous orbit. The Space Sextant unit has been flight qualified; however, the need for precise telescope measurements makes the instrument rather heavy and therefore limits its usefulness in many space applications.

Microcosm Autonomous Navigation System

The *Microcosm Autonomous Navigation System (MANS)* uses observations of the Earth, Sun, and Moon from a single sensor to provide real-time position and attitude data [Tai and Noerdlinger, 1989; Anthony, 1992]. These objects were chosen principally because they can be unambiguously identified with high reliability and low cost and observations can be done with minor modifications to attitude sensors already on most spacecraft. The MANS flight software can also make use of, but does not require, data from a GPS receiver, star sensors, gyros, and accelerometers. The addition of other data sources provides added accuracy but is not required by the system. In addition to orbit and attitude, MANS provides ground look point and Sun direction information (even when the Sun is not visible to the sensor). It can work at any attitude and any orbit from LEO to beyond GEO. MANS was flight tested on the TAOS mission launched in 1994 [Hosken and Wertz, 1995].

Stellar-Horizon Systems

A number of approaches for orbit and attitude determination have been proposed, based on the interaction of starlight with the Earth's atmosphere [Hummel, 1984]. Specifically, as stars approach the edge of the Earth as seen from the spacecraft, refraction will cause their position relative to other stars to shift, producing an effect which can be measured with considerable accuracy. Theoretical accuracies for such systems are projected to be in the vicinity of 100 m. However, none of these systems has been fully developed for flight as yet.

Satellite Crosslinks

A number of proposals have been made for using satellite crosslinks to provide orbit determination [Chory et al., 1984]. This is of interest because it can be done with crosslink equipment used for intersatellite communication, and, therefore, requires minimal additional hardware. Crosslink tracking has been proposed for a number of constellations, but tends not to be implemented because of several practical problems. One problem is that satellite-to-satellite tracking provides only the relative positions of the satellites in the constellation. This means that if the absolute position is needed for any purpose such as mission planning or data reduction, then an additional system must be provided to establish the orbit relative to the Earth's surface. A second problem is that the satellites become interdependent, so satellite-to-satellite tracking may not work well for the first satellites or may degrade if a satellite stops working. Therefore, an alternative system not based on satellite-to-satellite tracking is required. If additional systems must be provided, there is less benefit from the satellite-to-satellite tracking.

Landmark Tracking

Landmark tracking has also been proposed for orbit determination [Markley, 1981]. This has been established as feasible by using data returned from satellite payloads. However, it has not been used as a normal method for satellite navigation, due in part to the difficulty of establishing automatic, unambiguous identification of landmarks to ensure that tracking accuracy can be maintained in the presence of adverse weather or poor seeing conditions.

Earth and Star Sensing

The combination of Earth and star sensing works similarly to sensing the Earth, Sun, and Moon [Wertz, 2001]. The direction and distance to the Earth are sensed relative to the inertial frame of the fixed stars. This is then used to directly determine the direction and distance to the spacecraft. The Earth and stars are available nearly continuously in any Earth orbit and star identification is becoming less of a problem with the introduction of substantially better computers for space use.

11.7.3 Orbit Maintenance and Control

Chapter 6 presented relevant ΔV equations for orbit maintenance and control. This section discusses when orbit control is necessary and what options are available to do it. Section 2.1.2 discusses autonomous orbit maintenance as part of a fully autonomous spacecraft.

Most small spacecraft do not require orbit control and have no onboard propulsion. This has the advantage of eliminating one spacecraft subsystem and, therefore, reducing the spacecraft's cost, weight, and complexity. However, once the spacecraft has separated from the launch vehicle or upper stage, no further control of the satellite orbit is possible and the satellite will be subject to drag and orbit decay in low-Earth orbit and to cumulative secular perturbations in all orbits. Usually, this is only acceptable for satellites that will last from 1 to 3 years but may be suitable for longer periods in some cases. For example, the Voyager spacecraft, having left the solar system, will be uncontrolled indefinitely. The only real objective, however, is to maintain communications and, to some degree, sample the interstellar medium.

We can also adjust a spacecraft's orbit using other means than onboard propulsion, such as the Orbiter or an orbit-transfer vehicle or tug. For example, with the Space Telescope, the Orbiter is used for both instrument replacement and to return the Telescope to a higher altitude.

Orbit control is needed when any of the following are required:

- *Targeting to achieve an end orbit or position*—as in satellite rendezvous or interplanetary missions

- *To overcome secular orbit perturbations*—such as altitude maintenance in low-Earth orbit or geosynchronous stationkeeping

- *To maintain relative orientations*—as in constellation maintenance

Each of these is discussed briefly below. An orbit lifetime of more than 1 to 3 years usually demands some type of orbit maintenance or control.

Targeting to achieve a particular orbit or location in space is the most common reason for orbit control. Typically, we achieve the orbit objectives with one or two large maneuvers, using several small maneuvers in between or for final adjustments. For example, in transfer to geosynchronous orbit, an initial large maneuver occurs at perigee in low-Earth orbit. A second large maneuver follows at apogee near geosynchronous altitude. Finally, several small orbit maneuvers over an extended period place the satellite in its final position. This has been traditionally been done by using large, high-thrust engines for the major maneuvers and smaller engines for orbit adjustments. However, as described in more detail in Sec. 7.5, low-thrust engines can often be used efficiently for large ΔVs. This normally means smaller, lighter, less-expensive engines and much smaller, simpler control systems. Propulsive maneuvers usually are the largest attitude disturbance on the spacecraft and, therefore, affect the size of the required attitude control system. Small thrusters can reduce the weight, complexity, and cost of other components as well as the propulsion system itself.

We often associate major orbit changes with the early phases of a mission, but they can occur throughout the life of the spacecraft. For example, most geosynchronous spacecraft can be shifted so that the longitude of the spacecraft is adjusted to meet changing needs. Spacecraft can also be retargeted to achieve new objectives, such as the retargeting of the ISEE-C spacecraft to rendezvous with Comet Giacobini-Zinner in 1985. The need for maneuvers of this type may arise after the spacecraft has been launched. They are not planned in advance but simply take advantage of existing resources. Finally, as described in Sec. 21.2, it is becoming more critical to use an end-of-life maneuver for spacecraft disposal either in low-Earth orbit or in geosynchronous orbit. These maneuvers are used either to have the spacecraft reenter in a location that is not hazardous, or put the spacecraft in an orbit where it will not harm other spacecraft. While not important during the early years of space exploration, the requirement to maintain a clean space environment will become much more stringent in the future.

We must also maintain the orbit to overcome long-term secular perturbations, as described in detail in Chap. 6. All geosynchronous spacecraft require orbit maintenance in the East-West direction to avoid interference and possible collisions with other spacecraft. Furthermore, most use orbit maintenance in a North-South direction to maintain a near-zero inclination. In low-Earth orbit, altitude maintenance is used to overcome atmospheric drag and achieve a longer working life. Other orbit types, such as Sun-synchronous or repeating ground track, may also require orbit maintenance.

Nearly all constellations require some type of orbit maintenance to prevent collisions between satellites and maintain the constellation pattern over time. In principle, we could use *relative stationkeeping* in which we maintain the relative positions between satellites but not their absolute position. In practice, however, this will make orbit maintenance more complex and will not save propellant or reduce the number of computations. In a low-Earth orbit constellation with relative orbit maintenance, we would, in principle, maintain all satellites in the constellation to decay at the same rate as the slowest-decaying satellite at any time. But the entire constellation would still decay in this process. Therefore, it would slowly change its altitude and need to be reboosted at some later time.

The alternative is *absolute stationkeeping*, shown in Fig. 11-51. Here, we maintain each spacecraft within a mathematically defined box moving with the constellation pattern. As long as we maintain the constellation's altitude, absolute stationkeeping is just as efficient as relative stationkeeping. All in-track stationkeeping maneuvers are done firing in the direction of motion to put energy taken out by atmospheric drag back into the orbit. We put in more or less energy at any given time, depending upon the amount of drag and the atmospheric density, as described in Sec. 8.1.5.

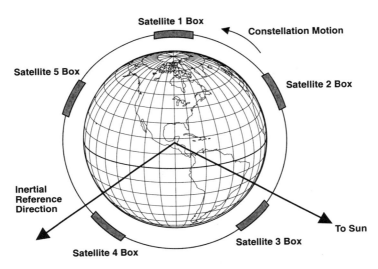

Fig. 11-51. **Constellation Stationkeeping Maintains Each Satellite in a "Box" Rotating with the Constellation.**

The amount of drag makeup for any satellite in a constellation depends on the satellite's observed drift relative to its assigned box. At the forward edge of the box, the applied ΔV is increased, thus increasing the orbit altitude and period and sliding the satellite rearward in phase relative to the box. Similarly, at the trailing edge, the applied ΔV is decreased, thus decreasing the altitude and period and sliding the satellite forward in phase. Because of the high spacecraft velocities in low-Earth orbit, timekeeping is critical to maintaining the satellites' relative positions. A 1 sec difference in the time corresponds to a 7 km difference in in-track position. Maintaining the same time throughout the constellation is important, but not difficult.

Although the perturbative forces are different, constellation maintenance in low-Earth orbit is analogous to stationkeeping in geosynchronous orbit. In-track and cross-track orbit maintenance in low-Earth orbit correspond to East-West and North-South stationkeeping, respectively, in geosynchronous orbit. Because the forces involved are different, the correspondence is not exact. Altitude maintenance is necessary in low-Earth orbit to overcome the effect of atmospheric drag. In geosynchronous orbit, this would correspond to maintaining the mean drift rate relative to the surface of the Earth at zero, so the stationkeeping box stays over a fixed location. In low-Earth orbit, this corresponds to maintaining the box at a mathematically fixed position in the constellation.

Ground-Based vs. Autonomous Orbit Control

Traditionally, orbit maintenance and control are executed from the ground. The required orbit adjustment will be computed at the ground station, and a series of commands will be generated which are uploaded to the spacecraft and then downlinked for verification. Finally, the ground station sends a command to execute the control commands, and the spacecraft carries them out based on an onboard clock. This sequence protects the spacecraft against communications errors and allows maneuvers when the spacecraft is out of sight of the ground station. The spacecraft is often out of sight because orbit maneuvers are usually 180 deg out of phase from the desired result. For example, in landing on the Moon, the principal burns bring the spacecraft down to the Earth-facing side of the Moon and take the returning astronauts back to Earth. Both burns occur on the far side of the Moon, entirely out of sight of the Earth.

In normal geosynchronous stationkeeping the spacecraft is allowed to drift to one side of the stationkeeping box. A maneuver then changes its velocity so it will drift across the box and back before needing another maneuver. This is similar to keeping a ball in the air by continuously hitting it upward with a paddle. The main reason for this sequence is that it maximizes the time interval between maneuvers and, therefore, minimizes the amount of ground operations, which carry the potential for communications errors or command mistakes.

In the past, there was no realistic alternative to orbit control from the ground. Now, however, autonomous navigation systems have made autonomous orbit maintenance possible, economical, and safe. Autonomous orbit maintenance drives down the cost and risk of missions by having a major part of the day-to-day operations on board the spacecraft. As described in Sec. 2.1.2, autonomous orbit maintenance is a key component in a fully autonomous spacecraft bus, which can further reduce mission costs. (For further discussion of autonomous orbit control, see Wertz [1996, 2001], Wertz, et al. [1997], Königsmann, et al. [1996], Collins, et al. [1996], or Glickman [1994].) Microcosm has developed a commercial, autonomous Orbit Control Kit flight software system scheduled for flight validation on UoSAT-12 in 1999. This and other approaches are also scheduled to be flight tested later on EO-1 which is intended to do formation flying with LandSat-7.

Orbit and attitude control are analogous, with several important differences. Typically, we must control attitude moderately often or continuously if we are to avoid serious consequences. A satellite that loses attitude control will usually tumble and then lose the payload function, power on the solar arrays, and contact with the ground. It also may point sensitive instruments at the Sun or have substantial thermal problems. Even a brief attitude control failure can destroy the mission. In contrast, orbit control maneuvers occur infrequently, and any computer that can do autonomous navigation will easily accommodate the necessary additional computations. As long as we control the orbit with low-thrust systems (Secs. 7.5, 17.4), a short-term failure will cause no damage. Gravity takes care of short-term orbit control very well. If the orbit-control system fails, the ground or the onboard system will determine that the satellite is slowly drifting from its assigned slot and a warning can be issued with adequate time to fix the problem or implement a back-up before adverse consequences occur.

With autonomous orbit maintenance we can optimize characteristics other than the time between maneuvers, such as the size of the control box. Just as we would not ordinarily implement attitude control to minimize the number of commands sent to the momentum wheel, we do not have to minimize the number of thruster firings so long as the duty cycle is within the range of the thrusters and the thruster pulses are long enough to achieve high efficiency. Ordinarily using many short pulses is not a problem, as thrusters are frequently used for attitude control and small thrusters have lifetimes far in excess of the required number of on-off cycles.

For some missions, autonomous orbit control is required simply because of the nature of the mission itself. This is the case for planetary flybys or a mission to the far side of the Moon in which either the spacecraft is out of contact or the communications delays are too long for normal ground control. Generally, ground stations do the orbit-control computations and upload them to the spacecraft for execution at a later time. This is a semi-autonomous approach which can meet mission requirements while maintaining ground control.

In geosynchronous stationkeeping, the main reason for autonomous orbit maintenance is to reduce operations cost and risk, rather than a specific technical requirement.* For a constellation at any altitude, the overall process of orbit determination and control represents a major operational cost. It also represents a significant risk element in which any operational error or failure of the ground system could damage or destroy the constellation. The orbit maintenance operation is necessarily carried out on board the spacecraft by firing thrusters. Performing the control computations on board the spacecraft can reduce both cost and risk. First, it eliminates the potential for operator error in a very repetitive function. Second, it reduces communication errors or failures frequently associated with operational activities.

The argument has been made that autonomous orbit control is a range-safety concern because having a spacecraft adjust its orbit without operator oversight could endanger other spacecraft or people. This is a reasonable concern for firing high-thrust, large ΔV engines, but it is not a problem for low-thrust maintenance. A watchdog timer can easily limit the propulsive burn time to keep burns small. Even if the watchdog system fails, the amount of propellant in most low-thrust systems is too low to endanger other space systems or people on the ground.

The principal reason for not undertaking autonomous orbit maintenance and control for future missions is tradition. It has not been done that way in the past, and there is a very strong desire in expensive space missions to maintain those procedures that have worked previously. A mechanism for overcoming this potential risk is *supervised autonomy* in which orbit maintenance maneuvers are computed on board the spacecraft and verification from the ground is required before they are executed. This allows mission personnel to gain confidence in the onboard computations before permitting fully autonomous operation.

A second alternative is to implement "autonomous orbit maintenance" from the ground. In this case, the computations would be done autonomously, but would be

---

* A technical requirement for autonomous orbit maintenance can arise in geosynchronous stationkeeping when there is a desire to place additional satellites in a narrow orbit slot. In this case, autonomous orbit maintenance is needed to reduce the size of the orbit-control deadband, which would be impractical if we had to do frequent commanding from the ground.

done at the ground station and then sent to the spacecraft for execution. This has the advantage of maintaining some characteristics of traditional orbit maintenance and also minimizes the amount of hardware on board the spacecraft. Unfortunately, this approach can add significant complexity and risk to the mission. If the navigation data is obtained on board the spacecraft, it would need to be communicated to the ground for processing. Then, the results and commands would go to the spacecraft and be verified for later execution. This makes the process much more complex and increases the potential for communications errors and transmitting the wrong data to the wrong spacecraft. Most likely, these disadvantages would outweigh any advantage of doing the small amount of command processing on the ground. If the spacecraft has enough computing power on board, a reasonable alternative might be to compute the orbit control on board the spacecraft and send it to the ground for verification and approval before actually executing the command. This allows full ground override. It also allows the system to use an on-orbit process with less cost and higher reliability, whenever operators are confident that the system is working smoothly.

If the spacecraft navigation is done with the traditional approach of tracking from the ground, doing the commanding from the ground and uploading commands to the spacecraft is an efficient alternative and minimizes the overall communication problem. In this case, the traditional cost and complexity of orbit maintenance from the ground remain, although it may be possible to reduce this somewhat with more automation.

Autonomous navigation and orbit control can significantly reduce the cost of space operations. For most space missions, this is a major cost. It should also reduce risk because there are fewer failure modes than with operator-driven or ground-based systems. For modern spacecraft, almost no one would recommend doing attitude determination and control from the ground, even though this is a more complex task with more serious adverse consequences than orbit control. The main argument against autonomous orbit determination and control is tradition—we have not done it that way in the past and, therefore, should not do it in the future.

11.7.4 Sizing Autonomous Guidance and Navigation

The implementation of a purely ground-based guidance and navigation system is both expensive and straightforward. This is the traditional approach, and is described in detail in Chap. 14. The process is well established for geosynchronous spacecraft, low-Earth orbit spacecraft, and interplanetary missions.

The main reason for considering autonomous guidance and navigation is to reduce mission cost and risk. But we can also extend mission life, put more spacecraft into a geosynchronous slot, or undertake missions which we could not realistically do without some autonomy.

Orbit and attitude sensing and control are strongly interrelated. In many cases, they will use the same sensors and the same actuators. The attitude control sensors and actuators will need to control the spacecraft during orbit maneuvers, which will probably be the largest source of disturbance torques during the spacecraft's operating life. In addition, orbit and attitude budgets are often combined to produce pointing and mapping budgets to satisfy mission requirements (see Fig. 4-6 and Sec. 5-4). Our goal should be to reduce the total cost and risk of attitude and orbit determination and control. (For an extended discussion of this objective, see Wertz [2001].)

In most cases, we would meet our objective of minimum cost, weight, and risk by combining elements of orbit and attitude determination and control. Unfortunately, this may be difficult to achieve in practice. Traditionally, attitude determination and control is the responsibility of a controls organization, and is a well-defined element of the spacecraft bus. Orbit determination and control, on the other hand, is usually the responsibility of a systems organization which will ultimately implement this function on the ground. The guidance and navigation function is ordinarily not allocated weight or power on board the spacecraft and is not considered a normal spacecraft function. This chapter contains separate sections on the attitude control system and the guidance and navigation system because these are normally separate functions. Getting these functions to work together is perhaps the single largest hurdle in developing autonomous navigation and orbit control and, therefore, fully autonomous spacecraft.

TABLE 11-67.   **Size, Weight, and Power of Alternative Autonomous Navigation Systems.** All systems are evolving, so size, weight, and power will probably decrease in the future.

System	Manufac-turer	Orbit/ Attitude	Size (cm³)	Weight (kg)	Power (W)	Other Requirements/ Comments
GPS	Honeywell	Orbit	4,000	4	35	
GPS	RI	Orbit	4,700	4	12	
MANS	Microcosm	Both	(1)	(1)	(1)	Uses spacecraft clock and spacecraft computer
Space Sextant	Martin Marietta	Both	4×10^5	25	50	Can eliminate other attitude sensors
Stellar Refraction	N/A	Both	(1)	(1)	(1)	(4)
Landmark Tracking	N/A	Both	(2)	(2)	(2)	(4)
Satellite Crosslinks	N/A	Orbit	(3)	(3)	(3)	(4)
Earth and Star Sensing	N/A	Both	(1)	(1)	(1)	(4); similar to MANS

(1) Uses sensors normally used for attitude only. Very small (or no) added weight and power for navigation sensing.
(2) Intent is to use observation payload sensor.
(3) Uses crosslinks on board for intersatellite communications. Would need secondary system until sufficient number of satellites are in place.
(4) Conceptual design only. Would probably use spacecraft computer.

The problem of combining orbit and attitude systems is illustrated by Table 11-67, which provides size, weight, and power for alternative autonomous navigation systems. Several of the systems, such as the GPS receivers, are orbit-only, although in the future they may be expanded to include attitude as well. Other systems, such as MANS and the Space Sextant, do both the attitude and orbit function. In these cases, one should look at the difference in weight and power required to achieve navigation from that required for attitude alone. Even this assessment is complicated because orbit determination requires computer resources which may be regarded as a part of the attitude control system, the navigation system, the command and data handling system, or the general spacecraft computer. For many of the systems, such as landmark tracking or crosslink navigation, the computer is the principal element associated with autonomous navigation and control. The main message is to avoid "double booking" components for guidance and navigation and to look at the joint implementation of orbit and attitude determination and control when beginning to optimize system performance.

Autonomous navigation and orbit maintenance is too new to have a standard implementation in terms of where computations are done. But the nature of the computations themselves and the data used suggest a natural configuration: using a single spacecraft processor for determining and controlling the attitude and orbit. These functions will probably use either the same or similar sensors and may use the same actuators. Most of the computing is associated with sensor processing, data handling, and anomaly resolution. The orbit and attitude computations themselves are normally much smaller. The implementation of either orbit or attitude control algorithms represents by far the smallest part of the throughput requirement. Thus, control adds little burden for any processor which is already determining the orbit or attitude.

A reasonable initial design would incorporate all of these functions in a single spacecraft processor. Actual implementation may vary, depending upon the specific hardware and software. For example, star-sensor processing may be incorporated within the star sensor itself, or may be done in the same processor as other orbit or attitude functions. The overall objective, however, should continue to be to minimize the cost and risk of determining and controlling the orbit and attitude for the entire mission.

References

Agrawal, Brij N. 1986. *Design of Geosynchronous Spacecraft*. Englewood Cliffs, NJ: Prentice-Hall, Inc.

Ahern, J. E., R. L. Belcher, et al. 1983. "Long Life Performance of Satellite Thermal Control Surfaces." Paper presented at the 18th American Institute of Aeronautics and Astronautics Thermophysics Conference, Montreal, Canada, and the 21st American Institute of Aeronautics and Astronautics Aerospace Science Meeting, Reno, NV.

Anthony, Jack. 1992. "Autonomous Space Navigation Experiment." Paper No. AIAA 92-1710 presented at the AIAA Space Programs and Technologies Conference. Huntsville, AL, March 24–26.

Battin, Richard H. 1999. *An Introduction to the Mathematics and Methods of Astrodynamics*. New York: AIAA Education Series.

Booker, R. A. 1978. "Space Sextant Autonomous Navigation and Attitude Reference System—Flight Hardware Development and Accuracy Demonstration. Rocky Mountain Guidance and Control Conference." Paper AAS 78-124.

Brown, M. J. and L. J. Gagola. 1965. *Thermal Properties Handbook*. ER 13997. Baltimore, MD: Martin Marietta Aerospace Group (internal publication).

Bruhn, E. F. 1973. *Analysis and Design of Flight Vehicle Structures*. Tri-State Offset Co.

CCSDS 201.0-B-1. January 1987. *Blue Book Telecommand Channel Service Part 1 Architectural Specification*.

Chao, C.C., H. Bernstein, W.H. Boyce, and R.J. Perkins. 1992. "Autonomous Station-keeping of Geosynchronous Satellites Using a GPS Receiver." AIAA-92-4655-CP, A92-52106.

Chetty, P. R. K. 1991. *Satellite Technology and Its Applications (2nd Edition).* New York: McGraw-Hill.

Chobotov, Vladimir, ed. 1991. *Orbital Mechanics.* Washington, DC: AIAA.

Chory, M. A., Hoffman, D. P., Major, C. S. and V. A. Spector. 1984. "Autonomous Navigation—Where We are in 1984." Paper presented at AIAA Guidance and Control Conference, Seattle, WA. August 20–22.

Chory, M. A., Hoffman, D. P. and J. L. LeMay. 1986. *Satellite Autonomous Navigation—Status and History.* TRW Internal Report.

Collins, John T., Simon Dawson, and James R. Wertz. 1996. "Autonomous Constellation Maintenance System." Presented at 10th Annual AIAA/USU Conference on Small Satellites, Logan, Utah, September 16–19.

Conley, Peter L. 1998. *Space Vehicle Mechanisms.* New York: John Wiley & Sons, Inc.

Escobal, Pedro R. 1965. *Methods of Orbit Determination.* Malabar, FL: Robert E. Krieger Publishing Co.

European Space Agency, 1989. *Spacecraft Thermal Control Design Data.* ESA PSS-03-108, issue 1.

European Space Agency, 1998. *Thermal Control Standards* ECSS-E-30-00 Part 2-1.

Fortescue, Peter, and John Stark. 1995. *Spacecraft Systems Engineering.* New York: John Wiley & Sons, Inc.

Gilmore, D.G. and Mel Bello. 1994. *Satellite Thermal Control Handbook.* Los Angeles, CA: The Aerospace Corporation Press.

Glickman, Ronald E. 1994. "TIDE: The Timed-Destination Approach to Constellation Formationkeeping." Paper AAS 94-122, presented at the AAS/AIAA Spaceflight Mechanics Meeting, Cocoa Beach, FL, February 14–16.

Hofmann-Wellenhof, B., H. Lichtenegger, and J. Collins. 1997. *GPS Theory and Practice (4th Edition).* New York: Springer-Verlag Wien.

Hosken, R.W. and J.R. Wertz. 1995. "Microcosm Autonomous Navigation System On-Orbit Operation." Paper AAS 95-074, presented at the 18th Annual AAS Guidance and Control Conference Keystone, CO, February 1–5.

Hummel, S.G. 1984. "Spacecraft Autonomous Navigation Using Stellar Refraction Measurements." Paper presented at AIAA Guidance and Control Conference, Seattle, WA. August 20–22.

IEEE. 1972. *IEEE Standard 100. IEEE Standard Dictionary of Electrical and Electronic Terms.*

IRIG, Telemetry Group. 1986. *IRIG Standard 106 Telemetry Standards.* Inter-Range Instrumentation Group, Range Commanders Council.

Jordan, J. F. 1981. "Deep Space Navigation Systems and Operations." Paper presented at the ESA International Symposium on Spacecraft Flight Dynamics. May.

Kaplan, Elliott D. 1996. *Understanding GPS Principles and Applications*. Boston, Artech House, Inc.

Kaplan, Marshall H. 1976. *Modern Spacecraft Dynamics and Control*. New York: John Wiley & Sons.

Königsmann, Hans J., John T. Collins, Simon Dawson, and James R. Wertz. 1996. "Autonomous Orbit Maintenance System." *Acta Astronautica* 39:977–985.

Kreith, F. and M. S. Bohn. 1996. *Principles of Heat Transfer*. PWS Publishing Co.

Leick, Alfred. 1995. *GPS Satellite Surveying (2nd Edition)*. New York: Wiley-Interscience Publication, John Wiley & Sons.

Long, Cappellari, Velez, and Fuchs, eds. 1989. *Goddard Trajectory Determination System Mathematical Theory, Revision 1*. Goddard Space Flight Center and Computer Sciences Corporation. NASA/Goddard Space Flight Center, Flight Dynamics Division, Code 550.

Markley, F. Landis. 1981. "Autonomous Satellite Navigation Using Landmarks." Paper No. 81-205 presented to the AAS/AIAA Astrodynamics Specialist Conference. Lake Tahoe, NV, August 3–5.

Martin Marietta Aerospace. 1977. *Design and Laboratory Testing of Self-Contained High Altitude Navigation System, Phase 1—The Space Sextant Autonomous Navigation Attitude Reference System (SS-ANARS)*, Phase 1 Final Report, MCR-77-196. Denver, CO.

MIL-A-83577. 1988. *General Specification for Moving Mechanical Assemblies for Space and Launch Vehicles*.

MIL-HDBK-17B, Change Notice 1, 16 June 1989, *Polymer Matrix Composites*. Philadelphia: Naval Publications Forms Center.

MIL-STD-1582. 1984. *Satellite Data Link Standard (SDLS)*.

Miller, M., H. Hanover, K. Patel, and E. Wahl. 1990. "Neptune Encounter: Guidance and Control's Finest Hour." Paper presented at the 13th Annual AAS Guidance and Control Conference. Keystone, Colorado. February 3–7.

Morgan, Walter L. and Gary D. Gordon. 1989. *Communications Satellite Handbook*. New York: John Wiley & Sons.

National Aeronautics and Space Administration. 1972. *Phase-Change Materials Handbook*. B72-10464. Marshall Space Flight Center, Huntsville, AL: National Aeronautics and Space Administration.

National Aeronautics and Space Administration. 1975. *Astronautics Structures Manual*, Marshall Space Flight Center, Huntsville, AL: National Aeronautics and Space Administration.

National Aeronautics and Space Administration. 1982. *Space and Planetary Environment Criteria Guidelines for Use in Space Vehicle Development*. Technical Memorandum 82473. Vol. 1.

National Aeronautics and Space Administration, 1983. *Space and Planetary Environment Criteria Guidelines for use in Space Vehicle Development.* Technical Memorandum 82501, Vol. 2.

National Aeronautics and Space Administration. 1988. *TDRSS Users Guide.* NASA Goddard STDN 101.2.

National Aeronautics and Space Administration, Jet Propulsion Laboratory. 1989. *Deep Space Network Interface Design Handbook.* Vol. I and II. JPL-DSN 810-5 Revision D.

Noton, Maxwell. 1998. *Spacecraft Navigation and Guidance.* London: Springer-Verlag.

Parkinson, B.S. and S.W. Gilbert. 1983. *NAVSTAR: Global Positioning System—Ten Years Later.* Proceedings of the IEEE. Vol. 71, No. 10. October 1983.

Parkinson, Bradford W., and James J. Spilker, Jr., eds. 1996. *Global Positioning System: Theory and Application.* Vols. I and II. Washington, DC: AIAA.

Pisacane, Vincent L. and Robert C. Moore. 1994. *Fundamentals of Space Systems.* New York: Oxford University Press, Inc.

Porter, J. P. and W. A. Hite. *Overview/Current Status of the NAVSTAR Global Positioning System,* IEEE PLANS. 1984.

Roark, R. J., and W. C. Young. 1975. *Formulas for Stress and Strain (5th Edition).* New York: McGraw-Hill.

Rolfe, S. T., and J. M. Barsom. 1977. *Fracture and Fatigue Control in Structures.* Englewood Cliffs, NJ: Prentice-Hall Inc.

Roy, A.E. 1991. *Orbital Motion (3rd Edition).* Bristol and Philadelphia: Adam Hilger.

Sandia National Laboratory. 1981. *Solar Heating Materials Handbook: Environmental and Safety Considerations for Selection.* DOE/TIC-11374 and UC-11-59-59C. Albuquerque, NM: Sandia National Laboratory.

Sarafin, Thomas P., ed. 1995. *Spacecraft Structures and Mechanisms: From Concept to Launch.* Torrance, CA, and Dordrecht, The Netherlands: Microcosm Press and Kluwer Academic Publishers.

Schildcrout, M. and M. Stein. 1928. *Critical Combinations of Shear and Direct Axial Stress for Curved Rectangular Panels.* NACA TN 1928. Washington, DC: National Advisory Committee for Aeronautics.

Siegel, Robert and John Howell. 1968. *Thermal Radiation Heat Transfer.* Vol. I, *The Black Body Electromagnetic Theory and Material Properties.* Lewis Research Center, Cleveland, OH: National Aeronautics and Space Administration.

Singer, Fred S. 1964. *Torques and Attitude Sensing in Earth Satellites.* New York: Academic Press.

Stimpson, L. D., and W. Jaworski. 1972. "Effects of Overlaps, Stitches, and Patches on Multilayer Insulation." Paper No. 72-285 presented at the American Institute of Aeronautics and Astronautics 7th Thermophysics Conference, San Antonio, TX.

Tai, Frank and Peter D. Noerdlinger. 1989. "A Low Cost Autonomous Navigation System." Paper No. AAS 89-001 presented at the 12th Annual AAS Guidance and Control Conference, Keystone, CO, Feb. 4–8.

TOR-0059(6110-01)-3, Reissue H. 1987. *Air Force Satellite Control Facility Space/Ground Interface.*

Tsai, S. W. 1987. *Composites Design (3rd Edition)*, Dayton, OH: Think Composites, Inc.

U.S. Air Force Materials Laboratory. 1987. MIL-HDBK-5E, *Metallic Materials and Elements for Aerospace Vehicle Structures.* Wright-Patterson AFB, OH: Air Force Materials Laboratory.

Vallado, David A. 1997. *Fundamentals of Astrodynamics and Applications.* New York: McGraw-Hill.

Wertz, James R. ed. 1978. *Spacecraft Attitude Determination and Control.* Dordrecht, The Netherlands: D. Reidel Publishing Company.

Wertz, James R., and Wiley J. Larson, eds. 1996. *Reducing Space Mission Cost.* Torrance, CA: Microcosm Press, and Dordrecht, The Netherlands: Kluwer Academic Publishers.

Wertz, James R., John T. Collins, Simon Dawson, Hans J. Koenigsmann, and Curtis W. Potterveld. 1997. "Autonomous Constellation Maintenance." Presented at the IAF Workshop on Satellite Constellations, Toulouse, France, Nov. 18–19.

Wertz, James R. 2001. *Mission Geometry; Orbit and Constellation Design and Management.* Torrance, CA: Microcosm Press and Dordrecht, The Netherlands: Kluwer Academic Publishers.

Wong, H.Y. 1977. *Handbook of Essential Formulae and Data on Heat Transfer for Engineers.* London and New York: Longman.

Chapter 12

Spacecraft Manufacture and Test

Emery I. Reeves, *United States Air Force Academy*

12.1 Engineering Data
12.2 Manufacture of High-Reliability Hardware
12.3 Inspection and Quality Assurance
12.4 The Qualification Program
12.5 Spacecraft Qualification Test Flow
12.6 Launch Site Operations

This chapter presents an overview of the spacecraft manufacturing, assembly, and test process and the underlying test theory. A system designer must understand this process because hardware manufacture and test heavily influence the program's cost and schedule. In addition, testing technology and special facility requirements may affect program feasibility. This chapter is oriented principally toward the construction of a single satellite or the first satellite of a production run. See Sec. 19.1 for a discussion of the "production line" approach applicable to building satellites in large numbers such as those which are a part of a large constellation.

Table 12-1 lists the names we use in this chapter to describe the parts of a spacecraft. For example, *piece parts* are individual parts, including transistors, integrated circuits, or mechanical parts such as housings, panels, bearings, and gears. A *component* is a complete unit or black box such as a transmitter, receiver, computer, or electromechanical actuator. Sometimes a functional group of parts, or *assembly*, is manufactured or tested together. An assembly as used here may be part of a component or may be integrated directly into the spacecraft. *Subsystems* consist of groups of components. They may be assembled and tested as subsystems or integrated into the spacecraft as components.

TABLE 12-1. Hardware Nomenclature. Spacecraft are built up from subsystems, which are composed of components.

Piece Part	Individual part such as resistor, integrated circuit, bearing, circuit board, or housing
Component	Complete functional unit such as a control electronics assembly, an antenna, a battery, or a power control unit
Assembly	Functional group of parts such as a hinge assembly, an antenna feed, or a deployment boom
Subsystem	All of the components and assemblies that comprise a spacecraft subsystem
Spacecraft	Complete vehicle

Methods for the manufacture and test of spacecraft and spacecraft component: derive from the aircraft and electronics industries. But spacecraft hardware is les: plentiful and less accessible for maintenance. In spacecraft production, a run of 10 i: high volume and spacecraft recalls are extremely rare. Furthermore, environmenta: forces severely stress the hardware during launch.

The theory of *type test* (see Table 12-2) is a basic principle affecting the manu facture and test of spacecraft hardware. Type test theory depends on preparing an controlling complete and exact engineering data (drawings, specifications, and proce dures). If the engineering data controls the hardware construction completely, all item built to the same data are equivalent and the results of any single-item test are valid fo all like items. In particular, if a representative article (*type test article*) passes sequence of *qualification tests*, all other articles built to the same engineering dat: should also pass. In other words, the design is *qualified*. We simply have to make sur articles are identical by controlling the engineering data and manufacturing processes Less severe *acceptance tests* then certify proper workmanship.

TABLE 12-2. Theory of *Type Testing*. The type test theory is the basis for qualification testing

Basis	Engineering data is complete and exact.
	Engineering data completely controls manufacture.
	All items manufactured to the same engineering data are identical.
Therefore	Results of qualification test for one article are valid for all articles.

Table 12-3 lists the steps in manufacturing, integrating, and testing a spacecraft. A system designers, we must determine how long each step will take and identify an: test or facility requirement that is risky or peculiar to the program. Above all, w should schedule qualification tests to qualify the spacecraft completely before launch The following sections address these steps and describe important aspects of system design.

TABLE 12-3. Steps in Manufacturing, Integrating, and Testing.

Step	Description	Comments
Prepare Engineering Data	Complete drawings and all supporting information such as material and part specifications and processing methods	Engineering data will consist of several hundred drawings for each component, specifications for each piece-part type and process, assembly drawings, and test equipment data
Manufacture Component	Stages: 1. Manufacture planning 2. Parts procurement and test 3. Component assembly 4. Component acceptance test	Typical timing: 1. In parallel with engineering-data preparation 2. Mechanical parts and materials 1–6 mos Electronic parts 3–18 mos 3. 1–3 mos 4. 1–3 mos; acceptance test includes functional test and environmental exposure
Qualify Component	Functional test and environmental exposure	Takes 1–6 mos, depending on complexity and fragility of component, severity of environment, and number of failures
Integrate and Test Spacecraft	Mechanical assembly, functional test, and environmental exposure	Takes 6–18 mos

12.1   Engineering Data

Engineering data (see Table 12-4) is drawings, specifications, and procedures. Standard formats and contents for military contracts are in MIL-STD 100 (drawings)[*] and MIL-STD 490 (specifications). Most aerospace companies use similar standards. Various documents combine to control the hardware manufacture. Each piece part, assembly, or component is described by its own individual drawing, and drawing call outs describe materials and processes. Drawing numbers identify parts. In the same way, drawings control assembly of parts into higher-level assemblies and identify integration hardware. Drawings also control interfaces, size, shape, and mounting provisions.

TABLE 12-4.   Defining and Controlling Engineering Data.

Engineering Data	Drawings, specifications, and procedures
Role of Engineering	Produce engineering data
Role of Manufacturing	Build hardware to meet engineering data
Role of Quality Control	Ensure that the hardware is built and tested to meet engineering data
Configuration Management	Systems and procedures that identify, account for, and control engineering data

Configuration management is the process of controlling engineering data. It includes identifying the engineering data required for manufacture (*configuration identification*), controlling changes, maintaining the engineering database (*configuration control*), reviewing and auditing the engineering data (*configuration audit*), and verifying that the hardware is built as designed (*configuration verification*). Changes to the engineering data (*engineering change orders* and *procedure change orders*) are also tracked. For military contracts, MIL-STD 483, MIL-STD 1521, and DOD-STD 480 itemize configuration-management requirements. Most aerospace contractors have systems that conform to these standards.

12.2   Manufacture of High-Reliability Hardware

The first step in manufacturing spacecraft hardware is to translate the engineering data into manufacturing plans, flows, instructions, manufacturing aids, and tooling. This occurs as the design matures and manufacturing personnel may influence the design for ease of manufacture. Engineering data is formally released at a *Critical Design Review*. Typically, we review manufacturing plans at this time or in a *Manufacturing Readiness Review* shortly afterward.

The manufacturing planning starts with subassemblies by generating parts kits, preparing detailed procedures for assembly, and identifying inspection and test requirements. We must also identify special facilities (such as clean rooms); manufacturing methods, precautions, and controls; and training and certification levels of personnel. Based on these plans, we then call out the manufacturing steps in detailed procedures. A copy of the procedures travels with the hardware and is checked off as

---

[*] Military specifications and standards are no longer being maintained and are due to be superseded by industry standards which are not yet in place.

the steps are completed. This paper becomes part of the permanent record of the assembly. Inspection and test results are also part of the record, and some installations include photographs of each assembly.

Manufacturers buy most raw materials from certified vendors, who certify the material quality. They buy electronic piece parts to meet individual specifications which call out performance and quality requirements. Under current practices, high-reliability electronic parts are constructed for a particular program. Such procurement takes a long time, and most parts undergo extensive testing before combining into components. Flight hardware usually requires the highest level of reliability, which in the past has been called *S-level*. Lower reliability parts may work for prototype and qualification units not intended for flight. In addition to vendor lead times, each part order must be bid and negotiated, which adds weeks to the procurement time. The current practice is to have the part supplier test each part for performance (*group A tests*) and a sample from each lot to extremes (*group B and C tests*). The manufacturer burns the parts in while monitoring their performance and does a destructive physical analysis (*DPA test*) on a sample.

The industry is currently undergoing a major revision in the way that parts are specified and procured. In the Department of Defense this is called "Acquisition Reform" and is characterized by elimination of government specifications and standards for parts and processes. Industry is expected to replace these specifications and standards with their own controls or use Professional Society Standards. A consistent set of such standards is not yet in place although efforts are under way to produce them (see for instance *AIAA Recommended Practice for Parts Management* AIAA R-100-1996). Extensive parts information is also available on the Internet at Air Force and NASA sites.

To contend with the lead times for parts and materials, manufacturers must order them before they see component engineering data. Thus, we establish a *Project Approved Parts List* (*PAPL*) and *Project Approved Materials List* (*PAML*) early in the program. Because designers must use the preferred part or identify and justify new parts, these approved lists reduce the number of part types and allow early procurement.

Manufacturing facilities cover mechanical manufacturing, electronic manufacturing, spacecraft assembly and test, and special functions. *Mechanical manufacturing* includes standard machine shops, plus locations for mechanical assembly, plating and chemical treatment, composite manufacture, adhesive bonding, and elevated temperature treatment. Although most aerospace facilities are quite clean, mechanical manufacturing does not normally need controlled cleanliness. But electromechanical and optical manufacturing, as well as the tailoring of thermal blankets, do need controlled clean rooms—normally separate from conventional mechanical manufacturing. Table 12-5 shows cleanliness requirements for various operations.

Electronic manufacturing facilities include areas for building printed circuits and clean rooms for building and testing electronic assemblies. Test facilities may include anechoic chambers and screen rooms containing various types of general-purpose test equipment as well as special purpose testers for circuit boards and components. Component tests may also require environmental test facilities.

Spacecraft assembly and test operations are normally conducted in controlled-cleanliness facilities, which are often high-bay hangars. Spacecraft functional tests are conducted with special purpose test sets. Spacecraft environmental testing requires large vibration and thermal-vacuum equipment.

TABLE 12-5. Facility Cleanliness Requirements (FED STD 209). Class 10,000 means less than 10,000 particles per cubic foot.

Facility/Operation	Cleanliness
Mechanical Manufacturing	Not controlled
Electronic Assembly	Class 10,000
Electromechanical Assembly	Class 100
Inertial Instruments	Class 100
Optical Assembly	Class 100
Spacecraft Assembly and Test	Class 100,000

Many times a spacecraft will need a special facility to protect sensitive equipment or prevent outside interference. Because optical equipment is especially sensitive to contamination, it must be ultra-clean. Payload instruments that require cryogenic temperatures, absence of magnetic fields, or RF isolation call for special facilities which may increase program cost and schedule.

12.3 Inspection and Quality Assurance

Quality assurance verifies that the manufacture and testing of the spacecraft and its components conform to engineering data. MIL Q 9858A describes quality assurance for military programs. Table 12-6 lists its elements: quality program management, facilities and standards, control of purchases, and manufacturing control.

A key element in the quality program is establishing points in the production flow where we can make sure the hardware construction complies with its engineering data before the next step keeps us from inspecting it. In some cases, controlling the process on the production line will ensure the hardware quality. Process inspection and control thus substitutes for direct test.

We must also verify that vendors supplying spacecraft hardware have satisfactory quality assurance programs by certifying their programs and periodically auditing their performance.

Test surveillance involves certifying test equipment and procedures, witnessing tests, approving test records, and reviewing results. Test equipment normally conforms to a reduced set of controlled engineering data such as mechanical assembly drawings, parts lists, wire lists, panel photographs, and calibration test procedures. Test surveillance personnel certify the construction and calibration of the test set. They approve hardware tests, and they prepare and control the test data package. They also formally review test results before the next operation proceeds. Generally, the test conductor chairs the test review board, but test surveillance provides the records and documents the results for the archives.

Quality assurance must maintain data records for formal tests and failures. Often quality-assurance people keep all failed parts and record failure and anomaly results, so they can identify repetitive failures and correct the design weaknesses. Although formal procedures govern tests, troubleshooting may deviate from them. For such operations, quality assurance approves and maintains records of the exact steps involved.

TABLE 12-6.   Elements of *Quality Assurance* (MIL Q 9858A).

• **Management**
– Organization
– Initial quality planning
– Skill requirements, training, personnel certification, records
– Work instructions
Manufacturing inspection and test program
Workmanship
Visual aids
– Records
– Corrective action
– Costs
• **Facilities and Standards**
– Drawings, documentation, and changes
– Measuring and test equipment
– Production tooling and inspection media
– Inspection equipment
– Special metrology
• **Control of Purchases**
• **Manufacturing Control**
– Materials and material control
– Production processing and fabrication
– Completed item inspection and test
– Handling storage and delivery
– Nonconforming material
– Inspection status

12.4 The Qualification Program

Qualification tests of all flight-type hardware and software show that a spacecraft design is suitable. These tests certify that the hardware and software work properly and that the hardware can survive and operate in the prescribed environment. The qualification program must test each component and the complete spacecraft. It may also include functional or environmental testing of selected assemblies or subsystems. Table 12-7 lists ways to qualify a spacecraft design.

We qualification test a component by checking how it performs (*functional test*) and testing its ability to survive the anticipated environment. Each component is powered and exercised by its own special-purpose test set during test. Component environmental tests include vibration, shock and thermal vacuum. If the component must survive nuclear weapons effects, it may also undergo a flash X-ray test, which simulates the prompt radiation dose. Sometimes, it may need a test of electromagnetic compatibility (EMC). Figure 12-1 shows a typical qualification sequence.

TABLE 12-7. Qualification Methods.

Method	Characteristics
Dedicated Qualification Hardware	A separate set of qualification components is constructed and tested at qualification levels. This set of components or a second set of qualification components is assembled into the qualification spacecraft and tested as a spacecraft at qualification levels.
Qualify the First Set of Flight Hardware	The first set of flight components is tested at qualification levels, then assembled into a spacecraft which is tested at qualification levels. This spacecraft is then launched. This is the *proto-flight* concept.
Qualify by Similarity	Demonstrate that the component and the environment are identical to previously qualified hardware.

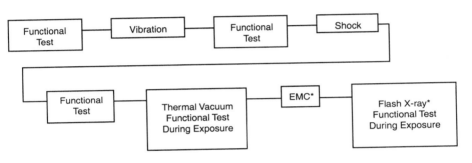

*May not be required

Fig. 12-1. Flow of Qualification Testing for Components. A component is qualified by a series of functional tests and exposure to environmental conditions.

A component must withstand vibration caused when launch vehicle acoustics and engine rumble couple to it through its structural mount. Vibration is a random-signal spectrum of frequencies from 20 Hz to 2,000 Hz. Chapter 18 gives vibration data for current launch vehicles. Figure 12-2 shows the vibration level for the Atlas-Centaur, as well as acceptance and qualification spectra [General Dynamics Space Systems Division, 1988]. The acceptance spectrum envelops the expected environment and is higher than the conducted level specified by the launch-vehicle contractor to account for structural resonances and acoustic input. The qualification spectrum is uniformly 6 dB higher than the acceptance spectrum. To vibrate a component, an electromechanical shaker drives its base at a specified level of acceleration.

Components experience shocks from explosive release devices such as aerodynamic fairing separation or spacecraft separation bolts. The shock pulse is a complex wave which induces mechanical response over a wide band of frequencies. The pulse is specified by the peak acceleration response it excites in a mechanical system with 5% damping—as a function of the mechanical system's resonant frequency. Chapter 18 shows shock-response spectra for various launch phenomena. Figure 12-3 shows the response spectra for acceptance and qualification of an Atlas-Centaur device. We can use an electromechanical shaker to produce a shock pulse for component testing, or we can mount the component on a test structure and hit the structure with a calibrated hammer blow.

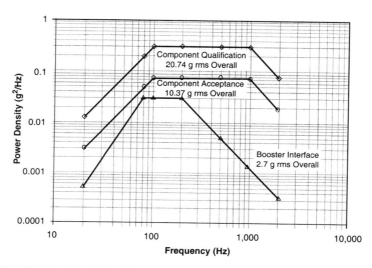

Fig. 12-2. Vibration Levels for Components on the Atlas-Centaur. Component qualification vibration levels are specified by a power density spectrum which is higher than the expected spectrum.

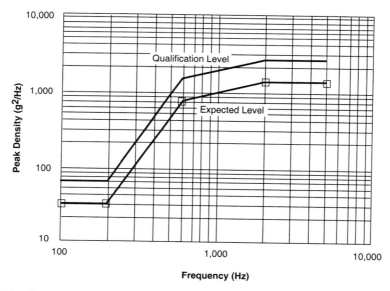

Fig. 12-3. Shock Levels for an Atlas-Centaur Device. Qualification levels are designed to be greater than the expected design levels.

Component temperature requirements derive from the spacecraft's thermal design. To test a component under thermal vacuum, we mount it to a temperature-controlled base plate inside a vacuum chamber. The vacuum chamber's walls have thermal shrouds for cooling or heating. Using conductive coupling to the base plate and radiative coupling to the chamber walls, we can cycle the component through its specified temperature range. The component must work at temperature extremes and during transitions. The temperature extremes equal or slightly exceed expected temperatures for the acceptance test and exceed expected temperatures by a margin (typically 11°C) for the qualification test. Sometimes, the thermal-vacuum qualification test may include a cold soak to demonstrate survival at low temperatures. Figure 12-4 shows a typical temperature cycle.

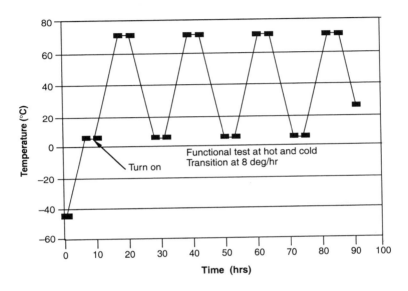

Fig. 12-4.   Typical Temperature Cycle for Thermal-Vacuum Test of Components. The key issues on the thermal-vacuum test are temperature extremes, pressure level and number of cycles necessary.

We do not always know what a spacecraft "component" is. For example, we may define the solar array and individual antennas as components and qualify them separately, or we may test them environmentally as a part of the complete spacecraft. The structure is tested by static loading but receives its environmental exposure as a part of the spacecraft. Mechanical assemblies, such as deployment joints or hinges, may be tested at qualification environmental levels separately as components or as part of the spacecraft.

The qualification program should evaluate each spacecraft function and all environmental effects the spacecraft may encounter. Table 12-8 lists the steps for designing a qualification program. First, we must prepare a list of all spacecraft and payload functions, as well as equipment (including redundancy), all modes of operation (including failure modes and backups), and software code. The system and subsystem specifications should provide this information. Second, we identify environmental effects, including those on the launch vehicle (acoustics, vibration, and shock) and in orbit (temperature, vacuum, and radiation). Third, we check to see which spacecraft functions must be tested with each environment. Fourth, we identify the major spacecraft configurations that require qualification, typically including a boost configuration and one or more on-orbit configurations. Fifth, we devise functional tests for each of these major configurations which completely evaluate the spacecraft. Sixth, we lay out the sequence of tests and environmental exposures. Finally, we must identify program span time and test equipment requirements.

TABLE 12-8. **Steps in the Design of a Spacecraft Qualification Program.** This is a typical approach for a high reliability program.

1. Identify Spacecraft and Payload Functions	Test each spacecraft and payload function for proper operation. Identify the top functional requirements of the spacecraft in the top system specification, and glean subsystem functions from the subsystem specifications.
2. Identify Environments	Environments for transportation and storage, launch, and orbit include vibration, shock, temperature, vacuum, and radiation.
3. Correlate Functions and Environments	During transportation the spacecraft is off, although sensitive components may be powered. During launch, some equipment will be in standby and some will be operating. Test the operating equipment during spacecraft vibration and check all modes of on-orbit operation.
4. Identify Main Configurations	Include boost configuration and one or more orbital configurations.
5. Devise Functional Tests for each Major Configuration	Test each function appropriate to a particular configuration, including all equipment and software.
6. Lay Out the Sequence of Functional Tests and Environmental Exposures	See Fig. 12-5 for a typical sequence.
7. Identify Span Times and Special Facility Requirements	

Some spacecraft functions are best demonstrated by testing a group of components or assemblies under special conditions or with special test equipment or facilities. These *design verification tests* (see Table 12-9) can even be run on nonflight (breadboard or engineering model) equipment. They demonstrate proper equipment function and allow simplified functional testing at the spacecraft level. For instance, a closed-loop tracking test helps us evaluate performance of an antenna pointing control system. During spacecraft test, the antenna feed can be stimulated by RF signals simulating those encountered during the closed-loop test, thus simplifying the test setup and conduct.

TABLE 12-9. **Typical Design Verification Tests.** These tests demonstrate functionality of components, subsystems, and systems.

• **Structural Tests**
  – Static loads test
  – Modal survey
• **Deployment Tests**
  – Solar array
  – Antennas
  – Experiment booms and appendages
	• **Separation Tests**
	• **Antenna Tests**
	  – Pattern tests
	  – Closed-loop tracking tests
	• **Attitude Control Tests**
	  – Closed-loop functional tests

12.5 Spacecraft Qualification Test Flow

The qualification test of a spacecraft is a lengthy and demanding process. Besides proving the design, it may also be the first chance to evaluate the entire spacecraft in operation. Some parts of the qualification test may provide data to validate the engineering design and verify proper interaction of the equipment. Assembling the first unit is particularly difficult because components are seldom available in the best sequence and unexpected interference and test peculiarities always occur. All operations must be planned and executed under control and quality surveillance. Records must be maintained and the results of each integration and test step reviewed before proceeding to the next operation. Typically, tests to qualify a spacecraft last more than a year.

Spacecraft integration normally starts with delivery of an assembled structure or structural frame. Liquid propulsion parts are assembled to the structure first because the fields used to braze them together are not compatible with electronic components. The spacecraft wiring harness can go on before or after the propulsion components.

Electronic assembly of the spacecraft commonly starts with electric power and command and data handling subsystem components. Integration of the remaining subsystems depends more on the particular design. Although the subsystems may have been preassembled and functionally tested, it is more common to integrate components one at a time and test the subsystems after they are integrated on the spacecraft.

To test a spacecraft, a crew normally uses central control and display equipment, with a computer for command generation, telemetry decoding, logging, display, and automated test sequences. They operate the equipment by keyboard and monitor status by video displays. These displays typically consist of an event screen that shows all signal activity and page displays devoted to particular subsystem parameters. The test set also provides stimuli and data-measurement equipment for end-to-end testing of the spacecraft and payload subsystems. A complete spacecraft test may take several days to run and involve subroutines for each subsystem or functional group. Such a test is called an *integrated system test (IST)* or a *comprehensive system test (CST)*. It uses subroutines that test each subsystem, so we need to rerun only the subroutine for a subsystem anomaly. We may also use a short version of this functional test at selected points in the integration flow.

The qualification test sequence normally matches the expected flight sequence: vibration, shock, and thermal vacuum. We also configure the spacecraft to match the operational sequence by folding the solar array and deployables during vibration test and deploying or removing them during thermal vacuum. Often the test facilities will limit the test article's size or the ability to perform particular tests. Figure 12-5 shows a typical sequence along with the special test facilities used [Reeves, 1979]. First, the mechanically integrated spacecraft goes through temperature cycling to verify proper assembly. A comprehensive system test and performance testing of the payload then take place in a special test facility. (For a communications satellite, this might be an anechoic chamber or screen room.) Separate performance testing of the solar arrays precedes their integration with the spacecraft. *Vibration tests* consist of acoustic testing and vibration by low-frequency sine waves. The system must also pass a *pyroshock test*—firing of the deployment ordnance—and a check of the mechanical deployments. Once the deployables are restowed, a thermal vacuum test of the folded configuration follows. After demonstrating the deployments again, the test crew then removes the deployables so the spacecraft will fit into the thermal vacuum (TV) chamber. The spacecraft is then thermal vacuum tested in its orbital configuration. The deployable units are tested in parallel with the spacecraft. An ambient comprehensive system test and mechanical verification series complete the qualification. *Integrated system tests* (shorter functional tests) supplement this sequence between each environmental exposure and during TV test.

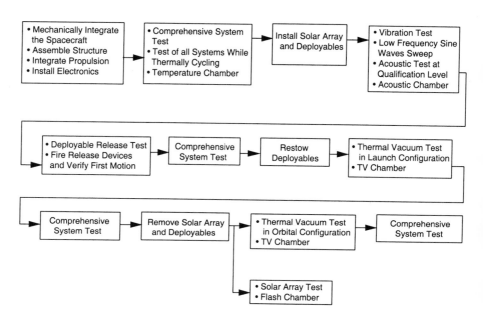

Fig. 12-5. Typical Flow of Qualification Test. Note that system tests are performed after each major activity.

During the qualification sequence, the test crew records all anomalies or out-of-tolerance measurements and formally resolves each discrepancy. Anomalies which result from operator error or malfunctioning test equipment and which do not damage the spacecraft are easy to resolve. But spacecraft malfunctions demand thorough investigation. If design errors have caused the problems, the design must be corrected and retested. Test rules normally require rerun of an integrated system test (or all affected subroutines) if any electrical, pneumatic, or hydraulic lines are disconnected.

12.6 Launch Site Operations

Spacecraft travel either by air or on air-cushioned trailers. Crews record vibration during transport, and spacecraft packaging conforms to the specified environment. Launch-site operations include installing and validating the test equipment (EAGE), testing the spacecraft's performance, installing propulsion (AKM), loading propellant, mating the spacecraft to its launch vehicle, installing ordnance, and monitoring. Crews may also install flight batteries at the launch site. Figure 12-6 shows a typical flow of launch-site activities.

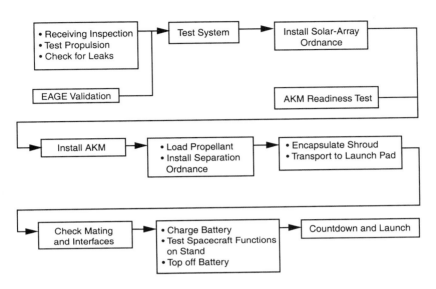

Fig. 12-6.   Typical Launch-Site Activities. The time required for launch site activity may vary from several months to several days.

Normally, one of the launch-site test hangars houses the spacecraft test set for performance testing nearby. Crews install propulsion in an explosive ordnance area. They mate the launch vehicle and spacecraft, load propellant, and install ordnance on stand. Trained people conduct these hazardous tasks, using appropriate equipment and safeguards. Some spacecraft components can be replaced at the launch site and even on stand, but restricted access to the spacecraft usually makes replacement difficult. Much time at the launch site goes simply to monitoring the spacecraft's state of health through hard-line connections or, during some integrated tests, through RF links. Spacecraft commanding is strictly controlled. Launch procedures include configuring the spacecraft for launch and removing hard lines.

References

General Dynamics Space Systems Division. 1988. *Atlas II DoD User's Mission Planning Guide*. San Diego, CA: General Dynamics Space Systems Division.

Reeves, Emery I. 1979. *FLTSATCOM Case Study in Spacecraft Design*. AIAA Professional Study Series. TRW Report 26700-100-054-01.

U.S. Air Force Rome Air Development Center. 1986. MIL Handbook 217E. *Reliability Prediction of Electronic Equipment*. Griffiss AFB, NY: U.S. Air Force Rome Air Development Center.

U.S. Air Force Standardization Office. 1985. MIL Std 1521B. *Technical Reviews and Audits for Systems, Equipments, and Computer Software*. Hanscomb AFB, MA: U.S. Air Force Standardization Office.

U.S. Army Armament Research Development and Engineering Center. 1978. MIL Std 480A. *Configuration Control—Engineering Changes, Deviations, and Waivers.*

———. 1983. DOD Std 100C. *Engineering Drawing Practices.* Picatinny Arsenal, NJ: U.S. Army Armament Research Development and Engineering Center.

———. 1985. MIL Q 9858A. *Quality Program Requirements.*

———. 1985. MIL Std 483A. *Configuration Management Practices for Systems, Equipment, Munitions, and Computer Programs.*

———. 1985. MIL Std 490. *Specification Practices.*

U.S. Army Missile Command. 1987. FED Std 209C. *Federal Standard Clean Room and Workstation Requirements, Controlled Environment.* Redstone Arsenal, AL: U.S. Army Missile Command.

Chapter 13

Communications Architecture

Fred J. Dietrich, *Globalstar L.P.*
Richard S. Davies, *Stanford Telecommunications, Inc.*

A *communications architecture* is the arrangement, or configuration, of satellites and ground stations in a space system, and the network of communication links that transfers information between them. This chapter discusses this arrangement of links, their operation, and their effect on system design. More detailed information on satellite communications is available in Morgan and Gordon [1989] and Sklar [1988].

Table 13-1 lists the steps required to specify the communications architecture. The first step is to define the mission objectives and requirements in enough detail to evaluate and compare alternative architectures. Section 13.1 describes alternative configurations and the criteria used in their selection.

The second step is to determine the data rates for each of the links identified in step 1. To do this it is necessary that the required data throughput accuracy of data transmission be specified, and whether or not there will be data processing on board the satellite. This process is described in Sec. 13.2.

The third step is to design each link in the network, as explained in Sec. 13.3. Principal factors are the availability of a radio frequency spectrum, coverage area of the satellite antenna beam, and path length between satellite and ground station. These

TABLE 13-1. Specifying a Communications Architecture. Evaluating alternative architectures may require designing the links and sizing the communication payload as described in Table 13-9.

Step	Process (= Detailed Step)	Reference

1. Identify Communication Requirements	• Develop mission data flow diagram based on mission requirements
	• Specify:
	  – Data sources, data end users, & locations
	  – Quantity of data per unit time
	  – Access time
	  – Transmission delay
	  – Availability, reliability
		Sec. 2.1
		Chap. 4

2. Specify Alternate Communications Architectures	• Identify links and ground station locations • Consider use of relay satellites and relay ground stations • Determine data processing location	Fig. 13-2 Fig. 13-2 Sec. 2.1
3. Determine Data Rates for Each Link	• Determine sampling rates, quantization levels • Specify bits per sample	Sec. 9.5.1 Sec. 13.2 Table 13-4
4. Design & Size Each Link	• Evaluate alternatives and compare	Tables 13-2,13-9
5. Document Reasons for Selection		

factors in turn determine antenna size and transmitter power—the major cost drivers in sizing the space system. The fourth step, described in Sec. 13.4, provides information to aid the reader in estimating the size and mass of the satellite antennas, and primary power and mass of the satellite transmitters. These parameters are inputs to the spacecraft design process described in Chap. 10, and the ground system design process described in Chap. 15.

13.1 Communications Architecture

A communications architecture is a network of satellites and ground stations interconnected by communication links. The term *ground station* is equivalent to *Earth station*, *ground terminal*, and *Earth terminal*, including land mobile, airborne, and shipborne terminals. All of these names refer to the same thing: the antenna, transmitter, receiver, and control equipment required to communicate with the satellite.

Communication links allow a satellite system to function by carrying tracking, telemetry, and command data or mission data between its elements. Figure 13-1 illustrates the ground station-to-satellite *uplinks*, satellite-to-ground station *downlinks*, and satellite-to-satellite *crosslinks*, or *intersatellite links*, that support a space system. Not shown are additional communication links which may be necessary to transfer data between the ground stations and a mission control center or users. For example, the Air Force's Satellite Control Network uses the DSCS-III communications satellite to relay data between remote tracking stations and the satellite control facility in California.

In space systems, the transmitter and receiver must be in view of each other, using frequencies high enough (above 100 MHz) to easily penetrate the Earth's ionosphere.

A satellite in a nongeostationary orbit is often out of view of its user's ground station. In this case a second satellite, usually in a geostationary orbit, may be used to relay data between the satellite and its ground station. The ground station-to-satellite link is the *forward link*, and the satellite-to-ground station link is the *return link*. As shown in the figure, both the forward and return links contain uplinks, downlinks, and crosslinks.

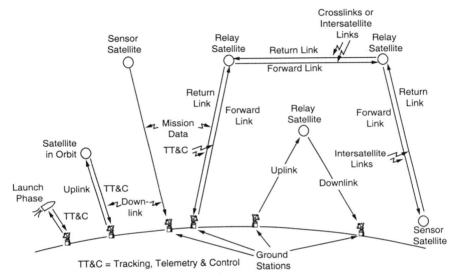

Fig. 13-1. **The Communications Architecture Consists of Satellites and Ground Stations Interconnected with Communications Links.**

Since 1990, several new classes of satellites in the orbital category of low-Earth orbit (LEO) and medium-Earth orbit (MEO) have been developed for various commercial uses. They will be used for both satellite-based cellular telephone service as well as for wideband data trunking (fiber optic type service where there is no fiber) and data distribution for Internet-type services.

In these new systems, the terms *forward link* and *return link* sometimes take on new definitions. As in terrestrial cellular, the *forward link* for a satellite-based cellular system is from the ground station (gateway) to the user terminal (UT) or radio-telephone, and the *return link* is from the radiotelephone to the ground station. In this terminology, both the forward and return links contain uplinks and downlinks to and from the satellite.

A constellation at an altitude of 1,000–1,500 km typically requires from 48 to 64 satellites to provide complete Earth coverage. A constellation at 10,000 km typically requires 12 satellites.

13.1.1 Communications Architecture Defined by Satellite-Ground Station Geometry

The geometry formed by satellite orbits and ground stations determines the basic communications architectures illustrated in Fig. 13-2. Table 13-2 lists the principal advantages and disadvantages of each.

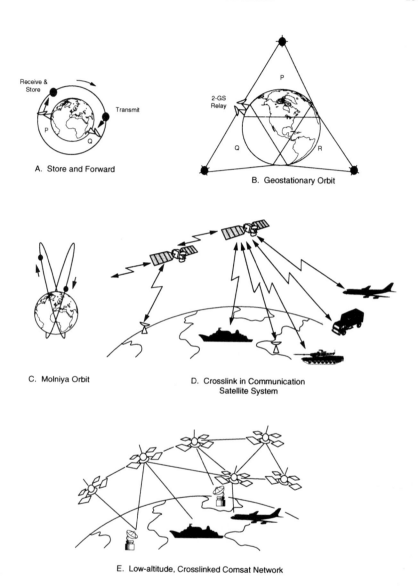

Fig. 13-2. Typical Communications Architectures Used to Satisfy Different Mission Requirements. Table 13-2 summarizes the characteristics of each architecture.

Store and Forward (Fig. 13-2A): The architecture for relaying communications by satellite appeared in 1960, when the U.S. Army launched the Courier satellite [Mottley, 1960]. In this configuration, the satellite orbits at low altitude (under 1,000 km), receives data, and stores it in memory. When it moves in view of a receiver ground station, the satellite transmits the stored data. This architecture permits the use of a low-cost launch vehicle due to the low-altitude orbit. The satellite cost is also lower

TABLE 13-2. Comparison of Five Example Communications Architecture.

Architecture	Advantages	Disadvantages
A. *Low-Altitude, Single Satellite, Store & Forward*	• Low-cost launch • Low-cost satellite • Polar coverage with inclined orbit	• Long message access time and transmission delay (up to several hours)
B. *Geostationary Orbit*	• No switching between satellites • Ground station antenna tracking often not required	• High-cost launch • High-cost satellite • Need for stationkeeping • Propagation delay • No coverage of polar regions
C. *Molniya Orbit*	• Provides coverage of polar region • Low-cost launch per satellite	• Requires several satellites for continuous coverage of one hemisphere • Need for ground station antenna pointing and satellite handover • Network control more complex • Need for stationkeeping

D. *Geostationary Orbit with Crosslink*	• Communication over greater distance without intermediate ground-station relay
	• Reduced propagation delay
	• No ground stations in foreign territory:
	   – Increased security
	   – Reduced cost
		• Higher satellite complexity and cost
		• Need for stationkeeping
		• Relay satellite and launch costs
		• No coverage of polar regions

E. *Low-Altitude Multiple Satellites with Crosslinks*	• Highly survivable—multiple paths • Reduced jamming susceptibility due to limited Earth view area • Reduced transmitter power due to low altitude • Low-cost launch per satellite • Polar coverage with inclined orbit	• Complex link acquisition ground station (antenna pointing, frequency, time) • Complex dynamic network control • Many satellites required for high link availability

due to the wider antenna beamwidth required to illuminate the Earth, which reduces the satellite antenna size and stabilization requirement. Usually satellite stationkeeping is not required. The principal disadvantage of this architecture is its long access time and transmission delay, perhaps hours, waiting for the satellite to pass into view of the user ground station.

There are several commercial systems planning to use store-and-forward communication for very low-cost service. They include ORBCOMM, Starsys, Vita, LEO-1, FAI, and ESat. They typically operate in the VHF portion of the radio spectrum at very low bandwidths. These are sometimes referred to as "little LEOs."

Geostationary Orbit (Fig. 13-2B): Virtually all communication relay satellite systems and many meteorological satellites use this architecture, in which the satellite is placed in a near-zero deg inclination orbit at 35,786 km altitude. The period of the orbit is exactly equal to the period of the Earth's rotation, making the satellite appear stationary when viewed from the ground (see Sec. 6.1). The cost of ground stations is usually less for this architecture because little or no antenna pointing control is required. A stationary network is far easier to set up, monitor, and control compared to a dynamic network containing nonstationary (relative to Earth) satellites. There is no need to switch from one satellite to another, for the satellite is always in view of the ground station. Principal disadvantages are lack of coverage above 70-deg latitude and the high launch cost. Furthermore, the delay time for propagation to and from the synchronous orbit is about 0.25 sec, which sometimes causes problems (echoes, acknowledgment protocols) in communications satellite systems.

Molniya Orbit (Fig. 13-2C): The Russian space program uses this architecture to cover the northern polar regions. The satellites are in highly elliptical orbits with an apogee of 40,000 km, a perigee of 500 km, and an inclination angle of 63.4 deg (see Sec. 6.1). The apogee is over the North Pole to cover northern latitudes. The period of the orbit is 12 hr, but because it is highly elliptical, the satellite spends about 8 hr of each period over the northern hemisphere. Two or more satellites orbit in different planes, phased so that at least one is always in view from all northern latitudes. Unfortunately, the Molniya orbit requires continuous changing of antenna pointing angles at the ground station and switching links between satellites as they move in and out of view.

Geostationary Orbit with Crosslink (Fig. 13-2D): When a geostationary satellite is beyond line-of-sight of a ground station, a second geostationary satellite relays data between it and the station. A relay satellite is better than a double-hop link using two adjacent ground stations as a relay (shown in Fig. 13-2B), because the relay ground stations must often be on foreign territory, which is more costly, less secure, and less survivable. The obvious disadvantage of this architecture is the added relay satellite and its crosslink, which increase the system's complexity, risk, and cost.

Low Altitude (Fig. 13-2E): This architecture places 20 or more satellites in low-altitude (500 to 3,000 km) orbits and sometimes connects them with crosslinks. The system divides digital messages into packets of a few hundred or thousand bits, labels each packet with time-of-day and its destination, and then transmits it in a short burst. Packets may arrive by different paths with different propagation times, depending on the satellite-ground station geometry at the time of transmission. The receiving station must sort and reassemble the packets in the correct order to obtain the original message. Because so many alternate paths are available, the system is highly survivable. The low-altitude orbit also improves immunity to jamming from the ground since the satellite sees a smaller Earth area. Finally, the uplink transmitter power is lower due to the shorter distance between ground station and satellite, making unauthorized reception less probable. On the other hand, this architecture needs complex network synchronizing and control functions because of relative motion between satellites and ground stations. Without stationkeeping, the satellites may drift together in bunches, leaving gaps in the coverage which lead to significant link outages.

Note: The satellite-based cellular telephone service mentioned in paragraph 13.1 and Table 13-2 and below do not use packet transmission as described above, but typically CDMA or TDMA waveforms.

There are several classes of nongeostationary, low- and medium-altitude satellites which have begun to appear since about 1990. These are:

- *Low-Earth Orbit (LEO) Cellular Satellites.* These are the systems such as Iridium, Globalstar, and Ellipso which provide cellular telephone-type service to hand-held and mobile phones from the low-Earth orbits.

- *Little Low-Earth Orbit (Little LEO) Satellites.* These satellites, including Starsys, Vita, LEO-1, FAI, and ESat, are commercial, low-powered, low-orbit satellites intended for store-and-forward communication at a very low price at VHF frequencies.

- *Medium-Orbit (MEO) Satellites.* These include ICO, Star Lynx, Pentriad, Ellipsat, and TRW Global EHF Satellite Network. They provide cellular as well as high-speed data transfer service.

- *Low-Earth Orbit (LEO) Data Satellites (also called "Big LEOs").* These include Teledesic, Boeing's Aeronautical Radio Navigation, Globalstar GS-40, and Iridium Macrocell. These satellites will provide digital data trunking as well as Internet access to many users.

Other architectures may meet particular mission requirements. For example, Chapell [1987] suggests a hybrid system which contains satellites at both low and high altitudes, thus combining their advantages. Lee [1988] compares three architectures for providing regional communications satellite service to the continental United States. These architectures contain four geostationary satellites, which are either isolated, interconnected by double-hop ground stations, or interconnected by crosslinks.

Some systems combine geostationary and low-Earth orbit satellites. Alcatel's Sky-bridge combined with Loral's Cyberstar and Globalstar's GS-2 systems both combine low-Earth orbit satellites for primary customer connection and geostationary satellites to act as relays between LEOs, as well as connecting to some customers. The service is typically data services such as Internet connectivity. Intersatellite links are widely used both between low-Earth orbit satellites and between low-Earth orbit and geosta-tionary satellites in order to perform the required communication functions.

13.1.2 Communications Architecture Defined by Function

Three types of communications architectures, shown in Fig. 13-3, are tracking, telemetry, and command; data collection; and data relay. A point-to-point network is used to provide a link to a single ground station. The broadcast architecture transmits data to multiple ground stations located in different areas. This architecture requires either a broad-beam satellite antenna, a narrow-beam antenna rapidly switching between ground stations, or a multibeam antenna (see Sec. 13.4). In communications-satellite applications, the satellite network control is often part of a larger telecommunications network containing thousands or millions of users. Thus the satellites may be shared among many nonsimultaneous users, using multiple-access techniques described in Sec. 13.5 (see Chap. 2 of Morgan and Gordon [1989]).

The satellite communications architecture may be affected by data processing on board the satellite. Usually data collected by a satellite is transmitted directly to the user via the downlink. However, the data often requires processing to make it useful. If the system contains many ground stations, processing the data in the satellite and then transmitting it may be more economical than having each user process it separ-ately. This processing often reduces the data rate, leading to further cost savings (see Sec.13.2). A satellite system using onboard processing is Europe's Meteosat, which stores and formats cloud-scanner data before transmitting it to several ground stations at a reduced data rate.

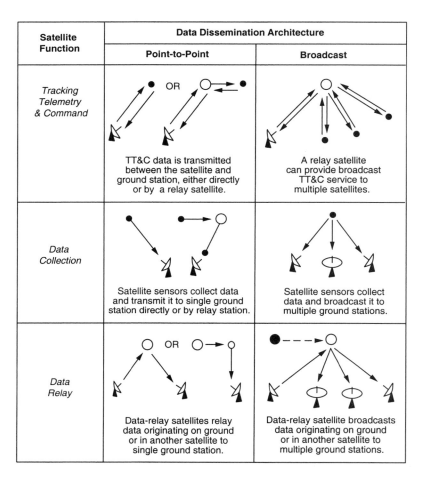

Satellite Function	Data Dissemination Architecture	
	Point-to-Point	Broadcast
Tracking Telemetry & Command	TT&C data is transmitted between the satellite and ground station, either directly or by a relay satellite.	A relay satellite can provide broadcast TT&C service to multiple satellites.
Data Collection	Satellite sensors collect data and transmit it to single ground station directly or by relay station.	Satellite sensors collect data and broadcast it to multiple ground stations.
Data Relay	Data-relay satellites relay data originating on ground or in another satellite to single ground station.	Data-relay satellite broadcasts data originating on ground or in another satellite to multiple ground stations.

Fig. 13-3. Communications Architectures may be Defined by the Function Performed.
Open circles represent a relay satellite.

Operators at ground (air, ship) stations usually control the mission in (near) real time by transmitting commands to the satellite. When the satellite is not in view of the ground-control station during part of its orbit, commands previously received and stored in the satellite are executed by an onboard timer. The advantages of this approach are flexibility to changing requirements, greater reliability, and a less complex, lower-cost satellite. The disadvantages are vulnerability to human error or failure of the ground control facility. Costs for the ground control segment, especially operations, can be high.

On the other hand, the satellite itself can control a mission by using onboard data-sensing and programmed decision-making processes. This arrangement replaces ground control, is highly survivable, has fast response time (communication link delays eliminated), excludes errors introduced by human operators, and reduces ground equipment and operations cost. But it is less responsive to changing or unanticipated requirements, and the satellite itself is more complex, more costly, and potentially less reliable. Even when using an autonomous control architecture, a ground station is generally required to collect data from the spacecraft and to serve as a backup to the onboard control system.

Usually, ground stations control unmanned satellites to simplify the satellite design. In the future we expect more functions, such as stationkeeping, to be performed in the satellite to reduce the dependence on the ground station control (see Sec. 16.1).

During the operation of a satellite system, the communication links may need to be reconfigured, or its parameters, such as power or bandwidth, adjusted to accommodate a change in requirements. The process for doing this is called network control. Communications architectures may require a number of control functions (Table 13-3). Early satellites, such as Sputnik, did not need these functions because their systems used only one satellite, a single satellite-to-ground link, and a broadbeam antenna. On the other hand, a communications satellite system such as the NASA ACTS [Naderi and Kelly, 1988] contains many narrowbeam satellite antennas with demodulators and switching circuits. This architecture requires a sophisticated system for network control. Network control can be centralized using a single ground station or satellite, or distributed with multiple ground stations or satellites. Distributed configurations use a control hierarchy, or set of priorities, to avoid conflicts. Distributed control makes the network less vulnerable to failure of a single control element (see Chap. 14 for further details).

TABLE 13-3. Network Control Functions.

Function	Example
Resource Allocation	Frequency channel, bandwidth assignment Time slot assignment Date rate assignment Modulation/coding assignment Antenna beam assignment Transmitter power control Crosslink assignment
Link Acquisition	Antenna pointing Frequency acquisition Time acquisition Acknowledgment protocols Crypto synchronization
Performance Monitoring and Redundancy Switching	Spectrum analysis Signal-to-noise ratio reduction (due to rain, etc.) Interference identification Fault identification Redundancy switching Reallocation of resources
Time/Frequency Standard	Provide universal time
Tracking, Telemetry, and Command (TT&C)	Range, range-rate measurement Command signal formatting, verification, execution Telemetry signal demultiplexing, processing, display
Stationkeeping	Ephemeris prediction Thrust control

13.1.3 Criteria for Selecting Communications Architecture

Individual users will assign different priorities to the criteria for selecting a communications architecture. For example, a commercial company will try to reduce cost and risk, but the military may make survivability the top priority. The factors which affect the criteria are explained below:

Orbit: The satellite orbit determines how much time the satellite is in view by the ground station and the potential need for intersatellite links. The satellite altitude determines the Earth coverage, and the satellite orbit determines the delay between passes over a specified ground station. Together, orbit and altitude set the number of satellites needed for a specified continuity of coverage (see Sec. 7.2). Transmitter power and antenna size depend on the distance between the satellites and the ground stations (see Sec. 13.3). Satellite view time determines the signal-acquisition and mission-control complexity (see Chap. 14).

In the satellite-cellular systems described above, intersatellite links are not necessarily used. Instead, the constellation is designed so that at least one satellite is in view by the gateway and every user at all times, so that there are no "outages." Coverage is determined by the number of satellites, the inclination of their orbits, the latitude of the gateway and user, and the number of gateways located around the world, if intersatellite links are not used.

If intersatellite links are used, then the number of gateways and their location becomes much less critical, as many satellites can connect to a single gateway through intersatellite links. Various systems proceeding now have used different philosophies with respect to intersatellite links, which can have great effect on the capital cost of the system. Intersatellite links make the satellites more expensive, but eliminate the need for many fairly expensive ground stations (gateways), for example.

There are many systems proposed in various frequency bands which use not only the geostationary orbit, the low-Earth orbit discussed above, and also what is called a medium-Earth orbit (MEO), which ranges in altitude from about 10,000 to 20,000 km. These are typically inclined with respect to the equator as the LEOs are, and can address users with small, hand-held UTs, but can see a much larger portion of the Earth at one time, so that only 10 or 12 of them are required to give nearly complete Earth coverage.

RF Spectrum: The RF carrier frequency affects the satellite and ground station transmitter power, antenna size and beamwidth, and requirements for satellite stabilization. In turn, these factors affect satellite size, mass, and complexity. The carrier frequency also determines the transmitter power needed to overcome rain attenuation (see Sec. 13.3). Finally, it is necessary to apply for and receive permission to use an assigned frequency from a regulatory agency such as the International Telecommunication Union, the Federal Communications Commission, or the Department of Defense's Interdepartmental Radio Advisory Committee, and every nation's regulatory agency. These agencies also allocate orbit slots for geostationary satellites (Chap. 21).

Data Rate: The data rate is proportional to the quantity of information per unit time transferred between the satellite and ground station (see Sec. 13.2). The higher the data rate, the larger the transmitter power and antenna size required (Sec. 13.3). Processing the spacecraft-generated data on board the satellite reduces the data rate without losing essential information, but makes the satellite more complex (see Sec. 13.2).

Duty Factor: The fraction of time needed for operation of a satellite link is the *duty factor*, which is a function of the mission and the satellite orbit. A low duty factor enables a single ground station to support more than one satellite (usually the case for telemetry and command). Alternatively, several users may share a single satellite link (see Sec. 13.5).

In the case of LEOs used for cellular service, one gateway will typically have several antennas communicating simultaneously with several satellites, each of which may be carrying 1,000 or more individual circuits. In this case, the ground station duty factor will be nearly 100% as antennas switch from satellite to satellite; the UT use factor will be quite small, however, as is the use of a telephone.

Link Availability: *Link availability* is the time the link is available to the user divided by the total time that it theoretically could be available. It depends on equipment reliability, use of redundant equipment, time required to repair equipment, outages caused by rain, and use of alternate links. Typical goals for link availability range from 0.99 to 0.9999, the latter value applying to commercial telephone networks. (See Chap. 19 for a discussion of reliability.)

Link Access Time: The maximum allowable *link access time*, or time users have to wait before they get their link, depends on the mission. For example, we usually demand access to a voice circuit in seconds. Meteorological data is needed in less than an hour to be useful in weather forecasting. On the other hand, X-ray data from a scientific satellite can be stored and transmitted later. Tracking, telemetry, and command links are often required in near real-time (a few seconds), especially if a problem requires an immediate response from the satellite-control operator. Link access time depends strongly on orbit selection, which determines when a satellite is in view of the ground station. Note that a real-time response is impossible for deep space missions, because the radio propagation time is minutes or hours long.

Threat: Various kinds of threats may influence system design. For military applications, choices of frequency, antenna, modulation, and link margin need to be evaluated for susceptibility to jamming. At the same time, a high-altitude nuclear detonation can disturb the propagation of radio signals. A physical threat to the satellite might dictate multiple satellites or a hardened design (see Chap. 8). A physical threat to a ground station might demand a data-relay satellite with crosslinks to allow the ground station to be relocated in safe territory.

The FireSat sample mission uses low-altitude satellites with limited coverage. If a ground station is near the forest area under surveillance, a store-and-forward or crosslink architecture is not required. The communications architecture is then simply a single satellite operating when in view of its ground station. A separate ground station is required for each major area under surveillance.

13.2 Data Rates

In designing a communications architecture for space missions, we must ask: what is the information to be transferred over our communication links? How fast must the transfer rate be? Keeping in mind that higher rates of data transmission mean higher system costs, we need to decide how we will transfer information to the user.

Satellite links originally used analog modulation techniques to apply the data onto the RF carrier for transmission over the link. Since 1980, however, most space-ground communication links use digital modulation. To implement a digital system, we must first sample the amplitude of the analog signal at a rate equal to at least twice the highest frequency in the signal spectrum, f_m. In 1928 Nyquist[*] showed that if we meet this condition we can theoretically reconstruct the original analog signal from the samples (see Sklar [1988], Sec. 2.4). For example, the normal human voice has a frequency spectrum range of about 3.5 kHz. Thus, to reproduce it digitally, the sampling rate must be at least 7,000 samples/sec. However, practical considerations, such as realizable filter limitations, suggest that the sampling frequency should be at least 2.2 times the maximum input frequency [Sklar, 1988]:

$$f_s \geq 2.2 f_m \qquad (13\text{-}1)$$

Using Eq. (13-1), our 3.5-kHz voice signal must be sampled at a rate of 7.7 ksamples/s. In fact, the sampling rate of commercial digitized voice systems is 8 ksamples/s. Another example is the sampling rate of the audio compact disc player which is 44.1 ksamples/s—about 2.2 times 20 kHz, the maximum source frequency of interest for high-quality music.

The analog amplitude sample is next converted to a *digitized word* composed of a series of *bits*. Consider the *analog-to-digital converter* process illustrated in Fig. 13-4, where three bits designate one of eight amplitude levels. For example, a 6.3 V amplitude converts to a 3-bit word—110. At the receiver, a *digital-to-analog converter* converts this word according to the algorithm $2^2 + 2^1 + 0^0 + 0.5 = 6.5$ V, leaving a *quantization error* of 0.2 V. This quantization error can be reduced by increasing the number of bits in the word.

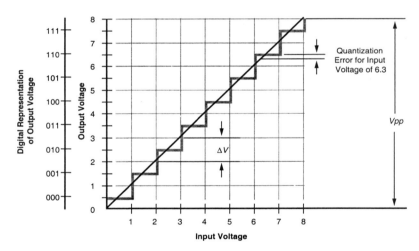

Fig. 13-4. Example of Analog-to-Digital Conversion for 8 Quantization Levels (3 bits).

The analog-to-digital converter divides the total amplitude range of the input sample into M quantizing levels, where $M = 2^n$, and n is the number of bits per sample. Assuming a uniform spacing of quantization levels, the maximum quantization error is $\pm 0.5\ \Delta V$ where ΔV is the *quantization step size*, equal to $V_{pp}\ /\ M = V_{pp}\ /\ 2^n$, where V_{pp} is the full-scale input signal voltage.

---

* Harry Nyquist set the stage for modern digital communications with his classic paper, "Certain Topics on Telegraph Transmission Theory," published in the *Transactions of American Institute of Electrical Engineers* in 1928 (vol. 47, pp. 617–644).

It can be shown [Panter, 1965] that the mean-squared noise power due to quantization is $(\Delta V)^2/12$. The *signal-to-quantization noise power ratio* is $(M^2 - 1)$, which is approximately equal to M^2 or 2^{2n}, assuming a uniform distribution of the input signal amplitudes over V_{pp}. See Table 13-4.*

TABLE 13-4. Required Bits Per Sample. The number of bits per sample is determined by the maximum quantization error and quantization noise allowed.

Number of Bits Per Sample	Maximum Quantization Error (%)	Signal Power to Quantization Noise Power Ratio* (dB)
3	6.25	18
4	3.13	24
5	1.56	30
6	0.79	36
7	0.39	42
8	0.20	48
9	0.10	54
10	0.05	60
11	0.02	66
12	0.01	72

* Assumes signal amplitudes and quantization errors are uniformly distributed.

The effect of quantization noise may be reduced by varying the size of the quantization steps, without adding to their number, so as to provide smaller steps for weaker signals. For a given number of quantization levels, coarser quantization is applied near the peak of large signals, where the larger absolute errors are tolerable because they are small compared to the larger signal amplitudes. These techniques are commonly applied to voice transmission, where the speech waveform is compressed at the transmitter end prior to digitization, and expanded at the receiver. Taken together, the compressor and expander are called a *compandor*.

The number of bits per sample is determined by the mission requirements. The Jet Propulsion Laboratory typically uses 256 shades of gray (i.e., $n = 8$) in differentiating voltages from its imaging sensors, and 32 ($n = 5$) is normally used for low-definition, black-and-white TV transmission. The length of the digital word selected depends on the precision one requires for the application. For example, we do not need this precision for satellite equipment temperatures, whereas scientific measurement may require a highly precise data link, if justified by the sensor accuracy.

The *data rate* is the number of samples per second times the bits per sample, or the number of bits per second, abbreviated *bps*. Thus a digitized voice circuit in a commercial telephone network requires 8 ksamples/s times 8 bits/sample for a data rate of 64 kbps. Table 13-5 lists other examples.

The cellular-satellite systems are using methods for modulation, demodulation, and forward error correction that greatly reduce the data rates needed for transmission. For example, the Globalstar system uses code division multiple access with Reed-Solomon coding for forward error correction and Viterbi decoding. It sends voice through a variable-rate encoder/decoder at an average rate of 9.6 kb/s, and can reduce to a 4.8, 2.4 or 1.2 kb/s rate under stress. These rates are much less than those discussed above, and in Table 13-5, and represent new methods of coding digital data. With these

---

* In dB, the signal-to-quantization noise ratio is $6n$. See Sec. 13.3 for the definition of dB.

TABLE 13-5. Bit Rate Required to Transmit Analog Information Over a Digital Communication Link. The bit rate of digitized, or *pulse code modulated* (*PCM*), voice can be reduced by transmitting only the changes in amplitude between consecutive samples. This technique is known as *Delta PCM*.

Analog Information	Max Input Freq., f_m (Hz)	Sampling Frequency (Samples/s)	Number Bits Per Sample n	Data Rate R (bps)
Voice (PCM)	3,600	8,000	7	64 k*
Voice (Delta PCM)	3,600	8,000	6	56 k*
Cellular Voice	4,800	4,800	1	4.8 k
DS1 Multiplexer 24 Voice Channels	—	—	—	1,544 M
Original Picturephone®	900 k	2 M	3	6 M
Color Television (commercial quality)	4.0 M	8.8 M	5	44 M
Color Television (broadcast quality)	4.2 M	9.25 M	10	92.5 M

* After 1 bit per sample added for signalling and supervision.

new codes it is possible to achieve a *bit error rate* (BER) of about 10^{-10} at a *bit energy to incremental noise* (E_b / N_o) ratio of only 5 dB.

Digital communication techniques are used instead of analog for a number of reasons. First, digital signals can more precisely transmit the data because they are less susceptible to distortion and interference. Second, digital signals can be easily regenerated so that noise and disturbances do not accumulate in transmission through communication relays. Third, digital links can have extremely low error rates and high fidelity through error detection and correction. Also, multiple streams of digital signals can be easily multiplexed as a single serial-bit stream onto a single RF carrier. Other advantages are easier communication-link security and implementation by drift-free miniature, low-power hardware, including microprocessors, digital switching, and large scale integrated circuit chips. In this chapter we will consider only digital communications.

Using the formulas developed in Chap. 5, we can easily determine the relationship between the quantity of data, D, the data rate, R, and the parameters for a single ground station pass from Sec. 5.3.1. Specifically,

$$D = R(FT_{max} - T_{initiate}) / M \tag{13-2}$$

$$F = (1 / \lambda_{max}) \quad acos(cos\lambda_{max} / \cos \lambda_{min}) \tag{13-3}$$

where T_{max} is the maximum time in view (i.e., the pass duration when the satellite passes directly overhead) from Eq. (5-52), F is the fractional reduction in viewing time due to passing at an Earth central angle λ_{min} away from the ground station, λ_{max} is the maximum Earth central angle from Eq. (5-36), $T_{initiate}$ is the time required to initiate a communications pass, and M is the margin needed to account for missed passes due to ground station down time, sharing of ground resources, transmission of other data, or conflicts on board the satellite or within the communications process. A reasonable value for $T_{initiate}$ is about 2 minutes. M is conservatively estimated at a value of 2 to 3 unless it is a dedicated ground station with a specified value for the percentage of pass time that will be used for collecting data. For the fraction of time in view, we may wish to use mean values rather than one for a specific ground station pass in Eq. (13-3). As discussed in Sec. 5.3.1, the average value of F is about 80% for satellites in a circular low-Earth orbit, and 86% or more of all passes will have F greater than 0.5.

With this background on digital techniques, we now consider the data rate requirements for the three types of architectures discussed in the previous section: telemetry, tracking, and command (TT&C); data collection; and data relay.

13.2.1 TT&C

The number and accuracy of functions being monitored in the satellite determines the telemetry data rate. Several hundred functions such as voltages, temperatures, and accelerations may require monitoring to determine if all satellite subsystems are operating correctly, and, if not, to determine where a failure occurred. Sampling each telemetry sensor in sequence with a *multiplexer* combines all telemetry data into a single bit stream. The sampling rate is usually low, perhaps once every second or once every 10 sec, because the monitored parameters vary slowly. For example, suppose we want to monitor 50 temperature sensors and 50 voltages once every 10 sec with an accuracy of 1.5%. The data rate required is 100 samples per 10 sec times 5 bits per sample, or 50 bps. Some applications require precise time or amplitude resolution of the data. In these cases, the data may be transmitted in analog form by frequency modulation of one or more subcarriers [Morgan and Gordon, 1989].

The rate needed to transmit commands to a satellite is usually quite low—perhaps only one per second. A command message may be 48 to 64 bits long, consisting of a synchronizing preamble (a set series of bits), an address word that routes the command to its satellite destination, the command itself (often a single on-off digit), and some error detection bits to make sure the command was correctly received. Some commands can cause irreversible functions or damage the satellite if performed at the wrong time. These commands are usually first transmitted and stored in the satellite. Correct reception by the satellite is verified by telemetry, after which a second command is transmitted to execute the function. If the command is to be executed later when the satellite is out of the ground station's view, a time of execution is added to the command word and stored in the satellite. The command is executed later when the time contained in the command word coincides with the satellite's clock time.

To track a satellite, the ground station measures range or range rate for computing and updating the orbit ephemeris. For example, the Air Force adds a one Mbps pseudorandom (*PN*) code to the command link. The satellite command receiver extracts this code. It is then retransmitted as part of the telemetry downlink signal. The ground station measures the arrival time of the code relative to its uplink transmission time to determine the round-trip delay, from which the range is computed. NASA's Goddard Range and Range Rate system operates the same way except it uses several harmonically related sinusoidal tones plus a pseudorandom code. Intelsat uses only four ranging tones.

In most cases we would want to use an existing TT&C ground station network. Table 13-6 summarizes the key parameters of four networks (see Chaps. 11 and 15 for additional details). The ratio of downlink-to-uplink frequencies listed in the table applies when the satellite transmitter is phased-locked to the received uplink carrier. This mode allows the Doppler frequency shift of the RF carrier to be accurately measured at the ground station to determine the range rate. The United States has NASA's Deep Space Network and the Air Force's Satellite Control Network. Intelsat and other communications satellite operators use their own TT&C system, which eliminates the need to pay for the services of a larger network. The TT&C requirements for FireSat are quite modest and can easily be handled by its own system, except during the launch phase.

TABLE 13-6. Parameters of Existing Satellite TT&C Systems.

Network	Command (Uplink)		Telemetry (Downlink)		DL/UL Carrier Freq. Ratio	Range Measurement
	Freq. (GHz)	Data Rate (bps)	Freq. (GHz)	Data Rate (bps)		
Air Force SCN (SGLS)	1.76–1.84	1,000 2,000	2.2–2.3	125– 1.024M	256/205	1 Mbps PN Code
NASA DSN	2.025–2.120 7.145–7.190	1.0–2,000	2.2–2.3 8.4–8.5	8–6.6M	240/221 749/880	PN Code at 1 Mbps + 8 Ranging Tones, 8 Hz to 500 kHz
Intelsat/ COMSAT	5.92–6.42 14.0–14.5	100–250 100–250	3.9–4.2 12.2 or17.7	1,000–4,800 1,000–4,800	Not applicable	4 Ranging Tones: 27.777 kHz, 3,968.25 Hz, 283.477 Hz, 35.431 Hz
TDRS*** (user satellite altitude below 12,000 km)	*MA S-Band 2.1064 **SA S-Band 2.025–2.120 **SA K-Band 13.775	10 kbps max 300k max 25M max	MA S-Band 2.2875 SA S-Band 2.2–2.3 SA K-Band 15.0034	1k to 1.5M 1k to 12M 1k to 300M	(S) 240/221 (S) 240/221 (K)1,600/1,469	3 Mbps PN Code
*MA—Multiple Access, up to 20 users simultaneously **SA—Single Access ***Frequencies to and from user satellite						

Also shown in Table 13-6 is NASA's Tracking & Data Relay Satellite (*TDRS*), which provides an alternative to ground stations for supporting the TT&C link [Yuen, 1983]. The orbit of the user satellite must be below synchronous altitude to be in view of the TDRS antennas.

13.2.2 Data Collection

In the second type of network, a satellite sensor, such as an optical or radar scanner, collects data. This data is transmitted to the ground station for processing and viewing by the user. The pictures of global cloud cover on the evening television news come from a satellite sensor. Although we could include the sensor data as part of the telemetry data discussed above, we usually consider the sensor data separately when the sensor data rate is greater than 100 kbps or so. Data rate requirements for payload sensors are discussed in Sec. 9.5.5.

Table 13-7 lists the data rates for two satellite-sensor configurations. One example is a geostationary satellite with a radiometer which scans the entire Earth in 20 min with 1 km resolution. Here the data rate is 1.42 Mbps, similar to a meteorological satellite such as GOES.

On the other hand, using Eq. (9-23), we calculate the FireSat data rate to be 85 Mbps—too high for any practical, cost-effective system. Let us review the FireSat mission requirements. First, we need to scan 150,000 acres (about 25 km by 25 km) in 4 min with 30 m resolution. By limiting the sensor coverage to the 150,000-acre area under surveillance as the satellite passes overhead, the sensor output rate would be only 2,900 pixels per second,[*] for a data rate of 39 kbps (for 1.6 samples per pixel, 8 bits per sample and $q = 0.95$). This rate is far more attractive for practical designs. We could also use some form of data processing on board the satellite, as discussed below.

TABLE 13-7. Data Requirement for Two Example Sensor-Satellite Systems. For FireSat the satellite ground track velocity provides scanning of the ground.

Parameter	FireSat	Geostationary
h = Orbit altitude (km)	700	35,786
v = Ground track velocity (m/s)	6,800	0
d = Ground resolution (m)	30	1,000
S_w = Scan width (km)	2,700	Earth coverage
w = Scan width (deg)	57.9	18
z = In-track scan (deg)	Continuous	18
T = Scan time	145 μs/scan width	20 min/Earth image
s = Number samples/pixel	1.6	1.6
n = Number bits/sample	8	8
q = Frame efficiency	0.95	0.95
R = Date rate – bps	85×10^6	1.42×10^6

In many cases we do not need all data collected by the satellite's sensor. For example, our FireSat data is of no interest unless the sensor observes heat from a forest fire. Onboard data processing can be used to dramatically reduce the required data rates. For FireSat, onboard processing would consist of selecting and transmitting only those pixels receiving thermal energy above a specified temperature. The amount of data-rate reduction depends on the portion of the observed area that is burning. We must also insert extra bits to identify the position of the pixels or groups of pixels in the scan. Such data processing can reduce the data rate by a factor of 3 to 10 or more, depending on the nature of the data.

Another technique for reducing the data rate is to transmit only the *changes* in the amplitude of the data samples. For example, the amplitude of the first pixel in a frame of data could be transmitted by an 8-bit word. Changes in amplitude of subsequent pixels, relative to the previous pixel amplitude, are then transmitted as 3-bit words, thus reducing the data rate by 3/8.

Considerable effort has gone into reducing, or compressing, the data rate of a digitized voice channel [O'Shaughnessy, 1987]. One technique is *Adaptive Differential Pulse Code Modulation*, which transmits the difference between the actual voice sample and a predicted value based on several previous samples. Data rates have been reduced from 64 to 32 or 16 kbps using these techniques while maintaining commercial toll-quality voice. Even greater reduction in data rates have been achieved with *Vocoders and Linear Predictive Coders*. With this method, receivers use transmitted spectral or excitation parameters to control a voice synthesizer. It requires a data rate of only 600 bps to 2,400 bps, but the voice often sounds unnatural. Voice-excited vocoders combine the best features of the approaches described above, producing a reasonably natural-sounding voice channel with a data rate of 4,800 or 9,600 bps [Gerson, 1990].

---

* The calculation is $\left((25 \times 25) km^2 \right) \left(\dfrac{10^3 m}{km} \right)^2 \left(\dfrac{1\,pixel}{30\,m} \right)^2 \left(\dfrac{1}{4\,min} \right) \left(\dfrac{1\,min}{60\,sec} \right)$.

Reducing the data rate by processing or compression on board the satellite decreases the required transmitter power, significantly reducing satellite mass. (Data processing or compression uses VLSI circuits, which add little to satellite mass.) Instead of reducing the satellite mass, the ground stations can be made smaller with increased mobility and lower cost. A lower rate link can also better survive jamming. However, a compressed signal is less tolerant to bit errors, thus negating some of the advantages listed above. We expect increased use of data compression in the future as performance improves and cost decreases.

13.2.3 Data Relay

Most communications satellites and data-relay satellites simply retransmit the data received through a receiver-transmitter combination called a *transponder*. The total bandwidth capacities of three communication satellites are in Table 13-8. (See Table 13-6 for the TDRS capacities.) Transponder bandwidths of commercial geostationary communication satellites are usually 36 MHz or 72 MHz. (These transponders are repeaters. See Sec. 13.5 for a description of the processing transponder.

TABLE 13-8. Relay Bandwidth Capabilities of Representative Communication Satellites. The maximum data rate can be several times the bandwidth, depending on the modulation and ground station size. The first entry for Intelsat-V is read as 4 transponders at 36 MHz and 1 transponder at 41 MHz. The *total relay bandwidth* is calculated by multiplying the number of transponders by their bandwidth and adding them together.

Satellite	Band	Transponder Bandwidth (MHz)	Number Transponders	Total Relay Bandwidth (MHz)
Intelsat-V	C	36/41	4/1	2,137
		72/77	12/4	
	Ku	72/77	2/2	
		241	2	
DSCS-III	X	50	1	375
		60	4	
		85	1	
Globalstar	L, S, C	16.5	16	264
Generic Internet	Ka	100–1,000	20	4,000

The cellular satellites have much different bandwidths. For example, Globalstar has 16.5 MHz transponder channels to accommodate 13 of the CDMA 1.23 MHz channels. The wideband Internet relay satellites proposed at Ka-band may have transponders with 1 GHz bandwidths, to accommodate multiple 155 Mbps channels for trunking Internet data between computers. The satellite communication world is becoming much more complex.

13.3 Link Design

The overall process of link design and, subsequently, payload sizing, is summarized in Table 13-9 and described in detail in this and the next section. The process for developing the communications architecture and determining the link requirements was detailed in Table 13-1.

To understand link design, we need to define the relationship between data rate, antenna size, propagation path length, and transmitter power. This relationship is

TABLE 13-9. Link Design and Payload Sizing Process.

Step	Process (= Detailed Step)	Reference
1. Define Requirements for Each Link		Table 13-1
2. Design Each Link	(Input link geometry, data rate from Table 13-1.) A. Select frequency band B. Select modulation, coding C. Apply antenna size, beamwidth constraints (if any) D. Estimate atmospheric, rain absorption E. Estimate received noise, interference powers F. Calculate required antenna gains and transmitter power	Sec. 13.3.5 Sec. 13.3.3 Eq. 13-19 Sec. 13.3.4 Table 13-10 Secs. 13.3.2, 13.3.6
3. Size the Communication Payload Subsystem	A. Select payload antenna configuration B. Calculate antenna size C. Estimate antenna mass D. Estimate transmitter mass and power E. Estimate payload mass and power	Tables 13-14, 13-15 Table 13-14 Table 13-16 Fig. 13-15 Secs. 11.2, 11.3, + 3C and 3D above
4. Iterate Back to Table 13-1		Table 13-1

defined by a *link equation* or *link budget* which relates all of the parameters needed to compute the signal-to-noise ratio of the communications system. The basic equation used in sizing a digital data link is

$$\frac{E_b}{N_o} = \frac{P L_l G_t L_s L_a G_r}{k T_s R} \qquad (13\text{-}4)$$

where E_b/N_o is the ratio of received energy-per-bit to noise-density, P is the transmitter power, L_l is the transmitter-to-antenna line loss, G_t is the transmit antenna gain, L_s is the space loss, L_a is transmission path loss, G_r is the receive antenna gain, k is Boltzmann's constant, T_s is the system noise temperature, and R is the data rate. The propagation path length between transmitter and receiver determines L_s, whereas L_a is a function of factors such as rainfall density. In most cases, an E_b/N_o ratio between 5 and 10 is adequate for receiving binary data with low probability of error with some forward error correction. Once we select the orbit and determine the transmitter-to-receiver distance, the major link variables which affect system cost are P, G_t, G_r, and R. Rain absorption also becomes a significant factor at radio frequencies above 10 GHz.

Figure 13-5 illustrates the key relationships between power of the satellite's transmitter, diameter of the ground-station antenna, and data rate for the downlink. These parameters are nearly independent of frequencies between 200 MHz and 20 GHz under clear weather conditions. The required transmitter power is relatively independent of satellite altitude when the antenna beamwidth is set to just illuminate the coverage area indicated. At low altitudes the required transmitter power is reduced in the Earth coverage case because the area in view of the satellite is smaller.

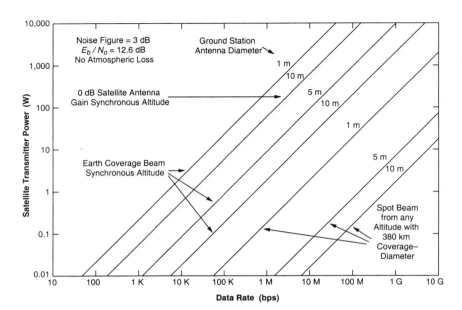

Fig. 13-5.   Satellite Downlink Characteristics. The required transmitter power is relatively independent of satellite altitude for constant coverage. At low altitude the required transmitter power is reduced in the Earth coverage case because the area in view of the satellite is smaller. The power is relatively independent of carrier frequency between 200 MHz and 20 GHz in the absence of rain and antenna-pointing error at the ground station.

13.3.1 Derivation of Link Equation

Consider a transmitter located at the center of a sphere of radius S, radiating power PL_l isotropically, and thus uniformly illuminating the surface of the sphere. The *power flux density*, W_f, received on the sphere's surface is the radiated power divided by the area of the sphere, that is, $PL_l/4\pi S^2$. The radiated power is the transmitter power, P, reduced by the line loss, L_l, between the transmitter and the antenna.

If the transmitting antenna has a narrow beamwidth, the power flux density is increased by the *transmitting antenna gain*, G_t, defined as the ratio of power radiated to the center of the coverage area to the power radiated by an isotropic (omni-directional) antenna. The received flux density is reduced by the *transmission path loss*, L_a, which includes atmospheric and rain absorption. The power flux density then becomes:

$$W_f = \frac{PL_l G_t L_a}{4\pi S^2} \equiv \frac{(EIRP)L_a}{4\pi S^2} \tag{13-5}$$

where W_f is typically expressed in W/m². $PL_l G_t$ is called the *effective isotropic radiated power,* or *EIRP*, in watts. In Fig. 13-6, the same *EIRP* and received power flux density is produced two ways, one using a high-power transmitter, the other a low-power transmitter. The difference between them is that the approach using a low-power, high-gain antenna illuminates only a limited coverage area, which may or may not meet the mission requirements.

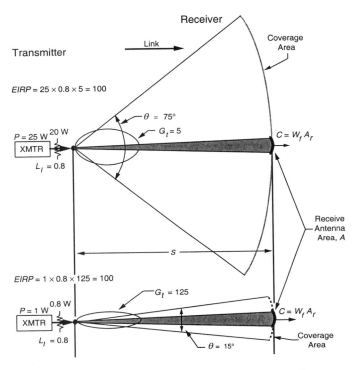

Fig. 13-6. Trade-off between Antenna Gain and Transmitter Power. These two communication links have the same *EIRP*, received power density, and received signal power. However, the 1-W transmitter with the high-gain antenna has only 1/25 the coverage area provided by the 25-W transmitter with low-gain antenna.

The received power, C, is W_f times the *effective receive antenna aperture area*, A_r. Here A_r is equal to the physical aperture area, $\pi D^2/4$, times the *antenna efficiency*, η. The efficiency, η, is a Figure of Merit between 0 and 1, and is a function of various imperfections in the antenna, including deviations of the reflector surface from theoretical, feed losses, and aperture blockage. A typical value for parabolic antennas is 0.55, though 0.6 to 0.7 often occur in high quality, ground antennas. Substituting for W_f, from Eq. (13-5), and A_r, we obtain

$$C = \frac{PL_lG_tL_aD_r^2\eta}{16S^2} \tag{13-6}$$

where D_r is the diameter of the receive antenna.

The antenna gain may also be defined as the ratio of its effective aperture area, A_r, to the effective area of a hypothetical isotropic antenna, $\lambda^2/4\pi$, where λ is the wavelength of the transmitted signal. For the receive antenna

$$G_r = \left(\frac{\pi D_r^2\eta}{4}\right)\left(\frac{4\pi}{\lambda^2}\right) = \frac{\pi^2 D_r^2\eta}{\lambda^2} \tag{13-7}$$

Substituting Eq. (13-5) into Eq. (13-4) we obtain

$$C = PL_lG_tL_aG_r\left(\frac{\lambda}{4\pi S}\right)^2 \equiv PL_lG_tL_sL_aG_r \equiv (EIRP)L_sL_aG_r \qquad (13\text{-}8)$$

where C is the received power and $(\lambda/4\pi S)^2$ is defined as the *space loss*, L_s.

In digital communications, the received energy per bit, E_b, is equal to the received power times the bit duration, or

$$E_b = C/R \qquad (13\text{-}9)$$

where C is in W, R is the data rate in bps, and E_b is in W·s or J.

The noise power at the receiver input usually has a uniform *noise spectral density*, N_o, in the frequency band containing the signal. The total received noise power, N, is then N_oB, where B is the *receiver noise bandwidth*. (B is determined by the data rate and the choice of modulation and coding, as discussed later in this chapter.) N_o and N are related to the *system noise temperature*, T_s, by:

$$N_o = kT_s \qquad (13\text{-}10)$$

and $\qquad (13\text{-}11)$

$$N = kT_sB = N_oB \qquad (13\text{-}12)$$

where N_o is in W/Hz, N is in W, k is Boltzmann's constant $= 1.380 \times 10^{-23}$ J/K, T_s is in K, and B is in Hz. By combining Eqs. (13-9) and (13-10) with Eq. (13-8), we obtain our original link equation, Eq. (13-4).

13.3.2 Link Design Equations

The link equation is a product of successive terms and, therefore, can be conveniently expressed in terms of *decibels* or *dB*. A number expressed in dB is just $10 \log_{10}$ of the number. Thus, a factor of 1,000 is 30 dB and a factor of 0.5 is –3 db. If the number has units, they are attached to the dB notation. For example, 100 W is 20 dBW. Antenna gain is the ratio of radiated intensity in a specific direction to that of an isotropic antenna radiating uniformly in all directions and, therefore, is a pure number which should, in principle, be expressed in dB. However, we use dBi (dB relative to isotropic) as the units for antenna gain to be consistent with standard practice in the industry.[*]

Eq. (13-4) can be rewritten in decibels as

$$E_b/N_o = P + L_l + G_t + L_{pr} + L_s + L_a + G_r + 228.6 - 10 \log T_s - 10 \log R \qquad (13\text{-}13)$$

where E_b/N_o, L_l, G_t, L_s, L_a, L_{pr}, and G_r are in dB, P is in dBW, T_s is in K, R is in bps, and $10 \log k = -228.60$ dBW/(Hz·K). This can also be written as

$$E_b/N_o = EIRP + L_{pr} + L_s + L_a + G_r + 228.6 - 10 \log T_s - 10 \log R \qquad (13\text{-}14)$$

---

[*] *Editor's Note:* Of course, modern computers are fully capable of multiplying real numbers instead of adding logarithms. Like much of astronautics, this peculiar nomenclature remains intact primarily to ensure the full employment of communications systems engineers.

where the *EIRP* is in dBW and sensitivity of the receiving station, $G_r/T_s = G_r -$ 10 log T_s, is expressed in dB/K. Eq. (13-14) is preferred when specifying the transmitter *EIRP* and receiver G_r/T_s separately.Note that in these equations, G_r and T_s must be specified at the same point, usually the junction between the receive antenna terminal and the Low Noise Amplifier.

We can find the *carrier-to-noise-density-ratio*, C/N_o by multiplying E_b/N_o by the data rate, R [see Eq. (13-9)]. Carrier power in W is the energy per bit in J times the number of bits per second.

$$C/N_o \equiv E_b/N_o + 10 \log R \tag{13-15a}$$

$$= (EIRP) + L_s + L_a + G_r/T_s + 228.6 \tag{13-15b}$$

From Eq. (13-12), the *carrier-to-noise-ratio*, C/N is $C/N_o B$ or, in dB, is $C/N_o -$ 10 log B. Combining this with Eq. (13-15b):

$$C/N = EIRP + L_s + L_a + G_r + 228.6 - 10 \log T_s - 10 \log B \tag{13-16}$$

where B is the noise bandwidth of the receiver in Hz. C/N also equals $E_b/N_o +$ 10 $\log(R/B)$.

The *Received Isotropic Power*, or *RIP*, is the power received if the receive antenna gain is 0 dB. If we substitute $G_r = 1$ (0 dB) into Eq. (13-8), then $C = RIP$. Combining this expression with Eq. (13-4), and converting to dB yields:

$$RIP = E_b/N_o - G_r/T_s - 228.60 + 10 \log R \tag{13-17}$$

where *RIP* is in dBW. A good way to specify the receiving system performance is to specify the *bit error rate* (the probability a data bit is incorrectly received) required for a given *RIP*. The designer then has the freedom to trade off his demodulator design (which determines the E_b/N_o required to meet the specified bit error rate) against the antenna gain and noise temperature (G_r/T_s) to meet the *RIP* specification at minimum cost.

We can similarly convert Eq. (13-7) to dB. Using the relationship $f = c/\lambda$, where c is the velocity of light in free space $\approx 3 \times 10^8$ m/s, we obtain the following equation for the antenna gain, G, in dB:

$$G = 20 \log \pi + 20 \log D + 20 \log f + 10 \log \eta - 20 \log c \tag{13-18a}$$

or

$$G = -159.59 + 20 \log D + 20 \log f + 10 \log \eta \tag{13-18b}$$

where G is in dB, f is in Hz, and D is in m.

For a circular antenna beam the half-power beamwidth, θ, is the angle across which the gain is within 3 dB (50%) of the peak gain. We may estimate θ from the following empirical relationship:

$$\theta = \frac{21}{f_{GHz}D} \qquad \theta \text{ in degrees} \tag{13-19}$$

where f_{GHz} is the carrier frequency in GHz, and D is the antenna diameter in m.[*]

---

[*] Equation (13-19) yields a beamwidth about 20% greater than λ/D (Chap. 9). This accounts for the nonuniform illumination of the parabola by the feed.

The antenna gain is approximately $27,000/\theta^2$, obtained by combining Eqs. (13-7) and (13-19), and assuming $\eta = 0.55$. A noncircular antenna has an elliptical beam with the half-power beamwidth along the major axis equal to θ_x and the half-power beamwidth along the minor axis equal to θ_y. The gain of the noncircular beam antenna can be estimated:

$$G \approx 44.3 - 10 \log(\theta_x \theta_y) \tag{13-20}$$

where θ_x and θ_y are in deg, and G is in dBi. For example, an antenna with a 1 deg by 2 deg elliptical beam has the same gain as a circular antenna with a beamwidth of 1.4 deg. Gain calculated in this manner is generally accurate to within 25% (1.2 dB) for beamwidths less than 150 deg. The beamwidths θ_x or θ_y can be estimated from Eq. (13-19) with D equal to the major axis or minor axis diameters.

The above gain equations are for peak gain. However, a receive antenna might not be located at the center of the transmitter antenna beam, or vice versa. With narrow beamwidths, small errors in pointing the antenna (introduced by wind gusts on the ground or satellite stabilization errors, for example) can lead to significantly reduced gain. The following equation estimates the reduction from peak gain, L_θ, in dB caused by a pointing offset from beam center:

$$L_\theta = -12 \, (e/\theta)^2 \tag{13-21}$$

where θ is the antenna half-power beamwidth, and e is the pointing error. For example, for e equal to $\theta/2$, the pointing loss is 3 dB. In calculating a link budget, we would subtract this pointing loss from the antenna gain.

From Eq. (13-6) the space loss, L_s, is

$$L_s = (\lambda/4\pi S)^2 = (c/4\pi Sf)^2 \tag{13-22}$$

Converting into dB, this gives:

$$L_s = 20 \log (3 \times 10^8) - 20 \log (4\pi) - 20 \log S - 20 \log f \tag{13-23a}$$

$$= 147.55 - 20 \log S - 20 \log f \tag{13-23b}$$

where S is the path length in m, and f is the frequency in Hz.

The system noise temperature, T_s, is the sum of a number of individual contributions from various sources. We have divided the noise sources into two groups. Those originating ahead of the antenna aperture (e.g., in the atmosphere) we call the *antenna noise temperature*, T_{ant}. These noise sources are external to the ground station, except for the antenna itself, and include:

- Galactic noise
- Noise radiated by clouds and rain in the propagation path
- Solar noise (either in the antenna main beam or sidelobe)
- Presence of the Earth (typically 290 K) in a sidelobe
- Man-made noise (either in the antenna main beam or sidelobe)
- Contribution of nearby objects, buildings, radomes, etc.
- Temperature of blockage items in antenna subsystem such as booms or feeds

Figure 13-7 shows the estimated noise temperature from various external sources as a function of frequency. Note the necessity of keeping the receive antenna from pointing toward the Sun when the beamwidth is narrow (< 5 deg). Otherwise the Sun will significantly increase the antenna noise temperature.

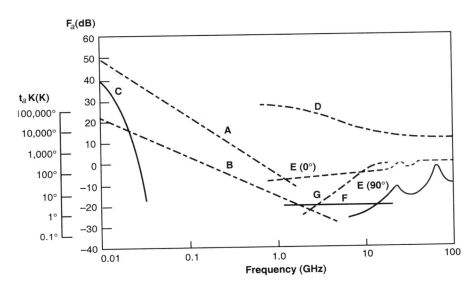

A: Estimated median business area man-made noise
B: Galactic noise
C: Atmospheric noise, value exceeded 0.5% of time
D: Quiet Sun (1/2 deg beamwidth directed at Sun)
E: Sky noise due to oxygen and water vapor (very narrow beam antenna);
   upper curve, 0-deg elevation angle; lower curve, 90-deg elevation angle
F: Black body (cosmic background), 2.7 K
G: Heavy rain (50 mm/hr over 5 km)

Fig. 13-7. Minimum Expected External Noise From Natural and Man-made Sources, 10 MHz to 100 GHz [Ippolito, 1986].

All noise sources between the antenna terminal and the receiver output are lumped together and called *receiver noise temperature, T_r.* Receiver noise originates from

- *Transmission Lines and Filters*—equal to $(1 - L)T$, where $L \equiv P_o / P_i$ is the ratio of output power (P_o) to input power (P_i) and T is the component temperature in K.
- *Low Noise Amplifier*—equal to $(F - 1)\,290$ K, where F is the noise figure from Eq. (13-24).

An additional contribution from subsequent amplifier stage noise exists, but is a small contributor because it is divided by the low noise amplifier gain.

The *noise figure, F,* of the receiver is defined as:

$$F = 1 + \frac{T_r}{T_0} \qquad (13\text{-}24)$$

where T_r is the noise temperature of the receiver itself, and T_0 is a reference temperature, usually 290 K. The noise figure is often expressed in dB (that is, 10 log F). For example, a cryogenically cooled receiver for reception of telemetry signals from a space probe may have a noise figure of 1.1 (0.4 dB) for a noise temperature of 29 K.

Adding the antenna noise and receiver noise gives us the *system noise temperature,* T_s. To find T_s we add the noise contribution of the transmission line and bandpass filter which connect the antenna to the receiver's low-noise amplifier. Thus:

$$T_s = T_{ant} + \left(\frac{T_0(1 - L_r)}{L_r} \right) + \left(\frac{T_0(F - 1)}{L_r} \right) \qquad (13\text{-}25)$$

where L_r is the line loss between the antenna and receiver, expressed as a power ratio. The second term in Eq. (13-25) is the noise contribution from the transmission line, and the third term is the contribution from the receiver. The receiver noise temperature is the sum of these two terms. These noise temperatures are referred to the antenna terminal by dividing by L_r. Continuing with our cooled receiver example, assume the line loss is 0.5 dB, making $L_r = 0.89$. Then the noise contribution from the line loss is 36 K and the receiver noise is 33 K, both referred to the antenna terminal. Then T_s is $T_{ant} + 69$ K.

Table 13-10 shows typical noise temperatures for satellite systems using uncooled receivers. When a narrow satellite-antenna beam looks at Earth, the uplink antenna noise temperature is the temperature of the Earth, about 290 K. In the future, improvements in design of low-noise amplifiers will reduce the receiver noise figures, especially at higher frequencies.

TABLE 13-10.   Typical System Noise Temperatures in Satellite Communication Links in Clear Weather. The temperatures are referred to the antenna terminal. [See Eq. (13-25)].

Noise Temperature	Frequency (GHz)					
	Downlink			Crosslink	Uplink	
	0.2	2–12	20	60	0.2–20	40
Antenna Noise (K)	150	25	100	20	290	290
Line Loss (dB)	0.5	0.5	0.5	0.5	0.5	0.5
Line Loss Noise (K)	35	35	35	35	35	35
Receiver Noise Figure (dB)	0.5	1.0	3.0	5.0	3.0	4.0
Receiver Noise (K)	36	75	289	627	289	438
System Noise (K)	221	135	424	682	614	763
System Noise (dB-K)	23.4	21.3	26.3	28.3	27.9	28.8

13.3.3 Modulation and Coding

Before we can complete our link design, we need to select a modulation and coding technique. *Modulation* is the process by which an input signal varies the characteristics of a radio frequency carrier (usually a sine wave). These characteristics are amplitude, phase, frequency, and polarization. *Demodulation* of the signal at the receiver consists of measuring the variations in the characteristics of the received carrier and deducing what the original signal was. Amplitude modulation, though common in terrestrial services, seldom appears in satellite systems because it requires larger (and more costly) transmitters. Phase or frequency modulation techniques are preferred, because the transmitter can operate at saturation for maximum power efficiency.

Figure 13-8 shows the most common modulation techniques used in satellite systems. *Binary phase shift keying (BPSK)* consists of setting the carrier phase at 0 deg to transmit a binary 0, and setting the phase at 180 deg to transmit a binary 1. *Quadriphased phase shift keying (QPSK)*, takes two bits at a time to define one of four *symbols*. Each symbol corresponds to one of four carrier phases: 0 deg, 90 deg, 180 deg, or 270 deg. Note that the symbol rate is one half the bit rate, thus reducing the spectrum width by one half.

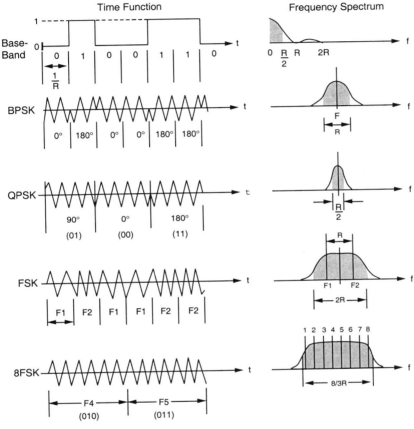

Fig. 13-8. **Modulation Types Commonly Used for Digital Signal Transmission in Satellite Communications.** *R* is the data rate. The shaded region is the required bandwidth.

Frequency shift keying (FSK) sets the carrier frequency at F1 to transmit a binary 0, and at F2 to transmit a binary 1. The separation between F1 and F2 must be at least equal to the data rate to avoid performance loss from mutual interference. Thus, the transmitted spectrum width is at least twice the width of the spectrum generated by BPSK. *Multiple frequency shift keying (MFSK)* sets the carrier frequency to one of *M* frequencies. For example, for *M* = 8, the first three binary bits, 010, determine that the transmitted frequency will be F5. The next three bits, 011, set the frequency at F6. The symbol rate is one-third the bit rate, and the transmitted spectrum width is about 8/3 the bit rate, where the separation between frequencies is the symbol rate. Less common spaceflight modulation schemes include minimum-shift-keying (MSK), offset-QPSK (OQPSK), 8PSK [Sklar, 1988], and BPSK/PM [Yuen, 1983].

These modulation techniques have different spectrum widths, and thus require different RF channel bandwidths, as illustrated in Fig. 13-8. An actual system may require wider bandwidths as a result of a tradeoff between performance and filter complexity. For example, the FLTSATCOM Experimental Package uses 8FSK as its modulation technique, with the tone frequency separation and total spectrum width twice as large as that shown in the figure to make the filters less complex [McElroy, 1988].

To demodulate a digital bit reliably, the amount of energy received for that bit, E_b, must exceed the noise spectral density, N_o, by a specified amount. Communication theory derives the E_b/N_o needed to achieve a required *bit error rate*, BER, at the receiver output (see Sklar [1988], Chap. 3). The BER gives the probability of receiving an erroneous bit. For example, a BER of 10^{-5} means that, on the average, only one bit will be in error for every 10^5 bits received.

FSK- and MFSK-modulated signals are usually demodulated by measuring the received power at each of the possible frequencies, and selecting the frequency with the largest power as the one transmitted. An advantage of this technique is that any variations in carrier phase introduced by the transmission channel (such as multipath) will not greatly degrade the link performance. For this reason, these modulation types are often used in military communications satellites (MFSK) and command links (FSK modulation of a subcarrier).

On the other hand, demodulation of either BPSK and QPSK requires us to measure the phase of the received carrier. Thus phase distortion caused by the transmission channel will significantly degrade performance. The effect of this distortion can be reduced using *Differential PSK (DPSK)*. This technique requires no change in carrier phase when transmitting a binary 0, and a 180-deg phase reversal when transmitting a binary 1. At the receiver the phase of the carrier received during each bit period is compared with the phase received during the previous bit period to determine whether a phase change took place. This modulation technique overcomes distortion effects provided the phase changes between successive bit periods are small (< 20 deg for < 0.5 dB degradation).

Unfortunately, both BPSK and QPSK modulation experience sudden phase transitions at the bit boundaries. These transitions generate sidebands outside the main signal spectrum which may interfere with an adjacent frequency channel. But if the carrier phase is *gradually* moved from one phase position to the next during the bit period (from 0 deg to 180 deg, for example), then the sideband power is greatly reduced. One version of this technique is *Minimum Shift Keying (MSK)* [Sklar, 1988, Sec. 7.9.2]. These *bandwidth efficient modulation* techniques are particularly useful in communications satellite systems with multiple channels closely spaced to fit the available frequency spectrum.

A modulation method that combines binary phase shift keying of a subcarrier with phase modulation of the carrier is *BPSK/PM* * modulation. Many TT&C links use this modulation because the carrier and data are transmitted at frequencies separated by the subcarrier frequency, which is made much larger than the data rate. This separation allows easy acquisition and tracking of the carrier, thus providing an accurate range-rate measurement. The disadvantages of this technique are additional power required for the carrier and increased bandwidth (to approximately twice the subcarrier frequency). For example, the SGLS TT&C downlink uses a 1.024 MHz subcarrier BPSK modulated by data rates up to 128 kbps. The subcarrier then phase modulates the

---

* FireSat will use this technique for the uplink.

carrier. For a phase modulation of ±1.0 radian, the transmitter power is divided between the carrier (60%) and subcarrier (40%). Figure 13-9 shows how the BER varies with E_b/N_o. Table 13-11 lists the types of digital modulation commonly used in space communication systems.

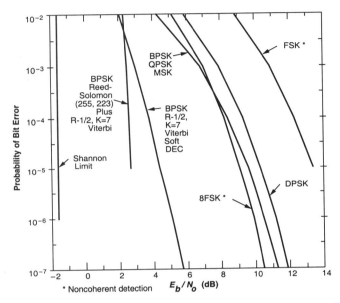

Fig. 13-9.   Bit Error Probability as a Function of E_b/N_o. The theoretical performance limit can be approached by use of error correction coding.

For the digital systems discussed in this chapter, the BER is used to evaluate the performance of the link. On the other hand, analog communication links are generally evaluated in terms of the output *signal-to-noise ratio, S/N*. The *S/N* is a function of the *C/N*, the modulation type, and how the *S/N* is expressed. For example, the *S/N* of a frequency modulated (*FM*) television signal depends on the modulation index and the pre-emphasis plus weighting factors in addition to the *C/N* (see Morgan and Gordon [1989], Sec. 3.4.5).

Forward error correction coding significantly reduces the E_b/N_o requirement, which in turn reduces the required transmitter power and antenna size, or increases link margin. Extra bits, called parity bits, are inserted into the data stream at the transmitter. These bits enable the receiver to detect and correct for a limited number of bit errors which might occur in transmission because of noise or interference. While complex, these techniques can be implemented at relatively low cost, using large scale integrated circuits with small size and low power consumption. This type of coding does have some disadvantages. The extra error correction bits increase the bit rate, and hence the transmission bandwidth, often a scarce resource. Also, a low E_b/N_o makes initial signal acquisition more difficult.

A common type of error correction technique is *convolutional coding with Viterbi decoding* [Viterbi, 1967; Sklar, 1988]. A *rate-1/2 convolutional code* is implemented by generating and transmitting two bits for each data bit. The data rate is therefore one-half the transmitted bit rate (hence "rate-1/2"). The receiver demodulates and stores in

TABLE 13-11. **A Comparison of Several Modulation and Coding Schemes Used in Satellite Communication Systems.** The theoretical values of E_b/N_o shown must be increased by 1–3 dB to account for filtering, timing, and frequency errors. See text for definition of modulation schemes. In the bottom row R = data rate and f_{SC} = subcarrier frequency.

Modulation	E_b/N_o for BER = 10^{-5} (dB)	Spectrum Utilization	Advantages	Disadvantages
BPSK	9.6	1.0	Good BER performance. Good use of spectrum.	Susceptible to phase disturbances.
DPSK	10.3	1.0	Not susceptible to phase disturbances.	Higher E_b/N_o required.
QPSK	9.6	2.0	Excellent use of spectrum.	More susceptible to phase disturbances.
FSK	13.3	0.5	Not susceptible to phase disturbances.	Higher E_b/N_o required. Poor use of spectrum.
8FSK	9.2	0.375	Good BER performance. Not susceptible to phase disturbances.	Poor use of spectrum.
BPSK and QPSK Plus R-1/2 Viterbi Decoding	4.4	0.5 and 1.0	Excellent BER performance.	Higher complexity. Reduced use of spectrum.
BPSK and Plus RS Viterbi Decoding	2.7	0.44	Best BER performance.	Most complex. Reduced use of spectrum.
8FSK Plus R-1/2 Viterbi Decoding	4.0	0.188	Excellent BER performance. Not susceptible to phase disturbances.	Poor use of spectrum. High complexity.
MSK	9.6	1.5	Low adjacent channel interference.	Higher complexity,
BPSK/PM ($\Delta\varnothing$ = 1.0 rad sinewave)	13.8	$\sim \dfrac{R}{2f_{sc}}$	Carrier transmitted for Doppler measurement.	Requires extra power and bandwidth.

memory a sequence (typically 62 bits long). As time goes on, additional bits are received and stored. These sequences are compared with coded sequences which could have been transmitted. The possible sequence which most closely resembles the received sequence is chosen as the most likely sequence transmitted. This process is repeated a number of times to reduce the probability of error. As shown in Fig. 13-9, Viterbi decoding greatly reduces the E_b/N_o required to obtain a specified BER. For example, a BER of 10^{-5} is achieved with an E_b/N_o of 4.4 dB. This represents an improvement, or *coding gain*, of 5.2 dB below the 9.6 dB required for uncoded BPSK.

There is a value of E_b/N_o equal to -1.6 dB, known as the *Shannon limit*, below which no error-free communication at any information rate can take place. This is derived from the *Shannon-Hartley theorem*, which states that the maximum theoretical data rate, R_{max}, which can be transmitted over a transmission channel with bandwidth, B, is

$$R_{max} = B \log_2\left(1 + \frac{C}{N}\right) \tag{13-26}$$

where C/N is the average carrier-to-noise power ratio in the channel (see Panter [1965], Sec. 19.2; or Sklar [1988], Sec. 7.4).

We cannot reach the Shannon limit in practice because the transmission bandwidth and coding complexity increases without bound. But we can approach this limit by using a double, or concatenated, coding scheme. The Pioneer deep-space communication link uses this technique to obtain the performance required to overcome the large space loss [Yuen, 1983, Sec. 5.4.4]. The binary signal is first block encoded using a 255-bit Reed-Solomon code with 32 parity bits. The block-encoded signal is then encoded using the rate-1/2 convolutional code. The data is recovered at the receiver with a Viterbi decoder followed by a Reed-Solomon decoder. As shown in Fig. 13-9, the BER performance for the concatenated code is only 4.2 dB above the Shannon limit at a BER of 10^{-5}. Longer and more complex codes can be used to push performance even closer to the Shannon limit. This technique is not presently used for data rates much above 100 bps due to the complexity and cost of the decoding process. However, we expect this limitation to disappear in the future with the development of high-speed, low-cost digital processing circuits.

Figure 13-9 demonstrates that the BER is sensitive to the E_b/N_o level. At error rates below 10^{-4}, a decrease of 1 dB in E_b/N_o will increase the BER about one order of magnitude. Thus, providing adequate link margin in our design (discussed later) is very important. Note that the E_b/N_o values given are theoretical, based on infinite bandwidth transmission channels and ideal receivers. In practice, we must account for band-limiting effects, deviations from ideal filter responses, phase noise and frequency drift in local oscillators, noise in carrier tracking loops, and bit synchronizing errors. Thus, we must add 1 or 2 dB to the theoretical E_b/N_o to allow for these losses.

Choosing which modulation and coding technique to use depends on cost, complexity, difficulty of acquiring the signal, limits on transmission bandwidth, and susceptibility to interference or fading. In the case of power-limited systems, where power is scarce but bandwidth is available, coding makes sense. Most systems today use some forward error correction coding to save transmitter power unless the data rate is greater than several hundred Mbps, in which case limits on both bandwidth and hardware speed become significant.

Table 13-11 showed that to obtain a low value of E_b/N_o for a BER of 10^{-5}, we must be less efficient in using the spectrum, which means the transmission bandwidth must increase. This is the result of adding forward error correction check bits to the signal before modulation, thus increasing the transmission rate through the channel. However, recent techniques can achieve significant coding gains **without** increasing the bandwidth [Ungerboeck, 1982; Sklar, 1988, Sec. 7.10.6]. The basic idea is to combine the coding and modulation process to generate a set of possible coded signal patterns at RF, each pattern corresponding to an n-bit word. The receiver knows the set of possible patterns and chooses the pattern that most closely resembles what it receives. For example, a four-state, rate-2/3, encoder combined with 8PSK modulation achieves a 3 dB coding gain over uncoded QPSK modulation with the same bandwidth. (8PSK is like QPSK except the carrier phase is set at one of 8 possible phases separated by 45-deg intervals.) Such schemes have been proposed for high-speed data communication on satellite channels and are likely to be used in future space systems [Deng and Costello, 1989].

13.3.4 Atmospheric and Rain Attenuation

The transmission path loss caused by the Earth's atmosphere, L_a, is a function of frequency as shown in Fig. 13-10. This figure gives the attenuation at 90-deg elevation, that is, zenith. To determine the loss at elevation angles, β, above 5 deg, divide the zenith attenuation by $\sin \beta$. At frequencies below 1,000 MHz, ionospheric scintillation can cause significant fluctuations in received amplitude and phase. These effects are most severe at frequencies below 200 MHz during periods of high sunspot activity, when they can disrupt communications (see Ippolito [1986], Chap. 8). Above 50 GHz, high attenuation occurs at frequencies corresponding to the oxygen absorption band. Virtually all Earth-space communications use frequencies between these two extremes, though some projects have considered the 90-GHz band. Of special note is the high absorption band of oxygen at 60 GHz. Intersatellite links often operate at 60 GHz, using the atmosphere to shield from interference or jamming originating on the Earth's surface.

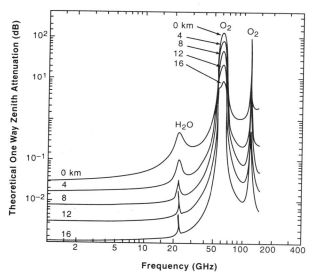

Fig. 13-10. **Theoretical Vertical One-way Attenuation from Specified Height to the Top of the Atmosphere.** Plot assumes 7.5 g/m^3 of water vapor at the surface, and does not include effect of rain or cloud attenuation [NASA, 1983].

Figure 13-11 shows that clouds and rain attenuation also add to losses, increasing with frequency, in the transmission path. This figure comes from the Crane model—a set of tables and equations based on observed climatic data used to estimate the rain attenuation (see Crane [1980], or Ippolito [1986]). This attenuation becomes an important consideration when designing systems employing satellite-Earth link frequencies above 10 GHz. Figure 13-11 also shows that the attenuation increases rapidly as the antenna elevation angle decreases below 20 deg. When operating at frequencies above 20 GHz, the minimum elevation angle to the satellite should be specified at 20 deg, especially in high rainfall areas. Note, however, that increasing the minimum elevation angle dramatically reduces the size of the satellite's coverage area (see Secs. 5.3 and 5.4).

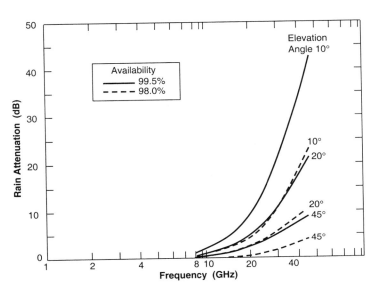

Fig. 13-11. Rain Attenuation Predicted by Crane Model for Rain Climate Typical of the Northern United States. Ground station altitude = 0 km, latitude = 40°. For other cases, see Ippolito [1986].

The percentage of time the link is available depends on the margin which the link has in clear weather. A higher link availability demands a greater margin. The availability numbers shown in Fig. 13-11 correspond to the climate of northeastern United States. The availability for a specified link margin will be higher in the Midwest and lower in the Southeast, compared to those shown in the figure. (Bear in mind that these availability numbers come from average weather conditions over many years; they do not account for the possibility that **this** year the rainfall may be higher or lower than a normal year.)

A by-product of rain attenuation is an increase in antenna temperature, T_a, given by

$$T_a = (1 - L_a)T_o \qquad (13\text{-}27)$$

where L_a is the rain attenuation given as a power ratio, and T_o is the temperature of the rain, usually assumed to be 290 K. The significance of this noise increase depends on the system noise in clear weather. For example, if a receiving system with a normal system temperature of 727 K experienced 10 dB rain attenuation, its noise temperature would increase to 988 K, increasing N_o by 1.3 dB. The decrease in C/N_o would then be the sum of the rain attenuation (–10 dB) and the increased noise (–1.3 dB) totaling 11.3 dB.

13.3.5 Frequency Selection

Regulatory constraints exist on the selection of frequency band, transmission bandwidth, and power flux density. For example, international agreements have allocated frequency bands for space communications, as listed in Table 13-12. These agreements originated with the *International Telecommunications Union* (*ITU*) and the

World Administrative Radio Conference (WARC). They are administered in the United States by the *Federal Communications Commission* (FCC) for commercial users, and by the *Interdepartmental Radio Advisory Committee* for military users. The system designer must apply for and receive permission from the appropriate agency to operate at a specified frequency with the specified orbit and ground locations. This is often a time-consuming procedure (see Chap. 21.1). For an excellent summary of this complex subject, see Morgan and Gordon [1989].

TABLE 13-12. Limitations on the Frequency Bands and Flux Densities Established by the ITU. Power density limits are for elevation angle >25 deg. They are about 10 dB less for lower angles.

Frequency Band	Frequency Range (GHz)		Service	Downlink Power Flux Density Limit (dBW/m^2)
	Uplink	Downlink		
UHF	0.2 – 0.45	0.2 – 0.45	Military	—
L	1.635 – 1.66	1.535 – 1.56	Maritime/Nav Telephone	−1.44/4 kHz
S	2.65 – 2.69	2.5 – 2.54	Broadcast, Telephone	−137/4 kHz
C	5.9 – 6.4	3.7 – 4.2	Domestic, Comsat	−142/4 kHz
X	7.9 – 8.4	7.25 – 7.75	Military, Comsat	−142/4 kHz*
Ku	14.0 – 14.5	12.5 – 12.75	Domestic, Comsat	−138/4 kHz
Ka	27.5 – 31.0	17.7 – 19.7	Domestic, Comsat	−105/1 MHz
SHF/EHF	43.5 – 45.5	19.7 – 20.7	Military, Comsat	—
SHF/EHF	49	38	Internet Data, Telephone, Trunking	−135/1 MHz
V	~60		Satellite Crosslinks	—

*No limit in exclusively military band of 7.70–7.75 GHz.

A criterion for frequency band allocation is the potential for one link to interfere with another. Extensive analysis is required when applying for a frequency band and orbit to avoid interference with, or by, existing services such as terrestrial microwave links and ground-based radar operations. Especially significant are the antenna sidelobe levels and the dynamic range of powers over which the system must operate. For ground-station antennas operating in the 4–6 GHz and 12–14 GHz bands, the FCC specifies the maximum sidelobe gain as $(32 - 25 \log \phi)$ dBi for 1 deg $\leq \phi \leq$ 48 deg; otherwise −10 dBi for 48 deg $\leq \phi \leq$ 180 deg, where *dBi* is dB relative to an isotropic radiator (0 dB gain) and ϕ is the angle in deg off the axis of the main antenna beam. Also shown in Table 13-12 are the maximum allowed power flux densities radiated by the satellite onto the Earth. These limitations, set by the ITU, are necessary to avoid interference to existing terrestrial services, such as microwave relay links.

Two geostationary satellites in approximately the same orbit location servicing the same ground area may share the same frequency band by: (1) separating adjacent satellites by an angle (typically 2 deg), which is larger than the ground station's beamwidth, and (2) polarizing transmitting and receiving carriers orthogonally, which allows two carriers to be received at the same frequency without significant mutual interference. Right-hand and left-hand circular polarization are orthogonal, as are horizontal and vertical linear polarization. Commercial systems use these frequency-sharing techniques extensively [Morgan and Gordon, 1989].

13.3.6 Link Budgets

The link budget provides the designer with values of transmitter power and antenna gains for the various links in the system. It is therefore one of the key items in a space system design, revealing many characteristics of the overall system performance. Table 13-13 presents link budgets for FireSat command and telemetry using the equations and figures given in this chapter.

TABLE 13-13.   **Link Budgets for FireSat.**

Item	Symbol	Units	Source	Command	Telemetry and Data
Frequency	f	GHz	Input parameter	2	2.2
Transmitter Power	P	Watts	Input parameter	1	20
Transmitter Power	P	dBW	$10 \log(P)$	0.0	13.0
Transmitter Line Loss	L_l	dB	Input parameter	−1	−1
Transmit Antenna Beamwidth	θ_t	deg	Input parameter	2.0	32.0
Peak Transmit Antenna Gain	G_{pt}	dBi	Eq. (13-20)	38.3	14.2
Transmit Antenna Diameter	D_t	m	Eq. (13-19)	5.3	0.3
Transmit Antenna Pointing Offset	e_t	deg	Input parameter	0.2	27
Transmit Antenna Pointing Loss	L_{pt}	dB	Eq. (13-21)	−0.1	−8.5
Transmit Antenna Gain (net)	G_t	dBi	$G_{pt} + L_{pt}$	38.2	−5.7
Equiv. Isotropic Radiated Power	EIRP	dBW	$P + L_l + G_t$	37.2	17.7
Propagation Path Length	S	km	Input parameter	2,831	2,831
Space Loss	L_s	dB	Eq. (13-23a)	−167.5	−168.3
Propagation & Polarization Loss	L_a	dB	Fig. 13-10	−0.3	−0.3
Receive Antenna Diameter	D_r	m	Input parameter	0.07	5.3
Peak Receive Antenna Gain (net)	G_{rp}	dBi	Eq. (13-18a)	0.74	39.1
Receive Antenna Beamwidth	θ_r	deg	Eq. (13-19)	150.0	1.8
Receive Antenna Pointing Error	e_r	deg	Input parameter	70	0.2
Receive Antenna Pointing Loss	L_{pr}	dB	Eq. (13-21)	−2.6	−0.1
Receive Antenna Gain	G_r	dBi	$G_{rp} + L_{pr}$	−1.9	39.0
System Noise Temperature	T_s	K	Table 13-10	614	135
Data Rate	R	bps	Input parameter	100	85×10^6
E_b/N_o (1)	E_b/N_o	dB	Eq. (13-13)	45.5	15.9
Carrier-to-Noise Density Ratio	C/N_o	dB-Hz	Eq. (13-15a)	65.5	95.2
Bit Error Rate	BER	—	Input parameter	10^{-7}	10^{-5}
Required E_b/N_o (2)	Req E_b/N_o	dB	Fig. 13-9	11.3	9.6
Implementation Loss (3)	—	dB	Estimate	−2	−2
Margin	—	dB	(1) − (2) + (3)	32.2	4.3

A detailed procedure for a downlink design is as follows:

1. Select carrier frequency, based on spectrum availability and FCC allocations. (Refer to Table 13-6 for TT&C, Table 13-12 for communication satellites.)

2. Select the satellite transmitter power, based on satellite size and power limits.

3. Estimate RF losses between transmitter and satellite antennas. (Usually between −1 and −3 dB.)

4. Determine the required beamwidth for the satellite antenna, depending on the satellite orbit, satellite stabilization, and ground coverage area (see Chap. 7).

5. Estimate the maximum antenna pointing offset angle, based on coverage angle, satellite stabilization error, and stationkeeping accuracy.

6. Calculate transmit antenna gain toward the ground station, using Eqs. (13-20) and (13-21). You might also want to check the antenna diameter, using Eq. (13-19), to see if it will fit on the satellite.

7. Calculate space loss, using Eq. (13-23a). This is determined by satellite orbit and ground-station location.

8. Estimate propagation absorption loss due to the atmosphere using Fig. 13-10, dividing the zenith attenuation by the sine of the minimum elevation angle (e.g. 10 deg) from the ground station to the satellite. (Consider rain attenuation later.) I would also add a loss of 0.3 dB to account for polarization mismatch for large ground antennas. Using a radome adds another 1 dB loss.

9. Select the ground station antenna diameter and estimate pointing error. If autotracking is used, let the pointing error be 10% of the beamwidth. Use Eq. (13-21) to calculate the antenna beamwidth.

10. Calculate the receive antenna gain toward the satellite. For FireSat we used antenna efficiency, η, of 0.55.

11. Estimate the system noise temperature (in clear weather), using Table 13-10.

12. Calculate E_b/N_o for the required data rate, using Eq. (13-14).

13. Using Fig. 13-9, look up E_b/N_o required to achieve desired BER for the selected modulation and coding technique. The downlink for FireSat is modulated with *BPSK* and the uplink is *BPSK/PM*. See Table 13-11.

14. Add 1 to 2 dB to the theoretical value given in Fig. 13-9 for implementation losses.

15. Calculate the link margin—the difference between the expected value of E_b/N_o calculated and the required E_b/N_o (including implementation loss).

16. Estimate the degradation due to rain, using Fig. 13-11 and Eq. (13-27).

17. Adjust input parameters until the margin is at least 3 dB greater than the estimated value for rain degradation, depending on confidence in the parameter estimates.

For communications satellites to evaluate a complete communication link (ground-to-ground), you must do the downlink shown above, and also calculate the uplink, and combine their E_b/N_o s in order to evaluate the communication link.

The downlink calculation described above provides the signal-to-noise at the ground station based on the assumed parameters for the downlink. In order to establish the performance of a communication link Earth-to-Earth, it is necessary to do the same calculation on the uplink from the ground station to the satellite. The overall link performance can then be predicted based on the design of the satellite communication payload. In a *bent-pipe* satellite, the signal-to-noise ratio established on the uplink is used as the "signal" input for the downlink, and a final signal-to-noise is calculated based on the noise already on the signal plus the noise gained on the downlink. For a *signal processing* payload (see Sec. 13.5.2), where signals are demodulated and re-modulated on board the satellite, then the overall signal-to-noise performance is only that of the downlink, because this is a "pure" signal generated on the satellite. Many of the new data satellites, and some of the cellular telephone satellites, use onboard processing.

When digital links are evaluated in terms of their bit error rate, the system for signal processing satellites gets more complicated, because there will be a certain bit error rate on the uplink which becomes the starting point for the downlink. The bit error rate will never be better than the weakest link. For these systems, it is desirable to make the uplink very robust so that bit error rates of 10^{-9} or 10^{-10} are achieved on the uplink, in order that, again, the downlink determines the overall performance. This is typical for most satellite links, as the satellite is limited in the amount of transmit power, whereas ground stations are relatively independent of that limitation, at least until frequencies of 30 GHz and above are used, in which case the cost of the transmitters becomes a limiting factor. Transmit power, transmit antenna gain, receiver noise figure, and receive antenna gain establish the maximum signal-to-noise that can be established on any link.

Many of the data handling satellites discussed elsewhere in this chapter expect to deliver 10^{-10} bit error rates on the entire Earth-Earth link. This is being achieved by using very powerful forward-error-correcting codes (see Sec. 13.3.3). Convolutional coding and Viterbi decoding (rate 1/3, K = 7) allow 10^{-10} bit error rates with E_b/N_o of only 5 dB for many newer, commercial data satellites.

The question often asked is, "How much margin is enough?" Clearly, too much margin is wasteful and costly, but not enough margin could occasionally lead to excessive bit error rates. Intelsat carries a 4 to 5 dB margin for their C-Band links. At frequencies above 10 GHz the margin should be 6 to 20 dB to accommodate atmospheric and rain losses, the exact amount depending on the required link availability and the amount of rainfall expected.

The order of the steps outlined above will depend on which parameters are specified. For example, one might start with link margin and solve for transmitter power. The uplink design is performed in the same way, except the receive antenna beamwidth may depend on the Earth-coverage requirement rather than size or pointing limitations.

Figure 13-12 illustrates how the downlink design, using a geostationary satellite, can vary with choice of carrier frequency. In this example, the satellite antenna's beamwidth is fixed at 6 deg to illuminate a specified Earth coverage area, and the ground-station size is fixed at 0.5 m for ease of transport. As the frequency decreases, the satellite antenna's diameter increases to maintain the specified beamwidth (and gain) until it reaches a maximum size (or mass) limit, which, in this example, is 2 m at 1.75 GHz. Reducing the frequency further requires more transmitter power to compensate for the loss in antenna gain [see Eq. (13-18a)]. On the other hand, going to higher frequencies requires more transmitter power to compensate for increasing receive antenna pointing loss and to provide a margin to operate through rain. The figure shows the preferred frequency is between 1 and 18 GHz.

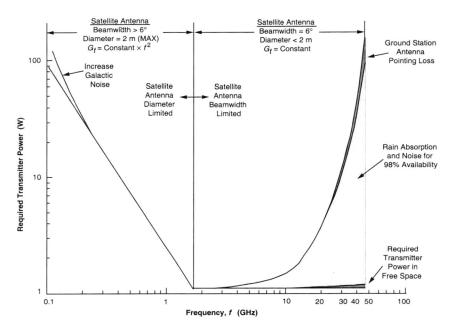

Fig. 13-12.   Example Downlink Design Showing Effect of Frequency Selection on Required Satellite Transmitter Power. A "window" exists between 1–20 GHz. The satellite is in a geostationary orbit and the ground terminal diameter is fixed at 0.5 m.

Table 13-13 shows we can satisfy the FireSat mission with a 20-W transmitter operating at S-Band (2.2 GHz) with a broad-beam antenna covering the entire Earth. A higher-gain antenna requires continuous steering to point toward the ground station, making the satellite far more complex. The diameter of the ground station antenna is 5.3 m.

13.4 Sizing the Communications Payload

We now have determined the satellite transmitter power and antenna aperture size required to support our links. These parameters have the greatest impact on satellite mass, and thus on the cost of the system. In this section we will describe these components and estimate their mass. This process is summarized in Table 13-9 in Sec. 13.3.

Up to now we have considered only the parabolic reflector antenna, which is best suited for applications where the peak gain is above 20 dB and beamwidth is less than 15 deg. For lower-gain, wider-beam applications, we may prefer other types of antennas with lighter mass and simpler design, especially at frequencies below 1 GHz (see Table 13-14). For example, an Earth-coverage satellite antenna has a beamwidth just big enough to illuminate the Earth. At geosynchronous altitude, this beamwidth is

TABLE 13-14. **Antenna Types for Satellite Systems.** (Formulas from Jasik [1961]) In these equations, C, D, L, and λ are in m and f is in GHz.

Antenna Type	Parabolic Reflector	Helix	Horn	Biconical Horn
Beam Shape	Conical	Conical	Conical	Toroidal
Typical Max Gain (dBi)	15–65	5–20	5–20	0–5
Peak Gain (dBi)	$17.8 + 20 \log D$ $+ 20 \log f$ $(\eta = 0.55)$	$10.3 + 10 \log (C^2 L / \lambda^3)$ $0.8 \leq C / \lambda \leq 1.2$ $(\eta = 0.70)$	$20 \log (C / \lambda)$ $- 2.8$ $(\eta = 0.52)$	$5 \log (h / \lambda) + 3.5$ $R > 2\lambda$ $\alpha = P(2\eta\lambda)$
Half-Power Beamwidth (deg)	$\dfrac{21}{fD}$	$\dfrac{52}{\sqrt{C^2 L / \lambda^3}}$	$\approx \dfrac{225}{(C / \lambda)}$	Typically $40° \times 360°$ for gain ≥ -1 dBi $70° \times 360°$ for gain ≥ -3 dBi
Peak Gain & Dimensions of 18° Beam at 400 MHz	$G = 19.1$ dBi $D = 2.9$ m	$G = 19.5$ dBi D: 0.19 m 0.24 m or L: 9.8 m 6.2 m	$G = 19.1$ dBi $D = 3$ m $h = 4$ m	—

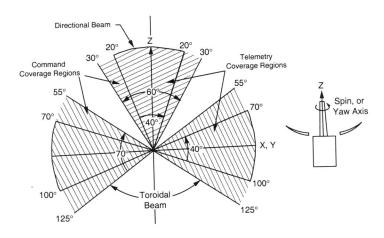

Fig. 13-13. **Typical Satellite-Antenna Requirements for Telemetry and Command Coverage.** The satellite is spin-stabilized during transfer orbit, and 3-axis stabilized in geostationary orbit [Lo and Lee, 1988].

18 deg. A simple horn antenna is often used at frequencies of 4 GHz or above (such as the Intelsat V C-Band TT&C antenna). When the frequency is below about 2 GHz, the helix often has lighter mass and is easier to mount on a satellite structure. Either a

single helix (FLTSATCOM UHF transmit), or quad-helix (Intelsat V L-Band) may be used. Table 13-14 compares these choices for an Earth-coverage antenna at 400 MHz. Wider satellite antenna beamwidths are required during launch and insertion into orbit because satellite maneuvers cause large variations in angle toward the ground station. During transfer orbit the satellite is often spin-stabilized, thus requiring a *toroidal beam* (omnidirectional in the plane perpendicular to the satellite spin axis) to provide continuous coverage. Figure 13-13 shows a typical antenna pattern. A biconical horn can generate this beam if it is mounted on a mast to avoid reflections from the satellite's body. The antenna gain is typically at least −3 dB ±35 deg from the spin axis. Once in orbit the satellite is stabilized toward Earth. A directional antenna, usually a simple conical horn, provides a gain of 9 dB or more over ±20 deg from the Earth vertical. As an example, the Intelsat V satellite combines the toroidal and directional (Earth-coverage) beams at a single port connected to the command receiver. The telemetry transmitters switch to either beam, depending on the mission phase.

High-gain antennas are required to support high data rates with low transmitter power. The basic antenna types used for this application, summarized in Table 13-15, are the reflector, lens, and phased array. The reflector is most desirable for satellites because of its low mass, low complexity and cost, and design maturity. The weight advantage of a reflector over a lens or phased array is especially significant for larger antenna diameters (> 0.5 m). Offset feeds can also simplify the satellite structure. The satellite structure contains the feed which is pointed at the reflector (Intelsat V, for example). Furthermore, an offset feed minimizes aperture blockage and therefore reduces the sidelobe levels.

In some missions it is necessary to change the direction the antenna beam is pointing. Steering the beam electronically is often preferred to mechanical methods, especially if the beam direction must be changed rapidly. The beam of a reflector antenna can be steered by switching to an off-axis feed. However, conventional reflector antennas have high losses when scanned off axis. A shaped secondary reflector can compensate for these losses, but scan angles greater than about 10 deg are difficult to achieve. We can design lens antennas for good scanning performance, but their mass is generally larger than the reflector plus feed when the diameter exceeds about 0.5 m. The lens or reflector antenna can perform beam scanning by switching between multiple feed elements or by using a phased array as the feed. Switching between feed elements or varying the amplitude or phase of each element electronically controls the feed. However, losses caused by the feed control (beam forming) network [L_l in Eq. (13-4)] can degrade the scanning antenna's performance.

A phased-array antenna may generate one or more beams simultaneously, forming these beams by varying the phase or amplitude of each radiating element of the array. This technique used for microwave radiometry is described in Chap. 9. We may also use an adaptive array to automatically point a null toward a jamming signal source to reduce the jamming-to-signal ratio.

To support high data rates with low satellite power, the antenna beamwidth should be narrow. However, as illustrated in Fig. 13-14, a narrow beam may not give us enough coverage to establish links between several widely separated ground stations. The DSCS III solved this problem by using waveguide lens antennas with variable beam-forming networks to generate a single beam with multiple lobes, each directed toward a ground station. An alternate approach is to generate multiple beams. Milstar uses a lens with a switched-feed array, and ACTS uses a reflector with an offset

TABLE 13-15.   **Six Antenna Configurations Used in Satellite Systems.** These antennas are suitable for beamwidths less than 20 deg, producing gains above 15 dB.

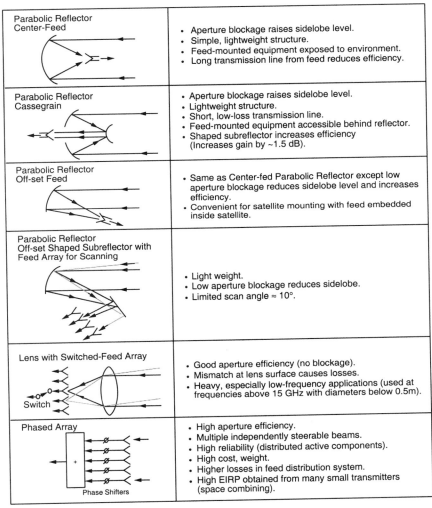

Parabolic Reflector Center-Feed	• Aperture blockage raises sidelobe level. • Simple, lightweight structure. • Feed-mounted equipment exposed to environment. • Long transmission line from feed reduces efficiency.
Parabolic Reflector Cassegrain	• Aperture blockage raises sidelobe level. • Lightweight structure. • Short, low-loss transmission line. • Feed-mounted equipment accessible behind reflector. • Shaped subreflector increases efficiency (Increases gain by ~1.5 dB).
Parabolic Reflector Off-set Feed	• Same as Center-fed Parabolic Reflector except low aperture blockage reduces sidelobe level and increases efficiency. • Convenient for satellite mounting with feed embedded inside satellite.
Parabolic Reflector Off-set Shaped Subreflector with Feed Array for Scanning	• Light weight. • Low aperture blockage reduces sidelobe. • Limited scan angle ≈ 10°.
Lens with Switched-Feed Array Switch	• Good aperture efficiency (no blockage). • Mismatch at lens surface causes losses. • Heavy, especially low-frequency applications (used at frequencies above 15 GHz with diameters below 0.5m).
Phased Array Phase Shifters	• High aperture efficiency. • Multiple independently steerable beams. • High reliability (distributed active components). • High cost, weight. • Higher losses in feed distribution system. • High EIRP obtained from many small transmitters (space combining).

switched-feed array. These antennas generate either simultaneous multiple beams or a single beam which is scanned or *hopped* over the Earth's surface using time multiplexing between channels (see Sec. 13.5). Thus high antenna gain and broad area coverage are achieved at the same time. Another advantage of the beamhopping technique is that the satellite coverage can be readily matched to the geographic traffic distribution by making the beam dwell time proportional to the traffic level.

LEO satellites such as Iridium and Globalstar form multiple beams using phased arrays covering the visible Earth. The multiple beams provide for spectrum conservation by frequency reuse. Iridium forms 64 beams which scan to continuously point to a point on the Earth as the satellite passes over that point. Globalstar has fixed beams which "sweep" over a point on the ground as the satellite passes over.

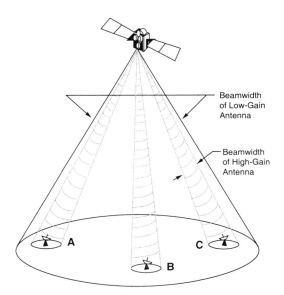

Fig. 13-14. Multibeam Coverage. To transmit simultaneously to A, B, and C, a multibeam antenna or a beamhopping antenna, with time-division multiplexing, will support higher data rates with lower transmitter power compared to a single, lower gain, broad-beam antenna system.

TABLE 13-16. Size and Mass of Typical Spacecraft Antennas. Antennas with shaped or multiple beams include the mass of complex feed systems. Lower gain values are at edge of coverage.

Type	Frequency Band (GHz)	Gain (dBi)	Beam-width (deg)	Mass (kg)	Satellite	Size (m)
Quad Helix	L (1.5)	16–19	18	1.8	Intelsat-V	0.4 × 0.4 × 0.47
Conical Log Spiral	S (2.2)	0–3	220	1.2	FLTSATCOM	
Parabola (fixed)	S (1.7)	16–19	18	3.9	GOES I, J, K	0.7 dia
Horn	C (4)	16–19	18	3.1	Intelsat-V	0.3 dia, 0.65L
Parabola w/ Feed Array	C (4)	21–25	*	29.4	Intelsat-V	2.44 dia
Parabola w/ Feed Array	C (6)	21–25	*	15.2	Intelsat-V	1.56 dia
Parabola—Steerable	Ku (11)	36	1.6	5.8	Intelsat-V	1.1 dia
Parabola w/ Feed Array	Ka (20/30)	45–52	*	47.1	SUPERBIRD	1.7 dia

*Beams shaped to illuminate specific land masses

The mass of a satellite antenna, including feed, depends largely on its size and the materials used in its construction. These factors are in turn a function of the frequency and beamwidth or gain. Table 13-16 lists some examples of satellite antenna weights, showing that the more complex shaped-beam antennas have relatively high mass because of their complex feed networks. Multiple and scanning beam antennas have comparable masses.

The power efficiency and mass of a satellite transmitter are often key factors in sizing a satellite. Figure 13-15 shows how the transmitter input power and mass varies with output RF power, based on actual satellite equipment. We can see that the solid-state transmitter has lower mass but requires more input power compared to the traveling wave tube amplifier. In general, solid-state transmitters are preferred for power outputs up to 5 or 10 W, except at frequencies below 2 GHz, where power outputs up to 80 W are achievable. Solid-state amplifiers are more reliable than the traveling wave tube amplifier, mostly because they require lower voltages. We expect improved solid-state amplifiers with higher powers and frequencies will become available during the next 10 years.

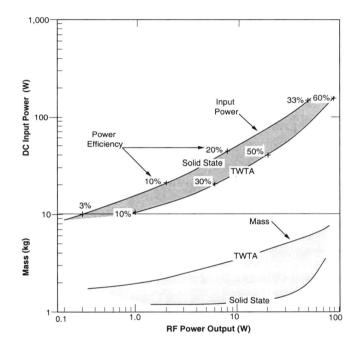

Fig. 13-15.   Satellite Transmitter Power and Mass vs RF Power Output. The curves derive from actual flight hardware. The data is relatively independent of output frequency.

Table 13-17 lists the payload parameters for FireSat, based on the link budgets in Table 13-13 and the payload characteristics found in Table 13-16 and Fig. 13-15. These parameters also enter into the total satellite power and mass budget discussed in Chap. 10.

13.5 Special Topics

This section discusses several special topics in selecting communications architectures.

13.5.1 Multiple Access: Sharing Communication Links

Some missions may require more than one uplink or downlink, especially where relay satellites have a number of satellites and ground stations interconnected in a sin-

TABLE 13-17. Communication Payload Parameters for FireSat. Parameters apply to both uplink and downlink.

Parameter	FireSat Up and Downlink
Frequency (GHz)	2.0/2.2
Transmitter Output Power (W)	20
Number Beams and Transmitters	1
Antenna Beamwidth (deg)	140
Antenna Diameter (m)	0.1
Antenna Mass (kg)	1.2
Transmitter Mass (kg)	2.5
Transmitter Input Power (W)	70

gle network. In such situations, it is cost effective to provide a means of sharing a limited amount of satellite link capacity between users.

Figure 13-16 shows three basic techniques for sharing link capacity. *Frequency division multiple access (FDMA)* assigns a single carrier frequency for each input signal. Bandpass filters at the receiver separate the individual carriers from each other. Most communications satellite systems use this technique with a repeater transponder operating in a linear (back-off) mode. The ground station is less expensive because the transmitter peak power is lower than the TDMA ground station described below.

Time division multiple access (TDMA) assigns a single time slot in each time frame to a single input channel. A digitized input signal is sampled and stored in buffer memory. These samples are then transmitted as short bursts within the assigned time slots. The bit rate during the burst is high, therefore requiring a high peak transmitter power. At the receiver the samples are sorted and stored, and then read out at the original rate. These samples are then converted to an analog signal and smoothed to obtain a replica of the original input signal. Some systems, including Intelsat and DSCS-III, use this method for some of their users.

If the satellite uses a multiple-beam antenna, a switching matrix on the satellite may be used with TDMA to route each time slot burst to the desired downlink antenna beam. The NASA ACTS [Naderi and Kelly, 1988] uses this technique, known as *Satellite Switched-TDMA (SS-TDMA)*. In some applications, the uplink uses the FDMA technique, and the downlink uses the TDMA [McElroy, 1988]. This method requires onboard processing (demodulation and remodulation) of the signal in the satellite, as discussed below.

A third technique, *code division multiple access (CDMA)*, consists of phase-modulating (BPSK or QPSK) a carrier with data, and then biphase-modulating the carrier with a pseudorandom noise (PN) code. (See Fig. 13-17.) The data rate is much lower than the PN code, or *chip* rate. Thus there are many code bits (or chips) per data bit. The receiver has a code generator which replicates the PN code of the desired signal on a carrier with frequency equal to the received carrier plus or minus an intermediate frequency (IF).

The PN codes are designed to have low cross-correlation properties (shown in Fig. 13-17B) so that two or more signals can be transmitted simultaneously at the same frequency with little mutual interference. Thus the received signals are essentially uncorrelated with the locally generated PN code and appear as noise to the receiver, except when identical received and locally-generated codes are aligned, or synchronized, in time. When this happens, the output of the mixer is a carrier containing only the narrow-band data modulation, which passes through the IF bandpass filter (BPF)

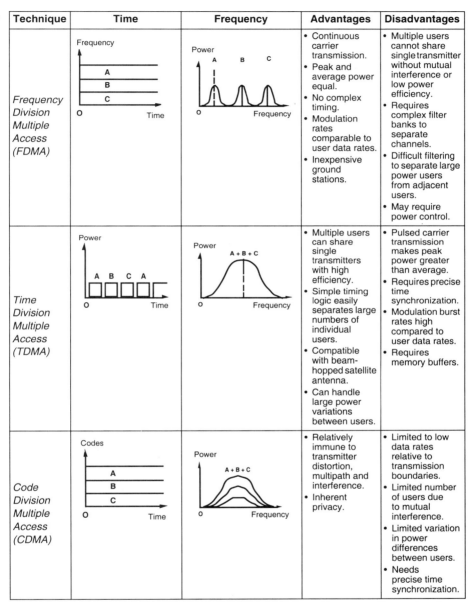

Technique	Time	Frequency	Advantages	Disadvantages
Frequency Division Multiple Access (FDMA)			• Continuous carrier transmission. • Peak and average power equal. • No complex timing. • Modulation rates comparable to user data rates. • Inexpensive ground stations.	• Multiple users cannot share single transmitter without mutual interference or low power efficiency. • Requires complex filter banks to separate channels. • Difficult filtering to separate large power users from adjacent users. • May require power control.
Time Division Multiple Access (TDMA)			• Multiple users can share single transmitters with high efficiency. • Simple timing logic easily separates large numbers of individual users. • Compatible with beam-hopped satellite antenna. • Can handle large power variations between users.	• Pulsed carrier transmission makes peak power greater than average. • Requires precise time synchronization. • Modulation burst rates high compared to user data rates. • Requires memory buffers.
Code Division Multiple Access (CDMA)			• Relatively immune to transmitter distortion, multipath and interference. • Inherent privacy.	• Limited to low data rates relative to transmission boundaries. • Limited number of users due to mutual interference. • Limited variation in power differences between users. • Needs precise time synchronization.

Fig. 13-16.   **Multiple Access Techniques Allow Different Users to Share the Same Transmission Channel.**

(Fig. 13-17D). The PN code modulation is completely removed, leaving the desired user signal "despread." At the same time, the signals from the undesired users appear at the output of the mixer as PN-coded signals with a spectrum width roughly twice the PN-code rate. Therefore, only a small portion of the undesired signal will pass through the narrow-band output BPF.

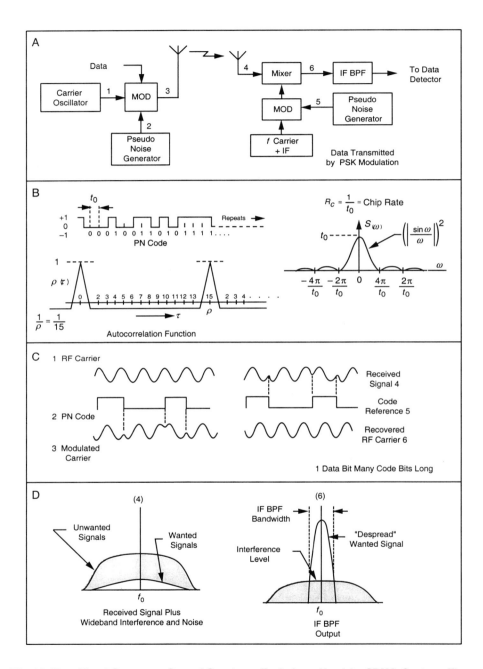

Fig. 13-17.　Direct-Sequence Spread-Spectrum Technique Used in CDMA System. The numbers in C and D above refer to the block diagram in A.

As shown in Fig. 13-17B, the power density of the PN-coded signal at center frequency is proportional to $1/R_c$ ($R_c = 1/t_0$), where R_c is the chip rate. The spectrum of the desired signal after despreading is proportional to $1/R$, where R is the data rate. Thus the power density of the desired signal is increased, relative to the power density of an unspread interfering signal, by the ratio R_c/R. This ratio is commonly referred to as the *coding gain* of the spread-spectrum system.

Note that in order to receive the desired signal, the locally generated code must be identical to the desired signal's code, **and** the two codes must be synchronized. We may accomplish this synchronization by scanning the locally generated code in time until synchronization takes place. To reduce the time to synchronize, a short code may be used for acquisition, such as the C/A signal in GPS.

The number of simultaneous users which the CDMA system can handle is limited by the noise generated by the undesired user signals. Suppose all CDMA user code rates and data rates are the same, and their carrier powers at the receiver input are equal. How many users can the system support? Eq. (13-28) estimates the maximum number of users, M, that the system can accommodate before the received energy-per-bit, E_b/N_o, drops below threshold.

$$M = 1 + R_c \left(\frac{1}{(E_b/N_o)R} - \frac{1}{C/N_o} \right) \qquad (13\text{-}28)$$

where C/N_o is the received carrier-to-noise-density ratio from one user (Eq. 13-15a), R_c is the PN-code chip rate, R is the data rate, and R_c is much greater than R. Note that in Eq. (13-28) C/N_o and E_b/N_o are power ratios, not dB.

An example of a CDMA system is the GPS navigation message on the C/A signal. Each satellite transmits a PN code with a different time phase. $R_c = 1.023$ Mbps, $R = 50$ bps, $E_b/N_o = 10$ dB $= 10$, and $C/N_o = 38.6$ dB-Hz $= 7,244$ Hz. Substituting into Eq. (13-28), we obtain $M = 1,906$ channels. This, of course, is much greater than the actual number of channels required for GPS (less than 28).

Remember that M is the number of equal-power users. Suppose one user is received with power 10 times greater than the other M-1 users. If this is an undesired user, it will generate interference equivalent to 10 undesired equal-powered users, thus reducing M by 9. For this reason an efficient CDMA system often requires some automatic means of controlling the transmitter power of each user.

The CDMA system is generally less bandwidth-efficient than FDMA or TDMA. On the other hand, the CDMA system is less susceptible to interference, including multipath caused by reflections from buildings or other objects. This makes CDMA of special interest to satellite communication systems with mobile terminals. For more information on multiple access techniques, see Morgan and Gordon [1989], Chap. 4.

Code division multiple access, CDMA, is used by the Globalstar cellular telephone satellite system in order to preserve spectrum. Each signal is 1.23 MHz wide, corresponding to the Cellular Telephone Industries Association Standard IS-95 for terrestrial cellular telephones. Up to 128 different signals can be "stacked" on the same frequency channel by virtue of being biphase modulated by different pseudorandom noise codes, or Walsh codes, as discussed above. The receiver recovers the signal by applying the same code to the incoming signal, demodulating it. If you divided this band by 64 kHz, the bandwidth of a typical digital telephone signal, you get only 19 signals in the same bandwidth. In practice, the number of signals on a given channel is limited to 30 to 50 because of satellite transmit power limitations.

13.5.2 Payloads with Onboard Processing

As discussed previously, communication relay satellites usually employ a *repeater transponder* (also called a *non-regenerative repeater* or *"bent pipe"*), which receives the uplink signals from all of the user transmitters operating in the assigned uplink frequency band, shifts the frequency to the downlink band, amplifies, and retransmits. This transponder is reliable and simple to use. But this technique has several disadvantages when more than one signal is received at the same time. First, the transponder's transmitter must operate as a linear amplifier to reduce mutual interference caused by intermodulation distortion. We must back off the input signals about 3 to 6 dB from maximum, thus reducing the transmitter's power output to roughly one-half to one-fourth of the maximum saturated power. We do so by controlling the transponder gain in the satellite (either automatically or by ground command), or by regulating each ground station's transmitter power. Second, the strongest uplink signal will tend to capture the satellite transmitter power, thus suppressing the weaker signals. This makes the transponder particularly vulnerable to uplink jamming.

An alternative to the repeater transponder is the *onboard processing transponder* or *regenerative transponder*. This transponder demodulates the received signal on board the satellite, and then routes the signal to the appropriate downlink modulator/transmitter and antenna beam (see Fig. 13-18). An example of this architecture is the Fleetsat Experimental Package developed for the Air Force by Lincoln Laboratory [McElroy, 1988]. The FDMA technique is selected for the uplink to allow use of low-power, ground-station transmitters as discussed above. The FDMA signals are demodulated by the satellite transponder, and the messages are reformatted into a single downlink TDMA data stream. TDMA is selected over FDMA for the downlink to allow the satellite transmitter to operate at saturation for high power efficiency. While more complex, the processing transponder overcomes the disadvantages of the repeater transponder.

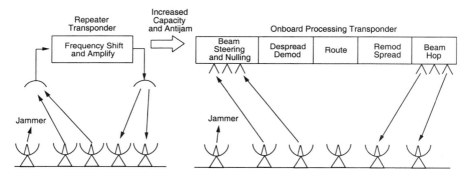

Fig. 13-18. **Two Payloads Commonly Used in Communication-Satellite Architectures.** The despread/spreading capability is used in military satellites to counter the effects of jamming.

A large number of 20/30 GHz synchronous (except for Teledesic) satellites were proposed to the FCC in August 1995. None has been licensed yet. They are listed in Table 13-18. Most of these satellites plan to use advanced onboard signal processing transponders. This allows onboard rerouting of the data, and improvement of the overall link performance because the noise of the uplink is not carried over to the downlink. Many of these proposals promised digital data transmission of the order of 10^{-10} bit error rate with E_b/N_os on each link of the order of 5 dB. They used coding scheme such as those discussed in Table 13-11 and the accompanying text in Sec. 13.3.3.

TABLE 13-18. 20/30 GHz FCC Proposals for Commercial Data Relay Satellites (Internet, Telephone Trunking).

Satellite Name	Owner	Satellite Name	Owner
Norsat	Norwegian companies	GE Americom	General Electric
Spaceway	Hughes	Motorola	Motorola
Teledesic	Teledesic	KaStar	
Cyberstar	Space Systems/Loral	Pan Am Sat	Pan Am Sat/Hughes
AT&T	AT&T	Net Sat 28	Net Sat
Lockheed Martin	Lockheed Martin	Morning Star	
Echostar		Cellular Vision	

Note: A total of 55 proposals have been filed with the International Telecommunications Union (ITU).

13.5.3 Antijam Techniques

Because the satellite is usually in view of a large segment of Earth, RF interference from Earth-based transmitters, either unintentional or deliberate, may occur. The frequency allocation procedures discussed previously minimize unintentional interference. Intentional interference, or jamming, is of particular concern in military applications. *Jamming* consists of transmitting a large modulated carrier to the receive terminal at approximately the same frequency, overwhelming the desired signal and thus disabling the link.

We can reduce the effects of jamming by using spread-spectrum modulation techniques [Dixon, 1984] to spread the transmitted signal in a pseudorandom manner over a bandwidth much larger than the data rate. The receiver takes advantage of the fact that he knows the code used to modulate the transmission while the jammer does not. A replica of the pseudorandom waveform is generated at the receiver and correlated with the received signal to extract the data modulation. Using this method, we can reduce the received jamming power relative to the desired signal by the ratio of the spread-spectrum bandwidth to the unspread signal bandwidth. For example, by hopping a BPSK-modulated signal of 100 bps over 1 MHz, the jamming power, on the average, is reduced by a factor of approximately 10,000, or 40 dB.

In a communication relay satellite, onboard processing is highly desirable to despread the received signal before retransmitting it on the downlink. Otherwise the uplink jamming signal will capture most of the satellite transmitter's power, leaving little for the signal. Another technique for countering uplink jamming, employed by the DSCS-III, is to generate a null in the antenna beam pointed toward the jamming source. This technique can lower the jamming power by 20 to 40 dB relative to the power of the received signal.

The satellite crosslink may also be jammed. The satellite can reject jammers located on Earth by using narrow antenna beams pointed away from the Earth. Operating at 60 GHz takes advantage of the oxygen absorption band, thus shielding the satellite from the Earth. Crosslinks may also use spread-spectrum and antenna-nulling techniques.

13.5.4 Security

A characteristic of space-ground communication is the ease with which the link can be intercepted by an unauthorized user, who may receive the data for his own use, or, even worse, take control of the satellite by transmitting commands to it. *Data encryption* techniques help us avoid these problems by denying access to the data and the satellite command channel unless the user has the correct encryption key. Recent developments have led to complete encryption and decryption devices being placed on single VLSI chips, thus adding little to satellite mass. The main issues are distributing the key and synchronizing time. To make sure a link remains secure over the life of the satellite, the encryption key must change at regular intervals, because others will monitor and eventually uncover it.

The receiver decryption device must be accurately synchronized with the transmitter encryption device to recover the original data. With some systems, both the satellite and ground station may need a very accurate atomic clock, especially if the data rate is high and one must acquire the signal within seconds. For command links to stationary satellites, the data rate is low and the acquisition time can be long. For this application, crystal oscillators are accurate enough. An alternative to atomic clocks is a GPS receiver, which automatically synchronizes itself to the GPS time standard.

13.5.5 Diversity Techniques

Diversity techniques consist of transmitting or receiving the same signal more than once to increase the probability of receiving the signal correctly. For example, a satellite may transmit a signal simultaneously to two ground stations. If the distance between these ground stations is greater than about 5 km, the probability of high rainfall attenuation existing at both locations at the same time is small. This technique is called spatial diversity and is an effective way to increase the availability of a satellite-ground station link operating at frequencies above 20 GHz, where rain attenuation can be large [Ippolito, 1986].

A second example of diversity is to transmit the same signal two or more times at different frequencies or time intervals. For example, multipath fading may be caused by reflections of the signal from parts of the aircraft structure in a satellite-to-aircraft link. The multiple reflected signals will interfere with the main signal, causing the amplitude of the received signal to vary with frequency and time as the aircraft moves. Use of frequency or time diversity will increase the probability that the message will be received correctly in at least one frequency or time interval. In military applications, frequency hopping over one data symbol provides frequency diversity protection against partial band jamming. Forward error correction coding followed by interleaving (e.g., scrambling the order in which the bits are transmitted) provides a form of time diversity protection against a pulse jammer. Time interleaving improves the decoding performance by randomizing a burst of errors caused by pulse jamming.

A third example of diversity is a technique used by Globalstar, "satellite diversity." In this case a ground station talks to a UT with circuits through two separate satellites. This is not to avoid rain outages as discussed above, but rather outages caused by blockage from buildings or trees for mobile UTs. If the path to one satellite is temporarily interrupted, the power is increased on the other link to maintain the contact. Also, when one satellite is "setting," this technique maintains the conversation through the "other" satellite while a new "rising" satellite can begin to carry a circuit. In this way, handover between satellites is transparent. This is accomplished by using a *RAKE receiver*. A RAKE receiver has the property of having several parallel digital processing channels which can correct for Doppler frequency offset and time delay so as to combine several digital signals in time alignment for the maximum signal-to-noise ratio. The technique requires two ground antennas, which are putting many circuits through each satellite simultaneously.

13.5.6 Optical Links

In recent years lasers generating narrow band energy at optical frequencies provide an attractive alternative to the microwave frequencies discussed above. Unfortunately, clouds and rain seriously attenuate optical links. Therefore, optical links have limited application in satellite-Earth communications. However, an optical link is well suited for communication between two satellites. Intersatellite links have been designed using optical links with capacities above 300 Mbps.

Optical crosslinks are superior to microwave crosslinks for high data rates because they can obtain extremely narrow beamwidths and high gains with reasonable size (see Fig. 13-19). Also, frequency allocation problems do not exist in the infrared or visual bands. On the other hand, the narrow optical beams are difficult to acquire and point accurately, requiring complex and sometimes heavy pointing mechanisms.

Figure 13-20 compares RF and laser crosslinks, demonstrating that RF links are generally better for data rates less than about 100 Mbps because of their lower mass and power. However, development of more efficient lasers and lighter steerable optics may soon make lower rate optical links attractive.

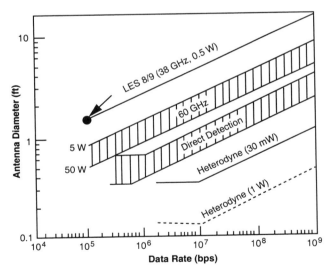

Fig. 13-19.   **Optical Systems (that is, Direct Detection and Heterodyne) Require Smaller Antenna Diameters Compared to RF Crosslinks.** [Chan, 1988].

One application of optical links between satellites and Earth is the *blue-green laser link* being developed by DARPA and the U.S. Navy for submarine communications [Weiner, 1980]. The laser frequency of 6×10^{14} Hz was chosen for its ability to penetrate sea water. Even so, the water loss can range from 5 to 50 dB or more, depending upon the actual depth of the submarine. In addition, loss due to cloud scattering is

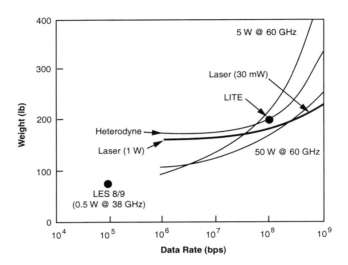

Fig. 13-20.   A Comparison of Crosslink Package Weights. At data rates greater than abou 100 Mbps, an optical system provides the lightest package [Chan, 1988].

between 4 to 14 dB. These losses are overcome by using low data rates of 10 to 10(bps, advanced coding techniques, and high-gain, narrow-beam optics.

Commercial satellites for both 2 GHz mobile satellite service and 38/49 GH trunking service have proposed use of optical intersatellite links between satellites fo efficient data transmission. They promise the advantages cited above in Figs. 13-1! and 13-20. A listing of these proposals is in Table 13-19.

TABLE 13-19.   Communications Satellite Systems.

System Name	Orbit	No. Satellites	No. Antenna Beams	Frequency (GHz)	Function
Star Lynx	GEO	4	204	38/49	Wideband data relay
	MEO	20			Wideband data relay
Expressway	GEO	14	204	38/49	Wideband data relay
SpaceCast	GEO	6	204	38/49	Wideband data relay
Orblink	MEO	7	100	38/49	Wideband data relay
CAI Satellite	GEO	1	5	38/49	Wideband data relay
Teledesic V-band Supplementary	LEO	72	64	38/49	Wideband data to the home/office
Celestri	LEO	72	64	38/49	Wideband data relay
	GEO	6			
Pentriad	Molniya	9	—	38/49	Wideband data to home/office
LEO-1	LEO	48	28	38/49	Wideband data relay
Cyberpath	GEO	4	—	38/49	Wideband data relay

TABLE 13-19. Communications Satellite Systems. (Continued)

System Name	Orbit	No. Satellites	No. Antenna Beams	Frequency (GHz)	Function
GS-40	LEO	80	30	38/49	Wideband data relay
GE*Star Plus	GEO	11	48	38/49	Wideband data relay
Lockheed-Martin	GEO	9	56	38/49	Wideband data relay
VStream	GEO	12	163	38/49	Wideband data relay
Aster Systems	GEO	25	48	38/49	Wideband data relay
Macrocell	LEO	96	228	2.0/2.2	Data relay to home/office
ICO Services	MEO	10	163	2.0/2.2	Data relay
Horizons	GEO	4	100–200	2.0/2.2	Data relay
Boeing Aero Nav	MEO	16	37	2.0/2.2	Aircraft data distribution
Ellipso 2G	Elliptical	—	127	2.0/2.2	Data relay
Celsat MSS HPCS	GEO	—	118	2.0/2.2	Data relay
PCSAT	GEO	—	23	2.0/2.2	Data relay
Globalstar GS-2	LEO	64	96	2.0/2.2	Data relay
	GEO	4	64/469		
Teledesic	LEO	288	64	18/30	Intelsat connection to home/business
Orbcomm	Little LEO	24	1	VHF	Store-and-forward comm relay
Starsys	Little LEO	24	1	VHF	Store-and-forward comm relay

References

Chan, V.W.S. 1988. "Intersatellite Optical Heterodyne Communications Systems." *The Lincoln Laboratory Journal*, 1(2):169–183.

Chapell, Paul M. 1987. "Implementation of SDI Resources for MILSATCOM User Support." In *Conference Record*, MILCOM 87: 20.6.1–20.6.5.

Crane, Robert K. 1980. "Prediction of Attenuation by Rain." *IEEE Trans. on Communications*. Com-28(9): 1717–1733.

Deng, Robert H., and Daniel J. Costello. 1989. "High Rate Concatenated Coding Systems Using Bandwidth Efficient Trellis Inner Codes." *IEEE Trans. Comm. Theory.* 37(5): 420–427.

Dixon, Robert C. 1984. *Spread Spectrum Systems (2nd edition)*. New York: John Wiley & Sons, Inc.

Gerson, Ira A., and Mark A. Jasiuk. 1990. "Vector Sum Excited Linear Prediction (VSELP) Speech Coding at 4.8 Kbps." In *International Mobile Satellite Conference*. 678–683.

Ippolito, Louis J., Jr. 1986. *Radiowave Propagation in Satellite Communications.* New York: Van Nostrand Reinhold.

Jasik, Henry, ed. 1961. *Antenna Engineering Handbook.* New York: McGraw-Hill.

Lee, Y.S. 1988. "Cost-Effective Intersatellite Link Applications to the Fixed Satellite Services." In *AIAA 12th International Communications Systems Conference.*

Lo, Y.T. and S.W. Lee, eds. 1988. *Antenna Handbook, Theory, Applications, and Design.* New York: Van Nostrand Reinhold.

McElroy, D. 1988. "The FEP Communications System." In *AIAA 12th International Communication Satellite Systems Conference Proceedings*, pp. 395–402.

Morgan, Walter L. and Gary D. Gordon. 1989. *Communications Satellite Handbook.* New York: John Wiley & Sons, Inc.

Mottley, T.P., D.H. Marx, and W.P. Teetsel. 1960. "A Delayed-Repeater Satellite Communications System of Advanced Design." *IRE Trans. on Military Electronics* April-July: 195–207.

Naderi, M., and P. Kelly. 1988. "NASA's Advanced Communications Technology Satellite (ACTS)." In *AIAA 12th International Communication Satellite Systems Conference Proceedings*, pp. 204–224.

National Aeronautics and Space Administration. 1983. *Propagation Effects Handbook for Satellite Systems Design.* Washington, D.C.: U.S. Government Printing Office.

O'Shaughnessy, Douglas. 1987. *Speech Communication, Human and Machine.* Reading, MA: Addison-Wesley Publishing Co.

Panter, Philip F. 1965. *Modulation, Noise, and Spectral Analysis.* New York: McGraw-Hill.

Sklar, Bernard. 1988. *Digital Communications Fundamentals and Application.* Englewood Cliffs, NJ: Prentice Hall.

Ungerboeck, G. 1982. "Channel Coding with Multilevel/Phase Signals." *IEEE Trans. Information Theory*, vol. IT28, Jan., pp. 55-67.

Viterbi, A.J. 1967. "Error Bounds for Convolutional Codes and an Asymptotically Optimum Decoding Algorithm." *IEEE Trans. Information Theory.* IT13:260–269.

Weiner, Thomas F., and Sherman Karp. 1980. "The Role of Blue/Green Laser Systems in Strategic Submarine Communications." *IEEE Transactions on Communications.* Com-28(9): 1602–1607.

Yuen, Joseph H. ed. 1983. *Deep Space Telecommunications Systems Engineering.* New York: Plenum Press.

Chapter 14

Mission Operations

John B. Carraway, Gael F. Squibb, *Jet Propulsion Laboratory*
Wiley J. Larson, *United States Air Force Academy*

Mission operations is the collection of activities performed by operations teams during the flight phase of the mission, together with the operations design activities they perform pre-launch, including development of a mission operations concept, policies, data flows, training plans, staffing plans, and cost estimates. The *mission operations system* is the integrated system of people, procedures, hardware, and software that must cooperate to accomplish these tasks. NASA, the DoD, industry, and other organizations have different requirements for mission operations and each organization has developed it's own philosophy and style for carrying out the mission.

Mission operations focuses on the period after launch, but substantial work must be done during all phases of mission design and development to prepare for operations. Failing to take operations into account in preliminary mission design will significantly increase both the cost and risk of the mission.

We must distinguish between the *mission concept*—how we conduct the overall mission and how the elements of the mission fit together (Chap. 1)—and the *mission operations concept*—how we do operations to carry out the mission objectives. In this chapter we define 13 key functions performed by mission operations (see Fig. 14-1) and discuss how they combine to meet the mission operations concept. Hardware, software, people and procedures operate together to complete these 13 functions. We must carefully trade automation against ground crew operations—on the ground and in space. Automating some of these functions can lead to lower operations costs and, in most cases, lower life-cycle costs. Organizations may group or name these functions differently, but we believe our list captures the tasks essential to mission operations.

The operations director must first define and negotiate the requirements on operations; then try to influence them to reduce cost and complexity. With requirements and constraints in place, the director must decide which functions to do, as well as their scope and how they can be done. Depending on the size and complexity of the mission, a director may even have to add functions to the list. In addition, the director must address organizational, hardware, and software interfaces between the functions.

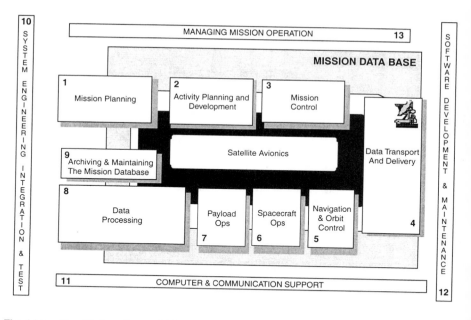

Fig. 14-1. The 13 Functions of Mission Operations System and How they Interact. Functions in the shaded area share data within the mission database. We briefly discuss them in Sec. 14.2. Functions 4, 8, and 9 are part of a broader category called data services.

Figure 14-1 overviews a mission operations system that carries out the mission operations concept. This system processes information and controls the ground and space assets so users and operators get needed information and services. Most missions today focus on providing information to users or customers—the downlink. But control drives much of the cost and complexity of operations—the uplink. We will see this in the following sections.

There is a fundamental difference in mindset between designers and operators of space systems. Designers expect things to work; operators expect things to go wrong! Both perspectives are necessary. Mission operators usually help develop the mission concept, but they often get involved too late—long after the initial design phase in which teams can make cost-effective choices. Operators should get involved early enough to influence planning and design, so they can reduce the life-cycle costs of space missions. They help develop flight rules, which govern operators' responses to anomalies and ensure timely responses consistent with mission success.

Figure 14-1 shows 13 mission operation functions. We can combine or eliminate some of them to reduce cost and complexity. Nine share data through the mission database and the space element's avionics and they require extensive data processing:

1. Mission planning—deciding what to do and when

2. Activity planning and development—creating operational scenarios and developing command loads

3. Mission control—managing daily activities

4. Data transport and delivery—establishing communication links and managing data flow

5. Navigation and orbit control—determining where the spacecraft is and planning maneuvers

6. Spacecraft operations—managing the spacecraft bus

7. Payload operations—managing the spacecraft's payload

8. Data processing—managing the data flow on the spacecraft and on the ground

9. Archiving and maintaining the mission database—managing all data generated by the mission

The order from upper left (mission planning) clockwise to lower left reflects the usual order in processing: analyzing uplinks, analyzing downlinks, and then planning new uplink activities. The other four functions provide support for all aspects of mission operations:

10. Systems engineering, integration, and test—maintaining the mission operations concept

11. Computers and communications support—planning for and maintaining the infrastructure

12. Developing and maintaining software—spread throughout all functions

13. Managing mission operations—maintaining the big-picture perspective, managing interfaces, and budget

A big question is, "How many people must we have to do mission operations?" The obvious answer is, "It depends." It depends on the organization's operational requirements and constraints, as well as the number and complexity of functions, complexity of the mission design, complexity of flight and ground systems, and how much risk the operations organization will take.

The number of operations people required can vary significantly between military, commercial, and scientific spacecraft and can vary significantly between individual missions within each category. The cost difference between a 5 person operations team and a 50 person operations team is 6 to 7 million dollars per year. This chapter will try and identify how differences in payload and spacecraft design, space environment, ground systems, the ops organization design, and risk policies can influence operations costs.

Section 14.1 describes a process for operations design and development. We emphasize the importance of defining an early operations concept to clarify requirements on the operations system. Armed with these requirements, we can meet goals for operations costs by negotiating with the other space mission elements to reduce their operational complexity, thus keeping costs acceptable across the full mission life cycle. Once we've iterated through the process, gotten costs within guidelines, and developed a workable mission operations concept and design, we can more effectively deal with the cost drivers for mission operations.

Section 14.2 defines mission operations in terms of 13 functions that are common across a wide variety of mission types and operations team sizes. Analyzing each function helps us understand recent trends in trying to reduce costs and in automation. Section 14.3 then discusses what determines the size and cost of space mission operations.

We're now converting functions that people have done into ones that hardware and software can do. We're also moving onto the spacecraft parts of operations functions traditionally done on the ground. Using software in place of people doesn't always save money but we need to assess this option. Section 14.4 provides some guidelines and insights about high payoffs from investments in operations technology and automation.

For a detailed discussion of the design of mission operations for scientific remote sensing missions, see Wall and Ledbetter [1991]. Because of the continuing pressure to contain cost, a number of authors has described methods for operations cost reductions on both individual missions [Bloch, 1994; Mandl et al., 1994; van der Ha, 1992] and various classes of missions [Cameron, Landshof, and Whitworth, 1994; Hughes, Shirah, and Luczak, 1994; Landshof, Harvey, and Marshall, 1994]. Much of this cost reduction strategy has been summarized by Boden and Larson [1995] and Marshall, Landshof, and van der Ha [1996].

14.1 Developing a Mission Operations Plan

For unmanned space missions the *Mission Operations Plan* (MOP) describes in operators' and users' terms the operational attributes of the flight and ground-based elements. The plan usually results from the cooperative work of several disciplines. Its development is similar to that of a space mission concept but the MOP is more detailed and emphasizes the way we operate the mission and use the flight vehicle (operational characteristics), crew, and ground operations team. It is generated in phases and becomes more detailed as our mission design progresses.

The MOP follows from, and must be consistent with, the mission concept. It is one of the most important deliverables from the mission operations organization before launch. A good MOP helps assure that the operations organization provides a tested and certified mission operations system that meets requirements at the lowest cost. The operations organization and management also use the MOP as their main tool to influence the design and operability of the mission and spacecraft. Its iterative development often changes the mission concept, flight vehicle, and software, both flight and ground. By combining the operations concept with assessments of operational complexity, we can determine the probable costs of operations early in the mission design. Table 14-1 outlines the steps needed to develop a useful MOP.

Step 1. Identify the Mission Concept, Supporting Architecture, and Key Performance Requirements

We begin developing the MOP by examining the mission concept and supporting architecture. By obtaining the information listed below, we can describe the mission in language that users and operators understand. Sometimes information isn't available. Sometimes, the operations design isn't very far along or isn't specified in the mission architecture. In these cases, we make assumptions and document them in the MOP. Of course, we update the MOP as the assumptions change or more data becomes available. We can then determine the cost of changes by modifying the operations concept and re-evaluating the cost and complexity of the mission operations.

Mission objectives identify what the ground crew, spacecraft, and payload must do for mission success. They help us define and describe how people will use the payload data. We also need to know the timeliness requirements for payload-processed data and the mission's overall criteria for success, such as the amount of data to be returned.

TABLE 14-1. **Developing a Mission Operations Plan.** Many items are detailed by Boden and Larson [1995]. See text for a discussion of each step.

Step	Key Items
1. Identify the mission concept, supporting architecture, and performance requirements (Chap.1)	• Mission scope, objectives, and payload requirements • Mission philosophies, strategies, and tactics • Characteristics of the end-to-end information system • Identify performance requirements and constraints
2. Determine scope of functions needed for mission operations (Sec. 14.2)	• Identify functions necessary for different mission phase • Functions usually vary for different mission concepts and architectures. Combine or eliminate if possible
3. Identify ways to accomplish functions and whether capability exists or must be developed (Sec. 14.2)	• Where functions are accomplished (space or ground) • Space-based crew capabilities • Degree of automation on the ground • Degree of autonomy on spacecraft and for flight crew • Software reuse (space and ground)
4. Do trades for items identified in the previous step.	• Try to define operational scenarios before selecting options. These trades occur within the operations element and include the flight software
5. Develop operational scenarios and flight techniques	• *Operations scenarios* and *flight techniques* are step-by-step activity descriptions. Identify key issues and drivers • Develop scenarios and flight techniques for functions from step 2 and options selected in step 4
6. Develop timelines for each scenario	• Timelines identify events, their frequency, and which organization is responsible. They drive the characteristics for each operations function
7. Determine resources needed for each step of each scenario	• Allocating hardware, software, or people depends on what, how quickly, and how long functions must be done
8. Develop data-flow diagrams (Sec 2.1.1)	• *Data-flow diagrams* drive the data systems and the command, control, and communications architecture
9. Characterize responsibilities of each team	• Identify organizations involved and their structure, responsibility, interfaces, and size. To be cost-effective, minimize the number of organizations and interfaces • Develop training plan for ground team and flight crew
10. Assess mission utility, complexity, and operations cost driver	• Refine development and operations costs each time you update the Mission Operations Plan
11. Identify derived requirements	• Identify derived requirements and ensure consistency with top-level requirements • Identify cost and complexity drivers • Negotiate changes to mission concept and architecture
12. Generate technology development plan	• If the technology to support mission operations doesn't exist, generate a plan to develop it
13. Iterate and document	• Iteration may occur at each step • Document decisions and their reasons

The mission description tells us the trajectory, launch dates and windows, trajectory profile, maneuver profile needed to meet mission objectives, mission phases, and the activities required during each phase. Observation strategies describe how we'll collect the mission data. We define and finalize the observation strategies and mission description before launch or adapt them to data gathered during the mission.

Sometimes, the operations organization develops the operating philosophies, strategies, and tactics. They may relate to the mission objectives or simply derive from the designer's background or experience. We must determine whether the philosophies, strategies, and tactics are mandatory, highly desirable, or just personal or organizational preference. Examples include

- Maximize real-time contact and commanding versus onboard autonomy and data-storage
- Maximize the involvement of educational institutions and teach students key aspects of issues like operations or space physics
- Make sure a central authority approves all commands
- Limit the image budget to 50,000 images
- Deploy a communications satellite early in the mission

The mission sponsor and the project manager may impose nontechnical constraints. Operators usually follow these constraints until the project manager learns they increase the mission's cost or make operations unacceptable or unsafe. Examples of program constraints include

- Limit mission cost and cost profiles
- Use a specific tracking network (for example, TDRS)
- Use existing flight hardware
- Use existing ground systems and design the spacecraft to be compatible with them
- Centralize or distribute operational teams
- Use multi-mission versus project-dedicated teams
- Involve educators and the academic community, including students

We must identify capabilities and characteristics of the end-to-end information system early on so we clearly understand the mission's information needs. These requirements at the system level include

- Using information standards
- Locating capabilities and processes (includes both space and ground)
- Characterizing inputs and outputs for the information systems

We must state the information systems requirements in terms that operators and users can understand, not in jargon used by computer scientists and programmers.

As discussed in Chap. 15, most missions are designed around a specific agency's ground system, such as the Air Force's Satellite Control Network, NASA's Satellite Tracking and Data Network (STDN), or the European Space Operations Center. Each ground system has standard services that support the mission at lower cost if the mission meets specified interfaces and standards. We follow these requirements on the flight system until they hinder our ability to meet mission objectives.

This first step is key to the overall success and cost of mission operations. Here we gain insight about what to do and why. We begin determining performance requirements and constraints that will drastically affect the mission's cost and complexity. If we get the requirements wrong, we get the system wrong.

Step 2. Determine Scope of Functions Needed for Mission Operations

Before deciding how functions must be done, we usually divide the mission into discrete, workable phases such as launch, early orbit, normal operations, entry, descent, and landing. These phases usually have distinct goals and objectives, so their operational requirements are different. The mission concept drives top-level functions, but abilities of the crew, spacecraft, and payload determine the detailed functions we must carry out. A completely autonomous payload requires few crew operations, whereas a spacecraft payload that can't compute or store enough data onboard may require more control or automation on the ground. Thus, to determine what we must do, we have to understand characteristics of the spacecraft bus and payload. Characteristics essential to the mission concept may become clear early in the conceptual design phase. Or we may develop them as part of the Mission Operations Plan. Through iterative discussions, the operators and developers define the characteristics described below.

Users often ask operators to support a wide variety and number of payloads during a mission. Including payload designers and mission planners while developing the mission concept leads to timely definition of the payload characteristics. For a multiple-payload mission, we must understand early how each payload's constraints interact with the operations of other payloads and the ground system.

To understand how a payload operates, we must describe what the payload does during an operational period by asking

- What are the payload attributes?
- What is the commanding philosophy—buffer use, micro-commands, tables?
- Does the payload use default values? Can ground operators change them?
- Can some commands damage the payload or endanger the spacecraft?
- Do some operations depend on previous commands?
- Does the payload use position commands or incremental commands to control rotating or stepping mechanisms?
- What command classes does the payload use—real-time, stored, 2-stage?
- What processing occurs within the instrument?
- How can we describe the payload in terms of
  - CPU/memory, closed-loop functions, and predictive commanding vs. event-driven commanding?
  - Instruments the space element must control?
  - Mechanical power and thermal attributes?
  - Avoidance areas (Sun, Earth, South Atlantic Anomaly, Venus, or Moon)?
  - Requirements for controlling the space element?
  - Safety constraints?

- Do the payload apertures drive the space vehicle's pointing control?
- What are the user-specified parameters for operation?
- How do these parameters convert into payload commands?
- What is the payload heritage?
- What ground processing and analysis do we need to support its operation?

The example below shows how planning payload operations helps us define the mission concept. It also shows how operational workarounds support mission success.

An instrument's aperture had a field of view of ten arc-sec. The spacecraft had pointing capability of 20 arc-sec. In this case, the operators could never be sure the object was in the instrument's field of view. The solution was to:

- Command the spacecraft to the desired position
- Observe with a different instrument having a wider field of view to see how far the spacecraft was off the desired position
- Generate attitude commands to move the spacecraft slightly (tweak commands) until the actual attitude corresponded to the desired attitude
- Verify the attitude errors were gone
- Select the instrument with the narrow aperture and observe as specified

This single design error caused real-time operations, such as decision making and commanding, to become nonroutine for this mission—a big cost driver. These operations required more ground software, controllers trained in commanding the attitude-control system, and more people whenever they used the instrument. The goal of generating a MOP is to identify early any incompatibilities and cost drivers like this one—before we design and build any hardware.

It's a good idea to ask the designer of a payload instrument how to go from an observer requirement to a set of commands for the instrument. Sometimes, the answer is simple, but it could be complex if instruments have been designed for a laboratory rather than for space operation.

As is true for the payload's capabilities and characteristics, timely definition of the spacecraft's characteristics depends on including spacecraft designers and mission planners in developing the mission concept. People working on the concept have to answer the following types of questions for the overall spacecraft and its subsystems:

- What are the spacecraft's operational attributes?
- How are the values of these commands determined?
- How many commandable states are required?
- Are engineering calibrations required? What are the purpose, frequency, and schedule constraints of the calibrations?
- How many engineering channels need monitoring?
- Do these channels provide subsystem-level information to the operators, or must operators derive information about subsystems?
- Are guide stars used? If so, how are they selected?
- How does the pointing-control accuracy compare to instrument requirements?

- What types of payload, ground and spacecraft system margins exist and which must be monitored and controlled in real time?

- What expendables need monitoring during flight?

- Does the spacecraft subsystem use any onboard, closed-loop functions?

- What are the attributes of the spacecraft's data system?

- What processing must we do on the ground to support spacecraft operations?

- What is the heritage for each of the spacecraft's subsystems?

Consider an example of how a simple design decision affects operations:

> The Galileo spacecraft was designed to take heat from the radio-isotopic thermoelectric generators and use it to warm the propulsion system. This design saved weight and power, and it cost less to develop. But the spacecraft's operational characteristics tied together subsystems for propulsion health and safety, thermal transfer, and power. Operators had to check each command load to see how it changed power states and affected the propulsion subsystem. Engineers from power, thermal, and propulsion had to check each activity, even if only the payload instrument's states changed. For example, turning an instrument on or off caused the heat output of the thermo-electric generator to change.

This example shows that highly-coupled subsystems can make operating the system more complex and costly.

We must work with the end-users or customers to determine how, and how often, they require data from the payload. We also need to identify key operator tasks to operate the payload successfully. By understanding these data products and required actions, we can start designing how to operate the payload, as well as to retrieve and process the data.

We must understand how confident the end-users are about the products. Often, they don't know what they want until they see how the payload works in flight and what it observes. In these cases, we must develop baseline processes before launch and refine them after launch. For attached and deployable payloads from manned space-craft, the payload must be mature enough to ensure no safety risks exist, either within a payload, or with another payload running simultaneously. We must develop early the procedures for payloads that we can maintain in flight, so we can use them during stand-alone and integrated crew training.

We also need to define the product's relationship to the payload data by answering the following questions:

- Is the product based on the payload's raw data or must we remove the payload instrument's signatures?

- Must the data be calibrated? How? Does it involve processing special calibration observations? At what rate do we expect the calibration files to change?

- Does the data need to be converted into geophysical units? How? Where do the algorithms for this conversion come from? Must the project generate them and update the mission database, as they become more refined?

- What are the formats and media of the payload's data products? Is there a community standard, such as the Flexible Image Transport System format used on all NASA's astrophysics missions?

- What ancillary data must we provide so the end-user can interpret the payload data—spacecraft position and attitude, ground truth, or crew commentary?

- Who processes the payload data—project or end user?

- How is the processed data archived—through the project or end user? How is quality of the product controlled?

- What, if anything, must the project archive after the flight phase is over? How long must information be stored?

Planners and payload engineers must consider these issues jointly to understand operationally what will occur during flight.

Step 3. Identify Ways to Accomplish Functions and Whether Capability Exists or Must Be Developed

At this level, many functions and the ways to do them are straightforward. For example, to track an interplanetary spacecraft, we use NASA's Deep Space Network. For other functions, we must identify options and describe them. For example, to determine a spacecraft's orbit, we might use the Global Positioning System and automated procedures on board, or we might track the spacecraft from the ground and calculate its orbit on the ground. Mainly, we must decide where the prime responsibility for each function lies—onboard autonomy, ground-based operators, or automated functions. Then we must review our decisions in light of the capabilities on board and on the ground, as well as how critical the action is.

To understand options, try building a table that contains the operations functions which apply to the mission's database and space element's avionics. Then, identify whether the avionics (automated) or the mission operations system are to do each function. If functions must be done on the ground, further determine whether the hardware, software, or operators should complete it. If a function could be done in more than one place, describe what to do in each place and options for doing it. Table 14-2 shows how such a table would look.

If the ground hardware and software do something, ask, "Could the avionics partially or completely do this function and lower the mission's life-cycle costs?" For example, if we were considering orbit determination, we'd ask, "Could we determine the spacecraft location within the avionics?" Then, we'd look at the accuracy of the GPS system, check the cost of GPS receivers that are flight qualified, and do a first-order estimate of the change in life-cycle costs compared to a more typical approach. Don't forget that the costs of tracking facilities are important in this type of trade. The longer the mission, the more ground systems cost.

Step 4. Do Trades for Items Identified in the Previous Step

For the options identified in step 3 that drive either performance or cost, a small group of operators, crew and designers need to do trades and decide how to carry them out. At this point these trades should involve areas that are very costly, controversial, or not well understood. In some cases, we may develop an operations scenario for each option to describe it in detail. Usually, these trades result in a traditional allocation of

TABLE 14-2. **Identifying Where to Carry Out Functions.** Using a table similar to this one helps us identify options for carrying out mission operations. We assume functions not included in table are done on the ground. As you evaluate each function, place a check mark in the table to show where you complete the function.

MOS Function	Where Accomplished	
	Spacecraft Avionics	Ground
Mission Planning		Operator augmented with automated tools is primary
Activity Planning and Development		Operator augmented with automated tools is primary
Mission Control		Operator is primary
Data Transport and Delivery	Many LEO telecommunications spacecraft implement much of this function onboard	Software and hardware provide primary capability
Navigation and Orbit Control	Software and hardware on spacecraft is an option	Software and hardware is primary
Spacecraft Operations		Short- and long-term planning by operators, augmented with automated tools
Payload Operations		Short- and long-term planning by operators, augmented with automated tools

functional responsibilities to the flight crew, avionics, and operators. The key here is to look for approaches that truly minimize life-cycle cost without jeopardizing the safety and reliability of the mission and systems.

Step 5. Develop Operational Scenarios and Flight Techniques

Operational scenarios are key to an operations concept. A scenario is a list of steps, and we can often describe a mission operations concept with several dozen top-level scenarios. Typically, we generate three types of scenarios:

- *User Scenario*—How the user interacts with the system elements and receives data.
- *System Scenario*—How systems and subsystems within an element work together.
- *Element Scenario*—How the elements of the space mission architecture work together to accomplish the mission.

During the early study phases of a mission, users develop a scenario to show how they want to acquire data and receive products from the payload. For a science mission, the user would be the principal investigator or science group or, for a facility spacecraft such as the Hubble Space Telescope, an individual observer.

We create a system scenario after we've developed the operations architecture. Here, we emphasize the steps within a process needed to conduct the mission. Finally, during element design, we expand these system scenarios to include more detailed information and subsystems. Once we have created scenarios for the user, system, and element, we must integrate them to understand what happens throughout the mission and make sure we've included all key activities and eliminated overlaps. The scenarios give us our first look at how our system operates as a unit to produce the mission data.

The mission concept, Mission Operations Plan, and design of the space and ground elements are closely related. As design proceeds, we should keep cost in mind and recognize that engineers normally focus on the technical challenge, with costs secondary. All participants in the conceptual design not only must keep cost in mind, but should view it as a design variable. As designs mature for all elements of the space mission architecture, we must do trade studies to get the most cost-effective mission design. If the budget is constrained, which it usually is, we must look for ways to reduce costs by changing the mission concept, mission requirements, and potentially, the mission's overall objectives.

Step 6. Develop Timelines for Each Scenario

Now we can add times needed to do each set of steps and determine which steps can run in parallel or have to be serial. This information becomes a source of derived requirements for the mission operations system's performance.

Many different timeline tools are used. None is standard, but many are modified from commercial, off-the-shelf software. Most missions use the same timeline tools for operational scenarios and activity planning. For many smaller payload users, asking for planning support from the MCC is less expensive than investing in a parallel planning system. These requirements must fit within the scope of the agreements among the program, operations, and the user. In many cases, the MCC has already required information that it can make available to you. In other cases, teams must generate ancillary data or change plans to fit unique needs for payload planning. If these changes are extensive, a separate planning system may be the effective solution.

Step 7. Determine the Resources Needed for Each Step of Each Scenario

Once we've developed scenarios, we may assign machines or people to do each step. This choice is obvious for many steps, but people or machines may do others, depending on performance requirements and available technology. The trend is toward automation on the ground and autonomy in space. We must be careful to identify, as specifically as possible, which tasks the flight crew should do.

Having allocated resources (hardware, software, or people), we must assign steps to hardware and software within data-flow diagrams. For steps assigned to people, we select an existing organization or develop an operational organization and assign steps and functions to teams.

At this point, we examine each step to which we've assigned an operator and ask, "Can this process be automated to eliminate the operator?" How? Allow a person to do something only when the complexity or flexibility or life-cycle costs mandate human involvement. Don't accept the idea that we need a person because we've always done it that way. Technology is advancing so rapidly that a machine may very well do now what a person had to do on the last mission. Through this process, we'll also discover areas on which to focus research and development funding for possible automation in future missions.

Step 8. Develop Data-flow Diagrams

System-engineering tools can convert machine steps into *data-flow diagrams* showing processes, points for data storage, and interrelationships. They also generate a *data dictionary* that ensures a unique name for each process or storage point in the data flow. These computer-aided systems engineering tools then generate information you can use for development. One of the most effective actions you can take to reduce overall life-cycle cost is to understand where data originates and the flow it takes throughout your system to the end user or customer and to your archive. Figure 2–2 in Sec. 2.1.1 shows a top-level data-flow diagram for FireSat.

Step 9. Characterize Responsibilities of Each Team

Once we've defined functions and processes, gathered the people-related steps, and formed an organization around them, we can assign teams to the steps and analyze the organization to establish operational interfaces. Generally, the more inputs we get from different teams, the more complicated, costly, and slow the operations organizations are. The goal of this step is to identify the number of people required to do mission operations during each mission phase. We'll use this information to estimate the cost of operations. We should be open to approaches and trade-offs that reduce overall life-cycle cost.

Step 10. Assess Mission Utility, Complexity, and Operations Cost Drivers

To assess mission utility, we follow the overall process defined in Sec. 3.3. To do so early in design may require flexible simulations, in which input parameters and system parameters have a range of values. Operational simulations that produce outputs which meet mission objectives are candidates for the Mission Operations Plan. As the plan matures, assessing mission utility yields confidence in the design or highlights shortcomings for re-design.

To assess operations complexity and how it drives mission operations cost, we use a complexity model that shows the relationship between operational parameters and full time equivalent operators. This model requires us to evaluate each of our operational activities as low, medium, or high complexity. Then, based on previous missions of the same class, the model produces the number of operations personnel. During design, this model gives us rules for trade studies to reduce operations costs. Boden and Larson [1996, Chap. 5] describe this model in detail. See Sec. 14.3 for a summary.

Step 11. Identify Derived Requirements

At this point, we've updated the Mission Operations Plan with new information associated with the mission operations concept, requirements, existing and new capabilities, scenarios, timelines. and the anticipated life-cycle cost. Now, we can identify new or derived requirements necessary to reduce cost and complexity and enhance the safety and reliability of operations. We should document these derived requirements in the Mission Operations Plan, being careful to identify what is to be done—not how—and why the requirements are necessary. Then, we should communicate the derived requirements to the group that develops the mission's conceptual design and negotiate them into the requirement baseline.

Step 12. Generate a Technology Development Plan

The technology to support a mission operations concept may not exist or may not be focused and prototyped appropriately for mission approval. With each iteration of the Mission Operations Plan, we must identify needed or risky technology, so someone can develop the technology or create work-arounds.

Step 13. Iterate and Document

The last, and usually most painful, step is to document the results of the iteration through the Mission Operations Plan to develop a baseline from which we can continue to improve and reduce life-cycle cost. The documentation should include at least these elements:

- Requirements and mission objectives
- Key constraints—cost, schedule, and technical performance
- Assumed mission and flight rules
- Scenarios—described in terms of functions
- Timelines for each key scenario
- Ground and flight crew tasks
- Organization and team responsibilities and structure
- Hardware and software functions
- Data-flow diagrams
- Payload requirements and derived requirements

Remember that this document is the basis for communicating the overall operations approach to users, operators, and system developers. If done and used properly, it can help save millions of dollars in the design and operation of space systems. We strongly recommend keeping the Mission Operations Plan in electronic form and making it readily available to all members of the conceptual design team, so they can keep the big picture in mind as they further develop concepts, systems, and approaches. The earlier the first Mission Operations Plan appears, the greater the leverage for reducing life-cycle costs.

Now that we understand how to develop a Mission Operations Plan, we need a more detailed understanding of what mission operators do, so we can create a better and more detailed plan.

14.2 Overview of Space Mission Operations Functions

Although space mission types vary widely across military, scientific, and commercial applications, flight teams on the ground operate with fairly common tasks and activities. In this section we'll describe these common operational tasks and then show how mission differences can influence performance requirements and hence the styles and costs of accomplishing the tasks.

Figure 14-1 at the beginning of the chapter shows 13 functions for space mission operations. Table 14–3 summarizes the associated attributes of these functions. For small projects, members of a single team may do these tasks. For larger projects, subteams may form; each dedicated to one of the functions in the table. On some projects, a single subteam does several of the functions, whereas another subteam may do only a part of one function.

TABLE 14-3. Functions and Attributes for Space Mission Operations.

Function	Key Responsibilities	Inputs	Outputs	Ops Design Considerations
1. Mission Planning	• Coordinate science, trajectory, and engineering plans • Allocate and manage mission consumables	• Planning requests • Science objectives • Eng. & nav. constraints • Mission performance metrics	• Activity timelines • Activity Planning guidelines • Mission rules	• Number and complexity of flight rules? • Consumables, performance, and timeline margins?
2. Activity Planning & Development	• Integrate activity plan requests • Develop time-ordered, constraint-checked activities	• Flight rules • Guidelines from mission planning • Activity timelines • Activity Plan requests	• Activity Plan review products • Spacecraft command load • Ground activity schedule	• Activity Plan duration? • Quantity and quality of constraint checks? • Activity level: over vs. under-subscribed?
3. Mission Control	• Monitor in real-time • Command in real-time • Configure and control ground data system	• Telemetry alarm limits • Real-time command requests • Ground system availability times	• Pass reports • Real-time commands • Ground system schedule	• Around the clock staffing? • Joystick vs. stored seq. ops?
4. Data Delivery — See Function (8)				
5. Navigation & Orbit Control	• Design trajectories or orbits • Determine position and velocity • Design maneuvers	• Radiometric data • Optical navigation data • Ephemerides • Spacecraft propulsion performance	• Trajectories or orbits • Prop. maneuver designs • Antenna projects	• Trajectory accuracy reqs? • Propellant margin? • Disturbance force complexity?
6. Spacecraft Operations	• Ensure spacecraft safety and health • Calibrate the spacecraft and establish engineering performance • Analyze anomalies • Maintain flight software	• Spacecraft engineering data • Activity Plan review products • Flight system testbed data	• Spacecraft constraints • Consumables status • Activity Plan requests • Real-time command requests	• Fault response vs. fault prediction trending? • Spacecraft performance margins? • Significant post-launch software development?
7. Payload Operations	• Ensure payload safety and health • Calibrate the payload • Do quick-look payload analysis	• Payload engineering data • Activity Plan review products	• Payload reqs. and constraints • Activity Plan requests • Real-time command requests	• Payload complexity? • Payload interactivity? • Number and complexity of outputs?

TABLE 14-3. Functions and Attributes for Space Mission Operations. (Continued)

Function	Key Responsibilities	Inputs	Outputs	Ops Design Considerations
8. Data Serv. Includes: (4) Data Delivery; (8) Data Processing; (9) Archiving and Maintaining Database	• Receive, stage, transport, process, display, and archive data • Transmit commands • Manage computer, comm. and database	• Noisy telemetry data • Radiometric data • Ground system monitor data • Command files	• Detected telemetry data • Processed engineering data • Processed payload data • Transmitted command data	• Payload data volume? • Data products? • Sensitivity to lost data? • Error free vs. nonerror free data?
10. System Engineering & Integration and Test	• Manage external interfaces • Manage system performance and internal interfaces • Recover from system failures	• System technical requirements and constraints • Optimization goals • Engineering change requests	• Operations concept • Interface agreements • I&T plans • Test results	• System optimization criteria? • Peer vs. hierarchical system engineering? • Test policies?
11. Computer and Communication Support	• Maintain the hardware • Maintain data links	• Anomaly reports • Outage reports	• Maintenance and upgrade plans	• Distributed vs. centralized • PC's vs. workstations or mainframes
12. Software Development & Maintenance	• Maintain the system • Upgrade the system • Train and certify operators	• New req's • Approved eng. change requests • Continuous improvement initiatives	• Development plans • Training plans • New, improved capabilities	• Dedicated dev. staffing? • Other dev. resources? • Training tools/aids? • Operator cert. policies?
13. Management	• Manage the overall mission • Work with sponsors and users	• Mission reqs. and constraints • Budgets • Ops concept • Ops component status reports	• Mission goals • Project policies • Allocated budgets • Status reports to sponsors	• Keep or delegate technical decisions? • Approval policies

These 13 operations functions in are useful for collecting and organizing require-ments on operations and to break out work for estimating, bookkeeping, and compar-ing operations costs. You can design an operations system by defining the interfaces between these 13 functions—specifying the flow of data products, interface formats, time-critical processes, and hardware and software tools. A thorough understanding of each function is important for efficient design, estimating costs, and managing space mission operations.

1. Mission Planning

Mission planning starts before activity planning. It defines how to use resources best in accomplishing mission goals. Mission planning produces rough activity time-lines across mission phases that identify the schedule and resources to complete major activities. The interface between mission planning and activity planning can be as gen-eral as operations based activity guidelines or as specific as a detailed timeline of activity requests that later gets converted into commands.

Resources that mission planners and designers are concerned with include the trajectory or orbit design, consumables over the spacecraft's lifetime, and long-range facility support. This task can be relatively easy for missions with low activity levels, simple trajectories, and reasonable margins for spacecraft consumables. It's also easy for supporting facilities with high availability that we can schedule far in advance. Mission planning can be very complex for missions that have oversubscribed timelines, a complex trajectory, tight budgets for spacecraft consumables, and over-used, shared resources. Timelines become oversubscribed whenever requests for engineering and payload activities overwhelm the time for the spacecraft to do them. Examples of complex trajectory design include planetary gravitational assists, rendezvous and docking, formation flying and planetary landers. Consumables that may have tight margins—requiring mission planning to allocate and manage budgets—include power (battery charge-discharge cycles), propellant, cryogen, total radiation dose, and cycles of mechanical devices, such as number of thruster firings, valve openings, and slewing of a scan platform. Examples of shared resource facilities that may be overbooked and complicate mission planning include the Deep Space Network's 70 m antennas and wide-band links on TDRSS. Complexity and costs for mission planning grow geometrically, rather than linearly, as each factor increases in complexity because their interactions increase.

Contingency planning is another task assigned to Mission Planning. We usually add contingency reserves to consumable budgets and design busy timelines to include a few planned, low-activity periods that are available for unplanned, contingency activities. Some missions demonstrate cost effectiveness by scheduling "bonus" payload activities they can drop to respond to contingencies, but which get executed if everything goes as planned. The most common form of operations contingencies we plan for are spacecraft or payload anomalies, degraded performance, and changes to facility schedules. Contingency planning for more adaptive missions sometimes include responding quickly to surprise opportunities or changing conditions.

2. Activity Planning and Development

This operations function produces the spacecraft's stored sequences discussed above. Activity planning converts requests for spacecraft and payload activities into a file of timed commands for uplink to the spacecraft. This command file is a sequence of commands that are integrated, time-ordered, constraint-checked, simulated, validated, reviewed, approved, and defect-free. Activity planning often generates real-time commands that may require simulation or constraint checking before passing them to Mission Control for uplinking.

Three factors influence the difficulty and cost of activity planning: complexity of the process for integrating requests, required accuracy of the constraint checking and performance models used to simulate and validate the activity plans, and duration for which the activity plan is to be active.

Missions with simple activity inputs and integration processes usually have a small number of users. They use an efficient format for activity requests, often electronic, with requests written by users in the command language. In addition, their mission resources are under-subscribed, so most requests can be scheduled and executed with few conflicts between requests. Examples of constrained mission resources that can complicate how we integrate the activity plan include limited onboard sequencing memory, limited time to carry out activities, limited power, limited onboard data storage or downlink bit rate, and limited tracking resources or downlink time. Missions that must develop highly constrained activity plans sometimes reduce costs by pre-allocating resources and then pushing conflict-resolution tasks back on the requesting teams from Mission Design, Spacecraft Engineering, Payload, or Operations.

Accuracy in spacecraft-performance models can also drive activity planning. For example, suppose we need to plan a spacecraft-pointing maneuver to move a sensor's boresight from target A to target B, using pre-planned, timed commands. This plan requires a predictive model of slew rates, turn times, and settling times. Models for propulsion, attitude control, thermal, power, and telecom can all require frequent tuning and calibration. That means designing and doing flight tests, as well as measuring and downlinking a spacecraft's performance parameters and then comparing them with predictions. After analysis, we would update the models. This process can consume a lot of operations resources, not just in activity planning but also in data collection and analysis. Missions with performance margins can use lower fidelity models and save money. But some missions, such as a planetary flyby, spend hundreds of millions of dollars to do a few hours of an activity plan for closest approach. They want to "optimize" spacecraft and instrument performance, so they're willing to invest in high-fidelity, accurately calibrated performance models.

Selecting the duration of a stored activity plan is a third factor that can significantly influence activity planning costs. Activity plans that last only a few hours require frequent uplinks, extra station coverage, and around-the-clock staffing. Short activity plans can be timed to correspond to mission events (such as orbit duration), don't require large memory storage on board, can respond to last-minute requests from users, and often can adapt to activity-request conflicts in the current sequence under development by postponing the activity to the next one. Activity plans designed to last for weeks or even months may require more onboard sequencing memory and more time for uplink transmissions. With longer activity plans, requesters must think over much longer planning epochs and, if an activity doesn't get scheduled in the current plan, it must wait until the next one. Requesters who miss getting commands integrated into lengthy activity plans often resort to real time commands, bypassing the stored sequence process and increasing the load on Mission Control. Missions that operate in this mode usually enjoy the advantage of being able to staff activity planning 5 days per week, prime shift only. We can reduce staffing by decreasing the number of activity plans and making each one last longer.

3. Mission Control

Mission Control handles activities that must be done in real or near real time. They do last-minute adjustments to pre-scheduled mission timelines. They manage the ground system's configuration and coordinate interface with operations facilities such as tracking stations and communication services. They also manage the spacecraft's configuration using real-time commands and telemetry monitoring. Mission controllers detect a spacecraft's telemetry alarms and anomalies, isolate problems, and call in appropriate experts on the spacecraft or ground system. They maintain a log of mission-significant events. They often work whenever the spacecraft is being tracked, so controllers may have to take other than prime shifts.

Mission control tasks may require significant staffing and high activity levels for missions that do many activities in near real time. This operations style is called *joystick* operations. It's often characteristic of missions that have short communication delays, continuous or nearly continuous coverage, and less elaborate software for onboard autonomy. These missions require rapid ground reaction to unplanned events. Joystick style missions must be able to tolerate occasional wrong commands or other operational mistakes common when people must respond rapidly.

Mission Control staffing is lower for missions that de-emphasize real-time decisions by ground operations. Instead, these missions rely on the ability to pre-plan activities and load the spacecraft with a timed command sequence that has a high probability of executing successfully with minimum ground interaction. What this *stored-sequence* style saves in real-time mission control, it usually more than pays for through increased costs for design and simulation tools. Missions often use it if they have long two-way light times, long periods without tracking or ground contact, and at least a modest investment in software for onboard autonomy. This software allows the spacecraft to react to unplanned events. Missions using stored sequences can't tolerate errors in critical events such as a planetary flyby or injection maneuver. Thus they rely on elaborate planning, simulation, and review to produce and store zero-defect plans. Software now handles most reflexive actions, such as monitoring and responding to faults in real time.

4. Data Delivery (see Data Services in Function 8)

5. Navigation and Orbit Control

This operations function delivers the flight system to the target or maintains an operational orbit. It designs trajectories to meet various requirements and constraints. Efficient designs for trajectories or orbits can save a lot in launch-vehicle costs and the spacecraft's required propellant mass. As described in detail in Chapter 7, techniques such as planetary gravity assists, aerobraking, and electric propulsion are some of the methods navigation uses to achieve efficiency goals.

As discussed in Section 11.7, navigation determines the flight system's location and velocity by using several types of data. Radiometric data collected during telemetry and command passes may include one- and two-way Doppler, as well as ranging codes. Some planetary spacecraft with imaging instruments use images of planets or asteroids and the star field background to give very precise spacecraft location data. Some Earth-orbiting spacecraft use onboard GPS receivers to locate the spacecraft. Navigation analyzes this position data to determine trajectory or orbit corrections needed to meet mission requirements. They also furnish position data to teams in Payload Operations who use it to interpret or calibrate instrument measurements. They also send trajectory and orbit data to Data Services, so these teams know when to schedule antenna tracks and where to point them.

Once the actual trajectory has been determined, orbit control function defines propulsive maneuvers needed for correction. Orbit control and Spacecraft Operations work together closely on this process, particularly the attitude-control and propulsion analysts. Desired changes to the spacecraft's velocity vector get converted to turns and propulsion burn times and then into spacecraft commands. Designing and analyzing a trajectory correction maneuver can get complicated. We may need to model the thrusters' performance, analyze gravitational effects and disturbance forces, and complete nominal and "clean-up" maneuver designs.

6. Spacecraft Operations

This spacecraft engineering function oversees the spacecraft's health and safety in flight and calibrates and maintains equipment to keep the spacecraft's performance within specifications. Engineers define detailed operational configurations and states necessary to support payload activities, such as modes for pointing, power, and data service. They also maintain flight software and analyze spacecraft failures.

Spacecraft Operations provides Mission Planning with requirements for calibrating and maintaining equipment to be integrated into mission activity plans. They also provide mission planners actual resource consumption data together with performance models to help them plan and manage mission resources.

Spacecraft Operations sends requests to Activity Planning for engineering activities to be converted into timed commands in an integrated sequence. They often review activity plans before uplinking to make sure they don't violate flight rules from spacecraft engineering and that they meet operational peculiarities and constraints of the subsystems. Reviews can be simple for missions that execute a routine daily set of standard commands and for missions that have invested in simulation software which automatically checks constraints. But they can be time consuming for missions that have a high percentage of new or unique activities every day, that can't simulate activities and constraints, and that have subsystems which strongly constrain operations.

Spacecraft Operations delegates to Mission Control the responsibility for real-time monitoring of the spacecraft's health and performance (such as monitoring telemetry alarm limits). This delegation allows Spacecraft Engineering to concentrate on more deliberative, non-real-time tasks, such as analyzing performance and predicting long-term trends. It also often allows them to staff only 8 hours per day, 5 days per week.

There are two styles of fault detection and anomaly analysis. The first—a *detect-respond* style—is less expensive. Operators watch for alarms or other unexpected failures, analyze them, and prepare strategies to correct or work them. The second—a *predict-prevent* style—costs more. Planners predict trends and performance, trying to forecast faults before they might occur and then plan preventative action. This task can involve elaborate models for telemetry, power, thermal, dynamics, and other systems and can require significant calibrations and performance tests during a flight. Detect-respond fault operations are characteristic of space missions that can handle occasional unplanned down times to recover from faults. Often these missions have single-string hardware configurations with limited options for recovering from failures. Predict-prevent fault operations are characteristic of space missions with engineering or science events, for which an unplanned down time could cause loss of a mission or loss of mission critical science data that we can't recover. Often these missions have block redundant hardware, allowing operators to select a back-up component when trending data suggests a possible future failure in the main on-line, component.

Failure analysis sometimes requires using test beds on the ground that allow analysts to simulate or recreate problems observed in flight. Spacecraft Operations runs and maintains them, which can become a significant operations cost for test-bed hardware (usually, engineering models for flight hardware) and managing and updating software.

Spacecraft Operations also maintains and fixes flight software, then continues to develop it after launch. Missions with long flight times, changing mission goals, or major anomaly recovery work-arounds will often require significant software re-development during flight. Some missions that have long flights to targets may even deliberately postpone developing target-related software until after launch. Software test and validation becomes another use for the flight-system testbed and another potential expense for Spacecraft Operations.

7. Payload Operations

This operations function has similar interfaces as the Spacecraft Operations function, but with responsibility for the payload instead of the spacecraft. Payload operators send measurement goals and activity requests to Mission Planning and Activity Planning respectively, which convert payload requests into mission timelines and integrated command loads. Some designs for mission operations try to simplify these interfaces by allowing payload controllers to directly command the payload's configuration and method of returning data, sometimes from sites located remotely from the

Mission Operations Center. This approach works best when individual payload instruments and sensors don't interact and when instruments don't contend much for spacecraft resources, such as power, pointing, data storage, and downlink bandwidth.

Payload operations use different data types. Operators monitor the payload engineering data to check the status of its health and safety. Often, Mission Control handles this task. Operators analyze data from the payload to verify that it is producing results in ranges of expected values. This is known as *quick-look* analysis. It doesn't replace the interpretation that becomes part of longer-term scientific discovery. Quick-look analysis happens quickly, so operators can discover and correct anomalies for instruments, sensors, and sequencing before they have collected, calibrated, and archived a lot of data. Quick-look analysis often requires concurrent recovery and analysis of a payload's ancillary data (such as where the spacecraft was located—orbit data), which direction it was pointed (attitude or scan-platform data), engineering states of instruments (gain settings or filter selections), and sometimes relevant environmental measurements (radiation field and temperature).

Complexity of instruments and investigations directly influence the size of the Payload Operations task. Instruments cost more to operate if they have many control states, are sensitive to interactions with other instruments or the background environment, must manage their own consumables (batteries or cryogen), and face many operational constraints (such as avoiding the Sun or Earth, warm up, cool down, radiator field of view). Investigations with complex targets, payload tasks (such as ground rovers or space weapons) and activity timelines cost more to operate. Some missions are designed to use correlative data from other sources (such as multiple spacecraft) or multiple sensors (such as measurements from ocean buoys). These missions may need more effort to plan operations and analyze data.

8. Data Services (Includes: (4) Data Delivery, (8) Payload Data Processing and (9) Archiving and Maintaining the Mission Database)

This function does various Data Service tasks including tracking and acquiring data, transmitting commands, transporting data, handling local computer and communication services, processing and displaying data, simulating data flow, maintaining the mission's database, building data products, and archiving data. IT also plans and analyzes performance, and coordinates operation of the ground data system. It supports all other operations functions. Well designed, well run data services can be vital to how efficiently other operations functions do their tasks. If a mission must return a lot of mission data, data services can dominate operations costs. New technologies, such as CD-ROMs and the worldwide web, have greatly reduced the cost of historical data products such as photographs and magnetic tape libraries.

Well engineered data systems apply existing data standards so we can use standard data services and tools. Handbooks from the Consultative Committee for Space Data Standards document standards for space data. International teams develop these handbooks with U.S. representation from DoD, NASA, and industry. The standards separate into layers for data timing, synchronization, transport, coding and representation. The direction has been away from data-stream services, such as the older systems using time-division multiplexed telemetry, and toward packet-based or file-based protocols that permit standard services independent of content.

Operations can be very efficient with well engineered data designs. These designs consider capabilities for ground processing and archiving while planning for onboard data capture, processing, and packaging. On the other hand, operating a poorly engineered data system can be costly. Such designs require operators to recover, extract, and correlate various separated data types. Other cost drivers include the robustness of payload data processing to data loss or drop-out and the project's policies on what percentage of data must be recovered for mission success. Many anecdotes describe missions that used 90% of their resources trying to recover the last few percent of data.

Operations design must choose between two competing approaches to data systems: multi-mission vs. mission-unique. Advocates of multi-mission data systems emphasize the efficiency of using a common design and a common set of operators to provide data services to more than one mission. We can amortize costs for parts of a ground data system that require large capitol investments, and this makes it more affordable. This approach also helps us use data standards and reuse systems. Cost-sharing efficiencies free up money to invest in data systems and to pay for overhead costs such as maintenance, planned upgrades, and new technologies. However, detractors point out that multi-mission, generic systems and organizations may be less efficient than those designed for, and focused on, a specific project's needs. Large heritage systems with significant capitol investments are expensive to change and can't respond to new technology innovations as rapidly as smaller, unique systems.

Usually, we decide on generic or unique data services for individual elements. Consider, for example, capturing telemetry data from deep space. This requires a network of large, expensive antennas—too expensive to build for each project. But calibrating payload data might require unique project software and processes, as well as operators dedicated to each project. Recent development of commercial off-the-shelf software that handles things like orbit analysis, command management, telemetry decommutation and display, and simulation has allowed us to take a cost-effective hybrid approach. These tools have made it easier to tailor commercial, multi-mission tools for unique data services.

Shared, multi-mission data services can present a problem when they are oversubscribed. Competition with other users can sometimes lead to complex scheduling that uses up a project's resources and complicates mission planning. Late schedule changes responding to other users' emergencies can require major replanning. Over-subscription can become significant whenever institutions fund multi-mission data services and provide them "free" to users. To help reduce this problem, NASA is starting *full-cost accounting*: it will allocate costs of multi-mission institutional services to individual users. The move to data-system architectures on workstations and personal computers also helps relieve over-subscription. Hardware dedicated to a project can host multi-mission software, and operators for that project can run it.

10. System Engineering and Integration and Test

This task considers the operations system as a whole, managing interfaces and adjudicating competing desires and requirements between operational teams, tasks, and functional elements. For example, activity planning is easier if Mission Planning produces a conflict-free timeline. Mission planning is easier if they only provide schedule requests and conflict-resolution guidelines to Activity Planning. System Engineering resolves issues like these to get the best performance from the system. They also define criteria for a best system, such as low cost, high reliability, minimum staffing, quick response, or all of these and more.

Just as System Engineering defines and manages interfaces between elements of a mission's internal operations, it does the same for elements of external operations, such as providers of institutional services or correlative data for a payload. System engineers focus on technical-interface agreements, such as data formats, procedures, and protocols. Management focuses on program interfaces such as funding agreements. System engineers develop and maintain operations concepts to help design and refine system architecture, team responsibilities, and operations interfaces. Operations concepts include scenarios, timelines, and product flows that capture the operations system performance requirements, design constraints, and requirements derived from the design.

System Engineering also plans for contingencies, resolves anomalies, and handles failure recoveries. System engineers coordinate anomaly analysis that involves interaction among several engineers or teams and provide technical approval of engineering change requests to develop fault repairs or work-arounds.

During operations, the Integration and Test task supports development by testing new or redelivered capabilities. It also supports testing of system interfaces when new institutional capabilities get delivered. Teams must develop test schedules to fit operational schedules. Engineers have to save previous versions of software to permit return to a known good system whenever testing shows the new version has problems. Independent verification and validation, i.e. assigning the Integration and Test task to engineers different from those in Development is a common practice to ensure independent, objective testing.

11. Computers and Communications Support

This function entails designing and buying or building hardware for the end-to-end information system. Because space missions typically produce so much data in electronic form, mission operations planners must ensure data moves efficiently. To do so, we prepare data-flow diagrams, requirements for computers or workstations, requirements for networking and data communication (within the control center and around the world), and requirements for voice communications. Having accurate data-flow diagrams (see Sec. 2.2.1) is the starting point for designing the hardware. From mission objectives, we learn how much data must flow, between which nodes, and how frequently. We use this information to diagram the data flow and allocate data-handling processes to software and hardware. From the diagrams, we list the numbers and types of computers, workstations, and other hardware. Knowing the communications architecture and organizational design helps us choose the hardware correctly. For example, a decentralized organization for mission operations requires us to connect dispersed staff members, which usually means preparing a communications network.

The networking requirements also come from the data-flow diagrams and usually require more support equipment. Other networking factors are availability, capacity, and security for mission data.The final piece of communication support is the voice and video-teleconferencing requirements. Early in design, we must establish any need for these special links so the operations organization can communicate efficiently during the mission.

Because of the proliferation of modern computing and communication equipment, planning for their support may be no more complex than ordering from industry catalogs. Designing and building unique hardware usually isn't cost effective, but special requirements may drive us to do trade-offs in this area. Finally, maintaining and administering the computer and communication hardware throughout the mission is a vital concern. Good designs allow us to repair and replace equipment and allocate staff for this activity.

12. Development, Maintenance and Training

Throughout a project's life cycle, we must develop and maintain software for:

- Creating normal space and ground system operations before launch
- Correcting spacecraft and ground system errors after launch
- Implementing changes in mission requirements after launch

One modern trend for interplanetary missions is developing operations software after launch. We program basic functions before launch, but create most of the operations software during the long cruise phase, then upload it in time for the planetary encounter. Similarly, some commercial communications constellations have been able to substantially shorten their deployment time by completing the onboard software subsequent to launch of the initial phase.

Another reason that flight software gets developed after launch is to deal with in-flight hardware failures. An example of an in-flight failure is the Galileo spacecraft's high-gain antenna. After launch it failed to deploy. So, programmers had to redesign the flight software during the transfer phase to provide data formats and enhanced onboard coding for the low-gain antenna to complete the mission.

To develop and maintain software for any mission, we need to know

- *System requirements*—use existing software from previous missions or develop new software for new requirements
- *Error reports*—error reports drive software maintenance
- *Change requests*—system capability changes usually require software changes
- *As-built documentation*—knowing the current software as "built" (not as designed), helps us develop and maintain it more effectively.

Several techniques for developing and maintaining software improve our probability of having software that works:

- Know the requirements and ensure they are testable and agree with the mission operations concept
- Use rapid prototyping to demonstrate system capabilities early
- Deliver the ground data system in increments
- Plan to develop and maintain software during operations (don't send the software design team home too early)
- Match software schedules to hardware schedules, so integration and testing aren't delayed
- Thoroughly test the software, usually incrementally
- Keep operations informed of software status so operators are ready when it is
- Use the new software in training, so operators know its capabilities and trust it
- Plan further software maintenance after delivery

Similar to computer hardware, software development and maintenance change rapidly. The software package that was unique on one mission may be commercially available for the next. New, highly flexible designs spring from unusual places, such as microsatellite designs and large constellations. Communicating what's available and what will work is a demanding process.

13. Management

Operations Management has overall responsibility for operations success. Managers must meet operations requirements within negotiated values for cost, schedule, and system performance. They provide the resources to make the other operations functions work. They focus on planning, monitoring, directing, and reporting of programmatic resources such as costs, schedules, and staffing.

Managers work with the sponsors and customers and negotiate mission goals vs. resources. They define the mission's operational policies and guidelines. An effective management technique is to allocate resources to each operations function or team and then not get involved in technical decisions or approvals unless they exceed the resources allocated. If management retains technical authority, managers also participate in and approve technical operations decisions. For any project, we have to think through how much we want management to participate in the daily approval of operations products in terms of operations response time, efficiency, team motivation, and value added. This operations element usually carries the budget for operational reserves and funding for other tasks, such as administrative support.

14.3 Estimating the Size and Cost of Mission Operations

How do you estimate the size of the operations task for a given mission? To some extent the size is influenced by the operations design itself which includes considerations such as the efficiency of how the teams are organized, how tasks are assigned, team member experience, how team members are trained and certified, and how the team is motivated and managed. But operations design efficiency typically can influence team size by only 10 to 20%. It does not explain why some missions can fly with only a few people, whereas others require an operations team of several hundred.

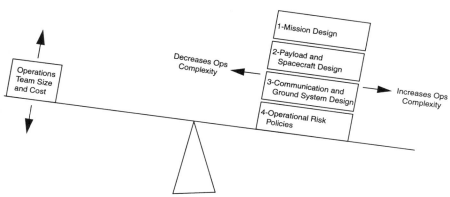

Fig. 14-2. Four Factors that Drive the Cost of Mission Operations.

Decisions made in designing missions, payloads and spacecraft, communication and ground systems, and policies on operational risk all affect operations cost and size. Operations often become expensive when we make these design decisions first, without considering how they affect operations. Instead, we should develop the operations concept concurrently with other elements, so we can select the best overall approach based on affordability and lifecycle cost. This section identifies important drivers of operations costs and organizes them into 4 factors you can use to estimate the size and cost of space mission operations. Figure 14-2 shows these 4 factors and their effect on an operations team's size and cost as operations complexity increases or decreases.

TABLE 14-4. How Mission Design Factors Affect Operational Complexity (Chaps. 1–8).

	Simple, Low-cost Ops ◀───────▶ Complex, High-cost Ops	
	Simple, Low-cost Ops (10 People or less)	Complex, High-cost Ops (50 People or more)
Mission Objectives	• Well defined • Single goal • Constant	• Ambiguous • Multiple competing goals • Changes with time and data
Mission Concept	• Few mission phases • Single spacecraft • Loose timelines • Low activity level • Repetitive activities • Repetitive opportunities	• Many mission phases • Constellation • Tight timelines • High activity level • Many one-time-only activities • One-time-only opportunities
Orbit/Trajectory Design	• Stable trajectory dynamics • Comfortable ΔV margin • Few nav data types • Loose accuracy reqs for nav. control and reconstruction	• Rapidly changing trajectory dynamics • Tight ΔV consumables budget • Many nav. data types • Tight accuracy reqs for nav. control and reconstruction
Space Environment	• Known environment • Stable environment	• New space environment • Rapidly changing environment

TABLE 14-5. How Design Factors for the Payload and Spacecraft Affect Operational Complexity (Chaps. 9–12).

	Simple, Low-cost Ops (10 People or less)	Complex, High-cost Ops (50 People or more)
Payload Design	• Single sensor/instrument • Independent sensor data • Simple calibration • Low rate data output • Mature technology • Small amount of software • Simple pointing • Instrument dedicated resources (power, pointing, data storage) • Few payload modes • Few data formats • Simple data processing	• Many sensors/instruments • Correlative data dependency • Complex calibration • High rate data output • New technology • Large amount of software • Complex targeted pointing • Instruments compete for shared resources • Many payload modes • Many data formats • Complex data processing, compression, and editing options
Spacecraft Design	• Single string • Mature technology • Comfortable margins (power, memory, thermal) • Spinner or gravity-gradient • Few flight rules and constraints • Few onboard engineering measurement routinely downlinked • Few telemetry rates and formats • Few articulating devices	• Block redundant • New technology • Low or negative margins • 3-axis stabilized • Many flight rules and constraints • Many onboard engineering measurements routinely downlinked • Many telemetry rates and formats • Many articulating devices

TABLE 14-6. How Design Factors for Communications and Ground Systems Affect Operational Complexity (Chaps. 13, 15, and 16).

	Simple, Low-cost Ops (10 People or less) ⬅➡	Complex, High-cost Ops (50 People or more)
Communication System Design	• Infrequent tracking required • Telemetry, command, and radiometric tracking reqs are compatible • Dedicated or undersubscribed tracking resource • Low data volume • High com link margins • Loose coupling between onboard and ground events • No station reconfiguration during pass • Highly repetitive tracking schedule	• Near continuous tracking required • Telemetry, command, and radiometric tracking reqs compete • Shared, oversubscribed tracking resource • High data volume • Low com link margins • Tight coupling between onboard and ground events • Frequent station reconfiguration during pass • Many uniquely scheduled tracks
Ground System Design	• Dedicated or undersubscribed ground system • Stable ground system and software design post-launch • Prime shift only ops team staffing • Project specialists, multi-tasked • Minimum number of operator tasks for data capture and alarm monitoring • Ops tasks done in single control center	• Multimission, shared, oversubscribed, ground system • Frequently changed ground system and software design after launch • 24-hour-per-day ops steam staffing • Task specialists, multi-project • Many operator tasks for data capture and alarm monitoring • Operations tasks distributed across many locations

TABLE 14-7. How Operational Risk Policies Affect Operational Complexity.

	Simple, Low-cost Ops (10 People or less) ⬅➡	Complex, High-cost Ops (50 People or more)
Operational Risk Policies	• Low investment in spacecraft and payload development • Few mission critical events • Spacecraft can safe itself without ground action • Performance analysis limited to fault detection • Command activity plans require simple simulation and constraint checking • High tolerance for "lost" payload or engineering data	• High investment in spacecraft and payload development • Many mission critical events • Spacecraft depends on ground to respond to anomaly • Performance analysis includes trending and fault prevention • Command activity plans require elaborate simulation and constraint checking • Low tolerance for "lost" payload or engineering data

- *Mission design* involves defining mission objectives, developing a mission concept, designing orbits and trajectories, and evaluating the space environment. Chapters 1 through 8 cover these design processes in detail. Table 14-4 shows how these design decisions affect operational complexity and the operations team's size.

- *Payload and spacecraft design*, is covered in Chaps. 9 through 12. It involves design choices that affect the flight system's operability. Table 14-5 shows how these design choices affect operational complexity.

- *Design of communications, operations and ground systems* is covered in Chaps. 13 to 16. It involves design and implementation choices that influence what tools and communication resources the operations team uses to support their tasks. Table 14-6 shows how design decisions here affect operational complexity.

- *Policies on operational risk* all effect cost and complexity. The more dollars and time we spend in developing a mission, the less operational risk we're usually willing to take. Operationally complex spacecraft require more cautious operations. Table 14-7 shows how these policies affect operations complexity.

After reviewing operations cost data for many government, commercial, and scientific missions, we have found that operations costs can be predicted reasonably accurately as a percentage of the total development cost of the spacecraft (i.e., payload plus bus). This makes sense because spacecraft that must perform complex missions and operate in severe environments tend to have higher development and operations costs while spacecraft with simpler missions and which operate in less severe environments tend to have lower development and operations cost. Organizations are willing to accept more operational risks (e.g. lower ops costs) for low cost spacecraft, while organizations that have a significant investment in an expensive spacecraft or payload, usually insist on more risk averse, higher cost operations. Table 14-8 shows an experienced-based model of this relationship between first unit spacecraft plus payload total development cost and first year ops costs.

TABLE 14-8. **Estimating Mission Operations Cost per Year.** First estimate the theoretical first unit cost from Chap. 20 and determine the category of spacecraft. Using the percentages in the second column estimate the low, high, and average mission operations costs per year. A more detailed ops cost model is available on-line from the Johnson Space Center home page, http://www.jsc.nasa.gov/bu2/SOCM/SOCM.html

Spacecraft plus Payload Theoretical First Unit Cost Category	% of Theoretical First Unit Cost Per Year for Mission Ops	Estimated Mission Operational Cost per Year for FireSat ($M FY99)
Traditional > $5M (FY99)	1–5 (average 3)	0.9–4.3 (average 2.7)
Low Cost < $5M (FY99)	3–12 (average 8)	

In order to refine the estimate obtained from Table 14-8, you can assess overall complexity of the mission systems using Tables 14-4 through 14-7. If your assessment results in a more complex mission, a higher percentage should be used from the second column of Table 14-8. A lower percentage would be used for a less complex mission.

One of the most important operational issues is whether support is required around the clock or only 40 hours per week. If the mission is designed with this in mind such that continuous operational support is not required, then personnel costs can be reduced by a factor of 4 or more. If continuous support is required to meet specific mission needs, we must then determine whether 4 or 5 operations teams are needed, based on the assessment in Table 14-9. Finally, we multiply the number of teams times the number of people per team to get the total operations personnel for the mission. This should be done separately for each of the major mission phases, since some phases may be much more operations intensive than others.

TABLE 14-9. Operations Concept Using Four and Five Teams. Typically we arrive at a staffing solution employing between four and five teams, giving duties besides spacecraft operations to the fifth team.

Number of Teams	Standard hr/yr/team	Required hr/yr/team	Result	Management Concern
4	2,000	2,190	190 hr/yr/team Overtime Wages	Highly sensitive to personnel absence and turnover
5	2,000	1,752	248 hr/yr/team Available Labor	Increased number of people to train and manage

TABLE 14-10. Relative Cost of Mission Operations Functions. The cost of each function is given as a percentage of the average annual operations cost for that mission. Data is from CSP Associates [1999]. Note that the functional breakdown differs slightly from that of Fig. 14-1 due to the categories in which data was collected in the CSP study.

Function	Development			Annual Operations		
	Low (%)	Typical (%)	High (%)	Low (%)	Typical (%)	High (%)
1. Management	14	70	194	0	8	22
2. Mission planning	0	78	169	0	4	12
3. Command Management	0	96	334	1	3	7
4. Mission control	4	146	410	9	22	45
5. Data capture	37	62	86	0	6	10
6. Navigation	8	78	212	3	9	26
7. Spacecraft planning and analysis	8	63	162	0	3	7
8. Science planning and analysis	0	87	662	0	17	72
9. Science data processing	0	181	480	0	17	42
10. Data archive	0	18	59	0	7	18
11. Systems engineering, integration, & test	20	197	437	0	3	17
12. Computers and communications	0	7	25	0	1	6
TOTALS	90	1,085	3,230	13	100	284

Other, more elaborate cost models have been developed to predict operations costs One model, based on estimating the value ranges of 91 operations complexity factor that correspond closely to the factors in tables 14-4 through 14-7, successfully pre dicted ops costs to within 25% for 13 out of 14 scientific mission case studies [Carraway, 1994, 1996]. NASA has also recently developed a Space Operations Cos Model that has operations cost estimation modules for planetary and Earth orbite missions, orbiting space facilities, launch system ops, and human spaceflight (lunar Mars) mission operations.

14.4 Automating Spacecraft and Ground Operations Functions[*]

As discussed in Secs. 2.1.2, 11.7 and Chap. 23, continuing technology advances in spacecraft computing and memory capacity, together with improved software capabil ities, are now making it possible to do things on board the spacecraft that people on the ground have done. Some examples are

- Monitoring alarms
- Managing spacecraft resources
- Analyzing performance and trends
- Onboard navigation and orbit control
- Adaptively planning activities
- Processing payload data
- Detecting faults—safing the vehicle—recovering from failures
- Adaptively capturing and downlinking data
- Constraint checking commands
- Mining data
- Archiving data

Migrating these functions from the ground to the spacecraft can save a lot of money by reducing ground operations tasks and the need for continuous staffing and fas response from the operations team. It can reduce the amount of routine data we mus transmit over expensive space communication links, process on the ground, and analyze. Processes performed on the spacecraft can use timely data, free from delay in space communication links and uncorrupted by transmission errors.

It's helpful to distinguish four levels of autonomy. The first level of autonomy i onboard closed-loop processes. Examples might be closed-loop thermal control momentum management, attitude control, and even navigation and orbit control. Th second level of autonomy is the ability of a spacecraft to execute planned events with out human intervention via a stored onboard sequence of timed commands. At the

---

[*] This section discusses general characteristics of autonomy, not necessarily related to reducing mission cost. Sec. 2.1.2 discusses the use of autonomy to reduce cost. Sec. 11.7 discusses the specific example of autonomous navigation and orbit control as a means of reducing both cos and risk. Finally, Sec. 23.3 discusses implementing the concept of "autonomy in moderation for general spacecraft functions to reduce cost and risk.

third level of autonomy, the spacecraft can react to unplanned events through event-driven rules. At the fourth level of autonomy, spacecraft react to unplanned events not just by executing rules but by using forms of onboard intelligence, inference engines, and planning agents.

Most spacecraft have level 1 autonomy and exploit onboard control loops. Most spacecraft have level 2 autonomy and can execute pre-planned sequences by using time-tagged commands referenced to their own clocks. Many spacecraft have level three autonomy in that they can sense and respond to unplanned hardware failure events by executing fault response rules that switch them to safe modes or that autonomously reconfigure the spacecraft to backup, redundant hardware components. Some spacecraft go beyond just hardware failure event response and have the capability to execute rule-based responses to payload-sensor events as well. Spacecraft are testing out level 4 autonomy by flying software that will perform autonomous onboard adaptive planning and resource management. Deep Space 1 has onboard intelligent agent software and Europa Orbiter will fly an adaptive, prioritized goal achieving execution engine.

Autonomy can enhance mission capabilities but it may not always reduce operational costs. Onboard control loops can require extra operations attention to the performance of flight software and management of flight computer resources. The creation of sequences or activity plans may require many people on the ground to plan, model, implement, constraint check, simulate, approve, uplink, and enable. If so, they'll end up being more expensive than a set of commands a small operations staff uplinks in joystick mode in real time. Fault protection rules onboard can require large teams to design and to then analyze the causes and results when they trigger. The cost of programming, "training", performance monitoring, and trouble shooting an intelligent agent can easily be underestimated. For a given mission, we can evaluate the cost/benefits of each of the four levels of autonomy in terms of several specific operational cost savings by answering the following questions.

Does it reduce the number or complexity of tasks that must be performed by ground operators? If the net number or complexity of tasks performed by operations goes down (after considering the additional operational tasks required to program, maintain, analyze, and trouble shoot the autonomy), then spacecraft autonomy can be justified in terms of ops cost savings.

Does it allow an increase in the time between spacecraft contacts? How long a spacecraft can go routinely unattended can depend heavily on the amount of onboard autonomy. The longer a spacecraft can go without ground interaction the more significant staffing cost savings may be by enabling single-shift rather than around the clock operations. The plan for the Pluto Express spacecraft during its nine year cruise to Pluto, is to have *lights out operations* (i.e., no activity) for 12 out of every 14 days.

Does it reduce the number of commands that we must routinely uplink? A spacecraft that can expand commands on board, use a high-order command language, or respond to goal-oriented commands can be cheaper to operate than a spacecraft that must be controlled by primitive, device level commands.

Does it reduce the amount of engineering and performance data that we must routinely downlink, process, and analyze on the ground? Spacecraft designed to do their own onboard monitoring, trending, compressing, summarizing, and archiving of engineering data can substantially reduce communication costs as well as the quantity of data that operations must deal with on the ground. Remember that spacecraft autonomy can add its own performance data to the set that must be routinely downlinked to the ground. The Pluto Express spacecraft is being designed to downlink only about 2% of the total engineering data that it will collect during cruise. The 98% that won't be routinely downlinked will be stored on the spacecraft in an onboard engineering data archive and will be downlinked only in the event of a spacecraft performance anomaly.

Does it reduce the amount of payload data that we must routinely downlink, process, and analyze on the ground? This question suggests that all things being equal, payload autonomy can reduce ops costs if it returns fewer telemetry bits while accomplishing the same mission goal. For FireSat, the least autonomous mode is to return all the data being captured by the payload sensor. A more autonomous mode would be to have the spacecraft decide when it was over the ocean and omit that data. An even more autonomous mode would include onboard detection of fires, and the data downlinked would be only the fire's location and extent.

Each mission will have different answers to these autonomy questions. You must analyze each mission's cost vs. benefit in deciding what spacecraft autonomy capabilities make sense. Autonomy can require additional spacecraft resources such as mass, power, memory, MIPS and data bus bandwidth, and the cost of these must be accounted for. Developing, testing, and validating flight software isn't cheap. You must compare cost increases for development and maintenance with operational cost savings. This trade may not be favorable for short missions or single spacecraft but may have high payoff for long duration missions or constellations. The cost and availability of space communication services and for transporting and managing data on the ground can also influence autonomy decisions.

References

Bloch, J., et al. 1994. "The ALEXIS Mission Recovery." Paper No. AAS 94-062, in *Advances in the Astronautical Sciences: Guidance and Control 1994*, R.D. Culp and R. D. Rausch, eds., vol. 86, pp. 505–520.

Boden, Daryl G. and Wiley J. Larson. 1995. *Cost-Effective Space Mission Operations.* New York: McGraw-Hill, Inc.

Cameron, G. E., J. A. Landshof, and G. W. Whitworth. 1994. "Cost Efficient Operations for Discovery Class Missions." In *Third International Symposium on Space Mission Operations and Ground Data Systems, NASA Conference Proceedings 3281*, Greenbelt, MD, November 15–18, pp. 809–816.

Carraway, John. 1994. "Lowering Operations Costs Through Complexity Metrics." Presented at Space Ops '94, Greenbelt, MD, November 15.

Carraway, John, and Gael Squibb. 1996. "Spacecraft Autonomy Metrics." Paper No. IAA-L-0302P, presented at the IAA Conference, JHU Applied Physics Lab, Laurel, MD, April.

Carraway, John, et al. 1996. "Application of Operations Concepts Methodology to Low Cost Planetary Mission Planning, Design, and Operations." Paper No. IAA-L-0504, presented at the IAA Conference, JHU Applied Physics Lab, Laurel, MD, April.

Carraway, John, and Gael Squibb. 1996. "Autonomy Metrics." Paper No. SO96.6.09, presented at Space Ops '96. Munich, Germany, September.

Carraway, John. 1995. "Assessing Operations Complexity." Chapter 5 in *Cost-Effective Space Mission Operations*, Daryl G. Boden and Wiley J. Larson, eds., New York: McGraw-Hill, Inc.

CSP Associates. 1999. *GSFC Ground Systems Cost Model Study; Final Results: Phases I, II, and III.* February 9. Cambridge, MA.

Davis, G. K. 1990. "NOAA Spacecraft Operations—An Operators Perspective." Paper No. 90-0322 presented at the 28th Aerospace Sciences Meeting, Reno, Nevada, 8–11 January.

Hughes, P. M., G. W. Shirah, and E. C. Luczak. 1994. "Using Graphics and Expert System Technologies to Support Satellite Monitoring at the NASA Goddard Space Flight Center." In *Third International Symposium on Space Mission Operations and Data Systems, NASA Conference Proceedings 3281*, Greenbelt, MD, November 15–18, pp. 707–712.

Jeletic, J. F. 1990. "Computer Graphics Aid Mission Operations." *Aerospace America.* 28(4): 28–32.

Landshof, J.A., R. J. Harvey, and M. H. Marshall. 1994. "Concurrent Engineering: Spacecraft and Mission Operations System Design." *Third International Symposium on Space Mission Operations and Data Systems, NASA Conference Proceedings 3281*, Greenbelt, MD, Nov. 15–18, 1994, pp. 1391–1397.

Ledbetter, K. W. 1995. "Mission Operations Costs for Scientific Spacecraft: The Revolution That is Needed." *Acta Astronautica*, vol. 35, pp. 465–473.

Mandl, D., J. Koslosky, R. Mahmot, M. Rackley, and J. Lauderdale. 1992. "SAMPEX Payload Operations Control Center Implementation." In *Second International Symposium on Space Mission Operations and Data Systems, Proceedings JPL 93-5*, November 16–20, pp. 63–68.

Marshall, M. H., G. E. Cameron and J. A. Landshof. 1995. "The NEAR Mission Operations System." *Acta Astronautica*, vol. 35, pp. 501–506.

Marshall, M. H., J. A. Landshof and J.C. van der Ha. 1996. "Reducing Mission Operations Cost." Chapter 6 in *Reducing Space Mission Cost*, James R. Wertz and Wiley J. Larson, eds. Torrance, CA and Dordrecht, The Netherlands: Microcosm Press and Kluwer Academic Publishers.

Mazza, C. 1987. "Data Processing Systems at ESOC." *Journal of the British Interplanetary Society.* 40 (6).

Roussel-Dupre, D., et al. 1994. "On-Orbit Science in a Small Package: Managing the ALEXIS Satellite and Experiments." In *SPIE Proceedings of Conference on Advanced Microdevices and Space Science Sensors*, San Diego, CA, July 28–29, vol. 2267, pp.76–89.

Squibb, Gael. 1995. "Developing a Mission Operations Concept." Chapter 4 in *Cost-Effective Space Mission Operations,* Daryl G. Boden and Wiley J. Larson, eds. New York: McGraw-Hill, Inc.

van der Ha, J. C. 1992. "Implementation of the Revised HIPPARCOS Mission at ESOC." *ESA Bulletin No. 69,* February, pp. 9–15.

Wall, Stephen D. and Kenneth W. Ledbetter. 1991. *Design of Mission Operations Systems for Scientific Remote Sensing.* London: Taylor & Francis.

Chapter 15

Ground System Design and Sizing

Gary G. Whitworth, *Applied Physics Laboratory*

A ground system (1) supports the space segment (spacecraft and their payloads), and (2) relays to users mission data generated by onboard instruments and received from the spacecraft. Table 15-1 summarizes these "functions" and corresponding options.

To support spacecraft and their payloads, the ground system must command and control them, monitor their health, track them to determine orbital position, and determine spacecraft attitude from sensor information. The ground system controls the spacecraft and its instruments or payloads by transmitting command data to the spacecraft. Except for passive echo tracking techniques such as radar or laser reflector, the ground system uses spacecraft housekeeping telemetry and mission data to carry out these functions. For example, the ground system may use instrument data from a spaceborne radar altimeter to refine knowledge of the spacecraft's orbit.

Ground stations acquire mission data from a spacecraft and its instruments and transfer it to the data users. The ground system also supplies any telemetry and tracking information the data users may need. Most space missions allow the user's evolving requirements to influence changes in the ground system's data relay and control functions.

Ground systems consist of ground stations and control centers working together to support the spacecraft and the data user. Figure 15-1 shows how these segments interact. Generally, the ground system commands and controls the spacecraft based on

621

TABLE 15-1.	Ground System Functions and Options.

Function	Options and Considerations	Key Criteria
Spacecraft/Payload Support		

Maintain RF communications links	Communications architecture
	  − Direct (spacecraft to ground)
	  − Relay spacecraft
	
	Number of spacecraft and orbital configuration
	
	Number and locations of ground stations
	  − Existing or new
	  − Dedicated or shared
	  − Fixed or mobile
	  − Data rates
	
	Ground station configuration
	  − Antenna size and track rates
	  − RF equipment frequencies and capabilities
	
	Intra-system communications required
	
	Simulation and verification	

Provide spacecraft and payload control
  − Issue commands
  − Determine orbital parameters
	Number of spacecraft and orbital configuration
	Tracking methods
	  −Range and range rate,
	    antenna viewing angles
	  −External tracking network
	  −Spacecraft autonomy (onboard navigation)
	Simulation and verification
		• Cost
		
		• Complexity
		
		• Timeliness
		
		• User Transparency
		
		• Availability

Process telemetry
  − Monitor spacecraft and payload health
  − Determine spacecraft attitude
	Number of spacecraft (determines processing load)	• Reliability
		
		• Survivability
		
		• Flexibility

Mission Data Relay		

Transport payload mission data
Transport spacecraft and payload telemetry as required
Provide data handling
Distribute to data user community
	Number and locations of data users
	Intra-system communications requirements
	Data handling capabilities
	  − Multiplexing and demultiplexing
	  − Encoding and decoding
	  − Encrypting and decrypting
	  − Data compression
	  − Data storage and archiving
	  − Timetagging
	  − Quality monitoring
	Simulation and verification	

Other		
Support mission operations	Space segment operations Ground system operations	
Maintain facility and equipment	Logistics Spare or repair	

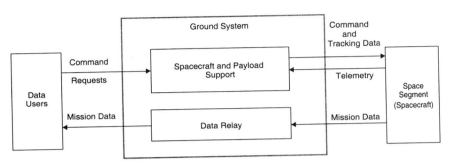

Fig. 15-1. Relation Between Space Segment, Ground System, and Data Users. Data users influence a mission by requesting commands through the ground system.

requests from the data user to the control centers. Except for communications satellites,* users do not send commands independently to the spacecraft, because its overall health depends heavily on the state of individual instruments and systems.

The ground system tries to provide highly available, high-fidelity access to the spacecraft while remaining transparent to both data users and ground controllers. In practice though, we must trade off transparency and cost. For example, we may accept some distortion or loss of mission data, as well as time delays between the spacecraft and the data users. These delays may range from subseconds to seconds for real-time data transfer, and from days to weeks for recorded data. In supporting the spacecraft and payload, we may need to balance length and number of opportunities to command or monitor the spacecraft with the risk inherent in being out of communication.

Because more complex ground systems are less transparent, we must design them as simply as possible, consistent with mission requirements. When designing a space mission, we should trade off space segment and ground system complexity through several iterations, until we produce best performance at lowest cost (see Chaps. 3 and 4).

15.1 The Ground System Design Process

Table 15-2 summarizes the ground system design process and references discussions pertaining to each step. This process is iterative because the steps interrelate and we must strike a balance in complexity between the spacecraft and the ground system. Each iteration must address:

- Ground station locations, based on spacecraft coverage and data user needs, balanced against cost, accessibility, and available communications. You will need new sites for a dedicated ground system, and suitable existing stations when using established ground systems.

- Link data rates, which establish the required gain-to-noise temperature ratios (G/Ts), and effective isotropic radiated powers (EIRPs). For dedicated ground stations, defer details of antenna and RF systems until you have established

---

* With communications satellites, "mission" data is really communications data being relayed between two or more "data users." Here, we simply expand the ground system's data-transfer function to include the path from the ground system to the satellite.

these two parameters. For an existing system, determine whether RF links are adequate and adjust data rates if necessary.

- Requirements for data handling, so you can determine where it will occur. Once you know the location, you can meet almost any need with available hardware and software, and at reasonable cost. A decision to perform data handling at a central facility instead of the ground station influences the location of the control centers.

- Appropriate communications between ground system elements and data users, for a dedicated system. When using an existing ground system, where most decisions are already made, confirm that data handling and bandwidth are adequate.

TABLE 15-2.   Summary of the Ground System Design Process. See text for discussion.

Step	Where Discussed
Establish number and locations of ground stations	Secs. 15.2, 15.5
Establish space-to-ground data rates	Secs. 13.2, 15.5
Determine required G/Ts and EIRPs	Secs. 13.4, 15.2
Determine required data handling	Secs. 15.3, 15.5
Establish data handling location	Sec. 15.5
Decide location of Spacecraft Operations Control Center, Payload Operations Control Centers, and Mission Control Center	Secs. 15.2, 15.3
Determine and select communications links	Secs. 13.3, 15.3
Evaluate complete or partial use of service-provided ground systems	Secs.15.4
Iterate as needed	

15.2 A Ground System's Basic Elements

Figure 15-2 shows that the ground system consists of mission elements and facility elements. *Mission elements* control the space segment or handle mission data. *Facility elements* support or are otherwise ancillary to mission elements. Both contain mixes of various hardware, firmware, and software. (In this chapter, I will use the most common names and acronyms.) The ground system staff uses and coordinates the operation of the physical components. Mission operations, discussed in Chap. 14, coordinates activities for the ground system and command and control of the spacecraft.

The ground station is the Earth-based point of communication with the space segment for control and, typically, user data. Figure 15-3 shows the basic ground station, which consists of the following components.

The single *antenna system* includes the antenna and mount, its associated electro-mechanical actuators, the consoles and servo circuitry which control the antenna, and the feeds and transmission lines which carry RF signals to and from the RF equipment. The antenna, along with the receive RF equipment, satisfies the required receive G/T at the frequency of the downlink carrier. It also works with the transmit RF equipment to provide the required EIRP at the uplink carrier frequency. Chapter 13 discusses these communications links. Antenna steering must provide the look angles required

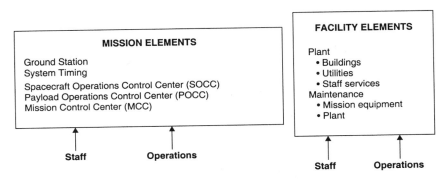

Fig. 15-2. **Mission and Facility Elements for a Ground System.** The staff provides the necessary operator input. Mission operations, discussed in Chap. 14, directs all activities of the mission.

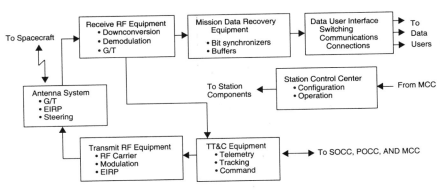

Fig. 15-3.   **The Basic Ground Station.** This figure displays only the minimum components needed to control a spacecraft and relay mission data to users.

by the mission. For low-Earth orbit, these can cover essentially all of the visible hemisphere. It must also provide the required steering modes, such as programmed computer steering and autotracking. *Autotracking* refers to the use of the received spacecraft signal itself to steer the antenna. In this case, the antenna system usually provides continuous pointing coordinates to the tracking component at the ground station.

The *receive RF equipment* is generally in suites of racks, located to minimize transmission-line losses to the antenna. This equipment accepts the downlink carrier frequency from the antenna system, downconverts it to intermediate frequencies, and demodulates it to baseband signals for the equipment devoted to mission data recovery and TT&C.

Also in racks near the antenna system, the *transmit RF equipment* accepts tracking and command signals from the ground system's TT&C component and modulates them onto the RF uplink, which it also generates. In the case of communications satellites, it also modulates user data onto an uplink carrier.

After the RF receive equipment demodulates the signals, the *mission data recovery equipment* conditions the mission data before relaying it to data users and ground system components. It typically has its own location in the system, but it may be intermingled with the receive RF suite for simple data streams.

The *data user interface* connects the mission data recovery equipment and the data user. If all parts of the ground system and the data user are colocated, this interface generally consists of no more than manual or electronic patching of the data lines between the ground station and the user facilities.

The *Telemetry, Tracking, and Command (TT&C)* equipment conditions and distributes received telemetry and tracking signals. It also electrically formats, authenticates, and times transmitted command and tracking signals. It usually processes these tracking signals and data on the antenna-pointing angle to inform users about range, range rate, and spacecraft position. TT&C functions are usually highly automated because of the need for speed, timeliness, and accuracy.

The *station control center* controls the configuration of, and the interconnects between, the ground station components. Operating under instructions from the ground system's mission control center, it keeps the ground station configured to support mission operations.

Ground system operations require time coordination, so one system element maintains a clock precise enough to meet mission requirements; it distributes clock time and reference frequencies to the other system elements (see Fig. 15-4), moving through the colocated elements of a ground system by wire or cable. It is accurate to within milliseconds or better. Its usual one-per-second timing pulses are synchronized to within a few microseconds or less to a world time scale, such as Universal Time Coordinated (UTC). Satellites able to transfer time even more precisely, such as GPS, permit us to synchronize well below a microsecond.

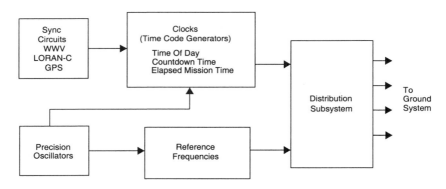

Fig. 15-4.   The System Timing Element. This synchronizes the elements of a ground system by distributing precision time and frequency.

Three types of control centers are generally found within ground systems. The *Spacecraft Operations Control Center (SOCC)* monitors and commands the spacecraft bus and common systems, as opposed to onboard instruments or payloads, which are controlled by the POCC, as discussed below. The SOCC analyzes spacecraft telemetry and, when necessary, telemetry and mission data from instruments which can affect the spacecraft's attitude and dynamics. As the only ground system element that directly commands the spacecraft, it coordinates and controls POCC access. Specifically, it approves the POCC's requests to command instruments after considering mission plans and schedules, spacecraft health, and the collective well being of the other instruments on board.

For simple spacecraft, the SOCC also serves as the POCC. On the other hand, we may need several SOCCs if the space segment is complex. An example might be one with several complicated spacecraft requiring intensive and near-continuous monitoring and control, such as ground systems which provide user services to other missions (see Sec. 15.4). We may also need more than one SOCC for backup or for security, survivability, or other political reasons.

Equipment and people make up the SOCC. The hardware includes data monitoring equipment and consoles, commanding facilities, and associated communications. This equipment is usually computer automated for quick response, but humans may intervene at any time to control the spacecraft.

The *Payload Operations Control Center* (*POCC*) analyzes telemetry and mission data from onboard payload instruments and issues commands to these instruments. Its commands depend on approval by the mission control center, with coordination from the SOCC. Interactive computer equipment also runs the POCC, with people standing by during communication with the spacecraft.

We may use multiple POCCs when several onboard instruments require careful independent supervision or when we need a backup for redundancy or survivability. We may also need or want to place the POCCs for some instruments near the manufacturer or the data user.

The *Mission Control Center* (*MCC*) plans and operates the entire space mission, including the configuration and scheduling of resources for both space and ground system. It computes and issues information needed by ground system elements and data users, such as data on the spacecraft's orbit, ground station pass times, and antenna pointing angles. In simpler systems, we may merge the MCC with the SOCC.

MCCs are best placed near the POCCs and SOCCs, but mission requirements or other considerations often call for placement elsewhere, thus greatly complicating the ground system. Location of the MCC depends on security, survivability, and political or administrative considerations. Sometimes, redundancy demands several MCCs —one as prime and the others as backups. If ground systems provide services to user missions, each user mission will have a dedicated MCC. These MCCs are frequently, but not necessarily, colocated. The host mission MCC may also, but again not necessarily, be near the user MCCs.

The software used for MCC and POCC activities is covered in more detail in Chap. 14. It is worth noting here, however, that there has in recent years become available commercial off-the-shelf (COTS) software for integration testing as well as mission and payload operations, analysis, and planning, and ground system operation. The best of these packages are very powerful, yet tailorable to a specific spacecraft, alleviating the need for a program to develop expensive mission-unique software. Most of these offerings are very efficient and versatile, permitting large reductions in the required operations team.

Plant facilities include buildings and grounds, utilities, services for the staff, and security. We normally would use commercial utilities with locally generated backup for emergencies. For security and survivability, utilities may be wholly self-contained. Because plants are expensive, we must decide whether to build unique, dedicated systems or to use existing alternative systems (see Sec. 15.4).

Availability is the percentage of time a ground system is available to support a mission. The availability we want in a system determines whether maintenance should use spare or repair. To achieve high availability often requires **hot** spares (powered on and ready for operation at a moment's notice).

Figure 15-5 shows the physical layout of a simple ground system with its data users colocated, eliminating the need for long-distance communications links between them. Such a system can provide low-coverage support of up to several spacecraft in low-Earth orbit on a timeshared basis, or virtually 100% coverage for a single space-craft in geostationary orbit. One of its advantages is that it can be dedicated to a particular mission, thereby eliminating schedule conflicts. Also, it is compact and self contained, allowing all communications between elements to be local and dedicated thus simplifying the system's operation and administration.

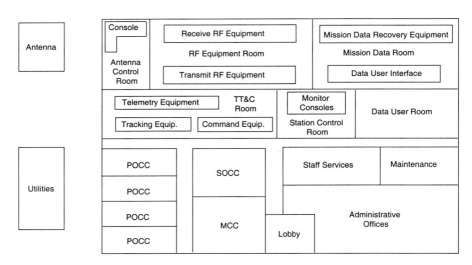

Fig. 15-5.   A Typical Installation for a Basic Ground System. Depending on the number of POCCs, the total area occupied may be between 100 and 500 m². The need for significant staff services can greatly increase the area for remote installations. See text for discussion.

Unfortunately, this simple ground system also has significant disadvantages. For example, the single ground station provides very low coverage for low-Earth orbit spacecraft because a pass (the time period that the spacecraft is in view of a ground station) lasts only a few minutes, and for other than near-equatorial orbits, does not occur often. As Sec. 15.5 points out, these conditions prevail because the maximum viewing angle of a low-Earth orbit spacecraft from any point on the Earth is 20% or less of the orbital path, and for inclinations above 25 or 30 deg, only 25 to 30% of the orbits are visible. An exception is polar orbits as seen from polar ground stations, where each orbit is visible. (See Chap. 5 for visibility formulas.) Another limit is the system's inability to support more than one spacecraft link at a time because of the single antenna and ground station. Also, the simple system can serve only on-site users. The system's most important disadvantage, however, is its lack of redundancy. If any element were to fail we would lose data, or possibly even a spacecraft. Thus, redundant equipment, spares, and maintenance, as well as geographical location, are particularly important. To be secure, a completely colocated ground system should be on domestic soil. Even so, it will not be very survivable, and it will take a long time to recover from any major damage. For this reason, the military would generally not want a simple or basic system.

15.2.1 GEOSAT—A "Simple" System

Though not as "simple" as the basic system described above, the GEOSAT's ground system shown in Fig. 15-6 is a good practical example. It supports only one spacecraft with one ground station and colocated components. Launched in the spring of 1985, this remote, ocean-sensing spacecraft is supported through an S-band telemetry and data downlink, a VHF command uplink, and VHF Doppler beacons for tracking. The ground system provides full telemetry, data, and command support of the spacecraft. It is located under one roof at the Johns Hopkins University's Applied Physics Laboratory in Laurel, Maryland, except for the Doppler tracking equipment in the worldwide system of TRANET Doppler tracking stations.

Figure 15-6 is a functional diagram of the ground system showing in solid lines its three major segments: the satellite test facility, the digital element, and the computer system. The figure and nomenclature reflect diagrams typically seen in descriptions of the GEOSAT ground system. The overlaid dotted boxes show how the functions fit into the basic components of a ground system with some functions overlapping. For example, the real-time frame sync and decommutator is involved in operations for both mission data recovery and TT&C, and the computer system supports many of the basic components. Further, as is often the case with simpler ground systems, the SOCC, POCC, and MCC are in one unit. Note that we have not yet discussed some of the functions, especially for mission data recovery equipment and the data user interface, which demand more than our simple ground system can provide.

15.3 The Typical Ground System

To support a realistic space mission, a ground system must usually provide high coverage simultaneously for several spacecraft in various orbits, with high levels of availability, security, and, for military missions, survivability. Such a system will usually include many elements in several configurations. The realistic system in Fig. 15-7 includes standard stations and less capable auxiliary stations, which may be on aircraft, ships, or land. These auxiliary stations fill gaps in coverage, using equipment similar to that of regular stations and providing radar tracking, telemetry, data reception, and backup command.

In a real ground system, we may also employ multiple control centers in separate locations. Thus, as shown, some POCCs are near the SOCC, whereas some are remote. In this generic system, we designate a SOCC as prime, colocate it with a prime MCC, and back up both prime centers with remote centers. Multiple control centers are redundant, survivable, and flexible, allowing responsibility for prime control to pass back and forth between the centers during various phases of the mission. For example, we may make the remote SOCC prime during launch and early checkout of the space segment because it is near the launch or simulation equipment.

Geographical dispersion and multiplicity of elements greatly complicate a ground system's design. For example, each location must usually have its own synchronized timing system, similar to that for the colocated ground system described earlier. Further, the distributed system requires several physical plants with different admin-

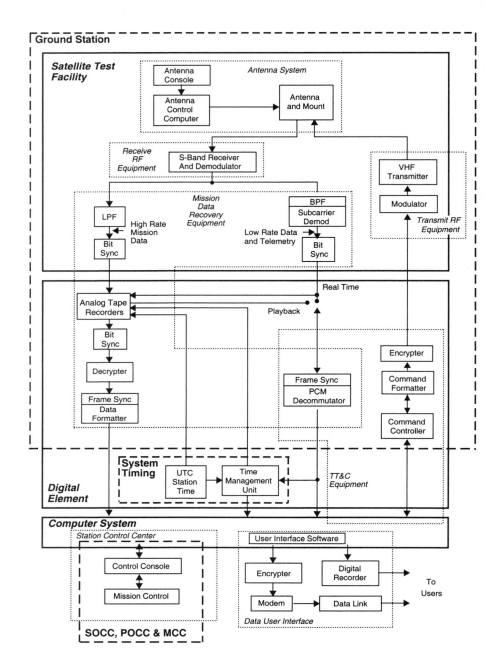

Fig. 15-6. Block Diagram for the GEOSAT Ground System. The large solid blocks (with underlined labels) show the system broken down into GEOSAT-unique *segments*. The dashed lines (with bold labels) show the division into *elements* as defined in Fig. 15-2. The dotted line (with italic labels) shows the division into *components* as defined in Fig. 15-3.

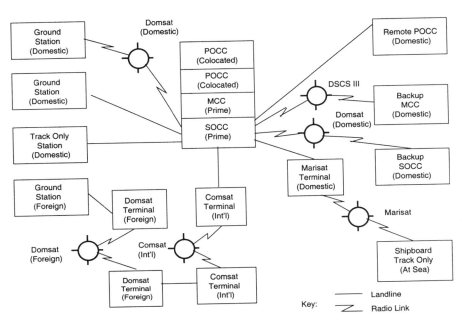

Fig. 15-7. Model of a Typical Ground System. Data is relayed between remote ground stations, backup control centers, and data users, by various communication links. (See text for discussion of these links.) Limited capability stations often supplement standard stations.

istrative structures, complicated maintenance and logistics networks, and reliable communications links—all difficult and expensive to implement for remote locations. Thus that remote tracking station in East Africa, for example, needed to fill a small but important gap in ground station coverage, may be disproportionately expensive.

15.3.1 Communications Links

Ground systems need long-distance communications links of sufficient bandwidth between their distributed elements and between them and the data users. These links mix landline (electrical, terrestrial microwave, and optical) and satellite connections. Unless the links are part of the ground system itself, they are usually subscribed to or leased.

We evaluate communications options while determining where to locate the ground system's components. For example, we may want to place the SOCC or the MCC near a metropolitan area to take advantage of its telephone system, but we would want to place ground stations in less populated areas to lessen radio-frequency interference. Yet installing new and dedicated communication facilities in a remote area may be quite costly.

As international, commercial domestic, and military satellite communications have become more available, we now prefer satellites to provide communications between remote ground system elements and data users. Because of their high capacity and performance, communications satellites (*comsats*) link nearly all intercontinental elements of ground systems as well as those lying far apart within continents. Worldwide systems may require more than one "hop," and more than one comsat. For example, a ground station located in India might communicate with a SOCC in the United States by accessing an international satellite through a domestic communications satellite (*domsat*). Table 15-3 shows several geostationary communications satellites typically used by ground systems with elements around the world. In general, portions or all of the available transponder bandwidth may be leased. Martin [1984] has summarized the technical details of these satellites, and user information is available from the operating agency. A good basic reference on satellite communications is Pratt and Bostian [1986].

Even when using comsats, we may need landlines to connect to the ground terminals which access the comsats. With many domsats and military communications systems, we may be able to use locally placed comsat terminals with user antennas that access the communication satellite directly. For international satellites, we must usually link in to a central ground terminal which accesses the satellite. This link is usually some form of landline, normally a leased telephone line with a high data rate.

To access these communications links, each element requires terminal equipment, whose complexity depends on the type of link. (Satellite links generally are most complex.) Domestic satellite links usually include options to build, lease, or buy the terminal equipment. Military communications systems provide qualified users links through small local terminals and centralized communications stations. Unfortunately, the small terminals often support relatively narrow bandwidths not suitable to relaying high speed data.

Thanks to the recent explosion in the ubiquity and versatility of communications products for networking, an increasing trend in moving data around a ground system is to use the Internet Protocol (IP) family of protocols. Physical connections can range from common low rate dialup lines to high speed dedicated channels. For low rate data, it is feasible to move data between system elements using real-time IP protocols. High rate data is generally stored initially at a ground station, and forwarded postpass by protocols such as FTP (File Transfer Protocol). With IP connections, the transport protocol may be either TCP or UDP.* Where dialup or dedicated lines are used, so that the potential for data loss is low, the ground system may use UDP, for the increased data efficiency it provides.

15.3.2 Optional Functions

Each element of the typical system may include functions beyond those in the basic elements. A good example is the ground station, which can be far more complex than the basic station we examined earlier. These stations may need several antennas to support more than one RF link or spacecraft at a time. The antennas may have multiple feeds to permit simultaneous links at different RF frequencies, or polarization diversity to permit simultaneous multiple links at the same frequency. Multiple RF links in turn require more RF and data recovery equipment, with enhanced performance needed to allow higher data rates and more sophisticated modulation techniques. Improved TT&C equipment permits increased rates for telemetry and command, more precise tracking, and coverage for several spacecraft at one time. We may also add to the ground station new functions such as simulation and verification systems.

---

* TCP (Transmission Control Protocol) provides a degree of assured data delivery, and adds overhead to the data transfer. UDP (User Datagram Protocol) uses less overhead but provides no assurance of delivery of any packet of data.

TABLE 15-3. Typical Geostationary Communications Satellites.

Satellite	Operator	Bandwidth at (Up/Down Frequency)	Longitude (deg)
INTERNATIONAL			
Intelsat V-A	Intelsat	36 to 72 MHz (6/4 GHz)	60, 63, 66 E
		72, 77, 241 MHz (14/11 GHz)	1, 18.5, 21.5,
Intelsat VI	Intelsat	36 to 72 MHz (6/4 GHz)	22, 27.5, 34.5 W
		72 to 159 MHz (14/11 GHz)	
Symphonie	CIFAS (Fr-Germ)	90 MHz (6/4 GHz)	11.5 W
DOMESTIC			
Anik C	Telesat Canada	54 MHz (14/12 GHz)	112.5 W
Westar	Western Union	36 MHz (6/4 GHz)	91 W
Telstar 3	AT&T	34 MHz (6/4 GHz)	87.95 W
RCA	RCA Americom	34 MHz (6/4)	66, 83, 119, 131,
			135, 139 W
Marisat	Comsat General	4 MHZ (1.6/1.5 GHz)	73, 176.5 E
		4 MHz (6/4 GHz)	15 W
		25, 500 kHz (UHF)	
L-Sat	ESA	40 MHz (30/20 GHz)	18 W
CS, CS2	NASDA (Japan)	200 MHz (6/4 & 30/20 GHz)	130, 135 E
		130 MHz (30/20 GHz)	
MILITARY			
NATO III	NATO	17, 50, 85 MHz (8/7 GHz)	18, 22.5, 50 W
FLTSATCOM	Navy Comm. Cmd	5, 25, 500 KHz (UHF)	23, 93,100 W
			72.5, 172 E
LEASAT	Navy Comm. Cmd	5, 25, 500 KHz (UHF)	5, 75, 176 E
			23, 100 W
DSCS II	Def. Comm. Agency	50, 125, 185 MHz (8/7 GHz)	66.8, 140, 175 E
			13, 130, 135 W
DSCS III	Def. Comm. Agency	50, 60, 85 MHz (8/7 GHz)	54, 175 E
		100 MHz (UHF)	13,135 W
MILSTAR*	50th Space Wing	44 GHz, 20 GHz	120 W, 4 E
		400 MHz, 225 MHz	

*Data from *MILSTAR SGLS Student Guide*, March 1997.

Data users' demands usually make ground systems more complex. The user interface must have versatile switching and interconnection options and connect to long-distance communications links. We may need to add data-handling equipment to distribute received data to different users. Data handling includes all processing of mission data between the ground station's data recovery equipment and the data user's communications interface. Sklar [1988] has rigorously defined various data handling operations, but I will summarize the most common functions in the following paragraphs.

Demultiplexing refers to the disassembly of composite data streams received from spacecraft into selected component data streams for routing to different users. With multiple POCCs, and possibly with multiple SOCCs, telemetry data also may require demultiplexing.

Classified or otherwise secure data from spacecraft is often *encrypted* on board before transmitting it to the ground where it is *decrypted*. The data then either flows to its users directly, without decryption, or it is decrypted before distribution preparatory to demultiplexing or other data-handling operations. Of course, some data may be re-encrypted before transmission to some users. To prevent unauthorized commanding by others, command data is often encrypted as well.

We may apply *encoding*, a technique which decreases errors in digital data because of noise, to data streams from spacecraft. We may *decode* this received data in the ground system and possibly reencode it before distribution. Command data is frequently encoded to ensure the spacecraft receives error-free commands.

Data compression refers to the increase in the information capacity of a data stream to permit delivering its information through a narrower band communications channel than would otherwise be possible. Alternatively, a compressed data stream can transmit more information over a given bandwidth medium. This technique permits us to combine one or more data streams into a composite data stream (that is, *multiplexing*, the reverse of the demultiplexing described earlier), and distribute it over a communications link which could not handle uncompressed data. We can also use data compression for a ground system with limited storage and to distribute received data more quickly.

Timetagging means adding timing information to data streams. If data is not time-tagged at the spacecraft, either the ground station or the data-handling equipment may apply time information to it before recording it or distributing it to users. Usually, we would add an epoch time reference to the data stream in one of several ways. We could reference it to the station timing system, by giving the time at which data was received at the ground station, with at least first-order corrections for known equipment delays. With more sophisticated systems, we could continuously correct the epoch with orbital information to estimate the actual time the spacecraft generated the data. To interpolate the timing of data between epochs, we can use the data clock itself as a time scale (for synchronous data only).

Instead of routing received data to users immediately, we may wish to record all or some of it, referred to as *data storage*, and transmit it later. Also, we may avoid wide-band communications links by first recording higher rate data and then transmitting it to users at lower playback rates. In all cases, storage protects against loss of data during distribution. Magnetic tape has been the storage medium of choice, but optical storage may eventually supersede it for very large amounts of data.

Data quality monitoring means examining the quality of the space-to-ground data link by checking predictable groups of bits or waveforms in the received spacecraft data stream. For example, with synchronous digital data, we can use frame-synchronization words to count the bit error rate and make sure the link is working properly. In this way, we can monitor all parts of the ground system used to relay mission data.

These data-handling operations may take place anywhere in the ground system. For small systems with only one or two ground stations, we would typically handle the data within the ground station, and transmit it directly to the user. In doing so, we can reduce bandwidths in communications links to users by separating high-rate composite data streams into their components and compressing the data at the ground station.

We can also control data security, preserve signal-to-noise quality in the data stream, and suppress accumulated distortions by processing data as quickly as possible after receiving and demodulating it. The ground station usually stores and timetags data as well because mission data has the smallest delay uncertainties there.

Although the ground station usually does the data-handling tasks best, a system with several ground stations would need links between each of the ground stations and each of the users to transfer the data—an impractical if not impossible arrangement. Thus, ground stations often transfer data directly to a central facility (the SOCC for example), handling only selected operations, such as recording, themselves. The central facility passes the data on to the users. This procedure minimizes ground station hardware, centralizes control, and gives more flexible service to data users. It usually requires dedicated communications links between the ground stations and the central facility, which can support higher data rates than might otherwise be necessary.

Simulation/Verification (Sim/Ver) systems test the ground system's readiness using realistic simulated signals and data. Tests may be at routine intervals, during prepass or postpass, or following system maintenance or upgrade. Sim/Ver also provides diagnostics for troubleshooting and calibrates equipment. When fully implemented, a Sim/Ver system not only can test individual ground system elements and components, but also can perform highly automated end-to-end tests of the entire ground system. But this type of Sim/Ver system is expensive, employed only within elements whose availability is critical.

15.3.3 Influence of Spacecraft Autonomy

Spacecraft autonomy could potentially simplify the tasks of the SOCC, POCC, and TT&C elements of the ground system. But unpredictable spacecraft upsets and malfunctions, including those in the autonomous systems themselves, will force us to use ground elements at the same level for some time. For example, onboard clocks may timetag data as it is generated, but the ground system's ability to timetag received data will probably be retained as a backup for the foreseeable future.

15.3.4 The DMSP Example System

The Defense Meteorological Satellite Program (DMSP) is an example of a typical distributed ground system. Using remote-sensing satellites in low-Earth orbit, it provides the Department of Defense important environmental information. Figure 15-8 shows its main elements. The spacecraft links are at L-band (1,750 to 1,850 MHz) for the uplink and S-band (2,200 to 2,300 MHz) for the downlink. Data rates for these links are 2 kbps (command) and 1,024 kbps (mission data), respectively. The DMSP ground stations are referred to as Command Readout Stations (CRS). They are supplemented by the Automated Remote Tracking Stations of the Air Force Satellite Control Network's (AFSCN) ground system.

Mission data is transferred from ground stations to DMSP central facilities by domestic satellite and landlines. The data is then relayed by similar communications links to the large data processing users, the Navy oceanographic, and Air Force weather forecasting centers. This system also is an example of a mission in which some data users receive mission data directly from the spacecraft. Shipboard and transportable landbased terminals throughout the world receive data on local environmental conditions directly for immediate use.

We might also note that with the current trend for commercial satellites which provide imaging and other forms of remote sensing, advances in receiver technology

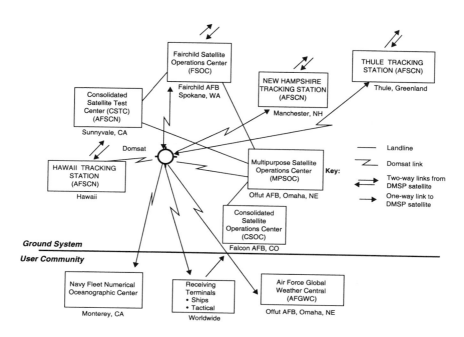

Fig. 15-8. The DMSP Ground System. It uses landlines and domsats to connect various ground stations to control centers and central facilities and for relay of data to large military data processing centers. Shipboard and mobile terminals receive data directly from the DMSP satellite.

and commercial processing software are now permitting private users to receive such data directly from these satellites at reasonable cost.

15.4 Alternatives To Building a Dedicated System

Instead of building a dedicated system, we can use existing ground support networks to supply part or all of the elements needed. A number of commercial and military ground systems can handle more than one mission and are available to support user missions. In this *service-provided* arrangement, the host ground system provides most of the elements of the user's ground system. But these existing configurations cannot satisfy all possible missions, so host systems are usually tailored to particular kinds. Where necessary, a user mission may have to provide some special equipment, but they can reduce the amount by designing missions to match the host's configuration wherever possible. In fact, users must usually meet severe constraints to make their missions compatible with the host system and other user missions.

Normally, full-service host systems provide all necessary elements of the ground system, which all users share. Major communication links connect ground stations to a central facility which houses the user mission's POCCs, SOCCs and MCC. Either the host or the user mission may supply the equipment and personnel for these centers. The central facility probably also contains the host's SOCC and perhaps, the prime or backup MCC.

Use of service-provided ground systems may follow a hybrid arrangement, combining some dedicated elements from the user mission with others from one or more host systems. Actually, most service-provided systems require some ground system construction and some significant elements, or parts of elements, from the user mission.

These alternative systems have some important advantages over dedicated ones. For example, they usually save a lot of money and have a defined and predictable cost schedule. Another advantage is high, predictable reliability and availability. Although not necessarily designed for it, most are highly survivable because they have many dispersed assets on the ground, making them important for military missions.

Alternative systems also have disadvantages. For example, matching the user mission with the system may make the overall mission less effective. Sharing host resources with other supported missions also demands coordination of activities and priorities, based on such things as the supported mission's relative importance, criticality of particular events, and the amount of control the host has over the ground assets. Contractual negotiations usually determine these priority agreements.

To evaluate potential alternatives to dedicated systems, we must begin by defining the requirements for key mission parameters and then matching them against the candidate host systems. As with a dedicated system, we would try to adapt the requirements to the host whenever possible through studies and discussions between planners of mission and host systems. To be efficient, we should evaluate all potential host systems at the same time. Further, if possible, we should examine service-provided systems while developing the preliminary design of a dedicated system, if we are considering one. The final configuration of our ground system would therefore be based on the best possible information.

In evaluating service-provided ground systems, we must determine how much users must tailor their missions, how much equipment users must provide, and how much access users have to ground stations or central distribution points. We would also need to consider the loading, lifetime, and upgrades planned for host systems. Another important comparison is between the well defined leasing costs of the host system, and the cost of building and maintaining a dedicated system, both evaluated over the mission's lifetime.

15.4.1 The Air Force Satellite Control Network (AFSCN)

An example of a service-provided ground system is the *Air Force Satellite Control Network*, *AFSCN*, a major ground system operated by the United States Air Force. It has 8 ground stations, called Automated Remote Tracking Stations (RTS), located throughout the world. Six of these are dual stations, able to support 2 spacecraft simultaneously. These stations communicate with 2 central facilities or nodes, called the Consolidated Space Test Center (CSTC) in Sunnyvale, California, and the Consolidated Space Operations Center (CSOC) at Falcon Air Force Base, Colorado, through an array of ground links and communications satellites, including DSCS, GE, and Intelsat. Each user mission's SOCC, POCC, and MCC are generally combined into a single MCC at one of the central nodes, with facilities including computers and software for both operations and planning.

Table 15-4 lists locations and key parameters for the AFSCN stations. To support spacecraft links, each station has 18 m and 14 m parabolic antennas, with RF and data-handling equipment for TT&C and mission data. The system of RF links with spacecraft is known as the S-band *Space-Ground Link Subsystem (SGLS)*.* The downlinks provide telemetry and mission data rates (signaling bit rates) up to 1.024 Mbps. The uplinks permit command rates of 2 kbps. Most RTSs also support non-SGLS downlinks at data rates up to 5 Mbps. The network has strong antijam and survival capabilities.

TABLE 15-4.   AFSCN Ground Stations. Six of the Remote Tracking Stations (RTS) are dual sites, capable of supporting two spacecraft simultaneously. See Chap. 13 for explanation of communications parameters G/T and EIRP.

Station and Antenna (1)		Location Deg : min	G/T		EIRP (3) (dBW)
			SGLS	Non-SGLS	
New Hampshire (Manchester)	18 m (2)	42:57 N	22.7	21.5	76.0
(NHS)	14 m	71:38 W	24.1	25.2	75.0
Vandenberg AFB (Lompoc, CA)	18 m	34:50 N 120:30 W	22.5	21.7	72.7
(VTS)	14 m		24.1	25.2	75.0
Hawaii (Oahu) (HTS)	18 m	21:34 N	22.5	21.7	72.7
	14 m	158:15 W	24.1	25.2	75.0
Guam (GTS)	18 m	13:37	22.7	21.5	76.0
	14 m	144:52 E	24.1	25.2	75.0
Diego Garcia (DGS)	10 m	7:27 N 72:37 E	18.1	17.3	72.0
Thule (Greenland)	4 m	76:31 N	7.7	—	61.5
(TTS)	14 m	68:36	24.1	25.2	75.0
Oakhanger (England)	18 m	51:07 N	25.0	—	76.0
(TCS)	10 m	00:54 W	18.1	—	72.0
Pike (CTS)	10 m	38:8 N 104:5 W	18.1	17.3	72.0

(1) Acronyms are AFSCN identifiers; parabolic antenna diameters are in meters.
(2) TT&C only.
(3) Nominal for 1 kW transmitter power; EIRP is 10 dB higher when using 10 kW transmitter.

A collection of detailed technical information on the AFSCN is available [Klements, 1987]. You may learn more about its use from the Air Force Systems Command, CSOC/50th OSS/Falcon AFB, Colorado.

---

* The SGLS incorporates 20 distinct and paired uplink and downlink channels. Uplink, or command channels, are in the 1,750 to 1,850 MHz range, and downlink channels are from 2,200 to 2,300 MHz. Each channel consists of a single uplink carrier and two downlink carriers which can be received simultaneously by an RTS to provide range and range rate, spacecraft telemetry, and mission data.

15.4.2 NASA Tracking and
Data Relay Satellite System (TDRSS)

Although it is not literally a ground support network, the TDRSS is the current support system for NASA satellites. Becoming operational in the mid-1980s, it replaced most of the Ground Space Tracking and Data Network. With its constellation of 3 geosynchronous relay satellites, the system can support up to 20 subsynchronous satellites with multiple-access S-band links, and two each single-access links at S-band and at Ku-band. The multiple-access links support satellites simultaneously at fixed frequencies of 2,106.4 MHz for the forward or command link (uplink) and 2,287.5 MHz for the return or downlink. The S-band single-access link supports one spacecraft at a time, with a frequency between 2,025 and 2,120 MHz for the forward path and between 2,200 and 2,300 MHz for the return path. Ku-band frequencies are 13.775 GHz forward and 15.003 GHz return.

The multiple-access links support spacecraft at lower data rates for extended periods. They use electronic beam forming by the TDRSS and signal separation by pseudorandom noise codes to discriminate between spacecraft. The single-access links provide users high data rates for short periods. With any of these links, the relay satellites pass signals between the user spacecraft and a single ground station located at White Sands, New Mexico. The system provides full TT&C and mission data, but the user spacecraft must match standard communications requirements and have standard TT&C hardware on board.

Figures 15-9 and 15-10 show the communications system performance required of user spacecraft. Figure 15-9 gives the spacecraft receiver G/T required for the forward (command) link to yield a bit error rate of 10^{-5}. Figure 15-10 illustrates how much EIRP a user spacecraft return link must have for the same bit error rate of 10^{-5}. In Fig. 15-10, "Power Received" refers to the power (in dBW) in the return signal at the TDRS spacecraft without the gain of the TDRSS receive antenna. In other words, it is the power in dBW of the user spacecraft's return signal after the space and absorption losses described in Chap. 13 are added to the user spacecraft's EIRP. "Encoded data" refers to data convolutionally encoded at rate 1/2. (See Sec. 13.3.3.) The user spacecraft's signal power received at the TDRS must be on the curves of Fig.15-10 to reach the Achievable Data Rate. Multiple access users must not exceed the curve by more than 3 dB—and then only by arrangement with the TDRSS. For both figures, the extent of the curve shows the permitted range of data rates.

The TDRSS ground station acts as a bent pipe for user command, telemetry, and mission data. It accepts previously formatted command data and routes telemetry and mission data to the user mission's POCC, SOCC, and MCC through the NASA NASCOM communications network. Although the TDRSS does not supply control centers for user missions, we can arrange to use standardized control centers (referred to by NASA as POCCs) at either Goddard Space Flight Center in Greenbelt, Maryland, or at Johnson Space Center in Houston, Texas. (These POCCs essentially combine the ground system's three basic control centers.) A mission may also have unique control centers remote from the NASA sites, but it must provide the required communications links with the NASCOM nodes at these sites.

Technical details and information on contracting to use the TDRSS are available in a NASA document [Goddard Space Flight Center, 1984, as revised]. To learn more about how to use the TDRSS, contact the Project Manager for Space Network, Code 452, Goddard Space Flight Center, Greenbelt, Maryland.

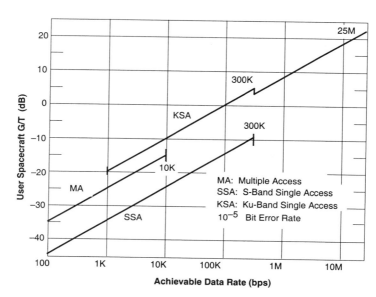

Fig. 15-9. TDRSS Forward Link G/T vs. Achievable Data Rate. These are typical G/Ts required for user spacecraft. See text for discussion.

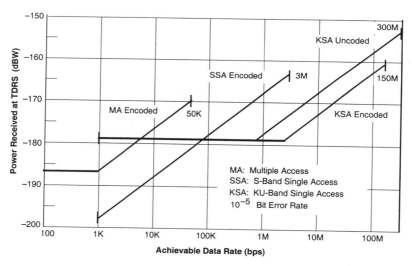

Fig. 15-10. TDRSS-Required Return Link Power Versus Achievable Data Rate. This is the effective power at the TDRS spacecraft antenna before the antenna gain is added. See text for further discussion.

15.4.3 Commercial Ground Systems

With the mid-'90s trend for the federal government to get out of the space-related services in general, and ground system services in particular, commercial ground system services are becoming available. The earliest of these ground systems, expected to be in operation in the late '90s, are fairly small, with initial assets consisting of two to four ground station sites and a central facility for control of the ground stations and for command, telemetry, and scheduling interfaces to the user. The ground stations are usually located at either or both low latitudes for spacecraft with low inclinations, and at very high latitudes, for polar orbiting spacecraft. The operators of some of these systems plan on providing links to government-owned or other private assets in the future.

Initially planned to provide services for scientific satellites, these commercial systems currently offer a mix of S-band and X-band downlink and S-band uplink services between the spacecraft and groundstation. The S-band downlinks provide housekeeping and payload data rates up to 10 or 15 MHz. Data rates to 150 MHz are served by the X-band downlinks. Uplink (command) data rates generally reach to 1 MHz. Ku-band operations are planned for the near future to accommodate the very high data rates of remote sensing spacecraft. These systems support both traditional TDM (Time Division Multiplexed) and CCSDS (Consultative Committee For Space Data Systems) spacecraft telemetry formats.

As a spacecraft program we might contract to use only the ground stations of one of these commercial ground systems. However, going beyond providing merely TT&C services, these private ground systems anticipate being able eventually to provide a full range of optional services including mission operations. After initial fixed charges, the costs of using these systems will be based principally upon use, generally calculated on a "per pass" basis. It is generally expected that these systems will serve not only the private space industry, but that Government agencies such as NASA will turn to them as well.

The user's primary communications with these commercial ground systems for command, telemetry, and scheduling are usually via Internet Protocol (IP). Physical connections can range from common low rate dialup lines to high speed dedicated lines. When using dialup or dedicated lines, where data security and delays are not serious problems, the ground system user may use the User Datagram Protocol (UDP) for efficiency. It is even feasible for spacecraft with low to medium downlink data rates, and where security and data latency for both command and telemetry are not critical, that the user's link with the ground system may simply be through the ubiquitous Internet "cloud." (It is advisable to use TCP here however, since even though it may slow down the average dataflow rate, it provides much more assurance that no data will be lost.) The transfer of high-rate spacecraft telemetry is generally stored locally at the station and transferred postpass by an efficient network protocol such as FTP, or shipped on a high density digital medium. Commanding may be either real time or store-and-forward at the system's central facility.

Two such commercial offerings are Universal Spacenet's Commercial Ground Network (CGN) and the Ground Network System (GNS) being developed by Allied Signal Technical Service Corporation. Spacenet plans to have X- and S-band stations in Alaska and Hawaii by 1999. Information on the use of the CGN is available from Universal Spacenet, 417 Caredean Drive, Suite A, Horsham, PA 19044. By the year 2000, Allied Signal will initially offer high latitude S- and X-Band stations in Greenland, Alaska, and Norway. Subsequent lower latitude stations, as well as ties to other existing stations are planned. Information on the GNS can be obtained from Allied Signal Technical Services Corporation, PO Box 5555, One Bendix Road, Columbia, MD 21045.

15.5 Key Design Considerations

Two key considerations in ground system design are the coverage required per spacecraft and the number, locations, and variety of data users. Together, they determine the number, location, and complexity of ground stations, of POCCs and SOCCs, and of the links which provide communications between the various system elements and data users.

15.5.1 Coverage Required per Spacecraft

The coverage required per spacecraft largely determines the number of ground antennas and ground stations necessary to support a mission. Coverage refers to how frequently and for what percentage of time a spacecraft must communicate with the ground system. For geostationary orbits, a single ground station can provide virtually continuous coverage. But low-Earth orbits may require many ground stations because each station has a limited time of view. This time limitation is described in Sec. 5.3 in which the time that the example satellite is in view of a ground station is calculated to be 12.3 min. Note that this is almost the maximum viewing time because this is nearly an overhead pass. The average viewing time will be considerably less than 12 min.

These short viewing times severely limit the time ground stations have to send commands to and receive data from spacecraft. Consequently, the data rates employed for these signals will depend on the amount of data to be transmitted during a pass. Most missions need only small amounts of command data and housekeeping telemetry, permitting low TT&C data rates, typically not more than a few kilobits per second. Usually, only a few seconds are needed to transmit commands. But we must often dump stored mission data to a ground station at high rates to transfer all the data in the available viewing time. Thus, limited ground station view times can create the most severe requirements on the performance of the space-to-ground communications link.

We can understand this requirement for high data rates for dumped data by assuming that the example spacecraft produces and stores on-board mission data at a 1 Mbps rate (a moderate rate for an Earth sensing spacecraft), for dump to the station once per orbit. The required rate for dumping data is the ratio of total stored data to available dump time of 12.3 minutes:

$$\frac{(105 \text{ min} \times 1 \times 10^6 \text{ bps})}{12.3 \text{ min}} = 8.54 \times 10^6 \text{ bps} \tag{15-1}$$

This is a rather high data rate. For any shorter viewing time, the rate will be even higher, requiring more performance of the space-to-ground link.

We can reduce this rate only by having enough stations symmetrically about the Earth to ensure a total dump time of more than 12.3 min per orbit. This would require many stations. Having only a few stations also increases the amount of data which must be stored on board, requiring more storage capacity to prevent loss of data from overflow. And, as we have seen, more stored data requires even higher dump rates to a limited number of ground stations.

Having only a few ground stations also significantly reduces spacecraft coverage for TT&C and for receiving mission data transmitted in real time. Because the amounts of data per pass are usually small, pass duration is not of much concern for TT&C. However, a mission's need for frequent TT&C contact, the lack of onboard command storage (spacecraft autonomy), and the requirement to receive real-time mission data typically demand more ground stations.

A dedicated ground system usually cannot have many symmetrically distributed ground stations, but we can increase their number by using hybrid configurations of various service-provided systems. In all cases, we must trade off the number of contacts needed with the spacecraft against the limited number of possible ground stations and the severe geographical and political restraints on their location.

15.5.2 Number, Locations, and Variety of Data Users

The data users' requirements determine the complexity of distribution systems. Data users frequently need different portions of spacecraft data streams with different requirements for handling, so data handling must be efficient and flexible. A typical approach is to sketch out the processing functions needed to supply data to users in a block diagram. The input to each diagram is the data stream from the data recovery equipment. We trace the data through each block, branch off the points where each user's data is available, and indicate any further handling or processing. We also show data rates at the input and output of each block. The final rate of the data stream to be sent to each user determines the bandwidth of the communication link.

Figure 15-11 shows the processing which might be applied to data from a scientific spacecraft, with data from the three onboard instruments combined into one data stream for transmission to the ground. We assume that three data users exist, each receiving all of the data from a particular instrument and "quick look" data—particular words—from each of the other instruments' data streams. For each user, the figure shows the data-handling functions and the rate at which the communication link must send data. Once we have selected the appropriate block diagram, we can fill in details at higher design levels.

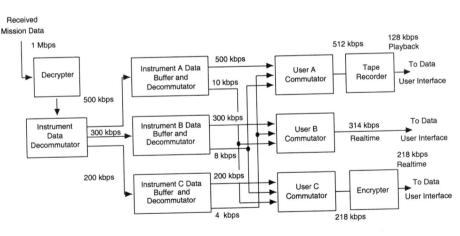

Fig. 15-11.   An Example of a Block Diagram for Data Handling. Composite data streams from a spacecraft must often be separated and recombined for relay to different users. The required communications link capacity may be seen explicitly. See text for discussion.

The users' location and variety also determine whether the individual ground stations or the central facility handles the data. Again, we must trade off the number, bandwidth, and complexity of the required data links against costs for procuring and maintaining the handling equipment. For smaller ground systems, it is often best to send data directly from the ground stations but in more complex systems, we usually prefer to distribute data from the central facility.

References

Klements, H. D. 1987. *Air Force Satellite Control Facility Space/Ground Interface.* TOR-0059(6110-01). El Segundo, CA: The Aerospace Corporation.

Martin, Donald H. 1984. *Communications Satellites, 1958 to 1986.* Report No. SD-TR-85-76. El Segundo, CA: The Aerospace Corporation.

National Aeronautics and Space Administration, Goddard Space Flight Center. 1984. *Tracking and Data Relay Satellite System (TDRSS) Users' Guide.* STDN No. 101.2.

Pratt, Timothy and Charles W. Bostian. 1986. *Satellite Communications.* New York: John Wiley & Sons.

Sklar, Bernard. 1988. *Digital Communications Fundamentals and Applications.* New Jersey: Prentice Hall.

Chapter 16

Spacecraft Computer Systems

L. Jane Hansen, *HRP Systems*
Robert W. Hosken, *The Aerospace Corporation*
Craig H. Pollock, *TRW, Inc.*

Mission-supporting computer systems include the computers onboard the spacecraft, as well as those on the ground, as illustrated in Fig. 16-1. On board the spacecraft, computers have become an integral part of the overall system, as well as being part of most spacecraft subsystems. Ground station computer systems are used to support daily operations after launch, and may be derived from systems originally used for developing and testing space-based elements. Thus, computer systems cross traditional subsystem and organizational boundaries.

In previous chapters we have described the various spacecraft subsystems. Through spacecraft evolution, most subsystems now contain elements of a computer system as shown in Fig. 16-2. This means that the computer system resource estimation process takes on a larger scope than in the past. In this chapter we discuss how to generate computer system resource estimates, refine the computer system requirements, estimate the effort in terms of resources, and define the tasks associated with developing computer systems onboard the spacecraft. Additionally, we will briefly examine the requirements for ground-based computer systems throughout the life-cycle development process.

As outlined in Table 16-1, we discuss the iterative process used to estimate computer resources, based on mission requirements. We will accomplish this by first discussing the computer system specifications and the task of creating a baseline computer system from top level requirements. Figure 16-3 shows that the computer

645

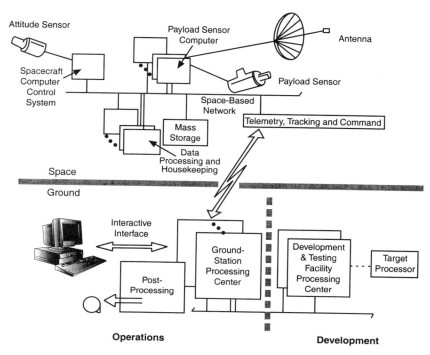

Fig. 16-1. Mission Computer Systems. Notice that there are many interfaces and managing their compatibility is critical to reducing cost and risk. Also, notice that the development tools and environment required to build, integrate, and test the computer hardware and software are included as part of the mission computer system.

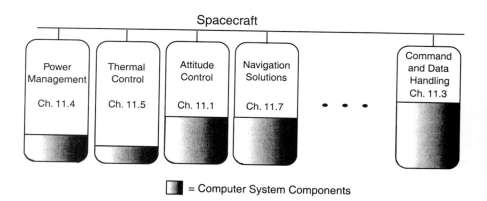

Fig. 16-2. Computer Systems Break Traditional Subsystem Boundaries. Today computer systems are an integral part of nearly every subsystem on board the spacecraft. In some cases, subsystems do not use computers if they are not required to meet mission requirements. However, in most cases, the task of defining computer system requirements and associated costs takes on a larger scope than in the past.

TABLE 16-1. Computer Systems Development Process. This iterative process defines top-level requirements during a program's concept through development phases.

Process Step	Where Discussed
Define Requirements	
– Evaluate Mission Objectives	Chaps. 1, 4, Secs. 16.1.2, 16.1.3
– Perform Functional Partitioning	Chap. 4, Sec. 16.1.1
Allocate Top-Level Computer Requirements	
– Evaluate Candidate Architectures	Sec. 16.1.2
– Perform Functional Flow Analysis	Sec. 16.1.2
– Evaluate Fault Protection	Sec. 16.1.5
– Establish System Baseline	Sec. 16.1.4
Define Computer System Requirements	
– Define Processing Tasks	Sec. 16.2.1
– Establish Computer Size and Throughput Estimates	Sec. 16.2.2
– Select Software Language	Sec. 16.2.2
– Select Hardware Instruction Set Architecture	Sec. 16.2.3
– Select Target Hardware and Supplier	Sec. 16.2.3
Define Development and Support Environment	
– Establish Development and Control Process	Secs. 16.2.2, 16.2.5
– Identify Required Support Tools	Secs. 16.2.4, 16.2.5
– Establish Test and Integration Approach	Sec. 16.2.4
– Estimate Life-Cycle Costs	Sec. 16.2.5
Document and Iterate	Sec. 2.1

system baseline includes hardware, software, and documentation. Next, we will evaluate the resources required to achieve the baseline system. This includes hardware and software, as well as life-cycle support equipment. Finally, we will use the FireSat example to clarify some of the key components and concepts of the estimation process. Table 16-2 provides definitions for terms frequently used in estimating computer system resource requirements.

In designing computer systems for space applications, we want to optimize the availability, capability, flexibility, and reliability of the system while minimizing cost and risk. Our objective is to meet the system and mission requirements, whether the resulting system is on the ground, in space, or distributed between the two. As mission objectives expand, we must blend complex hardware and software to meet them. An increase in system complexity leads to an exponential increase in the associated testing. We strive to keep the computer systems simple at the lowest level, while building up the capabilities to meet the top-level mission requirements.

The primary design drivers used to measure our success in optimizing the computer system design, are shown in Table 16-3. Mission requirements, shown on the left, typically dictate the system-level drivers, shown in the next column. These flow down to the subsystems where we establish driving requirements for computation. Finally, logistics support personnel set down the additional requirements which we feed back against the computer and system-level drivers, helping to manage the overall design process. We weight each of the design drivers based on mission objectives and constraints. Again, this iterative process requires multi-discipline participation and often crosses traditional subsystem and organizational boundaries.

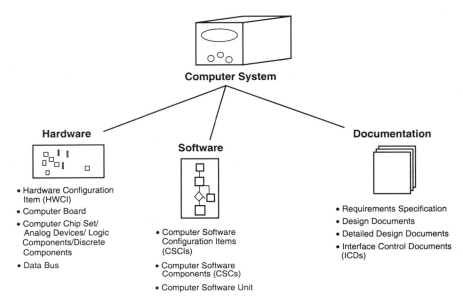

Fig. 16-3. Hierarchy of Elements in a Computer System. Computer systems consist of hardware, software, and their interface definitions and documentation. Hardware and software components are in a hierarchy, building to the final configuration item—either hardware or software. Documentation also has a hierarchy, but it starts with top-level requirements and leads to increased implementation detail.

TABLE 16-2. Definitions Associated with Computer Systems. Often when discussing computer system design and development we use terms which have a specific meaning to those involved in the discipline.

Embedded Systems	A built-in processor, providing real-time control as a component of a larger system, often with no direct user interface.
Real-Time Processing	Handling or processing information at the time events occur or when the information is first created. Typically, embedded or onboard processing is real-time.
Hard Real-Time	Requiring precise timing to achieve their results, where missing the time boundary has severe consequences. Examples include attitude control software and telemetry downlink. (For more information see Stankovic and Ramamritham [1988].)
Soft Real-Time	Requiring only that the tasks are performed in a timely manner, the consequences of missing a time boundary are often degraded, but continuous, performance. Examples include orbit control software and general status or housekeeping.
Operating System Software	Manages the computer's resources such as input/output devices, memory, and scheduling of application software.
Application Software	Mission specific software which does work required by the user or the mission rather than in support of the computer.

TABLE 16-3. Design Drivers for Computer Systems. These are factors that we evaluate throughout the design process. When flowing down mission requirements, including system level processing requirements, we must be careful to design hardware and software with the "ilities" in the fourth column in mind.

Mission Requirements	System Level Processing Requirements	Computer Level Requirements	Additional Requirements

• Customer Needs
  – Military
  – Scientific
  – Commercial
• Number of Satellites
  – Single
  – Multiple
  – Constellation
• Number and Location of Ground Stations
• Level of Autonomy
• Security Requirements
• Programmatic Issues
  – Cost
  – Schedule
  – Risk
	• Functional Capabilities
	• Processing Partitioning
	  – Payload vs. Spacecraft
	  – Onboard vs. Ground
	• Physical Characteristics
	  – Size
	  – Weight
	  – Power
	  – Temperature
	  – Radiation
	• Command Protection / Encryption
		• Throughput
		• Memory
		• Radiation Hardness
		• Development Tools
		• COTS Software availability
		• Emulator / Engineering Model availability
			• Testability
			• Feasibility
			• Usability
			• Reliability
			• Flexibility
			• Maintainability
			• Interchangeability
			• Replaceability

16.1 Computer System Specification

Chapter 4 discusses how to determine system requirements and allocate them to subsystems. Through that process, we identify operational modes for the spacecraft bus and payload, allocate top-level requirements to the computer system (among other spacecraft elements), and define the subsystem interfaces. Defining requirements for the computer system begins with these results. To arrive at a baseline computer system, we:

1. Allocate mission and system requirements to computer systems, detailing the computer system requirements

2. Define the computer system's operational modes and states, based on the computer system requirements

3. Functionally partition and allocate the computational requirements to space or ground, payload or spacecraft, individual subsystems, and to hardware or software

4. Evaluate the internal and external interfaces (analyze data flow), while evaluating the candidate architectures iteratively

5. Select the baseline architecture

6. Form the baseline system specification from the architecture, modes and states, and system level requirements

The first four steps typically occur before the System Requirements Review. We usually complete steps 5 and 6 by the Preliminary Design Review.

Revisiting and iterating between steps occurs frequently because requirements are often contradictory, especially in the early design stages. Requirements can also be unreasonable or too narrow in scope. For example, if we determine that the star tracker must have a specified level of accuracy on its own to meet pointing or mapping requirements, without synergy that might come from other ACS sensors such as Earth sensor or gyro, we might overspecify the star tracker. By using an iterative process, we can correct contradictory computer requirements and question the validity of others. An assumption made by one subsystem to reduce their complexity may increase the complexity and cost for another subsystem dramatically. Often, a compromise solution is needed.

16.1.1 Requirements Definition

As with all subsystems, poor computer system requirements definition results in an inferior product; erroneous requirements are very expensive to correct. (See Kane et al. [1993] for more information.) Thus, requirements have high leverage—a small improvement early avoids many problems later [Vick and Ramamoorthy, 1984, Boehm, 1984]. Defining system requirements is difficult, subjective, and time consuming. One approach to doing this is to study a set of questions, such as those shown in Table 16-4, which will motivate needed trade studies.

TABLE 16-4.   An Approach to System Requirements Definition. General questions which we ask in all aspects of life can be directly applied to computer system requirement derivation by evaluating the specific parameters listed below. (These questions are based on work by R. Holmes, S. Jacobs, and R. Lane of TRW.)

Questions to Ask	Parameters to Review
What must the system do?	Evaluate and establish basic functional requirements.
Why must it be done?	Establish traceability from functions to mission objectives. Be sure to challenge the requirements and assess their validity.
How can we achieve it and what are the alternatives?	Evaluate candidate architectures and understand the implications of interfaces in the data flow diagrams.
What functions can we allocate to parts of the system?	Perform functional partitioning to development block diagrams.
Are all functions technically feasible?	Determine if the value of state-of-the-art technology outweighs the risk. Look for data flow bottlenecks and reallocate functions to evenly distribute the data flow. Review baseline block diagrams for potential holes.
Is the system testable?	Develop nonintrusive testing which will ensure that the system will meet mission objectives. Are test points available outside the system for easy "black-box" testing?

To define requirements for a computer system, it is convenient to develop a computer system *state diagram*. The state diagram shows valid states of the system (such as "off" or "initialized") and the conditions required to achieve them, often based on mission requirements. The computer system states and state transitions must be consistent with its allocated requirements and the mission concept of operation (See Chap. 14.) Figure 16-4 is an example of a simple state diagram for a computer system, showing the general states and the source of their transitions. On and off are the obvious first choices for system states. Even when a system must be on at all times, we should consider having an off state to allow graceful degradation if the system were to shut down for some reason. Other states relate to what the system must do and can include related transitions. For a specific mission and mission requirements we might have several substates in place of one state shown in Fig. 16-4. Or we might not have a state shown in the figure if it's not applicable to our mission. For example, several fail-safe conditions will be associated with the error contingency state shown. However, it will be implemented differently for each specific spacecraft based on mission requirements and mission phase.

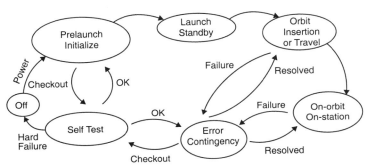

Fig. 16-4.   Typical State Transition Diagram for an Onboard Computer System. The state diagram shows the valid states of the system and the conditions needed to achieve each state.

When developing the state diagram for a spacecraft computer system, we must keep in mind implications for the ground system. Complex state transitions influence ground station software that deals with the spacecraft's limitations and constraints. Other organizations which define the spacecraft and the ground station need to help diagram the states for the spacecraft computer system.

Functional Partitioning

Functional partitioning is a structured methodology which begins with decomposing requirements into their lowest functional component and ends in the creation of multiple candidate architectures. This method allows us to group similar functions in subsystem definitions without unnecessary influence from traditional subsystem or organizational boundaries. The processing for a spacecraft system is usually partitioned between various processors in space and on the ground. This allocation of processing or functional partitioning is performed after the major processes have been determined and estimates of the processing time lines, or at least the time dependencies, are available.

The top-level considerations which determine where the processing will be performed to meet the system performance requirements (both technical and programmatic) are presented in Table 16-5.

16.1.2 Processing Architecture

An *architecture* is a framework for developing a computer system. We mold it to meet mission requirements and operational needs, creating a baseline system. The architecture shows us the system's parts and how they interact through a block diagram. Architecture studies for computer systems must address the top-level

TABLE 16-5. **Functional Partitioning Requirements for Spacecraft Computer Processing.** We partition functions in a general sense using mission timelines as the starting point. We must partition functions to various processors and subsystems so that each element of the system maintains an acceptable level of complexity. A large increase in the complexity of any element will greatly impact our test requirements.

Perform Processing in Space	Perform Processing on Ground
• When processing delays would be intolerable • When needed to make downlink bandwidth feasible (This case is treated for the FireSat example in Sec. 16.3)	• When human interaction with processing is necessary • When the downlink bandwidth is satisfactory
Perform Processing in Hardware	**Perform Processing in Software**
• When very high performance is needed • When well-defined, inexpensive hardware for the process is available	• When processing complexity exceeds that available in hardware • When changes in processing need to be made after hardware is acquired • When expensive, custom hardware can be replaced by software • When there is considerable unused computer capacity
Allocate Processing Between Spacecraft Bus and Payloads	**Do Not Allocate Processing Between Spacecraft Bus and Payloads**
• When payload processing is distinctly different from spacecraft bus processing • When payload performance accountability is critical	• When payload processing is minimal
Allocate Processing Along Organizational Lines	**Do Not Allocate Processing Along Organizational Lines**
• When there are geographical or other impediments to effective inter-organizational communication • When there are standard subsystems and accurate interface control documents which are typically managed within a defined organization	• When the project is small enough that there is a single organization with strong top-down authority • When subsystems are so complex that specific disciplines and experienced organization personnel are required.

block diagram, the data architecture, the hardware architecture, and the software architecture. *Data architecture* addresses the physical structure of the data network or bus, as well as the protocol or logical interaction across the bus. A *protocol* is a set of rules for sending data between computers, or between computers and peripherals. The *hardware architecture* defines the instruction set architecture (ISA) and the functional elements that are available in hardware. (For more information on hardware architectures see Hennessy and Patterson [1995].) Finally, the *software architecture* defines how the processing instructions execute. Software processing can function as a single thread, executing from top to bottom, repetitively, or as scheduled modules, where processing order is based on major and minor frames. Alternatively, software processing can be event driven, where interrupt service routines preempt normal execution in a deterministic way when hardware interrupts occur.

Tables 16-6 through 16-9 illustrate various system level and data architectures which we can use in their entirety or combine into a hybrid to meet the mission requirements. Along with a block diagram and brief description of each, a short list of positive and negative attributes have been listed in the tables. By using a hybrid system architecture, we can combine the positive attributes of several of the architectures while eliminating or reducing the risk associated with the negative attributes.

TABLE 16-6. Centralized Architecture. A *Centralized Architecture* has point-to-point interfaces between processing units and a single management computer, or central node, or *hub*.

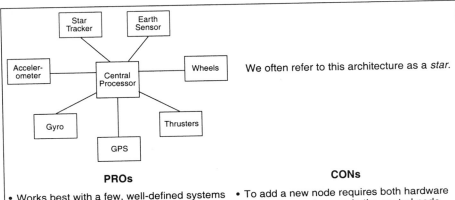

We often refer to this architecture as a *star*.

PROs	CONs
• Works best with a few, well-defined systems which all interface directly, and only, with the central computer.	• To add a new node requires both hardware and software changes in the central node.
• Highly reliable architecture where failures along one interface will not affect the other interfaces.	• Wiring harnesses become large because each node has duplicate transmission wires if data are sent to multiple receivers.

TABLE 16-7. Ring Architecture (Distributed). The *distributed ring architecture* establishes a way to arbitrate information flow control as the data are passed in a circular pattern.

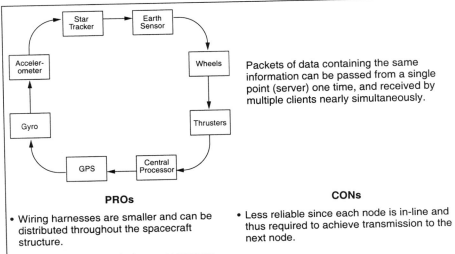

Packets of data containing the same information can be passed from a single point (server) one time, and received by multiple clients nearly simultaneously.

PROs	CONs
• Wiring harnesses are smaller and can be distributed throughout the spacecraft structure.	• Less reliable since each node is in-line and thus required to achieve transmission to the next node.
• Limited impact to central processor as we add new nodes.	

TABLE 16-8. Bus Architecture (Federated). A *federated bus architecture* uses a common data bus with all processors sharing the bus. This encourages the use of standard protocols and communication schemes for all nodes.

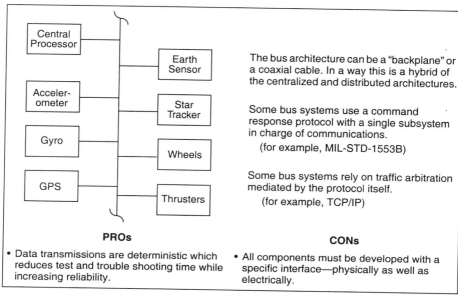

The bus architecture can be a "backplane" or a coaxial cable. In a way this is a hybrid of the centralized and distributed architectures.

Some bus systems use a command response protocol with a single subsystem in charge of communications.

(for example, MIL-STD-1553B)

Some bus systems rely on traffic arbitration mediated by the protocol itself.

(for example, TCP/IP)

PROs	CONs
• Data transmissions are deterministic which reduces test and trouble shooting time while increasing reliability.	• All components must be developed with a specific interface—physically as well as electrically.

TABLE 16-9. Bus Architecture (Distributed). A *distributed architecture* uses multiple "like" processors to execute all software on an as-needed basis.

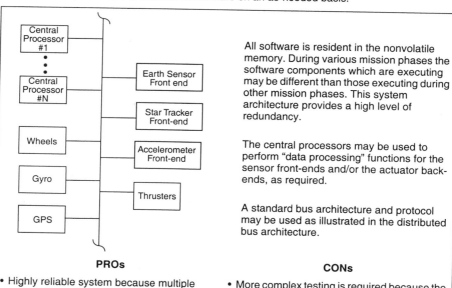

All software is resident in the nonvolatile memory. During various mission phases the software components which are executing may be different than those executing during other mission phases. This system architecture provides a high level of redundancy.

The central processors may be used to perform "data processing" functions for the sensor front-ends and/or the actuator back-ends, as required.

A standard bus architecture and protocol may be used as illustrated in the distributed bus architecture.

PROs	CONs
• Highly reliable system because multiple processing units can be used to execute software as needed.	• More complex testing is required because the system can reconfigure itself as software modules are allocated to processing resources.

We next analyze the flow of data to determine how to manage interfaces between components. We want clean, simple interfaces—a data path is inefficient and slows down the flow if it calls for data to pass through a component without being examined or used [Yourdon, 1989].

After partitioning functions, performing trades, evaluating the data architectures, and analyzing data flow, we can develop a block diagram for the computer system. The system *block diagram* illustrates how we implement an architecture, showing types and numbers of processors and networks, including topology and protocol when reasonable. It provides a point of departure for developing more detailed software, hardware, and interface requirements. We select an architectural baseline from among the candidates, shown in Table 16-6, by asking the questions listed in Fig. 16-5.

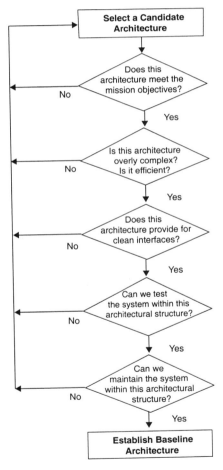

Fig. 16-5. Questions Used in Selecting the Architectural Baseline. These questions, when applied to the various architectures we are trading for a specific set of mission requirements, will lead us to an optimized architecture which effectively meets our specific needs. Once each of these questions has been successfully answered, we have selected a baseline architecture.

To clarify the hardware and software architecture issues, consider the example of a personal computer. Figure 16-6 shows the general components in the hardware architecture diagram. It contains a *central processing unit* (CPU), memory, and *input/output* (I/O) *devices*. The memory stores executable program code and data. *Random access memory* (RAM) does not retain information when we turn off the computer. *Programmable read only memory* (PROM) provides nonvolatile storage which retains information when not powered. In the simplest form, we input only from a keyboard and output only to a monitor. More complex forms may connect many input or output devices to the basic system.

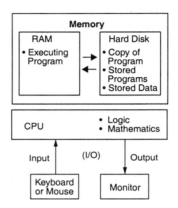

Fig. 16-6.   Personal Computer Architecture. The PC architecture shown here includes the various components discussed in the text. The inputs come from the keyboard or mouse and can be service either through an interrupt service routine (ISR) or through a polling scheme. The processing occurs in the CPU and uses the RAM as a "scratch pad" for computation. (This formulation is due to S. Glaseman of The Aerospace Corp.)

A personal computer used as a word processor illustrates the different software architectures which we might select, as well as the basic concepts for sizing computer systems and estimating throughput. The operator using the word processor will input data from the keyboard. The computer system software might recognize that an input has been made by receiving an interrupt from the keyboard each time a key is stroked. Alternatively, the computer system software might poll the keyboard at a specified rate, looking for a keyboard stroke to occur. The first approach illustrates *event driven software architecture* while the second is a *scheduled software architecture*. When using an event driven approach, we must be sure that we prioritize each event properly and define carefully the amount of time required to "service" the event or interrupt. When using a scheduled approach, we must be sure that the internal clock has enough resolution to accommodate the various scheduled rates. If the clock is set or reset by an external source, we must ensure that all software will continue to operate even with discontinuities in the time (such as, if time moves backwards when the clock resets).

Now, assume an operator types at most 100 words (600 characters) per minute. We can define 256 unique states on a typical keyboard, so a byte (8-bits) can represent all possible characters. Multiplying 8 bits per character by 600 characters per minute, we see that the system must input data serially (single stream) at 4,800 bits per minute to keep up with the keyboard operator.

Let's assume the computer system must prepare the input data for display. Also assume that each character received will require 10 instructions for processing and that each full word received will require 100 more computer instructions. We determine the computer's required processing rate by first multiplying the number of characters per minute by the computer instructions required per character. Then, we compute the product of the number of computer instructions per word times the number of words per minute. This example yields a processing rate of 16,000 instructions per minute, or 267 instructions per second, to prepare the input data stream for display. Now assume that each instruction requires five clock cycles to bring the data from memory and one more clock cycle for execution. Thus, to keep up with the typist, the CPU must provide at least 96,000 cycles per minute, or 1,600 cycles per second, requiring a clock rate of 1.6 kHz. This example illustrates how we can estimate computer throughput (instructions per second) and processor speed (cycles per second) requirements.

To store the input data and transfer the required 8-bit instructions to the CPU for processing, the computer must transfer data from memory to the CPU at approximately 2,216 bits per second (80 bits per second for input data storage and 2,136 bits per second for instruction fetch). We can size the memory by assuming the typist works continuously for 24 hours at the top rate. The resulting data would require 6.9 million bits of storage, plus 880 bits for the 110 instructions. The keyboard operator's typing rate limits the memory size requirements. Understanding such system bottlenecks is one key to defining requirements for computer systems.

16.1.3 Computer System Requirements

Once we have identified the top-level requirements, the state diagram with state transitions, and a basic system architecture we must evaluate the impact of mission requirements on the baseline computer system. This assessment begins the refinement process for establishing detailed computer system requirements. Information regarding the selected orbit, expected period of operation, and any high performance requirements such as large field of view with continuous coverage or tight pointing and mapping criteria will affect the fidelity of the hardware and software developed for the specific mission.

Several mission parameters drive the hardware selection. For example, the orbit we select will define the radiation environment. When we increase the required level of fidelity or include a requirement for autonomous operations, we often require a more capable computer system. Either of these conditions can develop if we have to perform rapid transitions between differing orbits, or travel to distant targets. These requirements may impact software development. More complex requirements lead to more complex software implementation, which requires a more robust design and more test cases to accomplish a desired level of preflight validation. Inadequate requirements definition may cause cost and schedule risk for both hardware and software, as modifications and last minute changes may be required.

When the mission is not as critical or multiple copies of a satellite will perform the same tasks, we can select computer hardware for the mission based on less stringent environmental testing. Often we can use commercial rather than space-qualified parts. Additionally, as the possibility for achieving unknown states goes down, we may reduce the software complexity and the level of preflight testing. As we increase the expected mission life or mission criticality, we should specify the use of more robust hardware, specifically space-qualified components and systems. We also increase the risk that software will operate in a manner that was not predetermined, and thus we should do more thorough preflight testing and system validation.

When we impose high performance requirements on the spacecraft computer systems we also increase the performance requirements of both the hardware and the software. High-data rate payloads such as imaging devices or communication subsystems impose the need for higher bandwidth data busses and often increase the CPU performance requirements. When we require tight attitude knowledge and control, we not only impose a requirement for increased accuracy on the sensors, we also increase the computational complexity of the software, which in turn affects the CPU performance requirements.

When we initially establish top-level mission requirements, we create a set of baseline computer system requirements, state transition diagrams, data flow charts, and system architectures. We can then perform trade-studies between the costs associated with our mission requirements and the costs associated with the hardware and software we selected to meet the mission requirements. Often when we iterate between the two, compromising when we feel we can on either side, we can reduce the overall mission costs.

16.1.4 Baseline Definition Expansion

If the initial analyses call for onboard processing, we should further partition functions between hardware, software, and firmware. *Firmware* is the software which resides permanently in nonvolatile memory. It reduces the susceptibility to upset, but we cannot modify it after launch. Certain elements of the system are clearly hardware: space-qualified computers and processors, the data bus, and so on. Software is for processing algorithms, which may change throughout the spacecraft's lifetime. Software typically executes out of random access memory (RAM). However, because RAM is susceptible to single-event upsets (Sec. 8.1.4), firmware is often the answer for critical processes such as initialization or contingency operations. Firmware often executes out of read only memory (ROM) or programmable ROM (PROM) where we can write once or some small number of times.

We next evaluate the *Instruction Set Architecture*, ISA. This is the machine code format used by a specific processor, such as the 80×86 family of processors, 68040, RH3000, and MIL-STD-1750A. The ISA defines the software developer's interface to the processor at the lowest level. To evaluate hardware architectures, we examine instruction sets, recognizing advantages and disadvantages of the two basic types: *general-purpose* and *custom ISA*. The former supports all kinds of processing but with only moderate performance. The latter supports specific algorithms or classes of functions very well but often supports varying applications poorly.

We should avoid custom architectures whenever possible because they are risky to develop, lack software support, and are hard to reprogram. General-purpose architectures allow us to modify algorithms more easily, but they slow down processing because they are not designed for a specific algorithm. In special cases, the faster speed of a custom ISA may drive us to select it despite the drawbacks.

In evaluating candidate software languages, we again have two basic options: assembly language and higher-level language. *Assembly language* contains the basic symbols and expressions used to program a specific computer, and the programmer must thoroughly know the computer being programmed. *Higher-level languages*, such as C, C++, or Ada, also have symbols and expressions, but they provide more sophisticated operations and add a level of abstraction. Assembly language software is more efficient and compact than software written in a higher-level language, but it often takes longer to generate. We prefer higher-level languages for maintenance, testability, and life-cycle costs. However, cross-compilers are often not available for custom machines, leaving assembly language as the only method of programming. A *cross-compiler* is one which resides on a standard host (such as SUN, DEC, SGI) and creates executable code for the target process (68040, 603e, and 1750A).

16.1.5 Methods for Tolerating Faults

Computer systems occasionally fail during operation. Since we know that this can happen, we can attempt to mitigate the risk by implementing a means of achieving graceful degradation, or fall backs to maintain some functionality. The most common are redundancy and distributed processing.

We use redundancy for flight critical components to assure that required data are always available to the system. We can implement redundancy in several ways: duplicate equipment, back-up capability using a different but comparable approach, perform the same tasks on the spacecraft and on the ground, use a bus network which allows for data to be sent to various applications or users, independently, or cross-strap equipment to various potential users.

Distributed processing allows us to allocate software functions to any one of a number of processors, depending on either mission phase, hardware availability, or subsystem failure. Each approach has pros and cons as outlined in Table 16-10.

TABLE 16-10.   Pros and Cons of Approaches for Providing Software Methods for Tolerating Faults. There are many methods for tolerating faults and no one is necessarily better than the next. However, when we evaluate our specific requirements against the pros and cons listed here, one solution may be more appropriate for our mission. (For more information on fault tolerance, see Magnus [1992].)

Redundancy	Distributed Processing
PROs: Provides backup which is identical to the original. • Does not require additional or special software to process the data. • Duplicate testing is straightforward since the back-up is identical.	PROs: Can reduce the system weight and power if the number of systems is optimized. • Provides a means of maintaining system performance until several failures have occurred. Then the system will operate in a degraded mode.
CONs: Additional weight, power and cost. Requires decision-making process to determine which to use.	CONs: Requires additional software to implement distribution methodology. Can be tricky to test and requires an extensive number of test cases

Hardware or software errors, as well as environmental effects (see Chap. 8) sometimes cause the computer to stop executing its intended program altogether or perform instructions in an incorrect sequence. We can mitigate this problem by designing special circuitry so that the computer restarts when it is hung in this condition. We can command this circuitry from the ground, or the computer subsystem can activate automatically. In this latter case, we call the circuitry a *watchdog timer*. The timer counts down from a given predetermined time and will reset the computer when it reaches zero. To prevent the reset during normal operations, the computer's operating system includes a function to reset the timer to its maximum time on a regular basis. The anomalous computer operation prevents this timer reset so the watchdog timer restarts the computer. The decision to include a watchdog timer usually does not depend on a subsystem trade-off because we have inadequate data on the rates of occurrence of this failure mechanism. It is often standard equipment on computers designed and marketed for space operations.

Spacecraft occasionally have a more serious computer problem in which the fault is permanent and resetting does not solve the problem. The designer must determine whether the reliability of the planned computer design is adequate, as discussed in Sec. 19.2. If not, we will need to provide a redundant computer and the special circuitry needed to switch from the primary to the secondary computer on failure.

When we determine that memory may be unusually susceptible to failure, we can mitigate risk by partitioning the memory into "blocks." We design memory to consist of independent blocks, where the failure of one block does not cause the failure of other blocks. In this case, the hardware designer may provide tolerance of failures of a memory block by creating a physical-to-logical map of the blocks at boot-up time. If failures are detected during operations, the system can work around bad blocks of memory. This is an example of a system architecture solution to a potential hardware problem.

Another redundancy approach is to duplicate critical programs stored in nonvolatile memory. For instance we often replicate the Start-Up ROM in physically isolated sections of nonvolatile memory. If this logic cannot execute, the computer can not run the operating system or the application programs. The rationale for providing this second version of the start-up ROM is to avoid the effect of such a catastrophic failure without including a second computer. Note that in this case, we need a hardware mechanism, possibly controlled from the ground, to initiate the execution at the secondary location because the computer is not under software control until the start-up program executes.

Error Detection and Correction (EDAC) circuitry is an example of hardware that provides tolerance of a bit error in a memory word. It is particularly important in the mitigation of single event upsets. EDAC corrects a single bit error in a word when it reads that word. If an upset has occurred, the EDAC will correct it. If a long time elapses before the CPU reads the word, a second bit in the same word may be upset. To prevent this from happening, the operating system executes a program known as a *scrubber* to read each word of memory. As it reads each word, it "scrubs" all the single bit errors. The design parameter available to the system engineer is the time between scrubs. If the time is too long, uncorrectable second upsets in words are likely to occur. If the time is too short, the scrubbing process will consume too much of the CPUs processing time. We can determine this necessary scrub time based on the anticipated rate of upsets and system probability requirement that a second, uncorrectable, upset will not occur. Table 16-11 provides an example of this calculation.

16.2 Computer Resource Estimation

The previous sections discussed how we define computer system requirements and generate a baseline architecture. In this section, we add detail to the baseline. With functional groups and a system architecture in place, we specify the needed processing tasks (in a general sense), determine the data requirements, estimate software size and throughput requirements, and identify computer hardware on which it will execute.

We use traceability analysis to make sure the requirements are complete. We must trace the computer system's requirements to parent requirements, which come from various sources. For example, we may derive them from top-level requirements, oper-

TABLE 16-11.   Example of Calculating the Time, *T*, to Perform EDAC Memory Scrub.

Scrub Memory to Prevent Second Single Event Upset
Given: • A single event upset (SEU) has occurred • The affected memory word has 22 bits (16 data bits plus 6 check bits) • We know the SEU rate is 10^{-7} upsets per "bit day" • We have a requirement that the probability of an uncorrectable second update is less than 10^{-6} per day • We assume a Poisson distribution for the bit errors then let λ = arrival rate of new bit errors in word = 21×10^{-7} upsets/day and P_{new} = probability of one or more new upsets in word $= P_1 + P_2 \cdots + P_{21}$ $$P_{new} = 1 - P_0 = 1 - \frac{(\lambda \cdot T)^0}{0!} \cdot e^{-\lambda \cdot T} \cong \lambda \cdot T$$ Therefore: $\lambda \cdot T \equiv P_{new} < 10^{-6}$ Or: $T < \dfrac{10^{-6}}{21 \cdot 10^{-7}} \approx 0.5 \text{ days}$

ational concepts, or launch-vehicle interfaces. We must also trace the flow of requirements to the components to reduce "gold-plating." Whenever a top-level requirement changes, good traceability allows us to examine the effect of this change on lower-level requirements and how we meet them. Often, we rethink a change when we see its effects. In any case, traceability allows us to identify all areas where we must evaluate the design to incorporate changes. If we flow the requirements properly during conceptual design, we can accurately run tests at each level during development.

16.2.1 Defining Processing Tasks

We document the requirements for processing tasks and system interfaces in Software Requirements Specifications and Interface Requirements Specifications. While establishing requirements for the spacecraft, we define processing tasks by classifying what the spacecraft must do. Software for onboard processing falls into four principal classes.

Control system software, such as attitude or orbit determination and control, requires an input stimulus and responds by changing the state of the system. This software is often mathematically intensive—requiring high accuracy and strict timeliness.

System management software includes such items as fault detection and correction, long duration event schedulers (such as reconfiguring the power system during eclipses), and payload system management. Software in this class manages control flow and is therefore logic intensive. Simple instruction sets are sufficient for this class because it requires few floating-point computations.

Mission-data software manipulates and compacts large quantities of data as they are collected. This function often demands special computer architectures, such as signal processors, as well as large storage capacity for collected data.

Operating system software directly manages computer resources and controls their allocation to spacecraft and mission tasks. This includes basic executive functions such as scheduling tasks for execution, time management, interrupt handlers, input/output device handlers and managing other peripheral drivers, carrying out diagnostics and built-in tests, and memory fault management. All computer systems must manage these processes. We often consider software for the operating system as overhead to application software.

After examining what the computer system must do, we can assess the nature of its processors and decide whether to use off-the-shelf processors or develop new ones. The same holds for the algorithms. We want to use established, proven algorithms because new ones involve technical risk, added costs, and a longer implementation schedule. Early emphasis on risk assessment and reduction is part of defining processing tasks.

16.2.2 Estimating Software Size and Throughput

We measure software size by words of memory, and processing time by *throughput*, usually expressed in thousands or millions of instructions per second (*KIPS* or *MIPS*, respectively). We estimate the size and throughput of onboard software for several reasons. When we begin defining a computer system, we use the software estimates in conjunction with requirements for spare processing to determine how much computing power we need to perform the mission. During system development, we revise the estimates to make sure hardware capacity is not exceeded. We also use software size estimates to estimate cost.

Processor throughput is a function of the instruction set and the clock speed. With only one instruction, a computer's throughput is proportional to clock speed. If it has two instructions, one (A) requiring two clock cycles and the other (B) requiring seven cycles, the computer's throughput also depends on the instruction mix. The *instruction mix* is the proportion in which the software uses the instructions. For example, if the software is 60% type A instructions and 40% type B, the throughput available with a 10 MHz clock is $(10/[(0.6 * 2) + (0.4 * 7)]) = 2.5$ MIPS. If the mix is reversed, the throughput available is 2.0 MIPS.

When we evaluate hardware architectures to determine their ability to meet our processing needs we use a *benchmark program* that contains a specified instruction mix so that various computers can be compared to a standard measure of performance. Typical benchmarks used for evaluating computer resources in space applications are shown in Table 16-12.

When we select a benchmark it is important that we use one that has a similar instruction mix to the one we expect in our operational flight software. For example, if our flight software will be mathematically intensive we could use Khornerstone, Linpack, or Whetstone. However, if our flight software focuses on integer math, the Dhrystone benchmark is a good match.

System Requirements Review is the milestone when we formally identify computer resource requirements. A good rule of thumb is to set the amount of computer memory and throughput at the System Requirements Review at four times the estimate of what is needed for software size and throughput. Empirically, initial software size and throughput estimates double from this review to launch because early requirements are uncertain, and changes in software are easier to make than changes in hardware during late stages of spacecraft development. We also want spare memory and throughput at

TABLE 16-12. Benchmark Programs Used to Evaluate Computer Performance. Listed below is a set of benchmarks and their strength in measuring the performance of a computer system [Beckert, 1993]. We use these benchmarks to evaluate the applicability of a specific computer to meet a specific software or mission objective [Santoni, 1997].

Benchmark	Measures
Dhrystone	A test designed primarily to measure a CPU's integer performance. An outcome of the Dhrystone test is the MIPS rating.
Khornerstone	A suite of 21 tests developed in 1987 by SRS/Workstation Laboratories to measure the overall performance of a computer's CPU, floating-point capabilities, and disk I/O.
Linpack	This test measures a CPU's floating-point performance. Its results are reported in mflops (millions of floating point operations per second).
MIPS	Millions of instructions per second. Refers to a CPU's ability to process integer operations. A Digital Equipment Corp. VAX 11/780—with its rating of one MIPS—is often used as a standard.
Whetstone	This test, written in Fortran and developed in the late 1960s at the National Physical Laboratory in Whetstone, England, measures floating-point performance. An optimized compiler can improve Whetstone performance, thus making the test somewhat misleading.

launch to correct anomalies or to increase performance after system calibration. Thus, we need to establish reserve capacity when initially defining requirements. A reasonable value for post-launch reserve is 100% spare (equivalent to 50% of machine capacity).

We should not attempt to use all of the available memory or throughput. Asynchronous processing, such as interrupt handlers, introduces a level of uncertainty in throughput. Costs also rise dramatically as we shoe-horn the software into existing memory [Boehm, 1981]. As a rule of thumb, we should use 70% or less of available throughput.

After System Requirements Review, we continuously update estimates for software size and throughput as requirements solidify. We plot them as *reaction curves* to ensure we can detect whether the software is growing too much. Figure 16-7 shows a typical reaction curve for software development, measured as a percentage of maximum use. As long as the estimates fall below the curve, no extraordinary action is required. When they exceed it, we must pare down the requirements, relax the restriction on reserve capacity, or increase the resources available.

Estimating Resource Needs for Application Software

Table 16-13 lists general categories of application software and estimates for size and throughput. It contains typical sizes and throughputs for several types of application software and is useful during conceptual design. Sizes for initial estimates are in words of memory, which are less sensitive to language choice. However, costing models typically use *source lines of code* (SLOC). Table 16-14 shows typical expansion ratios from words of memory to SLOC for various languages. Each of the functions in Table 16-13 is discussed below in terms meant to allow *sizing by similarity* for satellite applications as discussed shortly in Table 16-16.

Communications software includes processing external commands and collecting internal data for transmission to an external source. The information in Table 16-13

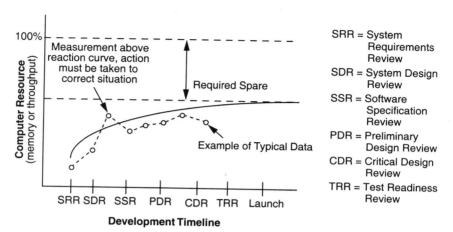

Fig. 16-7. Reaction Curve for Using Computer Resources. Reaction curves mitigate risk. They discipline our management of onboard computer resources. Whenever an estimate exceeds the reaction curve, we correct the situation.

assumes a modest number of ground commands (~100) and collection of data for telemetry to the ground. It does not include inter-processor commanding, but we can allow for this commanding by similarity of function.

Attitude-sensor software handles data from various sensors, compensates for sensor misalignments and biases, and transforms data from sensor to internal coordinates. Processing for gyros, accelerometers, Sun and Earth sensors, and magnetometers involves decoding and calibrating sensed data. Processing for star trackers involves identifying stars against a star catalog, which can require extensive data and memory resources. Mission-related or payload sensors typically require additional resources, which we should calculate separately.

The Attitude Determination and Control category covers various control methods. In *kinematic integration*, we estimate current attitude by integrating sensed body rates using gyros. Using *error determination*, we find how far the spacecraft's orientation is from that desired. For spin-stabilized spacecraft, we maintain attitude control using precession control. The precession-control size and throughput numbers reflect those of a thruster-based system. The thruster control function listed in the table is for a three-axis-stabilized control system using thrusters. Table 16-13 also lists control algorithms using reaction wheels, control moment gyros, and magnetic torquers. Also in this category are object ephemerides, which we can maintain using crude table look-ups and curve fits, or very complex algorithms. The orbit propagator integrates the spacecraft's position and velocity information.

Table 16-13 covers two levels of autonomy. Simple autonomy is for a simple system which requires little onboard support when not in contact with the ground stations. Systems will require complex processing if they need extensive management of onboard autonomy.

Fault detection is closely tied with autonomy. Monitors exist to identify failures or adverse conditions in onboard equipment. Size and throughput vary widely depending on the system. Processing for corrective actions usually depends on tables of pre-stored procedures and, therefore, requires considerable data.

TABLE 16-13. Size and Throughput Estimates for Common Onboard Applications. These values are based on 16-bit words and a 1750A-class Instruction Set Architecture and assume that software is developed in a higher-order language. Because the 1750A is a general-purpose processor, these numbers represent a good first estimate for other general-purpose ISAs. When estimating throughput by similarity, we should hold constant the ratio of throughput to execution frequency given in the table. Increased complexity will increase required size and throughput.

Function	Size (Kwords*)		Typical Throughput (KIPS)	Typical Execution Frequency (Hz)
	Code	Data		
Communications				
Command Processing	1.0	4.0	7.0	10.0
Telemetry Processing	1.0	2.5	3.0	10.0
Attitude Sensor Processing				
Rate Gyro	0.8	0.5	9.0	10.0
Sun Sensor	0.5	0.1	1.0	1.0
Earth Sensor	1.5	0.8	12.0	10.0
Magnetometer	0.2	0.1	1.0	2.0
Star Tracker	2.0	15.0	2.0	0.01
Attitude Determination & Control				
Kinematic Integration	2.0	0.2	15.0	10.0
Error Determination	1.0	0.1	12.0	10.0
Precession Control	3.3	1.5	30.0	10.0
Magnetic Control	1.0	0.2	1.0	2.0
Thruster Control	0.6	0.4	1.2	2.0
Reaction Wheel Control	1.0	0.3	5.0	2.0
CMG Control	1.5	0.3	15.0	10.0
Ephemeris Propagation	2.0	0.3	2.0	1.0
Complex Ephemeris	3.5	2.5	4.0	0.5
Orbit Propagation	13.0	4.0	20.0	1.0
Autonomy				
Simple Autonomy	2.0	1.0	1.0	1.0
Complex Autonomy	15.0	10.0	20.0	10.0
Fault Detection				
Monitors	4.0	1.0	15.0	5.0
Fault Correction	2.0	10.0	5.0	5.0
Other Functions				
Power Management	1.2	0.5	5.0	1.0
Thermal Control	0.8	1.5	3.0	0.1
Kalman Filter	8.0	1.0	80.0	0.01

* Notation here is both standardized and awkward. A lower case "k" is the metric prefix for 1,000; for example, a frequency of 5,000 Hz = 5 kHz. An upper case "K" in counting memory is used for 2^{10} = 1,024. Thus, 2 K words of memory = 2,048 words = 2,048 × 16 bits = 32,768 bits = 32 Kbits.

Power management and thermal control are support functions that often reside in onboard computers. Through power management, the computer controls battery charge and discharge and monitors the power bus. Active thermal control involves monitoring and controlling temperatures throughout the spacecraft.

TABLE 16-14. Converting from Source Lines of Code (SLOC) to Words of Memory. Software written in a high-order language is converted to assembly-level instructions by tools called *compilers*. The compilers usually offer optimization options: for minimum memory use, or for fast operation. The average number of instructions per SLOC is a function of this optimization. Single values are provided here for several common high-order languages. However, based on compiler optimization and designer style these values may shift as much as 25%. Some information was extracted from Boehm [1981]. Modern computers average about 1.5 words of memory per assembly instruction.

Language	Assembly Instructions per SLOC	Bytes per SLOC for 32-bit Processor
Fortran	6	36
C	7	42
Pascal	6	36
Jovial	4	24
Ada	5	30

Estimating Resource Needs for Operating-System Functions

When sizing the operating system software, we can use the numbers in Table 16-15 as baselines or averages. Because systems may require each component to do more or less than indicated, we must apply these numbers flexibly. (For more information on operating systems, see Lane and Mooney [1988] or Silberschatz and Galvin [1997].) For many hardware ISAs commercial operating systems are available and may be used in their entirety to reduce risk. However, this may add substantial memory requirements and increase the level of throughput required unnecessarily. The *executive* is the code that manages and schedules the application software and other operating-system functions. The executive provides interrupt services, schedules and manages tasks based on timers or interrupts, manages resources and memory, corrects single-event upsets, and detects memory faults. It also dynamically allocates memory and fault detection interfaces to the applications.

TABLE 16-15. Size and Throughput Estimates for Typical Onboard Operating System Software. The values below are based on 16-bit words and a 1750A-class ISA. Because the 1750A is a general-purpose processor, these numbers represent a good first estimate for other general-purpose ISAs. Operating-system overhead increases with added task scheduling and increasing message traffic.

Function	Size (Kwords) Code	Size (Kwords) Data	Throughput (KIPS)	Comments
Executive	3.5	2.0	0.3 n	n is the number of tasks scheduled per second. Typical: $n = 200$
Run-Time Kernel	8.0	4.0	see comments	Throughput is included in functions which use the features
I/O Device Handlers	2.0	0.7	0.05 m	m is the number of data words handled per second
Built-In Test and Diagnostics	0.7	0.4	0.5	Throughput estimated assuming 0.1 Hz
Math Utilities	1.2	0.2	see comments	Throughput is included in estimate of application throughput

Run-time kernel software normally supports higher-order languages. For example, it may represent, store, optimize, and pack data; drive input or output; handle exceptions or errors; and interact or interface with other programs, other devices, or even other mixed-language programs. The *I/O handler* controls data movement to and from the processor, as well as packing data for any specific interface. Likewise, the *device handler* or device driver software manages interfaces and data between the processor and any peripheral devices.

Utilities are software routines which several functions use. For example, different components of application software in a single processor might access a set of mathematical operations called *math utilities*. Their size and complexity vary directly with the application and its mathematical requirements. If the processor provides such utilities, they are referred to as *built-in functions*, as is the case in the MIL-STD-1750A ISA. Hardware specifications define these available functions.

Built-in test software provides initial, periodic, or continuous testing for computer elements under the control of software or firmware. Diagnostic software not only identifies faults or failures but also isolates them. We can make it sophisticated enough to recover from some of them. For built-in testing and diagnostics, we can write the software as firmware. The computer vendor may even supply them. If not, we would have to decide whether we need this added reliability, despite the processing and cost overhead.

For preliminary mission design, we recommend the estimation-by-similarity approach given in Table 16-16 to estimate the size and throughput for embedded software in an onboard computer, as well as its size. Another approach, *bottoms-up estimation*, is also discussed in Table 16-16. Both methods are demonstrated in Sec. 16.3, the FireSat example. For alternative approaches, see Rullo [1980].

16.2.3 Computer Selection Guidelines

Once the initial software size estimation process has been completed, we can begin identifying the hardware resources required. We must find a computer system which meets all of our basic needs, as well as the spare allocation, and has the required support environment. Each computer considered must have suitable system software (operating system or kernel and built-in functions such as mathematics).

Representative space computers are shown in Table 16-17. In almost all cases space computers should be purchased rather than developed to avoid paying the nonrecurring development costs, and incurring schedule risks. The exceptions are on opposite extremes. Large aerospace corporations have all the resources required to design and test a computer for a special application. Educational institutions, on the other hand, may make the same choice, not to save money, but because the process meets an educational objective and because reliability requirements are not stringent. Such an institution should verify the onsite availability of the hardware development environment, particularly a logic analyzer with the required pods.

The first performance criterion we can evaluate is the computation rate in MIPS. With Reduced Instruction Set Computer (RISC) this throughput rating is equivalent to 1.5 instructions times the clock cycle, typically expressed in MHz. The processor selected must meet the resource estimation we calculated based on the software functions required to meet the mission objectives. Nearly as important as the computational rate is the address space available. Each hardware address line can have two values: a one or a zero. Therefore, N address lines provide 2^N distinct addresses.

TABLE 16-16. Software Estimation Process. The *estimation-by-similarity* technique uses existing, well characterized functions and their relationship to functions under development to estimate processor memory and throughput needs for the new functions. The *bottoms-up estimation* process forces the estimator to break the functions into the smallest components, which are then evaluated based on experience.

Step	Notes
1. List all application functions allocated to the given computer.	Document any assumptions.
2. Break down the functions from step 1 into basic elements.	This often requires several iterative steps. We should continue to break down the functions until we have reached the lowest level we can identify.
3. Define the real-time execution frequency for each of the basic elements.	We only need to perform this step for those functions which are time critical. Execution frequency is not required for utility functions.
4. Estimate the source lines of code (SLOC) and memory needed for each function by: A. **Similarity**: Find a function from Table 16-13 with similar processing characteristics and known size. Compare the complexity of the known function with the new one and adjust the code size directly. Adjust the code estimate for differences in development language, such as assembly versus higher order language. Use Table 16-14 to determine the SLOC from the memory used.	**Similarity**—Requires knowledge of both the existing functions and the new functions in terms of specific requirements, complexity, and general implementation. Some rules of thumb are: • A 25% increase in complexity implies a 25% increase in code • If the known function is in assembly code, increase the code size by 25% for a higher order language. • If the known function is in a higher order language, decrease the code size by 20% for assembly code.
B. **Bottoms-Up**: Identify the SLOC for executable elements of the lowest level functions as well as the data structures and one time only initialization software. Sum the SLOC for all executable and nonexecutable functions separately. Use Table 16-14 to determine the memory requirements, based on SLOC.	**Bottoms-Up**—Requires knowledge of general elements of each function and how to implement the capability.
5. Estimate throughput requirements based on: A. **Similarity**: Find a function from Table 16-13 with similar processing characteristics and known throughput requirements. Compare the complexity of the known function with the new one and adjust the throughput value for differences in complexity directly. Based on the frequency of execution from step 3 above, compute the total throughput requirements for the new function.	Throughput should be expressed as instructions per second.
B. **Bottoms-Up:** Based on an average number of computer instructions executed per SLOC for a specific compiler and processor, multiply the number of instructions by the execution frequency. Define a "loop factor" for executable SLOC to represent internal loops within the function.	The internal loop factor for an ACS function associated with a three axis stabilized spacecraft will be 3— each function will be repeated for each axis. In order to estimate throughput from SLOC, the lines of code related to initialization and data must be separated from the executable lines of code. The loop factor is applied only to the executable SLOC
6. Determine the operating system and overhead requirements by **similarity** to other implementations. Compare the complexity of the known operating system and overhead functions to the new and adjust accordingly.	You must identify all application code prior to this step. Complexity for operating system functions is based on number of tasks scheduled per second, the number of interrupts to be handled, and the amount of I/O data.
7. Determine the margins for growth and on-orbit spare based on where you are in the development cycle and Fig. 16-7.	Growth and spare requirements are important and should be strictly calculated.

TABLE 16-17. Commercially Available Space Computers. These computers have been developed for use in a variety of general purpose space applications.

Supplier and Computer	ISA	Word Length (bits)	Memory (RAM + EEPROM)	Performance (MIPS)	Radiation Hardness	Connectivity	Heritage
Honeywell GVSC	1750A	16	16 MB*	1 to 3	1 MRad	1553B RS-232 RS-422	NEAR, ChinaStar, Clementine
Honeywell RH32	R 3000	32	4 GB	10 to 20	1 MRad	1553B RS-232 RS-422	GPS II SBIRS High
Honeywell RHPPC	603E	32	4 GB	20	100 KRad	RS-232 RS-422	VIIRS
L-M GVSC	1750A	16	16 MB*	1 to 2	1 MRad	1553B RS-232 IEEE-488	Cassini, Rapid I
L-M RAD 3000	R 3000	32	16 MB	10	Rad Hard	1553B RS-422	LM-900
L-M RAD 6000	RS 6000	32	16 GB	10 to 20	100 KRad	PCI Firewire HSS	Mars Pathfinder, Globalstar, Space Station, SBIRS Low, Mars 98
TRW	RS-3000	32	16 MB	10	Rad Hard	1553B RS-422	SSTI, T200b, Step-E
SWRI SC-2A	80C186	16	768 KB	0.3	10 KRad	Parallel RS-422	MSTI-2
SWRI SC-5	80C386	32	320 KB	0.6	10 KRad	Parallel RS-422	RADARSAT, SNOE
SWRI SC-7	TI320C30	32	640 KW	12	100 KRad	1553B	MSTI-3
SWRI SC-1750A	1750A	16	512 KB	1	10 KRad	RS-422 1553B	MSTI-1,2,3 New Millennium DS-1
SWRI SC-9	RS 6000	32	128 MB	20	30 KRad	RS-232	Space Station
SWRI MOPS	R 6000	32	128 MB	25	30 KRad	RS-422	Gravity Probe B, International Space Station Alpha (ISSA)
Sanders STAR-RH	R 3000	32	4 MB	10	50 KRad	1553B	CRSS
GDAIS ISE	603E	32	2 GB	25	Rad Hard	1553B	HEAO, AFAX
Acer Sertek	80186	16	512 KB	0.5	Rad Hard	1553B RS-422	ROCSAT

* Address Space   L-M: Lockheed Martin   SWRI: Southwest Research Institute

For example, a computer with 16 bit words would generally have a 64K word address space and a 32 bit word computer could address a 4 Gigaword address space. However, this may require a more in-depth examination. The Generic VHSIC Spaceborne Computer (GVSC) was developed by the Air Force from the MIL-STD 1750A, which defines one word as 16 bits, but it has an 8 Mword address space. This was achieved by built-in paging hardware. The GVSC is a special case because the Air Force developed it especially for space applications. We may find other exceptions as well.

If possible each candidate computer, or its engineering development unit equivalent, should be bench-marked against the relevant applications. This is rarely done because of time and lack of availability of the hardware. In the absence of real equipment or software benchmarks, the computer analyst should examine the individual instructions of each candidate and match them against the qualitative aspects of the computational requirements. For example, the processor we select to support the attitude determination and control function will perform many floating point and transcendental function computations. In this case it would be desirable for floating point and trigonometric instructions to be performed in hardware and not in software.

If the processor is to perform commutation or multiplexing at the bit level, the CPU should have instructions that support such processes.

On the other hand, the processor we select for control and data handling will need to support bulk moves of data. Thus it would be desirable if it had a *Direct Memory Access* (DMA) command or a block move command. DMA is very valuable for data handling processors with extensive I/O so that the range of data to be moved can be specified and letting the DMA hardware relieve the CPU of moving each word.

If we determine that several computers can meet the performance needs we outlined in the software resource estimation process, a computer's heritage can be a major selection criterion. We often select a computer previously used in space by NASA, ESA, the DoD, or a major commercial space venture. By starting with computers which have prior use in space, the major development, qualification testing, and documentation risks and costs will have been borne by the prior programs. However while older space computers are often highly reliable, they are typically more expensive and less capable.

16.2.4 Integration and Test

As Fig. 16-8 shows, testing usually begins at the lowest level and builds incrementally. By building our test scenarios from the bottom up, we can reduce the complexity and thus the risk. Testing must be rigorous at all stages from the unit level to the system level (Kaner et al. [1993]). Software and hardware testing follow the same general path, with the subsystem resulting when we integrate the hardware and software. At this level, we test the entire computer system. Finally, we test the whole spacecraft with computer systems becoming components of the subsystem as described in Chap. 12. Unfortunately, we cannot determine whether we have calibrated the computer equipment properly until it is in orbit. Once operational, the system needs general testing to ensure it continues to perform as required. Just as acceptance test procedures or inspections check systems for damage on delivery, retesting on-orbit checks for damage during launch. This testing is often referred to as *on-orbit check out and calibration.*

In general, integration and test pulls disciplines and subsystems into a configuration that meets top-level, system requirements. In this sense, as Chap. 12 suggests, integration and test is much the same as systems engineering. Testing includes all activities that increase confidence in the system's performance. It ensures that we have met requirements and that anything happening beyond these requirements does no harm to the system, while preserving specified functions. Testing for these "extras" is the most difficult because we do not always know what we are looking for. Testing, especially for the software-intensive computer resources associated with space systems, is a complex undertaking. It can consume up to half of the development cost and a significant percentage of support costs over the life-cycle.

16.2.5 Life-Cycle Support

Many issues associated with the development cycle affect conceptual design and long-term life-cycle costs. For example, we select hardware and software design concepts during requirements definition but we draw on implementation and development experience to do so. Because we often have to cost activities for developing a computer system before a complete design is available, we must take into account the many aspects of software and hardware development, testing, and integration. Software-based tools and standards such as MIL-STD-498, IEEE and SAE Specification Guide, and ANSI standards help us structure our methods and give us more manage

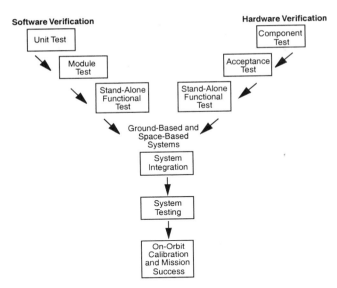

Fig. 16-8. Levels of Testing. Testing builds incrementally as the product develops from component to system. It begins at the lowest level building up into system and mission requirements verification as the elements are integrated.

ment and technical controls. The Software Engineering Institute (SEI) has established a rating system for the software development process. Companies can apply to SEI for increasing levels of performance ratings [Humphrey, 1995]. Also, the International Standards Organization has established a process-based rating known as ISO 9000.

In addition to the development process, long-term issues such as life-cycle costs require attention during conceptual design. The complete life-cycle costs include conceptual design, detailed design, implementation, system integration, test, and on orbit maintenance—plus the tools associated with each phase. There are many software development life-cycle models which we can use. However, these continually evolve to reflect the current state of the software development environment. According to Anderson and Dorfman [1991], "As the software development process evolves, so will these models to reflect new types of applications, tools, and design paradigms." Our experience has been that development support software will require 8 to 10 times as many lines of code as the flight software. We can use this to estimate the cost of the development software as discussed in Sec. 20.3.2. The cost estimate we generate using this method can be applied to either the development of support software or to procure COTS tools as discussed below. Table 16-18 summarizes the various life-cycle issues which we must address during the early phases of program development.

Building an operational system depends on the development philosophy and environment we select. Procuring *commercial off-the-shelf* (COTS) hardware and software is the easiest way we can build a capability. However, off-the-shelf items often don't meet our needs exactly. Thus, we must tailor the COTS products or adjust the need to match what is available. The break-even point is different for each program depending on the number and types of requirements, personal skills, our knowledge of the product, and whether the COTS products are maintainable and of high quality. If we opt for custom development, we must also consider the development and test tools

TABLE 16-18. Summary of Life-Cycle Cost Issues. Through the use of development and protocol standards, auto-code generation, and re-use of common software and hardware modules, as well as increasingly capable development tools we create a means of controlling and streamlining the costs associated with computer system development. (For additional information, see Boehm [1984].)

	Standards	Automation and Re-Use	Development Tools
Example	• DoD-STD-2167 • IEEE • ANSI • SAE • ISO • MIL-Specification • Structured Development • Object Oriented Development	• Commercial Off-the Shelf (COTS) Software • Computer Aided Software Engineering (CASE) • Automated Code Generation	• Compiler, Assembler, and Linker • Cross Compiler • Loader and Debugger • Code Analyzers and Optimizers • Test Case Generator • Code Management Software • Simulators and Emulators
Benefits	• Adds structure to development activities • Standardizes documentation between and among programs	• Reduces development time which implies lower costs	• Allows for more generic software to be developed and then compiled specifically for target hardware • Aids in configuration management • Simulation and emulation increase efficiency and effectiveness of testing
Notes	• Tailoring is critical • Must balance required amount of documentation with associated costs	• Need documentation which accurately reflects the implementation • MUST continue to test at the same level for all sources of software	• Be sure that a development environment is available for the hardware selected for the project

we will need to create the operational unit. We may need special hardware, software, or integrated systems. For single-unit or unique systems, developing support tools can cost as much as developing the operational units. We must evaluate these support tools during conceptual design, and include their cost in the overall development cost. As with the cost of other developmental tools, we may be able to amortize these costs over multiple products or projects.

When developing embedded flight software we must first determine if we want it in assembly language or a higher order language. If we have selected a higher order language, we must decide which tools to use. Depending on the target processor ISA and the language selected, we must evaluate the availability and quality of cross-compilers, linkers, assemblers, and the host processors on which they reside. A *host machine* is the computer where the development activity is to take place. The *target machine* is the embedded microprocessor or ground-based computer where the code will perform throughout its lifetime. A *cross-compiler* executes on the host, compiling software for the target processor. For both assembly language and higher order languages, the loaders and debuggers allow us to store software in the target processor and evaluate its performance based on either symbolic or physical information. Sometimes the host computer and the target computer have the same ISA. In this case the compiler is not the issue but library and other functions' availability on the target should be explored.

16.3 FireSat Example

The FireSat example can help us better understand the process of estimating computer resources. We show two examples of software estimation: one for the spacecraft attitude control capability, the other for payload control and data management.

16.3.1 FireSat Attitude Control Processing

To begin the estimation process, we need a list of allocated computer system capabilities. For this attitude control example, we assume that the spacecraft is a three-axis-stabilized vehicle using an Earth sensor, a Sun sensor, and rate gyros for sensing vehicle attitude and attitude rates. Reaction-control thrusters generate the control torques. Further, we assume that the highest frequency the system performs any function is 4 Hz (every 0.25 sec).

From this information we can decide what the system must do. For example, it must process data from the Earth sensors, Sun sensors, and rate gyros. To process the data and determine the current attitude, we need to perform kinematic integration, or something equally complex. To maintain the desired attitude, we need functions for determining attitude error and for thruster control. A function to estimate ephemeris allows us to keep track of orbital position. If the ground station regularly updates the ephemeris, we need only a simple function to propagate it. Remember to identify all assumptions whenever you estimate; this last assumption allocates functionality to a supporting ground station.

We can now estimate the memory and throughput requirements for the application functions. Assuming a 1750A-class host processor, apply the information in Table 16-13 directly to estimate memory for code and data. When using the table, if your particular function is estimated to be much larger or smaller than what is typical, adjust the numbers accordingly. Estimating throughput requires more effort. To use the throughput numbers in Table 16-13, we must first establish the execution frequency for the attitude control computations. Table 16-19 lists the assumed frequency for each application function. We estimate the throughput for a function executed at 4 Hz by using 4 Hz/10 Hz = 40% of the values in Table 16-13, which is based on a 10-Hz execution rate. Table 16-19 shows the estimated memory and throughput requirements by function.

We need to consider an operating system to complete the estimate for all the software in the attitude determination and control computer. We have assumed the software is in Ada; thus we will need a run-time kernel. A COTS Ada cross-compiler may include a kernel. We must also have a local executive to perform task scheduling and management. The number of scheduled tasks per second depends on the number of application functions and their execution rate. For this example, we estimate 80 tasks per second. Because of the four sensors being serviced and the connection with the data bus, we will need five data handlers. To estimate how much data the system must handle, first determine how much the sensors produce and how often they transmit. Then add external commanding and telemetry requests. For this example, the handlers will control the flow of 800 data words per second. The last functions to estimate are the built-in test and associated diagnostics. We determine memory and throughput for each operating-system function from Table 16-13, taking into account the assumptions above.

We have estimated memory and throughput requirements for all functions that the computer must support for attitude determination and control. But because early

TABLE 16-19. Estimating Size and Throughput for FireSat Attitude Control Software. We can use this general format for estimating the size and throughput requirements for any software application, based on the method of similarity. First list the anticipated applications functions. Using the mission requirements, we can establish a baseline frequency for the execution of each function. Then, based on either the estimation by similarity or by bottoms-up estimation, we determine the memory requirements for each function. Finally, using the estimation process and the estimated frequency of execution, we can determine the required throughput for each function. Notes for this table are on the following page.

Component	Estimation Source		Required Memory		Required Throughput (KIPS)
			Code (K words)	Data (K words)	
Application Functions		Frequency (Hz)			
Thruster Control	Table 16-13	4	0.6	0.4	2.4
Rate Gyros	Table 16-13	4	0.8	0.5	3.6
Earth Sensor	Table 16-13	4	1.5	0.8	4.8
Sun Sensor	Table 16-13	1	0.5	0.1	1.0
Kinematic Integration	Table 16-13	4	2.0	0.2	6.0
Error Determination	Table 16-13	4	1.0	0.1	4.8
Ephemeris Propagation	Table 16-13	1	2.0	0.3	2.0
(a) *Appl. Subtotal*			8.4	2.4	24.6
Operating System					
Local Executive	Table 16-15, with $n = 80$ [1]		3.5	2.0	24.0
Runtime Kernel (COTS)	Table 16-15		8.0	4.0	N/A
I/O Handlers (5)	Table 16-15, with $m = 800$ [2]		10.0	3.5	40.0
BIT and Diagnostics	Table 16-15		0.7	0.4	0.5
Utilities	Table 16-15		1.2	0.2	N/A
(b) Subtotal: COTS			8.0	4.0	N/A
(c) Subtotal: Non-COTS			15.4	6.1	64.5
(d) *O/S Subtotal*	(b) + (c)		23.4	10.1	64.5
(e) *Total Software Size & Throughput Est.*	(a) + (d)		31.8	13.5	89.1
Margin Calculations					
(f) Needed to compensate for requirements uncertainty	100% of non-COTS: $1.0 \times [(a) + (c)]$		23.8	8.5	89.1
(g) On-orbit spare	100% spare: $1.0 \times [(e) + (f)]$		55.6	22.0	178.2
Estimate of Computer Requirements	(e) + (f) + (g)		111.2	44.0	356.4

Assumptions:
A. Three-axis stabilized vehicle using Earth sensor, Sun sensor, and rate gyros.
B. Reaction-control thrusters used for attitude control.
C. No function needs to be performed faster than 4 Hz; Sun sensing and ephemeris propagation done at 1 Hz.
D. Ground station will update ephemeris frequently, allowing for a simple propagation mode.
E. 1750A-class target computer.
F. Software developed in Ada.
G. Target computer must have 50% spare capacity at launch.

Notes:
1. *Computation of n = number of scheduled tasks per second.*
   To determine *n*, first calculate the number of application functions times their frequency

                5 functions listed at 4 Hz = 20 functions per second
                2 functions listed at 1 Hz = 2 functions per second
                                Total ≈ 22 functions per second

   If no better information exists, assume 3 to 4 schedulable tasks for each major function. Thus, a conservative estimate for the number of schedulable tasks, *n*, in the FireSat ACS example would be 80.

2. *Computation of m = words/s*
   Rate gyros—12 @ 4 Hz, Earth sensors—20 @ 8 Hz, Sun sensors—10 @ 4 Hz   = 248 words per second
   Telemetry— 4 kbit telemetry stream   = 500 words per second
   Command—control commands   = <u>50 words per second</u>
                                                               Total ≅ 798 words per second

---

requirements are soft and the system will evolve, we must add substantial margin. The target computer in which this software resides must also have spare capacity to accommodate on-orbit growth in the software. Table 16-21 includes the margin we need. In this example, the ACS computer's minimum size and throughput are 155 K words and 356 KIPS.

To assess cost and develop a schedule, we need to estimate the number of source lines of code, or SLOC. This estimate should consider only the software that we must develop, so we exclude margins and off-the-shelf software. (The margin added for growth between the System Requirements Review and deployment is to reduce risk in determining processor requirements. Although not included in the calculation of source lines of code, we should consider it when we examine potential cost and schedule risk.) The amount of software to be developed is 23.8 K words of code and 8.5 K words of data memory. For costing, we use a factor of 25% to convert data memory to equivalent code words. We have to do so because developing a data word takes about one quarter of the effort for developing a word of executable code. Thus, for this attitude determination and control example, the total becomes 26.0 K words. Assuming the development is in a higher-order language such as Ada (using 1 line of code per 5 words of memory from Table 16-14), this translates to 5,200 source lines of code for the FireSat attitude control software.

16.3.2 FireSat Onboard Payload Processing

As part of defining the conceptual design for FireSat, we start with the assumptions for payload control & data management given in Fig. 16-9, including fire detection and reporting and fire parameter estimation. We show our orbit and sensor characteristic information in Table 16-20. We assume all FireSat candidate sensors operate in the single scan mode. The primary sensor differences for this example are in the scan width and the data sample sizes which, as we will see, are key drivers in both the system and the data flow architecture trades. The scan width options can imply different size satellite constellations to cover the same area per unit time, or imply multiple sensors per satellite, e.g., one satellite with a single A or B sensor can cover the same area as either one satellite with four C sensors, or four satellites with a single C sensor

In this example, we will focus on processor throughput. We also assume we will use a general purpose computer (identified as GP in Table 16-22) and that the best computer we can acquire has a throughput of 100 MIPS. (See Table 16-17 for a survey of currently available processors.) Assuming we want to launch with 50% spare, the

Payload Control & Data Management (PCDM) Interfaces & Capabilities

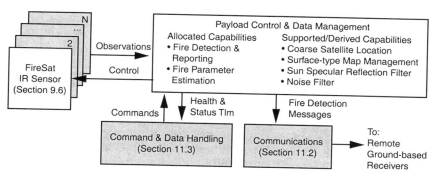

Fig. 16-9. Allocations to Onboard Computer System for FireSat Payload Management Capability. This figure delineates the various functional components of the FireSat payload management onboard capability. We can either host this software in a single onboard computer (OBC) or we can partition it between several OBCs. Likewise we might perform these functions using hardware if appropriate. How we partition the implementation depends on the computer system design and overall mission requirements.

available throughput is only 50 MIPS. As stated in Sec. 16.2, onboard software size and throughput historically double from initial conceptual design to launch. To account for this future growth, we set a maximum threshold of 25 MIPS per processor for this conceptual design phase exercise.

The primary software estimation method we use is Bottom-up Estimation (Table 16-16). We generate the needed software information from raw data and basic assumptions. Once we break the functions into manageable pieces, we expect to find elements that we understand and can estimate. Table 16-21 provides an example of the software elements and their characteristics that may be generated through the bottom-up estimation process. This example is not intended to represent optimum payload control and data management algorithms; it is presented as a set of sample algorithms that mission processing engineers may consider. Multiple algorithm sets are often considered for complex onboard systems, although we evaluate only one set here.

Several candidate architectures for this example are shown in Table 16-22. The first assumed architecture is a direct (point-to-point) connection between the IR sensor and the processing hardware in which the payload control & data management software will reside. The fire detection algorithm must first separate observations into "possible fires" and "noise". For first order calculations, we assume a simple thresholding algorithm to eliminate the noise; throw out samples below an intensity threshold, and throw out above-threshold samples for which there are no adjacent above-threshold samples. The clusters of continuous sample data are referred to as "events." Table 16-23 identifies the processing assumptions for a general purpose processor. For all three sensor candidates, the throughput calculations show a prohibitive situation. For sensor C with the narrow field of view, there appears to be some promise, but the throughput requirement still exceeds our 25 MIPS maximum. Additionally, the information in Table 16-23 doesn't include other software that we know must operate (such as the executive task scheduler or data I/O manager). We now must consider alternatives to our first architecture.

TABLE 16-20. Characteristics of Candidate FireSat IR Sensors. Based on the mission requirements, outlined in the top of this table, there is a variety of approaches to solving the problem. We must map out each alternative with quantifiable elements so that we can compare them fairly. No one solution is better than the next. However, one may meet our mission goals, requirements, and constraints better than the others.

Orbital and Mission Parameters (From Table 9-15, p. 287)			
Orbit altitude (*h*)	700	km (Defined in Sec. 7.4)	
Period (*P*)	98.8	min	
Ground Track Velocity (V_g)	6,762	m/s	
Min. Elevation Angle (ε)	20	deg	
Resolution (*d*)	30	m	
Max scan time (d/V_g)	4.437	msec (no overlap at Nadir)	
Min scan freq (V_g/d)	225.4	scans/s	
Peak fire density	40	detectable fires in 100 km path	
	1,600	detectable fires in 100,000 km^2	

Sensor Characteristics	A: Full FOV		B: Fewer Bits/Sample		C: Quarter FOV	
Scan Width (θ_x)	57.9	deg	57.9	deg	14.5	deg
Pixels per Scan	23,563	pixels	23,563	pixels	5,891	pixels
Samples per Pixel (*s*)	1.6	samples	1.6	samples	1.6	samples
Sample Rate:	37,702	/scan	37,702	/scan	9,425	/scan
	8,498,013	/sec	8,498,013	/sec	2,124,503	/sec
Bits/Sample (*b*)	8	bits	4	bits	8	bits
Frame Efficiency (*q*)	0.9		0.9		0.9	
Data Rate (DR)	76	Mbps	38	Mbps	19	Mbps
Bits/Scan	301,612	/scan	150,806	/scan	75,403	/scan
Bit Rate	67.98	Mbps	33.99	Mbps	17.00	Mbps
Earth Coverage Rate	5,327	km^2/s	5,327	km^2/s	1,203	km^2/s
"Design to" peak fire detection rate	85	/sec	85	/sec	19	/sec

One possible architectural change would be to include a separate processor, dedicated to the thresholding process. Again, the high throughput estimates in Table 16-23 indicate this is not a viable option, given the 25 MIPS throughput limit we have established.

Clearly now, we need to add a special-purpose sensor interface unit into the architecture to hold the scan information, perform the noise filtering function, and provide the events from each scan to the general purpose computer. Since a small change in a few parameters can lead to substantially different throughput estimates, we should perform additional analysis at this point to justify the need for this custom equipment. For example: doubling the ground resolution requirement (from 30 m to 50 m) will reduce the throughput estimate by a factor of four.

TABLE 16-21. Software Elements for Payload Control and Data Management. We can us this general format for estimating the size and throughput requirements for an software application, based on the bottoms-up method. The chart identifies a the elements of the Payload Control and Data Management function that will b implemented in software for the FireSat example. Additionally, it demonstrate how to calculate throughput, and what external environmental conditions driv the value. Cumulative throughput requirements are calculated in Table 16-25.

Functional Breakdown	Execution Frequency f		Exe. SLOC E	Inst./ Exe. A = 5E	Utility Inst. B	Loop Fact L	Throughput Est*
FIRE DETECTION							
Sensor I/O Control (assume DMA)							
– DMA control	per Sc	225 Hz	1	5	0	1	0.001 MIPS
– Data movement	per B		N/A	1	0	1	1 I/B
Noise Filtering (data intensity)							
– Check threshold intensities	per Smp		5	25	0	1	25 I/Smp
– Generate clusters†	per Smp		15	75	0	1	75 I/Smp
– Generate cluster "event"	per Ev	Table 16-25	30	150	0	1	150 I/Ev
Noise Event Filtering (geographic)							
– Orbit Propagator	per Sc	225 Hz		< By similarity:			4.5 MIPS
(Table 16-13)				0.020 MIPS * (225 Hz/1Hz) >			
– Convert to Lat/Long	per Ev	Table 16-25	10	50	250	1	100 I/Ev
– Filter out "ocean fires"							
– Build Surface map	per Sc	225 Hz	400	2,000	2,000	3	2.7 MIPS
– Filter out non-land events	per Ev		20	100	0	1	100 I/Ev
– Exclude Sun specular reflection							
– Locate specular region		2 Hz	150	750	0	1	0.002 MIPS
– Define exclusion zone	per Sc	225 Hz	30	150	0	1	0.034 MIPS
– Filter out pixels in E.Z.	per Ev	Table 16-25	5	25	0	1	25 I/Ev
FIRE REPORTING							
Determine Fire Lat/Long							
– Convert scan/pixel to Lat/Long	per Ev	Table 16-25	10	50	250	1	300 I/Ev
Message Generation							
– Correlate event to prior scans	per Ev	Table 16-25	15	75	0	2	150 I/Ev
– Generate containment ellipse	per F		150	750	1,000	2	3,500 I/F
– Generate average intensity	per F		10	50	0	2	100 I/F
– Format message	per F		20	100	0	1	100 I/F
Communications (Table 16-13)		225 Hz		< By similarity:			2.3 MIPS
				0.010 MIPS * (225 Hz/1Hz) >			
Supporting Math/Utility Functs							
– Square root	—	—	—	1,000	—	—	Utility thruput
– Trig Functions	—	—	—	250	—	—	factored in
– Inv. Trig Functions	—	—	—	1,500	—	—	above
Operating System	Assume 10% over-head on all functions						

Terminology:

I = Instruction
B = Byte
Smp = Sample— Raw digital intensity information from the IR sensor representing one measurement
Sc = Scan—One sweep across all pixels within the IR sensor
Ev = Event—A set of contiguous samples (2 or more) that are not removed by the Noise Filter
F = Fire—A set of events (may be only one) that are geographically connected scan-to-scan

* = L * (A + B)
† Eliminate single pixels as noise

ABLE 16-22. Onboard Computer System Architecture Evolution for FireSat Payload Control & Data Management Example. See text for discussion. Terminology: *GP* = general purpose computer, *SIU* = sensor interface unit.

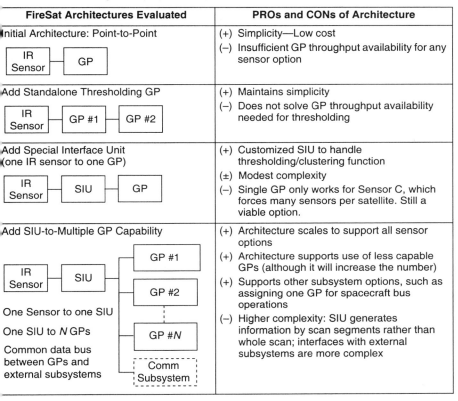

FireSat Architectures Evaluated	PROs and CONs of Architecture
Initial Architecture: Point-to-Point IR Sensor — GP	(+) Simplicity—Low cost (−) Insufficient GP throughput availability for any sensor option
Add Standalone Thresholding GP IR Sensor — GP #1 — GP #2	(+) Maintains simplicity (−) Does not solve GP throughput availability needed for thresholding
Add Special Interface Unit (one IR sensor to one GP) IR Sensor — SIU — GP	(+) Customized SIU to handle thresholding/clustering function (±) Modest complexity (−) Single GP only works for Sensor C, which forces many sensors per satellite. Still a viable option.
Add SIU-to-Multiple GP Capability IR Sensor — SIU — GP #1 / GP #2 / GP #N / Comm Subsystem One Sensor to one SIU One SIU to N GPs Common data bus between GPs and external subsystems	(+) Architecture scales to support all sensor options (+) Architecture supports use of less capable GPs (although it will increase the number) (+) Supports other subsystem options, such as assigning one GP for spacecraft bus operations (−) Higher complexity: SIU generates information by scan segments rather than whole scan; interfaces with external subsystems are more complex

We need to remain flexible, so we also assume that the sensor interface unit may operate as a one-sensor to many-processors interface. With the one-to-many approach, we assume that we can segment the fire detection and reporting capability using parallel processing. With a single-scan sensor, we can divide each scan into N contiguous segments. Then each processor will perform its fire detection algorithm on a swath of ground territory representing $1/N$ the width of the sensor field of view. Although the software functions do not change based on effective swath width, the special processor data management function changes with N, and we should size it accordingly. Note that we assume we use the same basic software in all the processors.

In order to proceed with our software estimation example, we need to identify the peak data rates from the sensor interface unit to each processor. Table 16-24 identifies the sensor interface unit assumptions. Using the throughput estimates of Table 16-21 and the event rate estimates in Table 16-24, we are now able to estimate overall throughput requirements for the payload control and data management software across the sensor type and number of processor combinations. Table 16-25 presents the computation for sensor A; the computations for sensors B and C are performed in the same manner. Figure 16-10 shows, for all candidate sensors, the throughput per processor as a function of number of processors supporting one IR sensor.

TABLE 16-23.   Noise Reduction Throughput Calculation. For each of the three example shown below, we calculate the resulting throughput requirement if we were t implement the noise reduction capability in software. Each solution exceeds ou original 25MIP limit imposed earlier in this example. Therefore, an alternativ approach must be found (for example, we can use hardware as shown in th candidate architectures in Table 16-22).

Throughput Calculations Using Noise Filtering Instruction Estimates (Table 16-21)			
250 instructions executed per above-threshold sample			
25 instructions executed per above-threshold sample			
	A: Full FOV	**B:** Small Sample Size	**C:** Quarter FOV
Peak % above threshold	1%	2%	1%
Samples above T	377 /scan	754 /scan	94 /scan
Samples below T	37,325 /scan	36,948 /scan	9,331 /scan
GP Instructions	1,027,367/scan	1,112,196 /scan	256,842 inst/scan
GP inst per second	232 MIPS	251 MIPS	58 MIPS
Implication	Won't Fit in GP	Won't Fit in GP	Exceeds Limit

TABLE 16-24.   FireSat IR Sensor-to-Processor Interface Unit Characteristics. This tabl shows how to calculate the throughput requirements for various sensor confi urations and differing numbers of general purpose computers.

SIU maintains information from the last 2 scans (allow for GP collection delays).
One "message" generated per scan (or scan segment).

	Message per Scan:	Number of events in scan, Time of scan	(60 bits)
	per Event:	Number of samples, first pixel location	(32 bits)
	per Sample	Measured Intensity	(8 bits)

For Single-Sensor to Multiple-GP Configuration
  Scan is segmented according to number, N, of general purpose CPUs in use.
  Dynamic thresholding performed over each segment of scan.

Maximum number of events per scan, based on Sensor scan width and number of general purpose processors in use:	**G.P. CPUs**	**IR Sensor** A	B	C	
	N = 1	189	378	48	events per scan
	N = 2	95	189	24	events per scan
	N = 5	38	76	10	events per scan
	N = 10	19	38	5	events per scan

Peak data transfer rate (Mbps), based on Sensor scan width and number of general purpose processors in use:	**G.P. CPUs**	**IR Sensor** A	B	C	
	N = 1	2.058	3.422	0.533	Mbps (SIU to GP)
	N = 2	1.041	1.718	0.273	Mbps (SIU to GP)
	N = 5	0.425	0.699	0.122	Mbps (SIU to GP)
	N = 10	0.219	0.356	0.068	Mbps (SIU to GP)

TABLE 16-25.   **Estimated Throughput Requirement for FireSat Payload Control and Data Management Software (Sensor A).** Using the individual throughput estimates for each basic software component (Table 16-21) and the event rates (Table 16-24), the overall throughput can be estimated. Note: For clarity of calculation, the event rates in Table 16-24 have been converted from events/scan to events/sec using a scan rate of 225 Hz.

Functional Breakdown for	No. Processors:	Throughput (MIPS)			
		1	2	5	10
Payload Control & Data Management Onboard Software	**Peak Events/Sec:**	85,050	42,525	17,100	8,550
	Peak Fires/Sec:	42,525	21,263	8,550	4,275
Data Transfer from SIU	**Throughput Formula**	— Sensor A —			
DMA Control	0.001 MIPS	0.001	0.001	0.001	0.001
Data Movement (data rate in Table 16-24)	1 Inst/Byte	0.257	0.130	0.053	0.027
Noise Filtering Geographic					
Orbit Propagator (Table 16-13)	4.5MIPS	4.500	4.500	4.500	4.500
Convert Scan/Pixel to LAT/LONG	300 Inst/Event	12.758	6.413	2.565	1.283
Filter Out "Ocean Fires"					
Build Surface Map	2.7 MIPS	2.700	2.700	2.700	2.700
Filter Out Non-Land Events	100 Inst/Event	4.253	2.138	0.855	0.428
Exclude Sun Specular Reflection					
Locate Specular Region	0.002 MIPS	0.002	0.002	0.002	0.002
Define Exclusion Zone for Scan	0.034 MIPS	0.034	0.034	0.034	0.034
Filter Out Pixels in E. Z.	25 Inst/Event	1.063	0.534	0.214	0.107
Fire Reporting					
Determine Fire LAT/LONG					
Convert Scan/Pixel to LAT/LONG	300 Inst/Event	0.026	0.013	0.005	0.003
Message Generation					
Correlate Event to Prior Scans	500 Inst/Event	6.379	3.206	1.283	0.641
Generate Containment Ellipse	3,500 Inst/Fire	0.298	0.149	0.060	0.030
Generate Average Intensity	100 Inst/Fire	0.009	0.004	0.002	0.001
Format Message	100 Inst/Fire	0.009	0.004	0.002	0.001
Communications (Table 16-13)	2.250 MIPS	2.250	2.250	2.250	2.250
Operating System	10% of above	3.428	2.195	1.447	1.198
Total Throughput Estimate per Processor (MIPS)		37.96	24.27	15.97	13.20

Based on the overall throughput requirement curves in Fig. 16-10, it appears that sensor A is viable if there are two processors, each handling half of the field of view. Sensor B requires a minimum of four processors. Sensor B has a smaller sample size (therefore a lower raw data rate), but the lower fidelity of the data does not allow sufficient noise reduction to lower the number of events being passed to the processor. This demonstrates the strong sensitivity between the noise reduction capability and the required processor throughput to detect fires. This noise reduction element has now

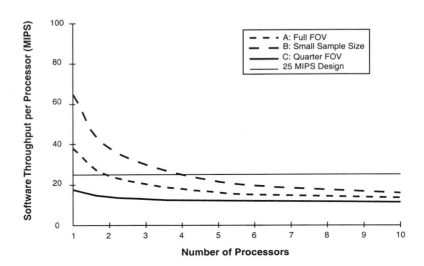

Fig. 16-10. Throughput Curves for FireSat Example. These curves illustrate how we can trade the number of onboard processors against the throughput and quantity of each processor to meet our overall mission requirements. Depending on size, weight and power constraints, as well as cost, we can determine which solution best meets our mission needs.

been identified to be a driver, given the algorithm suite provided in this example. It appears that sensor C can perform the mission with a single processor; however, the limited field of view of sensor C may imply the need for additional sensors per spacecraft.

Since processors are not 100% reliable, we need to establish a redundancy strategy. This strategy will involve the use of distributed vs. federated architectures. For a federated architecture, we will likely need 2-for-1 redundancy for each processor. A distributed architecture—if the SIU can support it—may allow more cost-effective collective redundancy such as 3-for-2. Given this, our preliminary conclusion is that sensor A in a distributed processing architecture will result in the fewest number of onboard processors for the complete FireSat system. Of course we recognize that this is sensitive to our assumptions and will continue to examine the trade space carefully.

16.3.3 Spacecraft and Payload Processing Consolidation and Effort Estimation

Although we have dealt only with selected elements of onboard processing for the FireSat space vehicle, areas of commonality have already emerged. As shown in Tables 16-19 and 16-21, both spacecraft and payload software requires mathematical library functions and some type of operating system. Some, perhaps significant, cost containment can be achieved using common elements: processors, software languages, utility libraries, and real-time operating systems.

Since the attitude control throughput requirement is small compared to the payload software, the spacecraft software may co-exist with the payload processing. Although merging the two may reduce the overall number of processors, the system development complexity may increase. Merging them induces tight coupling between the spacecraft and payload, and may increase system integration complexity. The hard-real-time deadlines associated with spacecraft and payload processing are unrelated, yet they will have to be arbitrated. A detailed trade study is needed to address the life cycle of the system, from conceptual design through development, integration, test, and orbit insertion.

For software cost, the most useful indicator is the number of source lines of code, or SLOCs. Various methods exist for determining overall effort and development duration from the total number of SLOCs and software productivity rates (See Table 20-10 in Sec. 20.3 or Boehm [1984]). When determining the total SLOC, consider only the software that we must develop. In the payload processing example, we concentrated on the executable SLOC. For overall costing, we must remember to add in the SLOCs for startup, system initialization, mode transition, fault responses, and other identified functions. In the attitude control example, our estimation method did not use SLOCs. For estimates made by similarity, the SLOC counts should be developed in the same manner as memory and throughput. If SLOC information is not available, then an order-of-magnitude estimate can be generated using the information in Table 16-10 to convert a memory estimate to equivalent SLOCS.

References

Anderson, Christine and Merlin Dorfman, eds. 1991. *Aerospace Software Engineering* Washington, DC: American Institute of Aeronautics and Astronautics.

Beckert, Beverly A. 1993. "The Lowdown on Benchmarks." *Computer Aided Engineering* 112(6):16(1), June.

Boehm, B.W. 1981. *Software Engineering Economics.* New Jersey: Prentice-Hall, Inc.

Boehm, B.W. 1984. "Software Life Cycle Factors," in *Handbook of Software Engineering.* ed. C.R. Vick and C.V. Ramamoorthy. New York: Van Nostrand Reinhold Co.

Hennessy, John L. and David A. Patterson. 1995. *Computer Architecture: A Quantitative Approach (2nd Edition).* San Francisco, CA: Morgan Kaufmann Publishers.

Humphrey, Watts S. 1995. *A Discipline for Software Engineering (SEI Series in Software Engineering).* Reading, MA: Addison-Wesley Publishing Co.

Kaner, Cem, Jack Falk, and Hung Quoc Nguyen. 1993. *Testing Computer Software.* Scottsdale, AZ: The Coriolis Group.

Lane, Malcolm G. and James D. Mooney. 1988. *A Practical Approach to Operating Systems.* Boston: Boyd & Fraser.

Magnus, Chris. 1992. *Software for Fault Tolerant Systems: A Development Handbook.* Prepared for The National Center for Dependable Systems.

Rullo, T.A., ed. 1980. *Advances in Computer Programming Management.* Philadelphia: Heyden & Sons, Inc.

Santoni, Andy. 1997. "Standard Tests for Embedded Processors to be Set by Industry." *InfoWorld* 19(31): 34(1).

Silberschatz, Abraham and Peter Baer Galvin. 1997. Operating System Concepts. Reading, MA: Addison Wesley Publishing Co.

Stankovic, John A. and Krithi Ramamritham. 1988. *Tutorial: Hard Real-Time Systems*. IEEE Catalog Number EH0276-6. Washington, DC: Computer Society Press of the IEEE.

Vick, C.R. and C.V. Ramamoorthy, eds. 1984. *Handbook of Software Engineering*. New York: Van Nostrand Reinhold Co.

Yourdon, E. 1989. *Modern Structured Analysis*. New Jersey: Yourdon.

Chapter 17

Space Propulsion Systems

Robert L. Sackheim, *TRW, Inc.*
Sidney Zafran, *TRW, Inc.*

17.1 Propulsion Subsystem Selection and Sizing
17.2 Basics of Rocket Propulsion
17.3 Types of Rockets
    Liquid Rockets; Solid Rockets; Hybrid Rockets;
    Electric Propulsion
17.4 Component Selection and Sizing
17.5 Staging

In the broadest sense, space propulsion systems do three things. They lift the launch vehicle and its payload from the launch pad and place the payload into low-Earth orbit. They transfer payloads from low-Earth orbits into higher orbits for mission operations or into trajectories for planetary encounters. Finally, they provide thrust for attitude control and orbit corrections. Table 17-1 lists the specific functions these systems perform during various mission phases and some typical performance requirements.

Performance requirements for propulsion systems include thrust, total impulse, and duty cycle specifications derived from mission profiles. Individual designs must meet other performance requirements, such as operating pressure, and internal and external leakage. Other specifications include physical characteristics, propellant, and mass properties. Configuration requirements include envelope dimensions, thruster locations, and alignment. In addition, plume efflux is frequently a design driver for payloads sensitive to contamination.

Once a payload is placed in low-Earth orbit by the launcher, an upper stage or onboard spacecraft integral propulsion system (see Table 17-2) is frequently used to transfer the payload to its operational orbit. These in-space propulsion system designs, especially their weight, size, and volume, are strongly driven by the performance and weight efficiency of the primary propulsion system. The specific impulse, propellant density, and overall stage mass fraction of the primary propulsion system are key parameters. Table 18-5 in Sec. 18.2 summarizes key features and performance characteristics of existing and planned upper stages for use with various launch vehicles.

With the heavy emphasis on driving down the cost of access to space, while still staying within the capabilities of the current fleet of launch vehicles, many space systems' prime contractors are designing and emphasizing higher efficiency and

TABLE 17-1.   **Typical Functions and Requirements for Space Propulsion.** The change in velocity required for orbit maneuvers is called ΔV. See Table 7-3 in Sec. 7.3 for specific ΔV requirements.

Propulsion Function	Typical Requirement
Orbit transfer to GEO (orbit insertion)	
• Perigee burn	2,400 m/s
• Apogee burn	1,500 (low inclination) to 1,800 m/s (high inclination)
Initial spinup	1 to 60 rpm
LEO to higher orbit raising ΔV	60 to 1,500 m/s
• Drag-makeup ΔV	60 to 500 m/s
• Controlled-reentry ΔV	120 to 150 m/s
Acceleration to escape velocity from LEO parking orbit	3,600 to 4,000 m/s into planetary trajectory
On-orbit operations (orbit maintenance)	
• Despin	60 to 0 rpm
• Spin control	±1 to ±5 rpm
• Orbit correction ΔV	15 to 75 m/s per year
• East-West stationkeeping ΔV	3 to 6 m/s per year
• North-South stationkeeping ΔV	45 to 55 m/s per year
• Survivability or evasive maneuvers (highly variable) ΔV	150 to 4,600 m/s
Attitude control	3–10% of total propellant mass
• Acquisition of Sun, Earth, Star	Low total impulse, typically <5,000 N•s, 1 K to 10 K pulses, 0.01 to 5.0 sec pulse width
• On-orbit normal mode control with 3-axis stabilization, limit cycle	100 K to 200 K pulses, minimum impulse bit of 0.01 N•s, 0.01 to 0.25 sec pulse width
• Precession control (spinners only)	Low total impulse, typically <7,000 N•s, 1 K to 10 K pulses, 0.02 to 0.20 sec pulse width
• Momentum management (wheel unloading)	5 to 10 pulse trains every few days, 0.02 to 0.10 sec pulse width
• 3-axis control during ΔV	On/off pulsing, 10 K to 100 K pulses, 0.05 to 0.20 sec pulse width

low-cost, onboard spacecraft *integral propulsion systems* (IPS). The functions of the onboard IPS are to provide much of the propulsion energy for ascent from low-Earth orbit to higher operational orbits and for controlled de-orbit/reentry at the end of mission life, in addition to the normal on-orbit attitude and velocity control propulsion functions. Spacecraft IPS have become almost routine for most designs. More than 85% of all spacecraft launched have operational orbits of greater than 500 km. In addition, new international regulations make controlled de-orbit a mandatory requirement which increases the onboard IPS energy requirements.

With this need for higher onboard IPS propulsion energy almost all spacecraft today employ one of three IPS designs or some combination: (1) storable bipropellant (i.e., employing N_2O_4 and monomethylhydrazine, or MMH, as the propellants); (2) Dual mode propulsion (employing N_2O_4 and hydrazine as high performance bipropellants for ΔV and monopropellant attitude control); (3) one of several different electric propulsion options, depending upon power available, burn time vs. weight requirements and considerations such as propellant commonality, volume limitations and electromagnetic issues. Typical spacecraft electric propulsion options are resisto-jets (delivering ~300 sec I_{sp} using decomposed hydrazine), arcjets (delivering ~500–700 sec I_{sp}, using decomposed hydrazine or ammonia), Hall effect thrusters (delivering ~1,500–2,000 sec I_{sp} using Xenon and derived from Russian technology applied to Russian satellites such as GALS and GONAS), and ion propulsion (delivers ~2,000–3,000 sec I_{sp}, also using Xenon). As more and more of the orbit raising impulse is incorporated on board the spacecraft instead of using a separate upper stage, modern designs are typically using an optimized combination of chemical and electric or hybrid onboard propulsion.

17.1 Propulsion Subsystem Selection and Sizing

The process for selecting and sizing the elements of the propulsion subsystem is shown in Table 17-2. We must carefully estimate the key performance requirements in steps 1 to 3 since they have the greatest impact on operation, weight, and cost. In step 4 we identify as many reasonable options as possible, and pare the list down to a few that have merit. We then proceed to step 5 to develop sufficient detail to estimate the performance, mass, and cost of each option.

TABLE 17-2. Propulsion Subsystem Selection and Sizing Process.

Step	Description of Process	Reference
1	List applicable spacecraft propulsion functions, e.g., orbit insertion, orbit maintenance, attitude control, and controlled de-orbit or reentry	Table 17-1
2	Determine ΔV budget and thrust level constraints for orbit insertion and maintenance	Sec. 7.3
3	Determine total impulse for attitude control, thrust levels for control authority, duty cycles (% on/off, total number of cycles) and mission life requirements	Table 17-1, Sec. 10.3, Chap. 11
4	Determine propulsion system options: • Combined or separate propulsion systems for orbit and attitude control • High vs. low thrust • Liquid vs. solid vs. electric propulsion technology	Secs. 17.1, 17.3
5	Estimate key parameters for each option • Effective I_{sp} for orbit and attitude control • Propellant mass • Propellant and pressurant volume • Configure the subsystem and create equipment list	Secs. 17.1, 17.3 Table 17-3 Tables 17-4, 17-6, 17-7
6	Estimate total mass and power for each option	Table 10-7
7	Establish baseline propulsion subsystem	
8	Document results and iterate as required	

Table 17-3 lists the primary options. Cold gas propulsion systems are inexpensive, low performance systems that are rarely used unless there is an overriding requirement to avoid the hot gases and safety concerns of liquid and solid systems. Solid propellant-based systems have been used extensively for orbit insertion, but the spacecraft propulsion subsystem must be augmented with another technology to provide orbit maintenance and attitude control. Liquid systems are divided into monopropellant and bipropellant systems with a third alternative, dual mode, that is a bipropellant derivative. Monopropellant systems have successfully provided orbit maintenance and attitude control functions, but lack the performance to provide high efficiency large ΔV maneuvers required for orbit insertion. Bipropellant systems are attractive because they can provide all three functions with one higher performance system, but they are more complex than the historic solid rocket and monopropellant combined systems.

TABLE 17-3. Principal Options for Spacecraft Propulsion Subsystems. See Sec. 17.3 for a definition of specific rocket types.

Propulsion Technology	Orbit Insertion		Orbit Maintenance and Maneuvering	Attitude Control	Typical Steady State I_{sp} (s)
	Perigee	Apogee			
Cold Gas			✔	✔	30–70
Solid	✔	✔			280–300
Liquid					
Monopropellant			✔	✔	220–240
Bipropellant	✔	✔	✔	✔	305–310
Dual mode	✔	✔	✔	✔	313–322
Hybrid	✔	✔	✔		250–340
Electric			✔	✔	300–3,000

Dual-mode systems are integrated mono and bipropellant systems fed by common fuel tanks. These systems are actually hybrid designs that use hydrazine (N_2H_4) both as a fuel for high performance bipropellant engines (i.e., N_2O_4/N_2H_4) and as a monopropellant with conventional low-thrust catalytic thrusters. The hydrazine is fed to both the bipropellant engines and the monopropellant thrusters from the common fuel tank. In this manner, high specific impulse is provided for long ΔV burns at high thrust (e.g., apogee circularization) and reliable, precise, minimum-impulse burns are provided by the monopropellant thrusters for attitude control. An additional capability to enhance dual-mode propulsion is the development of a bimodal thrust device which can operate either as a simple catalytic monopropellant thruster or as a high performance bipropellant thruster known as the Secondary Combustion Augmented Thruster or SCAT shown in Fig. 17-1.

Practical considerations may restrict the propellant choices to those that are readily available, storable, and easy to handle. Also, we must trade the lead time needed to develop new hardware against the combination of existing components or stages. Finally, limits on payload acceleration may dictate the maximum permissible thrust levels.

17.2 Basics of Rocket Propulsion

Two basic parameters of rocket engine design are thrust and specific impulse. *Thrust*, **F**, is the amount of force applied to the rocket based on the expulsion of gases:

$$\mathbf{F} = \dot{m}\,\mathbf{V}_e + A_e\left[P_e - P_\infty\right] \approx \dot{m}\mathbf{V}_e \qquad (17\text{-}1)$$

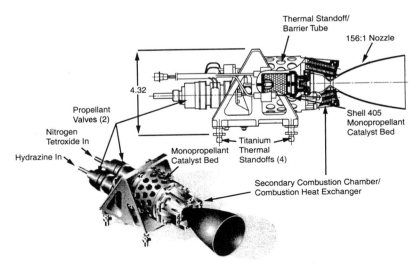

Fig. 17-1.   Secondary Combustion Augmented Thruster (SCAT). Bimodal spacecraft attitude and velocity control RCS thruster.

where A_e is nozzle exit area, P_e is the gas pressure at the nozzle exit, P_∞ is the ambient pressure, V_e is propellant exhaust velocity, and $\dot{m}$ is propellant mass flow rate. We simplify this expression by defining an *effective exhaust velocity,* **C**, as:

$$\mathbf{C} \equiv V_e + \frac{A_e}{\dot{m}} \left[P_e - P_\infty \right] \qquad (17\text{-}2)$$

Equation (17-1) then reduces to:

$$\mathbf{F} = \dot{m}\mathbf{C} \qquad (17\text{-}3)$$

At very high altitudes and in space, P_∞ is essentially zero; at lower altitudes, a rocket engine's thrust increases with altitude until the vehicle leaves the atmosphere.

Specific impulse, I_{sp}, is the ratio of the thrust, **F**, to the weight flow rate, $\dot{m}g$, of propellant.

$$I_{sp} \equiv \mathbf{F}/\dot{m}g \qquad (17\text{-}4)$$

I_{sp} is a measure of the energy content of the propellants, and how efficiently it is converted into thrust. For a chemical rocket, I_{sp} is directly proportional to the square root of the ratio of the chamber temperature, T_c, to the average molecular weight of the exhaust gases, M, as follows:

$$I_{sp} = K\sqrt{T_c / M} \qquad (17\text{-}5)$$

where K is a proportionality constant depending on the ratio of specific heats of the exhaust gas and the engine pressure ratio. This important relationship shows that we can maximize specific impulse by matching the highest possible total temperature with the lowest average molecular weight of the combustion products.

The primary measure of propulsion system performance capability is the velocity change, ΔV, that it can produce. We quantify this relationship by the *rocket equation:**

$$\Delta V = g\, I_{sp} \ln\left(\frac{m_o}{m_o - m_p}\right) \equiv g\, I_{sp} \ln\left(\frac{m_o}{m_f}\right) \equiv g\, I_{sp} \ln(R) \qquad (17\text{-}6)$$

where $m_f \equiv m_o - m_p$ is the final vehicle mass, m_o is the initial vehicle mass, m_p is the mass of the propellant consumed, and $R \equiv m_o / m_f$ is the *mass ratio*. This equation assumes zero losses due to drag and gravity, and is thus a limiting ideal case. In practice, the ΔV achieved will be somewhat smaller. Gravity and drag losses for launch vehicles are typically 1,500–2,000 m/s.

Another form of Eq. (17-6) provides the mass of propellant required for a given increment of velocity, ΔV:

$$m_p = m_f\left[e^{\left(\Delta V / I_{sp} g\right)} - 1\right] = m_o\left[1 - e^{-\left(\Delta V / I_{sp} g\right)}\right] \qquad (17\text{-}7)$$

This equation allows us to calculate the mass of propellant required based on either the initial or final mass of the rocket.

We obtain the highest thrust when exit pressure equals ambient pressure. Although Eq. (17-1) suggests that greater thrust can be obtained with an exit pressure greater than the ambient pressure, the exhaust velocity is reduced, resulting in a loss of thrust. As a result, we design rocket exhaust nozzles with an exit pressure equal to the ambient pressure whenever possible. The exit pressure is governed by the *nozzle-area expansion ratio*:

$$\varepsilon = A_e / A_t \qquad (17\text{-}8)$$

where A_e is the nozzle exit area and A_t is the area of the throat of the nozzle. Note that as the expansion ratio increases, the nozzle exit pressure decreases.

For launch vehicles (particularly first stages) where the ambient pressure varies during the burn period, trajectory computations are performed to determine the optimum exit pressure. However, an additional constraint is the maximum allowable diameter for the nozzle exit cone, which in some cases is the limiting constraint. This is especially true on stages other than the first, where the nozzle diameter may not be larger than the outer diameter of the stage below.

For space engines, where the ambient pressure is zero, thrust is always higher with larger nozzle expansion ratios. On these engines, we increase the nozzle expansion ratio until the additional weight of the longer nozzle (and increase in real nozzle internal boundary layer or drag losses) costs more performance than the extra thrust it generates.

Another important relationship for evaluating rocket performance involves two key Figures of Merit, the characteristic velocity of the combustion gases and the thrust coefficient. The *characteristic velocity*, **C***, is a measure of the energy available from the combustion process and is given by:

---

* The rocket equation was developed in the late nineteenth century and first published by Konstantin Tsiolkovsky [1903], a deaf Russian schoolteacher who was the first to develop much of the mathematical theory of modern rocketry. The most dramatic crater on the far side of the Moon was named *Tsiolkovsky* following its discovery in 1959 by the Soviet probe Luna 3.

$$C^* = \frac{P_c A_t}{\dot{m}} \qquad (17\text{-}9)$$

where A_t is area of the nozzle throat, and P_c is the combustion chamber pressure. Delivered values of C^* range from 1,333 m/s for monopropellant hydrazine (N_2H_4) to 1,640 m/s for Earth-storable bipropellants (N_2O_4/MMH) and up to 2,360 m/s for cryogenic LO_2/LH_2. The *thrust coefficient*, C_f, is a measure of the efficiency of converting the energy to exhaust velocity and characterizes the nozzle performance.

$$C_f \equiv \frac{F}{P_c A_t} \qquad (17\text{-}10)$$

Representative values for C_f are 1.6 (for nozzle expansion ratio, $\varepsilon = 30{:}1$) and 1.86 (for $\varepsilon = 200{:}1$). The product of these two Figures of Merit, divided by the gravity constant, gives the specific impulse:

$$I_{sp} = F/\dot{m}g = C^* C_f /g \qquad (17\text{-}11)$$

The physical parameters that interact to produce hot gases and associated thrust inside a chemical rocket chamber are illustrated in Fig. 17-2. Further information is given by NASA [1963].

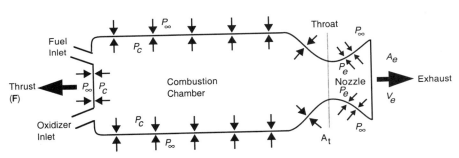

Fig. 17-2.   Simplified Diagram of a Chemical Rocket. Combustion of fuel and oxidizer in the combustion chamber produces expansion of the reaction product gases which are then expelled through the nozzle. The differential between pressure inside the chamber (P_c) and at the nozzle exit (P_e) produces a reaction that propels the chamber and the vehicle in a direction opposite to that of the exhaust gases. Atmospheric pressure (P_∞) reduces the effective thrust (F), so that a rocket engine actually works more efficiently in outer space than in the Earth's atmosphere. This simplified combustion chamber diagram is also representative of a monopropellant rocket (one fuel inlet only and the upper portion of the chamber is packed with catalyst), and of a solid rocket (where there are no inlets because both the fuel and oxidizer are preloaded in the chamber and burn only upon command for ignition).

17.3   Types of Rockets

Currently available space propulsion systems can be categorized as either *cold gas*, *chemical*, or *electric*. More exotic types, such as nuclear, solar sails or beamed-energy approaches may be feasible someday, but their consideration is beyond the scope of this book. Table 17-4 and Table 17-5 list the general characteristics of several types of propulsion systems.

TABLE 17-4. **Performance and Operating Characteristics of Propellants and Energy Sources for Spacecraft Propulsion Systems.** These propellant and energy sources start with the low-efficiency cold gas system and work up to the highest efficiency, space-qualified system, electromagnetic.

Type	Propellant	Energy	Vacuum I_{sp} (sec)	Thrust Range (N)	Thrust Range (lb_f)	Avg Bulk Density (g/cm^3)
Cold Gas	N_2, NH_3, Freon, helium	High pressure	50–75	0.05–200	0.01–50	0.28*, 0.60, 0.96*
Solid Motor	†	Chemical	280–300	50–5 × 10^6	10–10^6	1.80
Liquid:						
Monopropellant	H_2O_2, N_2H_4	Exothermic decomposition	150–225	0.05–0.5	0.01–0.1	1.44, 1.0
Bipropellant	O_2 and RP-1	Chemical	350	5–5 × 10^6	1–10^6	1.14 and 0.80
	O_2 and H_2	Chemical	450	5–5 × 10^6	1–10^6	1.14 and 0.07
	N_2O_4 and MMH (N_2H_4, UDMH)	Chemical	300–340	5– 5 × 10^6	1–10^6	1.43 and 0.86 (1.0, 0.79)
	F_2 and N_2H_4	Chemical	425	5–5 × 10^6	1–10^6	1.5 and 1.0
	OF_2 and B_2H_6	Chemical	430	5–5 × 10^6	1–10^6	1.5 and 0.44
	ClF_5 and N_2H_4	Chemical	350	5–5 × 10^6	1–10^6	1.9 and 1.0
Dual Mode	N_2O_4/N_2H_4	Chemical	330	3–200	—	1.9 and 1.0
Water Electrolysis	$H_2O{\rightarrow}H_2 + O_2$	Electric / chemical	340–380	50–500	10–100	1.0
Hybrid	O_2 and rubber	Chemical	225	225–3.5 × 10^5	50–75,000	1.14 and 1.5
Electrothermal:						
Resistojet	N_2, NH_3, N_2H_4, H_2	Resistive heating	150–700	0.005–0.5	0.001–0.1	0.28*, 0.60, 1.0, 0.019*
Arcjet	NH_3, N_2H_4, H_2	Electric arc heating	450–1,500	0.05–5	0.01–1	0.60, 1.0, 0.019*
Electrostatic:						
Ion	Hg/A/Xe/Cs	Electrostatic	2,000– 6,000	5 × 10^{-6}–0.5	10^{-6}– 0.1	13.5/0.44*/2.73* /1.87
Colloid	Glycerine	Electrostatic	1,200	5 × 10^{-6}–0.05	10^{-6}–0.01	1.26
Hall Effect Thruster	Xenon	Electrostatic	1,500– 2,500	5 × 10^{-6}–0.1	10^{-6}–0.02	0.22
Electromagnetic:						
MPD‡	Argon	Magnetic	2,000	25–200	5–50	0.44*
Pulsed Plasma	Teflon	Magnetic	1,500	5 × 10^{-6}–0.005	10^{-6}–0.001	2.2
Pulsed Inductive	Argon N_2H_4	Magnetic Magnetic	4,000 2,500	2–200 2–200	0.5–50 0.5–50	0.44 1.0

*   Gas densities at standard conditions of pressure and temperature
†   Several types in use: Organic polymers + ammonium perchlorate + powdered aluminum.
‡   MPD ≡ magnetoplasmadynamic

Cold gas propulsion is just a controlled, pressurized gas source and a nozzle. It represents the simplest form of rocket engine. Cold gas has many applications where simplicity is more important than high performance. The *Manned Maneuvering Unit* used by astronauts is an example of such a system.

Chemical *combustion* systems are the most common systems for space applications. They can be divided into three basic categories: *liquid, solid,* and *hybrid.* The terminology refers to the physical state of the stored propellants. Typically, rockets using solid propellants are called *motors* and rockets using liquids are called *engines* or *thrusters.*

TABLE 17-5. Advantages and Disadvantages of Propellants and Energy Sources for Spacecraft Propulsion Systems.

Type	Propellant	Advantages	Disadvantages
Cold Gas	N_2, NH_3, Freon, helium	Extremely simple, reliable, very low cost	Very low performance, heaviest of all systems for given performance level
Solid Motor		Simple, reliable, relatively low cost	Limited performance, higher thrust, safety issues; performance not adjustable
Liquid:			
Monopropellant	H_2O_2, N_2H_4	Simple, reliable, low-cost	Low performance, higher weight than bipropellant
Bipropellant	O_2 and RP-1	High performance	More complicated system
	O_2 and H_2	Very high performance	Cryogenic, complicated
	N_2O_4 and MMH (N_2H_4, UDMH)	Storable, good performance	Complicated
	F_2 and N_2H_4	Very high performance	Toxic, dangerous, complicated
	OF_2 and B_2H_6	Very high performance	Toxic, dangerous, complicated
	ClF_5 and N_2H_4	High performance	Toxic, dangerous
Dual Mode	N_2O_4/N_2H_4	High performance	Toxic, dangerous
Water Electrolysis	$H_2O \rightarrow H_2 + O_2$	High performance	Complicated, not developed, high power
Hybrid	O_2 and rubber	Throttleable, nonexplosive; nontoxic, restartable	Requires oxidizer fuel system; bulkier than solids
Electrothermal:			
Resistojet	N_2, NH_3, N_2H_4, H_2	High performance, low power, simple feed system	More complicated interfaces, more power than chemical; low thrust
Arcjet	NH_3, N_2H_4, H_2	High performance, simple feed system	High power, complicated interfaces (especially thermal)
Electrostatic:			
Ion	Hg/A/Xe/Cs	Very high performance	Very high power, low thrust, complicated, not well developed
Colloid	Glycerine	Moderately high performance	High development risk, high power, complicated
Hall Effect Thruster	Xenon	High performance, relatively high power/thrust density	High development risk, high power, complicated
Electromagnetic:			
MPD	Argon	Very high performance	Very high power, high development risk, expensive, complicated
Pulsed Plasma	Teflon	High performance	Low thrust, high power, contamination, complicated
Pulsed Inductive	N_2H_4 Argon	Very high performance, moderate thrust	High develop. risk, complicated, expensive, very high power

17.3.1 Liquid Rockets

In a *liquid rocket system* propellants are stored as liquids in tanks and fed on demand into the combustion chamber by gas pressurization or a pump. *Bipropellant engines* chemically react a fuel and an oxidizer, and *monopropellant engines* catalytically decompose a single propellant. Bipropellant engines deliver a higher specific impulse but involve additional system complexity and cost. Table 17-6 shows liquid rocket engines used for upper stages or integral propulsion systems on spacecraft.

TABLE 17-6.   Examples of Available Liquid Rocket Engines. For up-to-date and more detailed information, contact the developer.

Engine	Developer	Nominal Thrust (N)	Spec. Impulse (sec)	Propellants	Oper. Life (sec)	Engine Mass (kg)	Status
XLR-132	Rocketdyne	1.67×10^4	340	N_2O_4/MMH	5,000	51.26	In development
Transtar	Aerojet	1.67×10^4	330–338	N_2O_4/MMH	5,400	57.15	In development
Transtage	Aerojet	3.56×10^4	315	N_2O_4/A-50	1,000	107.95	Flown
Delta-II	Aerojet	4.36×10^4	320	N_2O_4/MMH	1,200	99.79	Flown
R-4D	Marquardt	4.00×10^3	309	N_2O_4/MMH	25,000	7.26	Qualified
OME/UR	Aerojet	2.67×10^4	340	N_2O_4/MMH	1,200	90.72	Modified Orbiter maneuvering engine
RL10-A	Pratt & Whitney	7.34×10^4	446	LO_2/LH_2	400	138.35	Flight qualified (Centaur)
DM/LAE	TRW	4.45×10^2	315	N_2O_4/N_4H_4	15,000	4.54	Flown
R4-D	Marquardt	4.89×10^2	310	N_2O_4/MMH	20,000	3.76	Flown
R42	Marquardt	8.90×10^2	305	MON-3/MMH	15,000	4.54	Qualified
MMBPS	TRW	4.45×10^2	302	N_2O_4/MMH	20,000	5.22	Flight qualified
RS-41	Rocketdyne	1.11×10^4	312	N_2O_4/MMH	2,000	113.40	Flight qualified (Peacekeeper)
ADLAE	TRW	4.45×10^2	330	N_2O_4/N_2H_4	28,000	4.50	In qual.
Chandra X-Ray Observatory	TRW	4.25×10^3	322.5	N_2O_4/N_2H_4	25,000	4.5	Flight qualified
HS 601 AKE	ARC/LPG	4.89×10^2	312	N_2O_4/MMH	10,000	4.08	In development
R-40A	Marquardt	4.00×10^3	309	N_2O_4/MMH	25,000	7.26	Qualified (mod. of Shuttle RCS engine)
HPLAM	TRW	4.45×10^2	325	N_2O_4/MMH	30,000	4.60	In advanced development

Liquid Bipropellant Engines. Figure 17-3 is a schematic of a bipropellant propulsion system used to change a spacecraft's velocity and to adjust its orbit. It is pressure-fed, uses Earth-storable propellants, and is designed for long life. This system contains one 100-lb$_f$, radiation-cooled liquid-fueled engine that uses N_2O_4 and monomethylhydrazine (MMH) propellants. There are two positive expulsion tanks for the fuel and two for the oxidizer. A pressurant tank stores helium at about 4,000 psia, and a quad-redundant regulator—coupled with a burst disc and relief valve—regulates

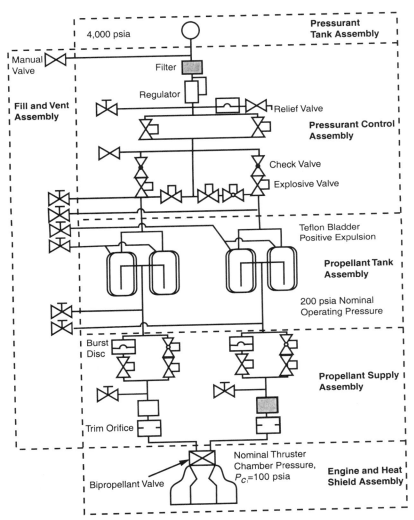

Fig. 17-3. Pressure-Fed Propulsion System Using Earth-Storable Bipropellant (N₂O₄/MMH). Propulsion system designers trade-off reliability and safety with complexity and mass.

flow. Together, they ensure 200 psia feed pressure to the propellant tanks, even after any single regulator failure. Both the fuel and oxidizer tanks can use propellant management devices to feed propellants to the 100 lb$_f$-engine on demand. Burst discs and pyrotechnically actuated squib valves isolate propellants from the engine (and high-pressure gas from the propellant tanks) until the system is ready for operation. Isolating the fluid enhances overall system reliability. This system also has manual fill and drain valves to load propellant and pressurant gas into the system, as well as additional manual valves for system leak checking on both sides of the pyro-isolation valves and regulators. Check valves ensure that fluid flow is only in the correct direction and that the fuel and oxidizer can never mix anywhere in the system, except in the engine. Finally, pressure transducers, filters, temperature sensors, and line and component heaters are provided to ensure proper subsystem operation.

Another type of system uses a cryogenic *topping cycle* engine. The fuel first travels through the thrust chamber's cooling jacket in a technique known as *regenerative cooling*. All of the fuel burns with part of the oxidizer in a high-pressure precombustor. The combustion products provide high-energy gas to drive the engine pump turbines. The total exhaust flow from the turbine is then injected into the main combustion chamber, where it burns with the remaining oxidizer. Because of the precombustor, the cycle lends itself to high-pressure operation, which results in a smaller thrust chamber. The extra pressure drop in the precombustor and turbines requires the pump-discharge pressure of both the fuel and the oxidizer to be much higher than with open-cycle engines. The topping cycle therefore needs heavier and more complex pumps, turbines, and piping. It can, however, provide the highest specific impulse for a given propellant combination.

The Space Shuttle's main engine uses a variation of the topping cycle by employing two separate precombustion chambers, each mounted directly on a separate main turbopump. The oxygen precombustor and turbopump burn an oxygen-rich mixture that expands through the oxidizer turbine to drive the pump. Then the mixture enters the main combustion chamber, where it burns with the fuel-rich mixture from the fuel-precombustor and turbine assembly. The Space Shuttle's main engine develops the highest specific impulse (455 sec at vacuum) of any flight-proven rocket engine using chemical propulsion.

The *expander cycle* is somewhat different, in that the engine pump turbines are driven by gaseous fuel which vaporizes in the thrust chamber cooling jacket. The expander cycle requires no precombustor. An expander cycle engine is shown in Fig. 17-4.

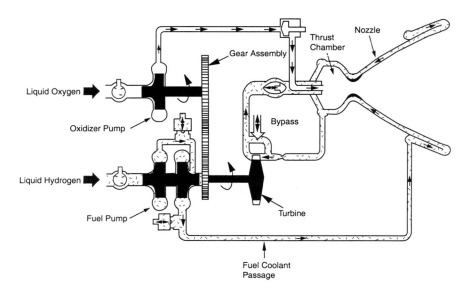

Fig. 17-4.   RL10 Expander Power Cycle. 16,500 lbf thrust engine used for the Centaur LO_2/LH_2 upper stage.

Liquid Monopropellant Engines. By far the most widely used type of propulsion for spacecraft attitude and velocity control is monopropellant hydrazine (N_2H_4). Its excellent handling characteristics, relative stability under normal storage conditions, and clean decomposition products have made it the standard. The general sequence of operations in a hydrazine thruster (Fig. 17-5) is:

- When the attitude-control system signals for thruster operation, an electric-solenoid valve opens, allowing hydrazine to flow. This action may be pulsed (as short as 5 ms) or long duration (steady state).

- The pressure in the propellant tank forces liquid hydrazine into the injector. It enters as a spray into the thrust chamber and contacts the catalyst beds.

- The catalyst bed consists of alumina pellets impregnated with iridium. The most widely used catalyst, manufactured by Shell Oil Company, is called *Shell 405*. Incoming liquid hydrazine heats to its vaporizing point by contact with the catalyst bed and with the hot gases leaving the catalyst particles. The temperature of the hydrazine rises to a point where the rate of its decomposition becomes so high that the chemical reactions are self-sustaining.

- By controlling the flow variables and the geometry of the catalyst chamber, a designer can tailor the proportion of chemical products, the exhaust temperature, the molecular weight, and thus the enthalpy for a given application (Fig. 17-5). For a thruster application where specific impulse is paramount, the designer attempts to provide 30–40% ammonia dissociation, which is about the lowest percentage that can be maintained reliably. For a gas-generator application, where lower temperature gases are usually desired, the designer provides for higher levels of ammonia dissociation.

- Finally, in the space thruster, the hydrazine decomposition products leave the catalyst bed and exit from the chamber through a high expansion ratio exhaust nozzle to produce thrust.

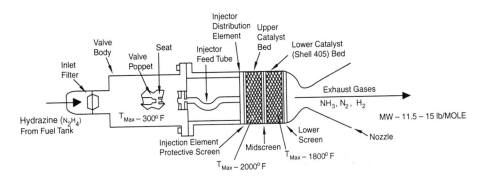

Fig. 17-5. Typical Hydrazine (N_2H_4) Rocket Engine.

Figure 17-6 shows a schematic of the monopropellant-hydrazine system used for the GRO spacecraft. One of the largest hydrazine systems ever built, it contains about 1,800 kg of hydrazine, and is the first such system designed to be refueled in orbit. It operates in a blowdown mode. For this operation, the propellant and pressurant gas are stored in the same tank. As propellant is expelled from the tank, the pressure level drops. The system has eight thrusters operating at a maximum thrust of 30 N for reaction control and four thrusters operating at a maximum thrust of 535 N to adjust the orbit and control altitude. Both sets of hydrazine thrusters are completely redundant for all functions. Both use Shell 405 catalyst and operate as described above. The system also has four large positive-expulsion tanks with elastomeric diaphragms, each holding about 450 kg of hydrazine.

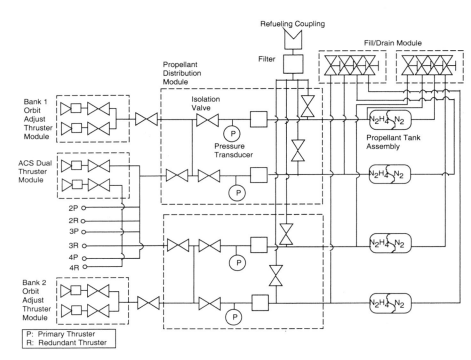

Fig. 17-6.   GRO Propulsion System Schematic. Monopropellant hydrazine fuel. Blowdown fuel system going from 400 to 100 psia tank pressure.

The system is completely two-fault tolerant (a Space Shuttle launch safety requirement) and therefore has three solenoid-latching isolation valves in series for every flow "leg" (between tanks and any one thruster) for a total of 18 latching isolation valves. They are called *latching valves* because they stay in whatever position they were last commanded (that is, open or closed), therefore only requiring power for the command period of approximately 200 ms. The system also has eight manual fill-and-drain valves. Each tank has one propellant and one gas valve to load the hydrazine and GN_2 pressurizing gas on their respective sides of the positive-expulsion tank. The GRO propulsion system also has miscellaneous components such as filters (to keep from contaminating the system), pressure transducers, temperature sensors, and thermostatically controlled catalyst bed and line heaters. These heaters increase thruster lifetime and prevent the propellant from freezing.

17.3.2 Solid Rockets

Solid rockets store propellants in solid form. The fuel is typically powdered aluminum, and the oxidizer is ammonium perchlorate. A synthetic rubber such as polybutadiene holds the fuel and oxidizer powders together. Though lower performing than liquid rockets, the operational simplicity of a solid rocket motor often makes it the propulsion system of choice. Table 17-7 gives representative examples of solid rocket motors.

TABLE 17-7. Representative Solid Rocket Motors. The firm United Technologies/Chemical Systems Division supplies the IUS SRM-1 and 2, as well as the current MinuteMan third-stage motor version of Leasat PKM. Thiokol Corp. supplies the STAR rocket motors as well as the LEASAT PKM.

Motor	Total Impulse (N·s)	Loaded Weight (kg)	Propellant Mass Fraction	Avg. Thrust (lbf)	Avg. Thrust (N)	Max. Thrust (N)	Effective I_{sp} (sec)	Status
IUS SRM-1 (ORBUS-21)	2.81×10^7	10,374	0.94	44,610	198,435	260,488	295.5	Flown
LEASAT PKM	9.26×10^6	3,658	0.91	35,375	157,356	193,200	285.4	Flown
STAR 48A	6.78×10^6	2,559	0.95	17,900	79,623	100,085	283.9	Flown
STAR 48B(S)	5.67×10^6	2,135	0.95	14,845	66,034	70,504	286.2	Qualified
STAR 48B(L)	5.79×10^6	2,141	0.95	15,160	67,435	72,017	292.2	Qualified
STAR 62	7.12×10^6	2,459					293.5	In develop.
STAR 75	2.13×10^7	8,066	0.93	44,608	198,426	242,846	288.0	In develop.
IUS SRM-2 (ORBUS-6)	8.11×10^6	2,995	0.91	18,020	80,157	111,072	303.8	Flown
STAR 13B	1.16×10^5	47	0.88	1,577	7,015	9,608	285.7	Flown
STAR 30BP	1.46×10^6	543	0.94	5,960	26,511	32,027	292.0	Flown
STAR 30C	1.65×10^6	626	0.95	7,140	31,760	37,031	284.6	Flown
STAR 30E	1.78×10^6	667	0.94	7,910	35,185	40,990	289.2	Flown
STAR 37F	3.02×10^6	1,149	0.94	9,911	44,086	49,153	291.0	Flown

Figure 17-7 is a schematic diagram of a typical rocket motor using a solid propellant. This motor, used for geosynchronous spacecraft, provides a circularizing burn when the spacecraft is at apogee, thus placing the vehicle into its operating orbit. The internal grain is shaped in a star configuration, so the grain's surface area will remain relatively constant as the motor burns. We often desire a constant burning area because it produces relatively constant pressure (and thrust) over the full burn period for a predominantly radial burning motor. An igniter in the forward or head end of the motor starts the burn on command from a control system. The igniter, when lit, sends burning particles into the main motor grain. These burning particles fully ignite the rocket motor. A solid motor typically operates with a single start and burns until the propellant is gone [Timnat, 1987].

17.3.3 Hybrid Rockets

A *hybrid rocket* is one in which the propellants are stored in different forms. Normally, the fuel is a solid and the oxidizer is a liquid or gas (Fig. 17-8). Hybrid rockets

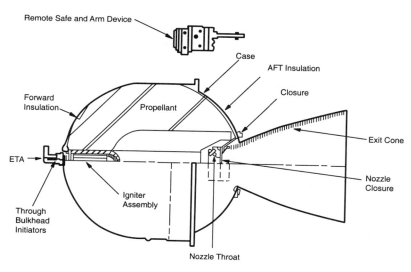

Fig. 17-7. Typical Solid Propellant Rocket Motor. ETA is the Explosive Transfer Assembly.

have several attractive features. These include: (1) **safety**—it is impossible to create an explosive mixture of fuel and oxidizer; (2) **throttling**—we can throttle the engine by modulating the oxidizer flow rate (useful for load alleviation during maximum dynamic pressure and for trajectory shaping). We can idle the engine also (10% thrust) to ensure system operation prior to launch commit; (3) **restart**—we can shut it off and restart it; (4) **storability**—the fuel is storable, as are many oxidizers; (5) **environmentally clean**—unlike solid rockets, hybrids can be made which produce no hydrochloric acid or aluminum oxide exhaust.

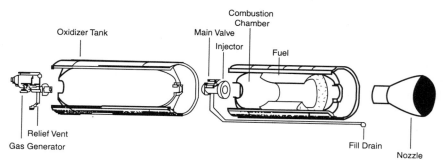

Fig. 17-8. Schematic Drawing of a Hybrid Rocket.

The California Rocket Society built hybrid rockets in the 1930s. The California-based Pacific Rocket Society also conducted research in the mid-1940s, using liquid oxygen with various fuels including wood, wax loaded with carbon black, and rubber. The LOX-rubber combination was the most successful, and was test-flown to an altitude of 9 km in June, 1951 [Altman, 1981].

ONERA in France developed a hybrid sounding rocket using amine fuel and nitric acid oxidizer. The 3.3-m long, 75-kg rocket first flew in 1964. These flights continued for three years, ultimately reaching altitudes in excess of 100 km [Salmon, 1968]. At the same time, United Technologies and Beach Aircraft developed a high-altitude supersonic target drone using *HTPB* (hydroxyl-terminated polybutadiene) fuel and *IRFNA* (inhibited red fuming nitric acid) oxidizer. This air-launched *Sandpiper* missile (later renamed HAST) first flew in 1968 and had a range in excess of 150 km [Altman, 1981].

Hybrid rocket technology has progressed slowly since the 1960s. As of 1990, engines with thrust levels of 75,000 lb$_f$ have been demonstrated in ground tests [Guthrie and Wolf, 1990]. Engines with thrust of 3,000,000 lb$_f$ have been proposed, but significant technical hurdles remain before such large hybrids will be feasible [Jensen, 1990]. Table 17-8 gives characteristics of several hybrid rockets.

TABLE 17-8. Representative Hybrid Rockets.

Motor	Average Thrust (lb$_f$)	Average Thrust (kN)	Burn Duration (sec)	Fuel	Oxidizer	Comments
American Rocket Company						
H-500	75,000	333	70	HTPB	LOx	Qualified for flight
H-250	32,000	142		HTPB	LOx	In development
H-50	10,000	44		HTPB	LOx	In development
U-50	6,500	29		HTPB	LOx	In development
U-1	100	0.44		HTPB	LOx	In development
United Technologies						
	40,000	178	300	HTPB	IRFNA	Flown on Firebolt air-launched target drone, 1968
StarsTruck						
	40,000	178		CTBN	LOx	Flown on Dolphin water-launched sounding rocket, 1984
USAF Academy						
H-1	55	0.25	2.3	HTPB	GOx	Flown on 4-ft tall rocket for student project, 1991

17.3.4 Electric Propulsion

Electric propulsion uses externally provided electrical power either from the Sun (converted through photovoltaic solar arrays—100% to date) or from nuclear or thermodynamic conversion thermal engines, to accelerate the working fluid to produce useful thrust. For example, in an *ion engine*, an electric field accelerates charged particles which exit at high velocity. Alternatively, in a *magnetoplasmadynamic*, or *MPD thruster*, a current-carrying plasma interacts with a magnetic field resulting in a Lorentz acceleration to expel the plasma.

There is no fundamental limit (other than the speed of light) to the exhaust velocity that we can obtain with an electric rocket. However, the power required may grow to the point where further acceleration is pointless. There is therefore an optimum exhaust velocity, and hence an optimum specific impulse, which depends largely on the electric power subsystem [Stuhlinger, 1964].

The propulsion system weight varies with the specific impulse (exhaust velocity), thrust level, and total impulse. Propellant weight clearly drops off as specific impulse increases. The power source requirements, however, are proportional to I_{sp}. Thus, the weight of the power source increases with specific impulse, leading to a minimum weight of the combined system (fuel and power source) at a particular value of I_{sp}. We may usually obtain cost savings by operating slightly below the optimum specific impulse, since propellant is usually cheaper than more power supply.

For an electric propulsion device, efficiency is defined as the ratio of kinetic energy expelled to the input energy. For small time intervals, where mass flow is constant, we express efficiency, η, as a power ratio:

$$\eta = \frac{\dot{m}v^2}{2P} \qquad (17\text{-}12)$$

where η is the overall efficiency, $\dot{m}$ is mass flow rate, v is exhaust velocity, and P is the input power. Recognizing that thrust $F = \dot{m}v$, and that specific impulse $I_{sp} = F/\dot{m}g$, where g is the gravitational constant, we rearrange Eq. (17-12) to get power required:

$$P = \frac{F^2}{2\dot{m}\eta} = \frac{FI_{sp}g}{2\eta} \qquad (17\text{-}13)$$

For a given I_{sp} and thrust, we use Eq. (17-13) to estimate power and mass flow rate of the working fluid. Typical values of η are included in Table 17-4, which also includes useful information on selected propellants and energy sources. See Sec. 11.4 for more on the power subsystem, or Clark [1975].

Electric Propulsion Systems Design Concepts

The five electric propulsion (EP) concepts shown on Tables 17-9 and 17-10 have achieved operational status and many programs are underway to increase the number and types of missions served by EP. The following will briefly highlight the characteristics of mature EP systems that have become operational, or for which near-term flight programs are firmly planned, and comment on the potentials of various classes of EP systems. Table 17-9 illustrates the three basic types of electrical energy thrusters. Table 17-10 and Fig. 17-9 show key characteristics of selected, mature EP systems. Figure 17-10 illustrates the fundamental concepts that enable operations for all electric propulsion systems. To mitigate the effects of mission specifics, the system specific mass (in kg/kW) only includes the mass of the thruster and power processor (The masses of the propellant subsystem, gimbals, and other mission specifics are not included.)

Resistojets have been used for North-South stationkeeping and orbit insertion of respectively, communications satellites in the United States and for orbit control and ACS functions on Russian spacecraft. Propellant temperatures are fundamentally determined by material limits in resistojets which implies modest (propellant specific) maxima for specific impulses of about 300 sec for the 0.5–1 kW-class resistojets

TABLE 17-9. Electric Propulsion; Three Classes of Accelerator Concepts.

	Electrothermal	Electrostatic	Electromagnetic
Mechanism	• Gas heated via resistance element or arc and expanded through nozzle • Resistojets • Arcjets	• Ions electrostatically accelerated • Hall effect (HET) • Ion • Field emission	• Plasma accelerated via interaction of current and magnetic field • Pulsed plasma (PPTs) • Magnetoplasmadynamic (MPD) • Pulsed inductive (PIT)
Power	0.4–2 kW	1–50 kW	50 kW–1 MW
Specific Impulse, I_{sp}	300–800 sec	1,000–3,000 sec	2,000–5,000 sec

TABLE 17-10. Characteristics of Selected Electric Propulsion Flight Systems.

Concept	Specific Impulse, (sec)	Input Power, (kW)	Thrust/ Power, (mN/kW)	Specific Mass, (kg/kW)	Propellant	Supplier
Resistojet	296	0.5	743	1.6	N_2H_4	Primex
	299	0.9	905	1	N_2H_4	Primex, TRW
Arcjet	480	0.85	135	3.5	NH_3	IRS/ITT
	502	1.8	138	3.1	N_2H_4	Primex
	>580	2.17	113	2.5	N_2H_4	Primex
	800	26*	—	—	NH_3	TRW, Primex, CTA
Pulsed Plasma Thruster (PPT)	847	< 0.03†	20.8	195	Teflon	JHU/APL
	1,200	< 0.02†	16.1	85	Teflon	Primex, TSNIIMASH, NASA
Hall Effect Thruster (HET)	1,600	1.5	55	7	Xenon	IST, Loral, Fakel
	1,638	1.4*	—	—	Xenon	TSNIIMASH, NASA
	2,042	4.5	54.3	6	Xenon	SPI, KeRC
Ion Thruster (IT)	2,585	0.5	35.6	23.6	Xenon	HAC
	2,906	0.74	37.3	22	Xenon	MELCO, Toshiba
	3,250	0.6	30	25	Xenon	MMS
	3,280	2.5	41	9.1	Xenon	HAC, NASA
	3,400	0.6	25.6	23.7	Xenon	DASA

* Thruster input power.
† Power dependent on pulse rate.

developed by Primex. Resistojets have several desirable features including values of thrust/power far higher than other EP options (due to their high efficiencies and modest specific impulses), the lowest EP system dry masses (primarily due to the lack of a requirement for a power processor), and uncharged/benign plumes. These features will continue to make resistojets attractive for low-to-modest energy applications, especially where power limits and/or thrusting times, and/or plume impacts are mission drivers. In addition, resistojets can operate on a wide variety of propellants which led to their proposed use as a propulsion/waste gas management concept on Space Station and, operated on hydrogen, for Earth-orbit insertions.

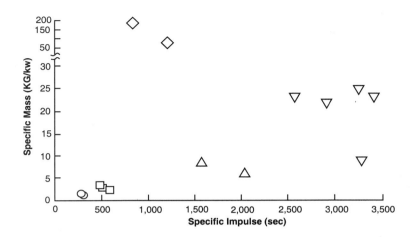

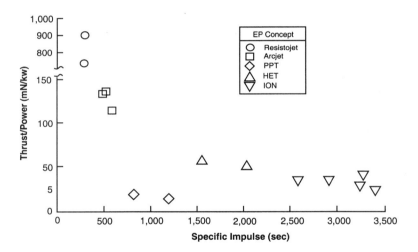

Fig. 17-9. Key Characteristics of Mature Electric Propulsion Systems. (A) Specific Mass and (B) Thrust/Power.

Table 17-10 shows three low-power arcjets. The two hydrazine concepts are produced by Primex and are being used for stationkeeping. The ammonia version was supplied by the Institut fur Raumfahrt Systeme and the Institut fur Thermodynamik und Technische Gebaudeausrustung in Germany and will be used for orbit raising and inclination change of a German amateur radio spacecraft. Arcjets have about twice the specific impulse of resistojets while maintaining some desirable features such as use of standard propellants and relatively low dry masses. The increased specific impulse coupled with relatively low efficiencies of about 0.3 to 0.4, lead to significant decreases (> 6X) of thrust/power relative to resistojets. In addition, as complex plasma/arc phenomena must be controlled, arcjets require relatively complex power conditioning which results in dry masses about twice those of resistojet systems.

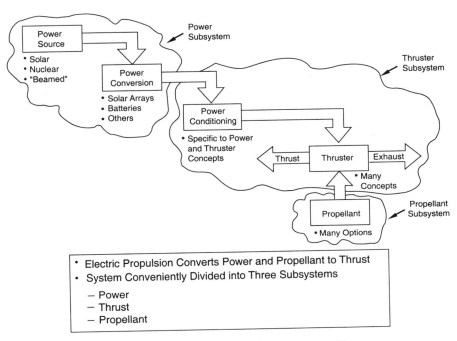

Fig. 17-10. Generic Electric Propulsion System. Functional Block Diagram.

Significant efforts including development of novel materials were necessary to define and validate the 600-sec, hydrazine arcjet. It is likely, therefore, that 600–650 sec represents the upper range of specific impulses that can be expected of low-power arcjets using storable propellants. Arcjets do provide major mass benefits for many spacecraft, are relatively simple to integrate, and are the least complex and costly of any plasma propulsion device. For those reasons, low-power arcjets can be expected to undergo evolutionary improvements and be used well into the future for a variety of medium-to-high energy propulsion functions. Figure 17-11 also shows a 26 kW, ammonia arcjet which operates at a specific impulse of 800 sec. The arcjet is part of a flight system called ESEX, built under a U.S. Air Force program by a TRW, Primex, and CTA Systems team, which includes the arcjet, supporting subsystems, and a diagnostic suite to evaluate plume and EMI effects. The increased specific impulse relative to low-power arcjets is largely due to reduction of losses associated with low Reynolds number flows that are fundamental penalties for low-power arcjets. The space test will represent a greater than tenfold increase in power level of flight-demonstrated EP devices and will address integration and mission issues critical to potential users of high-power (orbit-transfer-class) electric propulsion.

Pulsed plasma thrusters (PPTs) are inherently pulsed devices and versions which operate at about 847 sec specific impulse and were built by the Johns Hopkins University, Applied Physics Laboratory (JHU/APL) have successfully maintained precision control of three NOVA spacecraft for many years. PPTs feature very small ($\leq 4 \times 10^{-4}$ N·s) impulse bit capability, use of a solid propellant (Teflon), and the ability to operate at near constant performance over large power ranges. An improved

Fig. 17-11.   The 26 kW, ESEX Arcjet System.

version PPT which operates up to 1,200 sec specific impulse (Table 17-9), is being developed by a NASA/Primex team and a flight test in 1999 is planned on the Earth Observer 1 spacecraft to demonstrate propulsive ACS. The characteristics of PPTs will likely limit their power operating range to under a few hundred watts and, as suggested by Table 17-10, they have large dry masses. Within their operating capability, however, PPTs promise a combination of low-power, high specific impulse, and small-impulse bit that is unique. It is anticipated that PPTs will find uses for ACS and for modest energy ΔV applications for small spacecraft where the power and thrust limitations of PPTs are acceptable and/or desirable.

    Hall effect thrusters (HETs) and *ion thrusters* (ITs) represent the highest performance EP options and characteristics of mature versions of both concepts are shown in Table 17-10. HETs were developed and flown on dozens of Russian space missions for various functions and are under intense development for use on other nations' spacecraft. Flight-type HETs have been produced by Fakel Enterprise (Fakel), Keldysh Research Center (KeRC), and TSNIIMASH, all of Russia, and quite aggressive HET R&D programs are in place in Europe, Japan, and the U.S. Table 17-10 lists three HET concepts to illustrate the state-of-the-art. The 1,600 sec specific impulse concept was developed to flight ready status by a team including International Space Technology, Inc., Loral, and Fakel; the 1,638 sec I_{sp} device was built, qualified, and delivered for flight test (at reduced levels of power and specific impulse) by a NASA, Primex, TSNIIMASH team; and the high power HET is being built by Space Power Inc. and KeRC for a 1999 flight test on a Russian GEOSAT. In addition, two versions of

1.5 kW-class HETs traceable to the Fakel concept, are planned to provide stationkeeping for nine years on the French Stentor spacecraft. Table 17-11 summarizes the status of HETs.

TABLE 17-11. Hall Effect Thruster (HET) Status Summary.

Concept	Supplier	Power (kW)	I_{sp} (sec)	T (mN)	Demo Life (Khrs)	Maturity	Comment
SPT-100	Fakel (Russia)	1.35	1,500	83	>5.7	Flight	Most mature 1.5 kW-class concept. Multiple life tests >5 Khrs
D-55	TsNIIMASH (Russia)	1.39	1,638	88.6	0.64	Under Development	Several technical deifferences from SPT-100
T-100	KeRC (Russia)	1.29	1,650	80	0.63	Under Development	Nearly identical to SPT-100
SPT-140	Fakel (Russia)	3.0 5.0	1,579 1,929	177 263	–	Under Development	Operated from 1.5 to 5 kW
D-100	TsNIIMASH	3.0	1,849	184	–	Under Development	
T-160E, T-140, T-200	KeRC SPI (USA)	3.0 4.3	1,772 1,909	192 257	– –	Under Development	New high fidelity data from NASA

Five mature ion thrusters are also shown on Table 17-10. The 2,585 sec I_{sp} system was built by the Hughes Aircraft Company and is operational on a commercial COMSAT launched in 1997. The 2,906 sec I_{sp} concept was built by a team of Mitsubishi Electric Corporation and Toshiba Corporation of Japan and was flown on the ETS-VI spacecraft in 1994. An orbit insertion issue prevented the system from performing its planned stationkeeping function but in-space characterizations were performed in 1995 and an identical system will soon be flown on the Japanese COMETS spacecraft. The 3,250 and 3,400 sec I_{sp} systems were built in Europe by teams headed by, respectively, Matra Marconi Space (MMS) and DASA. These devices have been baselined for stationkeeping on the European Space Agency's Artemis spacecraft to be launched in 2000. The 3,280 sec, 2.5 kW device is the highest power, mature ion thruster for which data is available and is used on NASA's New Millennium DS-1 mission.

HETs and ITs are the highest specific impulse options available for mission planners, and many analyses have been conducted to evaluate their use for high energy missions. Comparisons of the two devices are difficult due to the relative lack of maturity of devices built to comparable powers and standards. ITs operate reliably at higher specific impulses than HETs and their performance and specific mass are deeply penalized by operation at specific impulses less than about 2,500 sec, due to the constraints imposed by the ion optics systems. On the other hand, HET systems perform at values of thrust/power 30% or more larger than those of ITs and are considerably lighter but HET operations above about 2,500 sec will pose major lifetime, or redesign challenges. Both concepts eject high-velocity, charged plumes and present approximately the same issues regarding spacecraft integration. Both HETs and ITs provide extreme benefits for emerging space missions and the choice between them will likely be quite mission specific. In general, however, ITs become increasingly beneficial as mission energies increase and HETs appear optimum for many time-constrained situations, typical of Earth-space missions.

17.4 Component Selection and Sizing

The simplest way to feed the propellant (or working fluid) to the thrust chamber on demand is to displace it from its storage vessels with a high-pressure gas. Alternative systems employ a pump, such as a piston or turbopump, which can be driven by turbines, gas pressure intensifiers, or directly by electric motors. Turbine-driven pump assemblies may obtain drive power either from a hot-gas cycle (the most common method) or, in a few cases, from electromechanical actuators that in turn receive power from batteries or solar arrays. A feed system using intensified gas pressure may soon find application in propulsion systems.

We typically use pressure-fed systems for rockets which deliver low to moderate levels of thrust and total impulse (see Fig. 17-12). The simplified propellant feed system reduces the overall weight of the propulsion systems. This simplicity, often resulting in increased reliability, is particularly attractive for spacecraft applications. However, as total impulse and/or thrust requirements increase, the weight of the propellant tanks may become prohibitive. This is because the propellant tanks for a pressure-fed system have to withstand a pressure somewhat higher than the engine combustion pressure, thus they tend to be heavier than those for pump-fed systems. Pump-fed systems are definitely lighter for applications using high thrust and longer total burn, such as for launch vehicles or large upper stages [Fritz and Sackheim, 1982]. Because of these variables, we must conduct design studies and trade-offs to select the best propellant feed approach. The next section includes schematic diagrams of typical liquid-propulsion systems using pressure or pump-fed systems. Sutton [1992] provides a more detailed discussion of turbopump assemblies.

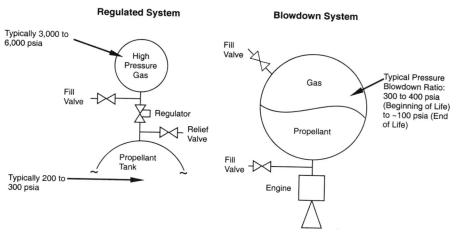

Fig. 17-12. Pressurization Systems.

For gas storage systems, only the pressure vessel, valving, and feed plumbing are required to direct the gaseous propellants under high pressure to the thrust chamber. Liquid storage systems are more complex, needing to manage the liquid propellant under zero gravity to ensure that liquid, rather than gas or vapor, is expelled from the tank. To manage the liquid propellant, we may use *artificial gravity* induced by a spinning spacecraft or by a settling burn from another small rocket, positive expulsion, or a surface-tension device. *Positive expulsion* systems use an active element (a bladder, diaphragm, piston, or bellows) to separate the pressurant gas from the liquid propellants under all dynamic conditions and to force the liquid from the tank into the feed lines on demand. The slightly higher differential gas pressure acting on the expulsion device forces the liquid to flow. Table 17-12 shows basic options using a positive expulsion tank.

TABLE 17-12. Available Options for Positive Expulsion Tanks. ΔP = pressure difference. After deciding that our spacecraft requires a liquid propulsion system, we must decide on system characteristics.

Tank Option	Advantages	Disadvantages	Typical Applications
Metal Diaphragm	• High volume efficiency • Good center of gravity control • No ullage volume • No sliding seals • Proven design	• High weight • High cost • High-expulsion ΔP • Optimizes only for special envelope	• Spacecraft control & maneuvering • Launch vehicles • Upper stages • Missiles
Rolling Diaphragm	• Light weight • Low cost • Low ΔP during expulsion	• Inspection of internal welds	• Missile interceptors • Maneuvering missiles
Piston	• Extensive database • Low ΔP during expulsion • Design adapts easily to growth	• High cost • Low volumetric efficiency • Critical tolerance on shell • Sliding seals possible blowby	• High acceleration missiles
Rubber Diaphragm	• Extensive database • Low ΔP during expulsion • Not cycle limited • Proven design • High expulsion efficiency	• Compatibility limits on propellants	• Spacecraft control & maneuvering • Launch vehicles • Upper stages
Metal Bellows	• No sliding seals • Good center of gravity control • Proven design • Good compatibility • Hermetically sealed	• High weight • High cost • Limited cycle capability • Low volumetric efficiency	• Missiles • Spacecraft • Launch vehicles

Surface tension systems passively manage propellants in a near zero-gravity environment by using vanes, screens, or sponges to wick the propellant into the propellant-tank outlets. In this manner the pressurizing gas bubble is always maintained in the center of the tank. All of these devices rely on surface tension forces to separate liquids from gases [Martin, 1986].

Depending upon the specific mission requirements, a spacecraft can employ from a few to as many as 30 thrust units. These thrusters may have to generate long, steady-state burns for velocity control maneuvers to adjust or maintain the orbit. They may also fire in short pulses of several milliseconds each to control attitude and manage momentum. In some applications, an attitude-control thruster may have to deliver in excess of a million pulses over its lifetime. This wide range of operating conditions places great demands on propulsion systems that control the attitude and velocity of spacecraft. Thus, designers of these thrusters place at least as much emphasis on operating life as on performance. In fact, in some cases absolute performance is secondary to requirements for extended lifetimes and reliability. Table 17-13 presents operating and performance specifications for typical flight-qualified attitude and velocity control thrusters.

Some means of directing the thrust vector may be required to ensure that the thrust vector points through the spacecraft's center of mass. This system must account for center-of-mass shifts as propellant burns and allow for necessary manufacturing tolerances. If the disturbance torques resulting from a misaligned thrust vector are small, the spacecraft's reaction-control system (that is, pulsing thrusters) can overcome them. In other cases, the entire vehicle may be spun intentionally during firings of the main engine to cancel out any misalignment of rocket thrust. If not, we may gimbal the large rocket or only its nozzle. Alternatively, several rocket clusters may provide the main thrust, with opposing rockets turned off briefly to compensate for the disturbance torques. Similarly, we can use a rocket engine's shallow-throttling ability to modulate its disturbances relative to any opposing rocket in the cluster.

Other techniques for thrust-vector control, such as exhaust-jet deflectors (vanes or tabs), are not as common as the ones mentioned above for exoatmospheric use. Table 17-14 lists some representative values for the range of control authority that we can achieve with various techniques for controlling the thrust vector.

To design a propulsion system, we must analyze requirements, trade off design features, and size the system iteratively until we arrive at the best configuration for a particular mission. Analyzing requirements includes performance, interfaces, and physical characteristics. Design trades investigate different types of propulsion systems, selection criteria, and design factors for the specific mission. Using sizing calculations, interface considerations, and safety criteria, we determine the types and quantities of propellants and pressurants, as well as how to configure the tanks, components, instrumentation, and power conditioning (if applicable). All choices must meet design requirements within suitable margins.

The total impulse that the propulsion system must deliver derives from the velocity increment (ΔV) it must impart to the spacecraft for maneuvers, the impulse required by the attitude-control system for limit cycling and managing momentum, and the spacecraft's weight. The allowable offset for center-of-mass affects the number and location of propellant tanks (or solid rockets), which usually occupy most of the propulsion system's volume. The ranges for maximum acceleration establish criteria for thruster sizing, whereas duty cycling determines the type, response time, and operating thermal characteristics of thrusters. Lifetime affects component selections. For

TABLE 17-13. **Flight-Qualified Thrusters Used to Control Spacecraft Attitude and Velocity.** Thruster life is expressed in terms of total impulse or number of pulses.

Nom. Thrust (N)	Nom. Thrust (lb$_f$)	Propellants*	Developers	Total Impulse (10³ N·s)	I_{sp} Range (sec) Steady State	Pulse†	Weight (kg)
0.09–0.22	0.02–0.05	Mono H	TRW, Olin/RRC ERNO, Marquardt, HAC	4–200	205–215	110–180	0.1–0.2
0.09–0.67	0.02–0.150	Mono H‡	TRW, Olin/RRC	90–800	285–320	250–290	0.5–0.9
2.22	0.5	Mono H	Olin/RRC, Marquardt	40–200	215–230	120–200	0.1–0.2
4.45	1.0	Mono H	TRW, SEP, HAC, Marquardt, Olin/RRC	40–1,100	210–230	120–210	0.1–0.2
13–18	3–4	Mono H	TRW, SEP, ERNO	40–1,100	215–235	150–210	0.2–0.3
22–36	5–8	Mono H	TRW, Olin/RRC, Marquardt, HAC	40–1,100	215–240	120–210	0.2–0.3
16–50	4–11	Bimodal	TRW (SCAT)	2.3 × 10⁶	305–325	310	2
45–67	10–15	Mono H	Marquardt, TRW	40–1,300	215–240	120–210	0.3–0.5
111	25	Mono H	Olin/RRC	40–400	215–240	150–220	1.5–1.6
133	30	Mono H	Marquardt	40–300	225–242	150–225	2–3
178–222	40–50	Mono H	Olin/RRC, TRW	40–200	220–245	150–220	1.4–1.8
445–689	100–155	Mono H	Olin/ RRC, Marquardt	1,100	225–245	150–225	1.8–2.3
1,335	300	Mono H	Walter Kidde, Olin/RRC	2,200	225–245	150–225	11.3
2,669	600	Mono H	Olin/RRC	200	225–240	N/A	8.2
11	2.5	N₂O₄/MMH	MMB	200	285	210	0.5
22	5.0	N₂O₄/MMH	Marquardt, ARC/LPG	200–400	290	220	0.7
111	25.0	N₂O₄/MMH	Marquardt	1,100	300	220	1.4
400–489	90–110	N₂O₄/MMH	ARC/LPG	9,000	308	N/A	4.1
450	100	N₂O₄/MMH	MBB	7,000	305	N/A	4.5
440	105	N₂O₄/MMH	TRW	11,000	325	N/A	4.5
450	110	N₂O₄/MMH	Marquardt	900	309	N/A	3.8
445	105	N₂O₄/N₂H₄	TRW	10,000	322	N/A	4.5
440	105	N₂O₄/N₂H₄	TRW	10,000	330	N/A	5.5
450	110	N₂O₄/N₂H₄	Royal Ordnance	7,000	317	N/A	4

* *Mono H* is monopropellant hydrazine (N_2H_4) and *MMH* is monomethyl hydrazine.

† For low duty cycles (<10%) and short pulse width (10–15 ms), use the lower pulsed I_{sp}. For duty cycles >10% and pulse widths greater than 50 ms, use the higher pulsed I_{sp}.

‡ Electrothermally augmented (resistojet/ EHT)

TABLE 17-14.   Representative Control-Authority Ranges for Some Typical Ways to Con
                trol the Thrust Vector.

Method	Typical Control Authority Range
Gimbals	±7 deg
Off-pulsing	20 to 40% of thrust
Shallow Throttling	±10 to 20% of thrust
Exhaust jet Deflectors (jet tabs and jet vanes)	±10 deg
RCS Thruster Control	Full range of attitude-control rates as determined by thrust level, moment arm (torque), and duty cycle

example, thrusters are designed and qualified for specific operating lifetimes. Mainte
nance, or maximum interval for resupply, affects propellant loads and tank sizing.

The propellant budget must include enough propellant to correct for errors stem
ming from injection dispersion. Both the number and accuracy of maneuvers affect th
reserves that must be carried. When a solid rocket performs a ΔV maneuver, especiall
a solid rocket restricted to a single firing, off-nominal performance will result in posi
tion errors and on-orbit velocity errors. For example, the Inertial Upper Stage has
three-sigma, geosynchronous position error of 43 km and a velocity error of 6 m/s.

For a typical design, we first size the propellant load based on the information pro
vided in Table 10-7 of Chap. 10. For initial sizing, we may consider only the first tw
items. Using the rocket equation presented in Sec. 17.2 (Eq. 17-6), we estimate th
propellant load, m_p, from total impulse requirements:

$$m_p = I_t / g \, I_{sp}$$ (17-14

where I_t is total impulse, I_{sp} is the specific impulse, and g is the universal gravity con
stant 32.17 ft/s^2. Knowing the propellant mass, we can determine the propellant vol
ume by dividing by propellant density. For gaseous propellant, density should be at th
storage pressure and maximum anticipated temperature.

Besides expelled propellant mass, many other things contribute to the weight c
propulsion systems. Some examples are thrusters, tanks, fluid components, instrumen
tation, lines, fittings, power conditioning equipment, dedicated power equipmen
pressurant, and residual propellants (depending upon propulsion type). Initial sizin
should take into account known hardware weights. Table 17-13 summarizes some rep
resentative, flight-qualified, spacecraft thrusters. We should use similar data to selec
flight-qualified, off-the-shelf hardware that are close to meeting mission requirements
A schematic or functional block diagram (see Sec. 17.2) builds on the basic concep
and size, as do the equipment list and weight summary. For a conceptual design, w
may use the methods outlined below.

We usually size solid propellant systems so that each motor in the system provide
the total impulse for a single maneuver. Thus, we need only know the mass fraction c
each motor to calculate total system weight, because the weight of the other hardwar
—such as the safe-and-arm device, explosive transfer assembly, and initiators—i
usually small by comparison. Solid rocket motors are typically over 90% propellar
by mass, including case and nozzle assemblies. A value from 91–94% is reasonabl
for present-day solid rocket motors with total impulse greater than 450,000 N·s

Improvements in materials of construction should increase this value to more than 95% by the mid-1990s.

For liquid propulsion systems, we size the tank after determining the propellant load because the tanks are the largest and heaviest components. For bipropellant systems, we determine the oxidizer and fuel requirements from:

$$O/F = m_{ox}/m_{fuel} \qquad (17\text{-}15)$$

where O/F is the *mixture ratio* needed to deliver the required specific impulse, m_{ox} is the oxidizer mass, and m_{fuel} is the mass of the fuel. We calculate tank volumes from propellant volumes loaded into each tank, plus reasonable allowances for *ullage* (gas volume in the propellant tank, approximately 5%), design margin, and propellant remaining in each tank because of trapped liquids or uncertainties in loading and performance. The mixture ratio is a critical parameter in propulsion system sizing; therefore, its selection is the first step in determining the propellant quantities for a given set of propulsion-system requirements (total impulse). Very often, we select the mixture ratio for system benefits other than just maximum I_{sp}. Consider the example of storable-bipropellant systems using N_2O_4/MMH for space systems. They almost always use an O/F ratio of 1.64 because this value results in tanks of equal size. Equal sizing simplifies tank manufacturing, packaging of propulsion systems (configuration layout), and integration.

The O/F ratio at which the engine is operating is defined as:

$$O/F = \dot{m}_{ox}/\dot{m}_{fuel} \qquad (17\text{-}16)$$

where $\dot{m}_{ox}$ is the oxidizer mass flow rate, and $\dot{m}_{fuel}$ is the fuel flow rate. The maximum theoretical value of the characteristic exhaust velocity, C^*, is achieved at a specific mixture ratio. This optimum O/F ratio depends on the particular propellant combination. Usually, we choose the O/F ratio so that the reaction products have the maximum achievable value of T_c/M and thus the highest possible specific impulse [Gordon and McBride, 1976].

In some situations a different O/F ratio results in a better overall system. For a volume-constrained vehicle with a low density fuel such as liquid hydrogen, we can significantly reduce vehicle size by shifting to an oxidizer-rich O/F ratio. In this case, the losses in specific impulse are more than compensated for by the reduced fuel tankage requirement, because combustion performance is not a particularly strong function of mixture ratio. A large orbital transfer vehicle transported to orbit by the Space Shuttle would use this type of mixture ratio, though there may be other situations as well.

Pressurant gas requirements depend on the type of pressurization system employed—regulated or blowdown (Fig.17-12), or some combination of the two. About 5% to 10% ullage is provided in the propellant tanks for pump-fed or regulated pressure systems. The total propellant tank volume for a blowdown system is the total of the propellant volume, V_p, and initial gas volume, V_{gi}, in the tank. They relate through the blowdown ratio, R, as follows:

$$R \equiv V_{gf}/V_{gi} \approx [V_{gi} + V_p]/V_{gi} \qquad (17\text{-}17)$$

where V_{gf} is the final gas volume, neglecting the propellant volume remaining at end-of-life as well as density changes with temperature. Design margin for liquid propellant loads depends on the mission but can be as high as 25% for early conceptual designs. Analyzing the residual propellant can get very detailed, involving statistical and deterministic error sources. A reasonable initial estimate is 5%.

For most systems, we can determine pressurant mass from the perfect gas law, but only when the propellant is withdrawn isothermally (in blowdown systems at low duty cycles). Otherwise, calculating requirements for pressurant mass of regulated systems can become complicated thermodynamically. Using a conservation-of-energy approximation to calculate the pressurant mass yields the following relationship:

$$m_{gi} = \frac{P_p V_p}{RT_i} \left[\frac{k}{1 - (P_g / P_i)} \right] \qquad (17\text{-}18)$$

where m_{gi} is the initial pressurant mass, P_p and V_p are instantaneous gas pressure and gas volume in the propellant tank, P_g (300–600 psia) and P_i (3,000–6,000 psia) are instantaneous gas pressure and initial gas pressure in the pressurant tank, respectively; T_i is initial gas temperature (275–300 K); k is specific heat ratio for a pressurant gas (1.40 for nitrogen, 1.67 for helium); and R is the pressurant gas constant (296.8 J/(kg·K) for nitrogen, 2077.3 J/(kg·K) for helium).

This equation does not apply to very high storage pressures, for which the compressibility factor becomes important. A more complete solution and derivation is in Sutton [1992].

We may estimate the propellant and pressurant gas tank weights using:

$$\sigma = \frac{pr}{t} \; (cylindrical) \qquad (17\text{-}19)$$

and:

$$\sigma = \frac{pr}{2t} \; (spherical) \qquad (17\text{-}20)$$

where σ is the allowable stress from Table 11-52,[*] p is the maximum expected operating pressure, r is the tank radius, and t is tank wall thickness. Usually, we select the material, estimate the tank size, determine the thickness from Eq. (17-19) or (17-20), and then compute the tank weight using the density of the material selected. For typical spacecraft propulsion, the tank weight will be about 5–15% of the propellant weight, depending on the basic design, safety factors, and construction materials. We must add 20–30% of the overall tank weight for mounting hardware and propellant management devices. See Sec. 11.6 for more detail.

We can estimate total weight of the liquid-propulsion system in a similar manner as for solid systems by estimating the fraction of the propulsion system mass that is propellant. Liquid propulsion systems are typically 85–93% propellant by mass, with the remainder consisting of pressurant, thrusters, tanks, fluid components, lines, fittings, and instrumentation. This fuel mass fraction, however, can be considerably lower in small systems. The higher fuel mass fractions are usually associated with large propellant loads and use of composite, overwrapped tanks. These tanks are fabricated with advanced materials which have higher strength-to-weight ratios. An example would be thin, metal-lined tanks using an overwrapping of high-strength carbon fibers, such as graphite-epoxy.

---

[*] Use the column labeled "F_{tu} = Allowable Tensile Ultimate Stress" plus a safety factor.

A significant part of the weight in cold gas propulsion systems is in the high-pressure tanks needed for storing the gaseous propellant. These systems most often appear in applications demanding low total impulse or extreme sensitivity to contamination. In this case, we can estimate tank weight with the aid of Eqs. (17-19) and (17-20). For example, the propulsion system which uses a nitrogen propellant for the Orbital Maneuvering Unit has a *mass fraction* (ratio of propellant mass to propulsion system mass) of 0.64. Thirty-five percent of its dry weight is in nitrogen gas tankage.

Thruster design requires specialized development to produce the best performance, life, chamber pressure, and expansion ratio. Often, existing thrusters can provide a basis for initial sizing. Sizes for fluid components (especially valves), as well as lines and fittings, develop from flow-rate requirements, which depend on the number of thrusters involved and the definition of specific impulse given in Sec. 17.2. Pressure drops across fluid components also depend on flow rates. A pressure schedule starts with requirements of the thruster's inlet pressure and works up through the pressure drops in each sequential component to the propellant and pressurant tanks.

To design a propulsion system, we must consider some additional special topics: interactions with the rocket exhaust plume, staging, maneuver accuracy, and thrust-vector control. Rocket exhaust plumes present three basic design issues. The first is plume heating of surfaces next to the rocket. The second concerns forces and moments that the plume places on the spacecraft. For example, thrusters that control inclination on geostationary satellites can lose about 10% of their delivered thrust because of plume drag on the solar arrays. To avoid this drag, we must mount the thrusters far from the solar array axis or cant them away from the arrays. The applied thrust vector is then degraded by the cosine of the thruster cant angle, because the applied thrust is no longer normal to the vehicle. A third issue concerns contamination by the plume. Depending on the propellants involved and the nature of their exhaust products, the plume may contaminate sensitive surfaces (such as optics or thermal control surfaces) of both the host and nearby spacecraft, such as the Space Shuttle or, eventually, the space station.

Rocket-plume exhausts divide into three regions: a continuous, forward-directed core flow; a transition region; and a rarefied backflow regime. By carefully placing thrusters on the spacecraft, we can usually avoid the first two regions. Avoiding back-flow effects is difficult, but we can reduce them with large separation distances, plume shields, covers on sensitive surfaces, and operational constraints on thruster firings. Determining plume effects in specific applications may require extensive analytical modeling supported by selected test data [Furstenau, McCoy, and Mann, 1980].

Optimizing weight for a given space propulsion system depends upon its type and specific operating parameters. Typically, for a system using a cold gas or pressure-fed liquid chemical, thrust-chamber pressure is the key to establishing its weight. Increasing the chamber pressure in a system design and trade study produces several effects. The engine will tend to be smaller and perform better because of increased area ratio and, to a much lesser extent, improved chemical kinetics (combustion efficiency). Therefore, the engine weight and propellant quantity will tend to decrease. But the propellant tank, components, and plumbing weights will increase because of the higher level of operating pressure required to force-feed propellants to the system, its components, and into the thrust chamber. The thrust chamber operates at some nominal pressure, which adds resistance by acting as a back pressure. Thus, for a given set of design and operating conditions, we can find an optimum design point for chamber pressure that results in a minimum weight for the propulsion system. For a turbopump-fed system, the weight of the feed system upstream of the pump inlet will not depend on the chamber pressure, so we can obtain the best size and pressure for the thrust chamber without considering the rest of the system. In this case, as mentioned earlier, we would trade the weight of a turbopump assembly against the weight of the pressure-fed components to determine which approach is better for a given propulsion system.

17.5 Staging

Multistage rockets allow improved payload capability for vehicles with a high ΔV requirement such as launch vehicles or interplanetary spacecraft. In a multistage rocket, propellant is stored in smaller, separate tanks rather than a larger single tank as in a single-stage rocket. Since each tank is discarded when empty, energy is not expended to accelerate the empty tanks, so a higher total ΔV is obtained. Alternatively, a larger payload mass can be accelerated to the same total ΔV. For convenience, the separate tanks are usually bundled with their own engines, with each discardable unit called a *stage*.

The same rocket equation as single-stage rockets describes multistage rocket performance, but must be determined on a stage-by-stage basis. The velocity increment, ΔV_i, for each stage is calculated as before:

$$\Delta V_i \equiv g\, I_{sp}\, \ln\left(\frac{m_{oi}}{m_{fi}}\right)$$ (17-21)

where m_{oi} represents the total vehicle mass when stage i ignites, and m_{fi} is the total vehicle mass when stage i burns out **but is not yet discarded**. We must always realize that the payload mass for any stage consists of the mass of all subsequent stages plus the payload. Then, the total velocity increment for the vehicle is the sum of those for the individual stages:

$$\Delta V_{total} = \Delta V_1 + \Delta V_2 + \cdots + \Delta V_n = \sum_{i=1}^{n} \Delta V_i$$ (17-22)

where n is the total number of stages.

We define the *payload fraction* λ as the ratio of payload mass, $m_{p/l}$, to initial mass, m_o. We do this for the overall vehicle:

$$\lambda = m_{p/l} / m_o$$ (17-23)

or for each individual stage:

$$\lambda = m_{p/l_i} / m_{oi}$$ (17-24)

where the subscript i indicates the stage number. Recall that the payload for each stage includes the mass of all subsequent stages. The overall vehicle payload fraction is then the product of those for the individual stages:

$$\lambda = \lambda_1 \cdot \lambda_2 \cdots \lambda_n = \prod_{i=1}^{n} \lambda_i$$ (17-25)

Another required definition is the *structure fraction:*

$$\varepsilon_{si} = m_{si} \, / \, m_{oi} = m_{si} \, / \left[m_{pi} + m_{si} + m_{p/li} \right] \qquad (17\text{-}26)$$

where $m_s \equiv m_f - m_{p/l} = m_o - m_{p/l} - m_p$ and m_p is the propellant mass.
Note that the mass ratio, R_i, from Eq. (17-6) is related to the payload and structure fractions as follows:

$$\frac{1}{R_i} = \lambda_i + \varepsilon_{si} \qquad (17\text{-}27)$$

Thus we can determine ΔV_i for any stage knowing I_{sp}, λ_i, and ε_{si}.

We say a multistage vehicle with identical specific impulse, payload fraction and structure fraction for each stage has *similar stages.* For such a vehicle, we maximize the payload fraction by having each stage provide the same velocity increment. We calculate the payload fraction for each stage by:

$$\lambda_i = e^{-\left(\Delta V_{total} / n I_{sp} g \right)} - \varepsilon \qquad (17\text{-}28)$$

and the overall vehicle payload fraction by:

$$\lambda = (\lambda_i)^n \qquad (17\text{-}29)$$

For a multistage vehicle with **dissimilar** stages the overall vehicle payload fraction depends on how we partition the ΔV requirement among stages. We reduce payload fractions when we partition the ΔV suboptimally. Techniques have been developed which yield an analytical solution for the optimal ΔV distribution [Hill and Petersen, 1970], or we may determine the optimal distribution by trial and error. In the latter approach, a ΔV distribution is postulated and the resulting payload fraction calculated as previously outlined. We have to vary the ΔV distribution until the payload fraction is maximized. Once we select the ΔV distribution, we size the vehicle by starting with the uppermost or final stage (whose payload is the payload) and calculating the initial mass of this assembly. This assembly then forms the payload for the previous stage and the process repeats until all stages are sized.

Results reveal that to maximize the payload fraction for a given ΔV requirement:

1. Stages with higher I_{sp} should be above stages with lower I_{sp}.

2. More ΔV should be provided by the stages with the higher I_{sp}.

3. Each succeeding stage should be smaller than its predecessor.

4. Similar stages should provide the same ΔV.

References

Altman, D.E. 1981. Presented at the AIAA Annual Meeting.

Clark, K.E. 1975. "Survey of Electric Propulsion Capability." *Journal of Spacecraft and Rockets.* 12 (11):641–654.

Fritz, D.E., and R.L. Sackheim. 1982. "Study of a Cost-Optimized Pressure-Fed Liquid Rocket Launch Vehicle." Paper No. AIAA-82-1108 presented at the AIAA 18th Joint Propulsion Conference, Cleveland, Ohio.

Furstenau, R.P., T.D. McCoy, and David M. Mann. 1980. "US Air Force Approach to Plume Contamination." in *Proceedings of the Seminar in Optics in Adverse Environments.* Society for Photo-Optical Instrumentation Engineers.

Gordon, S., and B.J. McBride. 1976. *Computer Program for Calculation of Complex Chemical Equilibrium Compositions, Rocket Performance, Incident and Reflected Shocks, and Chapman-Jouget Detonations.* NASA SP-273, Interim Revision. NASA Lewis Research Center: National Aeronautics and Space Administration.

Guthrie, Douglas M., and Robert S. Wolf. 1990. "Non-Acoustic Combustion Instability in Hybrid Rocket Motors." Paper presented at the JANNAF Propulsion Conference. Anaheim, CA.

Hill, Philip G., and Carl R. Peterson. 1970. *Mechanics and Thermodynamics of Propulsion.* Reading, MA: Addison-Wesley Publishing Co.

Humble, Ronald W., G.N. Henry, and W.J. Larson. 1995. *Spacecraft Propulsion Analysis and Design.* New York: McGraw-Hill.

Huzel, Dieter K., and David H. Huang. 1992. *Design of Liquid Rocket Propellant Rocket Engines.* Washington DC: AIAA.

Jensen, G.E., et al. 1990. *Hybrid Propulsion Technology Program Final Report.* United Technologies Chemical Systems. NASA/MSFC Contract NAS 8-37778.

Martin, J.W. 1986. "Liquid Propellant Management in Space Vehicles." *Quest Magazine* (TRW, Inc.) 9(1).

Sackheim, R.L. and D.H. Hook. 1996. "Development of a Bimodal Secondary Combustion Augmented Thruster (SCAT)." Penn State 8th Annual Symposium on Space Propulsion, State College, Penn., October 31.

Salmon, M. 1968. *World Aerospace Systems.*

Stuhlinger, Ernst. 1964. *Ion Propulsion for Space Flight.* New York: McGraw-Hill Book Company.

Sutton, George P. 1992. *Rocket Propulsion Elements (6th edition).* New York: John Wiley and Sons.

Timnat, Y.M. 1987. *Advanced Chemical Rocket Propulsion.* London: Academic Press.

Tsiolkovsky, Konstantin E. 1903. "Exploration of the Universe with Reaction Machines." *The Science Review #5* (St. Petersburg, Russia).

Chapter 18

Launch Systems

Joseph P. Loftus Jr. and Charles Teixeira,
NASA Johnson Space Center
Updated by Douglas Kirkpatrick,
United States Air Force Academy

The launch process can severely constrain spacecraft design. Primary restrictions are the launch vehicle's lift capability and the environment to which it subjects the satellite during ascent. A launch system consists of a basic launch vehicle incorporating one or more stages and an infrastructure for ground support. It alters velocity to place the spacecraft in orbit, creates a severe ascent environment, and protects the spacecraft from its surroundings. Ultimately, it places the payload into the desired orbit with a functional spacecraft attitude. In this chapter the term *payload* includes all hardware above the launch-vehicle-to-spacecraft interface, excluding the payload's protective fairing, which is usually part of the launch system. Thus, the launch-vehicle payload consists of the entire spacecraft above the booster adapter interface. For the Shuttle, it is customary to speak of the payload as the spacecraft to be deployed or the sortie mission payload to be operated from the payload bay with the Shuttle providing all of the support functions. Shuttle missions normally accommodate both types of payloads on each flight.

The aerospace industry has typically managed launch vehicles in one organization and spacecraft in another, reinforcing a tendency to budget and manage at the interface. This approach has been, and in many cases will continue to be, successful. However, we must continue to ensure it leads to the most cost-effective technical and management solutions. For example, traditional interfaces may not be appropriate as new generations of maneuverable spacecraft with their own inertial references become common. These spacecraft-borne inertial platforms could be used to guide and navigate launch systems, replacing similar systems in the launch vehicle.

Over the last 30 years, launch vehicle performance has improved tenfold while reliability has slowly increased from 0.85 to roughly 0.95. Planners often do not appreciate that launch-system reliability and cost are keys to a successful mission. Because the spacecraft usually costs more than the launch system, it can be very cost-effective to spend a bit more for a launch system with more reliability.

In this chapter we discuss some of the fundamental physical considerations of launch vehicles and upper stages, followed by a discussion of the launch system selection process as outlined in Table 18-1, and finally, a definition of spacecraft design envelopes and environments.

TABLE 18-1.　Steps in Selecting a Launch System.

Step	Comments and Required Information	References
1. Collect requirements and constraints, which depend on the mission operations concept. Consider the deployment strategy.	Number of spacecraft per launch Spacecraft dry weight Spacecraft dimensions Mission orbit Mission timeline Funding constraints	See text Chaps. 10, 11 Chaps. 10, 11 Chap. 7 Mission planning est. Mission planning est.
2. Identify and analyze acceptable configurations for the launch system.	Include the following information for each potential configuration: – Weight of spacecraft propellant – Orbit-insertion stage weight, if required – Weight of booster adaptor – Performance margin available – Boosted weight capability – Reliability	 Chap. 17 Chap. 17, Table 18-2 Secs. 11.6, 18.3 Sec. 18.2 Tables 18-2,18-4 Table 18-3
3. Select launch systems for spacecraft design. During conceptual design, identify several potential launch systems to make the launch more likely.	Criteria based on the following parameters: – Boosted weight capability – Cost – Performance margin available – Reliability – Schedule vs. vehicle availability – Launch availability	 Tables 18-2, 18-4 Chap. 20 Sec. 18.2 Table 18-3, 19.2 Mission planning est. Sec. 18.2, Eq. (18-5)
4. Determine spacecraft design envelope and environments dictated by the launch system selected.	Include the following information for each launch system, and include the worst-case environments for combined launch systems: – Fairing size and shape – Maximum accelerations – Vibration frequencies and magnitudes – Acoustic frequencies and magnitudes – Temperature extremes – Air cleanliness – Orbital insertion accuracy – Interfaces to launch site and vehicle	 Fig. 18-8 Table 18-8 Table 18-9, Fig. 18-10 Fig. 18-12 Sec. 18.3 Sec. 18.3,Table 18-7 Table 18-10 Fig. 18-8
5. Iterate to meet constraints on performance, cost, risk and schedule.	Document and maintain the criteria, decision process and data to support program changes.	

18.1 Basic Launch Vehicle Considerations

Space launch systems are unique forms of transportation since they are the only systems that accelerate continuously throughout their performance envelope. Consequently, *velocity* is the fundamental measure of performance for launch systems. A launch system's ability to achieve orbital velocity comes primarily from its propulsion efficiency, with vehicle weight and drag acting against it.

Figure 18-1 shows the forces acting on a launch vehicle and the associated free-body diagram. Note that the weight, W, of the vehicle acts at its center-of-gravity, cg, and the aerodynamic forces—lift, L, and drag, D, act at its center-of-pressure, cp. This configuration, with the center-of-gravity ahead of the center-of-pressure, is stable since the lift and drag forces cause restoring torques about the vehicle's center-of-gravity. Ideally, thrust, T, acts through the centerline of the vehicle but we can develop a control torque by gimballing the engine nozzle. We measure the vehicle's flight path angle, ϕ, from the local horizon to its velocity vector and we measure its angle-of-attack, α, from its velocity vector to its centerline.

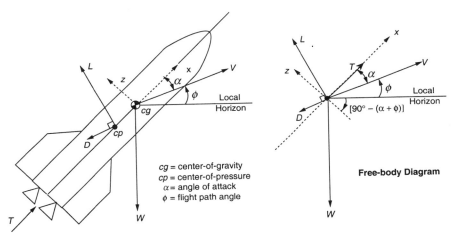

cg = center-of-gravity
cp = center-of-pressure
α = angle of attack
ϕ = flight path angle

Free-body Diagram

Fig. 18-1.   Forces Acting on a Launch Vehicle. By summing all the forces on a launch vehicle, we can compute its acceleration and velocity, using Eqs. (18-1), (18-2), and an integrating process.

Using the free-body diagram in Fig. 18-1, we can develop expressions for the acceleration along the axial, a_x, and lateral, a_z, body reference axes.

$$a_x = g \, [T/W - \sin(\phi + \alpha) - D/W \cos(\alpha) + L/W \sin(\alpha)] \qquad (18\text{-}1)$$

$$a_z = g \, [-\cos(\phi + \alpha) + D/W \sin(\alpha) + L/W \cos(\alpha)] \qquad (18\text{-}2)$$

We use Eqs. (18-1) and (18-2) to develop the vehicle's estimated acceleration shown in Table 18-8. We can calculate thrust and drag using Eqs. (17-1) and (6-21), respectively, and integrating Eqs. (18-1) and (18-2) in an inertial reference frame to yield the estimated velocity of the launch vehicle.

We can easily estimate the velocity that a launch vehicle should provide by

$$\Delta V_{design} = \Delta V_{burnout} + \Delta V_{gravity} + \Delta V_{drag} \qquad (18\text{-}3)$$

where $\Delta V_{burnout}$ is the velocity required for the desired orbit. We add the velocity losses, $\Delta V_{gravity}$ and ΔV_{drag}, to the burnout velocity to obtain the required design velocity. We also have to account for velocity losses from thrust vector control for trajectory shaping and other performance variables, such as solid rocket motor bulk temperature, which causes thrust-level variations. Fig. 18-2 shows values for gravity and drag losses for a typical two-stage vehicle. Note that these losses are sensitive to the initial thrust-to-weight ratio, T/W_0. A low thrust-to-weight ratio causes gravity losses to be high because the vehicle spends more time in ascent, while high thrust-to-weight causes drag losses to be high because of the higher velocities achieved in the atmosphere. The thrust-to-weight ratio is a key launch vehicle parameter because it dictates the vibration, acoustic, and dynamic load environment for the spacecraft. These environments are discussed in Sec. 18.3.

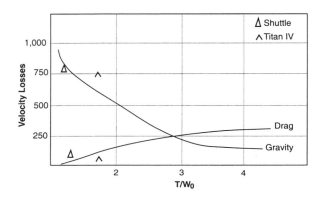

Fig. 18-2. Launch System Performance Losses. At the low end of T/W_0, gravity losses are higher because the launch vehicle spends more time ascending. At the high end of T/W_0, drag losses are higher because the launch vehicle reaches a higher velocity in the atmosphere.

If there were no atmosphere and no topographical variations, an optimum launch trajectory would be very similar to a Hohmann transfer and gravity losses would be minimized by thrusting normal to the radius vector. To accurately estimate gravity losses we need to know a precise ascent profile and time of flight. But for medium-to-large launch vehicles on nominal trajectories the velocity losses due to gravity fall between 750 and 1,500 m/s.

Aerodynamic drag forces acting on a launch vehicle are a function of the shape and size of the vehicle, speed, and angle-of-attack, α. We can manipulate Eq. (6-21) to get

$$D/W = C_d\ (A/W)\ q \tag{18-4}$$

where C_d is the dimensionless coefficient of drag (about 2.2), A is the vehicle's cross-sectional area perpendicular to its velocity vector, and the *dynamic pressure*, q, is one-half the product of the atmospheric density (at the vehicle's current altitude) and the velocity squared. For the current inventory of large, expendable launch vehicles, velocity losses due to drag are less than 3% of the total change in velocity required, about 20 to 40 m/s. The percentage decreases as the size of the vehicle decreases.

Once we know the required design velocity, ΔV_{design}, from mission requirements, we can estimate the mass of propellant required for the launch vehicle using Eq. (17-7) for single stage rockets and Sec. 17.5 for launch systems with multiple stages.

Several definitions are useful at this point. The *flight vehicle mass* is the sum of the propellant mass, structure mass, including mass of the fairing, and the mass of everything above the launch vehicle interface, including mass of the spacecraft bus, payload, and any upper stages. We use *mass fractions* to describe the portion of the flight vehicle devoted to certain sections. For example, the *propellant mass fraction* is the mass of propellant divided by the total flight vehicle mass; the *structure mass fraction* or *deadweight fraction* is the structural mass, including the mass of the fairing, divided by total flight vehicle mass; and the *payload mass fraction* is the payload mass divided by total flight vehicle mass. Typical values for propellant, structure, and payload mass fractions are 0.85, 0.14, and 0.01, respectively.

18.2 Launch System Selection Process

The first step in the launch system selection process is to establish the mission needs and objectives, since they dictate the performance, trajectory, and the family of vehicles which can operate from suitable sites. The mission need should be stated in terms of the specific return desired, e.g., Earth observation data over specific portions of the Earth, weather information, etc. The mission need may be very specific as in the case of a military objective, or as broad as a Presidential Directive to land man on the Moon within a decade. A clear understanding of the real mission need is extremely important since it can dictate the launch strategy. For example, large constellations of spacecraft may require periodic replenishment launches after the constellation is full. At the same time, a tactical satellite may require launch on demand within weeks or days. These drastically different requirements demand different performance from the launch system and its supporting infrastructure.

Another critical issue is whether the spacecraft will use a dedicated or shared launch system. The dedicated system may cost more, but it lessens the chances that a problem with another spacecraft will adversely affect the launch. Payload security may also demand a dedicated launch. On the other hand, shared launches are usually less expensive per spacecraft. Before deciding on a shared launch, we must consider the interaction between payloads in the shroud. If we mount them serially, for example, we must analyze the probability that the upper payload will not deploy and thus interfere with the lower payload's deployment. Examples of larger vehicles that can launch multiple payloads are the Space Shuttle, several Ariane 4 variants, Ariane 5, and Titan IV. We consider launching multiple spacecraft if their desired orbital altitudes and inclinations are compatible. This works especially well for the Space Shuttle, when primary mission payloads don't fill the payload bay, or when the spacecraft are deployable and sortie operations has room on the mission timelines.

Once we establish the mission need, then we determine specific mission requirements. For low-Earth orbit missions, these usually consist of orbit altitude, inclination, and right ascension of the ascending node. In addition, estimated payload weight and dimensions become requirements to the launch system. The mission concept specifies such parameters as number of spacecraft, anticipated lifetime and replacement strategy, and method of data retrieval and management. A required launch date also may become a selection parameter as it affects schedules, and the vehicle and launch site availabilities.

We allocate the mission requirements as functional requirements between the launch vehicle and payload. The basic question is, "What specific functions or operations must the payload accomplish, and which must the launch vehicle perform?" The two functions usually affected are propulsion, and guidance, navigation, and control. For example, can the launch vehicle achieve the final orbit, or must the spacecraft provide orbital maneuvering capability to raise the orbit to a higher altitude or change inclination? The launch vehicle may require an upper stage to achieve the final orbit, adding to the launch cost. The alternative is to provide sufficient propulsive capability on the spacecraft to perform the final propulsive maneuvers. We must carefully weigh the impact to the spacecraft in additional propellant and tankage, in terms of cost and complexity to the spacecraft design and trade against the potentially higher launch cost associated with an upper stage.

A similar trade is made in navigation, guidance, and control (Chap. 11.7). Spacecraft computer capabilities have grown by orders of magnitude over the past twenty years. Consequently, the spacecraft navigation, guidance and control subsystem can technically provide this function to the launch system during ascent. However, this approach results in a highly coupled payload and launch system which presents some negative attributes, including a more complex integration of the two, and makes it more difficult to manage at the spacecraft-to-launch-vehicle interface. The current trend is to separate the spacecraft and launch system functions at the interface, both functionally and physically, to minimize interface requirements and complexity.

We must assess each function required to achieve the mission objective through this process, and allocate functions based on cost, reliability, and risk. This is a classic systems engineering problem, which we must continuously evaluate as part of the vehicle selection process and as the spacecraft design matures.

Having established mission requirements, constraints, and the required information in Step 1 of Table 18-1, we must decide which launch-system configurations can deliver the spacecraft to its mission orbit. The launch systems selected during conceptual design should satisfy the mission's performance requirements and minimize program risk. We want to choose the launch systems early, so contractors for the spacecraft and launch system can negotiate requirements early, as well. Doing so decreases changes in design, cost, and schedule downstream. Recent experiences show that *we should design spacecraft to be compatible with several launch systems* to enhance launch probability, as well as to provide some leverage in negotiating launch cost. Fixing problems that may cause launches to fail takes months or years. If we change the launcher we may have to redesign the spacecraft and its interfaces with the launch vehicle. A redesign costs a lot of time and money. To solve this problem, we select an alternate launch vehicle, as a backup, early in the process, and design the spacecraft to be compatible with both.

Selecting a launch system depends on at least these criteria: the launch vehicle's **performance capability** to boost the necessary weight to the mission orbit, the required launch date versus **vehicle availability**, spacecraft-to-launch-vehicle **compatibility**, and of course, **cost** of the launch service. The launch system's performance capability must include factors such as performance margin, and a clear definition of weight and performance parameters as given in Table 18-2. Note that the payload performance quoted by launch vehicle manufacturers must be greater than the projected boosted weight. This difference (launch system performance capability minus spacecraft boosted weight) is referred to as *performance margin* and is an important selection criterion. Note that the performance margin is in addition to the allowances made for spacecraft weight growth (see Table 10-9).

Performance margin is an important parameter throughout a program, but it is particularly important early in a program. Ideally a spacecraft launches with a small positive performance margin, meaning that the spacecraft weighs (when it is completed and delivered) just what the launch system can place into orbit. For spacecraft with propulsion systems, we use propellant loading to trim the final weight.

TABLE 18-2. **Weight Parameter Definitions.** These are the key elements of a weight budget. Performance charts for launch vehicles usually list only payload performance capability.

Weight Parameters	Comments
1. *Spacecraft Dry Weight*	Weight of all spacecraft subsystems and sensors, including weight growth allowance of 15–25% at concept definition
plus Propellant	Weight of propellant required by the spacecraft to perform its mission when injected into its mission orbit
Yields	
2. *Loaded Spacecraft Weight*	Mission-capable spacecraft weight (wet weight)
plus Upper Stage Vehicle Weight	Weight of any apogee or perigee kick motors and stages added to the launch system
Yields	
3. *Injected Weight*	Total weight achieving orbit
plus Booster Adapter Weight	May also include airborne support equipment on the Space Shuttle
Yields	
4. *Boosted Weight*	Total weight that must be lifted by the launch vehicle
plus Performance Margin	The amount of performance retained in reserve (for the booster) to allow for all other uncertainties.
Yields	
5. *Payload Performance Capability*	This is the payload weight contractors say their launch systems can lift

Occasionally this happens, but more often the spacecraft's weight grows beyond projected weight growth allocations. Make sure the spacecraft weight is within the launcher capability and that the weight growth is within the allocated weight growth margin (see Table 10-10). The longer the wait to reduce weight, the more it costs. Do it early!

If the launch system does not have sufficient performance capability, discussions with the launch vehicle manufacturer can frequently result in some augmentation to the performance, selection of a higher performance launch system, or, if necessary, reevaluation of the spacecraft design and its requirements. These trades are part of the selection process whereby we continuously reevaluate cost, schedule, and risk.

We further evaluate the candidate launch systems which pass the performance "gate" based on their available payload fairings. The fairings must be physically large enough to house and protect the spacecraft during ascent, and the interfaces to the spacecraft, both structural attachment and other services such as cooling (see Sec. 18.3), must be acceptable to the spacecraft.

We must also consider the launch schedule and whether the preferred launch system will be available. Given the required launch date and window, we discuss availability with contractors of launch services. Schedule considerations should include the launch site's availability as well as the use of any unique facilities for ground processing. For example, on the requested launch date, the launch pad may be available but activities nearby may keep us from launching for safety or security reasons. Thus, we have to examine the entire infrastructure, including items such as ground-support equipment and networks for tracking and communications. Several off-site facilities are available for processing—commercially and through agreements with government agencies. Finally, if we intend to change the launch site, we must consider the effect on scheduling and cost.

Launch availability brings other dimensions to launch-system selection. The availability of a launch system depends on its reliability, production capacity, the ability of the launch operations system to support the desired launch rate, existing launch commitments, and its demonstrated stand-down time following a failure. The relationship

$$A = 1 - [L(1 - R)T_d / (1 - 1/S)] \tag{18-5}$$

allows us to discuss this concept. Here the *expected launch availability*, *A*, measured in percent of the time the launch system is available, depends on the vehicle's *reliability*, *R*, the nominal or planned *launch rate*, *L*, in units of flights per year, the demonstrated (or estimated) *stand-down time* following a failure, T_d, in units of years, and the *surge-rate capacity*, *S*, where $S = 1.5$ means the system can achieve a flight rate 50% higher than the planned rate *L*. Launch systems in the United States can typically surge to between 1.15 and 1.5 times the nominal launch rate. Figure 18-3 illustrates results from Eq. (18-5). Negative results mean that the system probably would not be available when needed. We must use Eq. (18-5) with caution since a singularity occurs for surge values approaching one, i.e., the system has no surge capacity. Commitment to two systems with poor availability means that some of the spacecraft so committed will not fly or may be delayed for several years. Table 18-3 provides estimated reliability and stand-down times for typical launch systems.

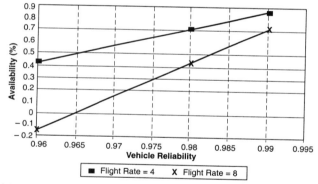

Fig. 18-3. Vehicle Availability. Even with a high surge capability (1.4), a launch system may have low availability (0.43) for a modest launch rate (4 per year) and a high reliability (0.96). For a high launch rate (8 per year), the same system may not be available when needed.

Using the above criteria, we can narrow the field of candidates to a few launch systems by evaluating them consistently and systematically. A risk analysis must also accompany these assessments by considering:

- Are the advertised cost and schedules reasonable?
- How do these numbers compare with past experience?
- Is the offeror likely to stay in business? Or in some cases, is the country providing the launch service stable?
- Are there any circumstances which are unique or new that could result in additional risk?

TABLE 18-3. **Reliability Experience of Launch Systems.** This table shows the reliability of international launch systems along with their stand-down times following a failure as of December 1998. *R* is the reliability. *Isakowitz* [1995] provides additional information.

Launch System	No. of Successful Launches	Total No. of Launches	R	Last Failure Downtime (months)	Average Downtime (months)	Launches Since Last Failure
Shuttle	93	94	0.989	32	32	69
Titan II (since 1970)	18	18	1.000	—	—	—
Titan IV	23	25	0.920	6	6	0
Atlas Centaur (since 1970)	142	155	0.916	8	10	45
Delta (since 1970)	170	179	0.950	4	4	1
Ariane 4	93	99	0.939	5	9	39
Ariane 5	2	3	0.670	17	17	2
Proton (since 1970)	216	232	0.931	4	3	41
Long March (CZ)	49	54	0.907	1	12	14
Zenit 2	23	27	0.852	14	14	2
Pegasus/XL	20	22	0.909	9	9	13
Soyuz/Molniya (since 1970)	1,225	1,293	0.947	1	unknown	24
Tsyklon	209	211	0.991	5	6	11
M-V	2	2	1.000	—	—	—
H-2	6	6	1.000	—	—	—

These are difficult questions to address, but a successful mission is predicated on careful consideration of these factors.

To identify the best combinations, we should examine several acceptable deployment strategies and staging concepts. There are three primary options for the ascent from Earth to the final mission orbit: direct injection by a launch system, injection using various launch and stage vehicle combinations, or injection using an integral propulsion system. Small payloads can usually use launch vehicles that insert directly into low-Earth orbit. For geostationary orbits, however, we typically need to augment the launch vehicle with an upper stage. The third method, *integral propulsion*, allows us to insert and maintain the spacecraft in its orbit and control its attitude with a single propellant system in the spacecraft. Because the system must operate for the entire mission with many restarts, it typically uses a liquid bipropellant or an ion engine.

Table 18-4 lists available launch systems, their ability to launch the boosted weight, compatible upper stages, available launch sites, and envelope dimensions for standard payload fairings. For more precise information on these systems, see the user manuals, listed just before the references at the end of this chapter. Figures 18-4 and 18-5 show performance curves for selected launch vehicles at various altitudes. Although options seem endless, usually only a few candidate systems meet requirements of payload weight injected into the desired orbit, weight margin, vehicle availability, cost, and reliability. In addition, Table 18-5 lists available stage vehicles and their performance. Tables 17-6 and 17-7 list typical solid rocket motors and liquid engines, respectively.

TABLE 18-4. Launch Systems Characteristics. The table shows characteristics for existing systems to 28.5 deg inclination, unless specified otherwise. Low-Earth orbit (LEO) is given here as approximately 185 km circular. GTO is geosynchronous transfer orbit, and GEO is geosynchronous orbit. Polar is 90 deg inclination and 185 km circular. See Isakowitz [1995] for details. Launch site letters are keyed to Table 18-6.

Launch System	Upper Stage (if any)	LEO (kg)	GTO (kg)	GEO (kg)	Polar (kg)	Launch Site	Payload Accommodations Dia (m)	Length (m)
ATLAS I	Centaur-1	—	2,255	—	—	B	3.3	10.4
ATLAS II	Centaur-2	6,580	2,810	570	5,510	B	4.2	12.0
ATLAS IIAS	Centaur-2A	8,640	3,606	1,050	7,300	B	4.2	12.0
DELTA II								
6920/25	PAM-D	3,990	1,450	730	2,950	A, B	2.9	8.5
7920/25	PAM-D	5,089	1,840	910	3,890	A, B	2.9	8.5
PEGASUS	—	375	—	—	—	Aircraft	1.3	4.4
PEGASUS XL	—	460	—	—	345	launch[4]	1.3	4.4
SHUTTLE		24,400	—	—	—	B	4.5	18.3[6]
	IUS	—	5,900	2,360	—			4.6[7]
	TOS	—	5,900	—	—			
	PAM-D	—	1,300	—				
	PAM-D2	—	1,800	—				
TAURUS[5]	STAR 37	1,400	450	—	1,060	A, B	1.4	2.8
TITAN II	NUS	—	—	—	1,905	A	2.8	3.7. 5.2, 6.7
TITAN IV	NUS	—	—	—	14,110	A,B	4.5[6]	9.7
	Centaur	—	8,620[1]	4,540	—	A,B		12.8
	IUS	—	6,350	2,380	—	A,B		15.8
	NUS[1]	21,645	—	—	18,600[1]	A,B		18.9
ARIANE 40 (France)	H-10	4,900	2,050	—	3,900	D	3.7	3.9 S 4.9 L 6.5 XL
42P	H-10	6,100	2,840	—	4,800			
42L[2]	H-10	7,400	3,380	—	5,900			
44P	H-10	6,900	3,320	—	5,500			
44LP[3]	H-10	8,300	4,060	—	6,600			
44L	H-10	9,600	4,520	—	7,700			
ARIANE-5	L9	18,000	6,800	—	12,000	D	4.5[6]	12.0
H-2 (Japan)	—	10,500	4,000	2,200	6,600	P	3.7, 4.6	3.5–5.0
MV	Numerous	1,800	1,215	490[7]	1,300	O	2.2	3.5
LONG MARCH (China)								
CZ1D	—	790	200	100[7]	—	K	1.5[6]	1.0
CZ3A	—	7,200	2,500	1,230[7]	—	L	3.0	4.0
CZ3B	—	13,600	4,500	2,250	—	L	3.8	6.0
CZ4	—	4,000	1,100	550	—	M	3.0	3.9
CZ2E	Star 63F	8,800	3,370	1,500[7]	—	L	3.8	6.0
PROTON	D1	20,900	—	—	—	H	4.1	15.6
(Russia)	D1e	—	5,500	2,200	—	H	4.0	7.5
PROTON K	DM	20,100	4,615	2,100	—	H	—	—
PROTON M	BREEZE M	22,000	5,100	2,500	—	H	—	—
ZENIT 2	—	13,740	5,180	1,535	11,380	H	3.4	5.9–8.4

[1] With solid rocket motor upgrade
[2] With two liquid rocket boosters
[3] With two liquid and two solid rocket boosters
[4] Carrier aircraft can stage from various locations
[5] Under development
[6] 4.5 m diameter and 18.0 m length allowing for dynamic clearance
[7] With perigee and apogee kick motors

NUS = No Upper Stage
IUS = Inertial Upper Stage
TOS = Transfer Orbit Stage
PAM = Payload Assist Module

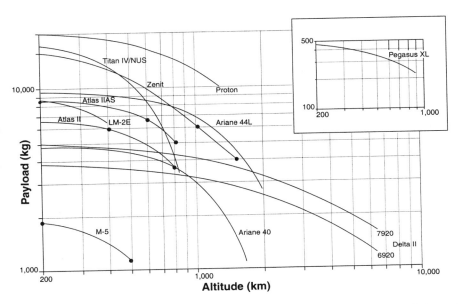

Fig. 18-4. Typical Launch-System Performance for Launches Due East from U.S. Launch Sites. Curves show delivery mass into circular orbit at the altitude indicated. The curves indicate development systems. Figures courtesy Capt. Marty France, U.S. Air Force Academy.

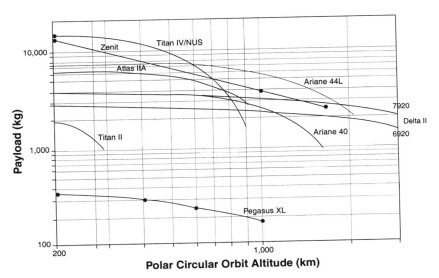

Fig. 18-5. Typical Launch-System Performance for Polar Launches from U.S. Launch Sites. For polar orbits there is no velocity assistance from the revolution of the Earth, so the launch vehicle has to furnish the additional ~450 m/s, thus reducing delivery mass.

TABLE 18-5. Orbital Transfer Vehicles. A number of upper stages are available to provide additional velocity beyond low-Earth orbit requirements. Both solid and liquid systems are shown and compatible launch vehicles identified.

Characteristics	PAM-D	PAM-DII	TOS	IUS		Centaur	H-10	D-M	L-9
Stage: Manufacturer	Boeing	Boeing	Lockheed Martin	Boeing		Lockheed Martin	Ariane-space	RSC Energia	Ariane-space
Length (m)	2.04	2.00	3.30	5.20		9.0	9.9	6.8	4.5
Diameter (m)	1.25	1.62	3.44	2.90		4.3	2.6	4.1	5.4
Engine: Manufacturer	Thiokol	Thiokol	CSD	CSD		Pratt-Whitney	SNECMA	Isayev	DASA
Type	(Star 48)	ISTP	SRM-1	SRM-1, SRM-2		RL 10A-3-3A	HM7B	11DM 58	Aestus
Number	1	1	1	1, 1		2	1	1	1
Fuel	Solid	Solid	Solid	Solid		LO₂LH₂	LO₂LH₂	LOX RP1	N₂O₄/ MMH
Composition	TP-H-3340	—	HTPB	HTPB		5.5:1	4.77	2.6	2.05
Total Thrust (N)	66,440	78,300	200,000	200,000	81,200	147,000	62,700	84,000	29,000
Specific Impulse (s)	292.6	281.7	294	292.9	300.9	442	444.2	361	324
Burn Time (s)	54.8	121	150	153.0	104.0	488	725	680	1100
Stage: Pad Mass (kg)	2,180	3,490	10,800	14,865		18,800	12,100	18,400	10,900
Impulse Propel. Mass (kg)	2,000	3,240	9,710	9,710	2,750	16,700	10,800	15,050	9,700
Burnout Mass (kg)	189	250	1,090	1,255	1,150	2,100	1,300	2,140	1,200
Airborne Support Equip. Mass (kg)	1,140	1,600	1,450	3,350		4,310	—	—	—
Illustration:									
Schedule: Start Date	1975	1980	1983	1978		1982	1986	New	1988
Operational Date	1982	1985	1986	1982		1990	1990	New	1996
Type of Development	Commercial	Commercial	Commercial	U.S. Gov't		U.S. Gov't	ESA	ILS	ESA
Sponsor	Boeing	Boeing	OSC	USAF		USAF	ESA	ILS	ESA

CSD—Chemical System Division, United Technologies; ESA—European Space Agency; ILS—International Launch Services; OSC—Orbital Sciences Corporation; RSC—Rocket Space Corporation; DASA—Daimler Chrysler Aerospace; SNECMA—Société Nationale d'Etudes et de Constructions de Moteurs d'Avion

The current version of the *Delta* launch system, Delta II, has two forms: the 6925 and 7925. The 6925 extended the earlier 3920's tanks to store more propellant and added Thiokol solid-rocket boosters for better performance. The 7925 uses an upgraded Rocketdyne RS-27A main engine and Hercules GEM solid rocket boosters. The 6920 and 7920 are two-stage versions of the three-stage 6925 and 7925, respectively. Both core vehicles use liquid oxygen and RP-1 propellant in the first stage and nitrogen tetroxide (N_2O_4) and Aerozine 50 (A50) in the second stage. They use inertial guidance and provide control moments from gimballed engines. Of the nine boosters, six ignite at lift-off and stage at 57 sec. The other three ignite at altitude during the first stage burn. The first and second stages are staged at 265 and 440 sec respectively. The third stage uses a spin-stabilized Star 48 B rocket motor in the PAM-D upper stage.

Atlas/Centaur launch vehicles include the Atlas I, Atlas II, Atlas IIA, and Atlas IIAS. The Atlas vehicle's core system uses liquid oxygen and hydrocarbons (RP-1) in a stage-and-a-half form, with three engines ignited at lift-off. Two of these booster engines jettison at 172 sec, and the third sustainer engine burns for another 111 sec. The Atlas is inertially guided and controlled in pitch and yaw by gimballed engines. Vernier engines on Atlas I control roll in the first stage. For Atlas II, IIA, and IIAS, a hydrazine roll control system attached to the interstage adapter controls the roll angle. The Centaur uses liquid oxygen and liquid hydrogen for propulsion. It has two RL-10A engines, which can start several times and burn 400 to 600 sec. Gimballed engines control the Centaur during burn, and 12 reaction-control engines control it when it coasts. On the IIAS version, four solid rocket motors improve the lifting capacity. Two motors ignite with the three main engines for lift off. When the first two motors burn out they drop off, then the other two burn until depleted and drop off.

The *Titan II* is a refurbished ballistic missile configured as a launch vehicle. It was essentially the core vehicle for the retired *Titan III* series. For fuel, the Titan II uses storable propellants: a nitrogen tetroxide oxidizer and Aerozine 50 (a 50/50 mixture of hydrazine and UDMH). Adding two solid rocket boosters and a third stage for the Titan III vehicle increased lift capability significantly. Fifty-five Titan II missiles were available to modify, eleven of which completed modification and launched success-fully, as of late 1998. The *Titan IV* vehicle adds solid rockets to the Titan core vehicle. The solid rockets are the "zero" stage ignited for lift-off, whereas the core vehicle's engines ignite when the solid-rocket motor's thrust tails off before separating. The Titan IV zero stage burns 138 sec, the first stage 164 sec, and the second stage 223 sec. A strap-down inertial system provides guidance using liquid injection (UDMH) to control the thrust vector in the zero stage and gimballed engines for the first and sec-ond stages. Titan IV is compatible with two upperstages, IUS and Centaur.

The *Space Shuttle* delivers, services, and recovers payloads. The Shuttle uses two solid-rocket boosters and three liquid-oxygen and liquid-hydrogen engines for propul-sion. The solid-rocket boosters burn out at approximately 123 sec. At first, the main engines burn in parallel with the solids, then continue to burn for 522 sec. For each flight, the Shuttle can carry up to three PAM-D payloads, one IUS, or one TOS.

The *Pegasus* air-launched booster and its -XL variant use wings to provide lift and three stages powered by solid-rocket motors. A strapped-down inertial system guides the booster. Aerodynamic fins control the first stage, and vectorable nozzles control the second and third stages. A cold-gas system for reaction control adjusts attitude while the spacecraft coasts and when the payload deploys and separates. Employing a Lockheed L-1011 as a launch platform reduces the propellant needed to achieve orbit and provides variable launch azimuths and locations for different orbital inclinations.

Space Data Corporation, a subsidiary of the Orbital Sciences Corporation, devel-oped a standard, small launch vehicle called *Taurus*. The first stage is essentially an MX missile first stage, and the upper stages are similar to those flown on Pegasus, without the attached wing. Two successful launches in 1998 placed seven satellites into low-Earth orbit.

Lockheed-Khrunichev-Energia International offers *Proton* for commercial launches. The original Russian D-1e version has four stages and delivers payloads to geostationary orbits. The first three stages use storable propellants (UDMH and N_2O_4), but the upper stage burns liquid oxygen and kerosene and has a multiple start capability. It has supported planetary and lunar launches and placed communication satellites in space. The 3-stage D-1 launches the MIR and other large payloads into low-Earth orbits. The new commercial K and M versions use the same first three stages, but add the DM and Breeze M upperstages, respectively.

China's Great Wall Industries builds, offers commercially, and operates the *Long March* vehicles. The Long March CZ3 is a three-stage vehicle, using storable propellants in the first two stages and cryogenic liquid-oxygen and liquid-hydrogen in the third stage. A new heavy-lift variant, the CZ3B, uses four liquid strap-ons and can carry 13,600 kg into low-Earth orbit.

The current versions of the *Ariane* launch system are the *Ariane 4* and *5* series. Arianespace operates these vehicles commercially. *Ariane 4* is a three-stage vehicle capable of using from zero to four strap-on solid or liquid rocket boosters. The liquid strap-ons and stages one and two use storable nitrogen tetroxide and UH-25 (a mixture of 75% unsymmetrical dimethylhydrazine and 25% hydrazine hydrate). The third stage burns liquid hydrogen and oxygen. *Ariane 5* is a new, heavy-lift vehicle, designed for reliability and cost effectiveness. It has a core stage that uses liquid hydrogen and oxygen, an upper stage that burns nitrogen tetroxide and monomethyl hydrazine, and two solid-rocket-motor strap-ons that are recoverable. Eventually, Arianespace will rate this vehicle for crewed launches. It has had two successful launches after an initial launch failure. Ariane launch vehicles are efficient boosters for low-inclination or geotransfer missions because of their Kourou launch site: just 5 deg north latitude. As a result of the location and launch successes, they have captured more than half of the launch traffic for commercial communication satellites.

Japan's space agencies operate two launch systems and are designing a third. The H-2 is a new heavy-lift vehicle that can place 10,500 kg into low-Earth orbit. This two-stage vehicle burns liquid hydrogen and oxygen in both stages and has two solid-rocket-motor strap-ons. Launching from Tanegashima Space Center, it has six successes in six attempts. The M-5 is a new three-stage vehicle, capable of launching 1,800 kg into low-Earth orbit. All stages use solid rocket motors, burning hydroxy-terminated polybutadiene (HTPB). This vehicle operates from the Kagoshima Space Center and has two successes in two attempts. A new design, J-1, uses the H-2 solid rocket booster as stage one and the two upper stages from the M-3SII as its upper stages. Because of these common parts, its design went quickly.

Table 18-6 lists the available launch sites for each launch system. Once we have identified the mission orbit and launch system, the appropriate launch site(s) become apparent. The key U.S. launch sites are the *Eastern Range* at Cape Canaveral Air Force Station, Florida, which serves the Kennedy Space Center next to it, and the *Western Range* at Vandenberg Air Force Base, California. The *Wallops Island* facility in Virginia can launch a number of smaller commercial launch vehicles and sounding rockets.

Figure 18-6 shows the location of the world's launch sites, and Table 18-6 identifies their coordinates. We can get the best performance from a launch vehicle into a direct orbit by locating the launch site at the equator to take advantage of the easterly velocity from Earth's rotation. Theoretically, we can attain any orbit inclination from the equator, but we may select other sites for convenient access, security, or political reasons. Launch sites at higher latitudes cannot directly access orbit inclinations much below their latitude, and trajectory profiles that go to higher inclinations sacrifice velocity and payload mass. An inclination change of one degree requires about 208 m/s of velocity in low-Earth orbit. As Eq. (6-38) shows, this number decreases as the altitude of the change increases. We can calculate the propellant mass required to achieve the plane change using Eq. (17-7).

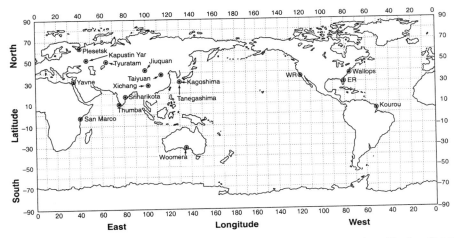

Fig. 18-6.   Launch Sites. These 17 sites have done or can do orbital launches. The location of the launch site and range safety considerations determine the acceptable launch azimuths.

TABLE 18-6.   Worldwide Launch Sites. The location of launch sites is useful when deciding on launch azimuth.

Launch Site	Map Designation	Country	Latitude (deg min)	Longitude (deg min)
Western Range	Vandenberg AFB, CA	United States	34 36 N	120 36 W
Eastern Range	Cape Canaveral AFS, Cape Kennedy Space Center	United States	28 30 N	80 33 W
Wallops Island	Wallops, VA	United States	37 51 N	75 28 W
Kourou Launch Ctr	Kourou	CNES/Arianespace	5 32 N	52 46 W
San Marco Launch Platform	San Marco	Italy	2 56 S	40 12 E
Plesetsk	Plesetsk	Russia	62 48 N	40 24 E
Kapustin Yar	Kapustin Yar	Russia	48 24 N	45 48 E
Tyuratam (Baikonur)	Tyuratam	Russia	45 54 N	63 18 E
Thumba Equatorial Station	Thumba	UN/India	8 35 N	76 52 E
Sriharikota	Sriharikota	India	13 47 N	80 15 E
Jiuguan Satellite Launch Ctr	Jiuguan	China	40 42 N	100 00 E
Xichang Satellite Launch Ctr	Xichang (Sichuan)	China	28 12 N	102 00 E
Taiyuan Satellite Launch Ctr	Taiyuan	China	37 46 N	112 30 E
Kagoshima Space Ctr	Kagoshima	Japan/ISAS	31 15 N	131 05 E
Tanegashima Space Ctr	Tanegashima	Japan/NASDA	30 24 N	130 58 E
Woomera Launch Site	Woomera	Australia/U.S.	31 07 S	136 32 E
Israeli Launch Complex	Yavne	Israel	31 31 N	34 27 E

Section 6.4 discussed the relationship between the desired orbital inclination and the required direction of launch from a specific launch site. The direction of launch, or launch azimuth, depends on range safety considerations that prohibit flying over certain land and ocean areas. Figure 18-7 shows the launch azimuths and inclinations directly available from the U.S. Eastern and Western launch sites. The Japanese have only two 45-day launch periods, one in the spring and one in the winter. At other times, the launch would threaten fishing fleets by dropping booster parts into the fishing area.

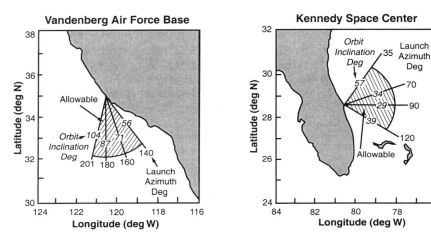

Fig. 18-7. Orbit Inclinations and Launch Azimuths Available from the Eastern and Western Ranges. In general, low-inclination orbits are possible from the East coast and high-inclination orbits are possible from the West. Note that orbital inclinations available from these launch sites depend on launch azimuth.

Given a choice of launch sites and launch dates, we must consider weather at the launch location. Bad weather can severely restrict chances to launch, thus costing time and money. The site's location and time of year determine surface weather and winds aloft. For example, Florida and French Guiana have many thunderstorms and frequent lightning during the spring and summer. Because the highly ionized gas in the launcher's exhaust plume can attract lightning, we must launch carefully in these seasons. During the winter, the winds aloft are severe because the jet stream moves to the south and often passes over both the Eastern and Western Ranges. As a result, we must study weather information for all possible sites.

Although only a few launch-system vendors operate in the U.S., we can choose intelligently from many types of launch vehicles. Several government organizations manage these systems. The Air Force Space Systems Division (*AFSSD*) developed the Titan, Atlas, and Delta launchers and matches launchers to users' requests. Air Force Space Command manages launch operations at DoD launch sites. NASA manages the Space Transportation System through NASA Headquarters. Vendors of commercial launch vehicles contract services directly with users. Contractors obtain license approvals to launch through the Department of Transportation, Office of Commercial Space Transportation.

18.3 Determining the Spacecraft Design Envelope and Environments

Once we've identified several launch systems for our mission, we must determine the configuration of the interfaces between the launch system and payload and understand the environments that the payload must withstand. We can use the parameters for the launch system we select; however, a preferred approach is to match environments and interfaces to the combined parameters of several launch vehicles. We should at least develop the information listed in Step 4 of Table 18-1. This process is referred to as *payload integration*, which means doing the management, program support, and analysis required to integrate the spacecraft (including upper stages) with the launch vehicle. We must consider this step early in mission design. The payload design must address launch environments and interfaces, whereas the launch system must support the payload during ascent. If we don't integrate these tasks carefully, problems in designing and manufacturing the payload can scuttle or seriously delay the launch. Payload integration must meet specific requirements of the program review, including interface control documents at both the Flight Feasibility Review and the Cargo Integration Review.

In the spacecraft design, we must consider the payload environment for the time the payload leaves the vendor's facility until the spacecraft completes its mission. In this section, we will address the environment from the time we install the payload on top of the launch vehicle until it's through the ascent. In many areas, the combination of high accelerations and vibrations coupled with the thermal environment and rapidly changing local pressures result in environmental conditions more severe than it would experience on orbit. Carefully considering these environments is critical to spacecraft design. Three areas require particular attention: usable payload volume offered by the available fairings, structural and electrical interfaces, and payload environments, as summarized in Table 18-7.

Fairings. Launch-vehicle vendors offer a wide range of payload fairings. They define the usable payload volumes and dimensions within the fairings for the spacecraft. These values account for payload and fairing deflections due to static and dynamic loads encountered during ascent. Table 18-4 provides approximate dimensions for various systems, and Fig. 18-8 shows a typical payload fairing.

The mission designer must ensure that the spacecraft will fit within the allowable envelope. The fairing protects the payload from aerodynamic loads and, in general, provides a benign environment, as discussed below, to lessen the impact on the payload. Generally the payload fairing is jettisoned late in the ascent when dynamic pressure and the heating rate are below acceptable levels specified by the launch vehicle vendor. Different times can be negotiated, if required. We do not need to analyze payload and upper-stage combinations which are similar to previously flown payloads as extensively as new or unique payloads, because analysis by similarity is acceptable. If available fairings are unsuitable, launch vehicle vendors can develop fairings to meet unique requirements.

The Space Shuttle's cargo bay is so large that it usually accommodates several payloads. To help planning, the bay is partitioned into four sectors, allocating a specified amount of electrical power, cooling, telemetry, and so forth, to each—comparable to that available from a Delta or Atlas vehicle. Larger payloads combine resources from several sectors to meet their requirements. The cargo bay doors and the fuselage provide protection equivalent to that of a shroud on an expendable launch vehicle.

TABLE 18-7. **Launch Vehicle Ascent Environments and Payload Fairing Constraints Must Be Factored into the Spacecraft Design Early in the Design Process.** Incompatibility found late in the design may require costly revisions or selection of another (less desirable) launch system.

Parameter	Typical Value/Comment	Reference
Payload Fairing Envelope	Consult user guide	Fig. 18-8, Table18-4
Payload to Launch Vehicle Interface	Specified bolt pattern	Launch vehicle user guides
Environments		
Thermal	10–35 °C	Launch vehicle user guides
Pad	188 BTU·ft^2/hr	
Ascent fairing radiant Aeroheating	100–150 BTU·ft^2/hr	
Electromagnetic	Consult range and launch vehicle user guides	
Contamination	Satisfy class 10,000 air	Sec. 18.3
Venting	Maximum of 1 psi differential	Fig. 18-9
Acceleration	5–7 g	Table 18-8
Vibration	0.1 g^2/Hz	Table 18-9, Fig. 18-10
Acoustics	140 dB	Fig. 18-12
Shock	4,000 g	Fig. 18-11

Structural and Electrical Interfaces. We must identify interfaces between the payload and the launch system early in the design process. For example, a payload adapter attaches the payload to the launch vehicle, and we have to determine whether the payload needs additional support. The adapters physically connect the payload and any required kick motors, spin tables, separation systems, or electrical interfaces. The launch-vehicle manufacturer usually provides them, if necessary, tailoring them to individual requirements. Adapter mass reduces available payload mass, so it is sometimes an important part of selecting the launch vehicle. Examples of booster-adapter masses are the 51-kg single-launch Type 1666A adapter for the Ariane 4, and, for dual-launch systems, the 440-kg Long SPELDA, also for Ariane. Section 11.6 provides a method to estimate booster-adapter weight, but actual weights are in the launch vehicle users' guide.

Launch-vehicle manufacturers must provide physical, electrical, radio frequency and optical access to the payload while the fairing encloses it. Effective operations demand the correct location of access doors and windows for radio frequencies and optics. In many cases, manufacturers must wire the launcher to command and safe the spacecraft. They also provide the mechanisms that separate the payload from the launch vehicle in orbit, typically by using redundant logic and circuitry to trigger redundant, ordnance-firing systems. High reliability in the payload separation mechanism is important.

The launch system and payload must match the desired communications architecture for launch operations. Communications requirements depend on the combined demands of the entire space mission: ground stations, payload, launch vehicle, range safety, and the user. We can adjust launch trajectories somewhat to provide redundant ground-station coverage during launch if needed. When we cannot get ground coverage, aircraft can cover critical events. We must check the entire communications

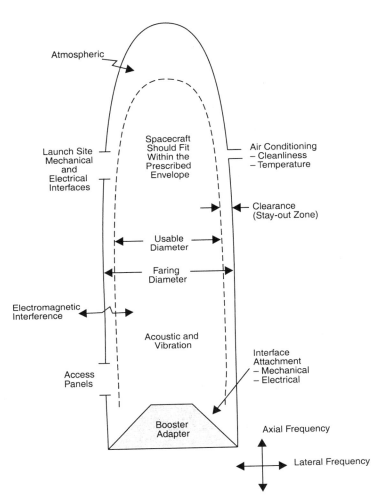

Fig. 18-8.   Typical Launch-System Fairing. This figure shows a typical payload fairing, illustrating the maximum dimensions and shape of the spacecraft allowed, as well as the separation plane between the spacecraft and launch vehicle.

architecture before launch if we expect the launch and early orbit operations to succeed. Chapter 13 discusses communications networks in more detail.

Payload Environments. We need to pay attention to the predicted payload environments so we can protect the payload during ground transportation, aircraft take-off and landing, hoisting operations, launch, and ascent. Table 18-7 lists the payload environments that we should assess for pre-launch and launch. The supply of conditioned air to the payload fairing controls the pre-launch thermal environment. Conditioned air typically moves through ducts into the top of the fairing, while vents near the fairing bottom maintain acceptable pressures and temperatures. Specifications normally call for static pressures of about 79 millibars, a temperature range of 9 to 37 $^{\circ}$C, relative humidity of 30% to 50%, and air filtration to class 10,000.

Contamination degrades the performance of solar panels, optical sensors, and surfaces used for thermal control. Where necessary, we need to control the handling of particles and molecules from the launcher's out-gassing materials. To model how contaminants move from source to removal, we analyze the thermal characteristics, compartment airflow, outgassing properties of materials, qualities and location of materials, and spacecraft components. We have to consider these analyses from ground processing through ascent.

Electrical signals must be compatible among the spacecraft bus, the payload, the launch vehicle, and the launch site. Electrical signals of different frequencies and powers can combine to form spurious radio transmissions and electric fields that spoof systems or fire ordnance devices. Shielding and design of ordnance circuits must conform to safety regulations at the launch site. For after lift-off, the launch-vehicle operator defines, documents, and integrates the flight electrical environment. The payload developer's analysis ensures that the spacecraft is compatible with these ascent conditions.

The following analyses of ascent environments concentrate on discrete events where flight experience shows the environment may drive the design. These design points include ignition and shutdown events, and periods of maximum dynamic pressure, maximum acceleration, peak heating rates, and heat loading.

Several analyses help us define and control the thermal environment for payloads during ascent. For expendable launch vehicles, we assess the thermal effects due to the radiant heat from the payload fairing internal surfaces. Maximum temperatures on the inner wall of a Delta-II vehicle range from 25 to 50° C. We also check radiated heat from the payload fairing or doors and free-molecular heating from rarefied air hitting the spacecraft after the payload fairing is jettisoned. When we use an upper stage, we have to consider the thermal effects from being exposed to the space environment during the parking and transfer-orbit phases. The Shuttle flights require similar analyses. We must consider any extended time the payload is aboard the Orbiter and the standard abort scenarios. We review results from these analyses to chart effects on mission operations and to constrain the time that the payload is exposed to the Sun or deep space while in the Orbiter bay.

Several static and dynamic loads affect the structures of the payload, adapters, and launch vehicle. These loads are either aerodynamic or they depend on acceleration and vibration. Aerodynamic loads are a function of the total pressure placed on the vehicle moving through the atmosphere. They consist of a static (ambient) pressure and a dynamic pressure (the pressure component experienced by a fluid when brought to rest). The relationship between altitude and velocity on the ascent trajectory determines these pressures. Payload fairings protect against dynamic pressure up to stated limits. (If strong winds on launch day result in excessive shear loads, the launch must be postponed.) At some point in the ascent trajectory the dynamic pressure will drop to levels which will not damage an unprotected payload; this results from the atmospheric density decreasing with altitude. At this point the fairing may be jettisoned to lighten the load on the booster. Typically payload fairings are not jettisoned until the dynamic pressure drops to 0.5 N/m^2.

During ascent a pressure differential occurs between the inside and outside of the fairing because the ambient atmospheric pressure continuously drops with altitude while the fairing contains higher-pressure air. Air trapped in compartments and crevices within the fairing and the spacecraft is at a higher pressure until the fairing vents to the outside. The venting rate depends on the pressure differential between internal payload compartments and the fairing, as well as the pressure differential between the volume enclosed within the fairing and the external environment. The venting rate also depends on the size and number of vent ports, on the spacecraft and on the fairing. Figure 18-9 gives typical data on pressure differentials (pressure inside the fairing minus external ambient pressure) for the Titan-III. Some values are ranges which cover trajectory dispersions and uncertainties in the venting rates. We may need to vent payload components to prevent damage during ascent.

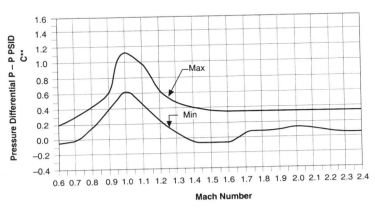

Fig. 18-9. **Pressure Differential in the Payload Compartment for Titan-III.** This chart shows the payload compartment's pressure relative to ambient air pressure during ascent. We must know this differential pressure to vent pressure adequately.

The acceleration loads, usually called *load factors*, experienced by the payload consist of *static* (steady state) and *dynamic* (vibration) loads. Table 18-8 shows typical values for several launch vehicles, measured in terms of *g levels* or multiples of sea-level, gravitational acceleration (9.806 65 m/s^2). Note that we must consider axial and lateral values. In addition to the normal ascent-acceleration and vibrational-load events, Shuttle payloads must look at events on orbit, during re-entry, and during emergency or nominal landings. Figure 18-10 shows the vibrational environments for several launch systems. These environments consist of launch-vehicle acceleration, variable combustion flows in the engines, aerodynamic drag and shear, acoustic pressures from the engines, and the mechanical response of the entire vehicle to these stimuli. We must design the payload and booster adapter to carry these loads. Table 18-9 specifies a lower limit for the fundamental frequency of the booster adapter and spacecraft combination, resulting in a structural stiffness requirement. However, if the combination is too stiff, the structure may interfere with the most energetic vibration frequencies shown in Fig. 18-10. Thus, we need to update analyses of static and dynamic loads, so they include more refined data on the design of the launch system, booster adapter, and payload.

To separate the launch vehicle from the spacecraft, or to deploy spacecraft components, we typically use pyrotechnic devices. These devices are light, highly reliable, and easily integrated into mating techniques that provide a high degree of stiffness. Unfortunately, when activated they generate a shock load that transmits through the structure to the payload. Figure 18-11 shows the shock characteristics for several launch systems.

TABLE 18-8. Launch System Acceleration. Steady-state and dynamic components for several critical ascent events are shown. We must design payloads to survive the sum of steady state and dynamic accelerations in the axial and lateral directions. When only dynamic load factors are given, they include steady-state load factors. All entries are in g's.

Vehicle	Lift-Off		Max Airloads		Stage 1 Shutdown (Booster)		Stage 2 Shutdown (Booster)	
	Axial	Lateral	Axial	Lateral	Axial	Lateral	Axial	Lateral
T34D/IUS								
Steady State	+1.5	—	+2.0	—	0 to +4.5	—	0 to +2.5	—
Dynamic	±1.5	±5.0	±1.0	±2.5	±4.0	±2.0	±4.0	±2.0
Atlas-II								
Steady State	+1.3	—	+2.2	+ 0.4	+5.5	—	+4.0	—
Dynamic	±1.5	±1.0	±0.3	±1.2	±0.5	±0.5	±2.0	±0.5
Delta (max* all series)								
Steady State	+2.4	—	—	—	—	—	—	—
Dynamic	±1.0	+2.0 to +3.0	—	—	—	—	+6.0	—
H-II								
Steady State	—	—	—	—	—	—	—	—
Dynamic	±3.2	±2.0	—	—	—	—	±5.0	±1.0
Shuttle with IUS								
Steady State	+3.2	+2.5	+1.1 to 3.2	+0.25 to - 0.59	—	—	+3.2	+0.59
Dynamic	+3.5	+3.4	—	—	—	—	—	—

* 2σ Values

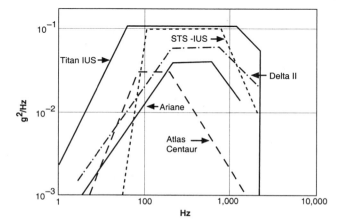

Fig. 18-10. Vibrational Environment for Launch Systems. The launch vehicle and fairing dictate the environment. The vertical axis is the power spectral density.

TABLE 18-9. Fundamental Frequencies for Spacecraft Design. The booster adapter and spacecraft structure should be designed for fundamental frequencies greater than or equal to those shown.

Launch System	Fundamental Frequency (Hz)	
	Axial	Lateral
Atlas II, IIA, IIAS	15	10
Ariane 4	*	10
Delta 6925/7925	35	15
Long March 2E	26	10
Pegasus, XL	18	18
Proton	30	15
Space Shuttle	13	13
Titan II	24	10

* 31 Hz for dual payloads, 18 Hz for single payloads.

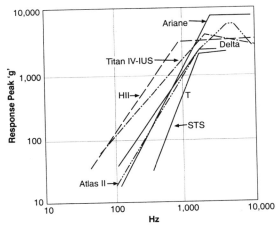

Fig. 18-11.   Shock Environments for Launch Systems Caused by Staging and Separation Events.

The acoustic environment is a function of the physical configuration of the launch vehicle, its acceleration time history, and the configuration of the propulsion system. The near-launch-pad acoustic environment depends on the reflected sound energy from the launch pad structures and facilities. The maximum dynamic environment is a consequence of the rate of acceleration and the aerodynamic smoothness of the launch vehicle shape. Solid-rocket boosters and first stages (often combined) usually provoke a more severe environment, and the smaller the launch vehicle, the more stressed the payload. In general, the closer the payload is to the launch pad, the more severe the acoustic environment prior to releasing the launch vehicle, and the more rapidly the vehicle accelerates, the more severe the environment at maximum dynamic pressure. Since the acoustic excitation is rapidly time-varying, we must design for the instantaneous values and the overall average. When the payload design is sensitive to the acoustic environment, it is common to add damping insulation to the fairing. Figure 18-12 shows the acoustic environments of several launch vehicles.

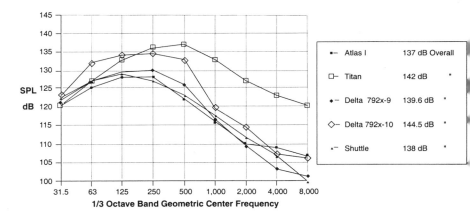

Fig. 18-12. Acoustic Environments of Typical Launch Configurations. These curves represent the *sound pressure level* (*SPL*) of the acoustic energy at one-third octave intervals for typical launch configurations. Customarily, the payload receives maximum energy at two points: at ignition before release of hold-downs and at transition through maximum dynamic pressure (*Max q*). Delta provides both a 2.9 and 3.2 m standard shroud for which the acoustics are different. Often shrouds are made to provide the desired environment.

Injection accuracy is also important to launch systems. Traditionally, the launch system and payload separate at a prescribed location and velocity. Launch-system vendors usually state an orbit-injection accuracy that depends primarily on the last stage of the vehicle. Table 18-10 gives injection accuracies for several systems launching a payload to geosynchronous transfer orbit.

TABLE 18-10. Injection Accuracy for Geosynchronous Transfer Orbits. These numbers depend on the last stage of the vehicle and type of guidance used. We may need more propellant on the spacecraft to correct errors in inclination and altitude at apogee.

Launch Systems	Injection Accuracies		
	Apogee (km)	Perigee (km)	Inclination (deg)
Delta II	337	0.25	0.12
Atlas/Centaur	82	1.7	0.01
Ariane 4	102	0.91	0.03
Shuttle/IUS	193	0.25	0.02

Early in the design, we must consider payload processing and integration procedures at the launch site. Commercial and government facilities are available for processing payloads at various launch sites. The actual flow depends on specific processing requirements, configuration of propulsion, ordnance elements, and available facilities. Payload processing includes receiving inspections, checking payload and ground-support equipment, installing hardware such as batteries and avionics, checking pressures and gas leaks, and testing functions and communications.

Potential integration functions at the launch site include mating the spacecraft with a stage vehicle, spin tests, loading propellants, mating with the launch vehicle, and prelaunch testing of all systems (integrated test). Two items to consider are the launch-site layout and how to integrate the payload and launch vehicle. A launch site may have several pads, so we can integrate several spacecraft at the same time. Others may have one pad that restricts the flow to one payload at a time. If the launch site uses a series approach, launch delays and failures in one spacecraft or launch system may adversely impact the next program.

Various launch systems use vastly different approaches to physically integrate the payload and the launch vehicle. The most common is vertical integration, which means erecting the booster and hoisting the payload on top of it. With this method, we must build platforms at various levels so we can get to the entire launch vehicle, stage vehicle, and payload. This approach works, but it is fraught with restricted access and safety problems. Another way is horizontal integration, which means securing the launch vehicle in a horizontal position and attaching the spacecraft. This approach eases access and lessens safety concerns. Actual payload processing varies according to the complexity of the spacecraft, and should be examined case-by-case. The time required at the launch site depends on its design, interface requirements, integrated test requirements, amount of shared resources, and the launch-site operator's philosophy and approach.

Following the process defined in this chapter, you should be able to identify several acceptable launch systems for a particular mission and establish preliminary design requirements for the spacecraft. The process is an iterative one, and you will probably need to use it several times prior to converging on the spacecraft design and launch vehicle.

Launch Vehicle User Guides

Ariane 4. Arianespace, Inc. 1747 Pennsylvania Avenue, N.W., Suite 875, Washington, DC 20006. Ariane Launch Vehicles, rue Soljenitsyns, 91000 Evry, France.

Atlas Launch System Mission Planners Guide. February 1995. Lockheed Martin Commercial Launch Services, Inc., 101 West Broadway, Suite 2000, San Diego, CA 92101.

Commercial Delta II User Manual. July 1989. McDonnell Douglas Commercial Delta, Inc., 5301 Bolsa Avenue, Huntington Beach, CA 92647.

Delta II Payload Planners Guide. October 1995. Boeing McDonnell Douglas, 5301 Bolsa Avenue, Huntington Beach, CA 92647-2099.

H-I and H-II Rocket. NASDA External Relations Department, World Trade Center Building, 4-1 Hamamatsu-cho-2-chome, Minato-Ku, Tokyo 105, Japan.

Long March Family of Launch Vehicles. China Great Wall Industry Corporation, No. 17, Wenchang Hutong Xidan, P.O. Box 847, Beijing, China.

Payload Users Guide Titan II Space Launch Vehicle. August 1986. Titan II Space Launch Vehicle Program, Space Launch Systems Division, Martin Marietta Corporation, Denver Aerospace, P.O. Box 179, Denver, CO 80201.

Pegasus Payload Users Guide Initial Release. December 1988. Advanced Project Office, Orbital Sciences Corporation, 12500 Fair Lakes Circle, Fairfax, VA 22033

Proton Launch Vehicle and Launch Services Users Guide LKE—(Lockhee Krunichev Energia). December 1993 (available from Lockheed Martin Commer cial Launch Services).

Shuttle Orbiter/Cargo Standard Interfaces, ICD 2-19001, NSTS Volume 14 Attachment 1, NASA L. B. Johnson Space Center, Houston, TX 77058.

Soviet Launcher Proton. Glavcosmos USSR, 103030, Moscow Krasnoproletarskay str., 9.

The Inertial Upper Stage Users Guide. January 1984. Boeing Aerospace Co., A Division of Boeing Company, P.O. Box 3999, Seattle, WA 98124.

Titan IV Users Handbook. June 1987. Titan IV Space Launch Systems, Marti Marietta, Denver Aerospace, P.O. Box 179, Denver, CO 80201.

Reference

Isakowitz, Steven J. 1995. *AIAA International Reference Guide to Space Launc Systems (2nd Edition).* Washington, DC: American Institute of Aeronautics an Astronautics.

Chapter 19

Space Manufacturing and Reliability

19.1 Designing Space Systems for Manufacturability
*Challenges for Manufacturing, Material, Test, and
Launch Processing; Creating the Manufacturing
Vision; Influencing the Design; Process
Development and Verification; Production*

19.2 Reliability for Space Mission Planning
*Design for Reliability, Design for Fault Avoidance,
Fault Tolerance, Test Techniques*

19.1 Designing Space Systems for Manufacturability

Wade Molnau, *Motorola Systems Solutions Group*
Jean Olivieri, *Motorola Advanced Systems Division*
Chad Spalt, *Motorola Satellite Communications Group*

Historically, satellite manufacturing, integration, and test has been a crafted and arduous process. Each spacecraft is essentially unique, and is manufactured and tested appropriately. Commercial satellites, while alleviating some of the major impedances to fast and efficient satellite manufacture, have fared only slightly better. The advent of commercial constellations of satellites forces us to seek and develop completely new strategies. We need to incorporate ideas and methods from other industries into satellite supply-chains to meet the cost and cycle-time requirements needed to make space systems compete effectively with their terrestrial counterparts.

This chapter describes a few of the vital changes that need to be addressed to manufacture and test multiple satellites efficiently. These methods and strategies apply to the whole satellite supply chain, and to piece-part, assembly, subsystem, and spacecraft levels. Chapter 12 detailed methods used to manufacture and test single satellites. This chapter augments Chap. 12 for multiple satellite systems.

Small satellite systems (< 10 spacecraft) may not fully benefit from the methods presented here, but some points will be applicable. Designing manufacturable satellites and associated production systems requires up-front investments in time, money, and capital. Each program needs to trade the benefits of these methods with anticipated investment costs. As the number of satellites grows, the benefits and the usefulness of these methods increases.

We begin with a short description of the goals and challenges of manufacturing, material, test, and launch processing organizations. The majority of the chapter then concentrates on four phases of the manufacturing and test of the spacecraft constellation: (1) creating the manufacturing vision, (2) influencing the design, (3) developing and verifying the process, and (4) producing the spacecraft.

19.1.1 Challenges for Manufacturing, Material, Test, and Launch Processing Teams

The challenges presented to the manufacture of satellite constellations are extremely different from those traditionally conceived for spacecraft. These challenges, however, are not much different from those for other commercial products. Spacecraft are designed and built to cost and schedule goals—just like video recorders or cars. Figure 19-1 compares a few of the challenges facing the manufacture of a constellation of satellites, while Fig. 19-2 shows manufacturing issues as they pertain to key metrics—quality, time, and cost.

Key Metrics	Traditional	Multi-Satellite
Quality	Build, Test, Rebuild, Retest . . .	Build, Verify, Ship, Shoot
Spacecraft Cost	66,000 $/kg	19,800$/kg
Cycle Time (Integration)	225 days	24 days
Cycle Time (Build to launch)	18 months	2 months

Fig. 19-1. Typical Constellation Manufacturing Challenges. Note the significant difference between traditional space vehicle manufacture and the newer manufacture process.

```
◆ Six-Sigma Quality
    – Satellites are not field serviceable!
◆ First to Market → Commercial
   Leverage the Market → Government/Military
    – Reduce benchmarked cycle time from 225 days to 24 days
      per satellite
◆ Fixed Budget, Commercial Pricing
    – Continuous profit improvement
```

Fig. 19-2. Constellation Manufacturing Issues. The quality, cycle time, and cost issue relating to the production of a constellation of satellites are different than those encountered while producing single satellites. While quality requirements are different—but still stringent—cycle time and cost issues are now paramount.

For instance, time to market often is not only a goal, but a competitive requirement for viability. Time becomes an overriding factor. This drives not only the overall time to market, but the time to complete each system element. For manufacturing, this translates to drastically reducing the cycle time required to produce a satellite.

Quality, as with all spacecraft, is still imperative. Satellites are not field serviceable and simple defects can render a satellite unusable. The traditional approach places emphasis on high-cost and time-consuming quality assurance methods that check and recheck, test and retest, and verify and reverify the hardware. Instead, we employ *six sigma* methods to verify that the processes are performing with high-quality results. Robust designs using in-control processes will produce high-quality products—without associated high costs and long cycle times. Six-sigma methods are a cornerstone for lean and agile manufacturing methods (see sidebar).

To meet the additional cycle-time and quality requirements, manufacturing personnel must participate concurrently during the creation of the design. They are responsible for influencing the design and tailoring the design into a final, producible product. Finally, manufacturing must effectively integrate the product during production.

In summary, the challenges presented to manufacturing are to recreate the value-added or useful elements of space hardware design and manufacture. We must eliminate non-value-added activities. We must add elements for quick, efficient, and high quality execution. We must drive out non-value-added or superfluous activities and requirements.

The Six Sigma program began in Motorola in 1985 as a means to measure and improve performance in all phases of its business. The ultimate target is virtually perfect execution.

Sigma (σ) is used to designate the distribution or spread about the mean or average of any process or procedure. A simple definition of *six sigma* is 3.4 *defects per million opportunities*, or 99.9997% perfect. Sigma level improvement is not linear. In fact, improvement rates escalate rapidly as the base sigma level slowly increases.

Six Sigma methodologies are a set of tools and procedures to improve processes. Fundamental methods include problem solving, process control and process characterization. They can be applied to any process from engineering to business. For more information, reference Harry [1997] or Harry and Lawson [1997].

Material Challenges

Traditional manufacturing drove spacecraft parts to the costliest and slowest-delivery grade—*S-level*. Unfortunately, these parts are not inherently better than commercial parts—they have just been screened for various properties to improve their overall expected performance or failure rates. The parts are mostly produced on commercial part lines and even from commercial part batches.

Our challenge is to use the *right* parts for the mission—not the highest level parts available. We need to target commercial parts, replacing them only when required. We need to discourage military grade or S-level parts, but allow them in situations where they *must* be used. A simple comparison of part costs, as shown in Fig.19-3, depicts the drastic difference for S-level, Mil-spec, and commercial parts. Significant cost savings can result from selecting lower-cost parts when applicable. As discussed in Sec. 12.2, significant time savings also can result from selecting lower grade parts. This yields a significant advantage of using Mil-spec parts over S-level parts. Furthering this comparison, we can receive commercial parts typically even quicker—often in less than a week.

Parts selection needs to consider producibility, radiation, out-gassing, and other requirements. We need to consider strongly parts that support efficient manufacturing methods. We should target parts that use the standard process flows—for instance, automated placement and mass reflow for electronic assemblies.

Finally, parts selection needs to be a value-added concurrent engineering activity. Success in the material area heavily influences overall program cost and schedule performance.

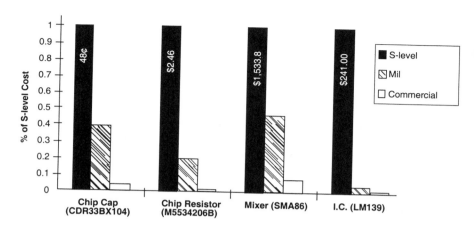

Fig. 19-3. Representative Parts Price Comparison. Commercial parts represent a significant cost savings over Mil-spec and S-level parts.

Test Challenges

The test function drives the overall cycle time of spacecraft delivery. This holds true not only for traditional satellites, but for constellations as well. The key is to design test as a process—this includes minimizing cycle time, maximizing throughput, and achieving high-quality results.

The purpose of the overall test program is to verify the adequacy of the design and assembly processes. The requirements to verify these two elements—design and processes—are inherently different.

Design verification, more commonly called *qualification testing*, requires detailed and thorough analysis of performance capabilities. This still is a time-consuming process that includes functional test, vibration and shock, thermal vacuum, electromagnetic compatibility, and other tests. This may include multiple iterations, and essentially involves testing the satellite to greater than expected levels. We should cut back and ultimately eliminate these qualification tests after we understand and proof the design. We can achieve significant savings in time and money by relinquishing the need for full-up testing on every unit.

Conversely, *process verification* continues throughout the production of the spacecraft. These tests only verify the performance of the production system and include only simple performance checks. The control of the individual processes throughout the production cycle ensures the high-quality results. Complete testing of the spacecraft would be redundant and non-value added—and not consistent with the principles of lean production. Figure 19-4 shows some key differences between traditional spacecraft production and lean production.

In regards to components, subassemblies, and subsystems, it is critical that they achieve their qualification testing and product testing as early as possible. We need to detect quickly issues with design or process. We can integrate high quality and well-understood components, subassemblies, and subsystems with no additional integration level testing—with high confidence. At integration, we need to verify only incremental processing—usually only shipping and connection.

TRADITIONAL	MULTI-SATELLITE
Craft Industry	**Lean Satellite Production**
• Optimize Locally	• Optimize Supply Chain
• Contracts Based	• Partnership Based
• Aerospace Practices	• Commercial Ingenuity
• Performance Only	• Process Driven Design
• Distributed Factory	• Assembly Line
• Unique Product	• Same Product

Fig. 19-4. **Lean Satellite Production Principles Compared to Traditional Craft Production.** Lean production methods originated in the automobile industry. These methods can be applied to the production of satellites as well. See Womack et al. [1990] for in depth discussion of lean manufacturing.

In summary, test provides possibly the greatest opportunity for significant cost and cycle-time improvements. Intelligent selection of reduced testing, clear distinction between design and process verification, and performance of lowest-level testing greatly enhances the cost-effective, high quality, low cycle-time production of spacecraft without significantly increasing risk.

Launch Processing Challenges

After finishing integration and test, we package and ship the satellites to the launch site for processing and preparation for launch. Traditionally, this has been a complex and time-consuming segment of the satellite delivery process. The requirements for building a multi-satellite constellation include quick and efficient processing of satellites for launch. These requirements may also force the use of multiple launch sites and launch-preparation areas, further driving the need for standardized and simplified launch processing.

We need to consider launch-site processing from the start of the program. We should target decisions and trades that support simplified processing. Changes and processing are much easier to handle while at the integration factory. This is where the facilities and skill levels are the most available. Launch-site processing typically takes place at a remote location—often not owned or operated by the satellite manufacturer—that makes processing more difficult.

The spacecraft are transported to the launch-site processing area via protected environment or container. The container environment must provide for temperature, humidity, vibration, and shock protection during the shipping process. In order to support the streamlined processing of the satellites upon their arrival, the shipping process is subject to basic process control. We deem the process to be successful as long as the required parameters—temperature limits, humidity limits, vibration levels, and shock levels—are not exceeded. If these parameters exceed their limits, we need to perform in-depth analysis or additional testing at the launch-site. But, if none of the parameters are exceeded, the satellite can continue with the streamlined process flow that contains little, or preferably no, testing.

Figure 19-5 depicts a streamlined launch-site process flow. The processing includes all the steps needed to prepare the spacecraft from shipping condition to flight condition. Launch team members receive, inspect, load onto the launch dispensers, fuel, package, and transport the space vehicles to the launch vehicle. They charge and monitor the flight batteries, which may be installed at the launch site, when the spacecraft are mounted on the launch vehicle.

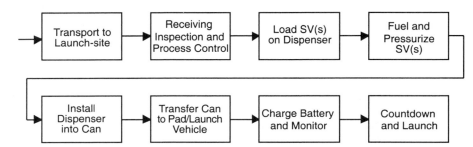

Fig. 19-5.   Streamlined Launch-Site Process Flow. Traditional launch-site processing can take several weeks. Streamlined processing can reduce the overall cycle time (transport to launch) to less than two weeks.

We remove all difficult tasks that can be driven to earlier processing. We minimize hazardous operations. For instance, we can replace deployment ordnance can be replaced with other mechanisms. We should minimize or eliminate test operations. If the shipping process was in control (e.g. no limits exceeded on shipping sensors), we need no test verification of the process. Well-designed tooling and process development can make the remaining processes quick and efficient with little threat of damage to the spacecraft.

19.1.2 Creating the Manufacturing Vision

One of the most important factors relating to a successful production program is the creation of a manufacturing vision. More and more customers today want the "better, cheaper, faster" solutions to their product needs. We should create the manufacturing vision based on the customer's and the market's expectations for the product. The entire project team must embrace this vision. All parties involved with manufacturing must share the manufacturing vision and we must also include everyone else involved with the program from the early design phases through production and test. We must create manufacturing goals to support the vision. Creating manufacturing goals for quality, cycle time and cost must directly relate to the manufacturing vision. For example, if there is a manufacturing vision to produce enough satellites to populate a constellation in a single year then the manufacturing goals must reflect this vision. These goals must be in the form of quantified short cycle times, quantified quality levels and quantified costs. Once the manufacturing goals are in place, the next step is to execute these goals and pursue the manufacturing vision. We must constantly measure the progress towards meeting these goals and share them with the team. We must evaluate progress and make necessary changes to achieve the stated goals. Changes may occur in the design or in the production and test processes. Changing the design and processes to meet the manufacturing goals is an iterative process. It is a give and take scenario where the sum of the changes equals achieving the manufacturing goals.

Establish Goals

We must establish goals in the areas of cycle time, quality and cost. In today's competitive environment, these areas are everything. Many times multiple companies have similar ideas and the only differentiating characteristic between their products is which one makes it to the consumer market first and captures the majority of the market share. Missing the time to market goals will cost a company market share, revenue and brand equity. Therefore we need to know up front when our product must be ready for our customer and then determine our cycle time goals to support this overall schedule.

As stated previously, one of the unfortunate aspects of building satellites is that they are usually not field serviceable. If there is a quality problem after we launch the product into space, we have to live with the degraded performance or, in an extreme case, accept that the product does not meet our mission requirements. We must establish quality goals so that we can design our product accordingly. When we choose components for our product and select processes to build and test our product, we determine our resulting quality levels. As we choose components and processes we need to calculate our cumulative quality predictions and track them to our goals. We will then have the visibility to see what components and processes have a positive or negative impact on the overall product quality. With this data we now make informed trade-off decisions. We can determine which components or processes to change to increase the product quality. Often quality drives cycle time. If a product is designed with low quality levels then we should expect to perform more rework and additional testing. This in turn adds cycle time. Hardware designed around robust processes which exhibit low defect rates result in products with fewer defects which leads to less test time, less rework and lower costs.

Most customers desire low-cost solutions. More and more of the traditional "cost plus" projects are being proposed as fixed price contracts. The days of cost-plus contracts and cost overruns being absorbed by the customer are quickly disappearing. Today the contractor carries the burden of performing to an agreed-upon fixed price. Because of this, the contractor must know the cost goals and understand how they are going to perform to these cost goals. Performance to cost goals directly relates to performance to cycle time goals and quality goals. The addition of unplanned cycle time and of effort required to correct quality problems result in increased cost. In general any unplanned work results in additional cycle time and in additional costs.

Involve Manufacturing Early

Manufacturing early involvement may not always be the accepted way of doing business. Early manufacturing involvement allows the design team to be informed regarding the effects of their choices on downstream manufacturing and test operations. It empowers the design team to make informed decisions and predetermine the expected cycle time, quality and cost performance of the assembly and test operations. It allows management to be aware of what to expect when the product reaches assembly and test.

Getting the design team to accept concurrent engineering can be a difficult task. Many times a company needs a culture change for everyone to embrace the early participation of manufacturing. Management must support concurrent engineering and must pay attention to the analysis results created by the manufacturing representatives. The manufacturing and test participants on the concurrent-engineering team must embrace the manufacturing vision and possess the tools to influence the design to meet this vision. They must show the other members of the design team that their participation is value-added and that early team involvement will save the program time and money in the long run. Everyone on the project team must share the manufacturing vision in order for it to become reality.

Select Parts and Processes Effectively

As was discussed earlier the sum of the project's individual process cycle times (both design and manufacturing/test) must support the overall cycle time goals and the manufacturing vision. Cycle time is inherent in the design. We predetermine the cycle time to assemble and test a product when we choose parts and processes for a particular product. Therefore, during the product design phase we must give appropriate detail to the proper selection of parts and processes. Overall the design team must realize that a design consists of parts and processes. Often the choice of a particular part will dictate the use of a particular process. This relationship is what gave rise to the phrase "pick a part, pick a process". Often design engineers choose parts that meet a limited set of criteria from which the design engineer is searching. They often do this without knowing the downstream impacts that these part choices have in assembly and test.

Therefore it's important to have manufacturing and test representatives involved in the part selection process. They can perform analyses to show how certain parts may require additional process steps which have lower process yields leading to more defects, higher cycle times and ultimately higher costs.

An electronic assembly example of this type of analysis is shown in Table 19-1. We can extend this analysis to satellite design and manufacture. Part X and part Y are equivalent in the areas of form, fit and function; therefore we would assume that the parts are interchangeable. However the parts vary greatly when we address associated processes, process yields and overall cost. Part X is an 84-lead *ceramic quad flat pack* that must have the leads formed, trimmed, and tinned. The part must then be loaded into a tray for presentation to the part placement (pick and place) equipment. Form, trim, tin and tray placement are non-preferred processes. We must perform these processes in-house, which contributes to increased process time for that part type. The low process yields associated with these non-preferred processes drive this part to contribute 0.273 *defects per unit* (DPU) which leads to increased cycle time and costs associated with rework and potential test failures. Part Y is an 84-lead plastic *quad flat pack* that comes to us formed, trimmed, tinned and loaded on tape and reel to be presented to the pick and place machine. This part uses only preferred processes which leads to lower process cycle times, lower defects per unit and lower rework costs. All of these factors are reflected in the lower total cost of Part Y. Therefore when selecting a part the design team must look at the downstream impact of their part selection and include the total cost impact of the part in their design tradeoffs.

By relating parts to processes we can determine the expected time to manufacture and test a product, the associated quality level and the overall cost. Once we make these estimates we can perform "what if" analyses to see if other part/process combinations will make attaining the cycle time, quality and cost goals more realistic. We perform this iterative process early in the design phase, so we can include many design tradeoffs between all of the disciplines.

Parts have many attributes that influence the effectiveness of manufacturing processes. Some of these attributes fall into the categories of cost, interchangeability, availability, reliability and simplicity. Early part-supplier involvement is critical to

TABLE 19-1.   Part/Process Cost Relationship

Part	Package	Part Cost	Processes Used	Process DPU	Process Time (Hr)	Assembly Cost	Rework Cost	Total Cost
Part X	84CQFP	$25.00	1. Lead form/trim 2. Lead tin 3. Pick & place —tray 4. Oven reflow	0.273	0.05	$3.20	$40.95	$69.15
Part Y	84QFP	$17.50	1. Pick & place —tape 2. Oven reflow	0.00306	0.002	$0.128	$0.459	$18.00

ensure acceptable levels in all of these categories. The suppliers need to know what our cost goals are for the parts being supplied. The supplier also needs to understand our product design and how the supplier part interacts with our design. The supplier can work with us to determine the appropriate variables to control to ensure that the part is interchangeable and reliable. The supplier also needs to understand our production schedule and our material management philosophy in order to make sure that their parts are available to us when we need them.

Part cost directly impacts product cost. It also has an impact on process development and process execution. Processes used to assemble high-cost parts are usually designed with additional controls and verifications. Typically this leads to higher stress levels for the people doing the assembly work and more engineering involvement for process verification. Use of higher cost parts drives the desire to perform more rework rather than scrap hardware in the event of unwanted process variations.

Part interchangeability is critical in a product design regardless of the production quantity. The ability to remove and replace a component quickly translates into less non-value-added time. Interchangeable parts have the same form, fit and function for all parts. The supplier of interchangeable parts must use process control to insure that the parts coming into the assembly area are truly interchangeable. If a part is not truly interchangeable, assemblers will have to spend additional time to alter mechanical characteristics of the part to make it fit or perform additional tests to adjust for the varying performance of the part.

Part simplicity directly relates to part cost, interchangeability, availability and reliability. Typically the more simple the part design the lower the cost. A simple part will have less variables associated with its design, therefore it will positively impact the component reliability and interchangeability. A simple part will require less production set up time and have fewer variables to cause scrap. Therefore a simple part should be more available for use.

Part availability is critical to maintaining product flow in the assembly area. We never want to stop the assembly process due to parts unavailability. Many times the risk of part unavailability is controlled by creating material inventories. However, there can be a great deal of cost associated with carrying part inventories. Typically we must allocate valuable factory floor space to material storage near its point of use. An alternative to storing material on the factory floor is to store it in a stock room and move the material to the assembly line when we need it. Both of these scenarios result in material being stocked at the assembly location and increase the risk of having large quantities of parts that may become obsolete or require rework/retrofit if there is a design change. The best alternative to either of these scenarios is to have *"just in time"* suppliers. Working with Just In Time suppliers requires less stocking activities and hence less chance of having vast amounts of product that has to be reworked or retro-fitted when there is a design change, but can add risk. The supply chain information—especially deliveries, cycle times, and quality—must be well understood to make Just In Time plausible. If not well understood and reliable, Just In Time is very risky.

Once material is ready for production it is time to use our manufacturing processes. In the typical aerospace manufacturing environment, processes were developed after the design, so the product could be manufactured and tested to meet the customer's needs. Before the implementation of concurrent engineering this meant that the design was "thrown over the wall" to manufacturing and manufacturing did "what ever it takes" to build, test and deliver the hardware to the customer. Typically this meant that the processes were highly flexible, required specially skilled workers, lacked good control and were seldom characterized. Today with the advent of *concurrent engineering*, we design in the controlled and repeatable manufacturing processes from the beginning and the manufacturing output is highly predictable.

Concurrent engineering has brought about a change of attitude regarding manufacturing processes in the aerospace industry. No longer do we perform processes just because "that is the way we've always done them." High-quality processes are characterized, controlled and repeatable. Many of these processes have been borrowed from the commercial manufacturing world. The mass reflow of electronic assemblies is an example of this exchange. Mass reflow—the mass soldering of parts on printed circuit boards using a single, controlled thermal cycle—was developed to efficiently produce electronic products for delivery to consumers. Although slow in gaining acceptance, we now use it for spaceflight hardware and other high reliability applications.

The concurrent manufacturing engineer should never be afraid to ask "why" we design or process something the way we do. An example of this is to look at the integration of deployables on a satellite. In the past certain deployables have always been integrated with the satellite in the vertical position. This was done to minimize the amount of stress that could be preloaded into the deployable mechanisms. Changing the satellite from the horizontal position to the vertical position and back to the horizontal position requires time, space, additional tooling and creates a lot of risk associated with handling the satellite. With today's modern analysis tools much of the stresses associated with preloading the deployable mechanisms in the horizontal position can be modeled and the risks analyzed. If the risks are determined to be low enough then the satellite plane change from horizontal to vertical can be eliminated. This will save time, space, and tooling and also reduce handling risk.

Typically satellite manufacturing has been a low-volume process. Currently many companies are proposing multi-satellite constellations. These multi-satellite projects drive manufacturing to create production lines for assembling and testing satellites. The "assembly line" concept of these satellite production lines has roots in the automotive industry. Many of the new satellite production lines take mass production assembly line technology one-step further and implement many of the *"lean" manufacturing* principles currently used in the automotive industry. Some of these "lean" manufacturing principles include Just In Time suppliers, supplier partnering, concurrent engineering, pull production,* and statistically controlled processes.

Other satellite manufacturers have taken the "lean" manufacturing philosophy to the next step—*"agile" manufacturing.* "Agile" manufacturing incorporates all of the principles used in "lean" manufacturing with the addition of flexibility. An "agile" manufacturing line can support multiple product lines on the same assembly line. It also can adapt and respond to product changes which may be directed by changing customer wants and needs.

One of the guiding principles behind "lean" and "agile" manufacturing is waste elimination. We must evaluate all of the processes associated with assembly and test in terms of value-added activities and non-value-added activities. During process development and implementation, we should maximize the value-added processes and the non-value-added activities should be minimized. Reduction of the non-value-added process steps will drive waste elimination.

Reduction in the number of inspection points is a good example of waste elimination. Typically satellite production programs have had a high content of quality inspection points. Inspection points are examples of non-value-added operations. In the past multiple inspection points were required because the assembly processes were highly variable and uncontrolled. By institutionalizing concurrent engineering, processes are now designed into the assembly flow that are controlled and repeatable. Because the output of these processes are predictable, we no longer need multiple inspection points. Therefore we eliminate waste by removing the non-value-added inspection steps.

Another non-value-added recurring process is test. Because test is non-value-added, we need to be minimize it to eliminate waste. Although some testing is required, we should minimize redundant testing. One of the benefits of using controlled and repeatable processes is that the output from these processes is predictable. Therefore we don't have to test something that has been tested earlier if we know that the processes used after the initial test have not injected variability into the product. A good example of this principle is the previously-mentioned minimization of launch-site processing. We test the spacecraft prior to shipping it to the launch site, and don't test it at the launch site unless shipping indicators or sensors show that it exceeded a shipping process limit.

Test is just another process. During the early phases of design, we should view the assembly and test efforts as processes that must be controlled and repeatable. Similar to satellite manufacturing processes, we should be establish goals for all test processes. These goals should relate to cycle time, quality (predicted test yields), and cost. We should question any non-value-added steps and, if possible, eliminate them. If a test continually exhibits a 100% yield rate then we need to determine if the test is necessary.

Reality is that test yields are not always 100% and finding defects at higher levels of integration can negatively impact cycle time and cost. The goal of performing a test is to find the defect at the lowest possible level of integration. This reduces the amount

---

* **Pull Production Systems**
    Traditional satellite manufacture and mass production in general is based on a push system. In other words, the material and work-in-process is generally pushed or processed through the system. The emphasis is on processing the hardware through the production line as quickly as possible. Unfortunately, this can create pockets of inefficiency and bottlenecks (throughput limiters). These problems require excess factory space for inventory of material and product, and can create excessive defects and cycle times.
    Lean production utilizes a pull production methodology. Material and work-in-process is not processed until upstream stations are ready. Downstream processes do not create excess inventory, because the production system will not allow production to continue unnecessarily. The system is tuned for a continuous process flow, with necessary materials being delivered Just In Time in small quantities.

of time and effort required to get to the problem area and fix it. Even after tests have been performed there are instances when the quality of the lower level hardware may still be suspect at the higher levels of integration. Therefore it is important to design in accessibility at the higher levels of integration. We may need this accessibility to repair a defect that has passed undetected through a lower level test operation or to incorporate late design modifications prior to sealing the satellite. Design for accessibility needs to be a goal of the program vision because it doesn't happen without conscious effort. For a detailed discussion of process design and production trends, see Shunk [1992].

19.1.3 Influencing the Design

Success of production depends on the manufacturing/production disciplines being involved throughout the product design. Through concurrent engineering, we develop the product involving various disciplines of the trade. For satellite design, this includes such expertise as electrical, mechanical, software, test, industrial, and reliability engineering; it also includes other less-technical but equally important functions such as contracts, finance, procurement and scheduling. The manufacturing role, in particular, acts as the main interface between many product teams due to their direct responsibility of ensuring a producible product. All experts must get involved with the design early and continually to the extent they can influence the product to meet their respective goals or to compromise with others. At the very least, if the experts can't reach an acceptable term, they can highlight the issue very early in the program so that they can continually work it into an acceptable term.

To influence the design, the program team must first establish a common vision for the enterprise. This common vision serves to create a working platform that allows the supply chain to collectively communicate. It provides the central theme to which the program will focus regardless of where anyone's portion of the business is located. The common vision creates the need for goals and guidelines in each concurrent design area within the program. Overarching goals must be established to provide incentive for the program; they should be difficult to attain, yet attainable within the program life.

To supplement the goals, the team must establish specifically defined rules and guidelines to bring the goals to fruition. The vision, goals, and guidelines provide the program with a standard, central direction that, in turn, provides the program with the opportunity to be efficient and united. The intention of a common vision is not to strip individual company identity, rather to converge on certain business ideas that alleviate challenges due to diverse mindsets, backgrounds, and corporate cultures.

Designing to goals and guidelines sounds simple but may be difficult in practice; few programs have these clearly outlined at the inception of the program. Implementing goals at the beginning of the program is where they are most effective. Studies have shown that approximately 70% of the cost of the system is fixed by the end of the conceptual design phase [Wade and Welch, 1996].

A concept becomes feasible through refinement and, in turn, producible through other iterations of refinement. If we initiate the guidelines of design at the conceptual level, it is less costly and less painful to iterate to a feasible design. The cost to correct problems occurring from a previous program phase increases by an order of magnitude, approximately (Fig. 19-7). A simple design change during the conceptual design phase will cost 10× during broadband phase and 100× during engineering model phase.

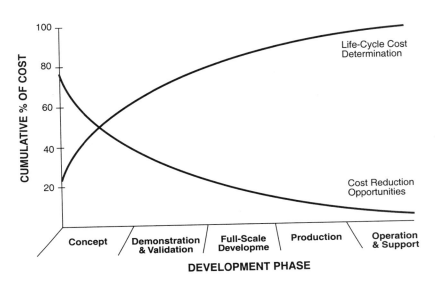

Fig. 19-6.   Project Phase/Cost Assignment. Adapted from Wade and Welch [1996]. The ultimate costs of a system are greatly influenced by decisions made early in the product life cycle. Based on model fit to empirical data.

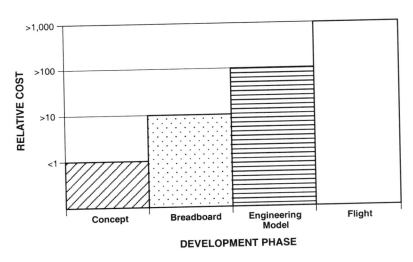

Fig. 19-7.   The Relative Cost of Modification / Corrective Action. Adapted from Wade and Welch [1996]. The relative cost of fixing a design problem escalates rapidly as the design matures.

The feasible design should already be largely configured to the goals and, therefore, iterating to the production level should be relatively smooth. The iteration process is smoother because the team knows what is expected through the long-term goals.

Production goals and guidelines should include all production-related information for the product. This information encompasses the expectations for both parts and processes as these two ingredients form a product. It also provides structure for the product to grow properly since it addresses quality, capabilities, production rates, facility requirements, procurement, cost, warranty, manufacturability, testability, etc. For example, The design dictates product quality. If the product has a 2-sigma design, the best we can expect is a 2-sigma product.

Knowing the expectations initially not only assists the product to mature into production, and assists the product team to develop. Having these goals gives the team a purpose to commonly focus their attention. Successful teamwork resolves issues seemingly otherwise unresolvable. Communication becomes clearer and less conflicting because the team uses common terms to explain the common goals. A good product team forms to produce a working product without relying upon functional titles and positions. Additionally, trust forms within the team. Yet, trust and competition are not mutually exclusive. Team members may and will form partnerships with competitors for different projects or products, but this practice must not hinder the primary business relationships. Lack of trust causes program inefficiencies.

Initially, the prime contractor must derive the goals and guidelines; however, the suppliers must then work closely with the prime contractor to further develop, understand and modify the expectations as necessary. In any event, if team members don't understand or believe the goals, the team's ability to influence the design is diluted, and the prime contractor must begin to educate them about the goals. Furthermore, it is especially important to derive the goals so they are meaningful. For example, assembly time of a high volume, high complexity product (indicative of the constellation satellite industry) typically comprises less than 35% of the overall cycle time while production test comprises the remaining 65% (Olivieri [1997]). Using the Pareto approach we find that the greatest opportunity to influence the design is in production test*.

Design influence is most effective through mathematical analysis. Personal bias and tradition are not the basis for influence. When we select a part for a product, we inherently select the process as well. This means that predicting behaviors (quality, reliability, cycle time, assembly process, etc.) of the product is possible. We do this by collecting data from suppliers but where data is not available we either run an experiment or locate like-product data. Key predictions include, but are not limited to: *defect per unit, defects per million opportunities*, statistical tolerance and variability assessments, *design to unit production cost*, solids modeling, cycle time, process modeling and mapping, design of experiments, and simulations.

Having the data equates to having the knowledge to influence design. The only way we achieve key goals in areas such as quality, manufacturability, and profit is if they are designed in.

Influencing the design is an enterprise challenge. Awareness of our up and down stream product map helps us make smart decisions for the overall program. Locally

---

* **The Pareto Principle**
     If we rank order contributors to a problem from greatest influence to least influence, the *Pareto Principle* states that only a few of the contributors will comprise the bulk of the influence. In general, 80% of the contribution comes from 20% of the factors. Pareto analysis is a technique for ranking items according to their contribution and analyzing the top contributors, or the "vital few." The other 80% of the factors that only influence 20% of the contribution are the "trivial many."

maximizing cycle time, quality, assembly ease and other areas of importance, does not always provide the same benefit to the overall program. Standardization of analysis tools, metrics, goals, and guidelines help to relate design impacts to the supply chain.

19.1.4 Process Development and Verification

Quality, and the resulting efficiency of manufacturing, is never an accident. It is the result of good design practices and sound process development methods–with the addition of management and execution diligence. Significant process development and verification activities must take place to manufacture high quality products. For satellite manufacturing, we need to apply these methods at all levels—suppliers, sub-tier suppliers, payload assembly and test, space vehicle integration, and launch operations.

In the manufacturing context, we define a process as any activity that changes or touches the product or requires resources. The objective is to use simple, uncomplicated processes that yield high quality results with short cycle times. Very few processes meet these general criteria without development. Therefore, sound development methods are key. The end goal is to have controlled and repeatable high-quality processes during the production phase of the program.

Development Methods

We can apply structured development methods to process and product development. Many methods exist, and most have merit. Here, we use a *four-phase process development*, a.k.a. *process characterization*, model. We should first apply this model to critical processes and then extend it to other processes as necessary.

The process characterization model consists of four phases.

Phase 1:   *Process Definition*—Map the process to understand the variables and characteristics involved.

Phase 2:   *Process Capability*—Establish current level of performance of the process. Does the process perform as needed?

Phase 3:   *Process Optimization*—Investigate the variables to determine which variables drive the process output. Determine the best levels for these variables to provide optimal process output.

Phase 4:   *Process Control*—Monitor the process and its important variables to determine when its performance has changed or is out of control.

We iterate the cycle to further improve the process as necessary. (See Fig. 19-8.) We characterize the process to identify and remove sources of variation. Ultimately, variation is the cause for product quality problems. We use statistical tools, such as *Design of Experiments* and *Statistical Process Control*, to reduce the variation, which improves quality—and subsequently reduces cost and cycle time. See Montgomery [1996] for detailed discussions about statistical tools.

Cycles of Learning

In addition to the iterations caused by characterizing the production process, we may need additional cycles of learning when developing satellite constellation production systems. Varying techniques are available, depending on the subject matter.

For instance, we use simulation techniques to model and predict performance. Discrete event simulation packages can model processes, factories, and whole supply

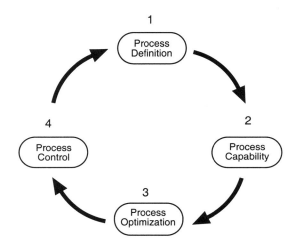

Fig. 19-8. Process Characterization Four Phase Model. We iterate this cycle to continually improve the product or process.

chains. We can use these models to determine resource requirements, facility size, inventory schemes, and delivery schedules. We can explore proposed changes to pre-established baselines without risk.

A second source of cycles of learning is the use of prototyping. Prototyping allows for physical evaluation—appearance and dimensional—while changes are easier and less expensive to make. Production changes are extremely costly, and potentially devastating. We can potentially avoid them through early prototyping. The use of *Dynamic Test Models* and *Engineering Development Models* are advanced functional examples of prototypes.

A third source is the use of *pathfinding activities*. The more complex a system and its logistics, the more beneficial pathfinding activities become. They revolve around building a scale representation, usually 1:1, of a product and using it to simulate all handling, manufacturing, and logistical activities. For satellites, this includes practicing and iterating every process, from subsystem receipt to integration processing to launch processing. The resulting benefits of pathfinding include handling streamlining, mechanical checkfit verification, documentation improvement, process development, tooling improvement, facility development, and logistical simplification.

In short, any activity that we can simulate or realize before we create an actual product can have significant payback, if enacted. These activities are important tools to gain significant process and product knowledge.

19.1.5 Production

All of the previous phases—creating the manufacturing vision, influencing the design, and process development—lead to the final phase—production. All previous goals are to make this phase easy to execute. Failure or lack of attention in the previous phases becomes readily apparent here. They surface as schedule slips, exceeded budgets, and poor quality.

Executing the overall plans and goals established in the earlier phases is paramount. While many production issues are similar for building one satellite or one hundred, some areas are significantly different. These areas include supply-chain management and production floor management.

Supply-chain Management

A supply-chain focus is key to the successfully executing the production phase of a multiple satellite program. The supply-chain is defined as all of the suppliers and activities involved with producing a satellite. This involves the major suppliers—bus, payload, and antennas—and the minor suppliers.

Communicating and coordinating across the supply-chain is critical. Fundamentally, all the suppliers need to have a common vision and sense of direction. In effect, the major suppliers become partners. Openly communicating goals, expectations, and problem solving is essential. To facilitate communication, we need to seek a common language and culture through common training and team building. The resulting program culture program can be built on the strengths of each partner.

Concurrent engineering across company lines must occur. All company sections must share the key design rules or characteristics early in the process and maintain them throughout the program. We must carefully define and fully understand the interfaces and handoffs. We should develop and use common tooling and processes. For instance, shipping containers carrying subassemblies to final integration should support integration activities without a tooling transfer. They should seamlessly integrate into the factory. Finally, common metrics across the supply-chain allow for meaningful summation of data, so managers can judge the overall system health. These metrics should center around quality, cycle time, and critical parameter performance.

Production Floor Management

Performance and executing the production phase uses principles similar to any production environment. In addition to supply-chain management, consistent performance takes lean manufacturing principles of production floor management, including good processes, qualified workers, and timely information.

Short-cycle, well understood processes and overall processing is fundamental to lean manufacturing. Simulation or other flow design tools help to determine optimum production flows that reduce the risks of production delays. The flows and processing should include Just In Time concepts to minimize Work In Process inventory and improve quality and cycle time. Just In Time hinges on delivering material items when the flow needs them in a pull fashion. We don't move items until they are ready for processing. Less hardware is in process, thereby reducing the potential for damage and reducing overall floor space requirements.

Qualified workers that understand the processing of space hardware are still required for satellite production. While still required, their roles and specific skills are very different than traditional programs. Highly specialized and narrowly skilled positions now become broader and more generalized. The design attributes resulting from a successful multi-satellite design team will require less intensive processing. While less difficult, the processing needs to be more repeatable. We can train or certify workers for each repeatable operation, so the can typically monitor their own work. We minimize quality inspections. Instead, the individual or the team is responsible for performing each operation correctly and safely.

The capture and use of timely production data is a third production floor management challenge. Producing multiple satellites with short cycle times requires us to collect, use, and archive data differently from traditional methods. Paper, the traditional documentation medium, is too cumbersome, untimely, and not available to multiple disciplines. Constellations based on a large number of satellites require volumes of data be available to multiple users. We need factory systems for work instruction and process planning; transmittal of product data between suppliers, integrators, and users; and for factory data collection. Factory data collection includes calibration, as-built, work-in-progress tracking, and quality data and/or information. An added advantage is using common tools for these functions across the supply-chain.

19.1.6 Summary

These methods apply exactly to multiple satellite constellations with communications payloads. These methods were applied to the Iridium® program, with outstanding cost and schedule results. Constellation reliability and performance data is starting to be collected at time of this publication.

Smaller constellations, non-communication payloads, and high-value missions may use subsets of these practices. In all cases, thoughtful selection of parts and processes and skillful application of concurrent engineering practices are paramount

More detailed information regarding the application of these methods and subsequent tradeoffs will become available as more constellations complete deployment. Currently, competitive issues restrict information flow as much data is viewed as proprietary.

The Iridium® Satellite Assembly Process

The Iridium®* Satellite Assembly Factory comprises 15 process stations. These stations complete specific functions. The satellites flow from station to station throughout the factory until Station 15. At Station 15, workers pack and send the satellites to the launch-site. During steady-state production, the total time to proceed from Station 1 to Station 15 is 24 days.

Station 0:　Material Receipt

All material and subsystems are received into the satellite factory. Major systems received into the factory include:

- Gateway antennas (4)
- Crosslink antennas, fixed (2) and moving (2)
- Panels (communications and gateway)
- Main mission antennas (3)
- Solar arrays (2)
- Satellite bus assembly

The satellite bus assembly, complete with propulsion and guidance systems, is received already mounted on a wheel-mounted dolly. This dolly, collaboratively developed for use throughout the satellite delivery process, is used for satellite assembly, shipment, and launch-site processing. The dolly also allows for access to all sides of the satellite via rotation, and can be easily maneuvered through the factory.

---

* Iridium® is a registered trademark and service mark of Iridium IP LLC.

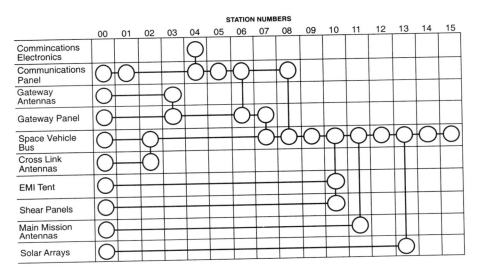

Station 1: Heat-pipe Bonding

The heat-pipe is thermally bonded to the communication panel. This is the only process that does not occur within the integration factory.

Note: Stations 2–4 operate in parallel

Station 2: NADIR Panel Assembly

The four crosslink antennas--two moving and two fixed--and the radiator plate are installed to the nadir bulkhead end of the bus.

Station 3: Gateway Panel Assembly

Gateway antennas (4), waveguides, antenna positioning equipment, cable harnesses, and other components are installed to the gateway panel. During this process, a trunion dolly holds and rotates the panel.

Station 4: Communications Panel Assembly

Motorola supplied units are installed on the communications panel. A rotating trunion dolly designed for the communications panel holds it.

Station 5: Communications Panel Test

The completed communications panel is tested.

Station 6: Communications Equipment Subsystem (CES) Test

The CES consists of equipment on the communications panel and the gateway panel. Cables are installed to temporarily link the two panels. This station initially verifies the functionality and performance of the CES hardware over temperature cycles, prior to being integrated in the spacecraft.

Station 7: Gateway Panel Assembly Integration

The completed gateway panel is installed onto the bus assembly. First, the panel is rotated to the vertical position, aligned to the spacecraft bus, and then mated to the bus. Cables and harnesses are then attached.

Station 8: Communications Panel Assembly Integration

The completed communications panel is installed onto the bus assembly. The panel is first rotated 180 degrees while in the trunion dolly. A strongback fixture is then attached to the panel for support and alignment. The panel is lifted over the spacecraft, aligned, and finally installed. The panel is securely fastened in place and all cables and harnesses are attached.

Station 9: Space Vehicle Test

(Or Integrated Communication Equipment Subsystem (CES) Test)
Testing is performed on the newly integrated communication panel, gateway panel, and spacecraft bus. This test verifies the CES functionality and performance.

Station 10: EMI Tent and Shear Panel Installation

An EMI tent is installed over the communications panel. Shear panels are then installed on the remaining two open sides of the bus structure.

Station 11: Main Mission Antenna Integration

The satellite is rotated horizontally on the dolly to attach the main mission antennas (MMA), one at a time. First, an attachment fixture is mounted on to the MMA. This fixture contains the lift points for using the lifting crane. The MMA is held vertically over the spacecraft and lowered to connect one end. RF, power, and signal cables are attached. The free end of the MMA is then lowered until flat against the spacecraft. The MMA deployment mechanisms are adjusted and set. Finally, this process is repeated for each of the three MMAs required.

Station 12: Space Vehicle (Integration) Test

Full-functional testing of the MMAs, communications payload, and bus is conducted The testing includes seven complete hot-cold thermal cycle tests.

Station 13: Solar Array Integration

Two solar array wing assemblies are integrated onto the spacecraft in a manner similar to MMA installation. First, an attachment fixture is mounted onto the solar array. This fixture contains the lift points for using the lifting crane. The solar array is held vertically over the spacecraft and lowered to connect one end. The cables are attached. The free end of the solar array is then lowered until flat against the spacecraft. The deployment mechanisms are adjusted and set. Finally, this process is repeated for the second solar array.

Station 14: Launch Confidence Test

Flight software is loaded and verified and solar array operation is tested. Finally, the vehicle is powered down and prepared to flight-ready condition.

Station 15: Space Vehicle Pack and Ship

The completed satellite is prepared for shipment to one of three launch-sites. Consent-to-ship authorizations and checkout procedures are completed. The satellite, still mounted on the dolly, is loaded into the shipping container. The shipping container provides for an environmentally controlled and shock protected delivery to the launch-site.

19.2 Reliability for Space Mission Planning

Herbert Hecht,* *SoHaR Incorporated*

When a bulb in our desk lamp burns out, it is easily replaced. When the switch that controls the bulb fails, the replacement is not quite as simple but still within the capabilities of most mechanically inclined teenagers and even some adults. We expect a higher reliability of the switch than of the lamp because it requires more effort to repair a failure. When a spacecraft command receiver fails on orbit, it takes an extraordinarily long screwdriver to fix it. You get the general idea: the command receiver has to be much more reliable than the light bulb or the switch on the desk lamp. This need for very high reliability in all parts and subsystems of a spacecraft is the basis for including a *reliability program* in most space projects.

Before describing the details of the reliability program, let us briefly discuss the meaning and metrics of *reliability* in the context of space missions. A common definition of reliability is "The probability that a device will function without failure over a specified time period or amount of usage." [IEEE, 1984] If the phrase "without failure" is taken to mean "without failure of any kind" it defines *basic reliability*; if it is interpreted as "without failure that impairs the mission" it defines *mission reliability*. In spacecraft that employ extensive redundancy there can be a significant difference between these two reliability metrics. Mission reliability is the more important concept, and when we use "reliability" without a qualifier, it always means mission reliability.

The elementary expression for the reliability of a single item, not subject to wearout failures, is

$$R = e^{-\lambda t} \tag{19-1}$$

where λ is the *failure rate* and t is the time. Here, R, is the probability that the item will operate without failure for time t (success probability). At this point we recognize only two outcomes: success and failure, and therefore the probability of failure, F, is given by:

$$F = 1 - R \tag{19-2}$$

More refined and practical methods of assessing mission success will be described later.

For a spacecraft made up of n nonredundant elements, all equally essential for spacecraft operation, the *system* (or *series*) *reliability*, R_s, or success probability, is computed as

$$R_s = \prod_1^n R_i = e^{-\sum \lambda_i t} \tag{19-3}$$

where R_i $(i = 1...n)$ is the reliability and λ_i the failure rate of the individual elements. For failure probabilities (λt) less than 0.1 or reliability greater than 0.9, the following

---

* Much valuable help in the formatting of this chapter was received from Emery I. Reeves of the U.S. Air Force Academy who also contributed the material for Table 19-6.

approximation is frequently used

$$e^{-\lambda t} \approx 1 - \lambda t \tag{19-4}$$

Most reliability computations, particularly prior to detailed design, use failure probabilities (which can be summed) rather than reliability values (that must be multiplied)

Where a system consists of n elements in parallel, and each of these elements can by itself satisfy the requirements, the *parallel* (or *redundant*) *reliability*, R_p, is given by

$$R_p = 1 - \prod_1^n (1 - R_i) \tag{19-5}$$

where the reliability of the parallel elements is equal, say R_a, the above simplifies to

$$R_p = 1 - (1 - R_a)^n \tag{19-6}$$

Examples of series and parallel structures are shown in Fig. 19-9 .

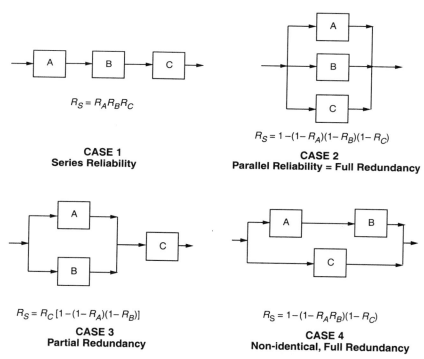

CASE 1
Series Reliability
$$R_S = R_A R_B R_C$$

CASE 2
Parallel Reliability = Full Redundancy
$$R_S = 1 - (1 - R_A)(1 - R_B)(1 - R_C)$$

CASE 3
Partial Redundancy
$$R_S = R_C [1 - (1 - R_A)(1 - R_B)]$$

CASE 4
Non-identical, Full Redundancy
$$R_S = 1 - (1 - R_A R_B)(1 - R_C)$$

Fig. 19-9. Series and Parallel Reliability Models. R_s is the system reliability. R_A, R_B, and R_C denote the reliability of the A, B, and C components, respectively.

In terrestrial applications it is customary to distinguish between active and quiescent (dormant) failure rates, the latter being about one-tenth of the active rates. This reduction accounts for the absence of electrical stress when a component is not energized. However the high reliability requirements of the space environment cause components to be derated so that the failure probability due to electrical stresses even in the active mode is quite small. The distinction between active and quiescent failure rates is therefore much less important for spacecraft applications.

When the spacecraft does not have any redundant elements, a plot of reliability vs. mission time will be concave toward the origin, but when redundancy is provided the initial part of the reliability plot tends to be convex to the origin. This accounts for the shape of the mission reliability curve in Fig. 19-10.

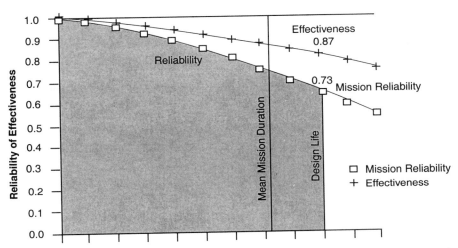

Fig. 19-10. Frequently Used Reliability Concepts. Design life is governed by wear-out and expendable stores. Mean mission duration is less than design life because failures can terminate a mission before end-of-life conditions are reached.

Design life, the intended operational time on orbit (see Section 19.2.1), is an important parameter for the reliability program. It determines the amount of consumables (control gas, etc.) that must be provided and establishes quality and test requirements for items subject to wearout, such as bearings and batteries. The mission reliability calculated at the design life is the *mission success probability*. Since this quantity is less than 1.0, the *expected life* is less than the design life. The customary measure of the expected life is the *mean mission duration* or *MMD*, given by:

$$MMD = \int T dR \qquad (19\text{-}7)$$

where T represents a horizontal (time) line in the shaded area of Fig. 19-10 and dR is the associated (vertical) increment in reliability. *MMD* is a frequently used figure of merit for spacecraft reliability improvement programs.

Most satellites perform multiple functions (e.g., weather observation and tracking ocean traffic), and performance of any of these functions may still be useful even when it is degraded. Under such circumstances the success criteria are no longer obvious. Should failure be defined as the event when the spacecraft fails to furnish any useful data, or should it be the point at which its performance deviates from any part of the specification? This problem can be addressed by establishing multiple reliability requirements. Assume that the specified weather observations are intended to use three frequency bands but that observations on even a single band are still useful. The success probability (SP) requirement may be formulated as shown in Table 19-2.

TABLE 19-2. Hypothetical Reliability Requirement for a Weather Satellite.

Condition	SP*
At least 1 band operational	0.96
At least 2 bands operational	0.92
All 3 bands operational	0.85

* Success Probability

During the development of the satellite it may be found that it will be difficult to obtain a success probability for all three bands operational greater than 0.80 but that the single band mission reliability can be raised to 0.99. If the requirements shown in Table 19-2 are part of a contractual document it may take lengthy negotiations to agree on the change, even though it may be technically quite acceptable. To avoid such situations, *mission effectiveness* rather than reliability may be specified. There are several definitions of mission effectiveness [MIL-HDBK-338B], but all aim at a single metric that represents reliability weighted by the operational capability level to which the reliability is applicable. In the above example assume that the 1, 2, and 3 band capabilities are assigned weights of 0.25, 0.35, and 0.4, respectively (it is desirable that the weights add up to 1.0). A simple mission effectiveness requirement may then be stated as $0.25 \times 0.96 + 0.35 \times 0.92 + 0.4 \times 0.85 = 0.902$. A similar calculation shows that raising the single band reliability to 0.99 will not compensate for reducing the 3 band reliability to 0.80, but that this reduction could be tolerated if the 2 band reliability is raised to 0.95 at the same time. Where there is a sound technical basis for specifying mission effectiveness it will generally reduce both cost and development time compared to specifying multiple reliability values. As indicated in Fig. 19-10, the effectiveness curve will lie above the reliability curve when the latter is constructed for an "all up" satellite.

19.2.1 Design for Reliability

The process for designing a reliable system is shown in Table 19-3. Evidence that simplicity makes for reliability can be seen in Table 19-4, which is excerpted from an earlier publication by the author of this section [Hecht and Hecht, 1985]. Complex functions had higher failure rates. In each case shown in the last columns of the table, there were probably compelling reasons for using the more complex implementation, but the reliability consequences of these decisions must also be recognized.

To make sure that the process leading to a reliable system is being carried out as intended, there must be an explicit assignment of responsibility for its implementation. Several alternatives for the assignment are listed in Table 19-5 in order of increasing direct cost. For very small satellites the low cost advantage of (A) or (B) will usually outweigh the benefits of the last two alternatives. When subsystems get more numerous and larger, alternatives (C) and (D) may result in overall savings, in spite of their higher immediate cost.

The activities listed below are suitable for alternatives (B) or (C). Where alternative (A) is adopted, at least a Reliability Program Plan should be prepared and a Failure Reporting System should be established. Alternative (D) implies that the program will be managed by reliability specialists who will generate their own process requirements.

TABLE 19-3.　Typical Steps for Achieving a Reliable System.

Process Steps	Details Found
Keep it simple (every additional function increases the failure probability)	See Table 19-4
Assure adequate strength (mechanical and electrical) of all parts, including allowance for unusual loads that may be imposed due to environmental extremes or failures in related components	Sec. 19.2.2
Provide alternative means of accomplishing the most essential functions where design for excess strength is not suitable (this includes most electronics)	Sec. 19.2.3
Plan a test program to assure that the above objectives have been achieved	Sec. 19.2.4
Collect and analyze of test and on-orbit failure data to guide future designs and mission plans	Sec. 19.2.1

TABLE 19-4.　Relationship Between Failure Rate and Complexity.

System	Simple		Complex	
	Type	Fail Rate*	Type	Fail Rate*
Telemetry	Hardwired	0.034	Programmable	0.190
Stabilization	Gravity Spin	0.038 0.216	3-axis active	0.610
Thermal	Passive	0.084	Active	0.320

* per orbit-year

TABLE 19-5.　Alternative Assignments of Responsibility

Alternative	Benefits	Disadvantages
(A) Designers responsible for reliability	No additional staff Familiarity with items Clear responsibility	Difficult to achieve uniformity; also limitations in (B)
(B) Designers responsible with policy guidance from management	All of the above plus some uniformity	No responsibility for subsystem interactions, opportunity for analytic redundancy may be overlooked, little awareness of reliability tools
(C) Designers responsible with guidance from reliability organization	Uniform procedures, above disadvantages largely overcome	Requires dedicated reliability function, possible confusion over responsibility
(D) Responsibility in reliability organization	Responsibility clearly defined, interactions likely to be identified	Lower motivation for designers, expense of reliability organization.

Reliability Program Plan

A *reliability program plan* adds little to the cost and is recommended even for the smallest spacecraft program. It specifies reliability objectives, assigns responsibility for achieving them, and establishes milestones for evaluating the achievements. It also serves as an agreement with other spacecraft functions regarding their responsibilities in support of reliability. The most significant interfaces usually are with quality assurance, test, configuration management, and thermal control.

Failure Modes Analysis

Failure Modes, Effects, and Criticality Analysis (FMECA) can provide valuable insights into how design decisions affect reliability. Typical benefits are

- Exposing single point failure modes in a subsystem assumed to be redundant

- Identifying opportunities for functional redundancy (see Sec. 19.2.3)

- Permitting components to assume a safe mode in the absence of required signals or power

Failure modes are usually recorded at the part level, e.g., for a capacitor the failure modes are open and short (sometimes a change in capacitance may also be recorded). Failure effects are assessed at the part level and also at higher levels such as assembly and major component or subsystem. The failure effect of a shorted capacitor may be a bias shift on a transistor (part level) which in turn cuts off the output of a demodulator (assembly level) and causes loss of voice communications (subsystem level). In addition, the FMECA forms usually contain fields for the method of detection and the means of failure alleviation. In our example, the detection may be by an output monitor included in the demodulator, and the failure alleviation may be use of an alternate voice channel.

A probability is associated with each failure mode, and the probabilities of all failure modes that cause a given effect are added in a summary section of the FMECA. Loss of modulator output cited in the example may be caused by several failure modes of the demodulator components, and loss of the voice channel may be caused by failure modes other than those in the demodulator. Although there may be considerable error in the estimate of a given failure mode, these tend to be evened out when arriving at estimates of failure effects at the subsystem and higher levels.

In digital microcircuits it is usually not possible to conduct FMECA at the level of primitive elements (gates or transistors) because there are too many of them and because causes of failure may affect multiple elements (e.g., voids in the oxide layer). The FMECA may then be conducted at the function level, where functions are timers, counters, and shift registers. Failure modes at the function level are generally not as well known as those at the part level, and a single cause of failure may affect multiple functions. In assessing the effects of failure at the higher level it may therefore be advisable to be conservative.

Sneak circuit analysis is usually considered a part of FMECA. This analysis establishes that explosive or other one-shot devices will not be accidentally actuated, and that they will always be actuated when intended. A good guide to FMECA is MIL-STD-1629.

Failure Reporting

Failure Reporting and Corrective Action (FRACAS) is a key element in any reliability program because:

a. It informs concerned parties that a failure has been observed

b. It furnishes a record through which trends and correlations can be evaluated at a future time (an example of a trend is that the probability of failure increases after x hours of use; an example of a correlation is that part y fails during a particular step in the test sequence).

c. It permits reassessment of the predicted failure rates and is the basis for consequent modifications of the fault avoidance or fault tolerance provisions.

Uses b and c require recording the operating time of all units in service; usually by means of an operating log maintained for each part number, with separate records for each serial number. Corrective action is typically also recorded on the failure reporting form. This facilitates configuration management in that it establishes at what point a failed component or subsystem has been returned to operational status. Reporting on the same form also facilitates future investigation of the effectiveness of the repair action. Corrective action frequently involves two steps: in the first, a failed part is replaced by a good one of the same design; the second step addresses the root cause, e.g., by tightening limits for the incoming test of this part. The results of retest are included in the corrective action report.

To establish a FRACAS the following must be identified:

- Scope of the activities (e.g., system test, field test, normal usage)

- Responsibility for cost and for report initiation

- Method and frequency of reporting (e.g., paper or electronic, each incident or by time interval)

The format used for reporting of failures and corrective actions is not standardized. The following are the most essential data:

1. Incident identification (e. g., report serial number)
2. Date, time and locale of the incident
3. Part no., name of the failed component, and its serial number.
4. Higher level part or system identifiers (subsystem or major component)
5. Lower level part or system identifiers (usually available only after diagnosis)
6. Operation in progress and environmental conditions when failure was detected
7. Immediate and higher level effects of failure
8. Names of individuals responsible for detection, verification and analysis
9. Diagnosis of immediate, contributory and root causes of the failure
10. Dates and nature of repair and results of retest.

Low Cost Methods of Reliability Assessment

Reliability prediction (usually by using a failure rate handbook) or estimation (based on experience with the component population) are routinely required for major satellite programs. Representative failure rates for reliability estimation are shown in Table 19-6.

These failure rates are abstracted from MIL-HDBK-217F and can be used for reliability predictions based on parts type and count. Lower reliability parts such as commercial parts have failure rates that are much higher (between 12 and 333 times). Values shown in the table are expected failures in 10^9 hours, they correspond to λ in Eq. (19-1).

TABLE 19-6. Representative Piece Part Failure Rates for High Reliability Parts. These failure rates are abstracted from MIL-HDBK-217F and can be used for reliability predictions based on parts type and count. Lower reliability parts such as commercial parts have failure rates between 12 and 333 times these failure rates. Values are failures in 10^9 hours, they correspond to λ in Eq. (19-1).

Part Type	Space Flight	Launch	Applicability
Bipolar Gate/Logic Array Dig	0.9–19	17–300	Min 1–100 gates; Max 60,000 gates
Bipolar Microprocessor	7–27	60–215	Min 8 bits; Max 32 bits
MOS Microprocessor	12–47	70–250	Min 8 bits; Max 32 bits
MOS Memory SRAM	2–11	24–75	Min 16 K; Max 1 M
Bipolar SRAM	2–8	30–75	Min 16 K; Max 1 M
Diodes General	1.3	170	
Transistors General	0.05	5	
Transistors RF Power	165	900	
Resistors	0.01	1	Composition/film
Capacitors	0.1	10	
Relays	40	6,000	

Parts count reliability estimation may be used for comparison of design alternatives, when test or on-orbit incidents indicate insufficient reliability, or when a more expensive payload is to be incorporated (which increases the cost of failure).

Software Considerations

Spacecraft operations are becoming increasingly dependent on software, in the spacecraft and in the ground segment. There have been spectacular launch and on orbit failures due to software faults, but many more missions have been saved by software (used for work-around) than have been lost due to it. Thus the way to more effective space missions may not so much lie in minimizing software as in making appropriate use of it. It should also be mentioned that most software used in spacecraft is actually *firmware* (programs furnished as read-only-memory chips).

Reliability assessment of software is not a precise science. The most widely accepted techniques depend on test (very expensive if high levels of reliability are required) or on *independent verification and validation* (even more expensive and less likely to yield a quantitative assessment). When software is simple and well structured its reliability is usually high and the cost of evaluation is low.

Cost effective software reliability assessment starts with an examination of the ways in which software can impact the mission. A convenient format for this is a fault tree, as shown in Fig. 19-11.

In this example, the software failures that are most likely to lead to mission failure are:

1. Complete halting of execution, e. g., due to being in an infinite loop or being directed to an incorrect memory location.

2. Faulty antenna orientation due to an antenna drive failure or faulty attitude determination

3. Failure to deploy the solar array.

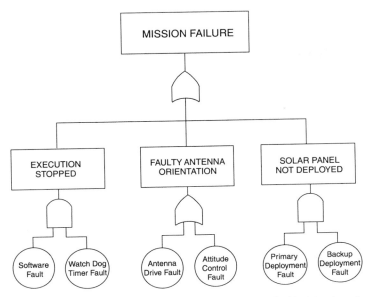

Fig. 19-11.   Fault Tree. Fault trees show which functions may be disabled due to a single failure.

Double failures are required to halt execution of programs and deployment of the solar array, because both of these have backup mechanisms. Faulty antenna orientation can be caused in two ways, and there is currently no backup provision for this hazard. Once this deficiency is recognized, alternatives for mitigation will be investigated. Re-orienting the spacecraft by ground command may be a worst case backup, provided that ground commands will be received regardless of initial antenna position.

Redundancy management is also frequently a very critical software function handled by the spacecraft computer.

A study of the space shuttle avionics software showed that causes of high severity failures were overwhelmingly associated with rare events, such as handling of exception conditions, management of hardware failures, and response to unusual or incorrect crew commands [Hecht and Crane, 1994]. These data, as well as others cited in the reference, indicate that testing of spacecraft software should demonstrate that unusual events in the environment (including hardware and software failures) are correctly handled.

19.2.2 Design for Fault Avoidance

Fault avoidance is most effective when there are only a very small number of significant failure modes. Common fault avoidance techniques are shown in Table 19-7, and application guidance is provided below.

Process Control

Control of the manufacturing process can only be exercised where parts are specifically manufactured for the spacecraft. In most cases this includes the structure, propulsion components, and segments of the environmental conditioning system. Control over the assembly process can usually be exercised for the entire spacecraft,

TABLE 19-7.   Representative Methods for Fault Avoidance.

Technique	Most Suitable for	Limitations
Process Control	Current deficiencies exist	User must be able to influence process
Design Margins, etc.	Known failure risk	Adds weight and cost
Coding Techniques	Memory upset	Digital components only
Part Selection and Screening	Modest improvement required	Critical measurements must be known

and this is where major emphasis for fault avoidance through tighter process control should be placed.

A good starting point for investigating the feasibility and effectiveness of tighter process controls for fault avoidance is to investigate past failures in test and operation, and, particularly for new components or processes, difficulties that have been experienced during engineering evaluations. These constitute the "current deficiencies" listed in Table 19-7, and for each the underlying cause of failure must be identified. Where dispersions in material or process characteristics are implicated, tighter process control can be expected to make a significant contribution to fault avoidance.

Design Margins, Derating and Environmental Protection

Design margins and derating accomplish the same goal: prevention of component failure due to higher than expected external stresses or other deviations from the nominal conditions. The term *design margin* is mostly used in structure and thermal subsystems and means that a component is designed to carry more than the expected load. *Propellant margins* in propulsion systems are an equivalent concept. The term *derating* is primarily applied to electrical and electronic components and involves the specification of a component that carries a higher rating than is needed for the application.

The reliability improvement by these practices is most significant if a part is initially used near its design strength or electrical rating. As an example (from MIL-HDBK-217F), the predicted base failure rate for a fixed film resistor at 40° C and used at 0.9 of rated power is 0.0022×10^{-6}/hour. Selecting a higher rated resistor, for which the dissipated power constitutes only 0.3 of rated power reduces this to 0.0011×10^{-6}/hour. But further reductions are hard to achieve. A resistor for which the dissipated power is 0.1 of the rating still has a failure rate 0.0009×10^{-6}/hour. Derated parts not only cost more, but are frequently larger and heavier than the ones that they replace. Derating only reduces the failure probability with respect to the stress that is being derated. In the example of the fixed film resistor, derating reduces the probability of failure due to power surges but it does not offer any protection against failures due to lead breakage or corroded connections.

Environmental protection can take the form of shock mounting, cooling or heating provisions, and shielding against radiation effects, Where derating reduces the failure probability by increasing the strength of the components, environmental protection reduces the failure probability by reducing the stress levels. In many cases, environmental protection adds considerable weight, and this, rather than cost, limits the amount of protection that can be provided.

Coding Techniques

Coding provides robustness by permitting continued operation in the presence of a defined spectrum of errors, primarily in memory and data transmission. Coding techniques are also available for detection or toleration of errors in arithmetic processors but are seldom used in this capacity in spacecraft computers.

The important coding techniques are *error detecting code* and *error correcting code*. The former is intended primarily for fault isolation (i. e., preventing an incorrect result from being used in subsequent operations). The latter is a fault tolerance mechanism that corrects a class of errors and permits operations to continue normally. All codes require the addition of *check bits* to the bit pattern that represents the basic information. If there is agreement between the check and information bits, the data is accepted. If there is no agreement, the data is rejected (for error detecting code) or corrected (for error correcting code).

The cost effectiveness of error correcting codes is shown in the following hypothetical example for a commercial earth observation satellite. The payload computer's 4 megabit dynamic random access memory has a mass of 400 gr. This memory (which does not incorporate error correction or detection) is expected to sustain two "upsets" per orbit-year. Upon detection of an upset by ground monitoring, the memory is reloaded, an operation that typically loses data from two orbits. The expected mission income is $1,000 per orbit. In the absence of error correction, the cost of memory upsets will therefore be $4,000 per year. The extra memory and coding/decoding chips will add 100 gr. to the mass of the memory and will cost $1,000. The cost/mass ratio for this satellite is $5,000 per kg, and thus the extra 100 gr will be equivalent to $500. In this example the cost of the error correction will be paid for in less than one-half orbit-year. As discussed in Sec. 8.1, the number of upsets to be expected depends on the size and type of memory, the orbit, and the amount of shielding provided by the spacecraft structure and the memory enclosure. In most cases coding is found to be very cost effective.

Part Selection and Screening

Screening (selection of parts by test) is a process that eliminates units that have a higher likelihood of failing in service than the other units in the lot [Chan,1994]. Whereas derating reduces the probability of failure by moving the average strength of the components higher, screening reduces the probability of failure by rejecting the lower tail of the distribution as shown in Fig. 19-12. A typical screening procedure for semiconductors is to measure the leakage current at elevated temperature.

The cost of screening is made up of two elements: the cost of the rejected product, and the cost of test. The cost effectiveness is high (1) for parts with an initially high failure rate, (2) for modest reliability improvements (generally those in which not more than 20% of the product is rejected), and (3) where the cost of test is small compared to the unit cost of the product under test (not over 10%)

Screening does not involve an increase in the size of the components and it is therefore preferred to derating for bulky or heavy parts. Screening is not very effective for reducing the failure probability in a mode for which components have been derated because the failure probability due to external stresses in that mode is already very low. Screening can be applied to assemblies, e.g., by subjecting them to combined temperature and vibration environments and is thus more versatile than derating. Screening at the assembly level is also likely to result in a lower ratio of test cost to product cost and thus produces a higher figure of merit.

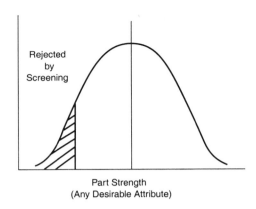

Fig. 19-12.   Attribute Control by Screening. Screening rejects parts likely to fail in service.

Screening is really a crutch that permits the use of products that do not, as delivered, meet all of the requirements of an given application. A more desirable reliability measure is to tighten the control of the process so that it can be relied on not to produce the outliers that must be screened out. This is not only a philosophical argument, but one with significant practical consequences as shown by the attribute distribution curves in Fig. 19-13.

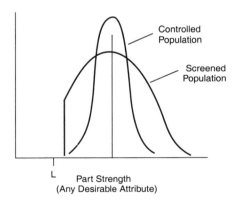

Fig. 19-13.   Attribute Control by Process Control. In a controlled population fewer parts are near the acceptance limit than in a screened population.

Both the screened population and the controlled population meet the acceptance criteria in that no parts fall below the lower limit, denoted as L in the figure. However, in the screened population a much larger fraction of the total population is near the lower limit than in the controlled population. Environmental effects and aging cause a dispersion of the attributes, and therefore will cause a larger fraction of the screened population to fall below the lower limit than in the controlled population. This effect can be compensated for by selecting an initial acceptance limit that is higher than the lowest value that can be tolerated in service. Where articles are procured specifically for space applications, costs will be reduced if the process can be improved so that only a small portion of the product will be rejected in screening.

19.2.3 Fault Tolerance

Fault tolerance protects against a wider spectrum of failure modes than fault avoidance. In most cases it also requires much more resources. A summary of the suitability and limitations of representative fault tolerance techniques is shown in Table 19-8.

TABLE 19-8. Representative Fault Tolerance Techniques.

Technique	Protection Against	Limitations
Same Design Redundancy	Random failures	High production cost, weight
Diverse Design Redundancy	Random and design failures	Same, plus design and logistic cost
k-out-of-n Redundancy	Random failures	Applicable only where multiple copies of an article are present
Functional Redundancy	Random and design failures	Diverse methods to accomplish a function must be available
Temporal Redundancy	Transient, intermittent failures	Time required for recovery

By *scope of redundancy,* we mean the size and importance of the entity that is being made redundant. Paralleling two relay contacts is a redundancy provision of very small scope, and dual telemetry systems represent redundancy of large scope. In Fig. 19-14 the single system line shows the reliability of a single element with MTBF of one time unit. The system redundancy curve shows the resulting reliability if two of these elements are operated in parallel. The partitioned redundancy curve refers to the original (single) system being divided into quarters, and then each quarter made redundant. The redundant and the quarter-partitioned curve involve (at least superficially) the same resources (in each case twice those of the single system), and yet the quarter configuration has a pronounced reliability advantage, particularly at the longer time intervals. It would therefore seem that redundancy of small scope is to be preferred over that of large scope.

In practice, the switching or voting provisions that are required for each partition can add considerably to the cost of the implementation, and since they are not likely to be 100% reliable they may also reduce the reliability benefits. System test can also be adversely impacted by redundancy of small scope. The system redundancy (middle curve in the figure) requires only two tests. The quarter redundancy requires 16 tests if all combinations are to be covered. Selecting the scope of redundancy must consider the means of failure detection, the ease of output switching or combining, the reliability gains that can be achieved, and the cost of implementation and test.

Same Design Redundancy

Same design redundancy involves installation of two or more identical components together with switching to make one of them active. In a few instances, particularly for power supplies, the outputs can be combined so that switching is not necessary. Voting can also be used for combining outputs of redundant units but this carries a high cost because at least three identical units have to be installed to make it practicable. Same design redundancy offers very high protection against random failures, and the

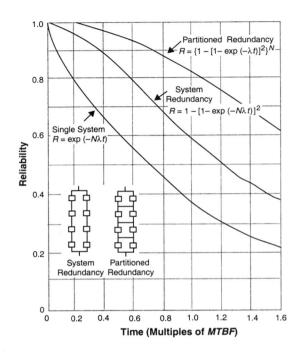

Fig. 19-14.   Effect of Partitioning on Reliability. t is the time from the start of the mission, R is the mission reliability or the probability that at least essential mission elements will survive (see Sec. 19.2.1), N is the number of individual blocks (4 in the figure), and λ is the failure rate of an individual block. For the elementary distributions considered here, $\lambda \equiv 1/MTBF$, where $MTBF$ is the mean time between failure for each block. For the system as a whole, $R_s = \exp(-\lambda_s t)$, where λ_s is $1/MTBF$ for the whole system.

benefits do not depend on knowledge of failure modes. It is not very effective against failures due to design deficiencies: if one component fails due to insufficient radiation hardness, the redundant one is very likely to fail soon thereafter. Because of the high cost, same design redundancy is used only sparingly in low cost satellites.

The cost of same design redundancy can sometimes be reduced by employing *k-out-of-n replacement*, in which a pool of spares can be assigned to replace any one of the pool of active units. An example is bulk memory (for data storage) which usually consists of multiple physically identical modules. Providing one or more spare modules on the same data bus permits replacement of any one failed module. The same scheme can be used for multi-cell batteries, solar panels, and other elements of the electric power supply.

Diverse Design Redundancy

Installation of two or more components of different design (called *diverse design redundancy*) to furnish the same service has two advantages: it offers high protection against failures due to design deficiencies, and it can offer lower cost if the back-up unit is a "lifeboat," with lower accuracy and functionality, but still adequate for the minimum mission needs. The installation of diverse units usually adds to logistic cost because of additional test specifications, fixtures, and spare parts. This form of redundancy is economical primarily where the back-up unit comes from a previous satellite design, or where there is experience with it from another source. Where there is concern about the design integrity of a primary component, diverse design redundancy may have to be employed regardless of cost.

Functional and Temporal Redundancy

Functional redundancy (sometimes called *analytic redundancy*) involves furnishing a service by diverse means. An example is the determination of attitude rate from a rate gyro assembly (direct), and from observation of celestial bodies (indirect). It is particularly advantageous when the alternate is already installed for another service, e.g., if a star sensor is provided for navigation. In these cases the only cost incurred is for the switching provisions and for data conversion. Both of these can frequently be achieved in an existing onboard computer, thus further minimizing the cost.

Functional redundancy can also take the form of a ground back-up for functions preferably performed autonomously, e.g., navigation, thermal control, or furnishing commands for sensor operation. In the communication subsystem there are frequently omni-directional and directional antenna systems. For some satellite missions these may also be considered a form of functional redundancy.

The chief benefits of analytic redundancy are: (1) it avoids the cost and weight penalties of physical redundancy, and (2) it is inherently diverse, thus providing protection against design faults. The major limitation is that the back-up provisions usually entail lower performance.

Temporal redundancy involves repetition of an unsuccessful operation. A common example is a retry after a failure within the computing process. The same technique is applicable to acquisition of a star, firing of a pyrotechnic device, or communication with the ground. This is obviously a low cost technique. It is most effective when the design incorporates an analysis of the optimum retry interval and of changes that may improve the success of later operations, e.g., switching a power supply, reducing loads on the power supply, or re-orientation of the satellite. The most important step is to plan ahead for retry of operations and to incorporate "hooks" that permit automatic or ground-initiated retry.

19.2.4 Test Techniques

The least expensive reliability test is one that is not run at all as a reliability test, but rather as a part of a qualification test, lot acceptance test for purchased parts, or as an acceptance test on the spacecraft as a whole or on a major subsystem. To use these activities for reliability assessment may require some additional instrumentation and sometimes an extension of the test time, but these are very small resource expenditures compared to those required for even a modest separately run reliability test.

Other alternatives to obtaining reliability data by test are:

- Using test data obtained by others (including vendors) on the same component
- Using test or experience data on similar components
- Stress-strength analysis (particularly for mechanical components)
- Reliability prediction by MIL-HDBK-217 or similar sources

One of the greatest problems with reliability tests is that the results are usually obtained after many months or even years of test. Once it has been decided that a reliability test is necessary, a suitable scope and test environment must be selected.

The scope designates the assembly level; parts and circuit boards usually being designated as small scope, while subsystem and system level tests represent a large scope. Advantages of small scope tests are:

- Low cost and small size of individual test articles permit testing of multiple items
- Inputs and outputs are easily accessible
- The test environment can be tailored to the requirements of the unit under test
- The test can be conducted early because it does not require integration.

Advantages of large scope tests are:

- Interactions between components can be observed
- Test results are easily translated to effects on the mission.

The test environment can be *quiescent* (room ambient) or *stressed*. Advantages of testing in a quiescent environment are:

- Low cost (no or only simple environmental chambers required)
- Articles under test are easily accessible
- No induction of failures due to unusually high stress

Advantages of a stressed environment are:

- Increased probability of failure (less test time required)
- Can identify environmental vulnerability of the unit under test.

The attributes of the approaches lead to the test recommendations shown in Table 19-9.

TABLE 19-9.   Typical Uses of Reliability Testing.

	Small Scope	Large Scope
Quiescent Environment	Suitability test Critical components—failure inducing stress unknown	High risk subsystems
Stressed	Critical components—failure inducing stress known	

Reliability testing at large scope and in a stressed environment is very expensive and is rarely warranted.

The time required for any reliability test can be significantly reduced if *testing by variables* is employed. This means that the value of significant attributes is numerically recorded (as contrasted with the commonly used pass/fail procedure). From these data, the distribution of parameters can be plotted, and the probability of dropping below an acceptance criterion can be assessed. The general technique is similar to the one described in the Screening Techniques section. For reliability assessment the parameter distributions are of interest, whereas in screening the attributes of individual units are the chief criterion.

The small component populations that are typical of space procurements preclude the use of conventional reliability demonstration methods. Even a modest subsystem requirement, such as reliability of 0.95 for 17,000 hours (approximately 2 years), corresponds to an MTBF of over 330,000 hours and will require over 1 million component-hours of test to arrive at a statistically meaningful assessment by conventional methods. Yet, experimental verification of the claimed reliability of a component or subsystem is frequently desirable and sometimes required. The following paragraphs explore low cost methods of accomplishing this.

The major causes of failures are workmanship and design. The first of these can be controlled by quality assurance. Design failures occur primarily because the strength of a component is not adequate for the environment in which it is used, or because the manufacturing process permits too much variability in component characteristics. This is most easily seen in mechanical spacecraft components where reliability depends on (a) the margin between the nominal (mean) strength of the component and the maximum service load and (b) the variability of strength about the mean in the delivered product. Since test can characterize the strength of mechanical components fairly easily, strength-load margins (design margins) have always played a major role in their reliability assessment. For electrical and electronic components the same relationship holds in principle, but it is usually much harder to define a single failure inducing stress or load. Nevertheless, test data can give valuable insights into potential reliability problems. Important requirements are (a) recording of test results in numerical form (not pass/fail), and (b) statistical evaluation of the probability of failure derived from the numerical test results, e.g. by applying the 6σ *criterion* (the mean value of a parameter is at least six standard deviations above the specified minimum or below the specified maximum) [Harry, 1997]. The tests from which the required data are obtained need not be specific reliability tests. Typically they are the qualification test and acceptance tests.

References

Chan, H. Anthony. 1994. "A Formulation of Product Reliability Through Environmental Stress Testing and Screening." *Journal of the Institute of Environmental Science*, March/April 1994, pp. 50–56.

Defense Systems Management College. 1986. *Integrated Logistics Support Guide (1st Edition)*. Fort Belvoir, VA: Department of Defense.

————. 1986. *Systems Engineering Management Guide (2nd Edition)*. Fort Belvoir, VA: Department of Defense.

General Accounting Office (GAO). March 1989. "Space Operations—NASA Efforts to Develop and Deploy Advanced Spacecraft Computers." GAO/IMTEC-89-17.

Harry, M. 1997. *The Nature of Six Sigma Quality*. Schaumburg, Illinois: Motorola University Press.

Harry, M. and L. Lawson. 1997. *Six Sigma Producibility Analysis and Process Characterization*. Schaumburg, Illinois: Motorola University Press.

Hecht, H. 1967. "Economics of Reliability Improvement for Space Launch Vehicles," Air Force Report No. SAMSO-TR-68-340.

————. 1970. "Interpretation of Test Data by Design Margin Techniques," in *Annals of the Assurance Sciences, 9th Reliability and Maintainability Conference*.

————. 1973. "Figure of Merit for Fault-Tolerant Space Computers." *IEEE Transactions on Computers*, C-22, No. 3: 246–251.

Hecht, H. and P. Crane. 1994. "Rare Conditions and Their Effect on Software Failures." *Proceedings of the 1994 Reliability and Maintainability Symposium,* January, pp. 334–337.

Hecht, H. and M. Hecht. 1985. "Reliability Prediction for Spacecraft," RADC Report RADC-TR-85-229. Rome Air Development Center, NY: Department of Defense.

IEEE. 1984. *IEEE Standard Dictionary of Electrical and Electronics Terms.* New York: Wiley Interscience.

Kececioglu, D. and D. Cormier. 1964. "Designing a Specified Reliability Directly into a Component," in *Proceedings of the Third Annual Aerospace Reliability and Maintainability Conference,* Society of Aerospace Engineers.

Lloyd, D.K. and M. Lipow. 1977. *Reliability: Management Methods and Mathematics (2nd edition).* Published by the authors.

MIL-HDBK-338B. 1998. *Electronic Reliability Design, Section 10, Systems Reliability Engineering.* Department of Defense.

MIL-STD-1629.1980. *Military Standard, Procedures for Performing a Failure Mode Effects and Criticality Analysis (FMECA) (rev. A).* November. Department of Defense.

MIL-HDBK-217F. 1991. *Military Handbook. Reliability Prediction of Electronic Equipment (rev. F).* (withdrawn as a DoD document)

Montgomery, D.C. 1996. *Introduction to Statistical Quality Control (3rd edition).* New York: John Wiley & Sons.

NASA and the Department of Defense. 1986. *National Space Transportation and Support Study.*

NASA Johnson Spaceflight Center (JSC). 1989. *Proceedings of the Satellite Services Workshop Number Four.*

Olivieri, J.M. 1997. "Commercial Satellite Manufacturing: Prerequisites for Success." Unpublished Master's degree thesis, Arizona State University, Tempe, Arizona.

Shunk, D.L. 1992. *Integrated Process Design and Development.* Homewood, Illinois: Business One Irwin.

Stoney, W.E. 1989. The Polar Platform and the SSS. Paper presented at the 20th Satellite Services System Working Group Meeting, 30–31 March.

Trivedi, Kishor S. 1982. *Probability & Statistics with Reliability Queuing and Computer Science Applications.* Englewood Cliffs, NJ: Prentice-Hall, Inc.

———. 1984. "Reliability Evaluation for Fault-Tolerant Systems," in *Mathematical Computer Performance and Reliability,* G. Iazeolla, P.J. Courtos and A. Hordijk, eds. The Netherlands: Elsevier Science Publishers V.V.

Wade, D.I. and C.S. Welch. 1996. "Spacecraft Manufacturing Implications for Volume Production Satellites." Paper No. IAF-96-U.4.08, presented at the 47th International Astronautical Congress, Beijing, China.

Womack, J.P., D.T. Jones, and D. Roos. 1990. *The Machine That Changed the World.* New York: Macmillan Publishing Company.

Chapter 20

Cost Modeling

Henry Apgar, *MCR International, Inc.*
David Bearden, *The Aerospace Corporation*
Robert Wong, *TRW, Inc.*

Cost is an engineering parameter that varies with physical parameters, technology, and management methods. A system's cost depends on its size, complexity, technological innovation, design life, schedule, and other characteristics. It's also a function of risk tolerance, methods for reducing risk, management style, documentation requirements, and project-management controls, as well as the size of the performing organizations. Analyzing and predicting program cost is becoming increasingly important, often critical, to determining whether a program proceeds. At the same time, sponsors, responding to budget reductions, and contractors, realizing that allowing technical performance alone to drive the design usually leads to a more expensive system, are systematically redefining the business of space, making it more difficult to accurately predict cost.

These trends dictate a changing role for cost estimation. In traditional, performance-only driven programs, cost modeling was primarily used to validate contractor cost estimates or give funding organizations an independent estimate of probable cost. Cost estimation was, to some extent, a self-fulfilling prophecy. Often, a space system would actually cost as much or more than what the budget allowed. However, this role is giving way to the more complex tasks of *design-to-cost* and *cost as an independent variable*, where performance is maximized subject to cost constraints. This entails a more proactive and interactive role for the cost estimator with involvement from the beginning of the process so that we may identify cost-effective solutions that meet a program's requirements. We can no longer just apply a general-purpose cost model to decide which programs to fund. Instead, we must develop a deeper understanding of cost-modeling methods, programmatic factors, and technological risks. As fiscal pressures continue to drive space budgets lower, cost estimates are being used at virtually all stages of space system procurement. Early in conceptual design, cost estimates help us assess whether development will succeed and identify key design decisions that will influence future costs. Project costs are monitored throughout the development cycle, and if they move much above budgeted amounts, we often must rescope or even cancel the program. Cost models must be flexible enough to evolve, from preliminary design to much later in the integration and test process, when we're deciding how to reallocate limited or diminishing resources.

In this chapter, we will

- Describe how to obtain cost estimates for space system elements
- Provide cost-estimating relationships useful for advanced system planning
- Describe how to assess the uncertainty (risk) in the cost estimates
- Show how cost and design may be integrated.

20.1 Introduction to Cost Analysis

20.1.1 Elements of Analysis

Figure 20-1 shows the relationships among key elements of cost analysis. The first step is to develop preliminary *cost analysis requirements descriptions* which identify the technical and operational parameters (*cost drivers*). These become "inputs" to cost models. We will develop just such a description in this chapter for FireSat. Each alternative concept specifies the configuration, number of units, orbits, equipment lists, hardware and software, and operational staffing needed for costing the system. The next step in the process is definition of a *Work Breakdown Structure* (*WBS*), an organizational table used to categorize and normalize costs. The WBS should cover all phases of the program. For example, the operations period follows *Initial Operating Capability* (IOC) and includes software maintenance costs and spare satellites.

Ground rules and assumptions should be laid out at the outset. These assumptions establish the foundation for understanding the costs and comparing them with those of other programs. Example assumptions include:

- Costs listed in constant-year dollars (fiscal funding information records costs in then-year dollars)
- Inflation rate forecasts (see Sec. 20.1.4)
- Exclusion of contractor fee and costs of the government project office
- Inclusion of government-furnished equipment
- Learning curve percentage (See Sec. 20.4.4)

Once input parameters have been specified and assumptions delineated, cost models support the preparation of estimates for each design alternative. The models combine cost data and cost-estimating relationships as described in Sec. 20.3. The cost estimates are organized by each segment's work breakdown structure, life-cycle phase, and schedule. Estimates prepared with cost models are not "end products"

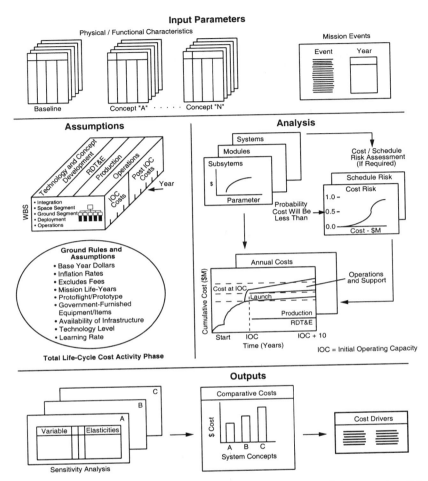

Fig. 20-1. Life-Cycle Cost Analysis Approach. This analysis approach is comprehensive and provides cost information necessary to properly evaluate system candidates.

themselves but, rather, means to an end. That end, typically, is a recommendation for the most cost effective design or the most affordable approach that meets mission requirements. Cost and schedule risk analysis is performed to capture uncertainties inherent in the analysis process and include effects of unusual requirements and beyond state-of-the-art technology. We distribute costs over the operational period according to a schedule of events and milestones. This schedule may not be necessary for early quick-look cost assessments and trade studies, but it is desirable for funding profile planning. Cost model outputs include:

- The major cost drivers—elements contributing most to total cost or mission requirements most affecting life-cycle costs

- A cost comparison of alternative systems or subsystems for trade studies

- The sensitivity of life-cycle costs to key assumptions and requirements

Figure 20-2 presents a more detailed WBS for a space mission which corresponds to the mission architecture elements of Chaps. 1 and 2, focusing on cost-related elements. It is very important because it helps us organize data, identify significant costs, and consistently compare one system to another.

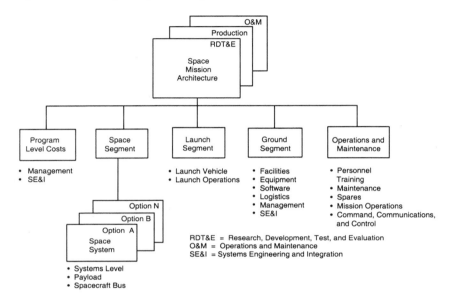

Fig. 20-2. Representative Work Breakdown Structure, WBS. The Work Breakdown Structure is an important tool for organizing cost information and ensuring consistency in comparing alternatives.

Life-Cycle Costs of a space mission architecture (i.e., the total mission cost from planning through end-of-life) are broken down into three main phases. The *Research, Development, Test, and Evaluation* (RDT&E) phase includes design, analysis, and test of breadboards, brassboards, prototypes and qualification units. Commonly referred to as the *nonrecurring* phase, RDT&E conventionally includes protoflight units and one-time ground station costs. This phase does not include technology development for system components. The *Production* phase incorporates the cost of producing flight units and launching them. A definition used to model costs is the *Theoretical First Unit* (TFU), which represents the first flight-qualified satellite off the line (for single-satellite missions the TFU is the flight article). For multiple units, production cost is estimated using a *learning curve* factor applied to the TFU cost as discussed in Sec. 20.4.4. Replacement satellites and launches after the space system *final operating capability* (FOC = full complement of on-orbit satellites) has been established are not considered as production units. The *Operations and Maintenance* (O&M) phase consists of ongoing operations and maintenance costs, including spacecraft unit replacements and software maintenance. Although the space, launch, and ground segments are usually the most important elements, O&M can sometimes be the system's most costly one (especially for constellations and reusable systems). For most space programs the primary ongoing operations and support costs are ground station operations and satellite spares; for reusable systems such as the Shuttle, this category consists of the ground crew and operations to support them.

20.1.2 Cost Estimating Methods

The work breakdown structure and system concept provide guidance for determining which cost methods apply. Three basic methods are used to develop cost estimates:

1. In *detailed bottom-up estimating* we identify and specify at a low level elements that make up the system. We then estimate the cost of materials and labor to develop and produce each element. This method has the advantage of being explicitly tailored to a specific program and contractor, but its basis of validity is the credibility of the experts called upon to estimate inputs, such as hours, labor rates, material costs, and indirect costs. Because this method is time consuming and because detailed design data is usually not available, this method is least appropriate for preliminary system studies. Bottom-up estimates are most commonly used during the production phase of a program, after design details are well known and a majority of technical uncertainties have been resolved during development.

2. In *analogy-based estimating* we use the cost of a similar item and adjust it for differences in size or complexity. We can apply this method at any level of detail in the system, but it is lower fidelity than a bottom-up estimate. This method also presumes that a sufficiently similar item exists and that we have detailed cost and technical data on which to base our estimate.

3. In *parametric estimating* we use a series of mathematical relationships that relate cost to physical, technical, and performance parameters that are known to strongly influence costs. An equation called the *Cost Estimating Relationship*, or CER, expresses the cost as a function of *parameters*. Cost drivers and function forms are selected based on a combination of engineering judgment and the statistical quality of the regression results. We may also apply complexity factors to the parameters to account for technology changes.

All three methods have advantages and disadvantages, depending on scope of the estimating effort and the amount of design and performance information (e.g. relevant historical data for analogy) available to estimators. Until recently, parametric models were avoided as the primary tool for developing a proposal bid and used only as back-up to validate another estimate. That situation is beginning to change, mainly due to the progress of the Parametrics Estimating Initiative led by several cost-estimating professional societies and endorsed by the DoD.[*]

The primary advantage of parametric models lies in their top-down approach. System requirements and top-level design specifications are all that are required to complete a cost estimate with a parametric model. Detailed hardware designs and development schedules do not have to be drafted to estimate costs of candidate system architectures. Incorporation of CERs within a system design/cost-engineering tool allows concept evaluation and technology studies to be conducted. Parametric cost models are therefore well suited for use in cost/performance trade studies that estimate how cost varies as a function of system requirements, in developmental planning and architecture studies, and in specific program assessments of cost vs. capability and

---

[*] Due to recent downsizing of estimating staffs in government and industry and general maturing of the parametric estimating process, the DoD has publicly advocated application of parametric estimating models as the **preferred method** of proposal estimating. See the DoD [1998] for guidance and examples.

individual parameter sensitivity. Since parameteric cost models are the most appropriate for trade studies, they are the focus of the remainder of this chapter.

Use of a parametric model implies several assumptions. First, because parametric models characterize historical cost trends as mathematical relationships, it is assumed that future costs will reflect these historical trends to some degree. Parametric models are applicable only to the range of historical data. In cases when major technology advancements are expected, or when fundamental paradigms in system architecture are shifted, parametric cost models based on old systems may very well not apply. One example of this problem is encountered when large-satellite-based cost models are used to estimate costs of today's smallsats. Such a paradigm shift, and perhaps a technology shift, requires a specialized model. Lacking new technology factors, CERs must be adjusted when applied to systems using beyond state-of-the-art technology. We derive the CERs from historical data, so their technologies may not reflect the advances usually considered in futuristic studies.

A second implicit assumption made when using parametric cost models is that program costs are random variables that cannot be predicted with 100% accuracy. Many more parameters influence costs than can be incorporated into a set of CERs, such as skill level of contractor engineers and technicians, occurrence of unforeseen technical problems, business base of all contractors involved, requirements changes, and test failures. Parametric cost models use a combination of parameters that explain historical cost trends while maintaining statistical integrity. Influence of all other variables manifests itself in estimating error, which is often quantified by using underlying data to calculate the *standard error, SE*.

There are three general sources for parametric cost estimating models:

- *Publicly available special purpose models*, such as the *Unmanned Space Vehicle Cost Model* (USCM), the *Small Satellite Cost Model* (SSCM), and the *Communications Payload and Spaceborne Electronics Cost Model* (CPCM). Such models are typically developed by the Federal Government and are usually available to the general public*. See the references at the end of this chapter for source information.

- *Publicly available general purpose models*, offered privately by commercial organizations such as PRICE Systems and SEER Systems. These general purpose models typically must be calibrated to the user's specific products and processes before they can be used for estimating space hardware and software components.

- *Private specific purpose models*, usually developed by a single organization from proprietary information, to estimate company-specific systems and components only.

In Sec. 20.3 we present CERs for computing the cost of space systems based on publicly available models. If these models do not apply to a specific concept, we may need to derive new relationships. The following section summarizes this process. Fisher [1970] provides a comprehensive discussion.

---

* Although we refer primarily to government programs and data, the models may be applied to commercial programs. The recommended cost models are appropriate for both large and small missions. (See Sec. 20.3.)

20.1.3 Cost Model Development

Figure 20-3 shows the procedure for developing a CER that represents how the cost properties of a system or subsystem vary with characteristic parameters. The first ingredient in defining a set of CERs is a historical database. Extensive research and data collection based on actual cost and technical data is needed. It's important to find out as much as possible about the origin of the data and the reference for measuring the characteristic physical parameters. We must also separate the cost elements into comparable physical subsystems or components, which become the costing elements we use to establish the entire system cost. Programs either already completed or awaiting launch within a year should be targeted as opposed to programs that are still in early stages of development. Most costs in the database should be actual program costs at completion. In cases where satellites are nearly complete but have not yet been launched, contractor estimates-at-completion costs may be used.

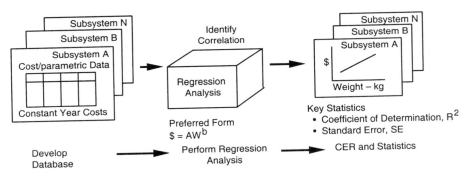

Fig. 20-3. Deriving CERs from Actuals. The key to a credible CER is an accurate database of relevant historical information from actual programs.

It is also important to *normalize* the historical data so it is consistent—correctly categorized between nonrecurring and recurring costs and in the same constant-year dollars. Then, we would compensate for economic differences in production quantities. The design or performance parameters to which costs will be related must be in the same units. These include programmatic, weight-based, and performance parameters for the satellite in general and each of the major subsystems. We typically exclude the prime contractor's fee but include all other direct, indirect, and general and administrative costs. Award fees and incentives are usually not included, nor are government costs (i.e., costs associated with the government procurement agency if one exists).*

After cost data are properly categorized and normalized, the task of CER development begins. To allow credible subsystem-level cost analysis of potential missions without requiring a detailed design we need to relate cost to technical characteristics. Other factors, named *wraps*, model nonphysical factors not included in the CER, such as system engineering, management, and product assurance, as well as the cost of integrating and testing the space system. Wraps typically account for about 30% of the development cost for space systems. Choosing cost drivers involves a combination of

---

* A fee should be added if we want to determine the purchase price. This fee is normally a negotiated value, but 10% is typical for a cost analysis. In addition, be aware of the need to include subcontractor fees according to the planned contractor tier arrangements.

statistics, engineering judgment, and often, common sense. We would hypothesize a relationship between costs and the explanatory parameters. For conceptual studies—the primary focus of this book—we would limit the parameters to one or two per subsystem.

For space systems, cost drivers would be primarily weight, power, and performance requirements, the parameters most likely to be available during an advanced system study. For example, in the derivation of the CER for the electrical power subsystem, we might consider the following: mass, beginning-of-life power, solar array area, on-orbit average power, design life, battery capacity, solar cell type, battery type, and payload power. As another example, if the attitude control subsystem is of interest, we would examine parameters such as pointing accuracy, knowledge, and required slew rate. We prefer to use a power law for relationships (see Fig. 20-3) because it allows for expected nonlinearities, and can be easily accommodated by standard regression packages. Also, since weight is often a key parameter for cost-estimating relationships, the power function with an exponent of less than one models the expected diminishing effects of increasing weight on costs. This highlights a danger not to forget: that all CERs are simplifications of the relationship they are emulating. Weight-driven CERs, for example, imply that lighter structures cost less. In fact, the opposite may be true. We need to recognize that when designers deliberately reduce weight, they may increase complexity, trade away ease of manufacturing an integration, or use inherently more expensive materials.

When deriving CERs we take note of statistical outliers and follow up to ascertain whether or not apparent discrepancies are attributable to numerical errors or nontraditional ways of accounting for certain costs. We evaluate the "goodness" of the relationship by evaluating the costs against their corresponding parameters using regression software. For further information on statistics and regression techniques, see, for example, Book and Young [1995]. The regression results in an equation between cost and the parameter or parameters, as well as statistics that indicate how well the relationship fits the data. Measures of the goodness of fit include the *coefficient of determination, R^2,* and the *standard error, SE. R^2,* is dimensionless between 0 and 1. *SE* is in units of cost or a percentage of the estimated cost depending on whether an additive or multiplicative error regression approach is used. A strong correlation is represented by R^2 near 1 and *SE* near 0. *SE* is important in evaluating the uncertainty in estimates as discussed in Sec. 20.4.

For additional information on space cost model development and space costing, see, for example, Apgar [1990], DISA [1997], Greenberg and Hertzfeld [1994], and Hamaker [1989]. For a discussion of cost modeling for low-cost missions see Wertz and Larson [1996] and Sarsfield [1998].

20.1.4 Types of Dollars

For consistency in referring to costs and to avoid confusion in the review of cost analysis results, constant-year dollars should be used. For examples in this chapter, Fiscal Year 2000 dollars (FY00$) are assumed. This simplifies the computations and interpretation of results, especially in making comparisons of alternatives. If project funding by year is required, then the costs should be spread first by year in constant dollars and converted to *real* or *then-year* dollars by multiplying each year's funding by an appropriate inflation factor. Table 20-1 provides a table of inflation factors for 1980–2020 relative to FY2000 as provided by the Office of the Secretary of Defense.

TABLE 20-1.   Inflation Factors Relative to the Year 2000 Based on Projections by the Office of the Secretary of Defense (January 1998). See text for discussion.

Fiscal Year (FY)	Inflation Factor to Base Year 2000	Fiscal Year (FY)	Inflation Factor to Base Year 2000
1980	0.456	2001	1.017
1981	0.510	2002	1.034
1982	0.559	2003	1.052
1983	0.610	2004	1.075
1984	0.658	2005	1.099
1985	0.681	2006	1.123
1986	0.700	2007	1.148
1987	0.719	2008	1.173
1988	0.740	2009	1.199
1989	0.771	2010	1.225
1990	0.802	2011	1.252
1991	0.837	2012	1.279
1992	0.860	2013	1.308
1993	0.883	2014	1.336
1994	0.901	2015	1.366
1995	0.918	2016	1.396
1996	0.937	2017	1.427
1997	0.958	2018	1.458
1998	0.970	2019	1.490
1999	0.984	2020	1.523
2000	1.000		

To convert costs from any year to fiscal year 2000 dollars (FY00$), divide by the inflation factor in the table. To convert from FY00$ to other year dollars multiply by the factor for that year. For conversion to years other than 2000, we calculate a new inflation factor as the ratio of the factors for the years of interest.

20.2 The Parametric Cost Estimation Process

Table 20-2 summarizes the procedure to compute parametric cost estimates using the cost models of this chapter. The information needed about the system design is shown in Table 20-3 along with values developed previously for the FireSat example. The steps shown follow the general process of Fig. 20-1. The procedure begins with developing the WBS and collecting the relevant space system characteristics. The next four steps develop the primary elements of life-cycle costs, computing the costs for the space, launch and ground segments, followed by the operations and costs. The CERs for estimating each cost item are given in Sec. 20.3. CERs are provided at the subsystem level for RDT&E and TFU costs. They have been adjusted to estimate fiscal year 2000 (FY00) costs. The TFU is the basis for computing the cost of multiple units in production as described in Sec. 20.4.4. Factors for program level costs and heritage are shown in Tables 20-7 and 20-8.

TABLE 20-2. Parametric Cost Estimation Process. This process provides an estimate of the total life-cycle cost.

Step	Reference
1. Develop Work Breakdown Structure – Identify all cost elements	Fig. 20-2
2. List Space System Characteristics – Identify advanced technology parameters	Table 20-3 Table 20-8

3. Compute Space Segment Cost
   – RDT&E cost
   – Software cost
   – Theoretical first unit cost
   – Subsequent unit costs
	Tables 20-4, 20-6, 20-9
	Table 20-10
	Tables 20-5, 20-6, 20-9
	Sec. 20.4

4. Compute Launch Segment Cost	Table 20-14

5. Compute Ground Segment Cost
   – First ground station
   – Software cost
   – Additional ground stations
   – Earth terminals
	Table 20-11
	Table 20-10
	Table 20-11
	Table 20-13

6. Compute Operations and Maintenance Cost
   – Space segment spares
   – Launch costs for spares
   – Ground system operations and support
	Sec. 20.5
	Table 20-14
	Table 20-12

7. LIFE-CYCLE COST	Sum of items 3–6 above

20.2.1 FireSat Cost Element Definitions

We include these items in each of the subsystems to be estimated:

- *Payload:* this includes communication systems and sensors (visible and IR). Some assumptions were made regarding electronic weights as these parameters were not available from the conceptual design.

- *Spacecraft Bus:* this is the spacecraft less the payload. Primary cost driver is mass.

- *Structure:* spacecraft structure items including enclosures, deployable components, supporting structure and launch vehicle interface. The spacecraft structure carries and protects the spacecraft and payload equipment through launch and deployment. Mass is the key metric that determines cost.

- *Thermal:* structure and devices for the purpose of maintaining all elements of a satellite system within required temperature limits. Thermal control systems may be classified as passive or active. An example of a passive system is paints, coatings and blankets, and a space radiator coupled to heat sources by conductive paths such as base plates. Active thermal-control subsystems include pumped-loop systems, heaters controlled by thermostats, mechanical devices (e.g. louvers) and refrigerators. In general, passive systems cost less than active systems.

- *Electrical Power Subsystem (EPS):* solar arrays, batteries, harness, and power management electronics. EPS mass, used in the cost model, is largely influenced by space radiation, which degrades performance of solar cells over time (FireSat has a 5-year design life). Choices for off-the-shelf solar-array cells include silicon or gallium arsenide. Battery choices include NiCd and NiH_2.

TABLE 20-3.   Space Mission Characteristics Required for Parametric Cost Modeling.

Characteristic	Reference	FireSat Example
1. *Constellation*		
No. of spacecraft in constellation	Secs. 7.1, 7.6	2
Orbit altitude	Secs. 3.3, 7.4	700 km
2. *External Communications Resources*		
TDRS	Secs. 13.1, 13.2	No
DomSat	Secs. 13.1, 13.2	No
3. *Space Segment*		
Payload		
Type	Sec. 9.1	IR
Weight (Communications)	Sec. 13.4	N/A
Aperture (IR, Visible)	Sec. 9.3	0.26 m
Spacecraft bus		
Dry weight by subsystem	Secs. 10.2, 10.3	Table 10-30
Volume	Sec. 10.5.1	1.7 m^3
Pointing accuracy	Sec. 11.1	0.1 deg
Pointing knowledge	Sec. 11.1	< 0.1 deg
Stabilization type	Sec. 10.4.2	3-axis
Flight software lines of code	Sec. 16.2	26 K
Average power	Sec. 10.6	110 W
BOL power	Sec. 10.6	NA
EOL Power	Sec. 10.4.2	NA
Solar array area	Sec. 10.6	8.5 m^2
Battery capacity	Sec. 11.4.2	17.5 A-hr
Data storage capacity	Sec. 11.3	
Number of thrusters	Sec. 11.1.4	4
Space segment design life	Sec. 10.4	5 yrs.
4. *Launch Segment*		
Launch vehicle	Secs. 18.1, 18.2	Pegasus
Upper stage	Secs. 18.1, 18.2	None
Launch site	Secs. 18.1, 18.2	N/A
No. of spacecraft per launch	Secs. 18.1, 18.2	1
5. *Ground Segment*		
No. of fixed and mobile sites	Secs. 13.1, 15.6	1 fixed
Software language	Sec. 16.3	Ada
Lines of code	Sec. 16.3	100 K
New or existing equipment and facilities	Sec. 15.2	New
Communications operating frequency	Sec. 13.3.5	S-band
6. *Mission Operations and Support*		
Mission duration from IOC*	Sec. 1.4	10 yrs.
No. of personnel	Sec. 14.3	10
No. of spare spacecraft	Sec. 19.1	0
No. of Shuttle support flights	Sec. 19.1	0

*IOC = Initial Operating Capability = time of launch of first satellite to provide operational data.

- *Tracking, Telemetry, and Command and Data Handling (TT&C/DH):* com‑ mand/telemetry electronics, onboard computers, transponders, transmitters receivers, data storage, antennas, and associated avionics. Primary cost driver of the TT&C/DH subsystem are subsystem mass, frequency of the uplink an‑ downlink, and data rate. In some cases, this is further divided into TT&C (th‑ communications system) and C&DH (the spacecraft computer plus other dat‑ storage and handling equipment).

- *Attitude Determination and Control Subsystem (ADCS):* stabilizes and orient‑ FireSat during its mission using sensors and actuators. ADCS is tightl‑ coupled to other subsystems, especially the propulsion subsystem. The prima‑ ry drivers are requirements for pointing knowledge, control stability, an‑ maneuvering (such as slewing or payload repointing). The FireSat missio‑ requires a 3-axis control system to point in a specific direction to within 0.2 deg; knowledge of spacecraft attitude is 0.10 deg.

- *Propulsion:* Provides thrust for attitude control and orbit correction. FireSa‑ uses a liquid system for on-orbit attitude correction and maneuvering (Tabl‑ 7-3). The relevant cost drivers are the propulsion subsystem bus dry mass c 8.4 kg (Table 10-31) and the spacecraft dry mass of 112 kg (Sec. 10.6). If orb‑ insertion requires an apogee kick motor, additional costs should be assessed.‑

- *Integration, Assembly, and Test (IA&T):* labor and material costs (primaril‑ testing) for integrating spacecraft and payload subsystems into an operation‑ space vehicle. Does not include costs for integrating components into subsystem (these costs are included in the subsystem CERs) or for integratin‑ the space vehicle with the launch vehicle. The total cost of IA&T for a satelli‑ includes research/requirements specification, design and scheduling analys‑ of IA&T procedures, systems test and evaluation, and test data analysis.

- *Program Level:* contractor costs for systems engineering, program manage‑ ment, reliability, planning, requirements flowdown, quality assurance, projec control, data preparation, and other costs which cannot be assigned to individ‑ ual hardware or software components. Program management includes effor‑ associated with planning and directing prime and subcontractor efforts an‑ interactions. System engineering includes activities required to ensure that a satellite subsystems and payloads function properly to achieve system goa‑ and requirements. Data and report generation is a program-level function th‑ includes efforts required to produce internal and deliverable documentation.

- *Ground Support Equipment (GSE):* test and support equipment needed fc assembly, development and acceptance testing and integration of satelli‑ subsystems and satellite to the launch vehicle. This equipment is required ‑ support the satellite and provide physical, electrical, and data interfaces wi‑ the satellite during IA&T. It is therefore classified as a nonrecurring cost.

- *Launch and Orbital Operations Support (LOOS):* planning and operation‑ related to launch and orbital checkout of the space system. These costs ar‑ those costs typically incurred by the spacecraft prime contractor involvin‑ prelaunch planning, trajectory analysis, launch site support, launch-vehic‑ integration (spacecraft portion), and initial on-orbit operations before owne‑ ship of the satellite is turned over to the operational user (typically 30 days They are generally categorized as recurring costs.

20.3 Cost Estimating Relationships

This section presents the specific CERs recommended for preliminary mission analysis and design.

20.3.1 Space Segment Costs

The CERs of Tables 20-4, 20-5, and 20-6 may be used to estimate costs in thousands of fiscal year 2000 constant dollars as a function of the specified parameters. (See Sec. 20.1.4 for conversion to other years.) These CERs are derived from historical data and, therefore, their validity is limited to a range of parameter values. The tables present the range of application; to preserve their validity, the equations should not be used further than 25% beyond the parameter ranges given. The CERs were derived using different satellite data, statistical frameworks, and error models as noted. Some CERs provide total subsystem cost while others estimate RDT&E and TFU separately. Differences between the RDT&E and TFU CER ranges are due to availability of data. For some subsystems, an alternate CER is presented to accommodate a different cost driver. Generally, TFU CERs show higher variability.

TABLE 20-4. CERs for Estimating Subsystem RDT&E Cost (FY00$K). Applicable range for a good estimate is 25% above and below this data range. CER represents contractor cost without fee.

Cost Component	Parameter, X (Unit)	Input Data Range	RDT&E CER[*] (FY00$K)	SE (%)
1. Payload				
1.1 IR Sensor	aperture dia. (m)	0.2–1.2	$356{,}851\ X^{0.562}$	$53{,}559$[†]
1.2 Visible Light Sensor	aperture dia. (m)	0.2–1.2	$128{,}827\ X^{0.562}$	$19{,}336$[†]
1.3 Communications	comm. subsystem wt. (kg)	65–395	$353.3\ X$	51
2. Spacecraft	spacecraft dry wt. (kg)	235–1,153	$101\ X$	33
2.1 Structure	structure wt. (kg)	54–392	$157\ X^{0.83}$	38
2.2 Thermal	X_1 = thermal wt. (kg)	3–48	$394\ X_1^{0.635}$	45
	X_2 = spacecraft wt. + payload wt. (kg)	210–404	$1.1\ X_1^{0.610}\ X_2^{0.943}$	32
2.3 Electrical Power System (EPS)	X_1 = EPS wt. (kg)	31–491	$62.7\ X_1$	57
	X_2 = BOL power (W)	100–2,400	$2.63\ (X_1\ X_2)^{0.712}$	36
2.4 Telemetry, Tracking & Command (TT&C)/DH‡	TT&C/DH wt. (kg)	12–65	$545\ X^{0.761}$	57
2.5 Attitude Determination & Control Sys. (ADCS)	ADCS wt. (kg)	20–160	$464\ X^{0.867}$	48
2.6 Apogee Kick Motor (AKM)	AKM wt. (kg)	81–966	$17.8\ X^{0.75}$	—
3. Integration, Assembly & Test (IA&T)	spacecraft bus + payload total RDT&E cost (FY00$K)	2,703 – 395,529	$989 + 0.215\ X$	46
4. Program Level	spacecraft bus + payload total RDT&E cost (FY00$K)	4,607 – 523,757	$1.963\ X^{0.841}$	36
5. Ground Support Equipment (GSE)	spacecraft bus + payload total RDT&E cost (FY00$K)	24,465 – 581,637	$9.262\ X^{0.642}$	34
6. Launch & Orbital Operations Support (LOOS)	N/A			

*   Taken from USCM, 7th edition (1994) using minimum, unbiased percentage error CERs.
†   Absolute error (in FY00$K), not percentage error.
‡   Includes spacecraft computer. If separate CERs for TT&C and C&DH are desired, use a 0.45/0.55 split.

TABLE 20-5.   CERs for Estimating Subsystem Theoretical First Unit (TFU) Cost.

Cost Component	Parameter, X (Unit)	Input Data Range	TFU CER* (FY00$K)	SE (%)
1. Payload				
1.1 IR Sensor	aperture dia. (m)	0.2–1.2	142,742 $X^{0.562}$	21,424†
1.2 Visible Light Sensor	aperture dia. (m)	0.2–1.2	51,469 $X^{0.562}$	7,734†
1.3 Communications	comm. subsystem wt. (kg)	65–395	140 X	43
2. Spacecraft	spacecraft dry wt. (kg)	154–1,389	43 X	36
2.1 Structure	structure wt. (kg)	54–560	13.1 X	39
2.2 Thermal	thermal wt. (kg)	3–87	50.6 $X^{0.707}$	61
2.3 Electrical Power System (EPS)	EPS wt. (kg)	31–573	112 $X^{0.763}$	44
2.4 Telemetry, Tracking & Command (TT&C)/DH‡	TT&C/DH wt. (kg)	13–79	635 $X^{0.568}$	41
2.5 Attitude Determination & Control Sys. (ADCS)	ADCS wt. (kg)	20–192	293 $X^{0.777}$	34
2.6 Apogee Kick Motor (AKM)	AKM wt. (kg)	81–966	4.97 $X^{0.823}$	20
3. Integration, Assembly & Test (IA&T)	spacecraft bus wt. payload wt. (kg)	155–1,390	10.4 X	44
4. Program Level	spacecraft + payload total recurring cost (FY00$K)	15,929 – 1,148,084	0.341 X	39
5. Ground Support Equipment (GSE)	N/A			
6. Launch & Orbital Operations Support (LOOS)	spacecraft bus + payload wt. (kg)	348–1,537	4.9 X	42

*   Taken from USCM, 7th edition (1994) using minimum, unbiased percentage error CERs.
†   Absolute error (FY00$K), not percentage error.
‡   Includes spacecraft computer. If separate CERs for TT&C and C&DH are desired, use a 0.45/0.55 split.

The CERs for the payload and spacecraft bus subsystems are primarily based on parameters available during the concept and mission design phase. The models cover both the subsystem and system levels with the user choosing the appropriate level. The models used here are publicly-available, special-purpose models, developed by the U.S. Air Force and NASA. These CERs were obtained from the *Unmanned Space Vehicle Cost Model*, Seventh Edition [SMC, 1994], the *Communications Payload and Spaceborne Electronics Cost Model* [MCR Federal, 1997], and a derivative of the *Small Satellite Cost Model* [Bearden et al., 1996]. Other models that could be used in lieu of those given in the tables are given by Burgess, Lao, and Bearden [1995], and Management Consulting and Research, Inc. [1986].

The primary categories of costs are hardware, software and program level (or *wraparounds*) used to indicate that the estimates for these functions are based upon percentages of hardware cost. *Wraps* are costs associated with labor-intensive activities where a level of manpower is allocated over some period of performance. The functions in this category are management, systems engineering, product assurance, and system tests. The CERs of Tables 20-4, 20-5, and 20-6 provide the overall program level costs and Table 20-7 provides an allocation of program level costs to the wrap components.

TABLE 20-6. Cost-Estimating Relationships for Earth-orbiting Small Satellites Including RDT&E and Theoretical First Unit. Total subsystem cost in FY00$M as a function of the independent variable, X.

Cost Component	Parameter, X (Unit)	Input Data Range	Subsystem Cost CER* (FY00$K)	SE (FY00$K)
1. Payload	Spacecraft Total Cost (FY00$K)	1,922–50,651	$0.4\,X$	$0.4 \times SE_{bus}$
2. Spacecraft	Satellite bus dry wt. (kg)	20–400	$781 + 26.1\,X^{1.261}$	3,696
2.1 Structure†	Structures wt. (kg)	5–100	$299 + 14.2\,X \ln(X)$	1,097
2.2 Thermal‡	Thermal control wt. (kg)	5–12	$246 + 4.2\,X^2$	119
2.3 Electrical Power System (EPS)	Average power (W)	5–410	$-183 + 181\,X^{0.22}$	127
	Power system wt. (kg)	7–70	$-926 + 396\,X^{0.72}$	910
	Solar array area (m²)	0.3–11	$-210{,}631 + 213{,}527 X^{0.0066}$	1,647
	Battery capacity (A-hr)	5–32	$375 + 494\,X^{0.754}$	1,554
	BOL Power (W)	20–480	$-5{,}850 + 4{,}629\,X^{0.15}$	1,585
	EOL Power (W)	5–440	$131 + 401\,X^{0.452}$	1,603
2.4a Telemetry Tracking & Command (TT&C)**	TT&C/DH wt. (kg)	3–30	$357 + 40.6\,X^{1.35}$	629
	Downlink data rate (Kbps)	1–1,000	$3{,}636 - 3{,}057\,X^{-0.23}$	1,246
2.4b Command & Data Handling (C&DH)	TT&C + DH wt. (kg)	3–30	$484 + 55\,X^{1.35}$	854
	Data Storage Capacity (MB)	0.02–100	$-27{,}235 + 29{,}388 X^{0.0079}$	1,606
2.5 Attitude Determination & Control Sys. (ADCS)	ADCS dry wt. (kg)	1–25	$1{,}358 + 8.58\,X^2$	1,113
	Pointing accuracy (deg)	0.25–12	$341 + 2651\,X^{-0.5}$	1,505
	Pointing knowledge (deg)	0.1–3	$2{,}643 - 1{,}364 \ln(X)$	1,795
2.6 Propulsion††	Satellite Bus dry wt. (kg)	20–400	$65.6 + 2.19\,X^{1.261}$	310
	Satellite volume (m³)	0.03–1.3	$1539 + 434 \ln(X)$	398
	Number of Thrusters	1–8	$4{,}303 - 3{,}903 X^{-0.5}$	834
3. Integration, Assembly & Test (IA&T)	Spacecraft total cost (FY00$K)	1,922–50,651‡‡	$0.139\,X$	$0.139 \times SE_{bus}$
4. Program Level	Spacecraft total cost (FY00$K)	1,922–50,651‡‡	$0.229\,X$	$0.229 \times SE_{bus}$
5. Ground Support Equipment (GSE)	Spacecraft total cost (FY00$K)	1,922–50,651‡‡	$0.066\,X$	$0.066 \times SE_{bus}$
6. Launch & Orbital Operations Support (LOOS)	Spacecraft total cost (FY00$K)	1,922–50,651‡‡	$0.061\,X$	$0.061 \times SE_{bus}$

* CERs based on the Small Satellite Cost Model [Bearden, Boudreault, and Wertz, 1996], adjusted for inflation as shown in Table 20-1, and broken into subsystem cost using the percentages from Table 20-9.
† Aluminum materials primarily with selected use of advanced materials (e.g. composites, magnesium).
‡ Thermal CER appropriate for passive systems only.
** CER applies to UHF/VHF and S-band LEO systems
†† Hydrazine monopropellant and cold-gas stationkeeping systems only. CER not appropriate for bipropellant or dual-gas systems. Costs of AKM are not included.
‡‡ Input data range for items 3–6 calculated using min and max values of input data range for spacecraft bus cost CER in item 2.

TABLE 20-7. **Allocation of Program-Level Cost.** We allocate the program level or wrap costs in Tables 20-4, 20-5, and 20-6 to their components as shown.

Program Level Component	RDT&E	Theoretical First Unit
Program Management	20%	30%
Systems Engineering	40%	20%
Product Assurance	20%	30%
System Evaluation	20%	20%

TABLE 20-8. **Heritage Cost Factors.** We apply these factors to the CERs and their standard errors from Tables 20-4, 20-5, and 20-6 as described in the text. Data from Hamaker [1987].

Multiplicative Factors for Development Heritage (Apply to RDT&E Costs Only)	
New design with advanced development	> 1.1
Nominal new design—some heritage	1.0
Major modification to existing design	0.7 – 0.9
Moderate modifications	0.4 – 0.6
Basically existing design	0.1 – 0.3

Recurring and Nonrecurring Factors

Nonrecurring costs include all efforts associated with design, drafting, engineering unit IA&T, ground support equipment, and a portion of program management and system engineering costs. This includes all costs associated with design verification and interface requirements (e.g. drawings, schematics, mockups, boilerplates, breadboards and brassboards). *Recurring costs* cover all efforts associated with flight hardware manufacture, IA&T, and a portion of program management and system engineering costs. The CERs in Tables 20-4 and 20-5 are already separated between RDT&E (nonrecurring) and TFU (recurring) costs. Since the CERs in Table 20-6 provide estimates of total subsystem cost, factors for the split between RDT&E and TFU are needed. To meet this need, we present a list of recurring and nonrecurring factors in Table 20-9. These factors can be applied to estimated total subsystem costs from CERs to obtain estimates of the recurring and nonrecurring portions. Total production costs for all flight units are computed by multiplying the TFU cost by the learning curve factor L described in Sec. 20.4.4.

Prototype vs. Protoflight Approach

A *protoflight* approach is one in which the qualification test unit is refurbished for flight. The CERs in Table 20-4 assume a *prototype* approach, i.e., include the cost of one qualification unit. The small-satellite CERs in Table 20-6 assume a protoflight approach. The protoflight approach saves on costs since no "dead end" hardware will result. For refurbishment of the qualification unit to become the protoflight unit, 30% of the TFU should be added to the RDT&E cost. The RDT&E estimate will then include the first flight article.

Heritage Factors

Table 20-8 presents factors for development heritage. These are multiplicative factors to be applied to the RDT&E CER for design maturity of a given subsystem. The difficulty in incorporating heritage information in cost estimation has been, and continues to be, the quantification of heritage; part of this problem stems from a lack of a

TABLE 20-9. **Breakdown of Small Satellite Costs.** These factors can be applied to overall small spacecraft costs to estimate the cost of constituent subsystems and the RDT&E vs. TFU costs. In the first column, 100%;Total spacecraft hardware cost, excluding payload and wraps. Data on cost breakdown by subsystem courtesy Microcosm; derived from Wertz and Larson [1996] and Sarsfield [1998]. Comparable factors can be found in Bearden [1999] and Sadin and Davis [1993]. Recurring vs. Nonrecurring estimates are from Bearden, Burgess, and Lao [1995]. The final row represents the total system cost exclusive of launch cost and operations.

Subsystem/Activity	Fraction of Spacecraft Bus Cost (%)	Non-Recurring Percentage (%)	Recurring Percentage (%)
1.0 Payload	40.0%	60%	40%
Bus Total	100.0%	60%	40%
2.1 Structure	18.3%	70%	30%
2.2 Thermal	2.0%	50%	50%
2.3 EPS	23.3%	62%	38%
2.4a TT&C	12.6%	71%	29%
2.4b C&DH	17.1%	71%	29%
2.5 ADCS	18.4%	37%	63%
2.6 Propulsion*	8.4%	50%	50%
Wraps			
3.0 IA&T	13.9%	0%	100%
4.0 Program Level	22.9%	50%	50%
5.0 GSE	6.6%	100%	0%
6.0 LOOS	6.1%	0%	100%
Total	189.5%	92.0%	97.5%

* Propulsion costs may be excluded if, as is the case with many small satellites, the spacecraft doesn't have a propulsion system.

standard definition for heritage. We define *heritage* as the percentage of a subsystem that is identical to one or more previous spacecraft, by mass. This has the appeal of being a measurable quantity, but has some obvious drawbacks. For instance, while mass may be a reasonable measure of design heritage for structures, it may not be appropriate for the TT&C subsystem. With this in mind, heritage is a value, from 0% (no heritage) to 100% (all heritage), which varies by subsystem and is best evaluated by subsystem design experts. For example, if the subsystem represents an existing design with 70% heritage, the developments costs will primarily be engineering interface and drawing modifications so that only 30% of the RDT&E cost is needed.

Commercial Missions

The CERs presented here were derived primarily from government procurements. We should apply a RDT&E factor of 0.8 when the CERs are used to estimate commercial satellite costs. This is an average factor based upon comparison of commercial communication satellites with government-procured communication satellites. Smith, Stucker and Simmons [1985] have performed a comprehensive study that indicates a 19% cost growth for military satellite contracts compared to 2% for commercial. The rationale for the difference is the higher level of uncertainty in mission requirements as more changes typically occur on the military programs. However, their findings indicate little difference in unit recurring costs.

20.3.2 Software Costs

Table 20-10 presents software costing relationships for flight and ground software. It also provides factors for various programming languages. Section 16.3.3 discusses software development costs further. Flight software is assumed to cost more (per KLOC) because there is more testing required to meet mission criticality. If software reuse is employed, the heritage factors in Table 20-8 apply.

TABLE 20-10.   Software Development Costs. RDT&E costs only (in FY00$K). See Sec. 16.2.2 for estimates of the lines of code.

Flight Software	435 × KLOC
Ground Software	220 × KLOC
KLOC = Thousand of Lines of Code; cost without fee	

FACTORS FOR OTHER LANGUAGES	
Language	**Factor**
Ada	1.00
UNIX–C	1.67
PASCAL	1.25
FORTRAN	0.91

20.3.3 Ground Segment and Operations Costs

Ground segment costs vary significantly depending upon the purposes of the ground stations. For most ground station cost estimates, we must state requirements for square footage of facilities, and an equipment list of specific items (computers, RF equipment, and so forth) which are typically not determined during the concept development stage of a program.

For this model, the costs for various elements of a ground station will be based upon typical distribution of costs between software, equipment, facilities and wraps, as Table 20-11 indicates. The distribution is fairly representative of a number of space projects. For preliminary mission design, this may be translated into estimated costs as follows. First, compute the software costs from Table 20-10. Then estimate other ground segment costs as a percent of software costs using the representative distributions of Table 20-11. A column to simplify this calculation has been added to the table.

The operations and support costs during the operational phase of the ground segment consist primarily of contractor and government personnel costs as well as maintenance costs of the equipment, software, and facilities. Table 20-12 presents expressions for these costs. The labor rates include overhead costs and other typical expenses associated with personnel. For smaller Earth terminals, Table 20-13 provides some typical costs of communications equipment for commonly used frequency bands.

TABLE 20-11. Ground Segment Development Cost Model. For preliminary mission design this should be used in conjunction with the software cost estimate from Table 20-10.

Ground Station Element	Development Cost Cost Distribution (%)	Development Cost as Percent of Software Cost (%)
Facilities (FAC)	6	18
Equipment (EQ)	27	81
Software (SW)	33	100
Logistics	5	15
Systems Level		
Management	6	18
Systems Engineering	10	30
Product Assurance	5	15
Integration and Test	8	24

TABLE 20-12. Operations and Support Cost in FY00$. See text for details.

Maintenance	$0.1 \times (SW + EQ + FAC)/year$
Contractor Labor	$160K/Staff Year
Government Labor	$110K/Staff Year

TABLE 20-13. Earth Terminals, Antennas, and Communication Electronics. Costs are for hardware only, and assume attachment to existing facilities [DCA, 1996].

Frequency	Cost (FY00$K)
SHF	$(50 \times D) + (400 \times P) + 1,800$
K, C Band	640
Ku Band	750
D = antenna diameter in m	P = RF power in kW

20.3.4 Launch Costs

The launch cost model includes vehicle costs and operations costs at the launch location (Table 20-14). For most missions, the launch represents a significant portion of the costs and thus the mission designer must consider concepts that are constrained by launch mass. For reusable vehicles such as the Space Shuttle, operations represent the predominant costs. The costs are presented in terms of a unit launch cost with the exception of the Shuttle, where usage cost is based upon a formula using either weight or length in the Shuttle bay, whichever results in larger costs. The chart also indicates the costs/kg of payload to LEO. This indicates the range of cost and payload size and provides guidance in extrapolating to costs for new launch vehicles to be competitive.

TABLE 20-14. Launch Vehicle Costs in FY00$M. The data assumes launch from the country's main site. Except where noted, LEO altitude is 185 km and inclination is 28.5 deg (5.2 deg for Ariane). Data from Isakowitz [1995].

Launch Vehicles	Maximum Payload-to-Orbit (kg)			Unit Cost (FY00$M)	Cost per kg to LEO (FY00$K/kg)
	LEO	GTO	GEO		
USA					
Atlas II	6,580	2,810		80–90	12.2—13.7
Atlas II A	7,280	3,039		85–95	11.7–13.0
Atlas II AS	8,640	3,606		100–110	11.6–12.7
Athena 1	800			18	22.5
Athena 2	1,950			26	13.3
Athena 3	3,650			31	8.5
Delta II (7920, 7925)	5,089	1,840		50–55	9.8–10.8
Pegasus XL	460			13	28.3
Saturn V	127,000			820	6.5
Shuttle* (IUS or TOS)	24,400	5,900	2,360	400	16.4
Titan II	1,905			37	19.4
Titan IV	21,640	8,620	5,760 (Centaur)	214 (270)	9.9
Taurus	1,400	450		20–22	14.3–15.7
ESA					
Ariane 4 (AR40)	4,900	2,050		50–65	10.2–13.3
Ariane 4 (AR42P)	6,100	2,840		65–80	10.7–13.1
Ariane 4 (AR44L)	9,600	4,520		95–120	9.9–12.5
Ariane 5 (550 km)	18,000	6,800		130	7.2
CHINA					
Long March C23B	13,600	4,500	2,250	75	5.5
RUSSIA					
Proton SL-13	20,900			55–75	2.6–3.6
Kosmos C-1	1,400			11	7.9
Soyuz	7,000			13–27	1.9–3.9
Tsyklon	3,600			11–16	3.1–4.4
Zenit 2	13,740			38–50	2.8–3.6
JAPAN					
H-2	10,500	4,000	2,200	160–205	15.2–19.5
J-1	900			55–60	61.1–66.7
GTO = Geosynchronous Transfer Orbit; GEO = Geostationary Orbit; LEO = Low-Earth Orbit					

* There is no official price for a Space Shuttle launch. Following the Challenger loss, only government payloads have been allowed. The GAO has assigned a price of $400 million per flight, but the actual cost depends strongly on the flight rate.

20.4 Other Topics

20.4.1 Cost Modeling Errors and Cost-Risk Analysis

Parametric cost modeling relies on a statistical analysis of past data to project future costs. Evaluating the statistical uncertainty associated with this projection is called *cost-risk analysis* because it represents the probabilistic risk that the program cannot be completed within a specified cost limit. This limit is usually set as the projected cost plus a cost margin or *management reserve*, typically on the order of 20% for major programs.

The basic tenet of parametric cost modeling is to base estimates of what satellites will cost next time on what they cost last time. If you're developing a space system under exactly the same circumstances as before (the same design, organizations, people, technology, requirements, and procedures) you'd expect it to cost the same. But this scenario never exists. The RDT&E cost models of Sec. 20.3 assume a relatively new design but proven technology. The more new technology is considered, the more the risk that added time and effort will be required to complete the development. Program cost is a nebulous quantity, heavily impacted by technological maturity, programmatic considerations, "normal" schedule slips, and other unforeseen events. Cost estimates derived from the CERs should therefore be accompanied by a cost-risk assessment to estimate potential effects of a level of complexity below or beyond average.

Cost-Risk Analysis provides an assessment of the ability of projected funding profile to assure that a program can be completed and meet its stated objectives. Although technical risks are often one of the biggest cost drivers for space systems, many cost-engineering processes and models ignore effects of cost risk in the interest of quick-turnaround estimates. Cost-risk analysis is important because single-point cost estimates, while meeting the top-level needs of budgetary planners, often do not meet the needs of those who want to perform more detailed trade-offs between cost and performance (see Book [1993]). The purpose of cost-risk analysis is threefold:

1. Translate qualitative risk assessments into quantitative cost impacts

2. Assist program managers in managing risk

3. Establish an empirical basis for estimating future programs with confidence

This section describes a method of assessing the uncertainty in cost estimates. This includes identifying the sources of uncertainty, combining them to arrive at a program level cost uncertainty, and interpreting the results. Cost-risk analysis views each cost element as an uncertain quantity that has a probability distribution and attempts to evaluate technical, programmatic, and schedule risks in quantitative terms. Qualitative measures of risks are then translated into cost-estimate adjustments. A key to making quick, consistent, and defensible assessments is reducing subjectivity by making assumptions about sources and magnitude of cost risk. The major sources of cost uncertainty we consider here are:

*   Cost-estimating uncertainty as quantified by the standard error, SE

*   Cost growth due to unforeseen technical difficulties

Examples of risk drivers include beyond state-of-the-art technology (e.g. cooling, processing, survivability, power, laser communications), unusual production requirements (e.g. large quantities, toxic materials), tight schedules (e.g. undeveloped technology, software development, supplier viability), system integration (e.g. multi-contractor teams, system testing), and unforeseen events (e.g. launch slip, need for redesign)

The objective is to quantify the sum of the contribution of these uncertainties to the overall system or program costs. We will use the *most likely estimates* (MLE), the sum obtained from the cost models and derive probability distributions that contain the impact of technology uncertainty and the uncertainty of the cost estimates. Cost-estimating uncertainty is quantified by computing the *standard error* (*SE*). For development of space-system hardware, CERs usually have *SE* between 30% and 50%. Cost-estimating uncertainty is therefore quantified by a distribution that has a mean (the estimate, *C*), and a variance (square of the *SE*). *

Cost risk due to technical difficulties is estimated using the *technology readiness level* (TRL), shown in Table 20-15, a NASA classification scheme for the level of technology development (the inherent development risk). A TRL of 1 or 2 represents a situation of relatively high risk. TRLs of 3, 4 and 5 represent moderate risks, and 6 through 8 are low-risk categories. Based upon related experience, the suggested cost uncertainties are also presented in Table 20-15. Thus, a low developmental risk sub-system would have a one standard deviation uncertainty of less than 10%, about the most likely estimate.

TABLE 20-15.   Technology Classification and Relative Cost Risk. Definitions are from NASA.

Technology Readiness Level	Definition	Relative Risk Level	Standard Deviation about MLE (%)
1	Basic principles observed	High	> 25
2	Conceptual design formulated	High	> 25
3	Conceptual design tested analytically or experimentally	Moderate	20–25
4	Critical function/characteristic demonstrated	Moderate	15–20
5	Component or breadboard tested in relevant environment	Moderate	10–15
6	Prototype/engineering model tested in relevant environment	Low	< 10
7	Engineering model tested in space	Low	< 10
8	Full operational capability	Low	< 10

---

* The terminology here can cause confusion. Cost models use the terms *most likely estimate, MLE,* and *standard error, SE,* because the statistical data in the CERs is not truly Gaussian. However, using a Gaussian probability distribution is often convenient for analyzing errors. Consequently, throughout this section we will use *MLE* from the CERs interchangeably with the *mean* or *estimated cost, C,* and *SE* interchangeably with the standard deviation, σ, even though means and standard deviations are applicable only if the statistical data has a Gaussian distribution.

Other adjustment factors correct for uncertainty in the development status of a specific technology. The technological risk related to developing a space system depends on how we use the technology and on its degree of "flight qualification." If an item has already flown in space, it's more likely to work again, so it represents less risk to the user.

To the technology-based risk, we need to add the contribution of cost estimating uncertainty in a probabilistic fashion. Since the technology uncertainty and the cost estimating uncertainties are independent (uncorrelated), these may be combined using the standard square root of the sum of the squares. (See standard probability texts.) For example, suppose a subsystem TFU MLE is $5 million. The risk level is moderate with a 20% standard deviation in uncertainty and the cost estimating standard error is assumed to be 15%. The standard error for the sum of the two sources of uncertainty is the root sum square

$$\varepsilon_t = (\sigma_t^2 + \sigma_c^2)^{1/2} = (0.2^2 + 0.15^2)^{1/2} = 0.25 \tag{20-1}$$

where σ_t is the cost standard deviation in technology and σ_c is the cost estimating uncertainty. Thus the uncertainty standard deviation is 25% of the MLE for the subsystem. The input to the above equation will be obtained from Table 20-15 for the σ_t and from Tables 20-4, 20-5, and 20-6 for the σ_c. This provides the uncertainty measure for a subsystem.

For an entire system, the probability sum is more complex. We cannot simply use the root sum square since there are correlations between subsystems. That is, interrelationships exist between the development of subsystems. To capture these interrelationships requires more advanced methods than will be treated here. An approximation to computing the system uncertainty measure is:

1. Sum the uncertainty standard deviations for each subsystem. This provides a system uncertainty assuming perfect correlation among subsystems.

2. Take the root sum square of the subsystem uncertainty standard deviations. This provides a completely uncorrelated solution.

3. Take the average of the two values in steps (1) and (2). This provides an intermediate solution which is a reasonable approximation for most conceptual analyses.

An example of this process for FireSat is shown in Table 20-22 in Sec. 20.5. Fig. 20-4 in that section shows the actual shape of the probability distribution, based on the assumption of Gaussian statistics. The integrated area under the curve up to a given cost estimate value X_E on the horizontal axis yields the probability, P, that the actual system cost will lie at or below the X_E estimated value. The curve is given by the Gaussian distribution:

$$P(X_E) = \int_{-\infty}^{X_E} F(x)dx, \quad 0 < P(X_E) < 1$$

$$F(x) = \frac{e^{-\frac{1}{2}\frac{(x-C)^2}{\sigma^2}}}{\sqrt{2\pi\sigma}}, \quad \int_{-\infty}^{+\infty} F(x)dx = 1 \tag{20-2}$$

Here, C is the estimated cost and σ is the standard deviation of the estimate.

The above analysis assumes normal distributions for the uncertainties. This is not usually the case, but does provide simple analytical solutions. A comprehensive method using Monte Carlo simulation is treated by Wong and Sheldon [1986] and Dienemann [1966]. The treatment of cost/reliability relationships (that is, the risk associated with the failure of flight units) is discussed in Sec. 19.2 and by Gupta and Altshuler [1989]. A more comprehensive treatment of analytical methods and the method of moments to address cost risk, as well as Monte Carlo simulation methods mentioned earlier, are given by Wilder [1978], Abramson and Young [1990], Book [1993], Burgess and Gobrieal [1996], and Young [1992]. The end product of cost-risk assessment in this framework is a total spacecraft cost-probability distribution, from which the mean, standard deviation, percentiles, and other descriptive statistics can be determined. (See Fig. 20-4 for an example cost estimate probability density evaluation.)

20.4.2 Time Spreading of Costs

Prior sections have focused upon developing total cost estimates. We now address how costs will be spread over time. The following analytical cost spreading method was developed by Wynholds and Skratt [1977] and approximates the experience of actual programs.

The spreading of the costs to determine funding profiles can be approximated by a function of the form:

$$F(S) = A\,[10 + S\,((15 - 4S)S - 20)]S^2 + B\,[10 + S\,(6S - 15)]S^3$$
$$+ \,[1 - (A + B)](5 - 4S)S^4 \tag{20-3}$$

where $F(S)$ is the fraction of cost consumed in time S, S is the fraction of the total time elapsed, and A and B are empirical coefficients.

The values for the coefficients A and B depend on the expected loading of costs over time. For instance, a typical period for RDT&E and two production units is 5 years. The costs are usually heavier during the first 2 years when design, develop-ment and testing occur. Typically, 60% of the costs will be incurred by the midpoint of the schedule. The coefficients for various spending splits are:

% Expenditure at schedule midpoint	Coefficients in Eq. (20-3) A	B
80	0.96	0.04
60	0.32	0.68
50	0	1.00
40	0	0.68
20	0	0.04

A 60% distribution is suggested for the RDT&E and production of the first several satellites. If more than two satellites are included, the 60% decreases toward a limit of 50%. A specific example is provided in Sec. 20.5.

Another schedule-related issue that we may address is *present value*, which is based upon the consideration of the time value of money. One dollar in 2000 is worth more than a dollar in 2005, since the 2000 dollar could be invested and earn a return so that its value in 2005 is more than one dollar (in constant 2000 dollars). This value increase would occur even without inflation.

To illustrate the present value concept, consider a cost comparison of two projects. Both projects have satisfied all technical requirements. Project *A* has a three year development consisting of funding requirements of $10 million, $5 million, and $1 million in constant 2000 dollars. Project *B* has a funding profile of $1 million, $5 million, and $10 million in 2000 dollars. Thus, the totals are the same for both. However, the buyer of this project, government or otherwise, will prefer the second project on an economic basis since it will expend most of the funds in the future rather than the near term. The buyer of Project *B* could invest the $9 million excess funds in the first year and have additional funds at the end of the project. The conventional way of handling a comparison is to compute the present value for both projects using a discount rate. The *discount rate* is the time value of money. The appropriate discount rate is controversial. If possible, treat the discount rate as a parameter and determine at what point, if any, there is a crossover in the discount rate where one project would be preferred over another. A 10% rate is a standard value for study purposes.

The present value, *PV*, is obtained by multiplying the funding for each year by the factor:

$$PV \equiv \frac{1}{(1+d)^{n-1}} \qquad (20\text{-}4)$$

where *n* is the year of project (relative to the constant dollar year), and *d* is the discount rate.

For project *A*, the present value at a 10% discount rate is

$$PV_A = 10 /(1 + 0.1)^0 + 5/(1 + 0.1)^1 + 1/(1 + 0.1)^2 = 15.4 \text{ in } \$ \text{ Millions}$$

For project *B*, the present value is

$$PV_B = 1/(1 + 0.1)^0 + 5/(1 + 0.1)^1 + 10/(1 + 0.1)^2 = 13.8 \text{ in } \$ \text{ Millions}$$

Thus, project *B* is less expensive than project *A* by $1.6 million or roughly 10% in present value terms.

20.4.3 Rough Order-of-Magnitude Cost Estimates

As concepts are developed, it is helpful to have an estimate of the anticipated costs. By making "sanity" estimates, we can develop some idea of cost bounds for various concepts. Table 20-16 gives estimated costs in constant 2000 dollars. These should be used to give rough order of magnitude costs for missions under consideration.

For example, we can estimate the cost of four communication satellites (TDRS-class), two in each of two orbit planes with launches from both the Eastern and Western Test Ranges using two Titan IV/Centaurs. Table 20-16 provides an order-of-magnitude cost of:

Satellites: 4 × $126M $504 million

and Table 20-14 provides an order of magnitude cost of:

Launches: 2 × $333M $666 million

Total $1,170 million

TABLE 20-16. Space Systems Costs. This table can be used to obtain quick order-of-magnitude estimates. All values are in FY00$M.

Manned Space Programs	Total Program Costs($M)		
Apollo	152,000		
Orbiter	45,000		
Gemini	4,400		
SkyLab	3,100		
Mercury	1,100		
Space Observatories	**Total Program Costs**		
Space Telescope	2,270		
GRO	640		
HETE	31		
Sampex	75		
Communication Satellites	**Dry Weight (kg)**	**Average Unit Costs ($M)**	**$K/kg**
Intelsat VIII (commercial)	1,200	133	111
TDRSS (NASA)	1,550	126	81
DSCS IIIB (DoD)	806	114	141
Westar (commercial)	500	78	156
ORBCOMM	33	11	333
Surveillance/Navigation Satellites			
DSP	2,200	314	143
GPS-2	839	57	68
Meteorological Satellites			
GOES	500	84	168
DMSP	514	88	171
Interplanetary Spacecraft			
Pioneer (S/C bus only)	231	38	165
Mars Observer	1,018	77	76
Clementine	232	57	246
Experimental Small Satellites			
RADCAL	92	5	54
PoSAT-1	49	1.2	24
AMSAT AO-13	84	1	12
Freja	214	19	89
Ørsted	60	15	243

A rule-of-thumb for satellite development costs is that RDT&E (nonrecurring) costs are 2 to 3 times the unit costs (for high-tech programs, factors of 5 to 6 are used). Adding development costs of 2.5 times unit satellite costs brings the total costs to around $1.5 billion. Such quick computations can also be used as a cross check for estimates obtained from the cost models of Sec. 20.3.

20.4.4 Learning Curve

Historically, the majority of satellites built and flown have been one-of-a-kind systems. However with the proliferation of constellations where recurring costs and learning rates dominate the cost equation, a breakout of developmental and theoretical first unit costs is necessary.

We can adjust the single-spacecraft CERs to correct for cost reduction from learning how to do the job better or the effects of changing technological status. CERs effectively predict costs for developing a prototype or first production unit, often called the *theoretical first unit*, but they poorly estimate the cost for even the most modest production line. The staff who manufactures a second or third similar spacecraft will learn to do the job better and use economies of scale, which quantifiably improve performance. The learning rate for the space and the aerospace industry is such that, on average, the Nth unit will cost between 87% and 96% of the previous unit.

The *learning curve* is a mathematical technique to account for productivity improvements as a larger number of units are produced. It includes all cost reductions between the first production unit and subsequent units. This includes cost reductions due to economies of scale, set up time, and human learning as the number of units increase. The total production cost for N units is modeled as:

$$\text{Production cost} = TFU \times L \qquad (20\text{-}5)$$

where

$$L \equiv N^B$$

$$B \equiv 1 - \frac{\ln((100\%)\,/\,S)}{\ln 2}$$

TFU is the theoretical first unit cost, L is the learning curve factor, and S is the learning curve slope in percent. This form of the learning curve was chosen because of its fit to empirical data, based on the theory of T.P. Wright [1936].

The learning curve slope S represents the percentage reduction in cumulative average cost when the number of production units is doubled. The learning curve slope S sets the value of B. For example: if $S = 95\%$ and the first unit costs $1 million, then doubling the number to 2 units reduces the average cost of both to 95% of the first unit. Thus, the two units cost $1.9 million. The second unit cost is $0.9 million. The learning curve exponent B is 0.926 for $S = 95\%$.

For less than 10 units, we recommend a 95% learning curve slope be applied. Between 10 and 50 units, a 90% learning curve and 85% for over 50 units is appropriate. These will vary with the application and how the manufacturing and assembly activities are set up. The cost models presented earlier provide first unit costs so that total production costs are determined by multiplying TFU costs by the learning curve factor.

The following example table illustrates the impact of a 95% learning curve on unit costs. The *unit* or *marginal cost* is the difference in production cost between N units and $N-1$ units. For example, the cost of the fifth unit is the difference in production cost between the fourth and fifth units, that is, $4.44 - 3.61 = 0.83$.

Effect of a 95% Learning Curve

Unit Number	Production Cost (TFU × L)	Average Cost	Unit Cost
1	1.00	1.00	1.00
2	1.90	0.95	0.90
3	2.77	0.92	0.87
4	3.61	0.90	0.84
5	4.44	0.89	0.83

20.5 FireSat Example

John T. Collins, *Microcosm, Inc.*

We will apply the above cost models to the FireSat example to compute life-cycle cost estimates for the entire mission. We assume that two satellites are needed initially and that both are launched into a 150-km orbit by a Pegasus XL launch vehicle. The WBS consists of the space, launch, and ground segments. The space element consists of two satellites with infrared sensor payloads. The launch segment is two vehicles. The ground segment will consist of a single ground control station. All of the necessary data is given in Table 20-3.

To illustrate the use of the CERs in Tables 20-4, 20-5, and 20-6, the cost estimates will be developed to the spacecraft subsystem level. The weight, beginning of life power, sensor aperture diameter, and other technical characteristics are the key parameters for the estimate. The specific values are in Tables 20-17 and 20-18. The hardware RDT&E cost is based upon the CERs of Table 20-4 and 20-6 (nonrecurring portion) modified by the design status factors listed at the bottom of Table 20-8. The production costs are the result of the TFU CERs of Table 20-5 and 20-6 (recurring portion) multiplied by a learning curve factor $L = 1.9$ for two units. The computation of L is described in Sec. 20.4.4. The subsequent or second unit cost is then the difference between the production cost of two units and the TFU. The results for the large satellite cost model (based largely on USCM 7.0) indicate the IR payload contributes most to overall system cost. Thus, the payload sensor should dominate attempts to reduce cost.

Program-level costs are added based on the CERs in Tables 20-4, 20-5, and 20-6. Ground support equipment costs are then computed from Tables 20-4 and 20-6 based on RDT&E and first unit hardware costs. The launch operations and orbital support costs are obtained from Table 20-5 and 20-6. Finally, the satellite software costs are based upon Table 20-10 using Ada. This yields a total space segment cost of $549M using the USCM 7.0 model and $44M using the smallsat model. The large discrepancy in total space segment cost is due largely to the vast difference in cost between the payload cost estimates for each model. The large satellite cost model yields an estimate of

TABLE 20-17.   FireSat Space and Launch Segment Costs in FY00$K. Cost estimates based on data in Table 20-3 Tables 20-4 and 20-5 for traditional satellite designs. Input data from Tables 10-31 and 20-3.

Cost Component	Parameter, Value, Data Source	RDT&E Cost (FY00$K)	1st Unit Cost (FY00$K)	2nd Unit Cost (FY00$K)	Total Cost (FY00$K)	SE ($K)
1. Payload						
1.1 IR Sensor	aperture dia. = 0.263 m	168,462	67,386	60,647	296,495	94,265
2. Spacecraft bus						
2.1 Structure*	wt. = 32.0 kg	2,784	419	377	3,580	1,368
2.2 Thermal*	wt. = 6.8 kg	1,337	197	177	1,712	830
2.3 Electrical Power System (EPS)*	wt. = 45.7 kg	2,862	2,067	1,860	6,790	3,360
2.4 Telemetry Tracking & Command and Data Handling (TT&C/DH)*	wt. = 6.8 kg	2,356	1,894	1,705	5,955	2,819
2.5 Attitude Determination & Control Sys. (ADCS)*	wt. = 18.3 kg	5,753	2,799	2,519	11,071	4,570
2.6 Propulsion	NA†					
Spacecraft Bus Total Cost		15,092	7,376	6,639	29,107	9,739
3. Integration, Assembly & Test (IA&T)	Spacecraft bus + payload RDT&E cost = $183,554	40,453	1,456	1,310	43,220	19,826
4. Program Level	Same as previous	52,450	25,494	22,944	100,888	37,773
5. Ground Support Equipment (GSE)	Same as previous	22,184	—	—	22,184	7,543
6. Launch & Orbital Ops Support (LOOS)	Spacecraft + payload wt. = 140 kg	—	686	617	1,303	547
7. Flight Software	26 KLOC	5,655‡	0	0	5,655	—
Total Space Segment Cost to Contractor		304,297	102,398	92,158	498,853	
10% Contractor Fee		30,430	10,240	9,216	49,885	
Total Space Segment Cost to Government		334,727	112,638	101,374	548,738	
8. Launch Segment	2 Pegasus XL Launchers		13,000	13,000	26,000	
Total Cost of Deployment					574,738	136,947

* Spacecraft bus subsystem masses shown include a fraction of the spacecraft mass margin of 11.2 kg.
† FireSat propulsion system is taken into account in ADCS CER (2.5). The 18.3 kg value for ADCS mass includes propulsion system hardware mass.
‡ Assumes a heritage cost factor of 0.5 (i.e., moderate modifications to existing flight software). Calculated using flight software CER in Table 20-10.

$296M for RDT&E, first unit, and second unit costs for the infrared sensor, while the small satellite model yields a total cost of $6.9M. This dramatic difference in payload cost estimates illustrates the difficulty in finding an appropriate cost-estimating relationship for some classes of spacecraft payloads. For this reason, payload cost estimation is in many cases best achieved by using bottoms-up or analogy-based methods rather than parametrics. The difference in these numbers is attributable to paramount differences in the database of programs on which the CERs are based including: prototype vs. protoflight approach, risk acceptance, operational vs. demonstration

TABLE 20-18. **FireSat Space and Launch Segment Costs in FY00$K.** Based on data in Table 20-3, Small Satellite CERs in Table 20-6 and ratios for other cost compo nents in Table 20-9.

Cost Component	Parameter, Value, Data Source	RDT&E Cost (FY00$K)	1st Unit Cost (FY00$K)	2nd Unit Cost (FY00$K)	Total Cost (FY00$K)	SE ($K)
1. Payload	Spacecraft total cost (RDT&E + TFU)	3,049	2,033	1,829	6,911	1,946
2. Spacecraft bus						
2.1 Structure*	wt. = 32.0 kg	1,318	565	508	2,390	1,393
2.2 Thermal*	wt. = 6.8 kg	221	221	199	642	173
2.3 Electrical Power System (EPS)*	wt. = 45.7 kg	3,271	2,005	1,804	7,080	1,221
2.4a Telemetry Tracking & Command (TT&C)*	TT&C + DH wt. = 6.8 kg	641	262	235	1,138	793
2.4b Command & Data Handling (C&DH)*	TT&C + DH wt. = 6.8 kg	868	355	319	1,542	1,077
2.5 Attitude Determination & Control Sys. (ADCS)*	wt. = 9.1 kg,	767	1,306	1,175	3,247	1,744
2.6 Propulsion	Satellite bus dry mass = 112 kg	453	453	408	1,314	450
Spacecraft Bus Total Cost		7,538	5,166	4,649	17,353	4,883
3. Integration, Assembly & Test (IA&T)	Spacecraft total cost (RDT&E + TFU)	—	1,766	1,589	3,355	945
4. Program Level	Same as previous	1,455	1,455	1,309	4,218	1,188
5. Ground Support Equipment (GSE)	Same as previous	838	—	—	838	236
6. Launch & Orbital Ops Support (LOOS)	Same as previous	—	775	697	1,472	415
7. Flight Software	26 KLOC	5,655†	0	0	5,655	—
Total Space Segment Cost to Contractor		18,535	11,194	10,074	39,803	
10% Contractor Fee		1,854	1,119	1,007	3,980	
Total Space Segment Cost to Government		20,389	12,313	11,082	43,783	
8. Launch Segment	2 Pegasus XL Launchers		13,000	13,000	26,000	
Total Cost of Deployment					69,783	7,552

* Spacecraft bus subsystem masses shown include a fraction of the spacecraft mass margin of 11.2 kg.
† Assumes a heritage cost factor of 0.5 (i.e., moderate modifications to existing flight software). Calculated using flight software CER in Table 20-10.

/validation, and programmatic oversight and required documentation. From Table 20-14 the launch vehicle cost for 2 Pegasus systems is $26M, giving a total cost to de ployment of $575M and $70M for large and smallsat models, respectively.

Table 20-20 gives the ground segment development and operations costs for FireSat. We begin with the ground software cost for 100 KLOC of Ada code and use Tables 20-10 and 20-11 to compute the development cost. Our initial assumptions or a 7-year life after IOC (5-year spacecraft life) and 10 contractor personnel are used with Table 20-12 to estimate a total operations and maintenance cost of $6M/year in Table 20-19. Finally, all costs, including a 10% fee, are summed in Table 20-21 to yield a total cost plus fee of $671M in FY00 dollars.

TABLE 20-19. FireSat Annual Operations and Maintenance Cost in FY00$M.

Operations and Maintenance	
10 Contractor Personnel ($160K/Yr) including fee	1.6
Maintenance	4.4
Total Annual Cost	6

TABLE 20-20. FireSat Ground Segment and Operations Costs in FY00$M.

Development	Cost
Software 100 KLOC (Ada) @ $220/LOC	22.0
Equipment	17.8
Facilities	4.0
Subtotal	**43.8**
Management	4.0
Systems Engineering	6.6
Product Assurance	3.3
Integration and Test	5.3
Logistics	3.3
Total	**66.3**

TABLE 20-21. FireSat Life-Cycle Cost Estimate. All costs in FY00$M, including fee. Launch segment estimate is based on two Pegasus launches.

Initial Deployment	
Space Segment - Table 20-17	548.7
Launch Segment - Table 20-17	26.0
Ground Segment - Table 20-20	66.3
Subtotal	**641.0**
Operations and Maintenance - (Table 20-19)	
Annual Ops. and Maint.	6.0
Total Ops. and Maint. for 5 years	**30.0**
Total Life-Cycle Cost for 5 years	**671.0**

Let us now assess the cost uncertainty in our estimate for the hardware using the technique of Sec.20.4.1. For the FireSat example, the satellite hardware consists of the spacecraft bus and an IR payload with CERs for the TFU from Table 20-5 and Table 20-6. The spacecraft bus is assumed to consist of proven technology with a TRL of 6. Some new component designs are necessary, but the technologies have had applications where engineering models were successfully tested. The IR payload requires infusion of new technology and is assigned a TRL of 5. The spacecraft dry weight (including payload) is 140 kg and the IR payload aperture diameter is 0.26 m. Using the corresponding CERs from Table 20-5 to compute costs and the risk methodology we arrive at the results in Table 20-22.

TABLE 20-22.   FireSat Cost Uncertainty

Element	TFU Cost ($M)	System Technology Level	System Technology Std. Dev. ($M)	Cost Estimate Std. Dev. ($M)	Combined Std. Dev. ($M)
S/C bus	7.4	6 (= 10%)	0.74	1.11	1.33
IR payload	67.4	5 (= 15%)	10.1	10.1	14.3
Total	74.8		Step 1: Sum		15.6
			Step 2: RSS		14.4
			Step 3: **Average**		15.0

The standard deviation for the system technology is obtained by using Table 20-15. For example, the $0.74M standard deviation for the spacecraft system technology is 10% of the $7.4M cost. The cost estimate standard deviation is assumed to be 15% and the combined standard deviation is the root sum square (RSS) of the two components. Thus, the standard deviation of the total satellite hardware cost ($74.8M) is $15 million or 20%.

The cost to build two FireSat satellites is $194.5 million (see Table 20-17). Assuming this is spread over 5 years with 60% of the costs to be incurred in 2-1/2 years, then the coefficients would be $A = 0.32$ and $B = 0.68$. Inserting these into the equation for $F(S)$ [Eq. 20-30], we obtain the distribution in Table 20-23 for satellite recurring costs.

TABLE 20-23.   Time Distribution of FireSat Costs. Based on a total recurring cost for the first two units of $194.5M (Table 20-17) and assuming 60% of costs spent in the first 2.5 years.

Year	Cumulative Cost (%)	Cumulative Cost ($M)	Annual Cost (%)	Annual Cost ($M)
1	12.3	23.9	12.3	23.9
2	42.8	83.2	30.5	59.3
3	75.6	147	32.8	63.8
4	95.8	186.3	20.2	39.3
5	100.0	194.5	4.2	8.2
Totals			**100.0**	**194.5**

Figure 20-4 displays cost estimate probability density for spacecraft bus RDT&E, first unit, second unit, wrap costs (IA&T, program level, GSE, and LOOS), launch cost, flight software cost, and contractor fee for the two cost models presented in this chapter. All costs are represented for the space and launch segment, except payload cost (due to the large discrepancy in payload cost estimate between the two models). The results illustrate the difference between these two models and also, where estimates derived from these two models are most likely to fall. The small satellite cost model yields a best estimate of $62,181K and a standard deviation of $6,401K. The large satellite cost model result, depicted by the wider curve at the right, yields a best estimate of $90,595K and a standard deviation of $15,159K.

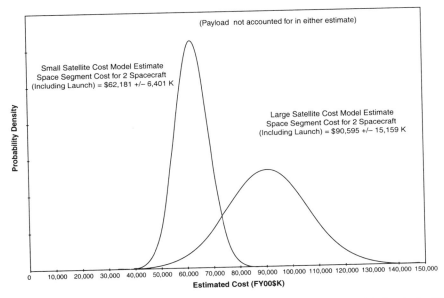

Fig. 20-4.   Probability Density of Estimated Cost Resulting from Cost Risk Analysis. The taller, narrower curve to the left is the result for the small satellite CERs from Table 20-6 and the wider, lower curve is the result for the large satellite CERs from Tables 20-4 and 20-5. Most missions of similar scope to FireSat will lie somewhere between these two extremes. The results of such statistical evaluations can be used to dictate reserves policy (e.g., "risk" dollars) or to guide technology investment.

20.5.1 FireSat Design Life Study

Using an integrated design-to-cost tool that uses design relationships previously presented and the small-satellite CERs found in Table 20-6, a trade analysis of design life was performed. The primary drivers for mass increase as a function of increasing design life are:

1. The requirement for additional propellant for on-orbit station keeping;

2. A need for more capable subsystems, for example the power subsystem where solar cell and battery degradation requires oversizing the system to meet a new higher beginning-of-life power requirement; and

3. Full redundancy assumed for 7-year lifetime.

We chose mass-based CERs and assume that payload requirements and performance parameters such as pointing accuracy, downlink data rate, data storage, and size remain fixed. The results are shown in tabular form in Table 20-24. Mass-based CERs are favored over the performance-based CERs.

TABLE 20-24. **FireSat Mass, Power and Cost as a Function of Design Life.** Data assumes fixed performance requirements. Redundancy and associated impacts to subsystem masses, and increased power and propellant requirements drive the results.

Design Life (years)	1	3	5*	7
Payload (kg)	20	24	30	35
Propulsion (kg)	7	8	9	18
ADCS (kg)	7	9	9	22
TT&C/C&DH (kg)	5	6	7	20
Thermal (kg)	5	6	7	9
Power (kg)	33	40	46	70
Structures (kg)	22	27	32	37
Spacecraft Dry Mass (kg)	99	120	140	211
Propellant (kg)	31	33	35	37
Launch Mass (kg)	130	153	175	248
Performance Pegasus (kg)	290	290	290	290
Launch Margin†	55%	47%	40%	14%
EOL Power (W)	140	140	140	140
BOL Power (W)	145	157	170	183
Space Segment Cost (FY00$M)	35	40	44	76
Launch Cost (FY00$M)	26	26	26	36‡
Operations Cost (FY00$M)	6	18	30	45
Total Cost (FY00$M)	67	84	100	157
Cost per Year (FY00$M)	67	28	20	22

* FireSat baseline design.
†25% margin required at conceptual design stage
‡Launch mass plus 25% margin exceeds payload capability of Pegasus XL. Two Athena 1 launches assumed at $18M per launch.

We include launch mass margin relative to estimated Pegasus performance to orbit, which is about 290 kg to a 700 km circular orbit at 55 deg inclination. Spacecraft power estimates, used only as intermediate results, demonstrate required growth in the power subsystem as a function of lifetime. Our required propellant mass increases as well. When the margin goes negative we are forced to launch Firesat on the more capable Athena 1 (costs $18M compared with the $13M Pegasus XL). Operations costs are not changed from our previous analysis of $6M per year.

Aggregate cost information along with cost per year is shown in Fig. 20-5. This can be compared with the notional representation in Fig. 1-5 and we have completed the requirements to design to cost cycle. Information like that shown in Fig. 20-5 is extremely useful for demonstrating how cost "pushes back" on requirements. When we specified a mission life of 5 years, we did not have enough information to know whether this was the proper choice from a cost-effectiveness standpoint. Now that we have completed the picture, we can see that (all factors considered) the optimal lifetime for FireSat is between 4 and 6 years where the curve is relatively insensitive to further changes in design life. This is also about the time when we would expect that the cost-effectiveness of the IR sensor payload and other components may become an issue. Even though amortized mission costs are lower, i.e., less dollars per year, there is proportionally less value.

FireSat Design Life Trade

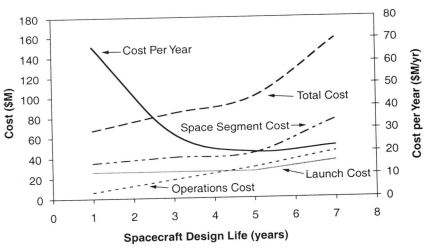

Fig. 20-5. **FireSat Design Life Trade.** Compare with Fig. 1-5 in Sec. 1.4. Total cost, the sum of individual spacecraft cost, launch vehicle cost, and operations cost, is shown as a function of spacecraft design life. On the right axis we plot cost per year which asymptotically approaches the yearly operations cost as design life goes to infinity. Optimal design life is chosen at a point on the cost-per-year curve that is relatively insensitive to changes in design life and while mission data still remains useful.

This analysis highlights an important role for shorter-lived small satellites over traditional large satellites. Since more rapid technology infusion may provide increases in cost-performance, we should be able to take advantage of state-of-the-art technology available 5-6 years from now, instead of continuing to operate an obsolete system. Therefore, we choose to utilize two FireSat payloads, each with a 5-year lifetime, but expect that the second version of the satellite will have higher performance per pound than what we designed with today's technology. We can launch a satellite every 5 years to assure that we don't experience an outage or design in extra mission life (with the mass and cost consequences shown in Fig. 20-5) if the second satellite is delayed.

References

Abramson, Robert L. and Philip H. Young. 1990. "FRISKEM—Formal Risk Evaluation Methodology," ORSA/TIMS Joint National Meeting, Philadelphia, PA, October 29–31, 1990.

Apgar, H. 1990. "Developing the Space Hardware Cost Model." Paper IAA-CESO-04 (90), AIAA Symposium on Space Systems Cost Methodologies and Applications, May.

Bearden, David A. 1999. "A Methodology for Technology-Insertion Analysis Balancing Benefit, Cost and Risk." Dissertation Paper, Department of Aerospace Engineering, University of Southern California, August 1999.

Bearden, David A., Richard Boudreault, and James Wertz. 1996. "Cost Modeling." In *Reducing Space Mission Cost*. Torrance, CA, and Dordrecht, the Netherlands: Microcosm Press and Kluwer Academic Publishers.

Book, S.A. 1993. "Recent Developments in Cost Risk." Space Systems Cost Analysis Group Meeting, Palo Alto, CA, June 7–8.

Book, S.A. and P.H. Young. 1995. "General-Error Regression for USCM-7 CER Development." Presented at the Society of Cost Estimating and Analysis National Conference, Tysons Corner, VA, 19-21 July.

Burgess, E.L. and H.S. Gobrieal. 1996. "Integrating Spacecraft Design and Cost-Risk Analysis Using NASA Technology Readiness Levels," The Aerospace Corporation. Presented at the 29th Annual DoD Cost Analysis Symposium, Leesburg, VA. February 21–23.

Burgess, E.L., N. Y. Lao, and D.A. Bearden. 1995. "Small-Satellite Cost-Estimating Relationships." Presented at the 9th Annual AIAA/Utah State University Conference on Small Satellites, Logan, Utah, September 18-21.

Defense Communications Agency (DCA). 1996. *Cost and Planning Factors Manual*. DCA Circular 600-60-1. Washington, DC: Defense Communications Agency.

Defense Information Services Agency (DISA). 1997. DISA Circular 600-60-1, *Cost and Planning Factors Manual*, version 1.2, 7 May. Washington, DC: Defense Information Services Agency.

Department of Defense. 1998. *Parametrics Estimating Initiative Handbook*.

Dienemann, Paul. 1966. *Estimating Cost Uncertainties Using Monte Carlo Techniques*. Report RM-4854-PR. Santa Monica, CA: The RAND Corporation.

Fisher, Gene H. 1970. *Cost Considerations in Systems Analysis*. RAND Report R-490-ASD. Santa Monica, CA: The RAND Corporation.

Greenberg, J.S. and H. Hertzfeld. 1994. *Space Economics*. Washington, DC: AIAA.

Gupta, A. K., and D. Altshuler. 1989. "Cost Effective Reliability Determination for Spacecraft Design." in *Institute for Cost Analysis/National Estimating Society Conference*.

Hamaker, J. 1987. "A Review of Parametric Cost Modeling Activities at NASA Marshall Space Flight Center." Paper presented to the AIAA Economics Technical Committee, Buffalo, New York.

Hamaker, J. 1989. "NASA MSFC's Engineering Cost Model (ECM)." Proc. Intl. Soc. Parametric Analysts, 10th Annual Conference, Brighton.

Isakowitz, Steven J. 1995. *AIAA International Reference Guide to Space Launch Systems (2nd Edition)*. Washington, DC: American Institute of Aeronautics and Astronautics.

Management Consulting and Research, Inc. 1986. *Space and Strategic Defense Cost Analysis: Cost Analysis Techniques Report*. TR-8618-2. Washington, DC: Office of the U.S. Secretary of Defense.

MCR Federal, Inc. 1997. *Communications Payload and Spaceborne Electronics Cost Model*. TR-9535/10. Arlington, VA: Air Force Cost Analysis Agency.

National Aeronautics and Space Administration. n.d. [ca. 1981] *Space Transportation System Reimbursement Guide*. Washington, DC: National Aeronautics and Space Administration.

Sadin, S.R. and R.W. Davis. 1993. "The Smallsat Revolution—Back to the Future?" Paper presented at IAF, Graz, Austria. IAF-93-U.5.570

Sarsfield, Liam. 1998. *The Cosmos on a Shoestring: Small Spacecraft for Space and Earth Sciences*. Santa Monica, CA: Rand Corp.

Smith, G. K., J. P. Stucker, and E. J. Simmons. 1985. *Commercial and Military Communication Satellite Acquisition Practices*. RAND Report R-3121-AF. Santa Monica, CA: The RAND Corporation.

Space and Missile System Center, Directorate of Cost, Los Angeles AFB, CA.1994. *Unmanned Space Vehicle Cost Model*, 7th ed. Los Angeles AFB, CA: Space and Missile System Center.

USAF. 1998. Raw Inflation Indices (Based on OSD Raw Inflation Rates), published by Secretary of the Air Force (SAF/FMCEE) and available on web site www.saffm.hq.af.mil; issued 14 January 1998. Washington, DC: U.S. Air Force.

Wertz, James R. and Wiley J. Larson. 1996. *Reducing Space Mission Cost*. Torrance, CA: Microcosm Press and Dordrecht, The Netherlands: Kluwer Academic Publishers.

Wilder, J. J. 1978. *An Analytical Method for Cost Risk Analysis* (Method of Moments), German Aerospace Corporation PDR-OP-B78-2. Bethpage, NY: Grumman Aerospace Corporation.

Wong, R. E. and D. Sheldon. 1986. "Cost/Schedule Risk Assessment of an Advanced Communications Satellite." Paper presented at 1986 ISPA Conference, Kansas City, Missouri.

Wright, T. P. 1936. "Factors Affecting the Cost of Airplanes." *Journal of the Aeronautical Sciences*. 3: 122-128.

Wynholds, H. W. and J. P. Skratt. 1977. "Weapon System Parametric Life Cycle Cost Analysis" in *Proceedings of the 1977 Annual Reliability Symposium.*

Young, P. H. 1992. "FRISK—Formal Risk Assessment of System Cost Estimates." The Aerospace Corporation. Presented at the AIAA 1992 Aerospace Design Conference, February 3–6, 1992, Irvine, CA.

Chapter 21

Limits on Mission Design

21.1 Law and Policy Considerations

William B. Wirin, *University of Colorado, Colorado Springs*

Why Worry About Law and Policy?

Engineers accustomed to precise answers often find that legal and political issues intrude on the space mission design process just when everything is going smoothly. However, I hope to shed some light on potential policy "show stoppers," and more importantly, provide some insight into legal thinking about space missions and valid, even critical, perils in the design process.

Policy results from balancing conflicting interests, so "valid" arguments may be rejected. Lawyers tend to give "answers" rather than an evaluation of political and legal risks. They unfairly believe that individuals from other disciplines will not understand and appreciate the balancing of interests. General James V. Hartinger, first Commander of Space Command, put it succinctly, "Lawyers are asked common sense questions so often that they begin to believe they have common sense." The mission planner should look to lawyers to evaluate policy and legal risks for various mission alternatives so they can be weighed along with technical factors.

Why worry about law and policy? The simple answer is that a perfect engineering solution is useless until it can be implemented. An example is the Apstar satellite. It was launched in July 1994 by the PRC without obtaining coordination from "owners" of nearby communications satellites as required by the International Telegraphic Union (ITU) regulations. Without the required consultations the satellite would not be permitted to transmit signals and therefore would be of little value. The result was a flurry of activity to conclude the negotiations quickly and this was accomplished a few months after launch. Had this been a U.S. launch it would have been postponed until the proper authorizations were accomplished, resulting in needless expense.

Another example is the Shuttle mission 51C to retrieve the Palapa and Westar VI satellites. These satellites remained in low-Earth orbit because the motors which should have transferred them to geostationary orbit failed. Hughes and NASA engineers worked out the technical solutions in a little more than 6 weeks. The lawyers spent more than 6 months resolving the legal questions concerning ownership, salvage rights, insurance coverages, and release of liability between the insurance carriers, Western Union Company and the Indonesian government. Also Hughes Aircraft Company and NASA had to agree on technical roles, liability, and compensation before the recovery effort began.

Without the implementing legislation in the Commercial Space Act of 1998 the DOT was not authorized to license private organizations to conduct space operations which included reentry of reusable space vehicles. This is a vital provision not only for NASA's X-33 and X-34 programs, but for companies building commercial reusable launch vehicles which will need to renter the Earth's atmosphere after delivering their payloads to orbit. Moreover the law is now clear that launch and recovery by definition are not an export or import. The first private launch of the Conestoga downrange to Matagorda Island off Texas required an export license in addition to other licenses. These additions to the law make way for commercial exploitation of new space opportunities and begin to strip away some restrictions on space activity.

21.1.1 Space Law

For a multinational project such as a space station, mission designers must recognize that lawyers from some nations approach policy and legal issues differently. England and her progeny are common law nations; all others are civil law nations. The latter seek to establish civil codes of law which in effect create all of the rules up front. Common law nations, in contrast, see law as evolving to solve particular problems. They look to past precedent and attempt to fashion a modern answer, while civil law judges apply the code and leave "changes" to legislators.

An interesting example of space law is that spacecraft have the right to pass over the territory and air space of other nations without their consent, whereas aircraft do not. But no one has defined where space begins. The Russians proposed 100–110 km because aircraft presently cannot fly at that altitude, and space objects burn up in the atmosphere below it. The U.S. position is to wait to see how technology will develop since there have been no particular problems. Equatorial nations have argued in vain that their sovereignty goes up to ~36,000 km (geostationary orbit) so they can control access and obtain compensation from space-faring nations who use slots above their territory.

Space missions are inherently an international activity, because space touches the sovereign territory of every nation on Earth. We must, therefore, consider diverse views on space issues. For example, space-faring nations may diverge from the feelings of developing nations, or Eastern and Western nations may disagree. Differences of opinions can arise from different cultures, economic status, political imperatives, and world view.

International law, including space law, evolves in part from treaties, including the U.N. Charter and U.N. resolutions plus organic documents of international organizations. It also depends on the practice of nations, as well as the writings of established authorities. Unlike U.S. domestic law, space law has no legislature to write the rules and no court to enforce them. Sovereign nations enter into treaties but may disregard them when they no longer serve their national security needs. The U.N. may pass resolutions, but they are not absolutely binding even when unanimous. Besides, treaties tend to have ambiguous language, permitting a nation to pursue its best interests. An example is Article IV of the Outer Space Treaty, which says that the Moon and other celestial bodies shall be used exclusively for "peaceful purposes," whatever that means.

The International Court at the Hague does hand-down decisions, but nations must usually agree to have the Court hear the case. Even when they do, there is no effective mechanism to enforce a decision. Having said all this, we must not conclude that space law has no value. There is a body of rules and a general understanding among nations as to what is proper and improper in space. Nevertheless, even though they may be ambiguous and no policeman enforces them, treaties are seldom if ever ignored.

Basic Do's and Don'ts: The Outer Space Treaty of 1967[*]

As the so-called *Principles Treaty*, this document lays down the basic philosophy and legal principles for outer space. In general, it says what should or should not be done, but does not spell out how to implement a given policy. Its preamble emphasizes that international cooperation is essential and recognizes that exploring space is in the common interest of all mankind. It goes further to declare that exploration should benefit all peoples. Provisions of the treaty include:

- All nations may scientifically investigate space, with international cooperation encouraged.

- No nation may claim sovereignty over outer space, including the Moon and other celestial bodies. As an illustration, Neil Armstrong claimed the Moon for mankind (not the United States) by stating that the mission had taken "One small step for a man, one giant leap for mankind." The American flag was present, but not to claim new territory, as Columbus and other explorers did when they came to the New World.

- The rules in space will follow the established principles and rules of international law and the U.N. Charter.

- No nation will place nuclear weapons or any other weapons of mass destruction in orbit around the Earth, or on the Moon, or on other celestial bodies. This restriction does not apply to nuclear power sources. The Treaty Banning Nuclear Weapon Tests in the Atmosphere, in Outer Space and Under Water (August 5, 1963) prohibits tests but not placement of nuclear devices in space.

- Nations must use the Moon and other celestial bodies exclusively for "peaceful purposes," but they may use military personnel in scientific research. The U.S. defines *peaceful* as "non-aggressive," thus permitting defensive measures.

- Astronauts are *envoys of mankind*. So long as they conform to accepted rules of activity in space, they have a form of immunity. Therefore, we must return them to their home nation promptly, and implicitly, may not charge to rescue them. The Agreement on the Rescue of Astronauts, the Return of Astronauts,

---

[*] The classic treatise on space policy is by Professor Carl Q. Christol, *The Modern International Law of Outer Space*, Pergamon Press, New York (1982). Other material is in the annual reports of the *Colloquium on the Law of Outer Space*, by the International Institute of Space Law, published by the American Institute of Aeronautics and Astronautics, Washington, DC.

and the Return of Objects Launched into Outer Space (April 22, 1968) could also apply.

- Recovered space objects must go back to the launching nation at its request and expense.

- Nations bear international responsibility for their activities in outer space, whether done by governmental agencies or private citizens. Thus, the U.S. must authorize and continuously supervise all space activities of its citizens. This requirement is unique to space activities, resulting from a compromise between the U.S. and the former USSR. The USSR had insisted that only governments should be permitted to go into space, whereas the U.S. insisted on permission for private entrepreneurs. The Commercial Launch Act of 1984 was enacted partly to carry out the U.S. obligations under the 1967 Outer Space Treaty. It has been amended from time to time to update policies and procedures.

- Launching nations are liable for damages to citizens of other nations caused by national and private launch activities. The Convention on International Liability for Damage Caused by Space Objects (March 29, 1972) could also apply.

- Nations must maintain a register of their launches. The purpose is to establish ownership, jurisdiction, and control over the spacecraft and its personnel. In essence, a U.S. spacecraft or space station is its sovereign territory. Other nations retain jurisdiction over their spacecraft and modules even when attached to another's station. Space debris falls in the same category, making clean-up politically difficult. Launches are also reported to the U.N. to provide an opportunity for the world community to learn about space activity. If a satellite or other space object were not registered, it would be considered a *rogue* and likely forfeit any legitimate status and protection under the Outer Space Treaty or international law. It would be awkward to put in a claim for damage to such a satellite under the liability convention. A nation that chooses for any reason not to play by the rules, cannot easily demand their enforcement. Of course, policy and politics would also restrain it from complaining.

- The Convention on Registration of Objects Launched into Outer Space (January 4, 1975) requires a launching nation to advise the U.N. Secretary General of the following: name of the launching nation, description or registration number, date and location of launch, basic orbital parameters (nodal period, inclination, apogee, and perigee) and general function of the space object. A nation need not update this information but may do so if it wishes.

- Nations must conduct space activities so as to avoid harming or contaminating the environment. *Project West Ford* influenced this provision. In 1963, the U.S. placed 480 million copper dipoles into orbit 2,000 miles above the Earth in an attempt to create an artificial ionosphere to enable radio messages to bounce from coast to coast. The in-house name for the project was *Needles*, which was an unfortunate choice. Complaints came from astronomers who were convinced the belt of copper needles would ruin their view of the universe. They also thought it might lead to worse experiments. The Soviets said, "What if one of the needles pierces the heart of a cosmonaut or puts out someone's eye when it reenters the atmosphere?" Although these objections were not reasonable, public ignorance made them real. Simply put, even though the project caused no significant problems, the planners had failed to consider fully the effects of adverse public opinion. A key question you won't find written anywhere is, "Does it pass the common sense test?" Project West Ford failed. Space law now requires that if a project might cause harm or interference with others, international consultations must occur before it proceeds (see Sec. 21.1.6 on Environmental Concerns). Planners should also review the Convention on the Prohibition of Military or any Other Hostile Use of Environmental Modification Techniques (May 18, 1977) if they anticipate large-scale effects.

Nuclear reactors in space create an environmental concern because they cause gamma-ray interference with some scientific satellites. NASA's Solar Maximum Mission is a case in point; its gamma-ray spectrometer suffered interference from reactors on Soviet radar satellites for ocean reconnaissance (RORSATs). If we plan a project such as the Gamma Ray Observatory, we must evaluate precautions and risks. At the same time, if we plan to use a nuclear power source for a mission, we must consider possible effects and protests. Some legislators in the U.S. have proposed a ban on the use of nuclear power sources in space. Although Congress is not likely to pass such a measure, the proposed ban illustrates environmental concerns.

- Stations on the Moon and other celestial bodies shall be open mutually to representatives of other nations after reasonable advance notice. Even this general language does not apply in space to space stations, shuttles, platforms, or satellites. Neither by treaty, international law, nor custom may anyone inspect sovereign facilities in space. Although the U.S. may agree to inspections on the Moon, Russia has taken a very stern stand against inspections; therefore, in my opinion, these inspections are unlikely. The Antarctic Treaty, in contrast, allows unannounced inspection of stations, installations, and equipment; therefore we must not confuse these provisions with those of the Outer Space Treaty. The Antarctic Treaty permits unconditional free access at all times. This could change if there were more than scientific interest in Antarctica. The Moon Treaty of 1979 has extensive inspection provisions but little force or effect, because no space-faring nation of the West or East ratified it. The "common heritage of mankind" provisions, which envision sharing profits with all nations even though they have not contributed, is a major sticking point.

Conflicts Between U.S. Law and International Law

Federal statutes and treaties are of equal authority. If a statute and a treaty conflict, the later in time controls (87 Corpus Juris Segundum Treaties 89). Therefore, statutes and regulations may control the activities of U.S. industry and government, even to the extent of revoking portions of an earlier international agreement or treaty.

Satellite Telecommunications. Communications have been the most commercially profitable use of space, with primary applications in telephone, telegraph, television, and data transmissions. The unique properties of the stationary orbit over the equator, which permits stationary positioning of a satellite, make it a limited natural resource. Space at 35,680 km above the Earth is ample to accommodate many satellites with little risk of collision. But the radio frequency and its bandwidth limit spacing to 2 deg in order to avoid interference. Reducing this spacing further will be expensive. The ITU authorizes and controls the effective use of geostationary orbits and frequencies. Under the International Telecommunication Convention and Final Protocol (Nairobi, 1982) and the ITU's radio regulations, the system uses a "first-come, first-served" procedure; once a satellite is registered with the ITU no subsequent system may interfere with its signal transmission. Generally, a similar replacement will retain the rights of a registered satellite.

A number of concerns over allocation of slots led to a call for a Space World Administrative Radio Conference in 1985, which reconvened in 1988 and 1992. In particular, developing nations worried that no slots would be left for them. A compromise left the conventional part of C and Ku radio frequency bands allocated only to the Fixed Satellite Service on a "first-come, first-served" basis. At the same time, all nations could obtain orbital slots and frequency channels on the expansion part of the Service's frequency bands. Policies set up at the first two WARCs began to be implemented at the most recent conference. However, it was also apparent that this implementation will continue to evolve as power shifts among participants. The nations also occasionally hold multilateral planning meetings to resolve area conflicts between themselves. Access to all other radio frequency bands allocated to all other satellite services is still "first-come, first-served" except the 12 GHz and 17 GHz bands for Broadcasting Satellite Service.

In exercising its registration function, the International Frequency Registration Board does not evaluate the particular nation's reasons for making a particular assignment. The Board seeks only to ensure that the national assignment conforms to identified services for the assigned frequency and that transmissions do not interfere with other broadcasts. Nations then determine how they will parcel out the allocated frequencies to their people.

If a proposed satellite could cause interference, the sponsoring nation resolves issues under negotiation procedures established by the ITU. The ITU conducts much of its technical activities through the International Frequency Registration Board and the International Radio Consultative Committee. Since 1973, orbital positions have been associated with particular frequencies and technical characteristics.

In the U.S., the FCC allocates orbital slots and frequencies. To obtain an orbital position or frequency, your legal department's FCC expert submits a formal application. The approval process may take 3 to 5 years for a new communications service or up to a year for an existing satellite.

An ongoing concern is the need to protect the GPS spectrum from disruption and interference. The Europeans at the World Administrative Conference in 1997 sought to overlay a commercial broadcast signal on the public service radionavigation band. This issue was addressed in the Commercial Space Policy Act of 1998 with the Congress encouraging regional agreements with foreign countries, like the 1998 agreement with Japan, in order to make the U.S. GPS signal a globally accepted standard.

Commercial Launches. Commercial launch planners must consider the Commercial Launch Act of 1984. As amended, it requires the DOT to approve and license all commercial expendable vehicles if launched from U.S. territory or by a U.S. citizen or corporation outside the U.S.

As the Federal Register indicates, before issuing a launch license the DOT requires both a mission review and a safety review. The first review focuses on the payload itself and the flight plan to ensure that the mission is in accord with U.S. policy and meets U.S. responsibilities under international treaties. Regulations require the DOT to review how missions affect national interests. Although some fear this authority will be used to block missions, Congress has declared privately conducted commercial launches to be consistent with the national security and foreign policy interests of the U.S. Time will tell what limitations will be imposed, but the Commercial Launch Act thus far gives government considerable power over commercial activity.

The DOT also reviews procedures for launch safety, including:

- Procedures for safety controls for launch sites and flight corridors
- Range safety expertise
- Procedures for ground and flight safety
- Range tracking and instrumentation
- Vehicle safety systems
- Proposed vehicle design

For the foreseeable future, I expect this review to be conservative. Missions using proven government launchers and national launching facilities should receive quick approval. But a mission proposing to use a new and untested vehicle or launch site will suffer close scrutiny.

DOT must coordinate commercial launches with all other interested agencies in the Federal Government, particularly the DoD and the Department of State. For current guidelines, contact the FAA.

The policy is now to require maximum utilization of commercial launch capability by the federal government. This includes planning missions in such a way so that they match the space transportation capabilities of our commercial launch providers. The main exceptions are national security or international collaborative efforts relating to science or technology. Congress additionally mandated in the Commercial Space Act of 1998 that DoD study Vandenberg AFB and Cape Canaveral Air Station to see what upgrades are needed to enable a flourishing commercial launch industry.

Military Space Activities. The development, testing, and deployment of a space-based defensive system must abide by the terms of the Anti-Ballistic Missile Limitations Treaty between the U.S. and the USSR (October 3, 1972). Administration policy will determine whether the U.S. will develop this system.

We also must acknowledge the possibility that international partners of the U.S. may object to legitimate national security activity. For example, our allies who signed up to help us develop the Space Station took exception to its potential undefined DoD activities, which they took to be SDI activities. Canada announced it would withdraw support for Space Station Freedom if the U.S. chose to conduct military activities, unless those activities were limited to research.

Patent Issues. Patent law and intellectual property rights in the U.S. stem from the first national patent law enacted in 1790. It was based upon the U.S. Constitution (1787), article 1, section 8, paragraph 8 that states, "Congress shall have power to. . . promote the progress of science and useful arts by securing for limited times to authors and inventors the exclusive right to their respective writings and discoveries." That limited period in the U.S. is now 20 years from the date of filing for the patent with the U.S. Patent and Trademark Office.

The U.S. law did not cover "space" so the U.S. Space Bill was enacted in 1990. It provides that, "Any invention made, used or sold in outer space on a space object of component thereof under the jurisdiction or control of the United States shall be considered to be made, used or sold within the United States." This legislation follows the "flagship" principle, which applies U.S. laws to ships on the high seas and aircraft flying over international water. With regard to the International Space Station, the International Governmental Agreement provides ". . . for purposes of intellectual property law, an activity occurring in or on a Space Station flight element shall be deemed to have occurred only in the territory of the Partner State of that element's registry, except that for ESA-registered elements any European Partner State may deem that activity . . . occurred within its territory."

In view of the U.S. Space Bill, the IGA, the Registration Convention and the provisions on registration in the Outer Space Treaty in appears that U.S. law is the law, which is most often applicable. In situations where U.S. law is not applicable there are no clear answers, and therefore considerable commercial risk to space missions from both a policy and economic perspective may be present.

The mission planner should be aware of two patent cases. Hughes Aircraft Company/Williams in 1973 obtained a patent relating to the spin stabilization of satellites to assure obtaining and maintaining satellite attitude on orbit. Hughes Aircraft Company sued the U.S. Government for infringement of its patent, because the government used the concept without paying for it. It was not until 1983 that multi-billion-dollar decision against the Government was handed down after the U.S. and others had launched 108 satellites, which infringed the patent rights of Hughes Aircraft Company. The lesson is clear; patent infringement may have very significant economic consequences

The second concern stems from a patent granted to TRW/Horstein et al. in 1995. Patent rights are based upon the "claims" in the patent application. The main claim of this patent in essence is an "orbital shell" above the Earth from 5,600 to 10,000 nautical miles reserved for TRW exclusively, for all communications applications to mobile handsets. Soon after the patent was granted TRW sued ICO Global Communications Ltd. to prevent them from launching satellites into the proscribed altitudes. The case was dismissed because there was no present infringement.

Nevertheless, this is a troublesome patent, because it flies directly in the face of the Outer Space Treaty provisions on nonappropriation of space by any nation including its private entities. In my opinion the patent is fatally flawed if it is interpreted to give any property rights to a spatial location. Were another nation to raise the issue in the U.N. or some other forum, the U.S., under its treaty obligations, would have no alternative but to "correct" the situation by legislation or cause a review of the original patent by the U.S. Patent and Trademark Office. A mission planner must, however, face the risks of litigation, which entail not only lost time but also the economic consequences.

21.1.2 U.S. Space Policy

A cohesive, consistent U.S. space policy does not exist. This conclusion follows from the lack of success of the program to get us into space. It has not reached new technological or cultural plateaus nor even maintained the initial pace.

What then must the mission planner consider from a U.S. policy prospective in mission design? Policy direction can change so quickly that the shorter the design and construction cycle, the better. Experience teaches us that government funding for projects can evaporate overnight. This being the case, get your money up front if possible, or at least attempt to obtain "fenced" funding.* For government projects you will need a single powerful backer or several who can provide funding. Political and economic support may be available from various sources. However, the advantage of securing interest and support from a number of sources may be more than offset by the necessity to serve various conflicting goals.

Policy edicts may be forthcoming from an administration, particularly the National Space Council, vocal committees of the Congress, and of course, industry groups. Do not be misled by the apparent credibility of the source or the clarity of the proposed goal. Words are cheap. All too often promises are made but are not backed up with budgeting and funding.

Even if you are doing a commercial project you will need an expert to advise you on national space policy sensitivities. The problem is you do not need to find friends—you only need to avoid enemies. The most likely enemies are potential competitors. While they probably cannot block your project directly, they may have friends in government who can. Accordingly, if it is possible, design the project in such a way that it is cooperative with others and not in direct competition. You should also be careful to assess what law and policy implications may arise from the proposed mission. These concerns may be national and international.

An example of this intergovernmental coordination was the effort by Space Industries, Inc., to secure government funding for the ISF (Industrial Space Facility). Hoping to get a fast start, they submitted the proposal to a congressional committee without adequate coordination with NASA and other interested governmental agencies. They were successful in getting support from the committee and the National Space Council. They were even mentioned specifically in President Reagan's 1988 space policy report. All of this withered and died because they had not made their peace with the supporters of the Space Station who felt that money spent on the ISF would be taken from Space Station funding and might ultimately result in the Space Station project being reduced or cancelled. The net result was that the ISF became tied up in red tape and did not get funding.

The bottom line is that space mission analysis and design must concern itself with the vagaries of policy and the multitude of concerns and interests that exist in the political arena.

Tighter constraints of federal budgets and an increasing demand for some form of economic return on federal investments has led to insistence by Congress that the federal government work with industry and buy from the private sector when possible. The 1998 legislation requires NASA to identify opportunities for commercial providers to participate in the International Space Station and to study the possibility of turning the space shuttle operation plus the space station over to commercial operators after it is assembled in orbit. Additionally space science data must be purchased from commercial providers to the extent possible.

21.1.3 Responsibility—Liability and Insurance

The Outer Space Treaty, Article VI, first established a nation's responsibility for its acts in space, as well as those of its citizens acting privately. This provision looks to the future and requires a nation to **authorize** and **continually supervise** all of its space activity to assure it conforms with treaties and international law. We must take into account this risk of regulation when we design a project and plan how to accommodate governmental supervision.

---

* *Fenced funding* is a term deriving from appropriations usage, meaning that monies allocated for a certain purpose cannot be otherwise allocated, even within the same agency or organization.

Liability, the other side of the responsibility coin, looks to the past. The Outer Space Treaty, Article VII, first specified liability for space activity. Then came the 1972 Convention on International Liability for Damage Caused by Space Objects. This Convention was intended to protect the nonspace-faring nations, so it held space-faring nations absolutely liable if they caused injury or damage on Earth. *Absolute liability* simply means that someone who is injured may claim compensation just by proving damage and who did it, without having to demonstrate negligence or fault. If damage occurs in space, however, space-faring nations must prove fault to recover damages.

The procedure for settling claims calls first for negotiations by the nations involved. The only claimant under the Convention is another nation, bringing a claim on its own behalf or on behalf of its citizens. If these negotiations fail, a three-member claims commission is formed. Each nation involved appoints one member. The two appointed members then appoint a third, the chairman.

The first case under this Convention involved the reentry of Cosmos 954 over the Northwest Territory of Canada on January 24, 1978. Negotiations between Canada and the USSR were successful. The nuclear reactor scattered radioactive debris over 124,000 km². The search and recovery effort (*Operation Morning Light*) cost Canada almost $14M Canadian dollars, but it settled for $3M. Canada elected not to claim the cost of $8M for its officials and employees nor for equipment. The USSR argued that Canada was too meticulous in the clean-up operations and that if this event had occurred in Siberia, she would not have expended that much effort. Additionally, the USSR asserted that she had offered to help clean up, but Canada rejected the offer. Had she been permitted to help, the costs would have been much less. This amicable settlement for 50% of the claim set a positive precedent for the future.

Under the Liability Convention, the U.S. must pay when her space activities injure citizens of other nations or their property. U.S. citizens may recover for personal injury or damage under the Federal Tort Claims Act. Thus, the government requires insurance before authorizing a commercial launch. Government launches, both military and civil, do not have formal insurance coverage, as the government is a self-insurer. If the government suffers a loss, it allocates more funds from the public coffers.

For commercial launches and potentially those conducted for foreign governments, we need to consider the following types of insurance:

- **Pre-Launch.** Pre-launch coverage insures against the risks from damage during shipment of the launcher and the satellite, as well as during integration and movement to the launch pad.

- **Government Equipment.** The government now requires launches on commercial expendable vehicles to provide insurance for loss to government launch equipment and facilities. The Commercial Space Launch Act (amended 1988) caps the amount at $100M. The Secretary of Transportation determines for each launch vehicle if less is appropriate; examples are $75M for the Atlas and $80M for either the Titan or Delta.

- **Third-Party Liability.** Launch vehicle operators and satellite owners may cause injury to others through the reentry. The risk is small because these objects generally burn up in the atmosphere during reentry. If they do not, they would typically strike the ocean or one of the uninhabited areas that make up most of the Earth's surface. Because of this low risk, coverage is less expensive though still required by DOT for launch vehicles and satellites and by NASA for commercial satellites.

When NASA was the only means of access to space for commercial satellites, their regulations established an insurance policy of $500M, with the government assuming the risk above that amount. This limit was a practical one because the insurance community could not insure a greater amount. Now that commercial launch vehicles will be the prime means of transportation into space, Congress has established a similar cap of $500M, until the world's insurance community can provide greater coverage. The purpose of this provision of the 1988 Amendment to the Commercial Launch Act was to attempt to level the commercial playing field as Arianespace requires an FF400M insurance policy. This amount was based upon the prior NASA policy of requiring $500M coverage.

Private operators of launch vehicles must also concern themselves with liability claims from citizens of the U.S. and around the world. The DOT regulations require coverage against this risk before issuing a launch license.

- *Launch Failure.* The most expensive insurance coverage insures the satellite value from the moment of lift-off to on-orbit checkout. The prices for this coverage have risen from 5%–10% of the satellite value to 25%–30% and then settled back to 16%–20%. Shuttle launches enjoyed about a 5% discount. Because it was a man-rated vehicle, the insurance community had a high confidence in its reliability. In the early days, insurance companies charged very low premiums for three reasons: they had few insurance losses; brokers knew little about space vehicles and the risks involved; and they competed to get into the business because space had pizazz.

  Premiums rose dramatically with the loss of the Shuttle Challenger in 1986 and several failures of expendable launch vehicles shortly thereafter. In fact, the space insurance community paid out 3 times more in loss claims than they took in through premiums. Once the U.S. and Ariane programs were back on track, premiums moderated. Still, as we plan commercial launches, we must try to choose launch vehicles and satellites with a history of success. Otherwise, insurance probably will be unavailable or prohibitively priced.

- *Reflight.* A corollary risk which the launch-vehicle manufacturer may cover is a guarantee to refly the mission or indemnification of the value of launch service. Whether a company will provide either at no cost depends on the launch vehicle's past experience and the competitive climate. A contract normally covers this risk, but I am sure the insurance community would cover it for a price.

- *Loss to Others on the Flight.* Under provisions of NASA launch contracts, the risk of damage to a payload by another payload or by NASA's actions falls under a *Hold Harmless Clause.* This provision means that all parties to the launch agree not to sue each other regardless of what happens. This practical remedy came about because it is very difficult to calculate the risk and determine the premium. The risk also seemed low, and experience bears out this assessment as no incidents have occurred. This solution is also true for companion satellites launched commercially.

- *On-Orbit Performance.* A further risk concerns whether a satellite will continue to perform. Communications satellites especially need coverage when the launch insurance terminates after on-orbit checkout. Satellite owners must have this "life" insurance coverage if they do not have sufficient resources to "self insure" the risks. Banks that finance communications systems usually demand this insurance. An alternative may be enough satellites on orbit to back-up the system.

In the past, buyers of space services (particularly communications service) contracted to build a satellite, sought a launch aboard the Shuttle or on Ariane, and then tried to secure one or more of the insurance coverages explained above. But competition in the satellite-construction and expendable-launch-vehicle industries has changed this procedure. Instead, the buyer puts out a request for proposals calling for on-orbit delivery with acceptance after checkout. This change shifts the burden and risks. Only time will tell whether the cost to the buyer of a space service will be higher. If competition is intense enough, which I predict it will be through the year 2010, the cost may decrease. If so, this *turnkey* package arrangement will be the norm rather than the exception.

One further note is appropriate. It might appear reasonable for one of the aerospace giants—manufacturer of satellites or expendable launch vehicles—to step up and provide insurance guarantees themselves. But the IRS has taken the position that a corporation may not set aside funds from successful launches to act as a reserve, without paying taxes on these funds. If companies could set aside untaxed reserves, they could gain a considerable economic advantage by offering insurance potentially unavailable from the insurance community, whose worldwide reserves are less than $20 billion. As an alternative, launch companies could acquire a block of insurance commitments for its launches from the insurance community and offer these to customers to sweeten a deal.

In a perfect world, actuarial risk would reflect the cost of risk plus overhead and profit. Cost of risk is simply the total value of all losses experienced divided by the total number of events or launches. Unfortunately, despite manufacturers' quoted success rates of 85%–98%, we do not have enough total launches to derive actuarially correct figures. Therefore, the price of insurance tends to depend on "feel" and various intangibles. As pointed out above, failures drive prices up, and successes decrease them. Interestingly, these variations occur whether or not a failure was insured. Space insurers paid nothing for the 1986 Challenger disaster; yet they were so shaken that insurance coverage was not available for a period of time. For up-to-date insurance coverage provisions and costs, contact Alexander and Alexander, Washington, DC; Coroon and Black, Bethesda, MD; or Marsh and McLennon, Washington, DC.

21.1.4 Remote Sensing

Anyone planning to launch a satellite that can sense the Earth needs to know the Principles Relating to Remote Sensing of the Earth from Space, a resolution adopted by the U.N. General Assembly on December 3, 1986. The USSR sought to prohibit the sensing of a nation without its consent and the developing nations sought to prohibit the release of information about a nation without its consent. Instead, this resolution was a victory for the "open skies" policy of the U.S. and the space-faring nations.

The key points are:

- It is a resolution, not a treaty, but as the world continues to follow its precepts it will become international law through custom.

- The resolution applies only to sensing "for the purpose of improving natural resource management, land use and the protection of the environment." It does not apply to applications related to national security.

- Data is divided into classes: primary or raw data, processed data, and analyzed information. *Primary data* is the same as *unenhanced data* under the Land Remote-Sensing Commercialization Act of 1984, so it must be released to all customers at reasonable cost. *Processed data* and *analyzed information* is data resulting from "value added activity" over which a company can assert property rights and refuse its release.

- The resolution states that nations will conduct remote-sensing activities with respect for other nations' sovereignty, not damaging the legitimate rights and interests of the sensed nation.

Legal arguments may affect how a reconnaissance or imaging satellite operates. Originally, in a classified memorandum, President Carter prohibited a U.S. company from disclosing remote-sensing information with a resolution better than 10 m. President Reagan's space policy rescinded this prohibition in February 1988, because the USSR was selling photographic images with 5 m resolution. France, Russia, Canada, China, Japan, and India provide commercially available remote sensing imagery; therefore if a nation or corporation wants to keep something secret, it had best hide it. The press will be the least of its worries. In my view, only cost will limit these activities.

Since 1982 the U.S. government policy towards remote sensing technologies has evolved from a very restrictive view of remote sensing as an intelligence gathering methodology which was classified to a broader view which promotes privatization, declassification and commercial exploitation. Efforts to expand this industry included requiring NASA to acquire remote sensing data from commercial providers. However, the industry has not developed as quickly as it might have because of lengthy coordination required by various federal agencies before a license to launch and operate a satellite system is issued. Unless the Department of State yields to the Department of Commerce and permits a reasonably unfettered development of a U.S. industry the market will be served by foreign competitors. In order to encourage the emergence of a competitive U.S. commercial remote sensing industry it will be necessary for the government to support investments in new remote sensing technologies, remove unnecessary restrictions on the dissemination of privately gathered data, and streamline the licensing process.

21.1.5 Import and Export Restrictions

We need not consider restrictions on imports into the U.S. unless a project has federal funding. In that event, the Buy America Act and domestic preference provisions in various other acts and regulations could affect purchases. In the past, an exploding number of statutes and regulations direct federal agencies to give domestic sources absolute or qualified preferences when they procure goods and services. These provisions, with exceptions and qualifications, appear in trade legislation, appropriation statutes, procurement laws and regulations, Executive Orders, treaties, and Memoranda of Understanding with foreign governments.

In general, the Buy America Act prohibits buying foreign goods rather than services, but NASA has held that a supplier of space services may not use foreign goods to provide the service. Under this Act, a product is not *foreign* if more than 50% comes from domestic sources, and the DoD may buy from NATO countries. The Act's provisions and rules will give American mission planners headaches and keep many contracting officers and lawyers fully employed.

Three mechanisms control exports from the U.S. First, commodities and technical data on the commodities control list fall under the licensing requirement of the Export Administration Act of 1979. The Commerce Department's Office of Export Administration runs this program and has published a series of detailed regulations. An example of this process in the space sector was the approval for Payload Systems, Inc. to fly a crystal-growth experiment aboard the MIR space station.

The second control applies when a patent involves military technology. The Departments of Defense and Energy review the export application, and either department may classify the information to keep it secret, thus preventing export.

The third restriction stems from the Arms Export Control Act. The Office of Munitions Control under the Department of State determines along with other Federal agencies what items fall under the Act's licensing provisions. This office has considerable discretion in determining whether a license should be issued. As a result, the United States Munitions List once included satellites and computer programs and many other items not traditionally thought of as weapons.

The International Traffic in Arms Regulations implement the Arms Export Control Act. The Munitions List specifically includes rockets, spacecraft, space electronics, and guidance equipment. *Defense services* are also included, defined as furnishing help to foreigners "in the design, engineering, development, production, processing, manufacturing, use, operations, overhaul, repair, maintenance, modification or reconstruction of articles." Included is the furnishing to foreigners any technical data whether in the U.S. or abroad. Finally, no one may provide foreigners technical data, meaning information classified or even related to defense articles.

Under these regulations, an export does not have to cross a border. For example, an export occurs when an individual discloses technical data concerning a spacecraft or rocket, even if the disclosure is part of a potential sale and within the U.S. These rules do not apply to NATO members, Australia, New Zealand, or Japan, but they still pose significant hurdles if we are seeking a world market. An example of this was the application to the Department of State by United Technologies Corporation to assist the Cape York Space Agency in developing a launch complex in Australia for the Soviet Zenit Launch Vehicle.

21.1.6 Environmental Concerns

The National Environmental Policy Act and its implementing regulations call for review of the Federal Government's actions to determine their effect on the environment. *Major actions* involve substantial time, money, and resources. They affect large areas or act strongly on small areas. Recommendations on these actions must consider their effects with special care.

When an action occurs in the U.S., we must take into account social, economic, and other environments. When it occurs outside the U.S., the requirements narrow. Whether environmental protection extends to activities in outer space depends on how we interpret the concept of *global commons*. In international law, "global commons" includes those territories outside the jurisdiction of any nation, such as the high seas, the upper atmosphere, the oceans, Antarctica, and particularly, outer space. Official U.S. documents which discuss "global commons" have not specifically mentioned outer space.

Whether the National Environmental Policy Act applies to outer space is not entirely clear, but the U.S. still has obligations under Article IX of the Outer Space Treaty which provides for international protection of the outer-space environment. Thus, we must comply with either the letter of the Act's provisions or their essence. If a proposed mission "has been done before," environmental questions may quickly be resolved. But if it includes something new, we may wait some time for approval, depending on how strongly the mission affects outer space or, especially, any celestial body.

For a government mission, the responsible agency must analyze the environmental effects. If more than one agency is involved, a lead agency supervises the preparation of the environmental documentation. For a private or commercial launch, the DOT would require an environmental impact assessment under its mandate from the Commercial Launch Act.

Under the National Environmental Policy Act, we must complete either an environmental assessment or an environmental impact statement. An assessment helps determine whether we need an impact statement for a particular action. It must include enough information to determine whether the proposed action is major and whether it significantly harms the environment of the "global commons." We must assess the environmental effects and the need for the proposed action plus available alternatives. Unless security restrictions intervene, the assessment is available to the public upon request. There is no need to obtain public comment. An assessment is less formal and rigorous than an impact statement. Typically, it applies to:

- Spacecraft development projects in space science and in space and terrestrial applications

- Specific experimental projects in space and energy technology

- Development and operations of new space transportation systems

- Advanced development of new space transportation and spacecraft systems

In contrast, we would need to file an impact statement when an action is "expected to have a significant impact on the quality of the human environment." Our draft statement should be thorough enough to permit analysis and comment. If it is complete enough after agency review, it goes to the public for comment. At this time, the Department of State, the Council on Environmental Quality, and other federal agencies also have the chance to comment. For the proposed action and reasonable alternatives, the environmental impact statement: (1) considers purpose and need; (2) provides a detailed description; (3) analyzes the environmental effects; (4) briefly describes the affected environment of the "global commons"; and (5) compares the alternatives' effects on the "global commons."

If we do not know something because it is unavailable or scientifically uncertain, we must say so in the impact statement. Public hearings are not required, but may be appropriate, depending on circumstances. After agencies and the public have commented, we redo the statement, further analyzing any issues they have raised. Then we publish it in the Federal Register before forwarding it to the decision maker for final action.

Normally, we would need an impact statement for R&D activities associated with developing and operating new launch vehicles, space vehicles likely to release large amounts of foreign materials into the Earth's atmosphere or into space, and certain systems for nuclear propulsion and generating power. Some listed exclusions do not require a statement, but we probably will have to do an assessment if any "significant environmental effects" are possible. Unless the project has been done before, we can expect a bureaucrat to require an assessment of environmental effects, so he or she can determine if an assessment is required! The first assessment should determine what is required: (1) no environmental assessment; (2) only an environmental assessment; or (3) an environmental impact statement. Although it is difficult to predict exactly how long this process will take, we should anticipate 3 to 6 months for a simple assessment, one to 2 years for a simple impact statement, and up to 5 years for international concerns or a suit in federal court.

In some cases, the action may affect the environment of a foreign nation or a resource designated as one of global importance. If so, we would use slightly different procedures. First, we would prepare an environmental survey or review. An environmental survey is a cooperative action and may be bilateral or multilateral. Whether or not we do a review depends on consultations which determine if the proposed action would do significant harm. The content is flexible, but generally includes (1) a review of the affected environment, (2) the predicted environmental effects, and (3) significant known actions that governmental entities are taking to protect or improve the environment against the proposed action. If the government is not acting, is this inactivity an oversight or a conscious decision?

The U.S. prepares an environmental review unilaterally. In effect, it is an internal action, by which one or more governmental agencies surveys the important environmental issues associated with the proposed action. It contains essentially the same information as an environmental survey.

Lasers and Particle Beams

Lasers and particle beams illuminating into space have caused almost no problems. The reason is fairly straightforward: in the U.S., such experiments are carried out by government laboratories or under government control. They have agreements with NORAD/USSPACECOM to search the space catalog and provide windows of opportunity, so illuminating does not interfere with operational spacecraft.

Nuclear Power Sources

Nuclear power sources have raised questions since they were first used. The U.S. program began using a radioisotope thermoelectric generator with SNAP-3A in the summer of 1961 and continued to 1977. In April 1965, SNAP 10 was the only U.S. space nuclear reactor. The U.S. safety program includes an Interagency Nuclear Safety Review Panel composed of three coordinators appointed by the Secretary of Defense, the Administrator of NASA, and the Secretary of Energy. The Nuclear Regulatory Agency, the Environmental Protection Agency, and the National Oceanic and Atmospheric Administration also participate in these reviews. The safety review ascertains whether the benefits of using nuclear power are worth the risks. The policy of the U.S. in using radioisotope thermoelectric generators following an aborted SNAP-A mission in April 1964 was to design the container so that all nuclear material would survive intact, regardless of the nature of the accident. The policy makers specifically envisioned reentry and impact on earth. On May 18, 1968, the range safety officer aborted the launch of NIMBUS-81 at an altitude of 30 km over the Santa Barbara Channel. The generator capsules were recovered without incident. Those who launched SNAP 10A with a nuclear reactor in 1965 launched the reactor in a subcritical mode, designed it to remain subcritical at or after impact should it reenter the atmosphere before start-up, and delayed its start-up until it had reached orbit. It is in an almost circular polar orbit, which has a decay life of 4,000 years. Additionally, this reactor package was designed to come apart on reentry.

Design for safety must include both system and mission design. The methods to reduce risks from nuclear materials include containing them within radioisotope thermoelectric generators, diluting and dispersing them with reactors, delaying their effects by boosting them into a higher orbit, and possibly retrieving them using a vehicle like the Shuttle.

On the international front, the untimely reentry of Cosmos 954 in 1978 (see Sec 21.1.3) caused great concern around the world and has been the subject of discussion in the U.N. Committee on the Peaceful Uses of Outer Space. A working group's February 1981 report on Cosmos 954 reaffirmed that nuclear power is safe in outer space if it meets safety requirements. Some people still suggest a ban on nuclear power, and Australia has specified solar rather than nuclear power in a request for proposals for surveillance satellites.

Only the former USSR routinely used nuclear power to run its Radar Ocean Reconnaissance Satellites (RORSATs). In the future, other nations will consider this power source to serve particular needs. For example, the U.S. had under discussion powerful reactors to provide power to space components of the Strategic Defense Initiative (SDI). The SP100s would generate 100 kW, some 25 times more powerful than the RORSATs. These developments pose two risks: return to Earth with much more fissionable material than on RORSATs and increases in radiation from gamma rays and positrons. The positrons from the reactor temporarily form an artificial radiation belt in the Earth's magnetic field. When they strike another spacecraft, the positrons produce penetrative gamma-rays. This radiation interferes with astronomers' readings of natural radiation from such phenomena as solar flares, neutron stars, and black holes.

In the late 1980s, such radiation from Soviet spacecraft occasionally overloaded the gamma-ray spectrometer on board the U.S. Solar Maximum Mission satellite. The Japanese X-ray satellite, GINGA, also suffered, and the U.S. Gamma-Ray Observatory may be affected.

Thus, as the need for nuclear power increases, we can expect greater pressure, particularly from the scientific and astronomical communities, to stop its use. Considering its unique abilities, I predict that nuclear power will not be banned from space, but its use will be closely regulated. Unless we must use a nuclear power source to complete a mission, we should select another form of power.

Space Debris

Space debris is the other significant environmental issue. It affects mission designers in two ways. First, we must plan the mission to prevent or abate space debris as much as possible. Second, we must design satellites to meet the risk of collision with debris. Aside from a few cries in the wilderness, space debris received little official notice until the late 1980s. Johnson and McKnight [1991] provide an extended technical discussion of this problem, and Reijnen and de Graff [1990] provide a discussion from a legal perspective.

In 1986, the U.S. Air Force Science Advisory board questioned its 1983 position that debris did not appear to be a problem. As a result, the DoD established in March 1987 a space debris policy. It states in part that the "design and operation of DoD space tests, experiments, and systems will strive to minimize or reduce accumulation of space debris consistent with mission requirements."

In a final statement on space policy, released in February 1988, President Reagan called for a review of the U.S. policy on space debris. As a result, a *Report on Orbital Debris by the Interagency Group (Space)* was released in February 1989. It reflected the Group's uncertainty as to the urgency for action. Its major recommendation was a joint study by NASA and DoD that would develop a comprehensive plan to improve the monitoring of debris, so debris predictions could be more accurate. The report identifies two critical areas in space for the near term. Low-Earth orbit requires attention because of the large masses of material and the high relative velocities. At the same time, the geosynchronous arc requires attention because of the number of spacecraft which will lose their maneuvering ability within the next few years. The report then echoes the DoD policy on managing debris.

This report stands in contrast with the report of the ESA's Space Debris Working Group, dated November, 1988. The Director General observed that clearly the present debris in the space environment poses little threat to either manned or unmanned missions. However, "we must adopt a conscious policy aimed at curbing the growth of debris." The report lists the same two areas of concern— low-Earth and geosynchronous orbits—and notes another concern for astronomical observers. It concludes by requiring preventive measures: observing and analyzing present debris, avoiding collisions, and minimizing collision effects and future space debris. It asks nations to reduce the number of pieces and the mass of space debris.

Strikingly, the ESA's report is a call to action. The agency is urged to start taking steps—organizational, technical, and institutional—to seek cooperation with others and thereby counter the threat to space flight. Why then does the U.S. simply want to study the matter further? The answer is both political and economic. Politically, the U.S. has had difficulty discussing technical issues with developing nations in the U.N. Time and again the U.S. has been outvoted on philosophical grounds and is not ready to hear political dialogues in the Committee on the Peaceful Uses of Outer Space. The U.S., therefore, looks to multilateral discussions with space-faring nations to solve problems.

The economic issue is equally sensitive. The U.S. is a space-faring nation which plays a major role in launching its own satellites and those of its western partners into space. Thus, it is loath to set a policy "prematurely" which could cost a lot of money to solve the space debris problem. The key words in the U.S. report are "where feasible and cost-effective." The U.S. recognizes that the problem demands multinational solutions but wants to avoid expensive "political solutions."

Advice: stay out of the politics if at all possible. Choose a launch vehicle that is at least no worse than average in terms of causing debris and plan the mission with space debris abatement features. If you minimize debris, you will get much faster approval for your space mission.

21.1.7 FireSat Legal and Policy Issues

At first blush it would seem that there could be no issues at all. However, the first question is whether FireSat is going to be a U.S. Government project or a project with private financing.

If this is a government project the mission planner needs to evaluate whether the Department of Agriculture or U.S. Forest Service is going to support funding or attempt to sabotage the project. Will they see it as a boon to their fire fighting mission or a threat, because there will be less need for fire fighting equipment and personnel, which will jeopardize Congressional funding? What agency will be responsible for the project after it is launched? Will their staffing and funding be increased to cover additional responsibilities? What international agreements cover firefighting support? Will foreign nations have a right to the information from FireSat? Does the U.S. face any liability from the release of the information on fires? In other words, would there be liability if FireSat did not report a fire? Under the U.N. Principles on Remote Sensing the U.S. is obligated to provide data to all nations at cost. Will the data supplied be raw data or enhanced data? Who will supply the equipment to analyze the data? Will the FireSat mission as planned support all requests for information? What infrastructure will be necessary to support the U.S.'s treaty obligations? Who will pay for it?

From a mission perspective how many satellites will be necessary? What is the replacement strategy? Who are the launch providers going to be? Can these satellites be launched with other satellites (size and position questions) or do they require dedicated launches? Are there any satellites using these altitudes? What is the space debris situation? Are any special shields required to protect the satellite? Do the FireSat sensors pose any threat of interference to other satellites? Are the mission and function of the FireSat satellites in total conformity with U.S. international policy? Might not Brazil, for example, object to the monitoring of their forests, fearing that the U.S. was trying to make a case to hold them responsible for their failure to control burning in the forests?

These are examples of the kinds of questions that need to be answered. Inevitably the answers cause more questions. From the limited information we have about the FireSat Mission, I believe that from a legal and policy perspective the mission is doable. I am aware of the saying, "If you cannot stand the answer do not ask the question." However, it is a foolish mission planner that refuses to at least know of the risks.

21.1.8 Asteroids

There has been recent interest in asteroids (Space Development Corporation) as a potential mining opportunity. Private businessmen have proposed launching missions to asteroids to bring back rare and precious metals. There is also interest in the Moon and it is said to contain aluminum, calcium, iron, silicon, and small amounts of chromium, magnesium, manganese and titanium. It also has oxygen and sulfur. Some have proposed to do this regardless of the legal and policy issues. The question of ownership of asteroids and the right to sever valuable ore is not totally clear.

The Moon Treaty of 1979 addresses the subject of mining the Moon, asteroids and other celestial bodies. In general it provides for the establishment of an International Regime to authorize and control any mining activity at such time as the exploitation of natural resources becomes feasible. A critical element of the Moon Treaty is the principle of the Common Heritage of Mankind. Under this principle there is the commitment that all nations must share in the management and benefits from such activity. It is this provision for sharing of benefits without contribution of investment that has caused the U.S. and other space-faring nations to refrain from joining the treaty. Only Australia, Austria, Chile, Mexico, Morocco, The Netherlands, Pakistan, The Philippines and Uruguay have ratified the agreement.

The argument could be made that as long as the U.S. is not a party to the treaty U.S. corporations should be permitted to conduct mining operations on the theory that what is not prohibited is permitted. Standing alone this logic is sound, but it overlooks the give and take in the international community on a vast number of issues. Many of the developing nations firmly believe that they are entitled to share in the profits from space activity and would deeply resent actions by U.S. or others from developed countries "stealing" from them. Accordingly it is a virtual certainty that the U.S. government would not issue a license to conduct mining activity on the Moon or other celestial bodies.

The United Societies in Space has proposed a Lunar Economic Development Authority and some modifications to the Moon Treaty so that mining operations could be explored. Creating legal certainty will be the first step in financing such a project.

There was a similar situation with the 1982 Law of the Sea Treaty which was resolved in 1994 by a U.N. General Assembly resolution. The provisions were modified to give the U.S. and other key nations a major say in the undersea mining and that has facilitated exploration and the beginning of operations. This appears to be the most viable solution to asteroid mining, but a first priority will be reducing transportation costs.

21.2 Orbital Debris—A Space Hazard

Ronald A. Madler, *Embry-Riddle Aeronautical University* Darren S. McKnight, *Titan Research and Technology*

About 20,000 tons of natural material consisting of interplanetary dust, meteoroids and asteroid/comet fragments filter down to the Earth's surface every year, with several hundred kilograms in LEO at any one time [Kessler, 1985; Zook et al., 1970]. This natural hazard has been recognized as a danger to space travel since the 1940s, but now human activities in space have created a hazard of even greater concern. Figure 21-1 portrays both the natural environment and the artificially created debris population [Adapted from NASA and NRC sources]. Millions of kilograms of artificial debris orbit the Earth and present a serious concern to continued safe access to space. The growth of orbital debris poses a series of difficulties for space mission designers. To control the growth of debris and its associated hazards, we should take a number of steps during the design process. As seen in Fig. 21-2, the debris mitigation process spans all phases of the mission profile. Similarly, NASA debris mitigation guidelines cover all aspects of the mission design process. These guidelines can be found in NASA Safety Standard 1740.14 [1995], while software to parallel the safety standard is also available from NASA.

21.2.1 Environmental Definition

Meteoroids have been a concern since the beginning of human spaceflight. Engineers performed a tremendous amount of work to understand the hazard posed to spacecraft by meteoroids. While meteoroid-effect studies and model improvements continue, researchers have understood the meteoroid background flux fairly well since the late 1960s. One meteoroid flux model is represented in Fig. 21-1. There has been a resurgence of interest in meteoroids due to the possible storm conditions associated with the Leonid meteor stream in 1998–1999 [Yeomans, 1998]. While most meteor-

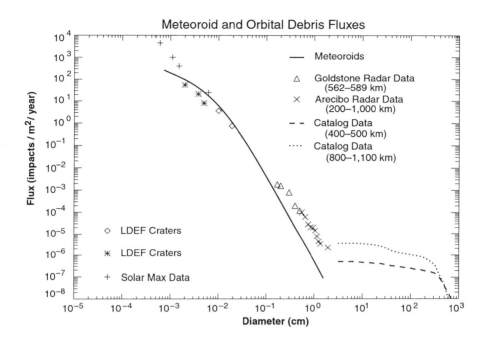

Fig. 21-1.   **Comparison of the Fluxes of Meteoroids and Orbital Debris for Low-Altitude Orbits (Adapted from the National Research Council and NASA sources).** This figure shows that human-generated orbital debris has much higher impact rates than meteoroids for large and very small debris diameters.

Fig. 21-2.   **Debris Mitigation Processes**.

oids have relative velocities of about 19–20 km/s with respect to the Earth, the Leonids have a relative velocity of approximately 70 km/s.* This higher velocity will produce more damage on impact, which could be of concern to spacecraft owners and operators. There were no significant anomalies associated with the 1998 Leonids.

We define *debris* as any nonoperational manmade object in space. These objects include nonfunctioning payloads, used rocket bodies, mission-related debris (e.g., lens covers or separation devices), debris from surface degradation (e.g., insulation or paint chips), and debris from on-orbit fragmentation. This derelict hardware accounts for 93% of the cataloged objects in orbit; only 7% are active payloads. Table 21-1 shows the breakdown of cataloged objects [Johnson et al., 2002]. Table 21-2 shows causes for these on-orbit fragmentation events. Debris from the more than 175 fragmentation events are by far the largest source of orbital debris.

TABLE 21-1.   Cataloged Orbital Debris Objects. We can help reduce the amount of orbit debris by acting responsibly.

Breakup Debris	40%	Spacecraft	30%
Rocket Bodies	18%	Operational Debris	12%

TABLE 21-2.   Causes for On-Orbit Fragmentation [Johnson et al., 2002]. We can reduce fragmentation debris by properly venting propulsion systems.

Deliberate Breakup	30%
Propulsion System Malfunctions	31%
Unknown Cause	28%
Battery	4.5%
Aerodynamics	6%
Collision	0.5%

The U.S. Space Command compiles the trackable debris tracked by its worldwide *Space Surveillance Network* (SSN) in a satellite catalog. This worldwide network of radar and optical facilities senses, tracks, identifies, and catalogs data on over 8,500 large orbiting objects. It senses objects with diameters as small as 10 cm in LEO and 1 m in GEO. This detection limit is due to the original design of the system for tracking large objects—debris detection was never envisioned as a task for the network. We know that there exists a much larger population of smaller objects. This has been confirmed by recent campaigns with more sensitive radars. Figure 21-1 shows that there are approximately an order of magnitude more objects in the 1 cm size than exists in the catalog. The 1 cm size is significant because it is the largest size fragment that we can effectively shield against. Thus, there are a significant number of objects that we cannot track, but which can cause substantial spacecraft damage.

We can quantify how crowded space has become by using the *spatial density,* i.e., the number of objects per volume of space. Figure 21-3 plots spatial density values out

---

* *Editor's Note*: The Leonids are remnants of Comet 55P/Tempel-Tuttle. They are in a retrograde orbit and collide with the Earth nearly head-on such that their velocity adds to the Earth's orbital velocity of about 30 km/s.

to 2,000 km altitude for cataloged objects of various sizes. The GEO curve represents the spatial density within 1 deg of the equatorial plane. The average density between 800 and 1,000 km is just above 10^{-8} km^{-3}. The smallest trackable objects may weigh tens to hundreds of grams. We cannot shield against objects of this size. In GEO the average spatial density is one to two orders of magnitude less than LEO. The **average** relative velocity in LEO between orbiting objects is 9 to 10 km/s with maximum values above 14 km/s due to eccentric and retrograde orbits. In comparison, the relative velocity between debris and satellites in GEO ranges between 100 and 500 m/s. The difference in relative velocities is due mainly to the lower orbital velocities and smaller inclination distribution of objects in GEO. This physical phenomenon couples with the lower spatial density values in GEO to make the collision hazard much smaller in GEO than LEO.

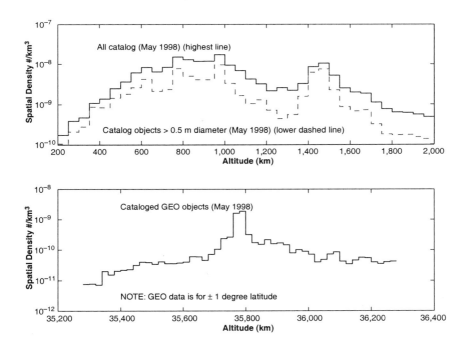

Fig. 21-3. Spatial Density Values. Densities in GEO are approximately 10 times less than in LEO. The probability of a spacecraft getting hit by something big is small, while the chance of getting hit by something small is big. See text for discussion.

In addition to the spatial density being different in LEO vs. GEO, the natural cleansing effects also differ. In GEO and geosynchronous transfer orbit the major perturbations are solar/lunar gravitational effects. For very small debris, solar radiation pressure may also significantly affect lifetimes, however, atmospheric drag at GEO has no measurable influence. Hence, a major breakup in geosynchronous orbit will affect all future operations in that regime. On the other hand, atmospheric drag greatly affects the lifetimes of objects in LEO. The smaller debris fragments have significantly larger area to mass ratios and thus drag affects them more. This natural removal of orbital debris is very important in the long-term definition of the environment. We have sensed decreases in the cataloged population only during periods of maximum solar activity (1979–80 and 1988–90). However, in the 2000–2001 period of high solar activity the overall cataloged population stayed fairly constant due to several significant breakup events.

The other major variables in debris growth are the number and types of satellite fragmentation events. As the debris population grows, the environment may become so severe that satellite or object fragmentations due to hypervelocity collisions with debris may occur [Kessler and Cour-Palais, 1978]. This type of event will mark a clear trend toward a worsening environment. The first confirmed collision between cataloged objects occurred in July 1996. The altitude of a breakup will foretell the effect on the long-term debris environment. Breakups below 500 km may have a major influence for less than a decade while fragmentations above 500 km may pose large hazards for many decades. Presently, there is no cost-effective method of removing debris already in orbit. The prudent design of future spacecraft will lessen the chance of debris generation and satellite fragmentation.

It is difficult to describe precisely the present status of the dynamic near-Earth debris environment, much less to accurately predict the future debris environment, due to uncertainties in traffic models and fragmentation rates. The trackable population grew at a nearly linear rate from 1960 to 1990. We don't know exactly how the undetectable population grew, but it is more strongly influenced by the fragmentation-event rate, satellite operational patterns, and the solar cycle [Johnson and McKnight, 1991]. Predictions of the actual number and flux of all debris rely on accurate modeling of its sources and sinks. We compare debris models with our best measurements of the environment: impact rates on returned spacecraft surfaces for very small debris and special radar measurements for detectable objects. We continually improve and update these models as our understanding of the debris environment grows. Spacecraft designers can look for the latest environmental models through one of the NASA, DoD, ESA, or other international space agency Space Debris program offices.

21.2.2 Design Considerations: Spacecraft Hazard and Survivability Analysis

Mission designers must address two main issues concerning space debris. First, we need to design debris protection for large, long-lived spacecraft to ensure mission success. Second, we must use debris mitigation methods to ensure that space activities will not continue to litter our near-Earth environment with more derelict hardware [NSS 1740.14, 1995].

We can approximate the probability, P_C, of a piece of debris impacting a space system using the kinetic theory of gases:

$$P_C = 1 - e^{(-SPD \cdot AC \cdot T \cdot VREL)} \tag{21-1}$$

where SPD is the spatial density of debris objects (i.e., average number of objects per volume in space), AC is the collision cross-sectional area, T is the mission duration, and $VREL$ is the relative velocity between the satellite and debris population. Using this simple equation, the rough order of magnitude approximation for collision probability for one year in orbit for a range of altitudes and satellite sizes is listed in Table 21-3. These values are approximate and only show the order of magnitude for the probability of collision. Detailed analysis should use a more accurate debris environment model [for example, Klinkrad et al., 1997 or Liou et al., 2002], as well as a more refined representation of the spacecraft's cross-sectional area. Also space satellite designers must take the meteoroid environment into account. Meteoroids dominate the hazard for sub-millimeter up to millimeter size, and can penetrate thin honeycomb structures.

TABLE 21-3. **Collision Probability per Year (in 1999).** The table values are approximated over all inclinations for a cross-sectional area range of 5 to 40 m^2. The cross-sectional area is defined as the area viewed from one orientation, and is approximately 1/4 of the total surface area for simple convex shapes.

Altitude (km)	Collision Probability per Year		
	Trackable	1 cm diameter	1 mm diameter
300	10^{-6}–10^{-5}	10^{-4}–10^{-3}	10^{-2}–10^{-1}
400	10^{-5}–10^{-4}	10^{-4}–10^{-3}	10^{-1}–1
500	10^{-5}–10^{-4}	10^{-4}–10^{-3}	10^{-1}–1
600	10^{-5}–10^{-3}	10^{-4}–10^{-2}	10^{-1}–1
800	10^{-4}–10^{-3}	10^{-3}–10^{-2}	10^{-1}–1
1,000	10^{-4}–10^{-3}	10^{-3}–10^{-2}	10^{-1}–1
1,200	10^{-4}–10^{-3}	10^{-3}–10^{-2}	10^{-1}–1
1,500	10^{-5}–10^{-3}	10^{-3}–10^{-2}	10^{-1}–1
2,000	10^{-6}–10^{-5}	10^{-5}–10^{-3}	10^{-2}–10^{-1}

The cataloged population presents a manageable debris hazard to even large spacecraft. However, incorporation of nontrackable objects into the hazard assessment produces much larger probabilities of collision. The flux rate increases several orders of magnitude for untrackable debris that can still cause damage. An encounter with a 1 cm fragment will likely produce significant amounts of debris, much of it trackable, while a 1 mm impact most likely will cause surface degradation, localized craters and small penetrations.

For GEO satellites, we may use alternate forms of probability of collision equations which are more convenient due to the physically and dynamically different environment [Johnson and McKnight, 1991]. The objects have lower orbital velocities and most reside in a narrow latitudinal band resulting in a distinct contrast to the more randomly distributed LEO environment. The hazard in GEO appears to be about a decade behind LEO, but the GEO population will grow quickly because there are fewer natural sinks.

While we can determine the expected rate of impact for certain sized meteoroids and debris relatively easily with one of the available orbital debris environment models, determining the hazard to a spacecraft is not as straightforward. Christiansen et al. [1992] outline this process as:

1. Determine failure or damage conditions for the mission

2. Determine impact conditions causing failure or damage

3. Determine likelihood of failure (integrating the step 2 equations over the flux, direction, velocity, and projectile characteristics from debris environment models)

4. Assess sufficiency of the design

5. Modify the design or requirements, if needed

Due to the dynamic nature of the debris environment, the best strategy is to integrate debris awareness into all phases of the design process. However, it will be increasingly expensive to actually implement any system changes the later in the design process. Special attention to the protection of mission critical systems may still be considered late in the design process.

The first step mentioned above entails determining what constitutes a failure or unacceptable mission degradation. This is something each spacecraft may define differently and will influence the hazard assessment. For the International Space Station, loss of a module or loss of life may be the unacceptable damage level, while for an unmanned spacecraft it may be loss of any critical system, such as attitude control.

The second step depends on the many unknowns of the projectile and the spacecraft structures. In order to proceed with the hazard analysis, we must derive some kind of equation relating damage to projectile characteristics. For pressure vessels, where penetration will have serious implications, a ballistic limit equation is determined which relates the projectile characteristics to the penetration ability for a specific spacecraft wall. Researchers have determined equations for single walled spacecraft and multilayered shielding configurations [Hayashida and Robinson, 1991; Christiansen et al., 1995; Christiansen, 1993]. Armed with the performance equations for the spacecraft surfaces, we can determine a probability of failure by integrating over all the expected projectiles from the debris models (step 3 above). The next step is to determine if the probability of success is sufficient for the mission. If the probability of success is not sufficient, then we need further protective measures or design modifications. The RADARSAT mission is an example of a spacecraft that did not have an acceptable level of risk after the preliminary design. Approximately 17 kg of shielding was added to reach a comfortable level [Warren and Yelle, 1994].

Debris Protection

We often describe the expected damage of a debris impact, to first order, by the relative kinetic energy of the impacting object. For a relative impact velocity of 10 km/s, a 100 g fragment (6–10 cm diameter) possesses the kinetic energy equivalent to 1 kg of TNT. On the smaller end of the scale, a 1.6 mm debris object has the same kinetic energy as a 9 mm pistol slug. The amount of energy absorbed by the structure and the level of damage is highly dependent on the impactor characteristics, satellite structure and location of impact. Thus, it is difficult to determine the effect of a hypervelocity debris impact on a satellite without a considerable amount of specific satellite and collision information. Nevertheless, a 100 g object impacting at 10 km/s will produce extensive damage on any satellite, and would destroy any small, compact satellite, given a center of mass collision.

Spacecraft **will** encounter micrometeoroids and orbital debris during their functional lifetime. However, there are passive and active means to protect them from most debris. *Passive* means include shielding and redundancy, while *active* generally refers to collision avoidance. Bumper shields are effective for passive protection against fragments smaller than 1 cm in diameter (mg range).

Figure 21-4 qualitatively shows how effective shielding systems are at defeating an impacting particle. The bottom curve is for a single sheet wall, while the other curves are for different shielding configurations which have the same weight as the single wall. Shielding systems have one or more outer bumpers and sometimes intermediate backup layers before the innermost wall. Bumper shielding systems work by fragmenting or vaporizing the projectile with the first layer when the projectile has a very high velocity. The resulting debris cloud expands and hits the next layers over a larger area, dispersing the energy of the projectile and the impulsive load. In the case of a pressure vessel, this shielding increases the probability that the pressure hull will survive impacts—without penetration, rupture or spall with a much thinner pressure wall. In other words, when we optimize bumper shields, the resulting two or more walls weigh less and provide better protection from particulate impact than a single wall design. The bumper system has three main design parameters:

1. Thickness and material of the outer wall (*shield* or *bumper*)

2. Spacing between the shield and the backup layers

3. Thickness and material of the backup layers

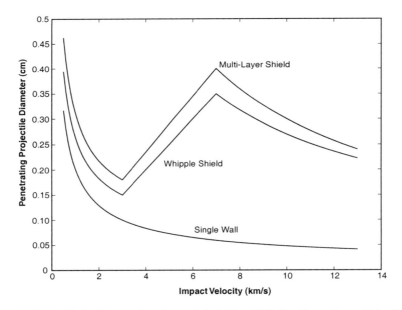

Fig. 21-4. Ballistic Limit Curves for Several Satellite Wall Configurations. This figure qualitatively shows the penetrating diameter of an impacting projectile vs. the impact velocity for a single wall and two shielding options. The three sections of the shielding curves correspond to projectile deformation, projectile fragmentation, and projectile melt or vaporization, respectively, as the impact velocity increases. This assumes spherical aluminum projectiles.

The design of the shield depends on the shield-to-projectile mass density ratio and the projectile's impact velocity. The debris environment parameters will vary by altitude and change over time as events alter the environment. For a given environment and spacecraft configuration (exposed area, structure and mission duration), we optimize the shield thickness, t_s, and wall spacing, S, to insure projectile fragmentation.

This highlights the main function of the shield: to break apart the impacting object so that its debris will hit the backup layer(s) over a larger area, causing less damage.

The ability of the shield to lessen the effect of the projectile's impact is directly proportional to the shield's ability to break it apart. The optimum situation occurs when the shield and projectile fragments vaporize or liquefy. In this state they present less hazard to the backup layer(s). Testing by NASA has shown that the ratio t_s/d, where d is the impacting particle's length, is around 0.1 to 0.2 for a functional shield [Cour-Palais, 1979]. This range is good for aluminum-on-aluminum impacts in the velocity range 5–12 km/s. We obtain more beneficial effects if the shield has a low melting point and high mass density, both of which will produce more damage to the projectile and a lower probability of solid particles striking the backup sheet [Cour-Palais, 1979].

The equation for determining the "optimum" backup sheet thickness, t_b, assumes that shield and projectile debris are mostly in molten or vaporized states. For a basic form of aluminum (7075-T6), the equation for the condition where the backup sheet will not deflect, rupture, or spall is given by Cour-Palais [1979]:

$$t_b = \frac{Cmv}{S^2} \qquad (21\text{-}2)$$

where t_b is the backup sheet thickness in cm, m is the projectile mass in g, v is the projectile velocity in km/s, S is the spacing in cm, and an empirically derived constant $C = 41.5 \pm 14.0$ (cm^{3}g^{-1}km^{-1}s).

The optimum design of a shield depends on the physical properties of the projectile and its velocity. A design which works at one speed may not be as effective at other speeds due to the characteristics of hypervelocity impacts. Also, the composite and sandwich construction methods common on many commercial spacecraft have different properties than simple aluminum structures [Taylor et al., 1998]. System designers should look for the latest references because the science and art of shielding progresses rapidly.

Structures such as the International Space Station must have extensive shielding to survive the particle environment. Multilayered shielding will greatly reduce the probability of a penetration to the pressure hulls of the Station. Other spacecraft also have vulnerable systems and components that must be protected to ensure a successful mission. Mission critical systems and components must either have special shielding from debris, or be shielded by less critical structures. An understanding of the debris hazard can help spacecraft designers to minimize the cost of the spacecraft by designing survival into the vehicle from the beginning. Table 21-4 lists some common sense design guidelines.

Debris Avoidance

Some satellites may face a significant hazard due to trackable objects. In this case, designers and operators may need to use passive and active avoidance techniques. *Passive collision avoidance* methods include minimizing the size (cross section) of the spacecraft, either by reducing the actual size or controlling the attitude to have a smaller profile with respect to the debris impact directions. For most LEO spacecraft, the normalized average relative impact rates by spacecraft surface are: 10 km/s for leading edge ± 45°, 2 km/s for surfaces 90° to the ram direction, 1 km/s for trailing edge, 0.1 km/s for space pointing, and 0.01 km/s for Earth pointing surfaces. Operators

TABLE 21-4.   Common Sense Design Guidelines.

• Design for end-of-life environment.	• Make critical systems redundant.
• Have a "debris expert" on the project.	• Don't create or leave debris in orbit.
• Make sure everyone on a project is familiar with the debris hazard.	• Shield or shadow sensitive surfaces and systems with less sensitive components when possible.
• Perform a cost-benefit tradeoff study of systems to allow for degraded performance due to debris (e.g., thermal control, power).	• Recognize that impacts of small debris make surfaces more susceptible to atomic oxygen damage and degradation.
• Orient sensitive objects that must be exposed to space on the spacecraft's trailing edge or facing the Earth to decrease particle strikes.	• Consider possible damage beyond perforation: spallation, high-velocity fragments from impact site, impulsive loading, and plasma from particle strikes on solar arrays (leading to possible electrical discharge).

could also place the spacecraft in a less populated orbital regime; however, this is not possible for most missions.

We may need *active debris avoidance* for critical space structures such as space stations or satellite constellations. This requires active tracking of co-orbiting objects, additional propellant, coordination with satellite tracking facilities (such as USSPACECOM), and coordination with co-orbiting spacecraft operators. At the present time, systems such as the International Space Station, the Space Shuttle, and LEO satellite constellations are the only missions using or planning to use active collision avoidance. The Shuttle has made several highly publicized maneuvers to avoid cataloged objects. Active avoidance is only practical for protection from the largest of fragments and may still be suspect since ground-based tracking accuracy is poor for most orbiting objects (unless the tracking system expends extra effort to update a specific satellite). Thus, it is not a cost-effective option for most satellite operators at this time.

Debris Mitigation

All space users have a responsibility to include debris mitigation techniques in their mission profiles. As identified in Fig. 21-2, there are three phases where users can minimize the addition of debris to the environment. While non-governmental missions do not have any debris mitigation regulations at this time, the U.S. Government requires its missions to minimize debris [NSS 1740.14, 1995; NMI 1700.8, 1993]. Additional regulations and guidelines are in work by both NASA and the UN.

Launch system designers should plan their scenarios to reduce launch-related debris such as protective shrouds, separation devices and expended rocket bodies. They should not allow this hardware to reach long-lived orbits. Historically, an average of three large pieces of debris are produced from each successful mission. However, the launch process probably generates many more nontrackable pieces.

As a satellite is inserted into its final orbit, we should take care not to produce debris during the last impulsive maneuvers. A major source of debris in the past was exploding rocket stages. Some of these propulsion-related explosions occurred during attempted burns while others resulted from inadvertent mixing of hypergolic fuels or tank overpressurization years after the rocket's last use. Designers corrected this problem on many rocket bodies by venting the fuel or conducting an idle burn. Before we declare a satellite operational, we must deploy its solar panels, uncover its instruments, and stabilize its orientation. Given the present design of satellites, all these activities may release hardware into space. As a satellite operates and ages, paint chips off and small pieces of hardware work themselves free, creating more orbital debris.

Once that satellite or rocket body ceases to perform a useful purpose, we may consider the entire system to be debris, and we should remove it for the sake of the remaining operational payloads. We should add **this mission termination phase to the mission profile,** using propulsive maneuvers to put it into a disposal orbit or re-entry trajectory. We use disposal orbits mostly for GEO satellites, while reentry is most economical for many LEO orbits. We must consider these maneuvers in the design process to control the growth of orbital debris. Regardless of our regime, all energy sources should be passivated (e.g., fuel and pressure tanks vented, batteries safed, and momentum wheels despun).

The design decision in response to orbital debris is basically a cost balance between design and risk costs. As discussed in Sec. 19.2, the *risk cost* is the expected cost of failure. This value is simply the probability of a debris encounter causing a failure times the cost to compensate for this failure. This may entail accepting degraded performance or launching a substitute. A worsening debris environment may increase the space segment failure probability (Sec. 19.2) which we can offset by changes in design such as shielding, redundancy, or avoidance maneuvers. Each of these counter-measures exact a financial burden of increased direct costs (design and manufacturing) and indirect costs (launch and maintenance), yet may result in greater reliability for all space systems.

References

Christiansen, E.L. 1993. "Design and Performance Equations for Advanced Meteoroid and Debris Shields." *Int. J. Impact Engineering*, vol. 14, pp. 145–156.

Christiansen, E.L., J.L. Crews, J.E. Williamsen, J.H. Robinson, and A.M. Nolen. 1995. "Enhanced Meteoroids and Orbital Debris Shielding." *Int. J. Impact Engineering*, vol. 17, pp. 217–228.

Christiansen, E.L., J. Hyde, and G. Snell. 1992. "Spacecraft Survivability in the Meteoroid and Debris Environment." Paper No. AIAA 92-1409, presented at the AIAA Space Programs and Technologies Conference, Huntsville, AL, March 24–27.

Christol, Carl Q. 1982. *The Modern International Law of Outer Space.* New York: Pergamon Press.

Cour-Palais, B.G. 1979. "Space Vehicle Meteoroid Shielding Design." Paper No. ESA SP-153 in *Proceedings of Comet Halley Micrometeoroid Hazard Workshop,* Noordwijk, The Netherlands: ESA Scientific and Technical Publications Branch.

Goldman, Nathan C. 1988. *American Space law, International and Domestic.* Ames, Iowa: Iowa State University Press.

Hayashida, K.B. and J.H. Robinson. 1991. "Single Wall Penetration Equations," NASA TM-103565, December.

Jasentuliyana, N. and R.S.K. Lee. 1979. *Manual on Space Law*. New York: Oceana Publications.

Johnson, N.L., A. Bade, P. Eichler, E. Cizek, S. Robertson, and T. Settecerri. 2002. "History of On-Orbit Satellite Fragmentations." 11th edition, July, JSC-28383, NASA Johnson Space Center.

Johnson, N.L., and D.S. McKnight. 1991. *Artificial Space Debris*. Malabar, FL: Krieger Publishing.

Kessler, D.J. 1985. "Orbital Debris Issues." *Advances in Space Research*. 5 (2): 3–10.

Kessler, D.J. and B.G. Cour-Palais. 1978. "Collision Frequency of Artificial Satellites: The Creation of a Debris Belt." *Journal of Geophysical Research*. 83:2637–2646.

Klinkrad, H., J. Bendisch, H. Sdunnus, P. Wegener, and R. Westerkamp. 1997. "An Introduction to the 1997 ESA MASTER Model." in *Proceedings of the Second European Conference on Space Debris*, Darmstadt, Germany, March 17–19, 1997, W. Flury, ed, ESA SP-393, May, pp. 217–224.

Liou, J.C., M.J. Matney, P.D. Anz-Meador, D. Kessler, M. Jansen, and J.R. Theall. 2002. "The New NASA Orbital Debris Engineering Model: ORDEM 2000." NASA/TP-2002-210780, May.

NASA. 1993. "Policy For Limiting Orbital Debris Generation." NASA Management Instruction (NMI) 1700.8, April 5.

National Research Council, Committee on Space Debris. 1995. *Orbital Debris: A Technical Assessment*. Washington, DC: National Academy Press.

National Research Council, Committee on International Space Station Meteoroid/Debris Risk Management. 1997. *Protecting the Space Station from Meteoroids and Orbital Debris*. Washington, DC: National Academy Press.

Natural Orbital Environment Guidelines for Use in Aerospace Vehicle Development. 1994. B.J. Anderson, ed. NASA TM 4527. June.

Office of Safety and Mission Assurance. 1995. "Guidelines and Assessment Procedures for Limiting Orbital Debris." NASA Safety Standard (NSS) 1740.14, August.

Reynolds, Glenn H. and Robert P. Merges. 1989. *Outer Space, Problems of Law and Policy*. Boulder, Colorado: Western Press.

Taylor, E.A., M.K. Herbert, D.J. Gardner, L. Kay, R. Thomson, and M.J. Burchell. 1998. "Hypervelocity Impact on Carbon Fibre Reinforced Plastic (CFRP)/Aluminum Honeycomb." In *Proceedings of the Institution of Mechanical Engineers Part G: Journal of Aerospace Engineering*.

Warren, H. and M.J. Yelle. 1994. "Effects of Space Debris on Commercial Spacecraft—The RADARSAT Example." In *Preservation of Near-Earth Space for Future Generations*, John A. Simpson, ed. New York: Cambridge University Press.

Yeomans, D.K. 1998. "Comet 55P/Tempel-Tuttle and the Leonid Meteors." LEONID Meteoroid Storm and Satellite Threat Conference, Manhattan Beach, CA, April 27–28.

Zhukov, G. and Y. Kolosov. 1984. *International Space Law*. New York: Praeger.

Zook, H.H., R. Flaherty, and D.J. Kessler. 1970. "Meteoroid Impact on the Gemini Windows." *Planetary and Space Sciences*. 18:953–964.

Chapter 22

Design of Low-Cost Spacecraft

Rick Fleeter, *AeroAstro*

22.1 Designing Low-Cost Space Systems
22.2 Small Space Systems Capabilities and Applications
Abilities of Small Space Systems; Emerging Miniature and Low-Cost Technologies; Potential Applications
22.3 Applying Miniature Satellite Technology to FireSat
22.4 Scaling from Large to Small Systems
22.5 Economics of Low-Cost Space Systems
22.6 Annotated Bibliography on Low-Cost Space Systems

Until about 1990, conventional satellite technology focused on relatively small numbers of highly capable, complex spacecraft. Recently, spacecraft have become more diverse, with the largest spacecraft now complimented by new systems using a larger number of smaller spacecraft in low-Earth orbit. While these are lower cost than their predecessors, this chapter focuses on the lowest tier of spacecraft cost to examine the particular methods and attributes characteristic of minimum-cost spacecraft. In certain applications these lower cost, smaller, simpler spacecraft are more effective. We will examine the tradeoffs between conventional technology and what is now referred to as *miniature satellite technology* or *microspace*. We will also consider the most successful applications for minimum cost spacecraft technology, as well as how engineering of low-cost and miniature spacecraft differs from that of conventional devices. Because modern, low-cost spacecraft design is a rapidly evolving technology, there are few references. We have included an annotated bibliography as a guide to further reading.

Ever since Sputnik in 1957 and Explorer in 1958, spacecraft developers have built small, simple systems alongside large, conventional satellites. Miniature satellites fill specific niches, especially for short-term missions with few users. A *miniature satellite* typically weighs less than 200 kg, has a shorter mission lifetime requirement, and is put together quickly by a small team. With less money invested, and because a minimum cost spacecraft has far fewer components and lower complexity, a user can sanction more liberal engineering designs and be more willing to use newer, less expensive technology, such as more contemporary electronic components or fabrication techniques.

The main advantages of miniature satellites are decreased costs and production times. Military, university, commercial, and institutional space programs sometimes cannot afford large, conventional spacecraft or take the time needed to build and launch them.

853

Since they provide fewer, more modest on-orbit abilities, miniature products have much simpler system architectures. They achieve reliability through simplicity rather than through expensive, redundant components. They can be much smaller and lighter than conventional products, thus reducing launch costs, which typically constitute half the cost of a satellite system on orbit. Very small devices can often fill small spaces available on large launchers. For ground transportation, one person can often carry a small satellite in a car or on an airline seat.

Military applications of larger spacecraft are vulnerable to failure during launch or on orbit, as well as to aggressive acts. We currently address these weaknesses mainly with redundant subsystems, highly reliable components, and defensive counter-measures, all of which add cost and size. A miniature technology approach would be to create many small, relatively vulnerable spacecraft, providing the same ability but a more difficult target.

Small size is not in itself a new feature. The earliest satellites were very small out of necessity, weighing 5 to 50 kg. In retrospect, it is remarkable how large satellites have grown, not how small a few satellites are. Table 22-1 surveys a sample of the small satellites launched from 1991 to 1995. Since 1957, larger systems, made possible by advances in the technology of satellites and launch vehicles, have absorbed the most engineering attention and resources. However, the continuity in their launch dates shows that small satellites have played a role throughout satellite history, and are still useful in specific applications.

TABLE 22-1. Selected Small Satellites from 1991–1995. These satellites are all under 425 kg. They performed their missions beyond their expected lifetimes, on the average.

Mo.	Company/ Sponsor	Satellite	Mission	Mass (kg)	Launch Vehicle
1991					
Feb	USSR	Kosmos 2125-2132	Military Comm. Sats	40 ea.	SL-8
Mar	SDIO/DSI	CRO	Research (3 sats)	70	Space shuttle
Jun	USAF/DSI	ISES/REX	Comm. Research	85	Scout
Jul	DSI	ASTP/Lightsat	Comm. (7 microsats)	23 ea.	Pegasus #2
Jul	U. of Surrey	UoSAT 5	Communications	49	Ariane 4
Jul	Tech. U. Berlin	TUBSAT-A	Communications	25	Ariane
Jul	Ball Aerospace	LOSAT-X	Research	75	Delta 2
Jul	DSI	ISES	Comm. Research	60	Scout
Jul	ESIEESPACE	SARA	Radio Astronomy	27	Ariane
Jul	OSC/CIT	ORBCOMM-X (VaSTAR)	Communications	17	Ariane 4
Jul	DARPA/ONR/DSI	ASTP/Lightsat 1-7	LEO Comms.	22 ea.	Pegasus #2
Aug	Japan/U.S.	Solar A (Yoko)	Research	200	M-3S2
Mar	DSI	CRO	Research (3 sats)	70	
Jun	USAF/DSI	ISES/REX	Comm. Research	85	
Dec	Czechoslovakia	Magion 3	Scientific	52	
1992					
		SARA	Planetary Geophysics	14	Ariane
May	India	SROSS C	Gamma Ray Detector	106	Indian ASLV
Aug	Matra/Surey	S80/T	Communications	50	Ariane 4
Aug	Korean Inst. of Tech.	Kitsat-A	Comm. Research	50	Ariane 4

TABLE 22-1. **Selected Small Satellites from 1991–1995. (Continued)**These satellites are all under 425 kg. They performed their missions beyond their expected lifetimes, on the average.

Mo.	Company/ Sponsor	Satellite	Mission	Mass (kg)	Launch Vehicle
			1993		
Feb	Brazil	SCD-1	Environmental Data	115	Pegasus #3
Mar	NASA/AF	SEDS 1	Tether Experiment	57	Delta II
Apr	DoE/LANL/ AeroAstro	ISES/ALEXIS	Research	109	Pegasus #4
Aug	Talspazio/Kayser Threde	Temisat	Ocean andTraffic Monitoring	50	Cyclone
Sep	South Korea	Kitsat-B	Comm. Research	50	Ariane 4
Sep	Portugal	Posat-1	Remote Sensing	50	Ariane 4
Sep	Intfrmtrcs/AMSAT	Eyesat		50	Ariane 4
Sep	Italian AMSAT group	Itamsat		50	Ariane 4
Sep		Healthsat		50	Ariane 4
Dec	DARA/OHB System	Safir R	Communications	50	Zenith
			1994		
	Pakistan	BADR-B	Remote Sensing	50	
	Italy	BARRESAT	Technology/Industrial	50	Ariane
	France	ENSAESAT 2	Research	50	Ariane
	France	Cerise	Miitary Eavesdropping	50	Ariane
	Energetics	Sattrack	Local Positioning	64	Soviet Proton
Jan	Tech. U. Berlin	TUBSAT-B	Research	40	Cyclone
Feb	Germany	Bremsat	Research	68	Space Shuttle
May	DARA/OHB System	Safir R	Communications	55	Zenit
May	India	SROSS-C2	Scientific	113	ASLV
Jun	UK DRA	Space Tech Res Vehicle	Test New Tech. (2 sats)	50	Ariane
Jun	UK DRA	STRV-1a, 1994-034c,	Component Testing	52	Ariane
Jun	UK DRA	STRV-1b, 1994-034b	Component Testing	53	Ariane
Aug	OSC	APEX, 1994-046a	Research	262	Pegasus
Dec	Russia	RS-15A, 1994-085A	Communications	70	SS-19 missle
			1995		
Jan	Russia	ASTRID, 1995-002B	Science and Tech.	28	COSMOS-3M
Jan	Russia	FAISAT-1, 1995-002C	Forward Communications	115	COSMOS-3M
Mar	Mexico	UNAMSAT	Research	12	Start-1
Mar	Russia	Techsat-1	Test Momentum Wheels	55	Start-1
	Israel	OFFEQ-3, 1995-018A	Astronomical Experiments	36	OFFEQ-3, 1995-018A
Apr	OSC	ORBCOMM FM1, 1995-01 & 1995-02	Global communications	47	Pegasus
Apr	NASA	MICROLAB-1, 1995-017	Weather	76	Pegasus
Apr	Germany	GFZ-1, 1995-020A	Passive Reflector	20	Progress M27 Rocket
Jun	USAF	STEP-3	Memory Experiments	268	Pegasus XL
Jul	France	Cerise, 1995-033B	Eavesdropping	50	Ariane
Jul	Spain	UPM-SAT	Comm. Research	44	Ariane
Aug	Chile	FASat-Alfa	Research	50	SICH-1

Early Space Systems

The earliest satellites were small because the first launch vehicles' payloads were limited. The satellites often had lifetimes limited by on-board battery power because they had no solar panels. They carried either simple analog transponders or simple beacons for researching signal propagation. The Echo series were passive reflectors formed of metallized polymer balloons.

Virtually all of today's satellite applications appeared in the first 10 years of satellite development. For example, the Telstar series first demonstrated television and telephone relay. In size, mass, power, and orbit these satellites resembled today's typical small satellites. They proved that satellite-linked TV was effective and desirable, thus blazing the trail for the large geosynchronous communications satellites in commercial and government service today. Several small Earth-surveillance systems had flown by 1967. The first weather satellite, Tiros 1, flew in 1960. VELA was one of the most successful small, Earth-observing satellites. Built by TRW for the Air Force and Department of Energy and weighing 152 kg, VELA satellites were the first to provide data from space concerning nuclear weapons testing on Earth. Though initially flown as an experimental satellite and developed by a small group in 18 months, VELA provided years of valuable reconnaissance service on orbit.

Radio amateurs were quick to see how they could apply satellites to communication and education. In 1961 the 5-kg OSCAR I (*Orbiting Satellite Carrying Amateur Radio*) was the first of a series of satellites radio amateurs built and operated worldwide. Now numbering about 30 and spanning over 35 years, almost all of these satellites have exceeded their operational design for on-orbit lifetime. The first commercial comsat, Early Bird, which weighed just 39 kg, was flown in 1965.

Continuing Applications of Miniature Satellite Technology

Between 1965 and 1985, space-faring nations deemphasized small satellites in favor of getting the most sophisticated performance from on-orbit resources. Because the United States was committed to crewed flight, including the lunar landing, we developed large boosters. Larger boosters also became available for placing large payloads into geosynchronous orbit, revolutionizing global communications and creating the infrastructure in place today.

As space systems rapidly grew in size, vigorous small satellite programs continued but without much attention from either the public or mainstream aerospace engineering. For example, several amateur radio satellites were developed using technologies and design approaches previously untried in spacecraft. They also continued to apply simpler, less expensive devices which were rapidly vanishing as satellites became larger and more complex. These small satellites employed photovoltaics to charge NiCd batteries for on-orbit lifetimes of several years. They carried VHF, UHF, and microwave transponders and had a range of operating modes controlled by ground command.

Small satellites often played a role behind the scenes in developing military systems. Very small devices were routinely flown to provide on-orbit targets and signal sources for tracking systems. These satellites usually carried active sensors which sensed their RF and optical environment and relayed the data back to ground stations.

Many nations have entered the space community through launches of small satellites, including Canada, France, Italy, England, Korea, Portugal, Sweden, Denmark, UK, Israel, Spain, Argentina, Chile, Mexico, Malaysia, Denmark, Japan, Germany, Czechoslovakia, India, and The Netherlands.

Modern Low-Cost Systems

By 1985, amateur satellites such as OSCAR 10 and UoSat 2/OSCAR 11 had demonstrated the value of reliable digital communications to small satellites. As Telstar clearly showed, we can use a single small satellite in low-Earth orbit only when the ground station can see it; its ability to communicate was limited. To overcome this problem, OSCAR 10 employed digital *store and forward* communications, using the small satellite platform as a mailbox or bulletin board. Users anywhere on Earth could transmit prepared messages to the satellite during its brief overhead pass. The satellite could then deliver these messages throughout the world. Consequently, a single, small, inexpensive satellite in low-Earth orbit provided global mail service—something not even a large geosynchronous satellite can do.

This new operating mode began at the same time that NASA extended its *Get Away Special* program to release small payloads into orbit. NASA's program had provided inexpensive transportation into space for small (< 68 kg), self-contained payloads. Replacing ballast used to balance the Space Shuttle's major payloads, the program had carried payloads fixed to the payload bay wall on roundtrips to orbit. Goddard Space Flight Center developed a Get Away Special container with an opening lid and a spring ejection system. This allowed orbital insertion of a 68-kg satellite for less than $50K. The European Ariane rocket had independently developed the *ASAP* (Ariane Structure for Attached Payloads). Up to 6 small satellites could be carried aboard the ASAP *ring*, each with mass up to 50 kg. Several Amsat and university satellites, as well as spacecraft developed under national R&D programs for a range of applications have since flown on ASAP. Due to the popularity and success of the ASAP program, it has been expanded for Ariane V, accommodating physically larger payloads with a mass of up to 100 kg. (For further details see London [1996]).

Thus, small, inexpensive satellites reentered the aerospace mainstream through the confluence of three developments:

- Low-cost access to space
- Highly capable digital communications systems whose weight, power, and volume were compatible with the Get Away Special
- Digital store and forward communications

Radio amateurs exploited the concept of a digital mailbox in a small, low-cost satellite to communicate with relief workers in remote parts of the world. The same technology was a fundamental feature of the University of Northern Utah's *NuSat*, which was designed to calibrate FAA air traffic control radars and to be an educational tool at the University. A similar application drove the *Global Low Orbiting Message Relay* (*GLOMR*) satellite's ability to carry data from remotely located sensors to a central command post.

The success of these low-cost satellites caused designers to reassess the roles of large and small space systems. Some payload organizations quickly recognized that the ability to fly a simple, small payload with low cost and fast turnaround was ideally suited to their needs. These first-generation users included scientific programs and smaller national programs that could not justify the resources required to fly a conventional technology project. Israel's Offeq satellite, launched in October, 1988, represents this facet of miniature technology. Today the small-satellite industry focuses on special niches, to satisfy missions that conventional satellite technology cannot cover with large spacecraft and multiple payloads, developed in programs spanning 5 to 15 years.

The ALEXIS (Array of Low-Energy X-Ray Imaging Sensors) 120 kg satellite, built in 1989 and launched in 1993, was representative of the increasing utility afforded by small, low-cost satellites. The payload required anti-Sun orientation with a slow 2-rpm roll about the Sun axis. This special stabilization requirement, plus the need for observation times of one month to one year, made this payload incompatible with larger spacecraft catering to multiple payloads. The ALEXIS spacecraft bus, developed for Los Alamos National Laboratory, weighs only 45 kg without its payload of scientific instruments and costs about $3M. Yet it supplies 55 W of continuous power to the payload, buffers 1 Gbit of science data between downlinks to a single ground station, provides all spacecraft guidance and position data, and offers telemetry down and uplinks at 750 and 9.6 kbit/s respectively. Built for a 6-month on-orbit mission ALEXIS has provided over 6 years of on-orbit science operations and remains in continuous operation as of mid-1999.

Chemical Release Observation canister (CRO), another advanced miniature satellite program, is a group of three small satellites, each carrying 25 kg of hydrazinic chemicals. Designed to eject the liquid chemical for optical observation from the ground and from the Shuttle, the satellites provide simple telemetry of the payload state (temperature and pressure) and respond to various ground commands. The satellites are aerodynamically stabilized to weathercock and fly oriented along their velocity vector. Built and flown for under $1M each, CRO was a highly successful application of a minimum cost spacecraft.

Since the success of the OSCAR 10 satellite in 1983, miniature technology has become an increasingly important element of hardware programs and systems architecture studies. OSCAR 10 provided analog and digital communications from a Molniya orbit to amateur radio operators for about 14 years. Since 1965, OSCAR-series satellites built by volunteer developers with limited budgets have demonstrated part of the potential of miniature satellite technology.

Figure 22-1 addresses one reason these small systems have become so important. The dramatic miniaturization of the electronic components composing most satellite payloads implies that the spacecraft could do as much or more while becoming smaller. But conventional satellites have increased their mass by three orders of magnitude despite the mass of some components shrinking by as much as four orders of magnitude. Miniature satellite devices built with advanced technology can do much more than the very large, costly devices of only a decade ago.

Computers are another instance where miniaturizing has provided highly capable, affordable machines. Miniaturizing has also changed the way we use these machines and greatly expanded their applications. The computer revolution came about because we thought of how to apply new technology in new ways. The technology is in place to create a new class of small, inexpensive, and highly capable space systems. Our challenge is to identify how to apply these new methods and products.

22.1 Designing Low-Cost Space Systems

Although we can design for low cost in different ways, some general rules apply based on space-system development, launch, and operations costs. Launch costs are often quoted simply as a linear function of mass on orbit for various orbit classes. This simple model implies that achieving very low mass automatically minimizes launch costs. But achieving low mass at low cost requires minimizing payload requirements, redundancy, and size. By including the mission performance requirements as part of

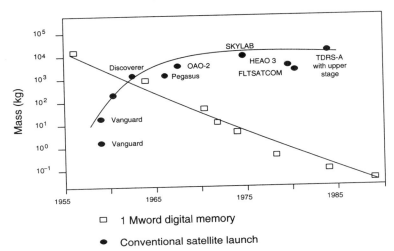

Fig. 22-1. Electronics and Satellite Mass Have Evolved in Different Directions Since Sputnik.

the design process, miniature technology tries to lower launch costs not by cost-saving measures applied to complex satellites, but rather by minimizing the requirements imposed on the spacecraft.

The single most significant factor driving ultimate mission costs will be the system performance requirements. Unfortunately, these requirements are often determined independently by the user organization before the design process begins. This practice is encouraged by engineers who complain that they can't design without specs to design to! The small satellite system attempts to provide a valuable capability within a severe set of volume, mass, power and complexity constraints. Only by tailoring requirements to that which can be realistically fit into this limited resource envelope can the small satellite design exercise come to a successful result. Typical tradeoffs which should occur in the early phases of the design process include reduced data storage and downlinking in favor of more onboard processing, reduced pointing and stability requirements vs. more adaptive sensors and actuators, and reduced power by careful design of payload instruments or reduction in duty cycle, leaving open the possibility of a multiple satellite system which in combination can provide a higher duty cycle and more frequent ground coverage.

Because launch resource is not a standard commodity, pricing in cost per unit mass is misleading. Very small launch resources, including the Ariane ASAP, Space Shuttle Hitchhiker and small payload space on most other major launch systems, often have a single price per payload up to a maximum mass. Hitchhiker and ASAP payloads are in general not mass but rather volume constrained. Thus, we need not spend payload development resources on very lightweight structures, as well as the detailed analysis and testing needed to design and verify them. The standard Hitchhiker carries 68 kg for as little as $150,000. The resulting cost of $2,200 per kg is about 10% of standard launch costs on a per unit mass basis. Maybe more significantly, the total cost is 0.1% to 1% that of larger spacecraft.

We must keep in mind that these systems are secondary payloads—flown "stand-by" and without a guaranteed launch date. If we build the payload to allow minimum support at the launch site, integration into a range of launch vehicles, and maximum flexibility in orbit requirements, we can minimize costs and program delays.

Dedicated small launch facilities, such as Pegasus or Shavit, price launch space per flight rather than per unit mass. Thus, we cannot save money by reducing mass below our allocation. In fact, because Pegasus and Shavit use solid propellant, they must carry ballast if the payload mass is too low. As a result, we should avoid radical design changes to shave mass, for these changes would require us to spend a lot of money unnecessarily on materials, fabrication and analysis.

Reducing costs is often erroneously associated with increasing mission risk. Conventional systems offer verifiable savings in expected cost of failure by lowering failure probability for an expensive mission. But if the program budget exceeds our available resources, we may have to abandon the program. This programmatic risk is mitigated by choosing missions achievable with a minimum cost approach.

One way to save money in designs using miniature technology is to dispense with full redundancy, thus risking a single-point failure. But net risk may in fact be lower for a small nonredundant system. This is because redundancy requires selection and switching mechanisms to arbitrate between elements. In miniature technology, we choose to back up only critical systems with known low reliability. In a system of n components each having a success (lack of failure) probability, R_o, the overall reliability of the system is

$$R = R_o^n \qquad (22-1)$$

Since the number of components, n, nominally scales with mass, a minimum cost small spacecraft may have n 1% as large as a larger spacecraft. Thus, the small system can use parts with 99.99% reliability and achieve the same reliability as a large system spending much more money to buy parts with 99.9999% reliability. Or, using the same part quality the little spacecraft will be more reliable. Thus the reliability gained in a larger system via redundancy is achieved in the smaller system through reduction in the number of parts. In fact, parts failure is no longer the major cause of spacecraft failure. Rather it is human errors in design and operation, which are in part the result of increased system complexity. Here too, a smaller, simpler system has a significant reliability advantage.

In practice, miniature satellite systems have an excellent record of success. The world production leader is *AMSAT*, the satellite organization for radio amateurs. Every one of their over 30 satellites has been successful over 35 years of development programs spanning many different development teams and missions. Their record stems directly from miniature technology's simplicity, its development process, and its organizational elements, as discussed below. Single-string design is riskier as it becomes more complex. The relatively simple design and subsystems in small satellites make them more reliable.

Low-cost space vehicles must be small enough to be assembled by a unified group with common goals. In a small design team, every member of the group has a direct link to every other member. They negotiate interactively and communicate efficiently. The team members take the minimum risk path to achieve the savings they mutually seek. Large, segmented development organizations have trouble working in this close, cooperative way.

The team designing a miniature satellite must consist of engineers with breadth and depth. Each specialist needs to understand the requirements driving the overall design.

A small team can produce unexpected innovations because each member can appreciate the problems of other members. Teams suffer if they lose a key person. This vulnerability can be even higher in small teams because every team member is essential. Managers of small programs need to ensure that each member works with a "buddy" who, though the buddy may be focused on another aspect of the program, understands to some extent what the engineer is doing. In the event of loss of that engineer from the program, the buddy can take on his or her tasks and responsibilities while the team redistributes work.

A small organization trades resources freely across disciplinary boundaries. No team member takes any resource for granted or "owns" an allocation. The program manager communicates concerns about cost throughout the group. The efficiency possible in a small group is so valuable that the individuals often work much harder rather than bring more communications and management burdens to the program. The spacecraft team can carefully explore and negotiate every systems-level requirement with the users in order to show them how to save money. Flexibility on both sides can cut costs dramatically. A small program should plan to invest in automated design and analysis tools. These tools reduce team labor hours, but more importantly they help to get the job done without team size inflating beyond the size (about 20 people) which allows close team interactions. Similarly the project should plan to subcontract specialized functions to keep the team as small, focused and interactive as possible.

A satellite program expends much of its resources on the ground station which may operate two or three times longer than the satellite development program and require expensive full-time staffing. To avoid the capital and maintenance costs for a large ground station, it is best located at the user's site—preferably in the user's office. Small satellites can be simple enough to be built around the user's personal computer and operated as a computer-controlled, laboratory apparatus.

Technologies employed in the ground station include computing systems from the consumer market, such as a PC-based station controller, read-write optical disc archiving, and commercial input-output cards. Because antenna costs increase rapidly with gain, we engineer the link based on the best antenna gain achievable at low cost. Sophisticated ground-station software eliminates most of the need for operator interaction and provides a simple, user-friendly interface requiring little special training. The computer is the only interface to the ground-station equipment, so the user need not understand any other equipment interfaces.

The program review used to develop conventional satellites is inappropriate for miniature satellites. For conventional program reviews, one person represents each department while other members keep working. But for the small satellite team, a department" may be one person. Thus, for best efficiency, teleconferences and smaller, less frequent Technical Interchange Meetings replace conventional reviews. Usually, the members directly involved in a given issue meet; rarely does the whole group need to attend.

Program costs depend on schedule about equally in large or small efforts. But because miniature satellite programs are short, delaying a few months can increase cost significantly with respect to the total development budget. Therefore program managers must take the schedule seriously, always looking for means to finish and launch the spacecraft as quickly as possible.

Like program management, quality assurance is a subfunction carried out within the team developing a miniature satellite. No single individual or discrete organization handles it. The team works case-by-case in deciding on fabrication standards, component qualification, or requirements for derating and previous flight experience. The same is true for applying specific fabrication standards or for inspections. This method allows the system engineer to reduce or eliminate quality requirements if their cost outweighs their contribution to the probability of success.

An example of selectively applying quality assurance standards is in the design of a spacecraft mass memory, which contains single controller and input-output (I/O) devices managing a large number of mass storage devices. Failure of one of the control or I/O devices is catastrophic. Because there are few of these devices, we can justify specifying highly reliable, highest-grade (and hence high cost) components. On the other hand, failure of one or even several of the many memory components is not particularly serious. Further, procuring many of these devices at very high standards is expensive. Thus, we procure memory for much less money at lower standards of quality and reliability.

Occasionally, a key component not previously used in spacecraft may improve performance or save money. With lower reliability requirements and the flexibility to consider parts individually, we can decide whether to apply the component, based on the following:

- Whether its failure brings catastrophe or merely degrades operation
- How well we can simulate its space operation
- How much not using the component would cost in performance and resources
- Whether we can meet system reliability goals with it

Saving money in testing does not necessarily imply higher risk of on-orbit failure. Here again the simplicity of miniature technology allows a different, more individual approach. The program manager may opt to eliminate subsystem and component tests in favor of a full system-level qualification. This decision does not reduce probability of mission success. In fact, simulating the space environment achieved in testing may be more accurate for the integrated system test approach because each part will be qualified while operating with actual flight interfaces. The tradeoff is in program risk vs. program cost. If many component or subsystem flaws exist, they are more easily corrected at the subsystem test level. On the other hand, if few failures occur, we should rely on the integrated system test, which eliminates long testing of subsystems and components.

Miniature satellite testing should take advantage of the system's simplicity. Wherever possible, we should use the actual flight and ground hardware for all tests, thus raising confidence that all system elements will actually work together in flight while lowering investment in simulations, test fixtures, and facilities.

Every mission develops with overt or subtle political pressure to satisfy the widest possible constituency. Conventional products typically carry a number of discrete, often unrelated payloads supporting many users. But programs using miniature technology cannot bear the complex payloads or engineering and management interfaces arising from this constituency building. To maintain the tight program staffing and focus needed for close communication, we must control complexity. Further, unless we limit spacecraft mass and volume, we may lose a launch niche. Increasing launch costs pressure us to provide higher reliability through redundancy, formal program controls, discrete quality assurance, and increased paperwork and subsystem testing. Instead, we should usually split demanding payloads into separate programs rather than lose the advantage of miniature satellite technology.

Earlier, we pointed out that trying to lighten structures beyond a practical minimum mass made designs more complex without reducing launch costs. A conservative, simple design helps us avoid complexity in the miniature satellite program. The program cannot support independent specialists analyzing various thermal, mechanical, stability, and other properties. Designing conservatively eliminates detailed analysis, thus maintaining the small development team and keeping the program within cost and schedule.

We can also justify less analysis because the small spacecraft's physical dimensions support smaller thermal gradients. Simultaneously, their vibration-resonant frequencies tend to be high whereas applied moments are low, owing to short unsupported structure lengths. Wall thicknesses typically depend more on screw thread depth requirements and machining tolerances. Thus, designs often include substantial structural and heat transfer margins. Finite element modeling can then be eliminated in favor of analyzing the overall system and a few critical parts. Therefore, we can save time and resources and need not hire more people to develop, run and maintain complex models and simulations.

22.2 Small Space Systems Capabilities and Applications

22.2.1 Abilities of Small Space Systems

Smaller space vehicles can support only one or two features of the most capable systems; larger systems dominate when we need many features. Thus, we will survey what small spacecraft can do with the caveat that we must normally customize them to each user's requirements. Although we can probably increase a parameter's performance if an application depended on it, we usually gauge what is possible against what has already been done or, at least, is in development.

Table 22-2 lists some of the common guidance and control techniques that have special merit for small-satellite programs. Chapter 11 discusses satellite stabilization in more detail. One option—no stabilization hardware at all—is a simple and therefore attractive alternative for small-satellite applications. To achieve downlink margin and adequate power in any attitude, satellite antennas need to have spherical coverage and solar cells must be distributed over the entire satellite surface. Many conventional satellites require gain antennas for two reasons. First, the satellite is in a high orbit and hence distant from the ground station. Second, the telemetry rates required by multiple on-orbit operations increase the load on the radio link. We do not need antenna gain or stabilization to operate small satellites only a few hundred kilometers above the Earth's surface with a single, low-data-rate mission.

Passive stabilization, either aerodynamic, magnetic or gravity gradient, is often used to minimize cost and complexity. Aerodynamically stabilized satellites are simple but must be in very low orbit to be effective. Because orbital decay shortens the mission life to less than one year, we should use it only in low-cost systems. Gravity gradient torques can passively stabilize a satellite in an Earth-pointed orientation. Small satellites have used this configuration for increased radio-link gain and imaging of the Earth's surface. Permanent magnets may also stabilize a small satellite by aligning it with the Earth's magnetic field. This technique often combines with spin maintenance schemes using solar radiation pressure. Completely passive and highly reliable, the technique can also allow about 3 dB of link gain.

TABLE 22-2. Guidance and Control for Small Satellites.

Technique	Typical Performance	Advantages	Disadvantages	Example
Unstabilized	NA	Simple	No gain in antenna or solar array	GLOMR
Aerodynamic	±10 deg	Aligned to velocity vector	Very low orbits only	CRO
Gravity Gradient	±10 deg	Earth-oriented	Damping and upset problems	UoSat1
Passive Magnetic	±30 deg	No active components	Limited to magnetic field alignment	OSCAR 4
Earth-oriented Spinner	±5 deg	Earth-oriented for part of orbit	Requires active control	OSCAR 13
Sun Spinner	±2 deg	Best use of Sun	Requires active control	ALEXIS

Because most small (and conventional) space vehicles are not oriented to the Sun, and because articulating solar panels are costly and complex, little electrical power is typically available in small satellites. But power has not typically constrained the design of small, low-cost spacecraft because they:

- Incorporate power management as described below

- Experience only intermittent contact with the ground station during typical LEO orbits

- Depend on the small satellite's large ratio of surface area to volume—i.e., since power consumption scales with volume (mass) but solar power by surface area, power requirements drop more quickly than the power available as size scales downward.

Typical power management measures include:

- Using low-power devices such as complementary metal-oxide semiconductors (CMOS) wherever possible

- Operating digital components at slow clock speeds to minimize power consumption

- Duty cycling all components not requiring continuous power

- Considering directional ground station antennas to reduce transmitter power requirements

Most power systems employ standard spacecraft solar panels and either NiCd or lead-acid batteries. (UoSat E was the first small satellite to incorporate GaAs solar panels as a means to increase available power in a miniature satellite.) Often, to control cost, we fly commercial-grade commercial batteries in several parallel stacks. To regulate charge, we can use simple current control or more advanced, digitally controlled, highly efficient circuits, depending on how much performance we must squeeze out of a system. Much of the risk of employing new, innovative solutions is unexpected effects elsewhere in a complex system. The simplicity of low-cost spacecraft lowers the risk associated with innovation. Thus, they often are first to employ new technologies such as was the case with full-time digital charge control.

Power available in contemporary low-cost satellites ranges from a few watts in unstabilized, Hitchhiker-sized spheres such as NuSat to over 60 W in Sun-stabilized vehicles like ALEXIS. While scaling arguments favor small vehicles, the absolute value of the steady-state power produced in LEO is typically small. Duty cycling usually satisfies a payload's need for more power. Where this is not possible, the power requirements can drive up costs, because the satellite must provide more solar panel area, and may need articulating solar panels when the spacecraft surface is not large enough.

Small satellites have provided superior information processing at a very low price. Small satellites with prices under $5 million are equipped with DSP and Pentium and Power PC microprocessors. Using static RAM or very low drain DRAM, the satellite can supply several Gbits of solid-state memory to buffer data between ground station passes while consuming less than 5 W of power for all systems combined. Satellites weighing under 20 kg have given us a full range of onboard signal processing: data compression, error encoding, as well as data checking and decoding or encrypting and deciphering algorithms.

Small, low-cost systems have flown with various telemetry and communications systems, including VHF, UHF, and microwave links using FM and digital transmission ranging from 300 baud to 1 Mbaud. To obtain digital rates up to about 9,600 baud duplex, we typically use nondirectional antennas on the ground and on the orbiting platform as well as NBFM (Narrow Band FM). The short slant range to a satellite in LEO has allowed us to develop ground units as small as a pocket calculator, which can uplink and downlink digital messages. At the same time, directional dishes operating at the S or X band can satisfy missions with requirements for very high data rates. Maintaining low program cost depends in part on the designer's ability to minimize data rate and bandwidth, which drive costs on the ground and on the satellite. As the data rate increases, we need more expensive components and new systems elements, such as a steerable dish capable of the high angular rates of LEO satellites observed from Earth.

As discussed in Sec. 22.1, launch costs do not change linearly with payload mass. In smaller vehicles, the actual cost will depend on negotiations with other payloads comprising the full launch payload, because even a very small payload is a significant fraction of the total. On larger vehicles, flying as a secondary, space-available payload may allow lower-than-standard pricing. Special launch services, such as the Shuttle, Hitchhiker and Ariane ASAP programs, fix prices up to a maximum allowable carrier capacity. To spend less money overall, we need to survey existing launch resources and build to match the largest number of candidate vehicles. In this way, we can get a quick, cost-efficient ride on the first launcher with available space.

Higher prices per kg are associated with small vehicles such as Pegasus and Shavit because economies of scale favor larger launch vehicles. However, because a small payload may occupy all or at least a major part of a small vehicle, the small payload operator has more control over launch schedule and final orbit than when purchasing secondary space on a larger rocket. Secondary payload accommodations being offered by the Hitchhiker and ASAP programs are priced mainly to cover basic administrative costs. The Space Shuttle programs offer an unusual value if we can live with the safety, mass, and orbit limitations of that program. Ariane and Delta offer adapters to carry many small payloads, and Ariane has launched several AMSAT satellites. These vehicles have well-established launch records and pricing structures well below small, dedicated launch vehicles. Both Ariane and Delta offer launches for commercial payloads.

22.2.2 Emerging Miniature and Low-Cost Technologies

Small satellites depend heavily on increased ability to compute and store data using low power. An 8.5-kg microsat's capability today was unavailable in the 1/2-ton satellites of 15 years ago. Digital communications and large data buffers ushered in the store and forward operating mode, which dovetailed with the communications architecture natural to LEO satellites. Advances in these same areas may further benefit small satellites, but our fear of risk slows applications of new technology to space. Particularly, concerns over the effects of radiation on integrated circuits in the space environment have retarded the transfer of digital technologies to miniature satellite devices. This partly explains why, as terrestrial machines transitioned to Pentium and Power PC microprocessors, the first 8086 derivative was not used in orbit until the 1990s. Thus, we have plenty of room to apply advances in integrated-circuit technology to miniature devices.

Whatever progress is made in data manipulation and storage devices, many payload devices will still have intrinsic power requirements which are hard to reduce. These include radio transmitters, optical beacons, active coolers, and guidance and control hardware such as magnetic torque coils and momentum wheels. Better conversion of solar to electrical power will enable advances in miniature technology devices now relying mainly on silicon-based photovoltaics and NiCd batteries. GaAs photovoltaics are becoming common on small spacecraft with even more efficient cells now appearing on the horizon. The major effect of improvements in technologies for photovoltaic and energy storage will be to increase the scope of orbital functions which miniature satellite devices can perform.

Often up to 20% of spacecraft mass is batteries. Lithium ion secondary batteries are now transitioning into microsatellites to reduce mass and improve performance. Momentum wheels are now scaling down in size, mass, power and cost to accommodate smaller satellites.

Advances in focal plane technology already allow very small satellites to carry digital imaging over a range of detector wavelengths. Sensitive, dense focal planes can enable high-resolution imaging with reduced objective lens diameter and poorer pointing stability. Advanced techniques for compressing image data, as well as increasing density of digital memory, will ease the burden of acquiring and storing image data for later transmission to Earth. Commercial organizations and countries now recognize that a low-cost satellite can perform meaningful imaging. This has spawned the startup of new commercial imaging companies serving specific market niches, as well as numerous new government-sponsored remote sensing satellite projects.

As we continue to reduce electronics size, the aperture requirements will increasingly determine satellite size, mass, and, ultimately, cost. Particularly, high gain antennas, very fast, high resolution optics, and solar power collection, require large apertures. Thus, a key technology will be development of low mass/low cost deployable optics, solar arrays and antennas.

A satellite on orbit is a type of robot. It is a device which carries out various physical activities under control of an autonomous, synthetic controller. As terrestrial robotic devices become more capable, it is logical that satellites will become more capable and more autonomous. This would enable new applications involving extended periods of autonomy (perhaps for missions outside of the Earth-Moon system), intelligent interactions with other space and terrestrial objects, and coordination of the behavior of large numbers of small satellites.

While the Defense Advanced Research Projects Agency, the Air Force, and NASA try to develop a new generation of launch vehicles for small payloads, no launch technology breakthroughs appear to be on the horizon. Organizations are working very hard to create marginal savings in launch costs, and conventional, large vehicles remain the most cost-efficient way to inject a satellite into orbit. We expect no major advances in launch technology in the next decade.

Until a few years ago, advanced designs for spacecraft architecture focused almost completely on conventional technology devices. The recent attention paid to systems composed of smaller, less costly components on-orbit is likely to produce new applications better suited to them, such as missions requiring a network of detectors distributed in orbit around the Earth or Sun. Previously, such concepts were dismissed because building and launching many conventional satellites into different, highly energetic orbits are expensive. Reducing launch mass by an order of magnitude or more may make these missions economically feasible.

In general, anticipated technological advances in information processing, solid state detectors, autonomous intelligence and power conversion will all increase applications of miniature technology. As we discover new systems architectures and develop new applications for satellite systems, miniature technology will take on heightened importance in future space programs.

22.2.3 Potential Applications

An array of satellites in random, low-Earth orbits can provide fully connected, continuous communications, as Fig. 22-2 shows. This fact has spawned numerous satellite programs which divide into the Big and Little LEOs. Big LEOs offer real-time connectivity, mainly for telephone but ultimately for very high bandwidth applications. Little LEOs are one- and two-way digital message carriers, used for asset tracking, monitoring and paging. All of the systems currently contemplated are constellations, meaning the spacecraft maintain constant position relative to each other. The number of satellites in the constellation ranges from 5 or 6 in the case of some Little LEOs servicing limited areas or with long *latency*—the time between passes over a user location. Big LEOs with constellations of 24 to over 240 satellites are in development. In general, larger numbers of satellites can provide higher bandwidth. Some of the systems use inter-satellite links to forward signals beyond their own footprint. Others immediately relay communication from the user to a ground station which feeds the data via the terrestrial network.

Unlike the constellations, satellite clusters consist of large numbers of satellite randomly distributed in their orbit planes without propulsion to maintain fixed relative to positions. These satellites are smaller and simpler than those in the constellations. They can be fit into available space on virtually any LEO launch. In time, such a network would become fully populated. Computer modeling of clusters shows that 400 satellites in random orbits provide 95% global coverage and 100% coverage from other orbiting platforms. Thus, the satellite cluster could support both point-to-point ground communications and satellite-to-ground links. One significant feature of satellite arrays, either constellation or clusters, to worldwide communications from the military viewpoint is its intrinsic survivability. Destruction of 1 or even 20 of the cluster's members barely affects the network's overall effectiveness. Presumably the small, simple, mass-produced satellites of the cluster would be less expensive than the weapon required to destroy them. Their small size would make them intrinsically difficult

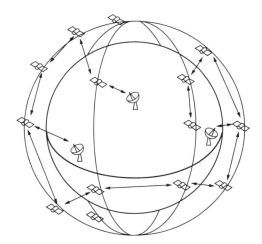

Fig. 22-2.   Satellite Cluster. Many small satellites in randomly distributed low-Earth orbits can provide point-to-point global communication.

to track and would also permit fabrication from materials transparent to radar. Their intrinsic redundancy and graceful degradation justify applying inexpensive, single-string design and fabrication techniques which will further reduce cost.

Another application of small satellites is for low-cost imaging. These systems use advanced focal-plane technologies to obtain fine optical resolution with relatively simple guidance systems. By using multiple satellites in clusters or constellations, frequent image updates can be combined with good ground resolution. Smaller, low-cost satellite can be optimized to specific applications such as agriculture, coastal zone management or land use and taxation. Several businesses and countries are developing systems that will eventually eliminate the monopoly of a few large government-owned systems on optical space surveillance of Earth-bound activities.

AMSAT and several commercial and government organizations are developing or using small satellites in geosynchronous orbits. AMSAT plans to supplement existing global digital communications from LEO satellites with real-time digital and analog communications. Indostar is the first of a new generation of geosynchronous satellites supporting only a few transponders. These small geosynchronous comsats could serve as on-orbit spares. Moreover, they appeal to smaller corporate and national users which cannot themselves support a conventional GEO comsat. These users must now either lease individual channels of a large satellite or become a member of a conglomerate such as Intelsat. Using miniature satellite devices with lower capability and price, smaller users can own and hence control the entire space asset, increasing their autonomy and security. Though GEO launch costs traditionally are high, small satellites can hold down those costs by piggy-backing on a GEO neighbor's launch, if available. However, few small satellites can take advantage of the GEO orbit.

Measurements of rapidly varying fields over astronomically significant baselines is impossible with a single satellite. By the time the satellite flies across the region of interest, temporal variations in the field distort the map. By flying tens to hundreds of very small satellites in varying trajectories, we can observe such phenomena as the charged-particle environments and magnetic-field variations of the Earth and Sun.

Particularly for solar observations, the desired trajectory energy is quite high, so we need low satellite mass. Small satellites are ideally suited for this class of mission—requiring many spacecraft inserted into energetic trajectories. Due to the scientific nature and high launch costs of such missions, they are best suited for NASA and other national science and research organizations.

22.3 Applying Miniature Satellite Technology to FireSat

To show how the small satellite conceptual process works, we will look at the application of miniature technology to the FireSat problem of detecting and monitoring forest fires either nationally or worldwide. This system is challenged to provide rapidly updated data on a firefront's genesis, topology and local progress to the Forest Service's central office. A network of low-flying reconnaissance spacecraft could resolve the firefront to a few meters through multispectral imagers. Then the satellite could find the spread rate by staring at the front during a pass and applying image processing techniques to the downlinked data. Direct broadcast to the field through a geosynchronous comsat would forward the derived data to the firefighting teams. Such a system would be highly complex and costly. We believe that this system does not exist because conventional satellite technology cannot meet the user's recognized requirements at an affordable price.

Serious forest and brush fires often begin with a simple match or a spark that is hard to see from an orbiting platform. One potential solution is to fly very low-resolution imagers filtered to the near IR. Data on nascent fires would be downlinked to a simple ground station which could reconstruct the low-resolution images and compare them with ground truth, such as the locations of large cities and other bright areas. The low resolution and subsequent low bandwidth will allow use of PC-based ground stations like those already built for earlier small imaging satellites. This solution parallels the conventional technology approach, but with minimum capability.

A more innovative solution would deploy simple thermocouple sensors from aircraft, with small nets to catch them in treetops, where forest fires spread. Fitted with a lithium primary battery and a 1-W uplink transmitter, the sensor could simply turn on (much like an Emergency Locator Transmitter) when its temperature reaches, say, 80 °C, broadcasting its digitally encoded serial number to a simple small satellite for digital messaging.

After discovering a forest fire, monitors would dispatch an aircraft to the firefront to drop a denser network of more sophisticated sensors. These sensors could transmit the local temperature and their own ID number to the aircraft, which would maintain a map of the sensor locations by recording them as each sensor is dropped. As the satellite overflew, it would receive the signals of the upgraded sensors, causing it to rebroadcast a map of sensor ID numbers and temperatures to the local firefighting crew. Their ground station would receive a map of sensor ID numbers and locations, so they could immediately derive a temperature contour map. Figure 22-3 shows the complete system.

A simple treetop sensor would be inexpensive because it consists of only a beacon, a battery and a thermocouple. Dispersing the sensors could occur during routine transportation and patrol flights because the sensors are merely scattered. The satellites serving this application would be very small, simple devices, which any sort of launch vehicle could launch. All firefighting crews would need small, portable ground stations, and the aircraft deploying sensors would use GPS to record location when

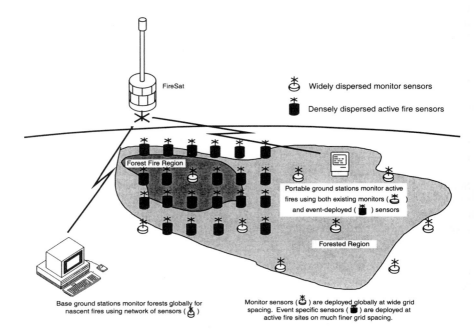

Fig. 22-3. FireSat System Concept. The miniature satellite concept for a FireSat system minimizes on-orbit cost by using an array of sensors for monitoring and for relaying information about particular fires to firefighters.

dropping each high-resolution sensor. A laptop computer would allow an operator to enter data as the sensors drop. The ground control stations consist of little more than the field units, a small computer, transmitter, receiver and omnidirectional antenna.

The system described above would provide the following:

- Global monitoring of new forest fires from a few ground stations located anywhere on Earth

- Ability to locate firefronts to treetop resolution after identifying a fightable blaze

- Communication of the fire progress directly to the field commander

With this approach, we benefit from minimal on-orbit requirements, cheap replacement, and built-in redundancy. With several satellites on orbit, if one fails, the system degrades only slightly. Any launch to LEO can potentially rebuild the system. Suppliers of miniature satellites can put needed spacecraft on orbit in under a year, and the system could be operating 18 months after contract go-ahead.

Note that in this example miniature satellite technology is considered a systems architecture discipline. This designation is accurate and intentional. Miniature technology is not a special set of technologies used for building conventional capabilities into small, cheap boxes. It is a new way of looking at an application to develop a solution which doesn't require conventional technology and which a closely-knit team of under 20 satellite engineers can handle. Using miniature technology architecture, a spacecraft needs only modest capabilities to detect and monitor forest fires. The satellites need only listen for beacons and crudely locate them to within a few kilometers or tens of kilometers. In the second mode of operation, at the firefront, the uplinked data from even a few thousand beacons is much less than 1 Mbit.

The 13-cm Microsat cube developed and flown successfully by AMSAT North America and shown in Fig. 22-4 is an example of the type of miniature technology device which can meet all of the satellite requirements for the FireSat mission. The microsats weigh only 8.5 kg and are fitted typically with 8 Mbits of RAM. The unstabilized satellites carry omnidirectional antennas providing sufficient link to recognize a low-bandwidth, 0.5-W beacon at the Earth's horizon from 800-km orbit.

Fig. 22-4. **MicroSat Cube Developed by AMSAT North America**. This 13-cm, 8.5-kg satellite first flown in 1990 could meet the space segment requirements for the FireSat concept shown in Fig. 22-3.

Even in small quantities these satellites can be produced for less than $400,000, because they are so simple. By using several satellites, the system itself is redundant, so each satellite can be simpler, less reliable, and of a single-string design. Lower reliability requirements allow us to procure commercial-grade components without special ordering, testing or quality assurance. With this method, we can buy parts cheaper and design more efficiently by using more modern, capable components. We can also cut engineering time spent finding qualified parts and working with vendors to meet program specifications.

Simple mission requirements allow the satellites to be quite small, thus greatly reducing launch and ground-support costs. In fact, we can transport them to a test or launch site by commercial airliner as carry-on luggage.

22.4 Scaling from Large to Small Systems

The fact that miniature technology devices tend to be physically small means that physical scaling laws will account for some fundamental differences between miniature and conventional satellite devices. A conventional device can also be physically small; this section addresses only the engineering differences which become significant when we develop a very small spacecraft.

As illustrated in Fig. 22-5, scaling up a simple cylindrical satellite by factor 5 in linear dimension increases its projected area by 25 and its volume by 125. If we assume that mass and power consumption are roughly proportional to the volume of electronics, this simple geometric scaling has implications for the systems designer. As satellite size decreases, its power requirement decreases faster than its projected area. So smaller satellites typically do not need deployable solar panels.

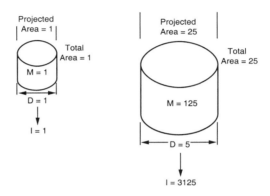

Fig. 22-5.   Effect of Scaling Up a Simple Cylindrical Satellite. *D* is the linear dimension, *M* is the mass, and *I* is the moment of inertia.

As we have mentioned, physically short thermal and load paths characteristic of smaller satellites usually allow less critical thermal and structural design. The small satellite of linear dimension 1 has to conduct 1 unit of heat flux over a maximum distance 0.5. The larger satellite must conduct 125 units of heat flux over a maximum distance of 2.5, requiring roughly 125 times more temperature difference between the satellite surface and interior. Presumably, thicker structural members in the larger satellite partly offset this difference, but satellite thermal considerations rarely affect structural design very much. Similarly, supporting the satellite from its edges (for example) results in much thicker structure relative to size for a large satellite than for a small one. This advantage to small size diminishes somewhat because wall thicknesses in very small devices often depend on machining and handling limits. Thus, we cannot build the small satellite to theoretical structural limits, so it tends to be heavier and a better heat conductor than it would be if designed optimally for strength.

The thermal and structural oversizing typical of small satellites tends to affect programs positively and negatively. Miniature technology devices tend to be built to exceed specifications, so they require very little analytical effort to ensure their structural and thermal integrity compared with conventional technology devices. Also they rarely require special, costly thermal and structural materials and devices such as high strength alloys, composites, or heat pipes.

On the other hand, miniature technology devices support the operational payload on orbit less efficiently because of design expediency and difficulty in manufacturing very thin, light, small structures. As Fig. 22-6 illustrates, OSCAR 13, a miniature technology comsat, has much more of its mass devoted to support functions such as structure, guidance, and propulsion than the geosynchronous comsat, Palapa B, which uses conventional technology. Only 21.5% of OSCAR 13's weight carries payload electronics compared with 34.5% of Palapa B's. We cannot precisely compare large and small satellites because their missions and operating conditions are so different. But Fig. 22-6 shows that the key to miniature systems is not a new technology or trick which allows its practitioners a special advantage over conventional systems. It is a systems discipline which allows its engineers to meet a user requirement cheaply and quickly despite its inherent disabilities.

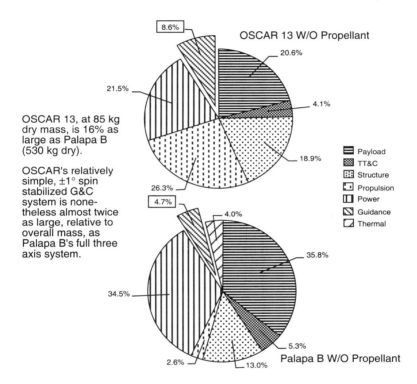

Fig. 22-6.   Comparison of Satellite Mass Efficiency. Comparison of OSCAR 13 with Palapa B illustrates the difficulty in scaling spacecraft G&C systems for small satellites.

22.5 Economics of Low-Cost Space Systems

To survive as an industry, low-cost space systems must be more cost effective than space- and ground-based alternatives. To confirm savings, we must verify that we can complete and fly hardware successfully at low cost. The value to the user must be at least as large in proportion to cost (including accounting for risk) as competing alternatives, such as conventional technology.

Experience in some of the more recent small satellites illustrates the range of costs which can be expected in a small satellite program. Table 22-3 lists approximate program costs for some small satellites (in 1990 dollars).

TABLE 22-3.   Cost Experience for Selected Small Space Systems.

Satellite	Mass (kg)	Stabilization	Developer	Approx. Cost	Comments	Year
OSCAR 10 & 13	100	Spin	AMSAT	$250,000	All volunteer engineering staff	1981
CRO	70	Aerodynamic	DoE/USAF	$1M		1980
GLOMR	68	None	DARPA	$1M	Cost includes ground station	1986
ALEXIS	45	Spin	DoE	$3.5M	Cost includes ground station	1993
UoSat E (typ)	61	Gravity-gradient	U. of Surrey	$1M	Commercial/University Cooperative	
Microsat	8.5	None	AMSAT/Weber	$200,000	Student and faculty support	
Indostar	600	3 axis	Orbital	$150M	Preliminary Cost Estimate	
SAMPEX	158	Momentum Bias	GSFC/NASA Small Explorer	$35M	1st Small Explorer Mission	1992
Clementine	424	3 Axis	NRL/LLNL/ BMDO/NASA	$70M	Lunar imager & first detection of lunar ice	1994
ASTRID 1 & 2	27	Spin	Swedish Space Corp./SNSB	$1.4M*	Earth & Space Physics	1995& 1998
HETE	120	3 Axis	MIT/AeroAstro	$12M**	Gamma ray burst detection. Launch vehicle failure. Awaiting reflight.	1996
NEAR	805	3 Axis	JHU/APL NASA	$150M	Asteroid Rendezvous	1996
SAPHIR-2	55	Gravity-gradient	OHB System (Germany)		Communications: messaging	1998

*Includes launch
**Includes Science Instruments, BUS and Multiple Ground Stations

We can easily survey small-satellite programs and demonstrate their low cost compared with conventional satellites. Understanding the basis for this cost difference is more complex. The cost pyramid in Fig. 22-7 is one way of explaining the dramatic difference between miniature and conventional costs for a small, astronomical satellite proposed as a university research program to NASA. The pyramid tracks cost growth in translating a hypothetical miniature satellite program to a conventional program with similar operational specifications. The pyramid topology is apt because each extra cost adds to the previous cost. Thus the cost growth is more geometric than arithmetic.

Starting with an initial satellite cost of $2M, we would establish a dialogue with the scientific user group to try to reduce program cost while maintaining the scientific value of the mission. This vertically integrated process leads us to modify operations by relaxing the pointing accuracy, thus saving considerable money at the second stratum of the pyramid.

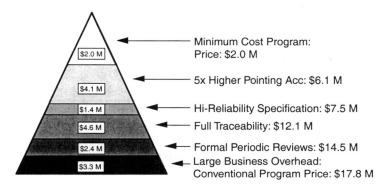

Fig. 22-7. **Conventional Satellites Impose Significant Costs over a Low-Cost, Small Satellite Program.** In this example based on design studies of a small astronomical satellite, the development cost is $2M compared with ~$18M for a conventional satellite for the same mission. Additional costs for test, ground station, and on-orbit operations will further widen the cost gap between the two approaches.

In this case, the user group had a fixed budget ceiling. Thus, as is often true in miniature satellites, the reliability trade off reduces to a choice between controlling cost (at increased risk) and cancelling the program. The choice to accept industrial-grade components saved ~$1.5M compared to space grade. Because small satellites have 10% to 1% as many parts as larger spacecraft, part failure is less significant to achieving mission reliability. Thus, small satellites routinely employ less expensive, more modern and more readily available commercial parts saving money via parts cost, shortened schedule and simplified design. In some cases, contractors cannot easily purchase individual space-grade parts, so they must buy parts in lots and test them, or use a more poorly suited component. Both of these options—in-house qualification or redesign around poorly suited components—cost much more money.

Component traceability strongly drives up costs, because it requires a program infrastructure, thus driving up direct and overhead costs. Direct costs increase partly because subcontractors must attend to traceability, thus increasing the cost of supplied components and assemblies. Traceability, like space-class specification, eliminates many potential components, because mass-produced devices not already made to space-class are not traceable. Industrial grade NiCd batteries are typical of this dilemma. Only MIL grade and above—ten times more expensive—are traceable.

Extensive periodic program reviews are of questionable value at all project scales. But a small team working on a miniature satellite project can save a lot of money by substituting regular communications for formal reviews. Of course, we can keep the team size down if we can avoid those earlier requirements, such as higher pointing accuracy, S-class parts, and traceability. A program review for any sized program has direct and hidden costs such as travel, time spent in meetings, preparation time, and misdirection of the hardware team towards generating paper rather than making hardware. But the penalty is more significant for a small program. Budget is often severely constrained such that airline tickets and hotel costs are significant. But more importantly, efficiency of small spacecraft development is in part rooted in the small size of the team. Additional burdens including numerous reviews and part tracking can inflate team size beyond that allowing very rapid, informal interactions or that encourage optimal resource allocations.

Although the exact numbers in the cost pyramid are all estimates derived from a study of system architecture, present experience with miniature and conventional programs and contractors supports them. The interest in missions using miniature satellites directly relates to this cost difference.

The above arguments notwithstanding, we still must show that particular missions can be performed more cost effectively with miniature technology than with conventional technology. Clearly, to achieve cost and schedule benefits, users of miniature technology must sacrifice some performance but also realize important gains in schedule. We cannot compare conventional and miniature products purely with numbers, because they differ qualitatively. But the following example illustrates how miniature technology can save money and increase reliability of an overall system.

A 50-kg satellite can be launched piggyback for between $50,000 and $1M, depending on the launcher. The Pegasus launcher has enough payload to launch about 10 satellites of 50 kg each for $1.5M each. When we have invested little in the launch, we can tolerate a less reliable satellite. A 90%-reliable satellite built for $1M plus a $1M launch investment (on-orbit cost of $2M) has an expected cost to 100% reliability of

$$\$2M \div 0.90 = \$2.22M$$

That is, to get 9 satellites working on orbit, we will need to build and launch 10, so the effective cost per satellite increases incrementally. As mentioned earlier, engineering to increase reliability in a single satellite costs a lot of money. Numerous subsystem assemblies and more complex control systems drive up these costs. The most reliable components can cost many times more than their commercial counterparts. A 28V stack of space-qualified NiCd batteries with 40 W-hr total charge costs about $40,000 and weighs 10 kg. The best commercial technology available from the same supplier costs $180 and weighs 4 kg. At $20,000/kg launch cost, the total savings is $159,820. Thus, the cost ratio between high and moderate reliability systems is 3:1.

Of course, we should not use small satellites when we cannot reduce the payload's size, mass, or support; the satellite will require a large fixed investment and a costly launch. For example, if the FireSat needs an imaging system with an optical objective of 1-m diameter, fitted to a steerable platform with arc-minute accuracy and stability, present technology would make the system's mass well over a ton. Launch costs will be near $40M, and payload costs could be equally large. With $100M invested in the system, the value of increasing reliability from 90% to 95% is $5.9M. Thus, adding redundant systems to roughly double reliability can be quantitatively justified if they increase costs less than this amount. Miniature technology is not appropriate for such a large mission, regardless of reliability. For one thing, a group of people each charged with understanding most of the total system could not fabricate such a large system. The more bureaucratic approaches of conventional technology will work better in this case.

We also cannot accurately measure the cost of failure simply by quoting orbit-system cost (satellite plus launch vehicle). Some payloads are worth more than their dollar price, the prime example being crewed vehicles. The value to society of preserving life is very high, particularly in a public government activity during peace time. Also, even if we accurately measure the space component's cost, we can still underestimate the real cost of a failure. Many missions require coordination with other valuable assets. Deploying and reconfiguring of space and ground assets for a space test can require more financial commitment than the space vehicle itself. Lost opportunities are also quite costly. If a single, large surveillance satellite fails on orbit, we may not secure continuous observation, thus losing politically vital data that is more valuable than the on-orbit asset.

Ultimately, we must associate a cost with on-orbit failure, assessing several levels of reliability in terms of program cost, complexity, and schedule. We can then decide whether to build to reliability requirements. In comparing a spectrum of candidates, we must remember that cost and reliability may not relate directly. A much smaller, simpler spacecraft, built by a very small group whose members are familiar with the whole system can be more reliable than a much more complex solution despite the latter's more reliable parts and redundancies.

Other non-economic factors also play a role in selecting conventional over miniature technology. Risks in program management often deter designers from applying miniature technology. The person who must deploy a one-of-a-kind space vehicle may know that the final price will be unimportant so long as the mission is a success. Or conversely, he or she understands that a failure, no matter how cheaply executed, is still a failure. Wise program managers use miniature satellite methods for small, relatively simple applications, not for highly complex missions relying upon large teams. Launch of many satellites also better follows probability distributions.

We should not, however, overestimate the importance of single string design to the cost advantages of small satellites. In fact, many small satellites have incorporated highly redundant architectures. ALEXIS includes a highly redundant digital system, a power system which is quadruply redundant and 3 parallel payload systems. DSI's MacSats were virtually fully redundant.

Also important is the application of low-cost approaches to subsystems. The example mentioned above of substituting commercial NiCd batteries for space-qualified ones can save up to $1M in the cost of a 200-kg small satellite. ALEXIS carries two custom Sun sensors built by AeroAstro, each for about 10% of the cost of existing space qualified units. Because they use less power, and are smaller and lighter than Sun sensors designed for larger satellites with more demanding performance and quality specifications, their cost savings ripple through the entire spacecraft bus design. The net savings in using simpler Sun sensors is estimated in the case of ALEXIS to have been several hundred thousand dollars.

Because the torques required to stabilize a small satellite are small, use of costly iron core torque coils is often not necessary. Instead, much simpler and lower cost air core coils can be used. Typically this can result in $50,000 in savings for a small satellite. When the design is correctly accomplished, a small satellite is not structure-limited. In fact, Martin Sweeting of Surrey Satellites has said that the difference between large and small satellites is whether a distinct structure is required. This is because small satellite characteristic lengths are short, and structure is usually over-built due to manufacturing constraints (metal needs to be thick enough to support fasteners and thereby becomes thicker than necessary for purely structural considerations) or because the mass savings of a weight reduction program aren't significant. In any case, there is a savings both in that higher cost materials, including composites, are generally not used, and little structural analysis is required.

Very significant cost savings are realized by specification of commercial, instead of military or space, grade components. Savings result from several benefits. Of course Mil-B and S-class parts are quite costly—sometimes 10 to 100 times more than the equivalent commercial part lacking the qualification inspections and paperwork. They are rarely available in small numbers, since they are built and tested in separate production runs. One integrated circuit, available for $50 commercially, may cost $500 in S-class, but may only be available in lots of 20—increasing the purchase cost to $10,000. Perhaps more importantly, the most modern technologies are seldom available in higher grade components, forcing a compromise to inelegant design solutions. The HETE spacecraft, built to commercial specifications, takes advantage of several modern, radiation-hard, semi-custom, integrated circuit components which eliminate hundreds of S-class integrated circuits from the parts count. Besides reducing parts costs by tens of thousands of dollars, the modern technology reduces the number of circuit boards, greatly reduces design, development and test labor, and requires much less power. The savings in spacecraft resources of power, space and mass, combined with the parts and labor cost savings, make the implementation of this commercial technology worth hundreds of thousands of dollars in savings to the $4M spacecraft budget.

Some components cannot be significantly altered for use on small satellites, compared with conventional designs. As an example, spacecraft photovoltaics (solar panels) have no market except spacecraft. In these cases, we try to work with traditional vendors to find ways to decrease costs. Paralleling the spacecraft design approach, these may include using flight spares developed for other programs, reducing non-hardware deliverables (meetings and paper) and interactive design to produce a design specification which is intrinsically inexpensive to build and test. Israeli Aircraft Industries' MLM division has succeeded in producing spacecraft photovoltaics at less than half the cost per installed watt of conventional spacecraft photovoltaic systems, without any decrease in product quality or performance through application of these steps.

Formal engineering guidelines institutionalize conventional satellite technology. The manager of a hardware development program that fails will not have to account for the failure if he or she documents the program thoroughly and builds the system to military specifications and Department of Defense guidelines. But program managers using miniature technology employ untried components when it is cost-effective and they perceive little risk. Yet, if their programs fail, they are accountable.

Thus far we have concentrated on two important motivators for application of small satellites—cost and unique capabilities. But small satellites, owing to their simpler architecture, the smaller team required for their development, and the smaller amounts of money required, can be built on very rapid schedules. AMSAT, motivated by the availability of a near-term launch slot, produced and flew a small satellite in 9 months. With typical development time as short as 18 months, the spacecraft development schedule is usually dominated more by the bureaucratic delays in getting a program started, than by the time to engineer, build, test, modify, retest and deploy a small satellite.

Figure 22-8 shows that besides the quantitative shortening of the development schedule, small satellite development is a highly interactive process. Activity begins when a particular mission is identified—in the case of Firesat this could be the mission to detect forest fires. In discussions between the user community and the development group in the ensuing 30 days, several very low resolution sketches of possible satellite configurations can be developed, each tailored to different launch vehicles, different size, mass, cost constraints, and different performance levels. This helps users to understand the impact of their budget on the ultimate capabilities they can achieve. If there is no way to get something of value for the user within the user budget, the program needs to be reconsidered and certainly there shouldn't be additional resource expended on detailed design until at least a tangency between the cost and utility curves is achieved.

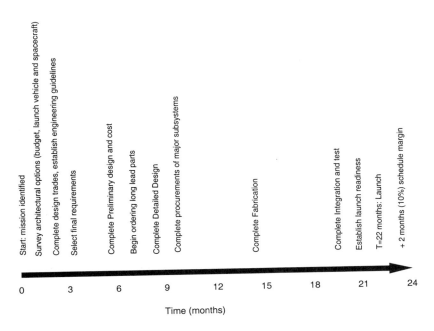

Fig. 22-8.   Small Satellite Development Program Timeline.

With a rough-order-of-magnitude budget in mind, planners can create a strategy of spacecraft size, launch vehicle and capability range. This loose set of guidelines then forms the basis for a conceptual-level, engineering trade study. This activity is generally carried out over a month or so with a team of 1, or at maximum 2, systems engineers with help from various subsystems specialists on an as-needed basis. The systems engineer should communicate frequently, no less often than weekly, with the user organization representative to review progress—particularly to discuss ways in which relaxations of requirements can reduce cost and to explore newly uncovered opportunities to provide additional capability without seriously impacting system complexity.

Only after the systems engineer and the user representative have explored this envelope of architectural options and the capabilities envelopes, do we draft a final requirements document drafted. This is important. We do not achieve the lowest possible satellite size and mission cost by a priori assignment of requirements to the satellite design. Just one especially difficult requirement, possibly not vital to the overall mission, can result in a costly and unwieldy design. Requirements and capabilities should be freely traded to reach a more global optimization of cost and complexity minimization coupled with achieving the best systems performance.

Equipped with an interactively defined set of specifications and detailed sketches of the spacecraft layout, along with weight and power budget estimates, the preliminary design process gets underway. Even at this very early stage, there will be components which everyone knows will be needed. These might include an oversupply of NiCd batteries for later selection or the photovoltaic cells necessary to build up solar panels. Also, bids can be solicited for guidance sensors and actuators based on estimated requirements. Also put on order any parts needed for building test fixtures or circuits. Even in a small, inexpensive satellite program, the value of beginning the hardware flow is high and the cost of parts ordered which end up unused is small—usually it is nearly zero. By ordering that which can be ordered as soon as it is identified, we spread the parts procurement process more evenly over the program lifetime. If certain components are unavailable or won't work in the proposed application, it's important to discover the problem as soon in the design and development process as possible.

We allocate the entire preliminary and detailed design process about 7 months. Fabrication formally begins, though by the Critical Design Review there should be a lot of working hardware already in house. This will give substance to the CDR and also allows the remaining fabrication of subsystems to be completed in an additional 4 months.

Six full months are allocated for integration and test. To have confidence in a system built with commercial components and possibly lacking full redundancy, test is extremely important and there is no substitute for uninterrupted time using the hardware in various modes—thermal vacuum, as well as desktop routine operation.

Launch readiness is the last opportunity to resolve issues which have come up in the development process. There will always be things which the team wishes would have happened or not happened. The flight readiness review team must be made aware of these concerns and must be ready to delay the launch if it determines some deficiencies are not acceptable. Ideally the launch will take place immediately after the review to minimize the temptation to "improve" the spacecraft.

A note on program delays may help. Interruptions in any program are always much more destructive of cost and schedule than managers appreciate. In a small program even more than a large one, the documentation is at a minimum and the focus of the small team creates efficiency. To withstand frequent delays and the start-stop-restart mode of many government-funded programs, a bureaucratic system is needed to document work and communicate progress as new individuals are assigned to roles. This will ultimately result in the small program taking on some characteristics of conventional, large programs with associated lengthening of the schedule, cost growth and increased difficulty in meeting requirements, particularly staying within the launch vehicle constraints.

In summary, how do we know when to procure a miniature technology system? The following questions will guide our decision:

1. Will miniature satellite technology enable flying a mission which will otherwise be shelved?

2. Can some give and take be allowed between requirements and capabilities?

3. Is reliability achieved through simplicity?

4. Can the flight hardware fly on a space-available basis?

   a: Is the flight hardware buildable at < 400 kg?

   b: Is there no special window required for orbit insertion?

5. Can the program management organization deviate from Mil-Specs and other norms of conventional satellite development?

6. Can hardware and software be built by < 20 people?

7. Is there a significant benefit in a rapid schedule?

8. Are many small systems preferred over a few large ones?

9. Do you want the user group to operate the ground facility?

10. Is the mission lifetime goal less than 5 years?

We mentioned earlier that ground operations can significantly affect overall program cost. Often, the satellite's orbital lifetime is longer than its development period, and staffing of a major ground station can be larger than that of a development program for a miniature spacecraft. Miniature technology reduces ground-station costs through its systems architecture and the ground station's design.

A low-cost satellite program cannot afford the luxury of ground stations staffed by a separate operations group. Also, we may not be able to operate costly remote ground stations. Thus, we need to build large spacecraft memories to buffer satellite data, so the satellite can store data over long periods (typically 12 to 18 hours) until it passes over a single ground station at the user's location. We also need to compress data on the satellite, because ground stations with limited antennas and RF links may not support high data rates.

We cannot simply design the satellite to rely upon highly expert ground controllers. The user organization typically does not know the special techniques of managing a spacecraft. Thus, we need to use a single small computer to control the ground station, so the user interface is a single machine. The ground station software should be simple and well structured, containing on-line and written support. Where possible, we should graphically represent the system status and the satellite orbit.

Whenever possible, the ground station should channel the user's activity toward the right solutions. Further, the ground station should be able to screen operator commands and activities that may adversely affect the satellite operations. When deciding how much effort to expend on the ground station, we must trade the real costs of the user group learning, input errors, and more people to manage the satellite against the cost of creating more capable software for the ground station.

Autonomy is another important factor in the cost of ground operations. Satellites in LEO pass the ground station a few times per day, every day of the year. While novel in the first few days and weeks, years of tending the satellite at roughly 6-hour intervals can be a tremendous drain on the user organization. The ground station should at least be able to buffer activities for a series of satellite passes. Thus, we should consider features allowing longer autonomous periods in the system design. An alternative to extended autonomy is remote operation. A dial-in system permitting limited control of the ground station from a remote terminal can be valuable. Security is an issue in a remote architecture, but we can devise various password and call-back schemes to protect the system.

22.6 Annotated Bibliography on Low-Cost Space Systems

The Logic of Microspace, by R. Fleeter, combines an overview of the technologies underlying small, low-cost spacecraft with the management and philosophy behind their development. It also provides a discussion of potential applications of small, low-cost space systems. Published by Kluwer Academic Publishers, Dordrecht, The Netherlands and Microcosm Press, Torrance, California.

AMSAT, the radio amateur satellite organization, publishes *The Amsat Journal* quarterly and the *AMSAT-NA Technical Journal* approximately once per year. These publications report program status and technical developments from AMSAT volunteers in North America. Write to: AMSAT-NA, P.O. Box 27, Washington, DC 20044.

The only textbook on small satellites, *The Radios Amateur's Satellite Handbook*, by M. Davidoff, is published by the American Radio Relay League (1998). Emphasizing amateur radio satellites, the book provides a useful foundation in the technical basis of small spacecraft. ARRL, 225 Main Street, Newington, CT 06111.

Also in the amateur satellite area, *QST*, the ARRL's monthly, occasionally highlights particular small-satellite technologies for amateur radio. Examples are June, 1988 (Vol. LXXII, No. 6); "Introducing Phase 3C: A New, More Versatile OSCAR" by Vern Riportella. Also see May, 1989 (Vol. LXXIII, No. 5) and June, 1989 (Vol LXXIII, No. 6); "Microsat: The Next Generation of OSCAR Satellites" parts 1 and 2.

A special supplement to Vol. 57, No. 5 of *The Journal of the Institution of Electronic and Radio Engineers* was devoted to University of Surrey's UoSat-2. It is a series of papers on the satellite's design which provide an excellent view into the engineering of a successful MST device. Write to IERE, Savoy Hill House, Savoy Hill, London, WC2R 0JD.

Proceedings of the Annual AIAA/USU Conference on Small Satellites contains almost all papers presented at the annual USU meeting. It is available from: Center for Space Engineering, Utah State University, UMC 4140, Logan, Utah 84322.

TRW Space Log is a comprehensive compilation of satellites launched to date. Write to: Editor, *Space Log*, Public Relations Department, TRW Defense and Space Systems Group, One Space Park, Redondo Beach, CA 90278.

Satellites of the World (Koredewa kara Sekai no Eisei) (in Japanese) by S. Shimoseko and T. Iida contains an excellent survey of both existing small satellite programs and several detailed concept studies. Available from the publisher, Nihon ITU Association, Nihon Kemigaru Building, 7th Floor, Nishi Shinbashi, 3 Chome, 15–12, Minato-Ku, Tokyo, 105, Japan. Phone (03) 3435-1931, FAX (03) 3435–1935.

The SPIE now includes, as part of its annual program on Planetary Exploration, a series of sessions on Small-Satellite Technology and Applications. Proceedings of the 1991 meeting, edited by B. Horais, include papers on remote sensing and supporting technologies. Volume 1495 available from SPIE, P.O. Box 10, Bellingham, Washington 98227-0010. Phone (206) 676-3290, FAX (206) 647-1445.

Space Almanac contains historical data on small satellites mixed with other general space system information: Arcsoft Publishers, P.O. Box 132, Woodsboro, MD 21798.

The best sources of information on small satellite launch vehicles are the manufacturers themselves. These include Orbital Sciences (Pegasus), Lockheed-Martin (Athena), NASA GSFC (Hitchhiker onboard Shuttle) and Arianespace (ASAP Secondary Payload Accommodation).

Chapter 23

Applying the Space Mission Analysis and Design Process

James R. Wertz, *Microcosm, Inc.*
Wiley J. Larson, *United States Air Force Academy*

Since the first edition of *Space Mission Analysis and Design* (SMAD) appeared in 1989, substantial progress has been made in applying this process more broadly to space missions, in part because of the continuing pressure to reduce mission costs. With the exception of launch, costs in all segments of space missions have been reduced. Launch costs, while not having succumbed as yet, are under attack from a wide variety of directions. A large number of communications constellations are being created that are bringing a more manufacturing-oriented methodology to bear, with an emphasis on both lower cost and high quality. Even interplanetary missions are becoming quicker and less expensive. The purpose of this chapter is to summarize how best to make the SMAD process work within the real environment of acquisition regulations, programs long past preliminary study phases, and the "new space" missions which we hope will be forthcoming in future years.

Much of this book ignores two of the major challenges to doing SMAD: organizational structure and the acquisition process. Both tend to introduce political and bureaucratic obstacles to our goal of developing cost-effective designs. As described briefly in Sec. 21.1, any space program exists within a broad and important context of law, policy, politics, and economics which **we must not ignore**. We are, after all, proposing to spend large amounts of someone's money designing and developing space missions. Even the best program has little chance of success if it does not fit into the political, economic, and policy context that must support it. Politics can include, for example, the need to have manufacturing distributed around the country for publicly funded programs or use of technology from specific companies for privately funded ones. In any case, we need to work within these boundaries.

Our goal is to learn from past space programs and apply that experience, along with our judgment of how to do it better, to new and ongoing programs. In practice, we often end up doing things the way we have always done them, perhaps because it's worked or is simply easier to do than trying to change the mindset of individuals and organizations. Introducing change is difficult in any venue, but we **must** change and, indeed, may need to take more calculated risks if we are to drive down the cost of space missions.

This book develops space mission analysis and design through a series of process tables indexed on the inside front cover. This process flow divides into three main areas: high-level processes, definition of the elements, and detailed design. The high-level processes develop concepts for the mission design. With these concepts in hand, we can begin defining the characteristic elements of the space mission architecture. Then we can design those elements in more detail. Table 1-1 in Sec. 1.1 gives you simplified steps to begin designing your particular mission. Remember that success depends on iterating and continuously improving the design.

We must caution you that SMAD does not work well whenever you need much more than typical estimation accuracy. Our high-level algorithms and sanity checks provide estimates for conceptual design. These estimates are enough in most cases, but whenever performance is critical, or if you are approaching a technology threshold, you should refer to more detailed information. Areas frequently needing more than typical estimation accuracy are satellite lifetimes in low-Earth orbit, computer systems sizing, and system cost modeling. Table 23-1 outlines ways to deal with these and similar problems.

TABLE 23-1. Dealing with Areas Needing Much More Than Typical Estimation Accuracy.

1. Estimate as realistically as possible by working through succeeding levels of the process tables outlined in these chapters.

2. Estimate both upper and lower bounds.

3. Use conservative values and substantial design margin.

4. If control of a parameter is critical:
   A. Identify drivers during design.
   B. Reestimate regularly as the design matures.
   C. Develop and maintain options.

5. Be prepared to trade with overall system performance.

23.1 Applying SMAD to Later Mission Phases

Nearly all space missions must respond appropriately to the economic and political forces acting on them throughout their design and operational life. At all stages of a program, we need to achieve two critical goals:

- Meet bonafide mission objectives and requirements while minimizing program cost and risk
- "Sell" the program and keep it sold, if it remains worth doing

The evolving political and economic environment may alter program objectives or make the program not worth doing. Even though a program continues to be worthwhile, it may still require a significant effort to demonstrate the mission utility within the ever-changing environment.

Minimizing Program Cost and Risk

Early conceptual design has a great impact on both cost and risk. However, costs fluctuate (usually upward) throughout a program. Increased cost is frequently justifiable, but can also be due to unnecessary engineering changes, schedule slips, and changes in the funding profile. Thus, we would like to apply the SMAD process throughout the mission life cycle to reduce, or at least contain, program cost and risk. This process is summarized below and described in more detail in Wertz and Larson [1996], Chap. 10.

Table 23-2 summarizes how to keep down cost and risk, but these steps can conflict with how space programs are typically run. Good mission engineering demands that we regularly review and consider revising mission objectives, requirements, and approaches. However, this review may make the program seem unstable. We might ask, "Why are we looking at our mission objectives and requirements again? We have already done that and now must build the system." We must do so because programs and mission designs must mature and adapt to new conditions. Thus our concept of operations, approach, and goals must also evolve.

TABLE 23-2.   Mission Engineering Activities Needed Throughout the Mission Life. These are key to containing or reducing cost.

Action	Comment
1. Maintain well-defined objective	Ensure all engineering personnel know the broad mission objectives
2. Revisit system-level trades regularly	Best done at beginning of each mission phase
3. Document reasons for choices	Does not occur in standard formal documentation, but is critical to maintaining options
4. Maintain a strong systems engineering group	Needed to continue developing lower-risk, lower-cost solutions and to keep program sold
5. Control system-level trades and budgets within the systems engineering group	Allows trade between elements and applying margin to different elements as needed
6. Update analysis of mission utility	Necessary to keep program sold in a strongly competitive, cost-conscious environment

A major part of doing our job better is to identify the right issues and look for key alternatives to investigate. Working the wrong issues can be costly. For example, early in the U.S. space program, scientists and engineers were trying to figure out how to handle the enormous heat generated during reentry into the atmosphere. At first the problem statement was, "Find a material that will withstand a temperature of 14,000 °F for 5 minutes." Much time and money were spent trying to solve this problem even though no known material could withstand the required temperature. Finally, someone recognized that the problem would be restated as, "Find a way to protect a capsule and the person inside during reentry." Ablative materials quickly solved this problem. The moral of this story is to make certain you are working the right problems and asking the right questions throughout the program.

Returning to see whether we are meeting basic mission objectives does not mean we will change the concept of operations 2 months before launch. The further a program has evolved, the more it takes to justify fundamental changes. Still, if a new technology or concept of operations can reduce mission cost, risk, and schedule, we should evaluate and use it if it is worthwhile. As the program evolves, the cost of implementing change becomes greater and the benefit decreases because other parts of the mission have been designed to work as originally intended. Nonetheless, changes late in the program may provide equal or better performance at less cost and risk.

Although reevaluating mission objectives may appear to jeopardize a program, the result becomes stronger and sounder. We could keep the same mission concept throughout, but a far better approach is to formally review mission requirements at the beginning of each mission phase. This review tells us where the program is going and why. At the same time, it allows us to resolve these issues by the end of each mission phase to maintain program stability within a highly competitive environment.

Secondly, we must keep the program flexible by assigning economic value to system flexibility. All too often, a program quickly becomes rigid by trying to optimize the meeting of narrow requirements for the current mission design. Thus, changing mission objectives or technology is difficult. Hewing to fixed requirements may reduce cost for a small, short-turnaround, unique mission, but is unlikely to succeed for larger-scale activities. For example, geosynchronous satellites should be, and usually are, designed to work in multiple slots so we can move them from one position to another. Expressed differently, the mission design as well as the design of the elements should be robust, i.e., stable in the presence of unknown perturbations. We need robust or flexible designs to help minimize the cost and risk of unforeseen yet inevitable changes in mission objectives and requirements.

One way to make a system more flexible is to carry out as many onboard tasks as possible in software and ensure we can reprogram the software from the ground. Now that we have more sophisticated processors and greater onboard memory, we can create very flexible satellites without driving up cost and weight. This allows us to both correct design flaws on orbit and respond to changing mission needs and conditions.

Too often, programs attempt to reduce risk by flying only components which have flown before. Many subcontractors change their design only when manufacturers stop making particular parts and components. A far better approach is be to allow some performance margin and then use modern technology to achieve the needed performance at lower weight, cost, and risk. As Chap. 20 outlines, we also make an informed assessment of the Technology Readiness Level of a particular approach, assign it a documented risk factor, and feed that information into the decision-making process.

An excellent approach to minimizing the risk of new technology has been used successfully by Surrey Space Technology Laboratory, Ltd. on a sequence of missions. As shown in Fig. 23-1, Surrey continually flies both an older, space-proven computer and a newer, more powerful one. This approach allows Surrey to have both a conservative flight-proven design to ensure mission success, while at the same time flying some of the newest, most capable processors in space.

Keeping the Program Sold

History has shown that space programs can be cancelled in virtually any phase—during conceptual design, during development and construction, after the first one or two launches, or in mid-life while the spacecraft is still operating on orbit. The message is clear: **if the program is worth doing, we must keep the program sold throughout its mission life**. Doing so is straightforward.We must analyze mission utility and trade cost against performance over the entire mission. We must also maintain a clear, easy-to-understand and consistent rationale for mission objectives, requirements and design decisions. At any time, we should be able to present a cost vs. performance analysis and demonstrate to those who are funding it the benefits of having the program proceed.

Maintaining an ongoing mission utility analysis capability means that we should always be aware of the fundamental mission objectives and how well they are being

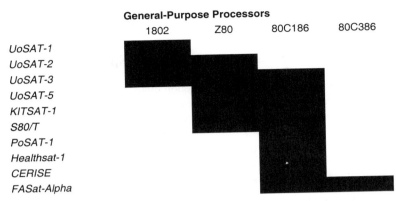

Fig. 23-1. Type of Onboard Computer for a Sequence of Surrey Space Technology Laboratory missions. Each generation of spacecraft typically flies the old standard, plus a new design which in turn becomes the next standard. (From Ward and Price, Sec. 13.4 in Wertz and Larson [1996].)

met. Thus, the mission objectives become important not only in the continuing design of the program, but also in keeping it alive.

System Engineering Over the Life Cycle

Strong system and mission engineering should be maintained throughout the life of the program: trading on requirements, reevaluating the basic mission concept, and evaluating alternatives and mission utility. Otherwise the program may not continue.

Overall, mission engineering later in the program should proceed as outlined in earlier chapters. At all levels, requirements should reflect **what** to do rather than **how** to do it. We want to let technology flow up and into the program where it can do the most good for cost, risk utility, or operations. We are unlikely to achieve the mission performance at minimum cost and risk by limiting ourselves to the technologies and approaches from the concept-design team.

A key element in mission engineering is to document the main system trades and, especially, the **reasons** for choices. Unfortunately, specifications demand only that you state what the system must do, how well it must be done, and how to verify that it meets the requirement. They do not ask you to say why you have chosen the require-ments. However, requirements are intended to quantify the trade between what we would like to achieve and what is possible within the established budget. Thus, we should document as fully as possible the reasons for these trades and requirements so we can effectively revisit them during mission utility assessments and regular require-ment reviews.

One way to initially reduce cost and schedule is to minimize the upfront engi-neering on the program. "Shoot the engineers and get on with the program," can get programs moving and contain growing engineering costs. This approach has some good features. We do not want to spend excessive time optimizing a strawman design that will change drastically at a later date. We also do not want to develop too much detail in the conceptual design. These activities divert resources from the more important, top-level mission trades. At the same time, we must have enough upfront system engineering as well as initial and continuing trades on objectives and mission utility. There is no substitute for these activities—ignorance does not improve system performance.

23.2 Lessons Learned from Existing Space Programs

For the most part, the existing space mission analysis and design process works well: trade-offs among technical alternatives in essentially all areas bring positive results. But in our experience there are four common pitfalls in the front-end planning of space missions:

- Disregarding or failing to understand the needs of operators, users, and developers during the early phase of design
- Failing to trade on requirements
- Constraining trades to too low a level too early
- Postponing or avoiding assessment of alternatives

Inappropriate results such as higher cost or reduced performance usually result from the acquisition process and organizational structure rather than from incorrect trade-off decisions. Occasionally, parochial views and politics keep the players from interacting in a way that fosters good upfront design and development of the space mission. Designing the mission operations concept, generating requirements, and performing early trades without taking into account the needs of the users, operators and developers can doom a program to failure from the beginning. A key to success is to foster continuous and open communications between the players. This can be an up-hill battle involving multiple organizations and possibly multiple nations, but it must be done.

The space mission concept and requirements drive any program, so users, operators and developers of the system must work together to identify credible, cost-effective mission approaches. Once requirements are defined, organizations often go to great lengths and expense to meet them because they do not want to be unresponsive or to say a given requirement is difficult. This can unnecessarily drive up cost. Table 23-3 shows steps that can help solve this problem.

TABLE 23-3. Steps to Help Alleviate the Problem of Failing to Trade on Requirements.

1. Specify functional requirements only—not **how** they should be accomplished
2. Identify and challenge difficult and costly requirements
3. Begin concept exploration with system requirements trades
4. Revisit top-level system requirement trades at each major milestone throughout the program
5. Maintain open communications among all groups involved

We often constrain trades to low levels too early. Early in a program we commonly do trades like altitude vs. resolution and coverage but ignore trades between system elements (see Fig. 1-3). We should look for cost-effective trade-offs among the elements such as shifting capability from the launch vehicle to the spacecraft or from the spacecraft to the ground segment. For example, we can move more computing and data processing from the ground station to the spacecraft for autonomous navigation. The question is, should we provide a spacecraft which can determine its position and velocity on board and remove the interface to the ground? This could increase the cost of spacecraft development but save a great deal in operations cost by reducing the ground system's need to track the spacecraft, process data, and provide spacecraft ephemeris. This is a politically sensitive trade because it takes money and responsibility away from the ground element and gives them to the space element. This can result in a turf battle where decisions are made on the basis of who controls the budget, rather than on what provides the best results at the lowest cost.

Two positive steps can help reduce these problems:

- Look explicitly for trades between system elements
- Do these trades as early as possible, before organizational structure and politics rule out major options

Look at the result of these trades carefully. For example, when someone presents an extremely low-cost spacecraft concept with minimal capability, look for higher costs in other mission elements.

We often postpone or avoid assessing major alternatives for space programs. We do so to meet tight budget constraints, to keep the program appearing stable, and to avoid reducing the momentum to proceed. As we know, change costs much more as a program proceeds. Change during concept exploration may do little or nothing to engineering cost, whereas changes during the definition phase involve more people and a reassessment of a more detailed design. Design changes involve formal paperwork and change controls. Changing anything once the hardware is in production can dramatically increase cost. It is more cost-effective to make changes as early as possible.

23.3 Future Trends

In several respects, we expect more changes in the way that space programs are conducted over the coming decade than have occurred since the start of the space program.* In the near term, three distinct trends are apparent:

- Increased political and economic constraints driving the industry toward lower-cost space missions and increased inter-organizational and international cooperation
- More reliance on advanced spacecraft processing
- The introduction of large constellations and, consequently, a more manufacturing approach to spacecraft design

Because of budget constraints on the space program as a whole, we expect a continuing strong emphasis on lower costs for spacecraft and space missions. Small spacecraft are a natural outcome, including single-purpose small satellites for data relay or materials processing and large constellations of small satellites for communi-

---

* The changing direction of future space programs has been discussed by a number of authors, including Butler [1990], Fogg [1995], Handberg [1995], Harvey [1996], Heiken, et al. [1991], Hoban [1997], Johnson-Freese [1997], Kay [1995], Launius [1998], Richard Lewis [1990], John Lewis [1997], Noor and Venneri [1997], Spudis [1996], Stine [1996], and Zubrin [1996].

cations, space surveillance, or tactical applications. Because of the accelerating use of onboard processors and miniaturization of many components, small, low-cost satellites will be as capable as older, larger satellites. (See, for example, Sarsfield [1998].) More people are building and working with *LightSats*—small, inexpensive spacecraft, traditionally in low-Earth orbit but moving rapidly outward. As described in Chap. 22, the technologies used here tend to be nontraditional and far less expensive. LightSats represent a new way of doing business in space that, if correctly exploited, can drive down costs by a factor of 2 to 10 with respect to more traditional programs [Wertz and Larson, 1996].

The biggest technical change in space missions will be the increased use of onboard processing. Space computers are emerging from infancy, with full-fledged, general-purpose processors now becoming available. We will use them much more as they become more economical and begin to duplicate some of the ground processing capability. As onboard processing increases, spacecraft will take on more complex tasks and become more autonomous.

Increases in spacecraft autonomy could drive costs either up or down, depending on the community's approach to autonomy and software development. We strongly recommend using software to help lower costs. As shown in Table 23-4, we believe *autonomy in moderation* can reduce cost and risk by automating repetitive functions on the spacecraft while leaving the higher-level functions to ground operations. Everyday tasks such as telemetry processing, attitude control, and orbit control should be done autonomously on board, with all of the relevant information attached to the payload data for downlink to the ground. On the other hand, trying to fully automate problem solving can be dramatically expensive. So long as the spacecraft is capable of putting itself in a fail-safe mode, this work is best performed on the ground where operators can do what they do best—use their intelligence to understand, resolve, and repair spacecraft problems and respond to changing circumstances.

TABLE 23-4. Development of Low-Cost, Autonomous Systems. The objective of "autonomy in moderation" is to automate repetitive tasks and spacecraft safing while allowing people to identify and fix problems and create long-term plans [Wertz and Larson, 1996].

Functions That Should Be Automated	Functions That Should Not Be Automated
• Attitude determination and control • Orbit determination and control • Payload data processing • Repetitive housekeeping, e.g., battery charging, active thermal control • Anomaly recognition • Spacecraft safing	• Problem resolution • Identification and implementation of fixes • One-time activities (e.g., deployment and check-out) • Long term operations planning • Emergency handling beyond safing

We do not know if proposals to drive down launch costs will succeed in the coming decade. If launch costs drop significantly, low-cost satellite systems will be possible. Reducing launch costs will itself drive down spacecraft cost because it will no longer be worth as much to optimize the spacecraft design or minimize weight [Wertz and Larson, 1996]. Costs for space exploration would become more consistent with aviation and other commercial activities.

We also expect to see significant changes in space payloads, due both to the increased level of processing and other factors. In observations and communications, the use of increased processing capability and array sensors for observations will greatly increase the capacity of small spacecraft. In addition, a number of new types of space missions will begin, most notably materials processing in space and other applications of a microgravity environment. This area is just now beginning and has enormous growth potential as low-gravity manufacturing processes become better understood. Much of this will require low-cost access to space and the return of payloads from space as well.

The commercial communications constellations have also had a major impact on the process of designing and manufacturing spacecraft [Logsdon, 1995]. Commercial geosynchronous spacecraft have been manufactured in an assembly line process for some time. Nonetheless, low-Earth orbit constellations require that the entire constellation be built and launched in a short time in order to allow the system to begin servicing customers. For example, the Iridium constellation of 66 satellites was launched in slightly over a year which required satellites to be coming off the assembly line at an average rate of approximately one per week. This significantly changes the focus of spacecraft design from performance optimization to high-reliability manufacturing. The two contrasting approaches are presented in Chaps. 12 and 19 above. To the extent that these manufacturing techniques can be applied to other space missions as well, this should continue to reduce the cost, risk, and schedule of spacecraft development. The application of this to spacecraft components is discussed by Zafanella [1998].

In addition, new constellations like Iridium, Celestri, Globalstar and even the existing GPS, provide an unprecedented infrastructure that can be used to create new and more cost-effective space mission architectures. For example having an abundance of bandwidth available in space can facilitate the movement of data around the world, thus alleviating the current problems we have with over-subscribed networks, and changing the way we do missions.

Since the first satellites, space has moved from a very novel activity to a routine commercial enterprise. Today, on the average, Americans use some form of space asset many times per day—weather, television, telephones, navigation, Internet, and more. Companies are looking for ways to use space and space-related assets to develop and provide highly marketable products and services. More than ever, companies are focusing on the end-to-end perspective—providing customers what they need in a more cost-effective way. For example, recent deregulation of electric utilities in North America has resulted in regional utilities trying to sell their electricity all over the continent. One severe problem they face is how to keep track of the customers' use of electricity, which has prompted the need for a method of automatically reading utility meters continentally—from space. The commercial space systems we propose must be relevant and financially viable, while government space systems must be relevant and cost-effective.

Today, space exploration depends more on policy, politics, and economics, than on technological limits. Virtually any mission is now technically possible. We can build a lunar base, send humans to Mars, develop space colonies and huge solar-powered satellites, or send manned or unmanned probes to the outer planets. Our limitations are set by what we choose to do or what we can afford. As environmental problems become more serious on the Earth's surface, intelligently using space will become more important. Yet the use of space resources cannot grow unless we can demonstrate that we are becoming more efficient and cost effective in our use of space. This is the challenge for space in the early 21st century.

References

Butler, George V., ed. 1990. *The 21st Century in Space*. San Diego, CA: American Astronautical Society.

Fogg, Martyn J. 1995. *Terraforming: Engineering Planetary Environment*. Warrendale, PA: Society of Automotive Engineers, Inc.

Handberg, Roger. 1995. *The Future of the Space Industry: Private Enterprise and Public Policy*. New York: Quorum Books.

Harvey, Brian. 1996. *The New Russian Space Programme: From Competition to Collaboration*. New York: John Wiley & Sons.

Heiken, Grant, David Vaniman, and Beven M. French. 1991. *Lunar Sourcebook: A User's Guide to the Moon*. Cambridge, MA: Cambridge University Press.

Hoban, Francis T. 1997. *Where do You Go After You've Been to the Moon?* Malabar, FL: Krieger Publishing Co.

Johnson-Freese, J. and Roger Handberg. 1997. *Space, the Dormant Frontier: Changing the Paradigm for the 21st Century*. New York: Praeger Publishers.

Kay, W.D. 1995. *Can Democracies Fly in Space? The Challenge of Revitalizing the U.S. Space Program*. New York: Praeger Publishers.

Launius, Roger D. 1998. *Frontiers of Space Exploration*. Westport, CT: Greenwood Publishing Group.

Lewis, John S. 1997. *Mining the Sky: Untold Riches from the Asteroids, Comets, and Planets*. Reading, MA: Addison-Wesley.

Lewis, Richard S. 1990. *Space in the 21st Century*. New York: Columbia University Press.

Logsdon, Tom. 1995. *Mobile Communication Satellites: Theory and Application*. New York: McGraw Hill.

Noor, Ahmed K. and Samuel L. Venneri, eds. 1997. *Future Aeronautical and Space Systems*. Reston, VA: AIAA.

Sarsfield, Liam. 1998. *The Cosmos on a Shoestring: Small Spacecraft for Space and Earth Sciences*. Santa Monica, CA: Rand Corp.

Spudis, Paul D. 1996. *The Once and Future Moon*. Herndon, VA: Smithsonian Institution Press.

Stine, Harry G. 1996. *Halfway to Anywhere: Achieving America's Destiny in Space*. M. Evans and Company, Inc.

Wertz, James R. and Wiley J. Larson, eds. 1996. *Reducing Space Mission Cost*. Torrance, CA: Microcosm Press and Dordrecht, The Netherlands: Kluwer Academic Publishers.

Zafanella, Carlo. 1999. "Reducing Cost and Increasing Volume by a Factor of 10 in the Space Component Business." *Journal of Reducing Space Mission Cost*, vol. 1 no. 2.

Zubrin, Robert. 1996. *Islands in the Sky: Bold New Ideas for Colonizing Space*. New York: John Wiley & Sons.

Appendices

Simon D. Dawson, *Microcosm, Inc.*

The fundamental physical constants and conversions factors based on them are those determined by the National Bureau of Standards using a least squares fit to the best available experimental data [Cohen, 1986]. Their intent is to create a set of constants which are mutually consistent to within the experimental accuracy. Other constants and conversion factors, such as the speed of light in vacuum or the conversion between feet and meters, are adopted as exact definitions of the units involved. For astronomical and astronautical constants, such as the values of *GM* for various objects in the solar system, values adopted by the International Astronomical Union are used. Many of these are quoted from *Astrophysical Quantities* [Cox, 1999], which kindly permitted use of proof copies for obtaining the most current data. We highly recommend this volume for those who need additional quantitative detail about the solar system or other astronomical topics.

References

Cohen, E. Richard and B.N. Taylor, 1986. *CODATA Bulletin No. 63*, Nov. New York: Pergamon Press.

Cox, A.N. ed. 1999. *Astrophysical Quantities*. New York: Springer-Verlag.

Appendix A

Mass Distribution for Selected Satellites

Table A-1 lists representative satellite masses with and without propellant for various types of spacecraft. Table A–2 further breaks down the dry mass by the percentage devoted to each subsystem. See Table 18-2 for definitions of the mass categories and Table 10-2 for definitions of the various subsystems. Statistical data for the various classes of missions are also provided. Table A–2 also includes subsystem masses as a percentage of the payload mass, since this may be the only mass known during early mission design.

The data here can be used as either a preliminary estimate or check on the reasonableness of more detailed methods (see Chaps. 10 and 11 for other methods). However, such historical data should always be used with caution and substantial margin should be applied. The first three categories are more traditional, older DoD spacecraft. The LightSats are newer, typically much less expensive systems from many developers, both US and international. As the satellite microminiaturization process continues, individual components will get smaller and percentages will shift, depending on both the component mix and the mission needs.

TABLE A-1.   Actual Mass for Selected Satellites. The propellant load depends on the satellite design life.

Spacecraft Name	Loaded Mass (kg)	Propellant Mass (kg)	Dry Mass (kg)	Propellant Mass (%)	Dry Mass (%)
Communications Satellites					
FLTSATCOM 1-5	930.9	81.4	849.6	8.7%	91.3%
FLTSATCOM 6	980.0	109.1	870.9	11.1%	88.9%
FLTSATCOM 7-8	1150.9	109.0	1041.9	9.5%	90.5%
DSCS II	530.0	54.1	475.9	10.2%	89.8%
DSCS III	1095.9	228.6	867.3	20.9%	79.1%
NATO III	346.1	25.6	320.4	7.4%	92.6%
Intelsat IV	669.2	136.4	532.8	20.4%	79.6%
TDRSS	2150.9	585.3	1565.7	27.2%	72.8%
Average	**981.7**	**166.2**	**815.6**	**16.9%**	**83.1%**
Standard Deviation	**549.9**	**179.9**	**389.4**	**7.3%**	**7.3%**
Navigation Satellites					
GPS Block 1	508.6	29.5	479.1	5.8%	94.2%
GPS Block 2,1	741.4	42.3	699.1	5.7%	94.3%
GPS Block 2,2	918.6	60.6	858.0	6.6%	93.4%
Average	**722.9**	**44.1**	**678.7**	**6.1%**	**93.9%**
Standard Deviation	**205.6**	**15.6**	**190.3**	**0.5%**	**0.5%**
Remote Sensing Satellites					
P80-1	1740.9	36.6	1704.4	2.1%	97.9%
DSP-15	2277.3	162.4	2114.9	7.1%	92.9%
DMSP 5D-2	833.6	19.1	814.6	2.3%	97.7%
DMSP 5D-3	1045.5	33.1	1012.3	3.2%	96.8%
Average	**1474.3**	**62.8**	**1411.6**	**4.3%**	**95.7%**
Standard Deviation	**660.9**	**66.8**	**604.5**	**2.4%**	**2.4%**

TABLE A-1.   Actual Mass for Selected Satellites. The propellant load depends on the satellite design life.

Spacecraft Name	Loaded Mass (kg)	Propellant Mass (kg)	Dry Mass (kg)	Propellant Mass (%)	Dry Mass (%)
LightSats					
Ørsted	60.8	no propulsion	60.8	n/a	100.0%
Freja	255.9	41.9	214.0	16.4%	83.6%
SAMPEX	160.7	no propulsion	160.7	n/a	100.0%
HETE	125.0	no propulsion	125.0	n/a	100.0%
Clementine	463.0	231.0	232.0	49.9%	50.1%
Pluto Fast Flyby '93	87.4	6.9	80.5	7.9%	92.1%
RADCAL	92.0	no propulsion	92.0	n/a	100.0%
ORBCOMM	47.5	14.4	33.1	30.3%	69.7%
AMSAT AO-13	140.0	56.0	84.0	40.0%	60.0%
AMSAT AO-16	9.0	no propulsion	9.0	n/a	100.0%
PoSat	48.5	no propulsion	48.5	n/a	100.0%
BremSat	63.0	no propulsion	63.0	n/a	100.0%
Average	**141.2**	**70.0**	**109.3**	**49.6%**	**77.4%**
Standard Deviation	**44.8**	**92.2**	**31.3**	**17.1%**	**18.4%**

Various algorithms and data are provided in Chaps. 10 and 11 to determine the mass of individual subsystems. Actual mass distributions are provided in Table A–2 to develop initial estimates or perform reasonableness checks. Be careful using this data. The percentages given do not necessarily represent the full spectrum of satellite mass and subsystem mass distributions.

TABLE A–2.   Mass Distribution for Selected Spacecraft.

Spacecraft Name	Percentage of Spacecraft Dry Mass						
	Payload	Structure	Thermal	Power	TT&C	ADCS	Propulsion
Communications Satellites							
FLTSATCOM 1-5	26.5%	19.3%	1.8%	38.5%	3.0%	7.0%	3.9%
FLTSATCOM 6	26.4%	18.7%	2.0%	39.4%	3.0%	6.8%	3.8%
FLTSATCOM 7-8	32.8%	20.8%	2.1%	32.8%	2.5%	5.7%	3.3%
DSCS II	23.0%	23.5%	2.8%	29.3%	7.0%	11.5%	3.0%
DSCS III	32.3%	18.2%	5.6%	27.4%	7.2%	4.4%	4.1%
NATO III	22.1%	19.3%	6.5%	34.7%	7.5%	6.3%	2.4%
Intelsat IV	31.2%	22.3%	5.1%	26.5%	4.3%	7.4%	3.1%
TDRSS	24.6%	28.0%	2.8%	26.4%	4.1%	6.2%	6.9%
Average	**27.4%**	**21.3%**	**3.6%**	**31.9%**	**4.8%**	**6.9%**	**3.8%**
Standard Deviation	**4.2%**	**3.3%**	**1.9%**	**5.3%**	**2.1%**	**2.1%**	**1.4%**
Navigation Satellites							
GPS Block 1	20.5%	19.9%	8.7%	35.8%	5.8%	6.2%	3.6%
GPS Block 2,1	20.2%	25.1%	9.9%	31.0%	5.2%	5.4%	3.3%
GPS Block 2,2	23.0%	25.4%	11.0%	29.4%	3.1%	5.3%	2.7%
Average	**21.2%**	**23.5%**	**9.9%**	**32.1%**	**4.7%**	**5.6%**	**3.2%**
Standard Deviation	**1.6%**	**3.1%**	**1.2%**	**3.3%**	**1.4%**	**0.5%**	**0.5%**

TABLE A–2.   Mass Distribution for Selected Spacecraft.

Spacecraft Name	Percentage of Spacecraft Dry Mass						
	Payload	Structure	Thermal	Power	TT&C	ADCS	Propulsion
Remote Sensing Satellites							
P80-1	41.1%	19.0%	2.4%	19.9%	5.2%	6.3%	6.1%
DSP-15	36.9%	22.5%	0.5%	26.9%	3.8%	5.5%	2.2%
DMSP 5D-2	29.9%	15.6%	2.8%	21.5%	2.5%	3.1%	7.4%
DMSP 5D-3	30.5%	18.4%	2.9%	29.0%	2.0%	2.9%	8.7%
Average	**34.6%**	**18.9%**	**2.1%**	**24.3%**	**3.4%**	**4.5%**	**6.1%**
Standard Deviation	**5.4%**	**2.8%**	**1.1%**	**4.3%**	**1.4%**	**1.7%**	**2.8%**
LightSats							
Ørsted	21.5%	38.3%	0.8%	15.8%	16.7%	6.8%	none
Freja	34.1%	22.7%	2.4%	19.0%	8.7%	6.0%	7.0%
SAMPEX	32.5%	23.1%	2.5%	25.0%	10.6%	6.3%	none
HETE	35.3%	16.0%	1.8%	20.3%	8.5%	18.1%	none
Pluto Fast Flyby '93	8.7%	18.1%	4.6%	24.1%	23.9%	8.3%	12.3%
RADCAL	22.5%	31.0%	0.3%	18.6%	9.4%	18.2%	none
ORBCOMM	25.3%	20.0%	2.5%	29.3%	8.8%	8.8%	5.3%
PoSat	12.2%	13.9%	0.0%	36.1%	17.5%	21.1%	none
BremSat	27.8%	20.6%	0.0%	33.3%	10.3%	7.9%	none
Average	**24.4%**	**22.7%**	**1.7%**	**24.6%**	**12.7%**	**11.3%**	**2.7%**
Standard Deviation	**9.3%**	**7.7%**	**1.5%**	**7.0%**	**5.4%**	**6.0%**	**4.5%**
All							
Average	**26.7%**	**21.7%**	**3.4%**	**27.9%**	**7.5%**	**8.0%**	**3.7%**
Standard Deviation	**7.5%**	**5.3%**	**3.0%**	**6.6%**	**5.4%**	**4.7%**	**3.2%**
Average % of Payload Mass	**100.0%**	**81.1%**	**12.7%**	**104.6%**	**28.2%**	**29.9%**	**13.9%**
Standard Deviation of % of Payload Mass	**0.0%**	**40.8%**	**16.4%**	**62.3%**	**35.5%**	**35.5%**	**30.2%**

   The average values and the associated standard deviation for each of the spacecraft subsystems are also listed in Table A–2. The final row lists the percentage of the **payload** mass devoted, on the average, to the individual subsystems. This information is useful in the beginning of a program when we only know the mass of the payload. When using these estimates, be sure to apply an appropriate margin for error (See Chap. 10).

Appendix B

Astronautical and Astrophysical Data

See Inside Front Cover for Fundamental Physical Constants and Spaceflight Constants. See Inside Rear Cover for tabular Earth Satellite Data.

TABLE B-1. Physical Properties of the Sun. (Data from Cox [1999]; Seidelmann [1992].)

Radius of the Photosphere	$6.95508 \pm 0.00026 \times 10^8$ m
Angular Diameter of the Photosphere at 1 AU	0.533 13 deg
Mass	$1.989\ 1 \times 10^{30}$ kg
Mean Density	1.409 g/cm^3
Total Radiation Emitted	3.845×10^{26} J/s
Total Radiation per Unit Area at 1 AU	1,367 Wm^{-2}
Apparent Visual Magnitude at 1 AU	−26.75
Absolute Visual Magnitude (Magnitude at Distance of 10 parsecs)	+4.82
Color Index, B-V	+0.650
Spectral Type	G2 V
Effective Temperature	5,777 K
Inclination of the Equator to the Ecliptic	7.25 deg
Adopted Period of Sidereal Rotation (L = 17 deg)	25.38 days
Period of Synodic Rotation Period (ϕ = latitude)	$26.90 + 5.2 \sin^2 \phi$ days
Mean Sunspot Period	11.04 years
Dates of Former Maxima	1968.9, 1980.0, 1989.6
Mean Time from Maximum to Subsequent Minimum	6.2 years

TABLE B–2. Physical Properties of the Earth. (Data from Cox [1999]; Seidelmann [1992]; Zee [1989]; McCarthy [1996].)

Equatorial Radius, a	$6.378\ 136\ 49 \times 10^6$ m
Flattening Factor (Ellipticity), $f \equiv (a - c)/a$	$1/298.256\ 42 \approx 0.003\ 352\ 819\ 70$
Polar Radius,* c	$6.356\ 751\ 7 \times 10^6$ m
Mean Radius,* $(a^2c)^{1/3}$	$6.371\ 000\ 3 \times 10^6$ m
Eccentricity,* $(a^2 - c^2)^{1/2}/a$	0.081 819 301
Surface Area	$5.100\ 657 \times 10^{14}$ m^2
Volume	$1.083\ 207 \times 10^{21}$ m^3
Ellipticity of the Equator $(a_{max} - a_{min})/a_{mean}$	$\sim 1.6 \times 10^{-5}$
Longitude of the Maxima	14.805° W, 165.105° E
Ratio of the Mass of the Sun to the Mass of the Earth	332,945.9
Geocentric Gravitational Constant, GM$_E \equiv \mu_E$	$3.986\ 004\ 418 \times 10^{14}$ m^{3}s^{-2}
Mass of the Earth	$5.973\ 7 \times 10^{24}$ kg

TABLE B–2. Physical Properties of the Earth. (Data from Cox [1999]; Seidelmann [1992]; Zee [1989]; McCarthy [1996].)

Mean Density	$5,554.8$ kg m^{-3}
Gravitational Field Constants (Data from JGM-2; the following constants should be used in conjunction with these data: $R_E = 6,378.136\ 3$ km; $GM_E = 398,600.441\ 5$ km^3/s^2; $\omega_E = 7.292\ 115\ 855\ 3 \times 10^{-5}$ rad•s^{-1})	$\begin{cases} J_0 & 1 \\ J_1 & 0 \\ J_2 & 0.108\ 262\ 692\ 563\ 881\ 5 \times 10^{-2} \\ J_3 & -0.253\ 230\ 781\ 819\ 177\ 4 \times 10^{-5} \\ J_4 & -0.162\ 042\ 999 \times 10^{-5} \end{cases}$
Mean Distance of Earth Center from Earth-Moon Barycenter	$4,671$ km
Average Lengthening of the Day	0.0015 sec/century
Annual General Precession in Longitude (i.e., Precession of the Equinoxes), at J.2000 (T in centuries from J.2000)	$50.290\ 966" + 0.022\ 222\ 6"$ T $-0.004\ 2"$ T^2
Obliquity of the Ecliptic, at Epoch 2000	$23° 26' 21.4119"$
Rate of Change of the Obliquity (T in Julian Centuries, T = (JD–2,451,545.0)/36,525))	$-46.815\ 0"$ T $-0.000\ 59"$ T^2 $+ 0.001\ 813"$ T^3
Amplitude of the Earth's Nutation	$2.556\ 25 \times 10^{-3}$ deg
Sidereal Period of Rotation, Epoch 2000	$0.997\ 269\ 68$ d$_e$ $= 86\ 164.100\ 4$ s $= 23^h56^m 04.098\ 9$
Length of Tropical Year (ref. = Υ), (T in Julian Centuries, T = (JD–2,451,545.0)/36,525)) T=(JD–2,451,545.0)/36,525	$365.242\ 189\ 669\ 8 +$ $0.000\ 006\ 153\ 59$ T $- 7.29 \times 10^{-10}$ T^2 $+ 2.64 \times 10^{-10}$ T^3 d$_e$
Length of Anomalistic Year (Perihelion to Perihelion), Epoch 1999.0	$365.259\ 635\ 4$ d$_e$ $= 31,558,432.5$ s
Mean Angular Velocity	$= 7.292\ 115\ 0 \times 10^{-5}$ rad s^{-1} $= 15.041\ 067\ 178\ 669\ 10$ arcsec•s^{-1}
Mean Orbital Speed	$= 2.978\ 48 \times 10^4$ m•s^{-1}
Mean Distance From Sun	$= 1.000\ 001\ 057$ AU $= 1.495\ 980\ 29 \times 10^{11}$ m

*Based on adopted values of f and a.

Phase Law and Visual Magnitude of the Moon

To determine the Moon's visual magnitude, $V(R,\xi)$, at any distance and phase, let R be the observer-Moon distance in AU and ξ be the phase angle at the moon between the Sun and observer. Then

$$V(R,\xi) = 0.21 + 5\log_{10}R - 2.5\log_{10}P(\xi) \qquad \text{(B-1)}$$

where the phase law, $P(\xi)$, for the Moon is given in Table B–3 [Hapke, 1974]. Note that the visual magnitude of the Moon at opposition (i.e., full Moon) at the mean distance of the Moon from the Earth is -12.73.

TABLE B–3.   Phase Law and Visual Magnitude of the Moon.

ξ (deg)	P(ξ) Before Full Moon	P(ξ) After Full Moon	V(R,ξ) – V(R,0) Before Full Moon	V(R,ξ) – V(R,0) After Full Moon	ξ (deg)	P(ξ) Before Full Moon	P(ξ) After Full Moon	V(R,ξ) – V(R,0) Before Full Moon	V(R,ξ) – V(R,0) After Full Moon
0	1.000	1.000	0.000	0.000	80	0.120	0.111	2.301	2.386
10	0.787	0.759	0.259	0.299	90	0.0824	0.0780	2.709	2.769
20	0.603	0.586	0.549	0.580	100	0.0560	0.0581	3.129	3.089
30	0.466	0.453	0.828	0.859	110	0.0377	0.0405	3.558	3.481
40	0.356	0.350	1.121	1.139	120	0.0249	0.0261	4.009	3.958
50	0.275	0.273	1.401	1.409	130	0.0151	0.0158	4.552	4.503
60	0.211	0.211	1.689	1.689	140	—	0.0093	—	5.078
70	0.161	0.156	1.982	2.016	150	—	0.0046	—	5.842

Geocentric and Geodetic Coordinates on the Earth

As shown in Fig. B-1, *geocentric coordinates* are defined with respect to the center of the Earth. However, latitude and longitude are frequently given in *geodetic coordinates* which are defined with respect to an oblate *reference ellipsoid* (i.e., a figure created by rotating an ellipse about its minor axis), with the height, *h*, measured perpendicular to a plane tangent to the ellipsoid. A triaxial ellipsoid is not generally used since the gain in representation is small. Although not normally used for space mission work, *astronomical latitude* and *longitude* are defined relative to the *local vertical*, or the normal to the equipotential surface of the Earth. Thus, *astronomical latitude* is defined as the angle between the local vertical and the Earth's equatorial plane. Maximum values of the *deviation of the vertical*, or the angle between the local vertical and the normal to a reference ellipsoid, are about 1 minute of arc. Maximum variations in the height between the ellipsoid and *mean sea level* (also called the *equipotential surface*) are about 100 m. Seidelmann [1992] provides an extended discussion of coordinate systems and transformations.

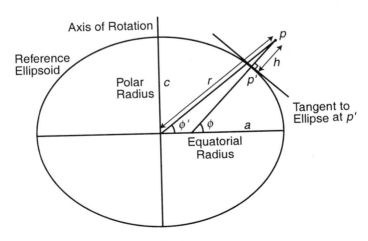

Fig. B-1.   Geocentric vs. Geodetic Coordinates.

The *geocentric latitude*, ϕ', of a point, p, on the surface of the Earth is the angle at the Earth's center between p and the equatorial plane. The *geodetic* or *geographic latitude*, ϕ, is the angle between the normal to the reference ellipsoid (at point p') and the equatorial plane. The ellipsoid is typically defined by the flattening, f, or the *inverse flattening*, $1/f$, given by

$$f \equiv \frac{a-b}{a} \approx 1/298.256\ 42 \text{ for the Earth} \tag{B-2}$$

where a and b are the *semimajor* and *semiminor* axes of the ellipsoid.

The *geocentric longitude*, λ, is defined by the angle between the *reference* (or *zero*) meridian and the meridian of point p (and p'), measured eastward around the Earth from $0°$ to $360°$. The *geodetic longitude* will be identical to the geocentric longitude assuming that the reference ellipsoid has the same axes and reference meridian as the geocentric system.

The *ellipsoidal height*, h, of point p is measured along the normal to the ellipse, and with respect to the point of intersection of that normal with the ellipsoidal surface, that is, from point p'. Other 'heights' are possible—for example, the *geocentric radius*, r, and the height above *mean sea level* or *geoidal height*, H.

We often express the position of a point, p, on or near the Earth in a right-handed geocentric cartesian coordinate system (X, Y, Z). Here the direction of the Z-axis is that of the rotational reference ellipsoid, the X-axis is perpendicular to Z through the reference meridian. The Y-axis completes the triplet in a right-handed sense.

Conversion from geocentric to geocentric cartesian coordinates is given by

$$\begin{bmatrix} X \\ Y \\ Z \end{bmatrix}_p = r \begin{bmatrix} \cos\phi' \cos\lambda \\ \cos\phi' \sin\lambda \\ \sin\phi' \end{bmatrix} \tag{B-3}$$

Conversion from geodetic to geocentric cartesian coordinates is given by

$$\begin{bmatrix} X \\ Y \\ Z \end{bmatrix}_p = \begin{bmatrix} \left(N_\phi + h\right)\cos\phi \cos\lambda \\ \left(N_\phi + h\right)\cos\phi \sin\lambda \\ \left(\left(1 - e^2\right)N_\phi + h\right)\sin\phi \end{bmatrix} \tag{B-4}$$

where e is the *eccentricity* of the ellipsoid

$$e \equiv \sqrt{a^2 - b^2}\big/a = \sqrt{2f - f^2} \tag{B-5}$$

and N_ϕ is the *ellipsoidal radius of curvature in the meridian*, given by

$$N_\phi \equiv a\big/\sqrt{1 - e^2 \sin^2\phi} \tag{B-6}$$

Conversion from geocentric cartesian to geocentric coordinates is given by

$$\begin{bmatrix} \phi' \\ \lambda \\ r \end{bmatrix}_p = \begin{vmatrix} \tan^{-1}\left(Z\big/\sqrt{X^2 + Y^2}\right) \\ \tan^{-1}(Y/X) \\ \sqrt{X^2 + Y^2 + Z^2} \end{vmatrix} \tag{B-7}$$

Note that $\phi' = \sin^{-1}(Z/r)$ may also be used to determine ϕ', and that these functions are evaluated such that $-90° \leq \phi' \leq +90°$, and $0° \leq \lambda \leq 360°$.

To convert from geocentric cartesian to geodetic coordinates is not as simple a procedure. An exact solution has been given by Borkowski [1989] based on using an expression for the reduced latitude in a solvable fourth-degree polynomial.

We first calculate intermediate variables as follows:

$$R = \sqrt{X^2 + Y^2} \qquad\qquad Q = 2\left(E^2 - F^2\right)$$

$$E = \left[bZ - (a^2 - b^2)\right]\big/aR \qquad D = \left(P^3 + Q^2\right)^{1/2}$$

$$F = \left[bZ + (a^2 - b^2)\right]\big/aR \qquad v = (D - Q)^{1/3} - (D + Q)^{1/3}$$

$$P = 4(EF + 1)/3$$

$$G = \frac{1}{2}\left[\sqrt{E^2 + v} + E\right]$$

$$t = \left[G^2 + (F - vG)/(2G - E)\right]^{1/2} - G$$

Finally, the latitude and ellipsoidal height are computed from:

$$\phi = \tan^{-1}\left(a\left(1 - t^2\right)/(2bt)\right)$$

$$h = (R - at)\cos\phi + (Z - b)\sin\phi \qquad\qquad\text{(B-8)}$$

To obtain the correct sign, set the sign of b to that of Z before beginning. This solution is singular for points at the Z-axis ($r = 0$) or in the XY-plane. For which:

$$\left.\begin{array}{l}\phi = \phi' = 90° \\ h = Z - b\end{array}\right\}\text{Z axis} \qquad \left.\begin{array}{l}\phi = \phi' = 0 \\ h = R - a\end{array}\right\}\text{XY plane}$$

$$\text{(B-9)}$$

Additionally, for points close to these conditions, some round-off error may be avoided and the accuracy improved slightly be replacing the value of v with

$$v = v^3 + 2Q/3P \quad \text{(near Z axis or XY plane)} \qquad\qquad \text{(B-10)}$$

Finally,

$$\lambda = \tan^{-1}(Y/X) \qquad\qquad\qquad \text{(B-11)}$$

References

Borkowski, K. M. 1989. "Accurate Algorithms to Transform Geocentric to Geodetic Coordinates," Bulletin Géodésique 63, no.1, p. 50–56.

Cox, A.N. ed. 1999. *Astrophysical Quantities, 1999*. New York: Springer-Verlag.

Hapke B. 1974. "Optical Properties of the Lunar Surface."

McCarthy, Dennis D., USNO. 1996. "IERS Technical Note 21." IERS Conventions.

Seidelmann, P. Kenneth, USNO, ed. 1992. *The Explanatory Supplement to the Astronomical Almanac*. Mill Valley, CA: University Science Books.

Zee, Chong-Hung. 1989. *Theory of Geostationary Satellites*. Netherlands: Kluwer.

Appendix C

Elliptical Orbit Equations

Argument of perigee: ω

$$\omega = \cos^{-1}\left[\frac{(\mathbf{n} \cdot \mathbf{e})}{(ne)}\right]$$

if $(e_z < 0)$ then $\omega = 2\pi - \omega$

See Table 6-2

Eccentric anomaly: E

$$\tan\left(\frac{E}{2}\right) = \sqrt{\frac{1-e}{1+e}}\,\tan\left(\frac{v}{2}\right)$$

$$\sin E = \frac{\sqrt{1-e^2}\,\sin v}{1+e\cos v}$$

$$\cos E = \frac{e+\cos v}{1+e\cos v}$$

See Table 6-3

Eccentricity: e

$$e = \frac{\left(r_a - r_p\right)}{\left(r_a + r_p\right)}$$

$$e = \frac{r_a}{a} - 1 \qquad e = 1 - \frac{r_p}{a}$$

$$e = \frac{r_2 - r_1}{r_1 \cos v_1 - r_2 \cos v_2}$$

See Eq. (6-3)

Eccentricity vector: $\mathbf{e}$

$$\mathbf{e} = \frac{1}{\mu}\left\{\left(V^2 - \frac{\mu}{r}\right)\mathbf{r} - (\mathbf{r}\cdot\mathbf{V})\mathbf{V}\right\}$$

See Eq. (6-9)

Flight path angle: γ

$$\tan\gamma = \frac{e\sin v}{1+e\cos v}$$

See Table 6-5

Inclination: i

$$i = \cos^{-1}\left[\frac{\hat{\mathbf{z}}\cdot\mathbf{h}}{|\hat{\mathbf{z}}|\,|\mathbf{h}|}\right]$$

See Table 6-2

Mean anomaly: M

$$M = n(t - t_0) + M_0$$

$$M = \left(\frac{\mu}{a^3}\right)^{1/2}\cdot(t - t_0) + M_0$$

$$M = E - e\sin E$$

See Table 6-3

Mean motion: n

$$n = \sqrt{\mu/a^3}$$

See Table 6-3

Nodal vector: $\mathbf{n}$

$$\mathbf{n} = \hat{\mathbf{z}} \times \mathbf{h}$$

See Eq. (6-8)

Period: P

$$P = 2\pi/n \qquad P = 2\pi\sqrt{a^3/\mu}$$

See Table 6-2

Radius of perigee: r_p

$$r_p = a(1 - e)$$

$$r_p = r_a\frac{(1-e)}{(1+e)}$$

$$r_p = 2a - r_a$$

$$r_p = \frac{r_1(1 + e\cos v_1)}{1 + e}$$

See Table 6-2

Radius of apogee: r_a

$$r_a = a(1 + e)$$

$$r_a = 2a - r_p$$

$$r_a = r_p \frac{(1 + e)}{(1 - e)}$$

$$r_a = r_1 \frac{(1 + e \cos v_1)}{(1 - e)}$$

See Table 6-2

Radius: r

$$r = \frac{a(1 - e^2)}{1 + e \cos v}$$

$$r = \frac{r_p(1 + e)}{1 + e \cos v}$$

See Eq. (6-3)

Right ascension of the node: Ω

$$\Omega = \cos^{-1}\left[\frac{\hat{\mathbf{x}} \cdot \mathbf{n}}{|\hat{\mathbf{x}}| \, |\mathbf{n}|}\right]$$

if $(\mathbf{n}_y < 0)$ then $\Omega = 2\pi - \Omega$

where $\mathbf{n}$ is the nodal vector not the mean motion
See Table 6-2

Semimajor axis: a

$$a = \frac{(r_a + r_p)}{2} = \frac{-\mu}{2\varepsilon}$$

$$a = \frac{r_p}{(1 - e)}$$

$$a = \frac{r_a}{(1 + e)}$$

See Table 6-2

Specific angular momentum: $\mathbf{h}$

$$\mathbf{h} = \mathbf{r} \times \mathbf{V}$$

$$h = r_a V_a = r_p V_p$$

See Eq. (6-7)

Time since periapsis: t

$$t = \frac{(E - e \sin E)}{n}$$

See Table 6-3

True anomaly: v

$$v = \cos^{-1}\left[\frac{(\mathbf{e} \cdot \mathbf{r})}{(er)}\right]$$

if $(\mathbf{r} \cdot \mathbf{V}) < 0$ then $v = 2\pi - v$

$$\cos v = \frac{r_p(1 + e)}{re} - \frac{1}{e}$$

$$\cos v = \frac{a(1 - e^2)}{re} - \frac{1}{e}$$

$$\cos v = \frac{\cos E - e}{1 - e \cos E}$$

$$\tan\left(\frac{v}{2}\right) = \sqrt{\frac{1 + e}{1 - e}} \tan\left(\frac{E}{2}\right)$$

See Table 6-2

Velocity: V

$$V = \sqrt{\frac{2\mu}{r} - \frac{\mu}{a}}$$

$$r_p V_p = r_a V_a$$

See Eq. (6-4)

Circular velocity: V_{circ}

$$V_{circ} = \sqrt{\mu/r}$$

See Eq. (6-5)

Escape velocity: V_{esc}

$$V_{esc} = \sqrt{2\mu/r}$$

See Eq. (6-6)

Appendix D

Spherical Geometry Formulas

This appendix provides a summary of basic rules. More detailed discussions are provided by Green [1985], Smart [1977], and Newcomb [1960]. Wertz [1999] provides a detailed discussion of global geometry and its application to problems of space mission analysis, including a discussion of "full-sky" techniques that eliminate the quadrant ambiguities which make automated spherical geometry solutions complex and inconvenient.

A *right spherical triangle* is one with at least one right angle. (Unlike plane triangles, spherical triangles can have 1, 2, or 3 right angles.) Any two of the remaining components, including the two remaining angles, serve to completely define the triangle. *Napier's Rules*, given in any of the above books, provide a concise formulation for all possible right spherical triangles. However, experience has shown that it is substantially more convenient to write out explicitly the rules for the relatively small number of possible combinations of known and unknown sides and angles. These are listed in Table D-1.

A *quadrantal spherical triangle* is one with at least one side which is 90 deg in length. As with right spherical triangles, any two of the remaining five components completely define the triangle. These are given by a corresponding set of Napier's Rules. Again, it is more practical to write out explicitly all possible relationships. These are given in Table D-2.

An *oblique spherical triangle* has arbitrary sides and angles. Sides and angles are generally defined over the range of 0 to 180 deg, although most of the spherical geometry relations continue to hold in the angular range up to 360 deg. A set of basic rules which can be applied to any spherical triangle are given in Table D-3. Finally, these general rules can be used to write explicit expressions for any of the unknown components in any oblique spherical triangle with any three components known. These are given in full by Wertz [1999].

References

Green, Robin M. 1985. *Spherical Astronomy*. Cambridge: Cambridge University Press.

Newcomb, Simon. 1960. *A Compendium of Spherical Astronomy*. New York: Dover.

Smart, W.M. 1977. *Textbook on Spherical Astronomy*, 6th ed. Cambridge: Cambridge University Press.

Wertz, James R. 1999. *Spacecraft Orbit and Attitude Systems* Vol. 1. Torrance, CA, and Dordrecht, The Netherlands: Microcosm, Inc., and Kluwer Academic Publishers.

Table D-1. Right Spherical Triangles

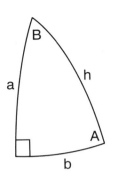

The line below each formula indicates the quadrant of the answer. $Q(A) = Q(a)$ means that the quadrant of angle A is the same as that of side a. "2 possible solutions" means that either quadrant provides a correct solution to the defined triangle.

Given	Find		
a, b	$\cos h = \cos a \cos b$ $Q(h) = \{Q(a) Q(b)\}^*$	$\tan A = \tan a / \sin b$ $Q(A) = Q(a)$	$\tan B = \tan b / \sin a$ $Q(B) = Q(b)$
a, h	$\cos b = \cos h / \cos a$ $Q(b) = \{Q(a)/Q(h)\}^{**}$	$\sin A = \sin a / \sin h$ $Q(A) = Q(a)$	$\cos B = \tan a / \tan h$ $Q(B) = \{Q(a)/Q(h)\}^{**}$
b, h	$\cos a = \cos h / \cos b$ $Q(a) = \{Q(b)/Q(h)\}^{**}$	$\cos A = \tan b / \tan h$ $Q(A) = \{Q(b)/Q(h)\}^{**}$	$\sin B = \sin b / \sin h$ $Q(B) = Q(b)$
a, A	$\sin b = \tan a / \tan A$ 2 possible solutions	$\sin h = \sin a / \sin A$ 2 possible solutions	$\sin B = \cos A / \cos a$ 2 possible solutions
a, B	$\tan b = \sin a \tan B$ $Q(b) = Q(B)$	$\tan h = \tan a / \cos B$ $Q(h) = \{Q(a) Q(B)\}^*$	$\cos A = \cos a \sin B$ $Q(A) = Q(a)$
b, A	$\tan a = \sin b \tan A$ $Q(a) = Q(A)$	$\tan h = \tan b / \cos A$ $Q(h) = \{Q(b) Q(A)\}^*$	$\cos B = \cos b \sin A$ $Q(B) = Q(b)$
b, B	$\sin a = \tan b / \tan B$ 2 possible solutions	$\sin h = \sin b / \sin B$ 2 possible solutions	$\sin A = \cos B / \cos b$ 2 possible solutions
h, A	$\sin a = \sin h \sin A$ $Q(a) = Q(A)$	$\tan b = \tan h \cos A$ $Q(b) = \{Q(A)/Q(h)\}^{**}$	$\tan B = 1/\cos h \tan A$ $Q(B)=\{Q(A)/Q(h)\}^{**}$
h, B	$\sin b = \sin h \sin B$ $Q(b) = Q(B)$	$\tan a = \tan h \cos B$ $Q(a) = \{Q(B)/Q(h)\}^{**}$	$\tan A = 1/\cos h \tan B$ $Q(A)=\{Q(B)/Q(h)\}^{**}$
A, B	$\cos a = \cos A / \sin B$ $Q(a) = Q(A)$	$\cos b = \cos B / \sin A$ $Q(b) = Q(B)$	$\cos h = 1/\tan A \tan B$ $Q(h) = \{Q(A) Q(B)\}^*$

* $\{Q(x) Q(y)\} \equiv$ 1st quadrant if $Q(x) = Q(y)$, 2nd quadrant if $Q(x) \neq Q(y)$

** $\{Q(x)/Q(h)\} \equiv$ quadrant of x if $h \leq 90$ deg, quadrant opposite x if $h > 90$ deg.

Table D-2. Quadrantal
Spherical Triangles

The line below each formula indicates the quadrant of
the answer. $Q(A) = Q(a)$ means that the quadrant of
angle A is the same as that of side a. "2 possible
solutions" means that either quadrant provides a
correct solution to the defined triangle.

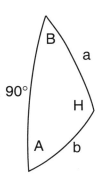

Given	Find		
A, B	$\cos H = - \cos A \cos B$ $Q(H) = \{Q(A) Q(B)\}*$	$\tan a = \tan A / \sin B$ $Q(a) = Q(A)$	$\tan b = \tan B / \sin A$ $Q(b) = Q(B)$
A, H	$\cos B = - \cos H / \cos A$ $Q(B) = \{Q(A) \backslash Q(H)\}**$	$\sin a = \sin A / \sin H$ $Q(a) = Q(A)$	$\cos b = - \tan A / \tan H$ $Q(b) = \{Q(A) \backslash Q(H)\}**$
B, H	$\cos A = - \cos H / \cos B$ $Q(A) = \{Q(B) \backslash Q(H)\}**$	$\cos a = - \tan B / \tan H$ $Q(a) = \{Q(B) \backslash Q(H)\}**$	$\sin b = \sin B / \sin H$ $Q(b) = Q(B)$
A, a	$\sin B = \tan A / \tan a$ 2 possible solutions	$\sin H = \sin A / \sin a$ 2 possible solutions	$\sin b = \cos a / \cos A$ 2 possible solutions
A, b	$\tan B = \sin A \tan b$ $Q(B) = Q(b)$	$\tan H = - \tan A / \cos b$ $Q(H) = \{Q(A) Q(b)\}*$	$\cos a = \cos A \sin b$ $Q(a) = Q(A)$
B, a	$\tan A = \sin B \tan a$ $Q(A) = Q(a)$	$\tan H = - \tan B / \cos a$ $Q(H) = \{Q(B) Q(a)\}*$	$\cos b = \cos B \sin a$ $Q(b) = Q(B)$
B, b	$\sin A = \tan B / \tan b$ 2 possible solutions	$\sin H = \sin B / \sin b$ 2 possible solutions	$\sin a = \cos b / \cos B$ 2 possible solutions
H, a	$\sin A = \sin H \sin a$ $Q(A) = Q(a)$	$\tan B = - \tan H \cos a$ $Q(B) = \{Q(a) \backslash Q(H)\}**$	$\tan b = -1/ \cos H \tan a$ $Q(b) = \{Q(a) \backslash Q(H)\}**$
H, b	$\sin B = \sin H \sin b$ $Q(B) = Q(b)$	$\tan A = - \tan H \cos b$ $Q(A) = \{Q(b) \backslash Q(H)\}**$	$\tan a = -1/ \cos H \tan b$ $Q(a) = \{Q(b) \backslash Q(H)\}**$
a, b	$\cos A = \cos a / \sin b$ $Q(A) = Q(a)$	$\cos B = \cos b / \sin a$ $Q(B) = Q(b)$	$\cos H = -1/ \tan a \tan b$ $Q(H) = \{Q(a) Q(b)\}*$

$*\{Q(x) Q(y)\} \equiv$ 1st quadrant if $Q(x) = Q(y)$, 2nd quadrant if $Q(x) \neq Q(y)$
$**\{Q(x) \backslash Q(H)\} \equiv$ quadrant of x if $H > 90$ deg, quadrant opposite x if $H \leq 90$ deg.

Table D-3. Oblique Spherical Triangles

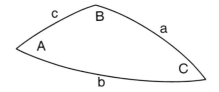

The following rules hold for any spherical triangle:

The Law of Sines: $\dfrac{\sin a}{\sin A} = \dfrac{\sin b}{\sin B} = \dfrac{\sin c}{\sin C}$

The Law of Cosines for Sides:

$$\cos a = \cos b \cos c + \sin b \sin c \cos A$$
$$\cos b = \cos c \cos a + \sin c \sin a \cos B$$
$$\cos c = \cos a \cos b + \sin a \sin b \cos C$$

The Law of Cosines for Angles:

$$\cos A = -\cos B \cos C + \sin B \sin C \cos a$$
$$\cos B = -\cos C \cos A + \sin C \sin A \cos b$$
$$\cos C = -\cos A \cos B + \sin A \sin B \cos c$$

Gauss's Formula:

$$\sin\left[\tfrac{1}{2}(A - B)\right] = \frac{\sin\left[\tfrac{1}{2}(a - b)\right]}{\sin c/2}\cos C/2$$

Useful Derived Formulas:

$$c = \tan^{-1}(\tan b \cos A) \pm \tan^{-1}(\tan a \cos B)$$

$$C = \tan^{-1}\left(\frac{1}{\tan A \cos b}\right) \pm \tan^{-1}\left(\frac{1}{\tan B \cos a}\right)$$

Appendix E
Universal Time and Julian Dates

James R. Wertz, *Microcosm, Inc.*

Calendar time in the usual form of date and time is used only for input and output because arithmetic is cumbersome in months, days, hours, minutes, and seconds. Nonetheless, this is used for most human interaction with space systems because it's the system with which we are most familiar. Even with date and time systems, problems can arise, because time zones are different throughout the world and spacecraft operations typically involve a worldwide network. The uniformly adopted solution to this problem is to use the local standard time corresponding to 0 deg longitude (i.e., the Greenwich meridian) as the assigned time for events anywhere in the world or in space. This is referred to as *Universal Time* (UT), *Greenwich Mean Time* (GMT), or *Zulu* (Z), all of which are equivalent for most practical spacecraft operations. The name Greenwich Mean Time is used because 0 deg longitude is defined by the site of the former Royal Greenwich Observatory in metropolitan London.

Civil time, T_{civil}, as measured by a standard wall clock or time signals, differs from Universal Time by an integral number of hours, corresponding approximately to the longitude of the observer. The approximate relation is:

$$T_{civil} \approx UT \pm (L + 7.5)/15 \qquad \text{(E-1)}$$

where T_{civil} and UT are in hours, and L is the longitude in degrees with the plus sign corresponding to East longitude and the minus sign corresponding to West longitude. The conversion between civil time and Universal Time for most North American and European time zones is given in Table E-1. Substantial variations in time zones are created for political convenience. In addition, most of the United States and Canada observe Daylight Savings Time from the first Sunday in April until the last Sunday in October. Most European countries observe daylight savings time (called "summer time") from the last Sunday in March to the first Sunday in October. Many countries in the southern hemisphere also maintain daylight savings time, typically from October to March. Countries near the equator typically do not deviate from standard time.

Calendar time is remarkably inconvenient for computation, particularly over long time intervals of months or years. We need an absolute time that is a continuous count of time units from some arbitrary reference. The time interval between any two events may then be found by simply subtracting the absolute time of the second event from that of the first. The universally adopted solution for astronomical problems is the *Julian Day*, JD, a continuous count of the number of days since Greenwich noon (12:00 UT) on January 1, 4713 BC*, or, as astronomers now say, –4712. Because Julian Days start at noon UT, they will be a half day off with respect to civil dates. While this is inconvenient for transforming from civil dates to Julian dates, it was useful for astronomers because the date didn't change in the middle of the night (for European observers).

TABLE E-1.   Time Zones in North America, Europe, and Japan. In most of the United States, Daylight Savings Time is used from the first Sunday in April until the last Sunday in October. In Europe, the equivalent "summer time" is used from the last Sunday in March to the first Sunday in October.

Time Zone	Standard Meridian (Deg, East Long.)	UT Minus Standard Time (Hours)	UT Minus Daylight Time (Hours)
Atlantic	300	4	3
Eastern	285	5	4
Central	270	6	5
Mountain	255	7	6
Pacific	240	8	7
Alaska	225	9	8
Hawaii	210	10	NA
Japan	135	−9	NA
Central Europe	15	−1	−2
United Kingdom	0	0	−1

As described below, there are four general approaches for converting between calendar dates and Julian dates.

Table Look-Up

Tabulations of the current Julian Date are in most astronomical ephemerides and almanacs. Table E–2 lists the Julian Dates at the beginning of each year from 1990 through 2031. To find the Julian Date for any given calendar date, simply add the *day number* within the year (and fractional day number, if appropriate) to the Julian Date for Jan 0.0 of that year from Table E–2. Day numbers for each day of the year are on many calendars or can be found by adding the date to the day number for day 0 of the month from Table E-3. Thus 18:00 UT on April 15, 2002 = day number 15.75 + 90 = 105.75 in 2002 = JD 105.75 + 2,452,274.5 = JD 2,452,380.25.

To convert from Julian Days to dates, determine the year in which the Julian Date falls from Table E–2. Subtract the Julian Date from the JD for January 0.0 of that year to determine the day number within the year. This can be converted to a date (and time, if appropriate) by using day numbers on a calendar or subtracting from the day number for the beginning of the appropriate month from Table E–3. Thus, from Table E–2, JD 2,451,608.25 is in the year 2000. The day number is 2,451,608.25 − 2,451,543.5 = 64.75. From Table E–3, this is 18:00 UT, March 4, 2000.

Software Routines Using Integer Arithmetic

A particularly clever procedure for finding the Julian Date, JD, associated with any current year, Y, month, M, and day of the month, D, is given by Fliegel and Van

---

* This strange starting point was suggested by an Italian scholar of Greek and Hebrew, Joseph Scaliger, in 1582 as the beginning of the current *Julian period* of 7,980 years. This period is the product of three numbers: the *solar cycle*, or the interval at which all dates recur on the same days of the week (28 years); the *lunar cycle*, containing an integral number of lunar months (19 years); and the *indiction* or the tax period introduced by the Emperor Constantine in 313 AD (15 years). The last time that these started together was 4713 BC and the next time will be 3267 AD. Scaliger was interested in reducing the astronomical dating problems associated with calendar reforms of his time and his proposal had the convenient selling point that it pre-dated the ecclesiastically approved date of creation, October 4, 4004 BC.

TABLE E–2. Julian Date at the Beginning of Each Year from 1990 to 2031. See text for explanation of use. The day number for the beginning of the year is called "Jan. 0.0" (actually Dec. 31st of the preceding year) so that day numbers can be found by simply using dates. Thus, Jan. 1 is day number 1 and has a JD 1 greater than that for Jan. 0. * = leap year.

Year	JD 2,400,000+ for Jan 0.0 UT	Year	JD 2,400,000+ for Jan 0.0 UT	Year	JD 2,400,000+ for Jan 0.0 UT
1990	47,891.5	2004*	53,004.5	2018	58,118.5
1991	48,256.5	2005	53,370.5	2019	58,483.5
1992*	48,621.5	2006	53,735.5	2020*	58,848.5
1993	48,987.5	2007	54,100.5	2021	59,214.5
1994	49.352.5	2008*	54,465.5	2022	59,579.5
1995	49,717.5	2009	54,831.5	2023	59,944.5
1996*	50,082.5	2010	55,196.5	2024*	60,309.5
1997	50,448.5	2011	55,561.5	2025	60,675.5
1998	50,813.5	2012*	55,926.5	2026	61,040.5
1999	51,178.5	2013	56,292.5	2027	61,405.5
2000*	51,543.5	2014	56,657.5	2028*	61,770.5
2001	51,909.5	2015	57,022.5	2029	62,136.5
2002	52,274.5	2016*	57,387.5	2030	62,501.5
2003	52,639.5	2017	57,753.5	2031	62,866.5

TABLE E–3. Day Numbers for Day 0.0 of Each Month. Leap years (in which February has 29 days) are those evenly divisible by 4. However, years evenly divisible by 100 are not leap years, except that those evenly divisible by 400 are. Leap years are indicated by * in Table E–2.

Month	Non-Leap Years	Leap Years
January	0	0
February	31	31
March	59	60
April	90	91
May	120	121
June	151	152
July	181	182
August	212	213
September	243	244
October	273	274
November	304	305
December	334	335

Flandern [1968] as a computer statement using integer arithmetic. Note that all of the variables must be defined as integers (i.e., any remainder after a division must be truncated) and that both the order of the computations and the parentheses are critical. This procedure works in FORTRAN, C, C++, and Ada for any date on the Gregorian calendar that yields JD > 0. (Add 10 days to the JD for dates on the Julian calendar prior to 1582.)

$$JD_0 = D - 32{,}075 + 1461 \times (Y + 4800 + (M - 14)/12)/4$$
$$+ 367 \times (M - 2 - (M - 14)/12 \times 12)/12$$
$$-3 \times ((Y + 4900 + (M - 14) / 12) / 100) /4 \qquad \text{(E-2a)}$$

Here JD_0 is the Julian Day beginning at noon UT on the given date and must be an integer. For a fractional day, F, in UT (i.e., day number D.F), the floating point Julian Day is given by:

$$JD = JD_0 + F - 0.5 \qquad \text{(E-2b)}$$

For example, the Julian Day beginning at 12:00 UT on December 25, 2007 (Y = 2007, M = 12, D = 25) is JD 2,454,460 and 6:00 UT on that date (F = 0.25) is JD 2,454,459.75.

The inverse routine for computing the date from the Julian Day is given by:

$$L = JD_0 + 68{,}569 \qquad \text{(E-3a)}$$

$$N = (4 \times L) / 146{,}097 \qquad \text{(E-3b)}$$

$$L = L - (146097 \times N + 3) / 4 \qquad \text{(E-3c)}$$

$$I = (4000 \times (L + 1)) / 1{,}461{,}001 \qquad \text{(E-3d)}$$

$$L = L - (1461 \times I) / 4 + 31 \qquad \text{(E-3e)}$$

$$J = (80 \times L) / 2{,}447 \qquad \text{(E-3f)}$$

$$D = L - (2447 \times J) / 80 \qquad \text{(E-3g)}$$

$$L = J / 11 \qquad \text{(E-3h)}$$

$$M = J + 2 - 12 \times L \qquad \text{(E-3i)}$$

$$Y = 100 \times (N - 49) + I + L \qquad \text{(E-3j)}$$

where integer arithmetic is used throughout. Y, M, and D are the year, month, and day, and I, J, L, and N are intermediate variables. Finally, again using integer arithmetic, the day of the week, W, corresponding to the Julian Date beginning at 12:00 on that day is given by:

$$W = JD_0 - 7 \times ((JD + 1) / 7) + 2 \qquad \text{(E-4)}$$

where W = 1 corresponds to Sunday. Thus, December 25, 2007 falls on Tuesday.

Software Routines Without Integer Arithmetic

While most computer languages provide integer arithmetic, spreadsheets such as Excel or MatLab typically do not. (See below for use of Excel and MatLab DATE functions.) Similar capabilities are available using integer (INT) or truncation (TRUNC in Excel, FIX in MatLab) functions. INT and TRUNC are identical for positive numbers, but differ for negative numbers: INT(-3.1) = -4, whereas

TRUNC(–3.1) = –3. It is the TRUNC or FIX function which is equivalent to integer arithmetic. Thus, using the same variables as above, we can rewrite Eqs. (E-2) for computation of JD from the date as:

$$C = TRUNC((M - 14)/12) \tag{E-5a}$$

$$\begin{aligned} JD_0 = D &- 32{,}075 + TRUNC(1{,}461 \times (Y + 4{,}800 + C)/4) \\ &+ TRUNC(367 \times (M - 2 - C \times 12)/12) \\ &- TRUNC(3 \times (TRUNC(Y + 4{,}900 + C) / 100) /4) \end{aligned} \tag{E-5b}$$

$$JD = JD_0 + F - 0.5 \tag{E-5c}$$

where again JD_0, Y, M, D, and C are integers and F and JD are real numbers. Applying the same rules to Eq. (E-3) gives the inverse formula for the date in terms of JD as:

$$L = JD + 68{,}569 \tag{E-6a}$$

$$N = TRUNC((4 \times L) / 146{,}097) \tag{E-6b}$$

$$L = L - TRUNC((146097 \times N + 3) / 4) \tag{E-6c}$$

$$I = TRUNC((4000 \times (L + 1)) / 1{,}461{,}001) \tag{E-6d}$$

$$L = L - TRUNC((1{,}461 \times I) / 4) + 31 \tag{E-6e}$$

$$J = TRUNC((80 \times L) / 2{,}447) \tag{E-6f}$$

$$D = L - TRUNC((2{,}447 \times J) / 80) \tag{E-6g}$$

$$L = TRUNC(J / 11) \tag{E-6h}$$

$$M = J + 2 - 12 \times L \tag{E-6i}$$

$$Y = 100 \times (N - 49) + I + L \tag{E-6j}$$

where the variables are the same as Eq. (E-3), except that D is now a real number corresponding to the date and fraction of a day. Finally, Eq. (E-4) for the day of the week becomes:

$$\begin{aligned} W &= JD - 7 \times TRUNC ((JD + 1.5) / 7) + 2.5 \\ &= JD - 7 \times INT ((JD + 1.5) / 7) + 2.5 \end{aligned} \tag{E-7}$$

where $1 \le W < 2$ corresponds to Sunday. The examples given above can also serve as test cases for Eqs. (E-5), (E-6), and (E-7).

Modified Julian Date

The Julian Date presents minor problems for space applications. Because it was introduced principally for astronomical use, Julian Dates begin at 12:00 UT rather than 0 hours UT, as the civil calendar does (thus the 0.5 day differences in Table E–2). In addition, the 7 digits required for the Julian Date did not permit the use of single precision arithmetic in older computer programs. This is no longer a problem with modern computer storage and number formats. Nonetheless, various forms of truncated Julian dates have gained at least some use.

The most common of the truncated Julian dates for astronomical and astronautics use is the *Modified Julian Date*, MJD, given by:

$$MJD = JD - 2{,}400{,}000.5 \tag{E-8}$$

MJD begins at midnight, to correspond with the civil calendar. Thus, in using

Table E–2, the MJD is given by adding the day of the year (plus fractions of a day, if appropriate) to the number in the table, with the ".5" at the end of the table-listing dropped. For example, the MJD for 18:00 UT on Jan. 3, 2002 = MJD 52,277.75. The definition of the MJD given here is that adopted by the International Astronomical Union in 1997. Note, however, that other definitions of the MJD have been used. Thus, the most unambiguous approach remains the use of the full Julian Date.

Spreadsheets such as Excel or MatLab

Spreadsheets, such as Excel or MatLab, typically store dates internally as some form of day count and allow arithmetic operations, such as subtraction. Thus, we can either subtract two dates directly to determine a time interval or convert them to Julian Dates by simply finding the additive constant, K, given by:

$$K = JD - I \tag{E-9}$$

where I is the internal number representing a known date, JD. Once this is determined, then the JD for any date is:

$$JD = K + I \tag{E-10}$$

Many versions of Excel use Jan. 1, 1904, as "day 0," such that $K_{Excel} = 2,416,480.5$. However, this should be checked for individual programs because other starting points are sometimes used and the starting point is a variable parameter in some versions of Excel. While this can be a very convenient function, Excel date routines run only from 1904 to 2078.

MatLab typically uses Jan. 1, 0000, 0:0:0 as "day 0." Thus, in the formula above, $K_{MatLab} = 1,721,058.5$.

Any of the day counting approaches will work successfully over its allowed range. However, systems intended for general mathematics or business use may not account correctly for leap years and calendar changes when historical times or times far in the future are being evaluated. Thus, the use of the full Julian Date remains the most unambiguous solution, particularly if a program or result is to be used by more than one person or program. For a more extended discussion of time systems, see for example, Seidelmann [1992] or Wertz [1999].

References

Fliegel, Henry F. and Thomas C. Van Flandern. 1968. "A Machine Algorithm for Processing Calendar Dates." *Communications of the ACM*, vol. II, p. 657.

Seidelmann, P. Kenneth. 1992. *Explanatory Supplement to the Astronomical Almanac*. Mill Valley, CA: University Science Books.

Wertz, James R. 1999. *Spacecraft Orbit and Attitude Systems*, vol. I. Torrance, CA and Dordrecht, The Netherlands: Microcosm, Inc. and Kluwer Academic.

Appendix F
Units and Conversion Factors

Robert Bell, *Microcosm, Inc.*

The metric system of units, officially known as the *International System of Units*, or *SI*, is used throughout this book, with the exception that angular measurements are usually expressed in degrees rather than the SI unit of radians. By international agreement, the fundamental SI units of length, mass, and time are defined as follows (see National Institutes of Standards and Technology, Special Publication 330 [1991]):

The *meter* is the length of the path traveled by light in vacuum during a time interval of 1/299,792,458 of a second.

The *kilogram* is the mass of the international prototype of the kilogram.

The *second* is the duration of 9,192,631,770 periods of the radiation corresponding to the transition between two hyperfine levels of the ground state of the cesium-133 atom.

Additional base units in the SI system are the *ampere* for electric current, the *kelvin* for thermodynamic temperature, the *mole* for amount of substance, and the *candela* for luminous intensity. Taylor [1995] provides an excellent summary of SI units for scientific and technical use.

The names of multiples and submultiples of SI units are formed by application of the following prefixes:

Factor by Which Unit is Multiplied	Prefix	Symbol	Factor by Which Unit is Multiplied	Prefix	Symbol
10^{24}	yotta	Y	10^{-1}	deci	d
10^{21}	zetta	Z	10^{-2}	centi	c
10^{18}	exa	E	10^{-3}	milli	m
10^{15}	peta	P	10^{-6}	micro	μ
10^{12}	tera	T	10^{-9}	nano	n
10^{9}	giga	G	10^{-12}	pico	p
10^{6}	mega	M	10^{-15}	femto	f
10^{3}	kilo	k	10^{-18}	atto	a
10^{2}	hecto	h	10^{-21}	zepto	z
10^{1}	deka	da	10^{-24}	yocto	y

For each quantity listed below, the SI unit and its abbreviation are given in brackets. For convenience in computer use, most conversion factors are given to the greatest available accuracy. Note that some conversions are exact definitions and some (speed of light, astronomical unit) depend on the value of physical constants. "..." indicates a repeating decimal. All notes are on the last page of the list.

To convert from	*To*	*Multiply by*	*Notes*
Acceleration [meter/second², m/s²]			
Gal (galileo)	m/s²	0.01	E
Inch/second², in/s²	m/s²	0.025 4	E
Foot/second², ft/s²	m/s²	0.304 8	E
Free fall (standard), g	m/s²	9.806 65	E
Angular Acceleration [radian/second², rad/s²]			
Degrees/second², deg/s²	rad/s²	$\pi/180$ $\approx 0.017\ 453\ 292\ 519\ 943\ 295\ 77$	E
Revolutions/second², rev/s²	rad/s²	2π $\approx 6.283\ 185\ 307\ 179\ 586\ 477$	E
Revolutions/minute², rev/min²	rad/s²	$\pi/1{,}800$ $\approx 1.745\ 329\ 251\ 994\ 329\ 577 \times 10^{-3}$	E
Revolutions/minute²	deg/s²	0.1	E
Radians/second², rad/s²	deg/s²	$180/\pi$ $\approx 57.29\ 577\ 951\ 308\ 232\ 088$	E
Revolutions/second², rev/s²	deg/s²	360	E
Angular Area [sr], book also uses deg²			
Degree², deg²	sr	$(\pi/180)^2$ $\approx 3.046\ 174\ 197\ 867\ 086 \times 10^{-4}$	E
Minute², min²	sr	$(\pi/10{,}800)^2$ $\approx 8.461\ 594\ 994\ 075\ 237 \times 10^{-8}$	E
Second², s²	sr	$(\pi/648\ 000)^2$ $\approx 2.350\ 443\ 053\ 909\ 289 \times 10^{-4}$	E
Steradian, sr	deg²	$(180/\pi)^2$ $\approx 3.282\ 806\ 350\ 011\ 744 \times 10^{3}$	E
Minute², min²	deg²	$1/3{,}600$ $= 2.777\ldots \times 10^{-4}$	E
Second², s²	deg²	$(1/3{,}600)^2$ $\approx 7.716\ 049\ 382\ 716\ 049 \times 10^{-8}$	E
Steradian, sr	rad²	1 rad²	E
Angular Measure [radian, rad]. This book uses degree (abbreviated "deg") as the basic unit.			
Degree, deg	rad	$\pi/180$ $\approx 0.017\ 453\ 292\ 519\ 943\ 295\ 77$	E
Minute (of arc), min	rad	$\pi/10{,}800$ $\approx 2.908\ 882\ 086\ 657\ 216 \times 10^{-4}$	E
Second (of arc), s	rad	$\pi/648\ 000$ $\approx 4.848\ 136\ 811\ 095\ 360 \times 10^{-6}$	E
Radian, rad	deg	$= 180/\pi$ $\approx 57.295\ 779\ 513\ 082\ 320\ 877$	E
Minute (of arc), min	deg	$1/60$ $= 0.01666\ldots$	E
Second (of arc), s	deg	$1/3{,}600$ $= 2.777\ldots \times 10^{-4}$	E

To convert from	*To*	*Multiply by*	*Notes*
Angular Momentum [kilogram · meter²/second, kg · m²/s]			
Gram · cm²/second, g · cm²/s	kg · m²/s	1×10^{-7}	E
lbm· inch²/second, lbm · in²/s	kg · m²/s	$2.926\ 396\ 534\ 292 \times 10^{-4}$	E
Slug · inch²/second, slug · in²/s	kg · m²/s	$9.415\ 402\ 418\ 968 \times 10^{-3}$	D
lbm· foot²/second, lbm · ft²/s	kg · m²/s	0.042 140 110 093 80	D
Inch · lbf· second, in · lbf · s	kg · m²/s	0.112 984 829 027 6	D
Slug · foot²/second, slug · ft²/s = foot · lbf · second, ft · lbf · s	kg · m²/s	1.355 817 948 331	D
Angular Velocity [radian/second, rad/s]. This book uses degrees/second as the basic unit.			
Degrees/second, deg/s	rad/s	$\pi/180$ $\approx 0.017\ 453\ 292\ 519\ 943\ 295\ 77$	E
Revolutions/minute, rpm	rad/s	$\pi/30$ $\approx 0.104\ 719\ 755\ 119\ 659\ 774\ 6$	E
Revolutions/second, rev/s	rad/s	2π $\approx 6.283\ 185\ 307\ 179\ 586\ 477$	E
Revolutions/minute, rpm	deg/s	6	E
Radians/second, rad/s	deg/s	$180/\pi$ $\approx 57.295\ 779\ 513\ 082\ 320\ 88$	E
Revolutions/second, rev/s	deg/s	360	E
Area [meter², m²]			
Acre	m²	$4.046\ 856\ 422 \times 10^3$	E
Foot², ft²	m²	0.092 903 04	E
Hectare	m²	1×10^4	E
Inch², in²	m²	$6.451\ 6 \times 10^{-4}$	E
Mile² (U.S. statute)	m²	$2.589\ 110\ 336 \times 10^6$	E
Yard², yd²	m²	0.836 127 36	E
(Nautical mile)²	m²	$3.429\ 904 \times 10^6$	E
Density [kilogram/meter³, kg/m³]			
Gram/centimeter³, g/cm³	kg/m³	1.0×10^3	E
Pound mass/inch³, lbm/in³	kg/m³	$2.767\ 990\ 471\ 020 \times 10^4$	D
Pound mass/foot³, lbm/ft³	kg/m³	16.018 463 373 96	D
Slug/ft³	kg/m³	515.378 818 393 2	D
Electric Charge [coulomb, C]			
Abcoulomb	C	10	E
Faraday (based on carbon–12)	C	$9.648\ 70 \times 10^4$	NIST
Faraday (chemical)	C	$9.649\ 57 \times 10^4$	NIST
Faraday (physical)	C	$9.652\ 19 \times 10^4$	NIST
Statcoulomb	C	$3.335\ 641 \times 10^{-10}$	NIST
Electric Conductance [siemens, S]			
Abmho	S	1×10^9	E
Mho (Ω^{-1})	S	1	E
Electric Current [ampere, A]			
Abampere	A	10	E

To convert from	To	Multiply by	Notes
Gilbert	A	$10/4\pi$	
		$\approx 0.795\ 774\ 715\ 459\ 5$	E
Statampere	A	$3.335\ 641 \times 10^{-10}$	NIST

Electric Field Intensity
[volt/meter $\equiv$ kilogram $\cdot$ meter $\cdot$ ampere^{-1} $\cdot$ second^{-3}, V/m $\equiv$ kg $\cdot$ m $\cdot$ A^{-1} $\cdot$ s^{-3}]

Electric Potential Difference
[volt $\equiv$ watt/ampere $\equiv$ kilogram $\cdot$ meter^2 $\cdot$ ampere^{-1} $\cdot$ second^{-3}, V $\equiv$ W/A $\equiv$ kg $\cdot$ m^2 $\cdot$ A^{-1} $\cdot$ s^{-3}]

Abvolt	V	1×10^{-8}	E
Statvolt	V	$299.792\ 5$	NIST

Electric Resistance
[ohm $\equiv$ volt/ampere $\equiv$ kilogram $\cdot$ meter^2 $\cdot$ ampere^{-2} $\cdot$ second^{-3}, Ω $\equiv$ V/A $\equiv$ kg $\cdot$ m^2 $\cdot$ A^{-2} $\cdot$ s^{-3}]

Abohm	Ω	1×10^{-9}	E
Statohm	Ω	$8.987\ 552 \times 10^{11}$	NIST

Energy or Torque
[joule $\equiv$ newton $\cdot$meter $\equiv$ kilogram $\cdot$ meter^2/s^2 , J $\equiv$ N $\cdot$ m $\equiv$ kg $\cdot$ m^2/s^2]

British thermal unit, Btu (mean)	J	$1.055\ 055\ 852\ 62 \times 10^3$	E
Calorie (IT), cal	J	$4.186\ 8$	E
Kilocalorie (IT), kcal	J	$4.186\ 8 \times 10^3$	E
Electron volt, eV	J	$1.602\ 177\ 33 \times 10^{-19}$	C
Erg $\equiv$ gram $\cdot$ cm^2/s^2			
= pole $\cdot$ cm $\cdot$ oersted	J	1×10^{-7}	E
Foot poundal	J	$0.042\ 140\ 110\ 093\ 80$	D
Foot lbf = slug $\cdot$ foot^2/s^2	J	$1.355\ 817\ 948\ 331\ 4$	E
Kilowatt hour, kW $\cdot$ hr	J	3.6×10^6	E
Ton equivalent of TNT	J	4.184×10^9	E

Force [newton $\equiv$ kilogram $\cdot$ meter/second^2, N $\equiv$ kg $\cdot$ m /s^2]

Dyne	N	1×10^{-5}	E
Kilogram-force (kgf)	N	$9.806\ 65$	E
Ounce force (avoirdupois)	N	$0.278\ 013\ 850\ 953\ 8$	D
Poundal	N	$0.138\ 254\ 954\ 376$	E
Pound force (avoirdupois),	N	$4.448\ 221\ 615\ 260\ 5$	E
lbf $\equiv$ slug $\cdot$ foot/s^2			

Illuminance [lux $\equiv$ candela $\cdot$ steradian/meter^2, lx $\equiv$ cd $\cdot$ sr/m^2]

Footcandle	cd $\cdot$ sr/m^2	$10.763\ 910\ 416\ 709\ 70$	E
Phot	cd $\cdot$ sr/m^2	1×10^4	E

Length [meter, m]

Angstrom, Å	m	1×10^{-10}	E
Astronomical unit (S1)	m	$1.495\ 978\ 706\ 6 \times 10^{11}$	AA
Astronomical unit (radio)	m	$1.495\ 978\ 9 \times 10^{11}$	NIST
Earth equatorial radius, R_E	m	$6.378\ 136\ 49 \times 10^6$	IERS
		$6.378\ 14 \times 10^6$	AQ
Fermi (1 fermi = 1 fm)	m	1×10^{-15}	E
Foot, ft	m	$0.304\ 8$	E
Inch, in	m	$0.025\ 4$	E

To convert from	To	Multiply by	Notes
Light year	m	$9.460\ 730\ 472\ 580\ 8 \times 10^{15}$	D
Micron, μm	m	1×10^{-6}	E
Mil (10^{-3} inch)	m	2.54×10^{-5}	E
Mile (U.S. statute), mi	m	$1.609\ 344 \times 10^3$	E
Nautical mile (U.S.), NM	m	1.852×10^3	E
Parsec (IAU)	m	$3.085\ 677\ 597\ 49 \times 10^{16}$	D
Solar radius	m	$6.960\ 00 \times 10^8$	AA
Yard, yd	m	$0.914\ 4$	E

Luminance [candela/meter$^2 \equiv$ cd/m^2]

Footlambert	cd/m^2	$\approx 3.426\ 259\ 099\ 635\ 39$	E
Lambert	cd/m^2	$(1/\pi) \times 10^4 \approx 3.183\ 098\ 862 \times 10^3$	E
Stilb	cd/m^2	1×10^4	E

Magnetic Field Strength, H [ampere turn/meter, A/m]

Oersted (EMU)	A/m	$(1/4\pi) \times 10^3$ $\approx 79.577\ 471\ 545\ 947\ 667\ 88$	E,1

Magnetic Flux
[weber $\equiv$ volt $\cdot$ s $\equiv$ kilogram $\cdot$ meter^2 $\cdot$ ampere^{-1}·second^{-2}, Wb $\equiv$ V $\cdot$ s $\equiv$ kg $\cdot$ m^2 $\cdot$ A^{-1} $\cdot$ s^{-2}]

Maxwell (EMU)	Wb	1×10^{-8}	E
Unit pole	Wb	$1.256\ 637 \times 10^{-7}$	NIST

Magnetic Induction, B
[tesla $\equiv$ weber/meter^2 $\equiv$ kilogram $\cdot$ ampere^{-1} $\cdot$ second^{-2}, T $\equiv$ Wb/m^2 $\equiv$ kg $\cdot$ A^{-1} $\cdot$ s^{-2}]

Gamma (EMU) (γ)	T	1×10^{-9}	E,1
Gauss (EMU)	T	1×10^{-4}	E,1

Magnetic Dipole Moment
[weber $\cdot$ meter $\equiv$ kilogram $\cdot$ meter^3 $\cdot$ ampere^{-1} $\cdot$ second^{-2}, Wb $\cdot$ m $\equiv$ kg $\cdot$ m^3 $\cdot$ A^{-1} $\cdot$ s^{-2}]

Pole $\cdot$ centimeter (EMU)	Wb $\cdot$ m	$4\pi \times 10^{-10}$ $\approx 1.256\ 637\ 061\ 435\ 917\ 295 \times 10^{-9}$	E,1
Gauss $\cdot$ centimeter^3 (Practical)	Wb $\cdot$ m	1×10^{-10}	E,1

Magnetic Moment [ampere turn $\cdot$ meter^2 $\equiv$ joule/tesla, A $\cdot$ m^2 $\equiv$ J/T]

Abampere $\cdot$ centimeter^2(EMU)	A $\cdot$ m^2	1×10^{-3}	E, 1
Ampere $\cdot$ centimeter^2	A $\cdot$ m^2	1×10^{-4}	E, 1

Mass [kilogram, kg]

γ (= 1 μg)	kg	1×10^{-9}	E
Atomic unit (electron)	kg	$9.109\ 389\ 7 \times 10^{-31}$	C
Atomic mass unit (unified), amu	kg	$1.660\ 540\ 2 \times 10^{-27}$	C
Metric carat	kg	2.0×10^{-4}	E
Metric ton	kg	1×10^3	E
Ounce mass (avoirdupois), oz	kg	$0.028\ 349\ 231\ 25$	E
Pound mass, lbm (avoirdupois)	kg	$0.453\ 592\ 37$	E
Slug	kg	$14.593\ 902\ 937\ 21$	D
Short ton (2,000 lbm)	kg	$907.184\ 74$	E
Solar mass	kg	$1.989\ 1 \times 10^{30}$	AA

To convert from	*To*	*Multiply by*	*Notes*
Moment of Inertia [kilogram · meter^2, kg · m^2]			
Gram · centimeter^2, gm · cm^2	kg · m^2	1×10^{-7}	E
Pound mass· inch^2, lbm · in^2	kg · m^2	$2.926\ 396\ 534\ 292 \times 10^{-4}$	E
Pound mass· foot^2, lbm · ft^2	kg · m^2	$4.214\ 011\ 009\ 380 \times 10^{-2}$	D
Slug · inch^2, slug · in^2	kg · m^2	$9.415\ 402\ 418\ 968 \times 10^{-3}$	D
Inch · pound force· s^2, in · lbf · s^2	kg · m^2	$0.112\ 984\ 829\ 027\ 6$	D
Slug · foot^2 = ft · lbf · s^2	kg · m^2	$1.355\ 817\ 948\ 331\ 4$	E
Power [watt ≡ joule/second ≡ kilogram · meter^2/second^3, W ≡ J/s ≡ kg · m^2/s^3]			
Foot · pound force/second, ft lbf/s	W	$1.355\ 817\ 948\ 331$	D
Horsepower (550 ft · lbf/s), hp	W	$745.699\ 871\ 582\ 3$	D
Horsepower (electrical), hp	W	746.0	E
Solar luminosity	W	3.845×10^{26}	AQ
Pressure or Stress			
[pascal ≡ newton/meter^2 ≡ kilogram · meter^{-1} · second^{-2}, Pa ≡ N/m^2 ≡ kg · m^{-1} · s^{-2}]			
Atmosphere, atm	Pa	$1.013\ 25 \times 10^5$	E
Bar	Pa	1×10^5	E
Centimeter of mercury (0° C)	Pa	$\approx 1.333\ 223\ 874\ 145 \times 10^3$	E
Dyne/centimeter^2, dyne/cm^2	Pa	0.1	E
Inch of mercury (32° F)	Pa	$3.386\ 388\ 640\ 341 \times 10^3$	E
Pound force/foot^2, lbf/ft^2, psf	Pa	$47.880\ 258\ 980\ 34$	D
Pound force/inch^2, lbf/in^2, psi	Pa	$6.894\ 757\ 293\ 168 \times 10^3$	D
Torr (0° C)	Pa	$(101325/760)$	
		$\approx 133.322\ 368\ 421\ 052\ 631$	E

Solid Angle (See Angular Area)

Specific Heat Capacity
[joule · kilogram^{-1} · kelvin^{-1} ≡ meter^2 · second^2 · kelvin^{-1}, J · kg^{-1} · K^{-1} ≡ m^2 · s^2 · K^{-1}]

cal · g^{-1} · K^{-1} (mean)	J · kg^{-1} · K^{-1}	$4.186\ 80 \times 10^3$	E
Btu · lbm^{-1} · °F^{-1} (mean)	J · kg^{-1} · K^{-1}	$4.186\ 80 \times 10^3$	E

Stress (see Pressure)

Temperature [kelvin, K]

Celsius, °C	K	$t_K = t_C + 273.15$	E
Fahrenheit, °F	K	$t_K = (5/9)\,(t_F + 459.67)$	E
Rankine °R	K	$t_K = (5/9)\,t_R$	E
Fahrenheit, °F	C	$t_C = (5/9)\,(t_F - 32.0)$	E
Rankine °R	C	$t_C = (5/9)\,(t_R - 491.67)$	E

Thermal Conductivity [watt · meter^{-1} · kelvin^{-1} ≡ kilogram · meter · second^{-3} · kelvin^{-1}, W · m^{-1} · K^{-1} ≡ kg · m · s^{-3} · K^{-1}]

cal · cm^{-1} · s^{-1} · K^{-1} (mean)	W · m^{-1} · K^{-1}	418.68	E
Btu · ft^{-1} · hr^{-1} · °F^{-1} (mean)	W · m^{-1} · K^{-1}	$1.730\ 734\ 666\ 371\ 39$	D

Time [second, s]

Sidereal day, d$_*$ (ref. = Υ)	s	$8.616\ 410\ 035\ 2 \times 10^4$	
		$= 23\text{h}\ 56\text{m}\ 4.100\ 352\text{s}$	AQ
Ephemeris day, d$_e$	s	8.64×10^4	AQ

To convert from	To	Multiply by	Notes
Ephemeris day, d_e	d_*	1.002 737 795 056 6	AQ
Keplerian period of a satellite in low-Earth orbit	min	$1.658\ 669\ 010\ 080 \times 10^{-4} \times a^{3/2}$	
		(a in km)	Table 6-2
Keplerian period of a satellite of the Sun	d_e	$3.652\ 568\ 954\ 757 \times$ $10^2 \times a^{3/2}$ (a in AU)	AA
Tropical year (ref.= Υ)	s	$3.155\ 692\ 597\ 47 \times 10^7$	AA
Tropical year (ref.= Υ)	d_e	365.242 198 781	D
Sidereal year (ref.=fixed stars)	s_e	$3.155\ 814\ 976\ 320 \times 10^7$	AA
Sidereal year (ref.=fixed stars)	d_e	365.256 363	AA
Calendar year (365 days), yr	s	$3.153\ 6 \times 10^7$	E
Julian century	d	36,525	E
Gregorian calendar century	d	36,524.25	E

Torque (see Energy)

Velocity [meter/second, m/s]

Foot/minute, ft/min	m/s	5.08×10^{-3}	E
Inch/second, ips	m/s	0.025 4	E
Kilometer/hour, km/hr	m/s	$(3.6)^{-1} = 0.277777\ldots$	E
Foot/second, fps or ft/s	m/s	0.304 8	E
Miles/hour, mph	m/s	0.447 04	E
Knot (international)	m/s	$(1852/3600) = 0.514444\ldots$	E
Miles/minute, mi/min	m/s	26.822 4	E
Miles/second, mi/s	m/s	$1.609\ 344 \times 10^3$	E
Velocity of Light	m/s	$2.997\ 924\ 58 \times 10^8$	E

Viscosity [pascal · second $\equiv$ kilogram · meter^{-1} · second^{-1} , Pa · s $\equiv$ kg · m^{-1} · s^{-1}]

Stoke	m^2/s	1.0×10^{-4}	E
Foot^2 · second, ft^2 · s	m^2/s	0.092 903 04	E
Pound mass· foot^{-1} · second^{-1}, lbm · ft^{-1} · s^{-1}	Pa · s	1.488 163 943 570	D
Pound force· second/foot^2, lbf · s/ft^2	Pa · s	47.880 258 980 34	D
Poise	Pa · s	0.1	E
Poundal second/foot^2, poundal s/ft^2	Pa · s	1.488 163 943 570	D
Slug · foot^{-1} · second^{-1}, slug · ft^{-1} · s^{-1}	Pa · s	47.880 258 980 34	D
Rhe	(Pa · s)$^{-1}$	10	E

Volume [meter^3, m^3]

λ ($1\lambda = 1\ \mu L = 1 \times 10^{-6}$ L)	m^3	1×10^{-9}	E
Foot^3, ft^3	m^3	$2.831\ 684\ 659\ 2 \times 10^{-2}$	E
Gallon (U.S. liquid), gal	m^3	$3.785\ 411\ 784 \times 10^{-3}$	E
Inch^3, in^3	m^3	$1.638\ 706\ 4 \times 10^{-5}$	E

To convert from	To	Multiply by	Notes
Liter, L	m^3	1×10^{-3}	E
Ounce (U.S. fluid), oz	m^3	$2.957\ 352\ 956\ 25 \times 10^{-5}$	E
Pint (U.S. liquid), pt	m^3	$4.731\ 764\ 73 \times 10^{-4}$	E
Quart, qt	m^3	$9.463\ 529\ 46 \times 10^{-4}$	E
Stere (st)	m^3	1	E
Yard^3, yd^3	m^3	$0.764\ 554\ 857\ 984$	E

Notes for the preceding table:

AA   Values are those of *Astronomical Almanac* [Hagen and Boksenberg, 1991].

AQ   Values are those of *Astrophysical Quantities* [Cox, 1991].

C   Values are those of Cohen and Taylor [1986].

D   Values that are derived from exact quantities, rounded off to 13 significant figures.

E   (Exact) indicates that the conversion is exact by definition of the non-SI unit or that it is obtained from other exact conversions.

IERS   Numerical standards of the IERS.

NIST   Values are those of National Institute of Standards and Technology [McCoubrey, 1991]

(1)   Care should be taken in transforming magnetic units, because the dimensionality of magnetic quantities (**B**, **H**, etc.) depends on the system of units. Most of the conversions given here are between SI and EMU (electromagnetic). The following equations hold in both sets of units:

$$\mathbf{T} = \mathbf{m} \times \mathbf{B} = \mathbf{d} \times \mathbf{H}$$
$$\mathbf{B} = \mu\mathbf{H}$$
$$\mathbf{m} = I\mathbf{A} \text{ for a current loop in a plane}$$
$$\mathbf{d} = \mu\mathbf{m}$$

with the following definitions:

$\mathbf{T} \equiv$ torque

$\mathbf{B} \equiv$ magnetic induction (commonly called "magnetic field")

$\mathbf{H} \equiv$ magnetic field strength or magnetic intensity

$\mathbf{m} \equiv$ magnetic moment

$I \equiv$ current loop

$\mathbf{A} \equiv$ vector normal to the plane of the current loop (in the direction of the angular velocity vector of the current loop about the center of the loop) with magnitude equal to the area of the loop.

$\mathbf{d} \equiv$ magnetic dipole moment

$\mu \equiv$ magnetic permeability

The permeability of vacuum, μ_0, has the following values, by definition:

$\mu_0 \equiv 1$ (dimensionless)   EMU

$\mu_0 \equiv 4\pi \times 10^{-7}$ N/A^2 SI

Therefore, in electromagnetic units in vacuum, magnetic induction and magnetic field strength are equivalent and the magnetic moment and magnetic dipole moment are equivalent. For practical purposes of magnetostatics, space is a vacuum but the spacecraft itself may have $\mu \neq \mu_0$.

Useful Mathematical Constants and Values

Constant	Value	
π	$\approx 3.141\ 592\ 653\ 589\ 793\ 238\ 462\ 643$	**(A)**
e	$\approx 2.718\ 281\ 828\ 459\ 045\ 235\ 360\ 287$	**(A)**
e^{π}	$\approx 23.140\ 692\ 632\ 779\ 269\ 006$	**(A)**
$\log_{10}x$	$\approx 0.434\ 294\ 481\ 903\ 251\ 827\ 651\ 128\ 9\ \log_e x$	**(A)**
$\log_e x$	$\approx 2.302\ 585\ 092\ 994\ 045\ 684\ 017\ 991\ \log_{10}x$	**(A)**
$\log_e \pi$	$\approx 1.144\ 729\ 885\ 849\ 400\ 174\ 143\ 427$	**(A)**

(A) are from *The Handbook of Mathematical Functions, with Formulas, Graphs, and Mathematical Tables* [Abramowitz and Stegun, 1970]

References

Abramowitz, Milton and Irene A. Stegun, eds. 1970. *The Handbook of Mathematical Functions with Formulas, Graphs, and Mathematical Tables*. New York: Dover.

Cohen, E. Richard and B.N. Taylor, 1986. *CODATA Bulletin No. 63*, Nov. New York: Pergamon Press.

Cox, A.N. ed. 1999. *Astrophysical Quantities*. New York: Springer-Verlag

Hagen, James B. and A. Boksenberg, eds. 1991. *The Astronomical Almanac*. Nautical Almanac Office, U.S. Naval Observatory and H. M. Nautical Almanac Office. 1992. Washington, D.C.: U.S. Government Printing Office.

McCarthy, Dennis D., USNO. 1996. "Technical Note 21." IERS Conventions.

McCoubrey, Arthur O. 1991. *Guide for the Use of the International System of Units (SI)*. National Institute of Standards and Technology (NIST), Special Publication 811, U.S. Department of Commerce: U.S. Government Printing Office.

Seidelmann, Kenneth P., USNO, ed. 1992. *The Explanatory Supplement to the Astronomical Almanac*. Mill Valley, CA: University Science Books.

Taylor, Barry N. 1991. *The International System of Units (SI)*. National Institute of Standards and Technology (NIST), Special Publication 811, U.S. Department of Commerce: U.S. Government Printing Office.

Index

—C—

—E—

—I—

—O—

Explanation of Earth Satellite Parameters

The following table provides a variety of quantitative data for Earth-orbiting satellites. Limitations, formulas, and text references are given below. The independent parameter in the formulas is the distance, r, from the center of the Earth in km. The left most column on each table page is the altitude, $h \equiv r - R_E$, where $R_E = 6{,}378.14$ km is the equatorial radius of the Earth.

1. *Instantaneous Area Access for a 0 deg Elevation Angle* or *the Full Geometric Horizon* $(10^6 km^2)$. All the area that an instrument or antenna could potentially see at any instant if it were scanned through its normal range of orientations for which the spacecraft elevation is above 0 deg [Eq. (7-6)].

2. *Instantaneous Area Access for a 5 deg Minimum Elevation Angle* $(10^6 km^2)$ = same as col. 1 but with elevation of 5 deg.

3. *Instantaneous Area Access for a 10 deg Minimum Elevation Angle* $(10^6 km^2)$ = same as col. 1 but with elevation of 10 deg.

4. *Instantaneous Area Access for a 20 deg Minimum Elevation Angle* $(10^6 km^2)$ = same as col. 1 but with elevation of 20 deg.

5. *Area Access Rate for an Elevation of 0 deg* $(10^3 km^2/s)$ = the rate at which new land is coming into the spacecraft's access area [Eq. (7-10)].

6. *Area Access Rate for an Elevation Limit of 5 deg* $(10^3 km^2/s)$ = same as col. 5 with an elevation of 5 deg.

7. *Area Access Rate for an Elevation Limit of 10 deg* $(10^3 km^2/s)$ = same as col. 5 with an elevation of 10 deg.

8. *Area Access Rate for an Elevation Limit of 20 deg* $(10^3 km^2/s)$ = same as col. 5 with an elevation of 20 deg.

9. *Maximum Time in View for a Satellite Visible to a Minimum Elevation Angle of 0 deg* (min) = $P\lambda_{max}/180$ deg, where P is from col. 52 and λ_{max} is from col. 13. Assumes a circular orbit over a nonrotating Earth [Eq. (5-49)].

10. *Maximum Time in View for a Satellite Visible to a Minimum Elevation Angle of 5 deg* (min) = same as col. 9 with λ_{max} for 5 deg taken from col. 14.

11. *Maximum Time in View for a Satellite Visible to a Minimum Elevation Angle of 10 deg* (min) = same as col. 9 with λ_{max} for 10 deg taken from col. 15.

12. *Maximum Time in View for a Satellite Visible to a Minimum Elevation Angle of 20 deg* (min) = same as col. 9 with λ_{max} for 20 deg taken from col. 16.

13. *Earth Central Angle for a Satellite at 0 deg Elevation* (deg) = *Maximum Earth Central Angle* = $\mathrm{acos}(R_E/r)$. Alternatively, Maximum Earth Central Angle is = $90 - \rho$, where ρ is from col. 49 [Eqs. (5-16), (5-17)].

14. *Earth Central Angle for a Satellite at 5 deg Elevation* (deg) = $90-\varepsilon-\eta$, where $\eta = \mathrm{asin}\,(\cos\varepsilon\,\sin\rho)$, ρ is from col. 49, and $\varepsilon = 5$ deg [Eqs. (5-26), (5-27)].

15. *Earth Central Angle for a Satellite at 10 deg Elevation* (deg) = same as col. 14 but with $\varepsilon = 10$ deg.

16. *Earth Central Angle for a Satellite at 20 deg Elevation* (deg) = same as col. 14 but with $\varepsilon = 20$ deg.

17. *Maximum Range to Horizon* = *Range to a satellite at 0 deg elevation* (km) = $(r^2 - R_E{}^2)^{1/2}$, where $R_E = 6{,}378.14$ km is the equatorial radius of the Earth.

18. *Range to a Satellite at 5 deg Elevation* (km) = *Maximum Range for Satellites with a Minimum Elevation Angle of 5 deg* (km) = $R_E (\sin\lambda / \sin\eta)$, where $R_E = 6{,}378.14$ km is the equatorial radius of the Earth, λ is from col. 14, $\eta = 90$ deg $- \lambda - \varepsilon$, and $\varepsilon = 5$ deg [Eq. (5-28)].

19. *Range to a Satellite at 10 deg Elevation* (km) = same as col. 18 with λ from col. 15 and $\varepsilon = 10$ deg.

20. *Range to a Satellite at 20 deg Elevation* (km) = same as col. 18 with λ from col. 16 and $\varepsilon = 20$ deg.

21. *Maximum Nadir Angle for a Satellite at 0 deg Elevation Angle* (deg) = *Max. Nadir Angle for Any Point on the Earth* = *Earth Angular Radius* = $\operatorname{asin}(R_E / r)$, where $R_E = 6{,}378.14$ km is the equatorial radius of the Earth [Eq. (5-16)].

22. *Nadir Angle for a Satellite at 5 deg Elevation Angle* (deg) = *Maximum Nadir Angle for Points on the Ground with a Minimum Elevation Angle of 5 deg* = 90 deg $- \varepsilon - \lambda$, is the Earth central angle from col. 14 [Eq. (5-27)].

23. *Nadir Angle for a Satellite at 10 deg Elevation Angle* (deg) = same as col. 22 with $\varepsilon = 10$ deg.

24. *Nadir Angle for a Satellite at 20 deg Elevation Angle* (deg) = same as col. 22 with $\varepsilon = 20$ deg.

25. *Atmospheric Scale Height* (km) = RT / Mg, where R is the molar gas constant, T is the temperature, M is the mean molecular weight, and g is the gravitational acceleration [inside front cover].

26. *Minimum Atmospheric Density* (kg/m^3), from MSIS atmospheric model [Hedin[*†‡], 1987, 1988, and 1991]. The solar flux value, F10.7, was chosen such that 10% of all measured data are less than this minimum (65.8×10^{-22} W·m^{-2}·Hz^{-1}). See Sec. 8.1.3. The MSIS model is limited to the region between 90 and 2,000 km. Below 150 km and above 600 km the error increases because less data have been used. All data have been averaged across the Earth with a 30 deg step size in longitude and 20 deg steps in latitude (-80 deg, to $+80$ deg). This over-represents the Earth's polar regions; however, satellites spend a larger fraction of their time at high latitudes. The solar hour angle was adapted to the individual location on the Earth with UT = 12.00 Noon.

27. *Mean Atmospheric Density* (kg/m^3) = same as col. 26 but with a mean F10.7 value of 118.7×10^{-22} W·m^{-2}·Hz^{-1}.

---

[*] Hedin, Alan E. 1987. "MSIS-86 Thermospheric Model." *J. Geophys. Res.*, 92, No. A5, pp. 4649–4662.

[†] ——. 1988. "The Atmospheric Model In The Region 90 to 2,000 km." *Adv. Space Res.*, 8, No. 5–6, pp. (5)9–(5)25, Pergamon Press.

[‡] ——. 1991. "Extension of the MSIS Thermosphere Model into the Middle and Lower Atmosphere." *J. Geophys. Res.*, 96, No. A2, pp. 1159–1172.

Explanation of Earth Satellite Parameters

28. *Maximum Atmospheric Density* = same as col. 26 but with a F10.7 value of 189.0×10^{-22} W·m^{-2}·Hz^{-1}. This is the F10.7 value such that 10% of all measured values are above it.

29. *Minimum ΔV to Maintain Altitude at Solar Minimum* (m/s per year) = $\pi(C_D A/m) \times \rho r v/P$, where ρ is from col. 26, v is from col. 41, P is from col. 52 expressed in years, and the ballistic coefficient, $m/C_D A$, is assumed to be 50 kg/m^2. ΔV estimates are not meaningful above 1,500 km [Eq. (6-26)].

30. *Maximum ΔV to Maintain Altitude at Solar Maximum* (m/s per year) = same as col. 29 with ρ from col. 28; Ballistic coefficient $m/C_D A$ = 50 kg/m^2.

31. *Minimum ΔV to Maintain Altitude at Solar Minimum* (m/s per year) = same as col. 29 with ρ from col. 26; Ballistic coefficient $m/C_D A$ = 200 kg/m^2.

32. *Maximum ΔV to Maintain Altitude at Solar Maximum* (m/s per year) = same as col. 29 with ρ from col. 28; Ballistic coefficient $m/C_D A$ = 200 kg/m^2.

33. *Orbit Decay Rate at Solar Minimum* (km/year) = $-2\pi (C_D A/m) \rho r^2/P$, where ρ is from col. 26, P is from col. 52 (expressed in years), and the ballistic coefficient, $m/C_D A$, is assumed to be 50 kg/m^2. Orbit decay rates are not meaningful above 1,500 km [Eq. (6-24)].

34. *Orbit Decay Rate at Solar Maximum* (km/year) = same as col. 33, with ρ from col. 28 and the ballistic coefficient, $m/C_D A$, assumed to be 50 kg/m^2.

35. *Orbit Decay Rate at Solar Minimum* (km/year) = same as col. 33 with ρ from col. 26, and the ballistic coefficient, $m/C_D A$, assumed to be 200 kg/m^2.

36. *Orbit Decay Rate at Solar Maximum* (km/year) = same as col. 33, with ρ from col. 28 and the ballistic coefficient, $m/C_D A$, assumed to be 200 kg/m^2.

37. *Estimated Orbit Lifetime at Solar Minimum* (days) = Data was produced using the software package SatLife. Ballistic coefficient, $m/C_D A$, assumed to be 50 kg/m^2.

38. *Estimated Orbit Lifetime at Solar Maximum* (days) = same as col. 37 with the ballistic coefficient, $m/C_D A$, assumed to be 50 kg/m^2.

39. *Estimated Orbit Lifetime at Solar Minimum* (days) = same as col. 37 with the ballistic coefficient, $m/C_D A$, assumed to be 200 kg/m^2.

40. *Estimated Orbit Lifetime at Solar Maximum* (days) = same as col. 37 with the ballistic coefficient, $m/C_D A$, assumed to be 200 kg/m^2.

41. *Circular Velocity* (km/s) = $(\mu_E/r)^{1/2}$ = $631.3481 r^{-1/2}$ [Eq. (6-5)].

42. *Orbit Angular Velocity* (deg/minute) = $360/P = 2.170\,415 \times 10^6 r^{-3/2}$, where P is from col. 52. This is the angular velocity with respect to the center of the Earth for a circular orbit. (See col. 47 for angular rate with respect to ground stations) [Eq. (5-31)].

43. *Escape Velocity* (km/s) = $(2\mu/r)^{1/2}$ = $892.8611 r^{-1/2}$ = $(2)^{1/2} \times v_{circ}$ [Eq. (6-6)].

44. *ΔV Required to De-Orbit* (m/s) = the velocity change needed to transform the assumed circular orbit to an elliptical orbit with an unchanged apogee and a perigee of 50 km [Eq. (6-32) and Sec. 6.3.1]. (Note that this a correction to the corresponding columns in SMAD I and SMAD II which were incorrect.)

45. *Plane Change* ΔV ((m/s)/deg) = 2,000 v_{circ} sin (0.5 deg), where v_{circ} is from col. 41. Assumes circular orbit and linear sine function; [Eq. (6-38)]

46. *ΔV Required for a 1 km Altitude Change* (m/s) = assumes a Hohmann Transfer with $r_B - r_A$ = 1 km; [Eq. (6-32)].

47. *Maximum Angular Rate As Seen from a Ground Station* (deg/s) = $2\pi r/hP$, where $h \equiv r - R_E$ is the altitude and P is from col. 52. This is the angular rate as seen from the surface of a non-rotating Earth of a satellite in a circular orbit passing directly overhead. (See col. 42 for the angular velocity as seen from the center of the Earth.) [Eq. (5-47)].

48. *Sun Synchronous Inclination* (deg) = acos ($-4.773\ 48 \times 10^{-15}\ r^{\,7/2}$); assumes circular orbit with node rotation rate of 0.9856 deg/day to follow the mean motion of the Sun. Above 6,000 km altitude there are no Sun synchronous circular orbits [Eq. (6-19)].

49. *Angular Radius of the Earth* (deg) = asin (R_E / r), where R_E = 6,378.14 km is the equatorial radius of the Earth [Eqs. (5-16)].

50. *One Degree Field of View Mapped onto the Earth's Surface at Nadir from Altitude h* (km) = The length on the Earth's curved surface of a 1 deg arc projected at nadir from this altitude. Note: This data is *very* nonlinear [Eqs. (5-26a), (5-26b), and (5-27)].

51. *Range to Horizon* (km) = same as col. 17 = $(r^2 - R_E^{\ 2})^{1/2}$, where R_E = 6,378.14 km is the equatorial radius of the Earth. For the range to points other than the true horizon (i.e., $\varepsilon \neq 0$ deg) use columns 18, 19, and 20 [Eq. (5-28)].

52. *Period* (min) = $1.658\ 669 \times 10^{-4}\ r^{3/2}$ = $(1/60) \times 2\pi\ (r^3/\mu)^{1/2}$. Assumes a circular orbit, r is measured in km, and μ = 398,600.5 km^3/s^2. Note that period is the same for an eccentric orbit with semimajor axis = r; [Eq. (7-7)].

53. *Maximum Eclipse* (minutes) = $(\rho/180$ deg)P, where ρ is from col. 49 and P is from col. 52. This is the maximum eclipse for a circular orbit. Eclipses at this altitude in an eccentric orbit can be longer. [See Example 1, Sec. 5.1]

54. *Revolutions per Day* (#) = 1,436.07/P, where P is from col. 52. Note that this is revolutions per sidereal day, where the *sidereal day* is the day relative to the fixed stars which is approximately 4 min shorter than the solar day of 1,440 min. [Note: This is a correction to SMAD I and II (printings 1 through 4) where the *revolutions per day* were defined as here but the data was produced using 1,440 min as the length of a day.]

55. *Node Spacing* (deg) = 360 deg $\times$ (P / 1,436.07), where P is from col. 52. This is the spacing in longitude between successive ascending or descending nodes for a satellite in a circular orbit [Eq. (7-13)]. Does not take into account node precession rate from col. 56.

56. *Node Precession Rate* (deg/day) = $-2.06474 \times 10^{14}\ r^{\,-7/2}\cos i$ = $-1.5\ n\ J_2$ (R_E/a)^2 (cos i) $(1 - e^2)^{-2}$, where i is the inclination, e the eccentricity (which is set to zero), n is the mean motion (= $(\mu/a^3)^{1/2}$), a the semimajor axis, and J_2 the dominant zonal coefficient in the expansion of the Legendre polynomial describing the geopotential. This is the angle through which the orbit rotates in inertial space in a 24 hour period. Assumes a circular orbit; r is in km in the first expression [Eq. (6-19)].

Earth Satellite Parameters

	1	2	3	4	5	6	7	8
	INSTANTANEOUS ACCESS AREA				AREA ACCESS RATE			
Alt. (km)	0 deg elevation (10^6km^2)	5 deg elevation (10^6km^2)	10 deg elevation (10^6km^2)	20 deg elevation (10^6km^2)	0 deg elevation (10^3km^2/s)	5 deg elevation (10^3km^2/s)	10 deg elevation (10^3km^2/s)	20 deg elevation (10^3km^2/s)
0	0.00	0.00	0.00	0.00	0.00	0.00	0.00	0.00
100	3.95	1.51	0.67	0.21	17.24	10.71	7.15	3.96
150	5.87	2.66	1.31	0.44	20.76	14.01	9.86	5.69
200	7.77	3.89	2.06	0.73	23.56	16.73	12.20	7.28
250	9.64	5.17	2.89	1.08	25.90	19.04	14.27	8.76
300	11.48	6.48	3.77	1.48	27.89	21.06	16.11	10.12
350	13.30	7.81	4.70	1.92	29.63	22.83	17.76	11.38
400	15.08	9.15	5.66	2.39	31.15	24.41	19.26	12.56
450	16.85	10.50	6.64	2.89	32.50	25.83	20.62	13.66
500	18.58	11.85	7.64	3.42	33.71	27.10	21.86	14.68
550	20.29	13.20	8.66	3.97	34.78	28.25	22.99	15.63
600	21.98	14.54	9.69	4.53	35.75	29.30	24.03	16.52
650	23.64	15.88	10.72	5.12	36.62	30.25	24.99	17.36
700	25.28	17.21	11.76	5.72	37.40	31.12	25.87	18.14
750	26.89	18.53	12.81	6.33	38.11	31.91	26.68	18.87
800	28.49	19.85	13.85	6.95	38.75	32.63	27.43	19.56
850	30.06	21.15	14.90	7.58	39.33	33.29	28.12	20.20
900	31.61	22.44	15.94	8.21	39.85	33.90	28.76	20.80
950	33.14	23.73	16.98	8.86	40.32	34.45	29.35	21.37
1,000	34.64	25.00	18.02	9.51	40.74	34.96	29.89	21.90
1,250	41.89	31.18	23.14	12.81	42.29	36.90	32.06	24.10
1,500	48.67	37.07	28.12	16.14	43.12	38.10	33.50	25.69
2,000	61.02	47.97	37.51	22.69	43.43	39.07	34.93	27.59
2,500	71.98	57.81	46.14	28.96	42.71	38.89	35.19	28.39
3,000	81.77	66.70	54.05	34.86	41.47	38.10	34.78	28.52
3,500	90.57	74.76	61.29	40.38	39.95	36.98	33.99	28.22
4,000	98.52	82.10	67.92	45.52	38.33	35.68	32.98	27.68
4,500	105.74	88.79	74.02	50.31	36.68	34.30	31.86	26.97
5,000	112.32	94.93	79.63	54.77	35.05	32.92	30.70	26.18
6,000	123.90	105.78	89.62	62.81	31.97	30.22	28.37	24.49
7,000	133.74	115.06	98.23	69.84	29.18	27.73	26.16	22.80
8,000	142.22	123.10	105.71	76.02	26.70	25.48	24.13	21.20
9,000	149.59	130.11	112.27	81.48	24.51	23.47	22.30	19.72
10,000	156.06	136.29	118.07	86.35	22.57	21.67	20.66	18.36
15,000	179.35	158.65	139.22	104.35	15.69	15.21	14.63	13.25
20,000	193.80	172.65	152.56	115.89	11.63	11.34	10.97	10.04
20,184	194.23	173.06	152.95	116.24	11.52	11.23	10.87	9.95
25,000	203.65	182.23	161.73	123.91	9.05	8.85	8.60	7.92
30,000	210.79	189.19	168.42	129.80	7.29	7.15	6.96	6.45
35,000	216.20	194.49	173.52	134.31	6.03	5.93	5.78	5.37
35,786	216.94	195.21	174.22	134.93	5.86	5.76	5.62	5.23

See Front of Table for Formulas and Sources.

Earth Satellite Parameters

	9	10	11	12	13	14	15	16
	MAXIMUM TIME IN VIEW				MAXIMUM EARTH CENTRAL ANGLE			
Alt. (km)	0 deg elevation (min)	5 deg elevation (min)	10 deg elevation (min)	20 deg elevation (min)	0 deg elevation (deg)	5 deg elevation (deg)	10 deg elevation (deg)	20 deg elevation (deg)
0	0.00	0.00	0.00	0.00	0.00	0.00	0.00	0.00
100	4.84	3.00	2.00	1.11	10.08	6.24	4.16	2.30
150	5.98	4.02	2.82	1.63	12.31	8.27	5.81	3.35
200	6.96	4.92	3.58	2.13	14.16	10.00	7.28	4.34
250	7.85	5.74	4.29	2.62	15.79	11.54	8.62	5.28
300	8.67	6.50	4.95	3.10	17.24	12.93	9.85	6.17
350	9.44	7.22	5.59	3.57	18.56	14.20	11.00	7.02
400	10.17	7.91	6.21	4.03	19.78	15.38	12.08	7.84
450	10.88	8.57	6.80	4.49	20.92	16.48	13.09	8.63
500	11.55	9.21	7.38	4.93	21.98	17.52	14.05	9.38
550	12.21	9.83	7.95	5.37	22.98	18.49	14.96	10.11
600	12.86	10.43	8.50	5.81	23.93	19.42	15.82	10.81
650	13.48	11.02	9.04	6.24	24.84	20.30	16.65	11.48
700	14.10	11.60	9.58	6.66	25.70	21.15	17.45	12.14
750	14.71	12.17	10.10	7.08	26.52	21.95	18.21	12.77
800	15.30	12.74	10.62	7.50	27.31	22.73	18.95	13.39
850	15.89	13.29	11.13	7.92	28.07	23.47	19.66	13.98
900	16.48	13.84	11.64	8.33	28.80	24.19	20.34	14.56
950	17.05	14.38	12.14	8.74	29.50	24.88	21.00	15.13
1,000	17.62	14.92	12.64	9.15	30.18	25.55	21.64	15.68
1,250	20.42	17.56	15.08	11.18	33.27	28.60	24.57	18.21
1,500	23.16	20.13	17.48	13.19	35.94	31.24	27.13	20.47
2,000	28.56	25.21	22.21	17.19	40.42	35.68	31.43	24.33
2,500	33.98	30.30	26.96	21.23	44.08	39.30	34.97	27.54
3,000	39.46	35.44	31.76	25.34	47.15	42.35	37.95	30.28
3,500	45.04	40.68	36.65	29.53	49.78	44.97	40.52	32.65
4,000	50.74	46.03	41.65	33.83	52.08	47.25	42.75	34.72
4,500	56.56	51.50	46.77	38.23	54.10	49.26	44.73	36.57
5,000	62.52	57.10	52.00	42.74	55.91	51.05	46.49	38.21
6,000	74.85	68.67	62.82	52.08	58.98	54.12	49.51	41.04
7,000	87.73	80.77	74.14	61.86	61.53	56.64	52.00	43.38
8,000	101.15	93.37	85.94	72.07	63.67	58.77	54.10	45.36
9,000	115.10	106.48	98.22	82.70	65.50	60.60	55.89	47.06
10,000	129.56	120.08	110.96	93.74	67.08	62.17	57.45	48.53
15,000	209.23	195.03	181.21	154.73	72.64	67.71	62.91	53.72
20,000	300.06	280.54	261.44	224.50	76.01	71.06	66.22	56.87
20,184	303.60	283.87	264.57	227.22	76.11	71.16	66.32	56.96
25,000	400.90	375.52	350.60	302.13	78.27	73.32	68.45	58.99
30,000	510.87	479.15	447.92	386.92	79.90	74.94	70.06	60.52
35,000	629.28	590.76	552.77	478.33	81.13	76.17	71.27	61.67
35,786	648.62	608.99	569.90	493.27	81.30	76.33	71.43	61.83

See Front of Table for Formulas and Sources.

Earth Satellite Parameters

	17	18	19	20	21	22	23	24
	MAXIMUM RANGE				MAXIMUM NADIR ANGLE			
Altitude (km)	0 deg elevation (km)	5 deg elevation (km)	10 deg elevation (km)	20 deg elevation (km)	0 deg elevation (deg)	5 deg elevation (deg)	10 deg elevation (deg)	20 deg elevation (deg)
0	0	0	0	0	90.00	85.00	80.00	70.00
100	1,134	707	477	277	79.92	78.76	75.84	67.70
150	1,391	942	671	406	77.69	76.73	74.19	66.65
200	1,610	1,147	846	530	75.84	75.00	72.72	65.66
250	1,803	1,331	1,009	649	74.21	73.46	71.38	64.72
300	1,979	1,500	1,160	764	72.76	72.07	70.15	63.83
350	2,142	1,657	1,304	876	71.44	70.80	69.00	62.98
400	2,294	1,805	1,440	984	70.22	69.62	67.92	62.16
450	2,438	1,944	1,570	1,090	69.08	68.52	66.91	61.37
500	2,575	2,078	1,695	1,193	68.02	67.48	65.95	60.62
550	2,705	2,206	1,816	1,294	67.02	66.51	65.04	59.89
600	2,831	2,329	1,932	1,392	66.07	65.58	64.18	59.19
650	2,952	2,448	2,045	1,489	65.16	64.70	63.35	58.52
700	3,069	2,563	2,155	1,584	64.30	63.85	62.55	57.86
750	3,183	2,675	2,262	1,677	63.48	63.05	61.79	57.23
800	3,293	2,784	2,367	1,769	62.69	62.27	61.05	56.61
850	3,401	2,890	2,469	1,859	61.93	61.53	60.34	56.02
900	3,506	2,994	2,569	1,948	61.20	60.81	59.66	55.44
950	3,608	3,095	2,667	2,035	60.50	60.12	59.00	54.87
1,000	3,709	3,194	2,763	2,121	59.82	59.45	58.36	54.32
1,250	4,184	3,665	3,221	2,537	56.73	56.40	55.43	51.79
1,500	4,624	4,102	3,648	2,932	54.06	53.76	52.87	49.53
2,000	5,433	4,905	4,437	3,673	49.58	49.32	48.57	45.67
2,500	6,176	5,645	5,167	4,368	45.92	45.70	45.03	42.46
3,000	6,875	6,342	5,856	5,032	42.85	42.65	42.05	39.72
3,500	7,543	7,008	6,516	5,671	40.22	40.03	39.48	37.35
4,000	8,187	7,650	7,154	6,291	37.92	37.75	37.25	35.28
4,500	8,812	8,274	7,774	6,897	35.90	35.74	35.27	33.43
5,000	9,422	8,883	8,380	7,490	34.09	33.95	33.51	31.79
6,000	10,608	10,067	9,558	8,649	31.02	30.88	30.49	28.96
7,000	11,760	11,217	10,704	9,779	28.47	28.36	28.00	26.62
8,000	12,886	12,342	11,826	10,888	26.33	26.23	25.90	24.64
9,000	13,993	13,448	12,929	11,981	24.50	24.40	24.11	22.94
10,000	15,085	14,540	14,018	13,061	22.92	22.83	22.55	21.47
15,000	20,405	19,856	19,327	18,339	17.36	17.29	17.09	16.28
20,000	25,595	25,046	24,512	23,507	13.99	13.94	13.78	13.13
20,184	25,785	25,235	24,701	23,696	13.89	13.84	13.68	13.04
25,000	30,723	30,172	29,635	28,619	11.73	11.68	11.55	11.01
30,000	35,815	35,263	34,724	33,700	10.10	10.06	9.94	9.48
35,000	40,884	40,332	39,791	38,760	8.87	8.83	8.73	8.33
35,786	41,679	41,127	40,586	39,555	8.70	8.67	8.57	8.17

See Front of Table for Formulas and Sources.

Earth Satellite Parameters

		25	26	27	28	29	30	31	32
			ATMOSPHERIC DENSITY			ΔV TO MAINTAIN ALTITUDE			
Alt (km)	Atm. Scale Ht. (km)	Minimum (kg/m^3)	Mean (kg/m^3)	Maximum (kg/m^3)		Solar Min 50 kg/ m^2 (m/s)/yr	Solar Max 50 kg/ m^2 (m/s)/yr	Solar Min 200 kg/ m^2 (m/s)/yr	Solar Max 200 kg/ m^2 (m/s)/yr
0	8.4	1.2	1.2	1.2		2.37×10^{13}	2.37×10^{13}	5.92×10^{12}	5.92×10^{12}
100	5.9	4.61×10^{-7}	4.79×10^{-7}	5.10×10^{-7}		8.95×10^{6}	9.90×10^{6}	2.24×10^{6}	2.47×10^{6}
150	25.5	1.65×10^{-9}	1.81×10^{-9}	2.04×10^{-9}		3.17×10^{4}	3.94×10^{4}	7.93×10^{3}	9.85×10^{3}
200	37.5	1.78×10^{-10}	2.53×10^{-10}	3.52×10^{-10}		3.40×10^{3}	6.72×10^{3}	8.51×10^{2}	1.68×10^{3}
250	44.8	3.35×10^{-11}	6.24×10^{-11}	1.06×10^{-10}		6.36×10^{2}	2.02×10^{3}	1.59×10^{2}	5.04×10^{2}
300	50.3	8.19×10^{-12}	1.95×10^{-11}	3.96×10^{-11}		1.54×10^{2}	7.47×10^{2}	3.86×10^{1}	1.87×10^{2}
350	54.8	2.34×10^{-12}	6.98×10^{-12}	1.66×10^{-11}		4.37×10^{1}	3.11×10^{2}	1.09×10^{1}	7.78×10^{1}
400	58.2	7.32×10^{-13}	2.72×10^{-12}	7.55×10^{-12}		1.36×10^{1}	1.40×10^{2}	3.40×10^{0}	3.50×10^{1}
450	61.3	2.47×10^{-13}	1.13×10^{-12}	3.61×10^{-12}		4.55×10^{0}	6.66×10^{1}	1.14×10^{0}	1.66×10^{1}
500	64.5	8.98×10^{-14}	4.89×10^{-13}	1.80×10^{-12}		1.64×10^{0}	3.29×10^{1}	4.11×10^{-1}	8.23×10^{0}
550	68.7	3.63×10^{-14}	2.21×10^{-13}	9.25×10^{-13}		6.59×10^{-1}	1.68×10^{1}	1.65×10^{-1}	4.20×10^{0}
600	74.8	1.68×10^{-14}	1.04×10^{-13}	4.89×10^{-13}		3.03×10^{-1}	8.81×10^{0}	7.58×10^{-2}	2.20×10^{0}
650	84.4	9.14×10^{-15}	5.15×10^{-14}	2.64×10^{-13}		1.64×10^{-1}	4.73×10^{0}	4.09×10^{-2}	1.18×10^{0}
700	99.3	5.74×10^{-15}	2.72×10^{-14}	1.47×10^{-13}		1.02×10^{-1}	2.61×10^{0}	2.55×10^{-2}	6.52×10^{-1}
750	121	3.99×10^{-15}	1.55×10^{-14}	8.37×10^{-14}		7.04×10^{-2}	1.48×10^{0}	1.76×10^{-2}	3.69×10^{-1}
800	151	2.96×10^{-15}	9.63×10^{-15}	4.39×10^{-14}		5.19×10^{-2}	8.63×10^{-1}	1.30×10^{-2}	2.16×10^{-1}
850	188	2.28×10^{-15}	6.47×10^{-15}	3.00×10^{-14}		3.97×10^{-2}	5.23×10^{-1}	9.94×10^{-3}	1.31×10^{-1}
900	226	1.80×10^{-15}	4.66×10^{-15}	1.91×10^{-14}		3.11×10^{-2}	3.30×10^{-1}	7.78×10^{-3}	8.25×10^{-2}
950	263	1.44×10^{-15}	3.54×10^{-15}	1.27×10^{-14}		2.48×10^{-2}	2.18×10^{-1}	6.19×10^{-3}	5.45×10^{-2}
1,000	296	1.17×10^{-15}	2.79×10^{-15}	8.84×10^{-15}		1.99×10^{-2}	1.51×10^{-1}	4.98×10^{-3}	3.77×10^{-2}
1,250	408	4.67×10^{-16}	1.11×10^{-15}	2.59×10^{-15}		7.69×10^{-3}	4.27×10^{-2}	1.92×10^{-3}	1.07×10^{-2}
1,500	516	2.30×10^{-16}	5.21×10^{-16}	1.22×10^{-15}		3.68×10^{-3}	1.95×10^{-2}	9.20×10^{-4}	4.88×10^{-3}
2,000	829	—	—	—		—	—	—	—
2,500	1,220	—	—	—		—	—	—	—
3,000	1,590	—	—	—		—	—	—	—
3,500	1,900	—	—	—		—	—	—	—
4,000	2,180	—	—	—		—	—	—	—
4,500	2,430	—	—	—		—	—	—	—
5,000	2,690	—	—	—		—	—	—	—
6,000	3,200	—	—	—		—	—	—	—
7,000	3,750	—	—	—		—	—	—	—
8,000	4,340	—	—	—		—	—	—	—
9,000	4,970	—	—	—		—	—	—	—
10,000	5,630	—	—	—		—	—	—	—
15,000	9,600	—	—	—		—	—	—	—
20,000	14,600	—	—	—		—	—	—	—
20,184	14,600	—	—	—		—	—	—	—
25,000	20,700	—	—	—		—	—	—	—
30,000	27,800	—	—	—		—	—	—	—
35,000	36,000	—	—	—		—	—	—	—
35,786	37,300	—	—	—		—	—	—	—

See Front of Table for Formulas and Sources.

Earth Satellite Parameters

	33	34	35	36	37	38	39	40
	ORBIT DECAY RATE				ESTIMATED ORBIT LIFETIME			
Alt (km)	Solar Min 50 kg/m² (km/yr)	Solar Max 50 kg/m² (km/yr)	Solar Min 200 kg/m² (km/yr)	Solar Max 200 kg/m² (km/yr)	Solar Min 50 kg/m² (days)	Solar Max 50 kg/m² (days)	Solar Min 200 kg/m² (days)	Solar Max 200 kg/m² (days)
0	3.82×10^{13}	3.82×10^{13}	9.55×10^{12}	9.55×10^{12}	0.00	0.00	0.00	0.00
100	1.48×10^{7}	1.64×10^{7}	3.70×10^{6}	3.96×10^{6}	0.06	0.06	0.06	0.06
150	5.30×10^{4}	6.58×10^{4}	1.32×10^{4}	1.57×10^{4}	0.24	0.18	0.54	0.48
200	5.75×10^{3}	1.14×10^{4}	1.44×10^{3}	2.67×10^{3}	1.65	1.03	5.99	3.6
250	1.09×10^{3}	3.45×10^{3}	2.72×10^{2}	7.99×10^{2}	10.06	3.82	40.21	14.98
300	2.67×10^{2}	1.29×10^{3}	6.67×10^{1}	2.95×10^{2}	49.9	11.0	196.7	49.2
350	7.64×10^{1}	5.44×10^{2}	1.91×10^{1}	1.23×10^{2}	195.6	30.9	615.9	140.3
400	2.40×10^{1}	2.48×10^{2}	6.01×10^{0}	5.50×10^{1}	552.2	77.4	1024.5	346.9
450	8.12×10^{0}	1.19×10^{2}	2.03×10^{0}	2.60×10^{1}	872	181	1,497	724
500	2.97×10^{0}	5.95×10^{1}	7.42×10^{-1}	1.28×10^{1}	1,205	393	2,377	3,310
550	1.20×10^{0}	3.07×10^{1}	3.01×10^{-1}	6.53×10^{0}	1,638	801	5,470	4,775
600	5.60×10^{-1}	1.63×10^{1}	1.40×10^{-1}	3.41×10^{0}	2,580	3,430	14,100	13,400
650	3.05×10^{-1}	8.83×10^{0}	7.64×10^{-2}	1.83×10^{0}	5,560	4,550	28,500	27,900
700	1.92×10^{-1}	4.92×10^{0}	4.81×10^{-2}	1.00×10^{0}	13,400	12,600	53,400	52,700
750	1.34×10^{-1}	2.82×10^{0}	3.36×10^{-2}	5.67×10^{-1}	24,400	24,300	98,500	97,700
800	1.00×10^{-1}	1.66×10^{0}	2.50×10^{-2}	3.30×10^{-1}	42,000	41,000	175,200	174,200
850	7.74×10^{-2}	1.02×10^{0}	1.93×10^{-2}	1.99×10^{-1}	76,600	76,200	307,400	306,700
900	6.12×10^{-2}	6.49×10^{-1}	1.53×10^{-2}	1.26×10^{-1}	127,000	128,000	521,000	520,000
950	4.92×10^{-2}	4.33×10^{-1}	1.23×10^{-2}	8.26×10^{-2}	21,1000	210,000	853,000	852,000
1,000	4.00×10^{-2}	3.03×10^{-1}	9.99×10^{-3}	5.70×10^{-2}	341,000	340,000	1,361,000	1,362,000
1,250	1.62×10^{-2}	9.00×10^{-2}	4.05×10^{-3}	1.59×10^{-2}	1,700,000	1,700,000	6,800,000	6,800,000
1,500	8.15×10^{-3}	4.32×10^{-3}	2.04×10^{-3}	7.17×10^{-3}	4,810,000	4,810,000	19,250,000	19,250,000
2,000	—	—	—	—	—	—	—	—
2,500	—	—	—	—	—	—	—	—
3,000	—	—	—	—	—	—	—	—
3,500	—	—	—	—	—	—	—	—
4,000	—	—	—	—	—	—	—	—
4,500	—	—	—	—	—	—	—	—
5,000	—	—	—	—	—	—	—	—
6,000	—	—	—	—	—	—	—	—
7,000	—	—	—	—	—	—	—	—
8,000	—	—	—	—	—	—	—	—
9,000	—	—	—	—	—	—	—	—
10,000	—	—	—	—	—	—	—	—
15,000	—	—	—	—	—	—	—	—
20,000	—	—	—	—	—	—	—	—
20,184	—	—	—	—	—	—	—	—
25,000	—	—	—	—	—	—	—	—
30,000	—	—	—	—	—	—	—	—
35,000	—	—	—	—	—	—	—	—
35,786	—	—	—	—	—	—	—	—

See Front of Table for Formulas and Sources.

Earth Satellite Parameters

	41	42	43	44	45	46	47	48
				VELOCITY-RELATED PARAMETERS				
Alt (km)	Circular Velocity (km/s)	Orbit Angular Velocity (deg/min)	Escape Velocity (km/s)	ΔV Req'd to Deorbit (m/s)	Plane Change ΔV (m/s)/deg	ΔV Req'd for a 1 km Alt Chg (m/s)	Max Ang Rate from Gnd Stn (deg/s)	Sun Syn-chronous Inclination (deg)
0	7.905	4.261	11.180	—	137.97	0.62	—	95.68
100	7.844	4.163	11.093	−15.2	136.90	0.61	4.49	96.00
150	7.814	4.115	11.051	−30.2	136.38	0.60	2.98	96.16
200	7.784	4.068	11.009	−45.0	135.86	0.59	2.23	96.33
250	7.755	4.022	10.967	−59.6	135.35	0.58	1.78	96.50
300	7.726	3.977	10.926	−74.0	134.84	0.58	1.48	96.67
350	7.697	3.933	10.885	−88.3	134.34	0.57	1.26	96.85
400	7.669	3.889	10.845	−102.3	133.84	0.57	1.10	97.03
450	7.640	3.847	10.805	−116.2	133.35	0.56	0.97	97.21
500	7.613	3.805	10.766	−129.8	132.86	0.55	0.87	97.40
550	7.585	3.764	10.727	−143.3	132.38	0.55	0.79	97.59
600	7.558	3.723	10.688	−156.7	131.91	0.54	0.72	97.79
650	7.531	3.684	10.650	−169.8	131.44	0.54	0.66	97.99
700	7.504	3.645	10.613	−182.8	130.97	0.53	0.61	98.19
750	7.478	3.606	10.575	−195.6	130.51	0.52	0.57	98.39
800	7.452	3.569	10.538	−208.3	130.06	0.52	0.53	98.60
850	7.426	3.532	10.502	−220.8	129.61	0.51	0.50	98.82
900	7.400	3.496	10.466	−233.1	129.16	0.51	0.47	99.03
950	7.375	3.460	10.430	−245.3	128.72	0.50	0.44	99.25
1,000	7.350	3.425	10.395	−257.4	128.28	0.50	0.42	99.48
1,250	7.229	3.258	10.223	−315.4	126.16	0.47	0.33	100.66
1,500	7.113	3.104	10.059	−370.1	124.14	0.45	0.27	101.96
2,000	6.898	2.830	9.755	−470.2	120.38	0.41	0.20	104.89
2,500	6.701	2.595	9.476	−559.6	116.94	0.38	0.15	108.35
3,000	6.519	2.390	9.220	−639.8	113.78	0.35	0.12	112.41
3,500	6.352	2.211	8.984	−711.9	110.87	0.32	0.10	117.21
4,000	6.197	2.053	8.764	−777.0	108.16	0.30	0.09	122.93
4,500	6.053	1.913	8.561	−836.0	105.65	0.28	0.08	129.86
5,000	5.919	1.788	8.370	−889.5	103.30	0.26	0.07	138.59
6,000	5.675	1.576	8.025	−982.8	99.04	0.23	0.05	—
7,000	5.458	1.403	7.719	−1,060.8	95.27	0.20	0.04	—
8,000	5.265	1.259	7.446	−1,126.4	91.89	0.18	0.04	—
9,000	5.091	1.138	7.200	−1,182.0	88.86	0.17	0.03	—
10,000	4.933	1.035	6.977	−1,229.3	86.10	0.15	0.03	—
15,000	4.318	0.694	6.107	−1,381.9	75.36	0.10	0.02	—
20,000	3.887	0.507	5.497	−1,453.8	67.85	0.07	0.01	—
20,184	3.874	0.501	5.478	−1,455.5	67.61	0.07	0.01	—
25,000	3.564	0.390	5.040	−1,485.7	62.21	0.06	0.01	—
30,000	3.310	0.313	4.681	−1,496.1	57.77	0.05	0.01	—
35,000	3.104	0.258	4.389	−1,494.2	54.17	0.04	0.01	—
35,786	3.075	0.251	4.348	−1,493.2	53.66	0.04	0.00	—

See Front of Table for Formulas and Sources.